Structural Geology

Principles, Concepts, and Problems
Second Edition

Robert D. Hatcher, Jr.

Department of Geological Sciences
University of Tennessee–Knoxville
and
Environmental Sciences Division
Oak Ridge National Laboratory

Prentice Hall
Englewood Cliffs, New Jersey 07632

Hatcher, Robert D., Jr.
Structural geology: principles, concepts, and problems / Robert D. Hatcher, Jr.—2nd ed.
p. cm.
Includes bibliographical references and index.
ISBN 0–02–355713–3
1. Geology, Structural. I. Title.
QE601.H35 1995
551.8--dc20

Cover Photo: Thin section negative and positive prints of a small fold in a garnet-rich (codicule) layer from the Einunnfjell region near Foldal, south-central Norway. Specimen courtesy of Elizabeth A. McClellan, Western Kentucky University.

Editor: Robert A. McConnin
Production Supervisor: Kelly Ricci, Spectrum Publisher Services
Production Manager: Aliza Greenblatt
Cover Designers: Robert D. Hatcher, Jr., Donald G. McClanahan, and Robert A. Freese

A Simon and Schuster Company
Englewood Cliffs, New Jersey 07632

Printed in the United States of America

10 9 8 7 6 5 4 3 2

ISBN 0–02–355713–3

PRENTICE-HALL INTERNATIONAL (UK) LIMITED, *London*
PRENTICE-HALL OF AUSTRALIA PTY. LIMITED, *Sydney*
PRENTICE-HALL CANADA, INC., *Toronto*
PRENTICE-HALL HISPANOAMERICANO, S.A., *Mexico, D. F.*
PRENTICE-HALL OF INDIA PRIVATE LIMITED, *New Delhi*
PRENTICE-HALL OF JAPAN, INC., *Tokyo*
SIMON & SCHUSTER ASIA PTE., LTD., *Singapore*
EDITORA PRENTICE-HALL DO BRASIL, LTDA., *Rio de Janeiro*

Dedicated to

My parents,
My uncle W. E. "Red" Harris and aunt Marie Harris,
My wife Diana and daughters Laura and Melinda,
Who have both tolerated me and influenced me
More than they realize;

and to

My teachers, particularly those who taught me
To think freely, clearly, and critically,
To synthesize,
To not be afraid to be wrong in the creation of new ideas,
and
To be a careful observer in the field:

Robert H. Barnes,
Douglas W. Rankin,
Richard G. Stearns,
and
George D. Swingle

PREFACE TO THE SECOND EDITION

THE SECOND EDITION OF *STRUCTURAL GEOLOGY: Principles, Concepts, and Problems* has undergone a complete revision from the first edition. It contains major changes in every chapter, many new line drawings and photographs, a new chapter on tectonic structures associated with plutons, and a new appendix containing a summary of available computer software for use in structural geology. Even with these additions, the total number of pages in the book is about the same as in the first edition because of efficiencies gained by sizing the art work and photographs myself based on a combination of the intended message and the nature of the art. The original goals and objectives of the first edition, however, remain the same: to present a balanced coverage of modern structural geology at a level appropriate for most one-semester elementary structural geology courses offered in the U.S. today.

Another primary goal remains the desire for students to enjoy their first structural geology course, despite the fact that only a few will become structural geologists. Hopefully, this text will kindle in them a lifelong interest in the architectural framework of the Earth.

This course probably involves more information from other courses than any other at the undergraduate level. It therefore should bring together many of the principles from stratigraphy/sedimentology, petrology and mineralogy, and even paleontology to aid in the understanding of how structures form and the way rocks behave as they are deformed.

This may be the first structural geology book produced almost entirely using off-the-shelf and public domain graphics and desk-top publishing software. All line art, except for a few diagrams that had to be reproduced photographically (as pickup), was drafted or scanned and redrafted on the Macintosh™ computer and composed with text and photographs to produce the final layouts of chapters.

ACKNOWLEDGMENTS

Aside from some preliminary reviews, most reviews of the second edition proceeded in small groups of from one to three chapters with the goal of focusing on technical content, and not the organization of the entire book. Reviewers include: Wallace A. Bothner (University of New Hampshire), Steven A. Boyer (University of Washington), Gregory A. Davis (University of Southern California), William M. Dunne (University of Tennessee–Knoxville), Steven G. Driese (University of Tennessee–Knoxville), Terry Engelder (Pennsylvania State University), Alexander E. Gates (Rutgers University), Lawrence B. Gillett (SUNY College at Plattsburgh), Richard A. Gilman (SUNY College at Fredonia), James P. Hibbard (North Carolina State University), Peter Hudleston (University of Minnesota), Raymond V. Ingersoll (University of California, Los Angeles), Kauko Laajoki (University of Oulu, Finland), Matthew Mauldon (Civil Engineering, University of Tennessee–Knoxville), Winthrop D. Means (SUNY–Albany), Charles M. Onasch (Bowling Green State University), Donald T. Secor, Jr. (University of South Carolina), James F. Tull (Florida State University), Jan Tullis (Brown University), Donald U. Wise (University of Massachusetts), and James E. Wright (Rice University). The efforts of all are very much appreciated. I must specifically thank Laurie Gillett for reading the first edition shortly after it was published and pointing out its many shortcomings, and then for agreeing to read the entire text of the second edition to identify any final corrections needed before it went to the printer. Bill Dunne, Terry Engelder, and Charlie Onasch also reviewed larger blocks of chapters, with Bill's reviews of the rock mechanics and joints chapters coming at the end. I am very grateful for all of their efforts. Several others who generously contributed photographs, unpublished line drawings, or data are acknowledged where these appear in the text.

Gina M. Keeling and Nancy L. Meadows assisted with preparation and revision of chapters. Gina Keeling also edited and proofed the chapters in various stages, obtained permissions, compiled the references cited, glossary, and indexes, and provided an extraordinary amount of order in an otherwise chaotic process. Donald G. McClanahan, graphics specialist and artist, prepared most of the diagrams and photographs for the book. This second edition would not have been completed without their assistance. Graduate students Donald J. Geddes and Mark W. Carter were very helpful at critical times by proofing the final text (Don) and drafting a number of diagrams on the computer (both). Diana S. Hatcher also helped by drafting a number of diagrams on the computer, in addition to her other duties at home. Freelance editor and long-time friend Jean Thyfault made my writing more clear and understandable. Finally, I very much appreciate the professional manner in which the process of completing this second edition has been conducted by the Macmillan and Spectrum staff, principally Aliza Greenblatt and Kelly Ricci, and orchestrated by Senior Editor Robert A. McConnin.

All of those acknowledged above contributed significantly to the second edition of *Structural Geology: Principles, Concepts, and Problems,* and I am very grateful for their help at various stages. I remain culpable, however, for all errors that may have found their way into print.

Bob Hatcher

PREFACE TO THE FIRST EDITION

STRUCTURAL GEOLOGY HAS PROBABLY CHANGED MORE IN last 15 years than in the previous 50 because we know more about the crust and the processes that deform it and because of the application of techniques developed both in structural geology and in the allied disciplines of geophysics and materials science. Also, more of the Earth's crust—and other planets—has been studied today by structural geologists than ever before. We have recently benefited, through the work of John Ramsay and a number of other geologists, from development and application of the techniques of fabric analysis that were invented in the early part of this century. We now routinely measure strain in deformed rocks, something done by a relative few a decade or so ago. We also today measure structures, such as kinematic indicators, that were not measured in the 1960s, to provide greater insight into the deformation and motion of rock masses. Moreover, most structural geologists today are familiar with seismic reflection imaging of the crust because of the resolution of structure by this technique, and many also routinely use paleomagnetism and other geophysical methods. Modern computer technology has made possible rapid development and widespread use of seismic techniques in the petroleum industry and more recent application in academic research. Few junior–senior level texts that were published in the 1970s and early 1980s considered all of these aspects of structural geology. Structural geology has also become more quantitative during the 1970s and 1980s. A goal of this book is to bridge the gap between older, less quantitative, texts and newer, but more advanced, mathematical structural geology books. An attempt is made herein to present concepts and discuss processes—many of which have a rigorous mathematical basis—with minimal use of advanced mathematics: mathematics used is mostly algebra and trigonometry. Use of higher mathematics is restricted to two or three places—where it is needed, and where understanding may be enhanced by incorporating it.

TO THE USER OF THIS BOOK

Structural Geology: Principles, Concepts, and Problems is intended to provide a one-semester junior–senior level course in modern structural geology and to present balanced coverage of the entire subject. Each chapter is designed to contain more material than can be covered during the allotted class time and may thus serve partly as a resource for material presented from a different point of view than the instructor's.

The twenty chapters are arranged in six parts. The introductory part is intended to provide background and review largely nonstructural, but closely related, topics. The "Rock Mechanics" part investigates the concepts of stress, strain, deformation mechanisms, and methods of strain measurement that form the basis for study of all the structures discussed in the remainder of the book. Several reviewers suggested that I introduce strain before stress, for good reason: structural geology is largely a study of strain—the structures we see clearly manifest strain features, with only rare vestiges of the stresses that produced them. But, because more structures and rock-mechanics experiments provide evidence suggesting stress produces strain, and therefore indicate a logical progression, I chose to introduce stress first. The chapters dealing with strain, and related topics (Chapters 4 through 7), may be taught independently of Chapter 3 on stress.

The part "Fractures and Faults" follows "Rock Mechanics" and stands largely independent of the later parts "Folds and Folding," and "Fabrics, Structural Analysis, and Geophysics." Consequently, any one can be taught independent of the others in the order preferred by the instructor. My own preference is to consider faults first, hence the placement of this part.

The last chapter—"Geophysical Methods"—is intended to introduce geophysical techniques useful in structural geology, not provide a course in geophysics. The techniques introduced here, principally seismic reflection and refraction, Earth magnetism, and gravity, have become important tools for understanding crustal structure because of the large amount of data available today. Several examples of structures imaged by seismic reflection and other geophysical techniques are scattered throughout the book, so I believe incorporation of this chapter is essential in an elementary structural geology text. This chapter also helps form a transition to study of large structures and tectonics.

Organization of individual chapters is intended to aid the learning process. Most terms and concepts are accompanied by illustrations. Problems are worked and examples are presented in most chapters to bring students closer to the subject. Words or phrases in ***boldface italic*** should, in my opinion, be understood and learned; those in *light italic* are terms that some instructors may not consider important enough to learn; other instructors may want to emphasize all italicized terms, bold and light. Most such terms are in the Glossary. The Glossary also contains a few terms that are not italicized in the text. Questions at the ends of chapters have a dual purpose: *review,* with answers readily accessible in the chapter, and *challenge,* to require additional thought about concepts presented in the chapter to reason about structural problems. Essays are intended to provide interesting nutshell-size sidelights to kindle further thought, provide different viewpoints, illustrate applications, or bring in controversial ideas. The interpretations in the Essays frequently represent my own, whereas attempts were made to balance all points of view in the body of the text. Appendices One and Two are intended to provide background information on fabric diagrams and structural measurements. Both will probably be covered in lab, but, because fabric diagrams and structural measurements are used throughout the text, some discussion is essential here. Appendix Three consists of a data set collected from an outstanding exposure of complexly deformed rocks at Woodall Shoals in the southern Blue Ridge. These data were plotted in a series of fabric diagrams and maps in the Essay for Chapter 19, and the structure there discussed further in the Chapter 16 Essay. Additional problems can be designed around this data set.

As a teacher, I want students to enjoy their structural geology course. The learning process need not be difficult or painful, but should be challenging, and can be approached as a game; the rules of the game should be spelled out by the instructor at the beginning of the course. The degree to which an instructor becomes involved in the course from the beginning largely determines the quality of the course and how much the students derive from it. The text chosen helps to determine the level at which the course is taught and the kinds of material to be covered. The balanced coverage in this text is intended to help with the involvement of both students and instructors in structural geology. I hope this book will kindle the interests of some of the students who use it so they will choose to become structural geologists.

ACKNOWLEDGMENTS

This text has been extensively reviewed at several stages of development at either the request of the publisher, or colleague reviews solicited by myself. Reviewers of the text at different stages include John Anderson (Kent State University), Tom Anderson (University of Pittsburgh), Ed Beutner (Franklin and Marshall College), Enid Bitner (Auburn University), Wally Bothner (University of New Hampshire), Donald Davidson (Northern Illinois University), Ian Duncan (Southern Methodist University), Jeremy Dunning (Indiana University), Terry Engelder (Pennsylvania State University), Eric Erslev (Colorado State University), Peter Geiser (University of Connecticut), Gary Girty (San Diego State University), Art Goldstein (Colgate University), Brann Johnson (Texas A&M University), Roy Kligfield (University of Colorado), Elizabeth Miller (Stanford University), Charles Onasch (Bowling Green State University), John Palmquist (Lawrence University), Donal Ragan (Arizona State University), Arthur Reymer (North Carolina State University), Jim Sears (University of Montana) Bruce Smith (Bowling Green State University), James Snook (Eastern Washington University), Rolfe Stanley (University of Vermont), John Tabor (University of Tennessee), Tom Tharp (Purdue University), Bill Travers (Cornell University), Daniel Tucker (University of Southwestern Louisiana), Brian Wernicke (Harvard University), Ian Williams (University of Wisconsin, River Falls), and Nick Woodward (University of Tennessee). Art Goldstein and Charlie Onasch did yeoman service by reviewing the entire text twice and exerted a very positive influence on the final version. Graduate students John Costello, Tim Davis, Teunis Heyn, Robert Hooper, Janet Hopson, Peter Lemiszki, and Beth McClellan reviewed several chapters, provided thin sections and hand samples of structures, and helped ferret out needed references. John Hanchar (Vanderbilt University) helped with photography of shattercone specimens. The efforts of all are very much appreciated.

Several colleagues and friends who provided key information, hand samples, interesting quotes, guidance to localities exposing spectacular structures, and needed conversation at critical times include Dick Campbell, Avery Drake, Bill Dunne, Steve Edelman, Hu Gabrielse, David Gee, André Michard, Bill Muehlberger, Bob Neuman, Alain Pique, Donald Ramsay, Brian Sturt, John Tabor, Bill Thomas, Rudolf Trümpy, Don Turcotte, George Viele, Rick Williams, Dave Wiltschko, and Nick Woodward. They might think their contributions small, but they improved the book, so these contributions cannot be small. Others who generously provided photos and diagrams that were used are acknowledged in figure captions.

Lynne E. Gaskin, Nancy L. Phalen, Karen M. Keener, and Nancy L. Meadows provided secretarial help by assisting with preparation and revision of chapters. Nancy Meadows helped additionally by editing some of the final chapters, obtaining permissions, and preparing the references cited, glossary, and indexes. Donald G. McClanahan, graphics specialist and artist, was also very helpful in skillfully preparing many of the diagrams and photographs for the book. This text would not have been completed without their assistance.

Free-lance editors Wendell Cochran and Jean Simmons Brown made my writing more clear and understandable. I learned a great deal from them about writing, yet remain culpable for all errors of scientific content or writing style. I appreciate the efforts of Merrill editorial, art, and production staff, principally Linda Bayma, Lorraine Woost, and Bruce Johnson, who paid close attention to detail and maintained high quality standards. I am grateful to the Merrill College Division Editors for their patience, their confidence in the project, and for shepherding it to completion.

Finally, I thank my wife Diana and daughters Melinda and Laura for their support while this and several other large projects were underway at the same time.

Bob Hatcher

CONTENTS

PART ONE

Introduction

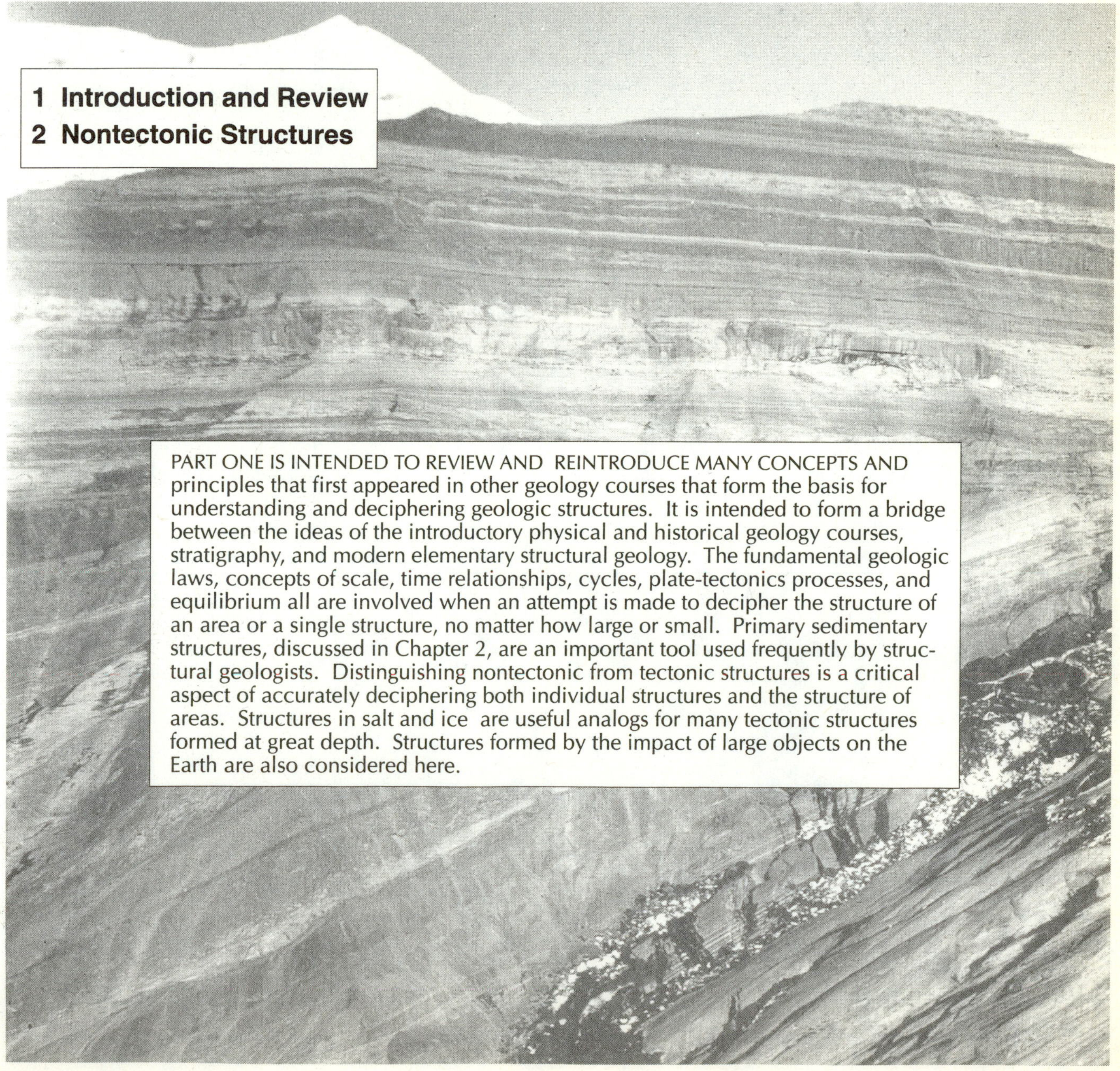

PART ONE IS INTENDED TO REVIEW AND REINTRODUCE MANY CONCEPTS AND principles that first appeared in other geology courses that form the basis for understanding and deciphering geologic structures. It is intended to form a bridge between the ideas of the introductory physical and historical geology courses, stratigraphy, and modern elementary structural geology. The fundamental geologic laws, concepts of scale, time relationships, cycles, plate-tectonics processes, and equilibrium all are involved when an attempt is made to decipher the structure of an area or a single structure, no matter how large or small. Primary sedimentary structures, discussed in Chapter 2, are an important tool used frequently by structural geologists. Distinguishing nontectonic from tectonic structures is a critical aspect of accurately deciphering both individual structures and the structure of areas. Structures in salt and ice are useful analogs for many tectonic structures formed at great depth. Structures formed by the impact of large objects on the Earth are also considered here.

1

Introduction and Review

A stone, when it is examined, will be found a mountain in miniature. The fineness of Nature's work is so great, that, into a single block, a foot or two in diameter, she can compress as many changes of form and structure, on a small scale, as she needs for her mountains on a large one; and, taking moss for forests, and grains of crystal for craggs, the surface of a stone, in by far the plurality of instances, is more interesting than the surface of an ordinary hill; more fascinating in form and incomparably richer in colour—the last quality being most noble in stones of good birth (that is to say, fallen from the crystalline mountain ranges).

JOHN RUSKIN, 1873, *Modern Painters*

FOR CENTURIES WE HAVE BEEN FASCINATED BY THE SHAPES OF continents and ocean basins, the linearity of mountain chains, the distribution of volcanoes, and the motions along large faults that produce earthquakes. We know today that most such features are the products of deep-seated processes that have been operating since the Precambrian to shape both past and present configurations of tectonic plates on the Earth (Figure 1–1). We are constantly reminded by the effects of earthquake and volcanic activity that the Earth is a dynamic planet, indicating the plates are driven by awesome forces. The lives of most of the Earth's population are influenced every day by tectonic activity; unfortunately, many are threatened by the potential for earthquakes and volcanic eruptions. Aside from the imminent danger and practical need to comprehend the processes that produce earthquakes and volcanoes, most geologists feel a basic scientific urge to understand these processes. Structural geologists are concerned with why parts of the Earth have been bent into smoothly curved shapes—***folded***—but others have been broken—***faulted***. They also deal with structures,

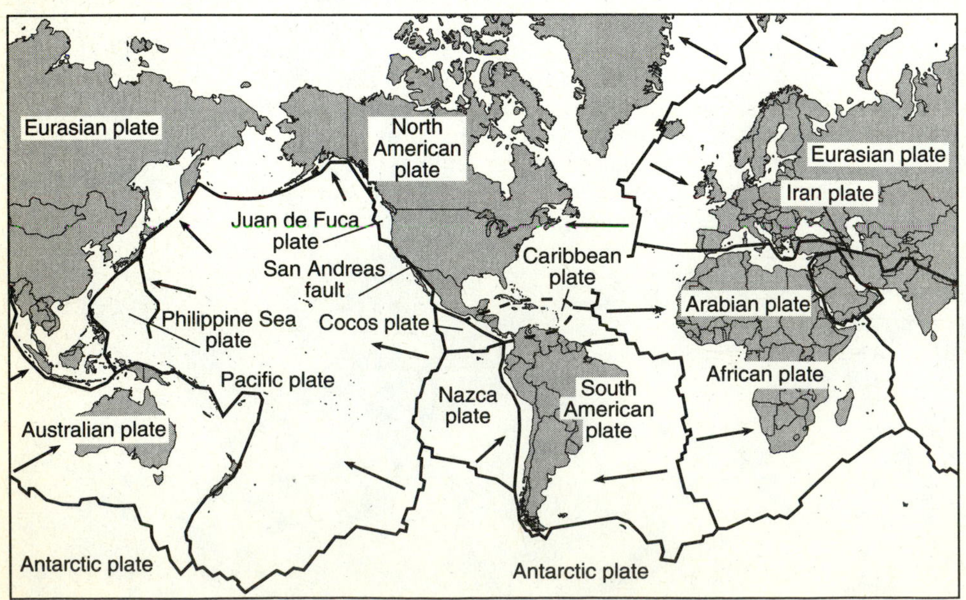

FIGURE 1–1
Distribution of tectonic plates on the Earth's surface. Arrows indicate principal directions of plate motion. Solid lines indicate plate boundaries.

such as faults and folds, on all scales and how they are related. Ruskin's statement about rocks illustrates how closely geologic processes are interwoven, whether we are dealing with parts of entire mountain ranges or structures we can observe in a hand specimen.

Tectonic structures are produced in rocks in response to stresses generated, for the most part, by plate motion within the Earth, and include different kinds of faults and folds, along with other structures. They make up the tectonic framework of the Earth. The kinds of structures that form in different parts of the crust are determined by (1) prevailing temperature and pressure, (2) composition, (3) layering, (4) contrast in properties with direction between and within individual layers *(**anisotropy**)* or the lack of contrast *(**isotropy**)*, and (5) amount and character of fluids within the rock mass. How rapidly the mass is deformed and the orientations of stresses applied to it also influence the kinds of structures produced. These factors determine whether deformation will be continuous *(**ductile deformation**)* or discontinuous *(**brittle deformation**)*, producing a great variety of structures both within and at plate boundaries in the rocks of the Earth and on other planets (Figure 1–2). *Continuous deformation* produces certain kinds of folds, ductile faults, cleavages, and foliations; *discontinuous deformation* produces other kinds of folds, brittle faults, and joints. Structures may also form as products of nontectonic processes, such as extraterrestrial impacts, landslides, and other structures produced by gravitational forces. We also recognize that many tectonic processes would not operate without a gravity component. It is useful to distinguish between tectonic and nontectonic structures (Chapter 2), because some nontectonic structures closely resemble—even mimic—structures formed by tectonic processes. Our main purposes here are to learn to recognize the various structures that are present in rocks and to understand how they form.

Structural geology deals with the origin, geometry, and kinematics of formation of structures. Structural geology is similar to architecture in that both require an ability to visualize objects in three dimensions (Figure 1–3). A close parallel exists between the shapes of geologic structures as they change through time and the physical conditions that formed them. In particular, the contrast in shape and type between structures that form near the Earth's surface and those that form at great depth under the weight of overlying rocks and at high temperature indicates profound differences in physical conditions. An appreciation of structural geometry thus provides greater insight into the origins of structures.

This discussion of structural geology may be your first encounter with visualizing objects in three dimensions. You might begin by attempting to think about familiar objects such as your car, house, or room, and then move on to less familiar tectonic structures. Keep in mind that most of us also had difficulty with this at the beginning, but we learned to visualize objects in three dimensions through practice. Much of our knowledge of geologic structures is derived from observing and attempting to understand structures in the field; thus, one of our goals here is to improve our abilities to recognize, describe, measure, and interpret both subtle and obvious structural features in rocks. Also, a better understanding of physical and chemical principles and the ability to use mathematics and computers are needed today to bridge the gaps between theoretical, field, and laboratory studies.

(a)

(b)

FIGURE 1–2
Continuous (ductile) and discontinuous (brittle) structures in rocks. (a) Folds produced by ductile flow in Precambrian gneiss near Central City, Colorado. Layers range from 5 mm to 10 cm thick. (b) Brittle deformation at the same locality as (a) produced several sets of joints. Older ductile structures are outlined by color banding. Scale indicated by small trees in foreground 3 to 5 m tall. (RDH photos.)

The link between field and laboratory studies is both essential and supportive, for structural geology is divisible into subdisciplines, most of which overlap. For example, laboratory studies determining fluid pressure that facilitates movement on faults are supported by field observations of evidence that fluid was present when a fault was active.

Rock mechanics is the application of physics to the study of rock materials. It deals with rock properties and the relationships between forces and the resulting structures, as well as with the study of structures produced in the laboratory in an attempt to duplicate natural structures (Figure 1–4). In the laboratory, we can simulate the higher temperatures and pressures thought to exist at great depths. Alternatively, very weak materials such as salt, gelatin, clay, putty, and paraffin, which behave like rocks being deformed at higher temperatures, may be used to produce experimental structures at room temperature. A severe disadvantage of laboratory experiments in rock mechanics, however, is that they cannot be run over geologic time—thousands to millions of years. They must be run on common rocks and minerals at temperatures and pressures far above those normally occurring in nature so that deformation rates will occur rapidly enough to be observed in a reasonable time. Artificial or natural materials deformed at reasonable rates that simulate the behavior of rocks must be scaled to approximate natural processes.

The study of ***field relations*** is an exceptionally important aspect of structural geology because it provides constraints for formulating models of regional deformation. In structural geology, we try to understand how small structures form and how they are related to larger structures and, ultimately, to crustal deformation. A geologist undertaking structural studies will (1) make accurate geologic maps of the geometry of structures, (2) measure orientations of small structures to provide information about the shapes and relative positions of larger structures in the field, (3) study the sequence of development and superposition of different kinds of structures in an area to determine the sequence of conditions of deformation, and (4) try to apply rock-mechanics data to relate structures to stresses that were present in the Earth at the times of deformation. Quite often, a structural geologist will also compare structures in one area with those elsewhere that may have been formed by similar mechanisms.

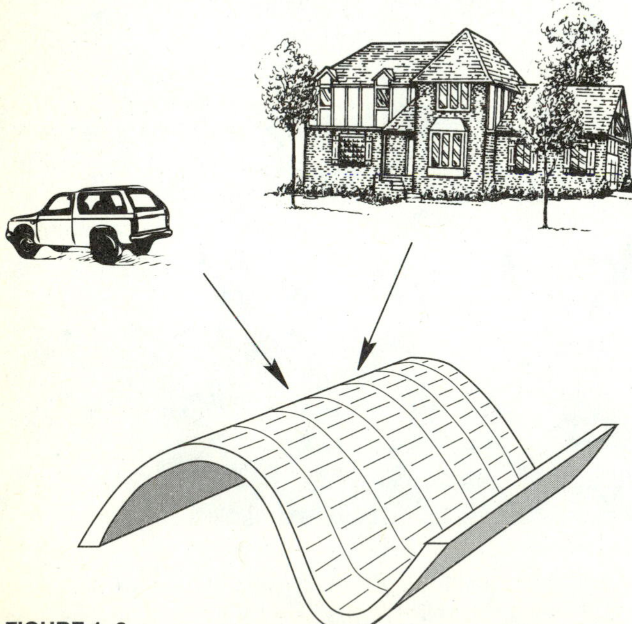

FIGURE 1–3
Visualizing objects in three dimensions. Start with a familiar object such as your house, car, or a familiar building, and then begin to think about geologic structures.

Tectonics and ***regional structural geology*** involve larger features. Studies of mountain ranges, parts of entire continents, trenches and island arcs, oceanic ridges, entire continents and ocean basins, and their relationships to stresses and tectonic plates are involved in these subdisciplines. ***Plate tectonics*** deals specifically with plate generation, motion, and interactions.

Separation of tectonics from regional structural geology is difficult. Regional structural geology is more concerned with continental structures and uses more data from detailed studies of small structures to reconstruct the deformational history and tectonics of a large region. Moreover, geophysical data and information derived from other disciplines of geology must be integrated with structural data for use in regional structural geology and tectonics. Use of geophysical data in structural geology is more common now because technology has made available more data of higher quality, especially gravity, magnetic, and seismic reflection data (Chapter 21).

It is easy to see that the many subdivisions of structural geology are related to other disciplines in geology as well as to the other sciences. Direct applications are made from physics to the origin of geologic structures. Isotopic data are frequently useful in working out the absolute time of formation of structures, and geochemical data may help to determine mobility of elements or isotopes during deformation. The chemical composition of highly deformed rocks may indicate the nature of the original material *(protolith)* and the environment of formation before deformation—a key factor in tectonics.

Structural geology can be applied to other fields. It is readily applied to engineering problems that involve bridges, dams, and power plants where large excavations are necessary, as well as highways where excavations extend for long distances. Studies of geologic structures beneath buildings, dams, and highway cuts are of great importance because of the potential for renewed motion along faults and other fractures, as

FIGURE 1–4
Experimental structures made in a centrifuge from viscous materials of different densities and fluid properties. Compare the shapes of these structures at this scale with those in Figures 14–19, 14–31, 14–32, 15–2, 15–11, 15–24, 15E–1, and 16–11(a). (From *Tectonophysics*, v. 19, H. Ramberg and H. Sjöström, p. 105–132, Fig. 15, © 1973, with kind permission from Elsevier Science, Ltd., Kidlington, United Kingdom.)

well as concern for the stability of slopes and geologic materials. Many power plants (both nuclear and conventional), large buildings, airports, and dams are under construction in different parts of the world. Siting these structures within active fault zones is not desirable, but sometimes it is impossible to build them in tectonically quiet areas. Therefore, geologists and engineers must work together from the design stage through construction to work out which structures are still active and might affect engineering works (Figure 1–5), as well as to minimize both cost and hazards.

Recognition of the importance of structural geology to environmental problems and land-use planning, such as earthquake hazard, waste isolation and disposal, and the controls of the distribution of ground water, provides additional applicability for this discipline. Documenting the antiquity or recent movement of faults is an important aspect that requires an understanding of structural geology. Location of sites for disposal of municipal, industrial, and radioactive waste requires application of structural and tectonic principles. Understanding the controls of large structures, such as folded layers of permeable and impermeable rocks that contain ground water, and small structures, such as fractures, on the distribution of ground water provides additional applications for this discipline.

Structural geology has long had a close working relationship with petroleum and mining geology. The geometric techniques of understanding and projecting fault surfaces, geologic contacts, and structures to depth have been used to great advantage by geologists and others who explore for valuable minerals. Similar applications of structural geology have been used for many decades in petroleum geology. The principles of tectonics have been applied to understanding larger trends and regional processes that control the concentration of mineral deposits and hydrocarbons.

The concept of ***scale*** is also of great importance in structural geology. Structures—such as geologic contacts and some foliations, faults, and folds—are commonly observed in the field in both hand specimens and at outcrop (or ***mesoscopic***) scale. Small structures that require magnification to be observed, such as many foliations and linear structures, are called ***microscopic***. Mountainside to map-scale structures of all kinds are called ***macroscopic*** structures. We must be constantly aware of the relationships between structures at all scales (Figure 1–6a). Scales and geometric perspectives of geologic cross sections must be maintained between the map from which the section is constructed and the section itself (Figure 1–6b).

Structures that occur as single features, for example, a fault or an isolated fold, are termed ***nonpenetrative structures***. These structures are not present on all scales, but others, such as slaty cleavage, foliations, and some folds, occur on any scale that we may choose for observation and are termed ***penetrative***. Some structures, such as joints (Chapter 8), may be penetrative on one scale but not on another. Joints are penetrative only on the macroscopic scale, rarely on the mesoscopic scale, and not on the microscopic scale.

FUNDAMENTAL CONCEPTS

The fundamental—almost simple—relationships to be discussed here provide us with the most powerful tools available to begin investigating the subject of structural geology. Without understanding them, we would be so severely handicapped that no technologically advanced equipment (for example, sophisticated computers, seismological equipment, or other analytical tools) could help solve our structural problems.

FIGURE 1–5
Complexly deformed metamorphic rocks exposed in the excavation for construction of Mica Dam, southern British Columbia. (RDH photo.)

Probably the most important foundation doctrine in geology is uniformitarianism. It was first stated by James Hutton, an eighteenth-century Scottish farmer and scientist. Because his writing style was obscure, his ideas did not become widely known until they were rewritten by John Playfair in the early nineteenth century. The ***doctrine of uniformitarianism*** states that *processes occurring today upon and within the Earth have probably gone on similarly in the past and will continue in the future;* stated more simply, *the present is the key to the past.* Hutton's conclusions were based on his observations along the coast of Scotland (Figure 1–7), where he could see waves wearing down rocks and producing pebbles that were further reduced to sand. He observed that sand bars and beaches were constantly being created and destroyed by storms, and were slowly rebuilt. Hutton also recognized that the sand in sandstone is the same as that moving about on a modern beach. He concluded that layers of sandstone turned on end were originally deposited horizontally and that an immensity of time must have elapsed since the sand grains were formed, became consolidated into rock layers, and were turned on end by crustal forces. His observations mark the beginning of modern geology—for the first time, it was recognized that a huge amount of time is both available and necessary to carry out geologic processes. Before Hutton and long afterward, the prevailing notion was that unknown catastrophic events were responsible for geologic processes and features. Uniformitarianism immediately led to conflict with religious dogma, resulting in numerous debates between scientists and theologians during the nineteenth century, particularly with the rise of other theories such as evolution.

Today, we recognize that most geologic processes require immense amounts of time. We have also realized that although the movement across a large fault may total many tens of kilometers, a large part of the motion may have occurred as relatively small displacements (perhaps nearly instantaneous movements producing earthquakes), not by continuous slippage through time. Study of active faults indicates that some segments move by continuous creep, but other segments undergo instantaneous catastrophic movement—earthquakes. Therefore, Hutton was probably correct when he recognized the immensity of time involved in geologic processes, and so uniformitarianism is the best means of thinking about geologic processes through time. *These long-term effects may, however, represent the sum of many instantaneous and even catastrophic events randomly distributed over the continuum of geologic time.*

The doctrine of uniformitarianism may break down if certain aspects of Precambrian geology are considered. The iron formations in the Lake Superior region are partly the result of a different atmospheric composition, and some geologists say that the smaller continental nuclei that existed during the Archean may show that the rules of plate tectonics may not have held during that early part of Earth history. Contrasts in the nature of the crust formed during Archean and Proterozoic time may further indicate fundamental differences

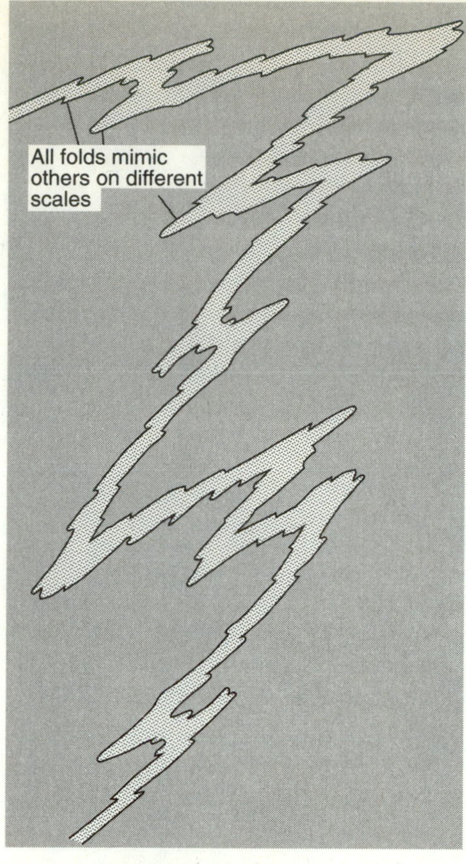

(a)

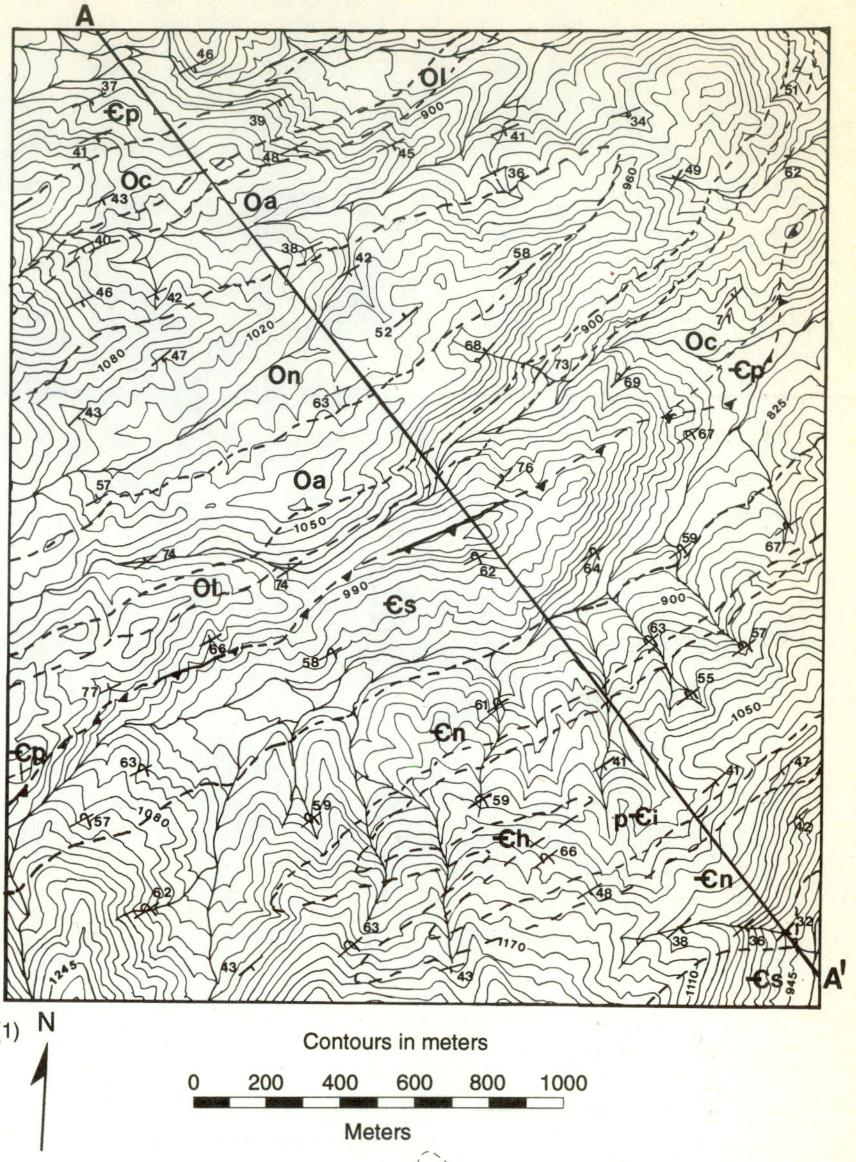

(b) (1)

N

Contours in meters

0 200 400 600 800 1000

Meters

FIGURE 1–6
(a) Relationships between small and large folds in the same structure. (b) Hypothetical geologic map (1) and cross section (2). Note that constructing an accurate cross section requires close attention to scale, strike and dip of layering (Appendix 2), and position of geologic contacts on the topographic surface. Vertical scale on the cross section is in meters; there is no vertical exaggeration.

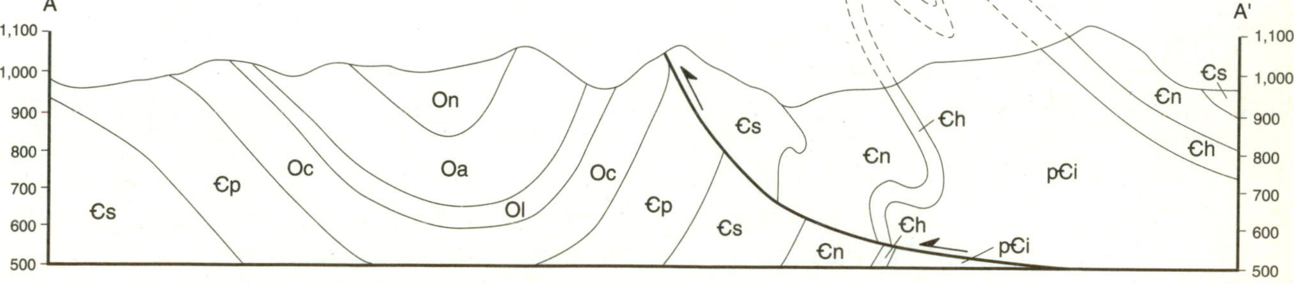

(b) (2)

in processes operating before and after about 3 billion years ago. Archean processes resulted in a crust dominated by greenstones intruded by large granitic batholiths, but Proterozoic crust involved appearance of the first platform sediments and cratonic basins, as well as the apparent addition of new crust around old nuclei, a process that continues today.

The ***law of superposition*** is another cornerstone of geologic thought. It states that *within a layered sequence, commonly sedimentary or volcanic rocks, the oldest rocks will occur at the base of the sequence and successively younger rocks will occur toward the top, unless the sequence has been inverted through tectonic activity.* Geology could not function as a science, and

FIGURE 1–7
Rocks along the coast of Scotland, like those in this scene near Loch Eribol in the footwall of the Moine thrust, enabled James Hutton to formulate the doctrine of uniformitarianism. (RDH photo.)

the understanding of many processes would be greatly impaired without this law and the doctrine of uniformitarianism. The first statement of the law of superposition was made during the seventeenth century by Nicolas Steno (Niels Stensen), a Danish physician with interests in geology. The law is of great importance in structural geology because it is necessary to determine whether the stacking order in a sequence is upright or has been tectonically inverted. The sequence may have been tilted, completely overturned, or repeated by folding or faulting (Figure 1–8a). Superposition is therefore an inviolate second principle in the study of structural geology.

The ***law of original horizontality*** is another fundamental geologic law stating that *bedding planes within sediments or sedimentary rocks form in a horizontal to nearly horizontal orientation at the time of deposition.* This law is fundamental in structural geology because bedding is the common initial reference frame (Figure 1–8b).

Another law that goes hand-in-hand with working out the structural history of an area is the ***law of cross-cutting relationships***, applied as either the ***law of structural relationships*** or the ***law of igneous cross-cutting relationships*** (Figure 1–8c). Both state virtually the same thing, *that an igneous body or a structure—that is, a fold or fault—must be younger than the rocks it cuts through.* In other words, the rocks that contain a structure or that form the host for an igneous body must have been there before the structure or igneous body formed. These laws provide a basis for placing structures in a time context. Truncation of an earlier structure or igneous body by a later structure, an unconformity (to be discussed below), or a younger pluton of known age provides a minimum age for the earlier features. Bracketing structures and igneous bodies in time is an essential part of understanding the geologic history of an area.

Initially, the ***law of faunal succession*** may seem far from useful to structural geologists. It states that *the fossil organisms should be systematically changed, possibly more advanced toward the top of a sequence.* This provides the basis for assignment of relative age to fossiliferous sequences and permits determination of whether a sequence is upright or tectonically overturned. It is therefore of major importance in unraveling the structural history of an area where the rocks are fossiliferous. Fossils have also played a key role in the

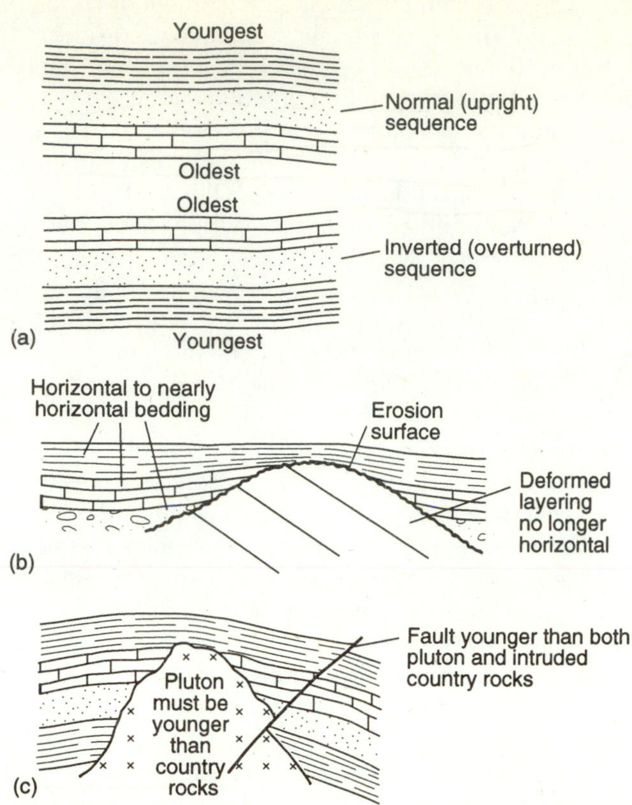

FIGURE 1–8
(a) Laws of superposition, (b) original horizontality, and (c) cross-cutting relationships.

identification of exotic terranes (see the next section, Plate Tectonics) because particular groups of fossil organisms were restricted to a particular region. Examples are the fusulinid fauna of the late Paleozoic Tethys and the different Cambrian trilobite faunas of North America and Europe.

The principle of ***multiple working hypotheses*** is a useful tool in structural geology. It enables us to formulate more than one possible explanation of the same data, to evaluate each, and to select the most likely hypothesis. Suppose you are working in a field area that lacks a critical exposure needed to correctly interpret a particular contact of an igneous body with the overlying sedimentary section (Figure 1–9). You have hypothesized that the contact can be (1) an intrusive contact, (2) a fault, or (3) an unconformity. Each hypothesis may be equally valid. You begin by sorting through your previous observations and ask: With regard to (1), have you observed baking or other alteration of the sedimentary rocks near the contact? With regard to (2), have you observed crushed rocks or other evidence of faulting near the contact? With regard to (3), have you observed clasts of the igneous rocks incorporated into the base of the overlying sedimentary sequence? One or two exposures of baked

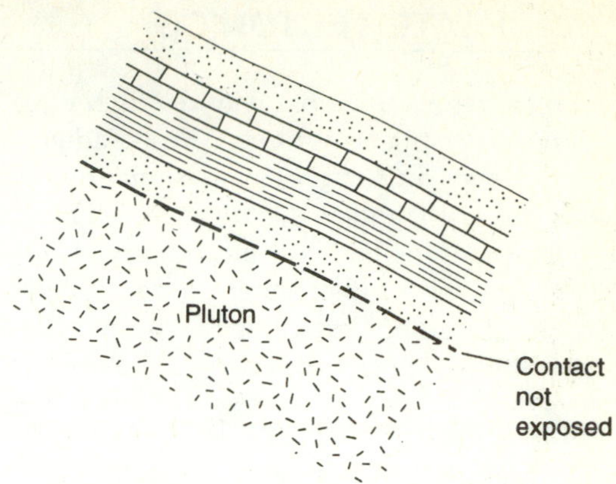

FIGURE 1–9
Use of the principle of multiple working hypotheses to interpret the contact between a pluton and an host rocks. Without the critical exposure, the contact could be an intrusive contact, an unconformity, or a fault.

rocks along the contact eliminates the second and third possibilities, but before discovery of the critical pieces of data, all working hypotheses were equally valid.

There is also much value in the ***outrageous hypothesis*** (W. M. Davis, 1926; Wise, 1963) as an alternative working hypothesis because it provides a focus for critical pieces of data toward a solution to the problem at hand. An outrageous hypothesis appears to be an impossible solution to the problem from the moment it is formulated. Considering the data, it may gain the position of a credible, alternative working hypothesis, or it may be quickly abandoned as other likely working hypotheses are formulated.

Another relationship of great importance in structural geology is ***Pumpelly's rule.*** It was first applied to geology by Raphael Pumpelly in the early twentieth century, and it states that *small structures are a key to and mimic the styles and orientations of larger structures of the same generation within a particular area.* (The quotation by Ruskin at the beginning of this chapter is an earlier statement of the rule.) Pumpelly's rule holds if all structures referred to were formed at the same time by the same stresses in rocks of similar properties and were deformed the same on all scales. Pumpelly's rule provides a basis for presuming that small and large structures of the same generation may be shown to be related within the same area. Because we are not always able to observe structures on all scales in an area, Pumpelly's rule allows us to assume similarity from hand specimen to map scale of structures formed at the same time (Figure 1–6). Pumpelly's rule may also be considered a statement of the *principle of self-similarity.* It thus enables us to visualize the configuration of a larger structure without ever directly observing the entire structure itself.

PLATE TECTONICS

Plate tectonics is the framework within which we assume all tectonic structures form. This paradigm is as fundamental to the earth sciences as atomic theory is to physics and chemistry and as evolution is to biology. Early formulation of the theory is attributed to Harry Hess, who during the 1930s conceived the *tectogene concept* of the subsiding crumpling crust driven by mantle convection. He later discovered that the sea floor is spreading apart at the mid-ocean ridges (Hess, 1962). Others, including Robert S. Dietz, W. Jason Morgan, Dan P. McKenzie, Xavier Le Pichon, Fred Vine, L. W. Morely, J. Tuzo Wilson, and Drummond Matthews, were also early contributors to different aspects of the theory; however, plate tectonics was first published as a unified theory by Bryan Isacks, Jack Oliver, and Lynn Sykes (1968).

The present surface of the Earth is divisible into seven major plates and several smaller plates that contain all of the continents and oceans. New oceanic crust formed at the oceanic ridges ultimately is consumed by subduction in the trenches (Figure 1–10). The thickness of plates corresponds to that of the ***lithosphere***, which is about 100 km, and includes all of the crust and part of the upper mantle (Figure 1–10). Plate generation and consumption are thought to be driven by convection in the mantle. Plate motion may be described using an Eulerian theorem that represents the motion of plates on a sphere in which displacement on the surface increases vectorially away from the spreading (rotation) axis. Angular displacement of a plate involves rotation about a line passing through the center of the Earth (Figure 1–11a), and rotation of a plate about the spreading axis may be expressed by the angular velocity (ω) on the sphere. Velocity increases away from the pole of the spreading axis; angular velocity remains constant. Consequently, along a spreading center (at a mid-ocean ridge), the spreading rate should increase to a maximum velocity 90° from the pole (at the "equator"). This variation in spreading rate has been demonstrated along the Mid-Atlantic Ridge (Le Pichon, 1968; Morgan, 1968). Each plate has an angular velocity on the sphere (determined by the absolute motion and position relative to the pole) with respect to other plates. Differences in displacement between plates are balanced by ***transform faults*** (Figure 1–11b). Plate boundaries where three plates meet are called ***triple junctions*** (McKenzie and Morgan, 1969). The change in locations of triple junctions may be predicted by spherical geometry and may be of several kinds, depending on whether they connect ridges, arcs, trenches, or combinations (Figure 1–11c).

A corollary to plate-tectonics theory is the generation of mountain chains as a result of either subduction (Cordilleran mountain chains—the Andes, the North American Cordillera) or continent-continent or continent-arc collision (collisional mountain chains—the Alps, the Himalayas), as described by John Dewey and Jack Bird (1970). Generation of mountain chains is much more complex than either of those latter mechanisms, for most collisional orogenic belts also had an earlier history of subduction. Before the landmark paper by Dewey and Bird, J. Tuzo Wilson (1966) had suggested that a proto-Atlantic Ocean had closed at the end of the Paleozoic, producing the Appalachian–Variscan mountain chain of North America and Europe, and then had reopened and produced the present Atlantic. This closing and opening of oceans has become known as the ***Wilson cycle.***

Another corollary, probably first recognized by Emile Argand (1925) and later by Wilson (1968), is that of ***accretionary tectonics,*** whereby *suspect* and *exotic terranes* of less than continental proportions are moved by plate motion to collision with each other or with continents. These were originally called ***microcontinents,*** or ***microplates,*** by D. P. McKenzie (1970) and by John Dewey, Walter Pittman, William Ryan, and Jean Bonnin (1973) on the basis of configurations of earthquake epicenters and tectonic units in the Mediterranean region. A ***suspect terrane*** is a rock mass in which original position is questionable with respect to the adjacent terrane or continent to which it is presently attached. An ***exotic terrane*** bears no resemblance to the mass to which it is attached, and the source may be on the opposite side of a major ocean. Boundaries of suspect and exotic terranes with the masses to which they are attached are always tectonic. Overlap sequences, deformational and metamorphic overprints, and plutons that cross-cut accretionary boundaries provide evidence of "docking" and frequently the time of docking of terranes. Warren Hamilton's (1979) compilation of the geology of the

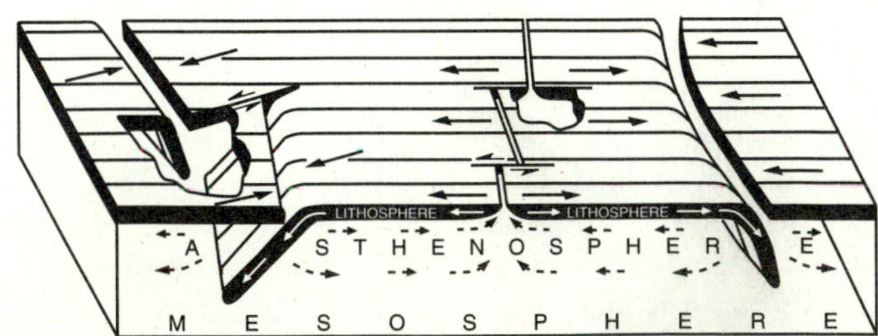

FIGURE 1–10
Generation of lithospheric plates at spreading centers and destruction at subduction zones. Differences in rate of motion or displacement between plates are taken up by transforms at ridges, trenches, and other boundaries. Arrows indicate direction of motion. (From B. Isacks, J. Oliver, and L. Sykes, Seismology and the new global tectonics: *Journal of Geophysical Research,* v. 78, p. 5855–5899, Fig. 1, © 1968 by the American Geophysical Union.)

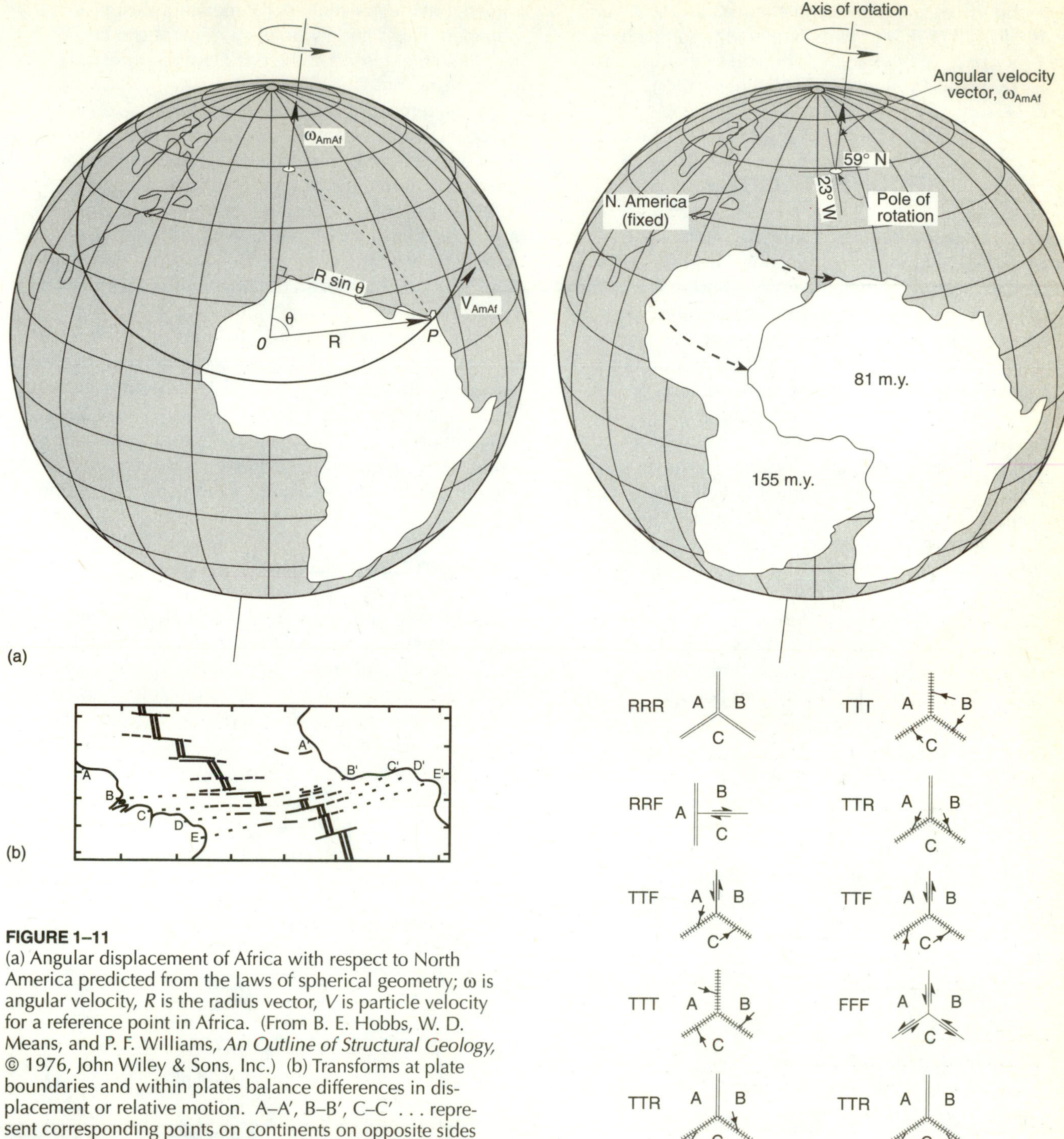

FIGURE 1–11
(a) Angular displacement of Africa with respect to North America predicted from the laws of spherical geometry; ω is angular velocity, *R* is the radius vector, *V* is particle velocity for a reference point in Africa. (From B. E. Hobbs, W. D. Means, and P. F. Williams, *An Outline of Structural Geology,* © 1976, John Wiley & Sons, Inc.) (b) Transforms at plate boundaries and within plates balance differences in displacement or relative motion. A–A′, B–B′, C–C′ . . . represent corresponding points on continents on opposite sides of the Atlantic. (From J. T. Wilson, reprinted by permission from *Nature,* v. 207, p. 343–347, © 1966, Macmillan Magazines, Ltd.) (c) Several kinds of triple junctions. (From D. P. McKenzie and W. J. Morgan, reprinted by permission from *Nature,* v. 224, p. 125–133, © 1969 Macmillan Magazines, Ltd.)

Indonesian region demonstrates that a complex of volcanic arcs, continental fragments, oceanic crust, and large continental blocks (such as Australia) are all in the initial stages of being accreted to Asia as Australia moves northward. Although the concept of accretionary tectonics has been applied to the plate-tectonic evolution of the Alps by Dewey and others (1973), it has been most fruitfully applied to the North American Cordillera by Peter Coney, David Jones, and James Monger (1980). It was later applied by Monger,

Raymond Price, and Dirk Templeman-Kluitt (1982) and David Howell (1985) to parts or all of the same chain and other parts of the Pacific Rim, and to the Appalachians by Harold Williams and Hatcher (1983).

GEOCHRONOLOGY

Radioactive decay is a process unaffected by heat, pressure, fluids, or chemical reactions. After a crystallized mineral cools sufficiently to prevent escape of the daughter products of radioactive decay, or the system is otherwise closed and the radioactive elements are locked into the rock, decay products are trapped in the lattices of minerals, and the age of the mineral or rock may be determined by several standard methods. The temperature below which a crystal lattice traps radioactive daughter products is the ***blocking*** or ***closing temperature***. The blocking temperature is different for each mineral and each daughter element because of the differences in crystal structures, as well as the size and physical state (solid, gas) of the daughter elements (Table 1–1).

Decay of an unstable parent radioactive element to a stable daughter element can be related to the number of atoms (N) present at the time the system closed, time (t), and a decay constant (λ) by

$$\frac{dN}{dt} = -\lambda N. \qquad \textbf{(1–1)}$$

The negative sign is needed because the number of parent atoms is decreasing. If this equation is treated as a definite integral between the initial (N_0) and final number of atoms (N) at times t_0 and t, the result is

$$\int_{N_0}^{N} \frac{dN}{N} = -\lambda \int_{t_0}^{t} dt,$$

and, because t_0 is zero, the integrated form is

$$\ln\frac{N}{N_0} = -\lambda t \text{ or } N = N_0 e^{-\lambda t}. \qquad \textbf{(1–2)}$$

Determination of radiometric ages of volcanic, plutonic, and metamorphic rocks, and of the detrital components of sedimentary rocks is an important aid in deciphering the structural history of an area, particularly if the rocks contain no fossils. The age of crystallization in plutonic and volcanic rocks is often determined, but the age highest T-P metamorphism is obtained with difficulty unless a mineral can be identified that closed at the peak temperature and pressure.

Refinement of sampling and analytical techniques in recent years has permitted reduction in the errors in most techniques for radiometric age determinations under optimum conditions to less than two percent. Rubidium-strontium ages are commonly reported with a possible error of ±2 million years if the age of the rock body is on the order of 300 to 500 million years. Similar results have been obtained with other techniques.

Radiometric ages can sometimes be determined for detrital minerals in sedimentary and metasedimentary rocks, thus providing an age of the source terrane from which the sediments were derived; this is particularly important in attempting to work out the nature and origin of accreted terranes. Next is a brief discussion of the principal methods used in radiometric age dating.

Uranium-Lead Method

Uranium-lead geochronology is probably the most reliable technique for rocks wherein ages exceed 10 million years. The most common way of applying the U-Pb method to date rocks is by analyzing zircon, monazite, and sphene separated from crushed rock samples. These minerals contain small quantities of the radioactive elements uranium and thorium. Neither parent nor daughter elements are easily released from the zircon lattice during deformation or metamorphism; consequently, it is possible to determine the ages of

TABLE 1–1
BLOCKING TEMPERATURES FOR DIFFERENT MINERALS AND SYSTEMS*

Mineral	System	Daughter	Blocking T °C
Zircon	U–Pb	$^{207,206}Pb$	>800
Monazite	U–Pb	$^{207,206}Pb$	700–725
Sphene	U–Pb	$^{207,206}Pb$	550–650
Garnet	U–Pb	$^{207,206}Pb$	>800
Rutile	U-Pb	$^{207,206}Pb$	400
Muscovite	Rb–Sr	^{87}Sr	
K–spar	Rb–Sr	^{87}Sr	
Biotite	Rb–Sr	^{87}Sr	300
Hornblende	K–Ar	^{40}Ar	480
Biotite	K–Ar	^{40}Ar	300
Muscovite	K–Ar	^{40}Ar	350

* Data compiled by Steven A. Goldberg, University of North Carolina-Chapel Hill.

many zircons by analysis of several uranium and thorium isotopes and the daughter products by using a mass spectrometer. These elements decay through a series of intermediate radioactive elements to stable daughter products:

$$^{238}U \rightarrow ^{206}Pb \text{ (half-life} = 4.5 \times 10^9 \text{ y)} \quad \textbf{(1–3)}$$

$$^{235}U \rightarrow ^{207}Pb \text{ (half-life} = 0.7 \times 10^9 \text{ y)} \quad \textbf{(1–4)}$$

$$^{232}Th \rightarrow ^{208}Pb \text{ (half-life} = 1.4 \times 10^9 \text{ y).} \quad \textbf{(1–5)}$$

Ratios of ^{206}Pb to ^{238}U, ^{207}Pb to ^{235}U, and ^{208}Pb to ^{232}Th are used to determine the age of the zircons by plotting them on a ***concordia curve*** (Figure 1–12). The curve shows the variation in ratios of parent and daughter elements through time. Zircons that plot on the curve are said to be *concordant;* those that do not are *discordant* and may have acted in part as open systems. A chord may be drawn through the points along which discordant zircons plot, yielding upper and lower intercepts on the concordia curve. The lower intercept yields a lesser age, indicating either *episodic lead loss,* commonly due to thermal resetting of the zircons, or ***inheritance*** (absorbing zircons into the magma from the country rock) of older zircons, in which case the lower intercept is the age of the rock. The upper intercept represents the original age of the crystallizing zircons, where there has been episodic lead loss, or the age of the inherited zircons.

The U-Pb method provides the most accurate ages for determining the time of crystallization of zircons in igneous and volcanic rocks. Sometimes metamorphic ages also may be obtained by this technique by analyzing sphene or monazite. Mafic rocks frequently contain baddeleyite (ZrO_2) instead of zircon, and yield the same or better quality results as zircons. Lower intercepts of discordant zircons are frequently metamorphic ages. Upper intercepts may either represent the original age of the body or show that the magma assimilated zircons from another rock mass. The latter are known as ***inherited zircons.*** Reset and inherited zircons (either detrital or assimilated) from an earlier tectonic cycle provide a major source of error in determining absolute ages of different rock materials.

The U-Pb method is particularly good because the blocking temperature for radiogenic isotopes to be locked into the zircon lattice is relatively high. Consequently, zircons are less subject to resetting by lower-temperature events, although other factors such as circulating water and radiation damage to the lattice by radioactive constituents may also reopen the system and cause loss of lead.

A technique invented by Thomas Krogh (1973) involves selective dissolution or abrasion of the outer layers of zoned zircon crystals to determine the composition and age of the outer layer. The remaining cores may then be dissolved and analyzed. An age greater than that of the younger rim may be obtained. Zoned zircons are very common. It is important in zircon geochronology not only to separate zoned from unzoned zircons but also to separate euhedral from anhedral or rounded crystals. Other variations in size, shape, or internal characteristics—such as found in clear or colored zircons—may indicate mixed populations that will yield different ages or, if not separated, discordant ages (Figure 1–13). This complexity creates analytical problems, but it may also provide important keys to the tectonic history of an area after the different zircon populations are separated.

Rubidium-Strontium Method

Rubidium-strontium age determination is also most applicable in rocks of ages that exceed 100 million years. The age is determined by analyzing ^{87}Rb, ^{87}Sr, and ^{86}Sr in a mass spectrometer. Ages of potassium minerals, primarily micas and feldspars, may be determined because the parent element Rb most commonly occurs in potassium minerals. Rb–Sr mineral ages are less reliable than ***whole-rock ages,*** where the Rb and Sr isotopes are determined in the entire rock. The decay scheme for this process is

$$^{87}Rb \rightarrow ^{87}Sr + \beta^- \text{ (half-life} = 48.8 \times 10^9 \text{ y).} \quad \textbf{(1–6)}$$

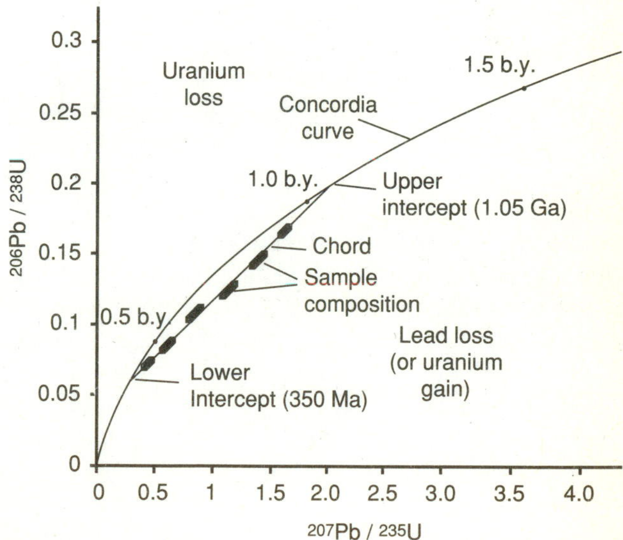

FIGURE 1–12
Concordia curve showing several hypothetical discordant zircon analyses with upper and lower intercepts at 1.05 billion years and 350 million years. Small elongate hexagons indicate margin of analytical error. Note that points that plot above the curve indicate uranium loss; points that plot below the curve indicate lead loss (or uranium gain). Relative size of hexagons indicates possible range of error.

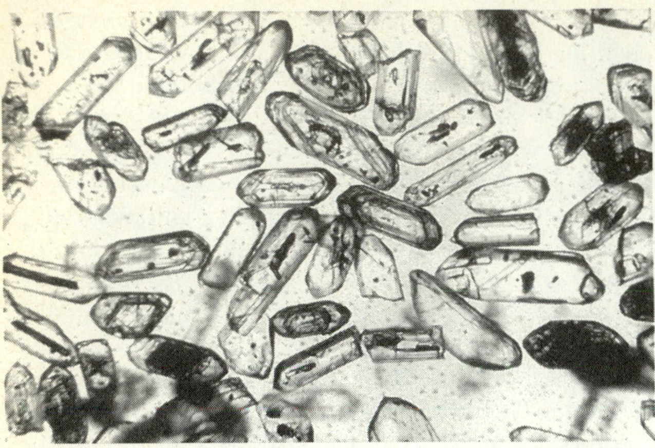

FIGURE 1–13
Zircon sample from the High Shoals Granite, North Carolina, illustrating various morphological types—euhedral, rounded, zoned. The largest zircons are approximately 1 mm long. Despite these variations in zircon morphology, the sample yielded a concordant age of 317 Ma. (From J. W. Horton, J. F. Sutter, T. W. Stern, and D. J. Milton, 1987, *American Journal of Science,* v. 287.)

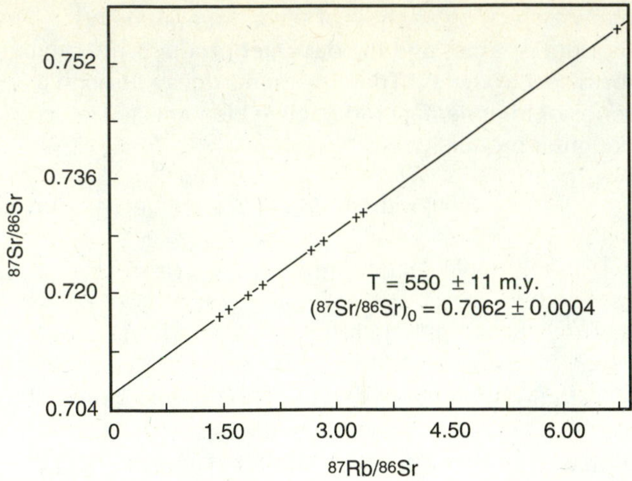

FIGURE 1–14
Rubidium-strontium isochron plot of a series of whole-rock analyses from a granitic batholith in south-central Libya yielding an age of 550 ± 11 Ma. The age of the rock is calculated from the slope of the isochron. (Isochron provided by P. D. Fullagar from *Earth and Planetary Science Letters,* v. 30, W. J. Pegram and others, p. 123–127, Fig. 3, © 1976, with kind permission from Elsevier Science, Ltd., Kidlington, United Kingdom.)

The extremely long half-life makes this method best suited to dating very old rocks. The age of the rock or mineral is obtained by plotting $^{87}Sr/^{86}Sr$ versus $^{87}Rb/^{86}Sr$ (Figure 1–14) and statistically fitting the points to the best straight line. The slope of the line yields the age of the rock or mineral. Points should plot very close to or on the line. If they do, the line is called an ***isochron*** and represents the age of the rock. If the points do not plot close to the statistically determined line, the line is called a ***mixing line,*** or *scatterchron,* and the age determined may represent an age between the real age of the rock or mineral and a more recent thermal or other event (such as hydrothermal activity) that has reopened the Rb-Sr system. The blocking temperature for the Rb-Sr whole-rock system is generally lower than that for U-Pb in zircons, and for minerals is about the same as for the K-Ar method (discussed in the next section). The principal advantage of the Rb-Sr system over K-Ar is that no daughter product that must be determined is gaseous, and so each has a better chance of being trapped in the rock, particularly if whole-rock analyses are made.

The isochron may be projected to the point where it intercepts the $^{87}Sr/^{86}Sr$ axis (Figure 1–14). This yields a value for $^{87}Sr/^{86}Sr$ that represents the value at the time the rock formed and is called the ***$^{87}Sr/^{86}Sr$ initial ratio***. Studies of isotopic compositions have shown that middle to upper continental crustal rocks have an initial ratio >0.706, but oceanic crust, Rb-depleted lower continental crust, and mantle rocks have initial ratios <0.706. This tool is very useful in determining the sources of magmas for plutons and volcanic rocks. It has also been used for reconstructing ancient continental margins by simply plotting the distributions of $^{87}Sr/^{86}Sr$ initial ratios on a map (Figure 1–15). If a clear-cut division appears between the values greater and less than 0.706, a line can be drawn that may represent the edge of the ancient continental crust.

The principal source of error for the Rb-Sr system is later metamorphism and hydrothermal alteration. Many rocks older than about 1,500 Ma have been sufficiently altered to produce errors of >200 m.y. in the Rb-Sr ages. Thus, zircon ages are more reliable.

Potassium-Argon Method

The potassium-argon method is based on the radioactive decay of ^{40}K into ^{40}Ar by a process called *branching decay*. Most of the ^{40}K actually decays into ^{40}Ca by release of a β^- particle, with a much smaller proportion decaying by electron capture into ^{40}Ar, but, fortunately, the small proportion of ^{40}Ar produced by this decay scheme is measurable because of the abundance of potassium in minerals. The decay process is

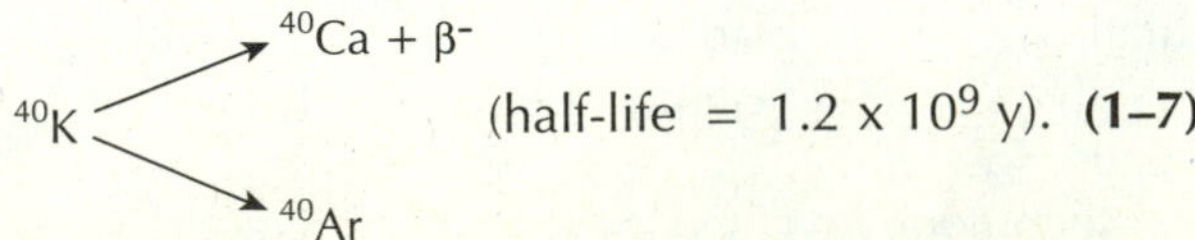

$$\text{(half-life} = 1.2 \times 10^9 \text{ y).} \quad \textbf{(1–7)}$$

This technique is generally used for determining the ages of minerals a million years old or more. The K-Ar method suffers from having a gaseous daughter product that must be trapped in crystals of the potassium-bearing mineral. Biotite, muscovite, and hornblende crystals retain argon better than do other minerals.

Potassium-argon age determinations provide information primarily on the time of uplift of a rock mass because the relatively low blocking temperatures of biotite (300° C), muscovite (350° C), and hornblende (480° C) are much below the crystallization temperatures of these minerals. This difference in blocking temperature can prove useful in determining the metamorphic age of fine-grained rocks (slates, phyllites) if the detrital and metamorphic micas can be separated mechanically. The ages of crystallization of unaltered Mesozoic and Tertiary volcanic rocks may also be determined using this method.

Argon-40/Argon-39 Method

A refinement of the potassium-argon technique employs samples irradiated with neutrons in a nuclear reactor. This converts ^{39}K to ^{39}Ar, and the ratio of ^{40}Ar to ^{39}Ar is determined in a mass spectrometer. This ratio is proportional to the $^{40}Ar/^{40}K$ ration in equation 1–7, and thus the age. Samples are heated incrementally from room temperature to 1000° C (or to fusion of the sample), and the amount of ^{39}Ar released is plotted versus calculated age (Figure 1–16). In the release spectrum for a sample with no excess argon, a plateau results for similar apparent ages for gas released with increasing temperature. This plateau represents the time at which the lattice became stable with respect to additional changes in temperature (below the blocking temperature) so that no Ar was allowed to escape. An age thus determined from the plateau is assumed to be the cooling age of the rock mass for the mineral being determined. An irregular *release spectrum* permits identification of samples containing excess argon and samples that have had an uneven history of opening and closing of the system to release or absorb argon. Studies using K-Ar and $^{40}Ar/^{39}Ar$ geochronology are useful for determining the time of uplift of an area and, under ideal circumstances, times of metamorphism and emplacement of large thrust sheets and other structures.

Samarium-Neodymium Method

A technique that gained extensive use in the 1980s is samarium-neodymium age dating. It involves determination of ^{147}Sm and ^{143}Nd in rocks with ages of several hundred million years or more. The half-life for the decay process is 106 x 10^9 y. The result is similar to that for the Rb-Sr method, where an isochron is plotted and the slope determines the age of the rock. The analytical technique for Sm-Nd requires even greater precision than that for determination of U, Th, and Pb isotopes, making age determinations using this method more tedious. The Sm-Nd method is best suited for determination of the ages of basaltic rocks, because sea water, which may come into contact with basaltic lava, contains almost no Nd and thus does not contaminate these rocks, and so the technique yields useful information on the age of the parent magma. On the other hand, Sr is abundant in sea water, and the Sr composition will vary widely with hydrothermal alteration, and so $^{87}Sr/^{86}Sr$ initial ratios are also commonly determined. The Sm-Nd technique is most commonly employed to provide a "model age" for the time of separation of a magma from the parent mantle and, in concert with Sr isotopes, data about sea water contamination and hydrothermal alteration of a mass of rocks.

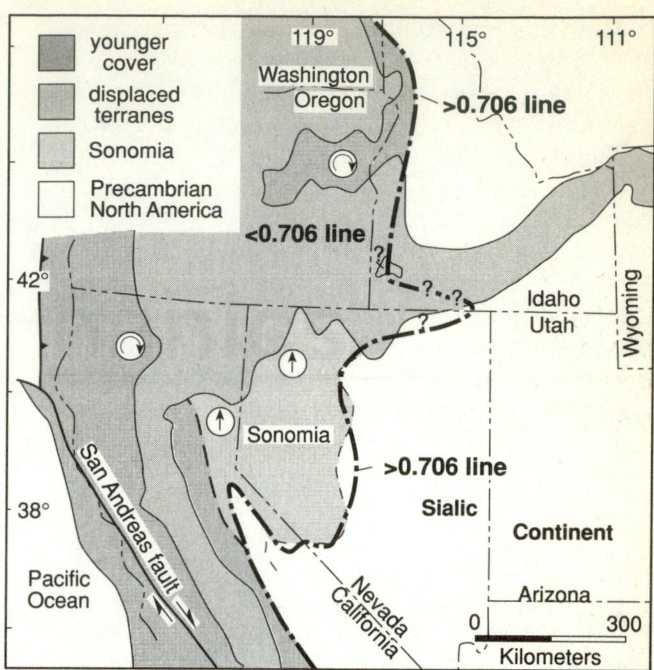

FIGURE 1–15
Map of the western United States showing the 0.706 line (heavy line). Plutons west of the line have initial ratios <0.706; those east of the line have ratios >0.706. Arrows indicate rotation sense determined using paleomagnetic data (Chapter 21) for different blocks. (Modified from R. C. Speed, AAPG *Memoir 34*, Fig. 2, © 1982. Reprinted by permission of American Association of Petroleum Geologists.)

EQUILIBRIUM

The Earth is a dynamic system. Energy from radioactive decay, the Earth's gravity field, and smaller components from our sun and moon drive processes within the Earth. Heat converted to work drives processes within the Earth that move plates, deform rocks of the lithosphere, and produce melts and metamorphism. Generally, some excess energy results that must be dissipated to restore a state of rest, or ***equilibrium***, to the part of the lithosphere where the excess occurs. The balance may be restored by volcanic eruption, breaking the crust along a fault, or some other process whereby heat may be converted into mechanical energy. The second law of thermodynamics predicts that a certain amount of energy is never available to do work and will be lost in any energy-consuming process, as long as the process is not 100 percent efficient. This amount of energy, called *entropy,* increases with time as more energy is expended. From a structural point of view, an increase in entropy is reflected in an increase in deformation relative to the undeformed state.

All processes in nature move toward a state of equilibrium. If heat is added to a system, the system will readjust to once again establish a state of equilibrium at the new temperature. The readjustment may be in the form of recrystallization, chemical reaction, change in deformation style from brittle to ductile (or vice versa), or some other process. Similar readjustments take place in response to changes in pressure. Striking a rock with a hammer produces elastic rebound if it is not struck hard enough to exceed the elastic strength of the rock (Chapter 6). If we strike the rock hard enough to break it, permanent deformation in the form of a fracture is produced, and any excess energy remaining is dissipated as a very small temperature increase in the vicinity of the fracture.

A large-scale attempt to restore equilibrium occurred in northern Europe and North America after melting of the Pleistocene ice sheets. When the ice sheets formed and loaded the continents with additional mass, the more rigid lithosphere sank to a lower level in the less rigid asthenosphere to attain a new equilibrium state. As the ice melted, the lithosphere was again forced out of equilibrium and accordingly began rebounding to restore a new equilibrium state. This process is continuing today. The greatest rebound occurs where the ice was thickest. The condition of balance, involving a state of equilibrium between blocks that occurs within the continents and between continents and the adjacent oceans, is called ***isostatic equilibrium*** (Figure 1–17). It is possible to calculate the viscosity of the mantle from the rate of isostatic rebound of the continents where information on the rate of uplift can be obtained. A good example of this is in the determination of rate of uplift of raised beaches from ^{14}C age determinations of wood fragments found in successive beach levels. The viscosity (μ) of the mantle beneath the uplifted beaches may be calculated from

$$\mu = t_r \rho g \lambda (4\pi)^{-1}. \qquad \textbf{(1–8)}$$

where t_r is relaxation (rebound) time, ρ is density, g is the acceleration of gravity (9.8 m/s^2), and λ is the wavelength of the displacement of the Earth's surface. (Equation 1–8 was derived in Turcotte and Schubert [1982].) Remember that the mantle is not an ideal

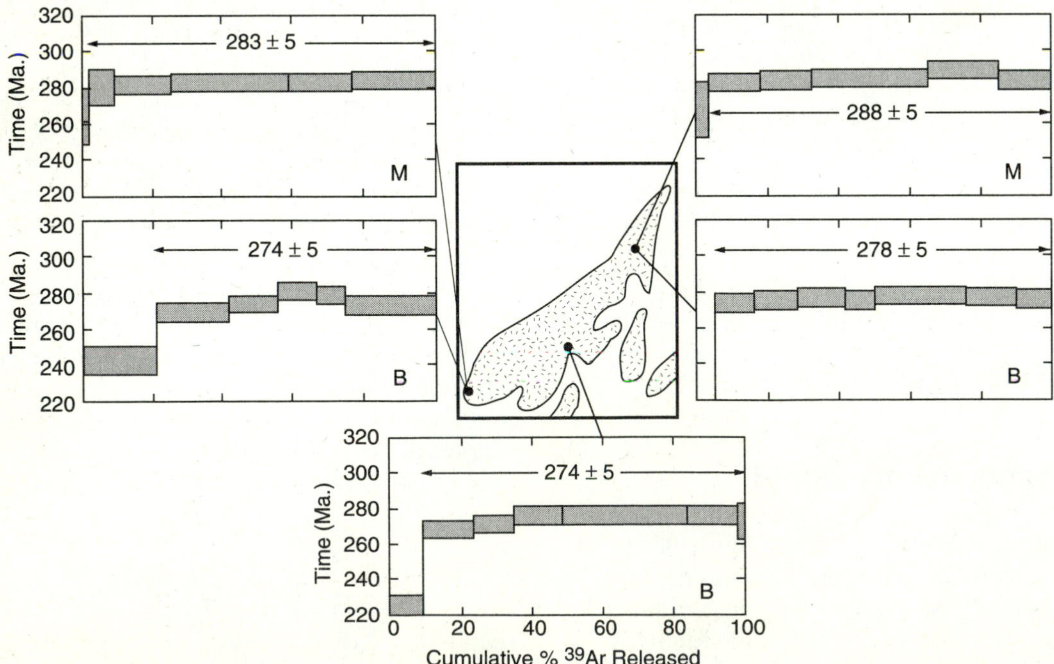

FIGURE 1–16
Incremental argon-release spectra from a New England pluton. B—biotite sample. M—muscovite sample. Arrows on both sides of the age in each diagram indicate the interval used for determination of the age. (From R. D. Dallmeyer and Otto Van Breeman, *Contributions to Mineralogy and Petrology,* v. 78, Fig. 7, © 1981, Springer-Verlag, Heidelberg.)

ESSAY

Capability of Tectonic Structures

The Code of Federal Regulations (Appendix A, Part 100) specifies that documentation of the antiquity of geologic structures in the foundations of critical buildings (such as dams and nuclear power plants) must be provided to show they have not moved during the past 500,000 years. As a result, faults discovered near or within excavations must be carefully studied to show whether or not they have moved during this time and if they are capable of moving during the projected useful life of the buildings planned. To some, it has been a very costly and difficult regulation, for it requires a level of study and documentation never before achieved. To many geologists, however, it has provided a wealth of new and useful information on the structural history of areas where these projects have been undertaken. Faults have been exposed that would not have been known otherwise, and details of their movement history have been brought to light. The techniques used in resolving these details have ranged from the applications of the classic geologic laws of cross-cutting relationships and superposition to modern isotopic and strain-analysis techniques.

In tectonically active areas such as the West Coast of the United States, documentation of antiquity of structures involves careful study of cross-cutting and overprinting relationships and age determinations using Holocene to Recent fossils, ^{14}C dating of organic material preserved in the fault zones, and dating undeformed sediments that truncate these zones. One such study was carried out by Kerry Sieh (1984) of the California Institute of Technology as part of his doctoral research. He studied the Holocene history of an excavation 50 m long by 50 m wide by 5 m deep along a segment of the San Andreas fault located 55 km northeast of Los Angeles. The excavation revealed evidence of repeated faulting in sediments deposited along the fault, indicating that 12 earthquakes occurred between A.D. 260 and 1857, with an average recurrence interval of 145 years. Of the 12 earthquakes he documented, 5 prehistoric earthquakes produced displacements comparable to the large-magnitude earthquake of 1857. Careful study of displacement sense and displacement amounts of faults, the overlap (superposition) of younger sediments (Figure 1E–1), and ^{14}C age

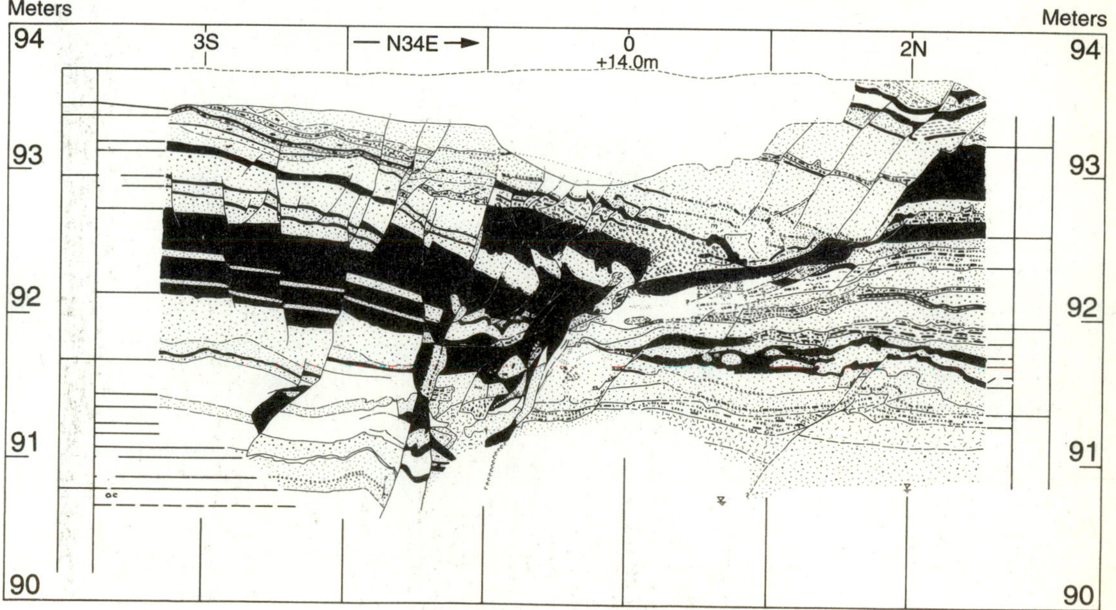

FIGURE 1E–1
Section in part of the trench along Pallett Creek on the San Andreas fault studied by Sieh (1984). It shows several overlapping sequences of sediment deposited during the Holocene and disrupted by recurrent movement along the fault. Most disruption events that produced cross-cutting faults were overlapped by deposition of younger sediments. (From K. E. Sieh, *Journal of Geophysical Research,* v. 89, Plate 4b, © 1984 by the American Geophysical Union.)

dates of wood fragments in sediments enabled reconstruction of the recent history of faulting here and led to a prediction of a 60 percent probability of another large earthquake in the area by the year 2000. Sieh used fundamental techniques to work out a very important interval of geologic history—the most recent prehistoric past—in one of the most populous areas in the United States, pointing out a major environmental hazard. Similar studies were carried out earlier in the region of the major New Madrid earthquakes of 1811–12 in southeastern Missouri and northwestern Tennessee by David Russ (1979) and more recently in the area near Charleston, South Carolina, site of the large 1886 earthquake, by Pradeep Talwani and John Cox (1985) and Steve Obermeier and others (1985). In both areas, it was possible to document earlier large earthquakes, but with much longer recurrence intervals—on the order of 1,000 years or more.

References Cited

Obermeier, S. F., Gohn, G. S., Weems, R. S., and Gelinas, R. L., 1985, Geologic evidence for recurrent moderate to large earthquakes near Charleston, South Carolina: Science, v. 227, p. 408–410.

Russ, D. P., 1979, Late Holocene faulting and earthquake recurrence in the Reelfoot Lake area, northwestern Tennessee: Geological Society of America Bulletin, v. 90, p. 1013–1018.

Sieh, K. E., 1984, Lateral offsets and revised dates of large prehistoric earthquakes at Pallett Creek, southern California: Journal of Geophysical Research, v. 89, p. 7641–7670.

Talwani, P., and Cox, J., 1985, Paleoseismic evidence for recurrence of earthquakes near Charleston, South Carolina: Science, v. 229, p. 379–381.

viscous material, but its behavior may be approximated as that of an ideal viscous material for our purposes. Consequently, calculations of this kind enable us to draw conclusions about the behavior of the mantle in areas that have undergone recent isostatic rebound. For example, we can calculate the viscosity of the mantle beneath the central Canadian shield by determining the uplift rate of beach terraces along the shore of James Bay in northeastern Ontario and by using estimated dimensions of the Keewatin ice sheet that covered this area during the Pleistocene. If the oldest beaches in that area are now 180 m above sea level and it is assumed (from gravitational data) that 20 m more uplift will occur from additional rebound, we can estimate the rate of uplift from the time of retreat of the glacial ice from this region, which was about 8,000 years ago. Moreover, sea level has already risen 125 m, and so 125 m must be added to the amount of uplift. Best estimates of the width of the Keewatin ice sheet give it dimensions of about 9,000 km, which is the wavelength of the displacement of the surface in equation 1–8. The density of the mantle is assumed to be 3,300 kg m^{-3}. Relaxation time (t_r) must first be calculated to take into account the decreasing rate of uplift with respect to the amount of uplift that has already occurred and the amount still to occur calculated from

$$w = w_m e^{-\frac{t}{t_r}}, \tag{1–9}$$

where w is uplift still to occur (~20 m), w_m is total uplift to date (200 m from beach data plus 125 m rise in sea level = 325 m), and t is time since uplift began (8,000 y). Rewriting equation 1–9 in logarithmic form, then solving for t_r,

$$\ln w = \ln w_m - \left(\frac{t}{t_r}\right),$$

$$t_r = \frac{t}{\ln w_m - \ln w}.$$

Substituting

$$t_r = \frac{8{,}000}{\ln 325 - \ln 20} = 2{,}869 \ y .$$

Calculating μ from equation 1–8,

$$\mu = [(2{,}869\ \text{y})\ (365\ \text{d y}^{-1})(24\ \text{h d}^{-1})\ (3{,}600\ \text{s h}^{-1})] \times 3{,}300\ \text{kg m}^{-3} \times 9.8\ \text{m s}^{-2} \times (9 \times 10^{6}\ \text{m}) \times (4\pi)^{-1} = 2.1 \times 10^{21}\ \text{Pa s.} \quad \textbf{(1–10)}$$

(The units of viscosity, here, Pa s, are pascal seconds. One pascal is 1 kilogram meter^{-1} second^{-2}.)

The calculation shows that the lithosphere responds to loads placed on it in relatively short periods of geologic time. The buoyancy of different crustal elements is fundamental and involves all parts of the lithosphere and asthenosphere.

The phenomenon of isostasy was first discovered in surveys in the flanks of the Himalayas, where the great relief led to an error in the calculations that could not be compensated by usual corrections. Early isostatic models based on either volume or density failed to satisfy the need for correction. Later, models incorporating both volume and density changes (Figure 1–17), along with flexural bending of the crust, best corrected the errors in the surveys and demonstrated the fundamental nature of the principle of isostatic adjustment.

The emplacement of large thrust sheets (Chapter 11) having areas of hundreds of square kilometers and thicknesses of 5 to 10 km would thicken the lithosphere in the immediate vicinity of the thrust sheet and would require profound adjustments in the asthenosphere beneath to accommodate the increase in lithospheric thickness. Similarly, crustal extension processes, such as those affecting the Basin and Range Province in the western United States (Chapter 14), serve to unload and thin the lithosphere, thus decreasing the amount of lighter material above the asthenosphere. Profound adjustments also occur here in both the lithosphere and asthenosphere. Because both the amount of light material and the total mass of the lithosphere are decreased, a significant amount of uplift does not occur after extension.

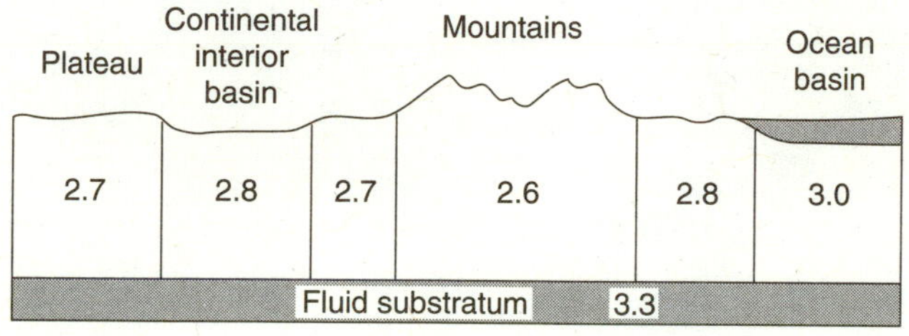

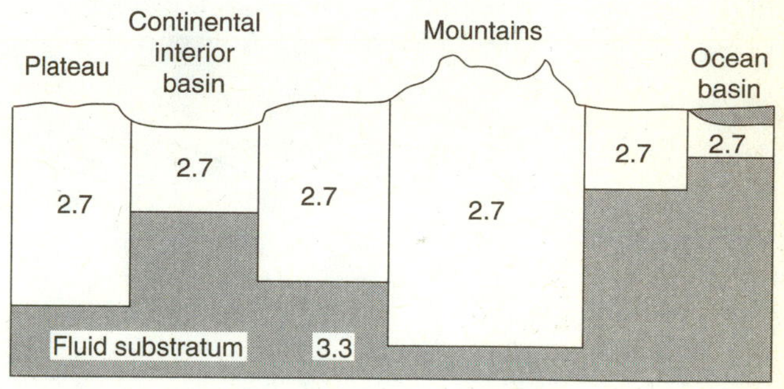

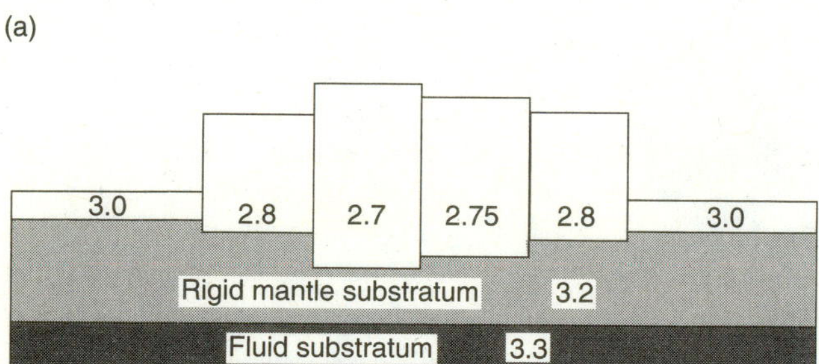

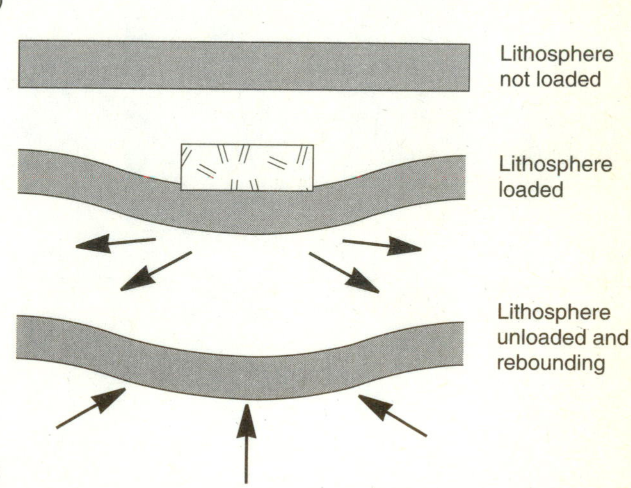

FIGURE 1–17
Isostatic equilibrium between crustal blocks of different densities and thicknesses, as well as between the continents and oceans; (a) and (b) are the early models of Pratt and Airy based separately on different density (in units of g cm^{-3}) and different size of blocks. We realize today that both density and size affect the isostatic equilibrium of the blocks (c), and are involved in isostatic compensation; (d) shows the effect of loading and unloading of a mass on the lithosphere. Arrows indicate directions of compensating flow in the asthenosphere during and after loading.

GEOLOGIC CYCLES

Most geologic processes are driven by cyclic changes of energy fluxes, often over millions of years. The ***rock*** or ***geochemical cycle*** is probably the most familiar of these geologic cycles (Figure 1–18). Each stage in the cycle, from the crystallization of magma to conversion of sedimentary or igneous rocks into metamorphic rocks, is in some way driven by thermal processes and, to a lesser degree, by changes in pressure. Inputs of heat or mechanical energy at particular places short-circuit the cycle. Chemical changes accompany deformation in several stages in the geochemical cycle. All stages tend to restore equilibrium to the system.

The ***Wilson*** or ***tectonic cycle*** (defined in the section on plate tectonics) involves plate motion, beginning with the opening of an ocean basin and producing a trailing plate margin like the present-day East Coast of the United States (Figure 1–19). The phase is terminated by formation of a subduction zone along the trailing margin which begins to subduct oceanic crust, generate heat and pressure, and form either a volcanic island arc offshore or a continental magmatic arc on the old continent. The Wilson cycle ends with continent-continent collision, closing the old ocean. Stages in the cycle reflect response to changing physical conditions in an attempt to restore a state of equilibrium to all or part of the plate system. Mountain building may be considered a direct consequence of a partial or completed Wilson cycle. The world's highest mountains formed in response to energy expenditure in a collision zone between India and Asia during closing of part of the ancient Tethys Ocean. The exceptionally rapid uplift of the Himalayas indicates an extreme isostatic imbalance in the crust because of the great thickness of continental crust there, indicated by the height of the chain. Erosion is rapidly reducing the elevations in this chain to levels that will be closer to equilibrium, but rapid uplift has thus far outstripped the erosion rate.

All processes operating upon or within the Earth act to achieve and maintain equilibrium. Energy is constantly being dissipated to keep the Earth in a dynamic state. Work is done to melt rocks within the Earth to restore equilibrium, and energy is used to drive several cyclic processes.

We continue our review and turn to a discussion in Chapter 2 of nontectonic structures, including a consideration of primary structures, many of which are useful to the structural geologist to determine the facing direction (top) of a sequence and to help distinguish tectonic from nontectonic structures.

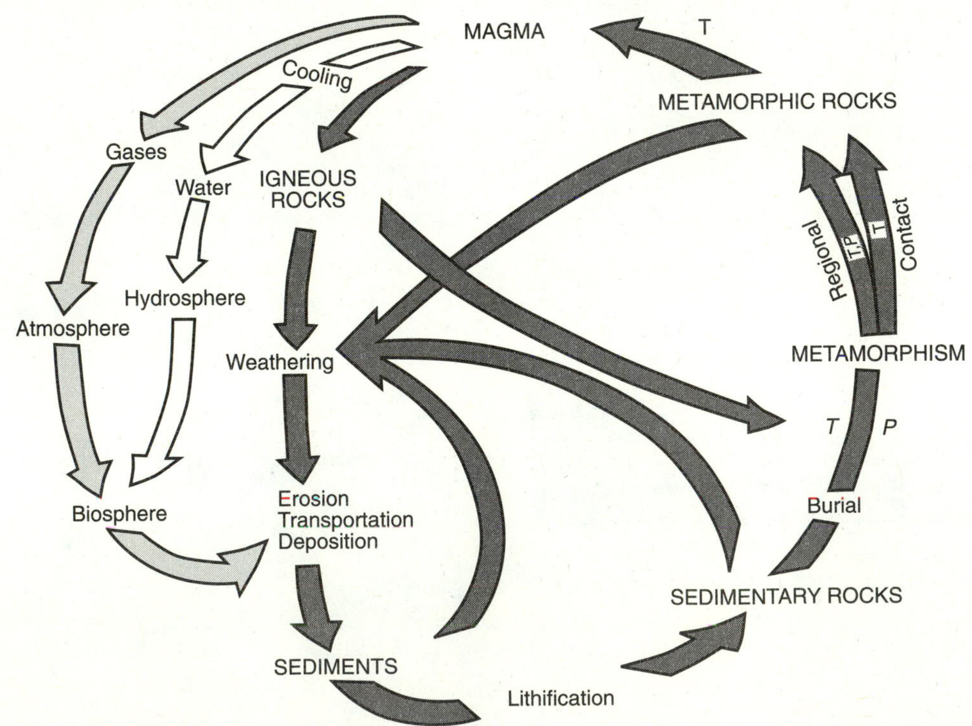

FIGURE 1–18
The rock, or geochemical- cycle—a thermally and mechanically driven equilibrium cycle involving many intermediate stages and shorter cycles.

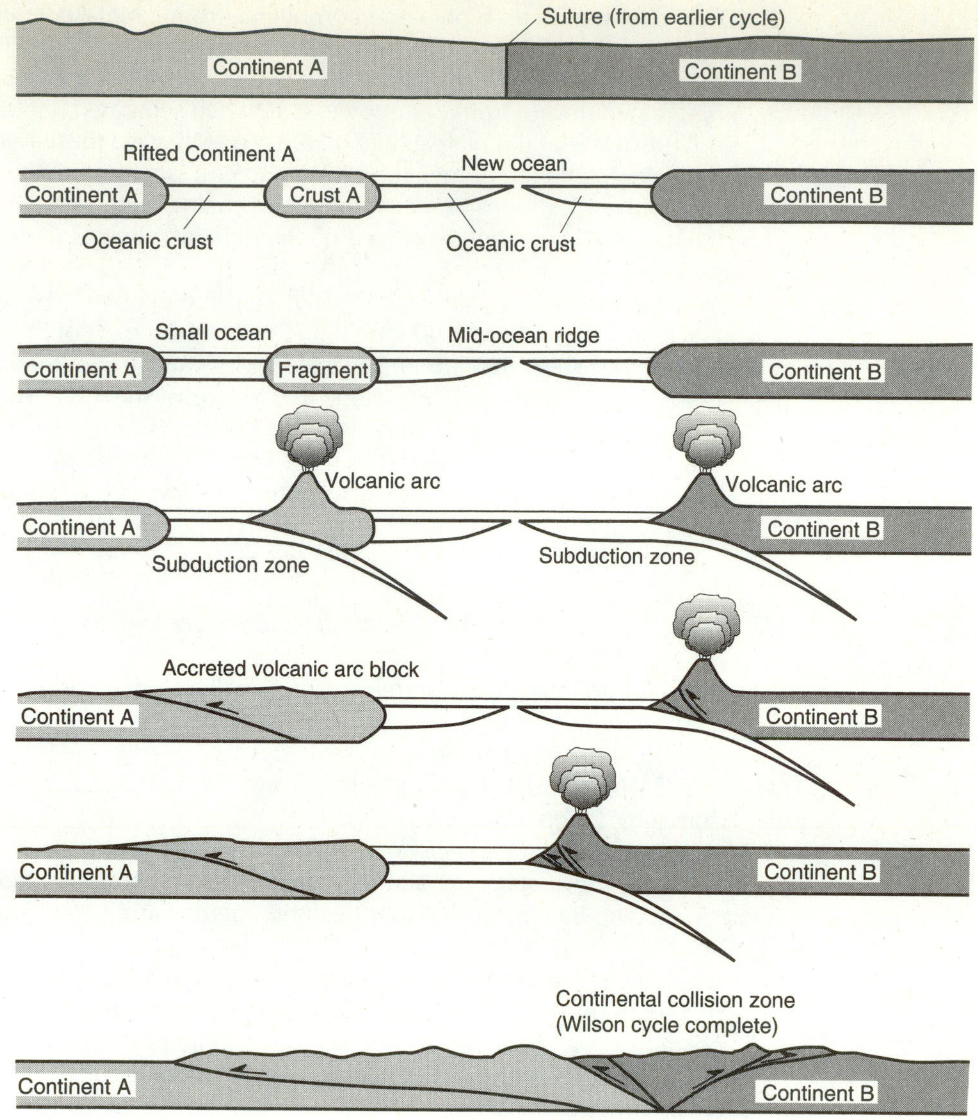

FIGURE 1–19
The Wilson cycle of the opening and closing of an ocean basin. The cycle may be complicated by the formation and movement of suspect terranes, partial closing of small oceans, and the lack of continent-continent collision to terminate the cycle.

Questions

1. In terms of energy distribution, why do geologic processes tend to reach a state of equilibrium?
2. How are the Wilson cycle and the doctrine of uniformitarianism related?
3. How can the laws of superposition, faunal succession, and cross-cutting relationships be used to date terrane boundaries?
4. Why was plate-tectonics theory not formulated in the nineteenth century, as were the unifying theories in physics and biology?
5. Zircons from a foliated granite are discordant, plotting on a chord that yields an upper intercept of 1870 Ma and a lower intercept of 480 Ma. What is the most reasonable interpretation of the two numbers?
6. An unfoliated granite sampled near the foliated pluton in Question 5 yields an Rb-Sr whole-rock isochron age of 365 Ma and an $^{87}Sr/^{86}Sr$ initial ratio of 0.712. What does the age tell you about the time of metamorphism of the rocks and the initial ratio about the source of the granitic magma?

7. $^{40}Ar/^{39}Ar$ studies on biotite and hornblende from a metasedimentary-metavolcanic sequence in a thrust sheet yield plateau ages of 320 Ma and 380 Ma. Slate interlayers from a limestone in the footwall of the thrust sheet contain fine-grained metamorphic muscovite that, when separated, yields a conventional K-Ar age and a plateau age of 460 Ma. What do these numbers mean relative to the timing of metamorphism and emplacement of the thrust sheet?
8. Why are the elevations of recent mountain chains, such as the Andes, Alps, and Himalayas, so high, but elevations of older chains, such as the Appalachians and Urals, so low?
9. Using the mantle viscosity of 2.1×10^{21} Pa s in equation 1–10, calculate the effects on a lithosphere (100 km thick, $\rho = 2{,}900$ kg m^{-3}) of the emplacement of a large thrust sheet 300 km long, 100 km wide, and 10 km thick and having a density, ρ, of 2,700 kg m^{-3}. Assume that the thrust sheet was emplaced during an instantaneously short period of geologic time.
10. How could you determine the timing of accretion of a suspect terrane?
11. What evidence would you use to reconfirm the law of original horizontality? superposition?

Further Reading

Adams, F. D., 1954, Birth and development of the geological sciences: New York, Dover Publications, 506 p.
Provides an interesting summary of the evolution of geological science from classical times through the beginnings of modern geology with Hutton, Lyell, Darwin, and others in the nineteenth century.

Cloud, P. E., 1970, Adventures in Earth history: San Francisco, W. H. Freeman and Company, 992 p.
A compendium of classic papers on the foundations of ideas on the origin of the Earth, the atmosphere and life, the geologic record, and geologic processes.

Dott, R. H., and Batten, R. L., 1981, Evolution of the Earth: New York, McGraw-Hill Book Company, 113 p.
A historical geology book and an excellent compendium of the fundamental principles upon which many observations in structural geology are based. Also contains a wealth of information on the tectonic evolution of North America.

Faure, G., 1986, Principles of isotope geology, 2nd edition: New York, John Wiley & Sons, 464 p.
Outlines the techniques of radiometric age dating with problems and discussions of the shortcomings of each.

Glen, W., 1982, The road to Jaramillo: Critical years of the revolution in Earth science: Stanford, California, Stanford University Press, 459 p.
Outlines the history of the development of plate-tectonics theory, emphasizing use of paleomagnetic measurements.

Hamilton, W. B., 1979, Tectonics of the Indonesian region: U.S. Geological Survey Professional Paper 1078, 345 p.
A synthesis of the geology of the Indonesian region, containing numerous maps showing the elements of a dispersed group of terranes ranging from Precambrian basement to recent volcanic-arc materials in the initial stages of being swept back into the Asian continent as Australia moves northward.

Howell, D. G., 1985, Terranes: Scientific American, v. 253, no. 5, p. 116–125.
Summarizes the distribution of microplates, or terranes, in and around the Pacific basin, presenting the background of plate tectonics and accretion concepts.

Wilson, J. T., 1966, Did the Atlantic close and reopen?: Nature, v. 211, p. 676–681.
This short paper set the stage for the concept of the Wilson cycle.

2

Nontectonic Structures

It has been said that stratigraphy is the basis of all geology (Weller, 1947). This is true of the metamorphic rocks as well as the sedimentary rocks. Cognizant of the importance of petrography, physical chemistry, and structural geology, the investigator of metamorphic rocks should nevertheless give more adequate treatment to the fascinating problems in stratigraphy, sedimentation and paleogeography that await solutions in the interesting rocks to which, perhaps foolishly, he has dedicated his scientific career.

MARLAND P. BILLINGS, 1950, Geological Society of America *Bulletin*

PRIMARY SEDIMENTARY AND VOLCANIC STRUCTURES FORM along with the rock mass of which they are a part, and have a nontectonic origin. They include bedding and features such as mud cracks, ripples, sole marks, vesicles, and others. For the structural geologist, primary structures, where present, are very useful for determining the ***facing*** (younging) ***direction*** in a sequence of rocks. Facing directions enable us to ascertain if a sequence is upright or overturned. Except for fossils, primary structures are probably the best tools for working out the structural geometry and history in deformed rocks. In the 1950s, Robert Shackleton used primary sedimentary structures to determine that a large part of the rocks in the southern Highlands of Scotland are upside down and concluded that they make up the inverted limb of a large overturned fold (Figure 2–1). He later published a paper outlining the significance of downward-facing structures (Shackleton, 1958).

In studies of deformed rocks, we need to know whether the observed structures have a tectonic or nontectonic origin (Figure 2–2). Many structures that formed in primary depositional environments may mimic structures in rocks that formed in response to tectonic deformation. Thus, it is important to make the distinction. Moreover, many sedimentary structures that form at or near the surface provide useful models to compare with tectonic structures that form at elevated temperatures and pressures. For example, structures formed by ductile flow (Chapter 6) in water-saturated silt, glacial ice, and evaporite (halite, gypsum,

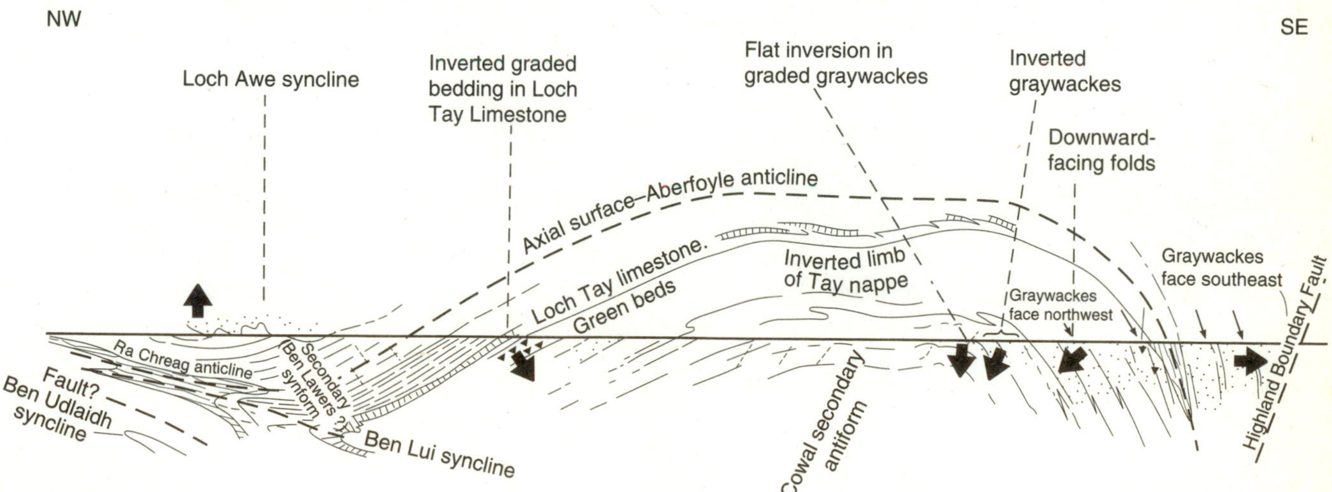

FIGURE 2–1
Geologic cross section of the Tay nappe in the Scottish Highlands. The resolution of overturned and upright structures was accomplished by R. M. Shackleton using facing directions of sedimentary sequences determined from sedimentary structures. Arrows indicate facing direction. (Reproduced by permission of the Geological Society, from Downward facing structures of the Highland Border, R. M. Shackleton, in *Quarterly Journal of the Geological Society of London,* v. 113, 1958.)

(a)

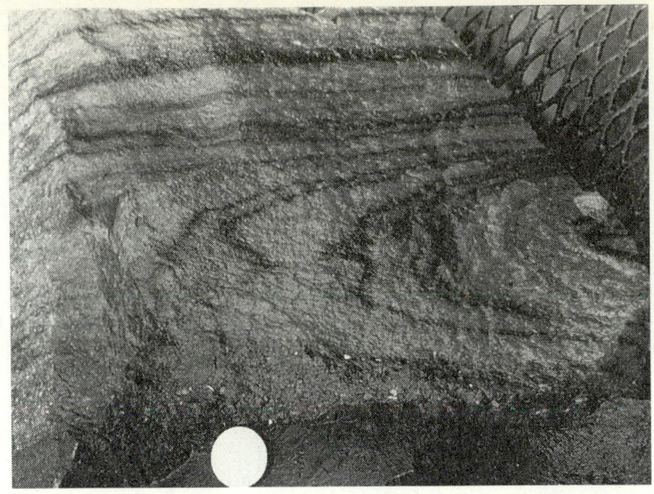

(b)

(c)

FIGURE 2–2
(a) Folds formed by slumping of nearly plastic, water-saturated sediment during freezing (from Byron Stone, 1976, *Journal of Sedimentary Petrology,* v. 46), and (b) soft-sediment folds in Upper Proterozoic Longarm Quartzite near Harmon Den, North Carolina, exhibit many of the characteristics of tectonic folds (c) that formed at high temperatures and pressures deep in the Earth. The coin in (b) is a U.S. quarter (2.4 cm in diameter). The example in (c) is sillimanite-grade gneiss in the Thor-Odin dome in the Shuswap complex, southern British Columbia. [(b) and (c) RDH photos.]

anhydrite) deposits at surface pressure and temperature (P-T) conditions are of the same styles as those formed at much higher P-T conditions in rocks deformed at depths of 15 to 20 km.

Distinguishing tectonic from nontectonic structures often requires observation of particular diagnostic features. Primary structures are mostly but *not always,* older than tectonic structures. So, if the structure in question overprints or cross-cuts an existing, obviously tectonic structure (such as a fault), it probably had a tectonic origin. If, on the other hand, tectonic structures overprint the structure in question, but cannot be shown to overprint earlier nontectonic structures, it may be difficult to determine if the origin was tectonic or nontectonic. Because of the crustal scale of deformation and the parallel orientation of forces over large parts of the crust, tectonic structures commonly have a similar orientation over wide areas. Small structures also mimic large (mappable) ones, both in style and orientation (Pumpelly's rule), but the orientations of primary (nontectonic) structures will be related to features (for example, currents) in depositional environments, not to large tectonic structures.

PRIMARY SEDIMENTARY STRUCTURES

Bedding

The most common characteristic and most diagnostic feature of sedimentary rocks is ***bedding,*** and because it forms mostly in a horizontal orientation, it is the first-order reference surface for most structural measurements. ***Bedding planes*** form in primary sedimentary environments and become mechanical zones of weakness as the sediment is lithified (Figure 2–3). They exist for several reasons, most commonly because of compositional or textural differences in sediment at the interfaces between adjacent beds. The differences in composition and texture reflect changes in sedimentary environment: sediment composed of particles of different size, shape, or (in many instances) composition may be deposited in various successions. For example, at a site where sand is being deposited, sand deposition may be interrupted for a short time, smaller clay and silt-sized particles are deposited, and then deposition of sand continues. The layer of fine-grained material forms a mechanical discontinuity in the mass of sand, and this discontinuity becomes a *bedding plane.* Repetition of this process produces a sequence of bedded sand.

Compaction of sediment already deposited may also provide a mechanical discontinuity that leads to formation of a bedding plane. When the next influx of sediment occurs, it is initially compacted less than that deposited earlier. Consequently, a bedding plane may form between newly deposited sand and sand deposited previously without the need for smaller particles to separate the two layers (Figure 2–3).

Graded beds contain a range of particle sizes from large at the base to small at the top. In some graded beds, particle size ranges from pebble or boulder at the base to clay at the top, and the entire thickness of the bed may range up to 2 to 3 m; in others, particle size may range from 1 mm or smaller at the base to clay at the top, and the thickness of the bed may be less than 1 cm. They form where a mass of sediment of contrasting particle size, commonly a *turbidity flow,* encounters a change in flow regime—from rapid to slow—producing rapid deposition (Figure 2–4). The largest particles, because of their mass, settle to the bottom first, followed by successively smaller particles. After lithification, the bottom of the bed may be conglomerate, the middle part sandstone, and the top of the bed shale, with a gradual transition between rock types. Graded beds form in both sedimentary and volcaniclastic deposits and are important tools for determining the ***facing direction***. Channels, shown in Figure 2–4, and other primary features frequently occur at the base of graded beds. *Reversed graded bedding* is relatively rare but may form as a primary structure during deposition of pumice fragments in water, or be a product of grain and debris flows. Larger pumice fragments will occur at the tops of beds, and smaller particles of other compositions will occur beneath, because pumice fragments initially float and thus settle more slowly than denser particles. The term has also been used (perhaps misused) to describe metamorphosed normal graded beds where the fine-grained top of the bed recrystallized to very coarse mica that is larger than the granular material at the original bottom of the bed.

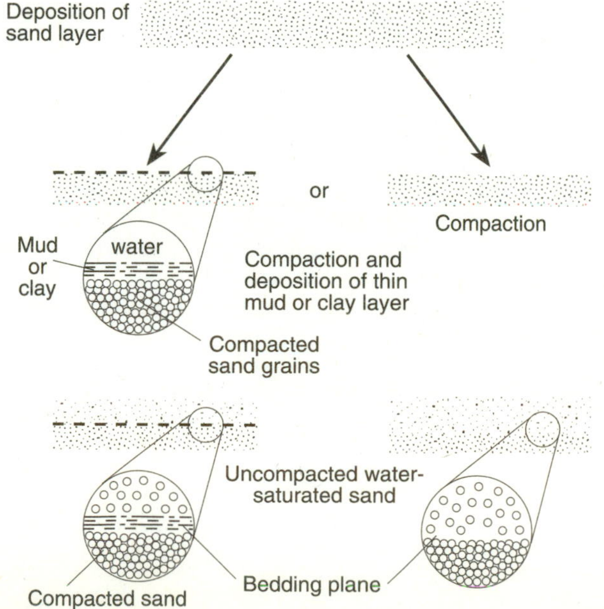

FIGURE 2–3
Formation of bedding planes.

Cross bedding, or cross stratification, is common in sandstones. It forms where sediment grains are transported by water or wind, producing migrating dunes, or mounds, with inclined (foreset) lamination planes within an individual bed, and is bounded by ordinary bedding. Common types of cross beds are *tangential, planar, trough*, and *festoon* (Figure 2–5). Tangential cross beds may be used for determining facing direction; planar cross beds, even though inclined, provide the same perspective whether overturned or upright and so cannot be used to determine facing direction. Two other types—*ripple cross laminations* and *hummocky laminations*—are small-scale cross beds that form in finer sediment, and may be useful for determining facing directions (Reineck and Singh, 1975; Harms and others, 1982). Those useful in determining tops (tangential) are generally concave upward, tangent to the bottom of the primary bedding plane, and truncated at the top (Figure 2–6).

Mud Cracks

Fine sediment deposited in water and later exposed to the atmosphere forms extensional shrinkage cracks called ***mud*** or *desiccation* ***cracks*** (Figure 2–7). In plan view, they form polygonal blocks bounded by cracks that taper downward from the surface of the sediment and terminate. Because of its fine texture, the surface layer may separate from the one below and curl

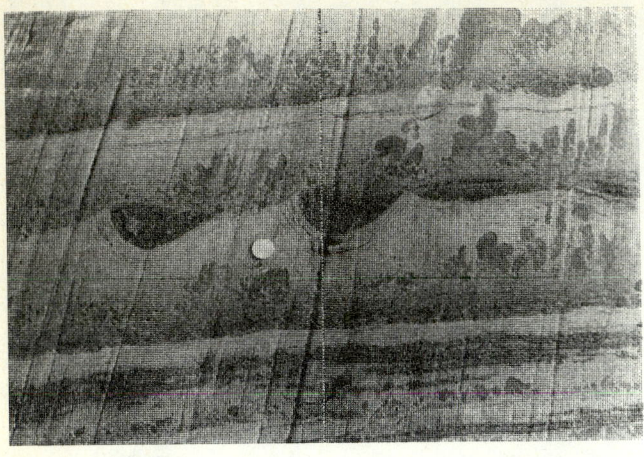

(a)

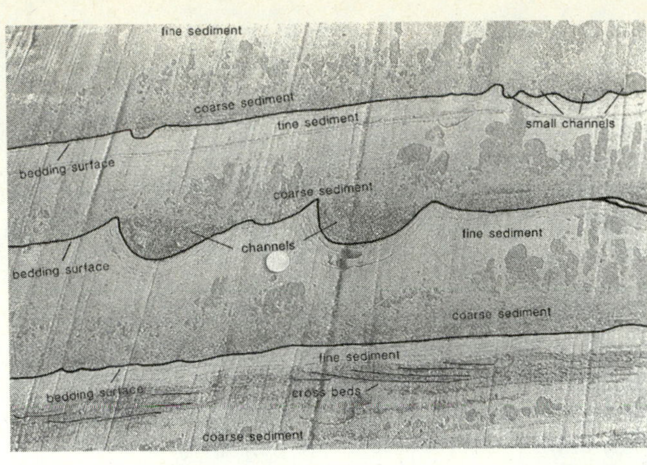

(b)

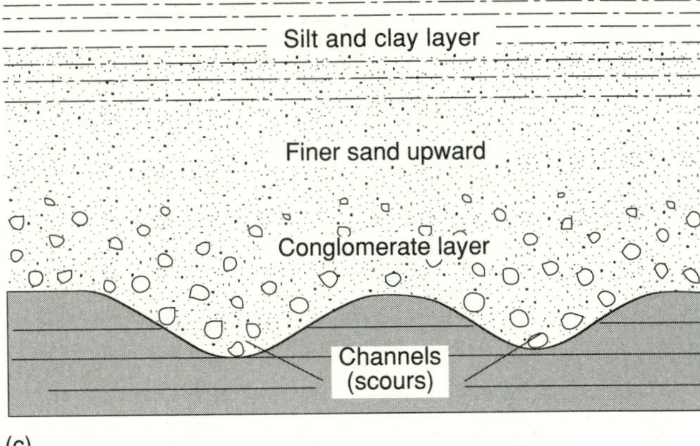

(c)

FIGURE 2–4
(a) Graded beds in Middle Silurian Ekeberg graywacke exposed on a glacially striated surface near Jämtängen, Sweden. Note there are three cycles of graded beds (with load casts or current scours at the base of the upper two) and fining-upward sequences. Cross beds appear in the lowest bed. (Dark circular areas are lichen.) (RDH photo.)
(b) Sketch of photo in (a). (c) General relationships of graded beds and associated textures.

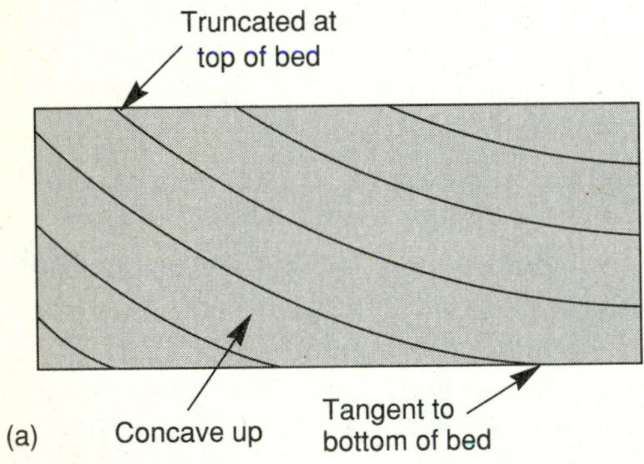

(a)

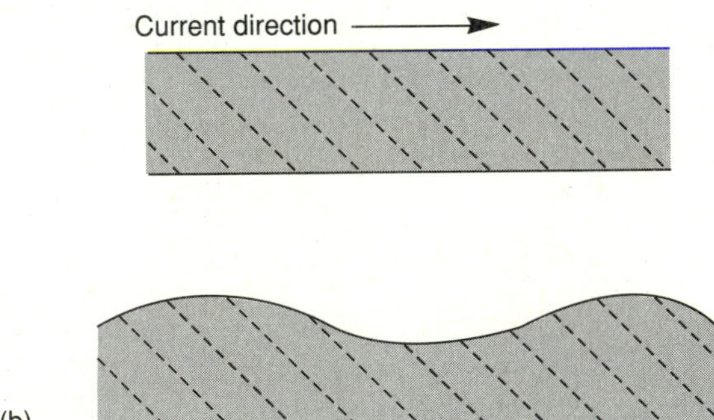

(b)

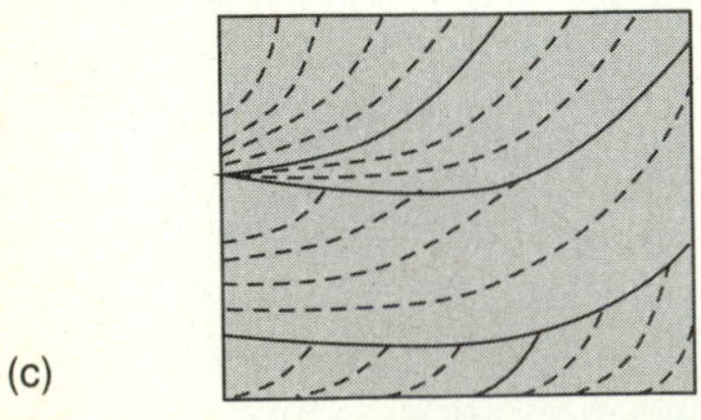

(c)

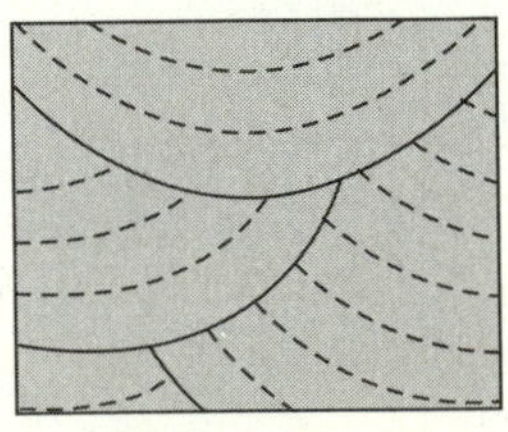

(d)

FIGURE 2–5
Tangential (a), planar abrupt (torrential) (b), trough (c), and festoon (d) cross beds. Facing direction can readily be determined in all but planar (torrential cross beds).

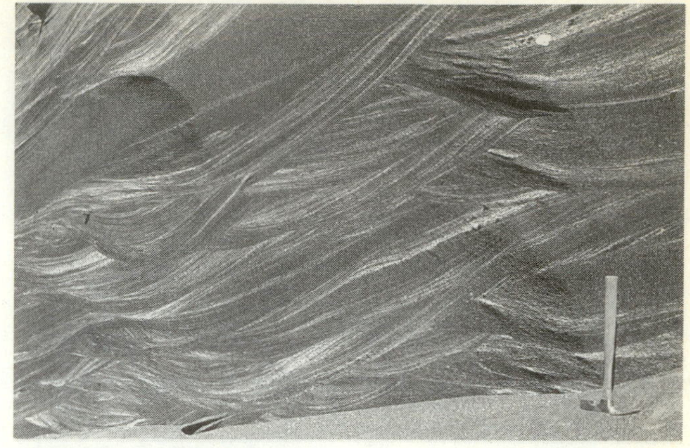

(a)

(b)

FIGURE 2–6
(a) Trough (festoon) cross beds allow determination of facing direction; Pleistocene sand, Elmore County, Idaho. (H. E. Malde, U.S. Geological Survey.) (b) Deformed cross beds in quartzite defined by thin dark laminae that intersect bedding at a low angle at Eidsvoll quarry, near Oppdal, southern Norway. (Locality courtesy of Allan G. Krill, Trondheim University. RDH photo.)

(a)

(b)

FIGURE 2–7
(a) Mud cracks in a dried-up shallow body of water near Salton Station, Riverside County, California. (G. K. Gilbert, U.S. Geological Survey.) (b) Middle Proterozoic mud cracks in siltstone, Isle Royale National Park, Michigan. (N. K. Huber, U.S. Geological Survey.)

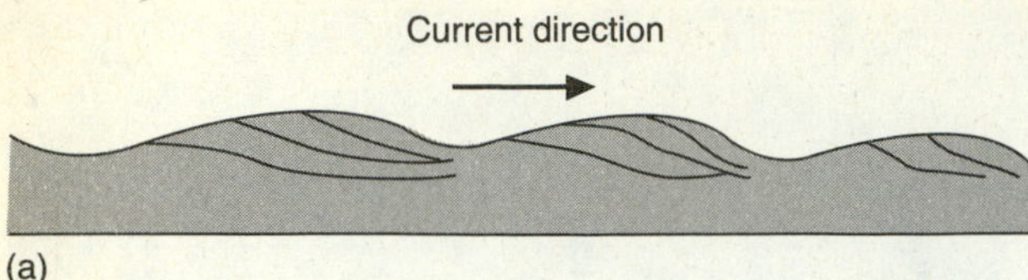

FIGURE 2–8
(a) Current (translation) ripple marks. (b) Deformed current ripples in silty carbonate and siltstone, Maddens Branch, Ocoee Gorge, Polk County, Tennessee. Bedding plane below (where the quarter is located) has minimal distortion, whereas the rippled top has been flattened into asymmetric fold-like structures. Flattening occurred parallel to slaty cleavage that dips approximately 30° to the right. (RDH photo.)

up, providing another indicator of facing direction. Mud cracks occur in the geologic record, but (because of the transient nature of the subaerial environment where they form) not as commonly as cross beds or graded beds. Where preserved, they are quite useful in determining the facing direction and may also be used as indicators of natural strain (deformation) in rocks (Chapter 5).

Ripple Marks

Ripple marks form where sediment is moved by a current or where the bottom sediment surface is otherwise disturbed by water moving above a threshold velocity. This setting is very common along beaches and streams, as well as in deeper water, where bottom currents or surface waves interact with bottom sediment. Two kinds of ripple marks have been recognized: ***current*** *(translational)* and ***oscillatory.*** Current ripples form where a prevailing direction of transport and deposition of sediment occurs; they are commonly asymmetric, and their steep sides face downstream in the direction of transport (Figure 2–8). Unless they contain tangential ripple cross laminae,

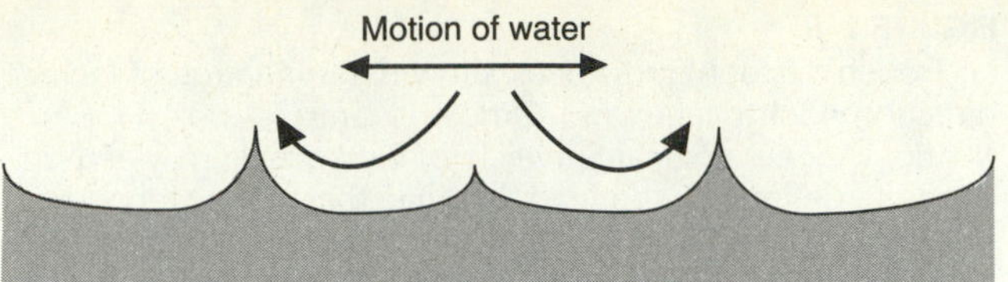

FIGURE 2–9
Oscillatory ripple marks.

current ripples have the same shape in cross section regardless of whether they are upright or overturned and therefore are not useful for determining facing direction. Oscillatory ripples are symmetrical and generally consist of high and low crests. They form where a back-and-forth motion exists in a body of water (such as a lake), or where there is no prevailing current direction. The crests have sharp peaks separated by rounded troughs (Figure 2–9). Oscillatory ripples may therefore be used for determining tops in sediments. Unfortunately, oscillatory ripples are less common than current ripples.

Rain Imprints

Rain imprints form where rain falls on fine sediment, and preservation in the sedimentary record depends on rapid, nondestructive covering of the imprints by another layer of sediment (Figure 2–10). Rain imprints can be used to determine facing direction, but because they are relatively delicate, are not preserved as often as many other primary structures.

Tracks and Trails

Tracks and ***trails*** left by organisms may be preserved and, under certain conditions, may be used to determine facing direction of beds (Figure 2–11). The weight of an organism produces a track or trail, and if the impression is filled with sediment during deposition of the overlying bed, a cast is formed and the facing direction may be determined. In some instances, partly filled burrows may be useful for determining tops if the fillings can be properly recognized and interpreted. The widely known cylindrical *Skolithos* tubes of the Cambrian of North America and the Scottish Highlands have also been used as strain indicators (Chapter 5).

Sole Marks, Scour Marks, Flute Casts

Marks formed as an object moves across a bedding surface or as currents scour a bedding surface are ***sole*** and ***scour marks*** (Figure 2–12). Sediment that fills

(a)

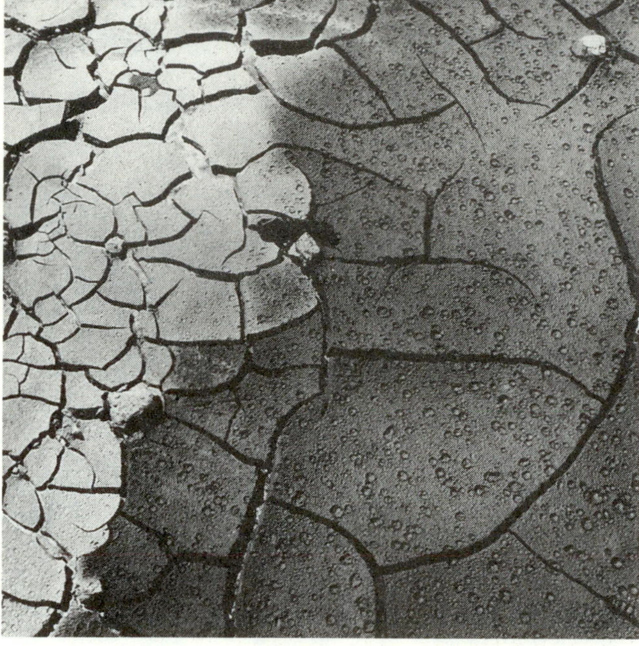

(b)

FIGURE 2–10
(a) Middle Proterozoic rain imprints in siltstone, Isle Royale National Park, Michigan. (b) Rain imprints on desiccated Recent mud cracks, Isle Royale National Park, Michigan. (Both photos by N. K. Huber, U.S. Geological Survey.)

FIGURE 2–11
Track left by an arthropod as it crawled across the Cambrian sea floor; Pyramid Shale Member, Titanothere Canyon, Inyo County, California. Tracks are approximately 1 cm wide. (A. R. Palmer, Cambrian Institute, Boulder, Colorado.)

scour or sole marks by subsequent deposition may allow the marks to be preserved, provide a sense of asymmetry from the bed below to that above, and thus indicate the facing direction. Flute molds consist of scoop-shaped structures formed where currents scour and erode a surface. The molds are filled with sediment to form *flute casts* as the bed above is deposited, producing similar asymmetry and another useful tops criterion. Flute casts are commonly found at the base of turbidite beds.

Dewatering Structures

Load casts form after deposition and dewatering of sediment as a result of gravitational instability at the interface between a layer of water-saturated sand and underlying mud. The weight of the newly deposited overlying sediment forces out the interstitial water. Compaction of the underlying sediment (commonly mud) beneath a sand bed commonly results in depressions in the bed below (Figure 2–13) as water is expelled from the mud during compaction. Load casts may be used with relative ease to determine facing direction. The broadly convex sides of load casts face toward the bottom of the layer. Load casts are common in any accumulation of sediment where there is a contrast in grain size, significant dewatering, and opportunity for differential compaction. Load casts commonly occur at the base of turbidite beds and in deltaic (flood-related) sediments. The most important factor in formation of load casts is the instability in sediment produced by rapid sedimentation that prevents some of the interstitial water from escaping.

Interesting analogs have been observed in soft sediment where deformation at hand-specimen or larger scale may approximate that which is occurring in metamorphic or igneous rocks at deeper levels within the Earth. Ductile, brittle, and fluid behavior have all been observed in soft sediment. An interesting phenomenon in soft sediment is dewatering: water trapped between individual grains in sediment may remain under pressure as long as the sediment is in the stable confined mass. If an abrupt change occurs in the plumbing within the mass of sediment and the water escapes suddenly, disruption of bedding and other primary structures results (Figure 2–14). Earthquakes have been known to cause dewatering in Recent sediments and have been suggested as the cause of disruption of layering in partly consolidated sands (Talwani and Cox, 1985).

(a)

(b)

FIGURE 2–12
(a) Sole marks on siltstone beds, Borden Formation, Bullitt County, Kentucky. (R. C. Kepferle, U.S. Geological Survey.) (b) Flute casts on the bottom of a turbidite bed near Ause Pleureuse, Gaspé Peninsula, Québec. Current was moving right to left. (L. B. Gillett, SUNY College at Plattsburgh.)

(a)

(b)

FIGURE 2–13
(a) Load casts on the bottom of a sandstone bed, which projects into underlying mudstone, Sunnyside No. 1 Coal, Carbon County, Utah. (J. O. Maberry, U.S. Geological Survey.) (b) Load casts on the bottom of a near-vertical sandstone bed, southwestern Montana. Top of bed faces into the road cut. (L. B. Gillett, SUNY College at Plattsburgh.)

Fossils

The fossilized remains of organisms preserved in the geologic record provide useful indicators of relative age (Figure 2–15). Fossils may also be used to determine facing direction in a sequence by studying the relative positions in the sequence where organisms of different ages are found. Sometimes, partly filled cavities also occur inside molds of organisms; other organisms, such as trees or bottom-living attached animals, may be preserved in the position in which they lived. If so, they may be used to determine the facing direction of a sequence. Fossils are also useful strain indicators (Chapter 5).

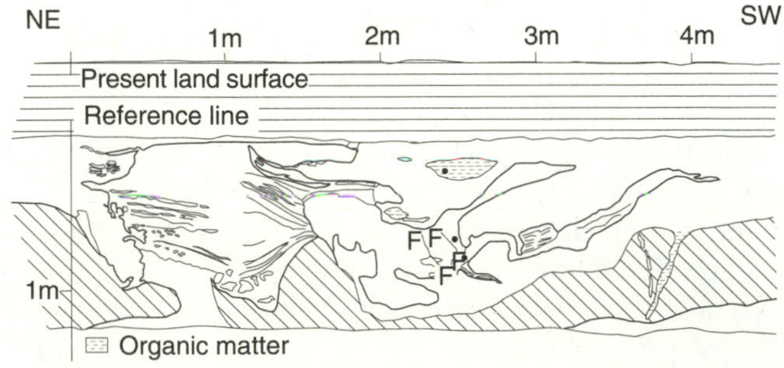

FIGURE 2–14
Dewatering structures in Holocene sediments, attributed to a large prehistoric earthquake. F marks faults. Horizontal and vertical scales are in meters. (From Pradeep Talwani and John Cox, *Science*, v. 229, p.379–381, © 1985 by the AAAS.)

FIGURE 2–15
Heads of several Middle Cambrian trilobites *(Paradoxides)* on a bedding surface in volcanic mudstone from near Batesburg, South Carolina. (Courtesy of Donald T. Secor, Jr., University of South Carolina.)

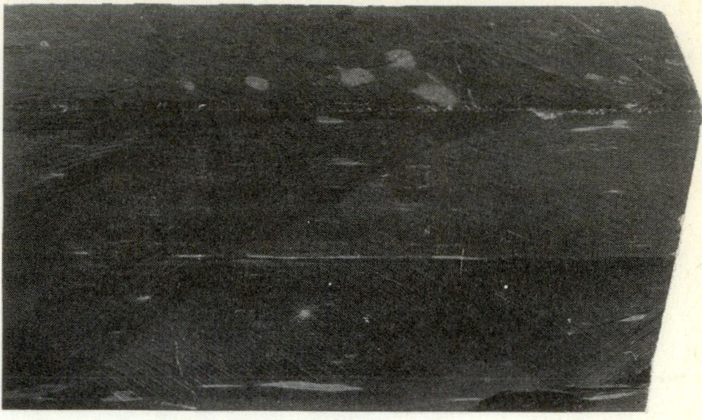

FIGURE 2–16
Deformed reduction spots in Eocambrian Metawee Slate from near Rutland, Vermont. The broken surface on top of the specimen exposes light-colored elliptical reduction spots, whereas the sawed surface in front displays thin vertical sections through the reduction spots. The sawed front of the specimen is 15 cm high. (RDH photo.)

Reduction Spots

Reduction spots are sedimentary structures produced by a small grain or fragment of organic matter or other material that is chemically different from the surrounding mass of sediment. The chemical difference may produce a nearly spherical area of reduction expressed as a color change in the immediate vicinity of the grain in the otherwise oxidized sediment (Figure 2–16). These features cannot be used to determine the facing direction of a sequence, but they serve as an important indicator of strain if the rock mass was internally deformed. Deformed reduction spots are common in slate and provide important constraints on the origin of slaty cleavage (Chapters 5 and 17). It is important to know when the reduction spots formed in order to study their role in the strain history of a rock mass, a subject to be discussed in greater detail in Chapter 5.

SEDIMENTARY FACIES

It was recognized early in the twentieth century that sedimentary (and volcanic) rock units vary laterally (and vertically) as paleoenvironments change. Such lateral and vertical differences of sediment (later rock) type are called ***sedimentary facies*** (Figure 2–17a), and each sediment (rock) type is called a *facies* or lithosome. One facies is separated from others by particular characteristics (composition, texture, sorting, physical and biogenic sedimentary structures, etc.) that set it apart from other facies within the same formation. Before this principle was recognized, numerous errors were made in interpreting the ages and nature of different rock types. It was assumed that each rock type was a separate formation (mappable unit), especially if there were no fossils to provide information on the ages of the rocks. The possibility of one rock type grading laterally into another within the same unit was not considered, nor was the possibility that both lateral and vertical changes in rock type were linked in transgressive and regressive sequences (Figure 2–17b). Careful studies of relations between interlayered and gradational sedimentary rocks and associated fossils showed that although rock types may change laterally as well as vertically, different *facies* of the same rock unit may be deposited at the same time and should not be considered different rock units solely because of lateral differences in rock type. The interrelationship between horizontal and vertical variation of facies is called *Walther's principle,* which states that only those facies (environments) that once existed side-by-side can be observed vertically juxtaposed in outcrop. Walther's principle does not apply to sequences containing unconformities (see below).

UNCONFORMITIES

A break in the sedimentary record, where part of a stratigraphic succession and thus history is missing, is called an ***unconformity.*** Unconformities are produced by erosion or nondeposition (or both), resulting in a lack of strata recording the history related to a segment

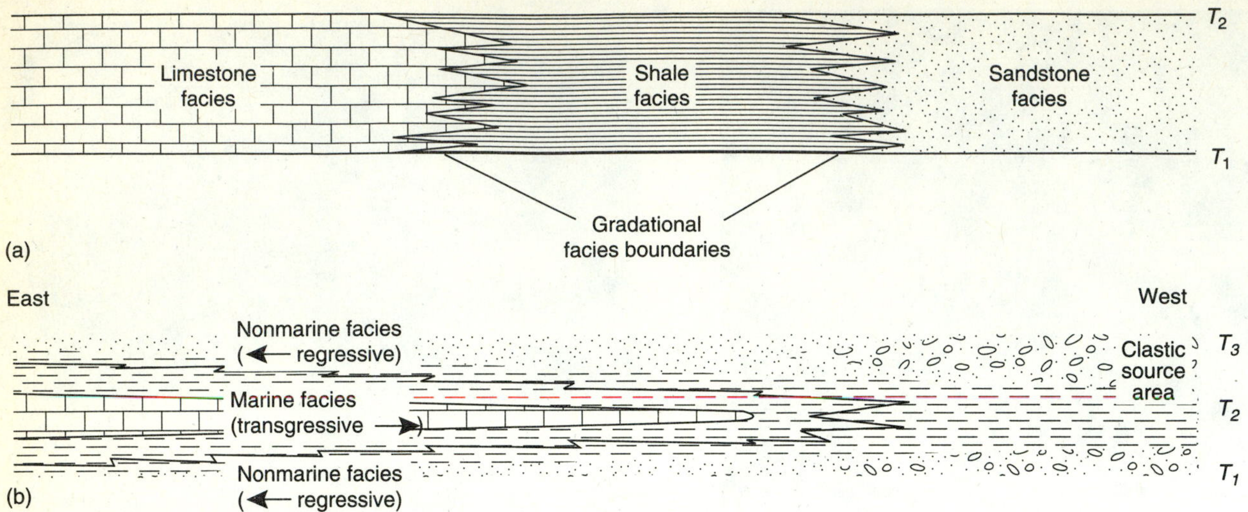

FIGURE 2–17
(a) Facies changes within the same rock unit in a sequence of sedimentary rocks. T_1 and T_2 refer to older and younger chronostratigraphic boundaries. (b) Lateral and vertical variations in facies related to transgression and regression of a shoreline with a clastic source to the west. T_1, T_2, and T_3 are again reference time boundaries.

of geologic time. Granted, the geologic record may be characterized more by missing than by complete history; unconformities, however, record major breaks on the order of millions of years.

The three fundamental types of unconformities are disconformities, angular unconformities, and nonconformities. A ***disconformity*** (Figure 2–18a) is produced by deposition of a sequence followed by erosion without tilting or deformation; then comes subsidence and renewed deposition. Bedding in the sediments above and below the unconformity remains parallel. Disconformities are recognized where part of a sequence is missing either because it was eroded or because it was never deposited. There may be topographic relief along the unconformity. The term *paraconformity* has been used in places where there is little relief on an unconformity, and bedding remains parallel on both sides.

An ***angular unconformity*** (Figure 2–18b) is produced where a sequence has been tilted as a result of slumping or the tectonic processes of faulting or folding. Erosion commonly accompanies or follows tilting. When deposition is renewed, an angular relationship exists between the rocks below and those above the unconformity (Figure 2–19). Although this process can occur rapidly, millions of years probably elapse in the formation of angular unconformities. The angular unconformity at Siccar Point in Scotland is probably the best known in the world, because there James Hutton gathered some of the evidence that yielded the principle of uniformitarianism.

A ***nonconformity*** (Figure 2–18c) is an unconformity in which igneous or metamorphic rocks (or both) occur below the erosion surface, and sedimentary (or metasedimentary) rocks occur above. Nonconformities indicate that a long time interval (probably millions of years) passed between formation of the igneous or metamorphic rocks at great depths in the Earth and deposition of sediment atop the exhumed and eroded ***crystalline rock*** mass. The characteristics of the rocks above help to identify the boundary as an unconformity rather than an igneous contact because

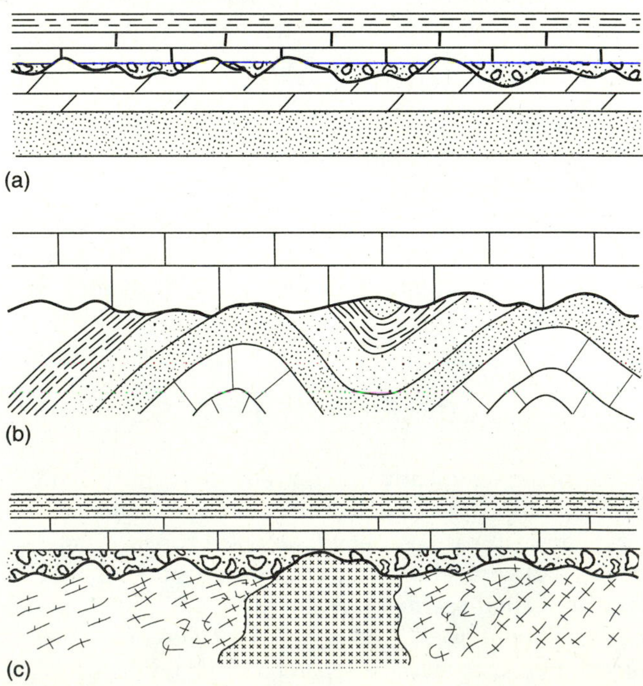

FIGURE 2–18
Types of unconformities. (a) Disconformity. (b) Angular unconformity. (c) Nonconformity.

ESSAY

Deciphering a Major Structure in the Southern Highlands of Scotland

The study by Robert M. Shackleton that was cited at the beginning of Chapter 2 provides an important key to later work in the southern Highlands of Scotland (Figures 2–1 and 2E–1) and illustrates the power of using primary sedimentary features to work out the structure of complexly deformed rocks.

To appreciate what Shackleton (1958) accomplished, it is important to know something about the deformed state of the rocks in this region. The sequence he studied consists of Upper Proterozoic to lower Paleozoic clastic sedimentary rocks that contain few fossils. At least one early Paleozoic Caledonian regional metamorphic event recrystallized the sequence to greenschist- and amphibolite-facies assemblages. During the thermal event, the rocks were ductilely deformed and subjected to several events of very tight folding, each of which overprinted the earlier episodes (see Chapter 16 for further discussion of complex folding). Ductile deformation also produced a strong foliation in most of the rocks.

Shackleton used the fairly abundant graded bedding and cross bedding that survived the rigors of Caledonian multiple deformation and metamorphism to determine that most of the rocks of the area (covering several hundred square kilometers) in the southern Highlands are downward-facing and therefore overturned. His conclusion must have been initially astounding, but it was reaffirmed by his many determinations of facing direction based on observation of primary sedimentary structures. This kind of observation should be made routinely in the course of the detailed structural study of any complexly deformed area. Other structural or stratigraphic criteria may help to determine if a sequence is upright or overturned, but primary structures may be the *only* criterion available to enable this important assessment of regional facing directions.

Reference Cited

Shackleton, R. M., 1958, Downward-facing structures of the Highland Border: Quarterly Journal of the Geological Society of London, v. 113, p. 361–392.

(a)

(b)

FIGURE 2–19
(a) Angular unconformity, southeastern Alaska. (D. J. Miller, U.S. Geological Survey.) (b) Tilted angular unconformity at Kingston, New York, between Ordovician Austin Glen (Normanskill) graywacke and Silurian Rondout Formation (sandy dolostone). (RDH photo.)

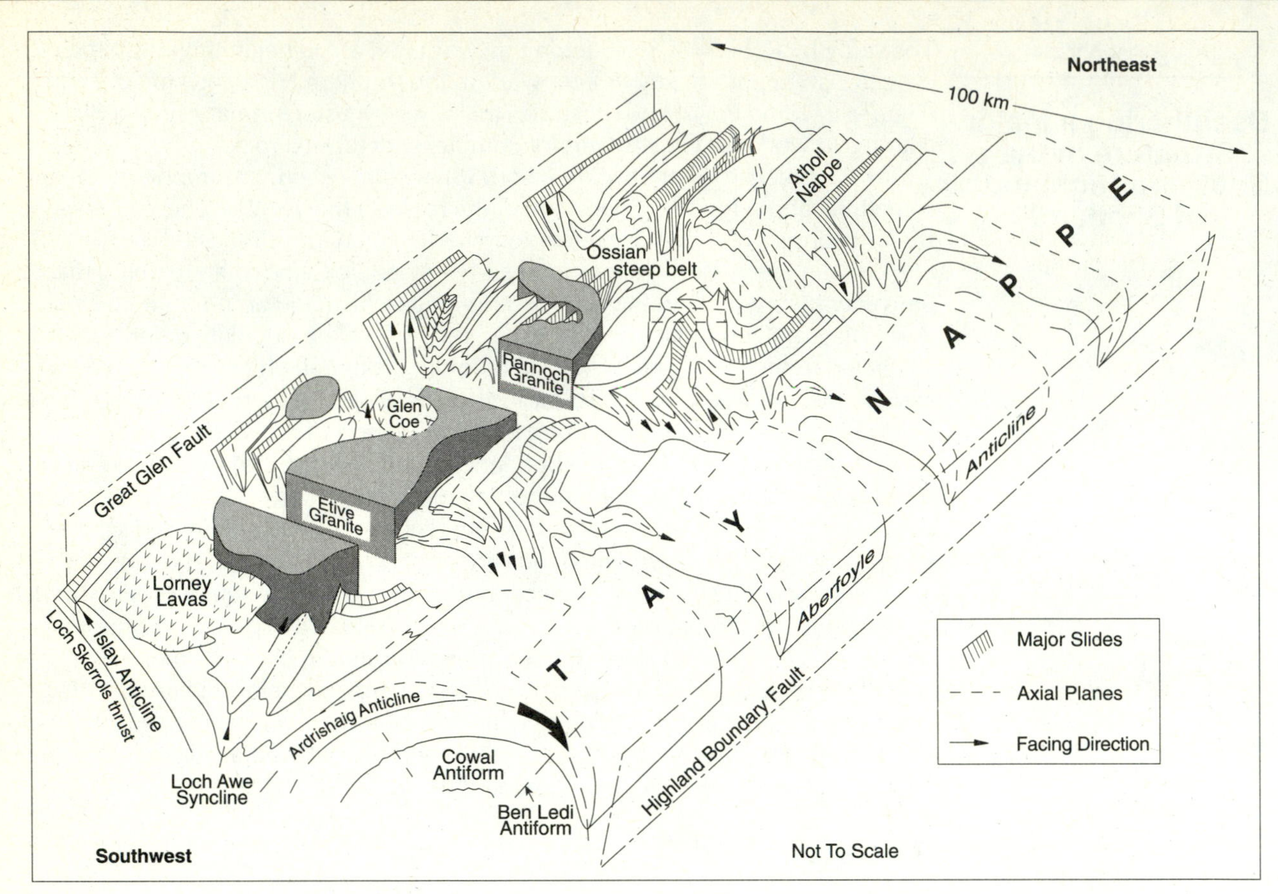

FIGURE 2E–1
Block diagram showing the major structures of the western Central Highlands of Scotland. Note the size of the Tay nappe relative to the other structures in this region. (Reproduced by permission of the Geological Society from *The Caledonides in the British Isles—Reviewed,* P. R. Thomas, *in* A. L. Harris, C. H. Holland, and B. E. Leake, eds., *The Caledonides in the British Isles—Reviewed,* Geological Society of London Special Publication 8.)

there is no contact metamorphic aureole. A contact metamorphic zone in the sedimentary rocks would be expected if the contact is intrusive. A basal conglomerate composed of clasts of the underlying crystalline rocks may also occur at the base of the sedimentary sequence (Figure 2–20).

Disconformities, angular unconformities, and nonconformities may all be represented at an unconformable surface if the rocks below the unconformity are metamorphic, or if the surface changes character along its extent. Layering or foliation in metamorphic rocks is commonly truncated at the unconformity, also producing an angular unconformity. Appreciable relief on the nonconformity surface suggests that it is also a disconformity. The classic unconformity at the base of the Precambrian sedimentary sequence in the Grand Canyon has all of those characteristics (Figure 2–21).

Another clue that a geologic boundary is an unconformity may be obtained from examination of the boundary: pebbles, boulders, and other debris accumulated from erosion may remain as part of the base of the unit deposited on the erosion surface (Figure 2–20). Not all unconformities have residual debris, either because no material was deposited at the site of observation or because all debris was removed from the area. Careful study of the rocks immediately beneath an unconformity may prove that the rocks along the old surface underwent a period of prolonged weathering before accumulation of erosional debris. Ancient soils (paleosols) have been recognized beneath many unconformities.

FIGURE 2–20
Basal conglomerate in Late Proterozoic Torridonian Sandstone, Assynt District, Scotland. Pebbles (light-colored) accumulated in the basal sand on the erosion surface on the darker Lewisian Gneiss that had been intruded earlier by light-colored felsic dikes, forming a nonconformity. (RDH photo.)

PRIMARY IGNEOUS STRUCTURES

Igneous plutons and lava flows commonly form in shapes ranging from approximately equidimensional to tabular. Less commonly, they contain features that result from flow within the magma that resemble sedimentary structures. Cross bedding, graded bedding, and other kinds of layering have been observed in igneous bodies. The most common primary structures in igneous bodies include flow foliation, phenocrysts, compositional banding, xenoliths, and vesicles. A foliation resulting from flow of crystallizing magma is difficult to distinguish from a tectonic foliation. Careful study of the structures in the enclosing rocks and texture of the igneous body may be necessary to document cross-cutting relationships or to show that the foliation is confined to the igneous body and is not a tectonic foliation in the country rock (Figure 2–22). Flow of magma also may produce folds in plutons that resemble tectonic structures in metamorphic rocks (Chapter 14). The tectonic aspects of pluton emplacement are discussed in Chapter 19.

Compositional banding in an igneous body (Figure 2–23a) may result from crystal settling, differentiation, fractional crystallization, and multiple parallel intrusions or flow processes that flatten xenoliths. Careful study of a layered pluton may reveal a differentiation or fractional crystallization sequence that resulted in the bottom of a pluton being more mafic and the top more felsic. If layering can be shown to have been produced by differentiation and gravitational settling, the upward change from mafic to felsic composition may be used to determine the top direction. Graded bedding or cross bedding may exist in rare instances, such as in the Skaergaard intrusion in Greenland (Wager and Deer, 1939). These features may be used in igneous bodies to determine facing directions in the same way as in a sedimentary sequence, except in a few places where they occur on the near-vertical walls of plutons.

Granophyric quartz toward the top of gabbroic sills has been used as a facing criterion in several places where differentiation has occurred. One example of where this texture occurs is in the upper part of the Palisades sill in New Jersey and New York (Walker, 1940; Poldervaart and Walker, 1962).

Vesicles are cavities left by gas bubbles that form in magma as the pressure decreases (Figure 2–23b). Vesicles form in lava flows and, less commonly, in plutons where magma has moved rapidly from deep to shallow levels in the crust. The bubbles move upward, being less dense, and make their way toward the top of a shallow pluton or lava flow, producing a cavity-filled zone that may be preserved and used as a facing criterion. Vesicles may become filled with secondary minerals and as such are called *amygdules* (Figure 2–23c). Their presence obviously would not affect determination of the facing direction in a flow or sill, but special care must be taken to not mistake the amygdules for phenocrysts. Amygdules and vesicles may occur in layers, thus revealing the horizontal plane. They may also be useful strain indicators (Chapter 5). Cracks in the tops of lava flows are also a useful indicator of facing direction.

Pillow structures form where lavas are extruded beneath or flow into water. Because of their shape, they can be used to determine facing direction (Figure 2–24). The vesicular, glassy curved tops and V-shaped nonvesicular bases are indicative of the top of a flow. Borradaile and Poulsen (1981) argued that the criterion is not useful in strongly deformed pillows, but others

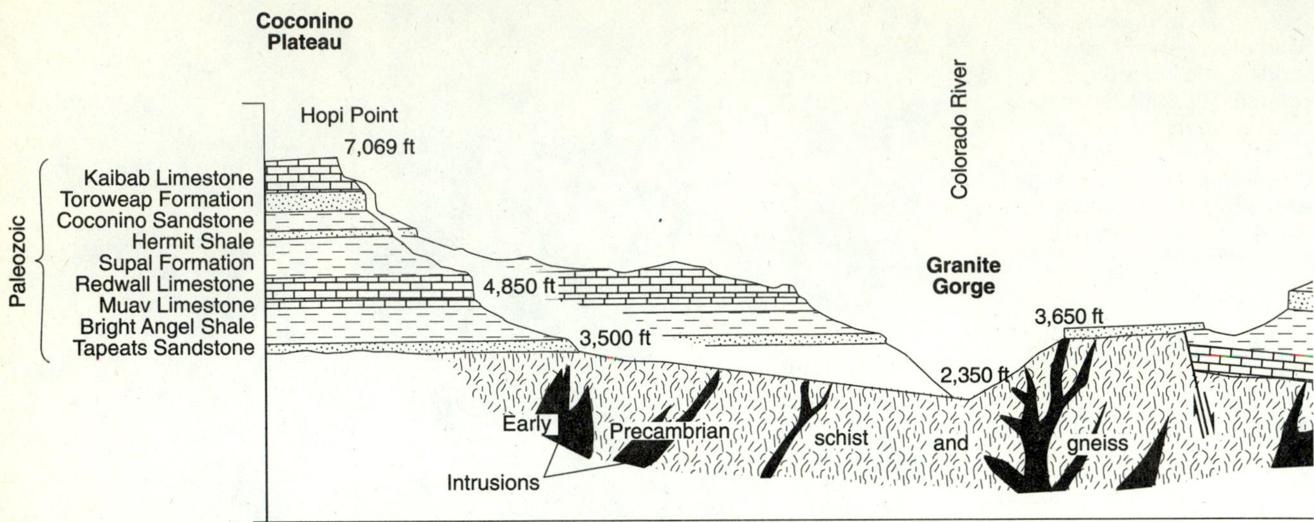

FIGURE 2–21
Cross section through the Grand Canyon showing the unconformity at the Precambrian-Paleozoic boundary. Note that the boundary is both an angular unconformity and a nonconformity, where it is underlain by tilted sedimentary rocks or metamorphic rocks, and has considerable relief. Elevations are in feet. (From F. E. Matthes, 1962, The Grand Canyon of the Colorado River: U.S. Geological Survey Bright Angel quadrangle, scale 1:62,500.)

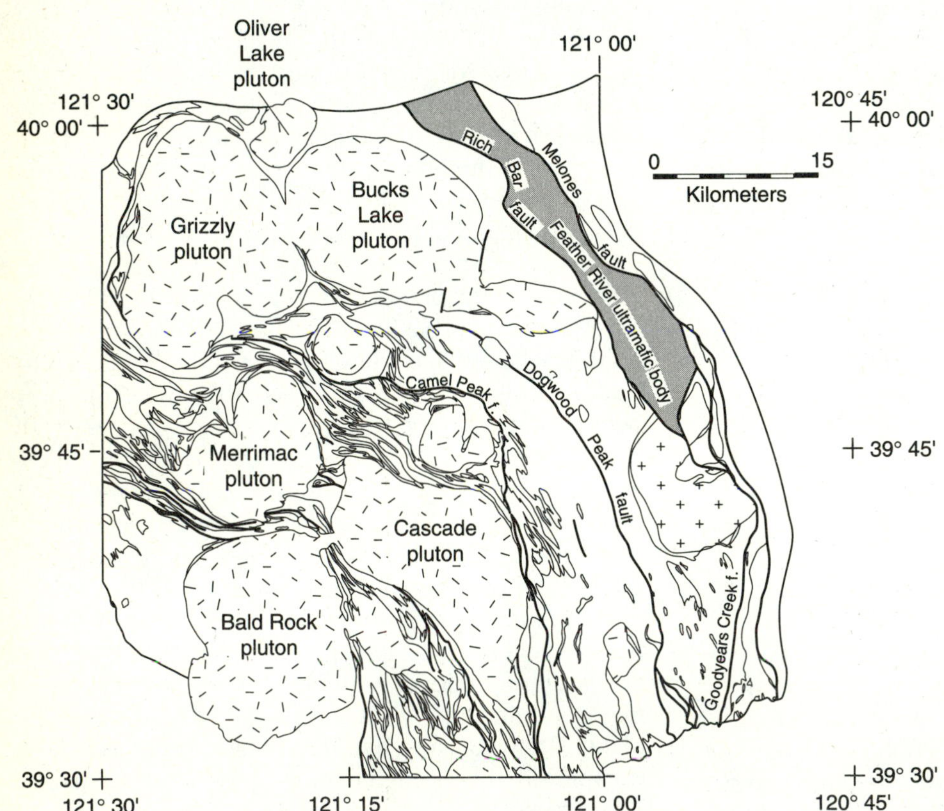

FIGURE 2–22
Geologic map of part of the Sierra Nevada showing cross-cutting of layering in country rocks by plutons. Granitic plutons shown by a random pattern, gabbro by a + pattern. (After Anna Hietanen, 1981, U.S. Geological Survey Professional Paper 1226-B.)

have employed it where the pillows are slightly to moderately deformed, particularly in areas where other criteria are scarce or absent. M. E. Wilson (1941, 1962) and G. O. Allard (1976) have used pillow structures to determine facing directions in the Noranda, Rouyn, and Chibougamau areas of Québec.

Pyroclastic rocks, ignimbrites, and water-laid tuffs *(volcaniclastic rocks)* frequently contain graded beds, reversed graded beds (with large pumice fragments at the top), cross beds, and other primary structures that can provide facing directions. Rapid facies changes also occur within volcanic assemblages.

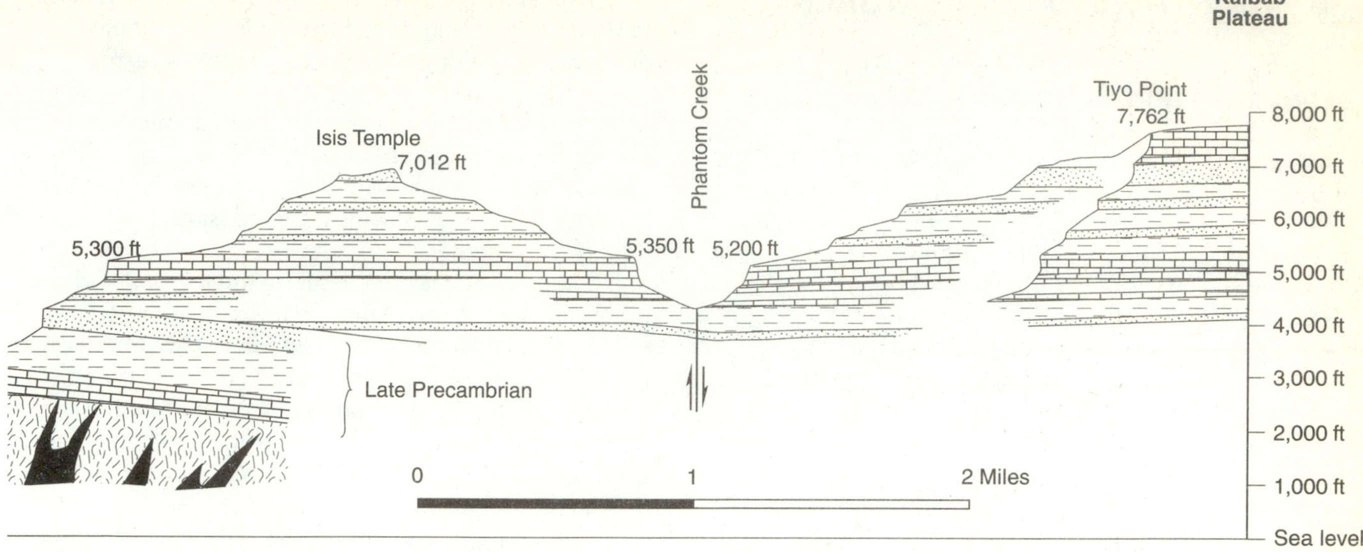

FIGURE 2–21 (continued)

(a)

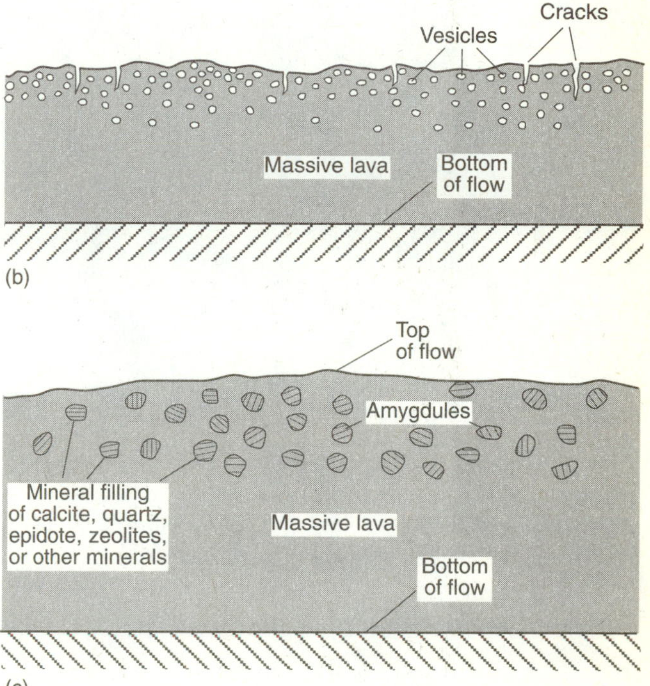

FIGURE 2–23
(a) Compositional banding in a layered gabbro (part of an ophiolite) south of Koppervik, island of Karmøy, southern Norway. Dark layers are rich in mafic minerals; light layers are rich in plagioclase. (Locality courtesy of Brian A. Sturt, Norwegian Geological Survey; RDH photo.) (b) Vesicles and cracks, and (c) amygdules, in the tops of lava flows.

Both rapid and gradual changes from volcanic to sedimentary assemblages and from one volcanic rock type to another characterize volcanic regions, making stratigraphic and structural studies more challenging. Resolution of the stratigraphy and younging directions of volcanic assemblages is frequently aided by geochronological data (see Chapter 1).

Contact metamorphic zones associated with sills and flows may be useful for determination of tops and in distinguishing sills from flows. Ideally, a sill (Figure 2–25a) could produce contact metamorphic zones on both the top and bottom contacts with the country rock. Therefore, the use of this feature for determination of tops is limited. A lava flow (Figure 2–25b), on the other hand, should metamorphose only the material below it, and there would be no metamorphism in a subsequently deposited overlying sequence. Thus, if a concordant tabular igneous body occurs in a sequence of sediments, and a contact metamorphic aureole occurs only below it, the body must be a lava flow and is useful for determination of facing direction.

GRAVITY-RELATED FEATURES

Landslides and Submarine Flows

Landslides occur both above and below sea level, and the results in either case may resemble tectonic structures. Slides may be triggered by earthquakes or other tectonic activity, either deep in the crust or on the surface many kilometers from the landslide.

The most obvious factor contributing to landslides is a slope, but not all sloping surfaces produce landslides. In addition to slopes (some of less than 1°), parallelism of bedding, foliation, or fractures with the surface slope may provide conditions favoring landslides. Weak material at or near the surface may also produce favorable conditions for a landslide, but not actually cause one.

Landslide-prone conditions exist in many places. Various phenomena trigger landslides: earthquakes, overloading of slopes, high precipitation, streams undercutting and oversteepening slopes, and human activities. Poorly consolidated water-saturated material—commonly of fine silt or sand size—may spontaneously liquefy when loaded and produce a landslide. An earthquake may provide the necessary shock to liquefy the material.

Landsliding may occur along the toe of an advancing thrust sheet or any other escarpment consisting of poorly consolidated or fractured material, particularly in submarine environments. Blocks may fall from the escarpment and be deposited in finer sediment. Masses of matrix-supported blocks such as this are known as *olistostromes* (bedded) and *diamictites* (no obvious bedding). *Pebbly mudstones* are matrix-supported masses containing rounded clasts. Tillites are a type of

(a)

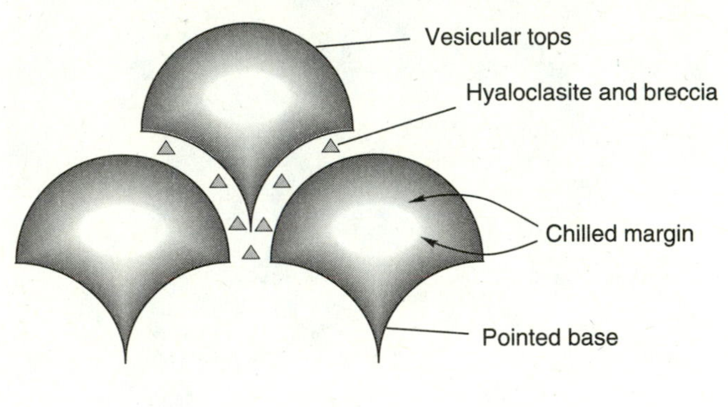

(b)

FIGURE 2–24
Pillow structures (a) in Perchas Lava Member, Barrio Gato, Puerto Rico. (H. G. Berryhill, Jr., U.S. Geological Survey.) Cross section (b) shows structure.

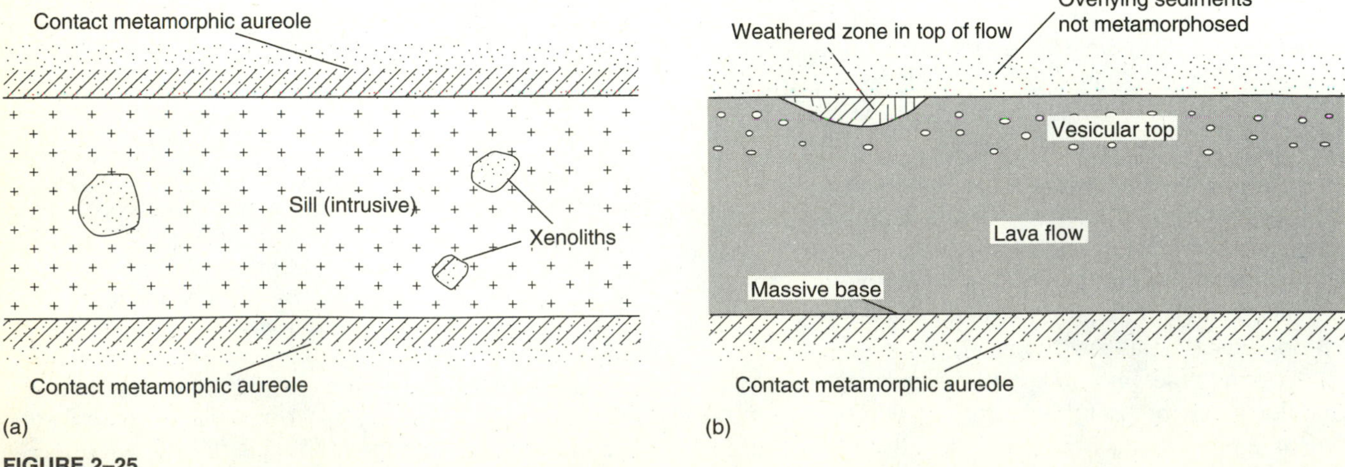

FIGURE 2–25
Distinguishing sills (a) from flows (b) by using contact metamorphic aureoles.

diamictite. Olistostromes may consist of exotic blocks (olistoliths) of any size, ranging from a few centimeters in diameter up to several kilometers, contained within a fine-grained matrix (Figure 2–26) (Abbate and others, 1970). Olistostromes are commonly interlayered with nonchaotic sediments and overall have a lenticular geometry.

Mélanges consist of mixtures of weak and strong rock materials—such as fragments of sandstone or basalt in a clay matrix—of either tectonic or nontectonic origin (Figure 2–27). E. Greenly (1919) first used the term mélange for rocks in Anglesey (northwestern part of Wales) that are characterized by fragments of stronger rock embedded in a sheared matrix of weaker materials (from Raymond, 1975). At present, we believe that tectonic mélanges commonly form in accretionary prisms along subduction zones and consist of coherent rock masses of different sizes of variously deformed materials contained in a pervasively sheared matrix (Hsü, 1968). Other mixtures of coarse and fine sediment form olistostromes from material broken off an escarpment of either a tectonic or nontectonic origin. Here, the matrix may not be sheared if the mass is preserved relatively undeformed, but the blocks in the mass will be both chaotic and of diverse lithology (Figure 2–28). The term *wildflysch* has been applied to a heterogeneous accumulation of blocks and smaller particles generally deposited in deep-water environments. Such an accumulation may result from either tectonic or nontectonic processes.

Turbidites are deposits produced by rapid flow of a sediment-laden turbidity current down a slope onto the sea floor or the floor of a large lake. They generally consist of an unsorted mass of sediment that cascades to lower levels, spreads out on the floor of the body of water, and settles as graded beds. Arnold H. Bouma (1962) recognized that these flows deposit graded, channeled, massive, and cross-bedded sequences that grade upward from the graded sequence at the base to the more massive and cross-bedded sequences near the

FIGURE 2–26
Olistostrome (wildflysch) near Albany, New York, composed of blocks of Ordovician Austin Glen Graywacke. The blocks were redeposited in Ordovician Normanskill Shale and were derived from an approaching thrust sheet. (RDH photo.)

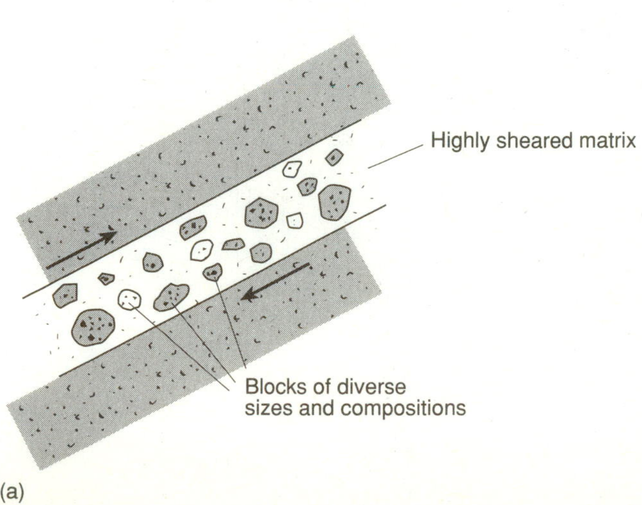

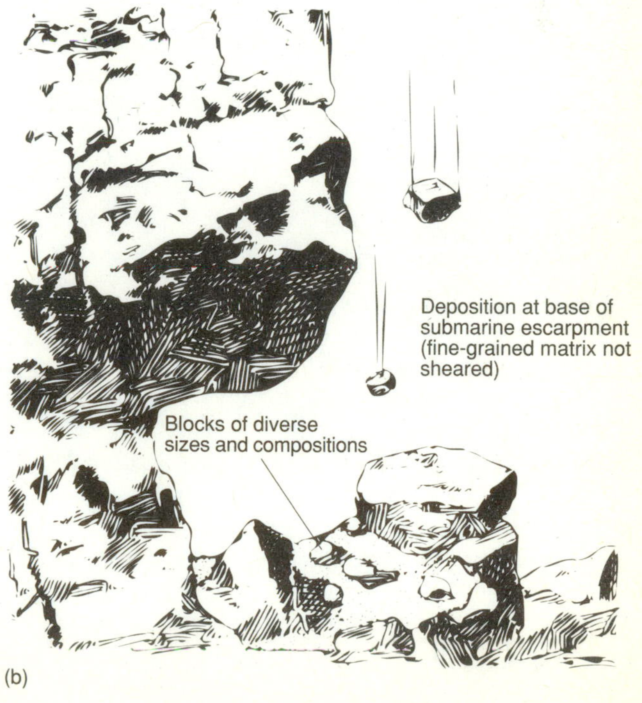

FIGURE 2–27
Formation of mélanges by tectonic (a) and nontectonic (b) mechanisms.

FIGURE 2–28
Mélange terrane near Tarnilat in northern Morocco. The high hills in the distance are held up by mélange blocks of limestone, and the lower hills and lowlands are underlain by shale. Rocks in the foreground are part of an underlying unit. (RDH photo.)

top (Figure 2–29), defining a sequence that now bears his name. Because of the cyclic asymmetry of these sequences and the units that they comprise, an intact or partial *Bouma sequence* is another facing criterion. Complete Bouma sequences are not preserved as commonly as are the lower (graded) parts because of the scouring that accompanies the arrival of the next overlying turbidity flow.

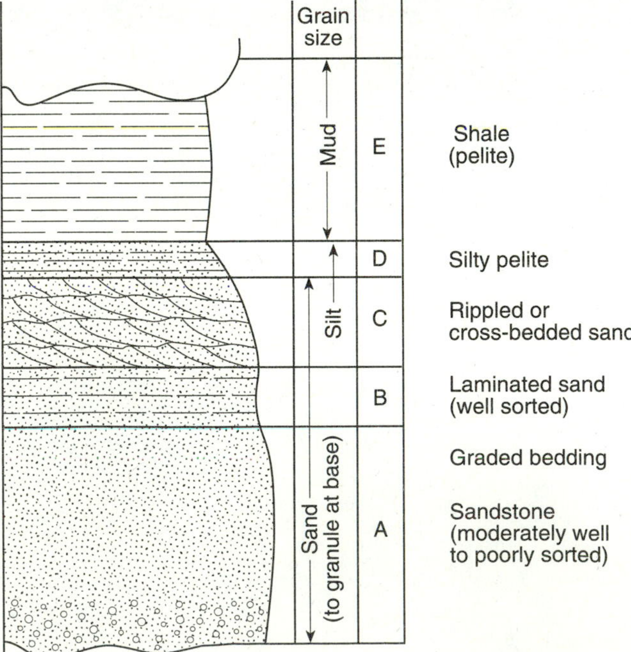

FIGURE 2–29
Bouma sequence. (From *Geologie en Mijnbouw,* v. 21, A. H. Bouma, p. 223–227, Fig. 8, © 1959, with kind permission from Elsevier Science, Ltd., Kidlington, United Kingdom.)

Salt Structures

Evaporite deposits occur in sedimentary sequences at shallow crustal levels in many parts of the world. Layers of anhydrite and gypsum undergo ductile deformation (Chapter 5) more readily than do the more common sedimentary rock types such as sandstone, dolostone, limestone, and shale. Rock salt, composed mostly of halite, flows more readily than does any other common rock type. Rock salt deposits, tens or hundreds of meters thick, formed by evaporation of sea water occur in the United States in the Texas-Louisiana Gulf Coast, West Texas, Kansas, Michigan, and New York; this is also the case in West Germany, Spain, Jordan, Iran, and elsewhere. Salt flows at surface conditions under the force of gravity. The density of salt contrasts with the greater density and strength of the enclosing sediments. A great variety of structures can be produced, ranging from salt glaciers on the surface (Figure 2–30a) to salt pillows, intrusive stocks, and domes at depth (Figure 2–30b). The internal structure of these salt features provides abundant evidence of plastic flow, with folds, foliations, and other structures resembling those formed at high pressures and temperatures in metamorphic rocks (Figure 2–31; Muehlberger, 1968; Jackson and Talbott, 1989; Jackson and others, 1990).

Salt or other materials—commonly water-saturated "mud lumps"—that move upward and gravitationally intrude the overlying sediments are called ***diapirs.*** They are common in the Mississippi delta and serve as important hydrocarbon traps (Figure 2–30b). Salt and mud diapirs provide useful analogs for deep crustal diapirs of lower-density rocks (such as granitic gneiss and granitic magma) that move upward through denser rocks and form domes (Figure 2–32).

The importance of salt deformation in controlling deformation of sediments and influencing formation of large hydrocarbon reservoirs in the U.S. Gulf Coast, the North Sea, Iran, and other hydrocarbon-producing regions has accelerated the rate of increase of knowledge about salt structures and the control that salt has on deformation throughout salt provinces.

Extensional processes in cover sediments related to flow of salt under the influence of gravity control the locations of diapirs and the geometry of structures that form in cover sediments (Vendeville and Jackson, 1992a, 1992b). The sizes and shapes of folds, displacements on faults, and formation of other structures are all related to flow of salt in the subsurface.

FIGURE 2–30
(a) Oblique aerial view of an eroded salt dome in Iran showing folded internal structure and boundaries of the dome. Layering in salt and country rocks is truncated at the contact. (Augusto Gansser, Eidg. Technisches Hochschule, Zürich, Switzerland.) (b) Shapes of large salt structures. Numbers indicate relative elevation above base. (From M. P. A. Jackson and C. J. Talbot, 1986, Geological Society of America *Bulletin,* v. 97.)

(a)

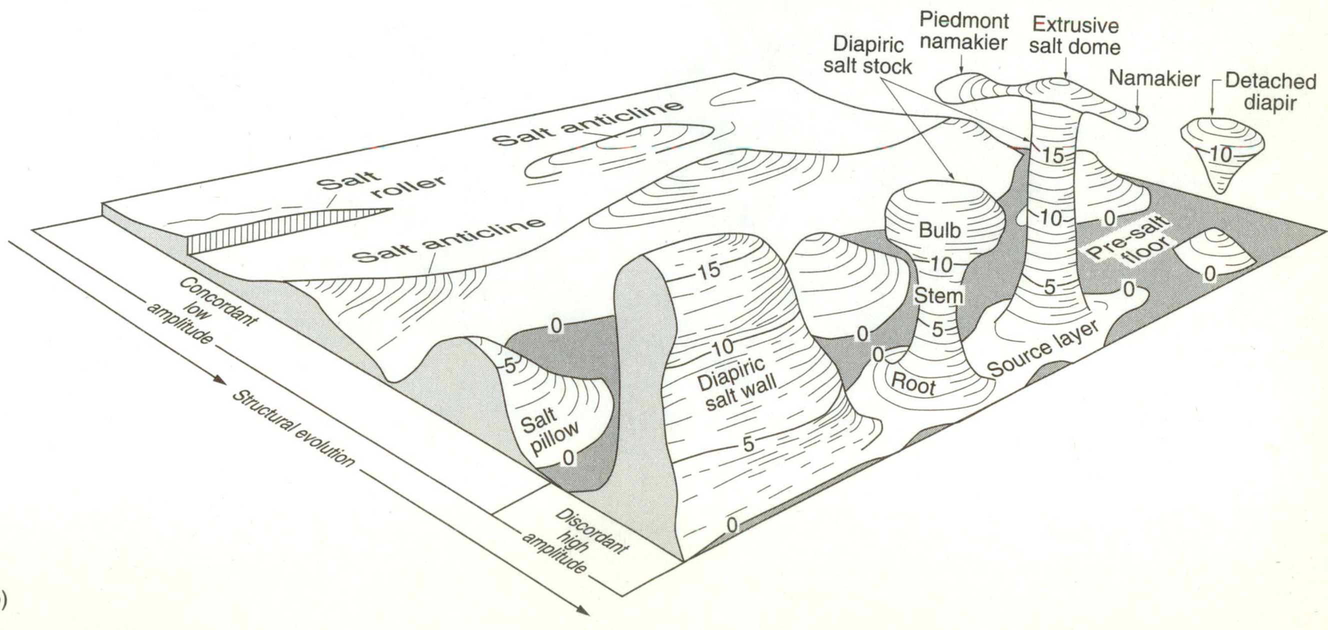

(b)

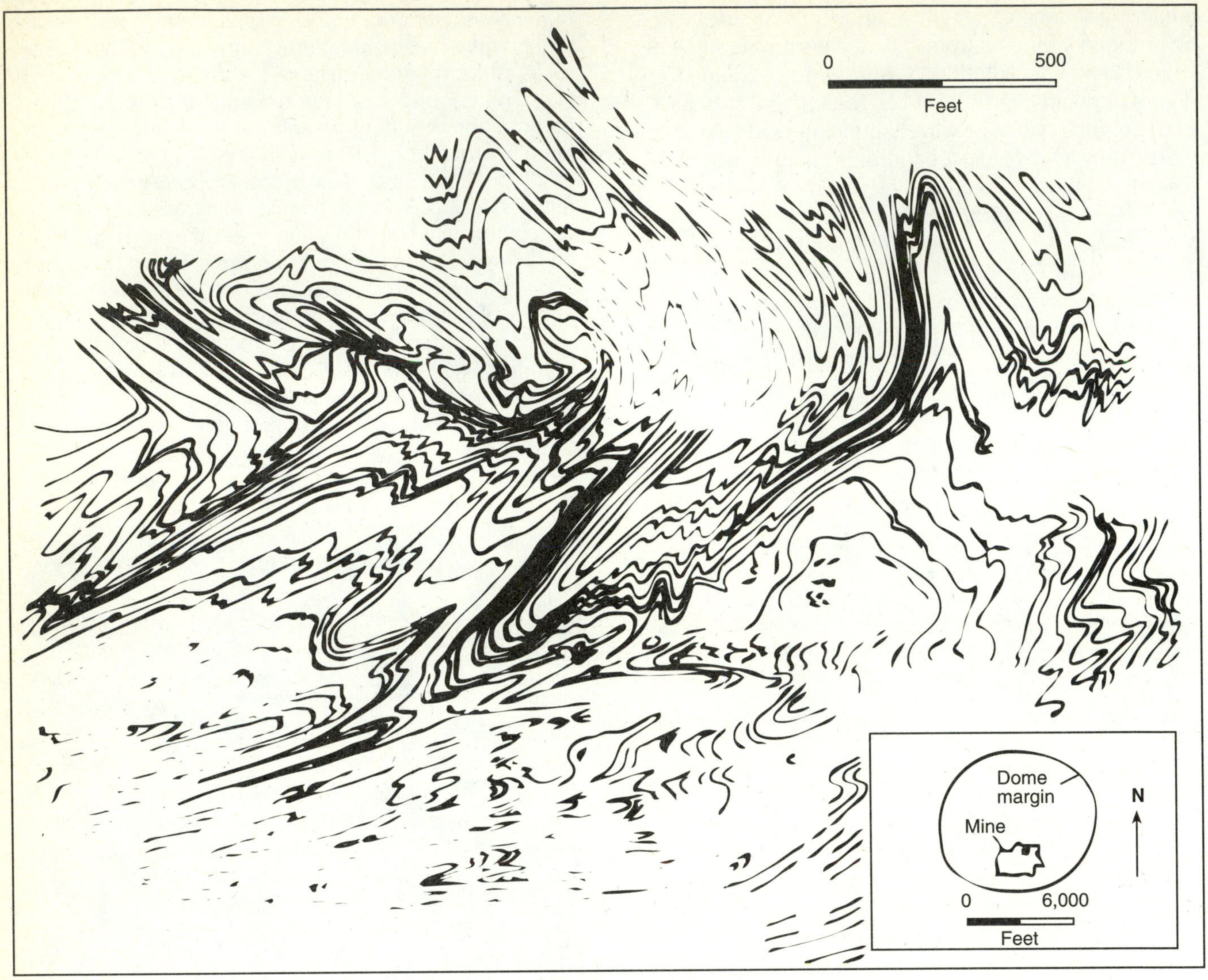

0 500
Feet

Dome margin
Mine
N
0 6,000
Feet

(a)

(b)

FIGURE 2–31
(a) Map of structures in salt in the Grand Saline salt mine, Grand Saline salt dome, Texas. (From W. R. Muehlberger, 1968, Geological Society of America *Special Paper* 88.) (b) Ductilely deformed, strongly foliated and folded salt inside the Grand Saline salt dome. Each dark mark in the salt is a shot hole spaced about two meters from the adjacent hole. (Courtesy of W. R. Muehlberger, University of Texas, Austin.)

FIGURE 2–32
Sections through diapirs produced in centrifuge experiments showing the shape of analog structures that resemble those in carefully mapped salt domes and some plutons. Black-and-white layers are markers with the same mechanical properties. (From *Journal of Structural Geology,* v. 11, M. P. A. Jackson and C. J. Talbot, p. 211–230, © 1989, with kind permission from Elsevier Science, Ltd., Kidlington, United Kingdom.)

IMPACT STRUCTURES

The surfaces of our moon and the planets with thinner atmospheres than that of the Earth reveal a history of impacts spanning billions of years. The Earth probably had a similar history, but evidence of such a lengthy history of impacts is lacking because of the dynamic surficial and tectonic processes that constantly change the surface of the Earth. Based on evidence of impacts of large objects on the other terrestrial planets throughout the history of the Solar System, large objects should have impacted the Earth over intervals of every few million years. Layered mafic intrusive bodies, such as the Sudbury (Ontario) Irruptive, have been interpreted as having been intruded following large impacts.

Structures generally having circular or elliptical outlines have been identified in different parts of the world but are not obviously related to tectonic processes. Unfortunately, many of these structures are deeply eroded, having formed before the Cenozoic. Others, for example, Meteor Crater in Arizona, have an origin more obviously extraterrestrial. Some of the older structures, including the Wells Creek structure in Tennessee, Jephtha Knob in Kentucky, Serpent Mound in Ohio, Kentland in Indiana, and Manson in Iowa, were once considered *cryptovolcanic* structures, related to explosive volcanic activity at depth that disrupted the surface rocks but did not produce surface evidence of volcanic activity—hence the prefix *crypto-* (hidden).

Direct evidence of meteorite impact has been found at Meteor Crater in the form of meteorite fragments and the high-pressure (shock-related), low-temperature silica polymorphs coesite and stishovite. The older structures mentioned above are more difficult to assess, but the cryptovolcanic interpretation has been questioned because none shows any vestige of the effects of volcanic activity at depth. One of the best-exposed older impact craters is the Wells Creek structure. It was first investigated by Bucher (1936), who cited it as an outstanding example of a cryptovolcanic structure (Figure 2–33), although it was discovered that Wells Creek and similar structures contain *shatter cones* (Figure 2–34), which are produced by brittle deformation like that observed where an explosive charge at the bottom of a drill hole produces a cone-shaped fracture propagating down and away from the explosion. Occasionally, shatter cones are obvious in highway cuts, radiating downward from the base of an exhumed drill hole. Where shatter cones can be seen in place, their apices mostly point upward (such as those in an exploded drill hole), suggesting impact from above. Careful measurement of the orientation of many shatter cones at Wells Creek indicates a common orientation (Figure 2–35), suggesting impact from above by a body with a trajectory inclined about 10° from the vertical. Estimates based on experiments and measurements of meteors traveling through the atmosphere suggest that the object that produced the Wells Creek structure was about 300 m in diameter, weighed 1.8 to 9 x 10^{10} kg, and was traveling at a velocity of about 40 to 50 km s^{-1} at the time of impact (Wilson and Stearns, 1968).

Studies of this and similar structures in North America, Europe, and in other continents led to the conclusion that they formed by meteorite impact. Some very large circular structures, such as the eastern part of Hudson Bay, have even been suggested to be impact structures (Dietz, 1960). The importance of large-body impacts on our moon and other planets is widely known and led to a controversy that began during the 1980s about the role of impacts in the extinctions of dinosaurs and other organisms (Silver and Schultz, 1982). Suggestions were made in the early 1990s that the impact that caused the extinction had been found, with the most likely candidate being the 65-m.y.-old Chicxulub crater in the Coastal Plain sediments in Yucatan (Swisher and others, 1992).

This concludes our review of related topics. We now will begin our discussion of structural geology via an introduction to rock mechanics.

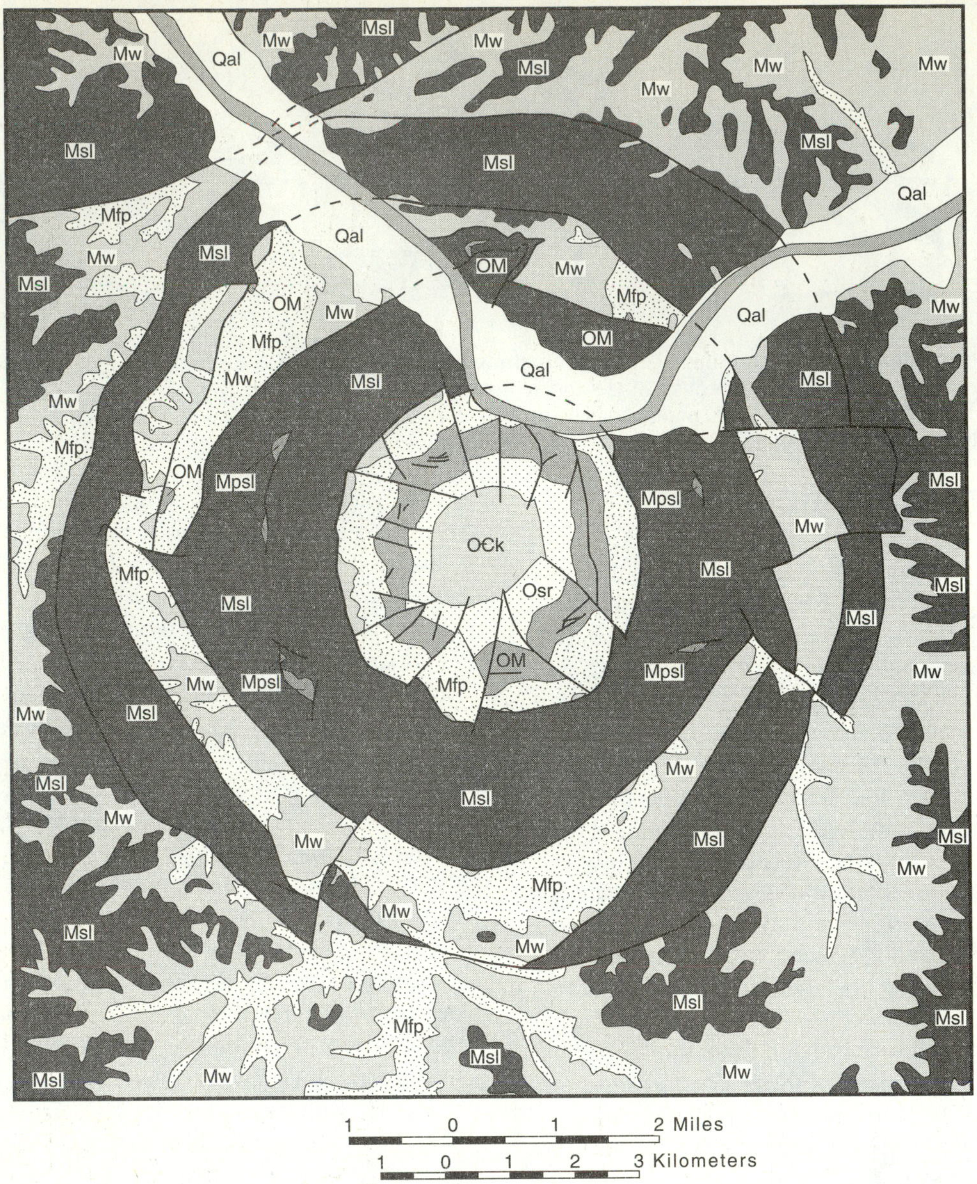

FIGURE 2–33
Geologic map of the Wells Creek structure in west-central Tennessee. Note the circular character of the outcrop pattern produced by doming and radial and concentric faults that produced steep dips in the core of the structure. Oldest units are exposed in the central part of the structure. O€k—Knox Group. Osr —Stones River Group. OM—Nashville Group through Chattanooga Shale. Mfp– Fort Payne Formation. Mw—Warsaw Limestone. Msl—St. Louis Limestone. Mpsl—Post-St. Louis Mississippian rocks. Qal—Alluvium. (From C. W. Wilson, Jr., and R. G. Stearns, 1968, Tennessee Division of Geology *Bulletin* 68.)

FIGURE 2–34
Shatter cones from the Wells Creek structure in Tennessee. (Specimen courtesy of Richard G. Stearns and John Hancher, Vanderbilt University.)

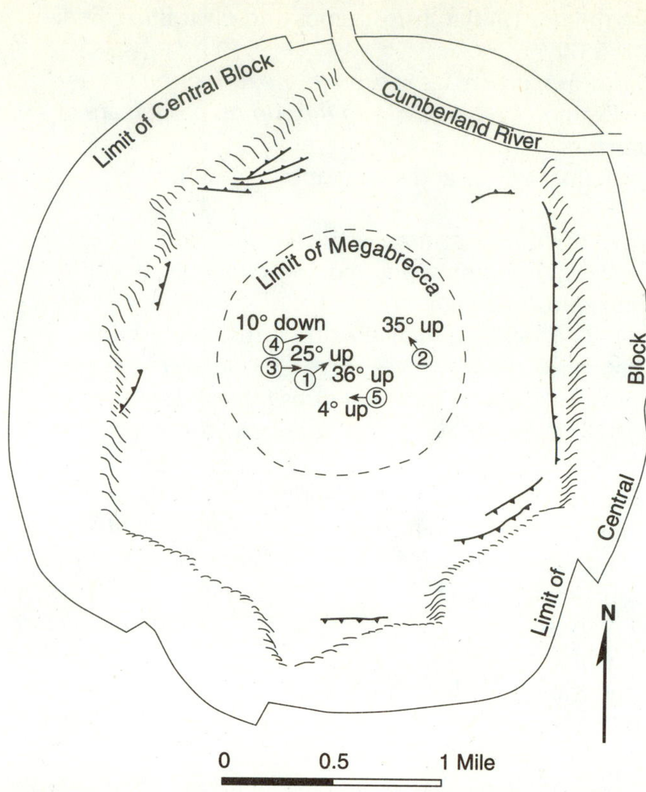

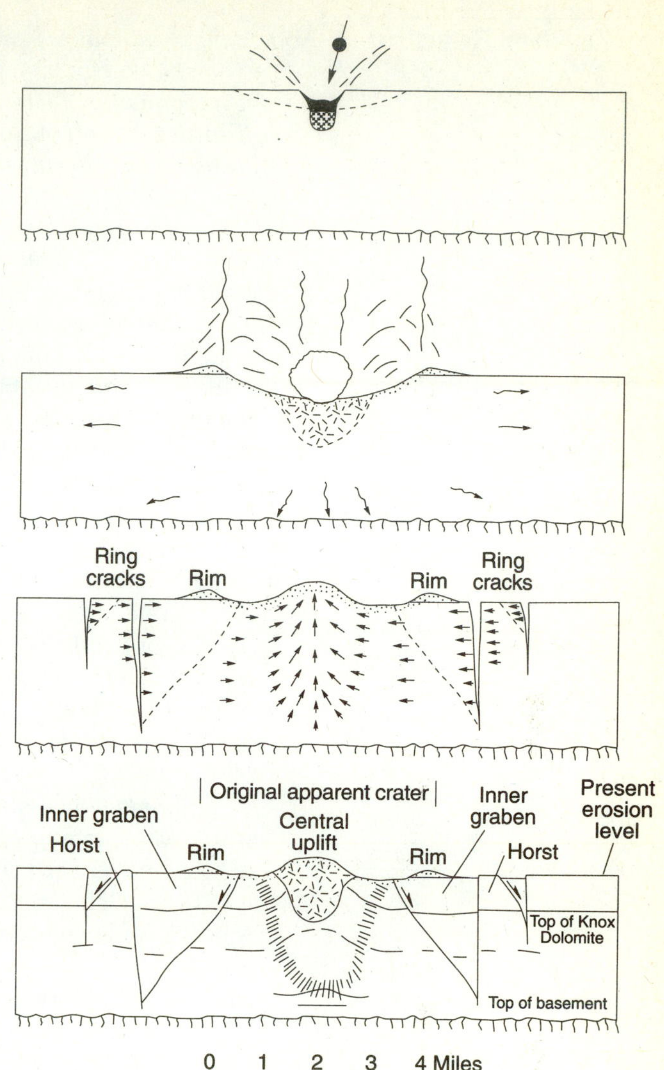

FIGURE 2–35
Reconstruction of the average orientations (numbered localities with arrows indicating directions and inclinations) of shatter cones at Wells Creek structure. Map (a) indicates an upward but not vertical orientation for the body at impact. Cross section (b) shows the sequential history of impact. (From C. W. Wilson, Jr., and R. G. Stearns, 1968, Tennessee Division of Geology *Bulletin* 68.)

Questions

1. How do we commonly distinguish tectonic from nontectonic structures? Why is it important to do so?
2. How does bedding form in sedimentary rocks?
3. Why are current ripple marks (excluding ripple cross laminations) generally not useful for determining facing directions?
4. Why are mud cracks and rain imprints less useful for top determinations than are graded beds and cross beds?
5. How can you tell the difference between a sill and a lava flow that has been buried in a sequence of sedimentary rocks?
6. Why do some areas with conditions that favor landsliding never have slides?
7. What is the difference between a mélange and an olistostrome?
8. Why are the lower parts of Bouma sequences most commonly preserved?
9. How can you tell an igneous intrusive contact from a nonconformity?
10. How can a nonconformity also be an angular unconformity?

Further Reading

Abbate, E., Bortolotti, V., and Passerini, P., 1970, Olistostromes and olistoliths: Sedimentary Geology, v. 4, p. 521–557.
Development of the terminology and processes related to the formation of olistostromes and olistoliths. Examples primarily from the Alpine orogen are discussed clearly and summarized.

Boggs, S., Jr., 1987, Principles of sedimentology and stratigraphy: New York, Macmillan, 784 p.
Contains an excellent treatment of primary sedimentary stuctures (Chapter 6), in-depth discussion of sedimentary environments, and modern concepts of stratigraphy, including seismic and magnetostratigraphy.

Collinson, J. D., and Thompson, D. B., 1989, Sedimentary structures, 2nd edition: London, Chapman & Hall, 224 p.
Outlines the kinds and origins of various primary structures in sedimentary rocks.

Reineck, H.-E., and Singh, I. B., 1975, Depositional sedimentary environments: Berlin, Springer-Verlag, 439 p.
Contains outstanding illustrations and photographs of sedimentary structures and textures and discussions of modern depositional sedimentary environments.

Shackleton, R. M., 1958, Downward-facing structures of the Highland Border: Quarterly Journal of the Geological Society of London, v. 72, p. 361–392.
The author employed sedimentary structures to decipher a major tectonic structure in the southern Highlands of Scotland.

Shanmugam, G., Damuth, J. E., and Moiola, R. J., 1985, Is the turbidite facies association scheme valid for interpreting ancient submarine environments?: Geology, v. 13, p. 234–237.
Raises fundamental questions related to use of the facies principle in the analysis of submarine fans. It discusses earlier works that apply the facies principle to both ancient and modern environments.

Silver, L. T., and Schultz, P. H., eds., 1982, Geological implications of impacts of large asteroids and comets on the Earth: Geological Society of America Special Paper 190, 528 p.
The controversy about extraterrestrial impacts raged throughout the 1980s after Luis and Walter Alvarez and others reported that a rare element, iridium, is concentrated in clays at the Cretaceous-Tertiary boundary. This publication is a compendium of papers from a meeting held to discuss various aspects of the problem, ranging from probability of impacts, through the geologic record of impacts, to other possible extinctions and the Cretaceous-Tertiary boundary problem.

PART TWO

Rock Mechanics

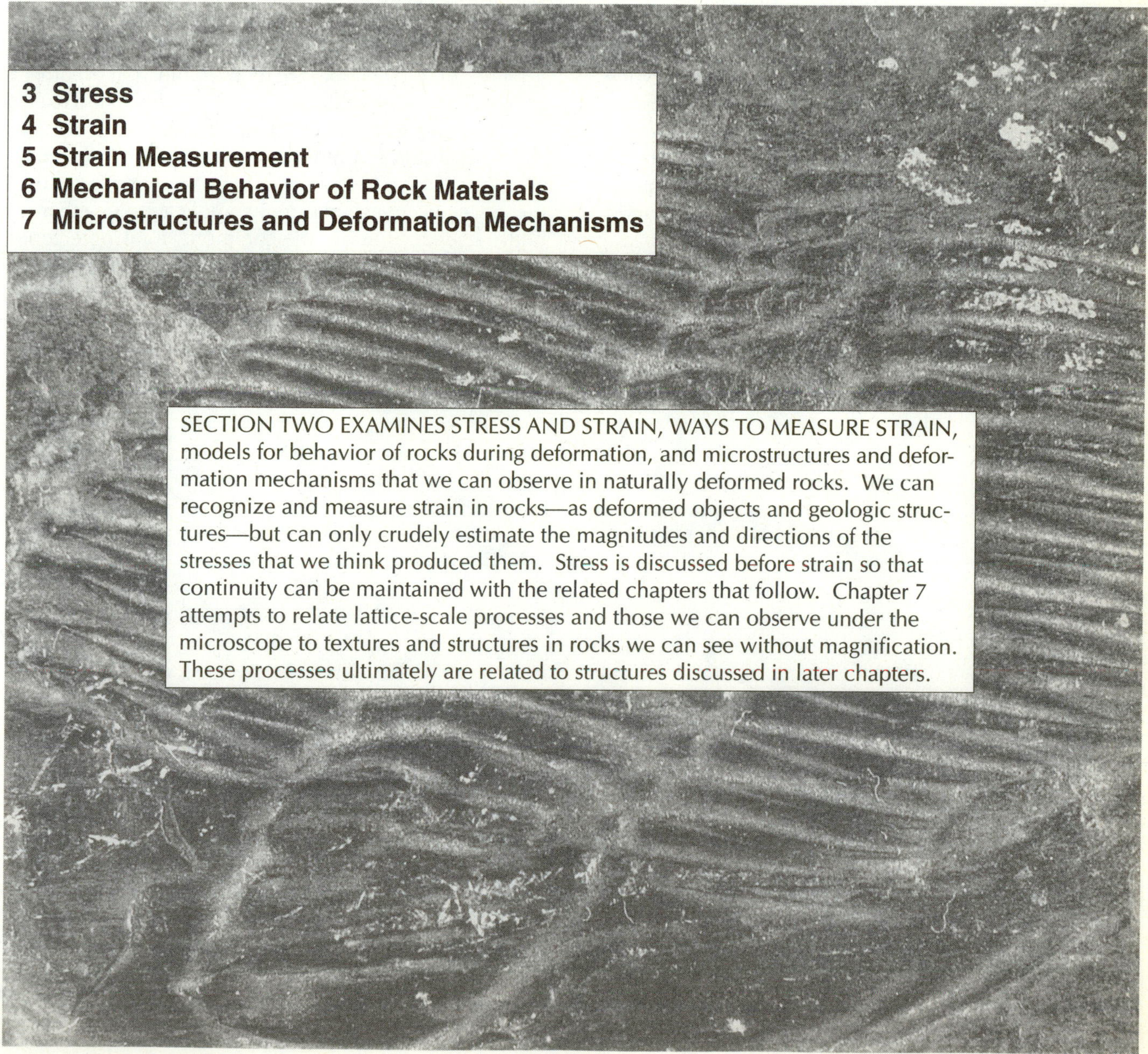

SECTION TWO EXAMINES STRESS AND STRAIN, WAYS TO MEASURE STRAIN, models for behavior of rocks during deformation, and microstructures and deformation mechanisms that we can observe in naturally deformed rocks. We can recognize and measure strain in rocks—as deformed objects and geologic structures—but can only crudely estimate the magnitudes and directions of the stresses that we think produced them. Stress is discussed before strain so that continuity can be maintained with the related chapters that follow. Chapter 7 attempts to relate lattice-scale processes and those we can observe under the microscope to textures and structures in rocks we can see without magnification. These processes ultimately are related to structures discussed in later chapters.

3

Stress

In all quantitative studies on the relationship between the original forces and the resulting deformations, an intermediate field of investigation enters, the condition of stress in the earth's crust. The original forces—whether primary or secondary, whether internal body or external boundary forces—set up a state of stress in the earth's crust which, in turn, produces the observed distortions.

W. HAFNER, 1951, *Geological Society of America Bulletin*

WE ARE ALL FAMILIAR WITH DAILY EFFECTS OF FORCES AND stresses. The most pervasive force affecting us is gravity: it holds the atmosphere, the oceans, and us to the Earth, and keeps the Earth and other planets from fragmenting. ***Forces*** change the velocity or direction of motion of a body. We use forces to open and close doors, ride bicycles and exercise machines, turn on lamps, and perform other everyday activities. In the Earth, forces act on rock bodies and drive certain kinds of deformation, but force is not as meaningful as stress in the study of rock deformation because we must consider the area affected by the force. Hafner's statement above about the direct connection between stresses and deformation can be demonstrated only for elastic deformation. There, a direct proportion exists between the amount of stress and the amount of deformation that results, and elastic deformation (Chapter 6) is, by definition, recoverable. Other forms of deformation, or strain, are not as easily related to forces, or stresses, particularly in the Earth.

Stress is force applied to an area, or force per unit area. It may also be considered a state within a solid that develops in response to the application of either body or surface forces. We commonly think of tectonic structures as products of stress (Figure 3–1), but ironically, we observe most of the effects of stress in tectonic structures without being able to measure the stress that produced the structure. A tectonic structure

FIGURE 3–1
The fault and accompanying folds exposed in this cliff in Alaska, like most others, are assumed to be a result of stress in the crust. (T. H. Moffitt, U.S. Geological Survey.)

is a manifestation of deformation that is assumed to result from stress imposed on a rock mass. Stress originates in processes that generate, move, and consume lithospheric plates, with the aid of gravity. Gravity alone induces the stresses primarily responsible for deformation in salt and impact structures. Thus, understanding the nature of stress is important to understanding tectonic structures. One of the goals of structural geology is to reconstruct the orientations and magnitudes of stresses that produced the structures we encounter in the field. Unfortunately, these stresses are usually no longer present. Experimental attempts to duplicate natural structures in rocks also help us to estimate the orientations and magnitudes of stresses needed to produce deformation in rocks.

DEFINITIONS

Several terms prove to be useful in our attempts to mathematically describe forces and stresses. A ***scalar*** possesses only magnitude; a ***vector*** possesses both magnitude and direction (Figure 3–2). A scalar is a number (for example, the price of oil, the score in a baseball game, temperature, or the thickness of a rock unit). A vector is a number with an indication of direction. For example, if we say a car is traveling northwest at 100 km per hour, we have defined a vector; if we say only that it is traveling 100 km per hour—with no indication of direction—we have defined a scalar. In this book, vectors in equations are indicated by **bold** type.

A ***tensor*** is a mathematical tool for defining and manipulating a group of quantities, where each quantity is represented by a magnitude, and most have an accompanying "direction." (Note that the direction does not have to be related to real coordinates in space or time.) Tensors have the same magnitude in any coordinate system, but the value of the components depends on the choice of coordinate system or "space." Tensors have different orders: a *zero-order tensor* is a scalar and has only one component; an example is the temperature at a particular time. A *first-order tensor* is a vector and has three components; a wind current is a first-order tensor if it is thought of as a single tensor quantity at a point. A *second-order tensor* relates sets of vectors to each other and has nine components. As we shall see, components of shear and normal stress at a point form a second-order tensor. The order of the tensor describes the minimum number of directions necessary to describe the dimensions under consideration. Thus the number of components for a particular order depends on the number of dimensions in the space being considered. For a first-order tensor, the number of components equals the number of dimensions in the space (2D=2, 3D=3, 4D=4, . . .).

A tensor may be used to describe a physical quantity by referring it to an appropriate coordinate system. The elastic constants relating stress and strain (Chapter 6) comprise a fourth-order tensor. The number of components in a tensor may be determined from 3^n, where n is the order, and the order equals the number of letters or numbers in the subscript for a particular element of the tensor. The symbol τ_{yz} for a shear stress tells us that it is part of a second-order tensor.

A ***force*** (Figure 3–3) is a vector that produces a change in the velocity or direction of motion of a body that may either may be stationary or may already be in motion. Newton's *second law of motion* states that

$$\mathbf{F} = m\mathbf{a}, \qquad (3\text{–}1)$$

where **F** is force (a vector), m is mass (a scalar), and **a** is acceleration (also a vector). *Body forces* act equally on all parts of a body. Examples are the effect of gravity or

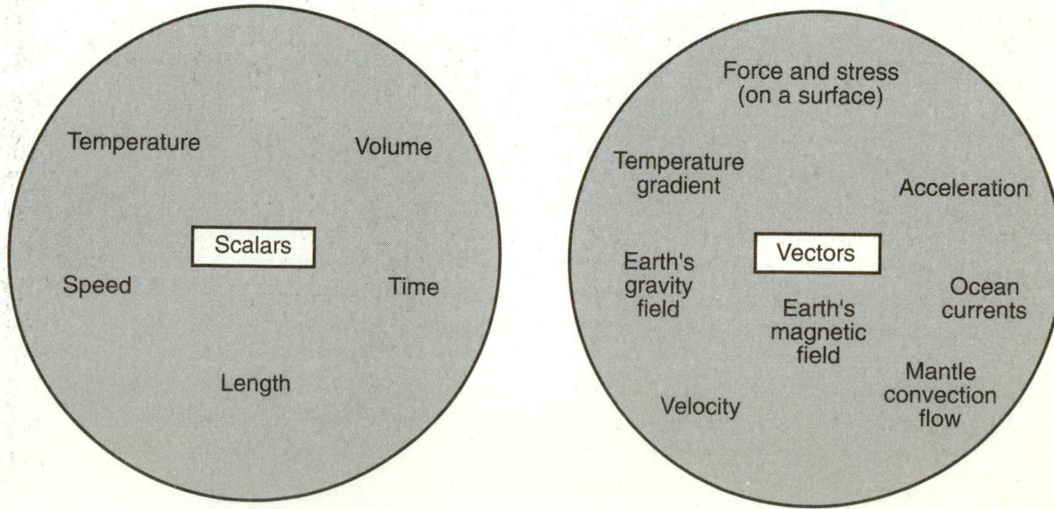

FIGURE 3–2
Scalars and vectors.

electromagnetic forces on a mass. Body forces must be considered if the behavior of fluid or ductile material is involved, but are not dependent on the rheology (state) of the material under consideration. *Surface forces* act on external or internal surfaces within rock masses and include forces acting along a fault or a major plate boundary. The magnitude of a gravitational body force is proportional to the amount of mass present, but the magnitude of a surface force is independent of the surface area affected. (For example, the same force is exerted by the heel of a shoe regardless of whether the person is wearing street shoes, field boots, or cowboy boots, or is a lady in spike heels, but the force on the heels is distributed over a smaller area beneath the lady in spike heels than the same lady wearing field boots.) Moreover, a surface force can be resolved into mutually perpendicular components: one normal to the surface, and one or two parallel to the surface. This relationship will reappear when we discuss normal and shear stress.

Force may be converted to stress by dividing by the area (*A*) affected by the force (Figure 3–4). Stresses may be ***tensional*** (pulling apart) or ***compressional*** (pushing together). Stress acting on a surface may be resolved into two vector components: ***normal stress***, σ_n, acts perpendicular to a reference surface; ***shear stress***, τ, acts parallel to a surface. Stress vectors acting across planes of zero shear stress are ***principal stresses***, commonly distinguished as three ***principal normal stress components***, σ_1, σ_2, and σ_3, of σ_n. The three principal normal stress components are oriented perpendicular to each other, and $\sigma_1 \geq \sigma_2 \geq \sigma_3$. Their directions are *principal directions of stress*. The planes that contain the principal stresses are *principal planes of stress*. Differential stress is the difference between the maximum (σ_1) and minimum (σ_3) principal normal stresses ($\sigma_1 - \sigma_3$); see equations 3–15 through 3–18 below, and related text. Mean stress is $(\sigma_1 + \sigma_2 + \sigma_3)/3$. *Deviatoric stress* is the nonhydrostatic component of stress, expressed as total stress with mean stress subtracted from the normal stress components. If the differential stress exceeds the strength of the rock, permanent deformation results. The *strength* of a material is the stress required to cause permanent deformation.

A *lithostatic state of stress* occurs where normal stress is the same in all directions in the Earth; a *hydrostatic pressure* is the confining stress acting on a body submerged in water at a known depth. In the case of a body buried in the Earth, the weight of the column of rock (instead of water) per unit area above it is called the *lithostatic pressure* (Figure 3–4). Under both lithostatic and hydrostatic conditions, one-third of the sum of all three principal stress components equals the mean stress. Under any deviation from lithostatic conditions, shear components will still exist.

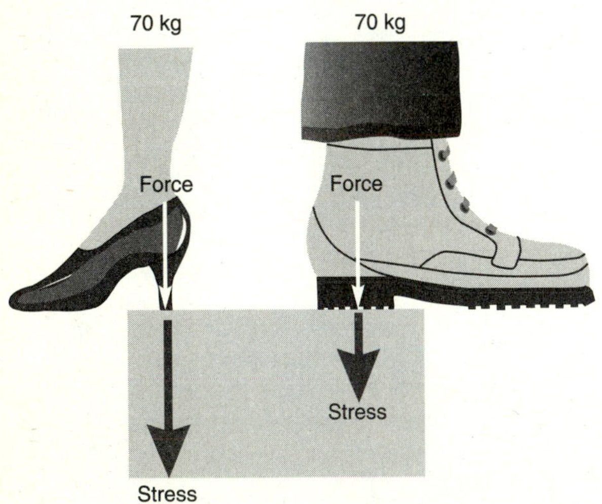

FIGURE 3–3
Forces and stresses. Two persons weighing 70 kg each stand on one foot on the edge of a box so that all of the weight is concentrated on the heel of the shoe. The person to the left is wearing a shoe with a narrow heel, but the shoe on the person on the right has a wide heel. Note that the same force is applied to the box by both people, but the applied stress is much greater with the person to the left because it is concentrated in a smaller area.

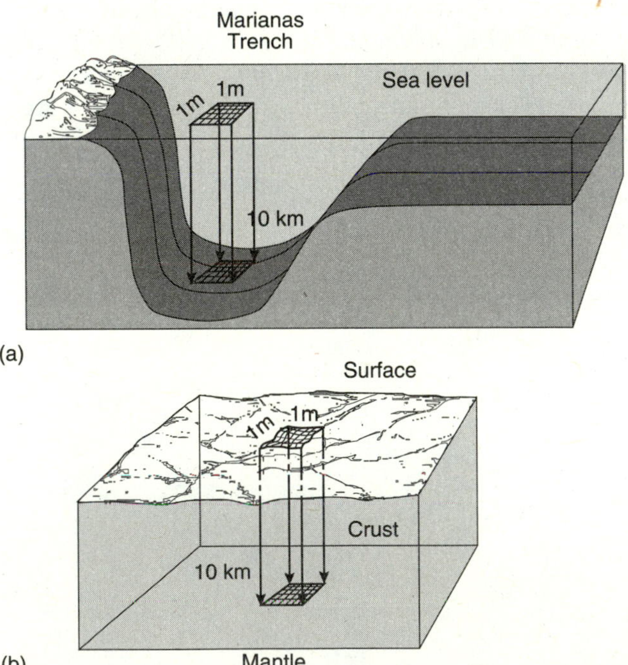

FIGURE 3–4
Stress on a body 10 km under water (a) or 10 km deep in the crust (b), assuming the stresses are distributed isotropically. The differences would arise principally from the weight of the column of water (a) or rock (b) above the 1-m^2 area at depth.

For example, the force on a body submerged at a depth of 10 km near the bottom of the Mariana Trench would equal the weight of the mass of water on it, expressed as

$$\begin{aligned} \mathbf{F} &= \text{weight of water} \\ &= \text{height x density of sea water} \\ &\quad \text{x area x acceleration of gravity} \\ &= 10^4 \text{ m x } 1{,}030 \text{ kg cm}^{-3} \text{ x } 1 \text{ m}^2 \text{ x } 9.8 \text{ m s}^{-2} \\ &= 1.009 \text{ x } 10^8 \text{ kg cm s}^{-2}. \end{aligned} \quad \textbf{(3–2)}$$

The hydrostatic stress is obtained by dividing by the area—that is, taking the water depth times the density of sea water times the acceleration of gravity, *g,* or

$$\begin{aligned} &10^4 \text{ m x } 1{,}030 \text{ kg cm}^3 \text{ x } 9.8 \text{ m s}^{-2} \\ &\quad = 100.9 \text{ x } 10^6 \text{ kg cm}^{-1} \text{ s}^{-2} = 101 \text{ MPa}. \end{aligned} \quad \textbf{(3–3)}$$

(1 Pa = 1 pascal = 1 kg cm^{-1} s^{-2}; 1 MPa = 10^6 pascals. 1 megapascal is the standard unit of stress in the Earth; it also is equal to 10 bars or 9.8 atmospheres of pressure.) If the same body were buried 10 km deep in the continental crust, the lithostat could be calculated by substituting the average density of the column of rock above the body, assumed to be 2,750 kg cm^{-3}. Substituting for sea water density in equation 3–3, we obtain the lithostatic value of stress

$$\begin{aligned} &10^4 \text{ m x } 2{,}750 \text{ kg cm}^{-3} \text{ x } 9.8 \text{ m s}^{-2} \\ &\quad = 269.5 \text{ x } 0^6 \text{ kg cm}^{-2} = 269.5 \text{ MPa}. \end{aligned} \quad \textbf{(3–4)}$$

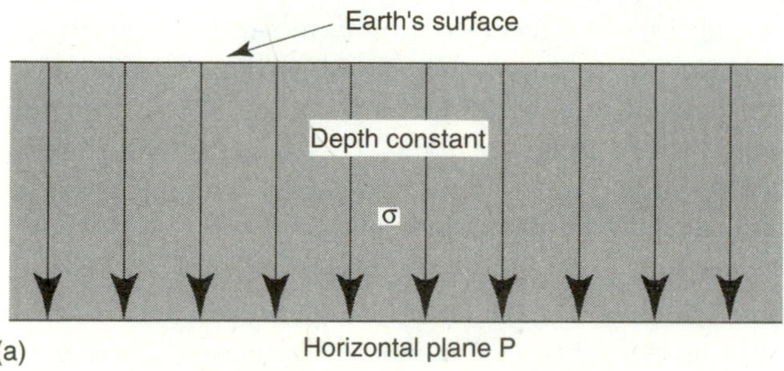

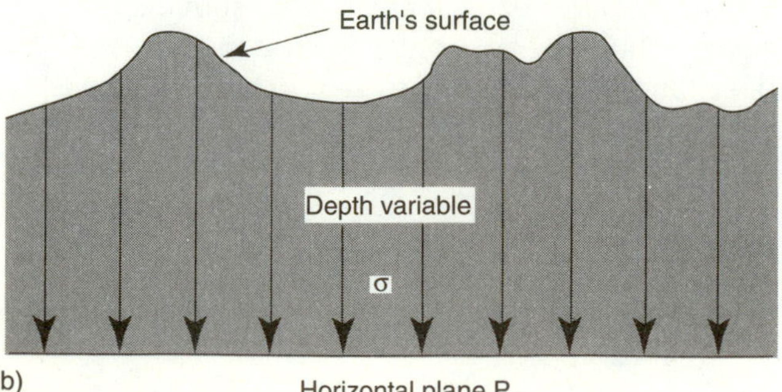

FIGURE 3–5
Stress vectors acting on a horizontal plane near the surface beneath smooth topography (a) and irregular topography (b). The inclined arrows represent a randomly oriented stress, σ.

STRESS ON A PLANE

Now that we have defined the terms used in describing and measuring stress, we should consider ways of using them. If we define a plane *P* in a mass of rock subjected to a vertical stress σ (Figure 3–5), the stress across a small part of the plane can be written as

$$\sigma = \frac{\Delta \mathbf{F}}{\Delta A}, \quad \textbf{(3–5)}$$

where *A* is area. If we assume this segment of the plane is infinitesimally small (equivalent to the statement of stress at a point, discussed in the next section),

$$\sigma_i = \lim_{\Delta A \to 0} \frac{\Delta \mathbf{F}}{\Delta A}. \quad \textbf{(3–6)}$$

or

$$\sigma = \frac{d\mathbf{F}}{dA}. \quad \textbf{(3–7)}$$

Equation 3–6 shows that stress on a plane σ is a vector quantity—from a vector (force) and a scalar (1/area)—and that the stress on each plane can be expressed as a unique set of normal and shear stress vectors (also known as *tractions).* Stress on any plane we choose in a mass of rock, particularly near the surface, is likely to vary from place to place on the plane (Evans and others, 1989), either because the amount of overburden varies or because the plane is inclined to the surface (Figure 3–6). At depths greater than a few kilometers in the Earth, stress on a horizontal plane is related to the density and height of the column of rock above it (ρ*gh* = density x gravity x height), although at depths of a few kilometers, variations in the amount of stress can occur, even within the plane. Determining the stress on an inclined plane is more difficult and is best accomplished using tensors (see Nye, 1957). The area of the plane, the density of the column of rock, and the height of the column are all involved, but the angles between the plane and the principal stress directions must also be known. Because the plane is inclined, the height of the column of rock varies along the plane. On the plane, the stress (σ) may be resolved into components of shear stress (τ) and normal stress (σ_n).

If we assume that our 1-m^2area in Figure 3–7 is a different plane now inclined 45° (for convenience here) to the horizontal, we should be able to calculate the vertical stress on the plane (Figure 3–7a). Because the plane is now inclined above the 1-m^2 area, the area of the plane is greater—1.41 m^2 (= $\sqrt{2}$ m^2). The vertical force, using **F** = m**a** = volume x density x acceleration of gravity, is

$$\begin{aligned} \mathbf{F} &= 10^4 \text{ m}^3 \text{ x } 2{,}750 \text{ kg m}^{-3} \text{ x } 9.8 \text{ ms}^{-2} \\ &= 2.7 \text{ x } 10^8 \text{ kg m s}^{-2}. \end{aligned} \quad \textbf{(3–8)}$$

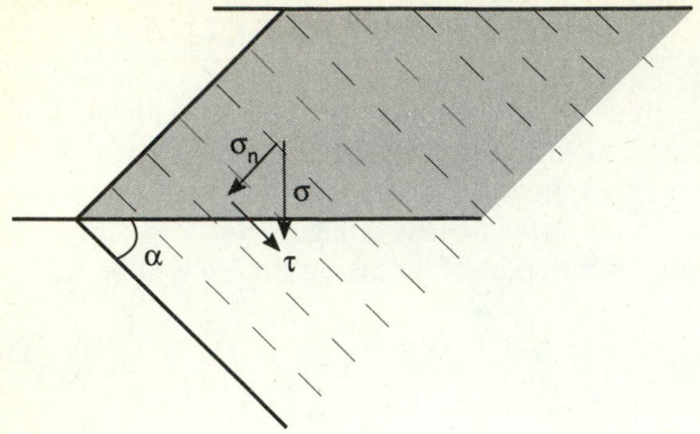

FIGURE 3–6
Stress vectors acting on an inclined plane, such as a fault surface, several kilometers deep in the crust. σ is the vertical component of stress resolved into normal (σ_n) and shear (τ) components.

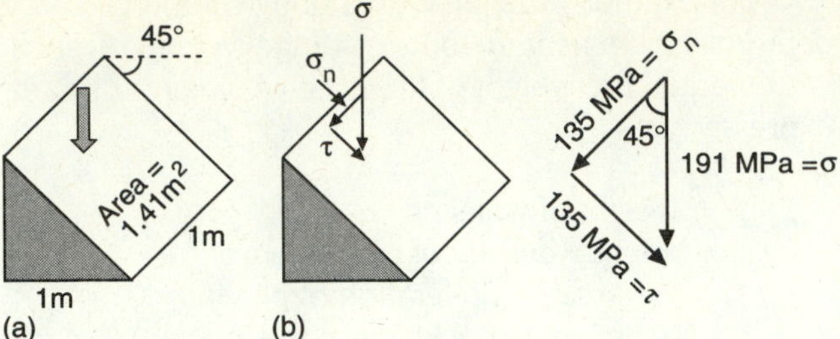

FIGURE 3–7
(a) Vertical stress acting on an inclined plane with an area of 1.41 m^2. (b) Resolution of normal and shear stress components of the vertical stress in (a) and calculation of the magnitude of stresses.

Conversion of force to stress involves dividing by the area of 1.41 m^2, which yields a stress on the inclined plane of 191 MPa. Note that the area has increased, and so the normal stress on the inclined surface has decreased, because the same force is distributed over a larger area. (Shear stress on the plane, however, must increase from zero to a nonzero value.)

Because any plane will have a unique stress vector resolvable into normal and shear components, the normal and shear components of stress (Figure 3–7b) can also be calculated from

$$\sigma_n = \sigma \cos 45° = 191 \text{ MPa} \times 0.707 = 135 \text{ MPa} \quad \textbf{(3–9)}$$

and

$$\tau = \sigma \sin 45° = 191 \text{ MPa} \times 0.707 = 135 \text{ MPa}, \quad \textbf{(3–10)}$$

where σ is the traction on the plane. Note that we are considering only a uniaxial vertical stress here; much greater complexity would be introduced if we considered a triaxial case including horizontal stresses that result from the horizontal resistance to expansion due to vertical loading.

Mohr Circle Derivation

It is useful to establish a mathematical relationship between the principal stresses in a rock mass, and between the normal and shear stresses on a plane or surface within the mass. The ***Mohr circle*** relationship is a simple and useful means of accomplishing this.

If we define an inclined plane *P* having unit area, and a randomly oriented stress σ acts on it (Figure 3–8), we can derive several expressions for the normal (σ_n) and shear stress (τ) on the plane. If a triangular segment ABC represents a small prismatic element with two sides, AC and BC, perpendicular to each other, and AC makes the angle θ with AB (which lies in the larger element *P*), AC and BC are also assumed to be perpendicular to the greatest and least principal (normal) stress components—σ_1 and σ_3. If A represents the area of the hypotenuse face *P* (one unit wide) of the prismatic element, and σ_n and τ are normal and shear stresses acting on this surface, we can derive several useful relationships from the prismatic element ABC. Assume that the prism ABC is infinitesimally small, so that the weight is negligible compared to the forces acting on it. We can also assume that the prism is in a state of static equilibrium: All forces acting to move the prism in any direction are countered by equal and opposite forces. The same is true for forces acting to move the prism in either direction horizontally. (These statements embody Newton's *third law of motion.*) We have already defined the area of the hypotenuse of the prism (plane *P*) as unity, and so the area of the left (vertical) face of the prism is 1 x sin θ, and the area of the base (horizontal face) is 1 x cos θ. The forces acting on the three faces are the stresses multiplied times the area of each face (force = force/area x area). If we sum the forces acting in both the horizontal ($\mathbf{F}_H$) and vertical ($\mathbf{F}_V$) directions (they must sum to zero in both directions),

$$\Sigma \mathbf{F}_H = 0 = \mathbf{F}_x - N \sin\theta - S \cos\theta, \quad \textbf{(3–11)}$$

and

$$\Sigma \mathbf{F}_V = 0 = \mathbf{F}_y + S \sin\theta - N \cos\theta. \quad \textbf{(3–12)}$$

N and S are normal and shear forces, respectively. The force on the left side of the prism (directed parallel to the *x* axis in an xy coordinate system) is $\mathbf{F}_x = \sigma_1 \cos\theta$;

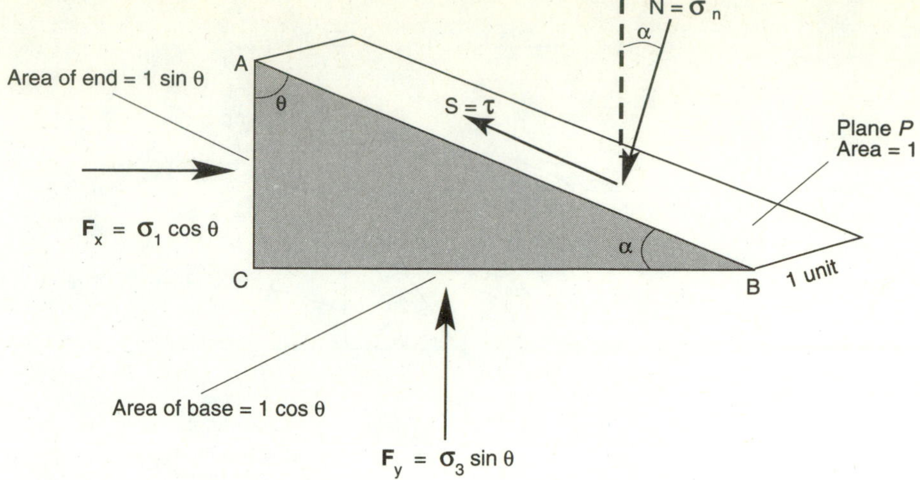

FIGURE 3–8
Relationships between a randomly oriented force is resolved into normal stress σ_n, and shear stress τ, components within the plane *P* having an area *dA* on which the force is applied, and resolved principal normal stress components σ_1 and σ_3, and the angles α and θ within the prism.

that on the base (directed parallel to the *y* axis) is $\mathbf{F}_y = \sigma_3 \cos\theta$. If the force terms are rewritten using components of stress, we can obtain expressions of the normal and shear stresses acting on a plane at an angle θ to σ_1 in terms of the principal stresses

$$\sigma_1 \sin\theta - \sigma_n \sin\theta - \tau\cos\theta = 0, \qquad \textbf{(3–13)}$$

and

$$\sigma_3 \cos\theta - \sigma_n \cos\theta + \tau\sin\theta = 0. \qquad \textbf{(3–14)}$$

If we rewrite equations 3–13 and 3–14 for σ_n and τ, we get

$$\sigma_n = \sigma_1 \cos^2\theta + \sigma_3 \sin^2\theta, \qquad \textbf{(3–15)}$$

and

$$\tau = (\sigma_1 - \sigma_3)\cos\theta\sin\theta. \qquad \textbf{(3–16)}$$

Otto Mohr, a German engineer who in 1882 developed the construction that bears his name, exhibited considerable insight when he realized that if equations 3–15 and 3–16 were rewritten to incorporate double angles, the determinations of σ_n and τ could be made for all cases graphically from a limited number of experimental determinations. He found a graphical solution to the problem much easier than one based on calculation—computers were not available then. Consequently, it was advantageous to introduce an imaginary double angle to replace the real angle.

Recall the trigonometric relationships $\sin 2\theta = 2\sin\theta\cos\theta$, $\cos 2\theta = 1 - 2\sin^2\theta$, and $\sin 2\theta = 2\cos^2\theta - 1$, and the identities

$$\sin^2\theta = \frac{1-\cos^2 2\theta}{2} \quad \text{and} \quad \cos^2\theta = \frac{1+\cos^2 2\theta}{2}.$$

Substituting in equations 3–15 and 3–16 in order to replace the single-angle with double-angle trigonometric terms, we obtain

$$\sigma_n = \frac{\sigma_1+\sigma_3}{2} + \frac{\sigma_1-\sigma_3}{2}\cos 2\theta, \qquad \textbf{(3–17)}$$

and

$$\tau = \frac{\sigma_1-\sigma_3}{2}\sin 2\theta \qquad \textbf{(3–18)}$$

Note that in these equations, θ is the angle between σ and the plane *P* (Figure 3–8 and 3–9). Equations 3–17 and 3–18 define the x and y coordinates of a circle–the ***Mohr circle***—centered on the *x* axis in a Cartesian coordinate system where the axes are the normal σ_n (*x*) and the shear τ (*y*) stresses. Here the general $(x, y) = (\sigma_n, \tau)$ coordinates for the circle are (*x*-axis value of the center + radius times cos 2θ, radius times sin 2θ). Therefore, the radius of the circle for this graphical construction is $(\sigma_1 - \sigma_3)/2$, and the *x*-axis value of the center is $(\sigma_1 + \sigma_3)/2$ (Figure 3–9). (Given θ, σ_1, and σ_3, values of σ_n and τ may be calculated for any plane *P* that is normal to the $\sigma_1\sigma_3$ plane.

A Mohr circle allows us to determine the normal and shear stresses across any plane that is normal to two of the principal stresses. Although we have done this here in two dimensions, some very important relationships will be produced from this type of analysis. In three dimensions, the three Mohr circles that

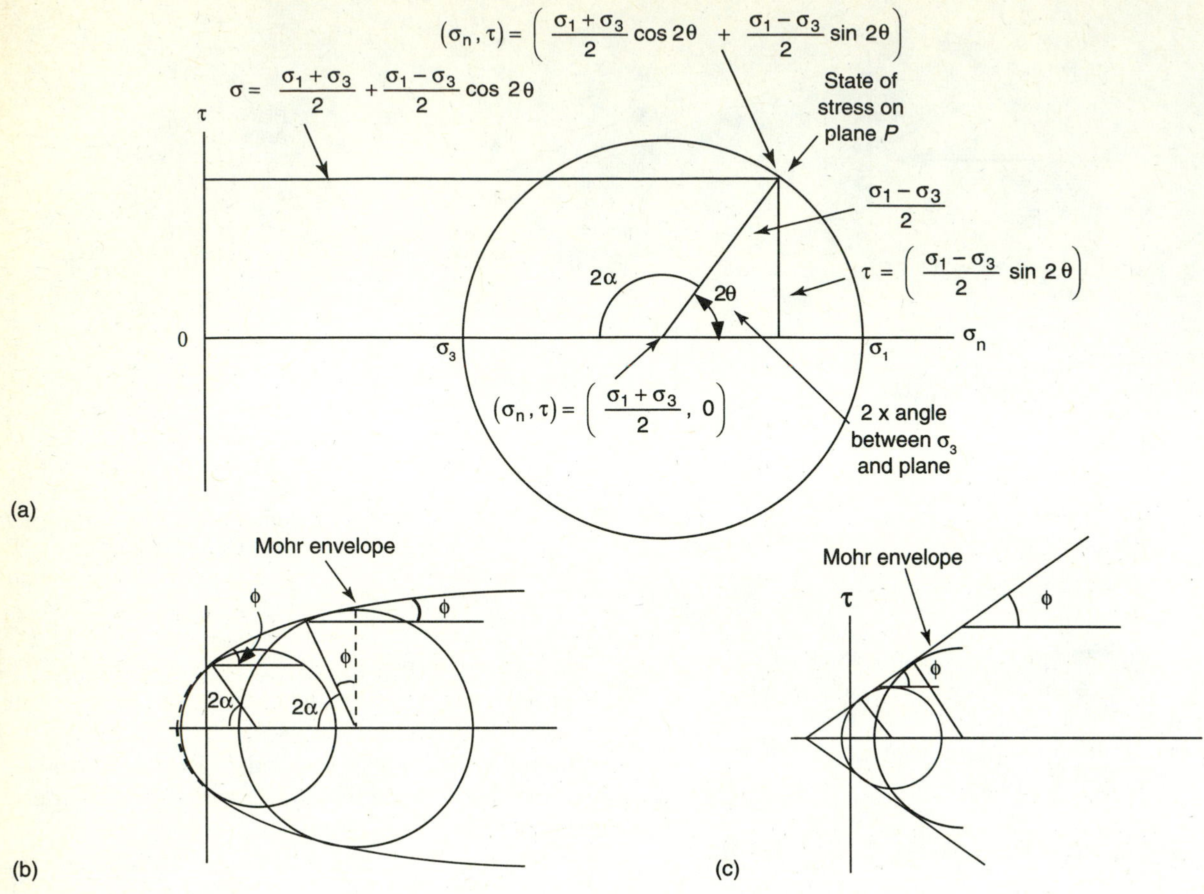

FIGURE 3–9
(a) Mohr circles plotted on σ and τ axes, the envelopes, and the relationships between 2α and 2θ. Note the differences in (b) and (c). With a curved envelope (b), ϕ is variable; with a straight envelope (c), ϕ is constant. ϕ is the angle of internal friction and tan ϕ is the coefficient of internal friction (see Equations 3–21 and 3–25 below).

define the normal and shear stresses on planes normal to two principal stresses bound all of the possible normal and shear stresses. Declan DePaor (1986) represented normal and shear stress with a graphic method employing an "orthonet." His technique provides a way to obtain values for σ_n and τ without trigonometry.

The angle θ is measured between σ_3 and the plane in the ***Mohr construction.*** The relationship may also be stated in terms of the angle α, which is the complement of θ, or the angle between σ_3 and plane *P* in Figure 3–9. These angles reappear in Mohr circles as 2θ and 2α.

STRESS AT A POINT

Having considered stress on a plane in two dimensions, we can now consider stress in three dimensions. Principal normal stresses, designated σ_1, σ_2, and σ_3 ($\sigma_1 \geq \sigma_2 \geq \sigma_3$), may be thought of as oriented parallel to coordinate axes *x*, *y*, and *z*, so that σ_1 corresponds to σ_x, σ_2 to σ_y, and σ_3 to σ_z (Figure 3–10). These normal stresses may also be considered parallel to the edges and normal to the faces of a cube. If the cube is reduced to an infinitesimal size, a stress σ applied to one of the planes that defines the cube may be considered as being applied at a point 0 (Figure 3–11a). We may also assume, using Newton's third law of motion, that the stresses on all sides of the cube balance and cause no rotation or translation, regardless of the size of the cube and regardless of whether the stresses are normal or shear stresses. The cube should be very small so that both shear and normal stresses balance. The stress σ defines a second-order tensor that may be expressed in both normal and shear components, σ_n and τ for any plane. In three-dimensional space, nine components (Figure 3–11b) are required to describe the stress σ at point 0. They include the three components of normal stress, σ_{xx}, σ_{yy}, and σ_{zz}, oriented parallel to the three coordinate axes, and six components of shear

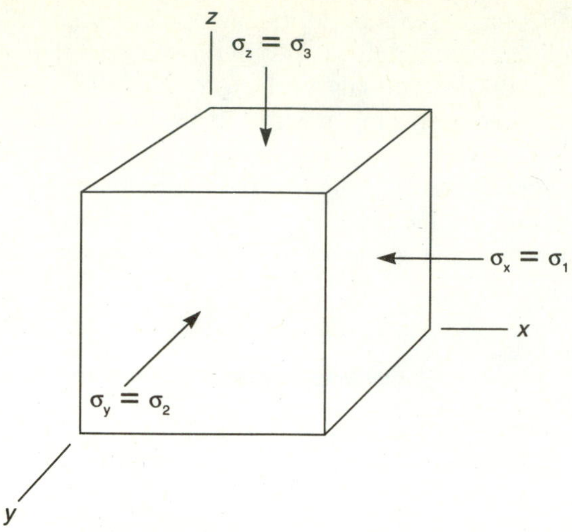

FIGURE 3–10
Principal normal stresses oriented parallel to the edges of a cube in *x-y-z* space. Note the position of the cube relative to the positive ends of the axes.

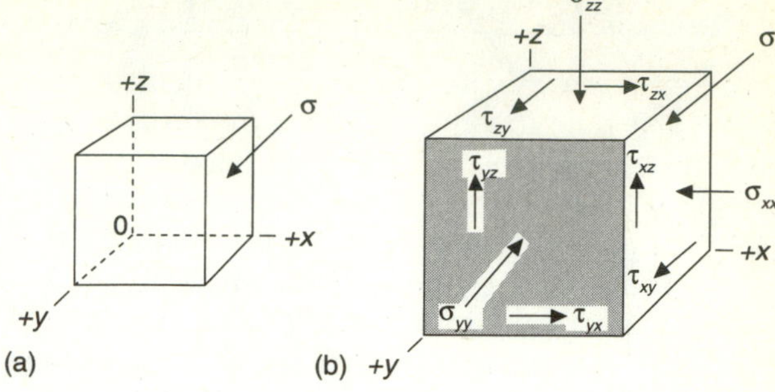

FIGURE 3–11
(a) Randomly oriented stress σ applied to an infinitesimally small (point size) reference cube in *x-y-z* space.
(b) Enlarged reference cube shows resolution of nine shear and normal stress components.

stress lying within the three faces of the cube defined as planes *xy*, *yz*, and *xz*. The symbol σ_{xx}, for one of the principal normal stresses, indicates that this is a normal stress acting perpendicular to the *yz* plane and parallel to the *x* axis. The components of shear stress are then expressed as τ_{xy}, τ_{yz}, τ_{xz}, τ_{yx}, τ_{zy}, and τ_{zx}. Any direction in the three-dimensional space is defined by the normal to the plane that contains the direction and a line in the plane. Thus, two lines define the orientation in three dimensions. Similarly for the cube, the notation *xz* means in the plane normal to *x* in the *z* direction; *yx* means in the plane normal to *y* in the *x* direction.

All nine components of the second-order tensor for stress at a point σ_{ji}, where *i* and *j* take the values of *x*, *y*, and *z*, may be arranged in matrix form as

$$\begin{pmatrix} \sigma_{xx} & \tau_{xy} & \tau_{xz} \\ \tau_{yx} & \sigma_{yy} & \tau_{yz} \\ \tau_{zx} & \tau_{zy} & \sigma_{zz} \end{pmatrix}. \quad \textbf{(3–19)}$$

This matrix contains the components of a *stress tensor* in terms of axes *x*, *y*, and *z*. The rows (horizontal) from top to bottom refer to stresses on a face normal to a particular coordinate axis, and columns (vertical) refer to stress directions parallel to coordinate axes (Figure 3–11b). The columns (vertical) from left to right represent the directions of stress components parallel to the *x*, *y*, and *z* axes. The left column represents stresses acting parallel to the *x* axis, the middle column the stresses acting parallel to the *y* axis, and the right column the stresses acting parallel to the *z* axis. If we designate the elements of the tensors *i* and *j*—again with *i* and *j* taking values of *x*, *y*, and *z*—we can state that the tensor in equation 3–19 is symmetric because $\tau_{ij} = \tau_{ji}$ (that is, elements *i* and *j* may be reversed without changing the values of the components, a requirement of the condition of equilibrium; the subscript *ij* also indicates that this is a second-order tensor). Symmetry is a requirement of the condition of force balance, or equilibrium, assumed at the beginning of this discussion. If the *x-y-z* coordinate system is oriented parallel to the principal stresses, no shear stress will be present on the faces (all values of $\tau = 0$). The same result can be accomplished by rotating the cube in the stress field until only normal stresses are present on the faces. The stress tensor for zero shear stresses may be written as

$$\begin{pmatrix} \sigma_{xx} & 0 & 0 \\ 0 & \sigma_{yy} & 0 \\ 0 & 0 & \sigma_{zz} \end{pmatrix}. \quad \textbf{(3–20)}$$

These three normal stresses are thus called the ***principal stresses.*** They act normal to planes so that shear stresses are zero.

STRESS ELLIPSOID

A graphic means of showing the relationships between the principal stresses is the ***stress ellipsoid*** (Figure 3–12). It is a triaxial ellipsoid in which the greatest, intermediate, and least principal axes are σ_1, σ_2, and σ_3. Principal planes in the stress ellipsoid contain the principal axes. The $\sigma_1 - \sigma_2$, $\sigma_2 - \sigma_3$, and $\sigma_1 - \sigma_3$ planes are principal planes. The $\sigma_1 - \sigma_3$ plane represents the maximum stress difference. All non-principal planes are shear planes.

MOHR CONSTRUCTION

Now we can use experimental results to calculate values of material properties. Plotting σ_n versus τ on two coordinate axes results in the Mohr construction (Figure 3–9a). Construction of the Mohr circle from values of σ_1 and σ_3 from laboratory measurements in a triaxial test apparatus (Figure 3–13) is remarkably simple. A cylindrical sample of rock is placed in the test cylinder. The specimen is jacketed in a thin sleeve of a soft metal such as copper, or thermal plastic, to protect it from a fluid (such as a low-viscosity oil) that is used to vary the confining pressure externally. In an axial compression test, the confining pressure is equal to both σ_2 and σ_3 because pressure is being applied equally around the cylinder. The load on the ends of the cylinder is usually increased until the material ruptures. This load gives rise to the axial stress σ_1. If the rock is loaded to the rupture point, a new cylinder of the same material is required for each run at increasing confining pressure σ_3. Table 3–1 contains values of σ_1 and σ_3 (in kilograms per square centimeter) derived from successive runs to failure on cylindrical specimens of limestone using an axial load apparatus. Plotting these data on the σ_n (horizontal) axis permits construction of a Mohr diagram for the limestone (Figure 3–14a). Plotting each pair of data on the σ_n axis determines the diameter of the circle for each run. The *Mohr envelope* can be constructed with ease, as the tangent to the circles. The Mohr envelope also separates *unfailed* (no permanent deformation) from *failed* (permanently deformed) regions in the diagram; the unfailed region lies within the envelope (Figure 3–14b).

A Mohr circle that does not intersect the envelope indicates that the sample did not rupture with the values of σ_1 and σ_3 employed in that run. As a result, the Mohr envelope can only be constructed using circles obtained when the sample actually ruptured.

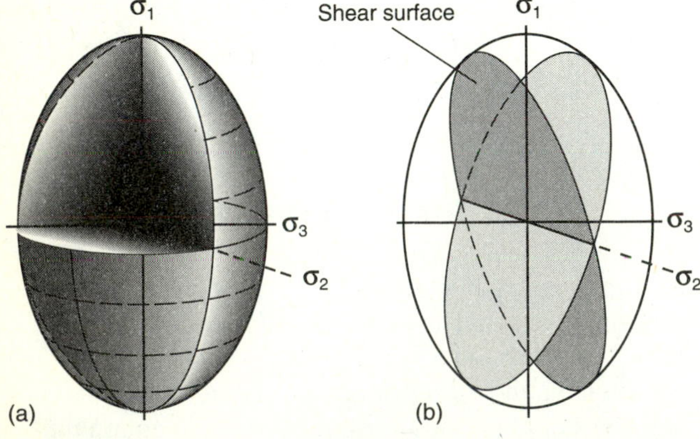

FIGURE 3–12
(a) The stress ellipsoid, a triaxial ellipsoid in which axes are the principal stresses σ_1, σ_2, and σ_3. (b) Planes of maximum shear stress are always parallel to σ_2 and at 45° to σ_1 and σ_3.

Curvature of the Mohr envelope is frequently related to inherent properties of the material (Figure 3–15). Brittle materials tend to produce relatively straight Mohr envelopes with steep slopes. With increasing confining stress (σ_3), even a brittle material behaves megascopically in a ductile manner. If the materials are ductile, they undergo some permanent, nonrecoverable strain before rupture (Chapter 6). With

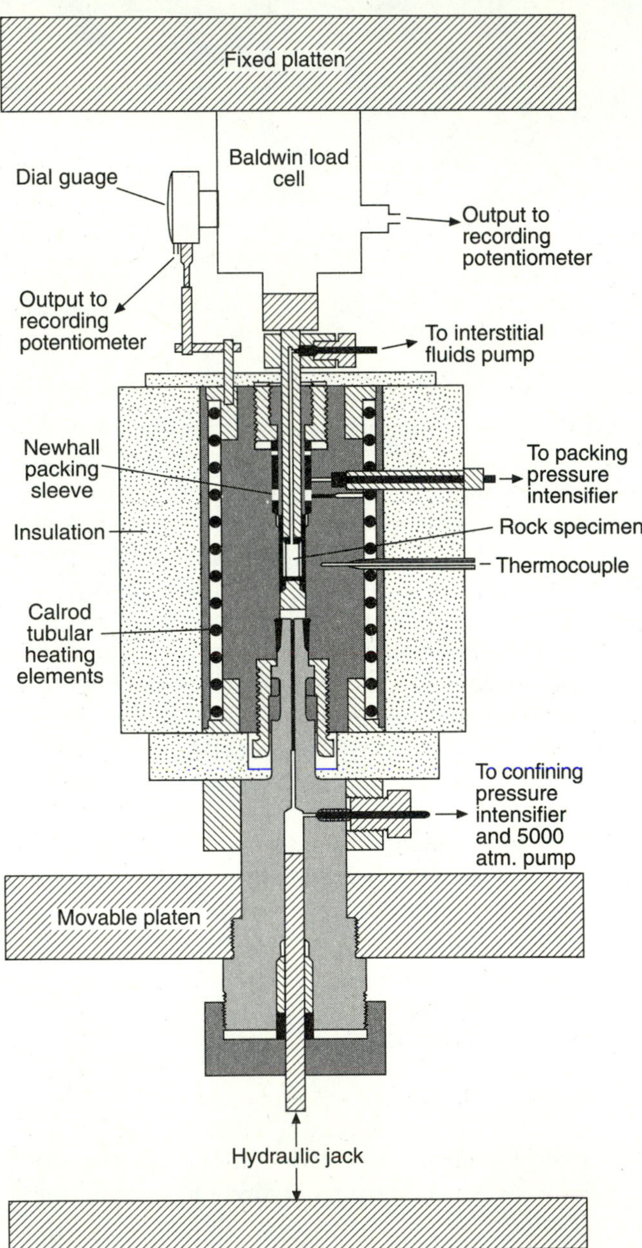

FIGURE 3–13
Triaxial test apparatus. Axial stress (σ_1) is applied with a hydraulic jack. The axial load is measured directly by the load cell. The dial gauge provides a direct measure of strain as percentage of shortening of the rock cylinder. (From D. T. Griggs, F. J. Turner, and H. C. Heard, 1960, Geological Society of America *Memoir* 79.)

TABLE 3–1

Test Run	σ_3 (kg cm^{-2})	σ_1 (kg cm^{-2})
1	0	750
2	250	1,750
3	500	2,400
4	1,050	3,550

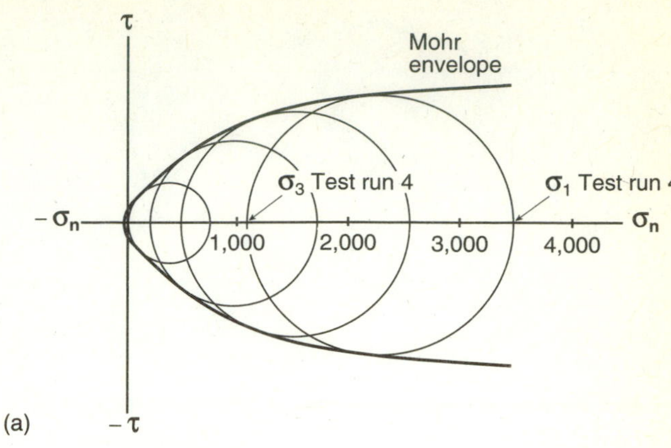

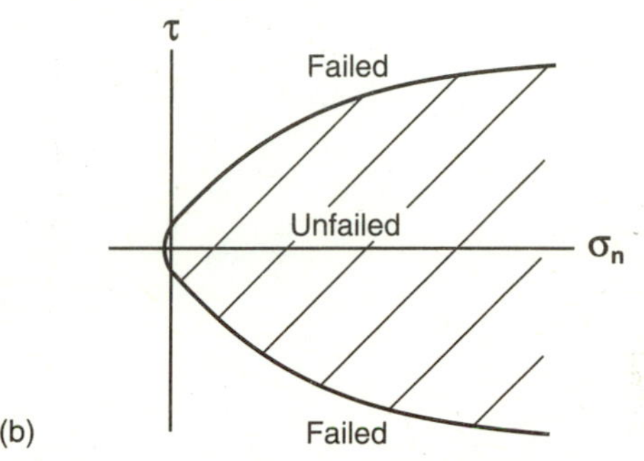

FIGURE 3–14
(a) Mohr diagram plotted from the values obtained from the experimental deformation of a limestone at different confining pressures (Table 3–1). (b) Mohr diagram showing unfailed (stable) and failed (unstable) regions.

an increase in ductility, the Mohr envelopes become flattened, resulting in an overall curvature. In general, the greater the curvature, the greater the amount of ductile strain before rupture, but Griffith materials (Chapter 10), which are actually brittle, produce curved envelopes. Their properties are related to the amounts of dilation of the preexisting microcrack population.

The angle 2α, termed the *conjugate shear angle,* is related to the coefficient of internal friction μ_i (or tan ϕ, defined below in equation 3–25), and α is the angle between σ_1 and the fracture that forms during rupture. The relationship is

$$90 - \phi = 2\alpha . \qquad \textbf{(3–21)}$$

(Here α is defined in degrees, as in Figures 3–9 and 3–15.) As the radii of the Mohr circles increase and approach a limit (with increasing stress difference and confining pressure), the slope of the envelope flattens as a function of increased ductility, and 2α approaches a maximum value of 90°. Therefore, the maximum shear angle α is 45°. This angle is most commonly attained (and sometimes exceeded) in ductile materials.

Several relationships within Mohr diagrams are summarized in Figure 3–16. Note that the τ axis separates compressional from tensional normal stress fields (Figure 3–16a). Materials with no tensile strength, such as dry sand, have an envelope that terminates at the origin (Figure 3–16b). Most geologic materials have compressive strengths (breaking strength under compressive stress) much greater than their tensile strengths (Figure 3–16c). Very few materials have tensile and compressive strengths even approximately the same, but a mild carbon steel (Figure 3–16d) is one that does, and a horizontal envelope that closes in the tensile field results.

AMONTONS' LAW AND THE COULOMB–MOHR HYPOTHESIS

A French physicist, Guillaume Amontons, suggested at a scientific meeting in 1699 that a direct proportional relationship exists between **F**, the shear force necessary for sliding along a contact surface, and W, the force perpendicular to the surface, expressed as an equality

$$\mathbf{F} = \mu_s W, \qquad \textbf{(3–22)}$$

and today known as Amontons' first law, where μ_s is the coefficient of sliding friction along the surface (Jaeger and Cook, 1976). The coefficient of sliding friction is a measure of the resistance of a material to sliding along a surface. If we divide both sides of equation 3–22 by the surface area (A), it becomes

$$\tau_s = \mu_s \sigma_n. \qquad \textbf{(3–23)}$$

Another French physicist, Charles A. Coulomb, recognized the linear relationship between shear and normal stress implied by Amontons' first law. From that, he derived another relationship that today bears his name, the *Coulomb criterion of failure,* proposed in 1773. It states that the absolute value of shear strength τ_s is the sum of the inherent shear strength S_0 and the coefficient of internal friction μ_i, or static friction, multiplied by the normal stress σ_n

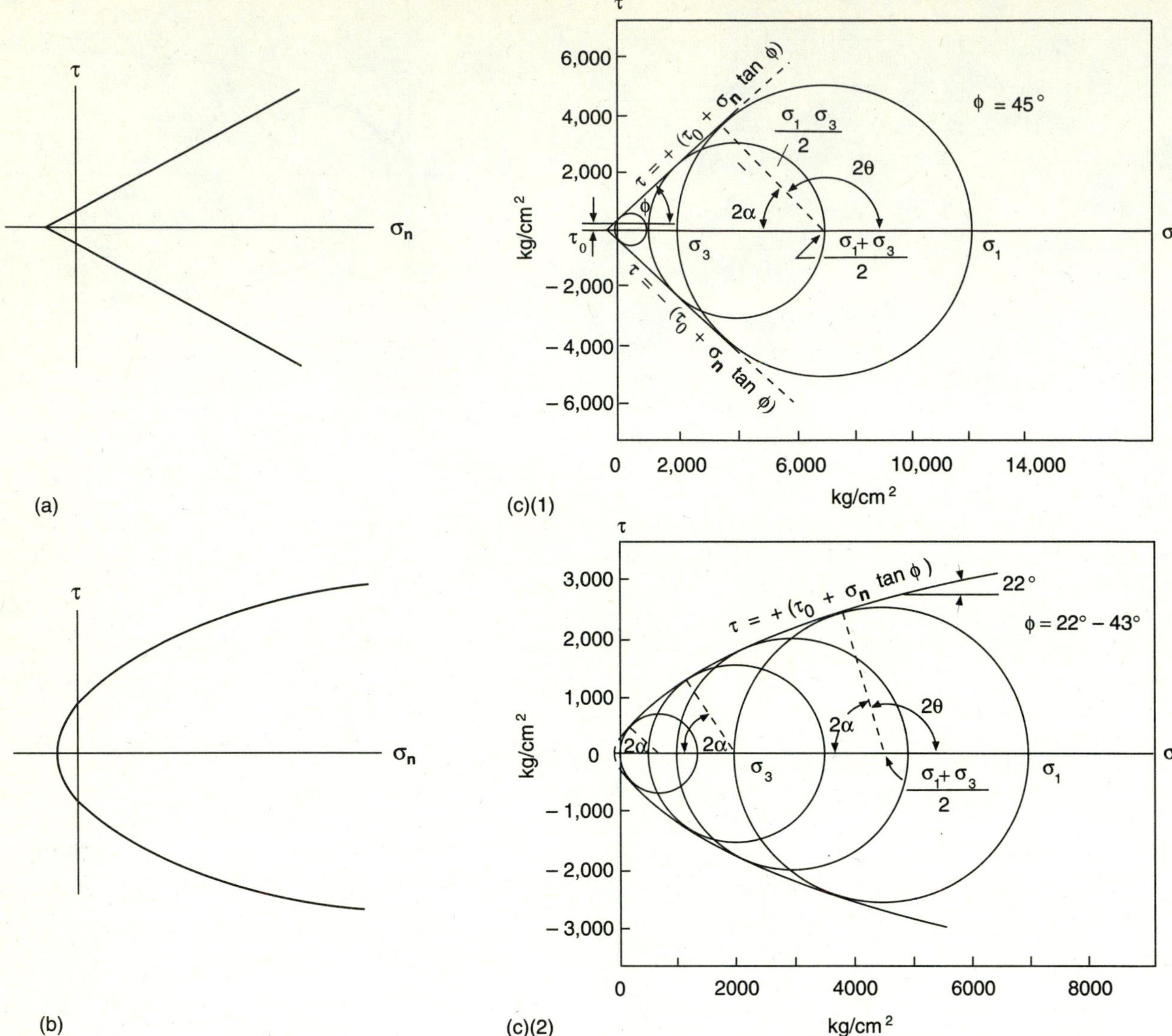

FIGURE 3–15
Mohr diagrams for brittle (a) and more ductile (b) materials. (c) Mohr diagrams for (1) Oil Creek Sandstone and (2) Blaine Anhydrite, deformed at 24° C and 0–2,000 atmospheric pressure. (From M. K. Hubbert and D. G. Willis, 1957, *American Institute for Mining, Metallurgical, and Petroleum Engineers Transactions,* v. 210.)

$$|\tau_s| = S_0 + \mu_i \sigma_n. \qquad \textbf{(3–24)}$$

Equation 3–24 predicts that fracturing will occur when the shear stress on a plane (such as a fault plane) reaches a critical value. The Coulomb equation is an equation of a straight line that approximates the Mohr envelope (Figure 3–9). It is commonly expressed today as

$$|\tau_s| = \tau_0 + \sigma_n \tan \phi. \qquad \textbf{(3–25)}$$

Here, τ_0 is the *cohesive strength* of the material at zero normal stress, and $\mu_i = \tan \phi$; ϕ is the angle of internal friction, and $\tan \phi$ is the *coefficient of internal friction.*

In 1900, Mohr generalized that shear strength is a function of normal stress,

$$|\tau_s| = f(\sigma_n)\,, \qquad \textbf{(3–26)}$$

where $|\tau_s|$ is the absolute value of *shear strength* (either a positive or negative number)—the resistance of a material to shear stress. The location of the Mohr envelope in $\sigma_n\tau$ space is directly related to the strength of the material. Mohr's hypothesis and the derivation of the equations for resolved shear and normal stresses on a plane (equations 3–15 and 3–16) are both incorporated in the Mohr construction (Figure 3–9).

We end our survey of stress and its mechanical basis. We turn in the next several chapters to a discussion of strain—the presumed effect of stress.

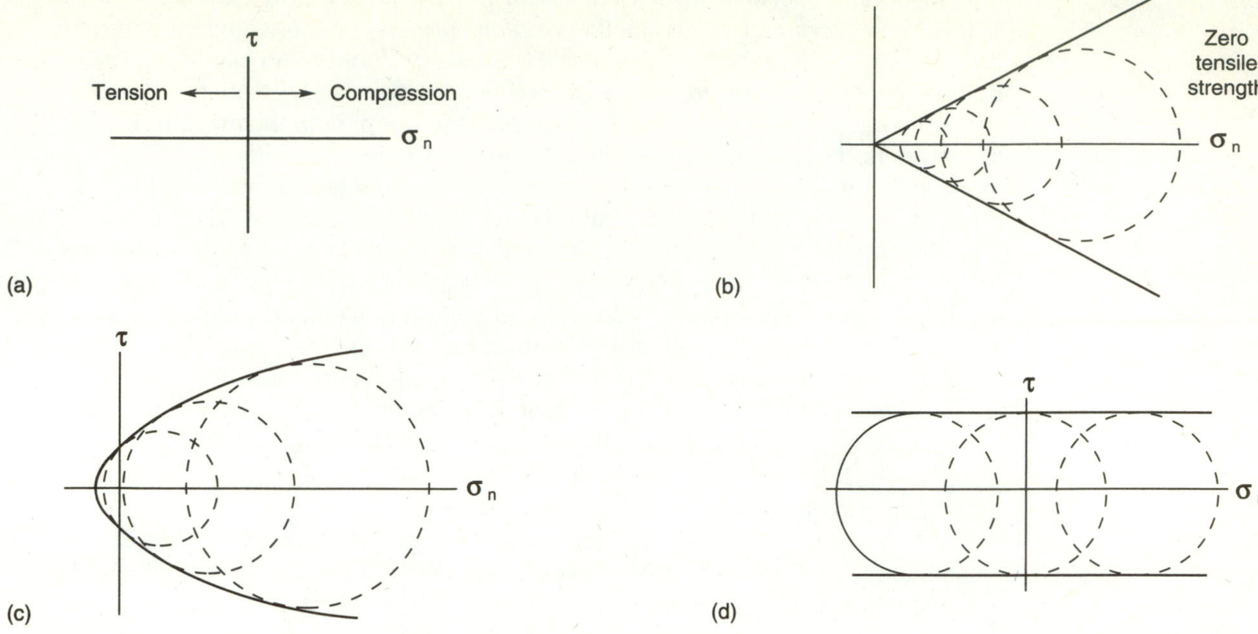

FIGURE 3–16
(a) Axes for Mohr diagrams showing extensional ($-\sigma_n$) and compressional ($+\sigma_n$) fields. (b) Mohr diagram for quartz sand, a material with no tensile strength. (c) Mohr diagram for an average rock that exhibits tensile strength at low stress, and greater compressive strength and possibly some ductility at higher stress. (d) Mohr diagram for a mild carbon steel, where tensile and compressive strengths are about the same, yielding an almost horizontal Mohr envelope that ultimately closes into the tensile field.

ESSAY

Measuring Present-Day Stress in the Earth

Knowledge of the orientation and magnitude of present-day stress in the Earth is important because it provides clues about the nature of active faults, information about what structures might become more active in the future, and insight into the kinematics of plate motion. Knowledge of these orientations bears directly on our understanding of the causes of seismic activity (Zoback, Tsukahara, and Hickman, 1980; Zoback and Zoback, 1980).

Present-day stress in the Earth can be measured by placing a strain gauge in rock and recording the *in situ* elastic strain. Such measurements are most frequently carried out by drilling a hole into bedrock, inserting the instruments, and recording the amounts and orientations of elastic strain. Two techniques are commonly used for measuring *in situ* stress: *overcoring* and *hydraulic fracturing.*

Overcoring involves drilling a hole 3 to 4 cm in diameter, then drilling a larger core (12 to 15 cm) with the same center as the smaller hole (Figure 3E–1; Hooker and Bickel, 1974). Before coring the larger hole, an instrument called a dilatometer (a stain gauge) is inserted into the smaller hole to permit measurement of the expansion (relaxation) of the rock mass as the larger hole is cored. Changes in shape of the small hole are recorded as it is overcored. The orientation of the small hole is known, so changes in shape are measured as the cross section relaxes from a circle to an ellipse. This provides a measure of the orientation of the present-day stress ellipsoid. If several measurements can be made at the same locality of overcores in different orientations—as is possible in a mine, tunnel, or quarry—a very good measure of the orientation of the stress ellipsoid may be obtained. Attempts have also been made to calculate the magnitude of the principal stresses using this technique.

Hydraulic fracturing involves drilling a vertical hole, sealing off a part of the hole with packers, and increasing the hydraulic pressure in the sealed-off part of the hole until the wall of the hole fractures (Figure 3E–2; Zoback and Haimson, 1982). Both the amount of pressure needed to produce fractures and the orientations of hydraulic fractures are measured to provide a measure of both the magnitude of minimum horizontal stress at the site and the orientation of maximum and minimum horizontal principal stresses. It is possible to make several measurements in the same hole if it is drilled to a minimum depth of 300 to 400 m. The hydraulic fracturing method requires that the hole drilled be nearly vertical, for it is assumed that one of the principal normal stresses is vertical and that the hydraulic fracture in a vertical borehole propagates perpendicular to σ_3 (Hubbert and Willis, 1957). Mary Lou Zoback and Mark Zoback (1980) concluded that field data also support the relationship between hydraulic fracture propagation and σ_3.

Numerous measurements of *in situ* stress using different techniques around the world have been compiled into a worldwide stress distribution map (M. L. Zoback, 1992). Such maps permit identification of areas or domains of common maximum principal stress orientation (Figure 3E–3).

In situ stress in a tectonically active area has been measured by Zoback, Tsukahara, and Hickman (1980) in a series of wells about 240 m deep drilled as far as 34 km from the San Andreas fault. They also made several measurements in a 1-km-deep well about 4 km from the fault. Their goal was to evaluate variations in the magnitude of the principal stresses with depth and to calculate the magnitude of shear stress along the fault. They used hydraulic fracturing to measure stress and obtained the magnitudes of $\sigma_{h_{max}}$ (maximum horizontal stress σ_1) and $\sigma_{h_{min}}$ (minimum horizontal stress) by modifying an equation originally derived by Hubbert and Willis (1957):

$$P_b = 3\sigma_{h_{min}} - \sigma_{h_{max}} - P_p + T, \tag{3E–1}$$

where P_b is breakdown (fracture formation) pressure, P_p is pore pressure, and T is the tensile strength of the rock mass. P_b is measured directly as the pressure necessary to hydraulically break the rocks. P_p is calculated from the pressure of a column of water (using ρgh) at the depth at which hydraulic fracturing was carried out, assuming that the pores in the rocks are interconnected and communicate with

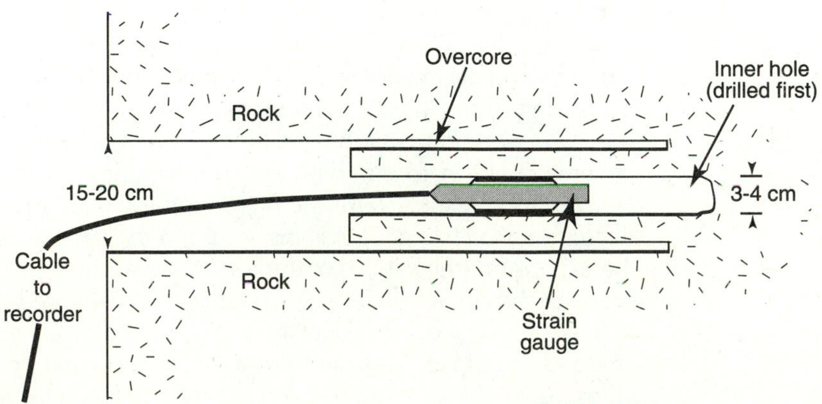

FIGURE 3E–1
Overcoring technique for measuring *in situ* stress. The strain gauge rests inside the small borehole and measures *in situ* stress—while the overcore is being drilled—by recording the amount and direction of change from a circular to an elliptical borehole.

the surface. The tensile strength of the rock mass can be estimated (as here) or measured on core samples obtained from the well. The magnitude of σ_3 is measured directly, and its azimuth is obtained from orientations of hydraulic fractures in the well. The value of σ_1 is then calculated from equation 3E–1 and is assumed to be horizontal.

Results of measurements in the two areas studied (Table 3E–1) indicate that the maximum principal horizontal stress, σ_1, is oriented about 45° from the strike of the San Andreas fault. Shear stress, determined by plotting the maximum and minimum principal stresses on a Mohr diagram, increases from about 2.5 MPa (25 bars) at depths of 150 to 300 m to about 8.0 MPa (80 bars) at 750 to 850 m. More recent work by Zoback and others (1987) based on *in situ* stress measurements in the Cajon Pass scientific borehole, other stress measurements along the fault nearby, and orientations of folds in Pleistocene sediments immediately adjacent to the fault indicate that σ_1 is oriented nearly perpendicular to the San Andreas fault northeast of Los Angeles. Thus, there is almost no shear stress along the fault near the surface in this area.

Plotting principal stress orientations on a map enables us to relate the present-day stress in the Earth to plate motion (Figure 3E–3). The nearly uniform N 70° E orientation of stress fields in the eastern United States is thought to be related to "ridge push" from the Mid-Atlantic Ridge, and the markedly different orientations in western states are related to interaction between the Pacific and North American plates (Figure 1–1; Richardson, Soloman, and Sleep, 1979).

In addition to direct measurements of stress, the orientation of maximum and minimum principal horizontal stresses can be obtained from well-bore breakouts (elongations) that form after a hole is drilled. A hole changes shape and fractures as

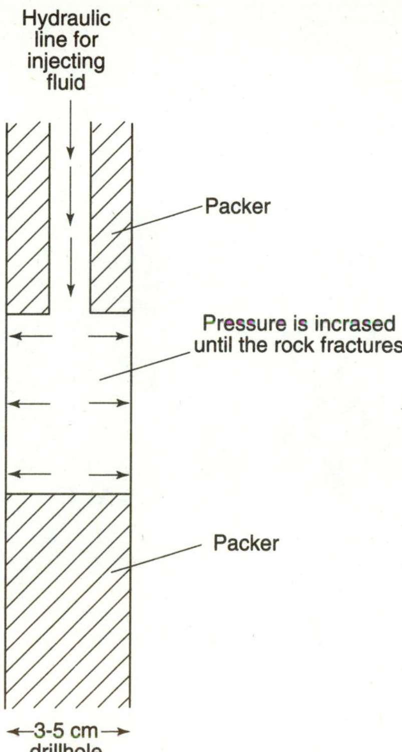

FIGURE 3E–2
Hydraulic fracturing method for measuring *in situ* stress. Larger diameter drill holes can be used but the pressures required to fracture the rocks increases as the radius of the drill hole increases by a power-law relationship. (Modified from R. O. Kehle, *Journal of Geophysical Research,* v. 69, © 1964 by the American Geophysical Union.)

the walls relax, producing curved fractures and causing rock to spall from the walls of the hole. These fractures form parallel to the orientation of the principal stress directions, but the magnitude of stress cannot be determined. Mount and Suppe (1987, 1992) used breakouts from drill holes near the San Andreas fault to independently conclude that σ_1 is normal to the fault.

TABLE 3E–1

HYDROFRACTURE DATA ALONG THE SAN ANDREAS FAULT

Well	Depth (m)	Distance San Andreas Fault (km)	Fracture Breakdown Pressure (bars)	Opening Pressure (bars)	Pore Pressure (bars)	σ_3	σ_1	σ_2	Tensile Strength (bars)	τ_{max} (bars)	Direction of Maximum Compression
1	167	4	200	69	17	73	133	45	131	30	N 4° W
1	196	4	209	74	20	77	138	53	135	31	N 1° E
2	338	4	109	63	34	74	125	91	46	26	N 43° W
2	561	4	163	130	56	150	264	152	33	57	N 20° W
2	787	4	192	124	78	183	346	213	68	82	N 19° W
3	80	2	144	24	8	23	38	18	120	8	N 20° W
3	185	2	250	73	19	56	73	43	177	6	N 23° W
4	167	4	139	47	17	51	89	45	92	19	N 83° E
4	230	4	164	85	23	83	140	62	79	29	N 14° W

(From Zoback, Tsukahara, and Hickman, *Journal of Geophysical Research,* v. 85, © 1980 by the American Geophysical Union. Used by permission.)

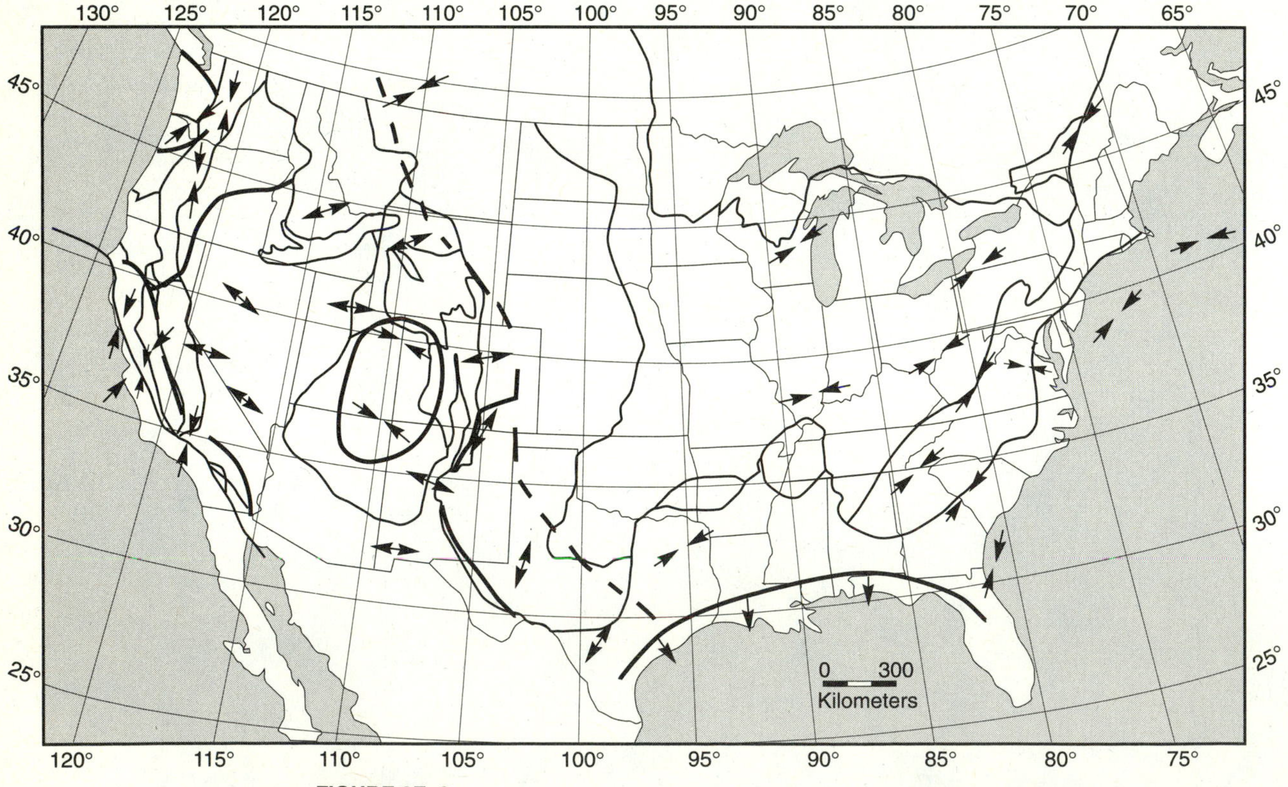

FIGURE 3E–3
Domains of common orientation of present-day maximum (compressional) and minimum (extensional) principal stress in the United States. Arrowheads indicate whether stress is extensional or compressional. (From M. L. Zoback and M. D. Zoback, 1989, Geological Society of America *Memoir 172.*)

References Cited

Hooker, V. E., and Bickel, L. D., 1974, Overcoring equipment and techniques used in rock stress determination: U.S. Bureau of Mines Information Circular 8618, 32 p.

Hubbert, M. K., and Willis, D. G., 1957, Mechanics of hydraulic fracturing: Journal of Petroleum Technology, v. 9, p. 153–168.

Mount, V. S., and Suppe, J., 1987, State of stress near the San Andreas fault: Implications for wrench tectonics: Geology, v. 15, p. 1143–1146.

Mount, V. S., and Suppe, J., 1992, Present-day stress orientations adjacent to active strike-slip faults: California and Sumatra: Journal of Geophysical Research, v. 97, p. 11,995–12,013.

Richardson, R. M., Soloman, S. C., and Sleep, N. H., 1979, Tectonic stress in the plates: Reviews of Geophysics and Space Physics, v. 17, p. 981–1019.

Zoback, M. D., and Haimson, B. C., 1982, Status of the hydraulic fracturing method for *in situ* stress measurements, *in* Society of Mining Engineers, Proceedings 23rd Symposium on Rock Mechanics: New York, American Institute of Mining and Metallurgical Engineers, p. 143–156.

Zoback, M. D., Tsukahara, H., and Hickman, S., 1980, Stress measurements at depth in the vicinity of the San Andreas fault: Implications for the magnitude of shear stress at depth: Journal of Geophysical Research, v. 85, p. 6157–6173.

Zoback, M. D. and 11 others, 1987, State of stress of the San Andreas fault system: Science, v. 238, p. 1105–1111.

Zoback, M. L., and Zoback, M. D., 1980, State of stress in the continental United States: Journal of Geophysical Research, v. 85, p. 6113–6156.

Zoback, M. L., and Zoback, M. D., 1989, Tectonic stress field of the continental United States, *in* Pakiser, L., and Mooney, W., eds., Geophysical framework of the continental United States: Geological Society of America Memoir 172, p. 523–539.

Zoback, M. L., 1992, First- and second-order patterns of stress in the lithosphere: The world stress map: Journal of Geophysical Research, v. 97, p. 11,703–11,728

Questions

1. Why do we commonly deal with stress, rather than force, in the earth sciences?
2. What does the shape of the Mohr envelope tell you about the strength of the material being deformed?
3. Why does the Mohr construction work using only values of σ_n?
4. What is actually happening in materials that produce straight Mohr envelopes?
5. Calculate the lithostatic value of stress on a fault plane inclined at 30° and located at a depth of 7 km in oceanic crust (ρ = 3,000 kg m^{-3}).
6. Does the limestone represented by the Mohr diagram in Figure 3–14 exhibit brittle or ductile behavior (units are kg cm^{-2})? Would values of σ_3 of 1,000 kg cm^{-2} and σ_1 of 2,500 kg cm^{-2} produce failure in a cylinder of this limestone? Explain.
7. Hydrofracture measurement of principal stresses along the San Andreas fault at a depth of >600 m in a drill hole by Zoback, Tsukahara, and Hickman (1980) yielded a value for σ_3 of 141 bars and for σ_1 of 258 bars. Determine the value of maximum shear stress at that point. (See Chapter 3 Essay for additional discussion of stress measurement techniques.)

Further Reading

DePaor, D. G., 1986, A graphical approach to quantitative structural geology: Journal of Geological Education, v. 34, p. 231–236.

A new graphical method is described for use in structural geology, determining normal and shear stress without trigonometry.

Engelder, T., 1994, Deviatoric stressitis: A virus infecting the earth science community: EOS, v. 75, p. 209–212.

A tongue-in-cheek discussion of deviatoric stress that is both educational and entertaining.

Hubbert, M. K., 1951, Mechanical basis for certain familiar geologic structures: Geological Society of America Bulletin, v. 62, p. 355–372.

This is a classic study relating experimentally produced structures to both theory and real tectonic structures as observed in the field. It also presents a clear, simple, and concise derivation of the Mohr circle equations.

July 30, 1992, Journal of Geophysical Research, v. 97, p. 11,703–12,013.

Contains a number of papers on stress patterns in different parts of the world, along with a new map by Mary Lou Zoback depicting the worldwide state of stress from measurements on all the major plates.

Means, W. D., 1976, Stress and strain: New York, Springer-Verlag, 339 p.

A clearly written introduction to the concepts of stress and strain through the principles of continuum mechanics. Mathematical concepts and derivations are presented understandably, using problems (with solutions) related to geologic situations. Emphasizes elastic and viscous strain.

Means, W. D., 1990, Review paper. Kinematics, stress, deformation and material behavior: Journal of Structural Geology, v. 12, p. 953–971.

An excellent summary of the relationships between stress and strain in experimentally deformed rocks. Mohr circles are presented as two-dimensional tensor quantities. Also includes a useful glossary of terms.

4

Strain

Stress conditions within the earth's crust change during the progress of time, and these changes often lead to the permanent deformation of crustal rocks. One of the prime aims of the structural geologist is to determine the nature and amount of these displacements.

JOHN G. RAMSAY, 1967, *Folding and Fracturing of Rocks*

STUDY OF STRUCTURAL GEOLOGY IS LARGELY A STUDY OF deformation. We assume that deformation is commonly the result of differential stress, but cases exist where deformations generate stresses.

Our ability to observe tectonic structures is restricted largely to a two-dimensional observation platform—the surface of the Earth—with the third dimension limited to a few kilometers of vertical exposure in mountain chains, deeply dissected high plateaus, and recently uplifted coastal regions. Most structures we can observe were formed long ago, kilometers to tens of kilometers deep in the crust. Fortunately, many tectonic structures we can observe on our two-dimensional platform do not parallel the Earth's erosion surface but are inclined obliquely to it (Figure 4–1). In such cases, we can sometimes observe an entire structure that formed over a range of depths in the crust and later was tilted and exposed by erosion.

One way we gain a greater understanding of tectonic structures and the conditions that formed them is by comparing their present *deformed state* with the original *undeformed state* for the same rock mass. This is sometimes difficult because we need an undeformed equivalent rock mass for direct comparison. Sometimes, parts of structures or objects within them help us to solve the problem. For example, deformed fossils are easily compared with their undeformed counterparts (Figure 4–2). An ideal study of deformation would consist of such a comparison of deformed and undeformed states, but the complexity of tectonic structures frequently frustrates our efforts. Recent work that describes deformation quantitatively has provided an important and useful aid in our understanding of deformation in rocks and also has helped us to construct more realistic models for restoring rock bodies to undeformed states. In this first discussion of the kind

FIGURE 4–1
Gently plunging folds in Middle Proterozoic gneiss near Central City, Colorado. These rocks were deformed at elevated temperature and pressure conditions that produced ductile behavior. Coin is a U.S. quarter. (RDH photo.)

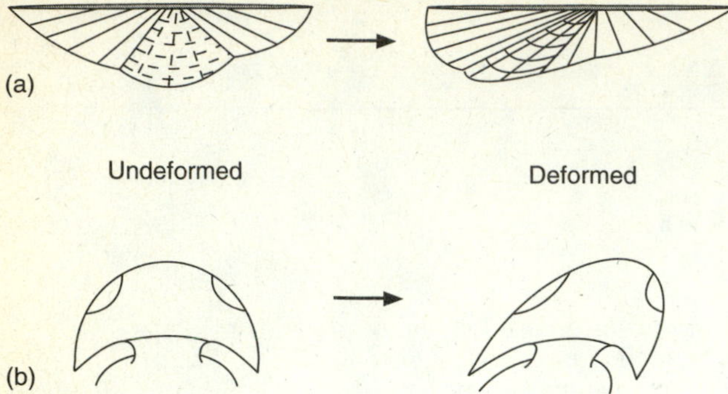

FIGURE 4–2
Undeformed and deformed fossils. Note changes in lengths and angles of material lines in most orientations in the undeformed fossils. (a) Brachiopod. (b) Trilobite cephalon.

and amount of strain affecting a mass of rocks we will consider only the initial and final states of deformation of the mass—not the *deformation path* or the stages of *progressive deformation* affecting the mass between the initial and final states.

DEFINITIONS

Deformation is the displacement field for tectonically driven particle motion(s) and involves the processes by which the particle motions are achieved (Figure 4–3). Deformation, unlike strain, encompasses both rigid-body rotation and rigid body translation. Deformation may be *continuous* (lines not broken) or *discontinuous* (lines broken). Terms that describe body motions (kinematics) include ***distortion*** that involves a change in shape, *rotation*, a change in orientation, and translation, a change in position. ***Strain*** may be described as aspects of shape change measured as changes in line length, changes in angular relationships between lines, or as volume changes. It is not uncommon for the term strain to be used loosely to describe displacements along faults, particularly in geophysical papers.

With *homogeneous strain,* lines that are straight and parallel before deformation remain straight and parallel after deformation (Figure 4–4). *Inhomogeneous strain* is the opposite: straight or parallel lines before deformation do not remain straight (or parallel) after deformation. Lines may also be broken during inhomogeneous deformation. Another factor influencing the kind of strain affecting a rock mass is scale: homogeneous strain affecting a rock mass on a scale of several kilometers may, on a scale of centimeters, be resolved into inhomogeneous strain.

Strain may occur in ***infinitesimal steps***, or as ***finite strain***, where a comparison is made between the present shape with some previous state for that shape.

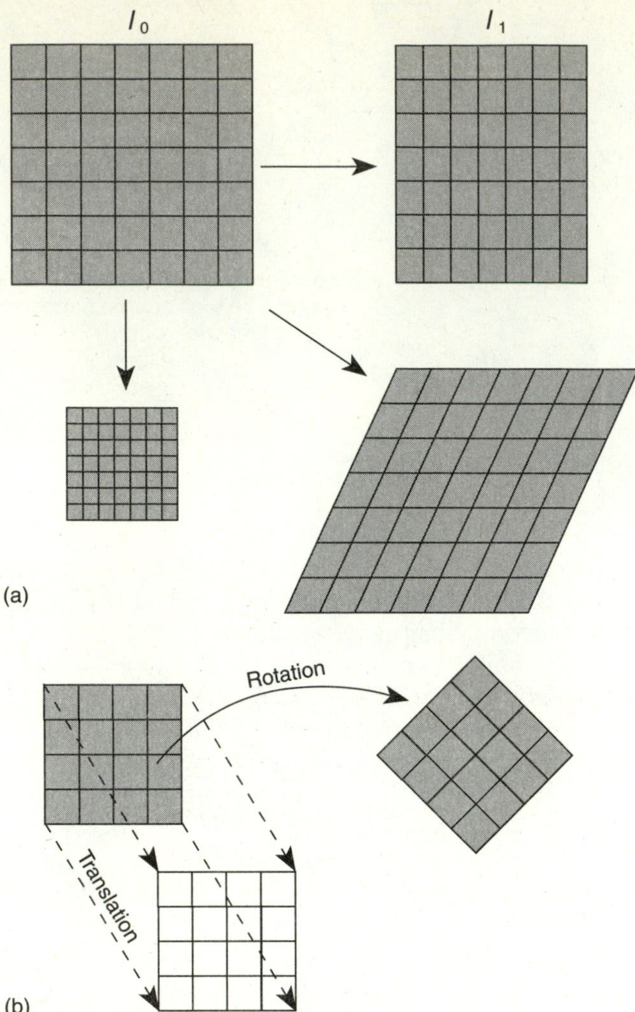

FIGURE 4–3
(a) Strain as distortion of initially parallel and perpendicular lines involving a change of length, shape, or volume of a mass. (b) Rigid-body translation or rotation of a mass without accompanying distortion.

For finite strain, the comparison is path independent. Strain that occurs during an event in the deformation history of the rock body is ***incremental strain.*** This incremental strain may be infinitesimal or finite. Most natural strain we observe in the field is finite strain and may or may not be separable into incremental strains.

MEASURES OF STRAIN

Strain may occur in several forms. It may be recognized as change in line length, angles between lines, or volume. If some fundamental measures are employed, several kinds of strain may be calculated for the same body.

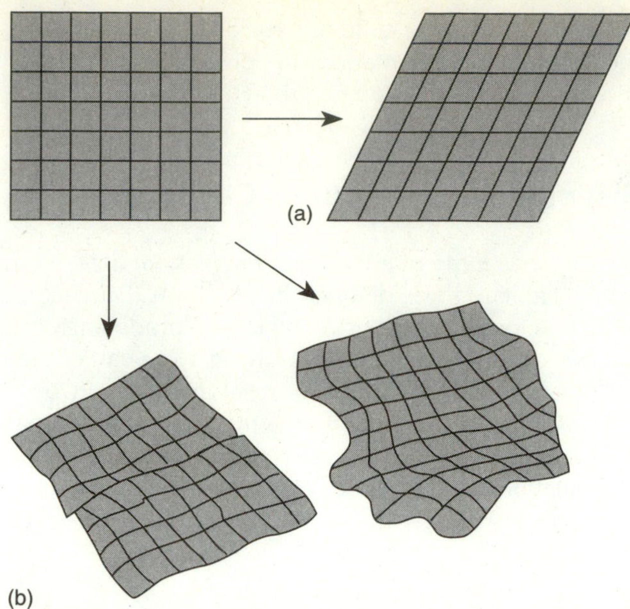

FIGURE 4–4
Homogeneous (a) and inhomogeneous (b) strain. Note that after deformation, originally parallel or straight reference lines within the body undergoing homogeneous strain remain straight or parallel. Similar reference lines in the body undergoing inhomogeneous deformation are either not parallel or straight or they are broken.

Linear Strain

Elongation. The *elongation (ε)* of a reference line in a rock mass (Figure 4–5) is the ratio of the length of the line in the deformed mass (l_1) minus the original length (l_0) to the original length; it is written mathematically for a finite strain as

$$\varepsilon = \frac{l_1 - l_0}{l_0} = \frac{\Delta l}{l} \qquad \textbf{(4–1)}$$

$$\%\ \text{elongation} = \frac{\Delta l}{l} = \times 100,$$

or for an infinitesimal strain, as

$$\varepsilon = \frac{dl}{l}. \qquad \textbf{(4–2)}$$

If $\varepsilon > 0$, or positive, an extension is indicated; if $\varepsilon < 0$, or negative, shortening is indicated. (Note that civil engineers use a negative sign to indicate extension and a positive sign for shortening.)

Stretch. The *stretch* (S)—or engineer's stretch—of a reference line in a rock mass is the ratio of the deformed line length (l_1) to the original length (l_0), or

$$S = \left(\frac{l_1}{l_0}\right) = (1 + \varepsilon). \qquad \textbf{(4–3)}$$

Quadratic Elongation. The square of the stretch is called the *quadratic elongation* (λ), or

$$\lambda = \left(\frac{l_1}{l_0}\right)^2 = (1+\varepsilon)^2 = S^2. \qquad \textbf{(4–4)}$$

All of the strains involving lines may easily be related to each other, so if one is known, the others can be calculated (Figure 4–5). All are dimensionless quantities because they are ratios.

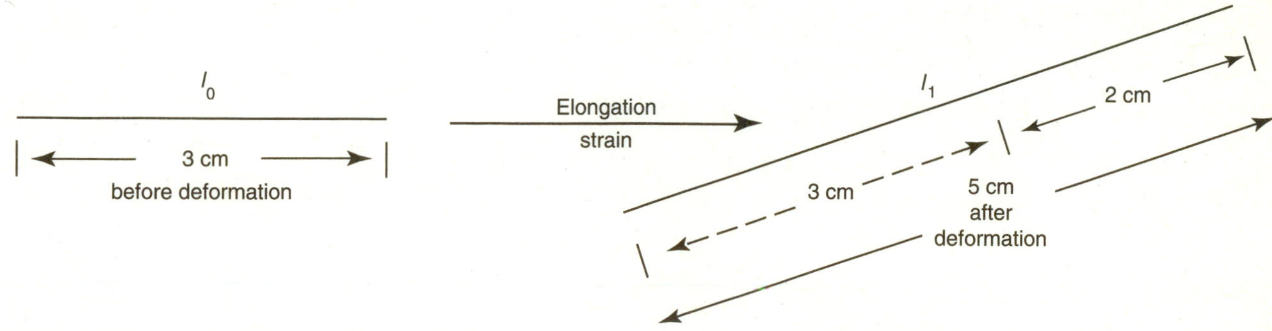

FIGURE 4–5
Magnitudes of linear strains. The elongation strain here is 0.67, calculated from equation 4–1. The stretch is 1.67 (from equation 4–3), and quadratic elongation is 2.78 (equation 4–4).

Shear Strain—Changes in Angular Relationships

Shear strain (γ) develops when differential movement occurs along a set of parallel lines. For example, in Figure 4–6, the movement is greater parallel to horizontal lines at the tops as compared to the bottom of the rock mass. As a result, a set of lines initially normal to the horizontal lines is rotated from orthogonality by an ψ, the angular shear. Shear strain is related to angular shear by

$$\gamma = \tan \psi. \qquad \textbf{(4–5)}$$

The history of differently oriented lines may be traced during progressive deformation for a shear strain applied to a reference cube in a rock mass (Figure 4–7). Lines in particular orientations undergo only extension; others undergo shortening and then extension.

Dilational Strain—Volume Changes

Changes in volume of a material, or *dilation,* may occur by at least three different mechanisms (Figure 4–8): (1) closing voids between grains—producing a negative volume change; (2) dissolving away part of the rock mass by pressure solution—resulting in negative volume change; and (3) fracturing the mass—producing positive volume change (Figure 4–9). A finite volumetric or dilational strain (Δ) is the ratio of the volume change to the original volume of the mass, expressed as

$$\Delta = \frac{(V_1 - V_0)}{V_0} = \frac{\delta V}{V_0}. \qquad \textbf{(4–6)}$$

If the strain is infinitesimal,

$$\Delta = \frac{dV}{V}. \qquad \textbf{(4–7)}$$

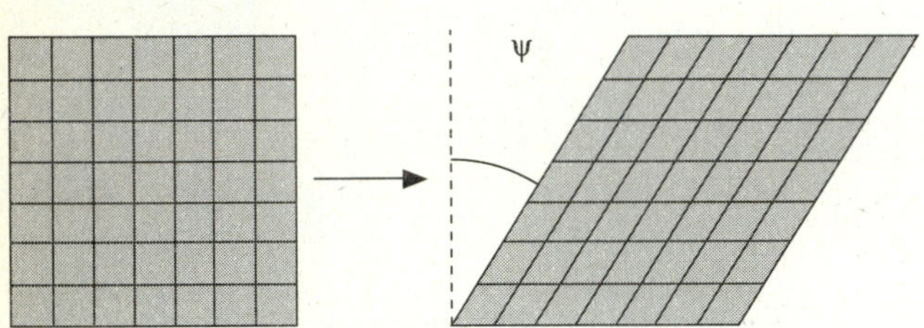

FIGURE 4–6
Shear strain. The rotation angle ψ is 32°, and the shear strain γ is 0.63 (from equation 4–5).

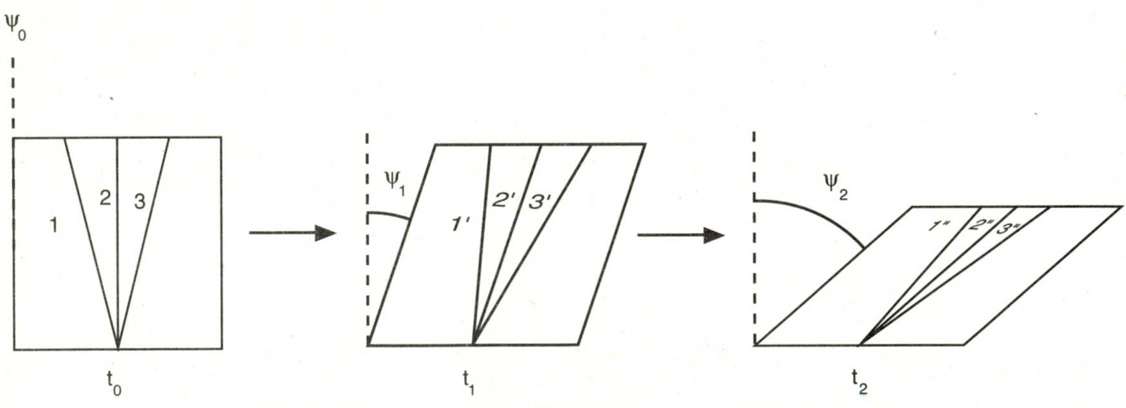

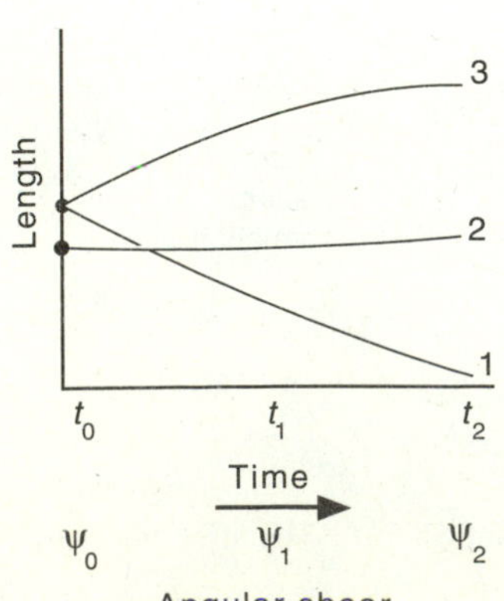

FIGURE 4–7
History of progressive deformation by application of shear strain ψ negative dilation in which reference line 1 is continuously shortened, reference lines 2 remains the same length, and reference line 3 is continuously lengthened. This kind of information is useful in attempts to decipher the progressive strain history in different parts of the same body, as well as the deformation of particular features as a function of orientation relative to strain.

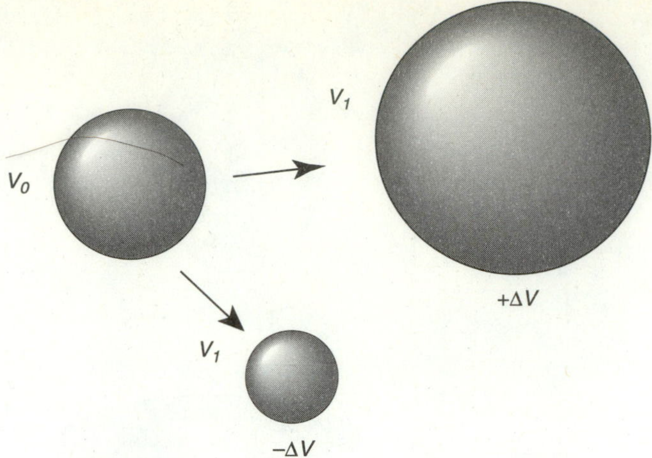

FIGURE 4–8
Positive and negative dilation, illustrated by soap bubbles that have expanded or contracted. If the radius of the bubble V_0 is 1.5 cm and that of the expanded bubble + V is 2.75, the dilation D is 5.16 (from equation 4–6). If the radius of the bubble − V is 1.1 cm, the dilation is –0.61 (from equation 4–6).

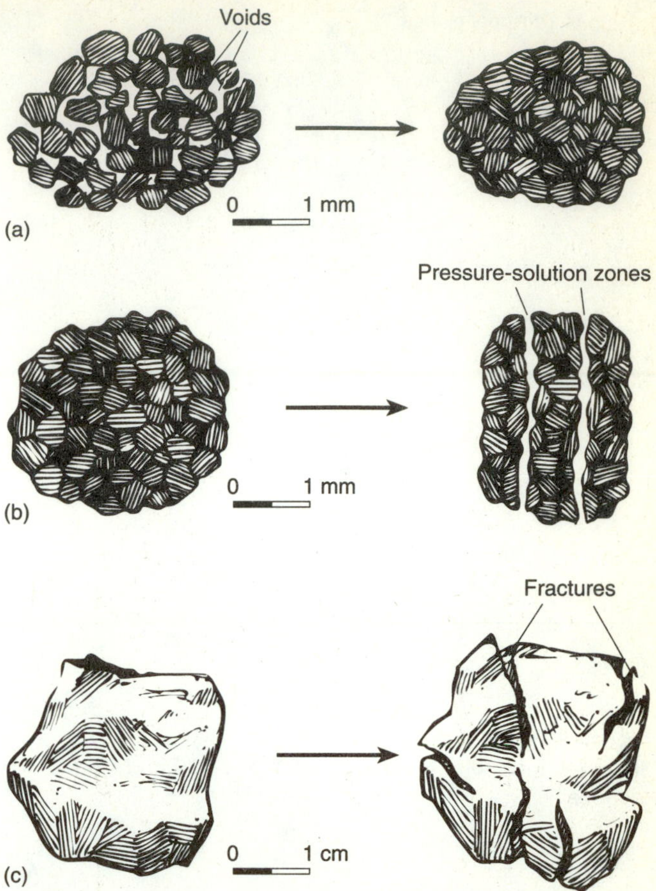

FIGURE 4–9
Three kinds of volume changes. (a) Closing voids between grains. (b) Dissolving part of the rock mass by pressure solution, here showing the pressure-solution zones pulled apart to make them more visible. (c) Fracturing the rock body—(a) and (b) produce negative volume changes, whereas an increase in volume occurs in (c).

STRAIN ELLIPSOID

The ***strain ellipsoid*** (Figure 4–10) is a graphical tool that provides a reference object for estimating shape change from an assumed initial sphere. Sections through it are ellipses that are occasionally printed on geologic maps and cross sections to indicate the distribution of shape changes as a function of geographic and geologic position. The strain ellipsoid consists of a triaxial ellipsoid generated by strain from an undeformed sphere of unit radius. The equation for the strain ellipsoid is

$$\frac{x^2}{X^2} + \frac{y^2}{Y^2} + \frac{z^2}{Z^2} = 1. \qquad \textbf{(4–8)}$$

The ellipsoid is referred to three mutually perpendicular axes x, y, and z, and has principal radii X, Y, and Z, in which the length of $X \geq Y \geq Z$. Therefore, X is the axis of ***greatest principal strain***; Y, the axis of ***intermediate principal strain***; and Z, the axis of ***least principal strain***. Three kinds of strain ellipsoids may be defined (Figure 4–11): the general case, where $X > Y > Z$ (Figure 4–10), and two special cases where $X = Y \geq Z$ (results in oblate spheroid with a pancake or hamburger shape) and $X \geq Y = Z$ (results in a prolate spheroid with a cigar or hot-dog shape). This distinction is immediately useful in diagnosing the kind of strain that may have affected a rock body—if we can find a natural strain marker (Chapter 5) such as deformed oöids.

Sections through the general strain ellipsoid that are parallel to any two of the principal strain axes—that is, the XY, ZY, and XZ sections—are elliptical (Figure 4–12). In the other special cases, XY (oblate) and YZ (prolate) sections are circular. In addition, any triaxial strain ellipsoid has two circular sections that are symmetrically inclined to the X and Z axes and contain the Y axis. An important property of circular sections is that the lines in them have undergone either no elongation, equal elongation in all directions, or they have returned to their original length so that the amount of lengthening equals the amount of shortening.

In the simplest case, the relationship of the strain ellipsoid to the stress ellipsoid is one of an inverse correspondence of axes. X, the axis of greatest principal strain, corresponds to σ_3, the axis of least stress. Likewise, the Z axis corresponds to σ_1; the two intermediate axes, Y and σ_2, are coincident. This relationship holds as long as the strain accumulates *coaxially* (parallel to principal axes and the strain is irrotational).

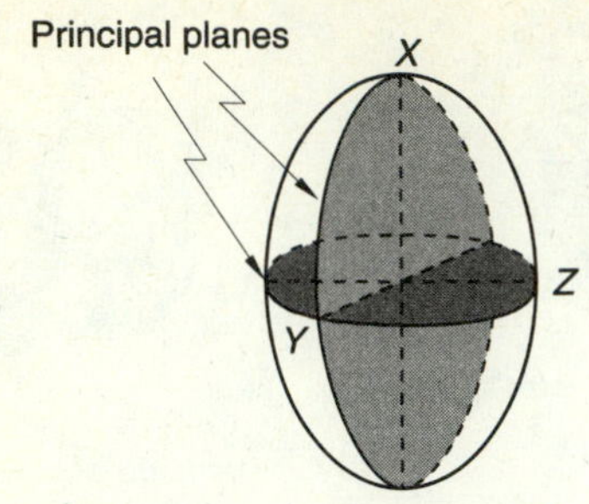

(a)

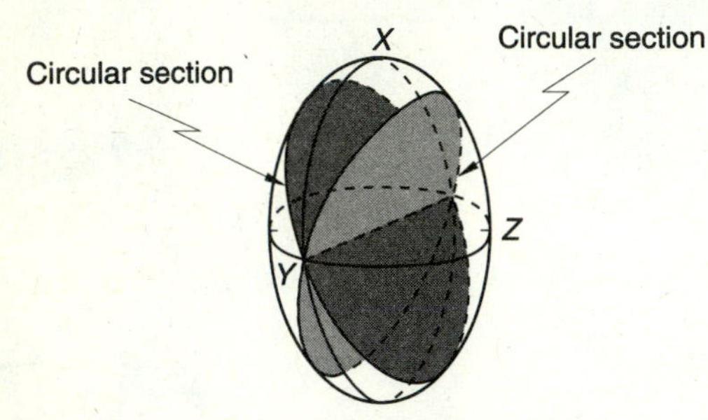

(b)

FIGURE 4–10
Strain ellipsoid with principal axes *X, Y,* and *Z* and circular sections.

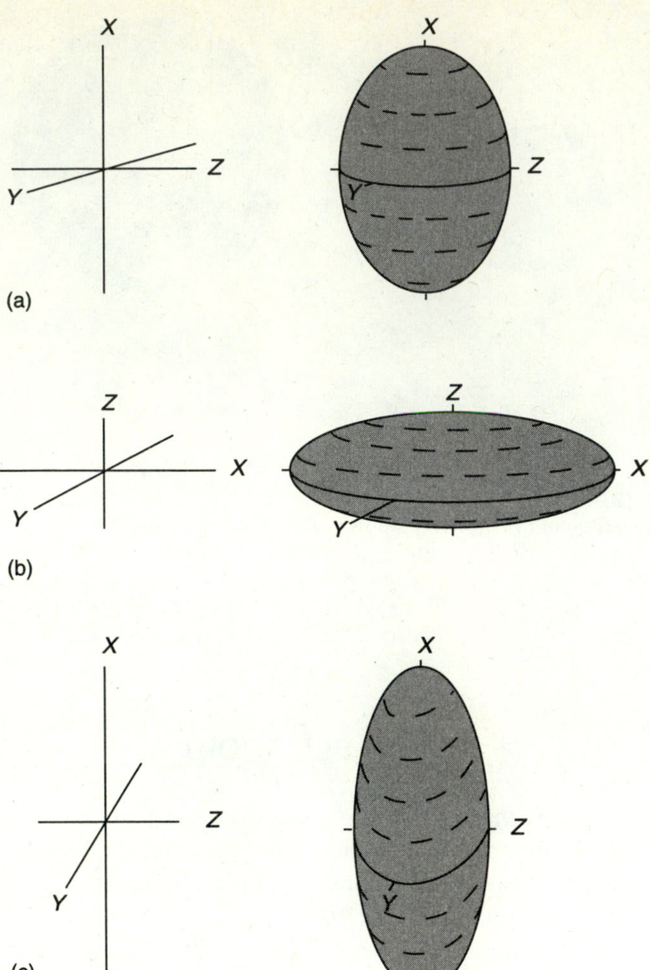

FIGURE 4–11
Three types of strain ellipsoids. (a) Triaxial ellipsoid, with $X > Y > Z$. (b) Oblate biaxial spheroid, with $X = Y > Z$ (axial shortening producing hamburger shape). (c) Prolate (biaxial) spheroid, with $X \geq Y = Z$ (axial elongation producing hot-dog shape).

The strain ellipsoid is particularly useful in the study of geologic bodies, for we can frequently relate it directly to individual structures. As we will see in subsequent chapters, fault planes (Chapter 10) correspond to shear planes in the strain ellipsoid, joints (Chapter 8) form parallel to the *YZ* plane, and slaty cleavage (Chapter 17) forms mostly parallel to the *XY* (perpendicular to *Z*) plane of the strain ellipsoid. Also, many fold axial surfaces (Chapters 14 and 15) form parallel to the *XY* plane.

MOHR CIRCLES FOR STRAIN

It is possible to construct Mohr circles for strain that are analogous to those used for showing the relationships between shear and normal stress (Chapter 3). The Mohr construction for finite strain involves plotting values of reciprocal quadratic elongation, $\lambda'(= 1/\lambda)$, on the horizontal axis and modified shear strain, γ' $(= \gamma/\lambda)$, on the vertical axis (Figure 4–13). These forms of elongation and shear strain are employed because they may be plotted as the locus of a circle (see equations 4–9 and 4–10 below) with a radius of $(\lambda'_1 - \lambda'_3)/2$ and an x- axis value for its center of $(\lambda'_1 - \lambda'_3)/2$. These equations have a form similar to equations 3–17 and 3–18 for the Mohr circle for stress. The mathematical basis for this graphical tool is more completely presented in Ramsay and Huber (1983). Values of λ' and γ' may be expressed in the form of equations of a circle (analogous to equations 3–15 and 3–16)

$$\lambda' = \frac{\lambda'_1 + \lambda'_3}{2} - \frac{\lambda'_1 - \lambda'_3}{2}\cos 2\phi' \qquad \textbf{(4–9)}$$

and

$$\gamma' = \frac{\lambda'_1 - \lambda'_3}{2}\sin 2\phi' \qquad \textbf{(4–10)}$$

(Ramsay and Huber, 1983). The angle ϕ' is the angle between a material line and the *X* direction. Values of

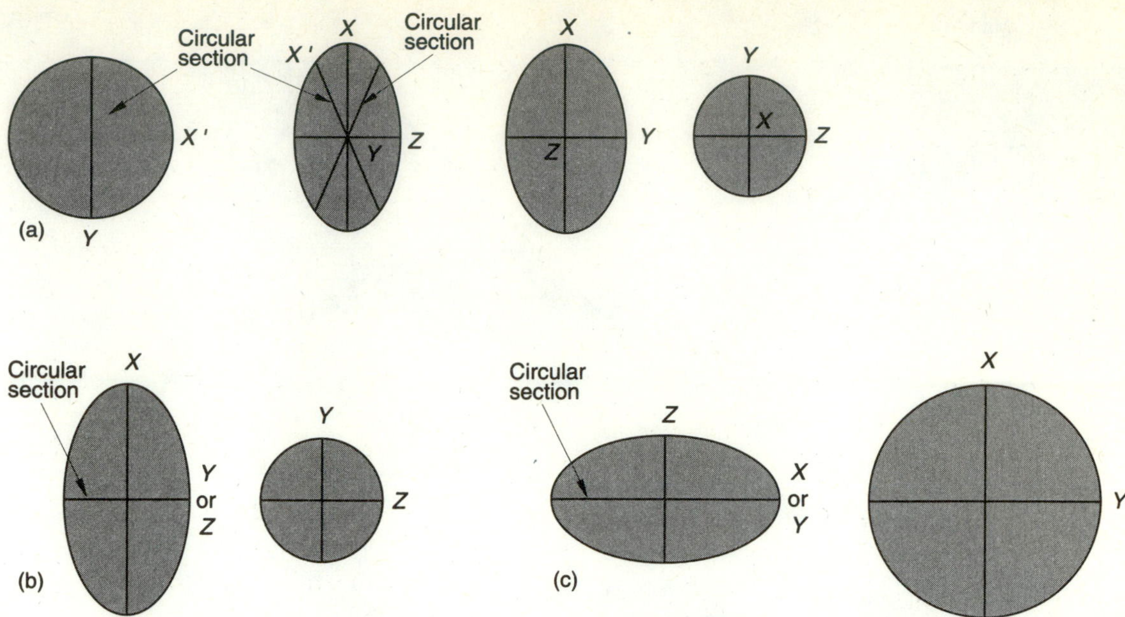

FIGURE 4–12
Sections through the strain ellipsoid: (a) triaxial, (b) prolate, and (c) oblate. Note that all sections, except the two special circular sections, are ellipses, but a section through the spheroids (Figure 4–11) normal to the unique axis is circular.

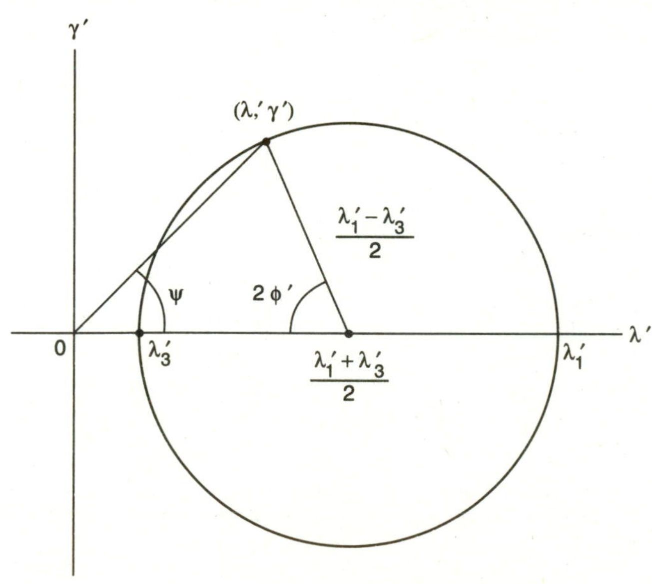

FIGURE 4–13
Mohr circle for finite strain. Note that $(\lambda'_1 + \lambda'_2)/2$ is the center of the circle. For a particular state of homogeneous strain, this circle describes the locus of all possible combinations of λ' and γ'.

γ' and λ' refer to this line. The angular shear ψ may be found for any value of ϕ' from

$$\psi = \tan^{-1}\gamma = \tan^{-1}\gamma'/\lambda'. \qquad \textbf{(4–11)}$$

We can determine λ' and γ' values for each ϕ' if we know values of the elongations ε_1 and ε_2. It should therefore also be possible to determine a Mohr strain circle from values of elongation determined from deformed objects in the field. (See Chapter 5 for more details of techniques for measuring strain; see also Ramsay and Huber, 1983.) In contrast to the Mohr circle construction for stress, which requires values of normal stress to obtain estimates of shear stress, the Mohr construction for finite strain provides an approximation of strain state in rocks deformed millions of years ago by stress long since dissipated. With such tools, we can construct models of former stress conditions in the Earth.

SIMPLE AND PURE SHEAR

Homogeneous deformation may be considered either rotational or irrotational. Rotational homogeneous strain is exemplified by the deformation observed when the cards in a deck are riffled so that each moves the same distance past the next (Figure 4–14). The relative motion of adjacent cards is parallel, with planes

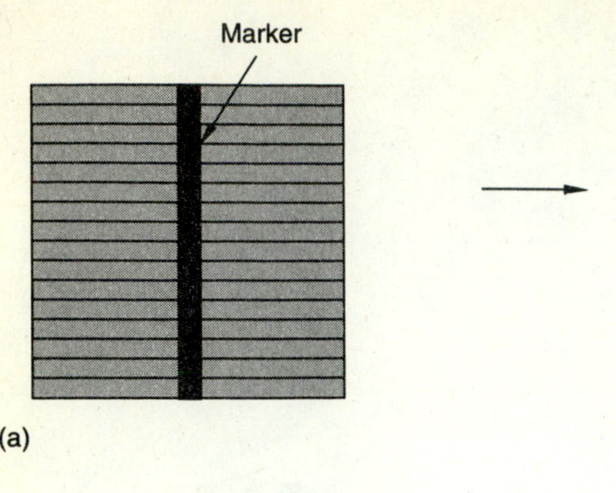

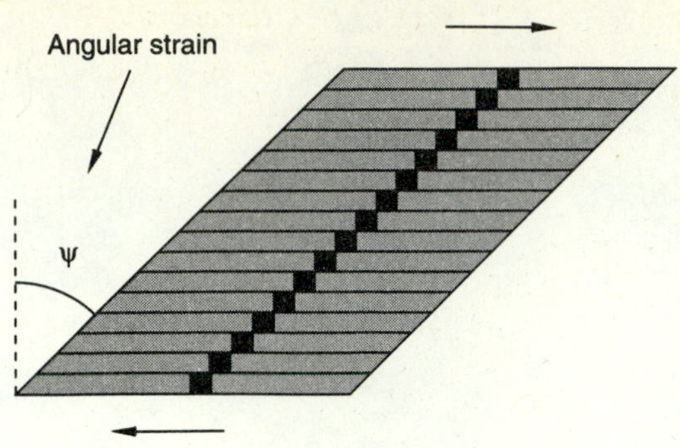

FIGURE 4–14
(a) Card-deck model of simple shear. (b) Deformation of a cube by homogeneous simple shear. *x*, *y*, and *z* here are coordinate axes, not the axes of the strain ellipsoid.

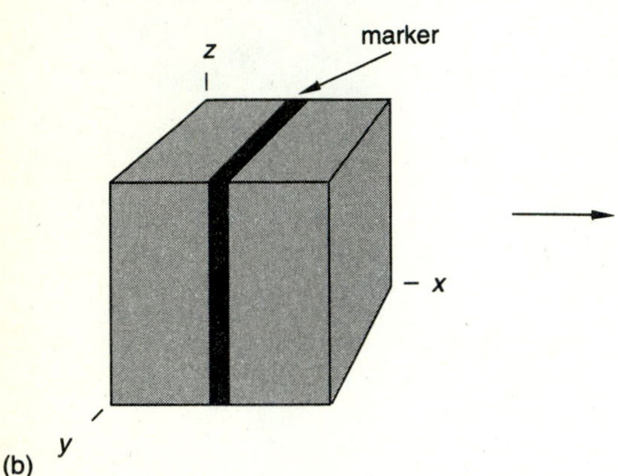

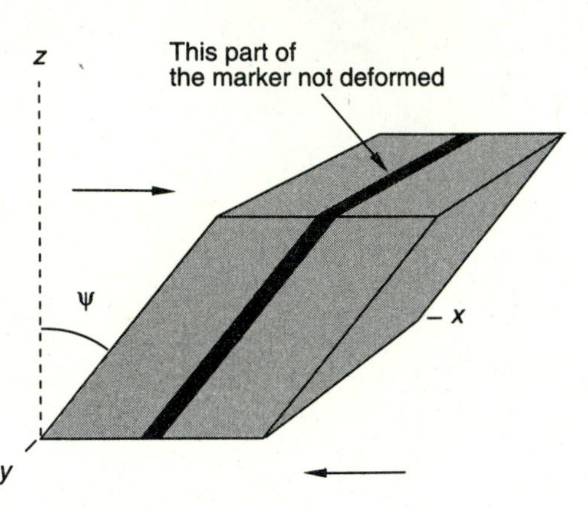

localizing displacement between each card. The net displacement of one card with respect to the one below it is a function of the amount of shear expressed as the angle ψ of rotation in equation 4–5 for shear strain. This kind of rotational homogeneous strain is called ***simple shear***, and it is accumulated by *noncoaxial deformation.* With progressive simple shear, the relationship between stress and strain axes changes continuously because strain axes rotate with respect to external reference axes. Thus, the inverse relationship between stress and strain ellipsoids described above does not hold.

Simple shear is an example of homogeneous plane strain where line lengths are unchanged parallel to the *y* axis during deformation. As a result, all strain is in the *XZ* plane and essentially is two dimensional. Deformation involving plane strains is easier to visualize and manipulate mathematically because the problem is reduced from three to two dimensions. For example, the second-order strain tensor only contains 4 components for plane strain, but 9 components are required for triaxial strain.

Pure shear results from distortion by homogeneous deformation in which the principal axes do not rotate (Figure 4–15). Pure shear may be thought of in terms of the ideal cases of axial flattening or elongation of the strain ellipsoid. Thus, by adding components of rotation and translation, pure shear may be transformed into simple shear as long as it is a plane strain pure shear. Thus any homogeneous strain may be represented by a combination of a pure shear, a rigid-body rotation, and a rigid-body translation.

Now that we can identify different kinds of strain, we can turn in the next chapter to the means we have to measure strain.

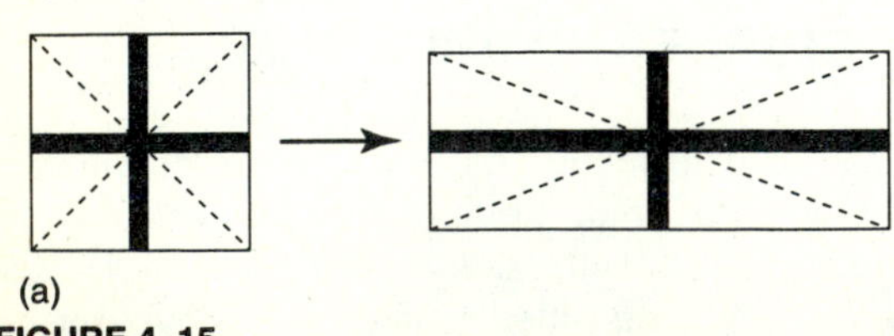

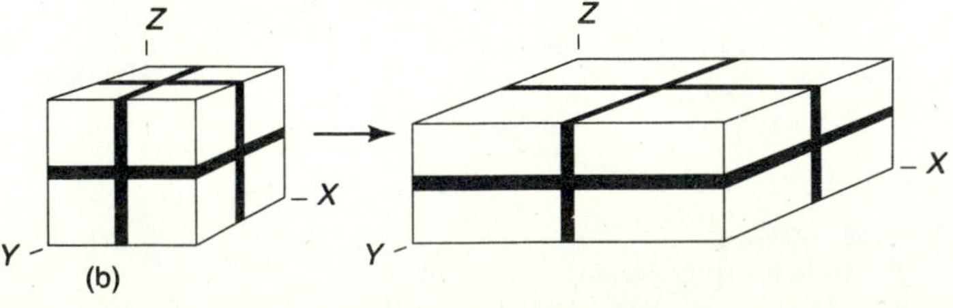

FIGURE 4–15
(a) Deformation by homogeneous pure shear. Note that the angles between the principal axes (heavy solid lines) remain unchanged, but the angles between other lines (dashed) do change. (b) Pure shear deformation of a cube.

ESSAY

Daubrée and Mead Experiment

A classic experiment involving fracture sets was made first by Auguste Daubrée (1879), a French geologist and mineralogist. He used plate glass for his experiment, but the experiment was repeated several decades later by W. J. Mead (1920), using a paraffin-coated rubber sheet in a square frame. The two men produced almost identical fracture sets within the glass or paraffin by applying a shear stress in the form of a couple (oppositely directed shears; Figure 4E–1). On shearing, a reference circle was deformed into a strain ellipse. In addition, four sets of fractures were easily recognized. One set paralleled the long axis of the strain ellipse (*X*), another paralleled the short axis (*Z*), and two conjugate sets formed about an acute angle with the *X* axis as a bisectrix.

Considering the work by Daubrée and Mead, it is easy to see that a fracture set parallel to the short axis of the strain ellipse results from extension. Extension fractures are known to form parallel to the *YZ* plane of the strain ellipsoid and normal to *X*. We can also see that these fractures formed normal to *X* and parallel to *Z* in the essentially two-dimensional strain model.

The conjugate set of fractures parallel to the edges of the glass plate are readily explained as shear fractures. If we consider that the stresses were applied parallel to the edges of the frame in Figure 4E–1, they must be considered shears, for they are oblique to the orientation of the normal stresses.

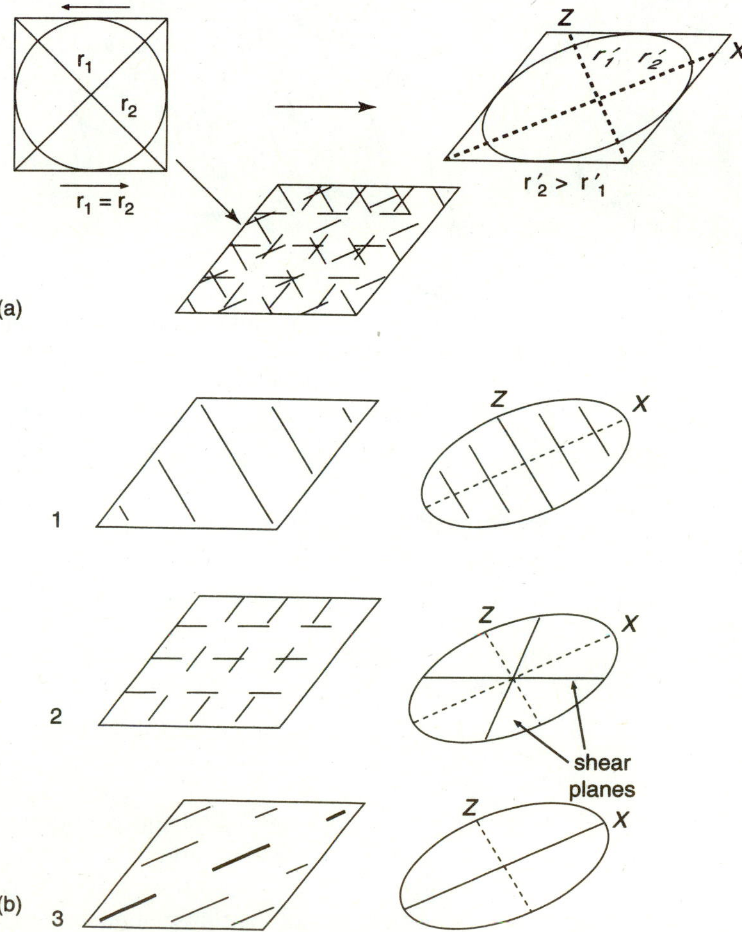

FIGURE 4E–1
Daubrée-Mead experiment (a) and interpretations of fracture sets in terms of the strain ellipse. (a) Four fracture sets in glass or paraffin. (b) Separating fracture sets. 1. Set parallel to *Z*, therefore, is composed of extension fractures (joints). 2. Possible shear planes. 3. Set parallel to *X* may(?) be thrust faults.

The fracture set parallel to the *X* direction is more difficult to explain. The set formed normal to σ_1 and *Z*—assuming that the stress field did not rotate during deformation—and formed parallel to a different oriented shear plane. If so, such fractures could be considered thrust faults (Chapters 9, 10, and 11), especially because they frequently form inclined to the surface of the glass or paraffin—*not* normal to σ_1. Otherwise, they are called "compression fractures". Any homogeneous material of approximately the same shape, strained in similar fashion, would yield about the same fracture pattern (Figure 4E–2).

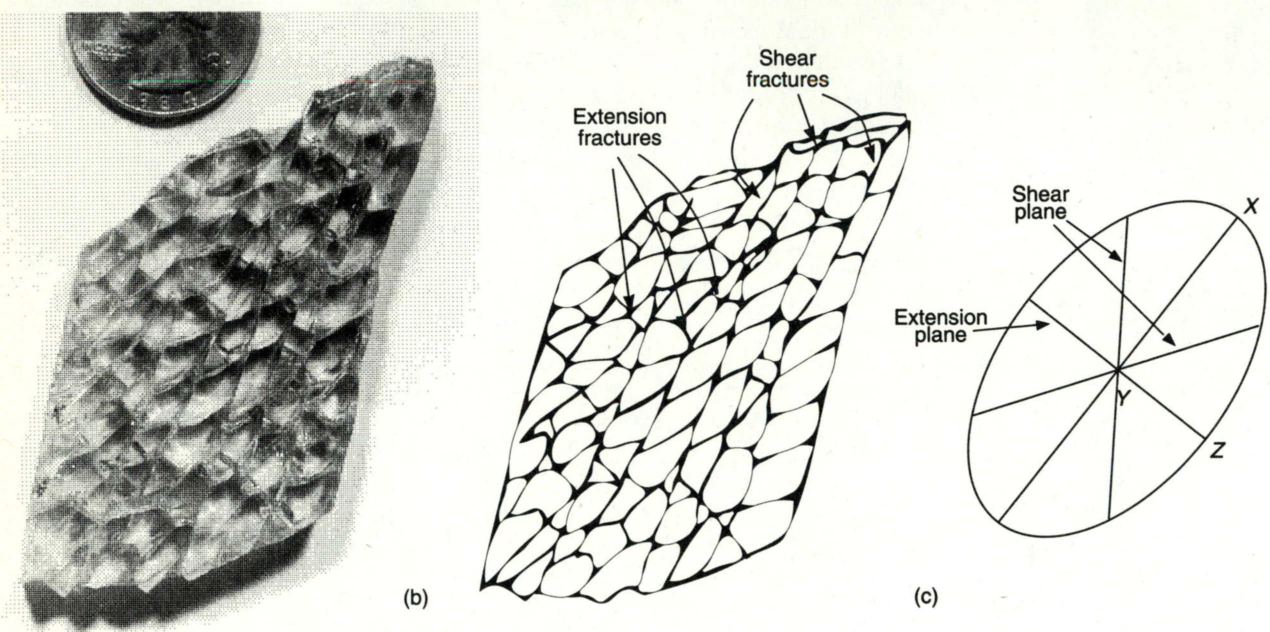

FIGURE 4E–2
(a) Fragment of safety glass from a wrecked motor car (unfortunately, Hatcher's) illustrating a fracture pattern almost identical with that produced by Daubrée and Mead. Deformation of the window glass probably occurred by homogeneous simple shear. Part of U.S. quarter indicates scale. (b) Sketch of the fracture pattern in (a). (c) Strain ellipse with principal and shear planes oriented parallel to most of the fractures in the glass.

References Cited

Daubrée, A., 1879, Etudes synthétiques de géologie expérimentale: Paris, p. 306–314.

Mead, W. J., 1920, Notes on the mechanics of geologic structures: Journal of Geology, v. 28, p. 505–523.

Questions

1. The angle between the foliation in metamorphic rocks and a shear surface was measured in a westward-dipping fault zone in eastern Connecticut. The angle ranges from 20° to 30°. Using information provided in Chapter 4, what can be concluded about the different amounts of strain in the rocks in this fault zone?
2. We have several ways to measure strain that involve change in line length. Why?
3. Why is the 45° angle between σ and the shear plane rarely attained in natural materials?

4. Construct a Mohr circle for finite strain in a deformed pebble, given the values $\varepsilon_1 = 0.6$ and $\varepsilon_2 = 1.2$ for elongation strain. To solve the problem, plot the values corresponding to λ' and γ' on the Mohr circle at intervals of 10° for $2\phi'$, corresponding to values of ϕ of 10°, 15°, 20°, etc.
5. Describe a real situation involving either natural or artificial materials where strain begins as pure shear (flattening) and then is transformed into simple shear.

Further Reading

Brace, W. F., 1961, Mohr construction in the analysis of large geologic strain: Geological Society of America Bulletin, v. 72, p. 1059–1080.
A very understandable discussion of the development of the Mohr circle for strain, and many of the pitfalls about strain that entrap structural geology students and even experienced geologists.

Hubbert, M. K., 1951, Mechanical basis for certain familiar geologic structures: Geological Society of America Bulletin, v. 62, p. 355–372.
This is a classic attempt to relate experimentally produced structures to both theory and real tectonic structures as we observe them in the field.

Means, W. D., 1976, Stress and strain: New York, Springer-Verlag, 339 p.
A well-written introduction to the concepts of stress and strain through the principles of continuum mechanics. Mathematical concepts and derivations are presented understandably, making problems (with solutions) related to geologic situations useful.

Means, W. D., 1992, How to do anything with Mohr circles (except fry an egg)—A short course about tensors for structural geologists: Geological Society of America Structural Geology and Tectonics Division Short Course Notes, 66 p.
An understandable introduction to the application of tensors to stress and strain. Emphasis is on Mohr circles for strain, but development of the Mohr stress circles for stress are also presented.

Ramsay, J. G., and Huber, M. I., 1983, The techniques of modern structural geology: Volume 1: Strain analysis: London, Academic Press, 307 p.
A detailed, well-illustrated treatment of strain and strain analysis using mathematics expected of junior or senior geology majors. Many concepts are presented by means of solved problems, using the illustrations in the book.

5

Strain Measurement

In 1966 we discovered that in addition to the relationship between lineation, oolite extension, and cleavage, the striations on slickensided surfaces, and growths of fibrous minerals in fractures were also related. If the ac [XZ] plane is normal to the fold axes that same surface contains the lineation a [X], the long axes of deformed ooids and the maxima for striations and mineral growths. This suggests a similar and uniform deformation plan for all elements, stratigraphically from the Precambrian at least to the Silurian. . . . To determine these facts it was necessary to sample the information and treat it statistically in graphs, charts, diagrams, and maps . . . .

ERNST CLOOS, 1971, *Microtectonics along the Western Edge of the Blue Ridge, Maryland and Virginia*

DISTORTED OBJECTS IN ROCKS REPRESENT DEFORMATION AND strain of objects of known original shape—an observation first reported by J. Phillips and D. Sharpe in the mid-nineteenth century. Henry C. Sorby (1856) first linked the distorted objects and the deformation (strain) ellipsoid. The first to measure strain using fossils were Albert Heim in 1878 and A. Wettstein in 1886 (Wood, 1973). Not until the mid-twentieth century was major use made of this early work; Ernst Cloos in 1947 published the first of his widely known works on the distorted oöids in the South Mountain fold in the central Appalachians of Maryland. All those studies paved the way for modern quantitative work on distortion of originally spherical and nonspherical objects, distorted fossils, growth of fibers, and other strain markers. Together, these methods permit us to measure strain in a rock mass. Today we can quantitatively assess the state of strain of a rock mass using relatively simple techniques, but suitable ***strain markers*** must be present. Strain markers are features—fossils, other primary sedimentary and volcanic structures, and some secondary features—where original shapes or distributions are known well enough so that their present shapes or distributions can be restored to measure the strain present in the rock (Figure 5–1). With enough strain markers, we can determine the distribution of strain as well as the amount of strain produced by different deformation mechanisms. Most structural studies investigate the effects of stress on rocks as manifested in the kinds of structures that can be observed. General observations of the direction of overturning *(vergence)* of folds (Chapter 14), orientations of fibers and slickenlines on fault or bedding surfaces, and the sense of offset on faults—all enable us to make qualitative and sometimes quantitative measurements of strain and displacement. Many other features can be used to measure the strain in rocks more precisely. Now we will look at some strain-measurement techniques and apply them in several examples.

KINDS OF STRAIN

Strain is change in shape or volume of a reference object between the initial undeformed state and the final deformed state. Strain may take the form of volume change (dilation), or changes in shape (distortion), or both (Figure 5–2). As previously defined, brittle strain occurs as a result of rupture. Brittle deformation results in loss of cohesion and development of fractures separating masses of undeformed material. In contrast, ductile strain—except for certain types of spaced cleavage (Chapter 17)—involves the entire mass, so that deformation is continuous from an initial unstrained state to a final strained state. Throughout a period of strain accumulation, any number of events may affect the state of strain in a rock body. The amount of strain in each event may vary considerably, but the total strain in the rock under study is always the result of a ***progressive deformation.***

The following terms, discussed in Chapter 4, are redefined here to reinforce their importance in measuring strain. ***Total finite strain*** relates the instantaneous shape of a rock mass at any one time to its initial

FIGURE 5–1
A rock mass subjected to homogeneous finite strain may be useful in determining the amount of strain—if it contains a useful strain marker such as undeformed brachiopods like these from the Devonian Hamilton Group, New York. (E. B. Hardin, U.S. Geological Survey.)

undeformed shape. Most deformation we observe in a rock mass represents a state of finite strain—the final or a cumulative state in a series of deformational events. ***Incremental strain*** involves separate steps that occur in small distortion or dilation events through progressive deformation. Finite strain is the sum of incremental strains. ***Infinitesimal strain*** is restricted to very small strain relative to any reference condition.

Observation of fractures and folds permits us to make qualitative or semiquantitative estimates of the amount of strain in a rock mass, but the actual amount of finite or incremental strain in the rock mass is impossible to determine simply by measuring fracture orientations or by measuring geometric features of folds. Sometimes we can determine the amount of strain if we assume the rocks have been folded by a relatively simple mechanism, but most fold mechanisms involve strains that are difficult to reconstruct to an undeformed state. Sometimes microscopic techniques, such as the *calcite strain gauge* in slightly deformed rocks (Groshong and others, 1984), or other methods involving microscopic features of individual grains, such as measurement of the numbers and character of subgrains, deformation lamellae, and other microstructures (see Chapter 7) help in determining finite-strain geometry.

It is useful to note whether the geometry of strain is ***homogeneous*** or ***inhomogeneous.*** Homogeneous strain represents strain occurring in a uniform manner so that straight and parallel reference lines in a material being deformed remain straight and parallel (Figure 4–4). Strain within one part of a homogeneously deformed rock body is the same as in another part. In contrast, inhomogeneous strain involves deformation in which strain is nonuniform throughout the body. The end result is that lines originally parallel are parallel no longer, angular relationships are changed in a nonuniform manner, and straight lines are no longer straight. Folding is a good example of inhomogeneous deformation.

Scale dependence of the kind of strain affecting a mass of rocks also exists. Strain may seem homogeneous on a scale of kilometers, but inhomogeneous on the outcrop or hand-specimen scale. For example, cleavage may seem to occur in all parts of a rock body on the map scale, but at the mesoscopic scale where measurements are made in the field, cleavage may be restricted to shales and may not occur at all in sandstones or limestones in a sequence.

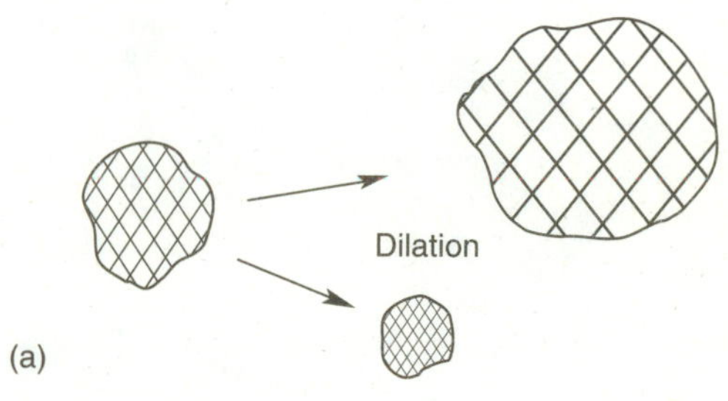

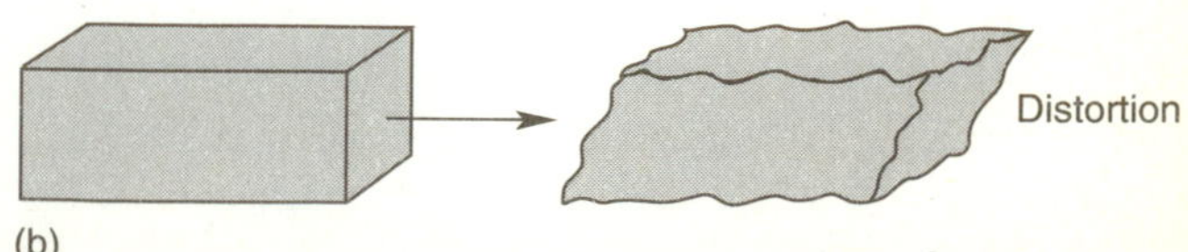

FIGURE 5–2
Example of pure dilation (a) and pure distortion (b). Many geologic deformations include both kinds of strain.

STRAIN MARKERS

Any deformed feature in a rock mass in which the original shape can be quantitatively compared with the present deformed shape may be used as a ***strain marker.*** Two characteristics are needed if a strain marker is to be used as a finite-strain indicator. First, the precursor of the feature should be identifiable and the original shape known or determinable. Second (for some strain-measurement techniques), the marker should have the same mechanical properties as the rest of the rock mass. Unfortunately, very few strain markers strictly meet this requirement. The simplest strain markers to use were originally spherical. Examples include oöids, reduction spots, and vesicles. Most natural strain markers, however, were not originally spherical, but techniques also exist for dealing with this problem.

Reduction spots are small, mostly spherical features in fine-grained sediments where the red to reddish-brown oxidized sediment has been chemically reduced to a greenish color (Figure 5–3a). Reducing conditions are produced locally by a grain or area of different composition. Ideally, these are perfect markers for determining finite strain because they probably have the same mechanical properties as the enclosing rock mass. The only difference is the color. Most reduction spots occur in shales and mudstones, and so their use for finite-strain measurement associated with slaty cleavage planes can indicate the amount of strain in slates. Interesting studies of finite strain have been made by John G. Ramsay and Dennis S. Wood (1972), who measured reduction spots in the Cambrian slates of Wales and Vermont. They concluded that elongations greater than 100 percent occurred by flattening in some samples in the *XY* plane of the strain ellipsoid and parallel to the slaty cleavage (Figure 5–3b). Use of reduction spots is complicated because not all were originally spherical, and their time of formation relative to deformation may not always be clearly known.

Pebbles are among the most frequently used strain indicators. Originally, most were ellipsoidal, but, if the amount of strain is large (greater than 10), an originally spherical shape may be assumed, and the resulting measurement error will not be large. Where the amount of strain is relatively small and the pebbles were not originally spherical, however, the assumption may lead to a large error in determination of finite strain (Figure 5–4). This can be compensated by measuring many ellipsoidal pebbles that were originally randomly oriented and employing a technique (R_f/ϕ) that permits determination of strain of initially ellipsoidal objects. Also, the mechanical properties of the pebbles and matrix may differ and introduce significant errors in measured strains.

Oöids and ***pisolites*** are good indicators of finite strain. They most commonly form in carbonate rocks and ironstone and are nearly spherical to slightly ellipsoidal before deformation. A classic study of oöid deformation was made by Ernst Cloos (1971) in the oölitic limestones around South Mountain in Maryland,

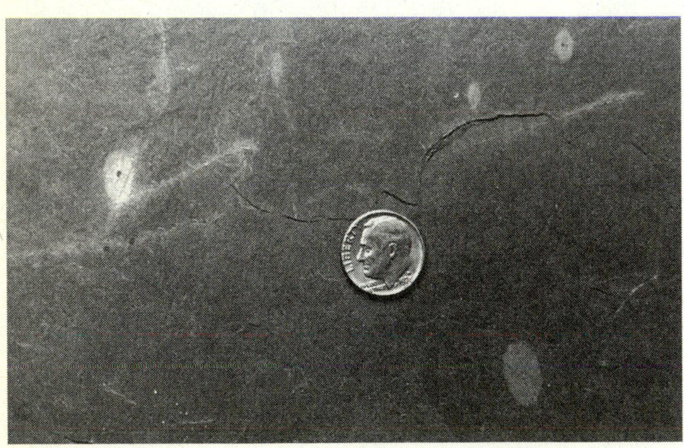

(a)

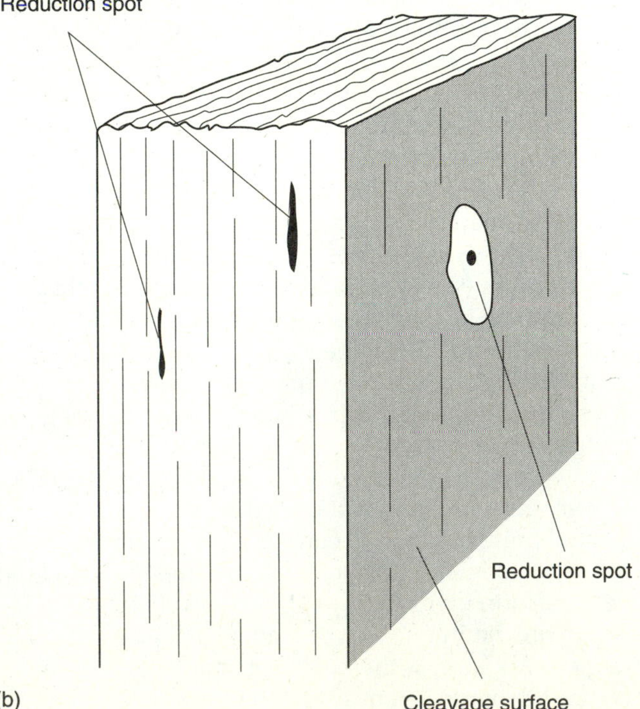

(b)

FIGURE 5–3
(a) Reduction spots in Metawee Slate near Rutland, Vermont, viewed normal to the cleavage plane. Also see Figure 2–15. (RDH photo.) (b) Relationship between reduction spots and slaty cleavage.

FIGURE 5–4
Pebbles in the Upper Proterozoic Bygdin conglomerate at Bygdin, southern Norway, locally exhibit very large amounts of strain that may have produced greater than 700 percent elongation, according to John P. Hossack (1968). Note that *YZ* sections of pebbles in the vertical face to the left of the knife appear much less deformed than *XZ* sections on the near-horizontal face to the right. Knife is 10 cm long. (RDH photo.)

Pennsylvania, and northern Virginia (Figures 5–5 and 18–13). After several decades and more than 10,000 measurements of strain in oriented limestone samples, Cloos resolved the strain accumulated during penetrative deformation of the Blue Ridge anticlinorium.

Fossils have been used to considerable advantage in determining finite strain in rocks (Figure 5–6). We usually know the original shape of an organism from specimens found in undeformed rocks, and we can readily determine if a fossil has been deformed; few organisms are originally spherical, however, and reconstruction of a deformed fossil to its undeformed shape may be geometrically difficult. Trilobites, belemnites, graptolites, *Skolithos* tubes, and brachiopods have been used as strain markers. Because we know their original shapes, many other fossils may also be used to determine shear strain (Chapter 4).

Vesicles (gas bubbles) in volcanic rocks may be used as finite-strain indicators—provided they were not appreciably deformed during initial outpouring of lava. Another problem with vesicles is the mechanical properties of gas in a vesicle are rather different from the mechanical properties of the bulk rock. Vesicles that become filled with minerals *(amygdules)* may also be used (Figure 5–7).

In submarine lava flows (commonly basaltic), ***pillows*** may be used to determine finite strain. Generally, these form as approximately equidimensional masses with a rounded glassy or vesicular top and a coarser-grained and more massive downward-pointing base (Figure 5–8). They have been severely deformed in many areas, and occasionally have been used as finite-strain markers. The usefulness of pillows in accurate determination of finite strain is limited, however, because their original shapes are not uniform. Where strains are large, pillows may enable us to estimate finite strain.

Indirect indicators of activity of organisms, such as **burrows**, may be used in determining finite strain. Most useful are almost-cylindrical burrows oriented normal to the surface of the sea bottom in clean sand environments. For example, the Cambrian Pipe Rock Sandstone in Scotland contains abundant *Skolithos* tubes. Deformation of the Pipe Rock Sandstone has been studied extensively (Figure 5–9); the tubes were initially cylindrical, and circular sections were deformed into ellipses. Thus, they may be used to determine finite strain in the plane of bedding (Coward and Kim, 1981; Fischer and Coward, 1982). They may

FIGURE 5–5
Deformed oöids in the Conococheague Limestone in the Great Valley, near Hagerstown, Maryland. (Charles M. Onasch, Bowling Green State University.)

FIGURE 5–6
Deformed trilobite *(Angelina)* from the Lower Ordovician of North Wales. (W. Stuart McKerrow, Oxford University.)

also be used to determine shear strain parallel to bedding—if the shear strain is penetrative—because the *Skolithos* tubes are characteristically oriented normal to bedding. *Skolithos* tubes are also commonly found in clean sandstones in Europe, North America, Asia, and Australia.

FLINN DIAGRAM

One of the most useful means of displaying constant-volume finite strain was invented by a British structural geologist named Derek Flinn (1962), who assumed that ellipsoidal objects such as reduction spots were initially spherical. Thus, the principal planes (*XY*, *YZ*) of the strain ellipsoid may be determined from principal axes and planes of the deformed objects.

Flinn showed that ratios of the principal axes of the strain ellipsoid could be used to determine a value *k* defined as

$$k = \frac{a-1}{b-1} = \frac{R_{xy}-1}{R_{yz}-1} = \tan q, \qquad \textbf{(5–1)}$$

where *a* is the length of the major axis divided by the length of the intermediate axis, and *b* is the length of

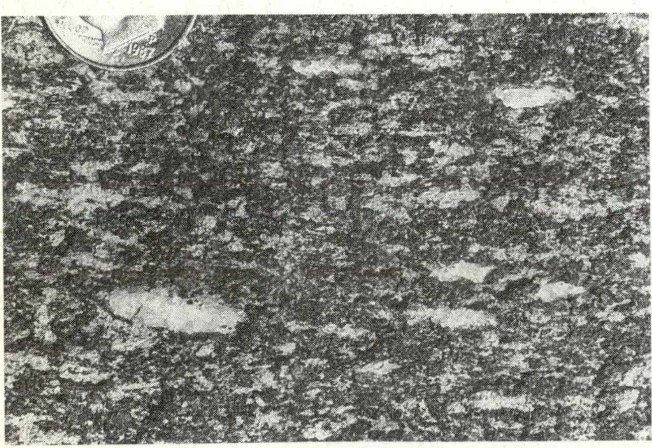

(a)

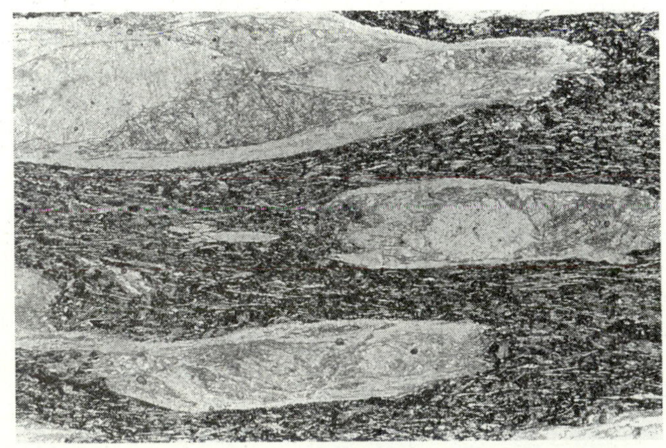

(b)

FIGURE 5–7
(a) Deformed amygdules filled with quartz and feldspar in amphibolite near Berner, Georgia. (Specimen courtesy of Robert J. Hooper, Conoco Research.) (b) Deformed amygdules filled with calcite in Eocambrian Sams Creek Formation near Union Bridge, Maryland. (Charles M. Onasch, Bowling Green State University.)

(a)

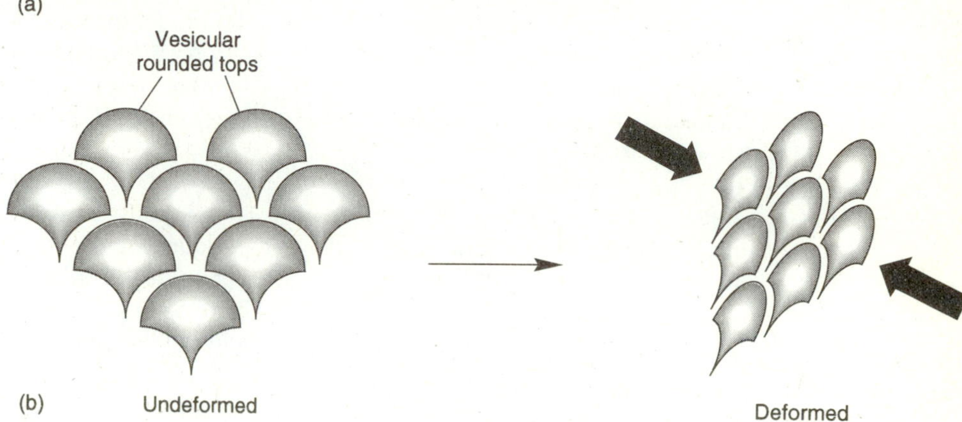

(b)

FIGURE 5–8
(a) Deformed pillow in the Chibougamou Lake area, Quebec. (G. O. Allard, University of Georgia.) This pillow was compressed from the bottom and top of the photo. The top of the deformed pillow is toward the right. Compare with undeformed pillows in Figure 2–24. (b) Comparison of pillows before and after deformation.

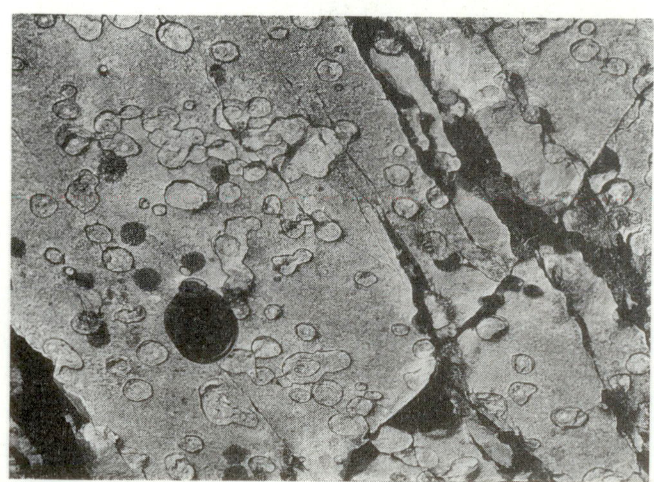

(a)

(b)

FIGURE 5–9
Deformed *Skolithos* tubes (pipes) in the Lower Cambrian Pipe Rock sandstone, Northern Highlands, Scotland. (a) Plan view on a bedding surface showing uniformly oriented elliptical sections of pipes. (b) Vertical section through a bed showing curving—deformed—pipes with greater shear deformation near the top of the bed. Unsheared pipes would be near vertical throughout, as in the thinner bed beneath. (Michael P. Coward, Imperial College of Science and Technology.)

the intermediate axis divided by the length of the minor axis. R_{xy} and R_{yz} refer to strains related to changes in the lengths of axes in the *XY* and *YZ* planes, expressed as

$$R_{xy} = \frac{1+\varepsilon_1}{1+\varepsilon_2}, \tag{5–2}$$

and

$$R_{yz} = \frac{1+\varepsilon_2}{1+\varepsilon_3}, \tag{5–3}$$

where ε is elongation strain (equation 4–1). The angle *q* is measured between the abscissa and the line connecting the arbitrary point *p* with the origin in the *Flinn diagram* (Figure 5–10). An ellipsoid plotting at point *p* in the diagram has a *k* value of tan *q*. Two-dimensional plots of R_{xy} versus R_{yz} yield a valuable relationship portraying three-dimensional strain. Along the horizontal axis, *k* equals 0; on the vertical axis, *k* is infinite. A line with a slope of 1 yields $k = 1$. As *k* approaches infinity, the shape of a deformed object begins to resemble a hot dog (prolate spheroid) and involves *axial elongation* (or constriction). Values near the horizontal axis indicate *axial flattening,* which produces a hamburger shape (oblate spheroid). *General strain* occurs in the region away from the coordinate axes because a deformed object has the shape of a triaxial ellipsoid, with axes of unequal length. Where volume change is zero, a value of $k = 1$ implies *plane strain* (where the *Y* axis has the same length as the diameter of the initial sphere) produced by either pure shear, simple shear, or a combination of these deformations. If plane strain can be proved, the Flinn diagram may also be used to estimate volume change.

STRAIN-MEASUREMENT TECHNIQUES

Wellman's Method

A simple geometric technique for determining both the orientation and shape of the strain ellipse was invented by H. G. Wellman (1962). His method is based on the angular distortion of reference lines originally aligned 90° to each other. The technique requires at least ten strain-marker objects, and all must be arranged randomly in the same plane. Accuracy of the method depends on the occurrence of enough fossils (or other strain markers with bilateral symmetry) in a small planar area. A distinct disadvantage is that a concentration of fossils must occur on a bedding surface; rarely do we find even two or three deformed fossils in a small area, and ten are very difficult to find frequently enough to characterize the strain of an area. Consequently, several techniques described in later sections have greater utility, even if they are not as easy to apply.

The most commonly used strain markers that meet the requirement of original 90° reference lines are brachiopods or tribolites lying on a bedding surface (Figure 5–11). Either a photograph or an accurate tracing of the surface containing the fossils is needed.

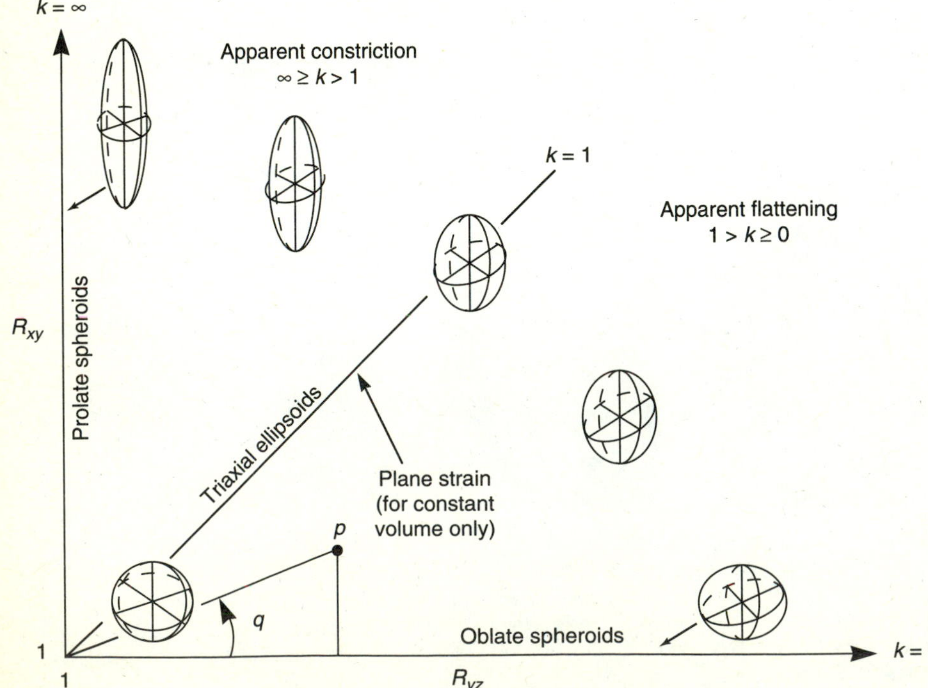

FIGURE 5–10
Flinn diagram. The *k* value for any ellipsoid, such as at point *p*, may be obtained from tan *q*. Note the difference in shape of ellipsoids that have low total strain and plot close to (1,1) or have high total strain and plot far from (1,1).

FIGURE 5–11
Deformed brachiopods on a bedding plane from the Ordovician Davidsville Formation, Gander Lake, central Newfoundland. (Robert B. Neuman, U.S. National Museum, and R. Frank Blackwood, Newfoundland Department of Mines; U.S. National Museum Specimen.)

Using brachiopods or tribolites as the strain marker, an arbitrary reference line *AB*, at least 10 cm long, is drawn in any orientation, not necessarily parallel to the hinge lines of any of the brachiopods, and the ends are labeled (Figure 5–12b). Lines are also drawn along the hinges of the brachiopods on the photograph or tracing and along the symmetry line (commonly marked by a sinus or fold) of each fossil.

A pair of lines is then drawn for each fossil parallel to both the hinge line and symmetry line passing through points *A* and *B*. If there is no strain in the bedding plane containing the fossils, the result is a rectangle. If the fossils are deformed parallel to the plane, a parallelogram results (Figure 5–12c). This procedure is repeated for the hinge and symmetry lines of each brachiopod in the plane. The corners of the parallelo-

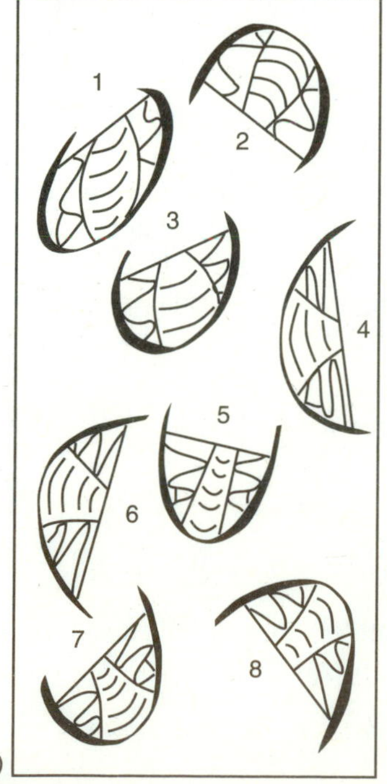

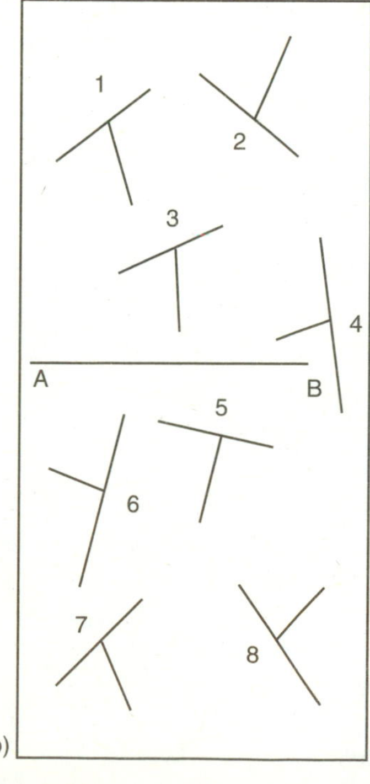

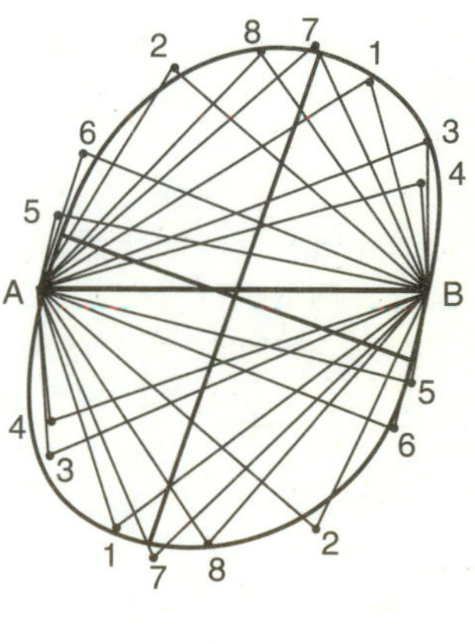

FIGURE 5–12
Determination of strain in fossils, using Wellman's method. Note that neither of the axes of the best-fit ellipse in (c) are parallel to the arbitrarily chosen line AB in (b). Using the ratio of long (*X*) to short (*Z*) axes in the ellipse in (c), the trilobite cephalons record a strain of 1.29.

grams, after being connected, outline an ellipse. The lengths and orientations of the axes of the ellipse indicate the amount of strain in the plane of bedding. Obviously, the more fossils available for construction of lines, the better defined the resulting strain ellipse will be.

Determination of Finite Strain of Initially Elliptical Markers

Strain markers with an initial elliptical shape, such as pebbles or sand grains, present another problem for determining finite strain. Three variables are involved: initial ellipticity, initial orientation, and orientation and magnitude of the principal strain axes. Pebbles, oöids, sand grains, and other initially elliptical markers may be randomly oriented during deposition. If so, the average initial shape can be considered spherical, and the initial ellipticity is less important. If such markers were deposited by a current, as in a stream channel, the long axes tend to parallel the current. Pebbles in a channel may thus be very strongly oriented and even imbricated upstream. A difficulty might arise in distinguishing a preferred orientation due to strain from a primary depositional orientation. Primary orientation does occur, but strain orientation is generally more consistent. The difficulty may be resolved by noting the relative parallelism of the present orientation of the objects to a known tectonic feature such as cleavage or foliation. Because most rocks commonly studied are marine, primary orientation is not so much of a problem, but it does occur as a primary bedding-plane orientation. Primary ellipticity is the rule.

The next few sections summarize the R_f/ϕ and several rapid methods for determining finite strain of initially elliptical objects. A more detailed and comprehensive discussion of these methods may be found in Ramsay and Huber (1983).

***R_f/ϕ* Method.** John Ramsay (1967) introduced the *R_f/ϕ method* (properly termed R_f/ϕ'), which was further developed by Dunnet and others in several papers and is comprehensively summarized in Ramsay and Huber (1983). Deformation of initially elliptical objects having an initial ellipticity R_i by a homogeneous strain produces objects that remain elliptical. The shape of the final ellipse (R_f) is determined by the initial shape and orientation of the starting ellipse (R_i) along with the shape or ellipticity (R_s) and orientation of the strain ellipse (Figure 5–13). The angle between the long axis of the ellipse and an initial reference orientation is designated as ϕ. The initial angles will change during deformation to ϕ', an angle more nearly parallel to that of the long axis of the strain ellipse. The only orientations that will not change will be those of ellipses that are oriented either parallel ($\phi = 0°$) or normal ($\phi = 90°$) to the long axis of the strain ellipse (using a reference line parallel to a principal axis). As a result, those parallel become *more* elliptical and those normal become *less* elliptical. Ellipticities of others with values of ϕ between 0° and 90° will lie between the two extremes, but most will attain greater ellipticity. As the amount of deformation increases, R_f increases and deviation from the mean value of ϕ' decreases. Curves that plot R_f versus ϕ' become more and more closed, and this deviation, termed the *fluctuation* (F), decreases so that curves become more closed and onion-shaped; the fluctuation is less than 90° (ranging from less than +45° to greater than –45°) for small to moderate values of R_i (1.5 to 3.0) and moderate to large values of R_s greater than 3.0 (Figure 5–14).

The ultimate objective of this analysis is determination of the relative contributions of initial ellipticity and R_s. An exercise in curve-fitting may determine the value of R_i that best fits the resulting R_f/ϕ' curve; this can be done manually or with a computer, and generally results in a curve relating initial elliptical shapes to high values of R_f. Curves are always symmetrical with respect to the long axis of the strain ellipse—unless there was an original preferred orientation of pebbles before deformation. Also, on each curve, points derived from a group of ellipses wherein initial orientation was random will cluster toward high values of R_f.

Ramsay and Huber (1983) described two situations that can arise: maximum R_i greater than strain R_s, and maximum R_i less than R_s (Figure 5–13). In the first case, the result will be data with a fluctuation of 180°, but points will concentrate about the maximum R_f value (Figure 5–14). The direction of maximum concentration of data will also indicate the orientation of the long axis of the strain ellipse, and data should be symmetrically distributed on either side of the maximum. If not, the initial distribution of ellipses was not random. The relationships are given by

$$R_{f\,\max} = R_s\,R_{i\,\max} \qquad \textbf{(5–4)}$$

and

$$R_{f\,\min} = R_{i\,\max}/R_s\,. \qquad \textbf{(5–5)}$$

Solving for R_s in both equations and substituting for R_s from equation 5–5 in equation 5–4,

$$(R_{f\,\max}\,R_{f\,\min})^{1/2} = R_{i\,\max}\,. \qquad \textbf{(5–6)}$$

Substituting for $R_{i\,max}$ from equation 5–5 in equation 5–4 and solving for R_s yields

FIGURE 5–13
Relationships between initial ellipticity (R_i) and orientation (ϕ), final orientation (ϕ') and ellipticity (R_f) in an array of elliptical markers. (a) Undeformed $R_s = 1.0$ with an initial ellipticity of 2.0. (b and c) Deformed, with vertical imposed flattening $R_s = 1.5$ and $R_s = 3.0$. F is the fluctuation. One data point on each curve corresponds to a pebble. (From J. G. Ramsay and M. I. Huber, *The techniques of modern structural geology,* Academic Press, v. 1, 1983.)

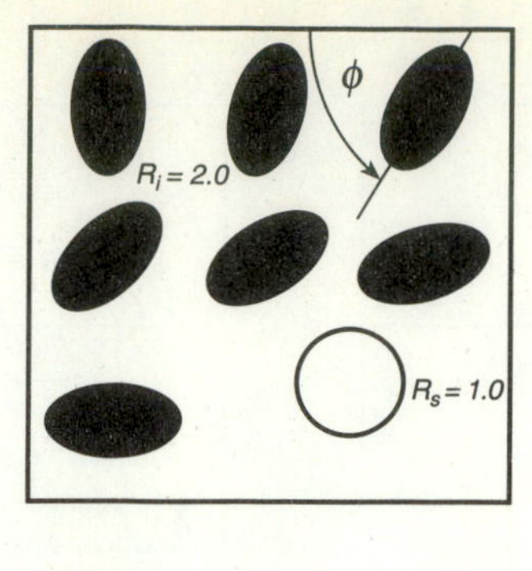

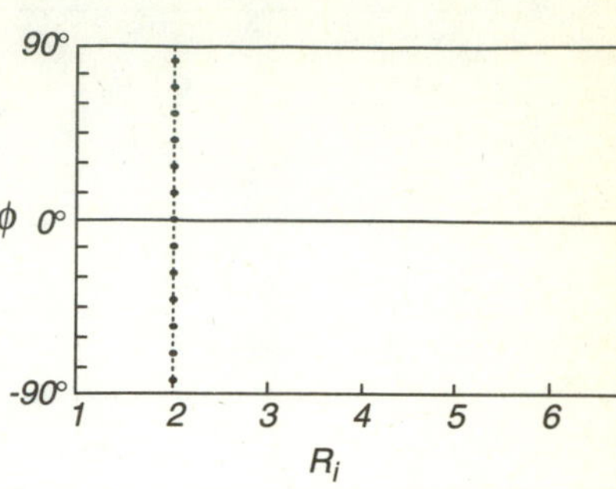

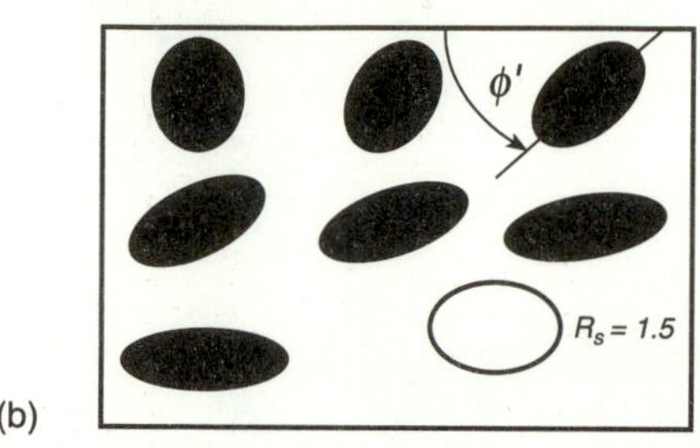

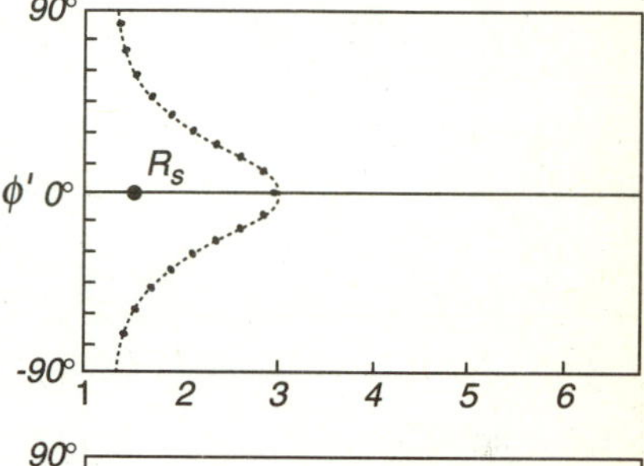

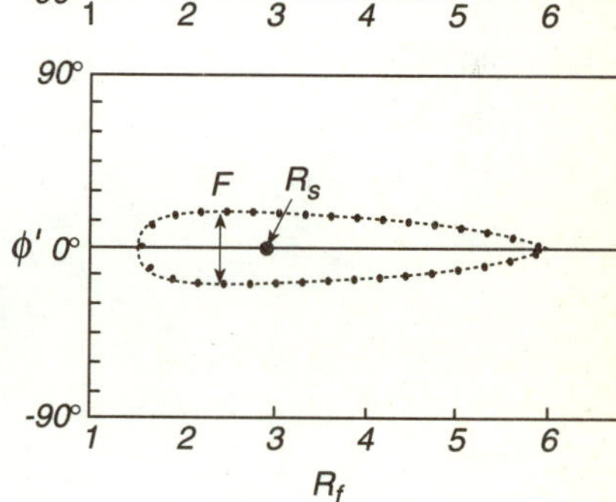

$$(R_{f\max}/R_{f\min})^{1/2} = R_s. \tag{5–7}$$

The second and opposite case, where R_i maximum is less than R_s, produces clustered data with closed envelope curves. Data should cluster toward $R_{f\max}$, be symmetrical about the long-axis orientation of the strain ellipse, and show minimal fluctuation (less than 90°). Relationships between $R_{f\,max}$ and $R_{f\min}$ are

$$R_{f\max} = R_s R_{i\max} \tag{5–8}$$

and

$$R_{f\min} = R_s / R_{i\max}. \tag{5–9}$$

Solving equation 5–9 for $R_{i\max}$, substituting in equation 5–8, and solving for R_s yields

$$R_s = (R_{f\max} R_{f\min})^{1/2}. \tag{5–10}$$

Solving equation 5–9 for R_s, substituting in equation 5–8, and solving for $R_{i\max}$ yields

$$R_{i\max} = (R_{f\max} / R_{f\min})^{1/2}. \tag{5–11}$$

Equations 5–6 and 5–7, and 5–10 and 5–11, enable calculations of the contributions of initial ellipticity and tectonic strain in a randomly oriented set of initially elliptical objects. If any initial preferred orientation of ellipses exists, an asymmetric distribution of data points will appear in the R_f/ϕ plot, and it will be difficult to obtain useful values of R_i and R_s.

Another problem common in using this technique is the difficulty in deciding on the best-fitting envelope curve, but this is an alternative procedure to making use of equations 5–4 to 5–11. Distribution of points often makes curve-fitting a tedious and even subjective part of the analysis, even with randomly oriented particles.

The R_f/ϕ technique is both tedious and time consuming without a computer and digitizer. Techniques discussed next are faster, frequently less accurate, and allow differentiation of R_i from R_s, and permit determination of tectonic strain; their reliability increases with increasing tectonic strain.

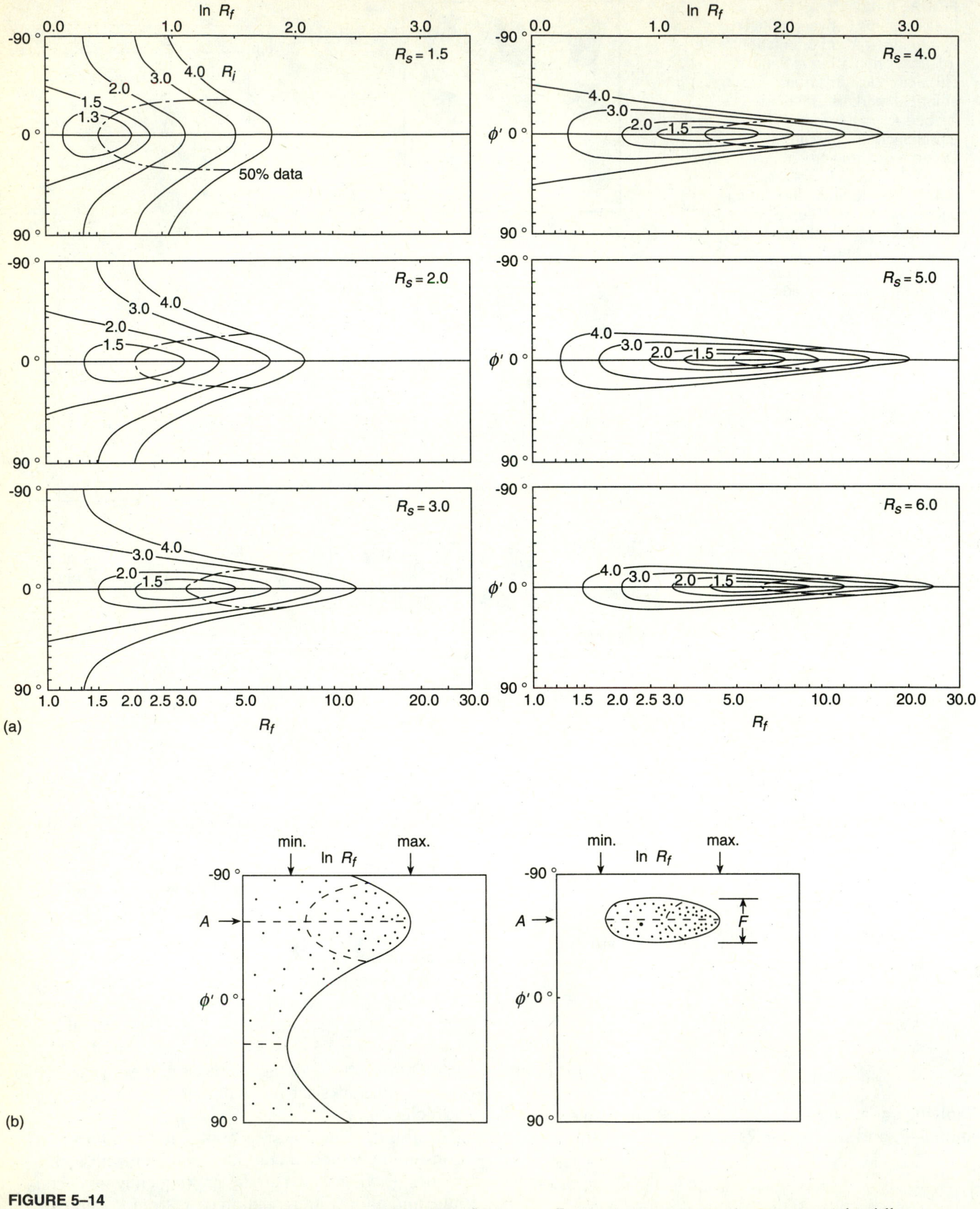

(a)

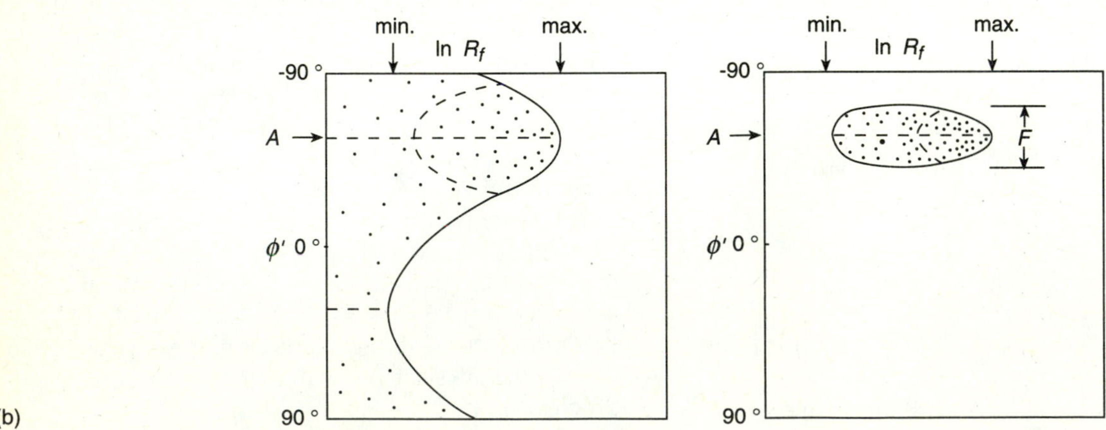

FIGURE 5–14
R_f/ϕ' plots showing different relationships between R_i and R_s and the resulting curves. (a) R_f/ϕ' reference curves for different values of initial ellipticity and the strain ellipse, R_s. (b) Features of R_f/ϕ' plots used for calculating strain. Note the symmetry of the fluctuation (*F*) about the orientation of the strain ellipse (A). In the left diagram, $R_i > R_s$; in the right, $R_s > R_i$. (From J. G. Ramsay and M. I. Huber, *The techniques of modern structural geology,* Academic Press, v. 1, 1983.)

Center-to-Center Method. The ***center-to-center method,*** also originally devised by John Ramsay (1967), is based on the principle that the distance and angular relationships between particles in an aggregate of objects (sand grains, pebbles, or oöids) with a statistically uniform initial distribution should help to determine the orientation of the strain ellipse in the deformed aggregate (Figure 5–4). The method involves measurement of the distances and angles between a reference grain and its nearest neighbors (Figure 5–15a). The technique is based on measurement of immediately adjacent grains only, so measurement of any tie line that crosses another grain should be discarded. A serious problem is that properly identifying nearest neighbors may be impossible.

Application of the technique is rather simple, but it is slow without a computer and digitizer. A transparent overlay is placed on an oriented photograph (at convenient magnification) of the objects to be measured. The overlay can also be placed onto a sawed slab of the rock if the objects to be measured (pebbles, pisolites, amygdules) are large enough. A reference line is drawn on the overlay. The center of each object is then marked with a dot, and tie lines are drawn connecting each point with its nearest neighbors. Distances (d') and angles ($+\alpha'$ or $-\alpha'$) between objects are then measured between the tie lines and the reference marker (azimuth) line.

A plot may be made of d' versus values of α' ranging from +90° to –90° (Figure 5–15b). The principal difficulty lies in constructing the best-fit curve and then determining the maximum and minimum values and the symmetry axes of the curve. Ramsay and Huber (1983) suggested determining the arithmetic means of the α' values at certain intervals (say every 10°) and using them to plot the best-fit curve. Maximum and minimum values of the curve will not represent exact values for tectonic strain, but a value for R (ellipticity of the strain ellipse) may be obtained from

$$R_s = \frac{d'_{max}}{d'_{min}} \quad . \tag{5–12}$$

Values of d'_{max} and d'_{min} are obtained directly from the $d' - \alpha'$ plot. The position of d'_{max} on the axis also represents orientation of the long axis of the strain ellipse.

The Fry Method. A technique invented by Norman Fry (1979) is a simpler version of the center-to-center method and works on the same principle. Angular relationships and distances between particles are modified according to the nature and amount of accumulated strain. The result of the Fry technique is a diagram containing a set of points with a circular-to-elliptical blank area of relative shape and orientation proportional to the shape and orientation of the strain ellipse. A circular area indicates that there is no strain in the rock.

To use the ***Fry method*** graphically, an overlay is made and a central reference point identified (Figure 5–16). A second overlay is made and the center of each particle is marked with a numbered point. (Centers may be marked directly on a photograph if additional prints are available.) The central reference mark on the first overlay is placed over a numbered point, and dots corresponding to all of the numbered centers of the particles are marked on the overlay. (It may be adequate to mark just the centers of the adjacent grains.)

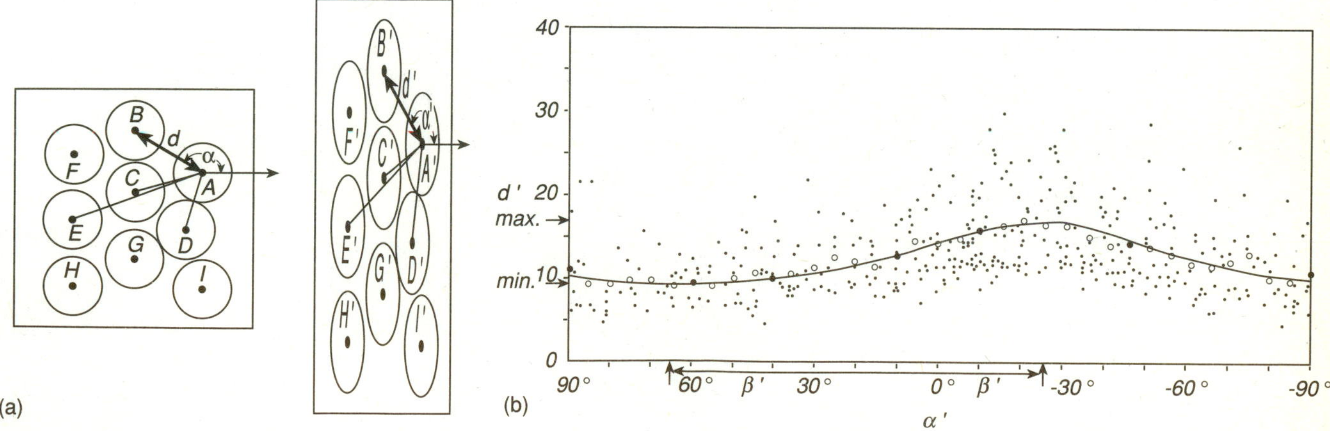

FIGURE 5–15
(a) Construction of tie lines in a deformed rock for determination of finite strain by the center-to-center method. (b) Plot of d' versus α'. The circles are averages of d' over 36 10° sectors. The distance AE should be discarded because it does not connect nearest neighbors. (From J. G. Ramsay and M. I. Huber, *The techniques of modern structural geology,* Academic Press, v. 1, 1983.)

The central reference mark is then moved (maintaining a constant orientation) to another numbered point in the array of particle centers, and the process of placing dots on the overlay is continued. The procedure is repeated until all of the numbered centers have been used, or until a vacant field emerges. If no vacant field emerges, initial (and final) particle distribution was Poisson random. If the field is circular, the rock is undeformed in the orientation used. An elliptical vacant area approximates the shape and orientation of the strain ellipse. Undeformed oöids and conglomerates—either of which may have a primary ellipticity—will yield mostly elliptical vacant fields.

Eric Erslev (1988) divided the center-to-center distances between grains by the sum of the average radii to eliminate variations related to object size and sorting. This results in better-defined vacant fields than those produced by the original Fry plots, and Erslev suggested these be called "normalized Fry" plots. Computer programs have been written for both the Macintosh and IBM™ personal computers to construct these plots with the aid of a digitizer (Appendix 3).

Discussion. The R_f/ϕ, center-to-center, and Fry methods all yield estimates of strain in elliptical objects, but each has advantages and disadvantages. The R_f/ϕ method is probably the most accurate, but it is very slow unless the parameters can be digitized for processing by a computer program. The center-to-center method has a severe disadvantage: the user may not be able to determine nearest neighbors accurately. The Fry method is probably the most rapid, and it can be used in conjunction with appropriate computer programs. All of these methods, however, are designed to determine two-dimensional homogeneous strain, if determination is made on only one surface. For complete determination of three-dimensional strain in a rock body, the technique should use three perpendicular sections, preferably oriented parallel to the three principal planes of the strain ellipse. Programs to run on desktop computers will transform two-dimensional strain to three dimensions and make three-dimensional analyses. McEachran and Marshak (1986) have pointed out that graphics capabilities of some small computers, particularly the Apple Macintosh™, permit writing simple programs that simulate strain and permit calculations to be made. Programs also exist for calculation of two-dimensional strain using the Fry, center-to-center, and R_f/ϕ methods.

FIBERS AS STRAIN INDICATORS

Growth of fibrous quartz, calcite, and other minerals in veins (Figures 5–17 and 5–18), pressure shadows, and movement surfaces make possible direct measurement of incremental strain. Veins that contain several layers of fibers indicate separate events of opening of the vein (crack-seal) and reveal part of the incremental strain history of the rock mass. Orientations of fibers also indicate the direction of opening of the vein, and curved fibers indicate rotation of the strain field or shear during formation.

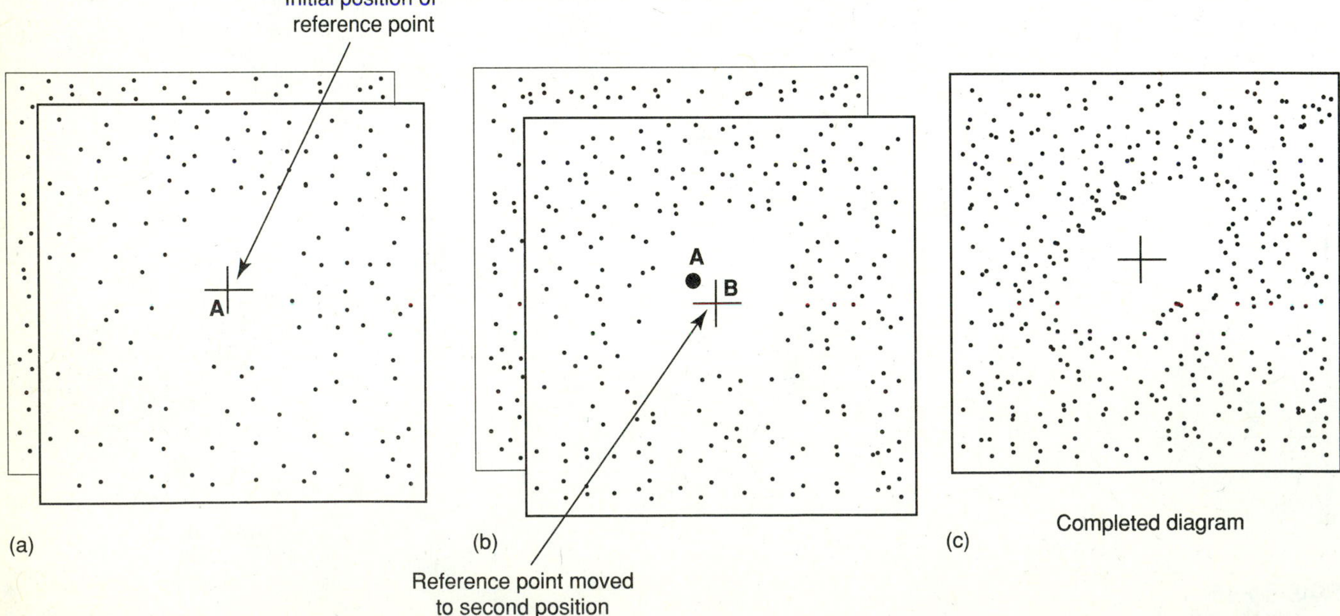

FIGURE 5–16
Stepwise plotting of a Fry diagram. The reference point is moved through successive positions (a and b) and the centers of grains in the photo below transferred until a vacant area appears (c), or until it is clear that none will appear.

Study of the multiple-opening history of extension veins yields details of progressive deformation as well as information about the deformation history of the rock mass. Most veins are filled with massive, randomly oriented crystals, but those nucleating fibrous growth during multiple events of progressive deformation provide a useful aid in studying the incremental strain history of an area.

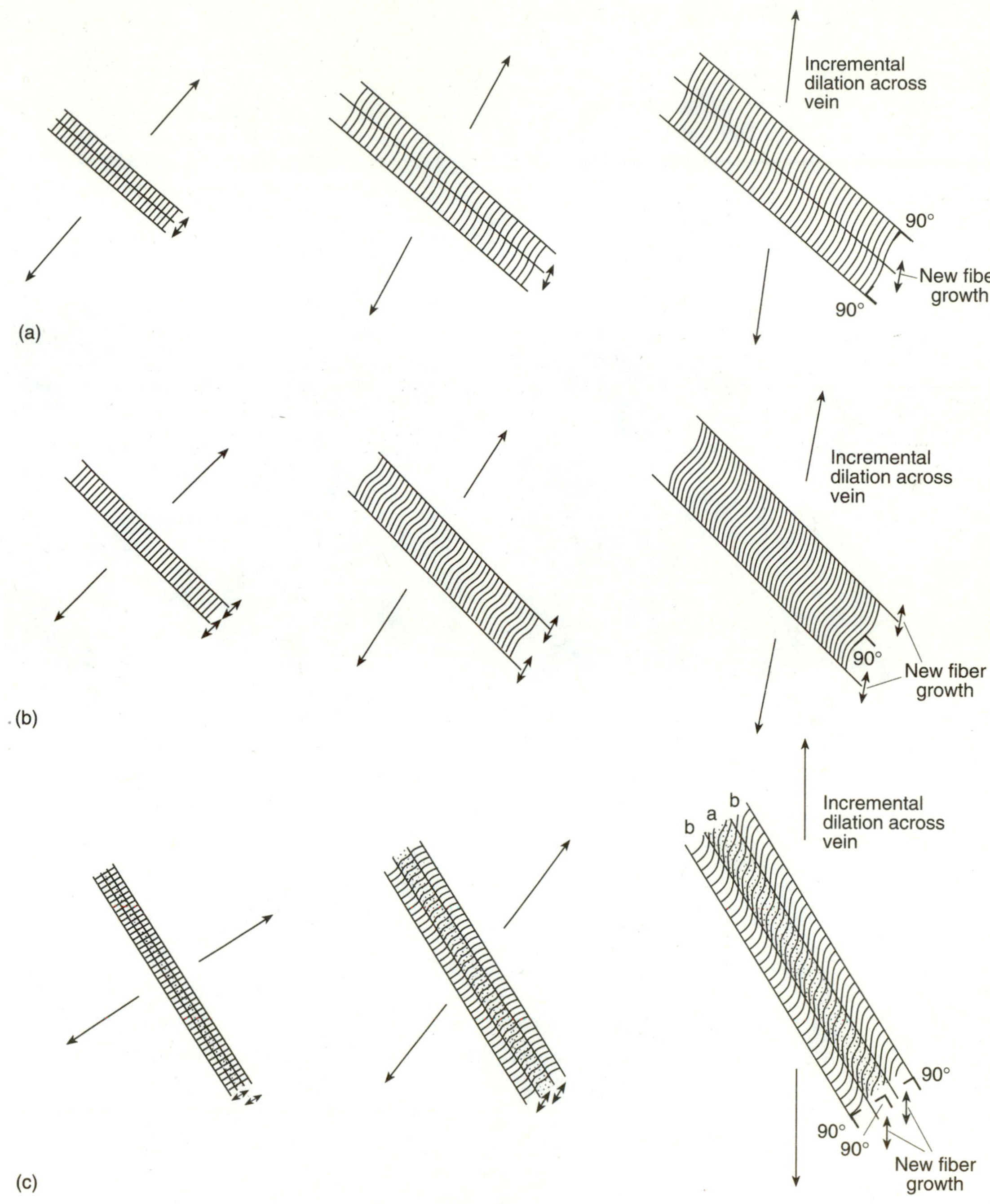

FIGURE 5–17
Several types of tectonic extension veins showing multiple stages of opening and filling. (a) Type 1—syntaxial crystal growth. New crystals in the vein form in optical continuity with wall-rock crystals. (b) Type 2—Antitaxial crystal growth. (c) Type 3—Composite new crystals b are in optical continuity with wall-rock crystals. Crystals a form last. (From D. W. Durney and J. G. Ramsay, *in* K. De Jong and R. Scholten, eds., *Gravity and tectonics,* © 1973, John Wiley & Sons. Reprinted by permission of John Wiley & Sons, Inc.)

DETERMINATION OF PRESSURE-SOLUTION STRAIN

Under ideal conditions, pressure-solution strain may account for 20 to 30 percent of the total strain in a rock body. This strain is commonly observed as stylolites, not all of which need be tectonic, and pressure-solution cleavage (Chapter 17). Direct effects are best observed in thin sections cut normal to surfaces formed by pressure solution (Figure 5–19). Estimates of the bulk strain in rock containing pressure-solution strain may be obtained by several means: (1) reconstructing partly dissolved clasts across several pressure-solution surfaces and calculating the relationship to a reference volume (actually, area in a thin section) in the rock, (2) unfolding folded veins produced as a product of pressure-solution shortening, and (3) determining the ratios of insolubles (such as clays and heavy minerals) in dissolved zones to those in the bulk undissolved rock mass.

Now that we have surveyed the different methods of measuring strain in rock, we can go on to Chapter 6, where we will consider the various models and examples of the mechanical behavior of rock materials.

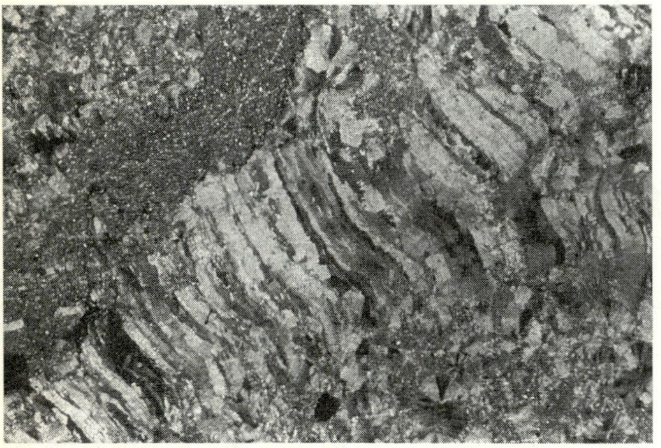

FIGURE 5–18
Vein filling showing curved fibrous calcite (syntaxial), indicating a change in orientation of strain axes as the vein opened. Oölitic limestone bed in Upper Cambrian Nolichucky Shale near Oak Ridge, Tennessee. Vein is approximately 2 mm thick; crossed polars. (Sample courtesy of Peter J. Lemiszki, Oak Ridge National Laboratory.)

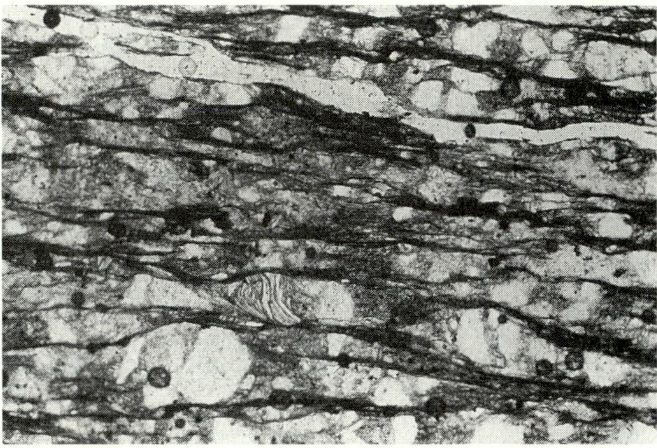

FIGURE 5–19
Pressure-solution cleavage in Martinsburg Slate near the Delaware Water Gap, New Jersey, showing quartz grains partly dissolved. Note the truncation of layers by pressure solution in the partly dissolved detrital muscovite grain in the lower center of the photo. Width of the long dimension of the photo is approximately 5 mm; plane light. (Thin section courtesy of Timothy L. Davis, North Carolina Geological Survey.)

Questions

1. How are finite and incremental strain related?
2. What are the desirable characteristics of a good strain marker?
3. What primary depositional processes can add to the difficulties of determination of finite strain of initially elliptical markers?
4. What characteristic of Flinn's *k* enables us to determine the kind of strain that distorted a marker?
5. Why does Wellman's method work?
6. With the strain markers in Figure 5–5, determine:
   (a) R_s, R_i, and ϕ', using the R_f/ϕ technique,
   (b) R_s and X using the center-to-center technique,
   (c) shape and orientation of the strain ellipse, using the Fry method.
7. Why does the Fry method work?
8. Why are fossils better than many other strain markers for quantitative determination of shear strain?

ESSAY PROBLEM

Finite Strain from Deformed Pebbles

The data in Table 5E–1 were collected by Baylus K. Morgan as part of a senior research project at Clemson University. For his study, Morgan measured deformed pebbles weathered from the Draytonville metaconglomerate in the Kings Mountain belt in the southern Appalachian Piedmont of South Carolina (Figure 5E–1).

TABLE 5E–1
FINITE-STRAIN DATA FOR DRAYTONVILLE METACONGLOMERATE

Sample #	*X* (cm)	*Y* (cm)	*Z* (cm)	Volume (cm^3)	Undeformed Diameter (cm)	Extension (%) *X*	Extension (%) *Y*	Extension (%) *Z*	*a* = *X/Y*	*b* = *Y/Z*	*k*
1	11.9	4.7	3.2	85	5.46	118%	-12%	-41%	2.5	1.5	3.0
2	10.4	5.1	3.0	60	4.86	114%	5%	-38%	2.0	1.7	1.4
3	8.4	4.2	2.4	30	3.86	118%	9%	-36%	2.0	1.80	1.2
4	10.3	3.8	2.2	35	4.06	154%	-6%	-44%	2.7	1.7	2.4
5	8.1	4.1	2.4	35	4.06	100%	1%	-39%	2.0	1.7	1.4
6	6.8	3.5	2.0	20	3.37	102%	4%	-39%	1.9	1.8	1.1
7	12.4	5.4	3.1	105	5.86	112%	-6%	-47%	2.3	1.7	1.9
8	9.2	4.0	3.4	60	4.86	89%	-16%	-30%	2.3	1.2	6.5
9	7.0	3.4	2.4	20	3.37	108%	1%	-27%	2.1	1.4	2.8
10	5.9	3.0	2.2	19	3.31	78%	-9%	-32%	2.0	1.4	2.5
11	8.9	3.3	1.9	25	3.63	145%	-9%	-46%	2.7	1.7	2.4
12	7.5	2.8	2.2	19	3.31	127%	-15%	-32%	2.7	1.3	5.7
13	8.2	4.5	2.5	39	4.2	95%	7%	-40%	1.8	1.8	1.0
14	8.7	4.6	2.7	48	4.5	93%	2%	-40%	1.9	1.7	1.3
15	8.6	3.1	2.1	20	3.37	155%	-8%	-36%	2.8	1.5	3.6
16	3.1	2.2	1.2	3	1.79	73%	23%	-31%	1.4	1.8	0.5
17	7.0	3.0	1.9	19	3.31	111%	-9%	-41%	2.3	1.6	2.2
18	4.2	2.2	1.9	8	2.48	69%	-11%	-23%	1.9	1.2	4.5
19	7.3	2.4	2.3	20	3.37	117%	-27%	-30%	3.0	1.0	50.0
20	7.8	3.3	2.0	25	3.63	115%	-9%	-43%	2.4	1.7	2.0
21	16.1	6.6	4.0	190	7.13	126%	-7%	-42%	2.4	1.7	2.0
22	7.1	3.1	2.1	20	3.37	111%	-8%	-36%	2.3	1.5	2.6
23	6.7	3.4	1.8	19	3.31	102%	3%	-44%	2.0	1.9	1.1
24	5.5	4.1	2.2	21	3.42	61%	20%	-34%	1.3	1.9	0.3
25	3.8	2.0	1.4	4	1.97	93%	2%	-27%	1.9	1.4	2.3
26	6.4	4.3	2.2	20	3.37	90%	28%	-33%	1.5	2.0	0.5
27	9.3	3.7	1.6	24	3.58	160%	3%	-55%	2.5	2.3	1.2
28	5.4	3.0	1.6	10	2.67	102%	12%	-40%	1.8	1.9	0.9
29	9.2	4.6	2.6	50	4.57	101%	1%	-43%	2.0	1.8	1.2
30	5.6	3.0	1.6	15	3.06	83%	-0%	-46%	1.9	1.9	1.0
31	9.7	4.1	2.7	40	4.24	129%	-3%	-36%	2.4	1.5	2.8
32	7.8	3.1	2.0	20	3.37	131%	-8%	-39%	2.5	1.6	2.5
33	8.1	5.3	2.8	58	4.8	69%	10%	-40%	1.5	1.9	0.6
34	6.9	3.5	1.9	25	3.63	90%	-2%	-46%	2.0	1.8	1.2
35	5.3	1.6	1.4	5	2.12	151%	-23%	-32%	3.3	1.1	23.0
36	5.6	2.9	1.8	10	2.67	110%	9%	-31%	1.9	1.6	1.5
37	9.4	6.1	3.0	80	5.35	76%	14%	-42%	1.5	2.0	0.5
38	5.4	2.5	2.0	10	2.67	102%	-6%	-25%	2.2	1.3	4.0
39	6.5	3.4	2.6	22	3.48	87%	-2%	-25%	1.9	1.3	3.0
40	8.1	4.1	3.0	45	4.4	84%	-5%	-30%	2.0	1.4	2.5
41	5.9	2.9	2.2	19	3.31	78%	-12%	-32%	2.0	1.3	3.3
42	5.7	3.7	2.2	20	3.37	69%	10%	-33%	1.5	1.7	0.7
43	4.7	3.2	1.9	15	3.06	54%	5%	-36%	1.5	1.7	0.7
44	5.4	2.4	1.3	8	2.48	118%	-3%	-46%	2.3	1.8	1.6
45	4.6	3.5	1.8	12	2.84	62%	23%	-35%	1.3	1.9	0.3
46	4.7	4.0	1.6	15	3.06	54%	31%	-46%	1.2	2.5	0.1
47	3.5	2.7	1.3	3	1.79	96%	51%	-27%	1.3	2.1	0.3
48	8.6	4.7	1.6	20	3.37	155%	39%	-51%	1.8	2.9	0.4
49	10.4	7.0	3.0	90	5.56	87%	26%	-46%	1.5	2.3	0.4
50	9.0	4.3	3.0	63	4.94	82%	-11%	-39%	2.1	1.4	2.8
51	6.8	3.0	1.5	15	3.06	122%	-0%	-49%	2.3	2.0	1.3
52	8.8	2.7	1.8	22	3.48	153%	-22%	-48%	3.3	1.5	4.6
53	5.4	2.7	1.4	10	2.67	102%	1%	-46%	2.0	1.9	1.1
54	7.5	3.2	1.8	15	3.06	145%	5%	-41%	2.3	1.8	1.6
55	13.6	3.9	2.2	50	4.57	198%	-13%	-50%	3.5	1.8	3.1
56	6.5	3.1	1.8	19	3.31	96%	-6%	-44%	2.1	1.7	1.6
57	5.3	2.0	1.6	8	2.48	114%	-19%	-35%	2.7	1.3	5.7
58	11.3	5.7	3.6	120	6.12	85%	-5%	-41%	2.0	1.6	1.7
59	4.4	2.2	2.0	8	2.48	77%	-11%	-19%	2.0	1.1	10.0

The Draytonville is a relatively clean metaconglomerate consisting of pebbles ranging up to 16 cm long in a matrix of predominantly recrystallized quartz grains. Most of the pebbles consist of recrystallized vein quartz, although a few (less than 2 percent) consist of other mineral or rock types. Microscopic grains within individual pebbles show no sign of the deformation evident in the shapes of the

TABLE 5E–1 (continued)

Sample #	*X* (cm)	*Y* (cm)	*Z* (cm)	Volume (cm³)	Undeformed Diameter (cm)	Extension (%) *X*	Extension (%) *Y*	Extension (%) *Z*	*a* = *X*/*Y*	*b* = *Y*/*Z*	*k*
60	6.5	3.8	2.1	20	3.37	93%	13%	-36%	1.7	1.8	0.9
61	4.3	1.8	1.3	4	1.97	118%	-7%	-34%	2.4	1.4	3.5
62	5.7	3.5	2.2	15	3.06	86%	14%	-28%	1.6	1.6	1.0
63	12.1	6.8	3.6	98	5.72	112%	19%	-37%	1.8	1.9	0.9
64	6.2	3	2.3	19	3.31	87%	-9%	-29%	2.1	1.3	3.7
65	4.5	3.2	1.7	15	3.06	47%	5%	-44%	1.4	1.9	0.4
66	4.5	2.1	1	8	2.48	81%	-15%	-58%	2.1	2.1	1.0
67	4.4	2.8	2.1	10	2.67	65%	5%	-21%	1.6	1.3	2.0
68	5.9	2.7	1.4	10	2.67	121%	1%	-46%	2.2	1.9	1.3
69	5.1	2.9	1.9	18	3.25	57%	-9%	-40%	1.8	1.5	1.6
70	7.9	4.2	2.8	36	4.1	93%	2%	-30%	1.9	1.5	1.8
71	5.1	2.4	2.1	10	2.67	91%	-10%	-21%	2.1	1.1	11.0
72	9.1	4.1	2.5	40	4.24	115%	-3%	-41%	2.2	1.6	2.0
73	5.9	2.4	1.5	10	2.67	121%	-10%	-42%	2.5	1.6	2.5
74	4.7	2.5	1.2	6	2.25	109%	11%	-45%	1.9	2.1	0.8
75	3.2	1.7	1	3	1.79	79%	-5%	-44%	1.9	1.7	1.3
76	5.4	2.9	1.6	8	2.48	118%	17%	-35%	1.9	1.8	1.1
77	9.6	4.2	2.2	35	4.06	136%	3%	-44%	2.3	1.9	1.4
78	6.9	2.9	1.5	15	3.06	125%	-5%	-49%	2.4	1.9	1.6
79	6.8	2.4	1.7	15	3.06	122%	-20%	-44%	2.8	1.4	4.5
80	8.3	4.3	2.3	40	4.24	96%	1%	-44%	1.9	1.9	1.0
81	5.3	2.6	2	12	2.84	87%	-8%	-28%	2.0	1.3	3.3
82	4.3	2.5	1.1	8	2.48	73%	1%	-54%	1.7	2.3	0.5
83	7.4	3.7	2.4	30	3.86	92%	-4%	-36%	2.0	1.5	2.0
84	8.5	4.3	2.9	55	4.72	80%	-7%	-37%	2.0	1.5	2.0
85	5.8	2.9	2.4	20	3.37	72%	-12%	-27%	2.0	1.2	5.0
86	3.9	2.7	1.6	12	2.84	37%	-3%	-42%	1.4	1.7	0.6
87	3.5	1.6	1.5	3	1.79	96%	-9%	-16%	2.2	1.1	12.0
88	4.1	1.6	0.8	2	1.56	163%	3%	-47%	2.6	2.0	1.6
89	9.8	5.3	2.7	80	5.35	83%	0%	-48%	1.8	2.0	0.8
90	7.9	4.4	2.5	50	4.57	73%	-2%	-45%	1.8	1.8	1.0
91	10.2	6.6	3.6	110	5.94	72%	11%	-39%	1.5	1.8	0.6
92	6.2	3.1	1.4	15	3.06	103%	1%	-54%	2.0	2.2	0.8
93	9.4	3.9	2.1	35	4.06	132%	-2%	-48%	2.4	1.9	1.6
94	8.1	3.9	1.5	20	3.37	140%	16%	-55%	2.1	2.6	0.7
95	12.6	6.1	2.9	95	5.66	123%	8%	-47%	2.1	2.1	1.0
96	3.6	1.8	1.6	3	1.79	101%	1%	-9%	2.0	1.1	10.0
97	4.1	2.3	1.3	4	1.97	108%	17%	-34%	1.8	1.8	1.0
98	5.9	3.4	2.9	20	3.37	75%	1%	-12%	1.7	1.2	3.5
99	5.9	3.2	2.4	19	3.31	78%	-3%	-27%	1.8	1.3	2.7
100	10.1	3.2	3	40	4.24	138%	-23%	-29%	3.2	1.1	22.0
101	8.5	2.9	2.6	23	3.53	141%	-16%	-26%	2.9	1.1	19.0
102	15.9	5	3.7	110	5.94	168%	-14%	-36%	3.2	1.4	5.5
103	12.5	5	3.4	110	5.94	110%	-14%	-41%	2.5	1.5	3.0
104	13	4.6	2.7	63	4.94	163%	-5%	-45%	2.8	1.7	2.6
105	11.5	4	2.6	60	4.86	137%	-16%	-45%	2.9	1.5	3.8
106	4.9	3.5	1.8	8	2.48	98%	41%	-27%	1.4	1.9	0.4
107	3.4	1.9	1	3	1.78	91%	7%	-42%	1.8	1.9	0.9
108	9.4	4.6	3.2	58	4.8	96%	-4%	-33%	2.0	1.4	2.5
109	8.6	4.1	2.2	38	4.18	106%	-0%	-47%	2.1	1.9	1.2
110	13.4	5.6	3.3	120	6.12	119%	-8%	-46%	2.4	1.7	2.0
111	3.3	1.8	0.8	2	1.56	112%	15%	-47%	1.8	2.3	0.6
112	2.9	1.3	0.6	2	1.56	86%	-15%	-60%	2.2	2.2	1.0
113	2.9	1.4	1	2	1.56	86%	-10%	-34%	2.1	1.4	2.8
114	3.4	1.9	1.1	3	1.78	91%	7%	-38%	1.8	1.7	1.1
115	5.1	2.4	2	10	2.68	90%	-10%	-25%	2.1	1.2	5.5
116	4	2.5	1.8	9	2.58	55%	-3%	-30%	1.6	1.4	1.5
117	5.1	2.7	1.4	10	2.68	90%	1%	-46%	1.9	1.9	1.0

pebbles today. Also, few of the pebbles bear evidence of pressure solution deformation. Long axes of pebbles are oriented parallel to the axes of upright folds that plunge gently northeast, parallel to a major syncline (Figures 5E–1, 5E–2).

Morgan collected more than 100 intact pebbles, weathered from the Draytonville metaconglomerate, and measured the major axes with a caliper to determine total strain. He also measured orientations of the long (*X*) axes of 143 pebbles in place to determine the finite-strain ellipsoid. Because the long axes of the pebbles all cluster about N 55° E, 15° NE (Figure 5E–2b), it is unlikely that any of the pebbles had appreciable original ellipticity unless their original orientation and that of the major elongation axis were the same. Localities where deformation is minimal confirm that the pebbles were mostly originally spherical.

Using any technique you think is appropriate and data from Table 5E–1 and Figure 5E–2, determine the orientation of the finite-strain ellipsoid and comment on the magnitude of strain affecting the rock mass. What additional sources of error may be present?

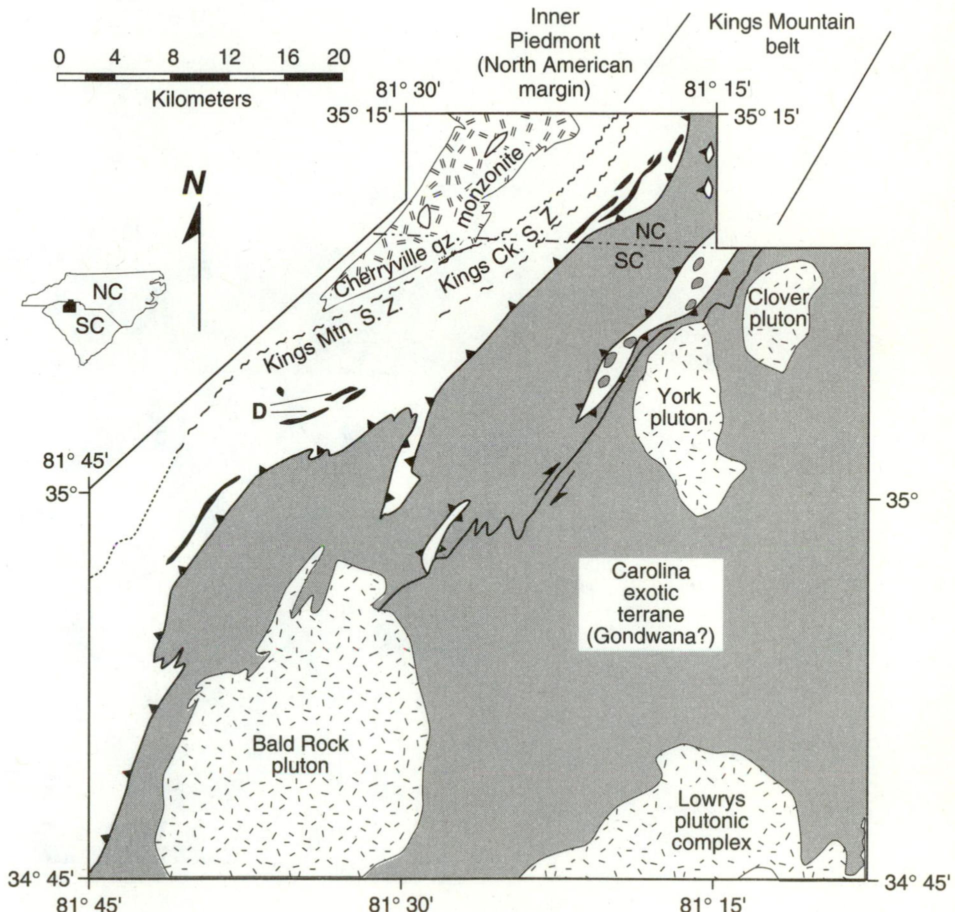

FIGURE 5E–1
Geologic map of the Kings Mountain belt in part of the Carolinas compiled from several sources showing the location of the study area in South Carolina. D—Draytonville metaconglomerate localities studied. Metaconglomerate bodies are patterned black. Barbed (teethed) line is probably a suture between an exotic terrane, the Carolina terrane (shown by a light shaded pattern), and North America, interpreted as the Central Piedmont suture, separating Laurentia and Gondwana. Accretion of the Carolina terrane occurred either during the early or middle Paleozoic. Dotted lines are traditional geologic subdivision boundaries labeled along the northeastern edge of the map. Wavy lines indicate shear zones (S. Z.).

One approach to solving this problem is to plot a Flinn diagram of the data and estimate the magnitude of strain. The center-to-center and Fry methods require observation and measurement of objects in their original positions relative to each other, and so we cannot use them here. In collecting the data in Table 5E–1, Morgan used an immersion technique to determine the volume of each pebble and back-calculated the original spherical dimensions. What sources of error could be important in this method of determining finite strain?

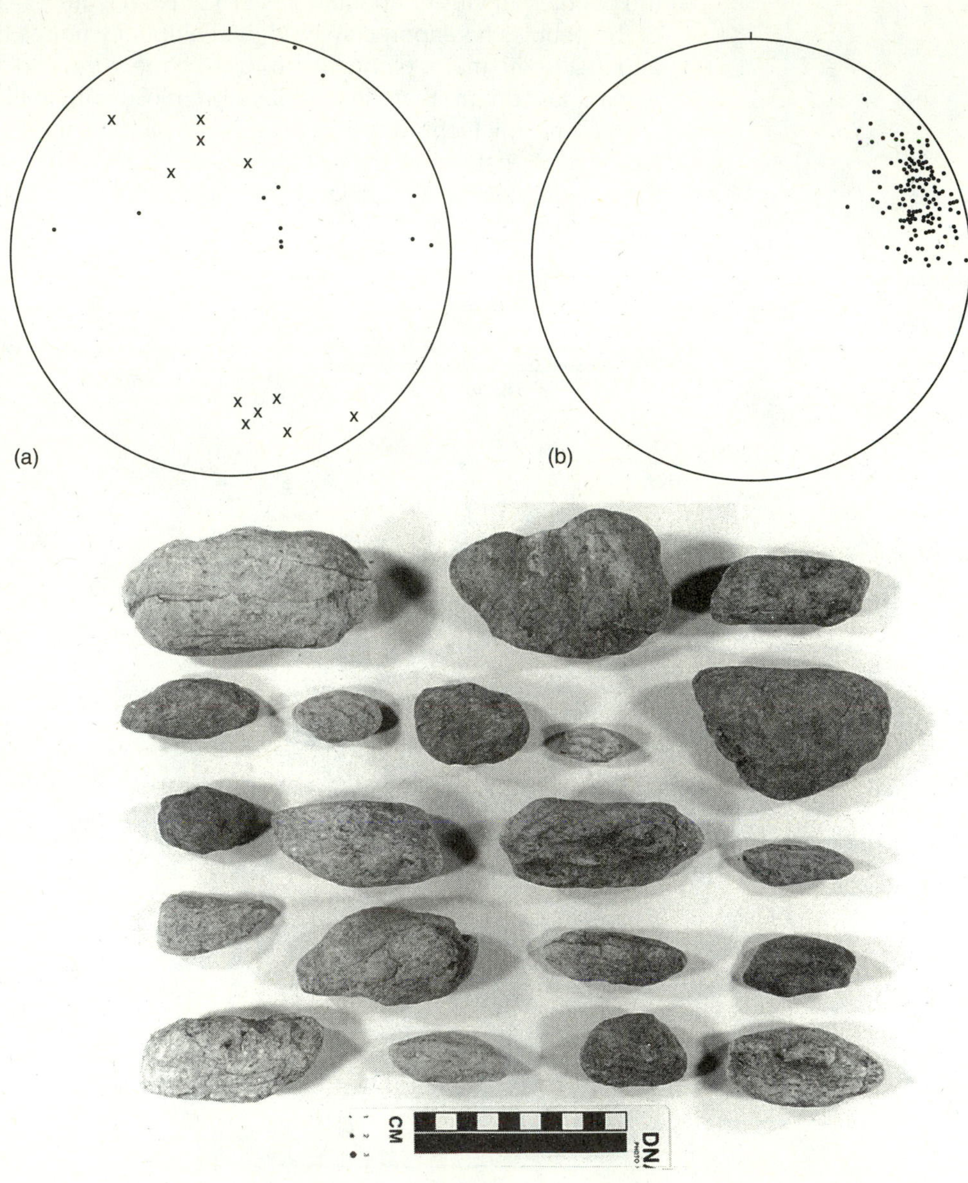

FIGURE 5E–2
Fabric elements in the Draytonville metaconglomerate, measured in place at the localities indicated in Figure 5E–1. (a) x—pole to the dominant foliation; dot—pole to a younger foliation. (b) Orientations of 143 long axes of deformed pebbles. (c) Representative pebbles, mostly composed of vein quartz.

Further Reading

Cloos, E., 1971, Microtectonics along the western edge of the Blue Ridge, Maryland and Virginia: Baltimore, Johns Hopkins University Press, 234 p.
Strain in a very large area in the central Appalachian Blue Ridge was documented by measuring dimensions of more than 10,000 oöids in a study lasting several decades.

Erslev, E. A., 1988, Normalized center-to-center strain analysis of packed aggregates: Journal of Structural Geology, v. 10, p. 201–209.
Modifies the original Fry method to permit more exact determination of strain in rocks by producing more sharply defined vacant fields in the Fry diagram.

Hossack, J. R., 1968, Pebble deformation and thrusting in the Bygdin area (S. Norway): Tectonophysics, v. 5, p. 315–339.
Determination of strain in pebbles in the Bygdin conglomerate and the relationships to thrust faulting are demonstrated. Some pebbles here exhibit very large elongation strains, and the rock mass containing them has been called a "walking stick" conglomerate because some pebbles are as much as 1 m long and only 1 to 2 cm in diameter.

Ramsay, J. G., and Huber, M. I., 1983, The techniques of modern structural geology, v. 1: Strain analysis: New York, Academic Press, 307 p.
Detailed summary of strain-analysis techniques. Undergraduate geology students should be able to work with the mathematics.

6

Mechanical Behavior of Rock Materials

The fact that rock does show marked differences in behavior in natural deformation provides the geologist with an unusual opportunity to learn more about the conditions of deformation if only he can relate these differences in some systematic manner to environmental and rock factors.

FRED A. DONATH, 1970, *American Scientist*

THE EFFECTS OF THE MECHANICAL BEHAVIOR OF MATERIALS are visible all around us, in deformed rocks and in the technology that produces many of the conveniences of everyday life. In any motor car, the steel making up the fenders and engine block displays contrasting physical properties produced by different compositions and physical conditions involved in manufacture. Steel is used for both, but for fenders a more ductile steel is rolled into sheets and stamped (deformed) into shape. The more brittle steel in the engine block is cast roughly in molten form and cooled under conditions that produce the needed strength and temperature resistance, and then it is machined into shape.

Plastics and ceramics are quite different from metals, but they, too, are pertinent here. What we now call plastics are mostly artificial polymers made from hydrocarbons. They are manufactured at temperatures high enough to soften—make them plastic—but not melt them. Ceramic materials in common use today—bricks, semiconductors, glasses, fine china, and ceramic internal-combustion engines—all exhibit elastic properties at room temperature but are made from different geologic raw materials at high temperatures, where they behave plastically or as viscous melts.

Much of the theory we use in structural geology to explain the behavior of rocks was originally developed to explain the behavior of metals under different conditions. Ceramic materials consist of silicates and oxides with properties more like those of minerals than the properties of metals. Because ceramic compounds, like minerals, form mostly low-symmetry crystals, in contrast with metals which form mostly high-symmetry crystals, structural geologists are now benefiting from theoretical developments in the fields of materials science as well as metallurgy.

Idealized mechanical models that simulate the behavior of rocks, soils, concrete, steel, and ceramics have been used for many years because they enable us to simplify and thereby understand the response of these real materials to stress (Figure 6–1). The three

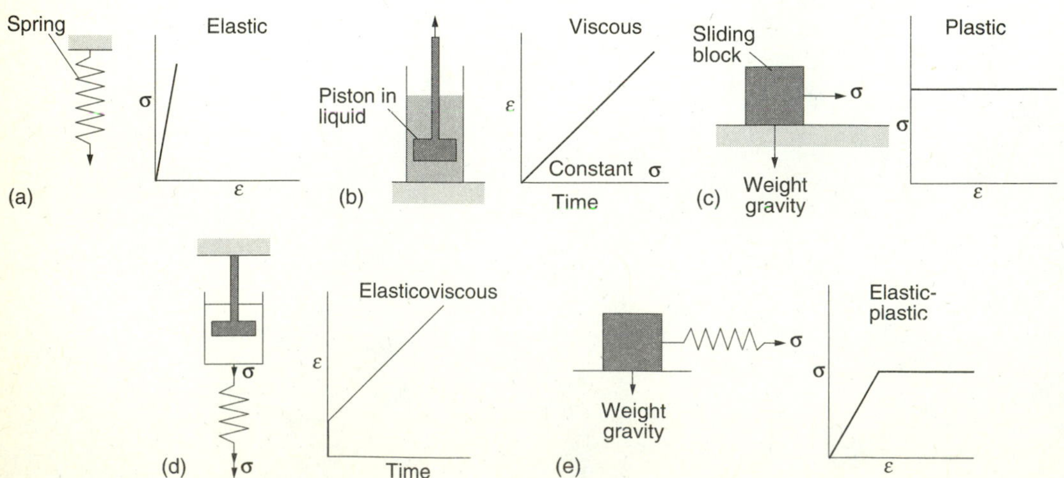

FIGURE 6–1
Ideal mechanical models and stress-strain or strain-time curves for material behavior. Think about the mechanical models in terms of how a spring, a piston in a liquid, and a sliding block behave independently or in combination. Strain is plotted versus time for viscous (b) and elasticoviscous (d) materials because strain in viscous materials (liquids) is time dependent.

ideal end-member behavior models—elastic, viscous, and plastic (to be discussed shortly)—also show us more clearly how to write mathematical expressions for ideal behavior. Some rocks approach ideal behavior under particular conditions; others exhibit more complex types of behavior under all conditions. Most rocks never exhibit ideal behavior but frequently respond to stress in ways that approach one of the ideal types. The discussions of stress and strain in Chapters 3 and 4 set the stage for us to look at the behavior of rocks under various conditions, thus enabling us to better understand the physical conditions that produce many geologic structures, as well as some of the bulk properties of rocks.

DEFINITIONS

Properties of ***homogeneous*** materials are the same throughout any sample of any size; those of ***inhomogeneous*** materials vary with location, either in a hand specimen or in a region. This inhomogeneity leads to *scale-dependent behavior* of rocks. ***Isotropic*** materials have the same properties in all directions, contrasting with ***anisotropic*** materials, wherein properties vary with direction. Some rocks, such as weakly bedded shale and strongly foliated gneiss, may behave as strongly anisotropic materials on the scale of a hand specimen, but they may behave isotropically on a scale of a rock mass of tens to hundreds of meters across. Large rock masses, such as many stocks or batholiths, may be homogeneous and isotropic in response to stress.

The uppermost crust of the eastern United States displays a very uniform distribution of present-day stress and thus is said to behave homogeneously with respect to the present-day stress field (Figure 3E–3); but earthquakes do occur in the East, and so there must be local inhomogeneities that interrupt the stress field, concentrate stress, or are relatively weak and allow rupture of the crust. Layered rocks are strongly anisotropic to stress, but the degree of expression of the anisotropy depends on the direction in which the stress is oriented. Rupturing of the crust producing eastern U.S. earthquakes probably occurs as a result of localized stress concentrations at suitably oriented anisotropies. These anisotropies may consist of old fault or fracture zones, rock bodies of contrasting composition and density (for example, granite and gabbro), and other features. Thus, an old fault zone may be inherently weak and provide a lower threshold for movement—lower strength than unbroken rock.

We commonly think of deformation in terms of three end-member behavior types: *elastic, plastic,* and *viscous* (Figure 6–2a). Strain that is recovered instantaneously on removal of stress is called ***elastic***. The object deformed elastically returns to the original undeformed shape after the stress is removed. No other behavior mode exhibits this reversible property or

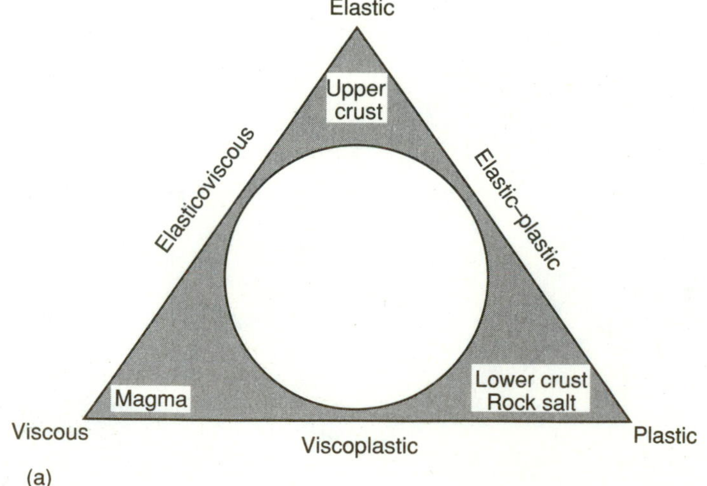

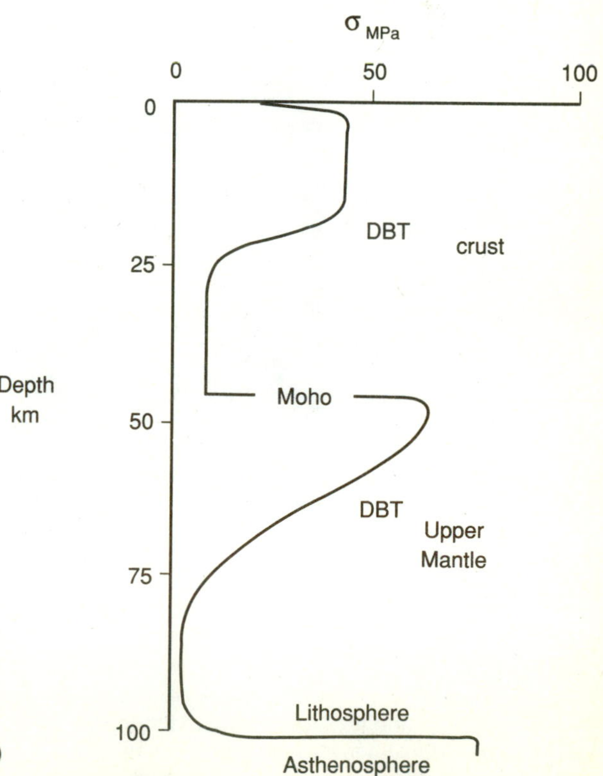

FIGURE 6–2
(a) Material behavior-modes triangle. Most rock materials do not exhibit ideal behavior, but most behavior falls outside of the circular vacant region of the diagram. (b) Stress vs. depth curve showing possible relationships of behavior of quartz (in the crust) and olivine (in the mantle) as a model for brittle-plastic behavior of crustal and mantle rocks. Variation in differential stress is directly related to the strength of the material. DBT—Ductile-brittle transition.

"material memory." ***Plastic*** behavior involves permanent strain that occurs without loss of cohesion and is the result of rearrangment of chemical bonds in crystal lattices in minerals by one (or more) of the dislocation creep mechanisms described in Chapter 7; a material behaves plastically if it strains only after exceeding a threshold or ***yield*** differential stress. Obviously, it is a pervasive strain that affects the entire rock mass. (NOTE: This definition of plasticity is the same as the frequently employed terms *crystal plasticity,* or *crystal-plastic behavior.*) ***Viscous*** behavior is the behavior of fluids such as water or magma or of any other substance with little internal structure. Viscous deformation is pervasive, permanent, and involves dependence of strain rate on stress. Although most rocks do not behave as viscous materials, some of their properties may be approximately explained by assuming viscous behavior.

Combinations of the three end-member types produce *elastic-plastic, viscoplastic,* and *elastico-viscous* behavior. Other combinations may approximate the behavior of most geologic materials. Fortunately, plots of the behavior of most rock types fall close to one of the end members or to an edge of the triangle—not in the middle—and so the end-member behavior models discussed in the next sections approximate the behavior of rock masses under a variety of conditions.

ELASTIC (HOOKEAN) BEHAVIOR

We say a material exhibits ***linear elastic behavior*** if it deforms in direct proportion to the applied stress and, after the stress is removed, immediately rebounds to its original configuration. ***Brittle behavior,*** as defined in Chapter 1, implies failure in the elastic range, actually at the *elastic limit* for the material. Linear elastic behavior in an isotropic homogeneous material is described by Hooke's law—named for an English physicist, Robert Hooke (1635–1703)—which states that the strain (ε) in the material is linearly proportional to the applied stress (σ) (Figure 6–3a):

$$\sigma \propto \varepsilon . \qquad \textbf{(6–1)}$$

Equation 6–1 is converted to an equality by inserting a proportionality constant:

$$\sigma = E\varepsilon . \qquad \textbf{(6–2)}$$

E is ***Young's modulus,*** an experimentally determined elastic constant in which

$$E = \frac{\sigma}{\varepsilon} = \frac{\frac{F}{A}}{\frac{\Delta l}{l_0}} . \qquad \textbf{(6–3)}$$

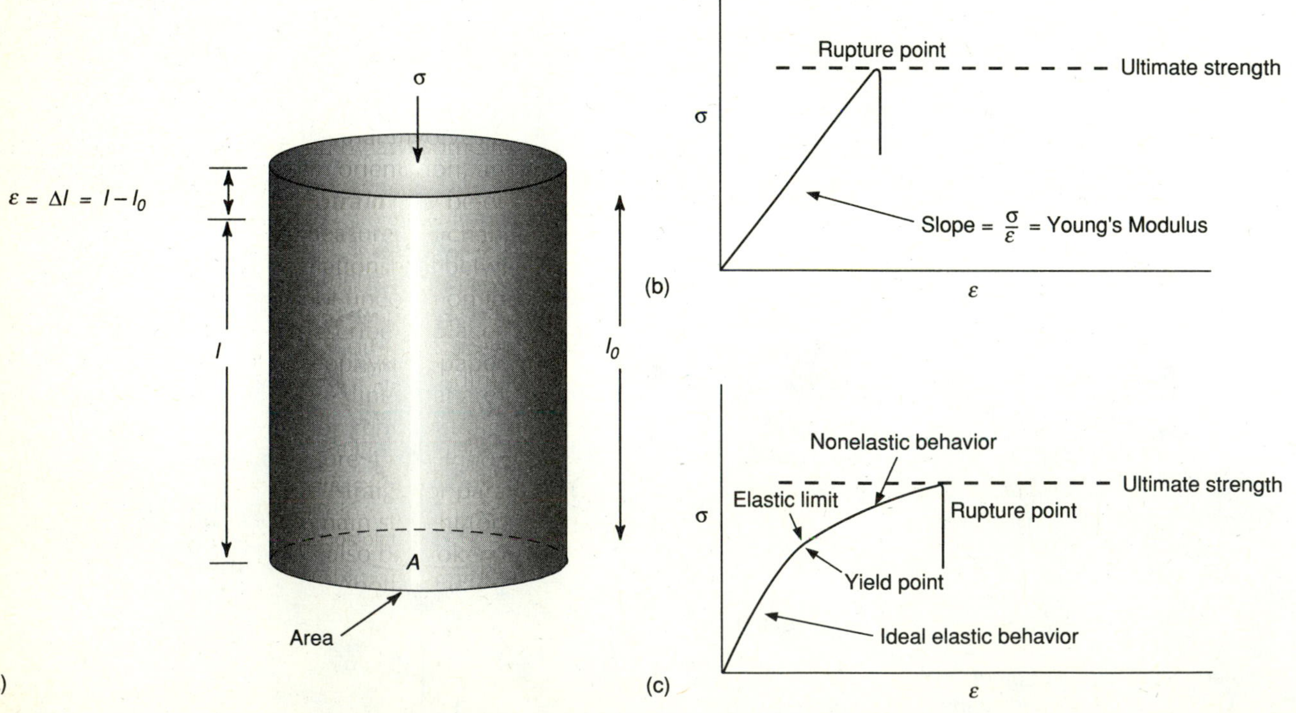

FIGURE 6–3
(a) Relationships between stress, σ, and elastic strain, ε, in a cylinder of rock in a test apparatus. The ends of the cylinder have a known area, A. (b) and (c) Stress-strain diagrams for an ideal elastic material (b) and a real elastic material (c). The real elastic material exhibits ideal behavior initially but begins to deviate and become nonelastic at higher strains.

Where F is force, A is area, Δl is the change in length of a reference line, and l_0 is the initial length of the line. Note that the strain in equations 6–1, 6–2, and 6–3 is the elongation strain (equation 4–1). For elastic behavior under shear stress,

$$\tau = G\gamma, \qquad \textbf{(6–4)}$$

where τ is shear stress, γ is shear strain, and G is another experimentally determined elastic constant, the ***rigidity***, or ***shear modulus***, defined as the ratio of shear stress to shear strain

$$G = \frac{\tau}{\gamma}. \qquad \textbf{(6–5)}$$

A linear relationship exists in ideal elastic behavior between the application of stress and the resulting strain. Below the elastic limit, an elastic material rebounds instantaneously to the original shape when the stress is removed. When the data are plotted on a stress-strain diagram (Figure 6–3), the slope of the line is determined by the elastic constant for the material used in the experiment. The stress-strain curve for an elastic material begins at the origin and abruptly changes slope at the ***yield point***, where Hooke's law no longer holds (nonelastic, permanent deformation behavior begins), or the material begins to fail and then ***ruptures***. Permanent strain occurs by rupture in an ideal elastic material. A change in slope of the stress-strain curve also indicates that direct proportionality between stress and strain no longer exists. The highest point on the curve is the ***ultimate strength*** for the material. The ***elastic limit,*** here the same as the yield point, is the point on the curve beyond which the material begins to undergo permanent deformation or ruptures (or both), although the elastic limit is reached in many rocks at about half the ultimate strength. Brace and others (1966) showed that microcracks begin to form, the rock begins to expand (dilate), and permanent deformation begins at a value of approximately half of the ultimate strength.

In a more realistic stress-strain curve for elastic materials (Figure 6–3b), the slope decreases before reaching the ultimate strength of the material. This point where the decrease in slope occurs is both the yield point and the elastic limit. Above the elastic limit, the material exhibits nonelastic behavior. Another frequently determined elastic constant in elastic materials, is ***Poisson's ratio***—a measure of compressibility named in honor of Simeon-Denis Poisson (1781–1840), a French mathematician. It is defined as

$$\nu = -\frac{\varepsilon_1}{\varepsilon_3}. \qquad \textbf{(6–6)}$$

where ν is Poisson's ratio, and ε_1 and ε_3 are linear elastic strains derived from changes of length measured

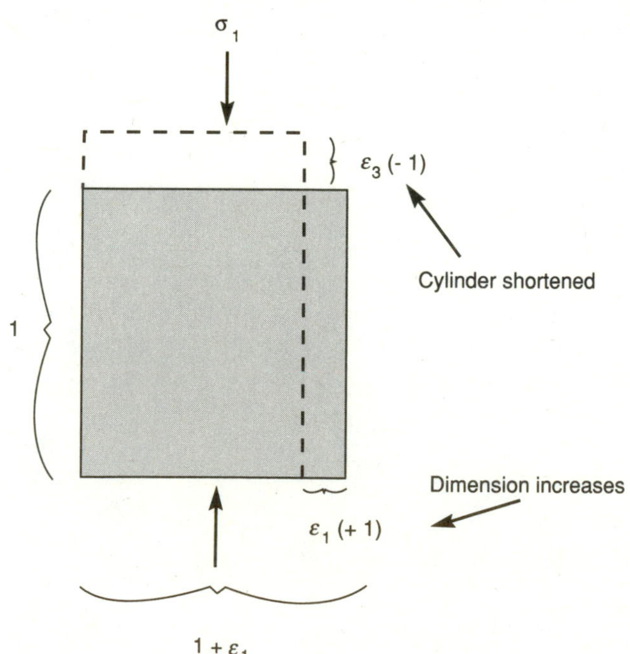

FIGURE 6–4
Definition of Poisson's ratio. The unconfined rock cylinder is compressed parallel to σ_1 and expands normal to σ_1. Poisson's ratio is the ratio of shortening parallel to σ_1 to the extension normal to σ_1.

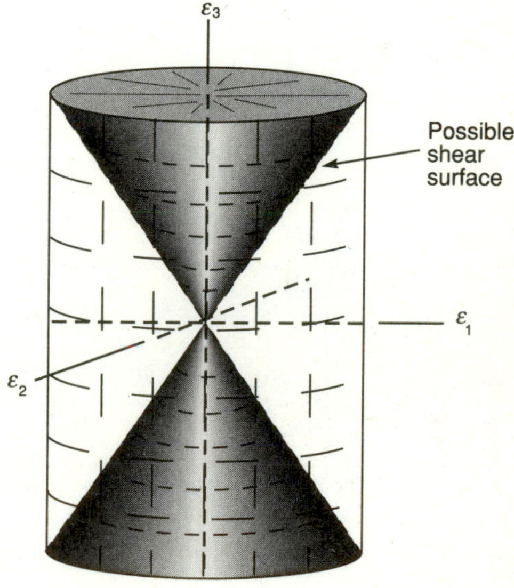

FIGURE 6–5
Cylinder of rock in an axial compression experiment showing orientation of strain axes, with $\varepsilon_1 = \varepsilon_2$ and $\nu = -(\varepsilon_1/\varepsilon_3)$. The conical areas inside the cylinder are locations of potential shear surfaces.

in uniaxial compression experiments (Figure 6–4). A test cylinder is shortened axially by ε_3, a principal strain, but the lateral dimension is simultaneously increased by another principal strain, ε_1, equal to ε_2 in the test cylinder (Figures 6–4 and 6–5). Values of ν are positive because one strain (ε_3) is negative and it thus cancels the negative sign in equation 6–6. Most rocks have a Poisson's ratio of less than 0.5, indicating that volume has decreased. Water, although not an elastic solid, is an incompressible material with a Poisson's ratio of 0.5; halite has a very low compressibility and has a Poisson's ratio of less than 0.5. Soft biological tissues and rubber have a Poisson's ratio of about 0.5; the ratio for cork is near zero; metallic lead, 0.45; aluminum, 0.33; most steels, 0.27; and polymers, 0.1 to 0.4 (Lakes, 1987). Table 6–1 contains representative values of Poisson's ratio for a variety of common rocks.

PERMANENT DEFORMATION—DUCTILITY

Permanent strain in the form of viscous or plastic deformation occurs in a material beyond the elastic limit. Viscous behavior occurs in fluids, whereas plastic deformation occurs in solids below their melting points.

TABLE 6–1
POISSON'S RATIOS FOR COMMON ROCK TYPES

Rock Type	Locality	Poisson's Ratio
Granite	Westerly, Rhode Island	0.25
Diorite	Mount Rainier, Washington	0.28
Gabbro	Duluth, Minnesota	0.30
Peridotite	Cypress Island, Washington	0.27
Dunite	Twin Sisters, Washington	0.28
Anorthosite	Adirondack Mountains, New York	0.31
Eclogite	Healdsburg, California	0.26
Quartzite	Baraboo, Wisconsin	0.10
Serpentinite	Burro Mountain, California	0.35
Slate	Poultney, Vermont	0.30
Amphibolite	Bantam, Connecticut	0.26
Quartz mica schist	Thomaston, Connecticut	0.31
Tonalite gneiss	Torrington, Connecticut	0.27
Felsic granulite	Saranac Lake, New York	0.26
Mafic granulite	Adirondack Mountains, New York	0.29
Sandstone	Berea, Ohio	0.26
Shale	Thorn Hill, Tennessee	0.26
Limestone	Thorn Hill, Tennessee	0.32
Dolostone	Thorn Hill, Tennessee	0.29

Poisson's ratios (ν) calculated from compressional (V_P) and shear (V_S) wave velocity measurements, where

$$\nu = \frac{1}{2}\left(1 - \frac{1}{(V_P/V_S)^2 - 1}\right)$$

at 200 MPa confining pressure.
(Data provided by Nicholas I. Christensen, Purdue University.)

Rutter (1986) maintained that *ductility* should reflect the capacity for large amounts of nonlocalized homogeneous strain. He would include all flow processes, including brittle mechanisms, in the usage of the term "ductile", and restrict the use of the term "plastic" to rocks that have been deformed by diffusion processes that deform crystal lattices (Chapter 7).

Scale-dependence is important when considering deformation that affects a rock mass, because the kind of deformation occurring in a few cubic centimeters of rock may not be the same as that occurring in several cubic kilometers of rock, even if the larger mass includes the smaller. Consequently, the term "ductile" will be used here for deformation in which plastic or viscous flow dominates over brittle processes. For example, in very fine-grained fault rocks formed under near-surface conditions, the hand-specimen to outcrop-scale deformation may appear to have involved ductile flow, but microscopic examination reveals that the fine-grained material still consists of uniformly pulverized original rock without any internal deformation of individual microscopic grains. The fault rock is still brittle (not ductile), but when the uniform very fine grain size is reached, the material will flow like wheat flour or dry cement.

Viscous Behavior

Viscous behavior is fluid-like behavior. Natural fluids, including water, magma, gases, and the molten part of the Earth's core, are all viscous materials. We can also argue, for purposes of simplification, that the mantle is a viscous material, using the example in Chapter 1 for calculating mantle viscosity from glacial rebound rates.

Stationary fluids will not transmit shear stresses and are called *perfect fluids.* Moving fluids undergo shear stresses on all planes of differential motion (nonzero shear-strain rate), including the fluid-rock boundaries. Fluids in which there is a linearly proportional relationship between differential stress (τ) and shear-strain rate are ***Newtonian fluids,*** expressed as

$$\tau \propto \frac{d\gamma}{dt} = \dot{\gamma} \qquad \textbf{(6–7)}$$

or

$$\tau = \eta\dot{\gamma}\ , \qquad \textbf{(6–8)}$$

where γ is shear strain and *t* is time, so that *dγ/dt*, or $\dot{\gamma}$, is the shear-strain rate, and η is the viscosity of the fluid and the proportionality constant. (The SI unit for viscosity is the *poise,* and 10 poise equals 1 Pa s.) The reciprocal of viscosity, 1/η, is called the *fluidity,* which is the ability of a fluid to move, rather than the resistance to motion measured by the viscosity. Stress-strain curves for viscous materials are not as useful as strain-time curves (Figure 6–6) because of the dependence of strain on time as well as stress. Shear stress exists in fluid only when it is in motion.

Rocks at high temperatures near their melting points behave in a nearly viscous fashion, but their behavior is better described as plastic (to be discussed). Useful models of crustal dynamics can, however, be formulated assuming viscous behavior.

Plastic (Saint-Venant) Behavior

Ideal ***plastic*** behavior involves permanent (nonrecoverable) deformation that affects the entire rock mass and begins at a yield stress. Ideal plastic behavior is also known as ***Saint-Venant behavior*** in honor of Adhemar-Jean-Claude Barre de Saint-Venant (1797–1886), a French physicist, who in the nineteenth century wrote mathematical expressions for plastic behavior in an attempt to relate various stress components in materials exhibiting plastic behavior.

In natural materials, plastic behavior is generally preceded by elastic behavior, with plastic behavior beginning at the elastic limit of the material being deformed (Figure 6–7). Total strain in a plastic material also depends on the *deformation path.* It thus depends on the sequence of stresses applied to the body rather than the final stress, as in an elastic material.

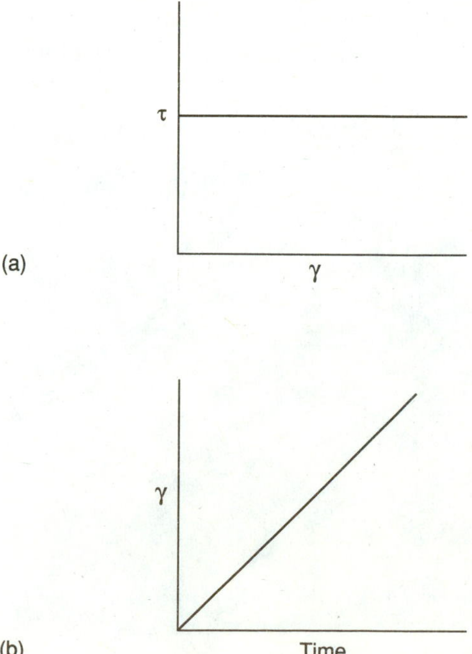

FIGURE 6–6
Stress-strain (a) and strain-time (b) diagrams for a viscous material at constant τ. This stress-strain curve is uninformative and even somewhat misleading because the strain rate depends on stress in viscous materials.

After the yield point is passed, the material flows at a constant stress unless one of two things occurs. The first is ***strain hardening*** (increased resistance to deformation as strain increases). The second, and opposite, is ***strain softening*** (Figure 6–8). Materials that either strain-harden or strain-soften should probably be described as non-ideal plastic because they exhibit a more general property of nonideal behavior. Stress-strain curves for the limestone in Figure 6–8b indicate that at relatively low confining pressure, the limestone deviates from ideal elastic behavior at moderate differential stress values before rupturing; at higher confining pressure, it still deviates but enters a realm of strain softening beyond the ultimate strength. At still higher confining pressure, the limestone exhibits almost ideal plastic behavior beyond the yield point. Plastic behavior is thought to dominate at depths in the crust and mantle where temperature reaches several hundred degrees Celsius and confining pressure reaches several kilobars. Presence of water or other fluid may lower threshold temperatures and pressures for ductile behavior, but also may promote brittle behavior if there is a rapid increase in fluid pressure. Recrystallization processes, grain-boundary sliding, calcite twinning, and some effects of pressure solution in metamorphic rocks (Chapter 7) are commonly associated with ductile deformation on the microscopic scale. The effects pervade the entire rock mass or, in some instances, are confined to tabular zones of ductile deformation (ductile shear zones; Chapter 10). Laboratory experiments designed to duplicate ductile deformation must be carried out at high temperature and pressure to attain strain rates sufficiently rapid to complete many experiments during a human lifetime. Structures in metamorphic and some fault rocks are thought to form under ductile conditions (Figure 6–9). In glacial ice

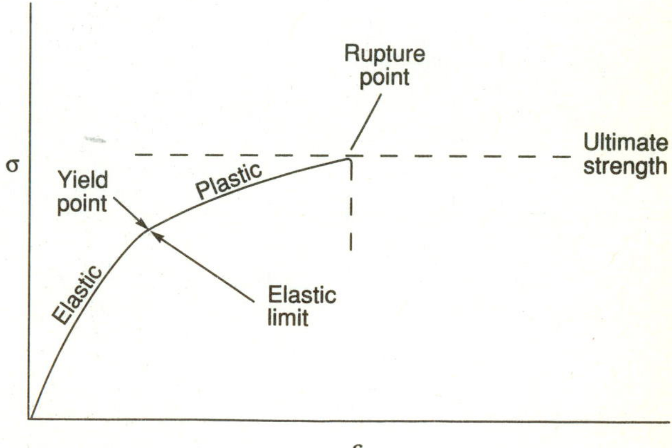

FIGURE 6–7
Stress-strain curve showing general properties of an elastic-plastic material (with strain hardening) and inflection points in the curve.

and rock salt, ductile behavior may be initiated under surface conditions, and so these rocks serve as useful analogs for rocks deformed deep in the crust.

Several models have been proposed to explain plastic behavior, but none fully succeeds in quantifying it because of the inherent complexity in the process. The mathematics describing plastic deformation is also complex, particularly that describing the plastic behavior of anisotropic crystalline solids like most rocks (Nicolas and Poirier, 1976). An example of a relatively simple flow law is

$$\dot{\varepsilon} = A\sigma^n , \qquad \textbf{(6–9)}$$

where $\dot{\varepsilon}$ is strain rate, A is a material constant, σ is normal stress, and n is another constant that depends on flow mechanism. A creep law for shear strain rate is

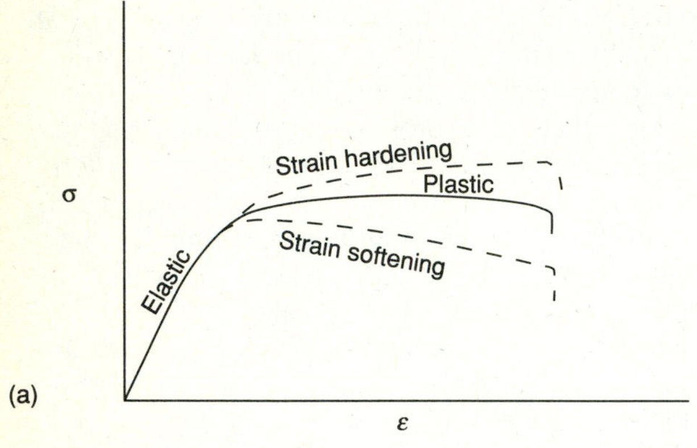

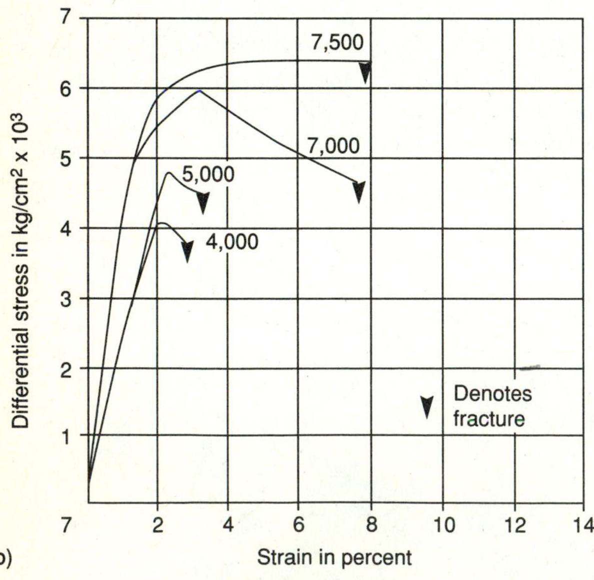

FIGURE 6–8
(a) Strain hardening and strain softening in an elastic-plastic material. (b) Experimental deformation of Solenhofen Limestone. Curves represent deformation at different confining pressures. Rupture occurs at differential stress indicated (in kg cm^{-2}) (From H. C. Heard, 1960, Geological Society of America *Memoir 79.)*

$$\dot{\gamma} = A\tau^n e^{\frac{\psi P}{RT}} , \qquad \textbf{(6–10)}$$

where $\dot{\gamma}$ is shear strain rate, τ is shear stress, ψ is another material constant, P is pressure in MPa, R is the gas constant (1.987 cal deg^{-1} mol^{-1}), and T is temperature. The two additional models outlined below were chosen because of their simplicity and historical significance.

Tresca's criterion states that plastic yield will begin when the maximum shear stress (τ_{max}) reaches a critical value (C_y), a specific constant for the material. This is, of course, half the initial stress difference ($\sigma_1 - \sigma_3$) when yielding begins, or

$$\tau_{max} = C_y = (\sigma_1 - \sigma_3)/2, \qquad \textbf{(6–11)}$$

where σ_1 and σ_3 are the principal stresses at yielding. It predicts that the material will yield when this maximum shear stress is reached regardless of the specific values of σ_1 and σ_3. This criterion was intended to account for some properties observed in metals where the constant maximum stress difference, $\Delta\sigma$ (= $\sigma_1 - \sigma_3$), indicates that yield stresses in both tension and compression have approximately the same magnitude. This relationship probably can best be used to explain the plastic behavior of metal rods in extension experiments (Figure 6–10), but it does not adequately describe the plastic behavior of rocks.

Another model is based on the *Von Mises criterion,* which proposes that yield will occur under any combination of principal stresses that will produce the

FIGURE 6–9
Strongly foliated Late Proterozoic biotite gneiss containing feldspar layers on the Yadkin River near Siloam, North Carolina. Quartz and micas were plastically deformed, and feldspar was deformed brittlely during the early Paleozoic. (RDH photo.)

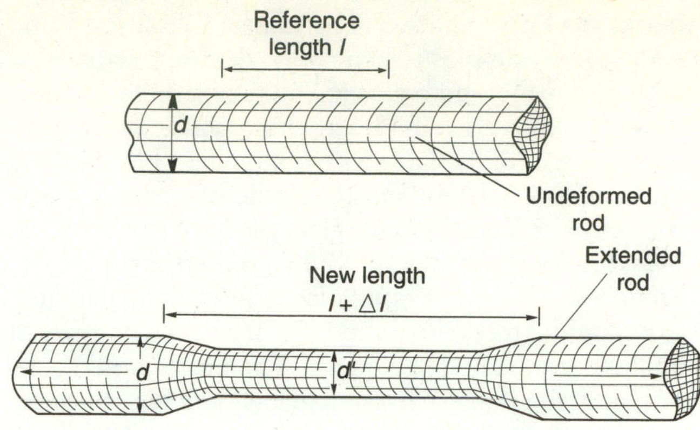

FIGURE 6–10
Extensional ductile necking of a metal rod.

same distortional strain energy that existed in a previous uniaxial test of the same material at the yield point. When the sum of the squares of the principal stress differences equals $2(\sigma_3)^2$, commonly expressed as

$$(\sigma_1 - \sigma_2)^2 + (\sigma_2 - \sigma_3)^2 + (\sigma_3 - \sigma_1)^2 = 2(\sigma_3)^2 = 6Cy, \quad \textbf{(6–12)}$$

where C_y is the yield constant, the criterion is satisfied. The Von Mises criterion does explain some plastic deformation in metals, but, again, not in rocks.

CONTROLLING FACTORS

Many factors affect and control the behavior of rock materials, including composition, texture, temperature, confining pressure, fluid (pore) pressure, rate of stress increase and strain rate, and character and spacing of anisotropies such as bedding or foliation. All affect the ways in which rocks deform, but particular variables, including temperature, composition, strain rate, confining pressure, and fluid content, have a greater effect than others. Given proper conditions, any of these variables may play a major role in the behavior of a rock mass being deformed. For example, most rocks deform elastically at low temperature but will deform ductilely at high temperature and low strain rate. At high strain rate, the same rocks may deform brittlely. Increased fluid pressure may produce ductile behavior at lower temperatures than those at which it would occur in a dry rock mass, although increased fluid pressure has also been shown by Handin and others (1963) to lower the threshold of brittle deformation.

BEHAVIOR OF CRUSTAL ROCKS

The ideal-mechanical behavior models discussed so far approximate the behavior of rock materials in the Earth. Some geologic materials exhibit almost ideal-model behavior (Figure 6–2). Thin sheets of muscovite and biotite exhibit elastic behavior as they are flexed at room temperature and spring back to their original shapes. Muscovite and biotite sheets also have an elastic limit that can be exceeded. Some fluids upon and within the Earth approximate ideal viscous behavior, and we have abundant evidence for occurrence of nonideal plastic deformation. Transitional behavior such as viscoelastic or elastic-plastic is also prevalent at proper temperature and pressure.

Ideal-behavior models (Figure 6–1) are defined in large part by the shapes of stress-strain or strain-time curves (or both). Further discussion of deformation mechanisms in Chapter 7 will help in understanding these modes of behavior in rocks.

Ductile-Brittle Transition

Brittle behavior generally dominates in the upper crust, with elasticity providing rigidity and permitting faulting and jointing to occur at the elastic limit. A transition zone occurs in the crust where brittle behavior is inhibited because increased ductility occurs as a result of increases in temperature and pressure (increase in confining pressure increases strength, though promoting ductile behavior) with depth. This is the ***ductile-brittle transition*** (DBT) or, according to Rutter (1986), the *plastic-brittle transition* (Figure 6–11). Direct evidence of ductile behavior in crustal rocks is shown by similar folds (Chapter 14) that can form only in the ductile realm. Laboratory studies of minerals in these folded rocks indicate that temperatures of several hundred degrees Celsius and pressures of several kilobars were required to form them deep in the crust. It would be incorrect to assume that there is no elastic behavior in the lower crust or the mantle well below the DBT. If the high pressure and temperature of the lower crust and mantle could suddenly be relieved, there would still be some elastic recovery, again illustrating the behavior of the elastic-plastic material (Figure 6–7).

The actual depth to the DBT in the lithosphere is determined by the thermal gradient (rate of increase of temperature with depth determined by the amount of radioactive heat-producing elements U, Th, and K, intrusion of magma into the crust, and mantle heat flux). It is also controlled by the quantity of fluid present and other variables that determine pressure gradients. In tectonically inactive parts of the continents such as the eastern United States, where thermal gradients are very low (15 to 25° C/km), depth to the DBT may be as much as 15 km, but in areas of high heat flow (30 to 40° C/km) in the continents, the depth to the DBT may range from 8 to 12 km. Depth to the DBT may be approximately estimated from the maximum focal depth of most earthquakes in an area (implying no elastic behavior here). In the eastern United States, most earthquakes occur at depths of less than 15 km; in the Great Basin in Utah and Nevada, where the thermal gradient is steeper, the foci of most earthquakes are within a few kilometers of the surface.

Large active faults that break the entire crust, such as the San Andreas in California, unquestionably exhibit brittle behavior in the upper crust, but probably involve movement in the fault zones by ductile flow below the DBT (Figure 6–11). Large faults also exhibit a kind of elastic-plastic behavior—the upper-crustal segments exhibit pure elastic behavior, and the lower-crustal segments exhibit nearly pure plastic behavior—but segments of the fault within the DBT (and sometimes below it) exhibit elastic-plastic behavior (Figures 6–7 and 6–8). High strain rates favor brittle behavior (Handin and Hager, 1957), but high fluid and confining pressures may permit rapid plastic flow along the fault.

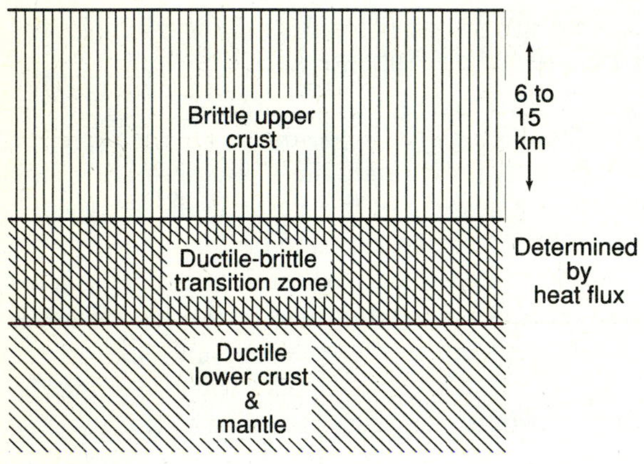

FIGURE 6–11
Simplified concept of the ductile-brittle transition in the lithosphere. The depth to the transition is determined by the amount of heat produced in that part of the lithosphere, the nature and amount of fluid present, and the pressure variables. Simple, and expected, variations in layered structure (e.g., massive vs. foliated rocks), water content, local strain rate, and local shortening and lengthening directions relative to structure can all produce a much more complex DBT.

Natural models for the DBT occur in glaciers, salt, and unconsolidated sediment. A definable brittle zone occurs in the upper parts of a glacier where crevasses and other brittle fractures form. The fractures close at depths rarely exceeding 60 m, giving way to a zone of ductile flow (Figure 6–12). Direct observations in the ductile zone have been made by excavating a shaft through the upper brittle zone into the lower ductile deformation zone. Observers can enter the shaft and make observations because ductile flow closes the shaft at a very slow rate. Salt structures afford similar opportunities to observe ductile behavior occurring near the surface in natural materials (Chapter 2). The unconfined parts of the salt yield brittlely; the confined parts yield ductilely (Jackson and Talbot, 1986). Other analogies of ductile or viscous behavior exist where folds have formed in slumped water-saturated sediments (Stone, 1976) and in glacial silts deformed by slumping or ice movement (Figure 2–2a; Stone and Koteff, 1979).

Elasticoviscous (Maxwell) Behavior

A combination of viscous and elastic behavior called ***elasticoviscous behavior*** was studied by a Scottish physicist, James Clerk Maxwell (1831–1879)—hence the alternative name; it depends on both stress and strain rates (Figure 6–13). Thus, equations 6–1 and 6–7 must be combined to express relationships between stress and strain in an elasticoviscous material. Total strain in the material is given by the sum of elastic and viscous strains,

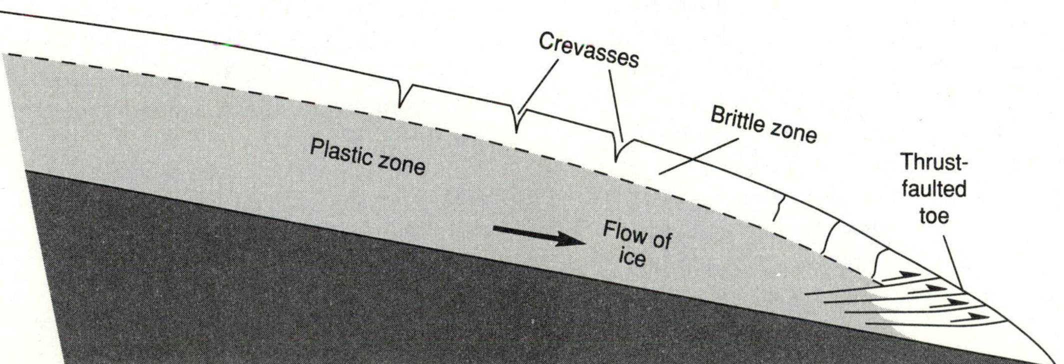

FIGURE 6–12
Cross section of a glacier showing deformation zones.

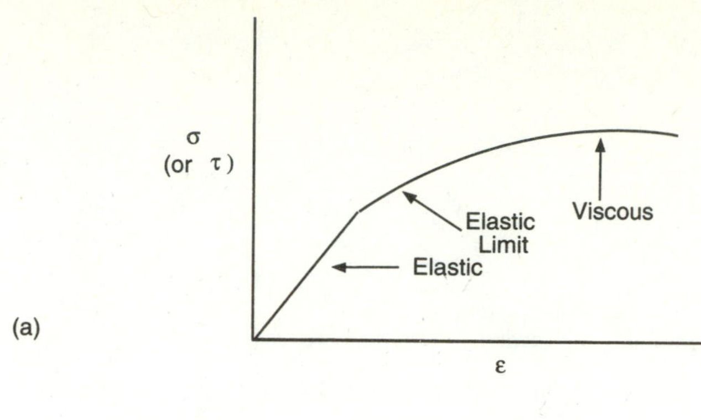

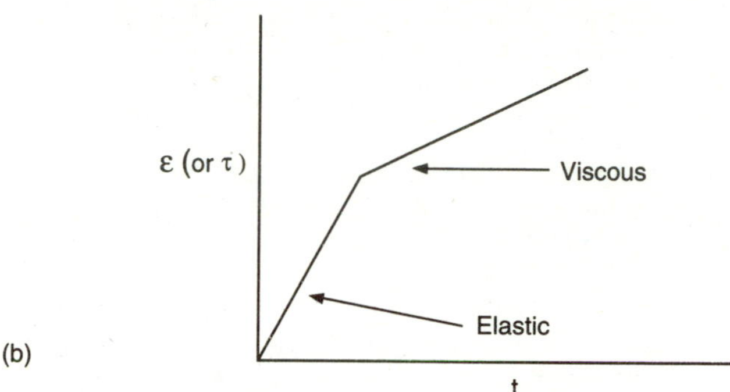

FIGURE 6–13
Stress-strain (a) and strain-time (b) diagrams for an elastico-viscous material. The same problems exist here with stress-strain curves as for ideal viscous materials (Figure 6–6).

$$\gamma = \gamma_e + \gamma_v , \quad \textbf{(6–13)}$$

where γ_e is the elastic shear strain and γ_v is the viscous shear strain. From equations 6–4 and 6–7,

$$\gamma = \left(\frac{\tau}{G}\right) + \left(\frac{\tau t}{\eta}\right). \quad \textbf{(6–14)}$$

Thus, the elasticoviscous strain (γ) equals the sum of the elastic strain and viscous strain components. G is the shear modulus, τ is shear stress, t is time, and η is viscosity.

Here, we must consider time-dependence of deformation, an important factor in geologic processes. Rocks that deform brittlely over a short time (milliseconds to a few years) will deform ductilely over periods of thousands to millions of years. The problem of why deep-focus earthquakes occur (see Chapter 6 Essay) is a good example of the time-dependence of geologic processes. Carey (1953) pointed out the relationship and suggested the term rheid for materials that deform like fluids (wherein the viscous strain component is at least 10^3 times the elastic component) at temperatures below their melting points. Most rocks are *not* rheids under strains of short duration, but more *are* rheids if deformation occurs over a geologically long time.

STRAIN PARTITIONING

Behavior of rocks may vary in space as well as in time; concentration of deformation into specific parts of a rock mass by different behavior or mechanisms is called ***strain partitioning***. This phenomenon may be related to differences in flow rate in a ductilely deforming mass (Lister and Williams, 1983), with the change in behavior type resulting from different physical properties or (occasionally) conditions (Figure 6–14). Plastic strain may be localized in narrow ductile deformation zones along deep crustal faults, leaving adjacent rocks mostly unaffected.

Different strains result from the bulk properties of the rocks being deformed. Relatively weak rocks (shale, salt, and schist, for example) commonly exhibit styles of deformation that contrast with those of stronger rocks (sandstone, gneiss, and amphibolite, for example) in the same deforming mass. Layers of different thickness in the same rock type may also cause partitioning of mechanical behavior. Shapes and wavelengths of folds are strongly influenced by layer thickness (Chapter 15). For example, folding of thinly bedded sandstone/shale layers in a sequence of massively bedded sandstone will result in strong contrasts in mechanical behavior. In massive layers, folds of longer wavelength and less curvature will be produced; in the thinly bedded weaker layers, folds will have very short wavelengths (Figures 6–14b and 6–15). This may be explained by differences in the original anisotropy of the sequence or by tectonically induced anisotropy (such as fractures and foliations) in the rock mass (Latham, 1985). Other factors, including local variations in fluid pressure, confining pressure, temperature, and strain rate, may create conditions under which strain will be partitioned in different parts of a rock mass.

The principles and problems of mechanical behavior of rocks have now been presented, and some understanding should have been gained of the application of these principles not only to ideal materials, but also to crustal rocks. In an attempt to round out our discussion of rock mechanics, we turn to an introduction to the theory and observation of microstructures in minerals.

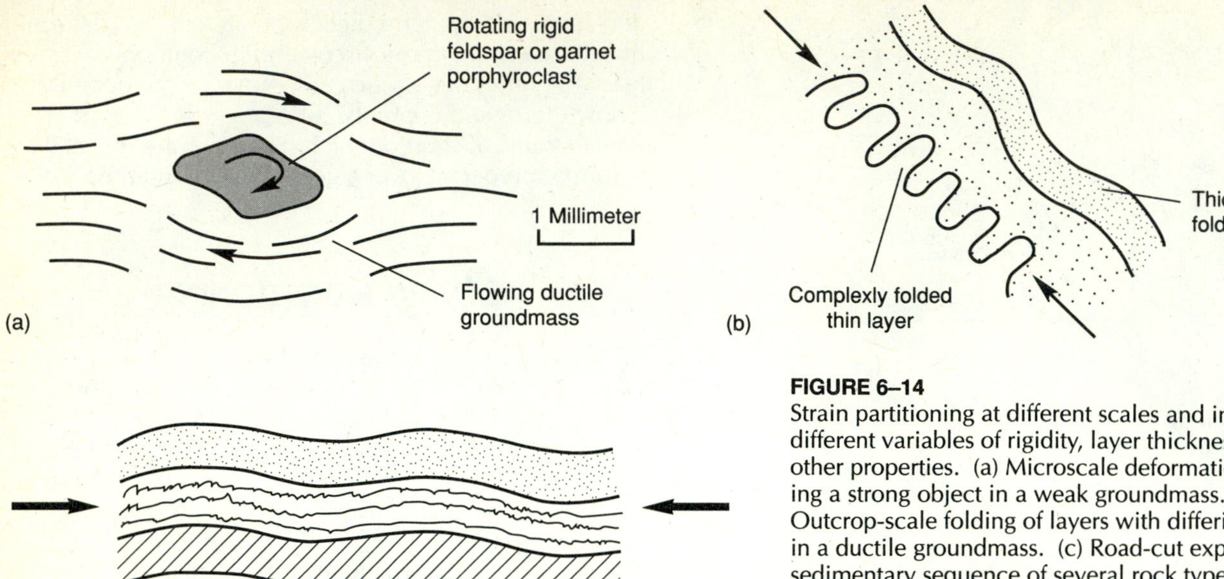

(c)

10 Meters

FIGURE 6–14
Strain partitioning at different scales and involving different variables of rigidity, layer thickness, and other properties. (a) Microscale deformation involving a strong object in a weak groundmass. (b) Outcrop-scale folding of layers with differing strength in a ductile groundmass. (c) Road-cut exposure of a sedimentary sequence of several rock types that exhibit different styles of behavior in different parts of the layered sequence.

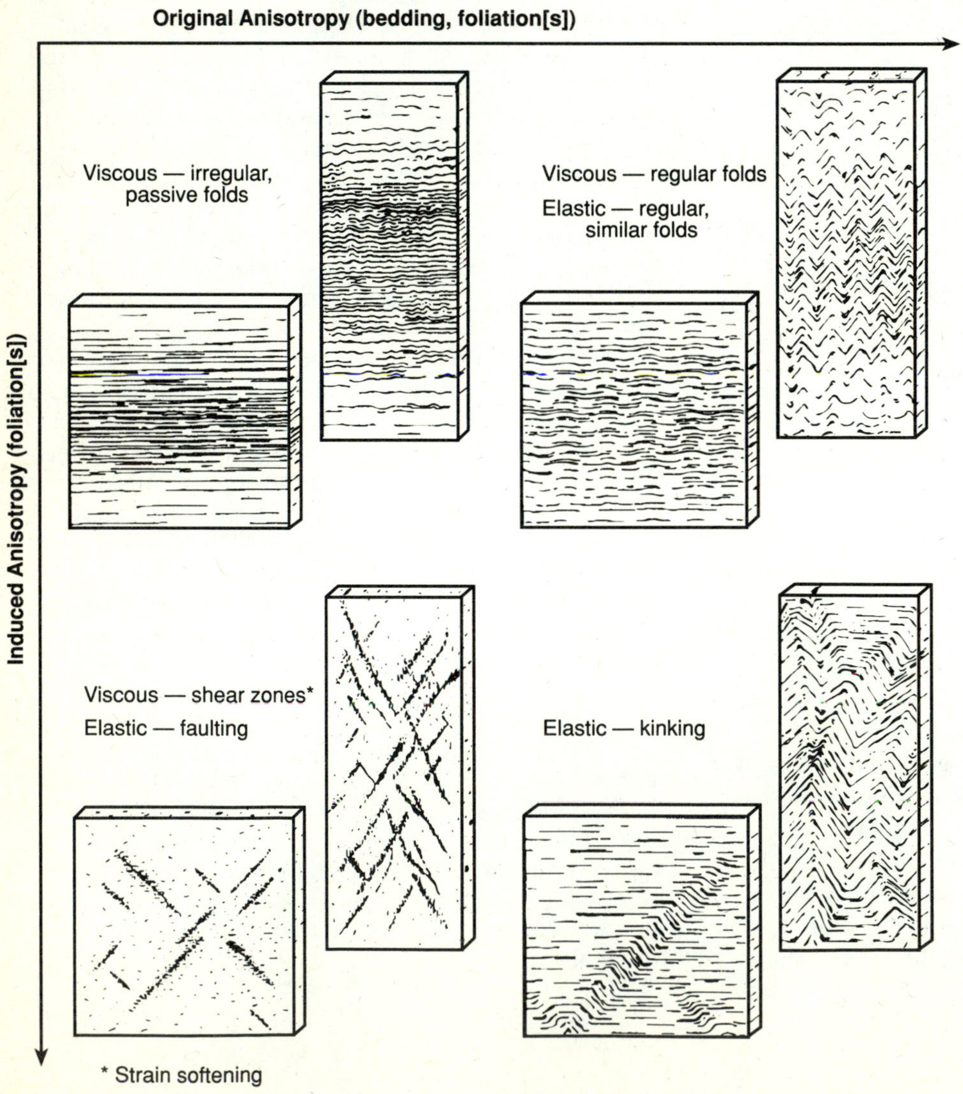

Original Anisotropy (bedding, foliation[s])

Induced Anisotropy (foliation[s])

Viscous — irregular, passive folds

Viscous — regular folds
Elastic — regular, similar folds

Viscous — shear zones*
Elastic — faulting

Elastic — kinking

* Strain softening

FIGURE 6–15
Relationships between original (intrinsic) anisotropy and tectonically induced anisotropy, producing different kinds of structures in rocks and an opportunity for strain partitioning. Return to this diagram after considering fold mechanics (Chapter 15) for a better understanding of both folds and strain partitioning. (Reprinted from *Journal of Structural Geology*, v. 7, J. P. Latham, p. 237–249, © 1985, with kind permission from Elsevier Science, Ltd., Kidlington, United Kingdom.)

ESSAY

Silly Putty™ and the Behavior of Mantle Rocks

An everyday model for elasticoviscous or elastic-plastic behavior is a silicone material known as Silly Putty (polydimethyl siloxane). It behaves as a ductile (probably viscous) material if it is deformed slowly and exhibits brittle-elastic behavior if it is strained rapidly (Figure 6E–1). Pulled slowly, it stretches; hit with a hammer, it shatters. Crustal rocks, except magma (even though magma has a finite yield strength and is thus not Newtonian viscous), probably do not exhibit viscous behavior, except perhaps over geologically long periods of time, but for simplicity, viscous behavior is used as an approximation for flow in the mantle (Chapter 1). A phenomenon in the mantle confounds any attempt to model the mantle as a homogeneous ideal viscous (or plastic) material: earthquakes occur here, at depths as great as 700 km (Figure 6E–2). Most geoscientists agree that earthquakes involve an elastic-rebound mechanism that produces a suite of surface and body waves (elastic waves that travel on the surface of and inside the Earth). The largest deep-focus earthquakes in the mantle never produce as much energy as the largest shallow-focus earthquakes generated in the upper crust, but the same general mechanism—elastic rebound—must account for their origin. We also know that some of the brittle crust may be subducted into the mantle as cold upper lithosphere. Even if subduction of cold brittle crust is a viable alternative explanation, generation of earthquakes requires sudden release of elastic strain energy, a rapid strain rate. Ductile flow may be occurring in the mantle much of the time, but rapid accumulation of strain energy may provoke transformation to elastic behavior that produces fractures and deep-focus earthquakes. Thus, the behavior of Silly

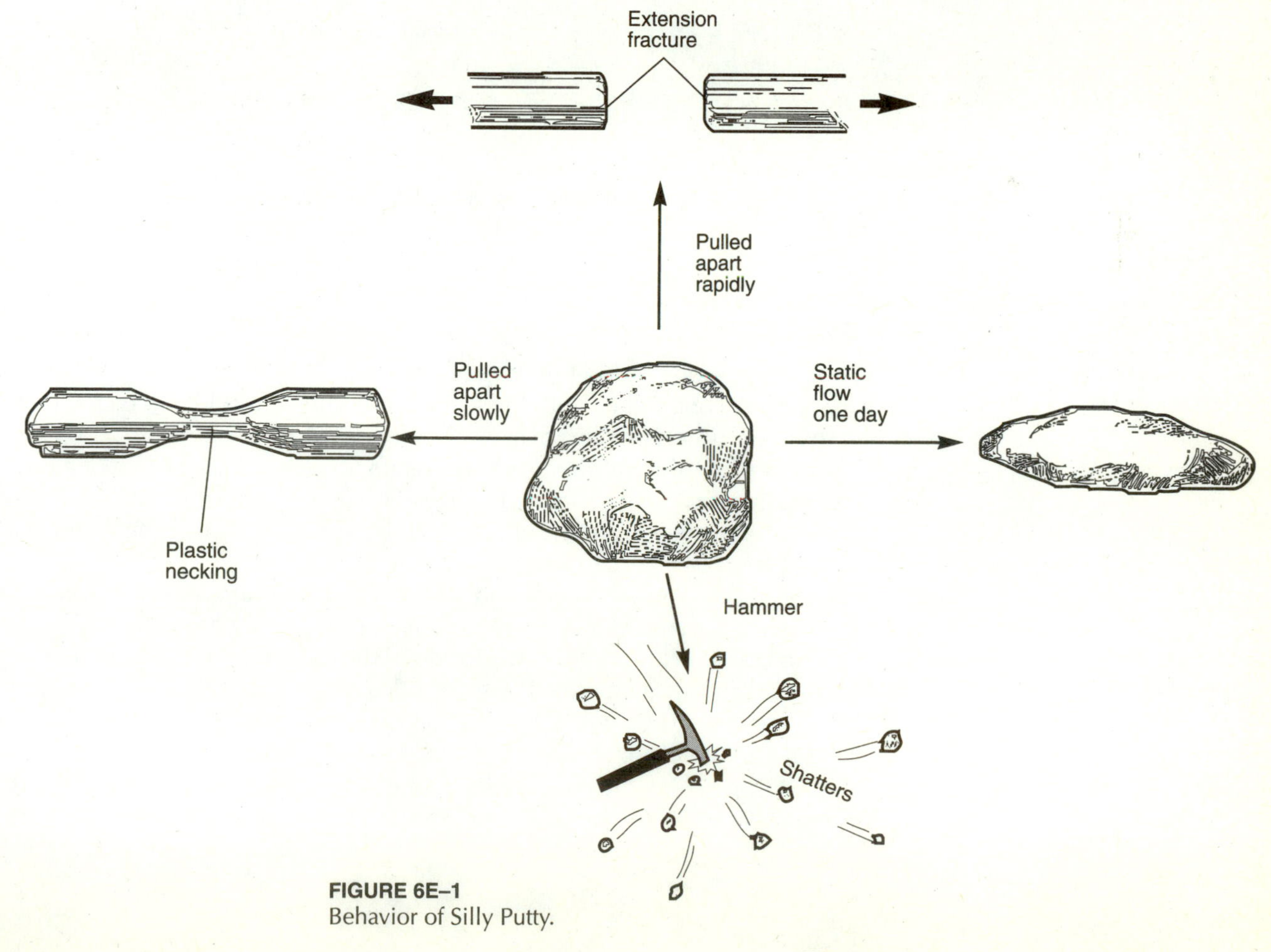

FIGURE 6E–1
Behavior of Silly Putty.

Putty subjected to markedly different strain rates may be a useful comparative model for understanding the behavior of mantle rocks, and it also helps to answer the question: Why do earthquakes occur there at all? Other possible answers include dehydration reactions that produce brittle rock and phase changes, in addition to the strain rate changes suggested by the Silly Putty analog.

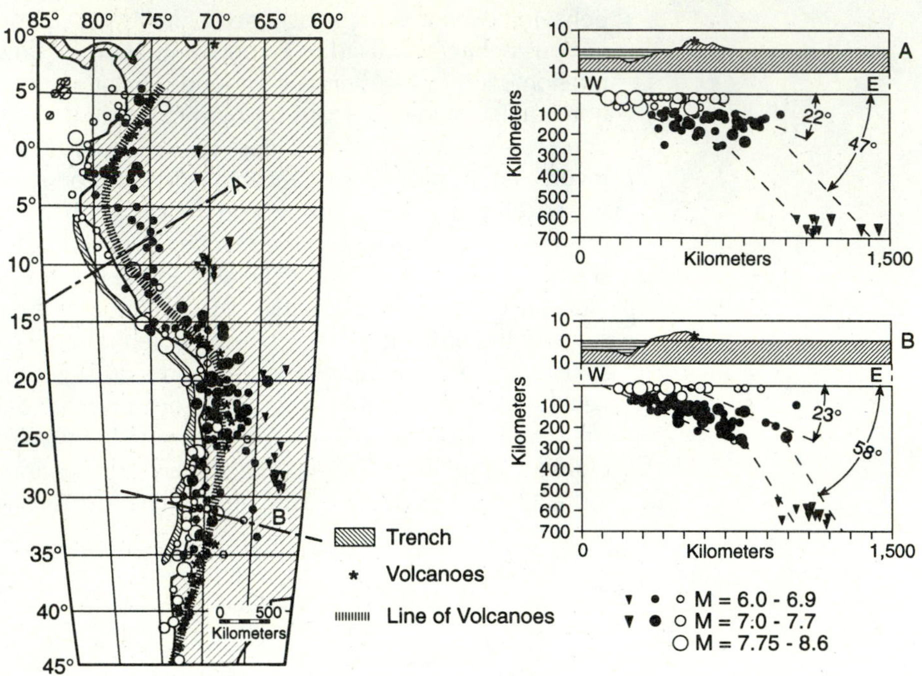

FIGURE 6E–2
Depth distribution of earthquakes beneath western South America. Note that few occur at depths greater than 250 km. (From Hugo Benioff, 1954, v. 65, Geological Society of America *Bulletin*.)

Questions

1. Why do many rocks exhibit nearly ideal behavior?
2. What factors control the depth to the ductile-brittle transition in the crust? How do these factors accomplish this?
3. Construct a hypothetical stress-strain diagram for inflation to rupture of a new (previously uninflated) balloon. If, instead of rupturing the balloon, you deflate it and then inflate it again, would the stress-strain curve be different before rupture?
4. Poisson's ratios for most natural and synthetic materials are positive. Predict the properties a material must have to exhibit a negative Poisson's ratio. See Roderic Lakes, 1987, *Science*, v. 235, p. 1038–1040, for a discussion of synthetic materials with negative Poisson's ratios.
5. Why do so many rock cylinders break along planar shear planes in uniaxial experiments although the shear surfaces in a cylinder compressed from the ends are expected to be conical? (Think about the way the experiments are designed, with $\sigma_1 \neq \sigma_2 = \sigma_3$. Are the principal stresses really so if planar shear surfaces form?)

6. The stress-strain data set tabulated here is from a loading experiment made by adding successive weights to the end of a 1-m length of steel wire 3 mm in diameter, producing an elongation strain (~1).

σ (kgcm⁻²)	Δl (mm)	σ (kgcm⁻²)	Δl (mm)
0.00	0.00	3.84	4.63
0.63	0.15	4.11	5.30
1.23	0.33	4.40	7.34
2.45	0.60	4.36	8.08
2.79	0.47	4.17	8.70
2.39	2.45	3.88	9.17
3.17	3.33	3.51	9.49
3.55	4.00		

Using the data, and assuming uniform strain along the entire wire, calculate the stresses in MPa, plot the data on stress-strain axes, and interpret the kinds of behavior exhibited by the wire throughout the experiment. Note that the load needed to produce the next increment of strain may actually decrease at some stages. Calculate Young's modulus for the wire from the elastic region of the curve.

7. Consider an experiment where standard axial compression experiments were run on shale, limestone, and sandstone at near-surface conditions to obtain stress-strain curves. Instead of discarding the broken rocks after failure, a second run was made on each confined rock to failure again. Would you expect the stress-strain curves to be different for each rock this time, or the same?

Further Reading

Bayly, B., 1992, Mechanics in structural geology: New York, Springer-Verlag, 253 p.

A concise book that presents the concepts dealt with in Chapters 3 and 4 in greater detail, yet understandably with many solved problems. Discusses fundamental strain principles, then presents stress and afterwards discusses various modes of behavior of Earth materials.

Donath, F. A., 1970, Some information squeezed out of rock: American Scientist, v. 58, p. 54–72.

Relationships between experimental rock mechanics and structures we commonly observe in the field are presented, clearly, at the level of elementary structural geology. Laboratory techniques are explained in a way that clarifies the operation of rock-mechanics experiments, the construction of stress-strain curves from experimental data, and the formation of many common structures.

Jaeger, J. C., and Cook, N. G. W., 1976, Fundamentals of rock mechanics: London, Chapman and Hall, 585 p.

An excellent summary of the mechanical properties of rock materials presented somewhat as seen by an engineer. It contains numerous practical examples as well as clear discussions of behavior models for rock materials.

Nicolas, A., and Poirier, J. P., 1976, Crystalline plasticity and solid state flow in metamorphic rocks: New York, John Wiley & Sons, 444 p.

This is a mathematically rigorous treatment of plastic behavior in rocks, but sections may yield some understanding of the plasticity of rocks without resort to the mathematical statements. A feeling will also be gained of the inherent complexity of plastic deformation processes.

Patterson, W. S. B., 1981, The physics of glaciers, 2nd edition: New York, Pergamon Press, 380 p.

Shumskii, P. A., 1964, Principles of structural glaciology, D. Kraus translation of 1955 edition: New York, Dover Publications, 497 p.

Both books survey glaciers and glacial ice, and each contains chapters on the ductile and brittle properties of ice. Shumskii includes photographs of glacial folds and other ice structures, the counterparts of which in rocks form only at great depths.

Turcotte, D. L., and Schubert, G., 1982, Geodynamics: Applications of continuum physics to geological problems: New York, John Wiley & Sons, 450 p.

Mathematical treatment of elasticity and viscous behavior, written at a suitable level and with many worked problems.

7

Microstructures and Deformation Mechanisms

Although the presence of dislocations in crystals was first proposed 30 years ago, it was only in the last 15 years that a general realization of their importance was developed. Today, an understanding of dislocations is essential for all those concerned with the properties of crystalline materials.

DEREK HULL, 1975, *Introduction to Dislocations*

OUR DISCUSSION OF IDEAL MECHANICAL MODELS FOR deformation in Chapter 6 provides a basis for exploring the microstructures and deformation mechanisms we actually observe in rocks (Figure 7–1) as well as some of the ideas that attempt to explain how observed mechanisms operate on the microscopic scale. Today we recognize the importance of grain-scale mechanisms in working toward a more complete understanding of crustal and mantle deformation. We have learned that rock deformation is largely scale dependent; for example, microscopic-scale brittle fractures produce small incremental displacements of a layer that appears on the map scale as smoothly curved folds that give the appearance of being the product of ductile deformation.

The theories and concepts of dislocations and deformation mechanisms developed in metallurgy, materials science, and solid-state physics enhance our understanding of rock deformation. The same deformation mechanisms identified in metals and ceramics

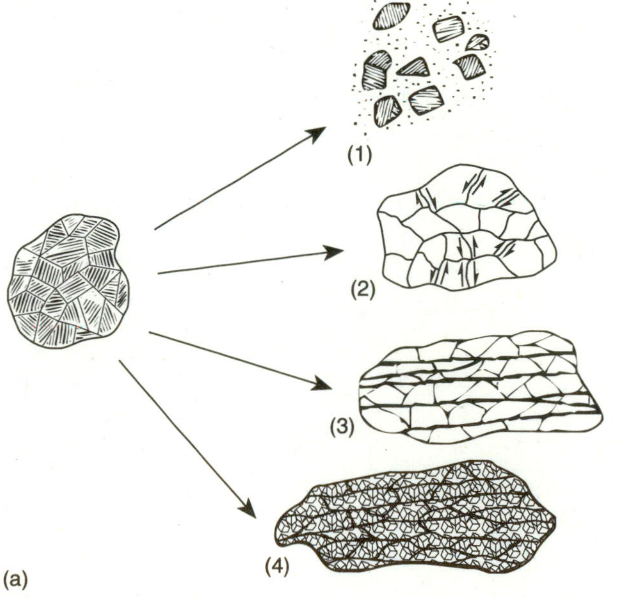

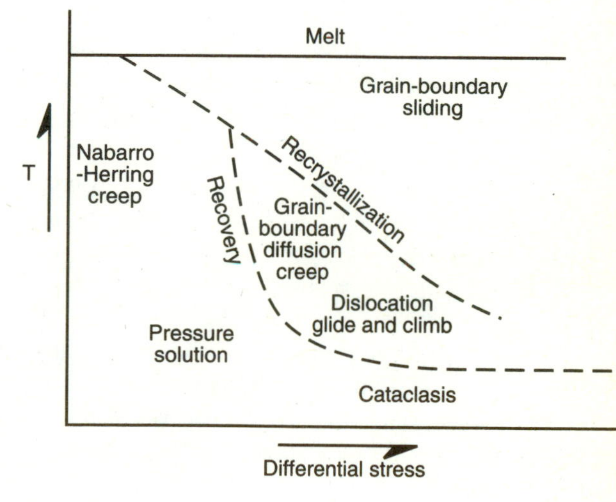

FIGURE 7–1
(a) Common modes of deformation in rocks. (1) Cataclasis, involving granulation and fragmentation of the original rock. (2) Grain-boundary sliding, producing deformation by grains sliding past each other. (3) Pressure solution, forming zones of residues where the rock has been dissolved. (4) Crystal-plastic deformation, involving partial to complete recrystallization of the rock, here producing preferred orientation of minerals that may preserve old (but deformed) external boundaries, but internally consisting of a mosaic of new unstrained crystals. (b) Relationships between temperature, T, stress, σ, and deformation modes. Strain rate is assumed to be constant.

have also been observed in rocks, indicating that these phenomena are present in all crystalline solids despite the fact that most metals form isometric crystals and are homogeneous. We have learned a great deal about deformation from studying simple monomineralic rocks (dunite, limestone, and quartzite), but most rocks are polymineralic. Rock mechanics structural geologists have the opportunity to add to the body of knowledge related to the behavior of dislocations in polycrystalline anisotropic materials as well as the conditions that influence the operation of deformation mechanisms, such as pressure solution.

In this chapter, we will see how rocks deform on a microscopic to submicroscopic scale, and then explore how microscopic-scale deformation is related to larger structures. When we understand the conditions that produce the variety of deformation mechanisms, we will be able to better deduce the conditions that affected an ancient rock mass that was deformed at great depth before erosion exposed it for us to study at the present surface. For example, some fault zones contain crushed and ground-up wall rocks (called ***cataclasite)***, whereas others may contain ductilely deformed wall rocks: the latter contain evidence of flow and reduced grain size ***(mylonite)*** and may even be recrystallized. What different physical conditions produced fault zones that differ so markedly, and what different behavior modes do these differences represent?

LATTICE DEFECTS AND DISLOCATIONS

We commonly think of crystals in terms of groups of spheres in a convenient packing arrangement or as a series of lines wherein intersections are occupied by atoms, ions, or groups of atoms such as CO_3 and SO_4. Regular geometric arrangement of atoms, ions, or groups in a compound is characteristic of a ***crystalline solid***. A ***crystal lattice*** is a hypothetical representation of the arrangement of the constituents of a crystalline solid that permits us to describe the symmetry of the crystal structure (Figure 7–2). It consists of a three-dimensional array of intersecting lines that portray the lattice as a series of parallelepipeds where the constituents (atoms, ions, groups) mostly occupy the intersections of lines (Hull and Bacon, 1984). These parallelepipeds form building blocks—repeat units—called ***unit cells***. Simple models like this relate some of the processes affecting crystalline solids on the atomic scale to larger structures, such as crystal faces, twinning, and zoning, that we observe in minerals and rocks. A crystal of a specific compound where all sites are filled with atoms, ions, or groups of the "correct" species, and which contains no *interstitial atoms* between lattice sites, is often called a ***perfect crystal*** and, in the context of the hypothetical lattice, is an orderly stack of unit cells. Ideally, perfect crystals exist only at 0° K (–273° C), where no thermal or other disturbances can affect the lattice. A halite lattice in which all the Na^+ sites are occupied by Na^+ ions and all the Cl^- sites are occupied by Cl^- ions at absolute zero is a perfect crystal. Such crystals are nonexistent in minerals and rocks; real crystals contain imperfections called ***defects***. Introducing ions of a different kind into the structure (such as K^+ into Na^+ sites or Br^- into Cl^- sites in halite), artificially making crystals or finding natural crystals with ***vacancies*** (unoccupied sites), changing the stacking order of layers, or otherwise distorting the lattice produces defects. Several kinds of defects have been described and categorized as *point, line, surface (planar),* and mixed defects. All these locally disturb the otherwise regular arrangement in a crystal.

In the past few decades, technological use of defects in crystal lattices has profoundly influenced the way we live. Mass-produced semiconductors made of high-purity silicon crystals manufactured (or "doped") with carefully added compositional defects (such as phosphorus or boron) have led to widespread use of transistors, microprocessor chips, and other components of computers, telephones, and other electronic systems. In the natural world, knowledge of defects helps us to understand the varied ways rocks deform on microscopic and larger scales.

Now we will examine crystal defects more closely.

Point Defects

Substitution or *interstitial* defects involving a foreign atom or ion in a lattice site, or a normal atom out of its proper place, may be introduced while a crystal is forming. Vacancies in a crystal may also be produced

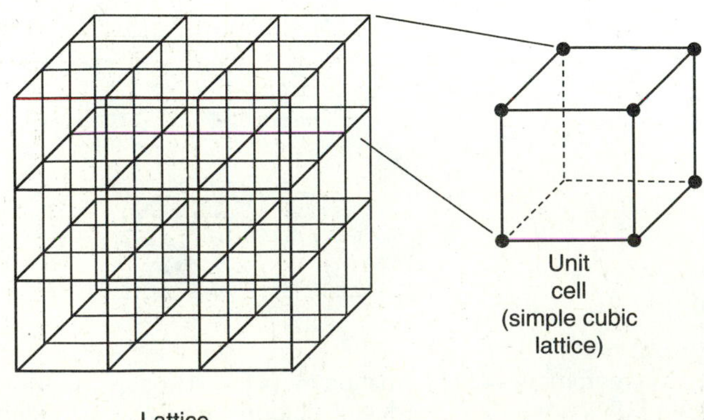

FIGURE 7–2
Crystal lattice (without atoms) and simple cubic unit cell.

during ductile deformation, by irradiation by high-energy particles, and by sudden cooling of a crystal from high temperature (Figure 7–3). An equilibrium concentration of vacancies will always occur at any temperature above absolute zero. In ionically bonded materials, substitution defects can be accommodated only if the charge balance is maintained, a factor relatively unimportant in metals (because of the nature of metallic bonding).

Line Defects—Dislocations

A ***dislocation*** is the line on a crystallographic slip plane that separates the slipped from the unslipped portion of a crystal (hence the name line defect). An ***edge dislocation*** is a line dislocation in which the slip sense across the crystallographic slip plane is *normal* to the dislocation (Figure 7–4a). A ***screw dislocation*** is a line dislocation in which the slip sense across the crystallographic slip plane is *parallel* to the dislocation (Figure 7–4b). Dislocations move sequentially and incrementally by displacing a single bond at a time in the crystal. A direct analogy with macroscopic faults exits where a fault propagates incrementally and recurrent motion occurs likewise. A comparison was may by Raymond Price (1988) between a particular kind of dislocation with the generation, propagation, and motion of a thrust sheet. Small, outcrop-size faults may involve motion along the entire fault surface, but earthquake distributions on large faults clearly indicate that this motion occurs on but a small part of the entire fault. The larger the earthquakes, the larger the increment of the fault surface is moved.

All dislocations in crystals evolve from lines to closed loops. These loops have different characteristics ranging from nearly pure edge dislocations, described above, to uniformly mixed, to nearly pure screw dislocations, depending on their propagation direction (Figure 7–5a). A dislocation begins as a line called a "Frank-Read source" that propagates with increasing shear stress until it forms a closed loop (Figure 7–5b). Loops formed in a common plane coalesce into edge, screw, or mixed dislocation and propagate through the crystal. A threshold length of a *Frank-Read source* exists at some critical stress τ_c and the source is said to be activated at values of $\tau > \tau_c$. Making a hypothetical incision in a hypothetical crystal, partly separating two layers, and then inserting an extra partial layer produces an ***edge dislocation*** (Figure 7–4a). The inserted layer is called an *extra half plane.* In real crystals, it is produced by subjecting a crystal to simple shear (which is frequently encountered in nature), leading to distortion of the crystal so that part of a layer of atoms is isolated halfway between two normal layers. The dislocation nucleates as a small loop and propagates outward as the crystal is progressively deformed by simple shear (Figure 7–5).

The term "line or edge dislocation" is derived from the location of the edge of the extra layer separating the dislocation from the undisturbed lattice. Edge dislocations are designated by $\perp$ if the dislocation is positive (the extra half plane lies above the reference line) and by **T** if the dislocation is negative (the extra half plane lies below the reference line).

The other end-member type, called a ***screw dislocation***, involves rotation about the dislocation line. The net result is that planes of atoms spiral around the dislocation line (Figure 7–4b). Screw dislocations are *right-handed* if the sense of rotation is clockwise (as seen looking down the screw axis; common wood screws are threaded right-handedly) and *left-handed* if the rotation is counterclockwise.

A convenient means of describing either edge or screw dislocations is by the ***Burgers vector*** and ***Burgers circuit.*** Imagine a layer in a perfect crystal where a

Vacancy

Interstitial atom (different species)

(a)

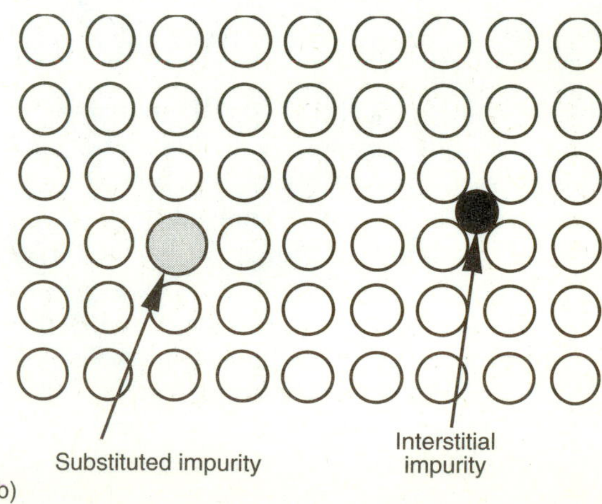

FIGURE 7–3
(a) Vacancies and interstitial atoms. (b) Impurities in crystals.

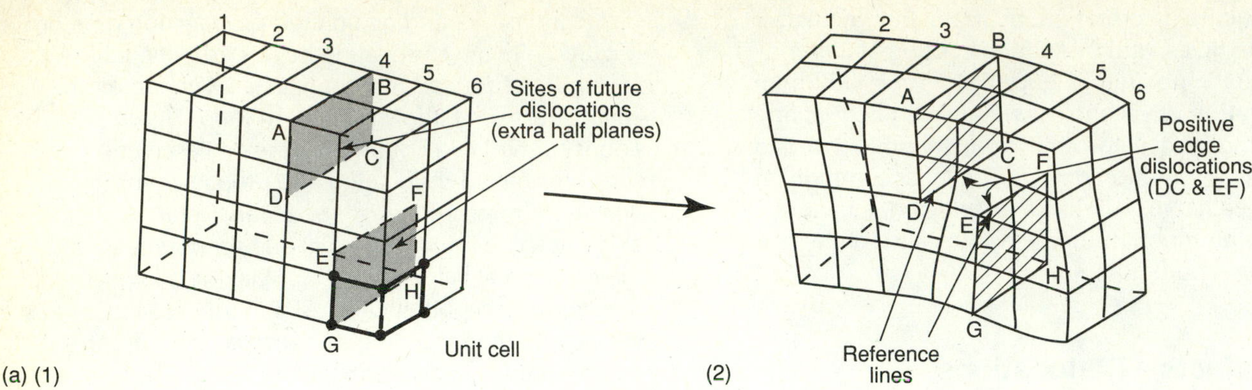

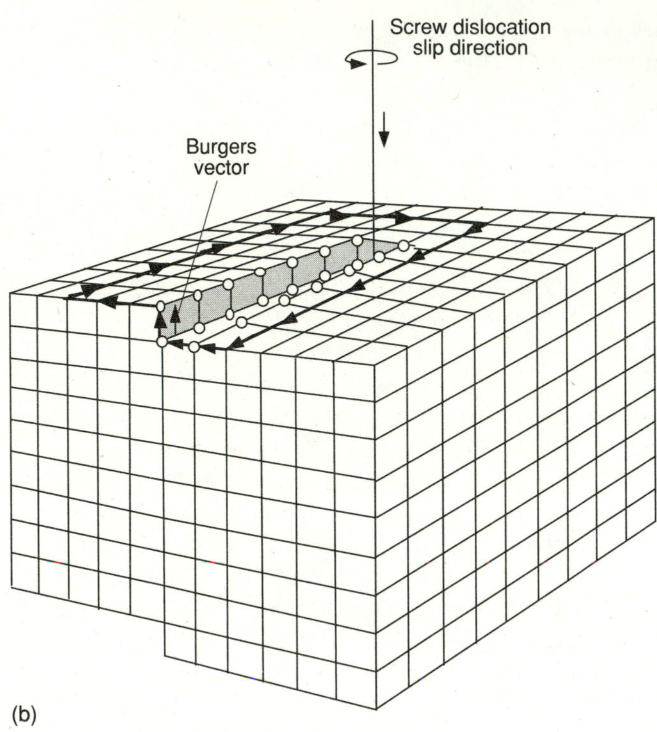

FIGURE 7–4
(a) Lattice before deformation (1), showing the site of the future edge dislocation and a unit cell. (2) Simple shear-driven propagation of an edge dislocation outward produces a displacement equal to Burgers vector as the dislocation moves out of the crystal. (b) Right-handed screw dislocation, propagating downward.

loop traverse is made from one site in the layer so that it returns to that site (Figure 7–6a). In a perfect crystal, the traverse should close precisely at the starting point. If the traverse encloses a dislocation, it will not close in the same distance as in the perfect crystal. The traverse, known as a *Burgers circuit,* requires that the traverse go an equal number of atomic units in each direction. The magnitude of the failure of the traverse to close in the same loop that surrounds a dislocation defines the unit shear or jump distance and direction of the dislocation, also defining a *Burgers vector,* **b** (Figure 7–6b). The Burgers vector of an edge dislocation is perpendicular to the line of dislocation; that of a screw dislocation (Figures 7–5a, 7–6, and 7–7) is parallel to the line of the dislocation (Hull, 1975). Most dislocation lines are located at an angle to the Burgers vector, giving them a mixed character, with an edge component and also a screw component (Figure 7–5a).

Dislocation lines can end at crystal (grain) boundaries, but never within a crystal. They must either join or branch into other dislocations, or form lines or closed loops in the same crystal. Three or more dislocations meet at a point called a *node,* and the Burgers vectors sum to zero at the node (Figure 7–8).

Dislocations in crystals may be best observed in extremely thin sections using the greater magnification of the transmission electron microscope (Figure 7–9). They were actually observed—but not understood—during the nineteenth century by examining etched surfaces of cleaved crystals with an optical microscope. Areas around dislocations are more soluble in acid because of the crystal unrecovered elastic strain near the dislocation. A combination of grinding and etching steps (called "decorating") made it possible to map the three-dimensional shapes of dislocations. Dislocations in some iron-bearing minerals, such as olivine, can be brought into view by heating at 900° C, long enough for some of the Fe^{II} to oxidize to Fe^{III} along dislocations, making them visible under a standard petrographic microscope (Mackwell and others, 1985).

Measuring ***dislocation density*** in a series of grains deformed by mostly dislocation glide (without climb) can be used to indicate the magnitude of differential stress affecting the crystal. Sometimes it is useful to construct *dislocation maps* of distributions and kinds of dislocations in crystalline solids. A saturation limit can be reached in most crystals; then dislocations from several different *slip systems,* or planes, become entangled, and no more dislocations can form in that part of the crystal (Figure 7–11). Slip systems consist of combinations of specific crystallographic planes with

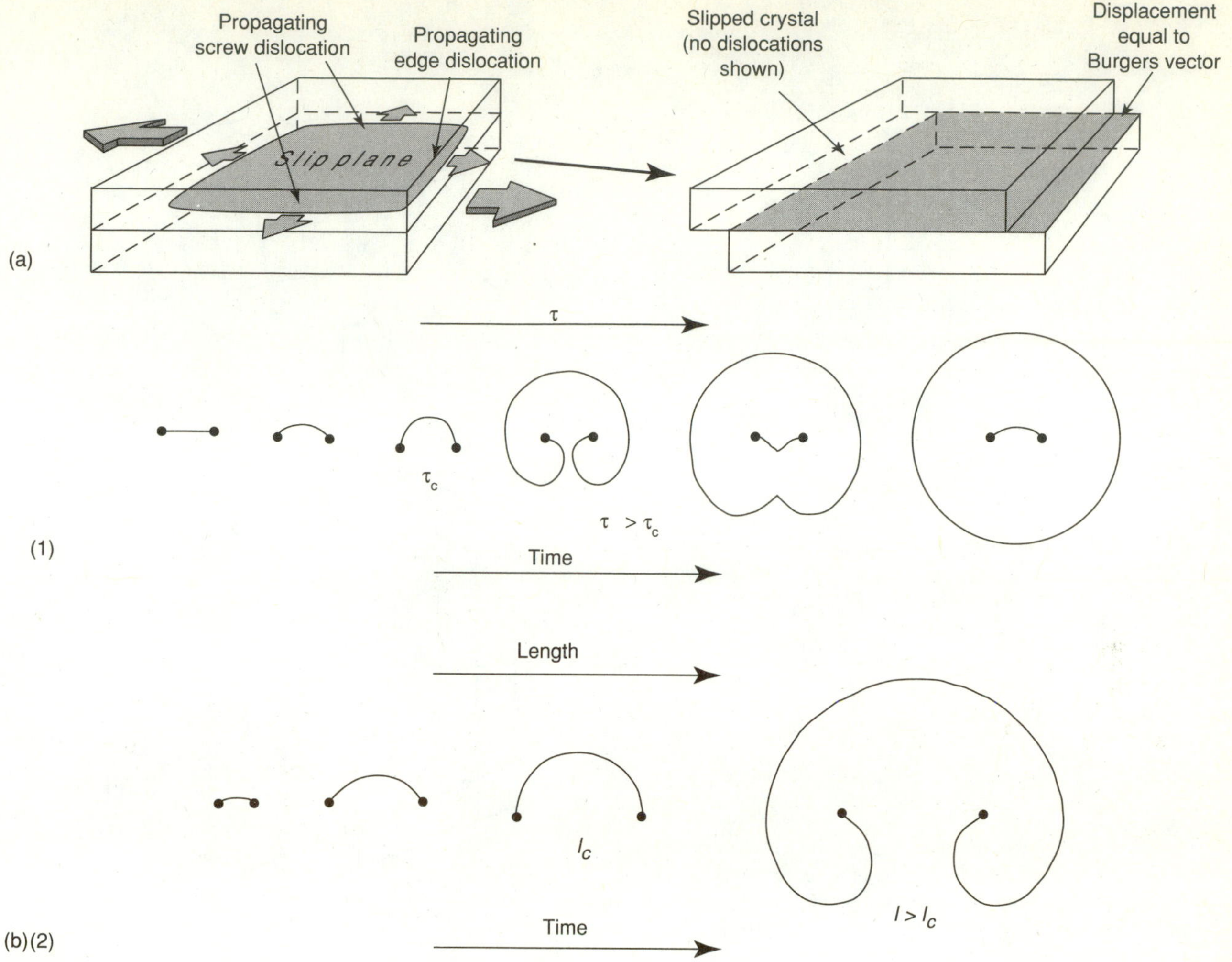

FIGURE 7–5
(a) Simple shear-driven propagation of slip in a mixed edge and screw dislocation that produces a displacement equal to Burgers vector as the dislocation moves through the crystal. The components of movement normal to the slip direction of the dislocation thus comprise the edge dislocation, whereas the components parallel to the slip direction are the screw dislocation. Once the dislocation propagates, the crystal structure has slipped and it displays no dislocations in that region. (b) Propagation of dislocations as Frank-Read sources. Think of this process as being analogous to blowing soap bubbles using a wire ring: the soap film stretches until it reaches a critical stress (τ_c) (1) or length (l_C) (2), the film forms an enclosed bubble that leaves the wire to become an independent feature, and the process is repeated. (Part b is modified from Jean-Paul Poirier, *Creep of crystals*, 260 p., © 1985, reprinted with permission of Cambridge University Press.)

crystallographic directions in each plane; active slip systems in crystals require minimum energy to cause displacement (Hobbs, Means, and Williams, 1976; Nicolas, 1987). Dislocations, after they are formed, may be thermally annealed out of a crystal by nucleation of new and unstrained crystals that replace the old strained configuration. Annealing is accomplished by the ***dislocation glide and climb, annihilation, and grain-boundary formation and migration*** mechanisms of recrystallization. Dislocation glide and climb, and annihilation may also be active without accompanying recrystallization. (Recrystallization mechanisms are discussed later in this chapter.) *Tangles* and *pileups* of dislocations at impurities in the crystal may inhibit but not prevent complete annealing (Figure 7–11).

Planar Defects—Stacking Faults

Among the most common planar defects in crystals with isometric symmetry are ***stacking faults,*** which consist of irregularities in the repeat order in a series of layers in a lattice (Figure 7–11). If a lattice has a repeat sequence between layers of ABAB (only two kinds of layers), the sequence can alternate in relatively few ways, and stacking faults can occur only if the repeat order is ABBA or BAAB. If the normal repeat sequence among layers is ABCABC (three kinds of repeat units), possibilities exist for locally changing the stacking order. If part of a layer is left out in a face-centered cubic lattice, so that part of the crystal has the stacking order ABCABABC. . . or ABCABACABC. . ., a local

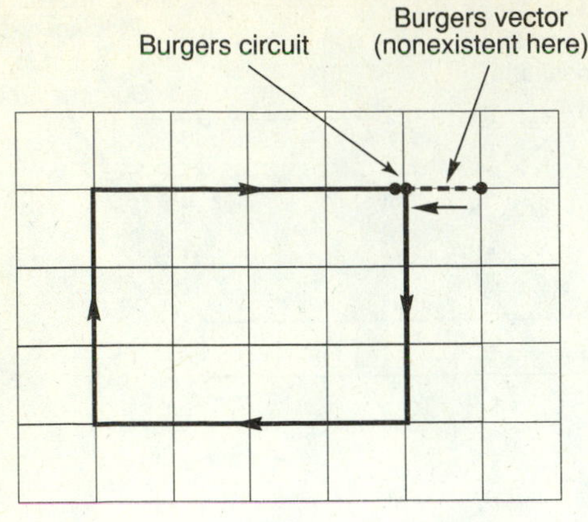

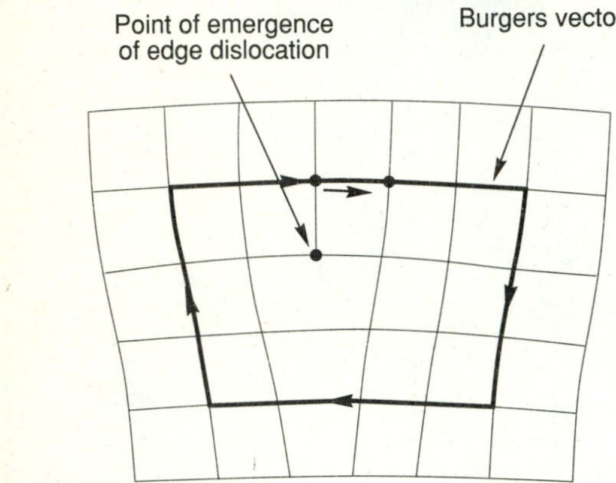

FIGURE 7–6
(a) A perfect crystal illustrating where Burgers vector would occur *if* the crystal was deformed. (b) Burgers vector in a deformed crystal with an edge dislocation. Note that to define Burgers vector, the traverse must involve an equal number of repeat intervals in each direction.

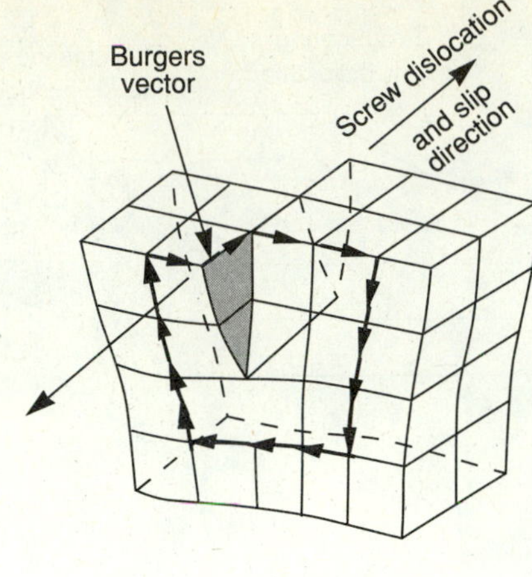

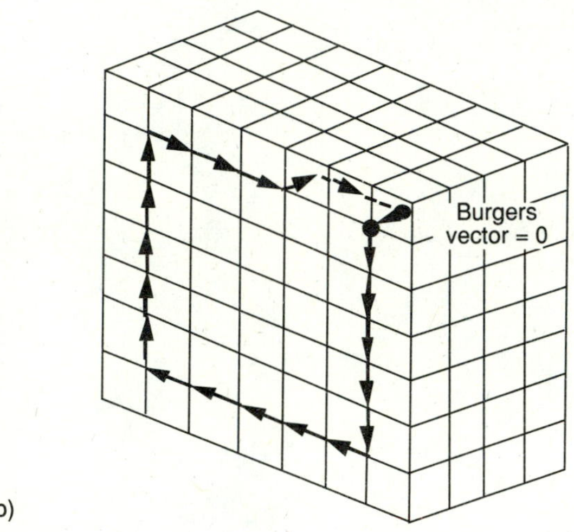

FIGURE 7–7
(a) Burgers vector for a screw dislocation. (b) Burgers vector for a perfect crystal for comparison (zero here).

discontinuity results, giving rise to two kinds of stacking faults (Figure 7–11). Additional possibilities for stacking faults also exist.

Dislocation and Twin Gliding

Deformation may occur in a crystal by homogeneous simple shear that causes movement of a dislocation located parallel to an existing crystallographic plane (Figure 7–12). This kind of deformation is called ***dislocation (translation) gliding***, or mechanical crystal plasticity (Knipe, 1989), and the specific planes along which offset occurs are called "slip systems." Slip may occur on the 0001, $10\bar{1}1$, and other planes in quartz (Christie and others, 1964; Carter, 1971); the $10\bar{1}1$, $02\bar{2}1$, and other planes in calcite (Turner and others, 1954); the 010 plane in plagioclase (Seifert, 1965); the 001 plane in the micas (Etheridge and others, 1973); and the 0kl, 110, and 010 planes in olivine (Raleigh, 1965; Carter and Avé Lallemant, 1970).

Deformation of crystals may also occur by homogeneous simple shear that produces deflection slip of

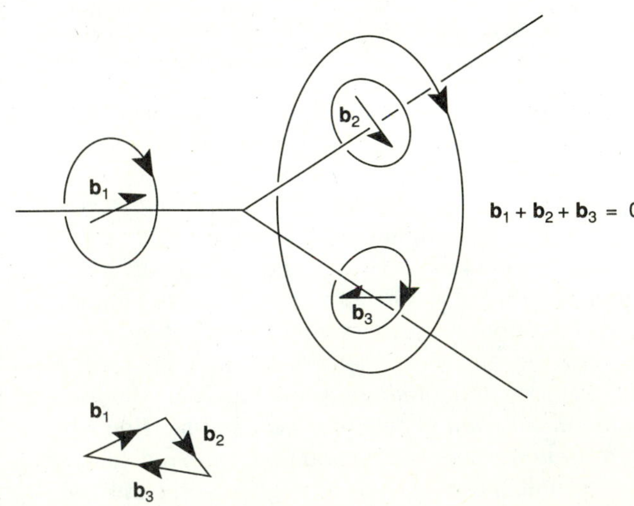

FIGURE 7–8
Several oppositely oriented Burgers vectors meet at a node and sum to zero. (Reprinted with permission from D. Hull and D. J. Bacon, *Introduction to Dislocations,* 2nd edition, © 1984, Pergamon Books Ltd.)

FIGURE 7–9
Dislocations in experimentally deformed synthetic quartz under the transmission electron microscope. Width of field is approximately 2 μm. Note the concentration (pinning) of some dislocations around fluid inclusions (arrow). (John W. McCormick, SUNY College at Plattsburgh.)

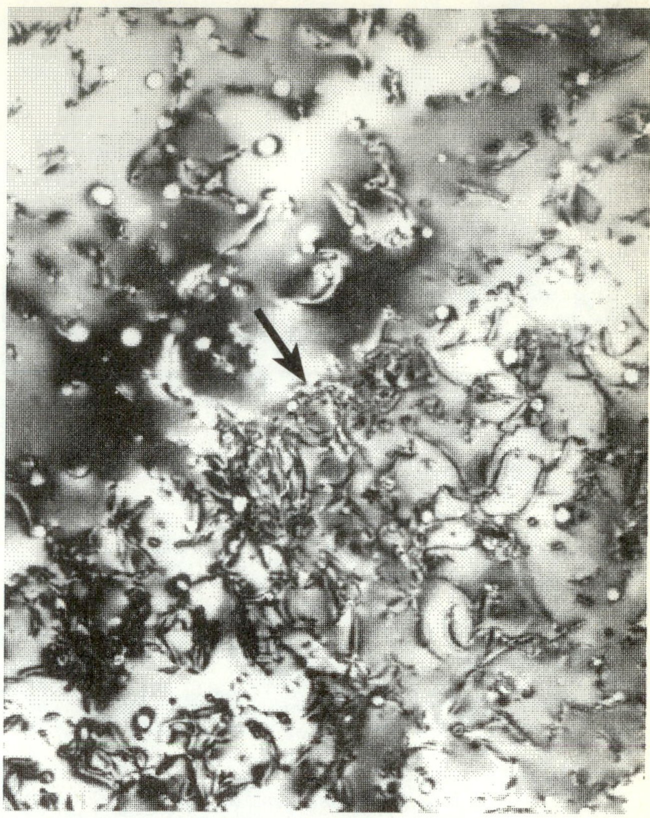

FIGURE 7–10
Dislocations piled up at a barrier (arrow) in experimentally deformed synthetic quartz under the transmitted electron microscope. Width of field is approximately 2 μm. (John W. McCormick, SUNY College at Plattsburgh.)

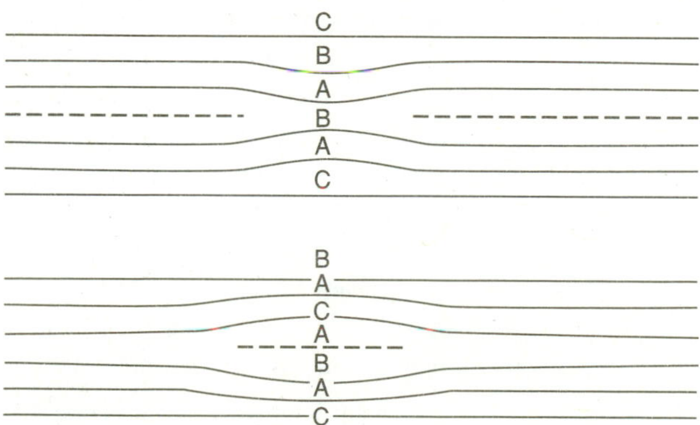

FIGURE 7–11
Two kinds of stacking faults (dashed).

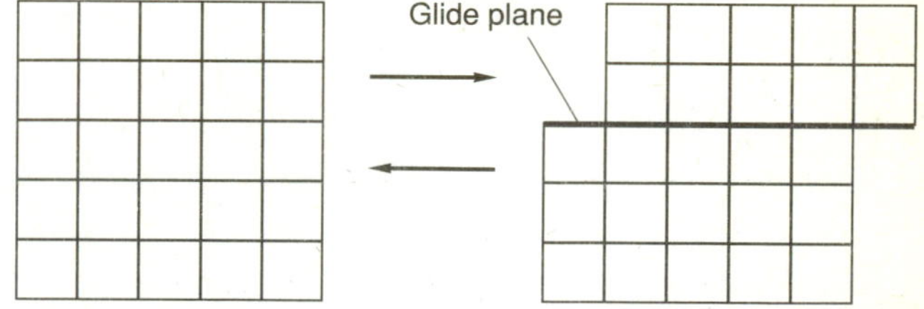

FIGURE 7–12
Translation gliding.

less than one repeat interval along crystallographic planes, producing shear-induced twinning of the crystal—another form of translation, called ***twin gliding*** (Figure 7–14). Twin gliding may be produced artificially in an Iceland spar calcite crystal by inserting a knife blade parallel to the $01\bar{1}2$ plane of the negative rhombohedron (Figure 7–14). Mechanical twinning also occurs in plagioclase twinned on the albite and pericline systems (Borg and Handin, 1966; Borg and Heard, 1970). For a much more complete discussion of mechanical twinning and slip systems, see Hobbs, Means, and Williams (1976), Nicolas and Poirier (1976), and Nicolas (1987).

DEFORMATION MECHANISMS

Factors that determine which deformation mechanism will dominate are temperature, total stress, differential stress, fluid pressure, composition, grain size, texture, and strain rate (not all are independent variables).

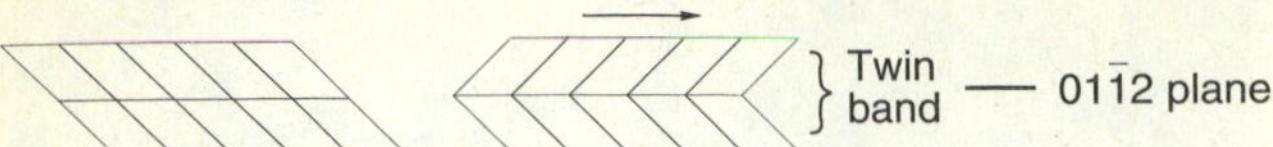

FIGURE 7-13
Twin gliding.

FIGURE 7–14
Mechanical twin of a cleavage rhomb of calcite produced by inserting a knife blade parallel to the 01 $\bar{1}$2 plane. The twin is the small upturned corner close to the pencil point. (Specimen courtesy of Theodore C. Labotka, University of Tennessee–Knoxville.)

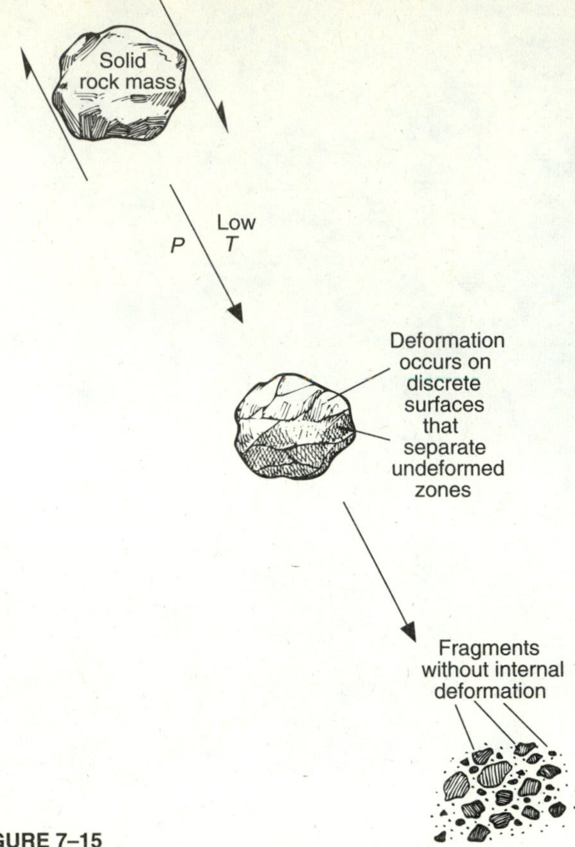

FIGURE 7–15
Cataclastic materials and behavior.

Deformation mechanisms (Figure 7–1) will be discussed here in order of dominance with increasing temperature.

Cataclasis

Brittle deformation is concentrated in a rock mass along movement surfaces that have no cohesion—surfaces that separate undeformed regions. The product of this process is the cohesive rock ***cataclasite***. This type of deformation, called ***cataclasis***, is characterized by megascopic breccias (breccias with visible fragments), gouge, fractures and faults, microbreccias, and microfractures (Figure 7–15). Cataclasis produces brittle granulation of rock at low temperature and low to moderate confining pressure (near-surface conditions) but may be facilitated by high fluid pressure. Also, where strain rate is high, failure occurs under brittle conditions by exceeding the elastic limit of the material. Cataclasis occurs mostly at low temperature within a few kilometers of the surface, but feldspars, garnets, and a few other minerals undergo cataclasis at high temperatures (500–1000° C).

In contrast, under conditions of moderate to low strain rate and temperatures of 300–800° C, feldspars and garnets deform by cataclasis, but quartz, calcite, and many other minerals deform ductilely. Total volume of rock may be reduced by closer packing of grains (that closes larger open spaces between grains) and by recrystallization that produces denser minerals. Fracturing and granulation, on the other hand, increase volume by creating pore space where none existed before, and also increase exposed surface area.

Some map-scale fault zones contain rocks that resemble ductilely formed fault rocks *(mylonite),* but microscopic examination reveals that deformation was brittle and that the rock was granulated frictionally to a very fine powder of unaltered grains. Apparent flow of fine-grained material produced by brittle deformation is called ***cataclastic flow***. The process may also describe large-scale changes in shape due to small-scale brittle fracturing. Apparent megascopic ductile behavior shown to be brittle (cataclastic flow) on the microscopic scale is sometimes called *triboplastic* behavior; it has been observed at shallow depth where rocks have been ground to fine grain size during motion along a fault (Figure 7–16). Dolomite or other rocks of high mechanical strength on both sides of a fault may produce triboplastic cataclasite at low temperatures.

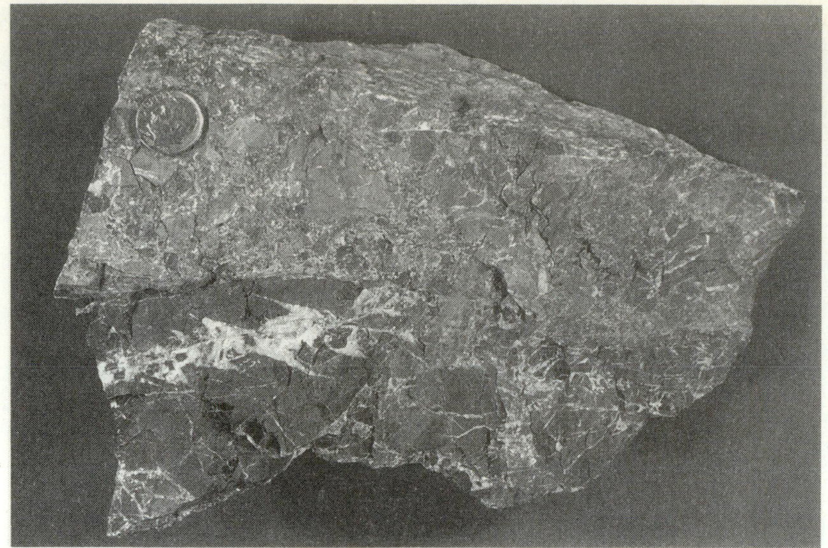

(a)

(b)

FIGURE 7–16
(a) Fine-grained triboplastic(?) carbonate gouge in a limestone breccia from cataclasite along the Saltville thrust in Knoxville, Tennessee. Small dark dolomite fragments were broken repeatedly and healed with white calcite.
(b) Photomicrograph of the same specimen as in (a) showing flow banding and stylolites in cataclasite that was broken initially, then healed with veins of white calcite, then broken again by brittle fractures. Width of photograph is approximately 5 mm. Plane light. (RDH photos.)

Creep Processes

Creep processes depend on strain rate, but the amount of strain is not limited by strain rate. These are also known as diffusional mass transfer processes (Knipe, 1989). Creep mechanisms may involve mass transport or diffusion of atoms or ions at grain boundaries, including pressure solution, climb of dislocations within a lattice, and diffusion of point defects through lattices. Each is a separate mechanism that is most efficient over a particular range of temperature, pressure, and grain size. These processes also overlap and compete with one another under the right conditions so that more than one mechanism may occur simultaneously, although one usually dominates. Creep is slow but may result in large strains if it occurs over a long time interval, as with mantle convection. Because these mechanisms occur over a range of conditions, it is useful to plot their occurrence on *deformation maps* (Figure 7–17). These maps reveal experimentally determined ranges of occurrence of several mechanisms for particular rock types or minerals together with the boundaries calculated using the appropriate flow law governing each process that can be expressed in terms of differential stress, strain rate, and temperature.

These mechanisms are sometimes referred to as *steady-state deformation mechanisms* because they operate above threshold values of pressure and temperature, and when initiated, continue unchecked until a competing mechanism becomes more efficient as temperature or pressure—or both—increase, or the fluid supply is exhausted (for pressure solution), or until the energy supply is exhausted. They are also considered to be steady-state processes because we do not yet

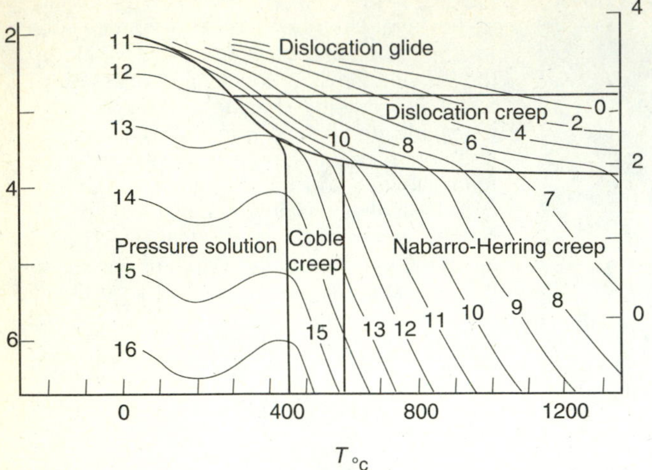

FIGURE 7–17
Deformation mechanism map for calcite, showing the different mechanisms that affect it at various temperatures and stresses. Contours are negative log strain rate per second. Vertical axis is in units of (negative log of) differential stress (Chapter 3) divided by the square root of 3 times the shear modulus (left axis) or (log of) differential stress divided by 10^5 Pa at 500° C (right axis). Thus the left axis numbers decrease upwards as differential stress increases (contrast with the right axis). Note that differential stress increases upward, so the highest-numbered contours represent the slowest strain rates. (After E. H. Rutter, 1976, *Philosophical Transactions of the Royal Society of London*, v. 283.)

know about critical changes in variables such as dislocation density, unit cell size, and grain size during deformation, so they are *assumed* to cancel—hence, steady state.

Pressure Solution. The phenomenon of ***pressure solution*** was first recognized in the mid-nineteenth century by Henry C. Sorby (1826–1908), the English geologist who pioneered microscopic petrography and made several important contributions to structural geology. Pressure solution consists of dissolution, under stress, of soluble constituents such as calcite or quartz, and is generally active at low to moderate temperature (<350° C) in the presence of water. The process is limited by the requirement of water between grains. After the water is lost, pressure solution ceases to occur. Pressure solution is commonly viewed as a stress-induced diffusion-flow process (Wheeler, 1991). Rutter's (1976) flow-law equation attempts to define most of the variables related to the process, and assumes Newtonian behavior:

$$\dot{\varepsilon} = \frac{K D_{gb} \Omega C_o}{x^3 k T \rho} \tau_d , \qquad \textbf{(7–1)}$$

where $\dot{\varepsilon}$ is strain rate, K is a material constant, D_{gb} is the coefficient of grain-boundary diffusion, Ω is molar volume, C_o is concentration of the solution outside the grain boundary, x is grain size, k is Boltzman's constant, T is absolute temperature, ρ is density, and τ_d is differential (shear) stress. Raj (1982) modified Rutter's flow law, by assuming the solute is transported by diffusion through a liquid film along grain boundaries. It employs an "island-channel" model for grain boundaries and the pressure solution process:

$$\dot{\varepsilon} = \frac{K \prod f h \Omega D \bar{c}}{x^3 k T b} \tau_d , \qquad \textbf{(7–2)}$$

where f is the area fraction of grain boundaries occupied by islands, h is the grain boundary thickness, D is diffusivity of solute in the fluid phase, and $\bar{c}$ is mole fraction of solute dissolved in fluid, and b is the "jump distance" in the diffusion process.

Parts of quartz, calcite, or other soluble mineral grains may be dissolved at points of greatest deviatoric stress, where one crystal touches another. Dissolution begins at these points of high stress (Figure 7–18a). Minerals dissolved by pressure solution are commonly precipitated in zones of lower pressure as overgrowths and fibers in pressure shadows, deposited in veins, or remain in solution and are completely removed from the region undergoing deformation. Parts of original grains, along with any reprecipitated material, may survive as evidence of pressure solution. In addition, residues of insoluble materials, including clays, micas, and opaque minerals, are commonly left behind on pressure-solution surfaces (Figure 7–18b). Pressure-solution surfaces appear to develop best in slightly impure carbonate rocks rather than in pure limestone (Marshak and Engelder, 1985). Fine (or mixed) grain size and impurities, because of greater surface area and compositional differences, initiate pressure solution more readily than coarser grain size and greater purity in carbonate or clastic rocks.

Pressure solution occurs during diagenesis, particularly during compaction and cementation of carbonate sediments where ***stylolites***—irregular surfaces coated with insoluble minerals—may form parallel to bedding (Figure 7–19), indicating that the maximum stress *(overburden pressure)* was vertical. Stylolites also form during deformation of carbonate rocks and sandstone normal to the direction of applied tectonic stress, with the orientation of the "teeth" of the stylolites normal to the plane of the stylolite, providing a more exact indicator of the orientation of σ_1. The dark irregular surfaces (lines resembling graphs in cross section) that mark stylolites are insoluble minerals that remain after soluble minerals have been dissolved by pressure solution. *Tectonic stylolites* (Figure 7–19c) may sometimes be distinguished from stylolites formed during diagenesis if one set is subparallel to bedding (generally diagenetic) and others are at moderate to high angles to bedding (generally tectonic).

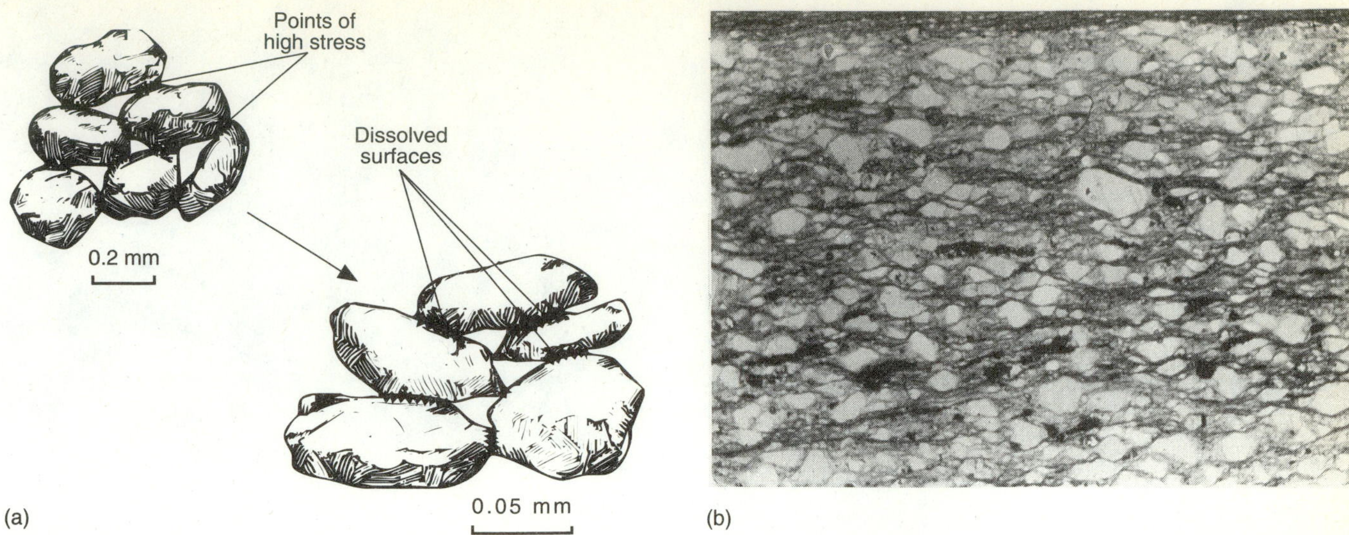

FIGURE 7–18
(a) Grains dissolve first at points of high stress. Note that the scale changes from before to after deformation. (b) Partly dissolved quartz grains along mica-rich slaty cleavage surfaces produced by pressure solution in Martinsburg Slate, Delaware Water Gap, New Jersey. Width of photograph is approximately 2 mm. Plane light. (Thin section courtesy of Timothy L. Davis, North Carolina Geological Survey.)

Pressure solution may proceed grain by grain without formation of stylolites if the rock mass is water-saturated and is sufficiently porous. Most porosity in a rock mass tends to be concentrated along bedding and fracture planes; pressure solution begins there, and stylolites form along existing surfaces. Preferential dissolution will occur along favorably oriented surfaces, normal to the maximum principal stress. Pressure solution may occur at high confining pressure and elevated temperature in both fine clastic and carbonate rocks where carbonate minerals and quartz are dissolved. It can be the dominant process up to 300° C (Rutter, 1976, 1983) and can reduce rock volume by as much as 50 percent if the system is open and dissolved constituents can escape (Wright and Platt, 1982). In the past, it has been difficult to understand the transport mechanism of dissolved constituents, given the low solubilities of quartz and calcite at surface conditions. It is now known, however, that during an orogeny, huge volumes of fluid are forced by deformation to move outward from the hotter inner parts of a orogen toward the cooler external parts. In the process, large quantities of quartz, calcite, metals, and hydrocarbons are dissolved and transported (Oliver, 1986). Veins composed of fibrous calcite, quartz, or other low-temperature minerals developed on bedding planes (Figures 5–17 and 5–18) or in vein arrays parallel and oblique to bedding may indicate excess fluid pressure during deformation (Cosgrove, 1993).

In many areas, pressure solution is the dominant mechanism forming slaty cleavage (Chapter 17). Irregular dissolution surfaces develop into discrete cleavage planes at increased pressure (20–40 MPa) and temperature (300–400° C) as deformation progresses. These surfaces form the well-developed slaty cleavage common in zones of low metamorphic grade. Direct evidence of pressure solution may be observed in thin sections as mica or quartz "beards" on larger grains (Figures 7–19 and 7–20), on partial grains (Figure 17–22), and in zones of insoluble residues developed approximately parallel to each other (Figure 17–22b).

Material dissolved in zones of compression is frequently precipitated in veins that fill fractures in zones of extension (Figure 7–19c). Fractures are commonly the sites for deposition of dissolved constituents, forming veins—particularly where fractures formed by extension remain open. Direct correlation generally exists between the composition of veins and the temperature—the metamorphic grade—of the enclosing rocks during deformation, except in rock masses uniformly affected by hydrothermal alteration and ore mineralization. Zeolite ($\pm$ calcite) veins (frequently with prehnite) occur at the lowest temperature. Calcite-quartz, then quartz-epidote and quartz-feldspar veins occur at progressively higher temperatures. At the upper end of the temperature range, deformation by diffusion processes (to be discussed) dominates in the rock mass, but water still moves and transports materials in solution that may be precipitated in zones of low pressure (Figure 7–21). Composite veins containing fibrous crystals and multiple mineral assemblages may document a history of repeated extension (Chapter 5) and a range of precipitation temperatures (although veins may also be the product of simple shear deformation).

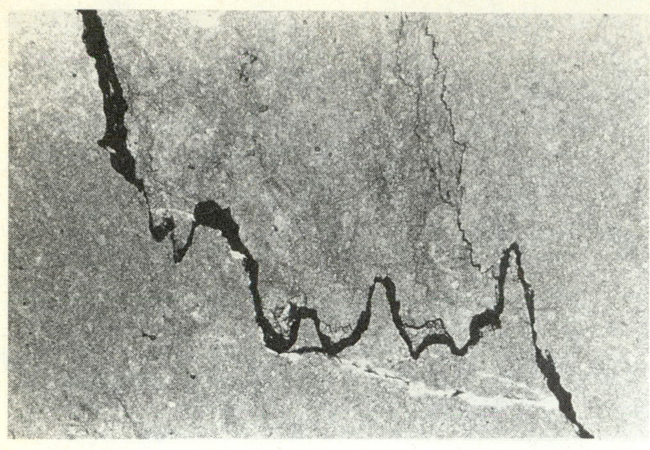

(a)

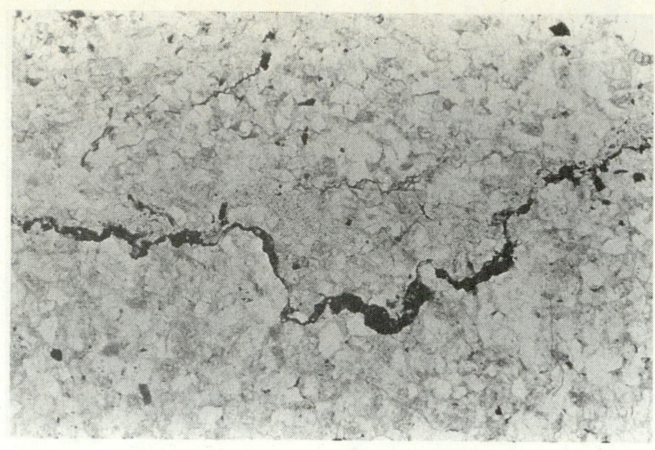

(b)

(c)

FIGURE 7–19
Stylolites subparallel to bedding in Tymochtee Dolomite, northwestern Ohio. Smaller stylolites are oblique to bedding (a) and parallel to bedding in Massanutten Sandstone, northwestern Virginia (b). Width of the photograph in both (a) and (b) is approximately 7 mm. Plane light. (Charles M. Onasch, Bowling Green State University.) (c) Negative print in plane light of a thin section of a folded oölitic limestone in the Nolichucky Shale (Upper Cambrian) from Melton Valley in the Oak Ridge National Laboratory reservation. Tectonic stylolites (white) are developed parallel to the axial surface of the fold (perpendicular to maximum compression), and calcite-filled tension veins (dark gray to black) formed mostly perpendicular to the stylolites. Veins that formed subparallel to the stylolites are later veins. Note the development of fine stylolites around the edges of oöids with minimal dissolution and thicker better-developed stylolites that have dissolved major parts of oöids. Width of field is 2.2 cm.

Grain-Boundary Diffusion Creep (Coble Creep). This process involves thermally driven diffusion mass transport along dry grain boundaries. Coble creep occurs at low to moderate temperature, but below 400° C is overwhelmed by pressure solution. Above 400° C, Coble creep becomes dominant where the diminished amount of water makes pressure solution less efficient and the additional heat energy accelerates this process. A flow law that describes Coble creep is

$$\dot{\varepsilon} = \frac{141 D_{gb} \delta \tau \Omega}{d^3 kT} \tag{7–3}$$

where $\dot{\varepsilon}$ is strain rate, δ is grain-boundary thickness, τ is shear stress, Ω is atomic volume, d is spacing of edge dislocations, k is creep rate, and T is absolute temperature (Poirier, 1985).

Dislocation Creep. Dislocations are a manifestation of strain through a combination of glide and climb motion of the dislocations through the crystal, driven by heat and pressure. This process occurs at moderate to high temperature as ***dislocation creep***. Dislocation creep is a two-step process that includes both strain and accommodation of the lattice to strain; the glide of disloca-

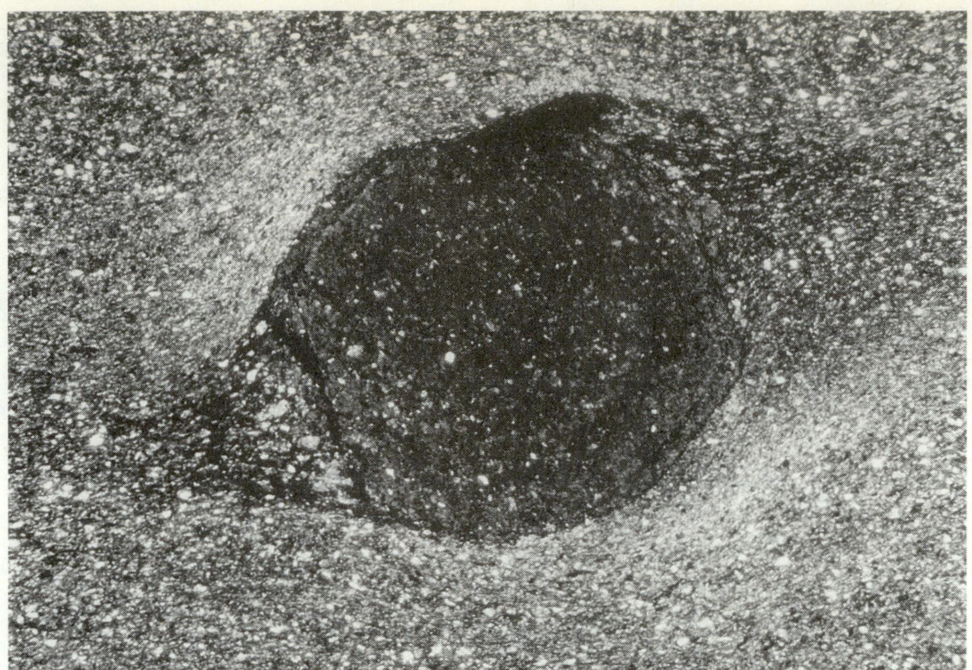

FIGURE 7–20
Mica beards formed parallel to slaty cleavage on an ankerite grain in fine-grained siltstone of the Wilhite Formation near Tellico Plains, Tennessee. Width of ankerite porphyroblast is approximately 1 mm. Crossed polars.

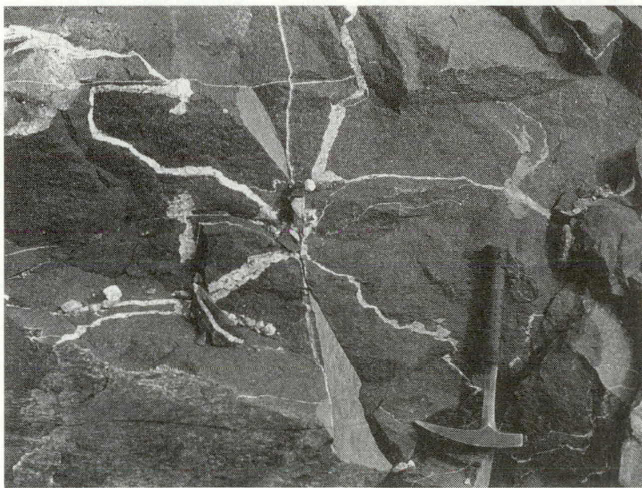

FIGURE 7–21
Veins in amphibolite in the Kapuskasing zone, south-central Ontario, filled with quartz and feldspar. Near-vertical vein in the center of the photograph probably formed later than the more gently inclined folded veins. (RDH photo.)

tions (strain) is accommodated by either recrystallization or dislocation climb (recovery). A flow law for dislocation creep is

$$\dot{\varepsilon} = \dot{\varepsilon}_0 \tau^n e^{-\left(\frac{Q}{RT}\right)} \qquad \textbf{(7–4)}$$

where $\dot{\varepsilon}$ is the initial strain rate for the crystal being deformed, n is the stress exponent, defined as $(d\ln\dot{\varepsilon}/d\ln\tau)$, Q is the apparent activation energy for the process, and R is the universal gas constant (1.987 cal/deg^{-1}/mol^{-1}) (Poirier, 1985). Dislocations are constantly produced and migrate through a grain as the rock is stressed, causing each grain to change shape as it deforms and as chemical bonds are rearranged. Dislocation glide does not require creep to occur, nor does climb of dislocations necessarily accompany glide. Minerals may be deformed by glide only, leading to formation of tangles of dislocations, deformation lamellae, and undulatory, patchy, or sweeping extinction under the microscope. Climb involves the reorganization of dislocations and may decrease dislocation density by formation of low-angle (subgrain) boundaries (see below). Climb of dislocations is thermally driven and thus is more efficient at higher temperature.

Dislocation creep produces grains that either contain fewer dislocations or that have a time-invariant (on average) dislocation density, which is related to the magnitude of the applied stress. At the low-temperature end of the process, because dislocation creep is temperature dependent, dislocations may become pinned at grain boundaries. As a result, the rock mass becomes strain hardened (Chapter 6), and greater stress is required to increase the amount of strain. (Strain hardening may also be produced by interference of dislocations.) In contrast, cataclastic flow may be accompanied by pressure solution that enhances sliding along irregular grain boundaries at low temperature, and, because it produces finer grains, it is a strain-softening process. At higher temperatures (greater than half the absolute melting temperature), thermal energy permits dislocations to bypass obstacles by climbing to a parallel lattice plane. Dislocations may be annihi-

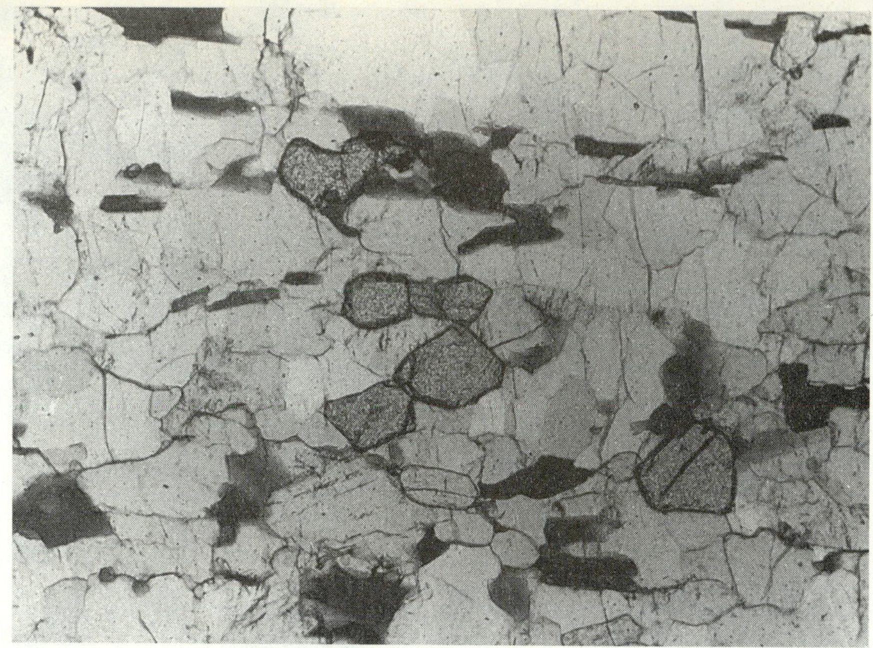

FIGURE 7–22
Annealed mineral assemblage in a thin section of highly strained Late Proterozoic(?) quartzite from near Franklin, North Carolina. The grains with high relief are garnet. Dark grains are biotite and magnetite (note alignment). Clear grains are mostly quartz with a few grains of plagioclase. Note that many quartz grain boundaries meet at 120° angles. Width of photograph is approximately 3 mm. Plane light. (RDH photo.)

lated by the same process, thereby reducing the number of dislocations and internal strain in the crystal. This is the most common plastic-flow process that leads to recrystallization in rocks at moderate to high temperatures (Kerrich and Allison, 1978) (Figure 7–22).

Volume-Diffusion (Nabarro-Herring) Creep. Another process involving diffusion of lattice defects is ***volume-diffusion*** or ***Nabarro-Herring creep***, which involves diffusion of point defects (vacancies) through crystals toward points of high stress and occurs at high temperature and low differential stress. It is a very slow process, even at high temperature, and is not considered an important deformation mechanism in crustal rocks. A flow law describing Nabarro-Herring creep is

$$\dot{\varepsilon} = \alpha \frac{D_{sd} \tau \Omega}{d^2 k T}, \tag{7–5}$$

where α is a factor determined by grain shape and the boundary conditions of deformation, and D_{sd} is the coefficient of self-diffusion of the crystal (Poirier, 1985).

Grain-Boundary Migration and Sliding. Movement of grain boundaries can occur either in the plane where they occur (sliding) or normal to the plane (migrating). The slip of grains past one another, analogous to the frictional process of sand grains or ball bearings moving aside when a stick is pushed into them, is an important deformation mechanism in rocks. It is a sliding process that implies no cohesion on grain boundaries. It is accompanied by other processes, such as Coble creep and grain boundary migration at moderate to high temperature, and is restricted to high-angle boundaries. This nonfrictional mechanism is called ***grain-boundary sliding***. A single grain in a rock being deformed by a creep mechanism will behave like the entire mass of grains; if the mechanism is grain-boundary sliding, a single grain will maintain its initial shape, even though the shape of the aggregate changes significantly. Cataclastic flow at low temperature is an analogous (but *not* identical) frictional mechanism occurring in fault zones where grains move past one another under a strong component of simple shear.

Grain boundary migration is a *fundamental kind of dynamic recrystallization* in rocks. It leaves behind grains containing few (or no) dislocation. Grains containing many dislocations and thus have higher free energy—the energy available to do work—that lie in the path of a migrating boundary are consumed. This describes the process of *strain-induced grain-boundary migration*. Other processes may produce *stress-induced* grain-boundary migration by an excess of elastic strain energy being stored in a crystal, or a *diffusion-induced* grain-boundary migration by differences in chemical potential between individual grains (Poirier, 1985).

Superplastic Flow

Superplasticity is a term used to describe a macroscopic behavior that results from sliding along grain boundaries and probably involves both Coble creep and grain-boundary sliding (Nicolas, 1987). It was first observed as a phenomenon of extreme ductility in certain fine-grained alloys undergoing deformation in high-temperature uniaxial tension experiments. Elongation strains of 1,000 to 2,000 percent have been observed in some alloys. ***Superplastic flow*** in metals is thought to occur by a combination of grain-boundary sliding (to provide the large strains) locally accommodated by diffusion creep (to provide a shorter diffusion path), because other slower processes would not permit the phenomenon to occur at finite rates (Poirier, 1985).

Superplastic flow is thought to occur at temperatures greater than half the absolute melting point in coarse-grained rocks (Boullier and Gueguen, 1975). In contrast, the mechanism has also been reported in fine-grained rocks (average grain diameter 10 μm or less) at low temperature under a strong component of inhomogeneous simple shear, as occurs in fault zones (White and White, 1980).

As indicated by the diversity of ideas just cited, superplastic flow of geologic materials may occur under a variety of conditions involving a combination of deformation mechanisms. Deformation producing a decrease in grain size at low temperature to the critical 10-μm size may enable strain rate to increase dramatically as grain-boundary sliding is initiated with no change in temperature. At high temperatures (>0.5 T_m), grain-boundary sliding and dislocation creep, or grain-boundary diffusion, may effect superplasticity (Kerrich and Allison, 1978).

UNRECOVERED STRAIN, RECOVERY, AND RECRYSTALLIZATION

We have examined deformation mechanisms and how they affect a rock mass; the remaining task is to clearly distinguish between a strained crystal and the processes of recovery and recrystallization that restore the crystal to a lower energy condition less strained to unstrained.

On the microscopic scale, deformation of a crystal is indicated by unrecovered strain. *Undulatory extinction* in quartz grains (Figure 7–23) is caused by a difference in optical-extinction properties, caused in turn by dislocations or microcracks distorting the lattice. ***Subgrains*** are small parts of grains with different lattice orientation in adjacent parts of the same grain. They begin to form as additional strain accumulates in the crystal through the mechanism of dislocation glide. Although the accumulation of dislocations in a crystal is a strain-hardening process, subgrain formation, once dislocations are no longer being produced, is a strain-softening process that represents the first visible stage of reorganization of the crystal to an internally less-strained or lower-energy state (Figure 7–24a). Subgrain boundaries consist of ***low-angle boundaries*** in which the lattice has a misorientation of less than 10° with respect to the nearest boundary of the crystal. This causes a change in optical properties so they are readily observed using crossed polars in a petrographic microscope.

As the crystal is deformed at higher thermal or strain-induced energy states, it will form sites for nucleation of entirely new unstrained crystals. In a crystal being deformed by dislocation glide and climb, steady-state flow may be accommodated by one of two thermally activated mechanisms—***dynamic recovery*** or ***dynamic recrystallization***. The term ***recovery*** is used to generally describe processes that reduce dislocation density, dislocation interaction, and increase the rate of dislocation glide and climb in crystals being deformed by dislocation creep (Nicolas, 1987). Recrystallization is a thermal- and stress-driven process that results in a decrease in dislocation density and creation of new grain boundaries through grain-boundary migration, and other deformation mechanisms discussed above. The net result is a decrease in strain in a crystal. A major difference between the two processes is that dynamic recrystallization involves strain softening, and dynamic recovery does not (Tullis and Yund, 1985). Recovery, which commonly produces low-angle boundaries, is the first stage in decreasing the amount of strain in deformed crystals; recrystallization, which may be initiated independent of recovery, completes the process of decrease in strain energy of a deformed crystal and produces high-angle grain boundaries (Figures 7–24b and 7–25). The newly formed grains contain lattices oriented at a high angle (>10°) to the lattice in the parent crystal, thereby producing ***high-angle boundaries.***

Recrystallization may occur as ***dynamic*** or ***syntectonic recrystallization*** while stress is being applied to the rock mass, or it may occur as ***static***

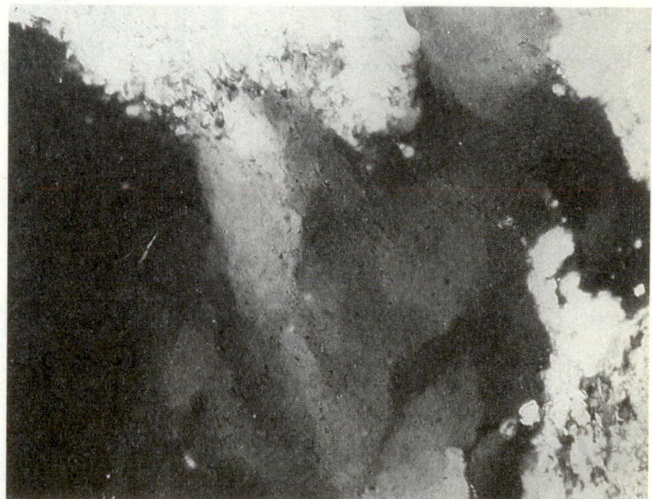

FIGURE 7–23
Subgrains indicated by undulatory extinction revealing subgrains in quartz. Cataclasite from the Homestake shear zone near Leadville, Colorado. Width of photograph is approximately 1 mm. Crossed polars. (RDH photo.)

(a)

(b)

FIGURE 7–24
(a) Plastically deformed quartz grains in Lower Cambrian Weaverton Quartzite from South Mountain near Harpers Ferry, West Virginia. Some grains have been strongly flattened into the foliation; a few maintain vestiges of originally round shape. All have very small unstrained grains at the edges of original sand grains with subgrain structure here outlined by undulatory extinction. Field is approximately 2 mm wide. (b) Completely recrystallized fabric in Setters Quartzite, Baltimore, Maryland. The entire rock is a mosaic of unstrained polyhedra; note 120° grain-boundary intersections and absence of subgrains. Width of field is approximately 3 mm. Both (a) and (b) were photographed under crossed polars. (Both photos courtesy of Charles M. Onasch, Bowling Green State University.)

recrystallization after stress has been removed (Sibson, 1977). Dynamic recrystallization is accompanied by a decrease in elastic strain energy and an increase in grain-boundary (surface) energy. Dynamic rotational recrystallization produces a "core-and-mantle texture" (Figure 7–24a) when dislocation climb continues to insert dislocations into subgrain boundaries at the perimeter of the grain. This continued insertion increases the magnitude of the angular mismatch of the lattice across the boundary until the point that the lattice breaks, converting the boundary from a low-angle subgrain boundary to a high-angle brain boundary. Core-and-mantle structure may also be produced by dynamic recrystallization where the concentration of dislocations or strain at the host grain boundary triggers the migration of a new grain boundary, forming small grains on the boundary of the larger host crystal. The original strained lattice may have formed subgrains to attain a lower energy state and then later recrystallized to form small new crystals with high-angle boundaries. The newly crystallized quartz or feldspar grains meet at angles near 120°, a low-energy state arrangement that therefore provides the greatest stability in an unstrained crystal (Figure 7–24b). As unstrained crystals form by dynamic recrystallization, they immediately are strained and again begin to form low-angle boundaries and subgrains that also eventually recrystallize.
Static recrystallization is driven by reduction in strain energy and a reduction of surface energy (Figure 7–26). If enough heat is available, recrystallization may thus be favored over high strain rate during continued deformation, even if the rock fabric is recycled several times during multiple episodes of recrystallization and progressive deformation. The end product of a texture dominated by strained grains would not provide enough clues that the grains were not original and are in fact the last of several generations of grains. Strain removed by recovery and recrystallization is the elastic strain of individual dislocations and defects.

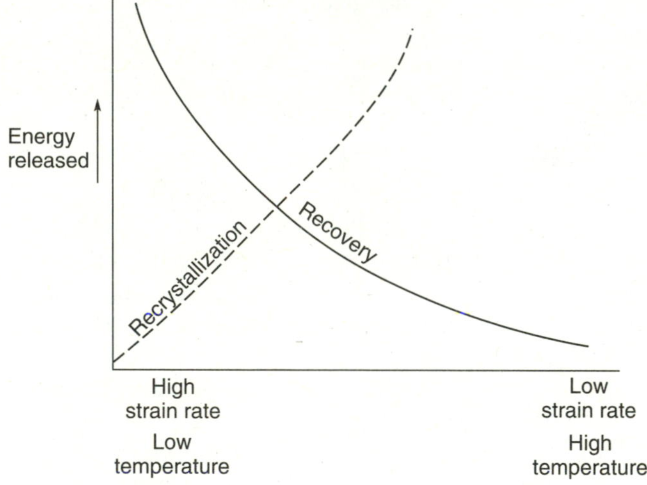

FIGURE 7–25
Relationships between strain rate, temperature, and energy released for recovery and recrystallization processes.

In order to better understand deformation and recrystallization processes in rocks, Greg Hirth and Jan Tullis (1992) have conducted a number of experiments using quartz aggregates (orthoquartzite and novaculite). As a result, they have recognized three regimes of dislocation creep. Regime 1 involves low-temperature

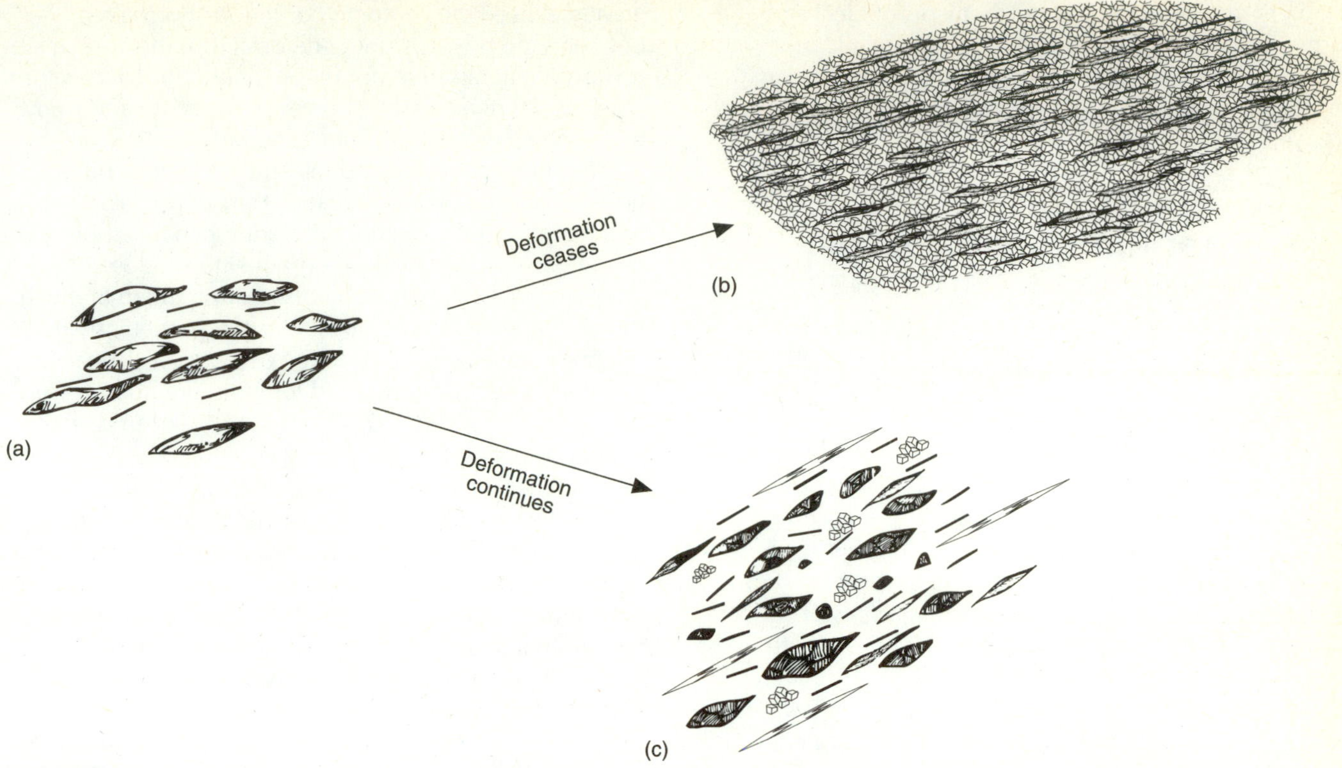

FIGURE 7–26
Relationships between static and dynamic recrystallization. (a) Deforming rock mass containing flattened original sand grains. (b) Static recrystallization; larger grains and groundmass are recrystallized. (c) Dynamic recrystallization; larger grains and groundmass recrystallize during deformation, and many grains record strain after deformation.

deformation at rapid strain rates wherein dislocation climb is hindered, and recrystallization occurs by strain-induced grain-boundary migration. Orthoquartzite (coarse-grained) samples undergo strain softening, but novaculite (fine-grained) samples do not. Quartz grains acquire a patchy undulatory extinction, and recrystallization begins at grain boundaries. Following nucleation of new unstrained grains, additional growth of new grains occurs by strain-induced grain-boundary migration. Regime 2 occurs at higher temperatures or lower strain rates. The rate of dislocation climb in Regime 2 becomes sufficiently rapid to involve recovery; steady-state flow is achieved at relatively low strain rates, and sweeping undulatory extinction outlining subgrains occurs in quartz grains. The low-angle boundaries continue to accumulate dislocations and thus evolve into high-angle boundaries—recrystallization. Recrystallized grains commonly form along grain boundaries. At still higher temperatures or lower strain rates, Regime 3 is characterized by sufficiently rapid dislocation climb to accommodate recovery, but with greatly increased grain-boundary migration and initiation of steady-state flow. Recrystallization occurs here by both grain-boundary migration and subgrain rotation. Regime 1 is the only one where strain weakening was noted. The addition of small amounts of water to the samples catalyzes the transition from Regime 1 to 2 and from Regime 2 to 3, permitting them to occur at lower temperature.

We have examined intracrystalline deformation mechanisms that function under different conditions, produce dislocations, and lead to recrystallization of individual grains. Most rocks that go through this process contain a fabric consisting of strongly oriented crystals—a foliation (Chapter 17). The question of how this crystallographic-preferred orientation of minerals is produced by deformation and recrystallization in rocks has been debated for many years. Metallurgists and geologists have also long been aware that dislocation glide results in rotation of crystallographic axes relative to the applied stress. The role of dynamic and static recrystallization in deformation processes has been resolved only recently. Obviously, formation of preferred orientations, cleavage, and foliations in metamorphic rocks, and in many fault rocks as well, depends on strain. Dislocation glide and rotation processes, in addition to recrystallization, are probably the most common mechanisms leading to preferred orientation. Preferred orientation has been produced experimentally without accompanying recrystallization (Tullis, 1971). In Chapter 17, we will return to this problem and examine the mechanisms of cleavage formation and their relationship to recrystallization, but meanwhile, we turn to a consideration of laboratory studies of recrystallization processes.

LABORATORY MODELS OF DEFORMATION PROCESSES

Many simple geologic materials have been deformed in the laboratory. Mixtures of quartz and feldspar (the common constituents of granitic rocks) and pure mineral aggregates such as quartz sandstone, pure marble (calcite), and dunite (olivine) have been studied extensively. Because of the geologically short time scale of human activities, mylonite, and other well-recrystallized and foliated rocks, can be produced experimentally in pure quartz sandstone only at temperatures in the range of 800 to 1000° C which permit high strain rates and plastic flow to be achieved in the range of 10^{-5} to 10^{-7} s^{-1} (Figure 7–27; Tullis and others, 1973). Both the temperature and strain rate needed to activate these processes in the laboratory greatly exceed the geologic conditions normally encountered by rocks being deformed in the core or toward the flanks of a mountain chain. Such laboratory conditions are necessary, however, to create and study these materials during the investigator's lifetime.

Mixtures of quartz and feldspar in various proportions are useful analogs because they permit study of two materials of markedly different mechanical properties in geologically realistic compositions (close to granite). Quartz deforms by dislocation creep at much lower temperature than does feldspar, particularly in the presence of a small quantity of water. Feldspar generally remains brittle until temperature and pressure are very high (900–1100° C, 1500 MPa), even with water present. Consequently, coexisting quartz and feldspars, as in granitic rocks, deform differently in the deforming rock mass. Experimental and field studies indicate that feldspar does recrystallize at higher temperature around the edges of large grains with the appearance of small and unstrained polygons, leaving the original large grains largely unstrained (Tullis and Yund, 1985). As recrystallization progresses, the size of the old grains decreases and new and smaller grains are replaced by recrystallized grains. Presence of micas in a quartz-feldspar aggregate also increases the rate of

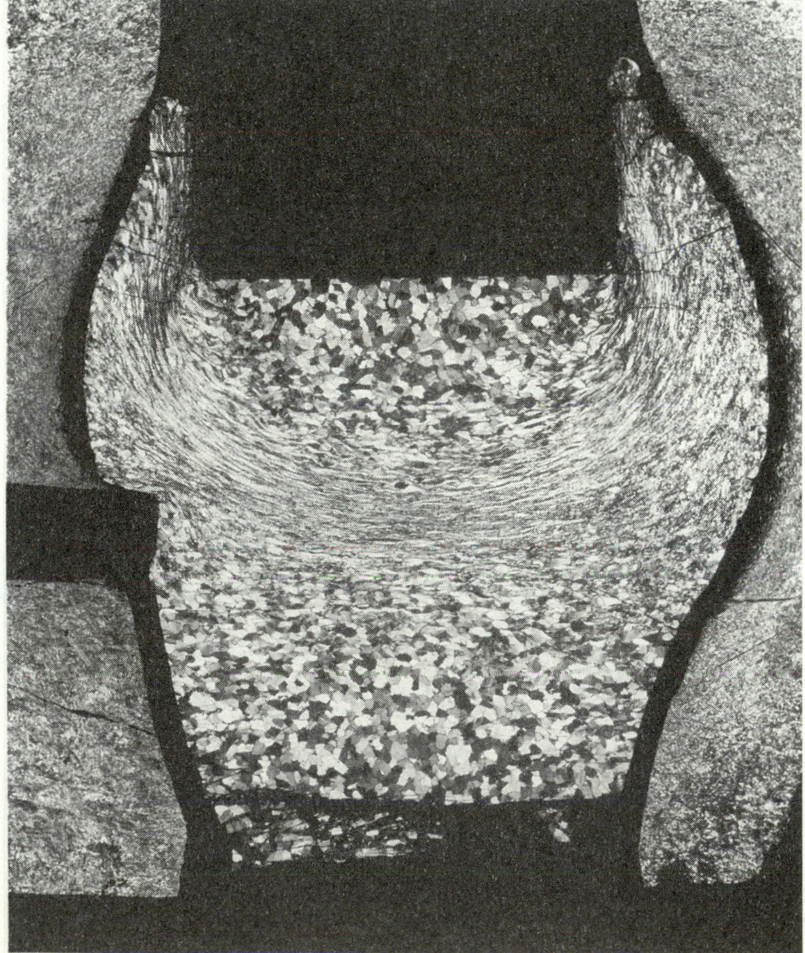

FIGURE 7-27
Thin section of an experimentally produced quartzite mylonite demonstrating that mylonite formed by both pure and simple shear. Curved light area of flattened quartz grains flowed around the sides of the dark obstacle, bulging the container as the ram compressed the sample from above. Undeformed original quartz grains are present on the ends of the sample immediately adjacent to the obstacle and the ram. This sample was deformed at 800° C at a strain rate of 10^{-7} s^{-1}, and the sample was shortened by 50 percent. Maximum width of the bulged part of the sample is ~6 mm. Crossed polars. (From J. Tullis, J. M. Christie, and D. T. Griggs, 1973, Geological Society of America *Bulletin.*)

deformation under the same temperature and pressure. Water, even in small quantities, increases the deformation rate and lowers the temperature threshold for most deformation mechanisms.

A series of experiments designed by Winthrop D. Means (1986, 1990) involves organic compounds with low melting points, such as monoclinic biphenyl, hexagonal octachloropropane, and triclinic paradichlorobenzene. These compounds strain and recrystallize under low stress at room temperature, because of their low melting points, and so their behavior can be observed directly under a petrographic microscope (Figure 7–28)—something impossible with most rocks, metals, and ceramics.

A number of intriguing surprises have arisen from this work. Means and his students have recognized a number of ways that subgrain boundaries can form and textures that appear to be the products of grain-boundary sliding from grain-boundary migration (and the converse). These experiments permit direct observation of both static and dynamic recrystallization as well as direct correlations of strain field with sequences of deformation mechanisms. These experiments have provided structural geologist with a greater appreciation of the complexity of these processes.

DISCUSSION

Understanding of the nature and kinds of microstructures should be an integral part of any investigation of geologic structure. We have now explored the different kinds of microstructures and deformation mechanisms and have gained some idea about the physical conditions under which they form.

Deformed rocks reveal a broad array of structures on all scales. Means (1993) has raised the questions, What kinds of deformation processes yield features that will have a memory of movement history?, and Do distinct deformation processes necessarily have distinct structural signatures? We know from our discussions in Chapter 7 that dislocations can be annihilated and the dislocation density reduced by recovery and recrystalli-

(a)

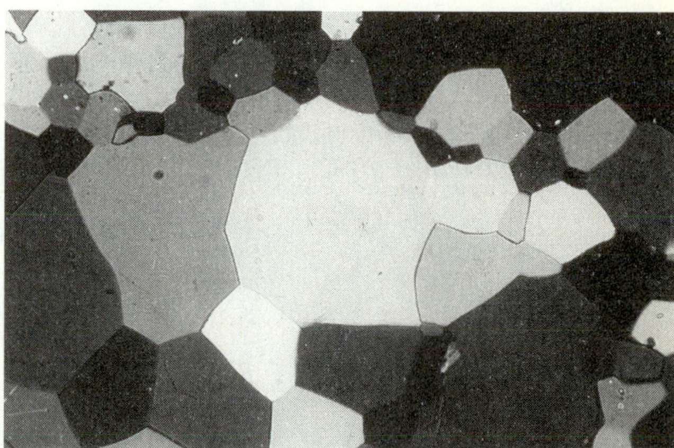

(b)

(c)

FIGURE 7–28
Low-melting-point organic compounds that deform under controlled conditions on the stage of a petrographic microscope and exhibit various recrystallization and deformation processes. (a) Deformation-induced twin lamellae (narrow lines) in crystals of paradichlorobenzene. Width of field is 0.7 mm. Crossed polars. (W. D. Means, 1986, *Journal of Geological Education,* v. 34.) (b) Undeformed (or annealed) mass of interlocking octachloropropane crystals. Note that junctions of most crystals make ~120° angles. Width of field is 0.5 mm. Crossed polars. (c) Abundant subgrains developed in crystals of (b) by horizontal, dextral shear of about 0.3. Also, note more irregular grain boundaries due to grain-boundary migration. Field width is 0.5 mm. Crossed polars. (Unpublished photos courtesy of Winthrop D. Means, SUNY Albany.)

zation. This process thus decreases the memory of part of the movement history within the rock mass. For sake of discussion, Means devised a scale-independent classification that attempts to differentiate between deformation mechanisms and processes that will retain some "memory" of movement history (conservative) and those that will not (nonconservative; Figure 7–29). It is based on whether the processes are active (velocity gradient across a structural element, such as a stylolite) or passive (no velocity gradient). Rate of local strain rate and migration rate of dislocations are additional variables. Taken together, these variables yield 12 classes, 6 each of conservative and nonconservative structures. In the classification, Means separated conservative from nonconservative processes—those that will retain memory of the movement history, such as dislocation glide, grain-boundary sliding, movement on faults, and fibrous veins preserving multiple growth history, and those that do not, such as dislocation climb, dike injection (retains memory of dilation direction, σ_3), static growth of porphyroblasts, and various chemical processes that erase earlier effects of movement. Means also recognized that composite processes are possible that involve more than one of his 12 classes. A stylolite in a limestone is a good example of a composite element. The entire stylolite surface is a nonconservative element that involves removal of material, hence some of the record, during formation of the surface but on a large scale suggests that it is not migrating. On a larger scale, however, undissolved calcite grains are migrating toward the boundary—the stylolite—and toward the insoluble materials that have accumulated along it as the grains are dissolved, making the stylolite a composite element. Means emphasized that the classification is far less important than the questions it raises and attempts to address. That many processes, when examined in detail, fit more than one of the 12 classes reveals the major problem with all classifications of natural phenomena: they are artificial attempts by scientists to divide nature into categories that really may not exist. We will encounter similar problems in Chapters 9 and 14 with fault and fold classifications.

We now turn from rock mechanics to a study of fractures and faults.

Deformation Processes					
$\dot{\varepsilon} \neq 0$				$\dot{\varepsilon} = 0$	
Active		Passive			
$\dot{M} \neq 0$	$\dot{M} = 0$	$\dot{M} \neq 0$	$\dot{M} = 0$	$\dot{M} \neq 0$	$\dot{M} = 0$
Dislocation glide **Class 1** Twin boundary migration	Grain-boundary sliding **Class 3** Fault slip	Dynamic grain-boundary migration **Class 5** Segregation veining	Passive grain-boundary deformation **Class 7** Folding with axial plane convection	Static grain-boundary migration **Class 9** Exsolution with lamella growth	Static polygonization **Class 11** Chemical exchange at fixed boundary
Dislocation climb **Class 2** Dissolution at stylolite margin	Cracking **Class 4** Dike injection	Dynamic phase-boundary migration **Class 6**	**Class 8** Dynamic chemical exchange with grain-boundary fluid	Static replacement veining **Class 10** Static porphyroblast growth	**Class 12** Non-conservative chemical exchange with grain-boundary fluid
Deformation mechanisms					

FIGURE 7–29
Means' 12 classes of deformation processes based on local strain rate ($\dot{\varepsilon}$) and migration rate ($\dot{M}$). Active structural elements involve a velocity discontinuity, passive elements do not. Note that only deformation mechanisms are active. (Reprinted from *Journal of Structural Geology,* v. 15, W. D. Means, p. 343–349, © 1993, with kind permission from Elsevier Science, Ltd., Kidlington, United Kingdom.)

ESSAY

Cataclasites, Mylonites, and Metamorphic Rocks

If dynamic recrystallization is occurring in a rock mass, the recovery and recrystallization processes in the crystals, which tend to remove strain, will compete with the tendency for renewed accumulation of strain in the grains. If this process is occurring in a fault zone within the ductile-brittle transition, brittle and ductile processes may compete as a function of strain rate. High strain rate may force the observed deformation from an otherwise ductile mode into a more brittle mode. Competition also exists between strain rate and recovery or recrystallization rate within the rock mass. The competing processes and the resulting products differ in subtle ways (Hatcher and Hooper, 1981). On the one hand, rocks are being deformed very rapidly but may not recover and recrystallize very rapidly—or may not recover at all. If deformation occurs at low temperature and low confining pressure, the deformation mechanism will be brittle fracture and frictional sliding, and the rock type produced will be a cataclasite (Figure 7E–1). On the other hand, if the temperature is somewhat elevated, fluid is available, and the strain rate is low enough (as along some fault zones), the rock will probably become a *mylonite.* Mylonites are strongly foliated metamorphic rocks that exhibit high ductile strain and incomplete recrystallization or recovery. Larger grains are flattened and extended into the foliation, and ribbon quartz (which may be internally recrystallized) is common. Reduced grain size characterizes mylonitization (Bell and Etheridge, 1973; Hatcher, 1978).

Factors that determine whether mylonites preserving a strongly deformed fabric forms, or if a metamorphic dynamically or statically recrystallized fabric forms, depend on both strain rate and temperature, as well as other factors such as quantity of fluid available and magnitude of shear stress. Sometimes, dynamic recrystallization takes place in a rock mass, but the mass cannot recover rapidly enough to keep up with the shear strain rate. If the mylonitic texture is to be preserved on the microscopic scale, it must be frozen in by rapid cessation of strain and cooling; otherwise, static recrystallization may coarsen the grains and eliminate high

(a)

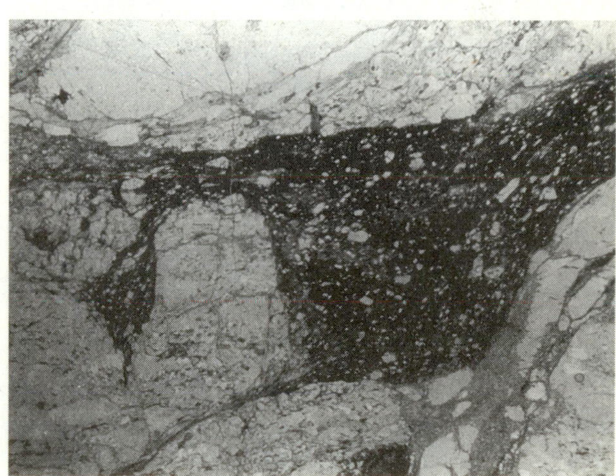

(b)

FIGURE 7E–1
(a) Cataclasite from the Brevard fault zone, South Carolina, formed at low temperature and pressure and containing highly strained constituents. The dark zone through the center of the specimen may have been a partially melted zone (called pseudotachylite) formed by rapid movement along the fault. (b) Thin section of cataclasite from the Homestake shear zone near Leadville, Colorado. Large, light-colored fragments of feldspar and quartz are fragmented and separated by very fine-grained zones of quartz and biotite. Width of field is approximately 16 mm. Plane light. (RDH photos.)

dislocation densities characteristic of mylonite (Figure 7E–2a). Finally, at moderate to high temperature and low strain rate, intracrystalline elastic strain is recovered or recrystallized faster than the strain rate can increase it by multiplication of dislocations (Wise and others, 1984). Enough thermal energy may remain after the stress is relieved so that the remaining strain is removed by static recrystallization, resulting in the microfabric of a "normal" metamorphic rock (Figure 7E–3) or an annealed mylonite. In the latter case, hand specimens preserve the mylonitic texture, but microscopically, it is entirely recrystallized (Figure 7E–2b).

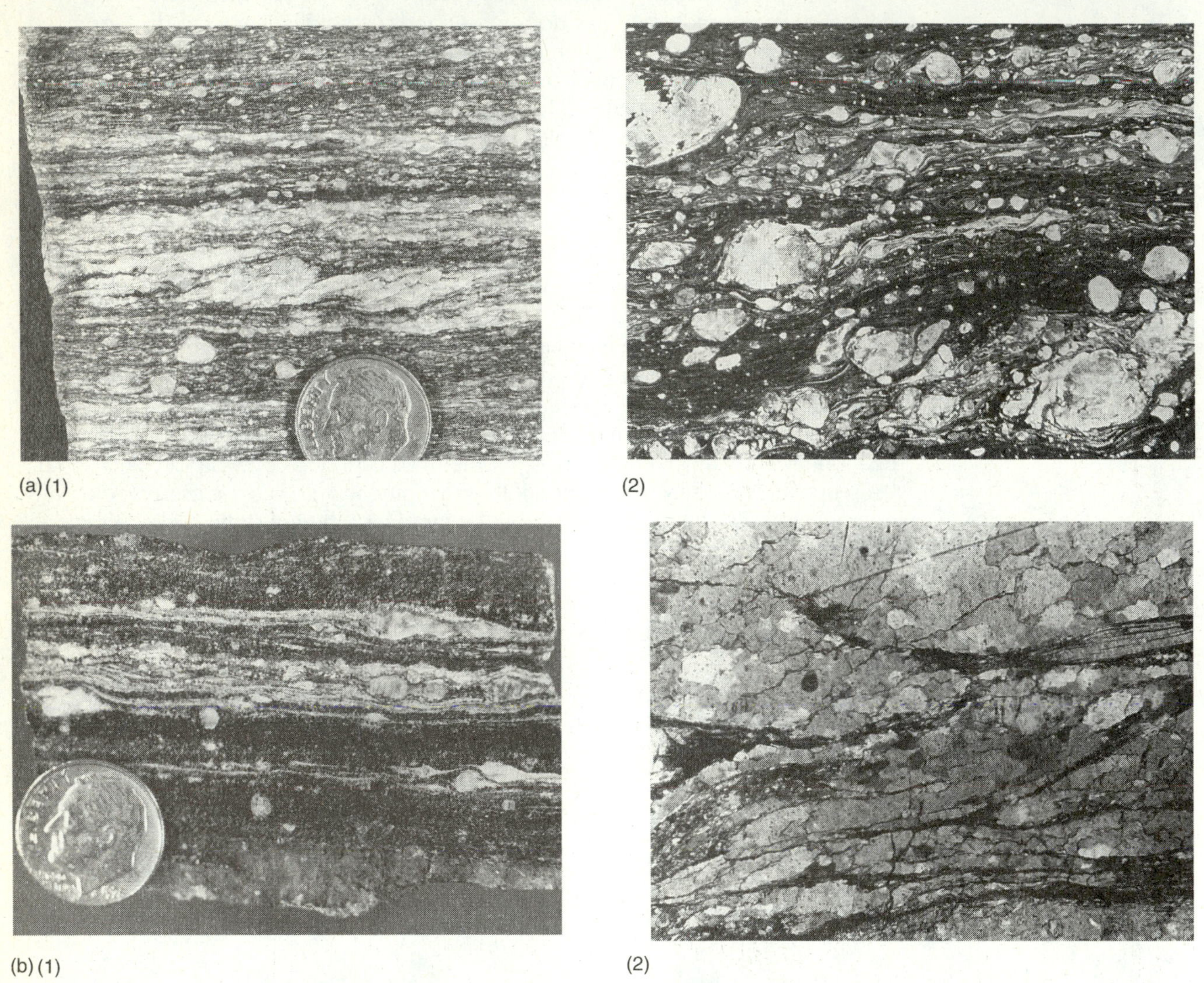

FIGURE 7E–2
(a) (1) Dynamically formed (nonannealed) mylonite from the Bartletts Ferry fault zone north of Auburn, Alabama. Note the S-C fabric that indicates dextral shear sense and the filamentous quartz ribbons. (RDH photo.) (2) Thin section of nonannealed mylonite from the Carthage-Colton mylonite zone, Adirondack Lowlands, New York. Field is approximately 2 cm wide. Plane light. (Thin section courtesy of Teunis Heyn, British Petroleum Company.)
(b) (1) Mylonite from the Goat Rock fault zone near Forsyth, Georgia, formed dynamically but statically annealed. The salt-and-pepper texture of the biotite-rich portion of this rock is a megascopic clue that the microstructure is annealed. Feldspars with tails preserve the original mylonitic texture. (RDH photo.) (2) Thin section of an annealed mylonite from the Box Ankle fault in central Georgia. Quartz ribbons are totally internally recrystallized, but the original shape of the ribbons is preserved. Width of field is approximately 3 mm. Partially crossed polars. (RDH photo.)

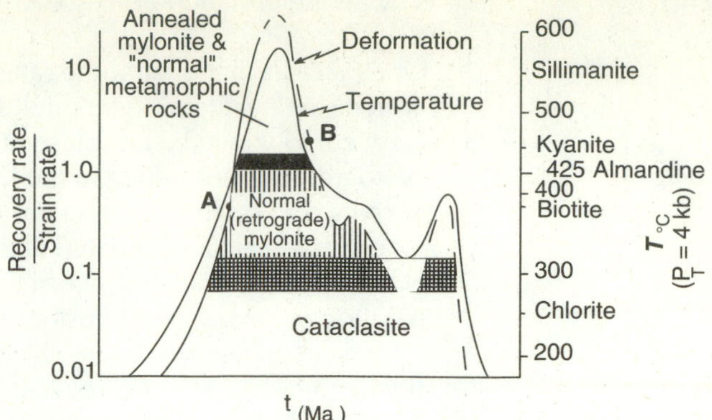

FIGURE 7E–3
Relationships between recrystallization-recovery rate, strain rate, temperature, and total strain and the products of deformation. Point A represents a rock that was deformed under conditions of increasing temperature. Mylonitic texture would probably be annealed following cessation of deformation, unless the rock mass was cooled without increasing temperature further so that the existing textures would be preserved. Point B represents a rock deformed following the thermal peak. Mylonitic (or metamorphic) texture would be preserved, and much of the strain in the rock may not be annealed. Black and cross-hatched shading represent transition zones.

Mylonite and cataclasite commonly occur along faults. In nature, the competing processes (just described) overlap to a great degree, and mylonites should not result from every occurrence of high strain rate. Mylonitic texture may develop in ordinary metamorphic rocks in areas where there are no faults, because of extremely localized development of high shear strain. Rocks in highly strained and thinned fold limbs that are undergoing ductile flow might contain mylonitic texture, because a large amount of strain can be concentrated there. On the other hand, faults that form before the metamorphic-thermal peak in a rock mass will produce mylonite, but the mylonite will probably be overprinted by metamorphism. Mylonite in the fault zone will exhibit recrystallized microtextures common to metamorphic rocks, rather than a mylonitic texture, but will preserve some of the megascopic textures of mylonite (Figure 7E–2b). This relationship is important because few clues remain concerning the origin and nature of these early-formed fault rocks in the internal parts of mountain chains. As a consequence, other criteria, such as stratigraphic data (Chapter 9), must be used to distinguish between stratigraphic and fault contacts separating rock bodies.

To summarize, products of strain—including strain recovery and recrystallization—and the competing processes that form cataclasites, mylonites, and normal metamorphic rocks are all gradational. They involve a delicate balance between the thermal regime of the rock mass, pressures and fluids within the rock body, the strain rate, and the timing relative to the thermal peak.

References Cited

Bell, T. H., and Etheridge, M. A., 1973, Microstructure of mylonites and their descriptive terminology: Lithos, v. 6, p. 337–348.

Hatcher, R. D., Jr., 1978, Eastern Piedmont fault system: Reply: Geology, v. 6, p. 580–582.

Hatcher, R. D., Jr., and Hooper, R. J., 1981, Controls of mylonitization processes: Relationships to orogenic thermal/metamorphic peaks: Geological Society of America Abstracts with Programs, v. 13, p. 469.

Wise, D. U., Dunn, D. E., Engelder, J. T., Geiser, P. A., Hatcher, R. D., Jr., Kish, S. A., Odom, A. L., and Schamel, S., 1984, Fault-related rocks: Suggestions for terminology: Geology, v. 12, p. 391–394.

Questions

1. Why are deformation processes in metals not good models for rock deformation?
2. Of all factors that control the rates and thresholds of deformation processes, which single factor do you think is most important? Why?
3. How could you tell a cataclasite from a mylonite in the field? With the aid of a petrographic microscope in the laboratory?
4. Given a thin section of a metamorphic rock, how could you determine the extent to which accumulated elastic strain in the rock has been relieved?
5. How would you determine whether a rock has been deformed by pressure solution or by volume-diffusion creep?
6. What is the difference between subgrain formation and formation of recrystallized grains in a strained rock?
7. Describe the sequence of deformation mechanisms that would occur with increasing temperature, in the presence of water, in an impure sandstone under stress from 100 to 450° C.

Further Reading

Groshong, R. H., Jr., 1988, Low-temperature deformation mechanisms and their interpretation: Geological Society of America Bulletin, v. 100, p. 1329–1360.
Reviews pressure solution and other low-temperature deformation mechanisms, and discusses how these are recognized in natural rocks.

Hirth, G., and Tullis, J., 1992, Dislocation creep regimes in quartz aggregates: Journal of Structural Geology, v. 14, p. 145–159.
Discusses results of experimental deformation of orthoquartzite and novaculite samples under varied temperatures and strain rates. They recognized three regimes in which dislocation creep and recrystallization processes are affected, presenting well-illustrated understandable examples of each.

Hull, Derek, 1975, Introduction to dislocations: New York, Pergamon, 271 p.
Defects and dislocations in crystals are explained clearly with good illustrations and without a great deal of mathematics.

Kerrich, R., 1978, An historical review and synthesis of research on pressure solution: Zentralblatt für Geologie und Paleontologie, Part 1, p. 512–550.
A useful review of our knowledge of pressure solution, from its earliest recognition in the nineteenth century to modern concepts of how it works.

Kerrich, R., and Allison, I., 1978, Flow mechanisms in rocks: Geoscience Canada, v. 5, p. 109–119.
A concise summary of rock-deformation processes with good examples.

Lloyd, G. E., and Knipe, R. J., 1992, Deformation mechanisms accommodating faulting of quartzite under upper crustal conditions: Journal of Structural Geology, v. 14, p. 127–143.
Discusses, with well-illustrated examples, the deformation mechanisms that accompany faulting and low-temperature deformation of pure quartzite.

Means, W. D., 1986, Three microstructural exercises for students: Journal of Geological Education, v. 34, p. 224–230.

Means, W. D., 1989, Synkinematic microscopy of transparent crystals: Journal of Structural Geology, v. 11, p. 163–174.
These two papers present techniques and well-illustrated examples for observing a variety of microdeformational processes using a petrographic microscope, simple glass-slide apparatus, and organic compounds that deform at room temperature. Means has shown that a number of static and dynamic recrystallization and other deformational processes can be modeled using this technique.

Means, W. D., 1990, Review paper: Kinematics, stress, deformation, and material behavior: Journal of Structural Geology, v. 12, p. 953–971.
Reviews basic continuum mechanics and discusses applications to deformation processes and material behavior. Mathematics is used, but at a level understandable to junior-senior geology students.

Means, W. D., 1993, Elementary geometry of deformation processes: Journal of Structural Geology, v. 15, p. 343–349.
Presents a classification that differentiates between deformation mechanisms and processes that will retain some "memory" of movement history and those that will not.

Nicolas, A., and Poirier, J. P., 1976, Crystalline plasticity and solid-state flow in metamorphic rocks: New York, John Wiley & Sons, 444 p.
A rigorous book providing a comprehensive survey of plastic-deformation processes affecting rocks.

Schedl, A., and van der Pluijm, B. A., 1988, A review of deformation microstructures: Journal of Geological Education, v. 36, p. 111–120.
Reviews the different grain-scale deformation processes, subgrain formation, recovery and recrystallization processes, and some shear zone terminology.

Schmid, S., 1983, Microfabric studies as indicators of deformation mechanisms and flow laws operative in mountain building, *in* Hsü, K. J., ed., Mountain building processes: New York, Academic Press, p. 95–110.
Discusses deformation processes that can be observed on a microscopic scale and how they are related to larger-scale structural features commonly observed in mountain chains.

Tullis, J., and Yund, R. A., 1985, Dynamic recrystallization of feldspar: Geology, v. 13, p. 238–241.

Tullis, J., and Yund, R. A., 1987, Transition from cataclastic flow to dislocation creep of feldspar: Mechanisms and microstructures: Geology, v. 15, p. 606–609.
These papers clearly distinguish between dynamic recrystallization and recovery processes and the deformation mechanisms that occur in feldspar under various physical conditions.

Wise, D. U., Dunn, D. E., Engelder, J. T., Geiser, P. A., Hatcher, R. D., Jr., Kish, S. A., Odom, A. L., and Schamel, S., 1984, Fault-related rocks: Suggestions for terminology: Geology, v. 12, p. 391–394.
Undertakes to clarify the confused terminology related to fault rocks and the processes by which both ductile and brittle faults produce a variety of rock types. Discusses the differences and similarities between cataclasites and mylonites and their relationships to metamorphic rocks.

PART THREE

Fractures and Faults

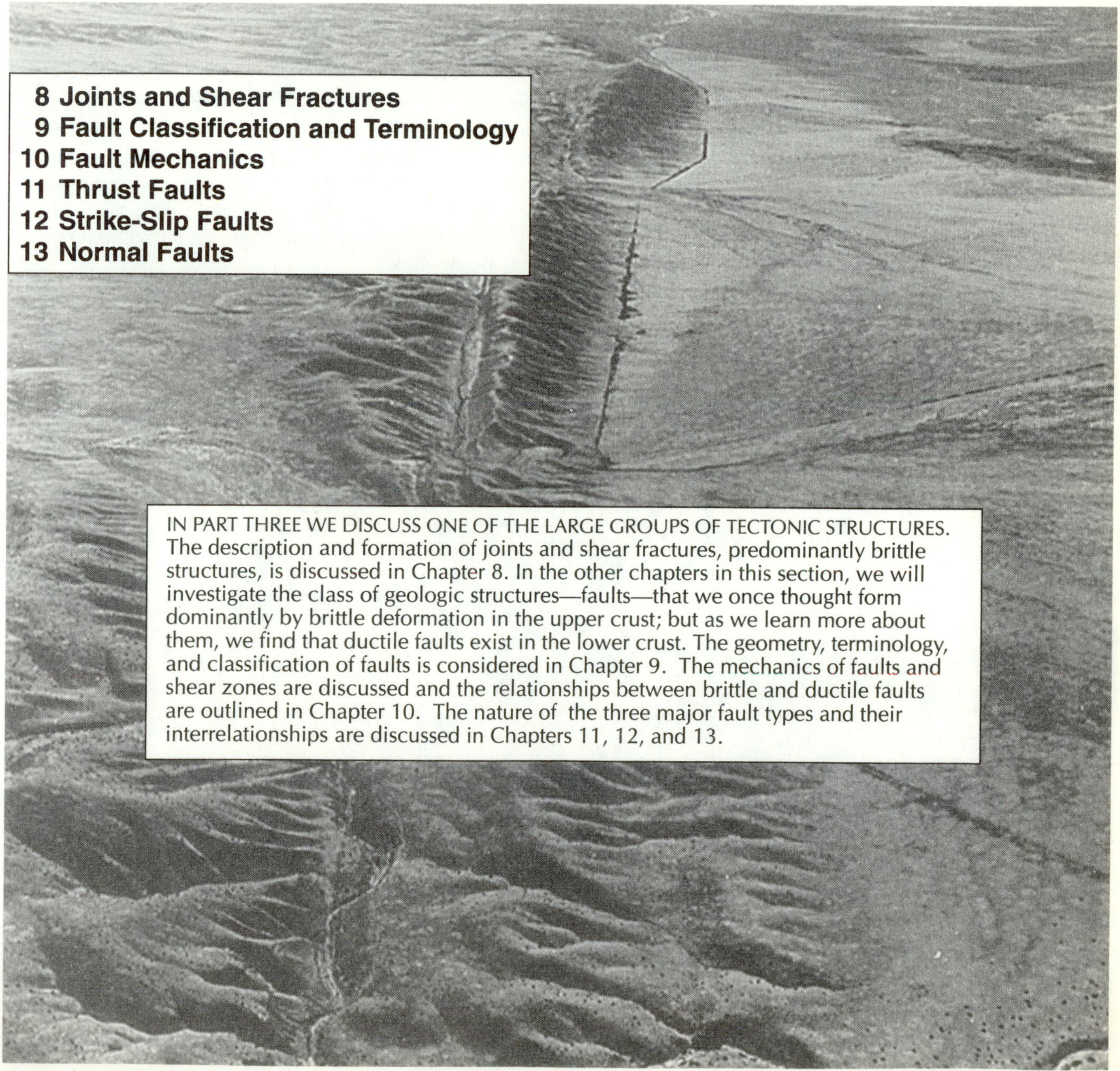

IN PART THREE WE DISCUSS ONE OF THE LARGE GROUPS OF TECTONIC STRUCTURES. The description and formation of joints and shear fractures, predominantly brittle structures, is discussed in Chapter 8. In the other chapters in this section, we will investigate the class of geologic structures—faults—that we once thought form dominantly by brittle deformation in the upper crust; but as we learn more about them, we find that ductile faults exist in the lower crust. The geometry, terminology, and classification of faults is considered in Chapter 9. The mechanics of faults and shear zones are discussed and the relationships between brittle and ductile faults are outlined in Chapter 10. The nature of the three major fault types and their interrelationships are discussed in Chapters 11, 12, and 13.

8

Joints and Shear Fractures

Joints are the most ubiquitous structure in the Earth's crust, occurring in a wide variety of rock types and tectonic environments. . . . They control the physiography of many spectacular landforms and play an important role in the transport of fluids. . . . Today there is no doubt that joints reveal rock strain accommodated by brittle fracture. Establishment of reliable relationships between joints and their causes can provide the structural geologist with important tools for inferring the state of stress and the mechanical behavior of rock.

D. D. POLLARD and ATILLA AYDIN, 1988, Geological Society of America *Bulletin*

FRACTURES ALONG WHICH THERE HAS BEEN NO APPRECIABLE displacement parallel to the fracture and only slight movement normal to the fracture plane are ***joints*** (Figure 8–1). This kind of fracture is extensional, with the fracture plane oriented parallel to σ_1 and σ_2 and perpendicular to σ_3. As such, joints form in a plane not subject to shear. Joints are the most common of all structures, present in all settings in all kinds of rocks as well as in partly consolidated to unconsolidated sediment. For example, joints occur in unconsolidated glacial sediments in New England and in unconsolidated sediments of the Gulf and Atlantic Coastal Plains. Fractures also form in association with faults and other shear zones. Many of these are extensional, and therefore are joints, but some are shear fractures (Figure 8–2). The kind of fracture that forms depends on the orientation and magnitudes of the principal stresses at the instant the fracture forms and on the mechanical properties of the rock (coefficient of internal friction, anisotropy, etc.).

Three types of fractures have been identified, with each type formed by a separate kind of motion (Figure 8–2a). Mode I fractures are joints formed by opening the fracture; mode II fractures form by sliding; and mode III fractures form by a tearing motion. Mode II and III fractures are ***shear fractures***, loose analogs of

FIGURE 8–1
Several orientations—sets—of near-vertical joints produced this irregular topography in flat-lying sandstone, Canyonlands National Park, Utah. (L. P. Arnsbarger, National Park Service, and S. W. Lohman, U.S. Geological Survey.)

strike-slip and dip-slip faults (Chapter 9) and more exact analogs of edge and dislocations. The same fracture can exhibit both mode II and III behavior in different parts of the slipped region; parts may even exhibit mixed-mode behavior (Figure 8–2b).

Study of joints and shear fractures is useful in both pure and applied science. Studying the sequence of fracturing and fracture filling in rocks to understand structures bracketed by fracturing events is applied directly in documenting the antiquity of deformation that produced fracturing (or lack of it) in the rocks beneath dams, bridges, and power plants (Figure 8–3). Fracture sequences help to explain the nature of brittle deformation and the relationships of joints to faults and folds. That valuable minerals may be found in joints and shear fractures has been known for centuries. Joints serve as the plumbing system for ground-water flow in most areas. They are the only routes by which ground water can move through igneous and metamorphic rocks at an appreciable rate, and they dominate as conduits in well-indurated sedimentary and volcanic rocks. Consequently, fracture porosity (and permeability) produced by joints is important for the water supplies in most areas, except those derived from glacial and stream deposits and the classic regional

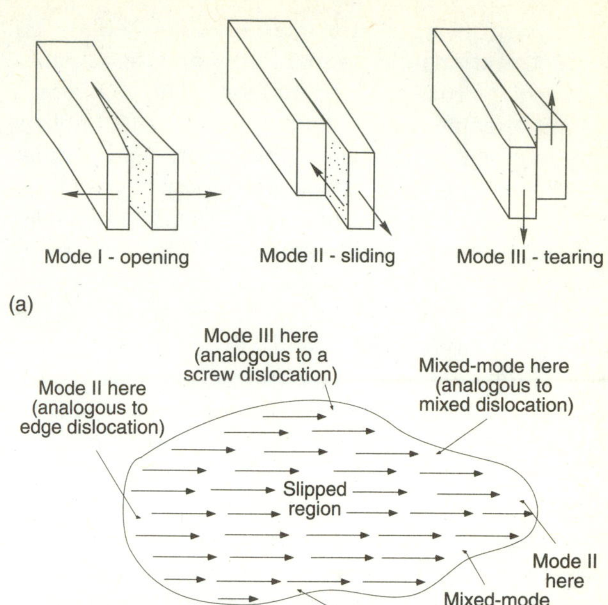

FIGURE 8–2
(a) Three modes of fracture formation. (b) Mode II and III behavior may occur during formation on different parts of the same shear fracture.

FIGURE 8–3
Near-vertical view of intersecting fracture sets exposed in the excavation for the foundation of a power plant in South Carolina. Before the foundation could be constructed, all fractures had to be demonstrated to have been inactive for the most recent 500,000 years and to be incapable of tectonic movement during the future life of the plant. Many of the fractures are filled with minerals that have not been deformed. Scale is indicated by construction workers. (Malcolm F. Schaeffer, Duke Power Company.)

aquifer systems, such as the unconsolidated sediments of the Gulf and Atlantic Coastal Plains and the Great Plains of the North American midcontinent.

Joint orientations in road cuts greatly affect both construction and maintenance. Those oriented parallel to or dip into a highway cut become hazardous during construction and later because they provide potential movement surfaces. Undercutting and oversteepening of slopes during construction may lead to sliding and increased maintenance costs. Rock bolts are frequently employed to stabilize large blocks that cannot be removed; cement grout, rock bolts, or both may be used if the rock is weak or the fracture density high. In mines, subhorizontal joints in the roof must be rock-bolted to prevent collapse.

Joints are planar or irregular surfaces that may be characterized as *systematic* or *nonsystematic*. ***Systematic joints*** have a subparallel orientation and regular spacing. Joints that share a similar orientation in the same area are referred to as a ***joint set.*** Two or more joint sets in the same area constitute a ***joint system.***

Joints that do not share a common orientation and those with highly curved and irregular fracture surfaces are ***nonsystematic joints.*** They occur in most areas but are not easily related to a recognizable stress field. Occasionally, it can be shown that systematic and nonsystematic joints formed at the same time, but nonsystematic joints usually terminate at systematic joints—indicating that the nonsystematic joints formed later (Figure 8–4).

Systematic joints may be *unfilled;* that is, the fractures may be open and devoid of minerals. Generally, they are the most recently formed fractures in an area, and their surfaces may be very smooth. Some joint surfaces are highly irregular; others are marked by low concentric ridges, and those with a feathered texture are termed ***plumose joints*** (Figure 8–5).

Veins are *filled joints* (and shear fractures; Figure 8–6), and the fillings range in composition from feldspar and quartz (high-temperature pegmatite and aplite veins) to quartz, calcite and dolomite, adularia, chlorite, and epidote (and combinations) as well as ore minerals (lower-temperature). Fractures may also be filled with combinations of zeolites, calcite, and other low-temperature minerals such as prehnite. Unraveling a sequence of cross-cutting relationships as well as the compositions of fillings, or lack thereof, and the orientations of fractures may lead to resolution of the chronology of brittle deformation.

Both filled and unfilled fractures may occur in ***conjugate*** (paired) systems. For paired sets to be conjugates, they must form at nearly the same time. Conjugate joint sets may be produced by tension or shear. Many intersect at an acute angle and are thus shear fractures. It is difficult to prove the conjugate nature of shear fractures unless intersections at outcrop scale emphasize their relationships to each other and to associated structures such as folds or faults (Figure 8–7). Many intersecting fracture sets are first taken to be conjugates, but detailed resolution of their movement histories shows that they formed at different times as joints. If conjugate shears can be proved, the acute angle will be bisected by σ_1 (and the *Z* axis of the strain ellipsoid) at the time of formation of the fractures. Curved fractures occur frequently and may be caused by textural or compositional differences within a thick bed or larger rock mass (Kulander and others, 1979) or by changes in stress direction or intensity. Many

(a)

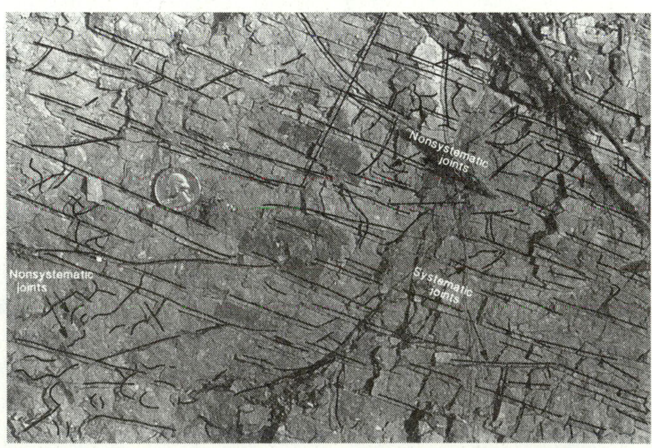

(b)

FIGURE 8–4
(a) Systematic and nonsystematic joints in Middle Cambrian Conasauga Shale, Oak Ridge, Tennessee. Note that systematic joints are linear and form parallel sets; nonsystematic joints are irregular. (RDH photo.) (b) Line drawing showing relationships in (a).

FIGURE 8–5
Plumose joints in Devonian siltstone near Watkins Glen, New York. Note the plumose structure on both layers in the center of the photo that indicates the fracture propagated right to left. (W. H. Bradley, U.S. Geological Survey.)

FIGURE 8–6
Several crossing sets of veins—joint sets later filled with calcite—in limestone at Highgate Springs, Vermont, probably formed at different times. The S-shaped veins may have had a component of shear, along with tension. (C. D. Walcott, U.S. Geological Survey.)

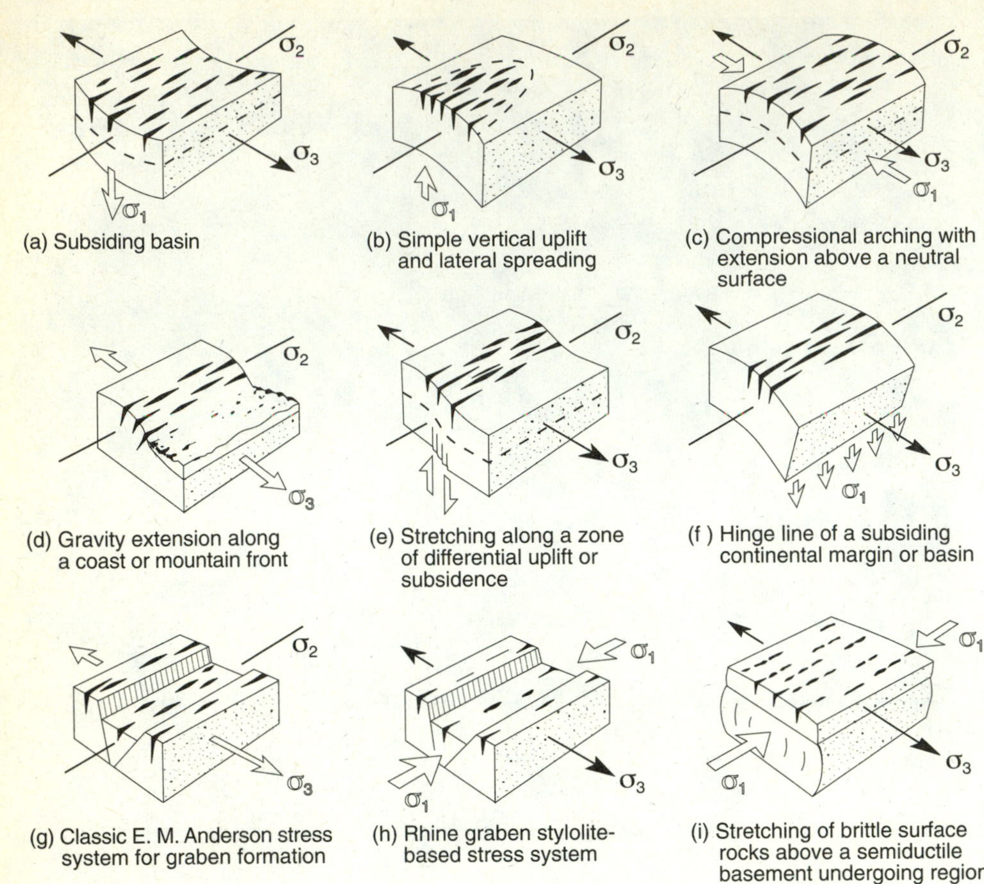

FIGURE 8–7
Possible joint orientations related to principal stress axes and larger structures. Note that joints form perpendicular to σ_3, which is tensile. (From D. U. Wise, R. Funiciello, P. Maurizio, and F. Salvini, 1985, Geological Society of America *Bulletin*, v. 96.)

systematic joints may be curved on the mesoscopic scale, but most are planar regionally, although they, too, may be curved regionally.

FRACTURE ANALYSIS

Study of joints in an area reveals the sequence and timing of formation—information with both pure and applied scientific value. Fracture studies in the field provide information on the timing and geometry of brittle deformation of the crust and the way fractures propagate through rocks. Laboratory study of fractures in rocks, artificial materials, and glacial ice enables comparison with natural fractures and leads to an understanding of the mechanism of brittle failure.

Significance of Orientation

Study of orientations of systematic fractures provides information about the orientation of one (or more) principal stress directions involved in brittle deformation.

Regional joint-orientation patterns may be determined by measuring strike and dip (Appendix 2) of mesoscopic-scale joints over a wide area. A good estimate of regional joint orientation is sometimes obtained by measuring the strike of linear stream segments on satellite imagery, topographic maps, or aerial photos (Figure 8–8). Data gathered either on the ground or from maps and photos related to orientation and spacing of lineaments may be analyzed to help us to understand the relationships between joints and their influence on drainage development and development of other topographic features. Data may be plotted using an equal-area net (Appendix 1) or rose diagrams (Figure 8–9), although the latter show only the strike and frequency of orientations. Rose diagrams are as useful as equal-area plots for joints in most areas because most joint sets dip steeply. Shallow-dipping sets are better shown on equal-area plots. Joint sets may also be analyzed statistically.

Studies of joint and fracture orientations from LANDSAT and other satellite imagery and photography have a variety of structural, geomorphic, and engineering applications. As one example, Wise and others (1985) have used LANDSAT data to interpret relationships between fractures and the Alpine tectonics of Italy

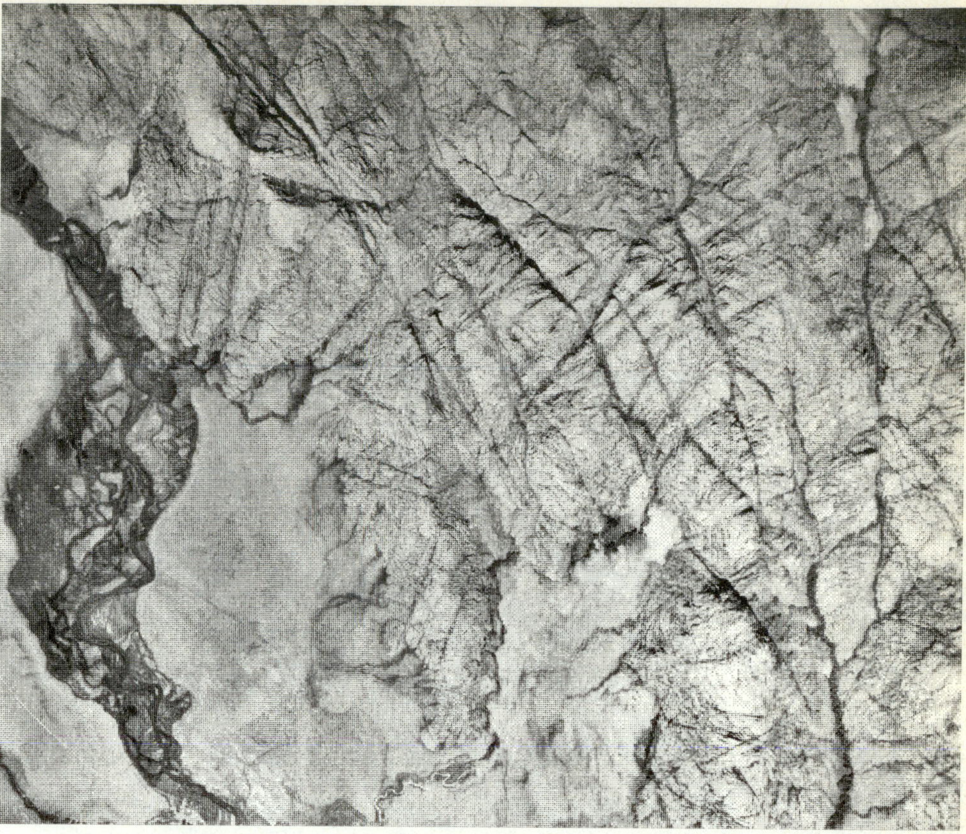

FIGURE 8–8
Aerial photograph of a highly fractured rock mass showing a joint system consisting of several crossing joint sets in Precambrian granitic rocks, Wyoming. (U.S. Geological Survey.)

(Figures 8–7 and 8–10). They were able to improve delineation of the boundaries of widely known structural features and, because the region is tectonically active, to draw conclusions about the orientation of the present-day stress field and the interrelationships of major structural features.

Strain-ellipsoid analysis of joints in an area may help to determine dominant crustal extension directions, keeping in mind that joints form in relation to the prevailing stress field. First, the fractures must be shown independently to be joints by finding evidence of extension across one or more sets. If it can be shown that two or more crossing shear fractures formed at the same time, and a third joint set bisects the acute angle between the shear planes, the joint set may be related to the *YZ* extension plane (Figure 8–11).

Fracture Formation in the Present-Day Stress Field

It has been suggested that joint orientation in the sedimentary cover may be controlled by stress in the crystalline basement beneath. Joints may propagate from older crystalline basement rocks into overlying, previously unfractured sedimentary rocks (Hodgson, 1961). The mechanism is poorly understood, but

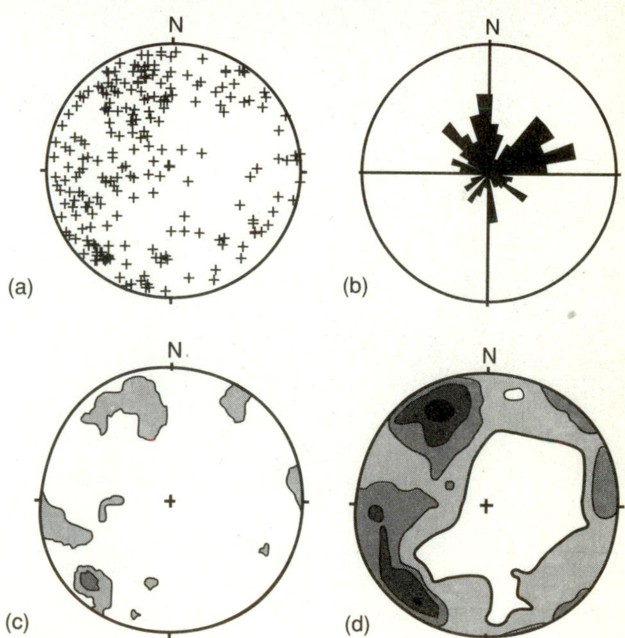

FIGURE 8–9
Lower-hemisphere, equal-area plots and rose diagrams of orientations of joints in the Blue Ridge Foothills near Pigeon Forge, Tennessee. (a) Point diagram of 256 poles to joints. (b) Rose diagram of strikes of joints. The circle diameter represents 10% of the population. (c) Percent per 1 percent area contour diagram of (a). (d) Kamb contour diagram of (a). Contour interval 2 sigma. (See Appendix 1 for plotting and contouring techniques.)

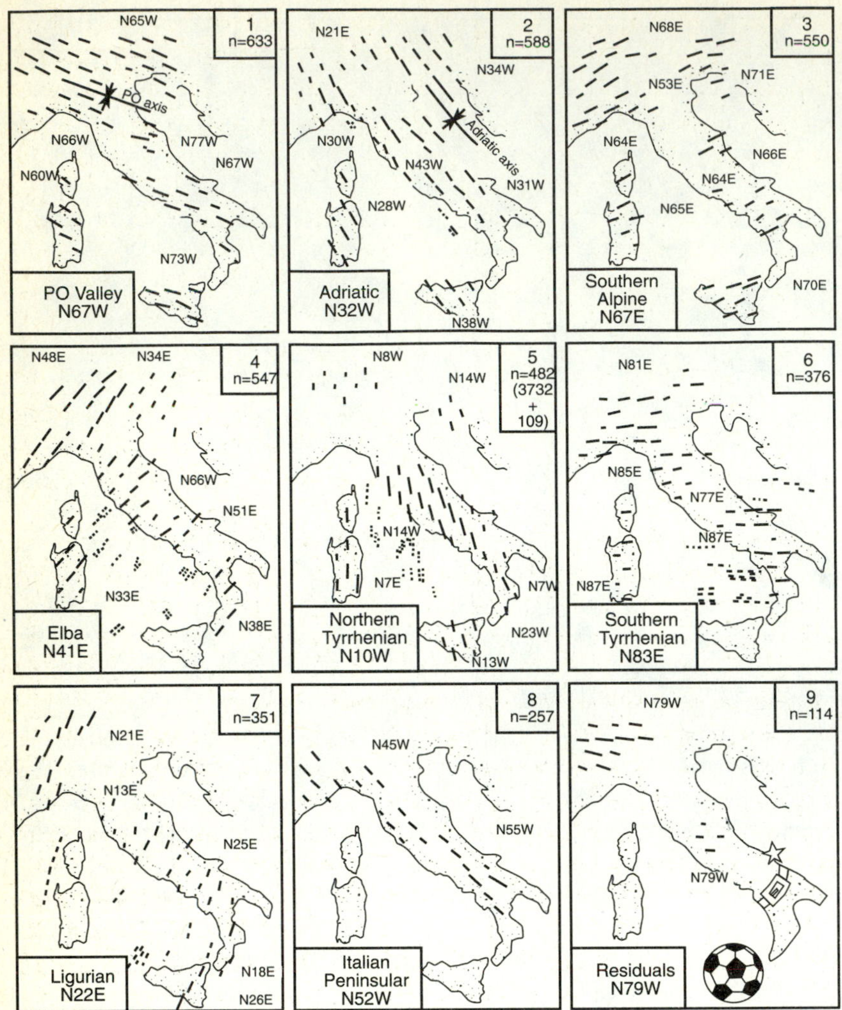

FIGURE 8–10
Fracture and lineament orientations in Italy. Frame 1 is oldest, 9 is youngest. (From D. U. Wise, R. Funiciello, P. Maurizio, and F. Salvini, 1985, Geological Society of America *Bulletin,* v. 96.)

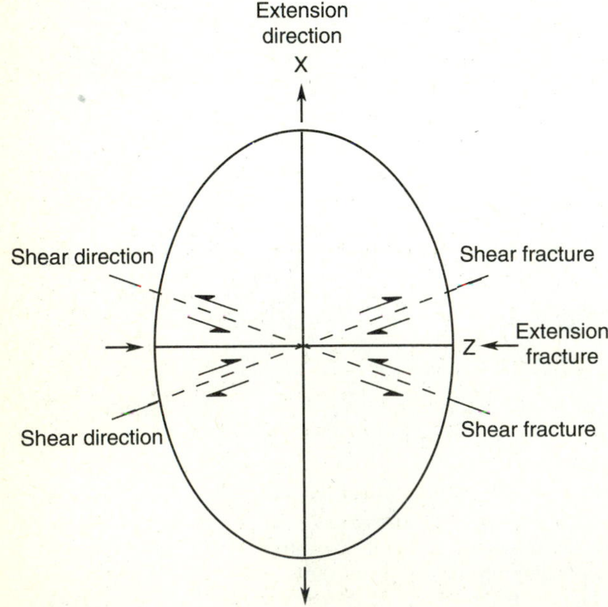

FIGURE 8–11
Jointing related to the strain ellipsoid.

propagation into cover rocks may occur during reactivation of basement fractures—provided the basement fractures and stresses are suitably oriented.

Engelder (1982) has hypothesized that some regional joint sets in eastern North America are a product of the present-day stress field—*not* inherited patterns from older crust. The joint sets thought to have formed in the contemporary stress field are those *not* related to any older stress regime. After considering their orientations, surface markings (such as smooth or plumose), and information about stress orientation derived from hydraulic-fracturing measurements and focal-mechanism solutions, Engelder concluded that joints form normal to present-day σ_3 (Figure 3E–3). Hancock and Engelder (1989) have investigated neotectonic joints in a variety of settings, including relatively young rocks. They concluded that neotectonic joints form in the upper 0.5 km of the crust where the stress difference is low, but where effective σ_3 is tensile and horizontal, and that these late-formed joints can in turn influence the orientation of the contemporary stress field in an area.

Fold- and Fault-Related Joints

Joints may form during brittle folding. They may form normal, parallel, and oblique to the fold axis and axial surface, depending on stress conditions (Figure 8–12).

In one practical application, Richard Nickelsen (1979) made a detailed study of a well-exposed series of deformed Pennsylvanian sandstone, coal, and shale layers at the Bear Valley strip mine in Pennsylvania. He was able to separate early jointing events from later faulting, cleavage formation, folding, later folding that produced new joint sets, and finally, more faulting, which also produced new joints.

Joints frequently form adjacent to brittle faults. Movement along faults commonly produces a series of systematic fractures in which spacing decreases closer to the fault zone, and the number of sets also increases. Most are probably joints, but some form as shear fractures.

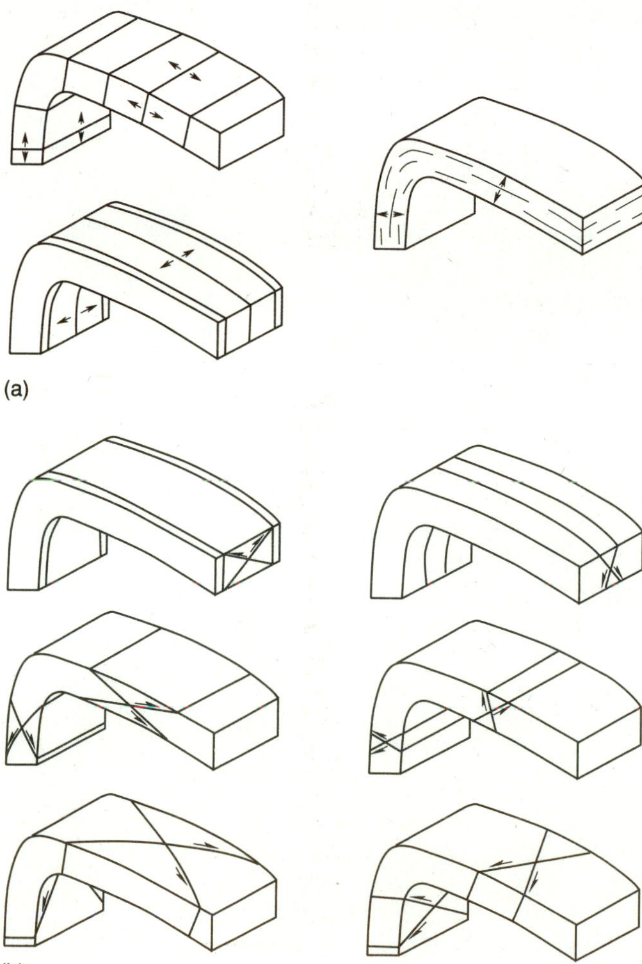

FIGURE 8–12
Fold-related joints and shear fractures. (a) Joints forming at different orientations in a fold. (b) Six possible shear-fracture orientations in a fold. (From *Journal of Structural Geology,* v. 7, P. L. Hancock, p. 437–457, © 1985, with kind permission from Elsevier Sciences, Ltd., Kidlington, United Kingdom.)

FORMATION OF FRACTURES: GRIFFITH THEORY

We can easily observe ruptures in the Earth's crust in the form of faults and joints. Many faults (Chapter 9) are traceable for hundreds of kilometers; joint sets are easily traced for tens of kilometers using the techniques of lineament analysis, along with measurement of joint orientations on the ground. For years, geologists have debated the origin of joints and other fractures in rocks. Much of the evidence favors origin of joints normal to the least principal stress, because many joints display no offset, except normal to the fracture surface, and thus should be related to a tensional stress field. Jointing and formation of shear fractures may be geometrically related by resolution of the same stress tensor (and brittle deformation) into normal and shear components, particularly in some folds and near some fault surfaces. This indicates that tensional stress may be present in rock masses undergoing simultaneous compression and shear. What is the ultimate source of these fractures? How do they begin? In 1920, Alan Arnold Griffith, an English engineer, suggested that elliptical microscopic cracks exist in glass (Figure 8–13)

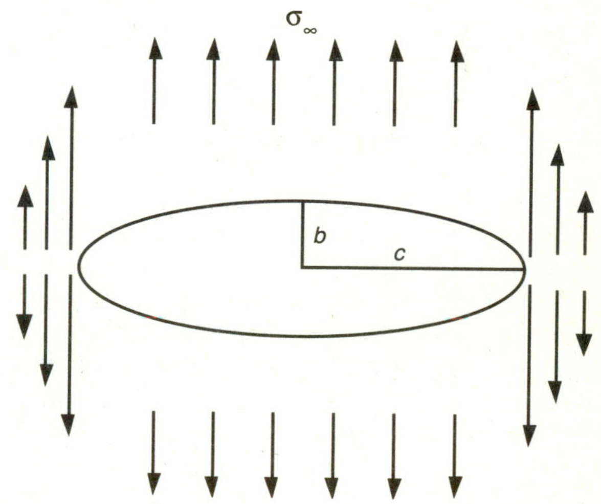

FIGURE 8-13
Stress distribution around an elliptical hole in a plate composed of a homogeneous material. The elliptical hole has semiminor and semimajor axes *b* and *c*, respectively. Arrows and length of arrows indicate orientation and local magnitudes of uniform stress, σ_∞, located far from the hole. (Modified from Figure 1.2(b) *in* C. H. Scholz, 1990, *The Mechanics of Earthquakes and Faulting,* Cambridge University Press.)

and hypothesized that stress is concentrated and magnified by several orders of magnitude at crack tips. This hypothesis may be originally attributable to Inglis (1913), with Griffith having built on these ideas and derived the energy balance relationships that determine whether or not a crack propagates. The following is summarized from a more detailed treatment of Griffith's theory in Scholz (1990). If an elliptical crack with semiaxes *b* and c (with $c >> b$, and the *c* axis of the elliptical hole perpendicular to one of the principal stress directions) occurs in a homogeneous solid, the stress concentrations at the ends of the crack increase in direct proportion to *c*/*b* as

$$\sigma \approx \sigma_\infty \left[1 + 2\left(\frac{c}{b}\right) \right] \quad \textbf{(8–1)}$$

or

$$\sigma \approx \sigma_\infty \left[1 + 2\left(\frac{c}{\rho}\right)^{\frac{1}{2}} \right] \approx 2\sigma_\infty \left(\frac{c}{\rho}\right)^{\frac{1}{2}}, \quad \textbf{(8–2)}$$

where ρ is the radius of curvature of the crack and σ_∞ is the remote principal stress that acts perpendicular to the *c* axis of the ellipse.

Griffith (1920) employed an energy-balance approach for crack propagation and assumed that the body containing the crack is perfectly elastic, the crack length is $2c$, and the crack is loaded by external forces. If the crack length exceeds $2c$ by an increment δc, the work, *W*, will be done by external forces and will result in a change in internal strain energy, U_e. Energy will be expended by creating the new crack surface, U_s. The total energy, *U*, for a static crack will be

$$U = (-W + U_e) + U_s . \quad \textbf{(8–3)}$$

If the effects of the cohesive strength between incremental extension surfaces δc are removed, crack propagation will proceed until a lower energy configuration is achieved. Surface energy increases as the crack grows because work must be done against cohesive forces to increase surface area. This results in lengthening (propagation) of the crack until this energy is expended.

Most fracture-forming environments are below the water table, and so an increment attributable to fluid pressure should also be considered rather than thinking of the fracture-forming process as taking place in totally dry rock. Donald Secor (1969) showed that work is also done on the walls of a crack by fluid pressure. If the dominant source of energy to drive crack propagation is the work done as a result of increased fluid pressure (hydraulic fracturing) during crack inflation, elastic strain energy will be stored around the inflating crack. When the crack propagates, energy is thus available for creation of new crack surface from both the stored elastic energy and the work done on the walls of the crack by the pressurized fluid as the crack opens incrementally during propagation. The available energy is usually a small fraction of the total surface energy of the crack, and so the crack must also propagate in small increments between times when the crack is being inflated (elastically) by flow of more fluid into it. At equilibrium, an expression for critical stress for crack propagation σ_f is

$$\sigma_f = \left(2E\ \gamma/\pi c\right)^{\frac{1}{2}}, \quad \textbf{(8–4)}$$

where *E* is Young's modulus and γ is the specific surface energy of the crack. If we consider the process of crack generation as one producing a set of multiple regularly spaced cracks, the theoretical tensile strength, σ_t, of a crack (actually is the strength of rock, because a crack cannot have any strength) can be expressed as

$$\sigma_t = \frac{E\lambda}{2\pi a} \quad \textbf{(8–5)}$$

and

$$2\gamma = \frac{E\lambda\sigma_1}{\pi}, \quad \textbf{(8–6)}$$

where λ is the wavelength of the sine wave stress (atomic forces) variation with separation of atoms by the crack, and *a* is the equilibrium crack spacing ($\lambda \approx a$). Combining equations (8–5) and (8–6),

$$\sigma_t = \left(\frac{E\gamma}{a}\right)^{\frac{1}{2}}. \quad \textbf{(8–7)}$$

The theoretical tensile strength will exist at the ends of a crack of length $2c$ when the following relationship to critical stress exists:

$$\sigma_t = 2\sigma_f \left(\frac{c}{a}\right)^{\frac{1}{2}}. \quad \textbf{(8–8)}$$

This equation relates theoretical crack strength to critical stress at equilibrium σ_f.

Griffith's equations describe behavior of the material under conditions of uniaxial tension, whereas Coulomb's equation for the Mohr envelope (Equation 3–25) describes behavior more exactly for compression (Figure 8–14; Secor, 1965). Griffith's criteria does not apply to all failure conditions because cracks are closed by compression, increasing rather than decreasing rock strength. Moreover, the stresses developed at the crack tips appear independent of crack size and are

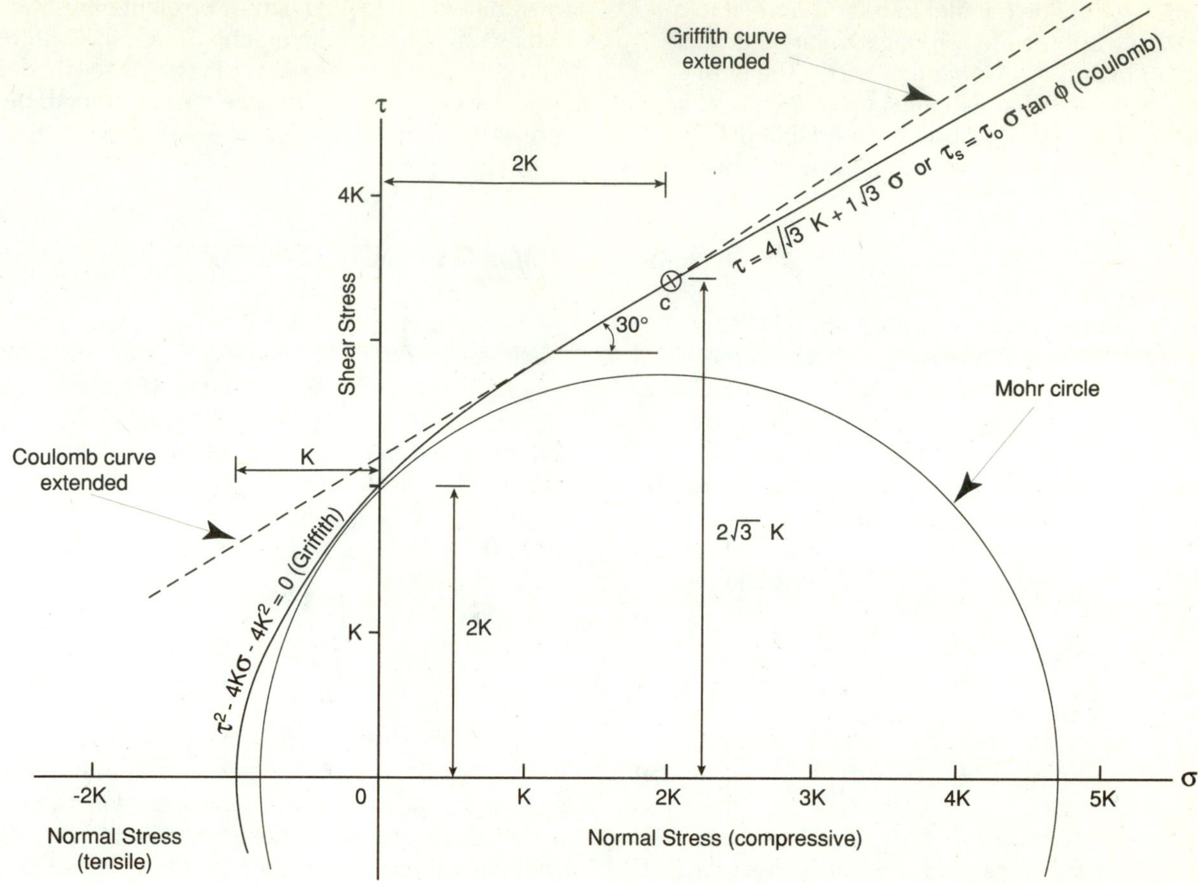

FIGURE 8–14
Mohr envelope combining both the Griffith and Coulomb failure criteria. Griffith's equation works better in the parabolic, mostly tensional region; Coulomb's equation works better in the straight-line part of the envelope. This Mohr diagram is plotted in units of tensile strength. The dashed lines are the projections of lines for the Coulomb and Griffith equations beyond the intersection point c (marked with a circled x). Mohr circle is added for reference. (After D. T. Secor, 1963, Geology of the central Spring Mountains, Nevada, [unpublished Ph.D. dissertation]: Stanford, California, Stanford University.)

determined by the shape of the crack although some have maintained that larger stress concentrations are a function of only crack length (Suppe, 1985). They also predict the orientation of the cracks (Jaeger and Cook, 1976).

Griffith cracks in glass have since been confirmed with the aid of the electron microscope and have dimensions on the order of 100 µm wide by 1,000 µm long. Their existence in other brittle materials, and the applicability of Griffith theory to propagation of fractures through rocks, have been debated for many years. Griffith cracks are generally accepted for convenience as a feature of rocks in order to apply the theory, despite the anisotropic nature of rocks and the other kinds of cracks that may be present (such as cleavage and grain-boundary segments). The size of Griffith cracks in rocks must be on the order of the maximum grain diameter.

Just how far Griffith theory is applicable to rocks is uncertain, but the theory helps us to understand the effect of the cracks on elastic properties of rock materials, particularly if the theory is applied to large rock bodies where the entire mass can be assumed to behave homogeneously (Jaeger and Cook, 1976).

A homogeneous-stress state exists in a rock mass if no Griffith cracks or discontinuities are present (granted, there may be other conditions that result in a homogeneous stress state). Griffith cracks concentrate stress at crack tips and edges, raising it by several orders of magnitude. The amount of increase depends on the orientation and dimensions of each crack. Larger and suitably oriented cracks will undergo greater stress magnification than smaller and less suitably oriented cracks. Cracks with greater stress magnification will propagate into larger fractures; those with less will remain as microcracks. Those suitably oriented for propagation would be parallel to σ_1 and normal to σ_3, least favorably oriented for propagation are cracks oriented normal to σ_1 and parallel to σ_3. The tensile stress, σ_t, at the tip of an elliptical crack (elliptical in the plane of the crack) in a brittle material under tension σ_∞, perpendicular to the crack, would be

$$\sigma_t \cong \frac{2}{3}\sigma_\infty \frac{(2c)^2}{2b}, \quad \textbf{(8–9)}$$

where $2c$ is the length of the crack and $2b$ is the width (Suppe, 1985). Solving this equation predicts that crack propagation may occur at low stress because amplification results from the relationship between length and width. For example, if the ratio of length to width of the crack is 10 to 1, stress amplification is about 66. If the ratio is 40 to 1, stress amplification is about 1,067.

Griffith theory predicts that, for principal stresses σ_1 and σ_3, failure will occur under extension in one direction if

$$(\sigma_1 - \sigma_3)^2 = 8K_0(\sigma_1 + \sigma_3)$$

or

$$(\sigma_1 - \sigma_3)^2 - 8K_0(\sigma_1 + \sigma_3) = 0 \quad \textbf{(8–10)}$$

or, alternatively solving equation 8–10 for the points of intersection of the straight line (Coulomb) and the parabola (Griffith) yields

$$\tau_0^2 = 4K\sigma_n - 4K^2, \quad \textbf{(8–10a)}$$

where K is uniaxial tensile strength and τ_0 is cohesive strength. If $\sigma_1 + 3\sigma_3 > 0$ in equation 8–10, the principal stresses required for failure are given by equation 8–10. If $\sigma_1 + 3\sigma_3 < 0$, failure is independent of the value of σ_1 and occurs when σ_3 equals the uniaxial tensile strength K. Equations 8–10 and 8–10a, along with the qualifiers, make up the *Griffith criterion for failure.* Griffith theory works well to predict the behavior of homogeneous materials under tensional stresses and will predict the shape of the fracture envelope in the tensional field of the Mohr diagram (Figure 8–14), but because there must be no cohesion between surfaces, it does not work well in the compressional field. A modified Griffith theory has been devised that permits fractures to form under compression with the necessity that they overcome friction.

The importance of Griffith theory is that it predicts that *the compressive strength (at zero confining pressure) of most materials will be eight times the tensile strength.* Measured ratios of seven rock samples range from 9 to 17, with an average near 14 (Jaeger and Cook, 1976, their Table 6.15.1). Another important conclusion to be drawn from this discussion is that the stress difference $(\sigma_1 - \sigma_3)$ must be eight times the compressive strength before the Coulomb-Griffith failure envelope in the Mohr diagram is reached in compression, and <4 times the tensile strength before it is reached in tension.

JOINTS AND FRACTURE MECHANICS

Development of the following concepts of fracture toughness and fracture energy are summarized from Scholz (1990). If a crack is planar, well defined, and there is no cohesion between the crack walls, a relationship exists between the stress σ_{ij} at a crack tip and the resulting displacement

$$\sigma_{ij} = K_n(2\pi r)^{\frac{1}{2}} f_{ij}(\theta) \quad \textbf{(8–11)}$$

and

$$U_i = \left(\frac{K_n}{2E}\right)\left(\frac{r}{2\pi}\right)^{\frac{1}{2}} f_i(\theta), \quad \textbf{(8–12)}$$

where r is the distance from the crack tip, θ is the angle measured from the crack plane to the point where the stress is σ_{ij}, and K_n is called the *stress intensity factor* and is specific for each fracture mode (Figure 8–2a), with the designations K_I for Mode I, K_{II} for Mode II, and K_{III} for Mode III. $f_i(\theta)$ expresses θ as a function of U_i, the energy term defined in equation 8–3. The notation σ_{ij} is tensor notation indicating the stress σ_{ij} is a second-rank tensor, and U_i, total energy, is a first-rank tensor; i and j can independently have values of 1, 2, 3, ... Values for K_I, K_{II}, and K_{III} depend on the geometry and magnitude of the applied stress and the determined intensity of the crack-tip stress field (Scholz, 1990). The stress intensity factor may be related to the Griffith energy-balance relationships by

$$\vec{G} = -\frac{d(-W + U_c)}{dc}, \quad \textbf{(8–13)}$$

where $\vec{G}$ is the energy release rate, or crack extension force, and U_c is elastic strain energy. Two different equations are needed for plane stress,

$$\vec{G} = \frac{K^2}{E}, \quad \textbf{(8–14)}$$

and plane strain

$$\vec{G} = \frac{K^2(1 - \nu^2)}{E}, \quad \textbf{(8–15)}$$

where ν is Poisson's ratio. For Mode III, the right side of equations 8–14 and 8–15 have to be multiplied by $(1 + \nu)$ for plane stress, and divided by $(1 - \nu)$ for plane strain. From equation 8–3 and the equilibrium condition $dU/dc = 0$ (recall c is equilibrium crack length), cracks will propagate when

$$\vec{G}_c = \frac{\vec{K}_c^2}{E} = 2\gamma, \tag{8–16}$$

where $\vec{G}_c$ is fracture energy when uniform stresses, σ_{ij}, are applied remotely (far field) from the crack, and $\vec{K}_c$ is the critical stress intensity factor (also called *fracture toughness*). $\vec{K}_c$ and E are both material properties that provide general failure criteria because they can be determined and related experimentally for individual rocks (Scholz, 1990). The relationships for the three modes (Figure 8–15) are

$$\begin{aligned} \vec{K}_I &= \sigma_{yy}(\pi c)^{\frac{1}{2}} \\ \vec{K}_{II} &= \sigma_{xy}(\pi c)^{\frac{1}{2}} \\ \vec{K}_{III} &= \sigma_{zy}(\pi c)^{\frac{1}{2}}, \end{aligned} \tag{8–17}$$

and, from equation (8–15), the crack extension forces for each corresponding mode are

$$\begin{aligned} \vec{G}_I &= (\sigma_{yy})^2 \frac{\pi c}{E} \\ \vec{G}_{II} &= (\tau_{xy})^2 \frac{\pi c}{E} \\ \vec{G}_{III} &= (\tau_{zy})^2 \frac{\pi c(1+\nu)}{E}. \end{aligned} \tag{8–18}$$

For plane strain, it is necessary to replace E with $[E/(1-\nu^2)]$ for both Modes I and II. Stress intensity factors permit useful inferences to be made about the propagation and behavior of joints. Using the expression for $\vec{K}_I$ in equation 8–17 for a Mode I fracture, $\vec{K}_I$ reaches a critical value,

$$\vec{K}_I = \vec{K}_{Ic} \tag{8–19}$$

where $\vec{K}_{Ic}$ is fracture toughness, a natural property, the joint will propagate. The most useful attribute of the stress intensity factor is that it is based on a heterogenous stress field at the crack tip. This property predicts, for example, that if two joints of unequal length exist in the same stress field, the longest will meet the criterion in equation 8–19 first. This property also predicts that a joint subjected to the greatest stress among a set of joints of equal length will propagate first in a stress field that varies with position in a rock mass (Pollard and Aydin, 1988).

Experiments can produce joints by forcing fluid into rock pores to lower the effective normal stress (normal stress minus pore pressure) until the tensile

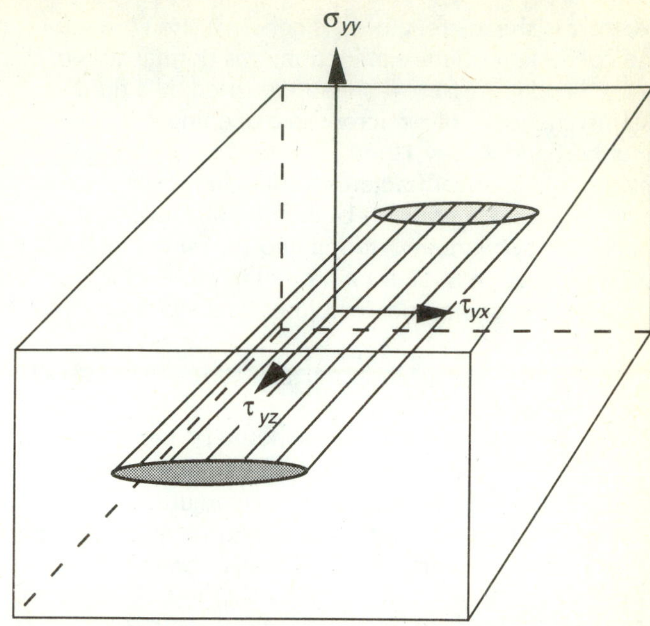

FIGURE 8–15
Resolution of normal (σ_{yy}) and shear (τ_{yx} and τ_{yz}) stress components around an elliptical crack in a plate subjected to a uniform stress field. (Modified from Figure 1.7(b) *in* C. H. Scholz, 1990, *The Mechanics of Earthquakes and Faulting*, Cambridge University Press.)

strength of the rock is exceeded. Theoretical and field studies indicate that fluid probably plays an important role in formation of many joints. Studies by Secor (1965), Pollard and Holzhausen (1979), and Segall and Pollard (1983) indicate that joints form more readily in the presence of fluid, generating fractures by hydraulic fracturing. Secor (1965) considered effective stress ($\sigma - P$, where P is fluid pressure) and concluded that joints can form in a compressive environment if fluid pressure is high. Fluid under pressure (compression) is forced into the fracture and promotes continued fracture propagation. This process occurs in the upper crust where water is abundant and, interestingly, may also be the mechanism for forcible dike emplacement, with magma as the fluid (Odé, 1957; Pollard and Muller, 1976; see Chapter 19). Fractures in rock may begin as Griffith cracks (microcracks in grains) and then coalesce into more continuous oriented fractures to ultimately form joints. One consequence of Griffith theory is the prediction that stress difference for jointing must be less than four times the tensile strength of the rock (Jaeger and Cook, 1976), although fracturing may occur whenever σ_3 is tensile and equal to the uniaxial tensile strength of the rock.

Terzaghi (1923) has suggested that shear resistance of soils (or rock) may be calculated by a modified version of the Coulomb criterion,

$$\tau = \tau_0 + \mu(\sigma - P), \tag{8–20}$$

where τ is shear strength, τ_0 is cohesive strength, μ is the coefficient of internal friction, σ is normal stress, and P is pore pressure—pressure exerted by a fluid against the walls of a microscopic opening in a rock (Hubbert and Rubey 1959). This idea has been applied extensively to shear fractures formed in rocks that contain a fluid (Figures 8–14 and 8–16). This must be taken into account in determinations of stress intensity factor. Equation 8–17 for a Mode I fracture should be modified to accommodate fluid pressure as

$$\vec{K}_I = \left(\sigma_{yy} - P\right)\left(\pi c\right)^{\frac{1}{2}}. \quad \textbf{(8–21)}$$

Estimates of fracture toughness should be modified accordingly, but the nature of predictability and usefulness of the stress intensity factor remain unchanged. While the simple Terzaghi modification of the Coulomb criterion can be readily verified under controlled conditions in the laboratory, in natural systems fluid pressure must be resolved into both vertical and horizontal components. This produces a net decrease in the diameter of the Mohr circle for effective normal stress. Nevertheless, the circle is still shifted toward the tensile field so that an additional increase in pore pressure may lead to failure at values of zero or negative (tensile) effective normal stress (Mandl, 1988).

Most joints form by extensional fracturing of rock in the upper few kilometers of the Earth's crust. The limiting depth for formation of extension fractures ideally should be the ductile-brittle transition (Chapter 6), which is influenced by rock type, fluid pressure, strain rate, and the stress difference at a particular time. High-temperature veins form below the ductile-brittle transition (e.g., pegmatites), and extension fractures commonly accompany ductile deformation in ductile shear zones (Hudleston, 1989; Chapter 10).

Features on joint surfaces, such as plumose structure (Figure 8–5), provide clues to the direction of joint propagation. Other features provide details of the mechanics of formation of extension fractures and information on the rate and direction (Figure 8–17) of propagation of individual joint planes. In Figure 8–17, *hackle marks* indicate zones where the joint propagated rapidly, while the *arrest line* is perpendicular to the direction of propagation and forms parallel to the advancing edge of the fracture (Hodgson, 1961; Kulander and others, 1979). Propagation of a joint always begins at a preexisting flaw in the rock mass, a grain of atypical size or hardness, a fossil, a concretion, a pore space, an irregularity in bedding, or other mechanical discontinuity. The joint may propagate from the flaw under ideal extensional conditions and form a symmetrical plumose surface with symmetrical twist hackle, Wallner lines, and arrest lines. Asymmetric plumose structure, twist hackle, and other structures require stress orientation to change as the joint propagates (Kulander and others, 1979). The explanation is reasonable if the asymmetric plume occurs on a fracture through a single bed and if the beds above and below have symmetrical plumes, but if the same sense of asymmetry occurs on plumes in successive layers, the fracture is likely to be a hybrid shear (mixed-mode fracture) and not a joint (Hancock and Engelder, 1989). Such observations are important to understanding the mechanisms of fracture formation and determining whether a series of fractures consists of joints, shear fractures, or hybrid shears.

Bedding and foliation planes in coarser-grained rocks constitute barriers to joint propagation. Bedding in uniformly fine-grained rocks, such as shales and volcaniclastic rocks, appears to be less of a barrier or none at all. Joints frequently propagate through a

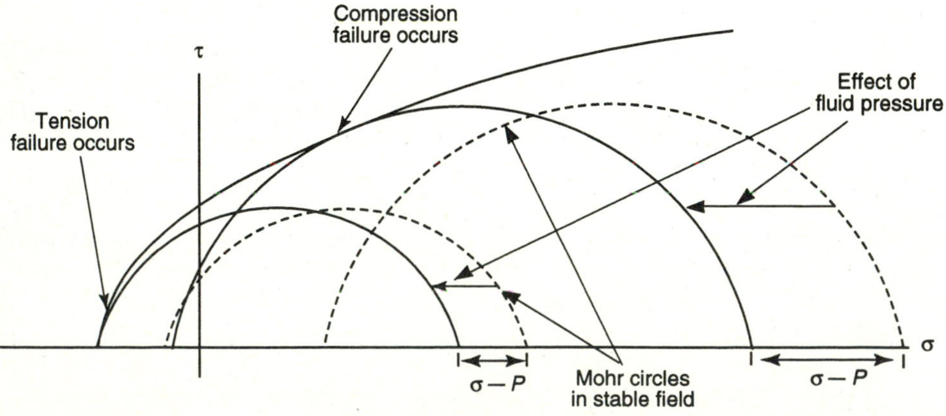

FIGURE 8–16
Mohr circles for stress, showing the effect of a fluid and pore pressure on effective normal stress. A dry material (dashed circles) remains in the stable field in contrast to the same material after adding fluid pressure. Failure would occur whenever the circles become tangent to the envelope. The envelope relative to the fluid-saturated material indicates failure because of decreased strength and effective normal stress.

sandstone bed and are slightly offset from those in the next layer above or below. Variation in bed thickness also affects propagation direction. Differences in stress magnitude between layers produce a barrier to joint propagation through a section of alternating sedimentary rock types. For example, assuming horizontal layering, joints will not propagate from sandstone into shale if the least principal horizontal stress in the shale is greater than that in the sandstone. Fractures will terminate at the contact between the two rock types (Engelder, 1985).

Terry Engelder (1985) characterized both the environment and the mechanisms of joint formation and described four categories of joints as end-member paths for increase of stress (Figure 8–18): *tectonic, hydraulic, unloading,* and *release joints.* All categories involve the assumption (based on many observations) that the failure mechanism is tensile. ***Tectonic*** and ***hydraulic joints*** form at depth in response to abnormal fluid pressure and involve hydrofracturing. Hydraulic joints form during burial and vertical compaction of sediment at depths greater than 5 km, where escape of fluid is hindered by low permeability, which creates locally abnormally high pore pressure. This process is static, whereas tectonic joints form essentially by the same mechanism but the stresses originate tectonically, and horizontal compaction occurs. Tectonic joints may form at depths of less than 3 km.

Unloading and ***release joints*** form near the surface as erosion removes overburden, and thermal-elastic contraction occurs. Unloading joints begin to form when more than half the original overburden has been removed from a rock mass. Contemporary tectonic or remaining ancient stresses may serve to orient these joints. For vertical unloading joints to form, the effective stress in the horizontal plane neces-

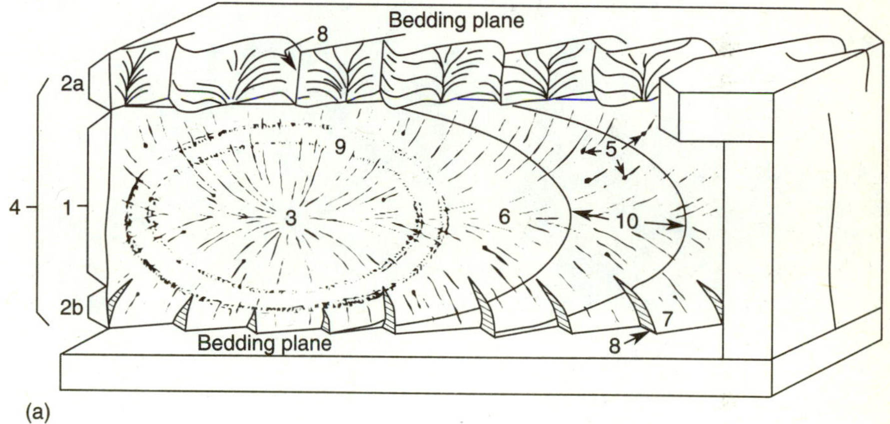

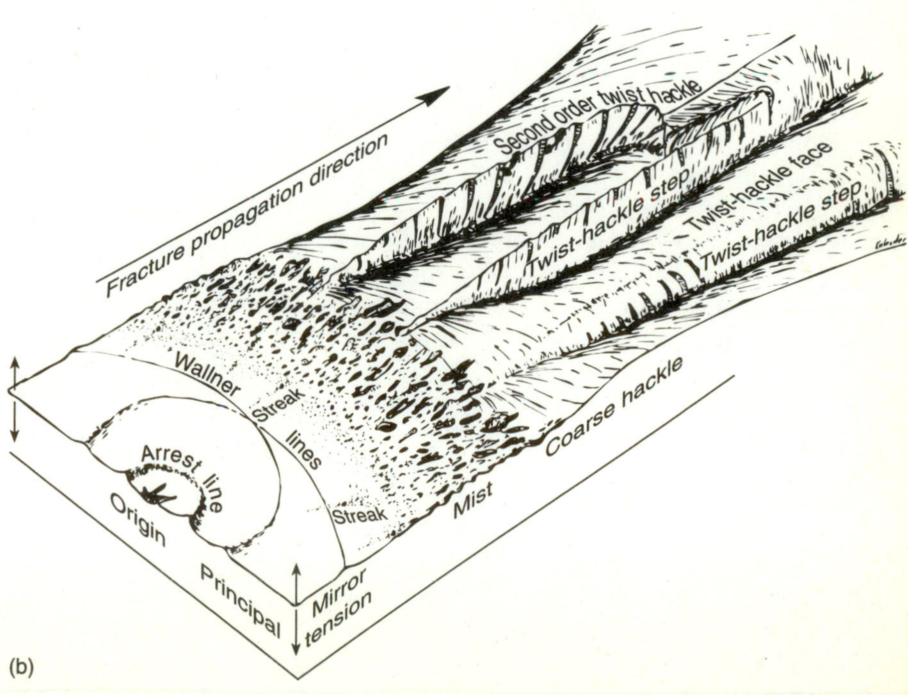

FIGURE 8–17
(a) Detail of plumose joint surface showing primary surface structures: 1. Main joint face. 2. Twist-hackle fringe. 3. Origin. 4. Hackle plume. 5. Inclusion hackle. 6. Plume axis. 7. Twist-hackle face. 8. Twist-hackle step. 9. Arrest lines. 10. Constructed fracture-front lines. (From B. R. Kulander and S. L. Dean, 1985, Proceedings of the International Symposium on Fundamentals of Rock Joints.) (b) Detail showing other features associated with propagation of the joint. (From Kulander, Barton, and Dean, 1979, *The application of fractography to core and outcrop fracture investigations,* U.S. Department of Energy METC/SP-79/3.)

FIGURE 8–17 (continued)
(c) Subhorizontal plumose joint in Middle Proterozoic Killarney Granite on Georgian Bay near Killarney, Ontario. The plumose pattern here indicates that the fracture propagated from top to bottom of the photo. The irregular pattern at the edge of the smooth fracture is twist hackle. This joint was created as a Pleistocene glacier moved over the rock surface and plucked a mass of bedrock. (RDH photo.)

(c)

sary to exceed the strength of the rock mass must become tensile. This is thought to occur during cooling and elastic contraction of a rock mass as it is exhumed by erosion and may occur at depths of 200 to 500 m.

Release of horizontal stress provides a similar opportunity for release joints to form. Orientation of release joints is controlled by the existing rock fabric, in contrast with the three other types recognized by Engelder, which are stress-controlled. Release joints form late in the history of an area and are ultimately oriented perpendicular to the original tectonic compression that formed the dominant fabric in the rock. Tectonic stress may further increase the stress normal to future joint planes as burial depth and degree of lithification increase. After erosion begins, and the rock mass begins to cool and contract, these joints begin to propagate parallel to an existing tectonic fabric, such as a prominent cleavage. In that way, release joints may also develop parallel to fold axes.

Alastair Beach (1980) has suggested that filled veins in low-grade metamorphic rocks are one of the best lines of evidence for hydraulic fracturing. Most of these fractures form by extension, but Beach showed that if the initial orientation of microcracks is parallel to shear directions, the cracks will then continue to propagate parallel to shear directions or perhaps terminate as asymmetrically forked veins. Shear fractures may be straight or slightly curved, but commonly they cannot be traced over as great a distance as can single extension fractures. Beach found that the length-to-width ratio of tensile veins is ordinarily greater than 500, whereas that of shear veins is less than 100. He also concluded that hydraulic shear fractures form where differential stress is low, but hydraulic tensile fractures may form over a wide range of effective stress values and therefore joints, not shear fractures, should be expected. His ideas remain somewhat controversial.

Paul Hancock (1985) has defined a joint as *any* fracture discontinuity in rock, even a shear fracture. He concluded that joints and shear fractures may form by extension, shear, or by a hybrid mechanism, and suggested that the history of jointing may be pieced together by using surface markings (smooth, plumose) and parallelism to nearby kinematic indicators (Chapter 10) if all can be shown to have formed by the same deformational event. Symmetry with respect to other structures, joint refraction, curviplanar joints (some-

(d)

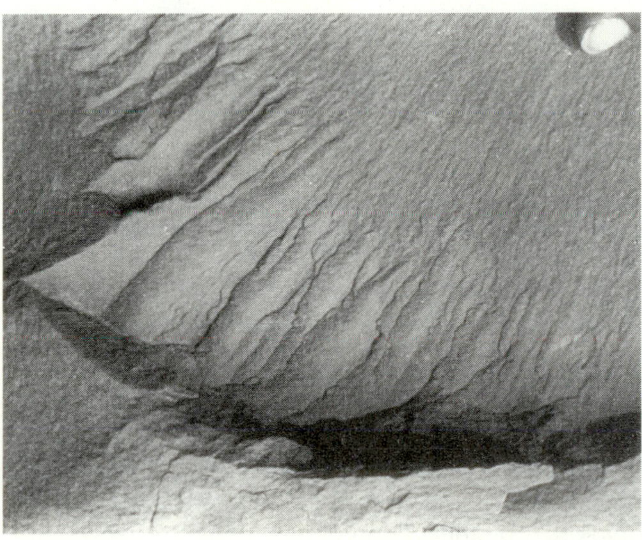

(e)

(f)

FIGURE 8–17 (continued)
(d) Glacially produced joint surface in Killarney Granite at the same locality as (b) with arrest lines that are concentric with the origin of the fracture. The joint therefore propagated right to left. Faint plumose structure is also visible on the joint surface. (RDH photo.) (e) Twist hackle at the edge of a joint in Jurassic Aztec Sandstone, southeastern Nevada. (f) Irregular plumose structure on a joint surface in Devonian Genesee siltstone near Watkins Glen, New York, indicating that either a change of propagation direction during fracturing or a component of simple shear was present. [Parts (e) and (f) courtesy of Atilla Aydin, Stanford University.]

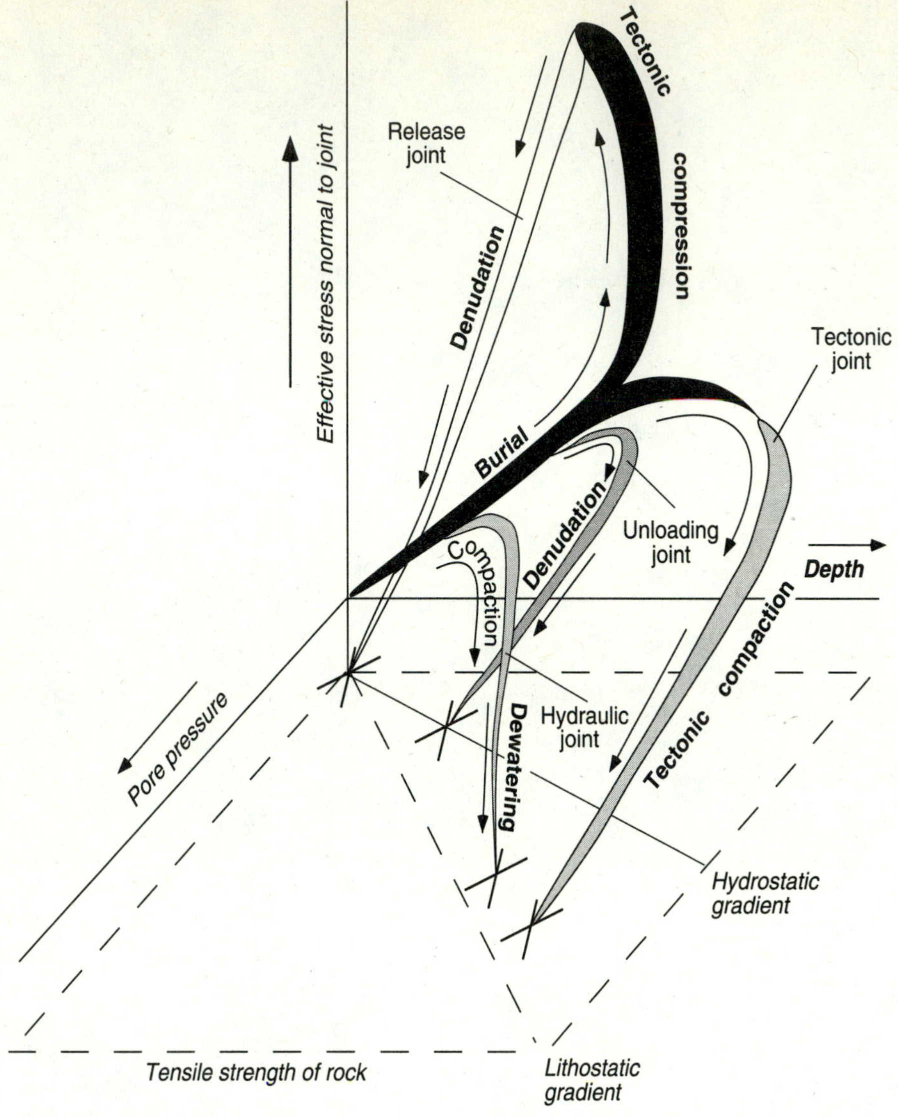

FIGURE 8–18
Loading paths for Engelder's tectonic, hydraulic, unloading, and release joints as a function of three variables: (1) pore pressure, (2) depth, and (3) effective stress normal to the joint. Arrows indicate expected directions of movement along pathways. (From *Journal of Structural Geology,* v. 7, Terry Engelder, p. 459–476, © 1985, with kind permission from Elsevier Sciences, Ltd., Kidlington, United Kingdom).

times indicating a transition from extension to shear), spatial relationships in the entire fracture system, and angular relationships among conjugate joints also may be used to decipher joint and shear fractures.

JOINTS IN PLUTONS

Fractures form in plutons in response to cooling and later tectonic stress (Figure 8–18). Orientations of joints that form in a cooling pluton may be influenced by the boundary of the pluton or by its general shape and internal structure. Some large stocks contain joints that are more or less concentric with the internal structure of the pluton and may also be related to the shape of the contact with the country rock. Fracture density and orientation may vary with changes in rock texture or fabric as well as proximity to the boundary of the pluton. Columnar joints (to be discussed below) form in flows and shallow plutons as a response to cooling and solidification of magma.

Robert Balk (1937) studied fractures in igneous bodies and recognized that fracturing occurs both as a result of flow related to emplacement and as a product of later regional stress fields. Early joints form in plutons during the final stages of crystallization of magma. Many joints are filled with hydrothermal minerals or crystallization products from late-stage volatile-rich magmas that crystallize pegmatite, aplite, and quartz veins. Balk recognized that many early fractures form in relation to flow banding in the pluton, either across, parallel, or diagonal to banding. He concluded that the cross fractures are joints and that the diagonal fractures, which form as conjugate sets, are shear fractures. The parallel, or longitudinal, fractures are more difficult to relate to an extensional or shear-stress field.

Balk also found that joints are more common and more closely spaced near the margins of a pluton. He suggested that processes related to magma emplacement and cooling of the margins may fracture both the early cooled parts of the magma and the country rock (Figure 8–19), forming both joints and faults.

Many joints in plutons, like in most other rock masses, form by tectonic and unloading forces superposed long after the pluton has cooled. Balk's analysis of fractures in plutons is useful where early formed fractures related to emplacement can be separated from later tectonic and unloading joints. One way to accomplish this is to study the joints present both inside and outside the pluton, taking into account differences in rock type.

NONTECTONIC AND QUASITECTONIC FRACTURES

Sheeting

A kind of joint that forms subparallel to surface topography, generally in massive rocks and corresponds to the unloading joints of Engelder (1985), is called ***sheeting.*** Most often, these fractures may be observed in igneous rocks, but they also form in metamorphic rocks. The spacing between sheeting fractures increases downward into the crust (Figure 8–20). Quarrymen use sheeting fractures in quarrying dimension stone to minimize the amount of blasting necessary to remove large blocks. Sheeting is thought to form by unloading over long periods of time as erosion removes large quantities of overburden from a rock mass. The mass expands normal to the Earth's surface (σ_3 vertical) so that extension fractures form normal to the expansion direction and parallel to the surface (Johnson, 1970). Many sheeting joints appear to form normal to a small compressive σ_3, an unexplained phenomenon that is not supposed to happen according to the Griffith failure criterion. Although Griffith recognized that tensile stress can still occur at a crack

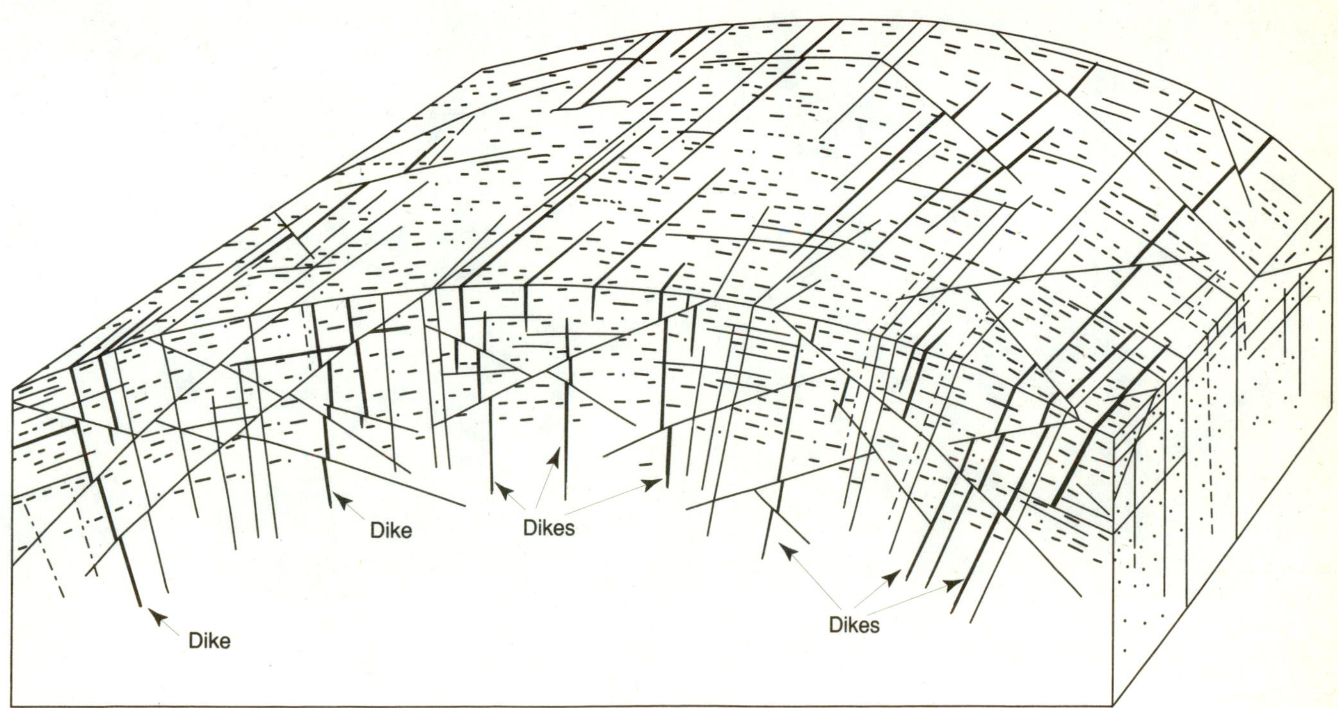

FIGURE 8–19
Joint patterns developed in a pluton. Note the greater joint density near the pluton margins (edge of block). The curved surface on top is the present erosion surface. Short, discontinuous lines are joints that form parallel to flow directions. Longer lines parallel to the shorter are sheeting (unloading) joints. Near-vertical joints form by regional bending of the crust, possibly related to intrusion of the pluton. Small faults displace the margin of the pluton, along with older joints, and dikes inside it. (Modified from Robert Balk, 1937, *Structural behavior of igneous rocks*: Geological Society of America *Memoir 5,* who redrew it from a drawing by Hans Cloos of structures in a pluton in the Strehlen massif, eastern Germany.)

tip even if far-field stress is all compressive (Price and Cosgrove, 1990). This appears to be a fundamental problem in need of further investigation.

Columnar Joints and Mud Cracks

Columnar joints form in flows, dikes, sills, and (occasionally) in volcanic necks in larger plutons in response to cooling and shrinkage of congealing magma (Figure 8–21) and are thus nontectonic structures. Thermal gradients and contraction processes in the magma control the orientation of columnar joints. The orientation of columns is generally normal to the sides of a pluton. They may become curved if the thermal gradients and contraction processes are nonuniform or if the magma is still moving slightly as the joints form. Columnar joints commonly have five or six planar sides. Ideally, the most efficient geometric resolution of fractures in a uniaxial contraction stress field is a set of hexagonal prisms if the stress field, cooling rate, and thermal gradient were perfectly uniform throughout the cooling magma body (Figure 8–22). A highly symmetrical cylindrical stress field will form in which jointing begins as the magma contracts. This forms hexagonal extension fractures in the congealing magma. Mud cracks (Chapter 2) form on the Earth's surface by a similar mechanism—shrinkage and evaporation of water in unconsolidated sediment (Figure 2–7); they appear as four-, five-, six-, or seven-sided polygons, which frequently curl up at the edges.

We now leave our discussion of joints and shear fractures and begin an exploration of faults and faulting processes.

FIGURE 8–20
Subhorizontal sheeting joints in a granite quarry in Massachusetts. Note that the joints are sub-parallel to the surface and that spacing increases with increased depth to a maximum of 6 m at the lower left. (L. C. Currier, U.S. Geological Survey.)

(a)

(b)

FIGURE 8–21
(a) Columnar joints in Quaternary basalt near Whistler, British Columbia. (b) Closeup of columnar basalt in (a). Columns are about 1 m across. Note fractures perpendicular to the long axes of the columns. (RDH photos.)

FIGURE 8–22
Formation of hexagonal fractures that form around uniformly distributed centers (axes) of contraction. Arrows indicate orientations of tensile stresses.

ESSAY

Mesozoic Fracturing of Eastern North American Crust—Product of Extension or Shear?

The present-day Atlantic Ocean probably opened along the coast of eastern North America during Early to Middle Jurassic time, following a period of rifting of the continental crust that provided block-faulted basins that formed the Triassic–Jurassic basins along the axis of the Paleozoic Appalachian orogen. The basins and the adjacent Piedmont were intruded by diabase dikes (Figure 8E–1) from Virginia southward, and they are oriented dominantly northwest; from Virginia northward, they trend more northerly to northeasterly (Ragland and others, 1983). A similar pattern occurs on both sides of the present Atlantic, interpreted by Paul May (1971) as indicating extensional deformation along the line where spreading of the continents began. An alternative explanation, by Jelle de Boer and Frederic Snider (1979), is that Mesozoic dikes in eastern North America formed as a product of doming over a mantle hotspot in the Carolinas. The dikes examined by May and by de Boer and Snider have since been shown to be cross-cut by a younger set of north–south-trending Mesozoic diabase dikes in the Carolinas and Virginia; this set converges toward a point near Charleston, South Carolina (Ragland and others, 1983; Figure 8E–1). Both sets of dikes appear to be extensional.

Fracture zones filled with quartz and then broken by reactivation several times and each time healed by quartz to form siliceous cataclasite occur in the same region. In the central and southern Appalachians, they are oriented approximately north–south and east–west (Conley and Drummond, 1965; Birkhead, 1973; Garihan and others, 1988, 1993); in some places, they appear to cut the diabase dikes, but in others, the siliceous cataclasites are cut by the dikes. Thus, the diabase dikes and siliceous cataclasite fracture zones may have formed about the same time in the Early Jurassic. If so, several extension directions may have existed at the same time, but that is mechanically unlikely. Note that the siliceous cataclasite zones are filled with remobilized quartz (with or without feldspar and prehnite). Many contain open spaces, boxwork structure, and vugs into which

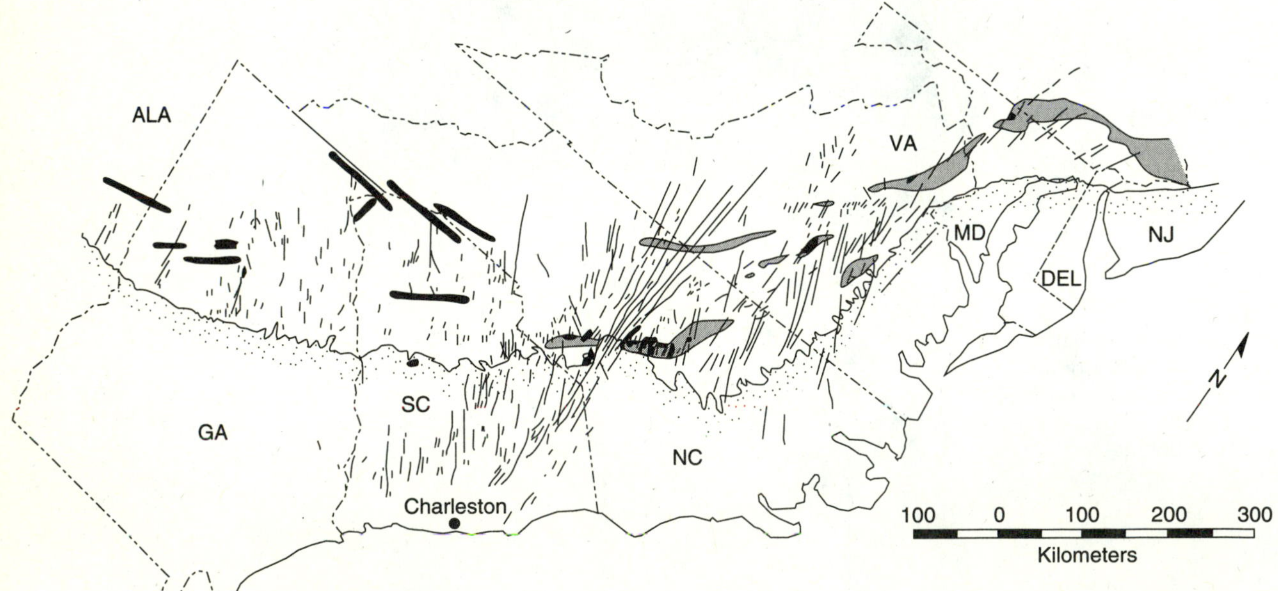

FIGURE 8E–1
Distribution and orientations of Triassic–Jurassic diabase (light lines) and some siliceous cataclasite dikes (heavy black lines) in the eastern United States. Triassic–Jurassic basins are represented by shaded areas. (Modified from P. C. Ragland, R. D. Hatcher, Jr., and D. Whittington, 1983, *Geology,* v. 11.)

quartz crystals have grown, further indicating their extensional nature. They also contain abundant evidence of reactivation, with ground-up and fragmented early quartz recemented by later vein quartz. These zones rarely exhibit evidence of offset parallel to their lengths: rock-unit boundaries between bodies that trend into them are only minimally offset from one side to another. One area in the southern Piedmont of Georgia contains siliceous cataclasite dikes that have a left-lateral displacement of a few kilometers (Hooper, 1989).

Is it possible that the early diabase dikes are purely extensional features and that the siliceous cataclasite zones are conjugate shears (Figure 8E–2)? If so, the siliceous cataclasite bodies are shears with no displacement parallel to the shear zone and after they were formed, were followed by repeated opening normal to the shears and filling with quartz. If not, simultaneous or alternating crustal extension in several directions would be required.

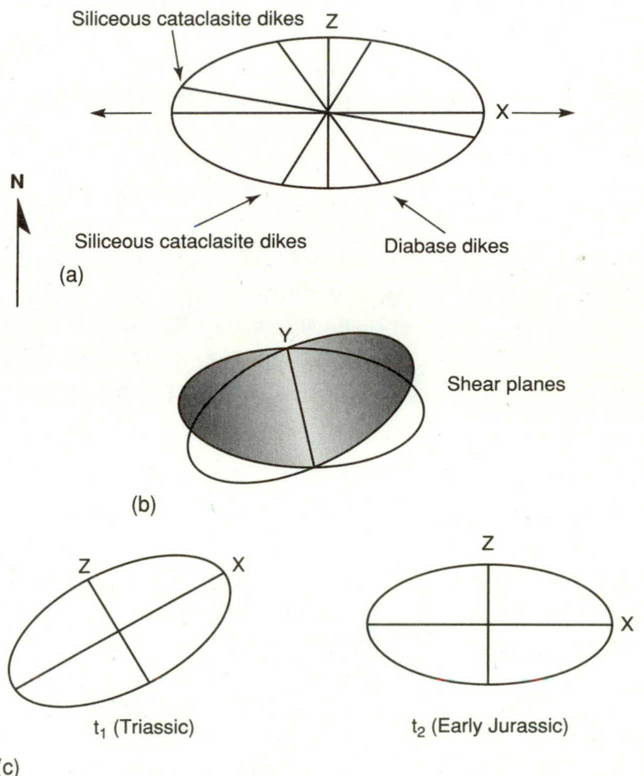

FIGURE 8E–2
Possible strain-ellipsoid orientations for siliceous cataclasite and diabase dikes in the eastern United States. (a) Model 1—Diabase dikes with a N 45° W orientation form by extension; siliceous cataclasite dikes form at the same time by shear. N–S diabases formed later by reorientation of the extensional stress field of the strain ellipse. *Against the model:* Because cataclasites contain open-boxwork quartz, extension is required for at least part of their history. They also have little or no displacement. (b) Model 2—All diabases and siliceous cataclasites formed by extension, produced by multiple orientations of the strain ellipsoid at about the same time. The N–S diabases formed later. *Against the model:* It requires nearly simultaneous and mechanically unlikely extension directions and is also unlikely over short periods of geologic time. (c) Model 3—N 45° W diabases formed by extension; siliceous cataclasites were initiated by shear without much accompanying displacement. Siliceous cataclasites were reactivated by later extension and filled with hydrothermal quartz. N–S diabases formed by later extension involving a strain ellipsoid with a different orientation.

References Cited

Birkhead, P. K., 1973, Some flinty crush rock exposures in northwest South Carolina and adjoining areas of North Carolina: South Carolina Geologic Notes, v. 17, p. 19–25.

Conley, J. F., and Drummond, K. M., 1965, Ultramylonite zones in the western Carolinas: Southeastern Geology, v. 6, p. 201–211.

de Boer, J., and Snider, F. G., 1979, Magnetic and chemical variations of Mesozoic diabase dikes from eastern North America: Evidence for a hotspot in the Carolinas?: Geological Society of America Bulletin, v. 90, p. 185–198.

Garihan, J. M., Preddy, M. S., and Ranson, W. A., 1993, *in* Hatcher, R. D., Jr., and Davis, T. L., eds., Studies of Inner Piedmont geology with a focus on the Columbus Promontory: Carolina Geological Society Annual Field Trip November 6-7, 1993, Field Guide, p. 55–65.

Garihan, J. M., Ransom, W. A., Preddy, M., and Hallman, T. D., 1988, Brittle faults, lineaments and cataclastic rocks in the Slater, Zirconia, and part of the Saluda 71/2 minute quadrangles, northern Greenville County, SC, and adjacent Henderson and Polk Counties, NC, *in* Secor, D. T., Jr., ed., Southeastern geological excursions: Columbia, South Carolina, Southeastern Section Geological Society of America, p. 266–312.

Hooper, R. J., 1989, Tectonic implications of a regionally extensive brittle fault system in the Piedmont: Evidence from central Georgia: Geological Society of America Abstracts with Programs, v. 21, p. 22.

May, P. R., 1971, Pattern of Triassic–Jurassic diabase dikes around the North Atlantic in the context of predrift position of the continents: Geological Society of America Bulletin, v. 82, p. 1285–1291.

Ragland, P. C., Hatcher, R. D., Jr., and Whittington, D., 1983, Juxtaposed Mesozoic diabase dike sets from the Carolinas: A preliminary assessment: Geology, v. 11, p. 394–399.

Questions

1. How can we distinguish between systematic and nonsystematic joints?
2. What does asymmetric twist hackle structure on a joint surface tell you about the way the joint formed?
3. How would systematic orientation of joints be established by a hydraulic-fracturing mechanism?
4. How could filled joints and unfilled joints form at the same time with different orientations?
5. What is the best evidence that most joints form by extension?
6. How do columnar joints form?
7. What evidence tells us that some regional joints in eastern North America may have formed in the present-day stress field?
8. Suppose joints formed at a locality where the earliest set was filled with quartz-feldspar (pegmatite), then cut by a set containing quartz-epidote, which was in turn crossed by a quartz-calcite-prehnite set, then by a set of calcite- and laumontite-filled joints, and finally by an unfilled joint set. (Each set has a different orientation.) What do you conclude about the conditions producing joints there?
9. Are sheeting joints and Engelder's release joints related?
10. How can tectonic stresses tens or hundreds of millions of years ago influence orientations of joints that formed in the past 5 million years?
11. Why are the tensile strengths of rocks almost always less by a large amount than the compressive strengths?

Further Reading

Engelder, T., 1985, Loading paths to joint propagation during a tectonic cycle: An example from the Appalachian Plateau, USA: Journal of Structural Geology, v. 7, p. 459–476.

Classifies joints and relates them to variables of effective stress normal to the joint, depth, pore pressure, tensile strength of the rock, and hydrostatic/ lithostatic gradients through time.

Engelder, T., 1987, Joints and shear fractures in rock, *in* Atkinson, B., ed., Fracture mechanics of rock: London, Academic Press, p. 27–69.

Reviews concepts of joint and shear fracture formation as well as features that occur on joint surfaces and information that can be derived from these surfaces.

Hancock, P. L., 1985, Brittle microtectonics: Principles and practice: Journal of Structural Geology, v. 7, p. 437–457.

Analyzes brittle structures as a way to solve tectonic problems. Joints are assumed to result from either extension or shear.

Nickelsen, R. P., 1979, Sequence of structural stages of the Alleghany orogeny, at the Bear Valley Strip Mine, Shamokin, Pennsylvania: American Journal of Science, v. 279, p. 225–271.

A classic study of a well-exposed series of rocks deformed during a single event, recording a sequence of extension followed by compression and possibly later extension. Nickelsen relates each stage of deformation to the strain ellipsoid and attempts to determine stress orientations for several stages.

Pollard, D. D., and Aydin, A., 1988, Progress in understanding joints over the past century: Geological Society of America Bulletin, v. 100, p. 1181–1204.

Reviews development of ideas on joint formation, including tectonic and nontectonic joints.

Secor, D. T., 1965, Role of fluid pressure in jointing: American Journal of Science, v. 263, p. 633–646.

Formation of joints under influence of pore fluids under pressure is a model that follows directly from Hubbert and Rubey's model for thrust faults (Chapters 10 and 11).

Wise, D. U., 1982, Linesmanship and the practice of linear geo-art: Geological Society of America Bulletin, v. 93, p. 886–888.

A tongue-in-cheek look at the ways lineaments are misinterpreted. Wise listed 32 rules derived from "logic" used in interpreting and misinterpreting lineaments.

Wise, D. U., Funiciello, R., Maurizio, P., and Salvini, F., 1985, Topographic lineament swarms: Clues to their origin from domain analysis of Italy: Geological Society of America Bulletin, v. 96, p. 952–967.

Lineaments in Italy from LANDSAT and other data were analyzed in an attempt to relate them to systematic joints in a tectonically active region.

9

Fault Classification and Terminology

The occurrence of faults and dykes must have been known from the earliest days of mining. It was only gradually, however, that they became an object of scientific study.

ERNEST M. ANDERSON, 1942, *The Dynamics of Faulting*

FAULTS ARE THE GENERATORS OF SEISMIC ACTIVITY; OUR interest in them is practical as well as scientific and aesthetic. Understanding faults is useful in the design for long-term stability of dams, bridges, buildings, and power plants, especially because of the devastating effects of active faults on populations who live near them. The study of faults helps us to understand mountain-building and deformation processes, and also yields results of practical value. Faults have produced some of the Earth's most spectacular scenery (Figure 9–1), including the Alps in Europe, the Grand Tetons in Wyoming, and the Canadian and Montana Rockies. Faults are largely responsible for uplifting blocks in the great mountain chains on the Earth.

FIGURE 9–1
Aerial view of the San Andreas fault, San Luis Obispo County, California. (R. E. Wallace, U.S. Geological Survey.)

A ***fault*** is a fracture having appreciable movement parallel to the plane of the fracture. Sometimes it is difficult to specify what "appreciable movement" is and what we may or may not choose to call a fault—a somewhat scale-dependent judgment. We have no difficulty at all identifying a very large fault that records many kilometers of movement; on the other hand, a fault having a few centimeters offset may be important in one study, but in another may be considered only a large fracture or a joint with slight offset because of the different scales of each study. Even small offsets become significant, however, when we weigh the impact of an active fault on nearby buildings or dams. Faults occur in many forms and dimensions. They may be hundreds of kilometers or only a few centimeters long. Their outcrop traces may be straight or sinuous. They may occur as knife-sharp boundaries or as *fault* or *shear zones,* millimeters to several kilometers thick (Figure 9–1). Fault or shear zones may consist of a series of interleaving, anastomosing brittle faults and crushed rock (cataclasite) formed near the surface, or of *ductile shear zones* composed of mylonitic rocks produced by faulting at great depth. Movement on ductile faults is distributed over a zone, in contrast with brittle faults, where movement may be confined to a single plane. Most of our discussion in this chapter will deal with the descriptive properties of brittle faults; we will discuss structures associated with ductile shear zones in Chapter 10.

ANATOMY OF FAULTS

We need to understand the basic anatomy of faults (Figure 9–2) before discussing the details of the behavior of specific kinds of faults. The most obvious feature related to faulting is the displacement of some marker, most commonly bedding. Displacement occurs along the actual movement surface—the ***fault "plane"***—commonly non-planar. If the fault plane is not vertical, the rock mass resting on the fault plane is called the ***hanging wall***, and the rock mass beneath the fault plane is called the ***footwall***—terms originated in mining. We can measure the dip and strike of the fault plane and make other measurements of structures (such as slickensides, to be discussed below) that indicate relative movement along the fault plane.

Two contrasting fault types exist: *dip-slip* and *strike-slip* faults (Figure 9–2). The ***slip*** along a fault describes the movement parallel to the fault plane. We speak of ***dip slip*** where movement is down or up

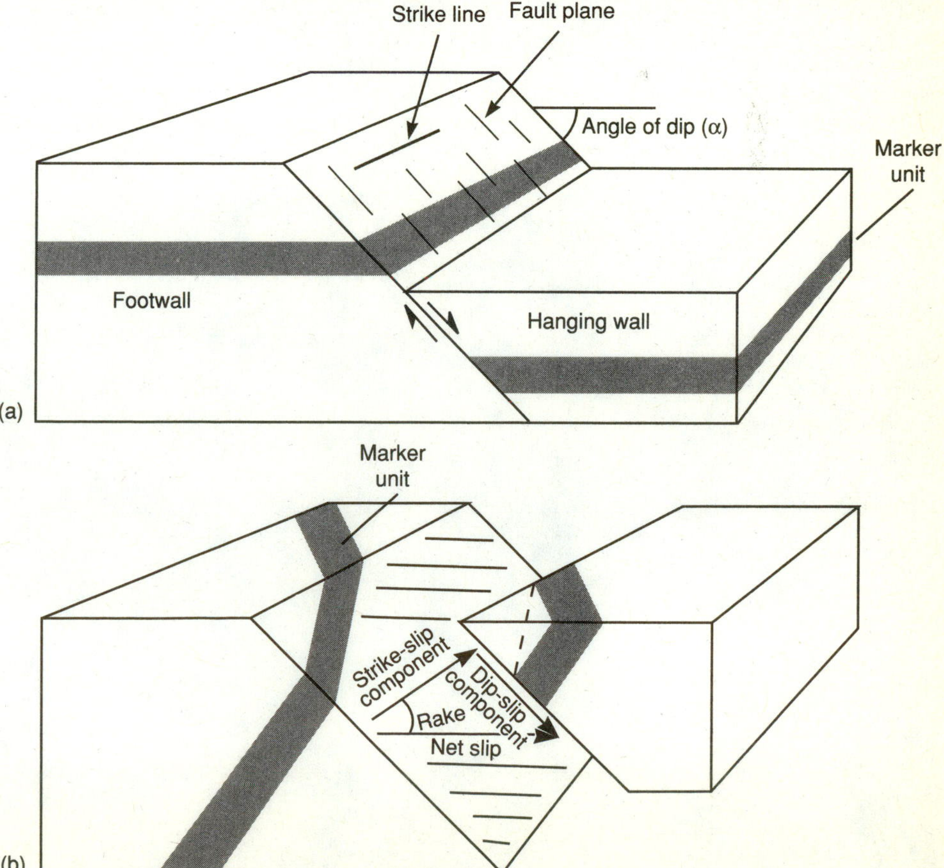

FIGURE 9–2
(a) Anatomy of faults. Arrows indicate sense of relative movement.
(b) Oblique-slip fault showing the components of net slip and the rake of net slip.

parallel to the dip direction of the fault. The term ***strike slip*** applies where movement is parallel to the strike of the fault plane. Most faults exhibit dominantly dip-slip or dominantly strike-slip motion. A combination is referred to as ***oblique slip***. The ***net slip***, or ***true displacement***, is the total amount of motion measured parallel to the direction of motion. We cannot actually determine the absolute motion sense on a fault without some primary criterion, such as by establishing benchmarks before an earthquake and surveying them afterward, but the *rake of net slip* is often a useful measurement (Figure 9–2b).

Another term is ***separation***—the amount of apparent offset of a faulted surface, such as a bed or a dike, measured in a specified direction (Figure 9–3). We can speak of ***strike separation***, ***dip separation***, and the total or ***net separation*** of a fault. The terms *heave* and *throw* are sometimes also used to describe horizontal and vertical components of dip separation. ***Heave*** describes the horizontal component of dip separation measured perpendicular to strike of the fault; ***throw*** is the vertical component measured in the vertical plane containing the dip (Figure 9–4).

Common features on fault surfaces are grooves and growth of fibrous minerals, both aligned parallel to the movement direction. They are commonly arranged in a series of steps, facing in the movement direction of the opposing fault block. Polished fault surfaces are called ***slickensides***, and the striations on them are called *slickenlines* (Figure 9–5). Aligned fibrous minerals on a movement surface are *slickenfibers*. All are considered a kind of lineation (Chapter 18) that indicates the trend of relative movement (such as north–south or northeast–southwest) along a surface.

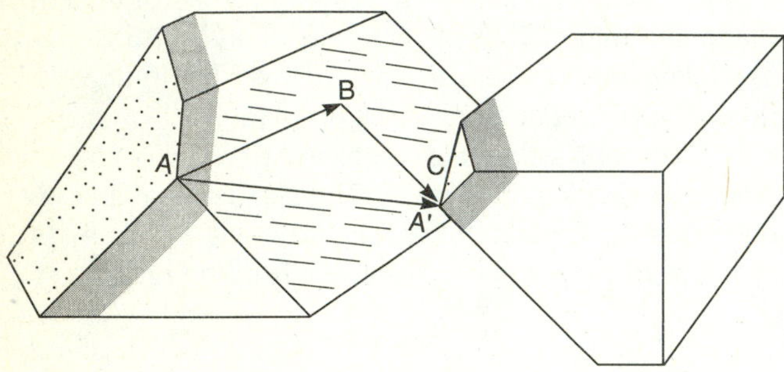

FIGURE 9–3
Oblique-slip fault showing the net separation (A–A′) of a marker layer and its components, strike separation (AB), and dip separation (BC), measured in the fault plane. These could also be termed the components of true displacement.

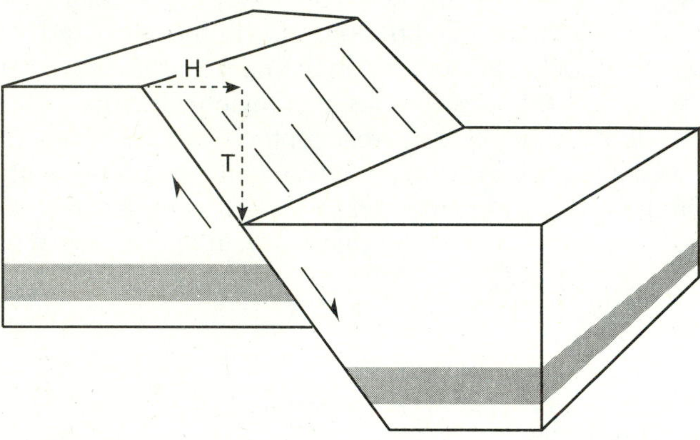

FIGURE 9–4
Components of dip separation—heave (H) and throw (T).

FIGURE 9–5
Calcite slickenfibers on a movement surface in the Lower Devonian Kalkberg Formation, Hudson Valley, New York. The field book provides the scale. (Stephen Marshak, University of Illinois.)

The small steps may also be used to determine movement direction. The direction of down-stepping is commonly the movement direction of the opposing wall (Chapter 10). Slickenlines most commonly record only the last movement event on the fault.

ANDERSONIAN CLASSIFICATION

The fault classification devised by Ernest M. Anderson (1942), a British geologist, defined three fundamental categories: *normal faults, thrust faults,* and *wrench (strike-slip) faults* (Figure 9–6). (Elements of his classification probably were in use before Anderson's work, but the history is difficult to trace.)

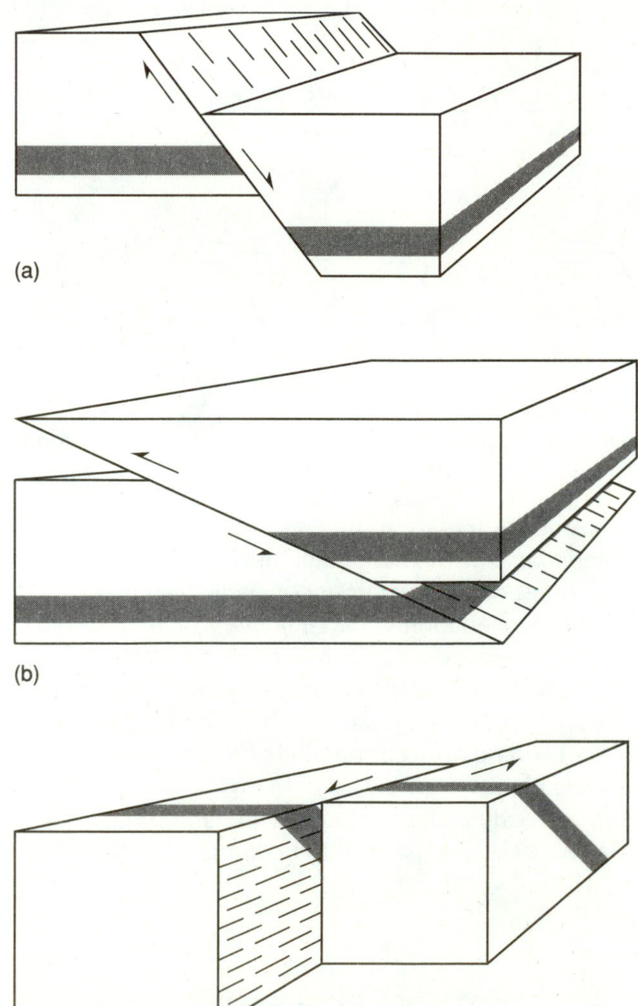

FIGURE 9–6
Anderson's fault classification. (a) Normal. (b) Thrust. (c) Strike slip.

Normal faults are dip-slip faults in which the hanging wall has moved down relative to the footwall. A ***graben*** consists of a block that has dropped down between two subparallel normal faults that dip toward one another (Figure 9–7a). The opposite is a ***horst***, consisting of two subparallel normal faults that dip away from each other so that the block between the two faults remains high (Figure 9–7b). Normal faults frequently exhibit ***listric*** (concave-up) geometry, so that they have steep dip near the surface but flatten with depth. The term *lag,* is less commonly used for listric normal fault.

The hanging wall has moved up relative to the footwall in *thrust* and *reverse faults* (Figure 9–6). Some geologists treat them separately, defining ***thrust faults*** as those with a low angle of dip (30° or less), and ***reverse faults*** as those with a moderate to steep dip (45° or more) and having the same sense of motion as thrusts. These two fault classes are mechanically similar, and we will assume them to be the same in our discussion here, although Anderson's theory of faulting (Chapter 10) does not predict the formation of reverse faults. Both high- and low-angle segments occur on many thrust faults; low-angle segments may likewise be found along many reverse faults—hence the similarity in treatment here.

Anderson described two categories of ***strike-slip faults*** on the basis of direction of relative motion as seen by an observer looking across the fault plane (Figure 9–8). Strike-slip faults, where the side opposite the observer moves to the right, are called ***dextral*** or ***right-lateral strike-slip faults***. Those in which the opposite side moves to the left are called ***sinistral*** or

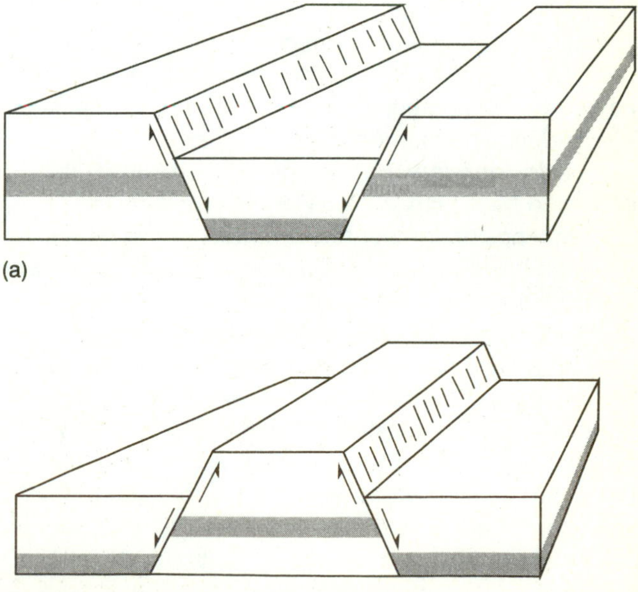

FIGURE 9–7
Graben (a) and horst (b) structures.

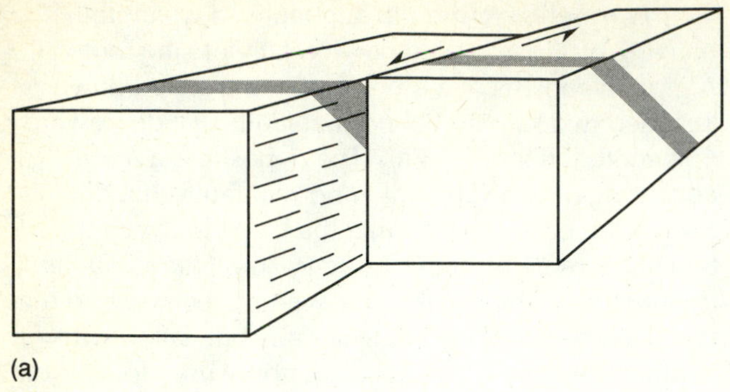

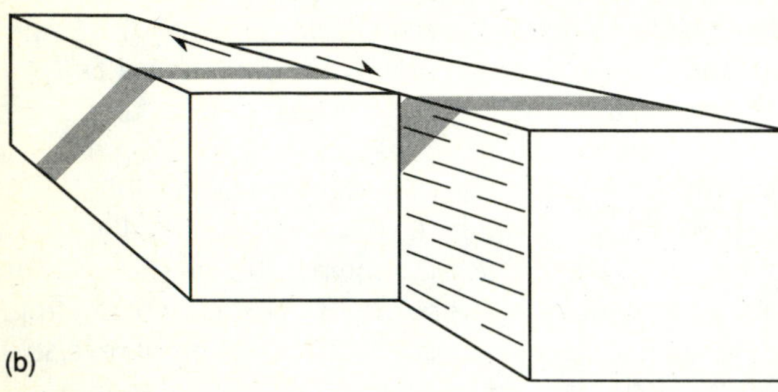

FIGURE 9–8
(a) Sinistral (left-lateral) and (b) dextral (right-lateral) strike-slip faults.

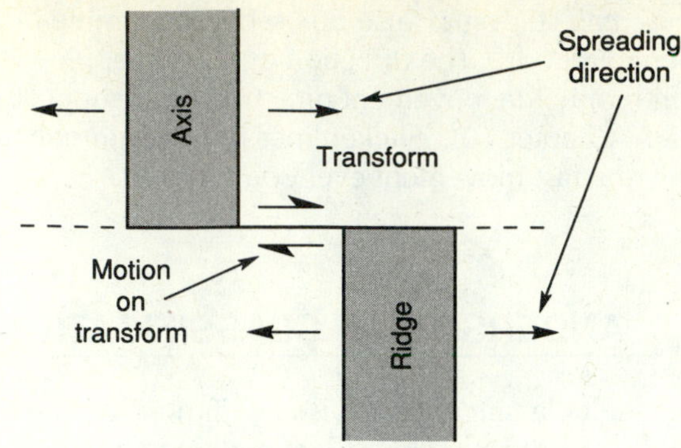

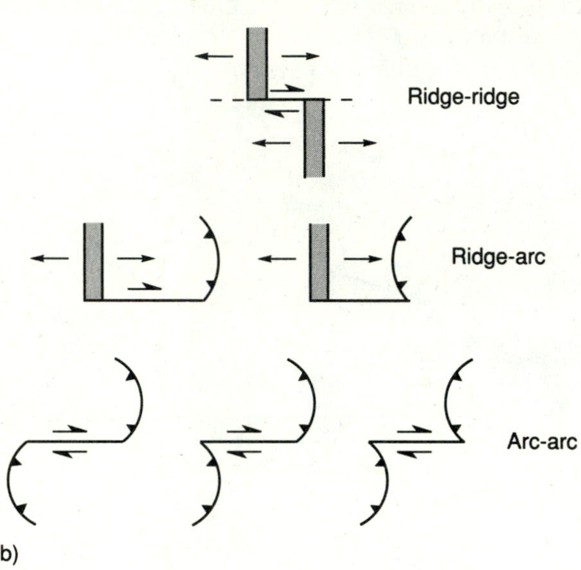

FIGURE 9–9
(a) Motion on transform faults. (b) Several kinds of transform faults.

left-lateral strike-slip faults. Note that if the observer moves to the opposite side of a hypothetical fault block from the side of first observation (and turns to face the fault again), the same sense of relative motion will again be observed. So, if the observer looking across the fault plane sees that the side opposite where he or she is standing has moved to the right, the fault is right-lateral. Strike-slip faults commonly have steep dip—near the vertical—but moderate and shallow segments have also been noted, as along the Alpine fault in New Zealand, and the San Andreas in California.

Transform faults are a type of strike-slip fault predicted by J. Tuzo Wilson (1965) as a necessary consequence of compensating differences in motion between lithospheric plates (Figure 9–9a). Unlike most faults, these are so named because they were first recognized in the oceans, linking areas where crust is "transformed" into mantle, or mantle is "transformed" into crust. More exactly, they occur where one type of plate boundary is transformed into another. Wilson's prediction was confirmed by seismologist Lynn Sykes (1967) from first-motion studies of earthquakes along the Mid-Atlantic Ridge. Because they connect parts of plates where crust is being transformed into mantle or vice versa, or are themselves plate boundaries, several kinds of transforms have been defined. They include ***ridge-ridge, ridge-arc***, and ***arc-arc*** transform faults (Figure 9–9b).

Among other terms that describe faults are *en echelon faults,* which approximately parallel one another but occur in short unconnected segments, sometimes overlapping; *radial faults,* which converge (or project) toward a single point; and *concentric faults,* which form concentric to a point (Figure 9–10; see also Figure 2–33). *Bedding faults* or *bedding-plane faults* follow bedding or occur parallel to the orientation of bedding planes (Figure 9–11). Many thrust and normal faults are bedding faults because the fault propagates along a weak zone parallel to bedding (Chapters 11 and 13).

CRITERIA FOR FAULTING

Existence of a fault at a particular locality is commonly demonstrated by the recognition of certain features and diagnostic characteristics. If all faults were active, we would need only two criteria: earthquakes or aseismic

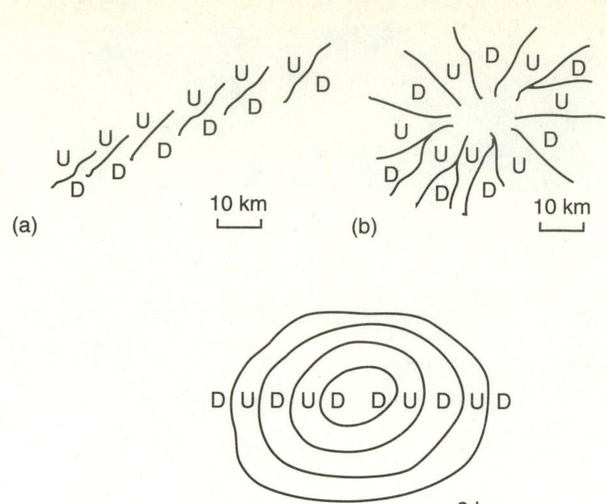

FIGURE 9–10
(a) *En echelon* faults with dextral steps, (b) radial , and concentric (c) faults in map view. U—upthrown side; D—downthrown side. (See Figure 2–33 for radial and concentric faults in an impact structure.)

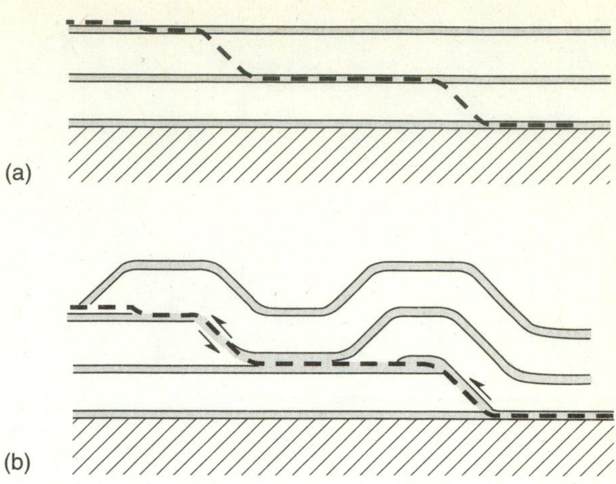

FIGURE 9–11
Bedding (plane) fault in cross-section view. (a) Before movement, showing the future path of the fault (dashed line) that will propagate parallel to some beds (weaker) and climb more steeply across others (stronger). (b) Fault formed by dip-slip motion of the hanging wall up relative to the footwall, producing a thrust.

creep and associated ground breakage. Most faults are inactive, however, and so we must use other means of recognizing them. Unfortunately, in many regions we cannot directly observe the fault surface because of younger deposits and soil or vegetation cover.

Probably the most fundamental way to demonstrate that a fault is present is ***repetition*** or ***omission of stratigraphic units***, or the ***displacement (offset) of a recognizable marker*** (Figure 9–12). Demonstrating that a rock unit or sequence has been repeated should prove the existence of a fault if it can be readily shown that repetition is not related to folding. Repetition by faulting commonly produces asymmetric repetition (units DCBA, A oldest, stacked onto units DCBA, with A or B of the sequence on top, resting on D of the sequence beneath, Figure 9–12a); repetition by folding commonly produces symmetric repetition (CDEDC in map view, Figure 9–12c). Likewise, units may be omitted from a sequence wherein continuity has been established elsewhere. Either repetition or omission can be demonstrated by recognition of marker units or by some other physical characteristic that allows us to make correlations and show that displacement or omission of rock units has actually occurred. Guide (index) fossils, color, texture, composition, or other distinctive features in a rock unit may set it apart so that it can serve as a marker unit.

The ***truncation of structures, beds***, or ***rock units*** against some feature is another criterion for faulting (Figure 9–12), but truncation must be used with care because structures may be truncated by unconformities and intrusions (Chapter 2). Therefore, a second criterion should be used to demonstrate that the boundary in question is a fault and not an unconformity. Truncation of structures on both sides of a discontinuity would most likely indicate that the boundary is a fault.

Occurrence of fault rocks, ***mylonite*** or ***cataclasite*** (or both), along a suspected fault zone is another good criterion for faulting. Those rocks may be used in conjunction with repetition or omission of stratigraphic units and truncation of structures to show that movement has taken place. Generally, either indicates that faulting has occurred, although to confirm the occurrence of faulting, it is generally best to use a displaced marker. Presence of S-C structures, rotated porphyroclasts, and other evidence of shear deformation (Chapter 10) probably indicates that faulting has occurred by formation of a shear zone and also yields the sense of displacement. Other criteria are required to determine the amount of displacement.

Abundant *veins, silicification,* or other *mineralization* along a fracture zone may suggest that faulting has occurred, but mineralization is a poor criterion because it may occur in fractures having no offset. It must therefore be used carefully and in conjunction with other evidence to prove whether a fault is present.

Drag is produced along a fault as units appear to be pulled into a fault during movement. The sense of motion may be determined because the drag folds exhibit asymmetry in the direction of movement. Drag may occur on thrust faults, so that the layers appear to have been pulled down-dip in the hanging wall or up-dip in the footwall toward the fault plane (Figure 9–13a), opposite the sense of relative motion. The opposite effect with drag folds occurs on normal faults. ***Reverse drag*** occurs along some listric normal faults

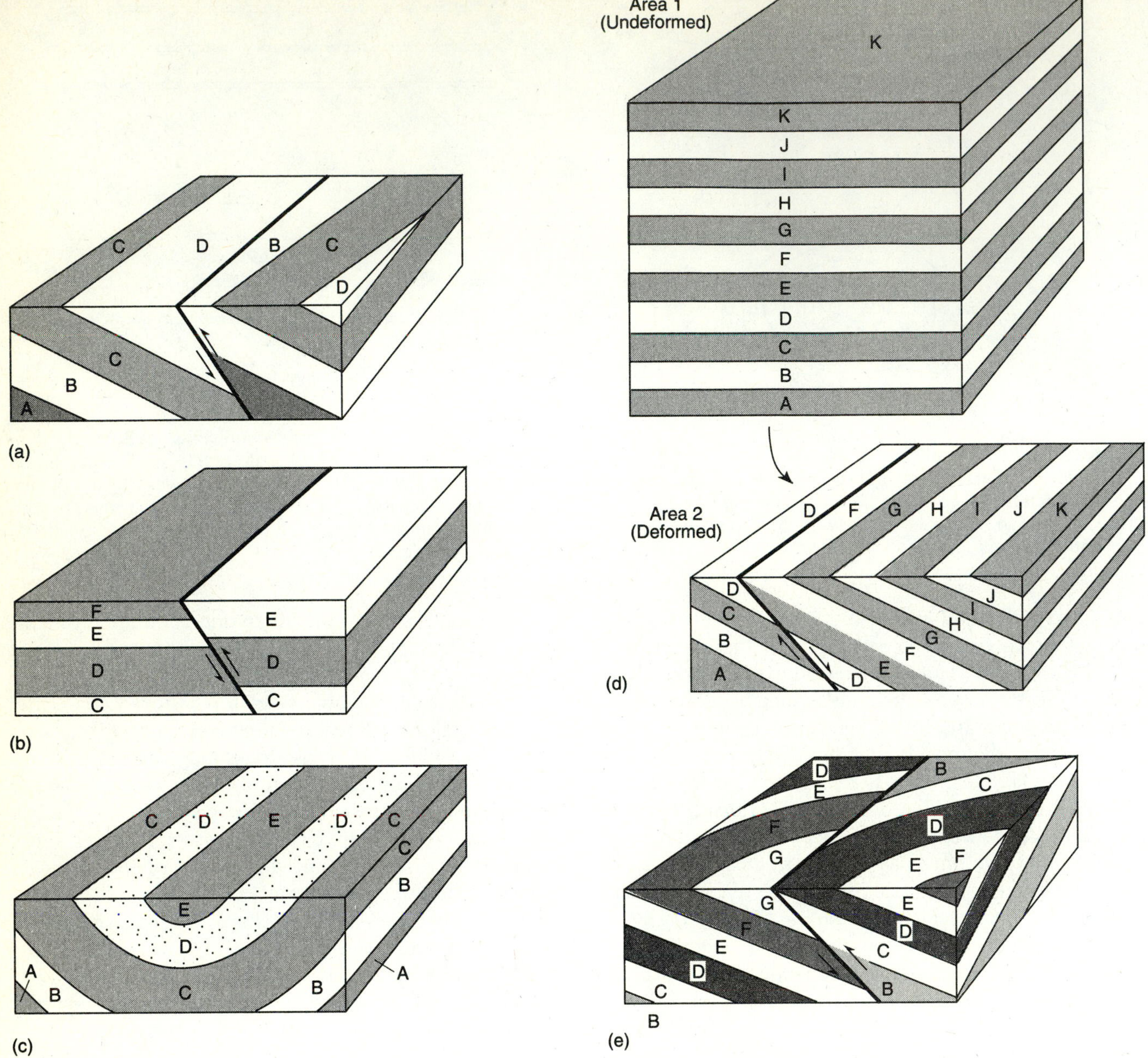

FIGURE 9–12
(a) Repetition or omission of stratigraphic units. (b) Offset of stratigraphic markers. (c) Repetition by folding. Note the symmetry in the outcrop pattern. (d) Omission of a unit by faulting in one area that can be shown to be present in a continuous sequence elsewhere. (e) Truncation of structures.

(Figure 9–13b) where the layering appears to have been dragged down-dip parallel to movement along the fault. Reverse drag is produced by movement of material into the void produced along the fault as it moves and the dip decreases, creating more of a rotational displacement up-dip on the more steeply dipping segment and apparent drag up dip.

Slickensides and *slickenlines* along a fault surface (Figure 9–5) often indicate movement but do not prove significant faulting because they may form on bedding surfaces during certain types of folding or may actually form on joints that have a small shear displacement. Both help to confirm only that movement has occurred, and the direction of last movement.

Certain characteristics of topography may indicate faulting has occurred. Drainage may be controlled by faults (Figure 9–14). If active faulting is present, drainage may be successively offset over a few thousand—or a few hundred thousand—years, thereby providing a primary criterion that can be observed on the surface as an indication of offset along a fault. ***Fault scarps*** form where a topographic surface is offset by dip-slip motion

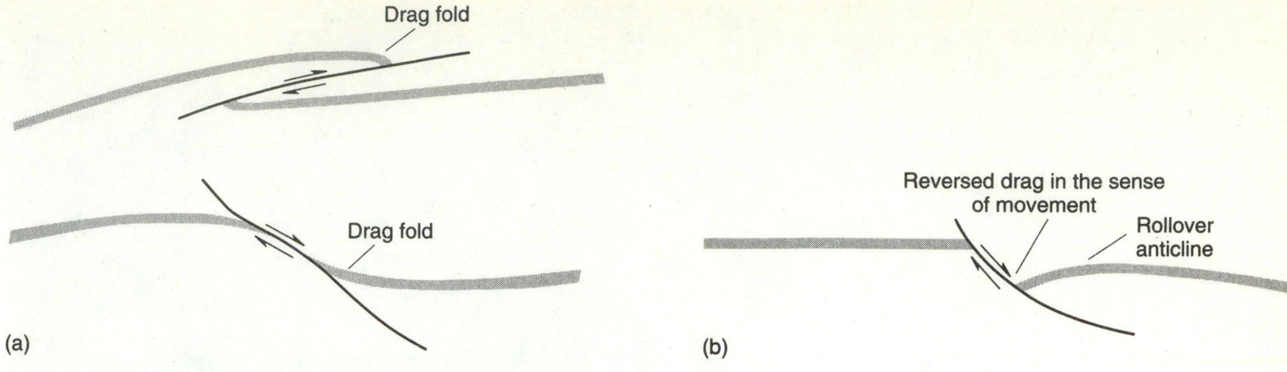

FIGURE 9–13
(a) Drag on thrust and normal faults. (b) Reverse drag on a normal fault. Cross-section view.

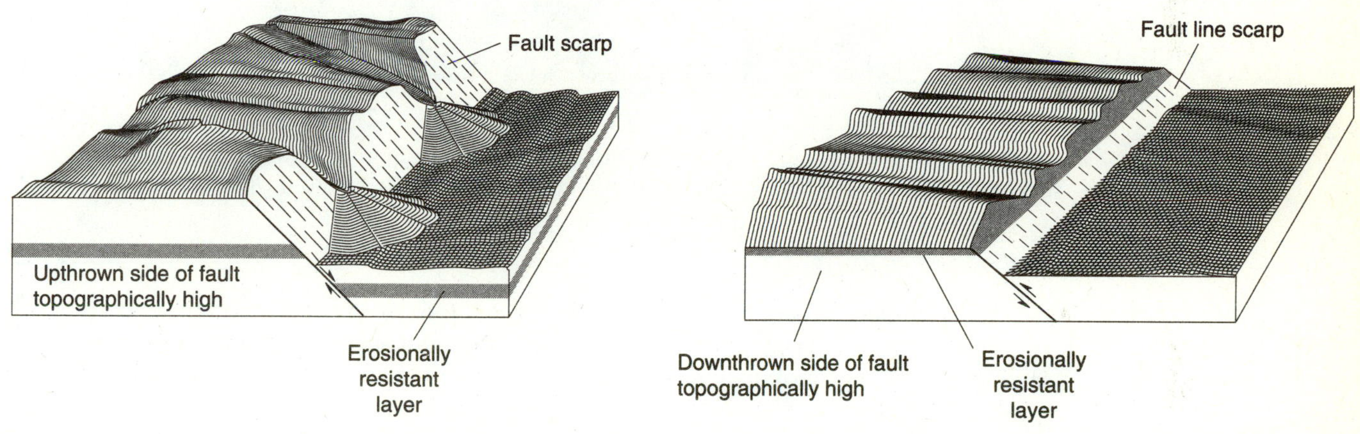

FIGURE 9–14
(a) Fault scarp. Note that the upthrown side of the fault is topographically high here. (b) Obsequent fault-line scarp with the downthrown side of the fault topographically high, because of the erosionally resistant layer.

along a fault and directly indicate the movement sense of the fault (Figure 9–14a; Figure 9-15). A fault scarp may evolve through time into a ***fault-line scarp*** by differential erosion along the fault. This will lower the level of the topographic surface and may remove a resistant layer in the hanging wall (Figure 9–14b). As long as erosion preserves the original motion sense of the fault, the fault-line scarp is called a *resequent fault-line scarp.* The resistant layer may later be etched into relief in the footwall, producing a false apparent sense of motion on the fault without subsequent movement, and is called an *obsequent fault-line scarp* (Figure 9–14b). The real sense of motion must be determined by using at least one other primary criterion.

The criteria for faulting just described may be listed as either primary, for identifying and proving existence of faults, or secondary, which must be used in conjunction with other criteria. Primary criteria include repetition and omission of stratigraphic units, displacement of a recognizable marker, and truncation of earlier structures—but the last criterion is best if truncation occurs on both sides of the fault.

Now that we have assembled the tools for recognizing and describing faults, we can proceed to Chapter 10 on fault mechanics and deal with the problem of how brittle and ductile faults form and move.

FIGURE 9–15
Fault scarp along the east side of Death Valley, California. The uniform slopes on the mountain front in the middle ground are actual movement surfaces on the fault. The right (east) side is upthrown; the left is downthrown. (Harold Drewes, U.S. Geological Survey.)

ESSAY

Existence and Displacement Sense of Large Faults

Some of the very large inactive faults in ancient mountain chains were discovered by recognition of one or more of the criteria for faulting discussed in this chapter. Yet the sense of motion and displacement on some of these faults has been debated for many decades because elements of the puzzle remain missing, or the history of movement is complex, or both.

The Lake Char fault (Char is short for Chargoggagoggmanchauggagoggchau-bunagungamaugg) in the southeastern New England Appalachians has been described by Roberta Dixon and Lawrence Lundgren (1968) as part of an extensive fault complex. It is a west-dipping low-angle fault in Connecticut that steepens along strike northeastward to join the more steeply dipping Clinton Newbury-Bloody Bluff fault system in Massachusetts (Figure 9E–1). The low-angle character of the Lake Char fault had been recognized earlier by the gentle dip along the outcrop trace, before Robert Wintsch (1979) suggested that the fault reappears encircling the core of the Willimantic dome farther west. The Lake Char fault has been assumed to be a thrust that moved rocks from the west in the hanging wall over the more easterly rocks now in the footwall. Unfortunately, there are no easily recognized marker units to help measure the displacement. Existence of the fault had been confirmed by Dixon and Lundgren primarily on the basis of mylonites along the trace that truncate earlier structures in the hanging wall and footwall. The sense of movement, as determined by Arthur Goldstein (1982), using several shear-sense indicators (Chapter 10), surprisingly turns out to be hanging-wall down toward the west, indicating that the last movement on the Lake Char fault was normal and thus may not be a thrust. The amount of displacement remains undetermined.

The Brevard fault zone in the southern Appalachians (Figure 9E–2) was first described by Arthur Keith (1907), who thought it was a syncline of younger rocks preserved within older rocks of high metamorphic grade. The sheared nature of the rocks within the fault zone was first recognized by Anna Jonas (1932), who concluded that the Brevard fault zone is a thrust and suggested it is correlative with a major thrust farther north—in Maryland and Pennsylvania—called the Martic thrust. The Brevard fault, like many others in the cores of mountain chains, does not offset easily recognized markers throughout its more than 300-km known length. Consequently, its displacement sense and magnitude remained obscure.

A more recent study by Jack Reed and Bruce Bryant (1964) used the orientations and displacement patterns of linear structures such as quartz *c* axes and mineral lineations along the North Carolina segment (see Chapter 18) and concluded that the Brevard is a sinistral strike-slip fault. In 1970, Reed, Bryant, and Myers suggested that the Brevard was a dextral strike-slip fault with a major thrust (dip-slip) component. Then, after studies in South Carolina and nearby Georgia, I reported exotic blocks of footwall rocks and a traceable stratigraphy within and southeast—but unfortunately not northwest—of the Brevard fault zone (Hatcher, 1971). My studies (Hatcher, 1971) also suggested that the

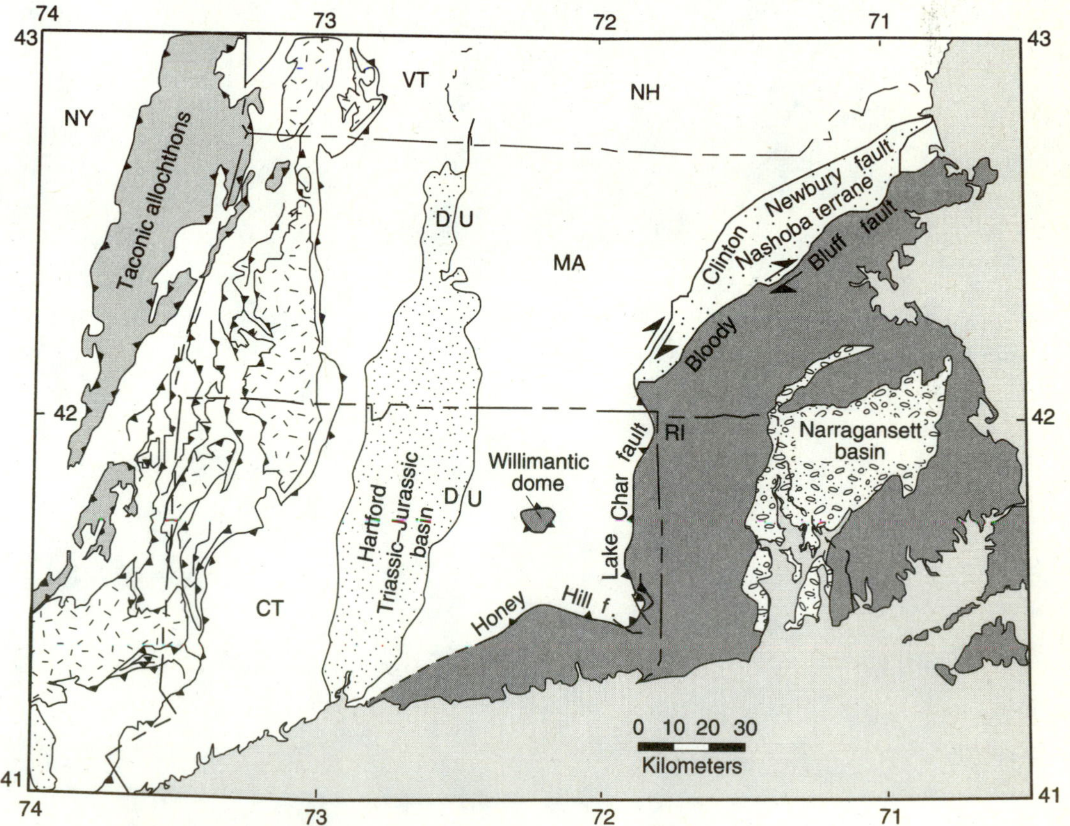

FIGURE 9E–1
Relationships of the Lake Char fault in eastern Connecticut to the Honey Hill and Clinton-Newbury and Bloody Bluff faults, as well as to the fault that frames the Willimantic dome window. The Narragansett basin contains Pennsylvanian-Permian rocks. The cross-hatch pattern represents ~1.1 Ga-old basement rocks of western New England. The dark pattern in southeastern New England is the Avalon terrane, an exotic terrane of volcanic and sedimentary rocks that was accreted to North America during the early Paleozoic. (From R. D. Hatcher, Jr., P. H. Osberg, A. A. Drake, Jr., P. Robinson, and W. A. Thomas, 1990, *Tectonic Map of the U.S. Appalachians,* Plate 1 *in* Geological Society of America, Geology of North America, v. F–2.)

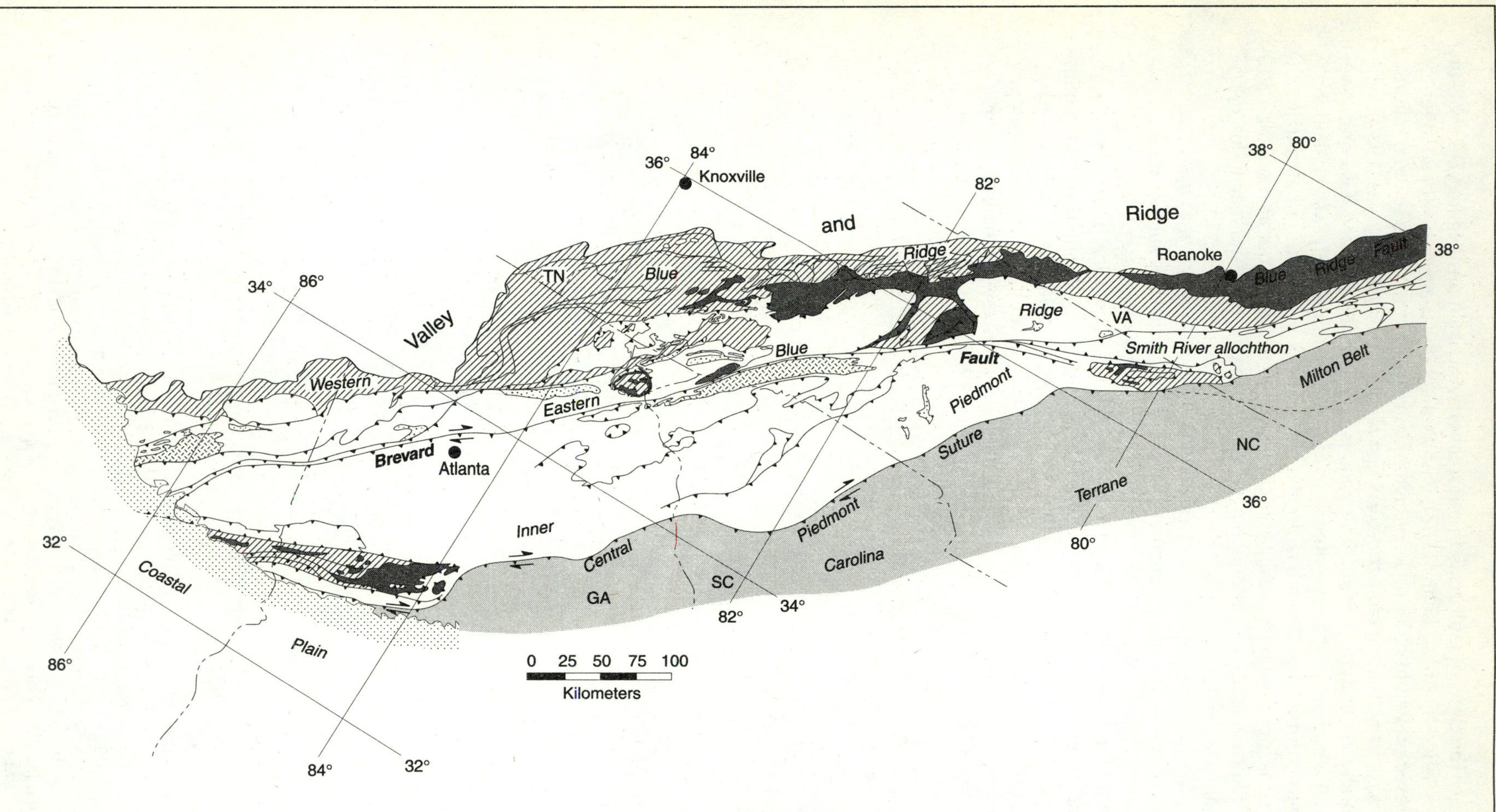

FIGURE 9E–2
Location of the Brevard fault zone relative to other major faults and tectonic units in the internal parts of the southern Appalachians. The Brevard fault zone is located at the boundary between the eastern Blue Ridge and the western Inner Piedmont. This region has no pattern because the same stratigraphic units occur on both sides of the Brevard fault, and so it cannot be a terrane boundary or a suture. The dark screen represents ~1.1Ga-old basement rocks. The striped pattern represents rocks that were deposited along the late Precambrian-early Paleozoic North American margin on the basement. The light stipple and cross-hatch patterns represent some of the Paleozoic plutons. (From R. D. Hatcher, Jr., P. H. Osberg, A. A. Drake, Jr., P. Robinson, and W. A. Thomas, 1990, *Tectonic Map of the U. S. Appalachians,* Plate 1 in Geological Society of America, Geology of North America, v. F–2.)

structure is a thrust, as originally concluded by Jonas. Since then, Andy Bobyarchick (1984) has found that the Brevard had a major dextral strike-slip component, and Steven Edelman, Angang Liu, and I (1987) concluded that the Brevard had an early thrust component, followed by later dextral strike-slip, and later still moved again by thrusting. This structure has also been interpreted as an Alpine root zone by Clark Burchfiel and John Livingston (1967) and as a transported suture by Douglas Rankin (1975).

Why so many conflicting interpretations for one fault? Part of the problem lies in the absence of critical marker units on both sides of the fault, while another lies in the complexity of the structure. It probably has undergone both dip- and strike-slip movement. Yet another difficulty is that different geologists, like the proverbial six blind men attempting to understand an elephant (see Essay in Chapter 17), have worked on different parts of a structure, exposed well in some areas and poorly in others. The underlying lesson is that large faults such as the Brevard and Lake Char, which bound major structural blocks, commonly undergo a long history of movement involving multiple displacements in different directions.

References Cited

Bobyarchick, A. R., 1984, A late Paleozoic component of strike slip in the Brevard zone, southern Appalachians: Geological Society of America Abstracts with Programs, v. 16, p. 126.

Burchfiel, B. C., and Livingston, J. L., 1967, Brevard zone compared to Alpine root zones: American Journal of Science, v. 265, p. 241–256.

Dixon, H. R., and Lundgren, L., 1968, The structure of eastern Connecticut, *in* Zen, E-an, White, W. S., Hadley, J. B., and Thompson, J. B., Jr., Studies of Appalachian geology: Northern and maritime: New York, Wiley-Interscience, p. 261–270.

Edelman, S. H., Liu, A., and Hatcher, R. D., Jr., 1987, The Brevard zone in South Carolina and adjacent areas: An Alleghanian orogen-scale dextral shear zone reactivated as a thrust fault: Journal of Geology, v. 95, p. 793–806.

Goldstein, A., 1982, Geometry and kinematics of ductile faulting in a portion of the Lake Char mylonite zone, Massachusetts and Connecticut: American Journal of Science, v. 282, p. 378–405.

Hatcher, R. D., Jr., 1971, Structural, petrologic and stratigraphic evidence favoring a thrust solution to the Brevard problem: American Journal of Science, v. 270, p. 177–202.

Jonas, A. I., 1932, Structure of the metamorphic belt of the southern Appalachians: American Journal of Science, 5th Series, v. 24, p. 228–243.

Keith, A., 1907, Description of the Pisgah quadrangle, North Carolina–South Carolina: U.S. Geological Survey Geologic Atlas Folio 147, 8 p.

Rankin, D. W., 1975, The continental margin of eastern North America in the southern Appalachians: The opening and closing of the Proto-Atlantic Ocean: American Journal of Science, v. 275-A, p. 298–336.

Reed, J. C., Jr., and Bryant, B., 1964, Evidence for strike-slip faulting along the Brevard zone in North Carolina: Geological Society of America Bulletin, v. 75, p. 1177–1196.

Reed, J. C., Jr., Bryant, B., and Myers, W. B., 1970, The Brevard zone: A reinterpretation, *in* Fisher, G. W., Pettijohn, F. J., Reed, J. C., Jr., and Weaver, K. N., eds., Studies of Appalachian geology: Central and southern: New York, Wiley Interscience, p. 241–256.

Wintsch, R. P., 1979, The Willimantic fault: A ductile fault in eastern Connecticut: American Journal of Science, v. 279, p. 367–393.

Questions

1. How does the strike of a plane constrain the orientation (direction) of the dip of the plane?
2. How would you make measurements of the orientation of planes in the Earth that have the two limiting special cases: horizontal and vertical surfaces?
3. What do slickensides tell us? What do slickenlines tell us?
4. If you found a limestone-sandstone contact in the field in which the contact and both units dip 40° in the same direction, how would you determine whether it was a fault, a stratigraphic contact, or some other feature such as an unconformity?
5. What actually determines whether a fracture is a fault or a joint?
6. Most faults—regardless of whether they are thrust, normal, or strike-slip faults—when studied carefully turn out to be oblique-slip faults with a dominant thrust, normal, or strike-slip component of motion. Why?
7. Why does drag occur on some faults? Reverse drag?
8. Which criteria are most reliable for recognizing faults? Why?
9. Would fault and/or fault-line scarps form where rocks on either side of the fault have equal resistance to erosion? Explain.
10. How are net slip and stratigraphic separation alike? How are they different?

Further Reading

Several other elementary structural geology textbooks discuss fault classification and terminology from a different viewpoint. Some, such as those by Davis, and Twiss and Moores, include fault mechanics in their treatment of faulting. It would be worthwhile to compare the approach in some of these other books to that employed here.

Billings, M. P., 1973, Structural geology, 3rd edition: New York, Prentice-Hall, 606 p.

Davis, G. H., 1984, Structural geology of rocks and regions: New York, John Wiley & Sons, 492 p.

Hills, E. S., 1963, Elements of structural geology: New York, John Wiley & Sons, 483 p.

Park, R. G., 1983, Foundations of structural geology: London, Chapman & Hall, 135 p.

Twiss, R. J., and Moores, E. M., 1992, Structural geology: New York, W. H. Freeman & Company, 532 p.

10

Fault Mechanics

If these principles hold in the case of rocks under pressure, rupture by shearing would be expected to occur along planes oblique to the axis of greatest and least intensity of compressive stress but (if the material is isotropic) inclined at angles of less than 45° to the axis of greatest stress. In such a case, two intersecting sets of planes of rupture may develop, cutting each other at an oblique angle, the greatest pressure bisecting the acute angle.

L. M. HOSKINS, U.S. Geological Survey Annual Report

WE DO NOT KNOW WHO FIRST RECOGNIZED A FAULT, OR when or where. Probably, it was many centuries ago, when someone noted ground breakage after an earthquake. The word "fault" may have been coined by eighteenth-century English coal miners when they recognized offset coal seams and considered this offset a defect or "fault" in the perfect deposition they expected in nature. Also in the eighteenth century, Coulomb formulated the failure criterion that bears his name (Chapter 3), recognizing that shear and normal stresses are interdependent with respect to the cohesive shear strength and the coefficient of internal friction of a material. In the 1800s, Playfair, Lyell, and other geologists wrote about faults much in the modern sense. In 1882, Otto Mohr related the Coulomb failure criterion to both shear stress and normal stress, and so the criterion is frequently referred to as the *Coulomb-Mohr criterion.*

During the late 1800s, Bailey Willis and H. M. Cadell experimented with some of the first geologic models (Figure 10–1), exploring the nature of faults and other structures, but it was not until E. M. Anderson (1942, 1951) systematically examined the three fundamental types of faults that modern study of *fault mechanics* began. Anderson's clear and simple—and rigorous—assumptions form the basis for most later

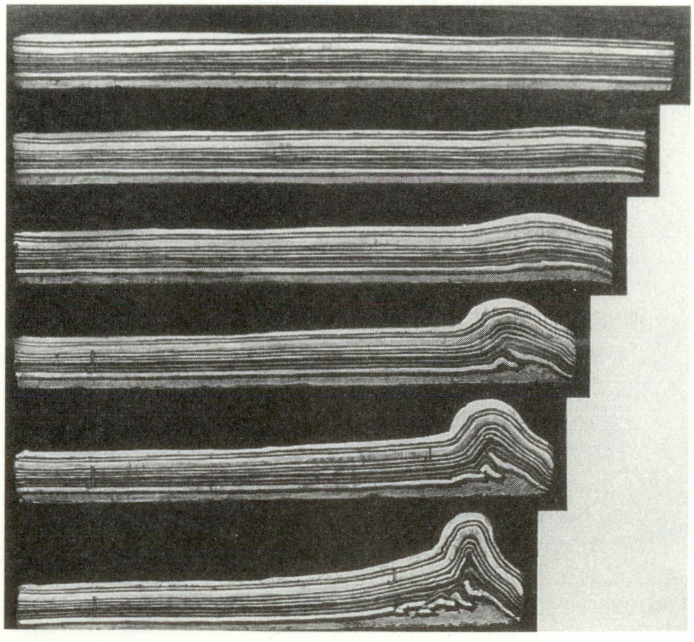

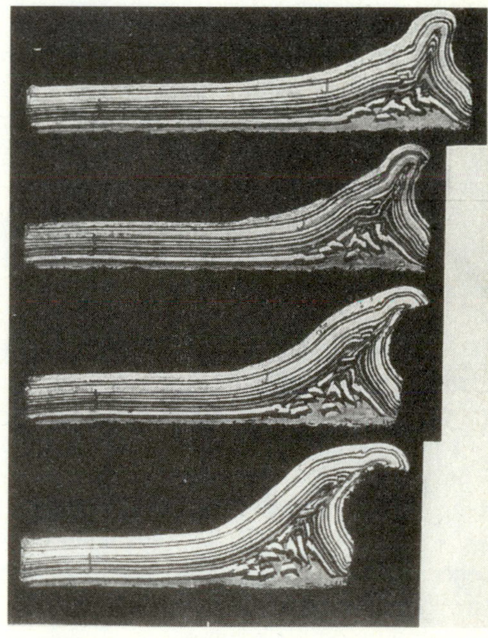

FIGURE 10–1
Bailey Willis's pressure-box model consisting of layers of differing strength that produced a faulted fold as they were compressed from the ends. The sequence of photos taken during the experiment begins at the top left, proceeds downward, then to the top of the group on the right and downward again. (U.S. Geological Survey Annual Report, 1893.)

studies of the mechanics of faulting. In this chapter we will outline the modern concepts of fault mechanics from the standpoint of both laboratory experiments and field observations. We will also look briefly at shear-zone mechanics and the criteria for determining shear sense in deformed rocks.

ANDERSON'S FUNDAMENTAL ASSUMPTIONS

E. M. Anderson in 1942 defined three groups of faults that may be related to principal and shear stresses in the Earth. In doing this, he first defined a "standard state" as

> a condition of stress which is the same in all directions at any point, and equal to that which would be caused by the weight of the superincumbent material [the overlying rock mass, lithostatic stress], across a horizontal plane, at the particular level in the rock. It is assumed in this definition that the surface is flat and that the strata are of uniform specific gravity.

Accordingly, his standard state represents a stress system made up of two parts: (1) the force of gravity and (2) a superposed horizontal stress that remains constant in any horizontal plane but increases with depth, because of the increase in the weight of the overlying column of rock (Hafner, 1951).

Anderson (1942) made several assumptions regarding stress in the Earth applicable to the other differentiated planets, assumptions still used in fault mechanics: (1) the crust was initially "intact" and unfractured; (2) one principal stress direction is vertical; (3) the two other principal stresses, which are perpendicular to the first and to each other, must be horizontal and probably maintain a constant orientation for thousands to millions of years; and (4) forces must balance each other in a body in equilibrium. Assumption 2 actually recognizes that the Earth's surface is a surface of no shear (that is, not subject to a shear traction); therefore, one principal stress must be vertical (at the Earth's surface—probably does not hold at depth). Anderson also recognized that stress orientation in the upper crust can differ from one region to another. His assumptions make possible more exact statements about fault mechanics, thus clarifying relationships between faulting and stress in the Earth.

Following up on his assumptions, Anderson suggested three possibilities for deviation from lithostatic stress conditions:

1. Principal stresses σ_1 and σ_2 are horizontal. The least principal stress σ_3 is vertical, and the dip on the fault plane will be $45° - \phi/2$ (about 30°).
2. Maximum and minimum principal stresses σ_1 and σ_3 are horizontal, implying that the intermediate stress σ_2 is vertical and that the fault plane will have a near-vertical dip.
3. Principal stress σ_1 is vertical. Therefore, the fault plane will dip $45° + \phi/2$ (about 60°).

Each of the three sets of conditions describes the stress in a major fault group: thrust faults, strike-slip faults, and normal faults. We will discuss them in greater detail in the next section and in the following three chapters.

Long after Coulomb formulated the theory of shear fracture in 1776, Navier observed in 1833 that rupture does not occur on the planes of maximum shear stress that bisect the angles between the greatest and least principal stresses. He suggested that "internal friction" was responsible for the difference and proposed a constant he called the *coefficient of internal friction* to account for it. The observation that σ_1 (maximum principal stress) bisects the acute angle between the conjugate shear planes of the stress ellipsoid is called ***Hartman's rule.*** This angle (2α) commonly is closer to 60° than to 90° because the shear angle depends on the angle of friction (ϕ) in the Mohr diagram (Figure 10–2), so that

$$\pm \alpha = 45° - \phi/2 . \qquad \textbf{(10–1)}$$

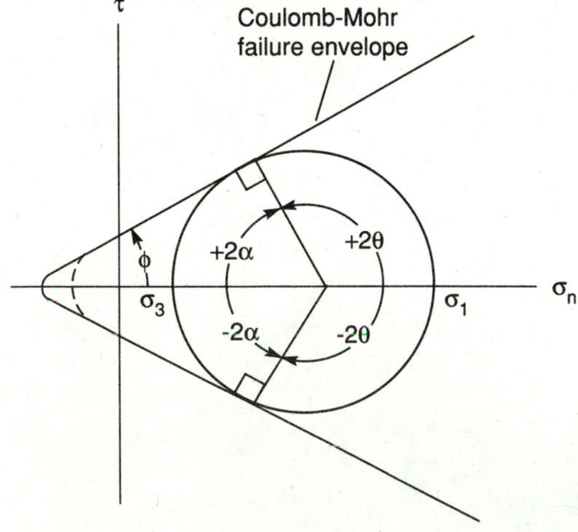

FIGURE 10–2
Mohr diagram showing relationships among ϕ, θ, α, and the Coulomb-Mohr failure envelope.

From the Mohr diagram in Figure 10–2, it is also evident that τ_{max} (the maximum shear) is at 45° to σ_1. The significance of this relationship lies in the ability to predict the angles produced by faults and other shear fractures, which occurs on planes of less than maximum shear stress (where frictional resistance is lower).

ANDERSON'S FAULT TYPES

Separation of stress regimes into the three fundamental possibilities just outlined gives rise to Anderson's three basic fault groups. The mechanics are easy to understand if his scheme is followed.

Type 1. Thrust Faults

Ideal thrust faults may be defined mechanically as faults produced by maximum and intermediate principal stresses oriented horizontally, with the minimum principal stress oriented vertically (Figure 10–3). A state of *horizontal compression* is thus defined for thrust faulting. The ideal strain ellipsoid corresponding to this stress configuration has fault planes oriented at 45° or less to the horizontal that strike parallel to the intermediate stress and strain axes. The fault planes commonly form at smaller angles because of inhomogeneities in the rocks and the complexity of the strain process (e.g., noncoaxial inelastic strain that precedes brittle failure). Strict adherence to Anderson's concept of thrust faults does not allow for the steep dips (>45°) of reverse and high-angle thrust faults, or for near-horizontal thrusts, because of the initial assumption of a homogeneous Earth. Variations in the relative strength and other properties of different rock types, however, permit more steeply and shallowly dipping segments to form in nature and do not violate Anderson's basic concepts.

The two shear planes are potential fault planes; ideally, a fault should form parallel to each shear plane, but commonly only one dominant shear plane becomes a fault. The alternate shear plane may form a fracture or series of fractures with minor displacement. The result is a thrust fault with the upper block (the hanging wall) moved up relative to the lower block (the footwall).

Mohr-circle analysis of thrust faulting (Figure 10–4) may involve successively increasing values of σ_1, producing circles of increasing radii until the failure envelope is reached. Alternatively, the value of a vertical σ_3 can be decreased by erosion to permit the failure envelope to be reached.

Type 2. Strike-Slip Faults

If the maximum and minimum principal compressive stress axes are horizontal and the intermediate principal stress axis is vertical, conditions are right for strike-slip faulting (Figure 10–5). As with all faults, the intermediate principal stress axis will lie within any fault plane that forms, and as a result, must here be vertical. The corresponding stress ellipsoid for strike-slip faulting also contains vertical planes of maximum shear stress, which ideally lie at 45° or less to the axis of minimum stress; shears commonly form at angles more acutely

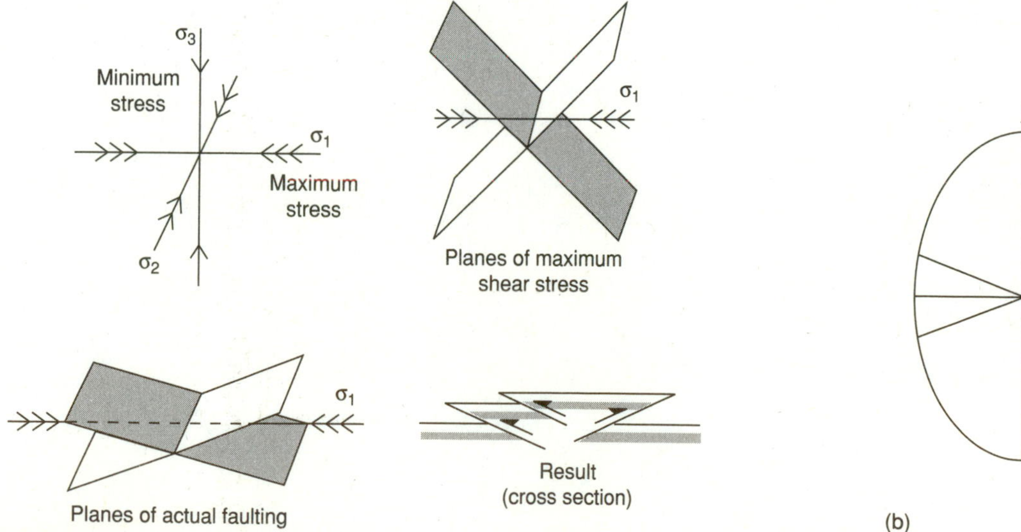

FIGURE 10–3
(a) Orientation of principal stresses for thrust faulting. (b) Strain-ellipse configuration for thrusting. Note that *Y* is horizontal and perpendicular to the plane of the page. View is parallel to strike of faults.

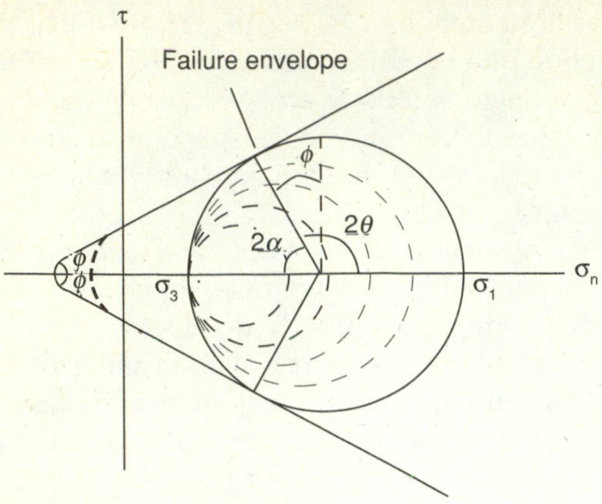

FIGURE 10–4
Mohr-circle analysis of thrust faulting. As σ_3 remains constant, σ_1 increases to failure (dashed circles).

and symmetrically about the axis of minimum strain (and maximum stress), mainly because of an optimum combination of high shear stress and low normal stress—and thus are found in isotropic rocks. In addition, inhomogeneities and the prerupture strain history in most rock masses can play an important role in formation of faults.

Mohr-circle analysis for strike-slip faults can be the same as for thrusts if sufficient differential stress exists to increase the size of the circles to reach the Mohr envelope for any of the Anderson fault types. With this criterion, even normal faults could be considered "compressional" (compression vertical).

Type 3. Normal Faults

Normal faulting involves extension in one horizontal direction, with the maximum principal stress (σ_1) vertical (Figure 10–6). Shear planes may be expected at 45° or less to the axes of maximum and minimum principal stress and strain respectively. They therefore form steeply dipping fault planes located symmetrically about the vertical axis. Both shear planes are frequently utilized in complex normal fault zones.

Mohr-circle analysis of normal faulting may be thought of as starting at a large confining pressure, with σ_3 decreasing (producing extension) until the size of the circles produces a stress difference ($\sigma_1 - \sigma_3$) large enough to reach the failure envelope (Figure 10–7). As with the other kinds of faults, other means may be used to increase the stress difference and cause failure. For example, depositional loading where σ_1 is vertical can lead to normal faulting.

ROLE OF FLUIDS

M. King Hubbert and William W. Rubey (1959) applied Terzaghi's observations from soil mechanics (Chapter 8) and demonstrated that fluid plays an important role in faulting. They recognized that the "lubricating" effect of fluid in a fault zone is really a buoyancy effect that reduces the shear stress necessary to permit the

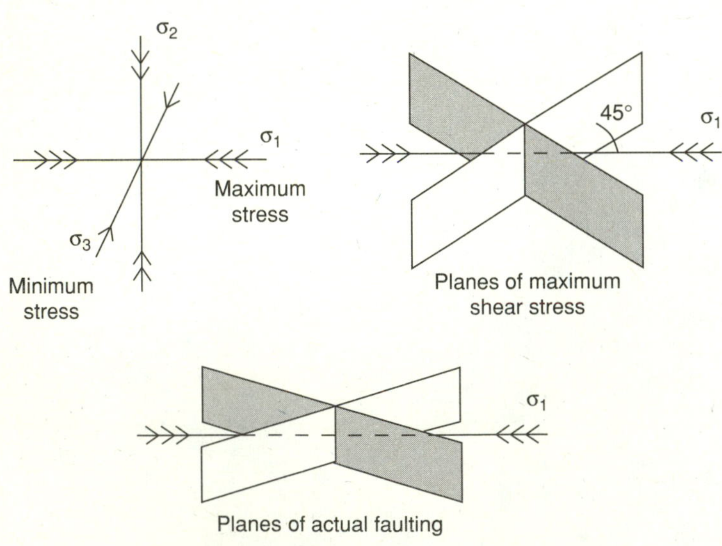

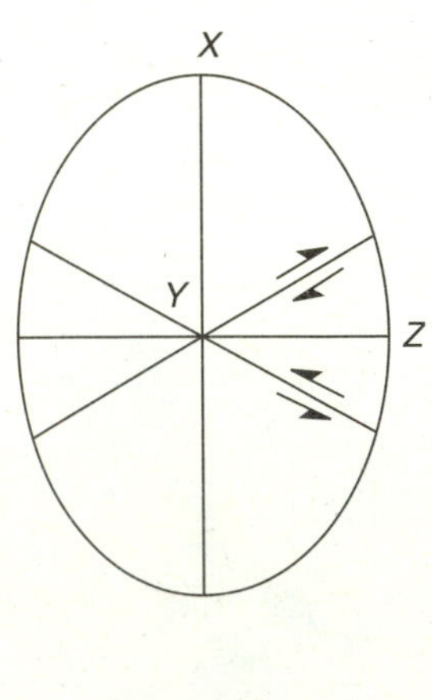

FIGURE 10–5
(a) Principal stress orientations for strike-slip faulting. (b) Strain ellipse for strike-slip faults. *Y* is vertical (in the crust) and oriented perpendicular to the page here. Note that this is a map view.

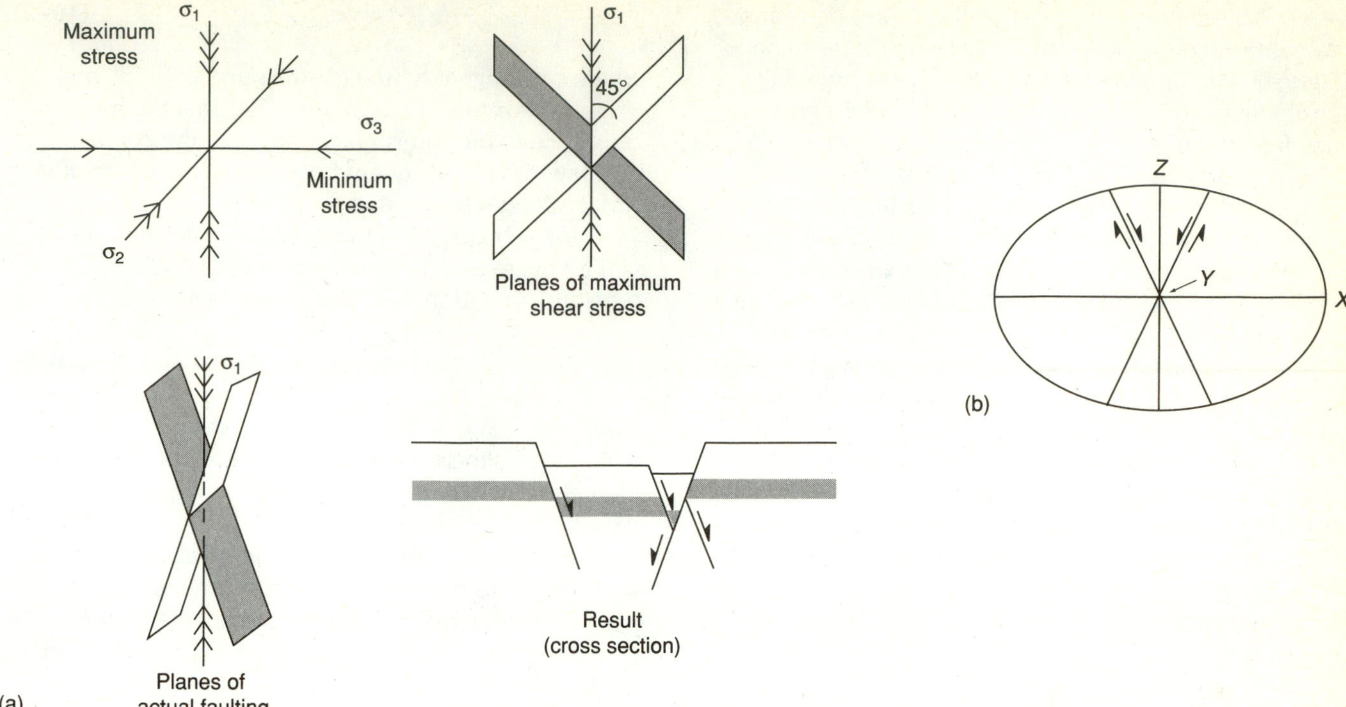

FIGURE 10–6
(a) Orientation of principal stress axes for normal faulting. (b) Strain ellipse for normal faults. *Y* is horizontal and perpendicular to the page. View is parallel to strike of faults.

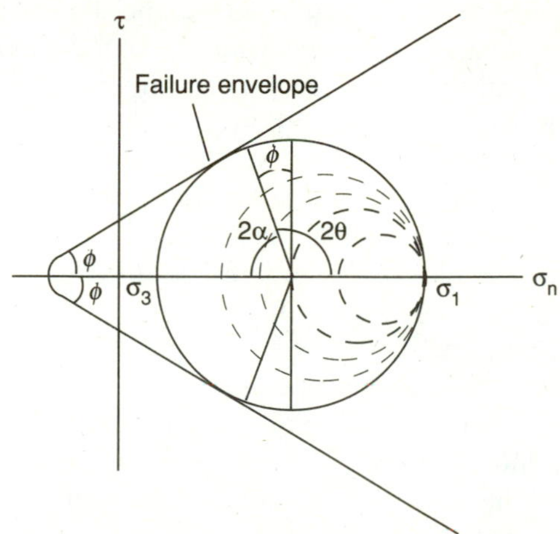

FIGURE 10–7
Mohr-circle analysis of normal faulting: σ_3 decreases with σ_1 constant (increasing the size of the Mohr circles, dashed circles) until the Mohr circle intersects the failure envelope.

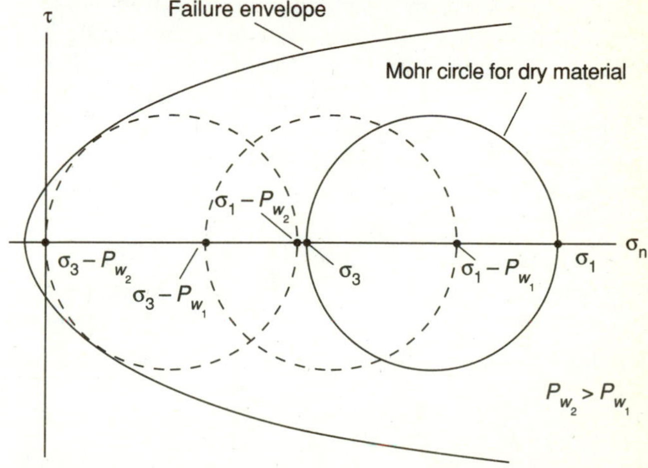

FIGURE 10–8
Mohr circles showing the effect of fluid pressure (P_W) on effective normal stress. Increasing fluid pressure reduces the strength of the material and forces it toward the unstable region outside the failure envelope. Note the curved envelope, indicating a component of ductile behavior.

fault to slip, as the fluid pressure reduces the normal stress on the fault plane. The resulting stress is *effective normal stress (S),* and we can determine it by subtracting the fluid pressure *(P)* from the normal stress (σ_n). Mohr circles are shifted to the left toward the failure envelope and then touch it, leading to rupture and movement along the fault (Figure 10–8). This concept is an oversimplification requiring resolution of both the horizontal and vertical components of effective normal stress. Horizontal effective normal stress is more easily assumed to be the same in all directions, whereas the vertical effective normal stress varies with the weight of the overburden minus the vertical component of fluid pressure. If higher pressure fluid from a nearby layer under higher pressure ("overpressured") is forced into a rock layer under lower fluid pressure, the values of σ_1 and σ_3 will be decreased by an additional amount

($-\Delta P$). This causes a decrease in the values of σ_1 and σ_3, but the size of the Mohr circle remains the same (Figure 10–8) and the circle is still shifted toward the tensile field, so that an additional rise in pore pressure may lead to failure at values of zero or negative (tensile) effective normal stress (Mandl, 1988).

The effect of fluid on movement is illustrated vividly by landslides and snow avalanches. Conditions favoring landslides may exist for centuries in an area without initiating slides (Chapter 2), then the influx of a large amount of water into the susceptible materials may reduce the effective normal stress within the mass and trigger a landslide. Snow avalanches travel at high velocity on a cushion of air—another fluid that decreases effective normal stress. In unconsolidated sediment in accretionary wedges and deltas, movement along faults is made easier in sections overpressured by fluid that cannot escape. Normal stress is borne by the confined fluid rather than by the sediment, facilitating movement.

FRICTIONAL SLIDING MECHANISMS

Some movement will occur on any fracture produced by shear. If appreciable movement is to occur on an existing fault surface, the coefficient of friction along the fault must be overcome. Two laws concerning friction were probably first discovered by Leonardo da Vinci and then were rediscovered by Guillaume Amontons (1663–1705), a French physicist who first presented them to a scientific gathering in 1699 (Suppe, 1985). ***Amontons' first law*** states that the tangential force parallel to a fracture surface necessary to initiate slip is directly proportional to the force normal to the fracture surface. The proportionality constant μ is the coefficient of friction. Dividing force by area, Amontons' first law may be stated as

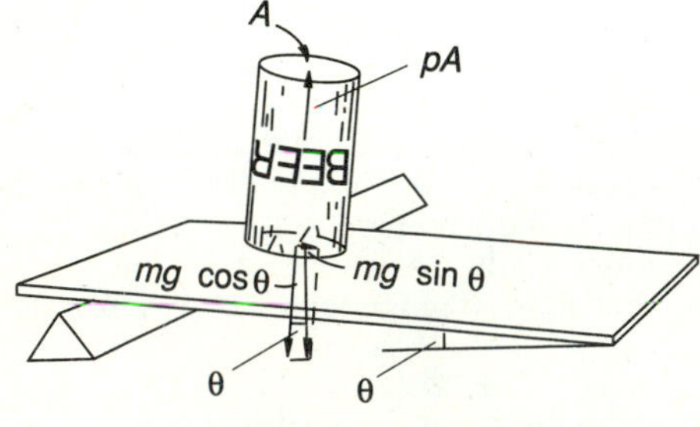

FIGURE 10–9
The Hubbert and Rubey beer-can experiment. If the glass plate is dry, $\theta > 15°$; if wet, $\theta < 2°$; m—mass of can, g—acceleration due to gravity, A—area of base of can, p—excess pressure of air inside can over that outside. (From M. K. Hubbert and W. W. Rubey, 1959, Geological Society of America *Bulletin*, v. 70.)

$$\tau = \mu\sigma_n, \qquad \textbf{(10–2)}$$

where τ is tangential (shear) stress and σ_n is normal stress. ***Amontons' second law*** states that the frictional resistance to motion is independent of the contact area. This law has an important bearing on the nature of fault surfaces and resistance to movement.

If water occurs in the fault zone, normal stress will be decreased by the quantity P_w, which is water pressure. Equation 10–2 then becomes

$$\tau = \mu\,(\sigma_n - P_w) = \mu S \qquad \textbf{(10–3)}$$

where $(\sigma_n - P_w)$ is effective normal stress, S. Therefore, water lowers the shear stress required for failure and fault motion and allows movement to begin at lower values of τ.

In 1959 Hubbert and Rubey illustrated the effective stress relationship, using only an empty beer can on a glass plate (Figure 10–9). First, they placed the can (open end down) on the dry plate and lifted one side of the plate until the can moved—an angle greater than 15°. Then they lowered the plate and wetted it with water, chilled the can, and again placed it on the wet surface. This time, when they raised one side of the plate, only a slight incline—about 2°—was needed for the can to begin sliding down the slope. As the can warmed and the air inside tried to expand, the buildup of air pressure along the wetted surface of the glass plate permitted movement of the can on a very low slope. The weight of the can corresponds to a vertical force (Figure 3–8), and the air pressure corresponds to the fluid pressure. A similar laboratory bench model used a concrete block weighing several hundred kilograms. The block was smooth and rested on a smooth surface. As long as the surfaces remained dry, the block could be moved only with great difficulty, but when a film of water between the two smooth surfaces counteracted the total stress and buoyed up the block, the block moved so easily that Hubbert and Rubey said that it was dangerous to stand near the heavy and unstable block.

MOVEMENT MECHANISMS

Movement on faults occurs in at least two different ways: by ***stick slip*** (unstable frictional sliding) and by ***stable sliding*** *(continuous creep)*. The stick-slip mechanism (Figure 10–10) involves sudden movement on the fault after long-term accumulation of stress (Scholz, 1990). This mechanism and the accompanying *elastic rebound* (Chapter 3) are what we think cause earthquakes. The stable-sliding mechanism involves uninterrupted motion along a fault, so that stress is relieved continuously and does not accumulate. The difference in behavior may be produced along segments of the same active fault undergoing stable sliding where ground water is abundant, but other segments, with less ground water, may move by stick slip. Some segments of the San Andreas fault in California exhibit stick-slip behavior; others, stable sliding. Other and more complex factors, such as curvature of the fault surface, may determine the mechanism involved.

Withdrawal of ground water may cause near-surface segments of active faults to switch mechanisms, from stable sliding to stick slip, thereby increasing the earthquake hazard. The converse may also be true, but there are difficulties. Pumping fluid into a fault zone has been proposed as a way to relieve accumulated elastic strain energy and reduce the likelihood of a large earthquake, but the rate at which fluid should be pumped into a fault remains unknown. Pumping fluid into a locked fault zone does raise the fluid pressure and lowers effective normal stress, but if pumping is too rapid, it could trigger earthquakes artificially by allowing the stick-slip mechanism to go to completion rather than slowly bleeding off accumulated stress by creep. Studies by Barry Raleigh, Healy, and John Bredehoeft (1976) on the fluid-induced seismic activity at Rangely, Colorado, and David Simpson (1986) on reservoir-induced and other man-initiated earthquakes, suggest that the Earth must be in a state of near failure (high *in situ* stress) for increases in fluid pressure to have a significant influence.

Fault Surfaces and Frictional Sliding

Fault surfaces between two large blocks are never perfectly smooth and planar, especially on the microscopic scale. Microscopic to megascopic irregularities and imperfections, called ***asperities*** (Figure 10–11), increase the resistance to frictional sliding because of the overall irregularly shaped fault surface; they also reduce the percentage of surface area actually in contact. F. Bowden, a metallurgist, first suggested that because of asperities, even the best-prepared surfaces will not fit together and that breaking asperities increases the contact area. The initial contact area may be as little as 10 percent, but when motion begins, irregularities on the surface are removed, and as the displacement increases, the contact area increases (Teufel and Logan, 1978). Because asperities provide the surface with frictional strength and resistance to movement, they must be broken through in order for movement to occur. Rock-mechanics experiments on fault surfaces demonstrate that resistance to frictional sliding increases after movement begins (strain hardening), suggesting an increase in strength or number of

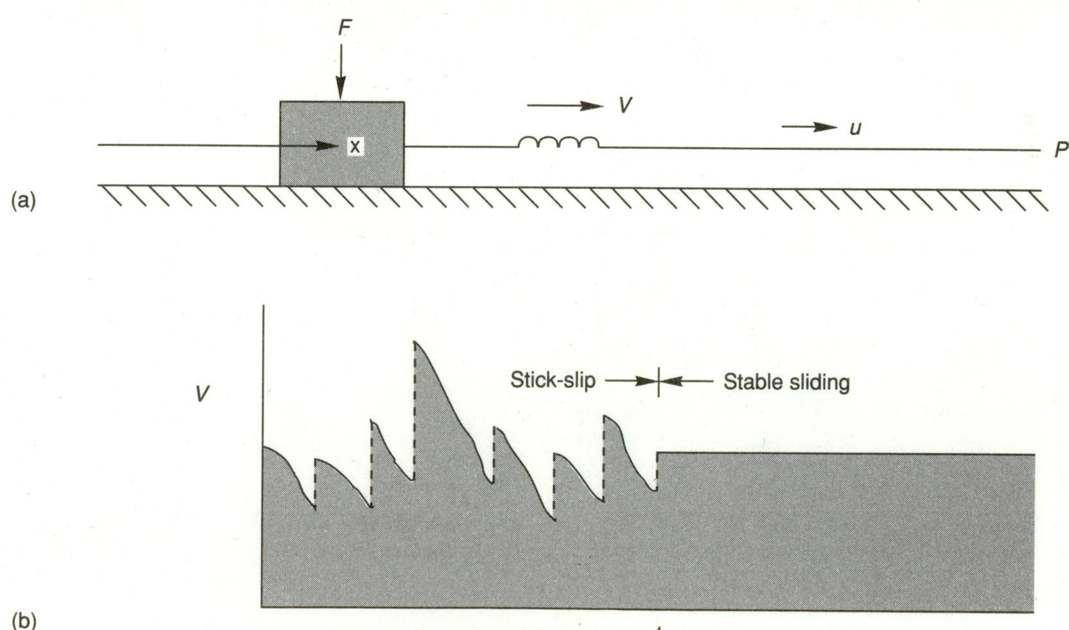

FIGURE 10–10
(a) To move an object connected by a spring by a line load (*P*) (e.g., a strong cable), friction (*F*) must be overcome before the object will move. The spring will stretch as the line is tightened at velocity (*u*) until friction with the surface is overcome, and the object will move at high velocity as the spring relaxes. If the spring remains tight and the block is not allowed to cease moving, the block will continue to move by stable sliding; if, on the other hand, the block stops even for a short time, friction will have to be overcome again, and the cycle involving stick slip will be repeated. (b) Diagram of time (*x* axis) versus velocity (*V*), illustrating stick-slip and stable-sliding mechanisms. (After James Byerlee, 1977, *Friction of rocks*, U.S. Geological Survey, Office of Earthquake Studies.)

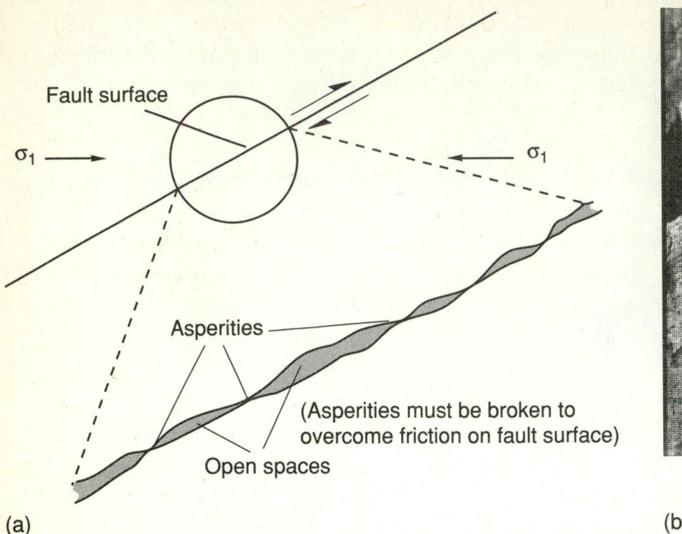

FIGURE 10–11
(a) Asperities and contact-surface area along a fault. (b) Holes and grooves in the footwall of a near-horizontal thrust fault at Vellerat near Choindez in the Jura Mountains, Switzerland. The holes were produced either by irregularities—asperities—on the footwall surface that were broken and removed during movement or by asperities in the hanging wall that gouged and plucked material from the footwall. Note the curved fractures above (south) of the large hole. They are extension fractures that indicate relative south-to-north motion of the hanging wall. Grooves (slickenlines) in the fault surface were also produced by irregularities on both the hanging wall and footwall surfaces. The orientation of the grooves indicate that motion was either south to north or north to south. Small steps on some of the grooves (not visible in the photo) suggest motion was south to north. (Locality courtesy of Hans P. Laubscher, University of Basel, Switzerland. RDH photo.)

asperities in the contact zone, but they also indicate dependence on both velocity and displacement (Scholz, 1990). The increase may result from the greater contact area, requiring that more asperities be broken to initiate movement, but the strength of the fault zone decreases again with continued movement. Thus, despite the increased contact area, additional movement ultimately reduces the asperity population and, hence, also the strength of the fault surface.

James Byerlee (1977, 1978) has shown experimentally (Figure 10–12) that for low effective normal stress, S (<0.2 GPa)—the usual condition in the upper crust—

$$\frac{\tau_{max}}{S} = 0.85 \,, \qquad \textbf{(10–4)}$$

where τ_{max} is maximum frictional shear stress, and 1 Pa, or pascal, is 1 kg m^1 s^{-2} (an MPa, megapascal, is 1 million pascals, and a GPa, gigapascal, is 1 billion pascals). This relationship has been called *Byerlee's law of rock friction* and is an important statement regarding faulting and the brittle behavior of most rocks under stress in the upper crust. That Byerlee's ratio for most rock types is 0.85 (Figure 10–12) indicates that the coefficient of friction in Amontons' laws is independent of rock type and depends solely on values of shear and normal stress. Byerlee also showed that for values of S between 0.2 GPa and 2 GPa,

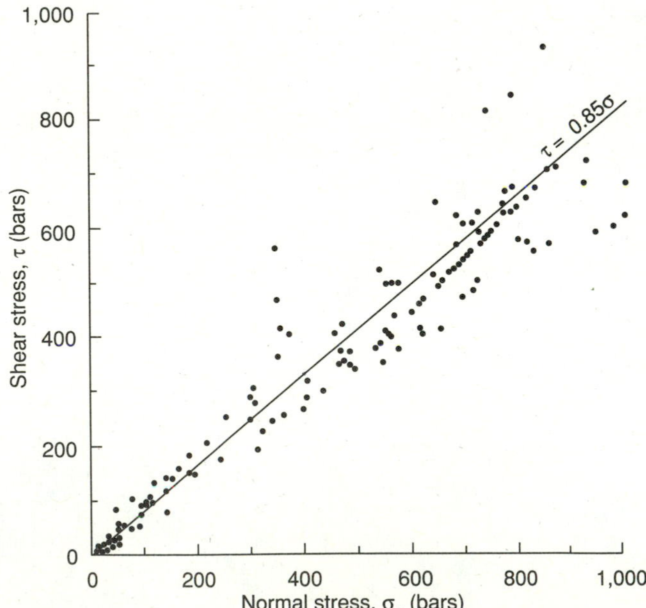

FIGURE 10–12
Least-squares fit of points derived from experimental measurement of maximum shear stress (τ) and normal stress (σ_n), producing a curve with the equation $\tau = 0.85\sigma_n$. Note that these measurements were made on sawed rock specimens, not fresh fractures. (From James Byerlee, 1977, *Friction of rocks,* U.S. Geological Survey Office of Earthquake Studies.)

$$\tau_{max} = 50 \text{ MPa} + 0.6\, S\,. \qquad \textbf{(10–5)}$$

These conditions probably occur at greater depth in the crust. There is some doubt that the 0.85 value for the coefficient of friction is valid—it is probably closer to 0.6—particularly because earthquakes occur in the shallow crust at ambient shear stress much below that suggested by Byerlee's experiments (Alsop and Talwani, 1984). Fluid, some clay minerals, or anything else that weakens rocks could cause failure at values lower than predicted.

SHEAR (FRICTIONAL) HEATING IN FAULT ZONES

The data gathered from frictional sliding experiments, the occurrence of veins of pseudotachylite (literally, false glass, Figure 10–13) in many deep-seated fault zones, the metamorphism along subduction zones, and the fact that mechanical work is accomplished during movement of faults led to the conclusion that frictional heat is generated along faults. The amount of work (W) accomplished in overcoming friction is related to the shear stress (τ) along the fault plane and the displacement (d) during a single movement event as

$$W = \tau d\,. \qquad \textbf{(10–6)}$$

The amount of heat (q) generated by the work done can then be estimated from

$$q = \frac{W}{J} = \frac{d\tau}{J}, \qquad \textbf{(10–7)}$$

where J is the mechanical equivalent of heat (4.186 joules calorie^{-1} gram^{-1} at 15° C). The heat generated can be related to an increase in temperature (T) by

$$q = mCT, \qquad \textbf{(10–8)}$$

where m is the mass of rock heated and C is the specific heat of the rock mass. Amount of heat (here heat flow) can also be related to movement velocity (v) and mean shear stress along the fault as

$$q \leq v\tau\,. \qquad \textbf{(10–9)}$$

According to Scholz (1990), equation 10–9 predicts that if the velocity of motion of a fault is greater than 1 cm/y and shear stress is greater than 50 MPa, shear heating as a consequence of faulting will be significant.

Scholz (1980) suggested that for the effects of shear heating to be recognized in geologic systems, temperatures must be increased by several hundred degrees Celsius within several kilometers of the fault and must be maintained long enough for metamorphic reactions to occur. The primary evidence for the effects of shear heating is twofold: (1) pseudotachylite, dark vein fillings of glass-like material thought to form by brittle failure and sudden release of elastic strain energy and friction-generated heat (Sibson, 1975); and (2) metamorphic reactions in subduction zones and accretionary complexes where greenschist- and blueschist-facies metamorphism in the fault zone can be directly related to heating of the overriding block by friction with the downgoing slab (Pavlis, 1986; Peacock, 1992). Frictional melts that produce pseudotachylite seem to be concentrated along very

FIGURE 10–13
Dark veins of pseudotachylite in lower-crustal gneiss (granulite facies stronalite) from the Ivrea zone, Valle dell Infierno near Ornavasso, Italy. Note that the large veins cross smaller veins that are perpendicular to the larger. (Locality courtesy of Hans P. Laubscher, University of Basel, Switzerland. RDH photo.)

thin fault zones (<1 cm, Sibson, 1973). H. Austrheim and T. M. Boundy (1994) have found pseudotachylite in the western gneiss region of southern Norway that formed at 800° C and 18 to 19 kb (≥ 60 km depth), demonstrating that they can form in the lower crust or upper mantle. In addition, metamorphic aureoles in footwall rocks beneath ophiolites (Williams and Smyth, 1973; Jamieson, 1986) have been cited as evidence for shear heating (e.g., Scholz, 1980), although it is difficult to prove that the heat was not acquired at greater depth in the crust before emplacement, and an already hot slab overthrust a cool footwall. Scholz (1980) has cited the Alpine fault in New Zealand as an excellent example of the direct relationships between a large crustal fault and metamorphic isograds that indicate increased grade closer and subparallel to the fault. Fault rocks within the Alpine and many other crustal fault zones, however, contain deformed retrograde (lowered or decreased grade compared with the surrounding rocks) hydrous minerals such as chlorite and muscovite. This suggests that the fault zone served as a conduit for rapid fluxing of large amounts of water and dissipation of heat during deformation, rather than as a locus for heat accumulation and metamorphism. This idea is supported by geochemical and isotopic evidence of large quantities of water moving through crustal fault zones to convert anhydrous minerals (e.g., feldspars, garnet, kyanite) to hydrous minerals and selectively remove particular elements and isotopes (Sinha and others, 1988). The parallelism of metamorphic isograds to the Alpine fault can also be explained without shear heating by upward tilting of the crust, exposing deeper levels—and metamorphic zones—near the fault by a dip-slip component of movement. Andrew Nicol and Donald Wise (1992) have shown that the Alpine fault underwent a complex movement history from the Late Cretaceous to the present involving both NW-directed (dip-slip) and SW-directed (dextral strike-slip) motion within the same plate-tectonic regime.

Friction-related heating along faults is a process that clearly occurs in the Earth, but it is difficult to demonstrate except along faults containing pseudotachylite and within subduction zones.

BRITTLE AND DUCTILE FAULTS

So far, most of our discussion of fault mechanics has been directed toward brittle faults in the upper 5 to 10 km of the Earth's crust. Faults in the upper crust consist of either a single movement surface or an anastomosing complex of fracture surfaces (Figure 10–14). An individual fault may have knife-sharp contacts (Figure 10–15), or it may consist of a zone of cataclasite (Chapter 7). The brittle-ductile (or brittle-plastic) transition (Chapters 4 and 6) is thought to occur 10 to 15 km deep in continental crust (Figure 10–16) but, remarkably, may occur locally very close to the surface where the combination of available water and increased heat permits the transition to occur. Faults that penetrate the entire crust, or most of it, must involve both ductile and brittle behavior. Large active strike-slip faults, such as the San Andreas fault in California, the Anatolian fault in Turkey, and the Alpine fault in New Zealand, probably penetrate most of the lithosphere, for they also are plate boundaries (Figure 10–17a). They consist of brittle fault zones at the surface but at great depth probably are generating mylonite. Direct evidence for the transition may be found in the variability of both brittle and ductile fault rocks at the present-day surface along some faults such as the Alpine in New Zealand (Reed, 1964) and the Borrego Springs–Santa Rosa mylonite zone in California (Simpson, 1984, 1985). Thrust faults that displace large blocks of crystalline rock may be generated within or below the ductile-brittle transition and propagate into the brittle upper crust (Figure 10–17b). Ultimately, the entire thrust sheet may be displaced onto upper-crustal rocks. Crustal extension, as in the Basin and Range

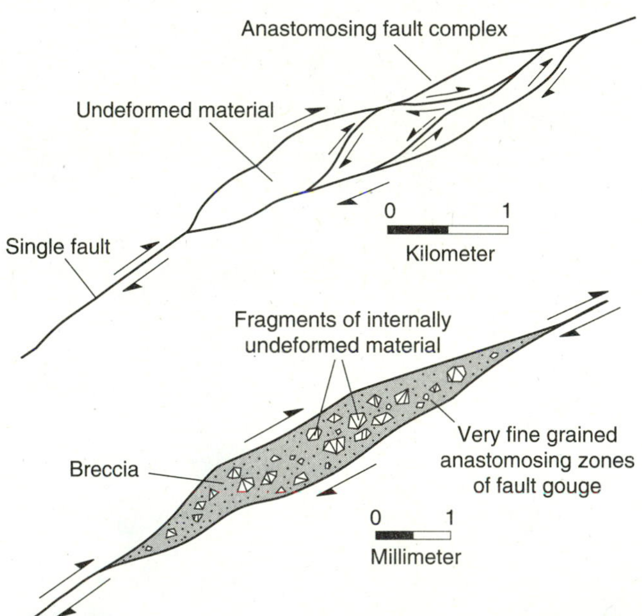

FIGURE 10–14
Characteristics of brittle faults in the upper crust. (a) Map view of a fault zone showing distribution of movement surfaces and the way a fault zone may change from a single fault to an anastomosing fault zone. (b) Close-up view of one of the faults in (a). Note that fragments of internally undeformed material survive to the microscopic scale.

(a)

(b)

(c)

FIGURE 10–15
(a) Knife-sharp fault contact along the Highgate Springs thrust (part of the Champlain thrust system) at Low Rock Point near St. Albans Bay, Vermont. Lower Cambrian Dunham Dolomite in the hanging wall is thrust over Middle Ordovician Iberville Shale. (RDH photo.) (b) Zone of cataclasite along a brittle fault, southern Wales. Note the dark rocks to the left of the lighter cataclasite in the fault zone (center of photo). (William M. Dunne, University of Tennessee.) (c) Cataclasite from the Homestake shear zone near Leadville, Colorado. An older ductile fabric of quartz and feldspar was fractured under brittle conditions. Width of field is approximately 7 mm. Plane light. (RDH photo.)

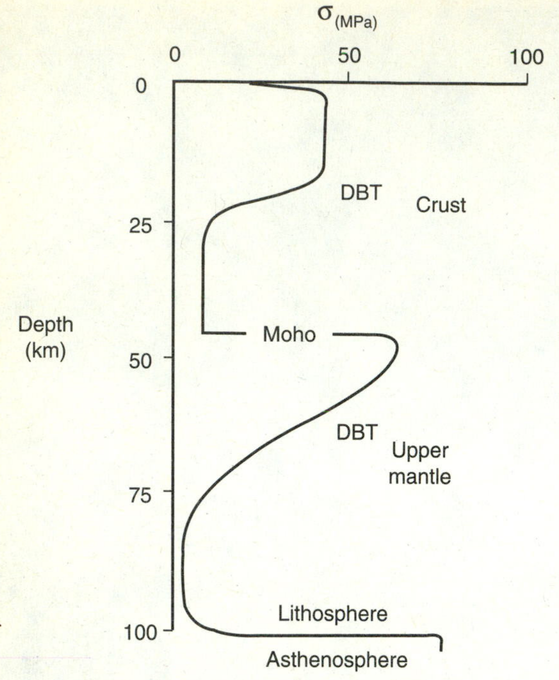

FIGURE 10–16
Model for the crust and upper mantle based on the mechanical behavior of dry and wet quartz for the crust and dry olivine for the mantle. DBT—ductile-brittle transition.

Province, may generate normal faults that on a regional scale propagate downward into the ductile-brittle transition (Eaton, 1979, Figure 10–17c).

SHEAR ZONES

Our previous discussions of ductile and brittle faults led directly to consideration of the characteristics of ***shear zones***. They are produced by both homogeneous and inhomogeneous simple shear, or oblique motion, and are thought of as zones of ductile shear (Figure 10–18a), although the term has been used to describe fault zones where brittle behavior dominates. John Ramsay (1980) has classified them as brittle, brittle-ductile, and ductile shear zones (Figure 10–18b).

Attributes of ductile shear zones have been listed by J. Carreras and Stanley H. White (1980), slightly modified here:

1. *Shear zones on all scales are zones of weakness and represent localized strain softening.*
2. *Strain softening in ductile shear zones is associated with formation of mylonite and may result from several processes.*

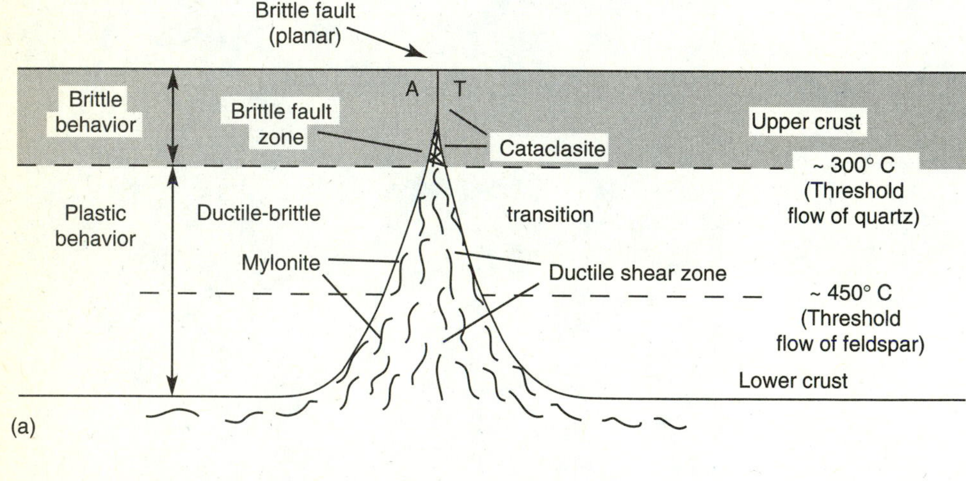

FIGURE 10–17
(a) Behavior of a large strike-slip fault in the upper and lower crust. A—away from the observer; T—toward the observer. (b) Generation of large crystalline thrusts by movement along the ductile-brittle transition zone and propagation into the upper crust. (c) Propagation of normal faults along the ductile-brittle transition zone in zones of crustal extension, and into the brittle upper crust.

(a)

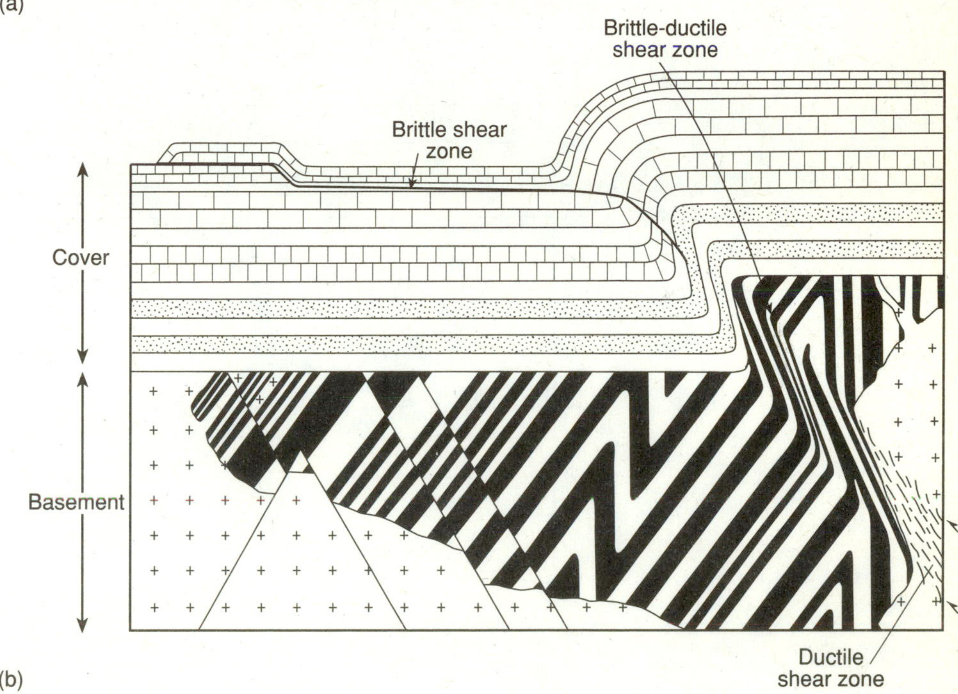

(b)

FIGURE 10–18
(a) Near-horizontal ductile shear zones in augen gneiss near Vålån in the Swedish Caledonides. Two well-defined ductile shears are visible just below the coin. Note how porphyroblasts are rotated dextrally (clockwise) by the shear zones. (RDH photo.) (b) Shear-zone development in part of the crust in a state of compression. Note the brittle shear zone in the cover undergoes a transformation at depth as it becomes more ductile and affects a larger part of the crust, then becomes a wide zone of ductile deformation deep in the basement where the crust is at a higher temperature. (Reprinted from *Journal of Structural Geology*, v. 2, J. G. Ramsay, p. 83–99, © 1980, with kind permission from Elsevier Science, Ltd., Kidlington, United Kingdom.)

3. *Transient brittle behavior (rate-dependent?) can occur in ductile shear zones.*
4. *Sheath folds [Chapter 15] are common in mylonite zones and apparently can form in both steady-state and non-steady-state flow regimes.*
5. *There is little evidence to indicate that shear heating is common in ductile shear zones.*
6. *Shear zones may act both as closed and open geochemical systems with respect to movement of fluids and elements, irrespective of the size of the zone.*

Ramsay (1980) added to these attributes the following boundary conditions as geometric features of ideal shear zones: (1) shear zones generally have parallel sides, and (2) displacement profiles along any cross section through a shear zone should be identical. His second condition implies that small-scale structural features and strain profiles across a shear zone in any cross section should be identical, except near the ends. This should permit study of finite strain and the geometry of any shear zone along several representative cross sections. Most real shear zones approximate

ESSAY

Artificial Earthquakes

Should we try to artificially relieve accumulated elastic strain energy on large faults otherwise certain to cause sizable earthquakes? The San Andreas fault is likely to produce an earthquake of magnitude 8 or larger in Southern California before the year 2020. One proposal is that we pump fluid into a locked segment of the fault zone at a suitable rate so that the strain energy will be relieved by very small earthquakes and thus minimize damage and loss of life. No one has ever successfully carried out such an experiment, however. Should Southern California cities be evacuated before pumping? Who will pay for damages produced by the small earthquakes? What if pumping triggers a large earthquake and devastates the region anyway? Who bears legal responsibility? Such questions discourage attempts to try the experiment.

Artificial earthquakes were produced unintentionally in the early 1960s as a by-product of disposal of toxic chemicals in a 12,045-foot well at Rocky Mountain Arsenal near Denver, Colorado. In 1965, geologist David M. Evans suggested that occurrence of earthquakes after disposal of fluid there from 1962 until 1965 showed that it was possible to influence the mechanism and timing of release of accumulated elastic strain (Evans, 1966). He plotted the occurrence of earthquakes from March 1962 to November 1965 and correlated them with the pumping of fluid into the well at the arsenal. His data showed that earthquakes began within a month after pumping started, decreased markedly from September 1963 through September 1964, when no fluid was injected—and resumed with greater frequency when pumping was restarted in September 1965 (Figure 10E–1).

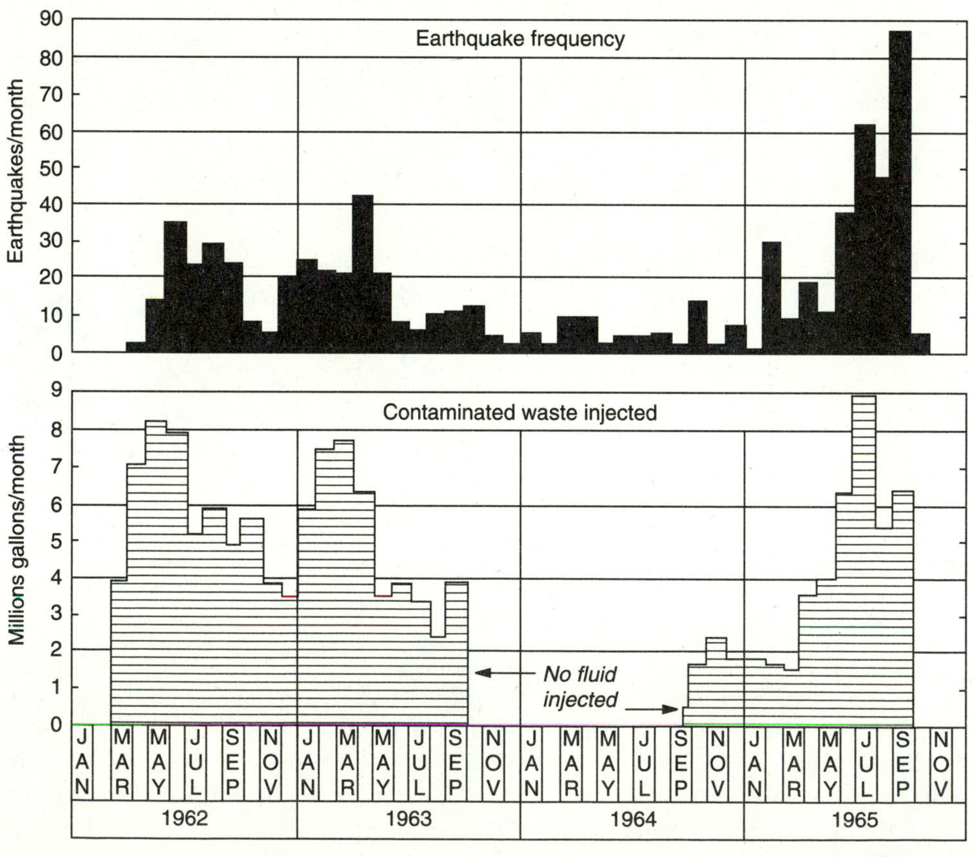

FIGURE 10E–1
Correlation of earthquake frequency with the injection of liquid waste at Rocky Mountain Arsenal from 1962 to 1965. (From D. M. Evans, 1966, *Geotimes*, v. 10.)

The waste fluid was pumped into Precambrian crystalline basement rock beneath the Paleozoic to Cenozoic sedimentary cover, with disposal in fractures that did not communicate with the present-day surface ground-water system. The fluid was initially injected under pressure, with the pressure relieved by flow along existing fractures enlarged in the crystalline rocks. New fractures were expected to form by hydrofracturing during injection.

Although seismic activity was not unknown, no earthquakes had been felt or recorded in the Denver area between 1882 and the beginning of pumping in 1962. From April 1962 until September 1965, local seismic stations recorded 710 earthquakes with epicenters in the vicinity of the Rocky Mountain Arsenal. Magnitude ranged from 0.7 to 4.3 on the Richter scale, and about 75 of the earthquakes were felt. Most foci were determined to be from 4.5 to 5.5 km deep (Evans, 1966)—about the same depth as the bottom of the well. Low-level elastic strain energy accumulated and was relieved by fluid pressure, reducing the effective normal stress so that elastic strain could be relieved by shear on appropriately oriented fracture surfaces.

No one doubts that the Denver earthquakes were produced by deep-well injection. Public opinion brought an end to the pumping, and contractors at Rocky Mountain Arsenal devised a chemical means of breaking down the toxic waste in order to dispose of it conventionally.

The events in Denver proved that we can induce earthquakes, but to enter a major active fault zone and attempt to relieve stored elastic strain energy artificially—before nature releases it suddenly and without warning— requires detailed knowledge of the fracture system associated with the fault at great depths. We do not have the data necessary to predict how much fluid should be pumped, where it should be pumped, and how rapidly it should be pumped into the system to lower the effective normal stress and bleed off the excess energy. Apparently, the crust in the Denver area did not contain a great excess of accumulated and stored elastic strain energy, as does the crust along the San Andreas fault in Southern California. Before we can attempt to drain off energy from a large and active fault zone, we must answer many questions—scientific, sociological, political, and legal.

Reference Cited

Evans, D. M., 1966, Man-made earthquakes in Denver: Geotimes, v. 10, no. 9, p. 11–18.

these features. By considering the various types of shear zones in light of these boundary conditions, Ramsay concluded that three kinds of displacement can occur in shear zones: (1) heterogeneous simple shear, (2) heterogeneous volume change largely by pressure solution, and (3) combinations of the first two.

SHEAR-SENSE INDICATORS

Determination of shear sense or direction of movement along ductile shear zones is an important factor that may later help to estimate displacement. Sense of shear may not be immediately obvious in many larger shear zones, and so indicators of shear sense on the mesoscopic and microscopic scales may be used ultimately to determine the sense of motion of shear zones that are kilometers wide. The shear sense commonly observed was produced by the *last* ductile event, although relicts of earlier events may sometimes be preserved.

Various criteria may help to determine shear sense, as summarized by Carol Simpson and Stefan Schmid (1983), Gordon Lister and Arthur Snoke (1984), Cees Passchier and Simpson (1986), Simpson (1986), and Simon Hanmer and Passchier (1991). The following discussion is partly based on their compilations.

Asymmetric Augen, Porphyroblasts, Porphyroclasts, and Pressure Shadows

Porphyroblasts are large grains that have grown in the rock mass during deformation and metamorphism, and *porphyroclasts* are relict earlier large grains that have survived ductile deformation. Porphyroblasts and porphyroclasts are commonly objects (e.g., feldspars, composite feldspar-quartz, garnet) that are harder than the matrix in which they reside, and so they behave much like ball bearings that allow the softer matrix to flow past. If they are rotated in a more ductile matrix—commonly finer grained—the rotation produces structures associated with these more brittle objects that provide clear indicators of shear sense (Figure 10–19). The associated structures are examined in the following sections.

Inclusions. Augen gneisses and mylonitic gneisses deformed ductilely by strongly asymmetric simple-shear commonly develop "tails" on porphyroclasts; the tails are asymmetric in the direction of ductile flow (Figures 10–19, 10–20a). Smaller grains making up the tails may be derived from both the porphyroclasts and the recrystallized groundmass, although compositional similarities to the porphyroclasts suggest that most of the material in the tails is derived from the porphyroclasts. Passchier and Simpson (1986) recognized that the asymmetric tails form a σ or δ shape, depending on the ratio of recrystallization rate to shear strain rate (Figure 10–21). They showed experimentally, using spherical porphyroclasts, that δ tails are formed at low ratios (0 to 0.033), whereas σ tails form at high ratios (0.066 to 0.13). Robert Hooper and I (1988), however, have recognized both σ- and δ-type porphyroclasts within 0.5 mm of each other in the same thin section, along with an additional type without tails that we called θ porphyroclasts. There may thus be other variables involved than the ratio of recrystallization rate to shear strain rate. Simpson and De Paor (1993) also attempted to account for the simultaneous formation of σ and δ porphyroclasts in a shear zone characterized by different flow (strain) rates where they have recognized sub- (low-) simple shear and super- (high-) simple shear flow regimes.

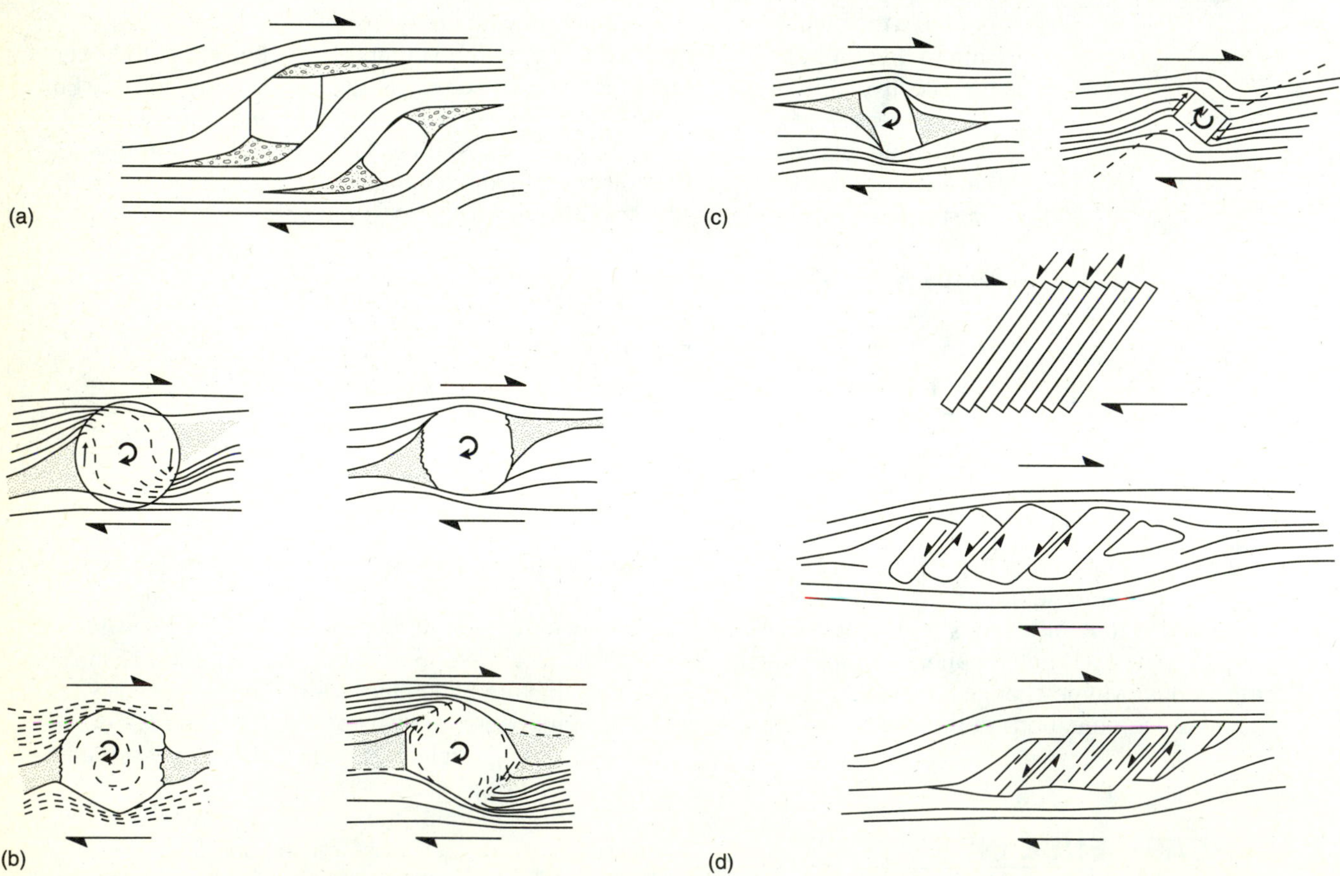

FIGURE 10–19
(a) Rotated porphyroclasts, (b) crystals, (c) pressure shadows, and (d) fractured grains. Arrows indicate movement sense. Note that fractured porphyroclasts frequently yield an opposite movement sense. (From Carol Simpson and Stefan Schmid, 1983, Geological Society of America *Bulletin,* v. 94.)

Hanmer and Passchier (1991) have termed the θ-type porphyroclasts "naked inclusions," and those with appendages (σ- and δ-type) "winged inclusions." They also suggested that both the deviation of the inclusion from perfect spherical shape (expressed as the aspect ratio) and the rotation rate of the wings relative to the inclusion can affect the shape of the wings attached to the porphyroclast, with more than one mechanism possible to achieve the same geometry. Despite the absence of tails, the shear sense can be determined using θ-type porphyroclasts because of the asymmetric flow of finer-grained matrix around the porphyroclast.

Boudins. Boudins (Figure 10–20b), which are sausage-shaped features formed by extension of a strong layer in a more ductile groundmass (Chapter 18), are sometimes rotated, producing pressure shadows or wrapped foliations that may indicate shear sense. They may also internally preserve an earlier foliation that can be used in concert with the enclosing foliation to determine the rotation sense. Boudins commonly rotate less than the groundmass.

Inclusion Trails. Inclusion trails in garnet, staurolite, cordierite, feldspars, and several other minerals may have a "snowball," an S, or a spiral asymmetric pattern (Figures 10–19b, 10–20c) that indicates shear sense. These crystals were rotated during crystallization in a more ductile matrix, recording the sense of rotation. Most inclusion trails record prograde deformation rather than retrograde. Care should be exercised in using inclusion trails in minerals such as garnet that

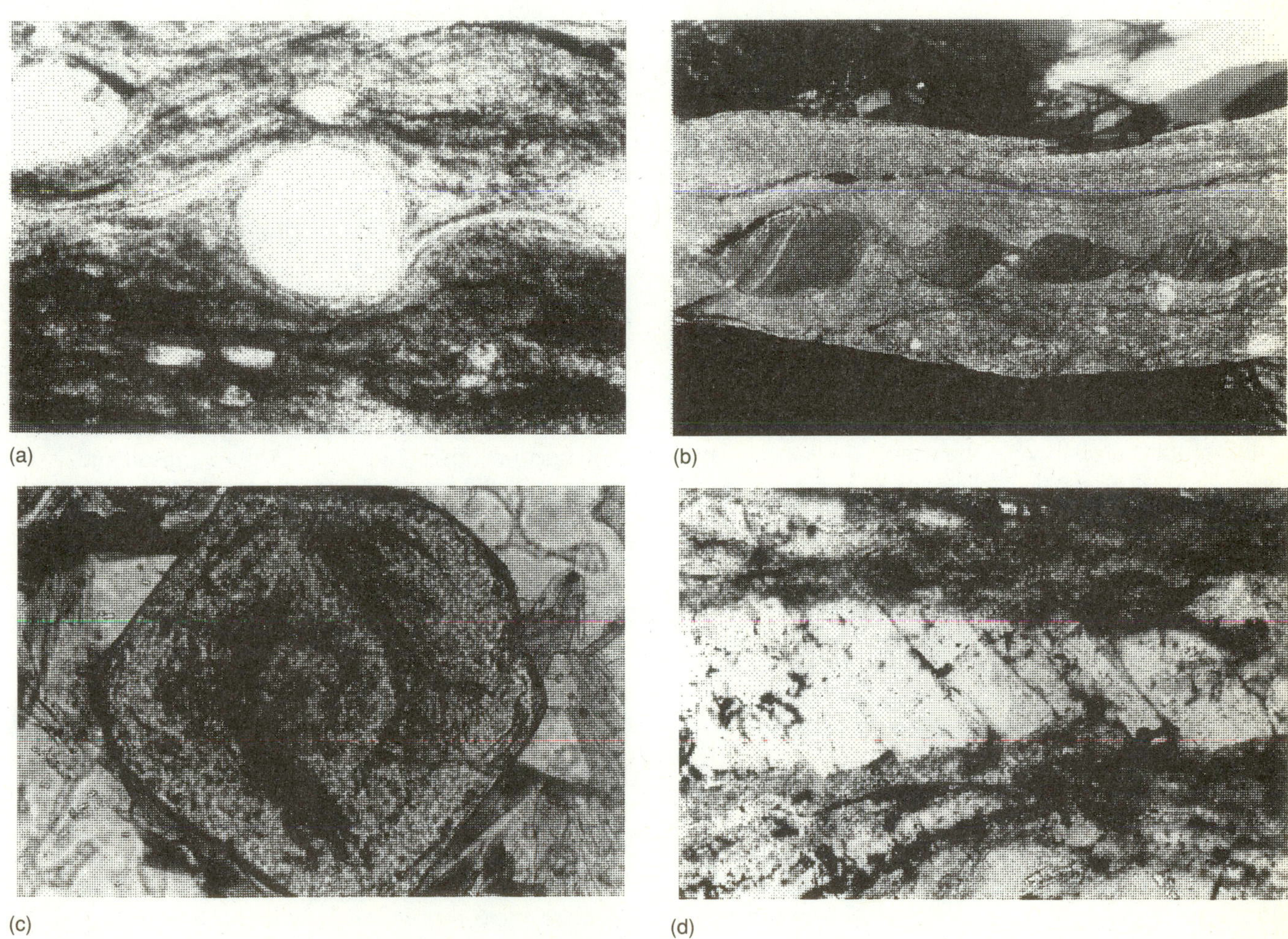

FIGURE 10–20
(a) Feldspar porphyroclast with asymmetric δ tails indicating (dextral) movement. Towaliga fault zone near Forsyth, Georgia. Width of photo is approximately 8.5 mm. Plane light. (From *Tectonophysics,* v. 152, R. J. Hooper and R. D. Hatcher, Jr., p. 1–17, © 1988, with kind permission from Elsevier Science, Ltd., Kidlington, United Kingdom.) (b) Amphibolite boudins in biotite gneiss showing sinistral rotation of original layering at Woodall Shoals, South Carolina. (RDH photo.) (c) Rotated inclusion trails of graphite in garnet in Poor Mountain metasiltstone near Walhalla, South Carolina, indicating dextral shear. This garnet has a spiral "snowball" structure and is approximately 0.7 mm wide. Plane light. (RDH photo.) (d) Fractured and rotated quartz porphyroclast from the Homestake shear zone near Leadville, Colorado, indicating apparent dextral, but actual sinistral shear. The quartz grain is approximately 1 mm long.

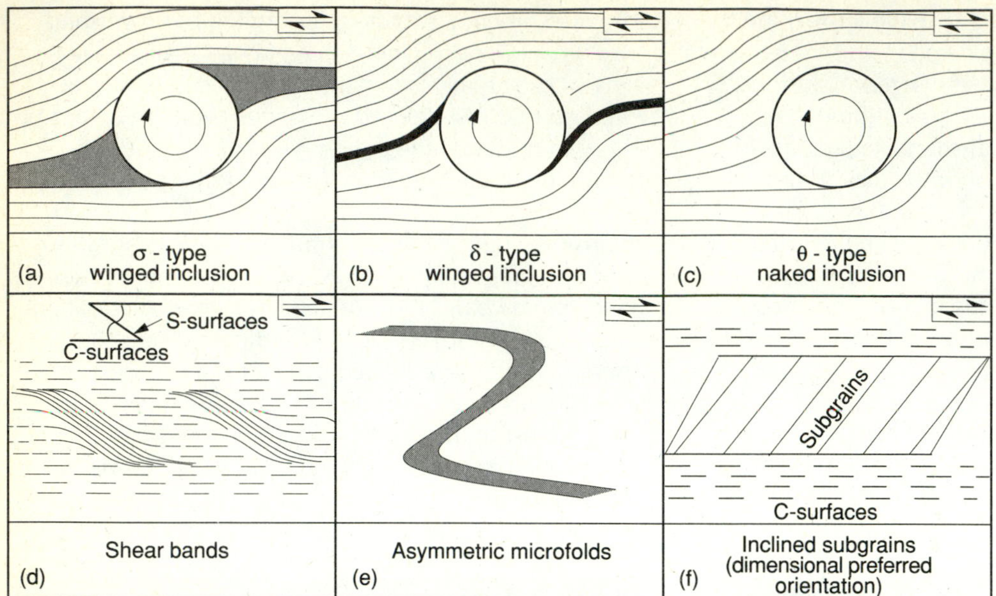

FIGURE 10–21
Several types of shear-sense indicators. Winged inclusions: (a) σ-type. (b) δ-type. Naked inclusions: (c) θ-type. Shear bands showing the orientation of C-surfaces (inclined) and relict S-surfaces (horizontal) (d). (e) Asymmetric microfolds. (f) Inclined subgrains (dimensional preferred orientation). Fine lines and dashed lines indicate shear (C) foliation. All rotation in each example is dextral (clockwise), as indicated by arrows. (Modified from *Tectonophysics,* v. 152, R. J. Hooper and R. D. Hatcher, Jr., p. 1–17, © 1988, with kind permission from Elsevier Science, Ltd., Kidlington, United Kingdom.)

record growth during a long time period because inclusion trails at a high angle to the enclosing foliation may indicate two directions of flattening and not simple shear (Bell, 1985; Vernon, 1988). According to Timothy Bell (1985), even asymmetric inclusion trails can be produced by different orientations of pure shear strains at different times, but spiral and snowball inclusion trails probably can be used to indicate shear sense.

Asymmetric Pressure Shadows. Pressure shadows formed adjacent to grains or crystals (e.g., pyrite, garnet) in a more ductile matrix may be used for determination of shear sense if the pressure shadows are asymmetric (Figure 10–19c). They function in the same way as asymmetric tails on porphyroblasts but display an opposite sense of asymmetry.

Distorted Layering. Geometry of distorted layering near porphyroclasts produces "drag fold" structures that indicate shear sense. The amplitude of the fold-like structures may also provide a measure of shear strain (Bjornerud, 1989). Peter Hudleston (1989) has recognized that paired folds and veins may form together in shear zones, and tension veins (dilational fractures) form about 45° to the shear zone boundaries. These veins are subsequently rotated into the shear direction, into an axial-planar orientation with respect to the folds, and are closed as the shear zone evolves. Paired hook folds adjacent to veins may serve as shear-sense indicators, provided they had a common origin.

Fractured Grains

Large grains of some minerals (for example, feldspars, amphiboles, pyroxenes, micas, and occasionally quartz) continue to deform brittlely, even in a ductilely deforming matrix. During progressive deformation in a zone of heterogeneous simple shear, they may fracture and become offset (Figure 10–20d). Simpson and Schmid (1983) have pointed out that these fractures indicate a sense of motion opposite that of actual movement (Figures 10–19d and 10–20d). In fact, the sense of motion depends on the original orientations of fractures.

Composite Foliations

S- and ***C-Surfaces.*** Foliations (parallel alignment of platy minerals or bands in a metamorphic rock) of several types develop in shear zones as products of progressive simple shear. A foliation that D. Berthé, P. Choukroune, and P. Jegouzo (1979) first called a ***C-surface*** (the C is from the French *cisaillement,* meaning shear) is related to shearing and forms in a shear zone during progressive simple shear (Figures 10–21d, 10–22). Depending on the initial orientation, slip may occur on the existing dominant foliation *(**S-surface**)* in the rock mass, but more frequently a new foliation made up of C-surfaces develops. We assume that the dominant simple-shear component is parallel to the walls of the shear zone and that C-surfaces form at an

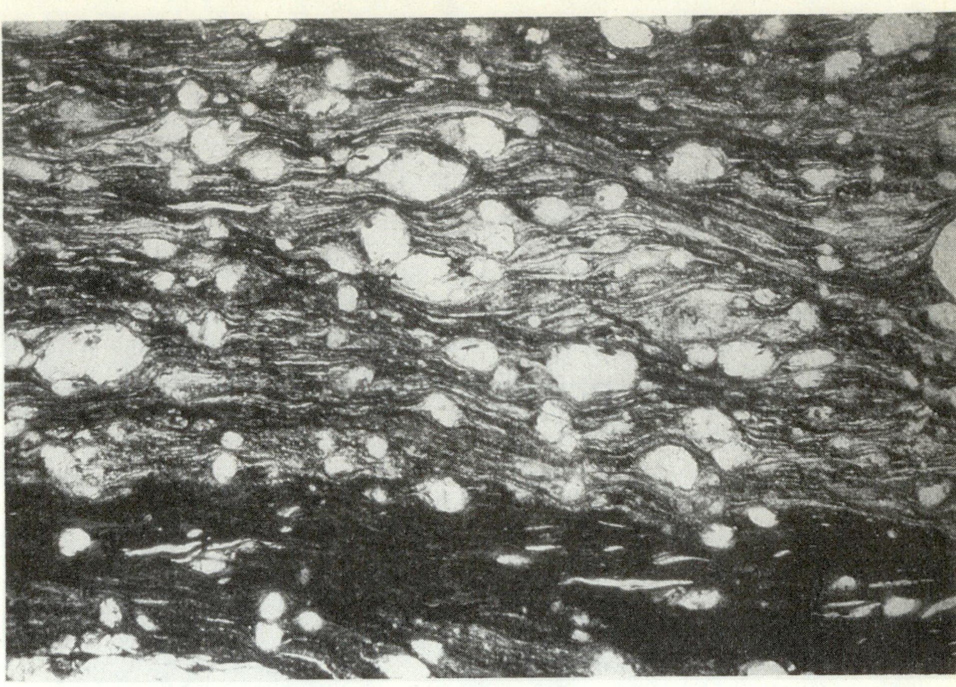

FIGURE 10–22
The acute angle between the older subhorizontal S- (or C-) surface (light-dark boundary near bottom of photo) and new inclined C-surfaces indicates the shear direction (dextral here). Mylonite from the Towaliga fault zone near Forsyth, Georgia. The C-surfaces parallel the sides of the shear zone. Width of field is about 4 mm. Plane light. (From *Tectonophysics,* v. 152, R. J. Hooper and R. D. Hatcher, Jr., p. 1–17, © 1988, with kind permission from Elsevier Science, Ltd., Kidlington, United Kingdom.)

angle of 18° to 25° to the shear-zone walls (Figures 10–22 and 10–24a). New C-surfaces continue to develop in this orientation and are rotated into parallelism with the shear zone walls as movement continues (Weijermars and Rondeel, 1984). S-surfaces may also form and rotate in the shear zone; sometimes they exhibit drag structure at the walls that indicate motion sense (Figure 10–23). The C-surface foliation in quartzofeldspathic or quartzitic rocks is the same as ***shear-band foliation*** (White and others, 1980) that forms in more micaceous rocks. Microscopically, both are recognized as thin layers of fine-grained recrystallized aggregates of minerals such as micas and quartz (Figure 10–24a). C-surfaces are thus transient structures that form and are superposed by younger C-surfaces until the shear zone ceases to move. After they are superposed and rotated, it may be possible to mistake older C-surfaces for S-surfaces or vice versa.

Lister and Snoke (1984) have called the S–C mylonites (just described) in quartzofeldspathic rocks Type I S–C mylonites. They defined another class, Type II S–C mylonites, as those resulting from mylonitization of quartz-mica rocks. Type II S–C mylonites contain flame-shaped buttons (or fish scales) of micas, which they called "mica fish." C-surfaces are defined by trails of mica, whereas S-surfaces are defined by an oblique foliation ("dimensional preferred orientation," DPO) in the adjacent quartz aggregate (Figure 10–21f).

Subgrains in quartz form at angles of 10° or less to the main lattice orientation and indicate shear sense in the same relationship as C- and S-surfaces (Lister and Snoke, 1984; Simpson, 1986). They appear as low-angle boundaries in a crystal having an extinction different from that of the larger host grain. Where the angle between the subgrain and host becomes greater than about 15°, recrystallization begins (Chapter 7).

Extensional Crenulation Cleavage. Another structure related to S- and C-surfaces originates by apparent extension and rotation as the shear zone moves; it is called ***extensional crenulation cleavage*** (Platt and Vissers, 1980; Figure 10–24a). Structures resembling intrafolial folds (folds lying within a foliation) appear between conveniently oriented active S-surfaces inclined opposite the shear direction and the C-surface orientation (Figure 10–25). They may result from shearing motion on an existing S-surface that becomes locked and then is forced to "ramp" to the next higher detachment S-surface on the hand-specimen to microscopic scale—a sequence similar to the detachment-ramp behavior of thrust faults at the road-cut to crustal scale (Figure 10–25). Formation of extensional crenulation cleavage requires a preexisting layering or layering produced during motion of the shear zone. An earlier foliation (S-surface), or new C-surfaces, provides the layering for formation of extensional crenulations (reverse-slip crenulations) (Dennis and Secor, 1987, 1990). Allen Dennis and Donald T. Secor (1987, 1990) noted that extensional crenulations permit the shear zone to shorten in the transport direction, whereas C–surfaces (structures they called normal-slip crenulations) permit material to extend in the transport direction (Figure 10–26).

Riedel Shears. Brittle shear fractures were first identified by W. Riedel (1929) in near-surface fault zones, and they have been produced experimentally in dry clay by J. S. Tchalenko (1970; Figure 10–27). ***Riedel shears*** and the oppositely moving structures called ***anti-Riedel shears*** form initially at very low strain, but movement is immediately transferred to shears that form subparallel to the shear-zone boundaries. Interesting comparisons can be made between the throughgoing shears that nearly parallel the Riedel and anti-Riedel shears and the C-surfaces that form in ductile shear zones. In contrast with normal and reverse-slip crenulations that form in ductile shear zones and function to thicken, thin, or maintain their thickness (Dennis and Secor, 1987), Riedel and anti-Riedel shears appear to form early in the movement history of a brittle shear zone and then are either destroyed or cease moving.

SHEAR-ZONE KINEMATICS

Regardless of whether a shear zone forms under brittle or ductile conditions, the kinematics appears similar and is also independent of scale. After simple shear is initiated and the rock mass begins to deform, the boundaries of the zone are defined. These boundaries expand as the zone evolves and the amount of shear strain in the zone increases. At advanced stages of movement, rotation of the existing dominant foliation (S-surfaces) or bedding approaches parallelism with C-surfaces, or shear bands, and the boundaries of the zone; Riedel shears and oppositely moving anti-Riedel shears form in brittle fracture zones. Depth, presence or absence of fluid, and P–T conditions determine to a

FIGURE 10–23
Truncation and rotation of an earlier foliation on the Chunky Gal Mountain fault near Shooting Creek, North Carolina. Foliation in the amphibolite to the right of the fault is rotated into near parallelism at the fault and is truncated by the near-vertical foliation in the biotite gneiss to the left of the fault. The rotation sense of the foliation is clockwise, indicating down-to-the-right motion on the fault. Despite this, the Chunky Gal Mountain is probably is a folded thrust, rather than a normal fault, based on the map relationships. (RDH photo.)

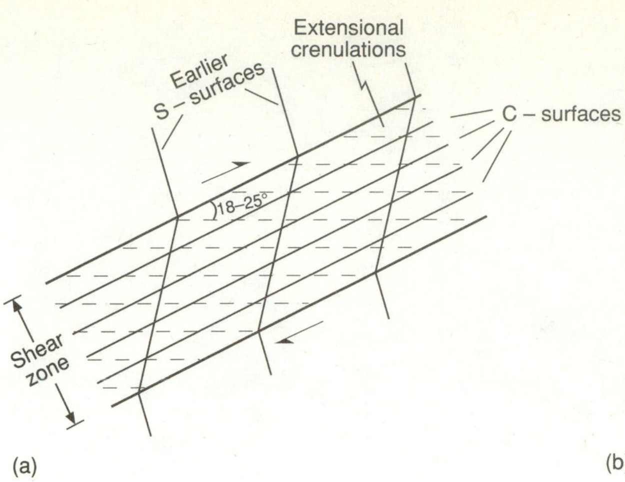

FIGURE 10–24
(a) Generalized shear-zone geometry showing development of foliations relative to the shear components and the walls of the shear zone. (b) Ductilely deformed Carboniferous megacrystic granitoid in the Pennine root zone of the Monte Rosa nappe in the Alps at Villadossola, Italy. The large fractured microcline phenocrysts are aligned parallel to the dominant foliation (S or an earlier C), and the younger dextral C-surfaces are inclined about 20° to the S-surfaces. This is a Type I S–C mylonite. (Locality courtesy of H. P. Laubscher, University of Basel; RDH photo.) (c) Phyllonite in the Brevard fault zone, Mountain Rest, South Carolina, a Type II S–C mylonite. (RDH photo.)

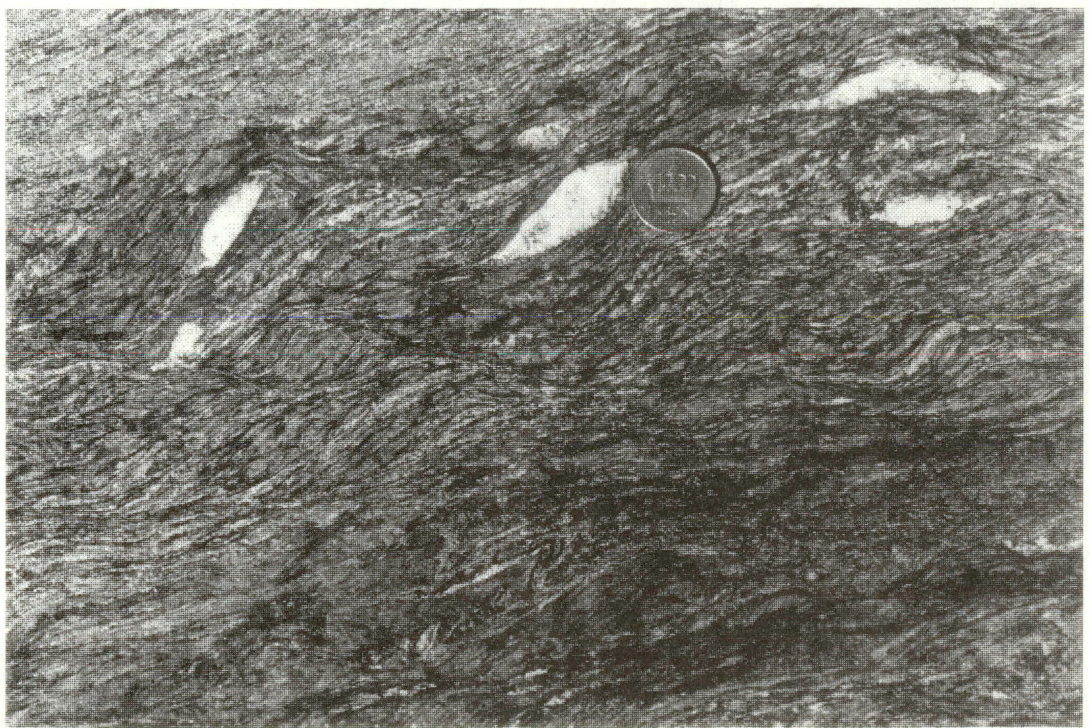

FIGURE 10–25
Shear band C-surfaces deforming earlier S-surfaces in a fine-grained Cambrian phyllonite from the fault zone at the base of the Seve thrust sheet on the west flank of Tronfjellet, south-central Norway. (RDH photo.)

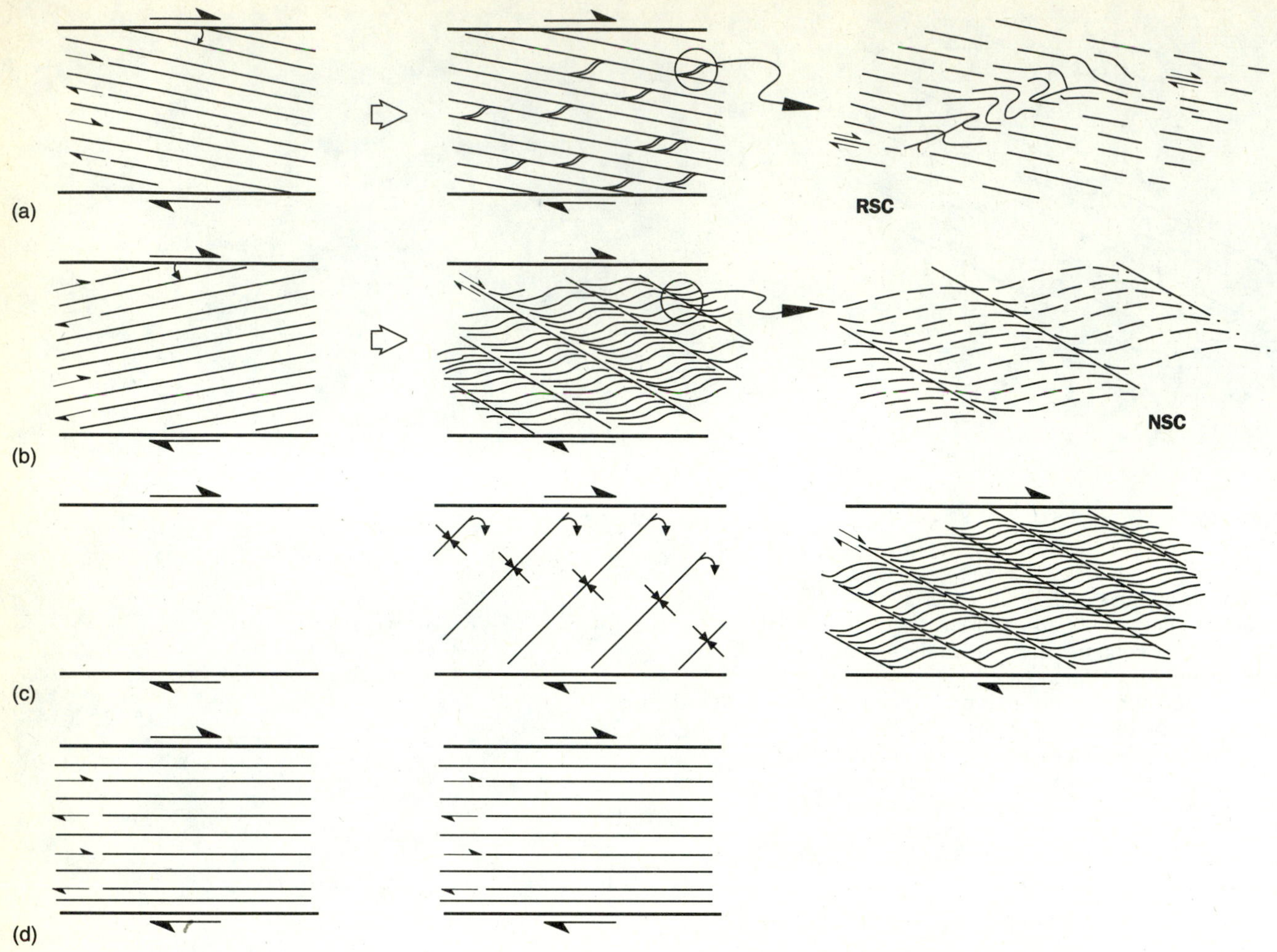

FIGURE 10–26
Development of C-surfaces and oppositely vergent structures related to conveniently oriented earlier S-surfaces that became reactivated. (a) Development of reverse-slip crenulations (**RSC**) compensates for movement normal to shear-zone walls where the original foliation is gently inclined clockwise to shear-zone movement direction. (b) Normal-slip crenulations (**NSC**) compensate for normal movement where the original foliation is inclined gently counterclockwise to the movement direction. (c) If shearing of the foliation perpendicular to the cumulative flattening direction occurs, so that the foliation is rotated into an easy slip orientation and slip is an important mode of strain in the shear zone, normal-slip crenulations will develop. (d) If the original foliation is parallel to the shear-zone walls, no compensating mechanism is required. (Reprinted from *Journal of Structural Geology*, v. 9, A. J. Dennis and D. T. Secor, p. 809–817, © 1987, with kind permission from Elsevier Science, Ltd., Kidlington, United Kingdom.)

large degree whether ductile or brittle shears will form, but strain rate is also important in determining whether ductile or brittle conditions prevail (Hatcher, 1978a; Wise and others, 1984). The deformation process that develops in mylonite and cataclasite is also determined by the ratio of strain rate to recrystallization rate (Chapter 7; Hatcher, 1978a; Hatcher and Hooper, 1981). Passchier and Simpson (1986) demonstrated the dependence on this ratio of flow processes in mylonites involving different kinds of porphyroclast systems that indicate shear sense. The shapes of the tails of δ porphyroclasts have been suggested as indicators of strain rate in experimental models of mylonite deformation. More curved tails indicate a power-law (plastic) rheology (higher strain rate), whereas lower strain rates that can be described by a more linear flow law (Newtonian) are indicated by δ porphyroclasts in which the tails become aligned with a plane that passes through the center of the porphyroclast (Passchier and others, 1993).

Strain-ellipsoid analysis of shear zones is initially unclear, because in a zone of heterogeneous simple shear, the ellipsoid has no constant orientation. Net flattening does occur within the shear zone, but at a high angle to the walls of the zone (Figure 10–24a).

Displacement of some marker by a shear zone may be estimated by a technique derived by Ramsay (1967). Lines making an original angle β with the shear-zone boundary (Figure 10–28) are displaced to a new angle α. The shear strain γ may be obtained from

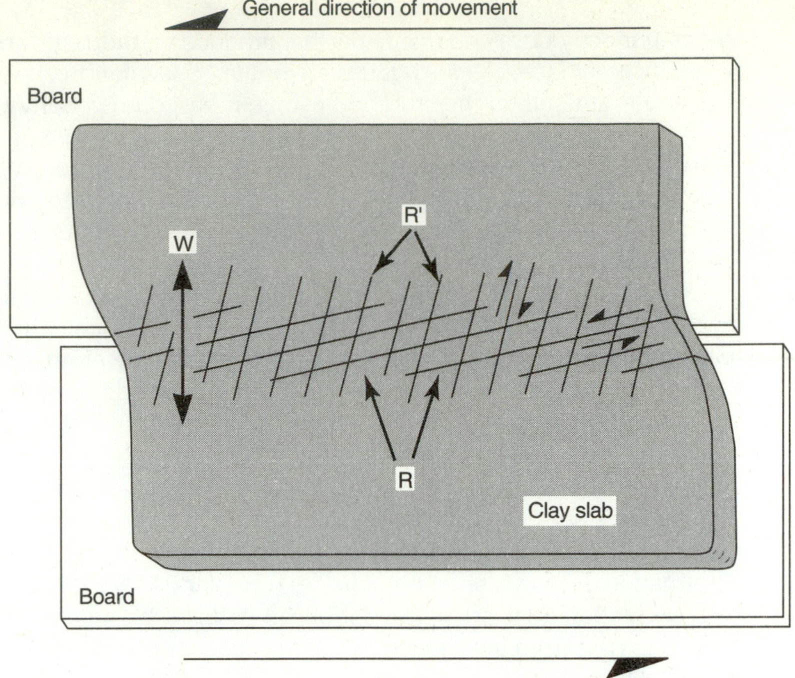

FIGURE 10–27
Experiment to produce Riedel shears in a clay layer. W—width of shear zone; R—Riedel shear; R′—anti-Riedel shear. (From J. S. Tchalenko, 1970, Geological Society of America *Bulletin,* v. 81.)

$$\cot \alpha = \cot \beta + \gamma \qquad \textbf{(10–10)}$$

or

$$\gamma = \cot \alpha - \cot \beta \,. \qquad \textbf{(10–11)}$$

The displacement D (actually a slip) on a shear zone may be estimated from the integral

$$D = \int_0^x \gamma \, dx \qquad \textbf{(10–12)}$$

derived by John Ramsay and Rod Graham (1970), where x is the width of the shear zone. If we can estimate the magnitude of shear strain on a shear zone from either equation 10–12 or equation 4–5, we can obtain an estimate for displacement by measuring the width of the zone in the field. For example, the Brevard fault zone in the southern Appalachians is about 5 km wide. Dextral shear strains, determined using equation 4–5, are near 1. Substituting and solving the integral produces $D = 1$ (5 km) = 5 km as the displacement estimated for the fault zone.

That ends our discussion of fault mechanics. Now we turn to more detailed consideration of each of the three kinds of faults, beginning with thrust faults.

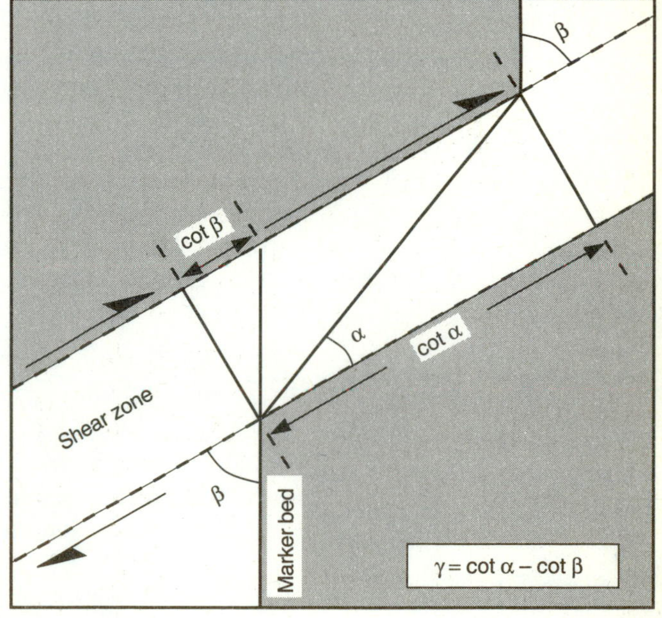

FIGURE 10–28
Displacement of a marker in a shear zone. The shear-zone thickness is one unit. (After J. G. Ramsay, 1967, *Folding and fracturing of rocks:* McGraw-Hill Book Company; and Ruud Weijermars and H. E. Rondeel, 1984, *Geology,* v. 12.)

ESSAY

Differences in Behavior of Crustal Faults

We can directly observe large faults that produce earthquakes and break the present-day surface. All such faults are brittle. Speculation about the behavior of large faults at depth in the crust and mantle is based on geophysical models and on direct observation of fault zones interpreted as having been formed at great depths but are now exposed after millions of years of erosion. We believe that faulting deeper in the crust (more than 10 to 15 km) occurs by ductile flow, which produces mylonite zones that range in width from less than a meter to several kilometers.

Studies of inactive broad mylonitic fault zones in present-day exposures in western Greenland led John Waterson (1975) and J. Grocott and Waterson (1980) to suggest that these broad zones may represent the lower-crustal equivalents of stick-slip brittle faults in the upper crust. The mylonite zones contain mineral assemblages, indicating that they formed at upper-amphibolite- to granulite-facies metamorphic conditions. Mineral assemblages in these fault zones, along with the textures of fault rocks, provide information about the conditions and depths where the zones formed. It would be useful if we could directly trace a fault from the lower crust, where it is now moving, to the surface, where it is undergoing brittle deformation. Seismic reflection profiles (Chapter 21) across active island arcs, such as the Barbados arc in the Caribbean (Westbrook and Smith, 1983), image faults that pass from the surface, where evidence for brittle deformation may be observed through the crust and into the mantle, where earthquake activity diminishes but the faults appear to continue. Ductile deformation would be expected at these greater depths, but direct observation is impossible. Thus, we must rely on indirect information from geophysical images and direct observations of ancient fault zones thought to have formed in the deeper crust.

References Cited

Grocott, J., and Waterson, J., 1980, Strain profile of a boundary within a large shear zone: Journal of Structural Geology, v. 2, p. 111–117.

Waterson, J., 1975, Mechanism for the persistence of tectonic lineaments: Nature, v. 253, p. 520–522.

Westbrook, G. K., and Smith, M. J., 1983, Long décollements and mud volcanoes: Evidence from the Barbados Ridge complex for the role of high pore-fluid pressure in the development of an accretionary complex: Geology, v. 11, p. 279–283.

Questions

1. Why are Mohr diagrams for normal and thrust faults constructed of nested circles of oppositely increasing radii?
2. How is Amontons' first law related to Byerlee's law?
3. How do Anderson's fundamental assumptions relate to real faults? Can you suggest an alternative scheme of orientation of principal stress (strain) axes that will work?
4. How does water affect the normal stress on a fault?
5. Describe the transformation of part of a fault from a stick-slip mechanism to a stable-sliding mechanism. What are the reasons for the transformation?
6. If a fault in a gneiss has a maximum shear stress of 68 MPa and an effective normal stress of 77 MPa, what is the coefficient of sliding friction for the fault?
7. A dextral fault zone, 2 km wide, exposed in the central part of the West African craton, consists entirely of mylonite. You can show (using zircon Pb-U geochronology) that it was formed 2.3 Ga in rocks that were in the upper-amphibolite facies (20 km deep, temperature above 500° C). What can you

deduce about the behavior of the fault in the upper crust before erosion removed the upper-crustal rocks?

8. Why do fractured grains commonly yield an opposite shear sense?
9. How can extensional crenulation cleavage form in a ductile shear zone that formed in a homogeneous granite?
10. How could you distinguish between a transposed C- (or C′-) surface truncated by younger C- surfaces and an earlier S-surface (similarly truncated)?
11. How could Riedel shears developed on a crustal scale be analogous to microscopically developed ductile shear bands? (For example, see S. Mosher, 1983, Kinematic history of the Narragansett Basin, Massachusetts and Rhode Island: Constraints on late Paleozoic plate reconstructions: Tectonics, v. 2, p. 327–344.)
12. Explain in terms of fault mechanics the mechanism by which earthquakes were produced in the Denver area soon after deep injection of fluid.

Further Reading

Anderson, E. M., 1951, Dynamics of faulting and dyke formation: Edinburgh, Oliver and Boyd, 191 p.
A classic and readable work discussing the characteristics and mechanics of the three major types of faults.

Flinn, D., 1994, Essay review: Kinematic analysis—pure nonsense or simple nonsense: Geological Journal, 29, p. 281–284.
An interesting short review that states some of the problems related to the use of many shear-sense indicators, along with a number of Flinn's opinions. This paper is worth a group discussion after it is read.

Hanmer, S., and Passchier, C., 1991, Shear-sense indicators: A review: Geological Survey of Canada Paper 90–17, 72 p.
A useful summary of the literature and of criteria for determining the sense of shear in rocks and of the literature.

Mandl, G., 1988, Mechanics of tectonic faulting: Amsterdam, Elsevier, 407 p.
Mathematical presentation of the mechanics of faults preceded by a discussion of the different geometric attributes of faults. A second section is a thorough discussion of the concepts of stress and strain and their application to faulting.

Scholz, C. H., 1990, The mechanics of earthquakes and faulting: Cambridge University Press, 439 p.
Contains a more in-depth summary of fracture and fault mechanics than herein; provides useful insight into the mechanics of faults leading directly to earthquakes as a consequence of faulting.

Simpson, Carol, 1986, Determination of movement sense in mylonites: Journal of Geological Education, v. 34, p. 246–260.
Discusses both shear-sense indicators and properties of fault rocks.

11

Thrust Faults

Eight great faults [occur in East Tennessee as] ribbon-like masses or blocks . . . crowded one upon another, like thick slates or tiles on a roof, the edge of one overlapping the opposing edge of the other.

JAMES M. SAFFORD, 1856, *A Geological Reconnaissance of the State of Tennessee*

THRUST FAULTS ARE FASCINATING STRUCTURES, CONSISTING of low-angle faults that transport thin sheets of rock—a phenomenon regarded as physically impossible when they were first discovered in the mid-nineteenth century. Many thrust sheets have huge areal extents and have moved horizontally tens to hundreds of kilometers. Chief Mountain, Montana, is a remnant of a formerly more extensive thrust sheet with a minimum displacement of 40 km (Figure 11–1). Thrusts form some of the largest structures in mountain chains, comparable in size to large accreted terranes or microcontinents—hundreds of kilometers long. Thrusts affect migration of ground water, oil and gas, and ore-bearing fluids during movement and also form traps for hydrocarbons and metallic minerals (Figure 11–2).

Thrust, or overthrust, faults are found in mountain chains—orogenic belts—and because of the stratigraphy present, are most easily recognized on the continentward side. A belt of thrust faults, called a ***foreland fold-thrust belt***, occurs between the undeformed craton and the metamorphic core of nearly every mountain chain (Figure 11–3). Most examples contain stratigraphic sequences from former stable trailing margins such as the present-day East Coast of the United States. Much larger thrusts also occur in the metamorphic cores of mountain chains, where some thrusts form under ductile conditions at moderate to high temperatures (>450° C), or at lower temperatures (300° C) if the rocks contain a small amount of water. Other thrust faults in orogenic interiors form later, at lower temperatures and under more brittle conditions. Thrusts also form in subduction zones as the overriding plate scrapes sediment off the descending sea floor, producing an ***accretionary wedge***.

The phenomenon of low-angle thrusting was, according to Bailey Willis (1923), first recognized in 1826 by Weiss near Dresden, Germany, where a Paleozoic granite occurs in horizontal fault contact above Cretaceous rocks. The idea of large-scale horizontal transport of great slabs of rock was conceived by Arnold Escher von der Linth, a Swiss geologist. During the 1830s, Escher first recognized thrust faults on the northwest side of the Aar massif in the Alps, and then in the 1840s was the first to correctly interpret the Glarus overthrust in Switzerland. He called these structures *Überschiebung,* which in German means "overpushing." Shortly afterward, James M. Safford (1856) recognized the Appalachian thrusts in Tennessee and visualized the fault geometry. In Switzerland, as related by Edward B. Bailey (1935), Roderick Impy Murchison reexamined the Glarus thrust in 1848 and confirmed Escher's interpretation, but he later refused to agree with James Nicol (1861)—a Scot known for the optical prism of Iceland spar that bears his name—that thrusts do indeed exist in the Scottish Highlands. In fact, some British geologists sought to disprove the existence of Scottish thrusts by testing a candidate structure in the Northwest Highlands near Loch Eribol and Assynt. First, Charles Lapworth, representing Her Majesty's Geological Survey, went to the Assynt District, where several months of study convinced him that the Moine thrust does exist. (He also recognized the importance of mylonite and coined the name.) Unfortunately, he fell mentally ill, believing the Moine thrust was still moving and threatening his cabin near the base of Knockan Crag, and became so disabled he never finished his work there. Later in the nineteenth century, several members of Her Majesty's Geological Survey, including B. N. Peach, J. Horne, C. T. Clough, and H. M. Cadell, mapped the same region in an attempt to dispose of any notion of large-scale horizontally transported thin sheets of rock in the British Isles. Instead, they published in 1884 compelling evidence that at least 16 km of transport had occurred along a low-angle movement surface. The usefulness of their geologic maps of the Assynt District

FIGURE 11–1
Chief Mountain, Montana—a klippe of the Lewis thrust that transported Middle Proterozoic Belt Series rocks over Upper Cretaceous sedimentary rocks. The Lewis thrust is the prominent inclined boundary traceable from the lower left of the mountainside and is truncated by the nearly horizontal dark boundary—another thrust carrying Precambrian rocks—near the top of the mountain. Boyer and Elliott (1982) concluded that the upper—nearly horizontal—boundary forms the roof of a duplex structure in the Belt Series rocks. (Bailey Willis, U.S. Geological Survey.)

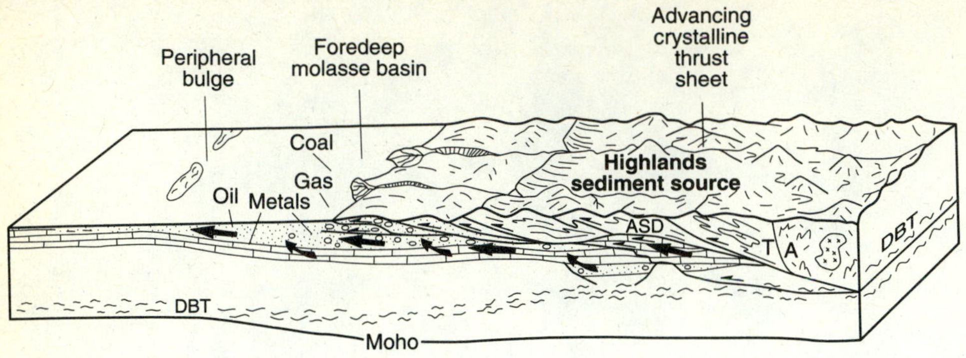

FIGURE 11–2
Effects of the arrival of a large thrust sheet on a continental margin. Fluids carrying hydrocarbons and metals (indicated by dark arrows) are expelled. Loading of the crust by the sheet produces a depression beneath and in front of it (foredeep basin) and a crustal arch (peripheral bulge) farther inland. DBT—ductile-brittle transition; ASD—antiformal stack duplex; T—movement toward observer; A—movement away from observer. (Modified from J. E. Oliver, 1986, *Geology,* v. 14.)

(Peach and others, 1888), including the Moine thrust (Figure 11–4), is likely to outlast the stone monument to their work that now stands weathering on a low hill near Inchnadamph.

As shown by studies of the Glarus and Moine thrusts, thrust faults have long provoked controversy. First, the very existence of horizontal sheets of rock that had been transported great distances was disputed; later, involvement of *basement* in foreland thrusts was debated. (Basement generally consists of crystalline rock that is the product of an earlier orogenic cycle and underlies a less deformed and less metamorphosed cover.) More recently, geologists have focused on whether the motive force for thrusting is gravity or compression. Great strides were made during the 1970s and 1980s toward understanding the geometry and mechanics of thrust faults, but even recent mechanical models—because they imperfectly describe nature—are not accepted by all. Thus, controversy and fascination with these structures continue.

This chapter is more detailed than the following chapters on strike-slip and normal faults partly because so much knowledge has accumulated on the geometry of thrusts. The petroleum industry has long explored thrust-faulted terranes, gathering extensive drilling and seismic reflection data to help us to understand the geometry and behavior of thrusts. Regions of normal and strike-slip faults are also much explored, but information—and controversy—about thrust faults remains more abundant.

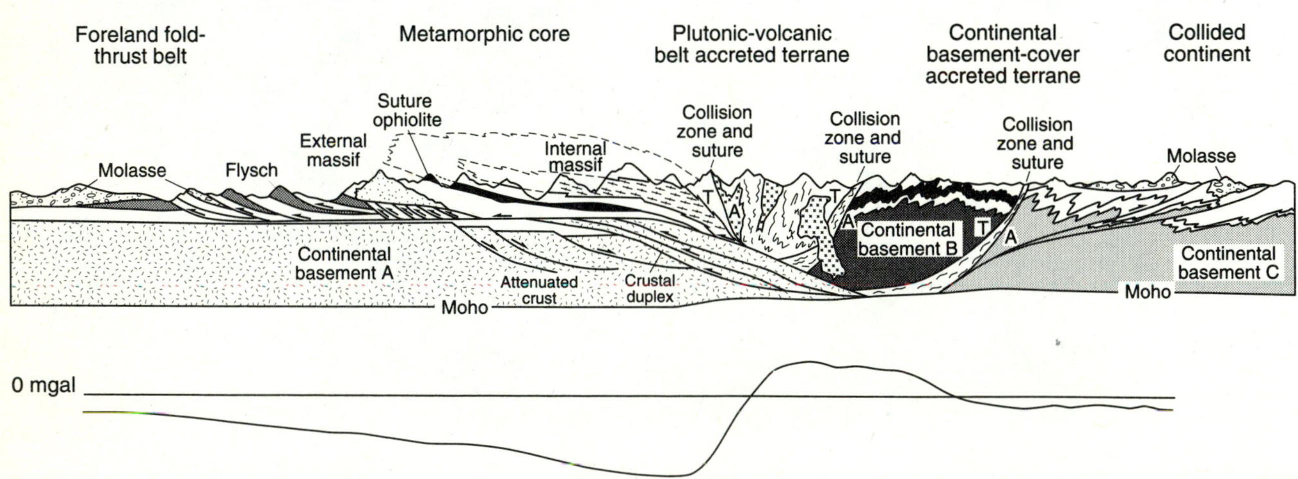

FIGURE 11–3
Ideal mountain chain. Masses of continental or oceanic crust or island arcs may be accreted by either head-on or oblique collision, producing a variety of deformational styles, metamorphism, and plutonism. The external massif is a block of basement caught up and transported in the thrust stack. Continental basements A, B, and C represent parts of different continents now amalgamated by collision. T—movement toward observer; A—movement away from observer. Curve is a hypothetical gravity profile of the kind observable in most mountain chains. (From R. D. Hatcher, Jr., and R. T. Williams, 1986, Geological Society of America *Bulletin,* v. 97.)

FIGURE 11–4
(a) The Stack of Glencoul at right center of photo (arrow) exposes the Moine thrust at the topographic break between the low hills on top and the flat at the base of the stack. This fault carried metasedimentary rocks of the Middle Proterozoic Moine Series over Paleozoic sedimentary rocks and basement in the foreland.
(b) Moine thrust on the Stack of Glencoul. White Cambrian Pipe Rock Sandstone (below the contact, see arrow) was overthrust and folded during the latter stages of emplacement of the darker-colored Middle Proterozoic Moine Series metasandstone.
(RDH photos.)

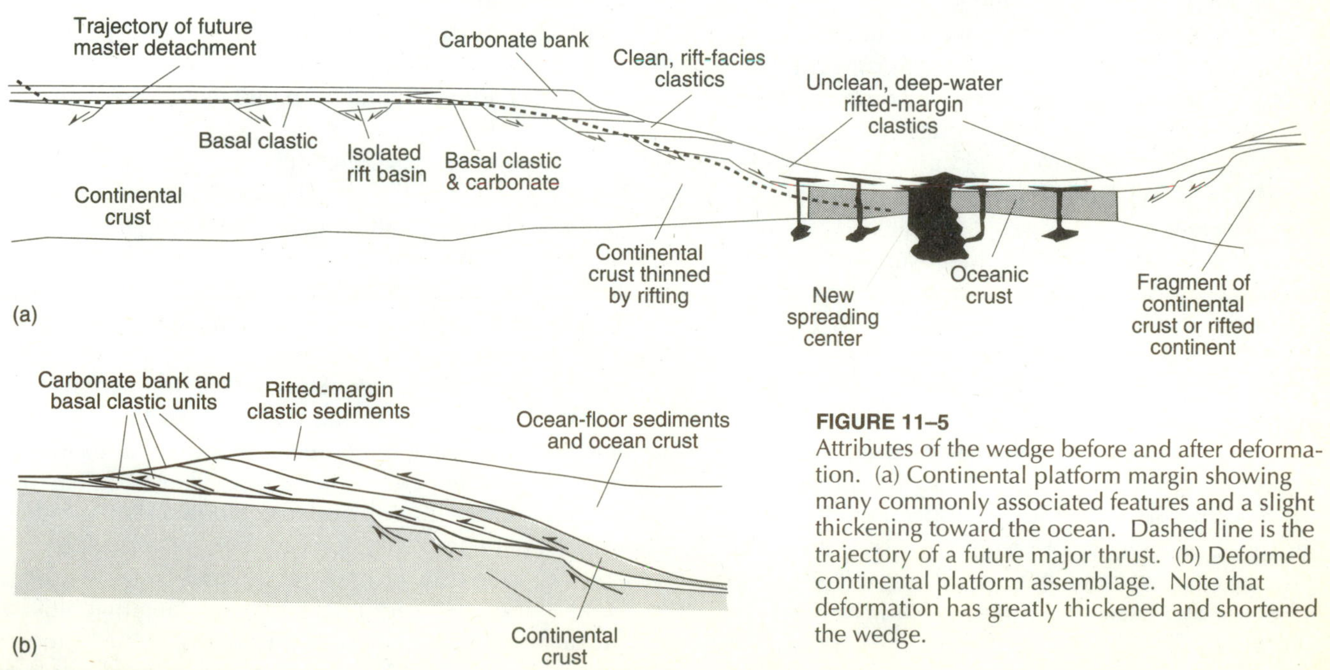

FIGURE 11–5
Attributes of the wedge before and after deformation. (a) Continental platform margin showing many commonly associated features and a slight thickening toward the ocean. Dashed line is the trajectory of a future major thrust. (b) Deformed continental platform assemblage. Note that deformation has greatly thickened and shortened the wedge.

(b)
(a)

NATURE OF THRUST FAULTS

Thrusts are gently dipping faults in which the hanging wall has moved up relative to the footwall. Several kinds of thrusts exist, and their behavior varies with the types and strengths of rocks involved, the ambient temperature at the time of formation, and the degree to which water was involved during movement.

Thrust faults are usually found in mountain chains or their eroded remnants. They occur most commonly in a continentward-thinning wedge of sedimentary rocks above an undeformed basement in a foreland thrust belt (Figure 11–5). The undeformed mass of continental-margin sediments has a characteristic wedge shape—a shape maintained throughout the deformational history of the belt—a feature inherited from the shape of the original continental margin on which the thrust belt is developed because trailing-margin sediments thicken oceanward. Wedge geometry in thrust belts was first recognized by David Elliott (1976a) as a fundamental property of deforming thrust belts. After studying structures in the still-active Taiwan thrust belt and formulating dynamic geometric models of them, John Suppe (1981) concluded that the shape is preserved during deformation through a combination of surface erosion and slumping. These processes tend to conserve the original angles—equilibrium angles—and thus the shape of the wedge. During thrusting, the principal change is that the wedge is shortened and thickened until it reaches a steady-state condition (Chapple, 1978; Suppe, 1981). Nicholas Woodward (1987) concluded that the wedge must be deformed internally as new material is added to the front if the shape of the wedge is to be preserved. Evidence is lacking for internal deformation in foreland-type wedges; the wedges probably become stronger overall during deformation, although fault zones may weaken them. Woodward suggested that the critical wedge models developed primarily from study of modern accretionary wedges may not survive rigorous application to foreland thrust belts because of differences in mechanical properties of the rocks and in the amount of fluid present. Despite this, we now know that the dip on the basement surface, and the mechanical properties of the basal weak unit (the detachment), strongly influence the amount of displacement that may occur in the deforming wedge and that a large amount of internal deformation within the wedge is not required.

Thrusts have historically been debated as either ***thin-skinned***, with no basement involved, or ***thick-skinned***, with basement involved (Figure 11–6). Because basement is not commonly observed at the surface in foreland thrust belts, many geologists came to believe that basement was also not involved in the subsurface. The thin-skin concept originated with A. Buxtorf (1916), who worked in the Jura Mountains in Switzerland. Later, John Rich (1934) outlined most of the principles of thin-skinned deformation during his classic study of the Pine Mountain block in the southern Appalachians (Figure 11–7a). Rich based his analysis on surface geologic mapping and a few shallow drill holes; data gathered since have modified his conclusions very little. Long after Rich, Shankar Mitra (1988) used seismic reflection and existing drill data to show that the subsurface structure was more complex than Rich had thought (Figure 11–7b), but he also reconfirmed Rich's fundamental conclusions.

Rich proved that thin-skinned deformation worked for the Pine Mountain overthrust sheet, and that set the stage for similar discoveries around the world. Even so, John Rodgers (1949, 1964) and Byron Cooper (1964) took opposing sides in a renewal of the thin- and thick-skinned debate in the Appalachians. A. W. Bally, P. L. Gordy, and G. A. Stewart (1966) used seismic reflection, drilling, and surface geologic data to prove that the deformation process in the Canadian Rockies fold-thrust belt, and others by analogy, was thin-skinned. Later studies, using seismic reflection profiling in other foreland thrust belts, confirmed Rich's conclusion that thin-skinned deformation is the dominant style. Today we know that thick-skinned thrusts (as originally defined) exist, but they are too commonly low-angle thrusts, suggesting that this argument is moot, and that we need to better understand how all thrusts form.

Metamorphic rock—either as basement slices or as progressively metamorphosed sedimentary rocks—appears on the thickened side of the wedge in the transition to the metamorphic core of a mountain chain (Figure 11–8). Thrusts involving crystalline (metamorphic and igneous) rocks are called *crystalline thrusts.* Those closer to the foreland form under brittle conditions; those farther into the metamorphic core form under conditions ranging from ductile to brittle, depending on temperature, availability of water, and strain rate during movement (Hatcher and Williams, 1986; Hatcher and Hooper, 1992). Thermal or strain softening of the crystalline mass may be required before initiation of faulting if low-angle thrust faults are to propagate through crystalline rocks. The thrust plane may be initiated by local ductile behavior, but the mass may be transported by brittle translation. The temperature during movement depends on when movement begins relative to one or more of the thermal-metamorphic peaks affecting the orogen. A source far inside an orogen may produce warmer thrust sheets than one on the outer flanks. Myself and Roy Odom (1980) observed a definite correlation between movement on large faults (mostly thrusts) and thermal-metamorphic peaks in the southern Appalachians, noting that older thrusts formed in the internal parts early in the history of the chain, but the youngest faults formed on the flanks, suggesting that the orogen had an inside-out deformation plan. A similar relationship exists in the Alps, the Canadian Cordillera, British and Scandinavian Caledonides, and—probably— other mountain chains. In contrast, we also noted that the Appalachians of New England and the Canadian Maritime provinces contain older thrusts along the western flank and younger faults toward the east, indicating an outside-in deformation plan. This also reflects a contrasting late Paleozoic plate-tectonic history for these regions—head-on collision in the southern Appalachians, oblique collision in New England.

Early concepts of thrust faults implied significant differences in the modes of origin, propagation, and motion of crystalline and foreland thrusts. For many decades since, we have been able to study their geometry in eroded surface exposures. The phenomenon of low-angle thrusting of crystalline rocks was first noted by Törnebohm (1872) in Norway and Sweden, and by Peach and others (1888), when they first mapped the Moine and Arnabol thrusts in Scotland. More recently, it has been shown that crystalline thrusts share many properties with foreland thrusts (Elliott and Johnson, 1980; Hatcher and Williams, 1986). The powerful combination of detailed surface geologic

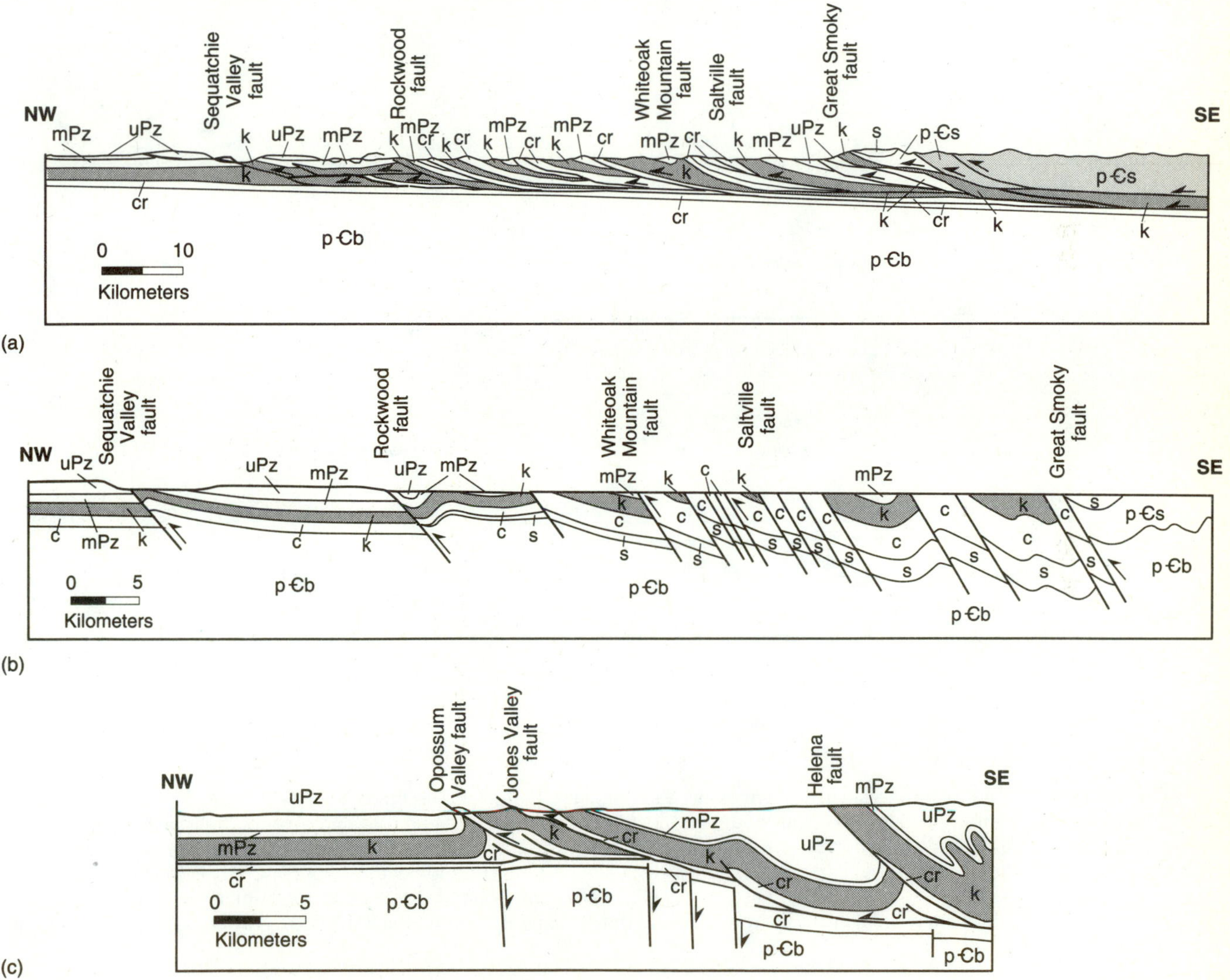

FIGURE 11–6
(a) Thin-skinned (no basement) thrusting versus (b) thick-skinned, basement-involved thrusting along nearly the same cross section through the Appalachians in Tennessee and western North Carolina. (Part a from a cross section by RDH; part b from John Rodgers, 1953, Kentucky Geological Survey Series 9, Special Publication 1.) (c) Cross section from the Appalachian Valley and Ridge in Alabama showing control of thin-skinned structures by older faults in the basement. (From W. A. Thomas, 1986, Virginia Tech Geological Sciences Memoir 3.) uPz—upper Paleozoic rocks; mPz—middle Paleozoic rocks; k—Knox Group carbonate rocks (Cambrian-Ordovician); c—Cambrian shale and carbonate; cr—Conasauga Group and Rome Formation (Cambrian); s—Shady Dolomite and Chilhowee Group rocks; p€ s—Precambrian sedimentary rocks; p€ and P€ b—Precambrian basement rocks.

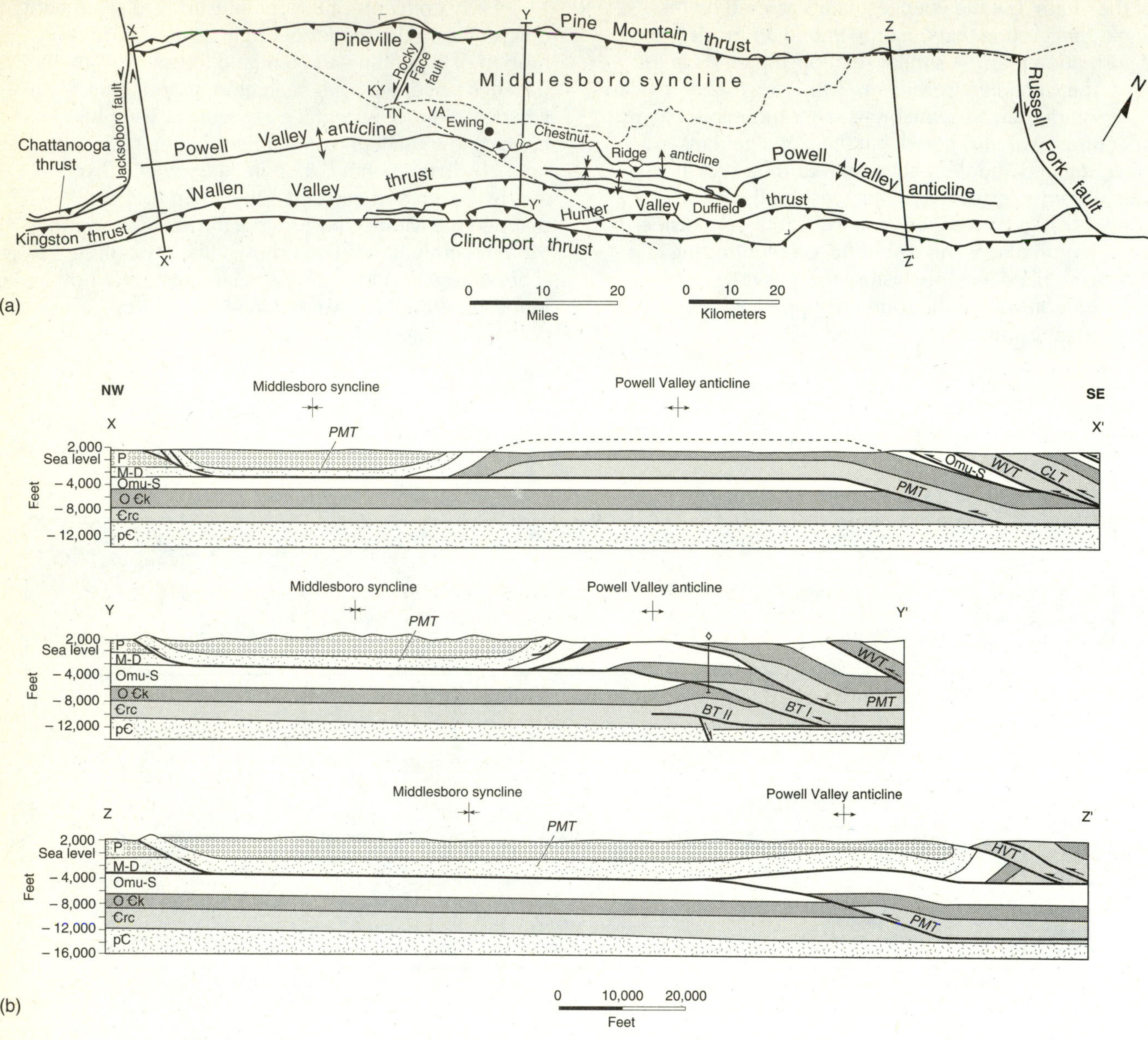

FIGURE 11–7
(a) Attributes of the Pine Mountain thrust sheet, Tennessee, Virginia, and Kentucky. The three lines X–X′, Y–Y′, and Z–Z′ locate the sections in (b). (b) Sections through the Pine Mountain block showing the fault-bend fold character of the Pine Mountain thrust (*PMT*). *WVT*—Wallen Valley thrust; *CLT*—Clinchport thrust; P—Pennsylvanian rocks. M-D—Mississippian and Devonian rocks; Omu-S—Middle to Upper Ordovician and Silurian rocks; O-Ꞓk—Cambro-Ordovician (Knox Group) carbonate rocks; Ꞓrc—Cambrian clastic rocks; pꞒ—Precambrian basement rocks. O-Ꞓk, Omu-S, and P are strong rock units; Ꞓrc and M-D are weak units. BT I and BT II in the middle cross section are blind thrusts I and II. (From Shankar Mitra, 1988, Geological Society of America *Bulletin,* v. 100.)

studies and geophysical data gathered in metamorphic cores of orogens, particularly in the Appalachians (Bryant and Reed, 1970; Hatcher, 1971, 1972; Cook and others, 1979; Hatcher and Zietz, 1980), the British Caledonides (Butler and Coward, 1984), and the Alps (Panza and Müller, 1979; Laubscher, 1983), has confirmed many of the similarities. Still, as the ensuing discussion will show, important differences remain.

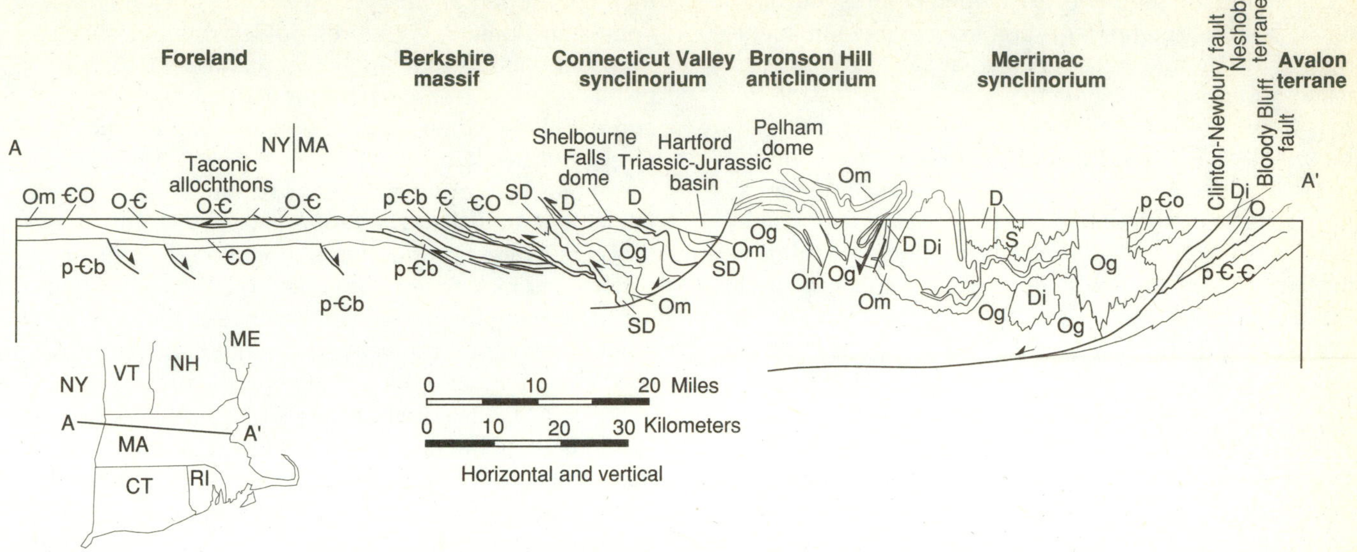

FIGURE 11–8
Formation of crystalline thrusts and basement involvement in thrusts at the edge of the New England foreland. ₸ J—Triassic and Jurassic; D—Devonian; Di—Devonian intrusive rocks; SD—Silurian and Devonian; S—Silurian; Om—Middle Ordovician; O—Ordovician; Oi—Ordovician intrusive rocks; Og—Ordovician granite; O-Є—Ordovician and some Upper Cambrian; ЄO—Cambrian and some Lower Ordovician; p-Є b—basement rocks; p-Єo—Precambrian rocks in eastern Massachusetts. (Reproduced by permission of the Geological Society from Thrust and Nappes in the North American Appalachian Orogen by R. D. Hatcher, Jr., in *Thrust and Nappe Tectonics,* Geological Society of London Special Publication No. 9, 1981.)

Many thrust faults form in a concave-up, spoon-shaped geometry and are called *listric* thrusts. Normal faults frequently exhibit the same geometry, indicating mechanical similarities to thrusts.

DETACHMENT WITHIN A SEDIMENTARY SEQUENCE

The concept of thin-skinned deformation is based on certain common properties of thrusts in foreland thrust belts. Foreland thrusts are characteristically low-angle faults that thrust older rocks over younger rocks. They may include high-angle segments and (very rarely) may bring basement rocks to the surface; they commonly repeat the same oldest rock unit across the entire belt. The oldest unit most commonly is a shale, evaporite, coal, or other weak *(incompetent)* rock type, or less commonly it may be a strong *(competent)* unit, one susceptible to strain softening. Thrusts generally propagate along these layers which are called *detachments* or *décollements.* Because they represent translation along a bedding plane, they are also called *bedding thrusts.* Décollements (French for "ungluing") may occur as the result of either folding or faulting (Figure 11–9), but the net result is relative displacement of the upper layers.

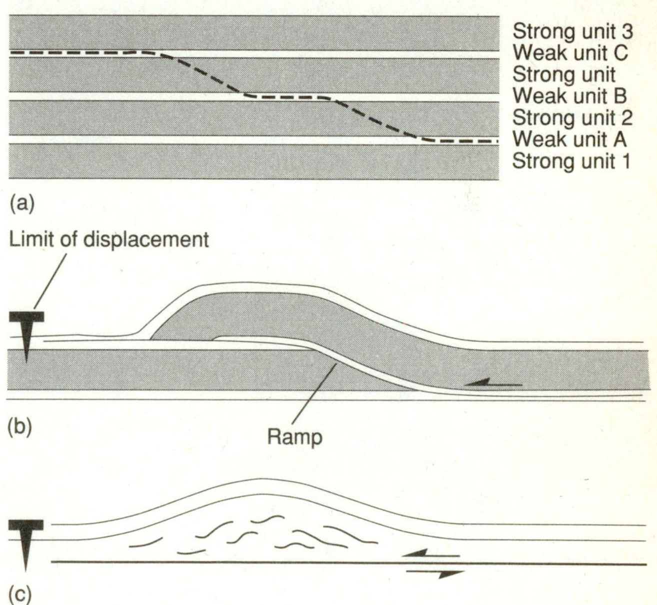

FIGURE 11–9
Formation of a décollement in a sequence of alternating weak and strong rock units. Dashed line in (a) is the path of a future décollement. Décollement follows weak units and ramps (or refracts) across strong units. Ramping occurs because of a change in the physical properties of a weak unit, as in weak unit A, B, or C in (a), or because of the thinning of a weak unit, as in weak unit in (b). The entire sheet moves over the undeformed footwall, forming a ramp anticline (or fault-bend fold; see Figure 11–11). (c) Folds may also form by this mechanism above a décollement.

High-angle segments exist where a thrust cuts across a strong (***competent***) unit from one detachment to another (Figure 11–10). A high-angle segment is called a ***ramp*** and may occur on all scales crossing units from a few centimeters to a kilometer or more thick. Massive limestone, dolostone, sandstone, or other competent rock types may compose strong units that cause thrusts to ramp. The thickness of strong units (also called a ***strut***) also limits the amplitude of folds formed in association with thrusting in a thrust-faulted sedimentary sequence.

Folds may form in association with thrust faults. Bailey Willis (1893), after studying the Appalachian foreland, distinguished *break thrusts*—which form by faulting the connecting limb of an anticline-syncline pair, overthrusting the *hanging-wall anticline,* and preserving a *footwall syncline*—from *shear thrusts,* formed independently of folding but not having a footwall syncline. John Suppe (1983) has defined *fault-bend folds* (Figure 11–11a) as those formed as the thrust surface changes from steeper dip to shallower in an up-dip direction. Fault-bend folds are similar to Willis' shear thrusts. Elliott (1976b) also discussed this mechanism but did not name it. A thrust sheet wherein dip flattens as it passes over a ramp would produce a fault-bend fold. Suppe (1985) also recognized *fault-propagation folds,* which form as layers fold during propagation of a thrust through a sedimentary sequence (Figures11–11b and 11–11c); they are similar to Willis' break thrusts, based on an analogy with kink folds, and thereby resurrect the earlier idea of Roger Faill (1973) connecting kinking with foreland deformation. Fault-propagation folds may evolve into fault-bend folds as displacement increases, perhaps accounting for the comparatively larger number of fault-bend folds in most foreland-thrust belts.

Folds that form near the thrust surface during movement are called ***drag folds*** (Figure 9–13a). Drag folds are a direct product of friction along the thrust surface and may range from open to tight folds (Chapter 14), depending on the competence of the rocks being deformed. The greater the amount of incompetent material in a section, the tighter the drag folds.

Thrust faults commonly break off and grind material along the fault surface. If the deformation mechanism is dominantly brittle, fracturing is more likely. Pieces of material called ***horses***, or ***slices***, measuring up to several kilometers in maximum dimension, may be transported within the fault zone

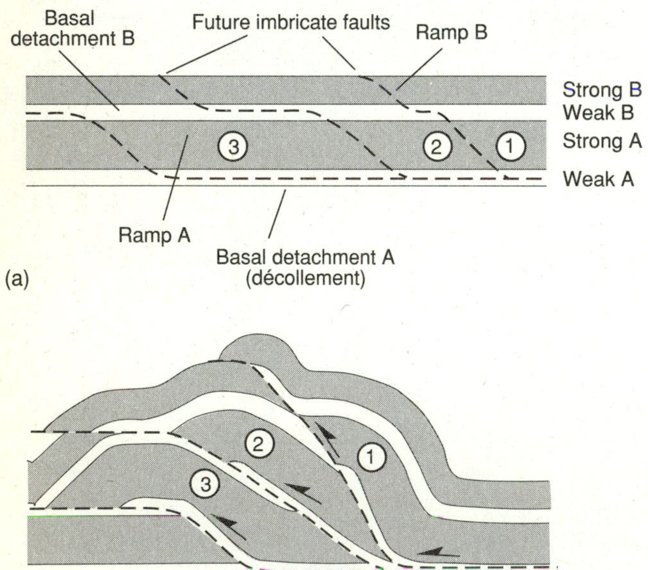

FIGURE 11–10
Properties and propagation of thrusts in cross section: (a) undeformed; (b) deformed. The fault labeled 1 forms first; that labeled 3 forms last. The dip of thrust sheet 1 increases as thrust 2 forms and moves; the dip of both 1 and 2 are increased as 3 forms and moves up the ramp.

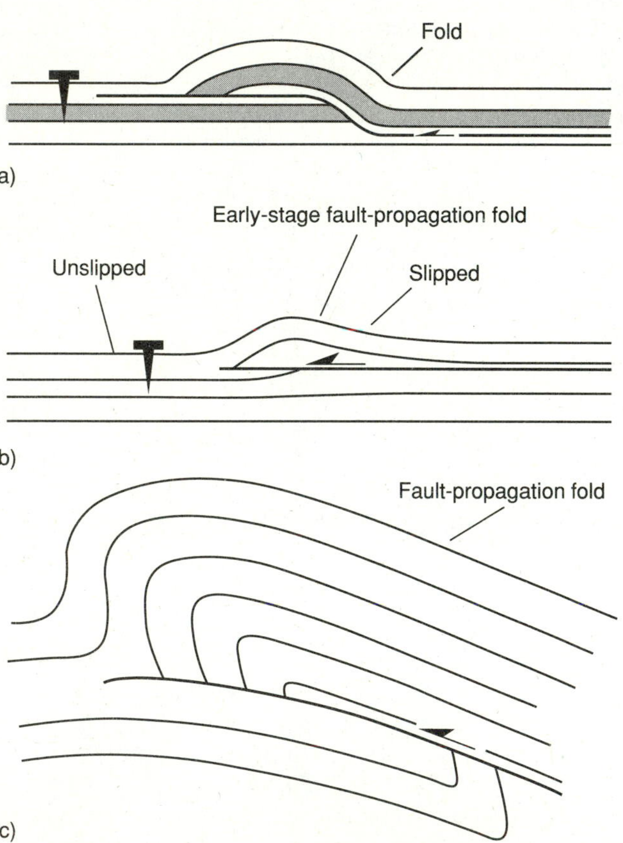

FIGURE 11–11
(a) Fault-bend folds. Rocks passing over the fault-bend (ramp) are ideally folded, and then unfolded. (b) Initial development of a fault-propagation fold as a blind thrust. (c) Well-developed fault-propagation fold containing a hanging-wall anticline and a footwall syncline. Displacement decreases toward the left, away from the hinge zone, and increases to the right of the hinge. Amplitude of the fold increases with increasing displacement.

beneath a thrust sheet. Here, a general rule applies: *the age of the material composing horses is usually intermediate between the ages of the hanging-wall and footwall rocks* (Figure 11–12). An exception occurs where the slice consists of crystalline basement. Slices may roll or slide along and, in the process, be folded and fractured internally. As a result of the mode of formation and transport, these blocks are commonly made of strong materials and are most frequently derived from the footwall, but horses derived from the hanging wall do exist. Obviously, fault surfaces completely surround a horse (Figure 11–13). Less commonly, they are composite strong/weak materials, with strong dominating. A *horse* may also be thought of as a slice that remains an integral part of the thrust sheet and is not transported far from its point of origin, but many geologists use the terms "horse" and "slice" synonymously. An intriguing variation on the term "horse" has been suggested by Sharon Lewis and Jerry Bartholomew (1989). They defined the term "orphan" as a complex of far-traveled horses and duplex imbricates along a thrust derived from a footwall syncline probably located on a former thrust ramp. Orphans contain rocks that are stratigraphically unrelated to the rocks in either the hanging wall or the footwall of the thrust system of which they are a part (Bartholomew and others, 1994).

Thrusts may occur as isolated faults, but many occur in groups consisting of a master fault with several associated smaller faults; the smaller ones are called ***imbricate thrusts*** (Figure 11–14a). The term *imbricate zone* may also be used to describe the series of thrusts making up such a group or the entire foreland thrust belt. Imbricates probably form by branching from a master décollement as the thrust sheet moves. As one segment "sticks," an imbricate propagates beneath it. In most cases, propagation is toward the foreland, but the term does not indicate propagation direction. A few imbricates may propagate in the opposite direction beneath a thin overburden. Repetition of the sequence results in an ***imbricate stack***. An ***accretionary wedge***, which forms as the sediments in a subduction zone are scraped off the descending slab, consists of an imbricate stack of thrusts (Figure 11–14b).

FIGURE 11–12
Exposure of the Champlain thrust at Lone Rock Point on Lake Champlain near Burlington, Vermont. The gently dipping fault zone contains a horse of white Lower Ordovician Beekmantown carbonate (right side of the photo) separating the hanging wall composed of gray Lower Cambrian Dunham Dolomite and footwall of black Middle Ordovician Iberville Shale. (RDH photo.)

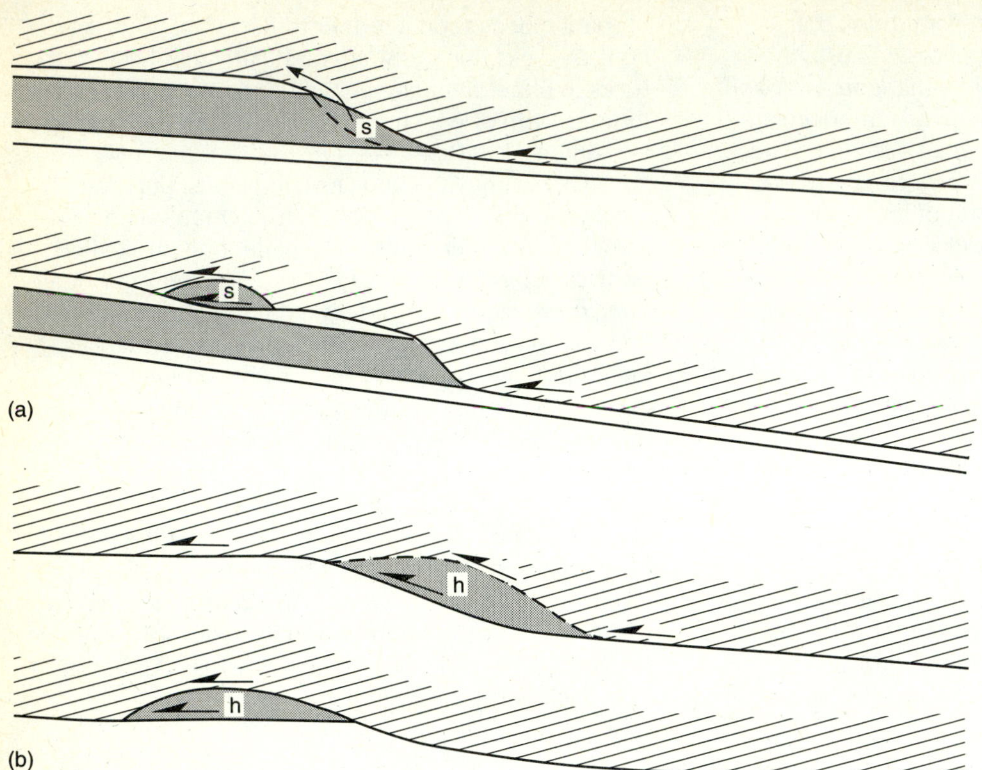

FIGURE 11–13
Formation of slices (s) and horses (h) in cross section. (a) Slices are commonly derived from the footwall of the fault; horses (b) may be derived from the hanging wall and remains as an integral part of the thrust sheet. The distinction is difficult and most blocks are simply called horses.

Structures called ***duplexes*** occur where two subparallel thrusts of approximately equal displacement are separated by a deformed interval that is thin relative to its total areal extent (Figure 11–14a). The upper of the two master faults is the ***roof thrust***, and the lower is the ***floor thrust***. Smaller imbricates connect the two master faults. The imbricates dip more steeply than the roof or floor thrusts and gradually merge with the master faults. Duplexes occur in most thrust-faulted terranes and are probably more common than once thought, even though they were recognized more than a century ago by Peach and others (1888)

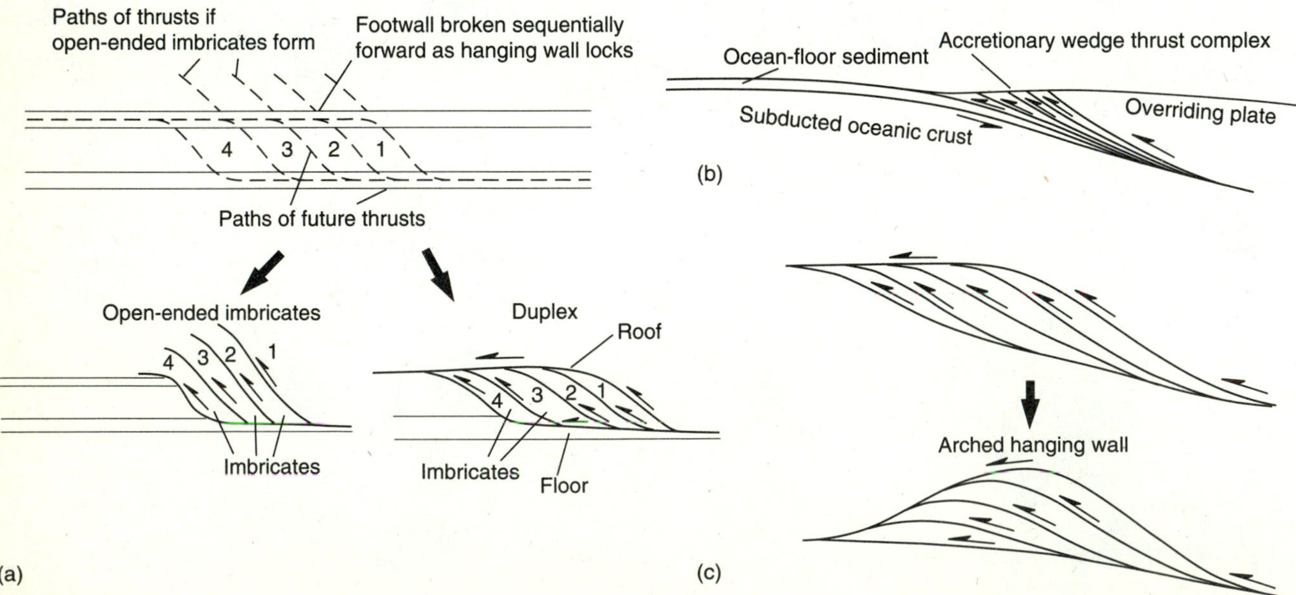

FIGURE 11–14
(a) Formation of imbricate zones and duplexes. Younger imbricates steepen the dip of older faults as they move. (b) Formation of accretionary wedges. (c) A hinterland-dipping duplex may evolve into an antiformal stack duplex by continued motion on the thrust sheet and imbricates, resulting in arching of the hanging wall.

along the Glencoul and Moine thrusts in Scotland, where they called them *schuppen.* They form as the roof thrust ramps and then locks, and successive imbricate faults form, ramp, and lock, allowing formation of the floor thrust and continued movement of the entire thrust complex.

Antiformal stack duplexes may be produced as the imbricate stack forms and rotates toward the horizontal (Figure 11–14c). Examples include the Mount Crandell structure in the Canadian Rockies (Dahlstrom, 1969; Price and Mountjoy, 1970) and the Limestone Cove window in the Appalachians of Tennessee (Diegel, 1986; Figure 11–15). Duplex development is a fundamental property of foreland thrust belts. It has been noted, using a combination of surface geologic and seismic reflection data, that a duplex, although formed in response to movement of a thrust sheet, frequently arches the thrust sheet as the duplex is built by duplication of rocks beneath it (Boyer and Elliott, 1982; Hatcher, 1991). Duplexes generally form beneath an appreciable overburden and with a ratio of strong rocks to weak between 1:1 and 4:1 (Costello and Hatcher, 1985). Mitra (1986) concluded that the geometry of a duplex is determined by a combination of ramp angle and height of the ramp, spacing between imbricates, and the total displacement in the duplex system. In some imbricate thrusts, a roof thrust exists with only minor displacement, but in others the roof thrust is easily identified and has considerable displacement. Banks and Warburton (1986) and Morley (1986) distinguished these as *passive-* and *active-roof duplexes.*

Synthetic and *antithetic normal faults* are normal faults that dip either subparallel to or opposite a larger fault respectively. They commonly form within a moving thrust sheet as the sheet passes over a ramp, and extensional stresses result as layers are extended and produce normal faults (Figure 11–16). They may also form as the entire sheet is extended as a single layer or by relaxation of the sheet. The strike of the faults generally parallels the strike of the more gently dipping thrusts. Dip of the normal faults is typically greater than 45°, but the dip may also become shallow, so that the faults are listric.

Using fold geometry and deformation mechanisms, Peter Geiser has classified thrusts into three major categories (Figure 11–17): *shortening, fracture,* and *fold thrusts.* Geiser's fracture thrusts are the same as Willis' shear thrusts and Suppe's fault-bend folds, and his fold thrusts are the same as Willis' break thrusts and Suppe's fault-propagation folds. Combinations are also possible. The layer-parallel shortening mechanism would account for only 15 to 25 percent strain if it depended solely on pressure solution as the dominant strain mechanism in the thrust sheet. Over a thrust belt 100 km wide, however, this mechanism could produce 15 to 25 km of displacement (decreasing with distance from the direction of thrusting). Other layer-parallel shortening mechanisms, such as contraction faulting (see next section), may result locally in 50 to 60 percent shortening. The mechanism related to pressure solution has been well documented by Geiser and Engelder (1983) in the Appalachian Plateau of New York and by Walter Alvarez, Terry Engelder, and

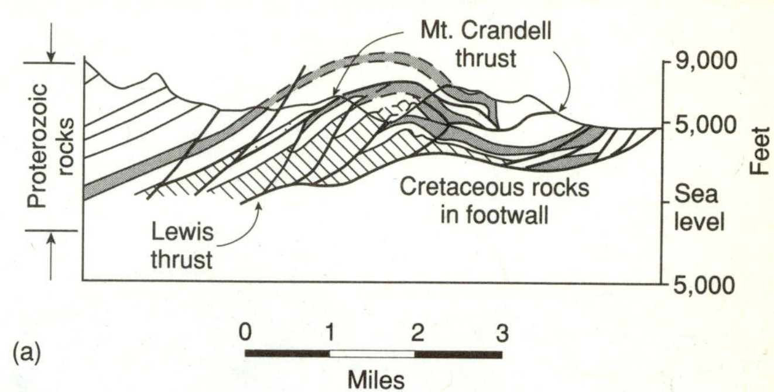

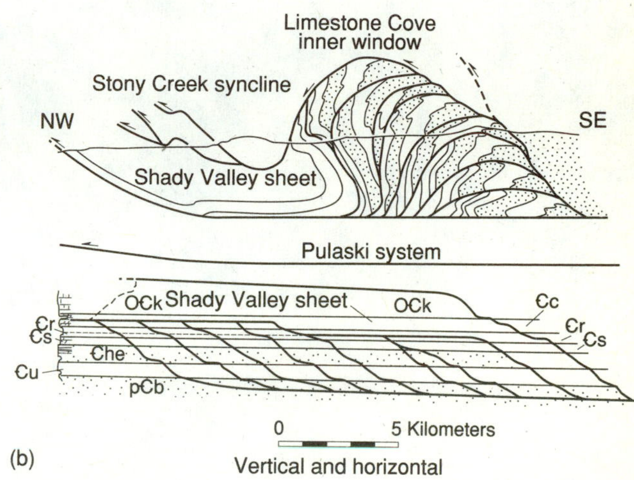

FIGURE 11–15
(a) Cross section of the Mt. Crandell structure in southern Alberta, a hinterland-dipping duplex. The Lewis thrust forms the floor, and the Mt. Crandell thrust forms the roof. (After R. J. W. Douglas, 1952, Geological Survey of Canada, *in* C. D. A. Dahlstrom, 1969, Canadian Society of Petroleum Geologists *Bulletin,* v. 18.) (b) Cross section of the Limestone Cove duplex, Tennessee, a foreland-dipping duplex. pꞒb—basement rocks; Ꞓu—Unicoi Formation; Ꞓhe—Hampton and Erwin Formations; Ꞓs—Shady Dolomite; Ꞓc—Conasauga Group; OꞒk—Knox Group. (From F. A. Diegel, 1985, Geological Society of America Southeastern Section Guidebook.) (c1) *(following page)* Exposure of a small duplex in Lower Devonian limestone in a quarry at Fuera Bush, south of Albany, New York. Dm—Manlius Formation; Dc—Coeymans Formation; Dk—Kalkberg Formation. (c2) Relationships between different elements in the duplex. (c1 and c2 courtesy of Stephen Marshak, University of Illinois-Urbana.)

FIGURE 11–15 (continued)

William Lowrie (1976) in the northern Apennines of Italy. Very large thrusts in the internal parts of foreland thrust belts may be initiated as a low-strain variety.

CONTRACTION (WEDGE) AND EXTENSION FAULTS

Several predominantly small-scale structures form in association with thrusts. Because they provide important clues to thrust mechanics, we will discuss them here.

Donald Norris (1958) described *extension faults,* in which bedding undergoes layer-parallel extension, and *contraction faults,* in which bedding is shortened, in coal mines of the foreland thrust belt of the Canadian Cordillera (Figure 11–18). Later, Ernst Cloos (1964) described contraction faults in the central Appalachians but called them *wedges* or wedge faults. They have since been reported in other thrust-faulted terranes. Frequently associated with extension and contraction faults is a *broken formation* zone. A broken formation generally develops beneath a detachment, perhaps by hydrofracturing under high pore pressure. It may also form where the ratio of competent material to incompetent is low (less than 1) and the rocks are too weak to support formation of a duplex. Steven Wojtal (1986) has interpreted broken formations as brittle shear zones.

Many other small-scale structures are associated with thrust systems, and either have already been discussed (cataclastic and mylonitic features, shear-sense indicators, drag folds) or will be introduced below.

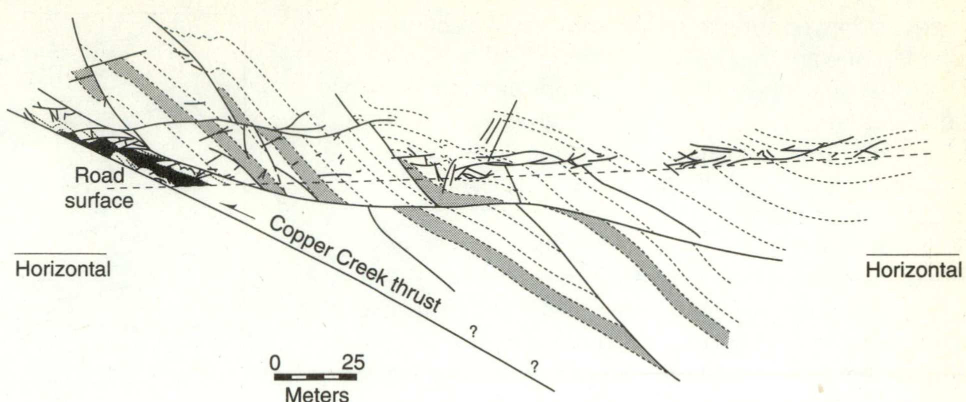

FIGURE 11–16
Synthetic and antithetic normal faults in sandstone and shale of the Lower Cambrian Rome Formation in the hanging wall of the Copper Creek thrust at Diggs Gap on Interstate 75 north of Knoxville, Tennessee. (Reprinted from *Journal of Structural Geology*, v. 8, Steven Wojtal, p. 341–360, © 1986, with kind permission from Elsevier Science, Ltd., Kidlington, United Kingdom.)

PROPAGATION AND TERMINATION OF THRUSTS

A thrust moves until the energy that causes it to move is spent or until something stops it from moving and the energy is transferred to another existing fault or to a new fault. A thrust belt rarely consists of a single thrust, and so there must be some reason for motion to be transferred from one thrust to another or for the unbroken sequence to form a new thrust. Movement along a thrust may become difficult or diminish where the thrust passes over a ramp. For that reason, the angle between the shear stress and the fault plane may differ from that along the main detachment or higher detachments. It can be shown mathematically that frictional resistance to movement is greater in the high-angle segment of the fault plane than along low-angle segments (Figure 11–19). If the geometry of the fault changes enough that the force necessary to overcome the frictional resistance exceeds the strength of the footwall rocks, a new thrust may form and propagate through the footwall from an existing detachment. New thrusts may form at any level, depending on where the older thrust surface becomes locked and a suitable detachment is present for continued propagation of the thrust. This is also part of the duplex-forming mechanism described above.

A thrust may be forced to ramp to a higher décollement by stratigraphic pinch-out of the detachment unit, by facies changes within the detachment (reducing the amount of weak material), or by basement irregularities. Basement irregularities change the angular relationships with the dominant shear strain, causing the fault to ramp to a higher detachment rather than continue to propagate along the lower detachment. Facies changes may also weaken the strong unit.

Thrust faults project down-dip until they join a larger fault. The line of intersection of two fault surfaces is called a ***branch line*** (Figure 11–20). Imbricates have only one branch line; a horse is terminated at each end by leading and trailing branch lines.

Thrusts may terminate up-dip in several ways. A thrust may terminate at the surface as an ***erosion*** (or ***emergent***) ***thrust*** (Figure 11–21). There, surficial materials may be caught up and overridden by the advancing thrust sheet. The amount of movement that occurs after the surface is reached depends on the available energy. Erosion thrusts have been documented by Declan DePaor and David Anastasio (1987) in the Pyrenees in Spain, where unconformities—formed beneath debris eroded from the advancing thrust sheet—had been deformed by the sheet. Smaller thrusts may terminate in the sedimentary section, either within a detachment or as a series of branching splays that distribute motion on the smaller faults (Figure 11–22). Several thrusts that branch and transfer displace-

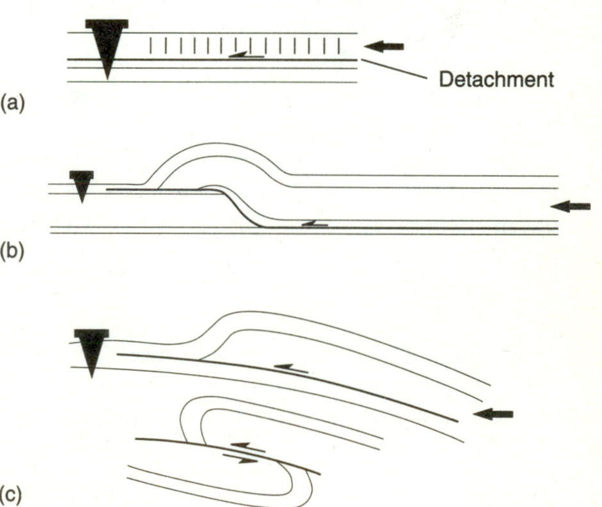

FIGURE 11–17
Geiser's thrust categories. (a) Shortening thrust. Pressure-solution strain in hanging wall permits displacement to occur. (b) Fracture thrust. Strong unit is broken through as fault propagates from lower to upper detachment. (c) Fold thrust. The thrust forms as a product of either folding above a detachment or along the common limb between an anticline and syncline. Note the similarities in (b) and (c) to Suppe's fault-bend and fault-propagation folds in Figure 11–11.

ment as they terminate in the sedimentary section or overlap one another have been described by Clark Burchfiel and others (1974) in the Spring Mountains of southern Nevada. The opposite of an emergent thrust is a ***blind thrust***, in which displacement decreases upward within the sedimentary section; it never reaches the surface.

Terminations of thrusts along strike sometimes yield information about subsurface up-dip behavior. A thrust may splay up dip or along strike and distribute movement among several smaller thrusts before terminating (Figures 11–22a and 11–22b). A single thrust or a set of splays may strike into the axial zones of anticlinal folds that plunge and die along strike away from the main thrust from which they were derived (Figure 11–22b). Folds and splays frequently strike at angles of 15° to 30° into the hanging wall of the thrust sheet from the strike of the main fault, a result predicted by orientation of the principal and shear planes in the strain ellipsoid.

Some thrusts terminate along strike and up-dip by decreasing displacement until the fault terminates in bedding. The *bow-and-arrow rule,* formulated by David Elliott (1976a), states that the length of a normal line at the bisectrix of the chord connecting the terminal ends of a thrust sheet is a measure of the maximum displacement of the thrust sheet (Figure 11–22c). Elliott's rule works only if displacement is approximately symmetrical to the ends of the sheet and reaches a maximum in the center.

The ends of a thrust sheet also terminate in strike-slip faults (***tear faults***), which at depth become part of the thrust sheet (Figure 11–22d). Along-strike changes in detachment level of a thrust may also occur where the thrust is forced to ramp to a higher level—a ***lateral ramp*** (Figure 11–22e).

Fault strike length and displacement are directly related, as noted by Elliott (1976b). He concluded that maximum displacement, u, may be approximated by

$$l \cong 14u \ , \qquad \textbf{(11–1)}$$

where l is fault length, obtained from his bow-and-arrow rule. As Elliott noted further, longer thrusts must be part of stronger thrust sheets. If so, we can predict that the largest of all thrust sheets—composite crystalline thrust sheets, to be discussed below—should be the strongest and have the greatest displacements. If we plot length versus known displacement for several thrust sheets (Figure 11–23), we find his conclusion to be correct. Elliott's equation may underestimate displacement by as much as 50 percent for very large thrust sheets unless independent criteria, such as seismic reflection data, are available to help to constrain cross sections.

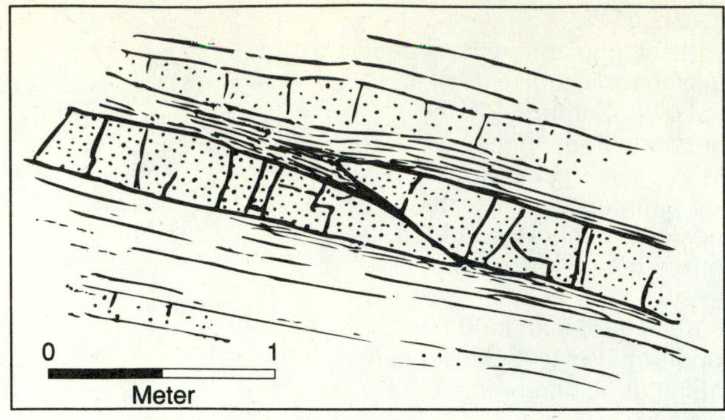

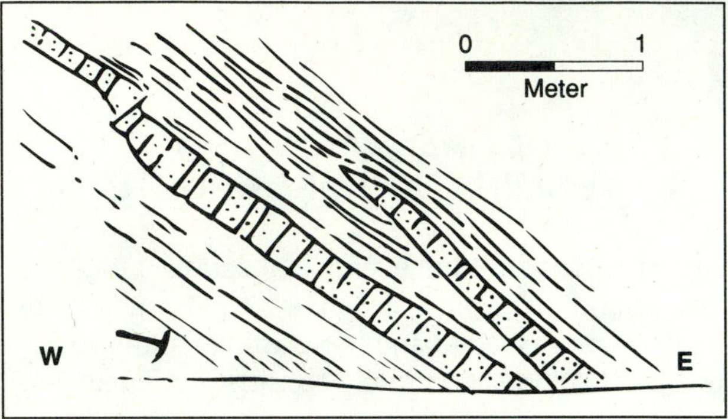

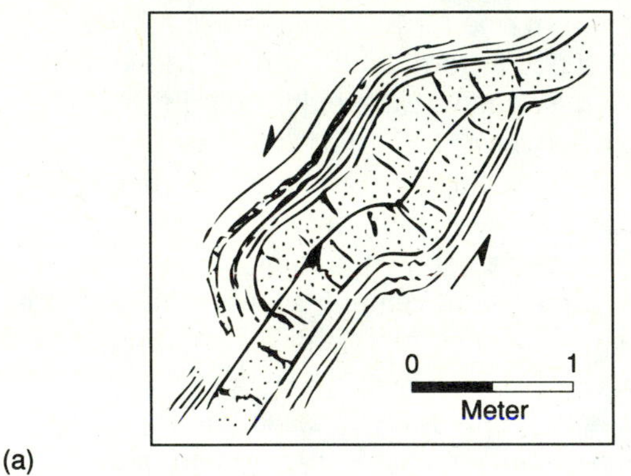

(a)

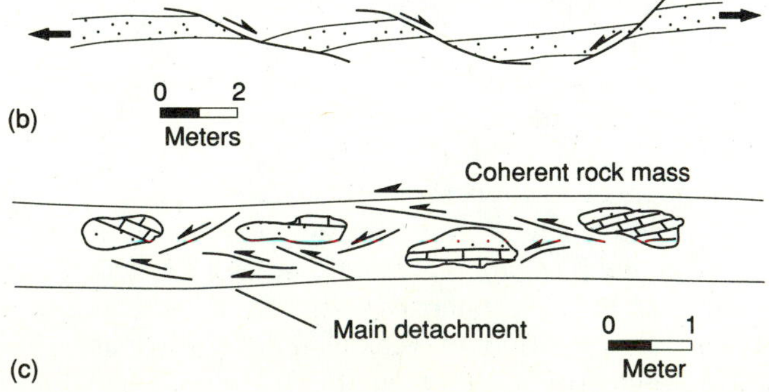

(b)

(c)

FIGURE 11–18
(a) Contraction faults shown in initial, intermediate, and advanced stages of development. Drawn from photographs by Ernst Cloos of structures in Maryland. (From Ernst Cloos, 1964, Virginia Tech Department of Geological Sciences Memoir 1.) (b) Extension faults. (c) Broken formation zone. All are shown in cross-section view.

Many geologists believe that thrust faults are produced by outward propagation from the internal parts of a mountain chain. These are *normal-sequence thrusts*. Sometimes a moving thrust sheet will lock or achieve a geometry that causes the hanging wall to break back of the leading edge, producing thrusts that are *out of sequence* as compared with normal inside-out geometry (Figure 11–24).

FEATURES PRODUCED BY EROSION

Thrust sheets characteristically have a large areal extent, and their fault surfaces commonly dip at low angles; also, many thrust sheets are folded after emplacement. This produces several important erosion-related features.

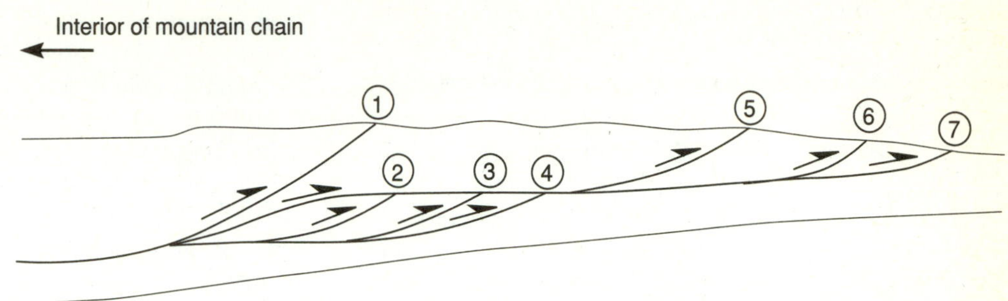

FIGURE 11–19
"Normal" propagation sequence in the development of a thrust belt (1, oldest; 7, youngest).

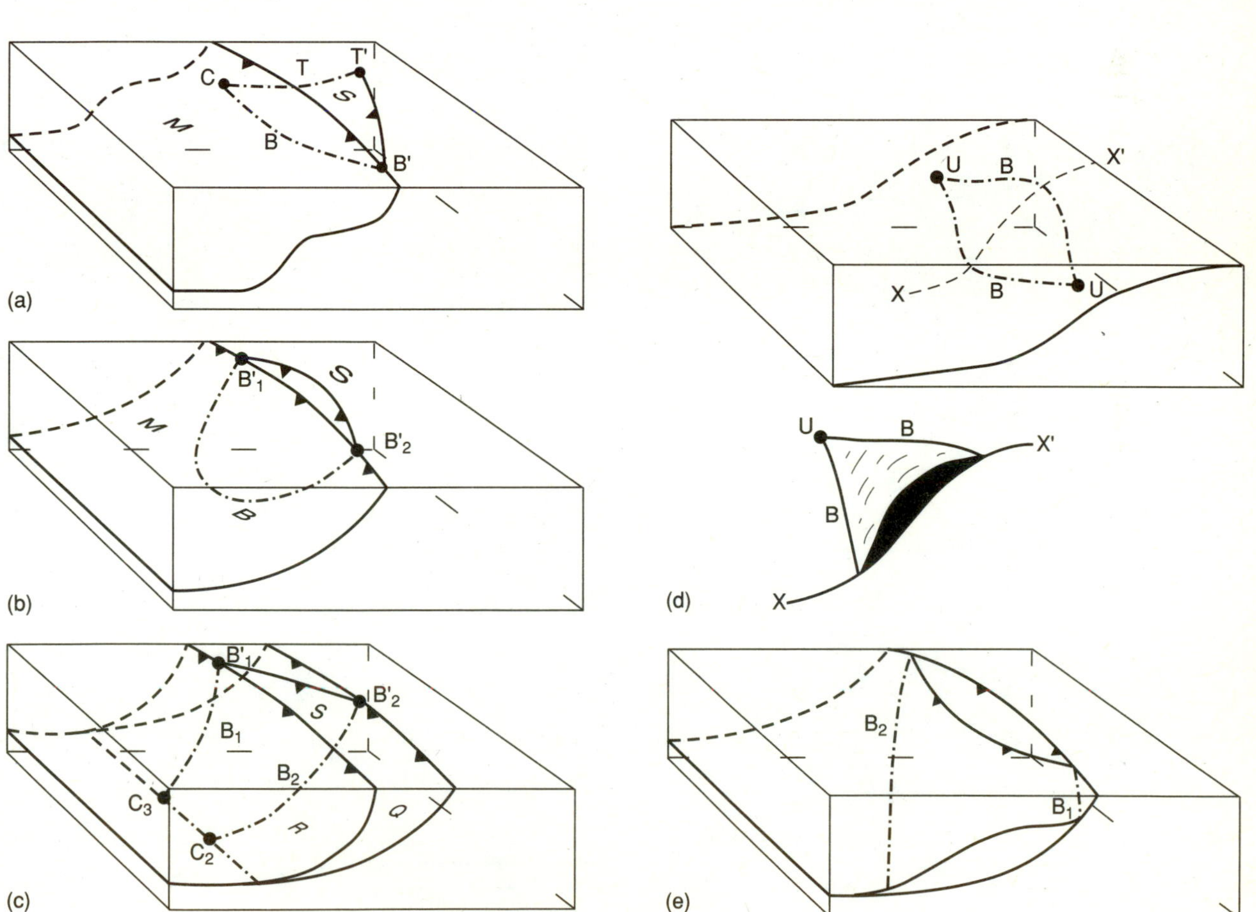

FIGURE 11–20
Branch lines. (a) Diverging splay S, which has a single tip line T and a map termination T′, and one branch line B that intersects the erosion surface at B′. (b) Rejoining splay with a single branch line B that intersects the map at two branch points B'_1 and B'_2. (c) Two major faults Q and R with connecting splay S. Two branch lines have surface terminations B'_1 and B'_2. (d) Horse surrounded by fault surfaces (above). Two fault surfaces meet at a single closed branch line B with two cusps U. Lower diagram illustrates half of horse cut along section X–X′. (e) Two fault surfaces meet along two branch lines B_1 and B_2; the map pattern resembles a diverging splay, and the cross section resembles a horse. (From S. E. Boyer and David Elliott, AAPG *Bulletin,* v. 66. © 1982. Reprinted by permission of American Association of Petroleum Geologists.)

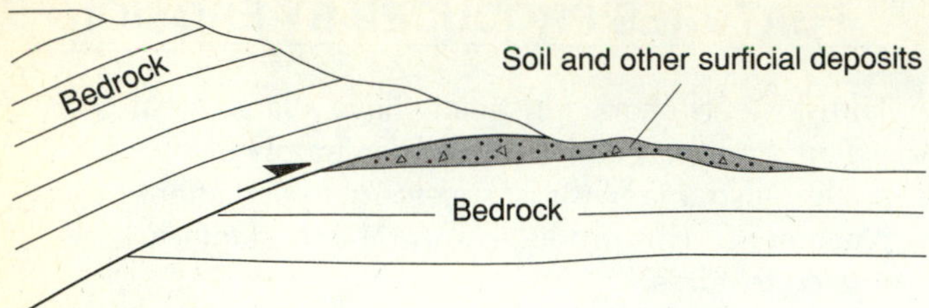

FIGURE 11–21
Emergent thrust breaks the surface and involves surficial materials.

Erosion may make holes in a thrust sheet, exposing the footwall rocks. Footwall rocks are thus completely surrounded by the hanging wall, producing a feature called a ***simple window***, or ***fenster***. A window usually develops where a thrust sheet has been antiformally folded (domed), causing part of the sheet to have a higher elevation than adjacent parts (Figure 11–25a). The higher part of the sheet may also have a greater fracture density, making it more susceptible to erosion.

A thrust sheet may undergo renewed movement during or after folding. An antiformal fold thus produced may prevent the entire thrust sheet from moving and lead to imbrication of the sheet along the crest of the antiform. Then erosion may expose a more complex window into the footwall rocks, but the window is opened along the trace of a fault and will appear to be partly closed because the elongate outcrop pattern parallels the fault trace (Figure 11–25a). Steven Oriel (1950) first used the term ***eyelid window*** to describe this complex geometry in the North Carolina Blue Ridge. A similar structure was described by James Gilluly (1960) on Goat Ridge in the Northern Shoshone Range, along the Roberts Mountain thrust in Nevada. From his study of the Goat Ridge window, Gilluly concluded that folding and faulting have to be synchronous for eyelid windows to form (Figure 11–25b).

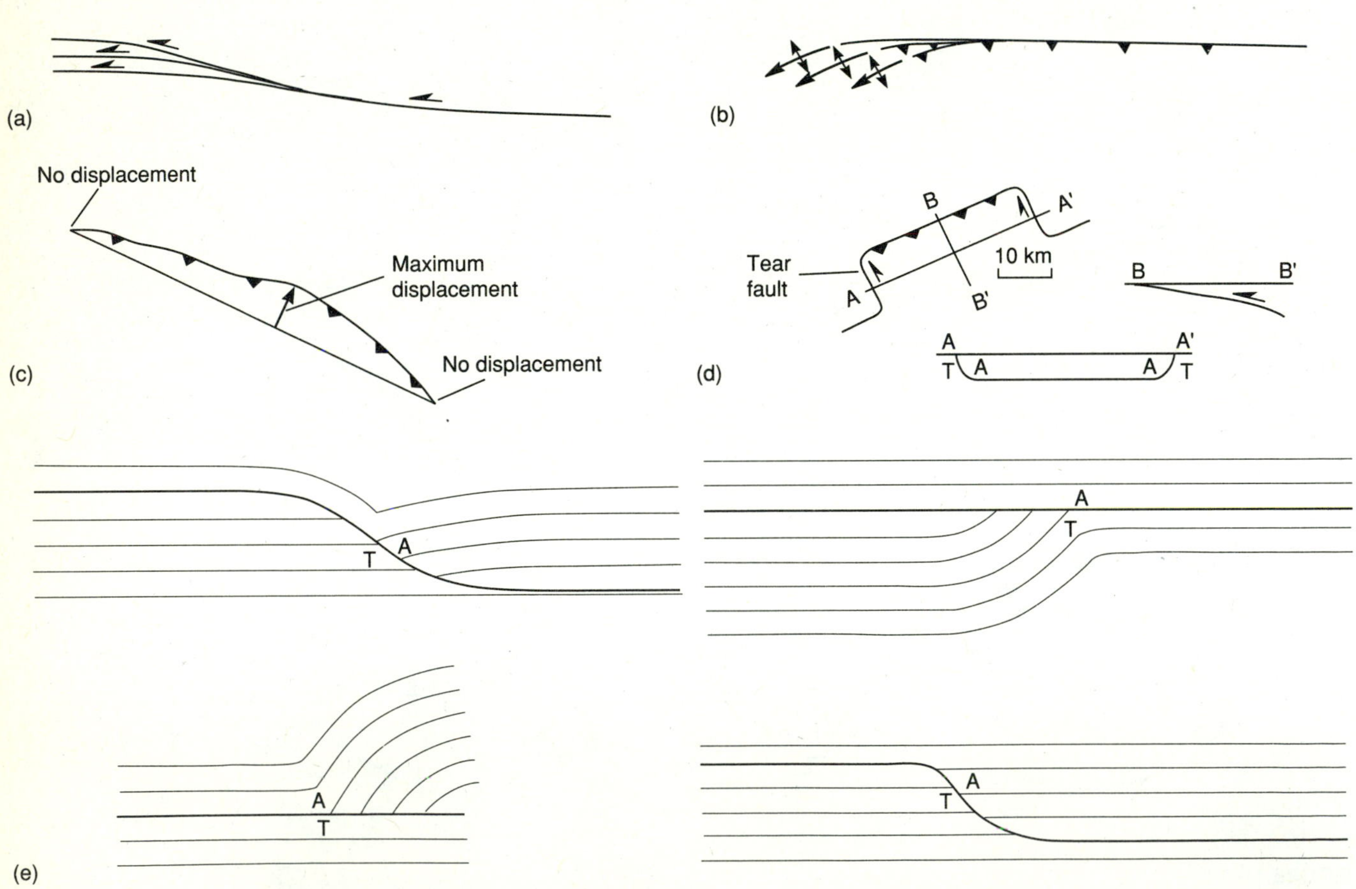

FIGURE 11–22
Termination of thrusts. (a) Branching splays, which take up displacement up-dip (cross section) or along strike. (b) Decreasing displacement and Elliott's bow-and-arrow rule. Displacement decreases away from the approximate center of the thrust sheet (map view). (c) Along-strike splays into plunging anticlines. Main thrust splays into smaller thrusts, and then into plunging anticlines (map view). (d) Tear faults (map view with cross sections a–a′ and b–b′). A—away from the observer; T—toward the observer. (e) Configurations of lateral ramps, viewed parallel to strike. All are cross sections.

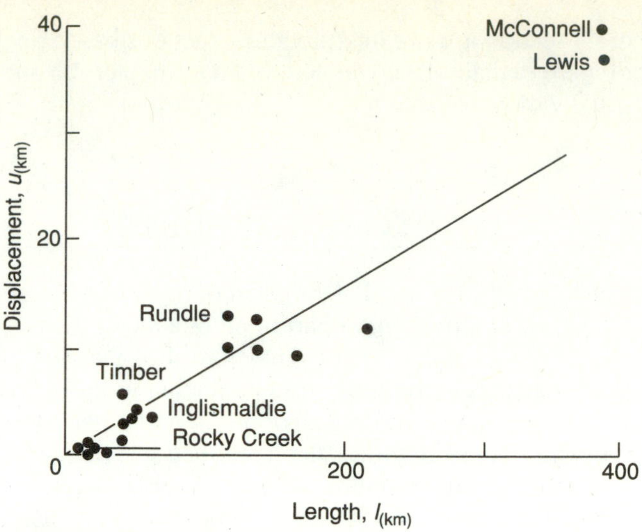

FIGURE 11–23
Plot of along-strike length versus displacement of thrust sheets in the Canadian Rockies, in Alberta. (Modified from D. Elliott, 1976, *Journal of Geophysical Research,* v. 81, p. 949–963. © American Geophysical Union.)

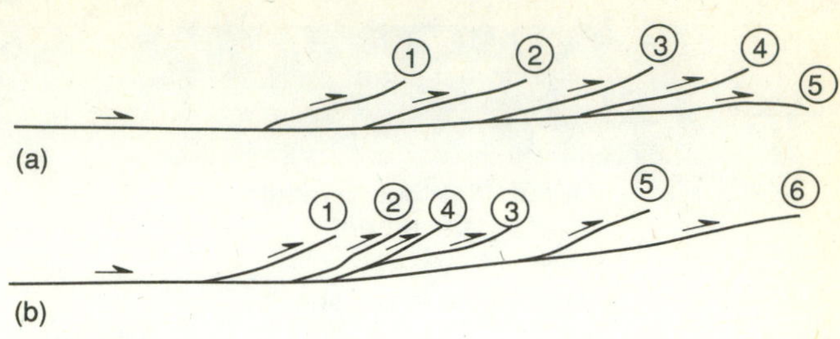

FIGURE 11–24
In-sequence (a) and out-of-sequence (b) thrusts. A normal sequence (1, oldest, through 5, youngest) involves progression of thrusts from one side of a deformed zone to another. Out-of-sequence thrusting (thrust 4) involves thrusts that form within a pile of already-formed thrusts.

Many windows in overthrust terranes expose footwall duplexes (Figure 11–15b), suggesting an interactive relationship between the growth of duplexes and the arching of a thrust sheet during thrust emplacement (Boyer and Elliott, 1982; Hatcher, 1991; Figure 11–26). Many examples of isolated domes occur in large thrust sheets; breached windows that expose duplexes in the footwall include the Grandfather Mountain and Mountain City windows in the southern Appalachians (Boyer and Elliott, 1982; Diegel, 1986) and the Assynt window in the northwestern Scottish Highlands (Elliott and Johnson, 1980).

Some geologists use the term "window" to describe any erosionally exposed rock unit that is surrounded laterally by the unit above it. If the contact between the two is not a fault but a normal contact or unconformity,

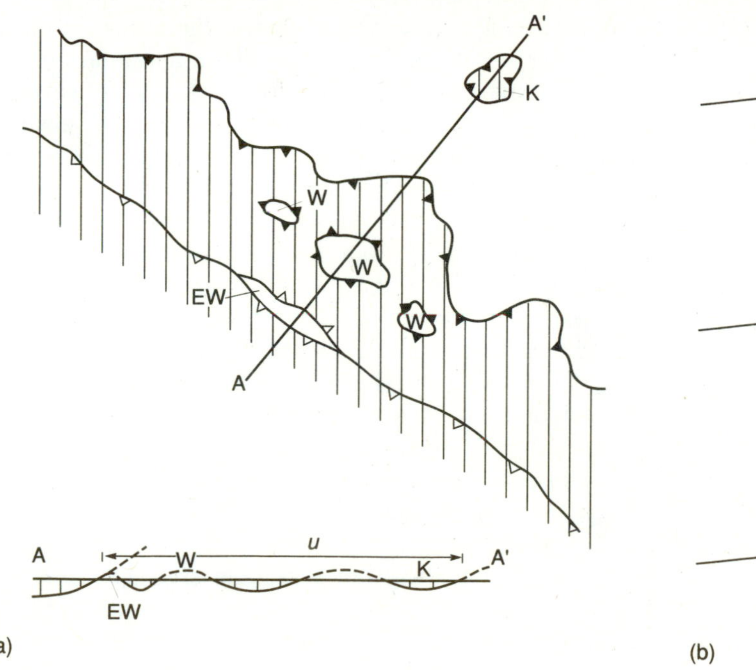

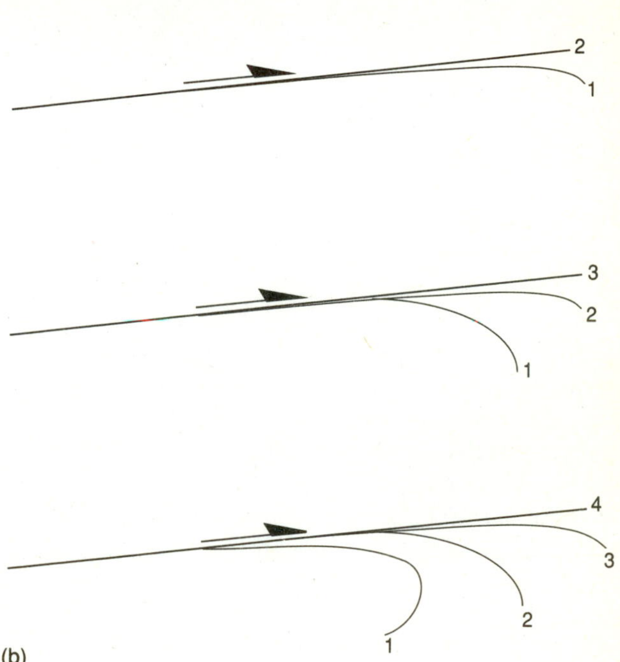

FIGURE 11–25
(a) Erosional features along thrusts: W—simple window; EW—eyelid window; K— klippe. Windows and klippes may be used in estimating minimum amount of transport (*u*). Rocks of the thrust sheet are striped. Younger thrust shown by open teeth on hanging wall. (b) Synchronous folding and thrusting in cross section, as envisioned by Gilluly (1960) for the Goat Ridge window, a complex eyelid window. Thrust 1 forms and is folded so that the hanging wall is broken at the crest of the anticline by continued movement, with the new up-dip portion of the thrust becoming thrust 2. The process is repeated to form thrusts 3 and 4, with each up-dip segment becoming inactive as it is folded.

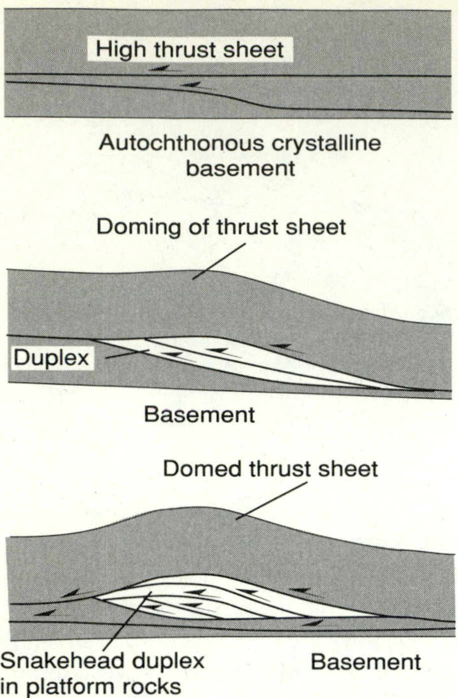

FIGURE 11–26
Possible interactive relationships between formation of a duplex during emplacement of a thrust sheet, arching of the thrust sheet by the growing duplex, and subsequent erosional unroofing of the domed area. The result is a window that exposes a duplex.

the proper term is *inlier,* because the rocks are still in stratigraphic order, with the oldest on the bottom. True windows involve younger rocks exposed by erosion through a sheet of older rocks.

Erosion also frequently isolates parts of a thrust sheet. Dismemberment of a thrust sheet by erosion may leave a remnant called a ***klippe*** (Figure 11–25a) or a larger remnant called an ***allochthon***. The term ***allochthon*** is also used for a large single thrust sheet. A klippe sometimes results from folding of the thrust sheet, where the preserved part remains in a synform, but erosional dissection of a nearly flat thrust sheet may also preserve parts of the sheet as klippes (or klippen). Klippes are *erosional outliers* of thrust sheets, but not all outliers are klippes (or allochthons).

Windows and klippes are useful in measuring minimum transport distance of a thrust sheet (Figure 11–25a). The minimum horizontal transport is the distance measured across strike from the outcrop trace of the thrust on the down-dip side of a window to the outcrop trace of the thrust on the up-dip side of a klippe, or leading edge of the sheet. This assumes the rocks in the window are younger than rocks in the klippe, that the thrust cuts stratigraphically upward, and that transport was across strike. Displacement is better measured on a map than on a cross section because a cross section is more interpretative; sections based on modern digital seismic reflection data (Chapter 21) are usually more reliable.

CRYSTALLINE THRUSTS

Crystalline thrusts involve transport of metamorphic or igneous rocks, or both, as part or all of a thrust sheet (Figure 11–27, Table 11–1). They have been known for many decades in the Alps, the Appalachians, the Scandinavian and British Caledonides, and other orogens. The large thrusts first identified in Sweden and Scotland are crystalline thrusts. Geologists once thought that their propagation and motion involved a process totally different from that in thin-skinned foreland thrusts in a sedimentary sequence, but many attributes of thin-skinned thrusts may also be identified in crystalline thrusts, leading to the conclusion that the processes may be nearly the same. That they are low-angle thrusts indicates an important similarity to thin-skinned thrusts, but how does a subhorizontal thrust surface propagate through a crystalline mass so that part of the mass becomes separated and moves as part of the thrust sheet? There must be important differences, such as the initially horizontal layering in sedimentary rocks, in view of the different material properties of sedimentary and crystalline rocks. After studying the Moine thrust zone, Elliott and Johnson (1980) concluded that thrusts follow zones of weakness in crystalline rock, as they do in a sedimentary section. Hatcher and Richard Williams (1986) came to a similar conclusion about propagation of thrusts in crystalline rocks and that the shared properties with thin-skinned thrusts lead to a general rule of crystalline thrusts: *The largest crystalline thrust sheet in an orogen is always larger than the largest foreland thrust.* The rule no doubt partly reflects the greater inherent strength of crystalline thrust sheets. The zone of weakness along which detachment occurs could be an appropriately oriented mechanical weakness, such as a preexisting fault or a strong foliaiton, but more likely it is the ductile-brittle transition in the large composite sheets discussed below.

Crystalline thrusts may form as large slabs wherein materials appear to exhibit brittle to semibrittle behavior. They may also form during ductile folding where large crystalline isoclinal recumbent folds (fold nappes) are produced. Hatcher and Hooper (1992) called the large slab-type thrusts Type C (for composite) crystalline sheets (Type 5 in Figure 11–27), and those related to ductile folding Type F (for fold-related) crystalline sheets (Type 3 in Table 11–1 and Figure 11–27b). These form because continued transport attenuates the overturned limb between an antiform and synform,

TABLE 11–1
TYPES AND PROPERTIES OF CRYSTALLINE THRUST SHEETS

Type	Name	Place of Emplacement in Orogen	T-P Conditions of Formation	T-P Conditions of Emplacement
1	Thin-skinned thrusts transporting basement	Inner foreland/outer metamorphic core	Low	Low (greenschist or below)
2	Ophiolites	Inner edge of foreland, metamorphic core, accreted terranes	Low to moderate; may be high in fault zone	Low; may be moderate T at base of sheet
3	Tectonic slides (Type F)	Metamorphic core	Low to moderate to high, depending on rheology; may form at low P-T in weak rocks	Low to moderate to high
4	Basement uplifts	Foreland; edge of craton	Low; faults initiated brittle in upper crust, ductile in lower crust	Low
5	Composite sheets (Type C)	Metamorphic core; may involve inner foreland	Moderate to high in fault zone, low in internal parts of sheet	Moderate to low higher along fault zone

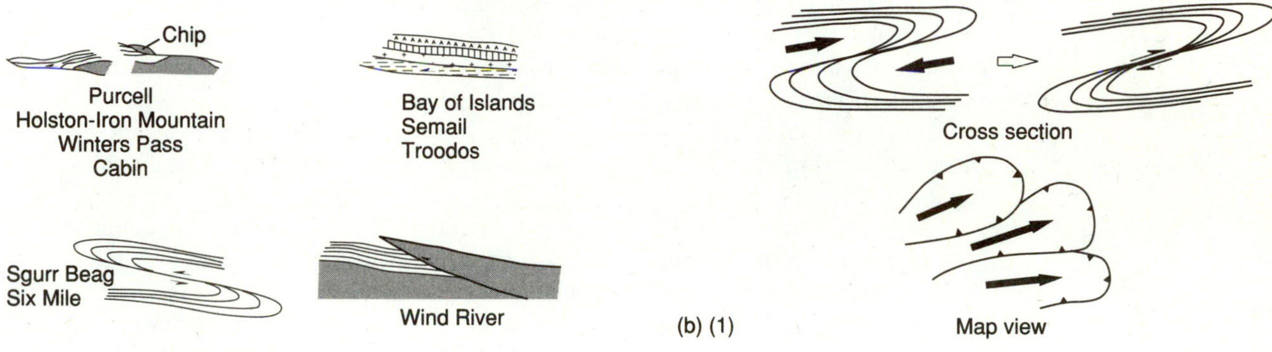

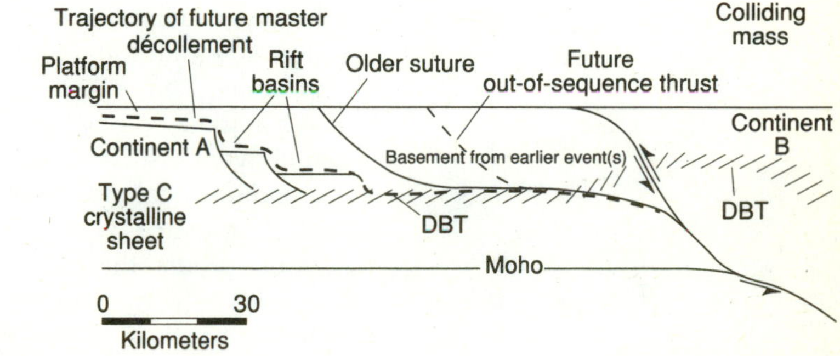

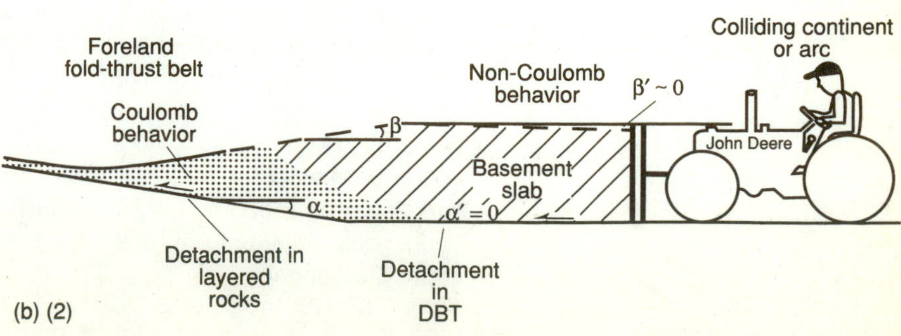

FIGURE 11–27
(a) Varieties of crystalline thrusts. (From R. D. Hatcher, Jr., and R. T. Williams, Geological Society of America *Bulletin,* v. 97, 1986.) (b) Types F (1) and C (2) crystalline thrust sheets. Type F (= Type 3 in a) sheets result from plastic folding and heterogeneous deformation and thus have lobate outcrop patterns in map view. Type C (= Type 5 in a) sheets are produced by detachment of a strong, largely brittle slab along the ductile-brittle transition (DBT) as a product of continent-continent, terrane, or arc-continent collision. The analog model below shows the elongate wedge shape of the crystalline sheet that pushes the deforming tapered wedge of the foreland fold-thrust belt in front of it. α (or α') represents the dip of the detachment. β is the surface slope angle. (From R. D. Hatcher, Jr. and R. J. Hooper, *in* K. R. McClay, editor, *Thrust Tectonics,* Chapman and Hall, London, 1992.)

ESSAY

The Paradox of Large Overthrust Faults

Thrust faults have been controversial ever since they were first discovered. Controversy has raged about their very existence, transport distance, the degree of basement involvement, and mechanics of emplacement. Several disputes have arisen because we find it hard to conceive of an enormous, thin sheet of rocks being detached from its roots and moving intact for many kilometers. The paradox was probably first stated clearly by M. S. Smoluchowski (1909) when he calculated the strength of thin sheets of dry rock and concluded that the maximum permissible width of thrust sheets is much less than the dimensions observed in nature. In search of better understanding of large thrust sheets, M. King Hubbert (1945) devised a model to simulate the strength of crustal rocks, depicting a hypothetical crane and bolt capable of lifting a 600-km-thick block of continental crust and mantle the size and shape of Texas. His calculation of the strength of the crust, assuming it was granite, showed that the bolt would pull out without lifting the block—that the block would not support its own weight. He concluded that "the good state of Texas is utterly incapable of self-support" and, equally important, that a large sheet of rock would not remain intact while moving for any appreciable distance. Large (Type C) thrust sheets (composite crystalline sheets of Hatcher and Hooper, 1992), however, do exist and have moved tens and even hundreds of kilometers. The Silvretta nappe in the Alps (Laubscher, 1983), several thrust sheets in the Scandinavian Caledonides (Gee, 1978), the large Yukon-Tanana sheet (Templeman-Kluitt, 1979), and the Blue Ridge–Piedmont thrust sheet (Hatcher, 1981) are all huge Type C thrust sheets that have moved long distances. Many smaller thrusts have moved lesser distances, even though their transport distances are proportional to the larger thrusts.

Raymond Price (1973) has outlined the problem of large overthrusts and discussed constraints placed by mechanics on the initiation and propagation of large overthrusts.

> "The mechanical paradox in overthrust faulting," Price stated in his abstract, "originates in the conceptual models chosen for mechanical analysis." He then summarized, "the notion that the maximum dimensions of a tabular mass of rock that may undergo translation along an overthrust fault are fixed by the relative magnitudes of: (1) the stress required to overcome the *total* frictional resistance to sliding (and the cohesion or adhesion?) over the *entire* fault surface and, (2) the strength of the rocks involved." That notion, Price said, "tacitly assumes that sliding is initiated, and occurs simultaneously, over the entire fault surface".

He went on to say:

> "A realistic model for overthrust faulting must be compatible with the nature of actual overthrust faults, the way in which displacements occur on active faults, and the focal mechanisms of earthquakes. Most large overthrusts are discrete slip surfaces within a coherent mass of rock that is physically continuous around the ends. The net slip, which changes along the fault surface, can be viewed as the cumulative effect of innumerable incremental displacements, each of which is initiated as a local shear failure that propagates as a dislocation on a scale and at a velocity that is small in comparison with the total area of the fault surface. The Coulomb-Navier failure criterion and the law of sliding friction, as modified for the effects of pore pressure, are empirical relationships derived from experiments on small rock specimens and may be applied over the area of a dislocation, but should not be extrapolated, in the same way, over a linear scale of up to six orders of magnitude to encompass an entire fault surface. The shape and dimensions of overthrust faults are controlled by the stress field, mechanical properties, and anisotropy of the rock mass, but not by the total frictional resistance to sliding over the entire fault surface."

Here, Price advocated application of physical laws to a complex structure by considering the real properties of the structure. For several decades now, we have been attempting to do so. In 1988, Price suggested, following studies of the nature of failure and generation of earthquakes along active faults (e.g., Aki, 1967), that propagation of thrust faults may be analogous to propagation of a Somigliana (smeared) dislocation, in which propagation of the fault occurs by increments rather than by overcoming friction over the entire ultimate extent of the fault. The Hubbert and Rubey (1959) suggestion for a solution to the paradox—involvement of fluid— is probably the best to date to overcome friction, despite questions about the ability to ignore τ_0, and involve gravity as the driving force for foreland deformation, among other objections.

One barrier to understanding why overthrust faults exist is our inability to make direct measurements on large active thrusts buried several kilometers in the crust beneath the surface. The barrier is likely to remain, but we can obtain a good idea of the shapes and properties of thrusts by using geophysical techniques (Chapter 21) to image the crust and by studying samples we can obtain from fault zones.

References Cited

Aki, K., 1967, Scaling law of seismic spectra: Journal of Geophysical Research, v. 72, p. 1217–1231.

Gee, D. G., 1978, Nappe displacement in the Scandinavian Caledonides: Tectonophysics, v. 47, p. 393–419.

Hatcher, R. D., Jr., 1981, Thrusts and nappes in the North American Appalachian orogen, *in* McClay, K. R., and Price, N. J., Thrust and nappe tectonics: Geological Society of London Special Publication 9, p. 491–499.

Hatcher, R. D., Jr., and Hooper, R. J., 1992, Evolution of crystalline thrust sheets in the internal parts of mountain chains, *in* McClay, K. R., ed., Thrust tectonics: London, Chapman and Hall, p. 217–233.

Hubbert, M. K., 1945, Strength of the Earth: American Association of Petroleum Geologists Bulletin, v. 29, p. 1630–1653.

Hubbert, M. K., and Rubey, W. W., 1959, Role of fluid pressure in mechanics of overthrust faulting: Part 1. Mechanics of fluid-filled porous solids and its application to overthrust faulting: Geological Society of America Bulletin, v. 70, p. 115–166.

Laubscher, H. P., 1983, Detachment, shear, and compression in the central Alps, *in* Hatcher, R. D., Jr., Williams, Harold, and Zietz, Isidore, Contributions to the tectonics and geophysics of mountain chains: Geological Society of America Memoir 158, p. 191–212.

Price, R. A., 1973, The mechanical paradox of large overthrusts: Geological Society of America Abstracts with Programs, v. 5, p. 772.

Price, R. A., 1988, The mechanical paradox of large overthrusts: Geological Society of America Bulletin v. 100, p. 1898–1908.

Smoluchowski, M. S., 1909, Some remarks on the mechanics of overthrusts: Geological Magazine, v. 6, p. 204–205.

Templeman-Kluitt, D. J., 1979, Transported cataclasite, ophiolite, and granodiorite in Yukon: Evidence of arc-continent collision: Geological Survey of Canada Paper 79–14, 27 p.

finally producing a ductile thrust. Such thrusts were first called *slides* by Sir Edward Bailey (1910), but Michael Fleuty (1964) has urged that they be called *tectonic slides* to prevent ambiguity. Tectonic slides are common on all scales in high-grade metamorphic rocks, but they may also form at lower grades in less-competent assemblages that have undergone isoclinal folding. Bailey (1934) distinguished *lags* as listric normal faults that cut out the upright limb of a recumbent anticline.

THRUST MECHANICS

Ever since their discovery in the nineteenth century, the mechanics of thrust faults have been studied intensively. Some of the early insights were derived from work with pressure boxes in which a piston deformed layers of different materials with contrasting properties and colors (Figure 11–28). In such experiments, the end opposite the piston is buttressed and the upper surface remains unconfined so that deformation may proceed internally. By using pressure boxes, H. M. Cadell (1890) in Great Britain and Bailey Willis (1893) in the United States made several contributions toward our understanding of thrust mechanics.

Attempts to apply the principles of continuum mechanics to thrusting began in the twentieth century with the work of M. S. Smoluchowski (1909), W. Hafner (1951), M. K. Hubbert (1951), and Jean Goguel (1965). Models of stress distribution in a thrusted block by Hafner are still used (Figure 11–29). Probably the most significant paper on thrust faulting published in mid-century was one by Hubbert and William W. Rubey (1959), who showed that fluid pressure along a zone of weakness facilitates motion with expenditure of much less energy than if the zone is dry (Chapter 10). Their calculations of the force required to move a thrust sheet (such as the Keystone thrust sheet in Nevada) always exceeded the inherent strength of the sheet if they assumed it slid on a dry fault—indicating that it would fragment before motion could occur and therefore could not exist in nature. They concluded that the buoyant effect of fluid in a mass of rocks would lower the frictional resistance to fracturing and subsequent motion—lower it enough to show how a very large thrust sheet can exist in nature, solving Smoluchowski's (1909) dilemma. After fracture is initiated, they suggested, the Coulomb–Mohr criterion may be modified to

$$|\tau_c| = \tau_0 + (\sigma_n - P_w) \tan \phi, \qquad \textbf{(11–2)}$$

where τ_c is the critical shear stress at failure, τ_0 is the inherent cohesive shear strength of the material, σ_n is normal stress, P_w is the pore pressure [$(\sigma_n - P_w) = \mathbf{S}$, effective normal stress], and ϕ is the internal friction angle. If τ_0 is small, equation 11–2 may be modified further to

$$|\tau_c| = (\sigma_n - P_w) \tan \phi. \qquad \textbf{(11–3)}$$

Effective normal stress is an important factor in movement along thrusts and other faults (Chapter 10), as well as in the general process of rock deformation. Pore pressure reduces the normal stress and in effect reduces the overall strength of the rock mass. Thus fluid on the fault plane promotes the development of large overthrusts, but fluid in the hanging wall inhibits their development. Fluid also enables renewed movement by lessening frictional resistance along existing faults. Hubbert and Rubey assumed that after movement

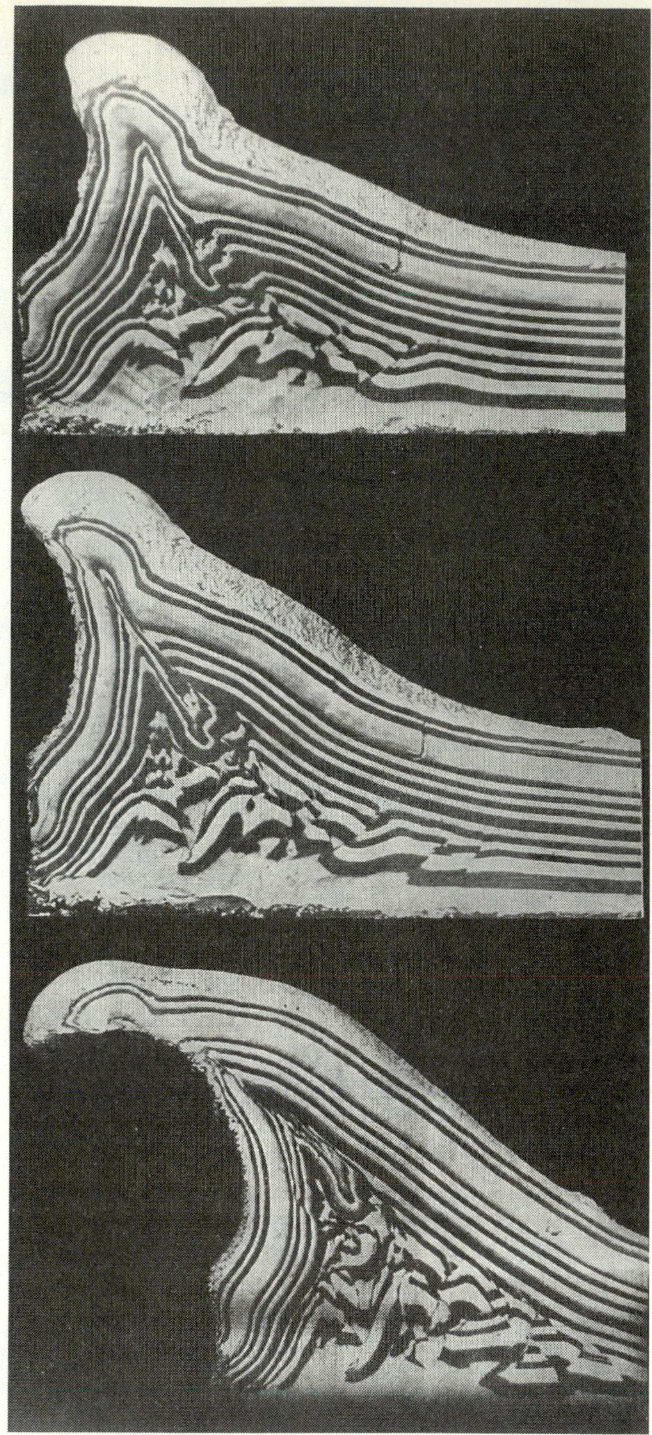

FIGURE 11–28
Thrusts produced in a pressure-box experiment. The sequence from top to bottom represents a fold formed in a pressure box that on continued deformation develops a thrust (a break thrust or fault-propagation fold). Layers are made of clays of slightly different properties. (Bailey Willis, 1893, U.S. Geological Survey, Thirteenth Annual Report, Part II.)

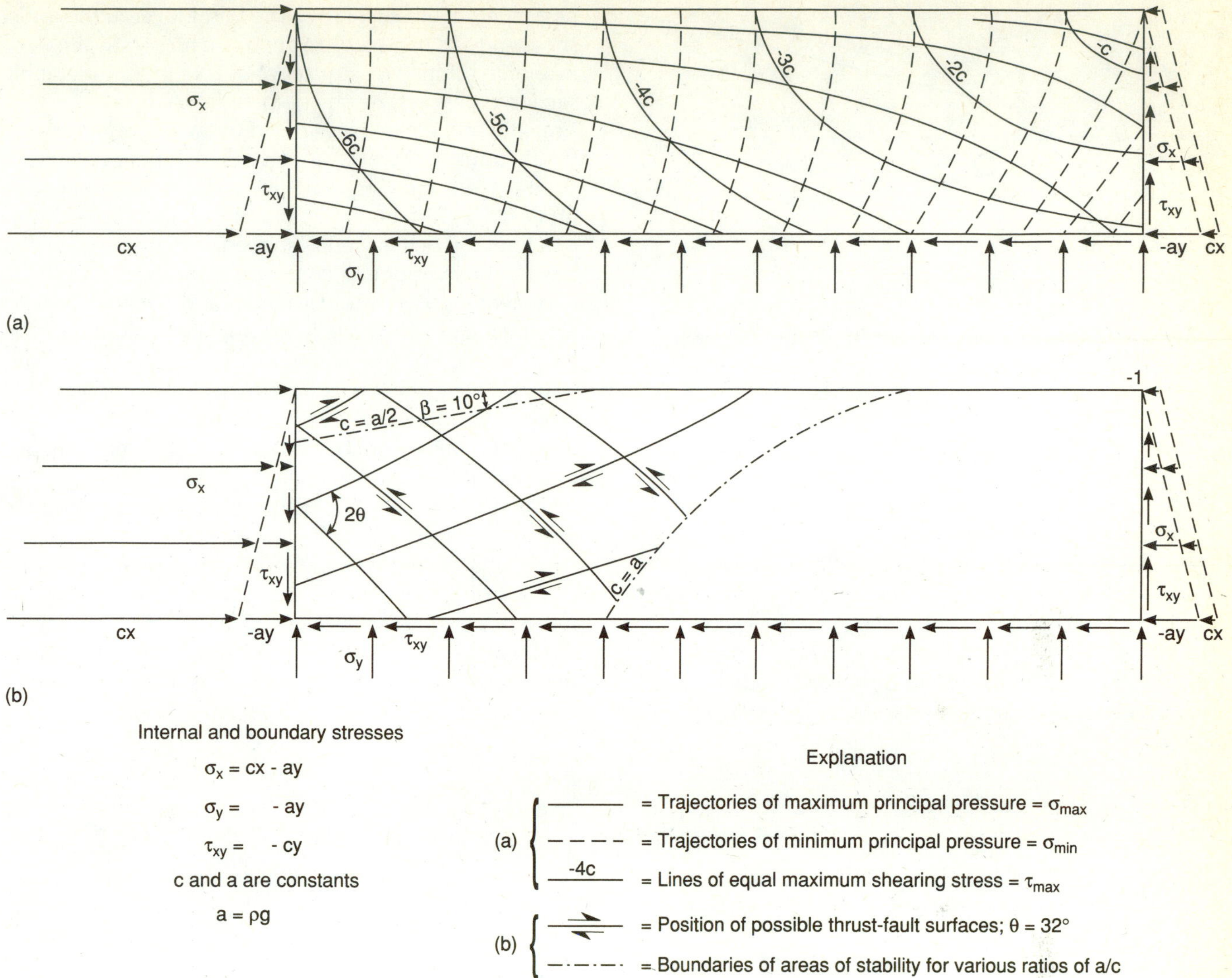

FIGURE 11–29
(a) Hafner model of stress distribution in a buttressed block. Stress is applied horizontally and is assumed to be constant with depth and with a constant lateral gradient. (b) Possible shear (fault) surfaces within and at the base of the block. (From W. Hafner, 1951, Geological Society of America Bulletin, v. 62.)

begins, cohesive shear strength along the fracture becomes negligible; stress is then transferred to the fluid, producing a buoyant effect and greatly enhancing motion. In the preceding discussion in Chapter 10, we saw a useful analog to the buoyant-fluid concept—Hubbert and Rubey's beer-can experiment (Figure 10–9). This experiment demonstrates the importance of buoyant force in decreasing effective normal stress.

As pointed out by Kenneth Hsü (1969), Hubbert and Rubey's simplification applies *only* if renewed movement occurs on an existing fracture. In all other situations, we must consider the cohesive shear-strength term (τ_0).

Study of thrust sheets in the Muddy Mountains of Nevada led William Brock and Terry Engelder (1977) to conclude that thrust sheets (or other fault blocks) may be too permeable to maintain fluid pressure of the kind Hubbert and Rubey postulated, except in a few places such as Taiwan and possibly the Gulf Coast of the United States. Fibers of calcite and other minerals on many fault surfaces also suggest that deformation occurs by a creep mechanism and may further indicate that the buoyancy mechanism is not viable for all thrusts. Fibers and mineral growth along faults indicate that fluid was present.

David Elliott (1976a, 1976b) suggested that a thrust propagates as fractures, with displacement related to the length of the outcrop trace. Elliott also suggested that much of the energy expended in emplacement is dissipated within the thrust sheet. In the McConnell thrust sheet in the Canadian Rockies, he concluded that energy was dissipated by frictional sliding only in the uppermost 5 km. Pressure-solution slip along individual surfaces and penetrative pressure

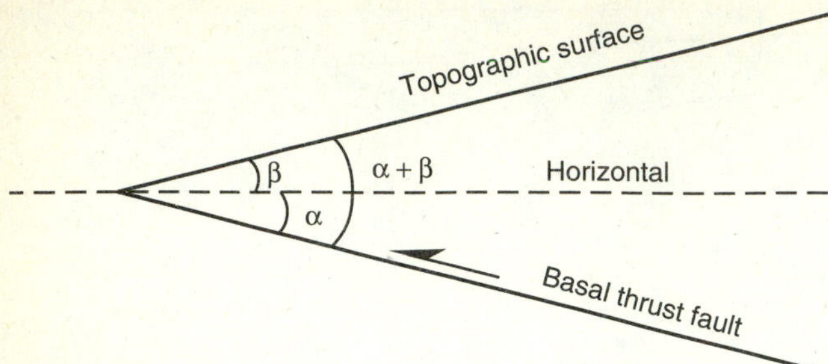

FIGURE 11–30
Ideal Coulomb wedge: α is the dip of the basal thrust; β is the topographic slope angle; $\alpha + \beta$ is the angle at the apex of the wedge.

solution with cleavage development dominate at greater depths but have also been found within hundreds of meters of the surface. Where pressure-solution slip has been observed, slip velocity appeared to vary with magnitude of shear stress.

The mechanical analysis of foreland thrusting by William Chapple (1978) may prove as significant as the earlier study by Hubbert and Rubey. Chapple recognized five characteristics now considered fundamental. (1) Foreland thrust belts are thin-skinned, and all deformation occurs above a particular layer without basement involvement. (2) The basal layer in the thrust sheet consists of a weak material such as shale, coal, or evaporite. If the weak material occurs in a stratigraphic section composed mostly of materials of higher strength, movement is restricted to the weak layer. (Many thrusts, however, involve detachment only in strong layers, creating difficulty in our basic concepts!) (3) Before deformation, the entire stratigraphic section of a foreland belt is wedge-shaped. The wedge thins toward the interior of a continent, with depth to basement increasing toward the sea (or an ancient sea); the surface of the wedge is initially flat or slopes gently away from continent before collision. (4) As deformation proceeds, the wedge is continuously lengthened as more faults form, shortened as thrusts override each other, and thickened, but it maintains its overall shape throughout deformation (Figure 11–30). (5) On a regional scale, the entire deforming foreland thrust belt behaves plastically.

Dynamic thrust models developed by John Suppe (1981) from study of the active accretionary wedge thrust belt in Taiwan suggest that an equilibrium angle (*critical taper*) is maintained throughout deformation of the wedge and that materials composing thrust sheets are Coulomb (brittle) materials (Figure 11–31). Thus,

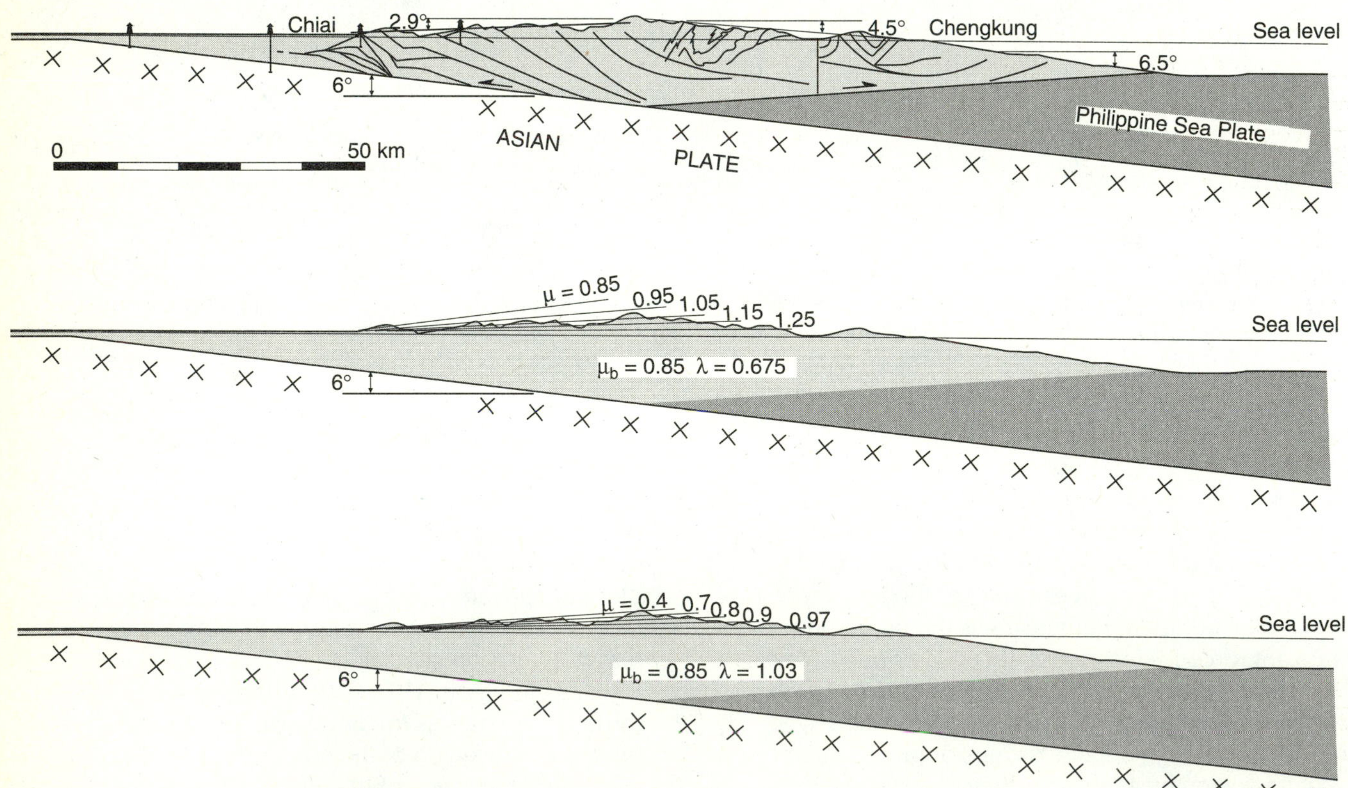

FIGURE 11–31
Propagation of thrusts in Taiwan maintaining the wedge shape and equilibrium surface angle of 2.9°. μ—coefficient of friction, μ_b—coefficient of friction along base of wedge; λ—pore-fluid pressure, obtained from a ratio of pore-fluid pressure to lithostatic load pressure. (From Dan Davis, John Suppe, and F. A. Dahlen, *Journal of Geophysical Research,* v. 88, p. 1153–1172, 1983, © American Geophysical Union.)

Dan Davis, Suppe, and Tony Dahlen (1983) have called this deforming wedge a ***Coulomb wedge***. The basal angle α is the dip of the basement surface, and the critical taper angle $\alpha + \beta$ is maintained by internal deformation within the wedge, out-of-sequence and synchronous thrusting, or by active erosion, which removes the steepened part of the shortened end of the wedge. These angles are also related to the coefficient of internal friction μ, which equals tan ϕ for the rock mass. In addition, the surface-slope angle is partly a function of climate; for example, in an arid climate the surface equilibrium angle would be maintained by increased landsliding, whereas in a humid climate, stream and slope erosion processes would maintain the angle. Even so, Davis and his colleagues concluded that the rock mass does not begin to move forward until the critical taper of the wedge is achieved.

To formulate an ideal mechanical model, we must see how materials involved in thrusting will behave under stress (that is, their *rheology*), as well as know their strengths. It is also essential that we understand the degree to which body (penetrative) deformation (such as cleavage formation) affects the thrust sheet during initiation and emplacement. Different ideal-behavior modes—such as elastic, viscous, and plastic—may be used in model experiments to duplicate natural behavior, but it may not be possible to describe material behavior exactly in terms of an ideal rheology. Consequently, we must make certain assumptions about rheology in order to devise a mechanical model for foreland thrusting.

The mechanical model formulated by Chapple (1978) embodies both elastic and plastic behavior. Chapple assumed that the basal weak layer and the total wedge behave as ideal plastic materials. As a result of his analysis, thrust faults are considered a manifestation and consequence of shortening of the wedge. A second conclusion is that plastic flow under compression is an attribute of foreland fold-thrust belts and that it provides most of the shear stress along the basal weak layer to drive deformation of the wedge.

Elliott (1976a) concluded from his mechanical analysis that gravity is primarily responsible for deformation of foreland thrust belts. He did not assume a specific rheology but did make preliminary assumptions about stress distribution within thrust sheets. He assumed that σ_1 is oriented at 45° to the basal layer in thrust sheets, and so he did not incorporate a weak basal layer with its attendant properties in his analysis. Thus, the fundamental strength of the entire thrust sheet must be the same as that of the basal layer. In contrast, Chapple assumed that the strength of the remainder of the thrust sheet was much greater than that of the weak basal layer. A consequence of Elliott's approach is that horizontal compression is not necessary to form a foreland thrust belt, leading to the additional consequence that high shear stress is not produced in the basal layer—contrary to Chapple's conclusion.

Several major thrusts, such as the Keystone in Nevada, appear to have involved only strong rocks, such as massive carbonates, and so the basal detachment may not have been involved or was not available when the thrusts formed. A surface slope, although present, is essential in Chapple's model, for critical taper must be achieved and maintained in the wedge, but it may not be necessary in Davis, Suppe, and Dahlen's model.

GRAVITY VERSUS COMPRESSION

Now let us compare gravity with compression as two possible primary motive forces producing foreland fold-thrust belts. Gravity is an appealing mechanism. Hubbert and Rubey (1959) argued that it would be very difficult to push a thin sheet of rock very far horizontally without breaking it up because its internal strength is too low; they reasoned that the magnitude of shear stress necessary to move a mass of rock would be greatly reduced if the base of the mass were under high pore-fluid pressure. They maintained, however, that body forces—gravity—were necessary to move a thrust sheet.

We can argue geometrically that gravity is the primary force responsible for moving thrust sheets. Robert Milici (1975) has pointed out that the pattern of overlap suggested by outcrop traces of thrust faults in part of the Appalachian foreland thrust belt yields a cross-cutting pattern in which the oldest thrusts form on the outer fringe of the belt, and successively younger thrusts form deeper in the core of the orogen (Figure 11–32a). The logical mechanism for producing this geometry is gravity, but the geometry may be a local aberration, produced during the latter stages of emplacement of out-of-sequence thrust sheets where the footwall sequence locks and the hanging wall breaks and moves over the deformed footwall (Morley, 1986).

The Bergamesc Alps of northern Italy were once considered a classic example of "gravitational gliding" tectonics, because of the syntheses of this region by Dutch geologist L. Ulbo deSitter (1949). Here the Paleozoic basement was thought to have been uplifted during the Alpine orogenies, and the cover slid (glided) gravitationally southward off the basement high along weak evaporites in the Triassic carbonate section (Figure 11–32b). Recent work by Swiss geologist, Hans Laubscher (e.g., 1990), and his graduate students, Gregor Schönborn (1992a, 1992b) and Markus Schumacher (1990), have shown that basement is

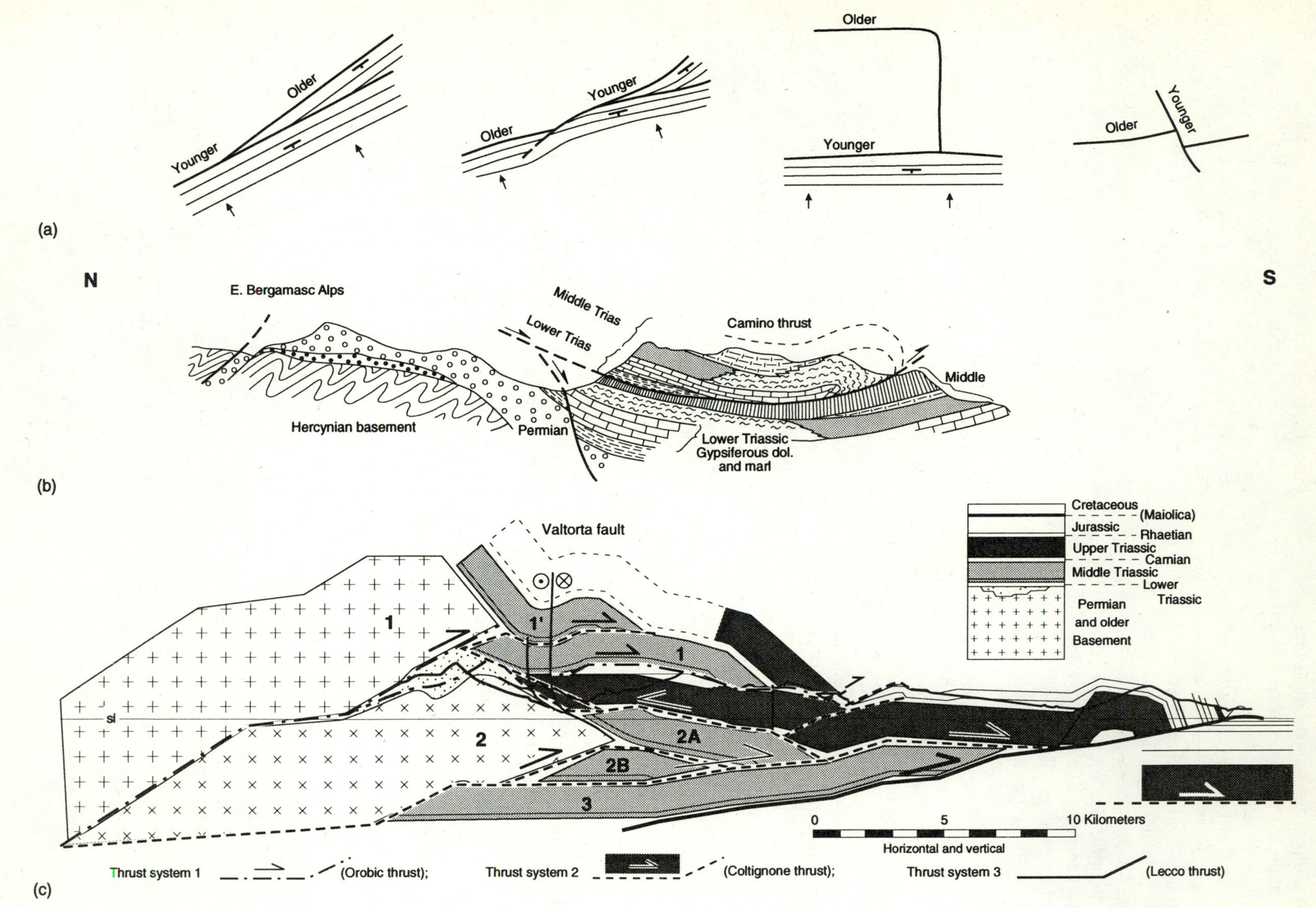

FIGURE 11–32
(a) Milici's geometric relationships suggesting development of thrusts by a break-back mechanism. (From R. C. Milici, 1975, Geological Society of America *Bulletin*, v. 86.) (b) Thrusts identified by de Sitter in the southern Alps of northern Italy as gravitationally produced (From L. U. de Sitter, *Structural Geology,* 1964 © McGraw-Hill, Inc., p. 243.) (c) Cross section by Gregor Schönborn through the southern Alps showing basement (x or + pattern) involvement with cover deformation (From G. Schönborn, Alpine Tectonics and Kinematic Models of the Central Southern Alps, *Memorie Di Scienze Geologiche,* 1992, v. 144, p. 386.)

indeed involved in deformation of the cover and probably drove the cover deformation south of the rising basement blocks (Figure 11–32c). This clearly makes primary compression the favored mechanism.

Model experiments by Hans Ramberg (1967) and field work, including studies of accretionary wedges in subduction zones (Stockmal, 1983; Dahlen, Suppe, and Davis, 1984), have shown that most thrust belts result from propagation of thrusts continentward from the core of the orogen. It has also been noted that similar map patterns exist in many foreland thrust belts where it can be shown independently that the deformation plan is inside-out. Independent evidence has also shown that the oldest thrusts—not the youngest—are in the interior of the orogen and that deformation there generally occurred long before deformation in the foreland. Such evidence led Hans Stille in the 1930s to suggest that an orogen, including the foreland thrust belt, is deformed from the inside out.

Another objection to the gravity mechanism has to do with the direction of dip of basement beneath the foreland. Nowhere in the world is there a foreland fold-thrust belt with the basement surface sloping *away* from the interior of an orogen: all are inclined *toward* the interior. The original wedge shape of the foreland wedge also argues against a reverse slope. It is difficult to envisage gravity pushing thrusts (or water) up-slope. Raymond Price (1974) attempted to resolve the problem with his alternative mechanism of *gravitational spreading,* based partly on earlier observations and studies of models by Walter Bucher and Hans Ramberg. Price suggested that as the core of an orogen is shortened and buoyantly uplifted during compression, metamorphism, and plutonism, the orogen compensates for the uplift by flowing laterally under the influence of gravity, as a large ice sheet may push lobes uphill (Price, 1974; Elliott, 1976a). Lateral spreading provides stress that deforms the foreland and pushes thrusts up the otherwise hinterland-dipping basement surface, but the amount of uplift actually needed to produce the observed deformation greatly exceeds most estimates. For this reason alone, Price's mechanism seems unworkable. Add the requirement of an internal zone of stretching and attenuation of structures (none has been discovered except in very special systems, such as the Gulf of Mexico, that involve movement down a slope with thrusting at the toe; see Chapter 13), and gravitational spreading becomes difficult to accept as a primary mechanism of foreland deformation. Price (1981), recognizing the problem, has recommended that the concept of gravitational spreading be abandoned.

The deformed wedge, the weak basal layer, and the overall geometry of the foreland thrust belt all suggest that Chapple and Davis, Suppe, and Dahlen are right. Their mechanical models—despite initial objections to compression as the primary mechanism of foreland deformation—best explain how foreland thrusts form.

MECHANICS OF CRYSTALLINE THRUSTS

The mechanics of crystalline thrusts remain a partly unsolved mystery. Crystalline thrusts share many properties with thin-skinned thrusts. They are transported along low-angle faults, occur in sheets up to a few kilometers thick, and may follow zones of weakness in the footwall sequence, but there are important differences. (1) They are among the largest structures in orogenic belts. (2) There is no obvious lithologic plane of weakness along which detachment may begin. (3) Crystalline thrust sheets may form as part of the thick foreshortened wedge but can also form as plastic structures, either as huge detached thrust sheets (Type C) or as products of folding (Type F) (Hatcher and Hooper, 1992).

Understanding the formation of detachments along which crystalline thrusts propagate is a difficult problem. Ronald Oxburgh (1972) has suggested that crystalline thrusts form as "tectonic flakes" when two pieces of continental crust collide. Richard Armstrong and Henry Dick (1974) have suggested that propagation and detachment of crystalline thrusts occur at the ductile-brittle transition in zones of high heat flow, as in back-arc basins. Oxburgh's mechanism may account for single, large crystalline-thrust complexes, such as the Type C Austroalpine sheets of the Alps, which can be directly related to a collision zone. Moreover, Armstrong and Dick's mechanism may apply to generation and emplacement of ophiolite thrust sheets, but many crystalline thrusts do not occur in the situations described by either of the two models. The majority of crystalline thrusts occur in the metamorphic cores of mountain chains. The five basic types of crystalline thrust sheets identified by myself and Richard Williams (1986) became elements of a mechanical model to explain formation of the composite, ophiolite, and thin-skinned types. This model was simplified mechanically to two basic types by myself and Robert Hooper (1992). Our model embodies Armstrong and Dick's idea of detachment within the ductile-brittle transition zone and also relates across-strike width to compressive stress, friction and dip on the basal thrust, and thickness of the sheet.

Most geologists at the present time agree that faults beneath crystalline thrust sheets (and deep-seated listric normal faults) propagate within the ductile-brittle transition, which occurs at a depth controlled by the thermal properties of the crust—thereby also controlling the thickness of these sheets.

ESSAY

Natural Model Gravity Foldbelt

An ideal fold-thrust system can form on a small scale under the influence of gravity, if all of the mechanical elements are present. In December 1974, I observed such a fold-thrust belt in a frozen, moss-covered road-cut in the North Carolina Blue Ridge (Figure 11E–1). The soil beneath the thin moss was water-saturated and had frozen overnight. As it froze, it expanded, forming needle ice that lifted the moss

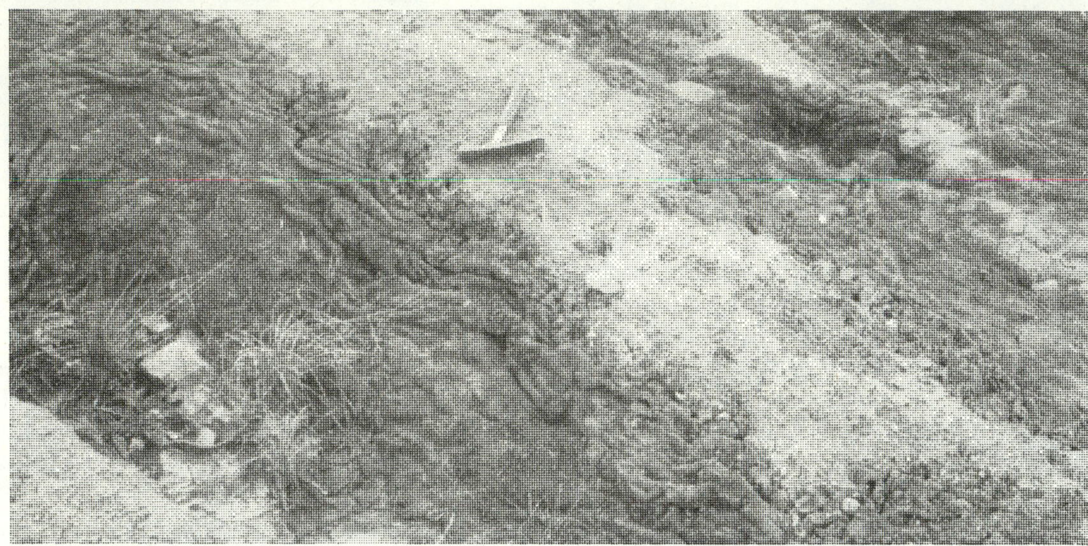

(a)

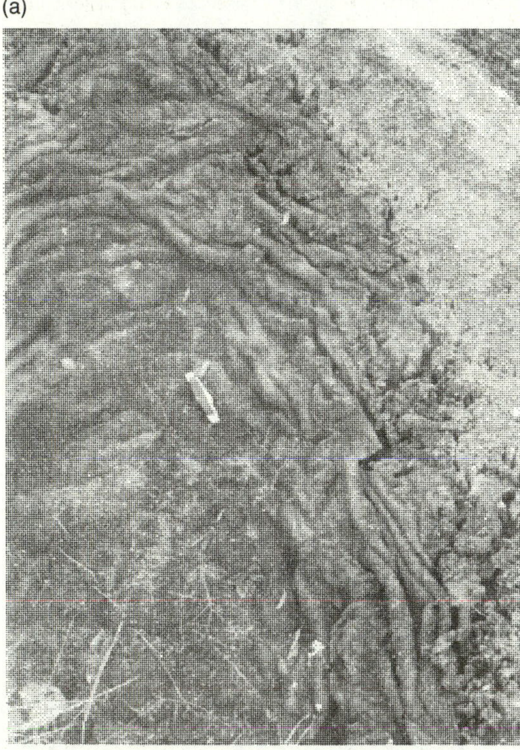

(b)

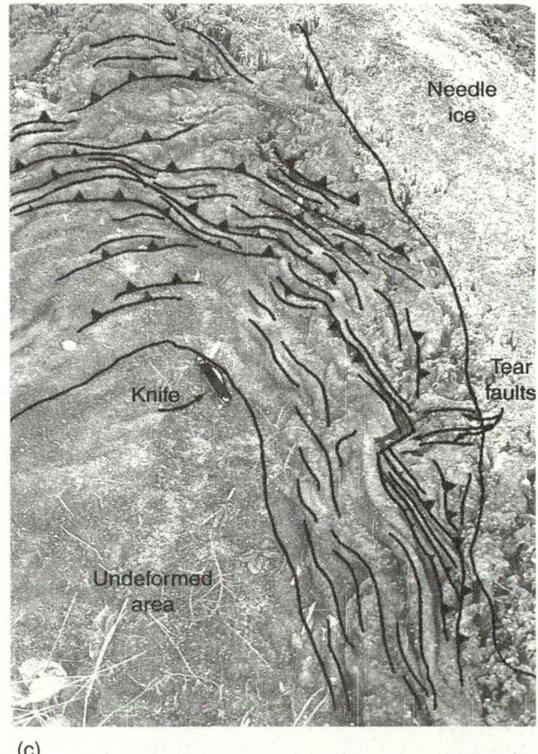

(c)

FIGURE 11E-1
Small-scale, fold-thrust system developed on a slope covered by a thin layer of moss that froze; the moss detached, moved downslope under the influence of gravity, and produced thrusts, fault-propagation folds, tear faults, and other features characteristic of a fold-thrust belt. (a) View of needle-ice-covered upper slopes from which a layer of moss has detached and slid downslope. (b) Close-up of the detached moss layers showing the variety of structures developed. (RDH photos.) (c) Sketch of structures in (b).

layer, increasing the slope angle and forming a detachment between the more rigid moss layer above and the more rigid (drier) soil below. The surface layer moved down the slope above the master detachment, perhaps lubricated by a film of water that had not yet frozen.

The sliding mass developed thrusts, drag structures, fault-propagation folds, transfer zones, and tear faults as it moved downslope above the detachment (Figure 11E–1c). Both folds and thrusts could be traced to termination points, with many terminations overlapping the terminal zones of other folds and thrusts. As the sheet moved, it began to fragment, but mostly remained intact.

The only motive force that created the mossy fold-thrust system was gravity—a phenomenon easily and frequently observed—and, until recently, downslope movement of a gravity-driven mass off a topographic high was the mechanism generally invoked to explain most foreland thrust belts in mountain chains. Edward Hansen (1971) has described a similar gravity-driven, small-scale, multiple-fold system in partly frozen surficial materials (solifluction lobes) in south-central Norway (Figure 14–17).

The Helvetic Alps and the Prealps in Switzerland and France were long thought to have slid northward off the Aar and Gotthard basement massifs and the more internal Pennine Alps (Trümpy, 1960). At one time, many geologists believed that the Appalachian foreland from Alabama to Pennsylvania had slid northwestward off the Blue Ridge (Gwinn, 1970) and that the Wyoming-Montana thrust belt had slid eastward from the internal higher parts of the Cordillera (Rubey and Hubbert, 1959). George Davis (1975), however, *has* documented gravity-driven folding and décollement from the gneissic basement of the Rincon Mountains near Tucson, Arizona. Also compare these structures with maps and cross sections of the Jura Mountains in Switzerland and France (Laubscher, 1992) or the Zagros Mountains in Iran (Alavi, 1994).

Compare the gravity-driven model described in Figure 11E–1 with the small thrust in northern Italy at a similar scale in Figure 11E–2. This fault formed in limestone at a time when the rock was strong enough to fracture and form a drag fold along the leading edge of the sheet as it moved. The displacement on this thrust sheet varies along it, indicated by the change in strike of the drag fold and the fault near the lower end and the internal dismemberment of the sheet by tear faults (fractures) oriented perpendicular to the fault plane and the fault trace. Could this thrust also have formed by gravity? Or, is it more likely that because the large thrust sheets in this region ultimately become involved in basement deformation (see Figures 11–32b and 11–32c; Schönborn, 1992a), the mesoscopic thrust here may be another product of the compressional regime that formed this part of the Alps?

References Cited

Alavi, M., 1994, Tectonics of the Zagros orogenic belt of Iran: New data and interpretations: Tectonophysics, v. 229, p. 211–238.

Davis, G. H., 1975, Gravity-induced folding off a gneiss dome complex, Rincon Mountains, Arizona: Geological Society of America Bulletin, v. 86, p. 979–990.

Gwinn, V. E., 1970, Kinematic patterns and estimates of lateral shortening, Valley and Ridge and Great Valley provinces, central Appalachians, south-central Pennsylvania, *in* Fisher, G. W., Pettijohn, F. J., Reed, J. C., Jr., and Weaver, K. N., eds., Studies of Appalachian geology: Central and southern: New York, Wiley Interscience, p. 127–146.

Hansen, Edward, 1971, Strain facies: New York, Springer-Verlag, 207 p.

Laubscher, H. L., 1992, Jura kinematics and the Molasse Basin: Ecolgae Geologicae Helvetiae, v. 85, p. 653–675.

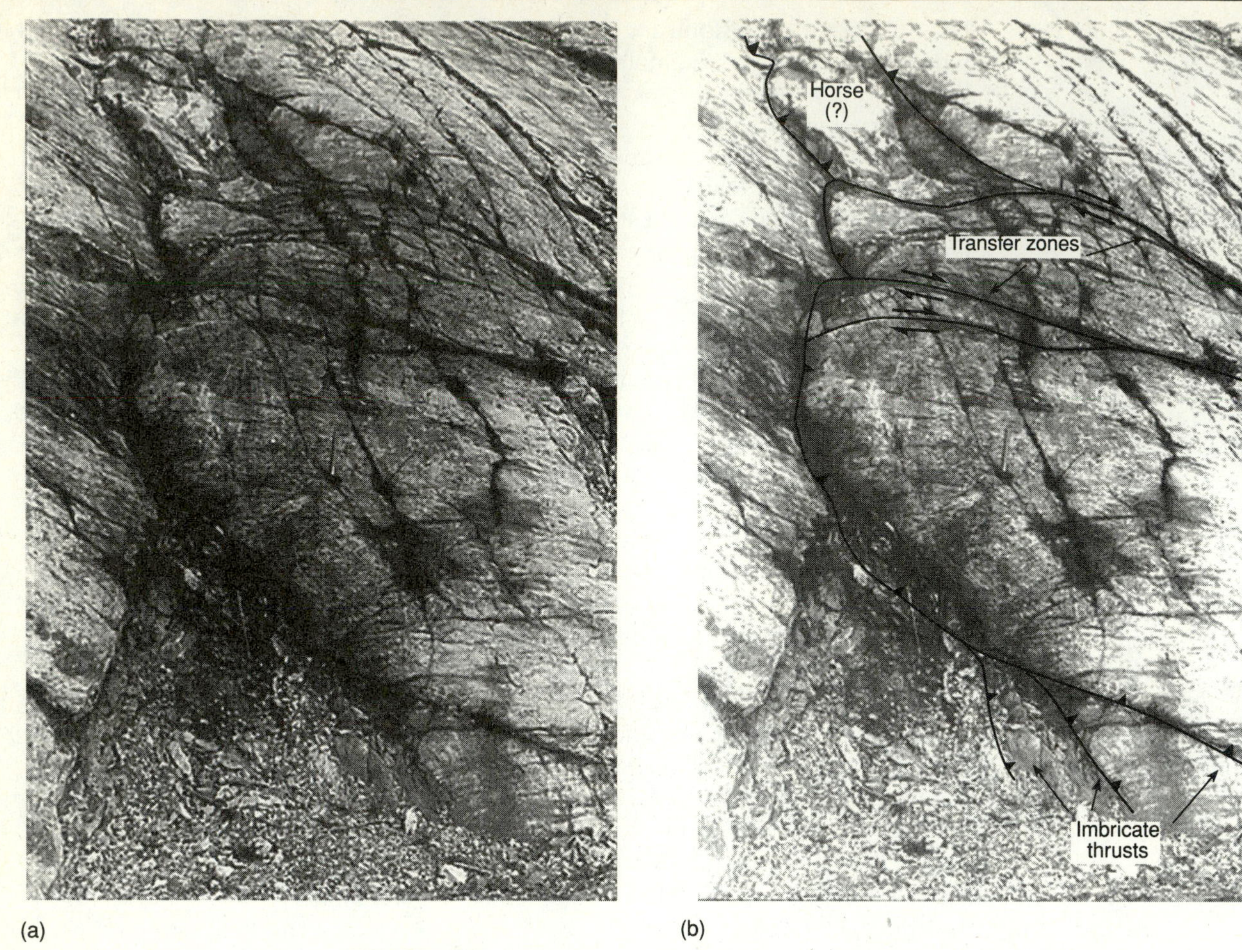

(a) (b)

FIGURE 11E–2
(a) Mesoscopic-scale thrust sheet in Lower Triassic Raibler limestone and shale between Monte Pora and Clusone, in the Italian Alps. Note the drag fold along the leading edge of the sheet, the decreasing displacement downslope along the trace of the fault, and the internal fractures (tear faultsor transfer zones) that compensate for additional variations in displacement within the internal parts of the sheet. (b) Sketch of relationships shown in (a). (Locality courtesy of Gregor Schönborn, University of Neuchâtel, Switzerland. RDH photo.)

Rubey, W. W., and Hubbert, M. K., 1959, Role of fluid pressure in mechanics of overthrust faulting, II. Overthrust belt in geosynclinal area of western Wyoming in light of fluid pressure hypothesis: Geological Society of America Bulletin, v. 70, p. 167–206.

Schönborn, G., 1992a, Alpine tectonics and kinematic models of the central Southern Alps: Memorie Di Scienze Geologiche, v. 144, p. 229–393.

Trümpy, Rudolf, 1960, Paleotectonic evolution of the central and western Alps: Geological Society of America Bulletin, v. 71, p. 843–908.

THE ROOM PROBLEM AND CROSS-SECTION CONSTRUCTION

Unfilled voids may initially seem possible in the Earth by students who are constructing their first cross sections in foreland fold-thrust belts. This may initially suggest not enough rock or too much space (or volume) is present—a statement of the ***room problem*** in structural geology and tectonics. Usually, the problem really lies in the skills of the person constructing the section because nature does not permit large voids in the crust at depths of several kilometers, where most structures form.

Many thrust faults form in regions where the local deformational style is governed by slip of layers past each other—flexural slip—(Chapter 15) and where ideally there is no ductile flow of material into voids, either potential or real. In a deformational realm dominated by horizontal compression, there should be no void space larger than pores or small fractures (openings of millimeters or less), even in rocks being brittlely deformed. Yet, given the geometric possibilities that exist in foreland thrust belts, voids of tens to hundreds of cubic meters might be expected; deep valleys, seismic reflection profiles, and drill data tell us that is *not* so.

Thrusts may also form in the cores of brittle (flexural-slip buckle) folds—generally, anticlines—as the folds tighten and a room problem arises in the more tightly folded inner layers. A single thrust fault may propagate, tighter folds may form, or two or more thrusts may dip in the same directions as the limbs of the fold and opposite each other (Figure 11–33b). The last comprise a ***triangle***, or ***delta***, ***structure.*** Triangle structures commonly form in the outer or upper parts of the foreland fold-thrust belt where the overburden is relatively thin, but this is not a hard and fast rule. They also form where the ratio of competent to incompetent rocks is high, on the order of 1:1 to 4:1. Duplexes may form at greater depths and serve the same purpose. Fold curvature, along with thickness and nature of the stratigraphic section involved, may alter the mechanism that solves the room problem.

Most folds in a foreland fold-thrust belt are rootless and are transported along thrusts as they form. As they tighten, thrusts may form in the core as a single imbricate, as multiple imbricates, as antithetic thrusts (triangles or deltas, just described), or as thrusts that work their way up through a fold as a series of branching splays or imbricates (Figure 11–33).

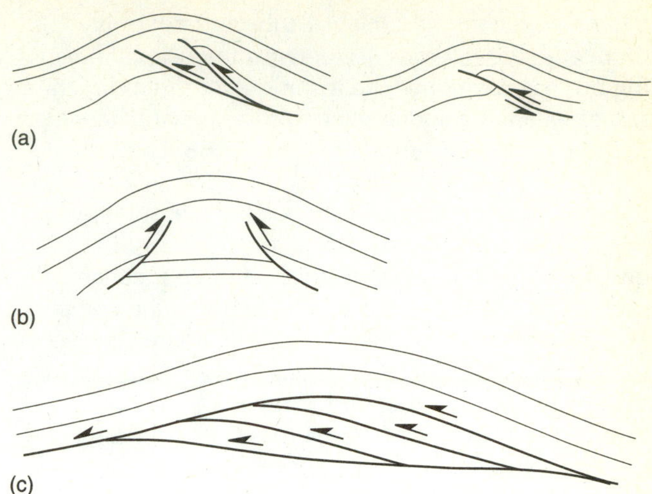

FIGURE 11–33
Possible solutions to the room problem in thrust-fold systems. (a) Single thrusts and splaying imbricates. (b) Triangle (delta) structures. (c) Duplexes.

DISCUSSION

The nature and mechanics of thrust faults is far from clear, as we have just seen. Techniques yielding many new kinds of data are being devised and employed to clarify our understanding of thrusts and other faults.

Seismic reflection data (Chapter 21), particularly the large amount of data produced in exploration for hydrocarbons, have greatly clarified our view of the subsurface geometry of thrusts, provided more-precise measurement of depths to basement, and confirmed the thin-skinned concept for deformation of foreland areas. At least one geometric term has arisen from study of subsurface geometry—*snake-head structure* (Figure 11–34)—for folds associated with ramps. For some time, snake-head structures had been known on the surface, but they were not so identified. Seismic reflection data from the COCORP (Consortium for Continental Reflection Profiling) Project at Cornell University have also helped to confirm existence of large crystalline thrusts in the Appalachians and regional listric normal faults through much of the upper crust in the Basin and Range in Utah (Cook and others, 1983; Allmendinger and others, 1983).

Seismic reflection data are essential when constructing palinspastic (balanced or retrodeformed) cross sections through brittlely (mostly nonpenetratively) deformed thrust-faulted terranes (Figure 11–35). The ***balancing of cross sections,*** as originally suggested by Hans Laubscher (1962) and later by Clinton Dahlstrom (1969), has become sufficiently mechanically iterative that it may be done with a computer. An oft-quoted rule is that "if a cross section will not balance, it cannot be correct; if it does balance, the section *may* be correct." Balancing may be accomplished by measuring the lengths of deformed layers and restoring the section to an undeformed condition (Figure 11–36; Woodward and others, 1985). If the section can be retrodeformed and reconstructed, the interpretation may be correct. Balancing by areas (thickness of rock units times length) is also possible by taking the areas of units in the section and restoring them to an undeformed condition. Major problems arise with construction and balancing of any section if internal deformation within rock units occurs and if there is strike- or oblique-slip deformation out of the plane of the section. The problem of internal deformation may be partly solved by area balancing. Deformation out of the plane of the section requires construction of sections in several directions—a form of three-dimensional analysis. Strongly curved segments of orogens, such as the Ligurian Alps in Italy, are impossible to balance, except by employing three-dimensional balancing techniques. Hans Laubscher not only pioneered the rigorous use of two-dimensional balancing but has developed the technique of three-dimensional balancing and applied this to crustal balancing of the Alps, (see Laubscher, 1983, 1988; Laubscher and others, 1992).

Accurate coupling of surface and subsurface data in cross-section analysis aids reconstructions that enhance our ability to construct accurate mechanical models. Note that in this kind of analysis the component of movement parallel to strike ultimately prevents balancing sections if the sections cross into the metamorphic core. Another problem with balancing sections in complexly deformed rocks lies in estimating the amount of penetrative strain (Woodward and others, 1986). A poor estimate may prevent balancing a section otherwise correctly drawn.

A technique borrowed from engineering has found widespread use in mechanical modeling of thrust faults—*finite-element studies.* It is a numerical method for dividing a continuum into triangles or parallelograms with nodes on the boundaries where displacements are calculated by a computer program and a structure is defined. Boundary conditions of the finite-element method are either nodal displacements or forces at nodes. The computer program simulates the deformation by applying stress of various magnitudes at various orientations to thrust or other fault-fold geometries, but it has some difficulty in handling discontinuities such as faults.

Photoelastic studies of pressure-sensitive plastics in polarized light (Figure 11–37), another technique from engineering, allow model studies of various stress distributions to be made with different fault, ramp, and fold configurations. Coupled with finite-element analysis, such studies have helped to elucidate simple foreland thrusts.

Now we can turn from thrust faults to the other class of faults that form under horizontal compression—strike-slip faults.

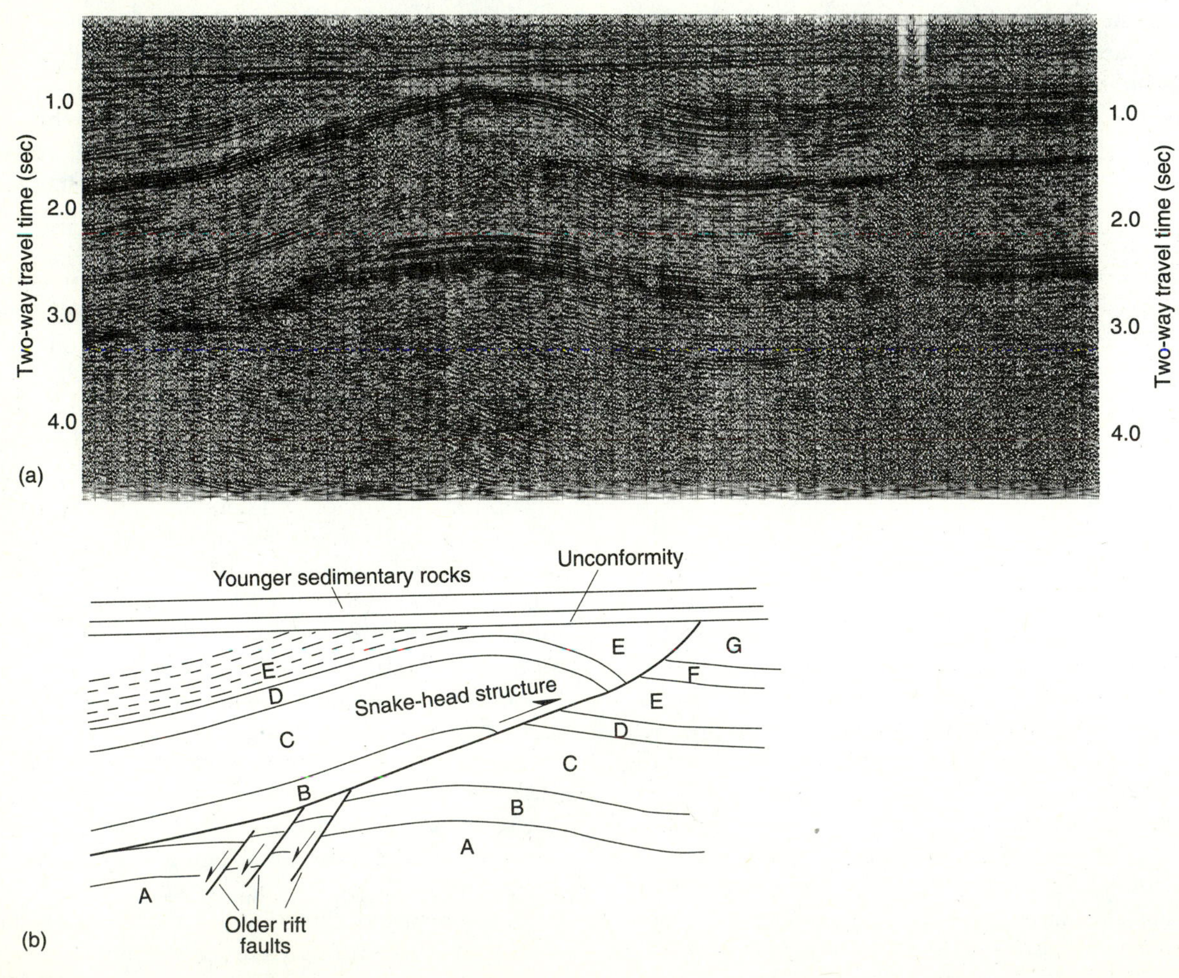

FIGURE 11–34
(a) Seismic reflection profile of snake-head structure from the Appalachians buried beneath the Coastal Plain of Alabama.
(b) Cross-section interpretation of the structure in the western half of the profile.

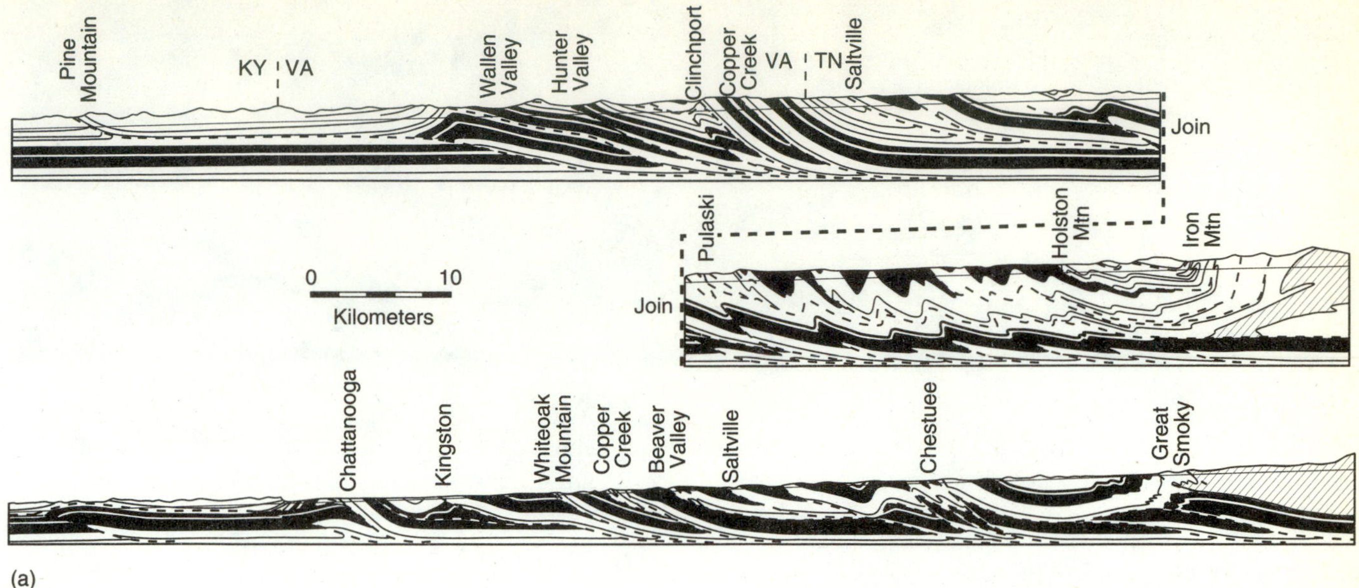

(a)

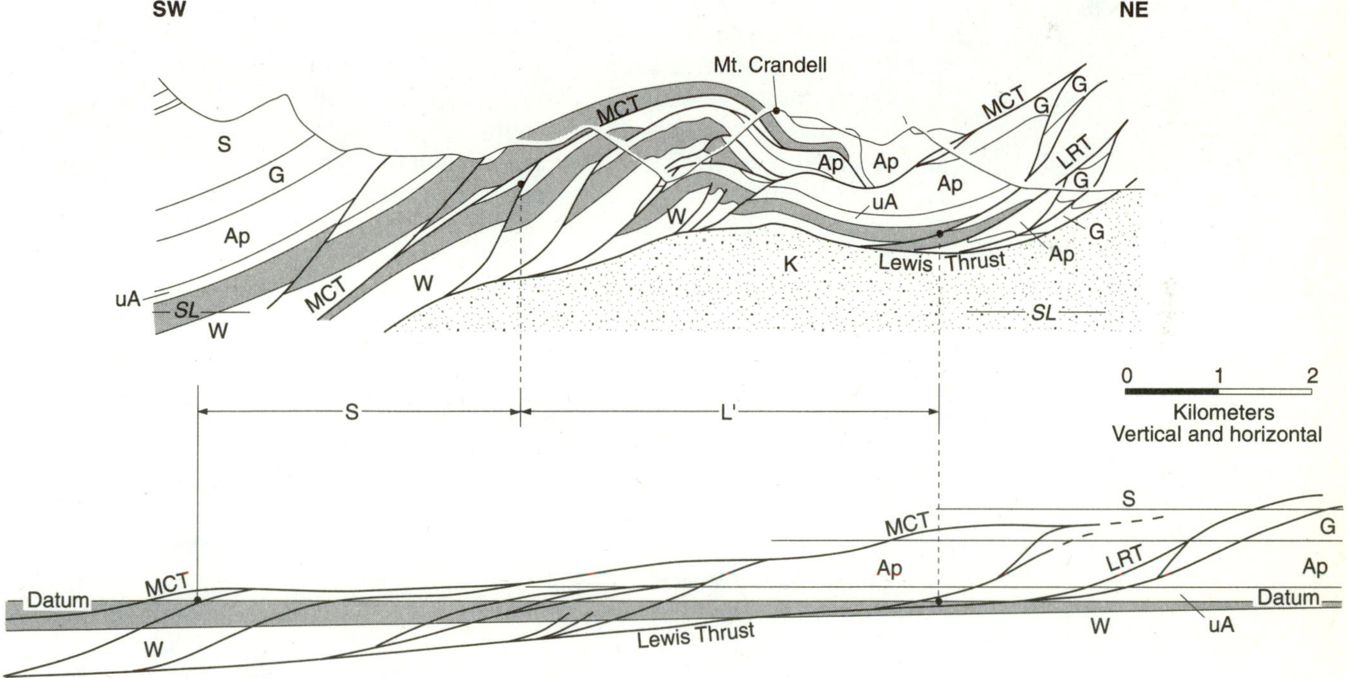

(b)

FIGURE 11–35

Cross sections through parts of a foreland fold-and-thrust belt. (a) Sections through the Appalachian Valley and Ridge. Black is the strong rock unit in the sequences. Names of faults appear above each. (From D. H. Roeder and W. D. Witherspoon, 1978, Palinspastic map of east Tennessee, *American Journal of Science*, v. 278, p. 543–555. Reprinted by permission of American Journal of Science.) (b) Section through part of the Canadian Rockies (above) and a restored (retrodeformed) version (below). MCT—Mt. Crandell thrust; LRT—Livingston Range thrust; S—Siych; G—Grinnell; Ap—Appekunny; fine stipple—lower Altyn; uA—upper Altyn; W—Waterton; K—Cretaceous. (From S. E. Boyer and David Elliott, AAPG *Bulletin*, v. 66, © 1982. Reprinted by permission of American Association of Petroleum Geologists.)

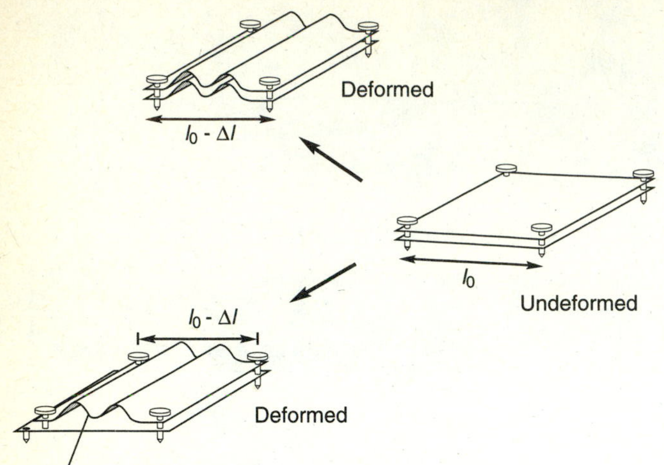

FIGURE 11–36
Retrodeforming a hypothetical block using the technique of line (or bed-length) balancing. (After C. D. A. Dahlstrom, 1969, *Canadian Journal of Earth Sciences,* v. 6.)

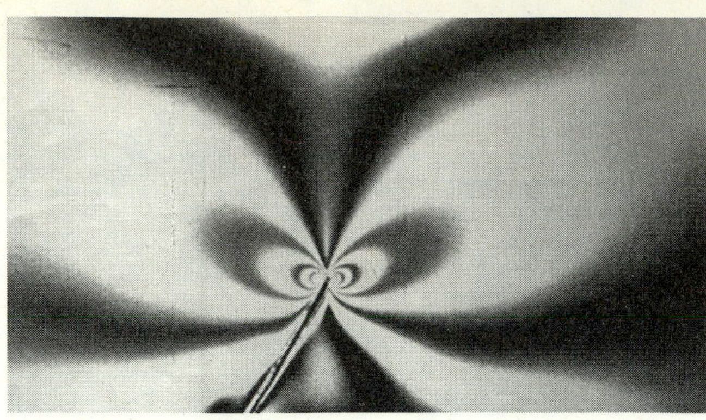

FIGURE 11–37
Photoelastic strain pattern developed around a slot in a stressed sheet of Plexiglas. Note pattern at crack and at crack tip. (Courtesy of John J. Gallaher, Jr., John J. Gallaher & Associates, Dallas, Texas.)

Questions

1. What evidence would you seek to prove that a large overthrust exists—or not—in an area where none has been found before?
2. Why does a foreland wedge maintain its shape throughout deformation?
3. Why do thrusts have steeper dips on ramps than within detachments?
4. How and why do slices (horses) form?
5. How does fluid pressure reduce the amount of energy needed to move a thrust sheet? If this is so, why do castles of wet sand stand (have greater cohesive strength) and those of dry sand form more gently sloping surfaces (have less cohesive strength)?
6. Why does Chapple's model for foreland thrusting require a weak basal layer?
7. How does an eyelid window form?
8. How does Price's concept of gravitational spreading partly dispel objections to a simple gravity model for thrusting?
9. How are potential voids filled beneath folds or in a stack of thrusts?
10. Why did Chapple choose an ideal plastic rheology for the foreland wedge even though evidence of brittle (elastic) behavior was abundant?
11. Why don't thrusts slide back down after they have been pushed up?
12. Estimate the displacement of the Little North Mountain thrust (in the central Appalachians), which has a strike length of 220 km.

Further Reading

The literature on thrust faulting is voluminous and growing. These works, and the papers they cite, will help to open the door to better understanding thrusts.

Boyer, S. E., and Elliott, D., 1982, Thrust systems: American Association of Petroleum Geologists Bulletin, v. 66, p. 1196–1230.
Describes the terminology and geometry of thrusts, with examples from several mountain chains, in both maps and cross sections.

Chapple, W. M., 1978, Mechanics of thin-skinned fold and thrust belts: Geological Society of America Bulletin, v. 89, p. 1189–1198.

Lists the properties of a foreland fold-thrust belt, then derives a mechanical model. The first part of the paper and the conclusions will be particularly useful to undergraduates.

Dahlstrom, C. D. A., 1969, Balanced cross sections: Canadian Journal of Earth Sciences, v. 6, p. 743–758.
Principles of balancing cross sections (mainly line balancing) are explained clearly and concisely, with examples mostly from the Canadian Rockies.

Hatcher, R. D., Jr., and Hooper, R. J., 1992, Evolution of crystalline thrust sheets in the internal parts of mountain chains, *in* McClay, K. R., ed., Thrust tectonics: London, Chapman and Hall, p. 217–233.
This paper outlines the properties of two major groups of crystalline thrust sheets and compares and contrasts their behavior with each other as well as with foreland thrusts.

Hossack, J. R., and Hancock, P. L., eds., 1983, Balanced cross sections and their geological significance: Journal of Structural Geology, v. 5, p. 98–223.
Special issue dedicated to David Elliott, containing 10 papers on balanced cross sections.

Laubscher, H. P., 1988, Material balance in Alpine orogeny: Geological Society of America Bulletin, v. 100, p. 1313–1328.

McClay, K. R.,1992, ed., Thrust tectonics: London, Chapman and Hall, 447 p.
A compendium of papers that updates the McClay and Price volume published in 1981 that contains examples of thrust systems from different parts of the world as well as up-to-date treatment of mechanics, geometry, and concepts of thrusting.

McClay, K. R., and Price, N. J., eds., 1981, Thrust and nappe tectonics: Geological Society of London Special Publication 9, 539 p.
This compendium ranges from thrust mechanics through rock products of faulting to studies of thrusts in different parts of the world. Papers by Bally, Ramberg, Ramsay, Price, and Coward and Kim, in particular, will interest undergraduates, but others also contain excellent examples and problems related to thrust faulting.

Mitra, S., 1986, Duplex structures and imbricate thrust systems: Geometry, structural position, and hydrocarbon potential: American Association of Petroleum Geologists Bulletin, v. 70, p. 1087–1112.
This study of duplex and imbricate thrusts provides more details than do Boyer and Elliott (1982) on this aspect of thrust faulting. Cites good examples from several fold and thrust belts.

Mitra, S., and Fisher, G. W., eds., 1992, Structural geology of fold and thrust belts: Baltimore, Maryland, The Johns Hopkins University Press, 254 p.
Another compendium of papers dedicated to David Elliott. It contains a number of new and useful papers on thrust faulting in different regions, as well as several papers on thrust mechanics.

Moores, E. M., 1982, Origin and significance of ophiolites: Reviews of Geophysics and Space Physics, v. 20, p. 735–760.
Summarizes the structure and processes that form and emplace ophiolite sheets, with examples from various settings.

12

Strike-Slip Faults

The powerful dislocation which intersects Scotland along the line of the Great Glen has, in the past, been regarded by most geologists as a normal or dip-slip fault with a predominant vertical downthrow to the south-east. A reconsideration of the entire problem now suggests that this view is no longer tenable and that the dislocation is, in reality, a lateral-slip [strike-slip] or wrench fault with a horizontal displacement of approximately 65 miles. . . .

WILLIAM A. KENNEDY, 1946, *Quarterly Journal of the Geological Society of London*

STRIKE-SLIP FAULTS ARE THOSE IN WHICH THE MOTION IS parallel to the strike of the fault plane. Both thrusts and strike-slip faults have produced very large earthquakes, and some have repeatedly caused heavy damage, killing or injuring millions of people. The 1960 Chilean and 1964 Alaskan earthquakes, largest of this century, were produced by thrusts, but the earthquakes near Kansu (1920) and Tangshan (1976), China, that killed 100,000 and 500,000 people, respectively, were produced by strike-slip faults. We now know that much of the destruction by moderate-to-large earthquakes is related to construction practices that do not include design for earthquake safety. Such disasters justify intensive study of active faults with the aim of improving our ability to predict earthquakes. Routine prediction of earthquakes is not yet possible, but we have gathered an enormous amount of data and greatly increased our knowledge of strike-slip and other active faults. Some of these large active strike-slip faults, such as the San Andreas in California and the Alpine in New Zealand, are located on the margins of the Pacific plate and function as an integral part of these dynamic plate boundaries. Others, such as the Great Glen fault in Scotland, described above by Kennedy, and the Brevard in the Appalachians, have been known for decades—not because of their activity—but because of their linear traces.

PROPERTIES AND GEOMETRY

It is relatively easy, using several different lines of evidence, to determine the sense of motion along a fault after an earthquake. One is accompanying ground breakage. Another is that topographic features, such as streams and ridges, may be offset, indicating movement (Figure 12–1). Strike-slip faults are characterized by their steep dips and the predominance of horizontally displaced markers (Figure 12–2). Most dip more than 60°, and many approach the vertical. Shallow-dipping segments also exist, particularly where a fault changes orientation.

Strike-slip faults may branch and splay into smaller segments and into other types of faults. They may contain *horses* plucked from either side of the fault and carried along in the same way that blocks are

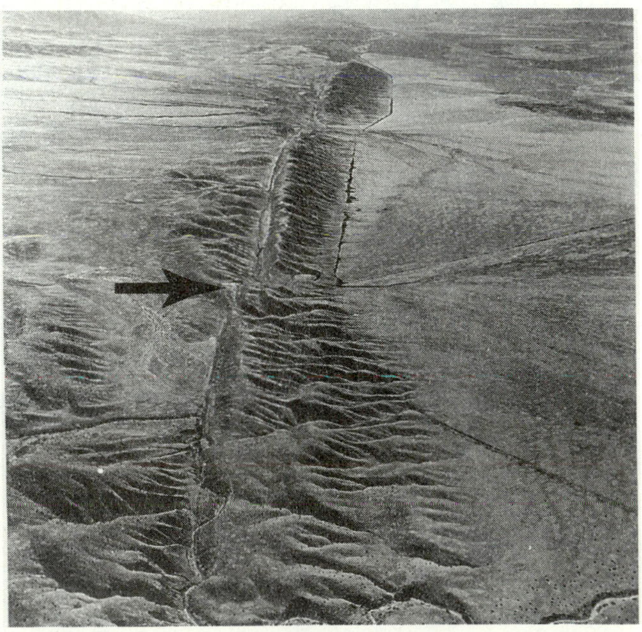

FIGURE 12–1
Dextral offset of streams (arrow) along the San Andreas fault, San Luis Obispo County, California. View is toward the south. (R. E. Wallace, U.S. Geological Survey.)

carried along thrust faults. Horses may be derived from either block, and so the general age-relationship rule applicable to thrusts (Chapter 11) does not apply here.

Categories of strike-slip faults are also called ***wrench, tear,*** and ***transcurrent*** faults. Arthur Sylvester (1988) has recommended the term "wrench fault" be abandoned because of the confusion generated by the name. The most recently named class of faults, called ***transforms*** by J. Tuzo Wilson (1966), was predicted by consideration of plate motion. ***Sinistral,*** or ***left-lateral,*** and ***dextral,*** or ***right-lateral,*** strike-slip faults were defined in Chapter 9 (Figure 12–2). Interestingly, the term "accident" is used by French geologists to describe large, steeply dipping faults with uncertain motion sense. No similar term appears in English terminology. Many faults called accidents by French geologists, such as the South Atlas fault in Morocco, are probably strike-slip faults.

Various things may happen to the dip and motion sense on a strike-slip fault where it abruptly changes strike. Both may change so that the segment with the markedly different strike may become a thrust or normal fault (Figure 12–3). Folds may also form along a strike-slip fault, either as a consequence of a change in strike, branching into either thrust or normal faults (which pass into folds as the displacement diminishes along the faults) or drag related directly to strike-slip motion. Shear sense on steeply plunging drag folds reflects the sense of motion of associated faults. Overlapping segments of *en echelon* (overlapping) strike-slip faults may indicate the sense of motion from the nature of the structures in the overlap zone (Figure 12–3a).

ENVIRONMENTS OF STRIKE-SLIP FAULTING

Many strike-slip faults (for example, the San Andreas in California, the Motagua in Central America, and the Alpine in New Zealand) occur at plate boundaries where one large block of crust moves past another; they are fundamental accommodation structures and consist of *transforms.* They also form the boundaries between many accreted terranes (Chapter 1) and the continent (or other terranes) to which accretion is taking place.

Strike-slip faults also occur behind oblique convergent margins within mountain chains, like several large faults in southeastern Alaska (the Fairweather fault) and adjacent parts of the Yukon and British Columbia (the Tintina fault), where strain in the upper plate is partitioned into strike slip parallel to the margin. Structures in these environments, such as folds and smaller dip-slip faults, are formed by margin-parallel compression. Other strike-slip faults include ***tear faults*** that bound the edges of thrust sheets. Classic examples include the Jacksboro and Russell Fork faults on the Pine Mountain thrust sheet in the Appalachians (Figure 12–3d).

NATURE OF FAULT ZONES

A strike-slip fault is commonly envisioned as a single brittle fault surface separating two oppositely moving blocks. Some do occur as simple planar faults, but planar faults are not common in upper-crustal rocks, and where they do occur are more related to rock type or some local variation in fluid content. Most strike-slip faults, like other large faults in the interior parts of mountain chains and at terrane and plate boundaries, are commonly *not* single, clean, brittle surfaces. They most frequently occur as cataclastic zones consisting of anastomosing faults (Figure 12–2) containing gouge and breccia or—if erosion has exposed the deeper parts of the zone—a considerable thickness of mylonite. These are the classic *shear zones* discussed in Chapter 10.

Evidence from inactive strike-slip faults and others eroded to expose lower-crustal segments suggests a continuous transition from brittle deformation near the surface to ductile deformation at middle- to lower-crustal depths (Figure 12–4). At the surface, elastic strain energy can be relieved by instantaneous strike-slip rebound (Chapter 10), as indicated by cataclastic rocks (breccia and gouge). At depth, this motion may occur as aseismic plastic creep, continuously producing mylonite and other ductilely deformed rocks (Sibson, 1983). Also, comparison of brittle strike-slip

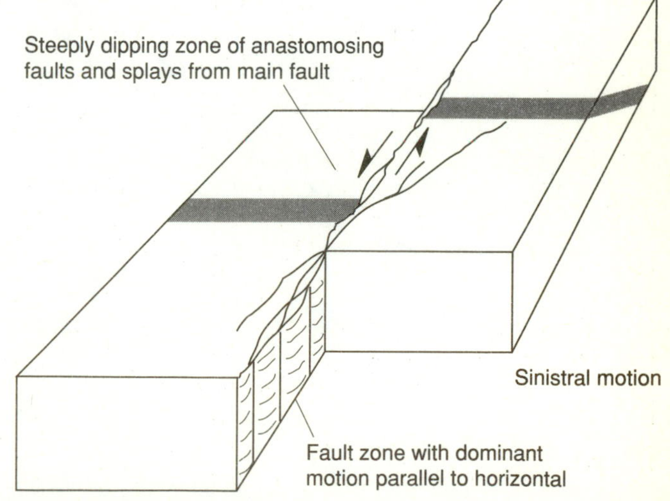

FIGURE 12–2
Geometry and anatomy of strike-slip faults.

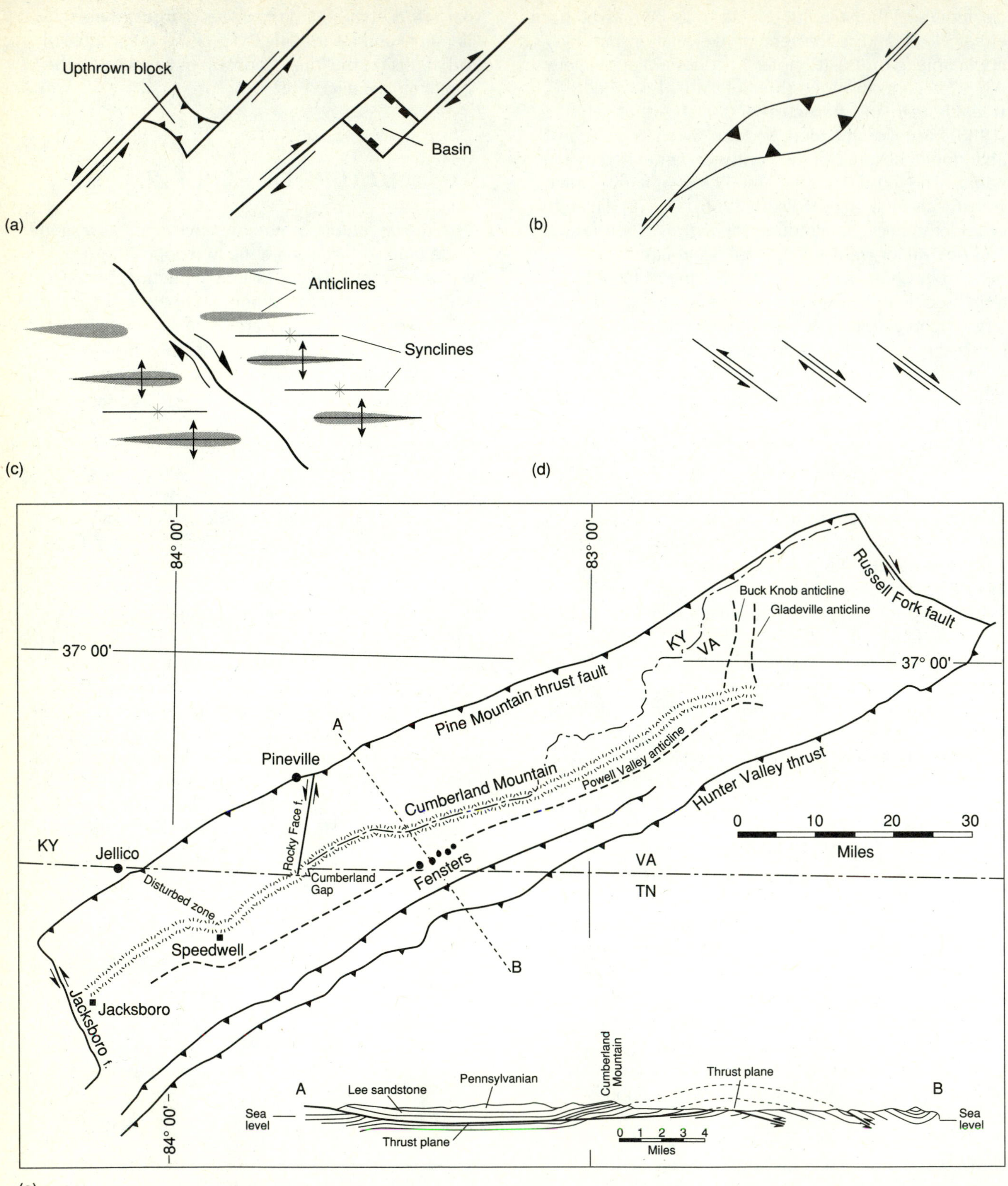

FIGURE 12–3
(a) Transfer of motion from strike-slip into thrust or normal slip motion. (b) Development of thrusts at a bend in a sinistral strike-slip system. Figure 12–10b is a cross section through a similar structure. (c) Gently plunging folds related to strike-slip motion. (d) *En echelon* segments of a dextral fault. (e) Map of the Pine Mountain block in Tennessee, Kentucky, and Virginia showing a thrust sheet bounded by tear faults. All are shown in map view. (e after J. L. Rich, 1934, AAPG *Bulletin,* v. 18. Reprinted by permission of American Association of Petroleum Geologists.)

faults and those in deeply eroded crust indicates that strike-slip fault zones near the surface are relatively narrow, but some ductile strike-slip fault zones in deeply eroded parts of high-grade Precambrian shields, such as western Greenland (Bak and others, 1975), may be several kilometers wide (Chapter 10). The mineralogy and texture in these zones exhibit abundant evidence of ductile deformation at high temperature and pressure—conditions found in the lower crust. Benjamin Page (1990) suggested that a broad zone of "pseudoviscous" strain exists along the San Andreas fault at depths greater than the depths where the fault generates earthquakes.

MECHANICS OF STRIKE-SLIP FAULTING

A fundamental attribute of strike-slip faults is that they form by horizontal compression. Supporting evidence is threefold: (1) the Andersonian solution of the strain ellipsoid for strike-slip faults (Figure 12–5a); (2) field evidence, including mesoscopic shear structures, drag folds, and the sense of motion determined by offset of marker units; and (3) first-motion studies of earthquakes produced by active strike-slip faults. The close association of strike-slip and thrust faults also shows that both form by horizontal compression.

Most strike-slip faults bear evidence of recurrent motion, and study of active faults indicates that recurrent motion is the rule over geologic time. As one example, the Ramapo fault in New Jersey and New York (See Chapter 13, Figure 13E–2) underwent recurrent strike- and dip-slip motion in the Paleozoic, Mesozoic, and Cenozoic over a period of about 400 m.y. (Ratcliffe, 1971). The Ramapo may even be active today, for many small earthquakes have been located along it (Aggarwal and Sykes, 1978). Moreover, geologic evidence indicates that the San Andreas fault that we see today has been moving blocks of continental crust past one another only since 5 Ma, and this motion has continued throughout the Holocene (Figure 1E–1; Sieh and Jahns, 1984; Powell and others 1993). Before that time, segments of nearby faults were active at least since middle Miocene time (~12–14 Ma). Various segments of the modern San Andreas have been active at different times. When accumulated elastic strain energy is relieved along a particular segment, other segments continue to accumulate strain to be relieved later, and so all segments of a large fault do not move at the same time. Geophysical evidence from the sea floor off California indicates that faults ancestral to the present San Andreas have been active throughout the past 30 to 35 m.y. (Atwater, 1970; Severinghaus and Atwater, 1990).

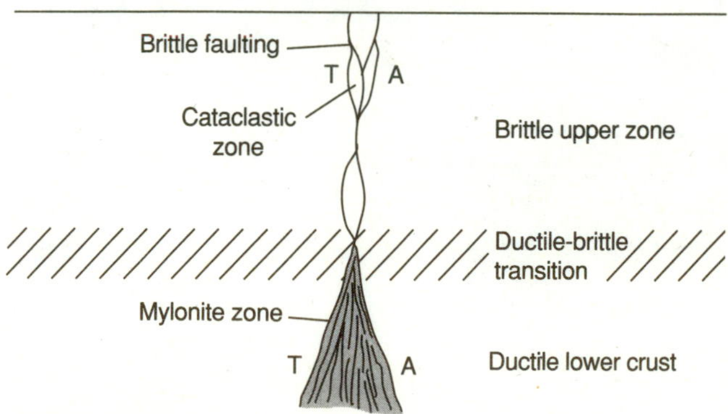

FIGURE 12–4
Along a strike-slip fault, the deformation type and kind of fault rock both vary with depth in the crust. T—motion toward the observer; A—motion away from the observer.

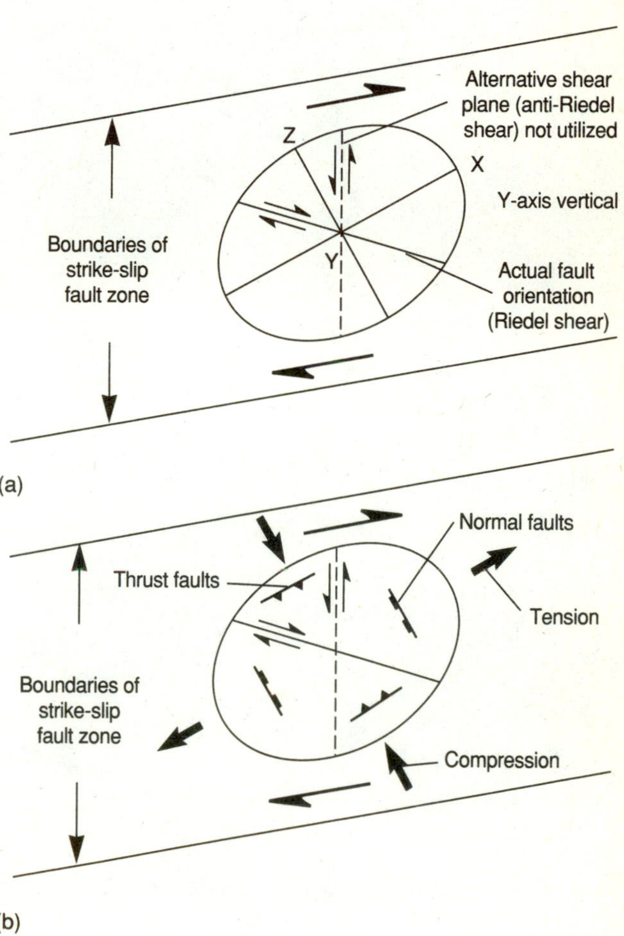

FIGURE 12–5
(a) Strain ellipsoid for strike-slip faults (map view). (b) Strain ellipse showing the possibilities for motion transferred to thrust or normal-slip domains. Motion transferred to thrust or normal faults must also occur on shear planes in appropriately oriented strain ellipsoids, not on the one shown here with a Y-axis vertical.

FAULT GEOMETRY RELATED TO OTHER FAULT TYPES

J. Tuzo Wilson (1965), Dan P. McKenzie and Robert L. Parker (1967), and Jason Morgan (1968), using the fact that the Earth is nearly spherical, recognized that because plate tectonics and spherical geometry govern the motion of both continents and ocean floors, major strike-slip and transform faults must follow small or great circles on the Earth (Figure 12–6). We can thus predict, using spherical geometry, that with any deviation of a fault plane from a small circle the sense of movement on the fault will include components of either normal or thrust motion (Figure 12–3). When the orientation of the fault plane deviates significantly from the primary orientation, motion along the fault passes from the realm of strike slip into the realm of normal or thrust motion (Figure 12–5b).

Relationships between *en echelon* or parallel segments of strike-slip fault systems that are not connected by segments of strike-slip faults have been described by Atilla Aydin and Amos Nur (1982), although the original observation of these relationships should probably be attributed to J. Tuzo Wilson (1965); they saw that motion may be transferred from one segment to another— by a *step-over*—through zones of extension or of contraction, depending on the relative motion of the fault segments and the sense of each step-over (Figure 12–7). As a result, an *en echelon* dextral strike-slip fault with right step-overs produces ***pull-apart***, or ***rhomb-graben*** (or *rhombochasm), basins.* The same result is produced by left-lateral strike-slip faults with left step-overs. Two sides of the basin will be bounded by segments of the strike-slip fault with a significant component of normal slip. The other sides will be bounded by normal faults oriented obliquely to the primary direction of horizontal strike-slip motion (Figure 12–7). The term "pull-apart" was first suggested by Clark Burchfiel and Jack Stewart (1966) to describe the structure of the Death Valley region. John Crowell (1974) and Thomas Dibblee (1977) later recognized the same relationships between step-overs and grabens and horsts along segments of the San Andreas fault.

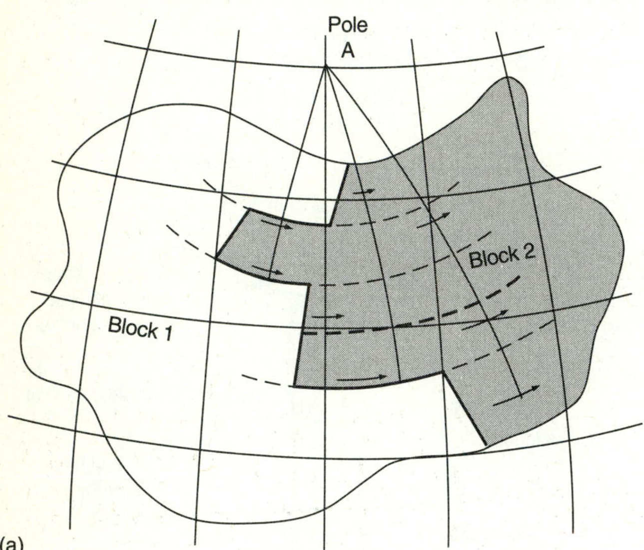

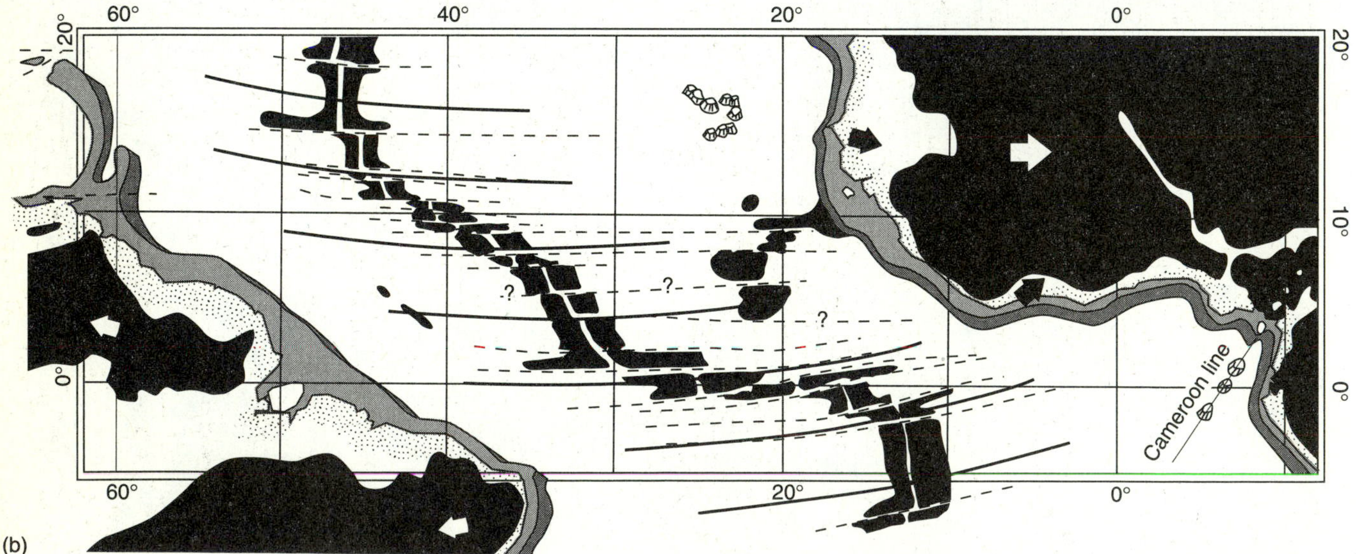

FIGURE 12–6
Small-circle geometry relationship of strike-slip and transform faults. (a) Generalized diagram showing the motion of two blocks on a sphere that must follow small circles on the sphere concentric to pole A. (b) Transforms (dashed faults) in the present-day Atlantic Ocean relative to small circles (heavy solid lines) concentric to a pole at 62° N 36° W. Dashed lines represent active and inactive transforms. (From W. J. Morgan, *Journal of Geophysical Research,* v. 73, p. 1959–1982, 1968, © American Geophysical Union.)

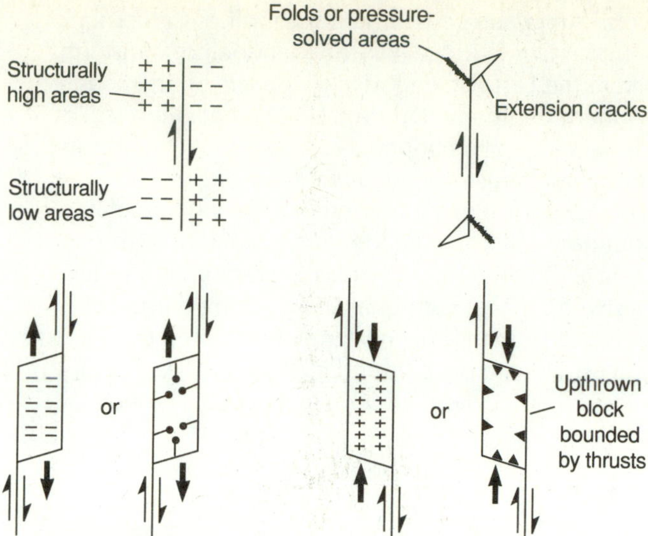

FIGURE 12–7
Transfer of motion between strike-slip fault segments producing rhomb-graben or rhomb-horst structures, depending on whether the step-over is in a right- or left-lateral sense. (After Atilla Aydin and Amos Nur, *Tectonics*, v. 1, p. 91–105, 1982, © American Geophysical Union.)

Dextral strike-slip faults with left step-overs or sinistral faults with right step-overs produce ***rhomb horsts****,* or *push-up ranges.* Strike-slip fault segments with significant thrust motion will bound two sides of a horst, and the other two sides will be bounded by obliquely oriented thrusts (Figure 12–3b).

Small rhomb grabens and horsts have been identified at many places on the Earth. The Dead Sea and the San Francisco Bay area have been interpreted by Atilla Aydin and Amos Nur (1982) and Aydin and Benjamin Page (1984; Figure 12–8) as rhomb-graben structures. Similar conclusions were reached earlier for the Dead Sea structure by Raphael Freund (1965) and Freund, I. Zak, and Z. Garfunkel (1968). The Ocotillo Badlands and Borrega Mountains along the Coyote Creek fault in California have been interpreted by Aydin and Nur (1982) as rhomb horsts (Figure 12–8b). Each of these structures formed as part of a complex of interacting plate boundaries where relative motion has changed through time.

Complex fault geometry frequently produces large rhomb grabens and horsts (Figure 12–9). Then, as noted by Aydin and Nur (1982), more faults form, producing smaller segmented grabens and horsts, which together form larger composite basins or uplifts. Aydin and Nur also noted a consistent ratio of length to width of pull-apart basins and push-up ranges varying from 2.4 to 4.3, with a mean of 3.2. The ratio appears to be scale independent, suggesting a fractal geometry, because length-width ratios of basins and uplifts of all sizes fall within this range.

TERMINATIONS OF STRIKE-SLIP FAULTS

Strike-slip faults may terminate by splaying along strike into smaller faults, by changing strike and becoming thrust or normal faults (which may further splay into smaller faults or terminate in folds), or by diminishing displacement along the main fault until none is detectable (Figure 12–10a). Splays may develop as Riedel and anti-Riedel shears off a main fault, reducing the total displacement.

In cross section, strike-slip faults frequently splay upward into outwardly branching segments that exhibit *flower (*or *palm-tree* or *horse-tail)* configurations (Figure 12–10b). These are more difficult to observe in the field unless vertical relief exposes a cross section of the fault zone. Existence of flower structures has been confirmed and observed as offset rock units in seismic reflection profiles. Some strike-slip motion involved in forming a flower structure must be resolved into dip slip. Faults associated with the Paleozoic aulocogen of southern Oklahoma exhibit flower and *inverted rift structure* (Figure 12–11; Harding, 1974).

Resolution of strike slip into significant contractional dip-slip motion is called *transpression,* which often results from oblique plate convergence where the dominant motion is strike slip. Brian Harland (1971) and James Lowell (1972) were perhaps the first to describe transpression, in the course of their studies of Caledonian structures in Spitzbergen. It has since been reported in many places, including the Mecca Hills near the Salton Sea in California (Sylvester and Smith, 1976). Transpression and its opposite, *transtension,* involve nothing more than complex oblique-slip motion resolved from dominant strike slip. It is surprisingly common, because most faults exhibit one form or another of oblique-slip.

TRANSFORMS

We have already discussed transform faults several times (see Figures 1–11, 9–9, and 12–6). Wilson (1966) first postulated that they exist to provide slip between adjacent plates, whether at a spreading center (such as an oceanic ridge), or between segments of a plate moving past one another, or connecting segments in a subduction zone where some segments are consumed faster than others. Shortly after Wilson's prediction, studies by Lynn Sykes (1967) of earthquake focal mechanisms along the Mid-Atlantic Ridge verified the hypothesis. Several kinds of transforms have been described, including ridge-ridge transforms, ridge-arc transforms, and arc-arc transforms (see Figure 9–9b).

The classic ridge-ridge transform has an apparently simple strike-slip offset, but the real motion on the fault is opposite the apparent offset sense. Transforms are clearly distinguishable from strike-slip faults (which are shears developed within a horizontal compressional field) because they are accommodation or transfer structures as well as strike-slip faults at plate boundaries. They form segments of convergent or divergent plate boundaries that are moving at different rates. The San Andreas fault fits this distinction between strike-slip and transform faults. It is an *intracontinental transform,* connecting the ends of the East Pacific Ridge, where it disappears beneath the North American continent in the Gulf of California, with the Gorda Ridge (via the Mendocino Transform) where the East Pacific Ridge reappears off the coast of northern California (Figure 12–12). It also is a strike-slip fault at a plate boundary. The connection through continental crust has not always existed, and the connection is regarded as temporary until the plate boundary shifts again either farther inland (to the Mojave Desert and Owens Valley region) or farther westward. Large earthquakes occurring during the early 1990s in unbroken segments of continental crust suggest that the plate boundary may shift eastward, transferring the Sierran block, now on the North

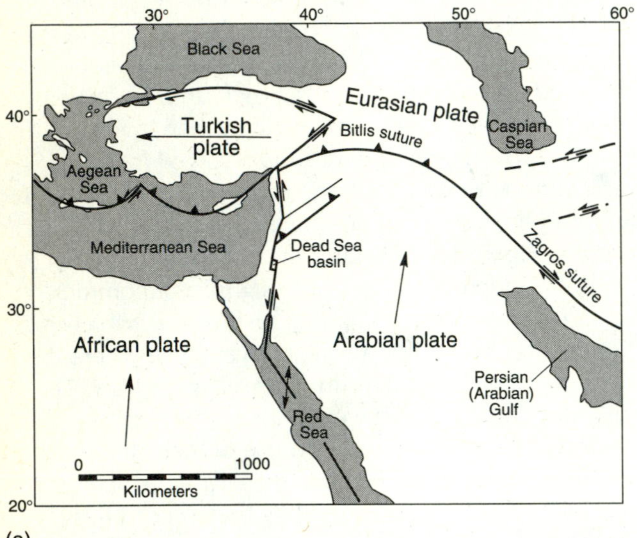

(a)

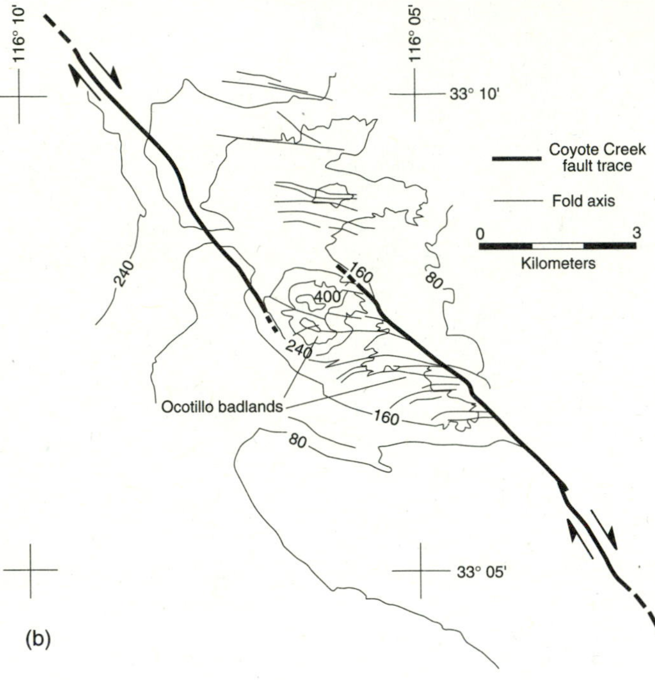

(b)

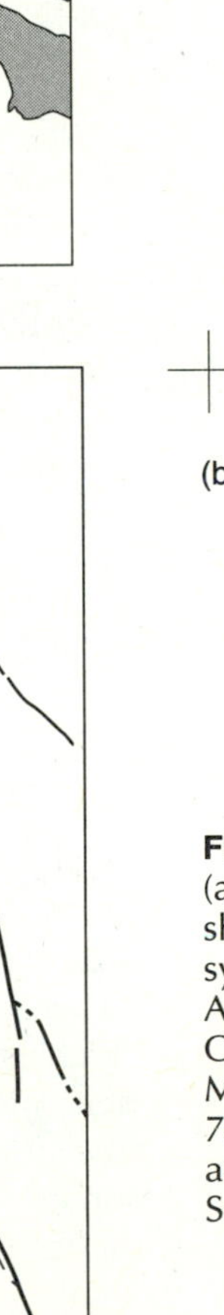

(c)

FIGURE 12–8
(a) Arrangement of plate boundaries in part of the Middle East showing the Dead Sea rhomb-graben and transform fault system. (After M. R. Hempton, 1987, *Tectonics,* v. 6, © American Geophysical Union.) (b) Rhomb horst along the Coyote Creek fault in California. (From R. V. Sharp and M. M. Clark, 1972, U.S. Geological Survey Professional Paper 787.) (c) Rhomb-graben fault system in the San Francisco Bay area. (After A. Aydin and B. M. Page, 1984, Geological Society of America *Bulletin,* v. 85.)

American plate, to the Pacific plate (Hauksson and others, 1993). Greg Davis and Clark Burchfiel (1973) earlier suggested that the Garlock fault east of the San Andreas in southern California is also an intracontinental transform: they pointed out that two different tectonic styles are present on each side of its trace. Consider too, the Dead Sea fault in the Middle East (Figure 12–8a), the Alpine fault in New Zealand, and

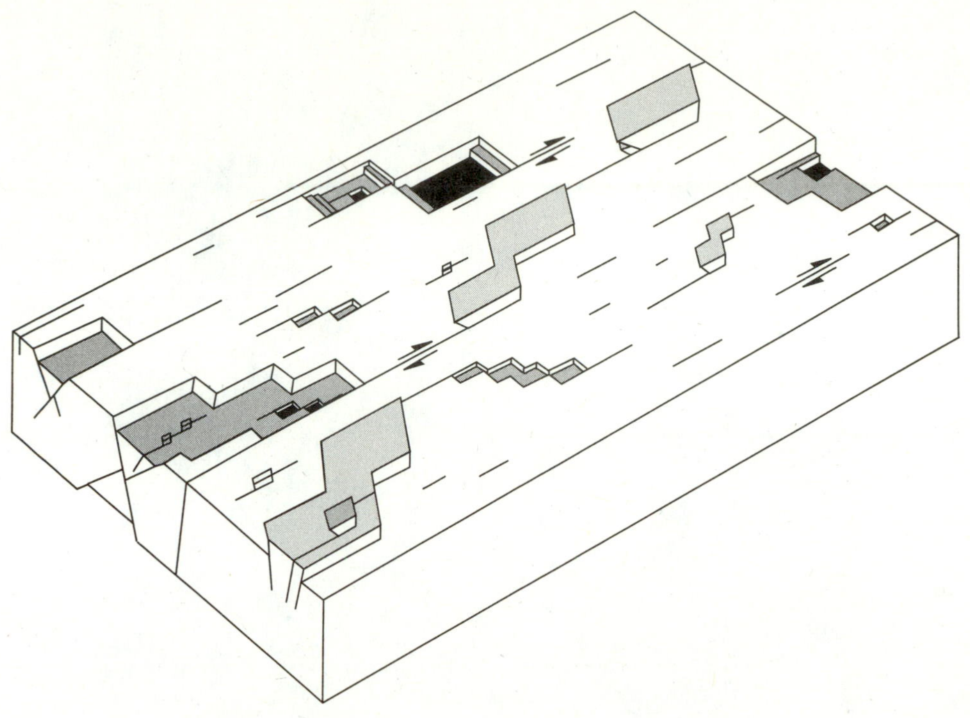

FIGURE 12–9
Complex rhomb-graben and rhomb-horst structure along a large fault system. (After A. Aydin and A. Nur, 1982, *Tectonics,* v. 1, p. 91–105, © American Geophysical Union.)

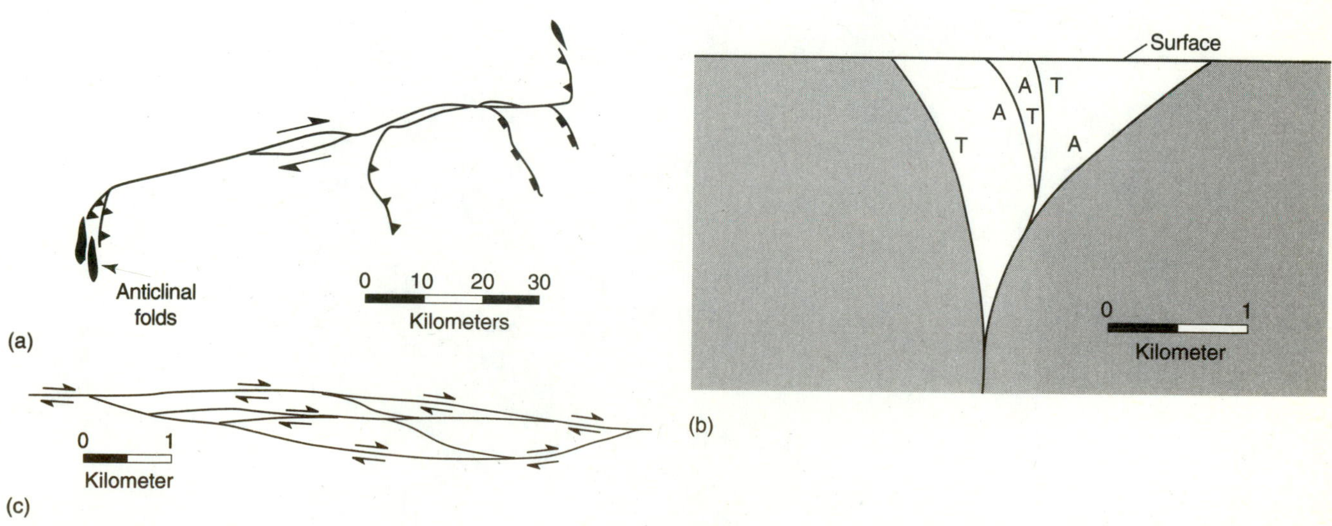

FIGURE 12–10
(a) Termination of strike-slip faults into thrust and/or normal faults, folds, and splays (map view). (b) Flower structure or horse-tail configuration in cross section along a strike-slip fault (A—movement away from observer; T—movement toward observer). (c) Flower structures may be a near-surface phenomenon, the flowers closing upward or downward in cross section into single fault zones, as convergence of faults along strike into a single fault is frequently observed in map view on the surface.

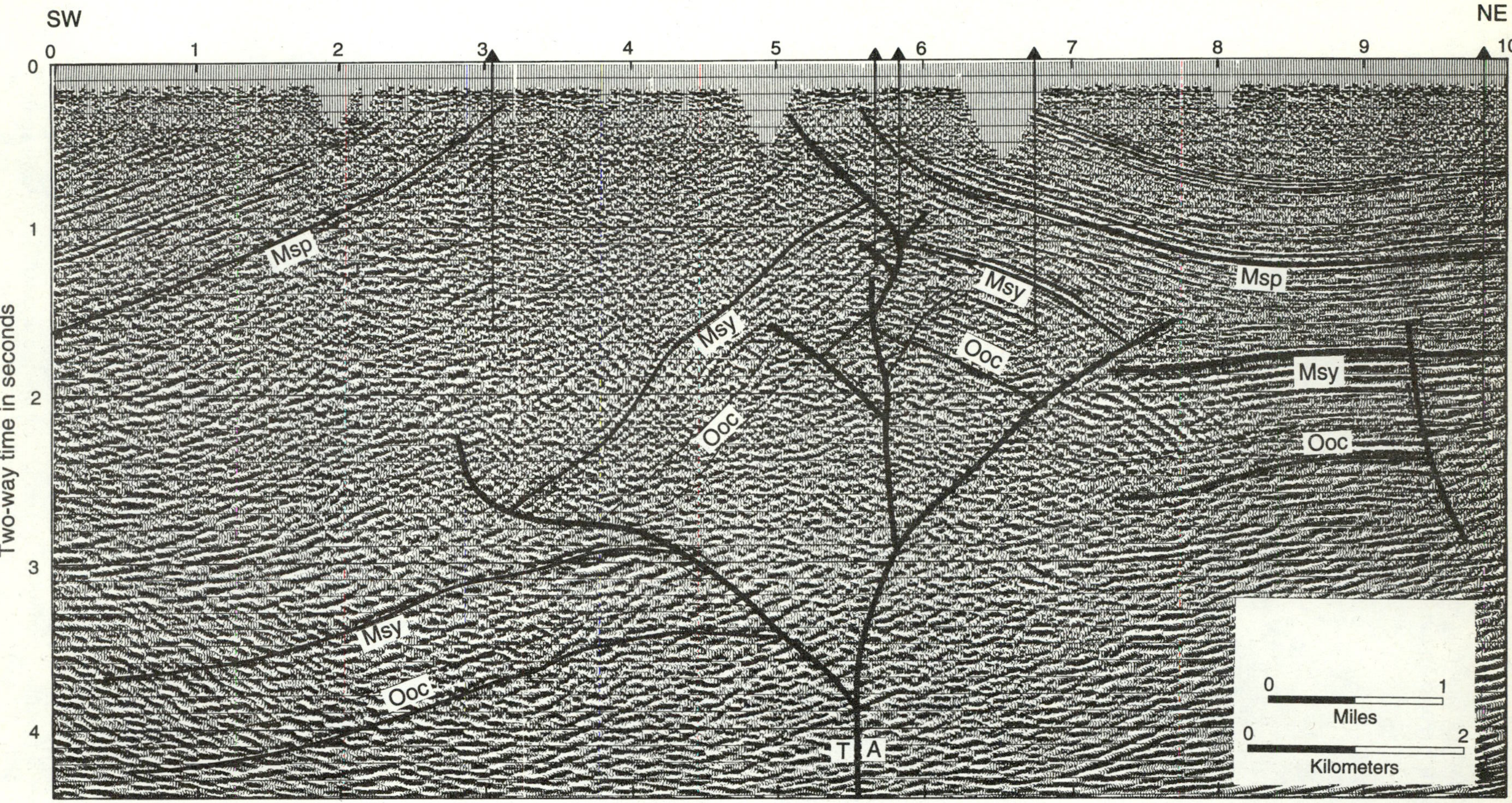

FIGURE 12–11
Structure section constructed on a seismic reflection profile and drill data through the Ardmore basin in the Oklahoma aulocogen, illustrating flower and inverted-rift structures. Msp—Springer, Msy—Sycamore, and Ooc—Oil Creek are Paleozoic rock units. (After T. P. Harding and J. D. Lowell, 1974, AAPG *Bulletin,* v. 58. Reprinted by permission of American Association of Petroleum Geologists.)

the Great Glen fault in Scotland. All may be intracontinental transforms but keep in mind that transforms are only a special kind of strike-slip fault, and that, by changing fault geometry slightly, motion may become either thrust or normal.

Here, we end our discussion of strike-slip faults with a note that even the largest and most widely known examples still deserve continued study, particularly because of the interrelated nature of all faults, and the seismic hazard posed by active faults.

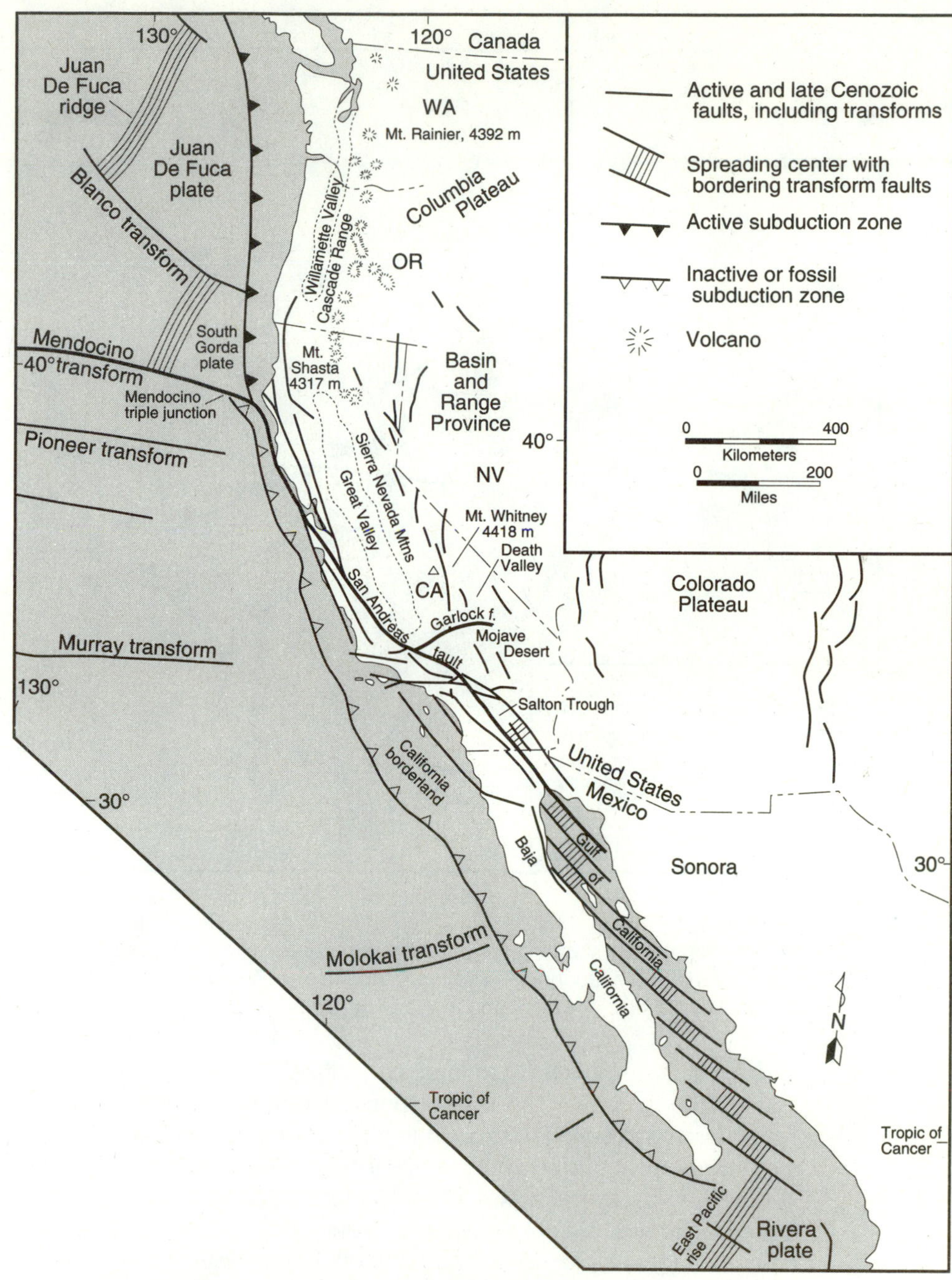

FIGURE 12–12
San Andreas and related fault systems in California, northern Mexico, and in the adjacent Pacific Ocean. (After J. C. Crowell, 1987, *Episodes*, v. 110.)

ESSAY

Rigid Indenters and Escape Tectonics

A large system of intraplate continental strike-slip faults was first described by Peter Molnar and Paul Tapponnier (1975) in central Asia, where they found that known strike-slip faults interconnect (Figure 12E–1). As a result, they suggested that India acted as a *rigid indenter* (Figure 12E–2) when it collided with Asia, and that strike-

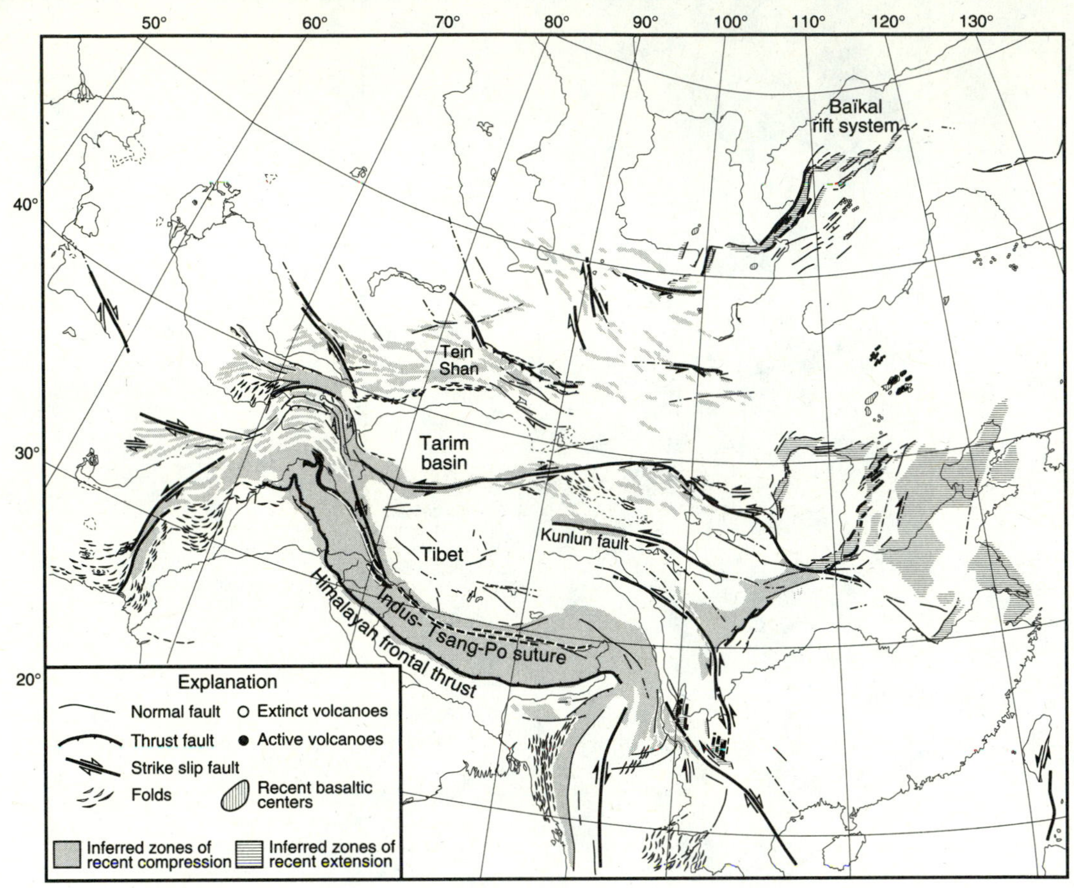

FIGURE 12E–1
Tectonic map of Asia showing major active faults (heavy lines). Arrows indicate sense of motion. (From Paul Tapponnier and Peter Molnar, 1977, *Journal of Geophysical Research,* v. 82, p. 2905–2930, © American Geophysical Union.)

slip deformation resulted when large parts of Asian crust moved laterally in response to the collision. The crust responded to the collision with the indenter by forming large strike-slip faults bounding blocks that moved laterally out of ("escaped" from) the collision zone, hence the term "escape tectonics." North of the indenter in Asia, dextral strike-slip faults ideally would have moved blocks westward from the west side of the collision zone, and sinistral faults (again, ideally) would have moved blocks eastward and out of the eastern part of the collision zone. We recognize, however, that both dextral and sinistral faults bound the escaping blocks because of the orientation of the more local stress fields and the shear planes that ultimately form faults. Deviation of actual motion from ideal behavior along the large faults was related to both the original shape of India as it collided with Asia (Figure 12E–1) which produced variations in stress trajectories, and also to the relative-motion vectors that brought Asia and India into collision. Tapponnier and others (1982) have built accurate scale models to better understand the process (Figure 12E–2).

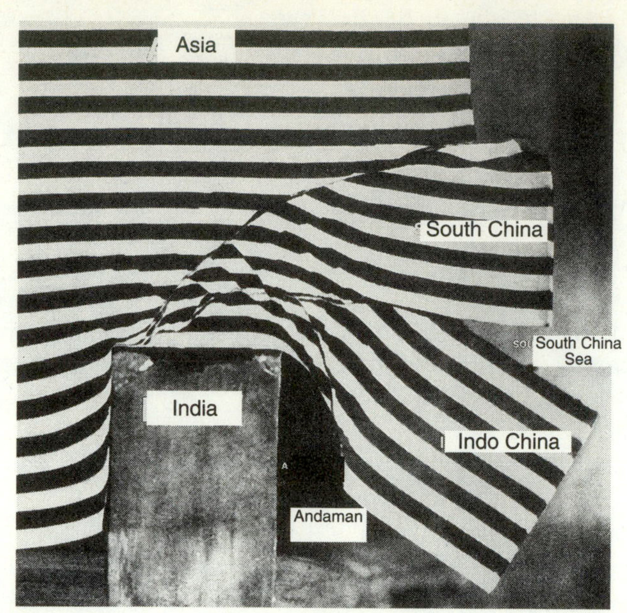

(a)

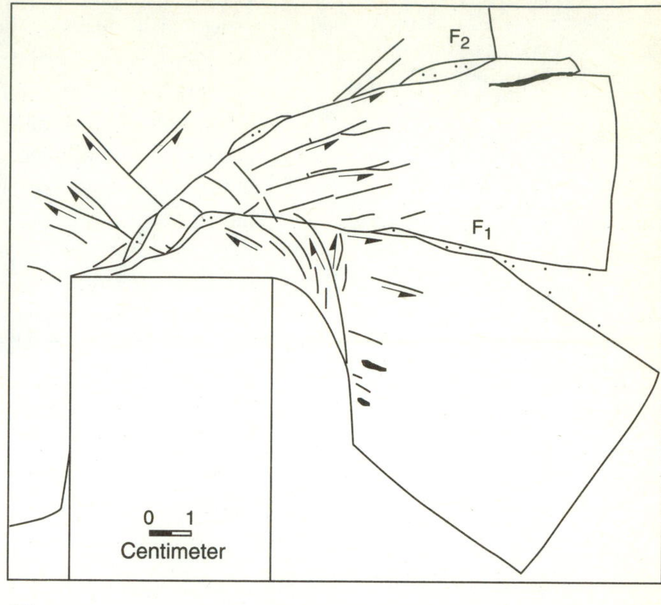

(b)

FIGURE 12E–2
(a) Plasticine model of escape tectonics involving a rigid indenter and several laterally moving blocks. (b) Sketch of the deformation in (a). Note the pull-apart basins along both faults (F_1 and F_2). (From P. Tapponnier, G. Peltzer, A. Y. Le Dain, R. Armijo, and P. Cobbold, 1982, *Geology,* v. 10, p. 611–616.)

The Asia-India collision zone is one of Earth's most populous regions, and so the earthquake hazard is of great importance. Thus, an additional incentive exists to study crustal deformation processes associated with the indenter mechanism and collision-related faults (both large and small) and to apply the results to earthquake prediction.

Since the indenter mechanism was originally proposed, Brian Davies (1984) has applied it to explain the origin of the Panafrican (Late Proterozoic) deformed belt in the Arabian shield, and Jean-Pierre LeFort (1984) has used it to explain the structure of large faults and the curvature of the central Appalachians. No less important is the westward escape of the Turkish Plate from the northward-moving Arabian Plate indenter (Figure 12–8a), along a number of active faults.

References Cited

Davies, B., 1984, Strain analysis of wrench faults and collision tectonics of the Arabian shield: Journal of Geology, v. 82, p. 37–53.

LeFort, J.-P., 1984, Mise en évidence d'une virgation carbonifère induite par la dorsale Reguibat (Mauritanie) dans les Appalaches du Sud (U.S.A.), Arguments géophysiques: Bulletin Société Géologique de France, v. 26, p. 1293–1303.

Molnar, P., and Tapponnier, P., 1975, Cenozoic tectonics of Asia: Effects of a continental collision: Science, v. 189, p. 419–426.

Tapponnier, P., Peltzer, G., Le Dain, A. Y., Armijo, R., and Cobbold, P., 1982, Propagating extrusion tectonics in Asia: New insights from simple experiments with Plasticine: Geology, v. 10, p. 611–616.

Questions

1. Why do strike-slip faults characteristically dip steeply?
2. How can strike-slip, normal, and thrust faults—as well as folds—all form in the same stress system?
3. Why do some strike-slip faults contain cataclastic material along their extent (or are single planar structures) but others contain mylonite and consist of zones several kilometers wide?
4. Why do some strike-slip fault zones contain mylonite that is cut through by later cataclasite?
5. What evidence supports the statement that some strike-slip faults are contractional structures?
6. Why must major strike-slip faults become some other type if they deviate from a small circle of the Earth?
7. How do strike-slip faults terminate?
8. Explain the origin of a rhomb graben.
9. Why does a ridge-ridge transform have a sense of motion opposite that of an ordinary strike-slip fault?
10. What distinguishes a ridge-ridge transform from a ridge-arc transform?

Further Reading

Aydin, A., and Nur, A., 1982, Evolution of pull-apart basins and their scale independence: Tectonics, v. 1, p. 91–105.
Describes strike-slip faulting and its relationships to the origin of pull-apart basins (rhomb grabens) and uplift (rhomb horsts), with numerous examples.

Biddle, K. T., and Christie-Blick, N., 1985, Strike-slip deformation, basin formation, and sedimentation: Society of Economic Paleontologists and Mineralogists Special Publication 37, 386 p.
Papers relating strike-slip faulting and formation of sedimentary basins. Several describe extensional settings; others, development of strike-slip faults and basins in compressional settings.

Davis, G. A., and Burchfiel, B. C., 1973, Garlock fault: An intracontinental transform structure: Geological Society of America Bulletin, v. 84, p. 1407–1422.
The Garlock fault—part of the San Andreas system—is interpreted as an intracontinental transform, in contrast with the San Andreas, which has been interpreted as a ridge-ridge transform as well as an intracontinental transform.

Powell, R. E., Weldon, R. J., II, and Matti, J. C., 1993, The San Andreas fault system: Displacement, palinspastic reconstruction, and geologic evolution: Geological Society of America Memoir 178, 332 p.
A series of papers that chronicles the geologic history, geometry, and deformation along the San Andreas fault system in California.

Sylvester, A. G., ed., 1984, Wrench fault tectonics: American Association of Petroleum Geologists Reprints Series 28, 374 p.
Contains several classic papers on strike-slip faults. Several papers by J. C. Crowell, J. Tuzo Wilson's paper defining transforms, J. S. Tchalenko's paper on shear zones, and the comparative paper by R. E. Wilcox and others on wrench-fault tectonics all make good reading.

Sylvester, A. G., 1988, Strike-slip faults: Geological Society of America Bulletin, v. 100, p. 1666–1703.
An excellent review of the nature of strike-slip faults written for students, structural geologists, and other geologists.

Tapponnier, P., Peltzer, G., Le Dain, A. Y., Armijo, R., and Cobbold, P., 1982, Propagating extrusion tectonics in Asia: New insight from simple experiments with Plasticine: Geology, v. 10, p. 611–616.
Models the strike-slip tectonics of Asia, based on experimental work and the concept of a rigid indenter. The results are striking.

13

Normal Faults

The labors of Pennsylvania geologists have rendered so familiar the structure of the Appalachians, that it has been accepted as typical of all mountains, and a comparison will facilitate an understanding of the basin ranges. Indeed, I entered the field with the expectation of finding in the ridges of Nevada a like structure, and it was only with the accumulation of difficulties that I reluctantly abandoned the idea.

GROVE KARL GILBERT, 1874, *Explorations and Surveys West of the One Hundredth Meridian*

G. K. GILBERT'S RECOGNITION OF THE BLOCK-FAULTED STYLE OF the Basin and Range Province was one of the great discoveries in nineteenth-century geology. Though chief geologist of the Wheeler reconnaissance, Gilbert was under the command of an Army officer and had only three field seasons to explore the geology of large parts of Utah and Nevada. He was also conditioned to expect Appalachian geology in the Far West. He employed the resources of an open mind and careful observation, hallmarks of a good scientist, together with hard work, that yielded the initial discoveries that provide some of the striking examples we will discuss in this chapter.

Normal faults are dip-slip faults in which the hanging wall has moved down relative to the footwall (Figure 13–1). In many parts of the world, they are important as structural traps for hydrocarbon accumulations. In the Houston area in the Texas Gulf Coast, they cause subsidence problems, affecting roads and buildings. (Excessive withdrawal of ground water from regional aquifers also contributes to the subsidence.) Normal faults are also the primary mechanism for extending and thinning the crust before the opening of a new ocean basin. Only now are we beginning to understand the similarities and differences between extensional processes that form symmetrical rifts (such as the Gulf of California) and those producing asymmetrical extension (for example, the Basin and Range Province and the East African Rift). These, too, involve normal faulting.

FIGURE 13–1
Normal faults in Pleistocene clay in northeastern Washington, rotated by landsliding, and later overlapped by terrace deposits. (F. D. Jones, U.S. Geological Survey.)

Normal faults have been called *gravity faults,* implying that the primary motive force is gravity. They have also been called *extensional faults* because they extend layering, or thin the crust. ***Listric faults*** (Figure 13–2) have concave-up surfaces, flatten with depth, and steepen toward the surface. They may be either normal or thrust faults, as "listric" describes the geom-

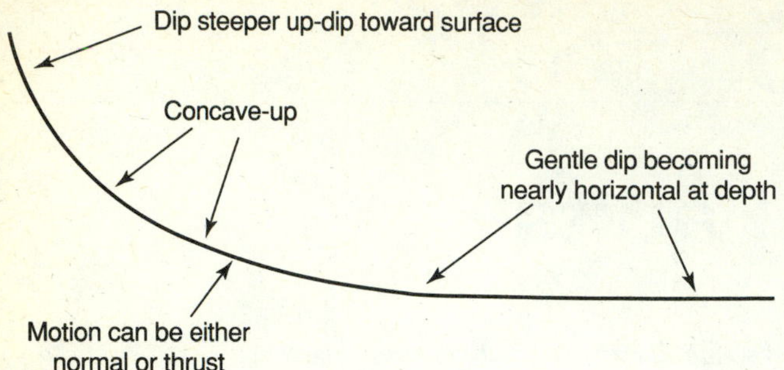

FIGURE 13–2
Characteristics of listric faults. Note that they apply to both normal and thrust faults.

etry and not the sense of motion. ***Growth faults*** involve simultaneous deposition of sediment and fault motion. Many growth faults are listric normal faults.

PROPERTIES AND GEOMETRY

Normal faults commonly have moderate to relatively steep angles of dip. They have long been thought to propagate directly into crystalline basement beneath the sedimentary cover and terminate deep in the crust without change in dip. With the accumulation of better field and geophysical data, it has been demonstrated that many normal faults flatten listrically into incompetent units at depth and display many of the detachment properties of thrust faults. Others pass from the sedimentary cover into the crystalline basement (Figure 13–3). Modern seismic reflection profiles, coupled with surface studies, suggest that many flatten in a crystalline basement into the ductile-brittle transition at great depth (Figure 13–4). The geometry and evolution of normal fault systems in the Gulf Coast region have been described by Ernst Cloos (1968), who modeled them in clay-block experiments. These experiments demonstrated the concave-up listric shape of these faults in three dimensions. Cloos also successfully modeled the map pattern of normal faults in Texas and Louisiana, where they are concave southward toward the Gulf of Mexico.

Normal faults display many of the branching characteristics of thrust and strike-slip faults (Figure 13–5a). ***Splays*** occur along normal faults. ***Synthetic faults*** dip in the same direction as the master fault and join the master fault at depth (Figure 13–5b). ***Antithetic faults*** likewise join the master fault at depth, but dip in the opposite direction.

Paired normal faults exist where related blocks are flanked by parallel-striking faults that dip either away from the blocks or toward them (Figure 13–6). A ***graben*** involves two normal faults that dip toward each other, with a down-dropped block in between. A ***half-graben*** is a block bounded by a normal fault on one side; on the other side, it passes into a gentle fold. A

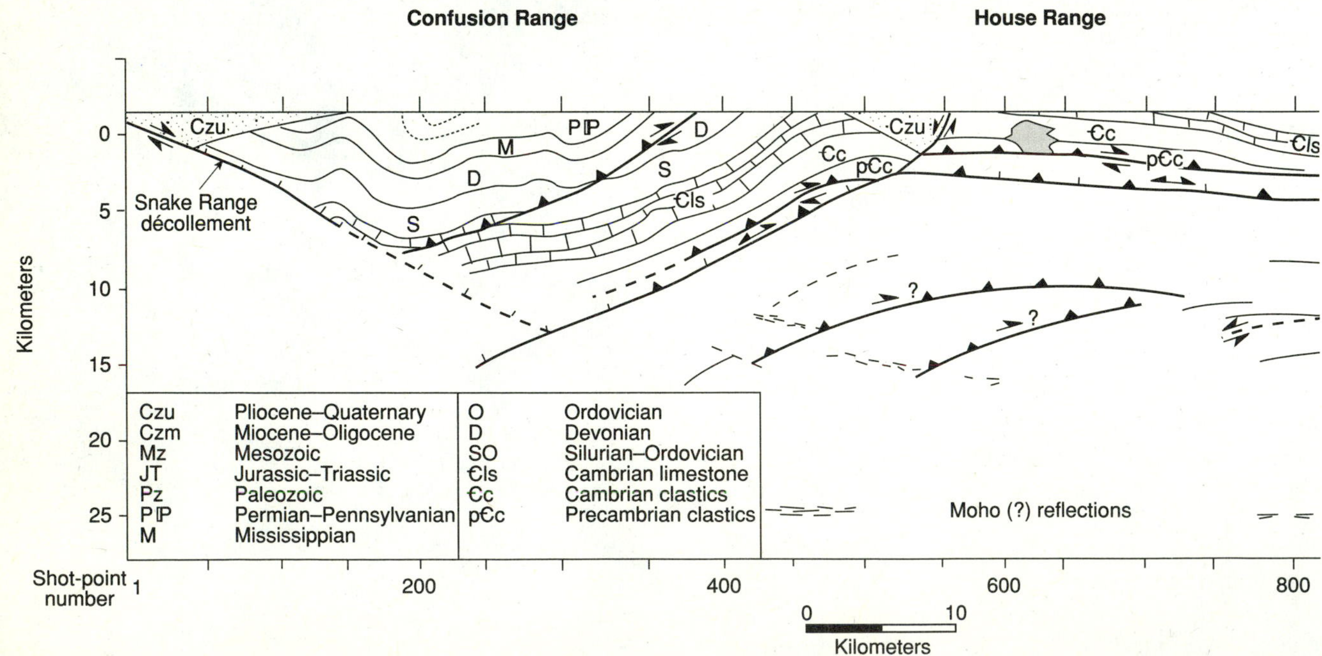

FIGURE 13–3
East-west cross section through the eastern Basin and Range Province in Utah showing flattening characteristics of listric normal faults (ticks), thrusts (solid teeth), and older thrusts inverted as normal faults (closed teeth with ticks). Section is based on the COCORP Utah seismic reflection line. (From R. W. Allmendinger and others, 1983, *Geology*, v. 11.) This section is derived from the seismic reflection profile in Figure 21–15.

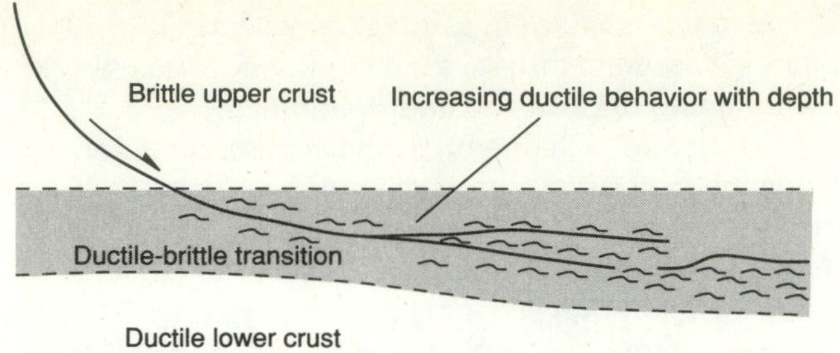

FIGURE 13–4
Listric normal fault flattens into the ductile-brittle transition.

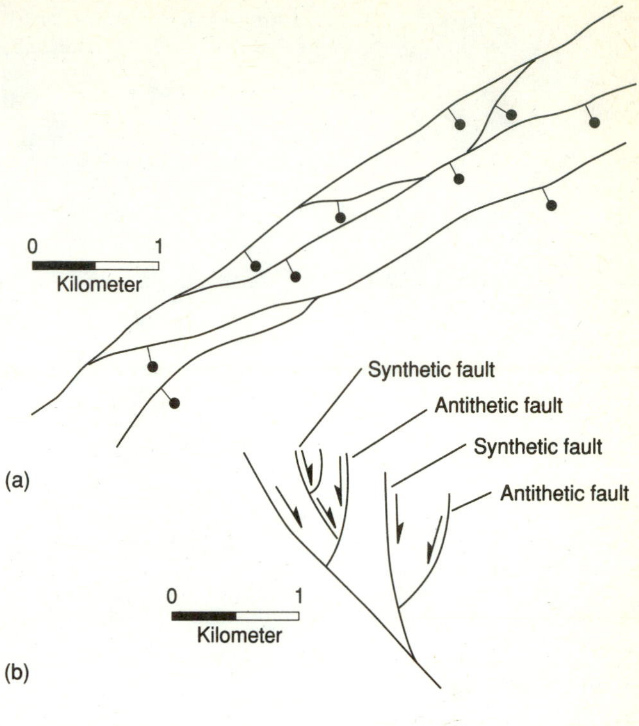

FIGURE 13–5
Branching characteristics of normal faults. (a) Anastomosing splays: ball-and-stick pattern on faults indicates the downthrown side (map view). (b) Synthetic and antithetic faults (cross section).

horst is a relatively upraised block flanked by symmetrical normal faults that dip away from the horst. The mountains and alluvium-filled valleys of the Basin and Range Province, which extend from southern Idaho into Mexico, may seem a classic graben-and-horst region, with the ranges as horsts and valleys as grabens, but crustal extension over geologic time has created a much more complex geometry of tilted fault blocks and half-grabens. The Rhine graben region of Germany and Switzerland is dominated by a single, large down-dropped crustal block, but within it are many smaller horsts and grabens.

Normal faults may also be related to folding. Some normal faults at depth pass into monoclinal folds near the surface or along strike (Figure 13–7). The shape of the fold is frequently related to the shape of

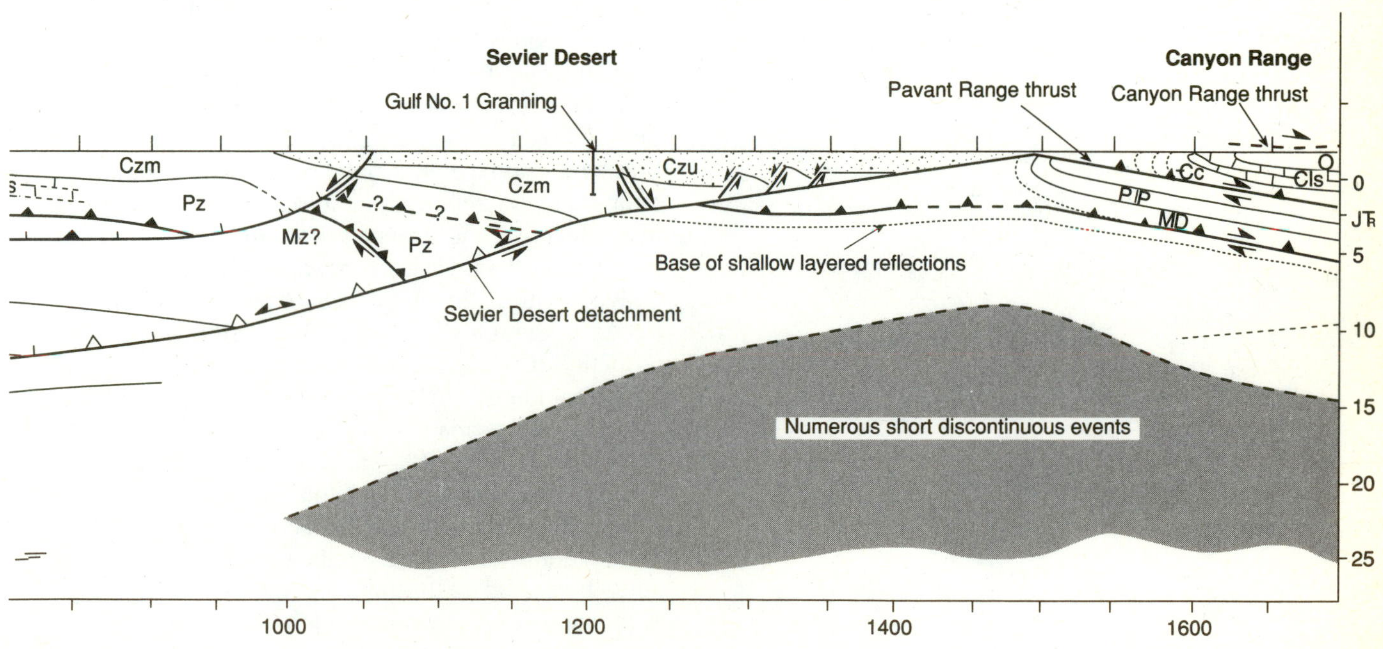

FIGURE 13–3 (continued)

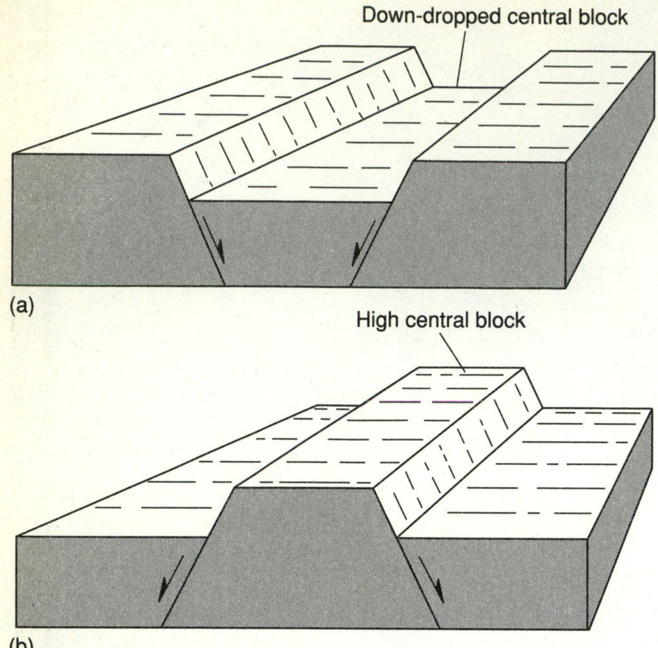

FIGURE 13–6
Graben (a) and horst (b) structures.

the fault surface. Good examples occur in the Grand Canyon region, where normal faults pass into monoclines along strike.

A normal fault may break and displace basement rocks but die out upward into the sedimentary cover, producing a ***drape fold*** in the sedimentary cover rocks.

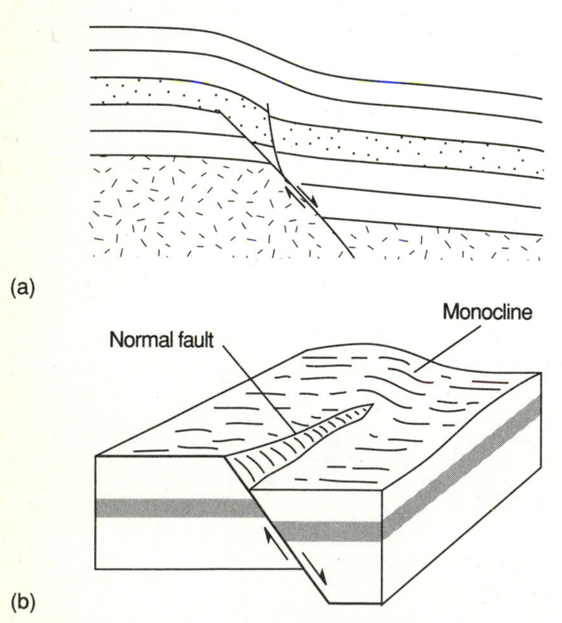

FIGURE 13–7
(a) Up-dip termination of a normal fault into a monoclinal drape fold (cross section). (b) Monocline grading along strike into a normal fault.

Drape folds also form in association with steeply dipping basement thrusts, as in the Rocky Mountain Front Ranges of Colorado and Wyoming.

Drag folds (Chapter 9) form because of friction along the fault surface and occur along normal faults; they provide evidence of motion sense on a fault through observation of the sense of shear of the folds along the fault. ***Reverse drag folds*** and *rollover anticlines* form along growth faults where the part of the downthrown block close to the fault is displaced downward more than the parts farther away (see Figure 9–13). These structures are related to the listric shape of the faults on which they commonly form. Rollover anticlines are targets for petroleum exploration, and in the Gulf Coast have proved to be highly productive.

ENVIRONMENTS AND MECHANICS

Normal faults are commonly described as brittle structures that develop only in the upper crust. Those formed near the surface generally have sharp contacts, cataclastic zones, breccias, and rotated blocks (or horses) characteristic of brittle fault zones; if formed deep in the crust, they exhibit ductile features—notably, mylonites—and develop thicker ductile shear zones. The large detachment-normal faults produced by crustal extension, as in the Basin and Range Province, probably formed in the ductile-brittle transition.

The extensional character of normal faulting may be examined with aid of the strain ellipsoid, assuming coaxial strain (Figure 13–8). Ideally, normal faults dip more than 50° and form on shear planes that are bisected by a vertical (Z) axis, with the maximum principal strain (X) horizontal. The Y axis lies in the fault plane, parallel to the intersection of the conjugate normal faults. The stress ellipsoid involves a vertical maximum principal stress (σ_1) and a horizontal σ_3, with σ_2 lying within the fault plane. As with other faults and fractures, the rules for fluid pressure, buoyancy, and reduction of frictional resistance (Chapter 10) all apply to normal faults.

Upward propagation of a normal fault—as argued by Neville Price (1977)—may result from hydrofracturing by overpressured fluid, generally water. When pore-water pressure exceeds the tensile strength of a potential glide zone within a part of a listric fault zone, hydraulic fracturing occurs. Frictional resistance to movement falls to zero, and the fault block tends to move, causing the fracture to propagate upward—still under high water pressure—until the entire block moves along the fault. The fault will steepen as the block moves down dip, forming a steeply dipping fault segment that propagates through the upper part of the

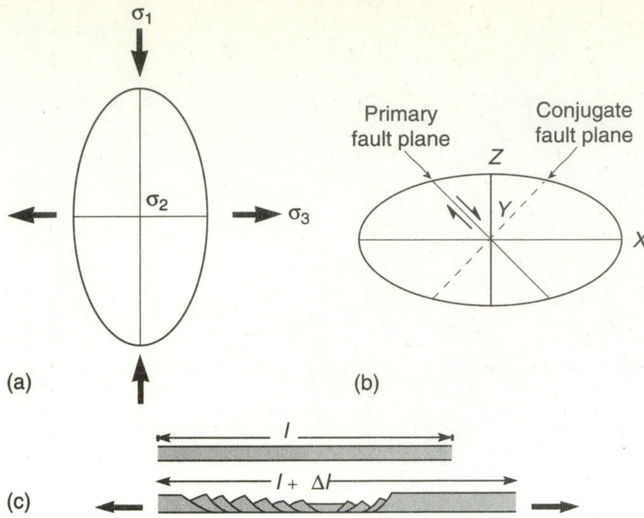

FIGURE 13–8
Stress (a) and strain (b) ellipsoids for normal faulting. (c) Sequential sections of the crust before and after normal faulting and extension, showing displacement taking place primarily as horizontal extension. A crustal segment of original length l is extended and thinned to a new length $l + \Delta l$.

block and permits it to move as a coherent unit. The block thus formed is bounded by a listric surface or is a *rotational slide block.*

Growth Faults

Growth faults commonly form in relatively unconsolidated sediments during deposition and produce thickened stratigraphic units in the downthrown block (Figure 13–9). They may be either normal or thrust faults, but normal faults are more common. The best-documented growth faults in North America (possibly the world) occur in the Gulf Coast region. There, drilling has revealed many large listric normal growth faults, some with vertical displacements of a kilometer or more. They have also been observed in the Niger delta, the Bay of Biscay, and in other basin margins. High-quality seismic reflection data have provided additional documentation of fault geometry and properties (Figure 13–10).

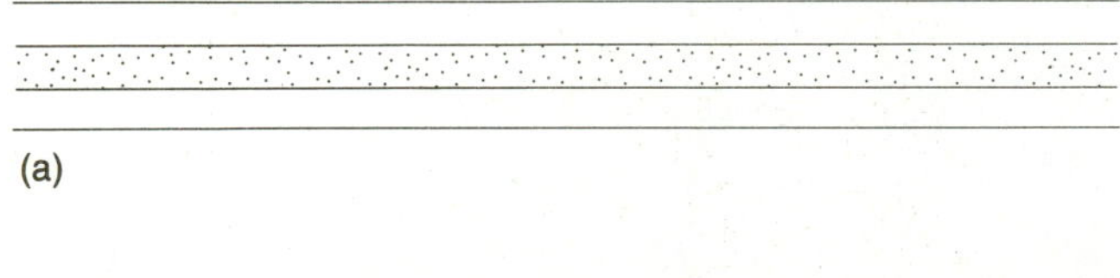

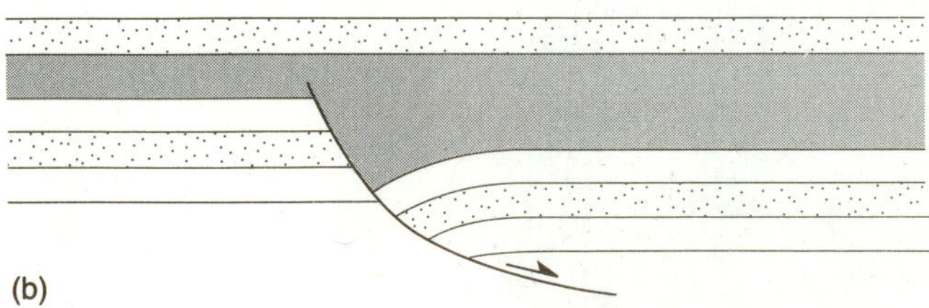

FIGURE 13–9
Geometry and characteristics of growth faults. (a) Uniform deposition before faulting. (b) Deposition continues during faulting, with excessive accumulation of sediment on the downthrown block of the fault. Movement ceases, but deposition continues on both sides of the fault. Compare with Figure 13–10.

Listric normal growth faults are among the best examples of gravity faults because they can move only under the influence of gravity. Their movement may be accelerated or slowed by the variations in load produced by different rates of sedimentation. Loading of the downthrown side—the *depocenter* for rapidly deposited sediment—may accelerate motion. Frequently, we can closely fix the time of motion of growth faults, provided motion has stopped and sediment has been deposited across the fault. Changes in relative thickness across the fault are therefore keys to timing of motion. The situation is unique because material is transported from one fault block to the other, both during and after movement on the fault. Hongbin Xiao and John Suppe (1986) have presented evidence suggesting that the curvature of listric normal growth faults is related to compaction of unconsolidated sediments after deposition on the downthrown blocks of growth faults. The actual motion along growth faults, and possibly other large normal faults, could also be attributable to block rotation of the hanging wall (Xiou and Suppe, 1992; Groshong, 1993).

In the Gulf Coast, overall movement along growth faults is toward the Gulf of Mexico basin. As a result, the faults tend to form arcuate, steplike patterns roughly concentric to the Mississippi delta and the Gulf Basin (Figure 13–11; Cloos, 1968). The term ***down-to-basin faults*** is frequently used because the downthrown side is always toward the Gulf Basin. These are also classic listric faults, flattening downward into a detachment in the sedimentary section. In the Gulf Coast, the detachments may be localized in either weak "geopressured" shales (actually unconsolidated shales—muds—containing excess trapped pore water), or they may propagate in the extensive Jurassic salt unit. Blocks thus formed are analogous to rotational landslides and slump structures. Compression and thrusting occur at the toes of both of these structures (Figure 13–12) because of the change during movement from extension at the rear to compression at the toe.

The entire process of down-to-basin faulting in the Gulf Coast and in other regions is intimately related to movement on either salt or other plastic materials (Figure 2–30b) and has been modeled as a thin-skinned extension process in which the cover sediments behave almost passively in response to deformation in the

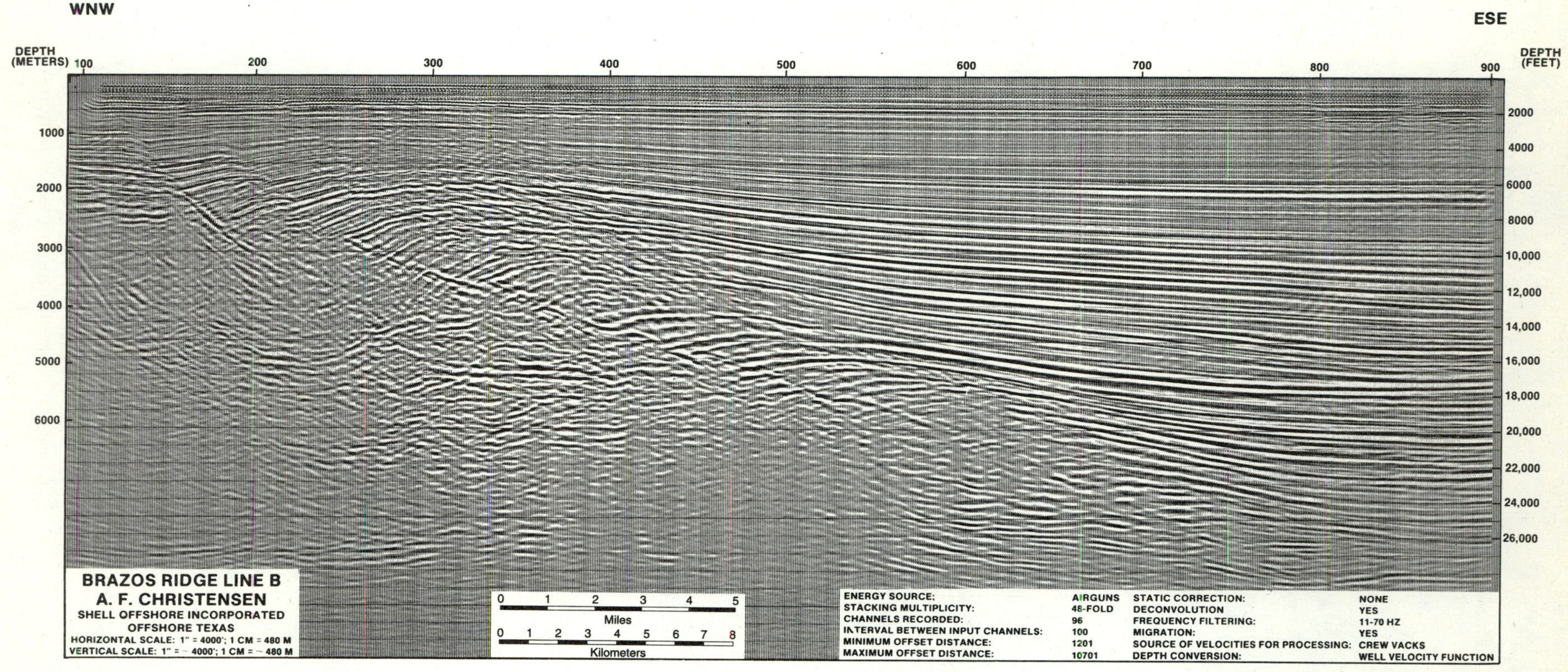

(a)

FIGURE 13–10
Seismic reflection profile (a) and interpretation (b) showing the geometry of the Corsair fault complex—a major growth fault with many smaller synthetic and antithetic normal faults, and accompanying rollover along the Corsair trend in the U.S. Gulf Coast. (Courtesy of A. F. Christensen and Shell Offshore, Inc.)

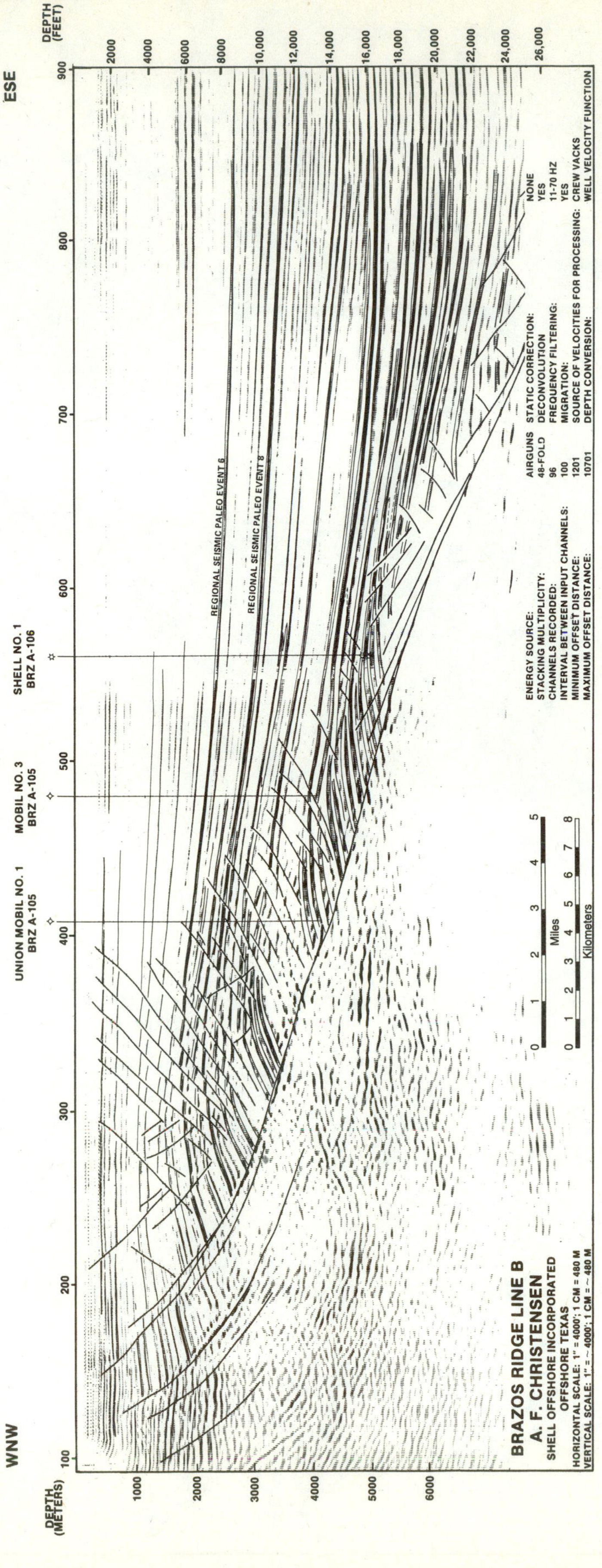

(b)

FIGURE 13–10 (continued)

FIGURE 13–11
Map of part of the U.S. Gulf Coast showing distribution of major regional growth faults (ticked lines) and salt-diapirs (black). The screened area has detached and moved toward the Gulf Basin. The southeastern edge has overthrust the sediments on the floor of the basin and produced the Perdido and Mississippi Fan foldbelts. (Modified from Plate 2, Principal Structural Features, The Gulf of Mexico Basin, compiled by T. E. Ewing and R. F. Lopez, *in* A. Salvador, *The Geology of North America*, Geological Society of America, v. J, 1991. Used by permission.)

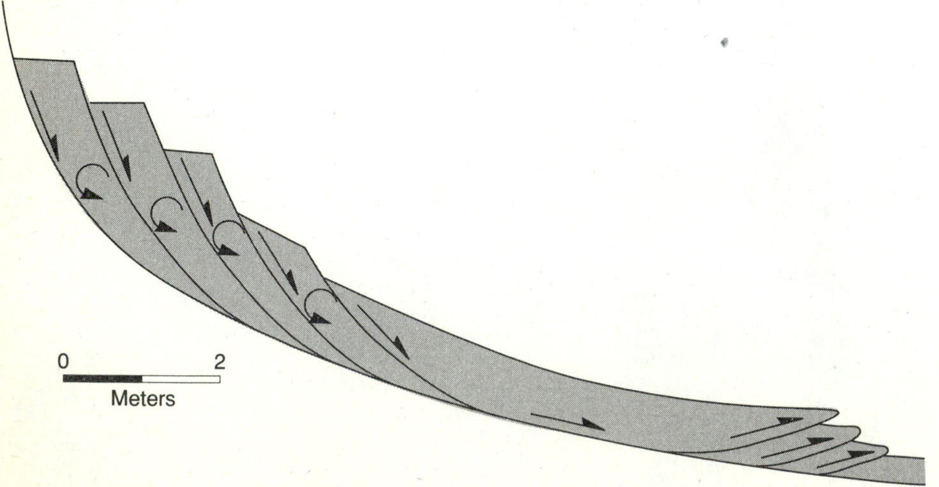

FIGURE 13–12
Cross section in landslide or slump structure showing rotational character and thrusting of the toes of listric faults in sediments. Curved arrows indicate counter-clockwise rotation of blocks.

underlying salt (Vendeville and Jackson, 1992a; Vendeville and Jackson, 1992b). Thickening and thinning of the cover, however, as products either of variations in the primary depositional framework and sediment dispersal patterns or of tectonic extension, provide zones where diapiric upwelling and piercement of salt can occur, forming most of the familiar salt structures (Chapter 2). B. C. Vendeville and Martin P. A. Jackson (1992a, 1992b), using both model experiments and natural examples, have recognized several stages of salt diapirism related to thin-skinned extension of the cover (Figure 13–13): (1) *reactive* stage, where diapirism begins as slow reactive piercement into a region where the cover has been thinned by normal faulting; (2) *active* stage, where fluid pressure at the crest of the diapir is high enough to lift and move the thinned cover aside, independent of regional extension; (3) *passive* diapirism (or piercement) stage that begins when the diapir reaches the surface, and so without additional sedimentation the diapir cannot continue to rise, but it can widen with additional regional extension; and (4) *fall* of diapirs which occurs with continued extension, so that the amount of salt available for intrusion and extrusion is exceeded by widening of the salt stock. The salt then sags downward toward the source (Figure 13–13b, stage 3), producing a low on the surface that acts as a depocenter.

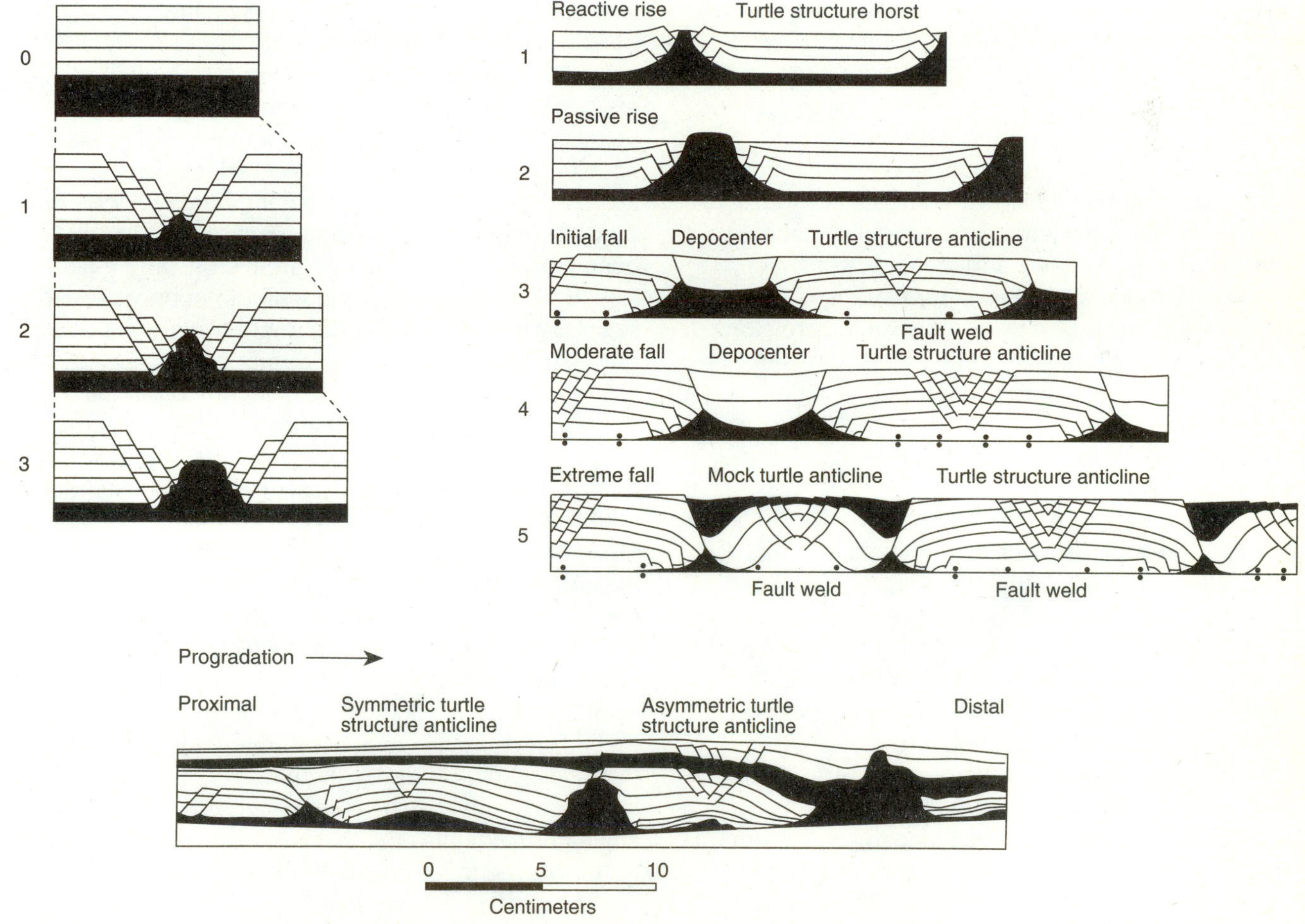

FIGURE 13–13
Salt tectonics in a region undergoing thin-skinned extension above a salt detachment, from experiments by Vendeville and Jackson (1992a, 1992b) using silicone (black) for salt and unornamented sand layers for brittle cover sediments. (a) (1) *Reactive* stage; (2) *active* stage; (3) *passive* diapirism (or piercement) stage. (b) *Fall* of diapirs follows the stages in (a) (1 and 2 here). The dots delimit areas that were formerly underlain by salt. (c) Regional relationships between the proximal sediment source, regional dip, and the distal edge of the deforming mass. Large blocks of isolated cover have been referred to as "rafts," and the process has been called "raft tectonics" (Jackson and Cramez, 1989; Duval and others, 1992). (From B. C. Vendeville and M. P. A. Jackson, *Marine and Petroleum Geology*, v. 9, 1992. Used by permission.)

Growth faults are aseismic. Areas where active growth faults are known to occur are commonly areas where no seismicity has been associated with faulting. Continuous and very recent movement on growth faults has been demonstrated in the Texas Gulf Coast, which is seismically one of the least active regions in North America. High fluid pressure in the fault blocks and the relatively unconsolidated nature of the sediments induces a state of stable sliding (Figure 10–10), and the blocks move aseismically toward the Gulf of Mexico.

Rift Zones

Narrow linear zones where the crust has been pulled apart, producing grabens, half-grabens, and other structures associated with normal faults, are called ***rifts*** (Figure 13–14a). Such zones may presage formation of a new ocean basin or a major zone of crustal extension. They have traditionally been thought to result from symmetrical thinning of continental or oceanic crust, and the structure of rift zones along mid-ocean ridges supports that assumption. After examining earthquake epicenter locations and making first-motion studies, Dan McKenzie, D. Davies, and Peter Molnar (1970) concluded that such a symmetrical spreading pattern exists in the Red Sea and the East African Rift. They also concluded that although spreading is symmetrical about a rift axis, it may decrease to zero along the axis to a pole of rotation. These ideas have since become known as the "McKenzie model" of rift formation, but newer evidence suggests that the Red Sea and East African Rift formed by asymmetric rifting (Figure 13–14c).

Aulocogens are tectonic troughs, bounded by normal faults and formed within continental crust at a high angle to a nearby continental margin. Kevin Burke and John Dewey (1973) have suggested that aulocogens result from failed formation of a ridge-ridge-ridge triple junction above a mantle plume where two spreading ridges have developed, but the third does not continue to develop and spread apart beyond the initial rifting stage (Figure 13–14b). The result is a ***failed rift***, or ***failed arm***, which receives a great thickness of sediment. The Paleozoic Oklahoma aulocogen was interpreted by Burke and Dewey as a failed rift, and Paul Hoffman (1973) similarly interpreted the Athapuscow aulocogen in the Proterozoic Wopmay orogen in the Canadian shield.

Rift basins, such as the Triassic-Jurassic basins of eastern North America, may form as marginal rifts before the opening of an ocean. Formation of the basins just mentioned also involved a major strike-slip component (Swanson, 1982). In contrast, the Gregory Rift of the East African system has been shown to be an asymmetric rift (Bosworth and others, 1986); there, the master fault occupies the east side of the rift and is down toward the west for more than 100 km along strike (Figure 13–14c) but then changes to the west side and becomes down to the east for more than 100 km north and south.

Regional Crustal Extension

Large-scale extension of continental crust occurred in the Basin and Range Province during middle to late Tertiary time (Eaton, 1979; Coney, 1980; Wernicke and Burchfiel, 1982; Wernicke, 1985). The crust there has been vertically thinned by at least 25 percent during lateral extension—but locally more than 500 percent in other places—in the Snake Range, in eastern Nevada (Miller and others, 1983; Gans, 1987). Phillip Gans (1987), using crustal balance calculations, estimated the eastern Basin and Range across central Nevada has been extended 77 to 140 percent.

Gently dipping faults in the Basin and Range, now exposed at the surface, were at first interpreted as thrusts. Thrusts do exist here, but many of these faults have been reinterpreted as segments of crustal-scale listric normal faults. They dip gently today because they either formed as low-angle segments that have been erosionally exhumed or are tectonically rotated steeply dipping faults. Steeply dipping upper crustal segments may flatten into the ductile-brittle transition during crustal spreading and thinning. Uplift and erosion expose old low-angle detachments at the surface. Because of a constant rate of heat flow, new levels of subhorizontal detachment form deep in the crust as the depth of crustal extension and detachment within the ductile-brittle transition remains constant. Uplift and erosion cause existing detachments to become inactive, and as extension continues, new detachments form beneath them (Eaton, 1979; Miller and others, 1983). In the Basin and Range, many older detachment zones are now exposed at the surface and have been cut by younger, active normal faults—faults that presumably flatten at depth into the present ductile-brittle transition. Alternatively, many of these low-angle faults may be reactivated thrusts, but low-angle normal faults exist that exhibit a brittle-to-ductile transition and that formed at relatively shallow depths.

Several models were proposed during the 1970s and 1980s to explain regional crustal extension (Figure 13–15). All are based on the fact that many low-angle fault surfaces in the Basin and Range are erosionally exhumed normal faults. Models have been developed involving one of these mechanisms: (1) symmetrical extension and pure shear, suggested by John Proffett (1977), Gordon Eaton (1979), and Warren Hamilton (1982) which is basically an application of the McKenzie model; (2) asymmetric extension involving

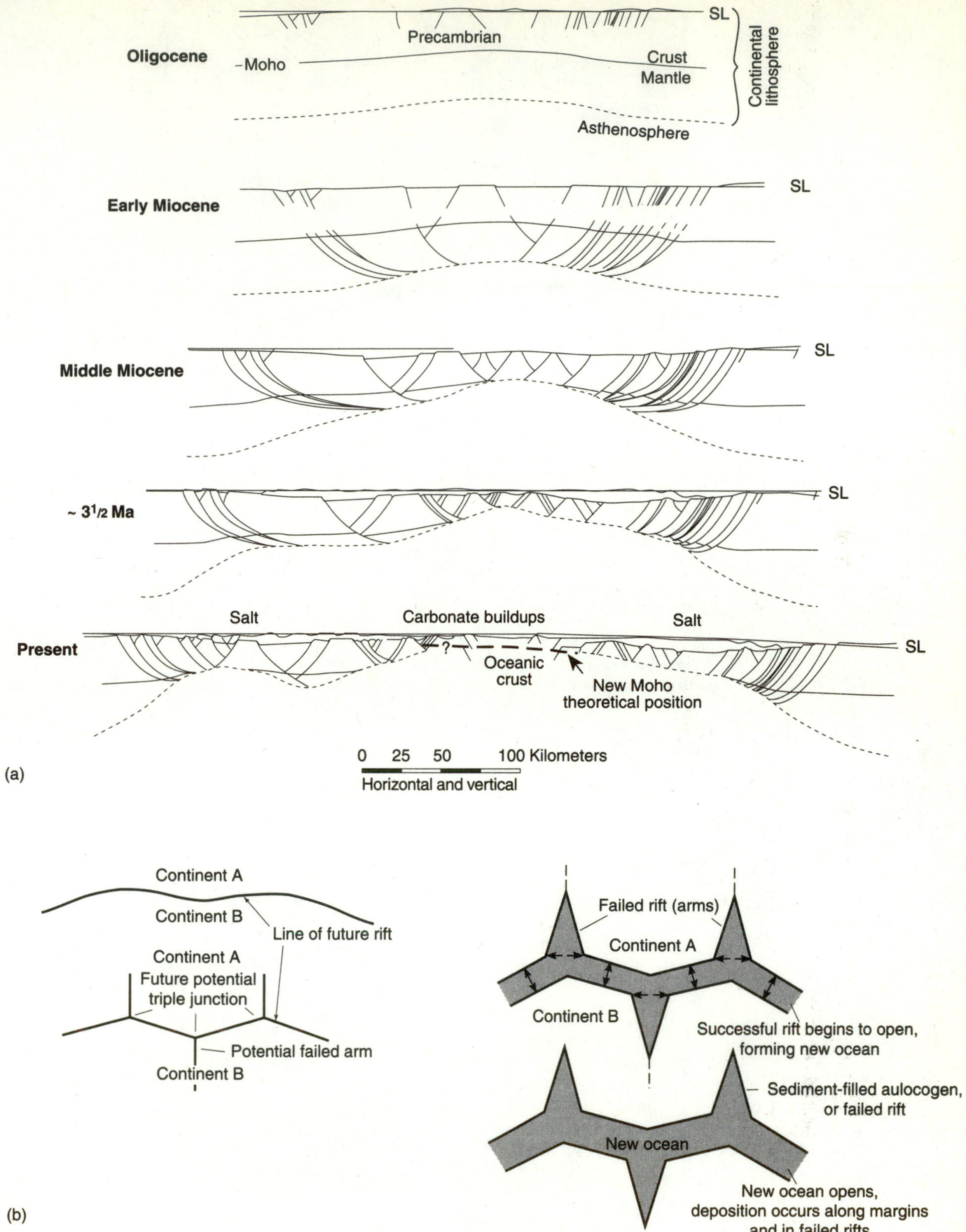

FIGURE 13–14
(a) Cross section through the Red Sea Rift. (From J. D. Lowell and G. J. Genik, 1972, AAPG *Bulletin,* v. 56. Reprinted by permission of American Association of Petroleum Geologists.) (b) Failed rifts and aulocogens. (c) (following page) Interpretive map of fault geometry in the Gregory Rift of the East African rift system made from LANDSAT images. Note asymmetry along the length of the rift, indicating that even supposedly classic symmetrical rifts may prove to have been formed by asymmetric extension. (From W. Bosworth, J. Lambiase, and R. Keisler, *EOS,* v. 67, © July 1986, American Geophysical Union.)

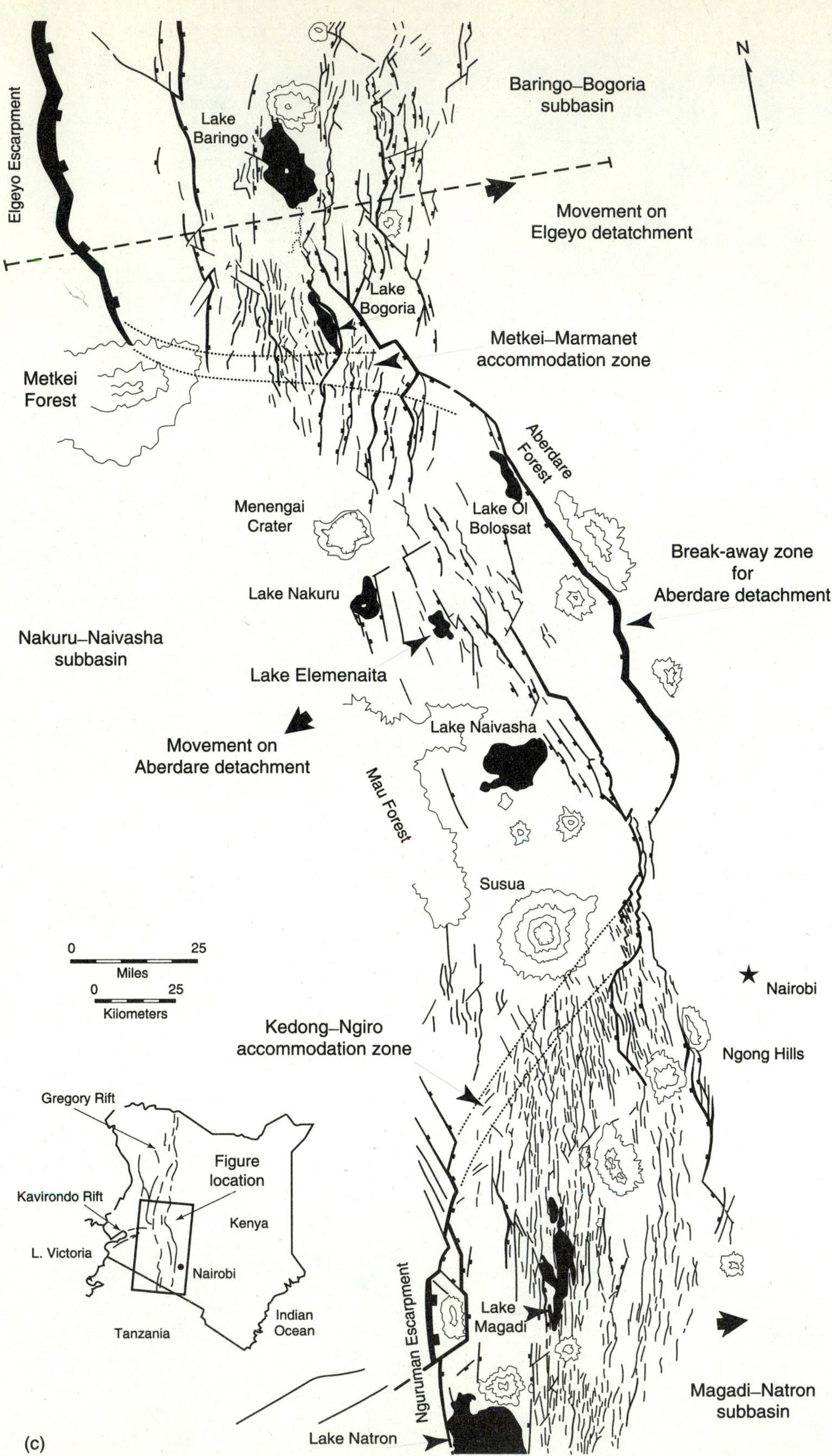

FIGURE 13–14 (continued)

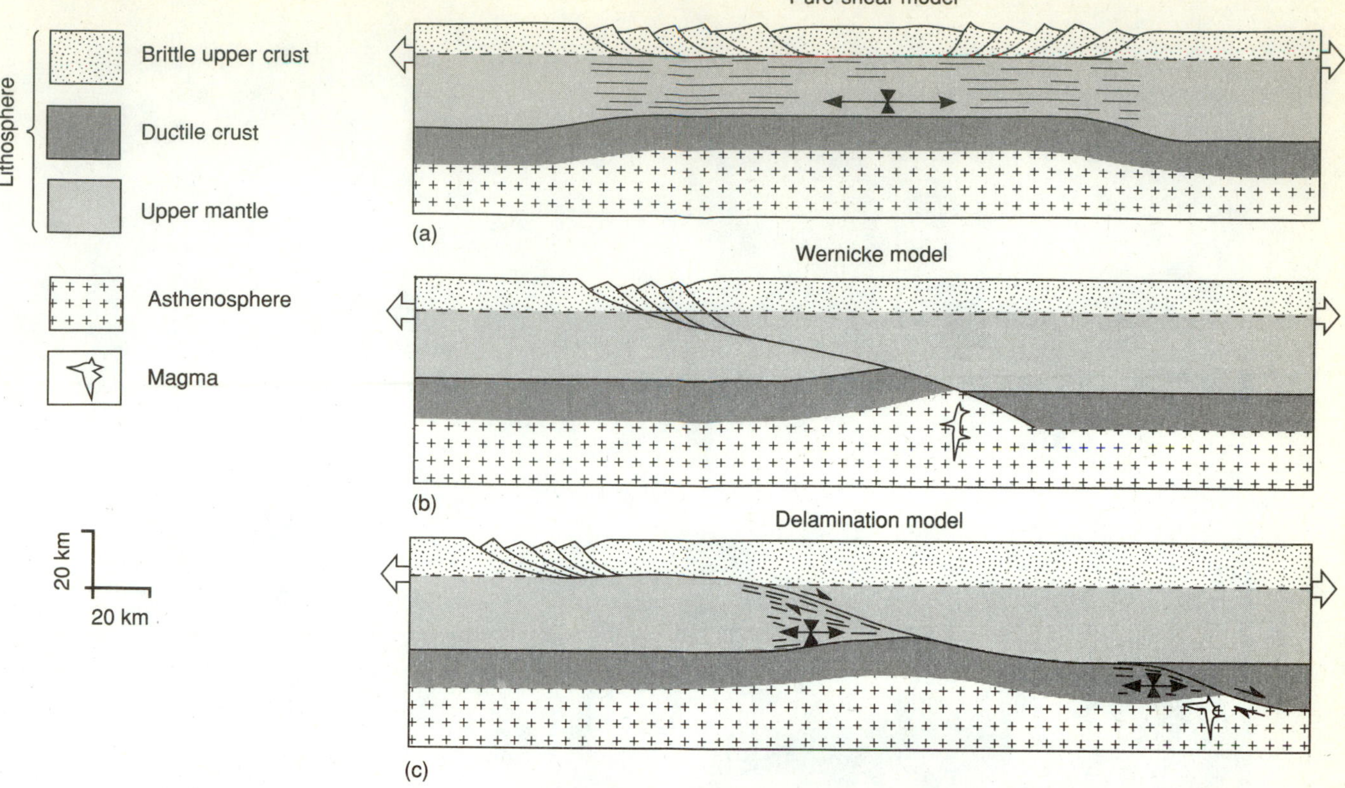

FIGURE 13–15
Three models of extension of continental crust. (a) McKenzie "pure-shear" model: symmetrical rifting and pure shear. b) Wernicke model: asymmetric rifting with simple shear. (c) Delamination model: asymmetric rifting with simple shear and delamination. (From G. S. Lister, M. A. Etheridge, and P. A. Symonds, *Geology,* v. 14, 1986.)

simple shear and a master detachment cutting through the entire lithosphere (Wernicke, 1985); and (3) asymmetric extension and simple shear involving delamination of the lithosphere (Lister, Etheridge, and Symonds, 1986). The symmetrical-extension model explains thinning of the crust but not the asymmetric geometry of several extended regions, such as the Basin and Range; this model was undoubtedly based on consideration of symmetrical rifts such as the Gulf of California and the mid-ocean ridges. Most geologists who work directly with extended regions, however, have accepted some form of Wernicke's asymmetric-extension model, analogous to the structure shown in Figures 13–4 and 13–15b. They regard symmetrical extension as having only historical interest or being applicable in very specific circumstances.

Extensional detachments are intriguing structures. Most that can be closely examined in the Basin and Range (Figures 13–16 and 13–17) consist of a fault surface underlain by a thin (several centimeters thick) layer of microbreccia (extremely fine-grained breccia). These breccias are developed across thicker (up to several hundred meters) chloritized breccias that formed during the shearing and retrogression of footwall mylonitic gneisses. The mylonitic gneisses were formed at depth along the detachment faults by extensional ductile shear and then were carried upward in the footwalls of the faults. Upper-plate rocks typically exhibit only brittle deformation, usually by brittle normal faults (Coney, 1980; G. H. Davis, 1980). An excellent smaller-scale example of this is at the

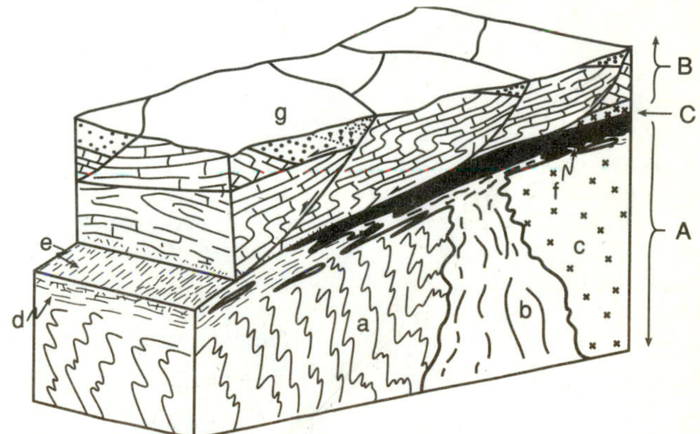

FIGURE 13–16
Typical Cordilleran metamorphic core complex. A—basement; B—cover; C—detachment; a—older metasedimentary rocks; b—older pluton; c—younger pluton (early to middle Tertiary); d—mylonitic foliation; e—mylonitic lineation; f—marble (black); g—lower to middle Tertiary sedimentary and volcanic rocks. (From P. J. Coney, 1980, Geological Society of America *Memoir* 153.)

(a)

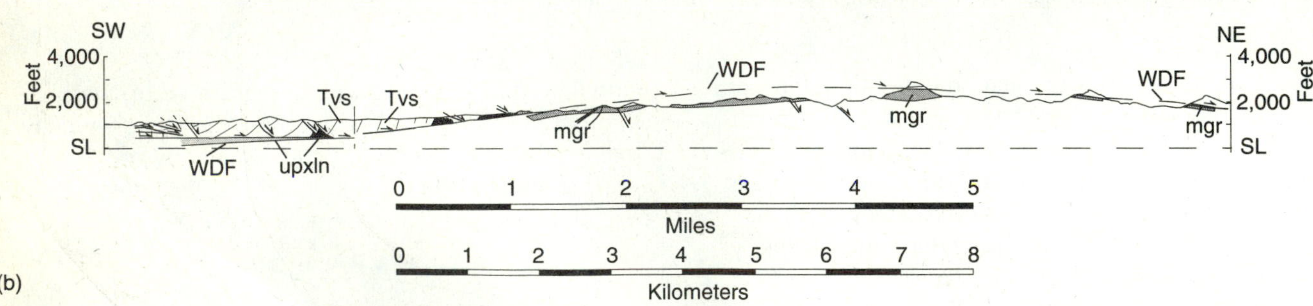

(b)

FIGURE 13–17
(a) View of the Whipple Mountains, California, looking southwest along the axis of the Whipple Peak antiform. The detachment fault is the light-dark contact that wraps around most of the range. Dark rocks in the upper plate consist of Miocene volcanic and sedimentary rocks, which were deposited on upper-plate crystalline rocks that form the light-colored area in the lowest portion of the photo. Light-colored rocks forming most of the range are lower-plate mylonitic gneisses that sit structurally beneath the nonmylonitic gneisses that compose the low-relief lower plate in the western part of the range (upper-right part of photo). (E. J. Frost, San Diego State University.) (b) Cross section across the Whipple Mountains, illustrating middle Miocene geologic relations before domal uplift and warping of the Whipple detachment fault (WDF). The cross section illustrates evidence for multiple phases of rotational normal fault displacement along the detachment surface. Tvs—Tertiary sedimentary and volcanic rocks; mgr—mylonitic granitic rocks; upxln—upper-plate crystalline rocks. (Unpublished section courtesy of G. A. Davis, University of Southern California.)

Turtlebacks along the eastern flank of Death Valley in California (Figure 13–18), part of the strike-slip pull-apart system first described in this area by Burchfiel and Stewart (1966) (Chapter 12). Here, semiconsolidated continental Plio-Pleistocene sediments are brittlely deformed above a detachment zone that underwent a transition from a brittle upper part to near-ductile deformation in the lower portion (formation of microbreccia and protomylonite) at and below the detachment at the contact with the 1.8-Ga-old Proterozoic basement. This transition is only about 2 to 3 m thick. The detachment had to be located within 5 km

FIGURE 13–18
The Turtlebacks along the eastern side of Death Valley National Monument, California. (a) Curved surfaces, such as the one here that gave the Turtlebacks their name, result from detachment faulting where Plio-Pleistocene sediments (light colored) have moved relatively down toward the valley in the foreground, leaving the erosionally exposed, 1.8-Ga Proterozoic basement gneisses (dark colored) visible in the footwall on the dome-shaped turtleback surface. (b) Closeup of a detachment zone. Ductile to semiductile deformation at the base of the fault zone in the Proterozoic rocks gives way upward to brittle deformation in the poorly consolidated overlying Plio-Pleistocene sediments. (Locality courtesy of John C. Crowell, University of California, Santa Barbara; RDH photos.)

(a)

(b)

of the surface when it formed, although the actual ductile deformation may have been brought up from greater depth along the footwall. The thermal gradient could also have been steep at the time the detachment was active, producing high heat flow. Evidence of abundant fluid is present at the exposed contact in the form of chlorite formed as an alteration product of biotite in the ductilely deformed basement gneiss. The Whipple Mountains detachment in southeastern California (Figure 13–17) is another excellent example: in many places, moderately to steeply dipping upper-plate structures have clearly been truncated by the underlying detachment (G. H. Davis and others, 1980; G. A. Davis and Lister, 1988).

Formation of crustal listric normal faults and resulting large-scale extension has been suggested to explain the opening of the Bay of Biscay between France and the Iberian peninsula (de Charpal and others, 1978). Seismic reflection profiling in the Bay of Biscay reveals large normal faults that dip steeply in the upper crust and appear to flatten into a major detachment zone in the lower crust. The detachment horizon is interpreted as the ductile-brittle transition. Such faults are thought to be extensional normal faults such as those in the Basin and Range Province.

J.-P. Brun and Pierre Choukroune (1983) have devised several models to explain the mechanics of extended crust (Figure 13–19). Their models take into account several boundary conditions and behavior modes, including (1) vertical zoning of the crust into ductile (continuous) or brittle (discontinuous) domains of deformation, (2) the nature of discontinuities in the brittle realm, (3) presence—or absence—of preexisting gently dipping mechanical discontinuities, and (4) existence of an extensional (stretching) gradient across the zone. To sum up Brun and Choukroune's models:

Model I—Crustal thinning occurs only by ductile processes, like the extensional necking of a metal rod.

Model II—A vertical downward transition from brittle to ductile deformation develops whereby near-surface, normal-fault displacement is related directly to thinning of the crust by ductile deformation at depth.

Model III—A preexisting mechanical discontinuity is assumed within the crust as a zone of movement. This discontinuity may be the ductile-brittle transition.

Model IV—All deformation is brittle.

Model V—Crustal extension may be accomplished by intrusion of magma, separating and isolating large blocks of crust.

Of the five models, II and III account for the most properties and known boundary conditions of zones of crustal extension. Geologic and geophysical evidence also favor models II and III.

Collapse Structures and Related Features

Some normal faults occur as collapse structures related to the subsidence of magma chambers, and they may deform volcanoes and the tops of plutons (Figure 13–20). They form calderas and other structures, and function in the overall process of magma emplacement (Chapters 2 and 19); they also provide conduits for magma to reach the surface.

The impact on the Earth of large bodies *(bolides)* from outer space produces rebound structures in which normal faults may develop concentrically or radially (see Figure 2–32). As a result, a series of concentrically arranged horsts and grabens are exposed in a deeply eroded *impact structure*. Such structures, also called *astroblemes*, or *impact structures*, are known in many parts of the world and are attributed to large meteors,

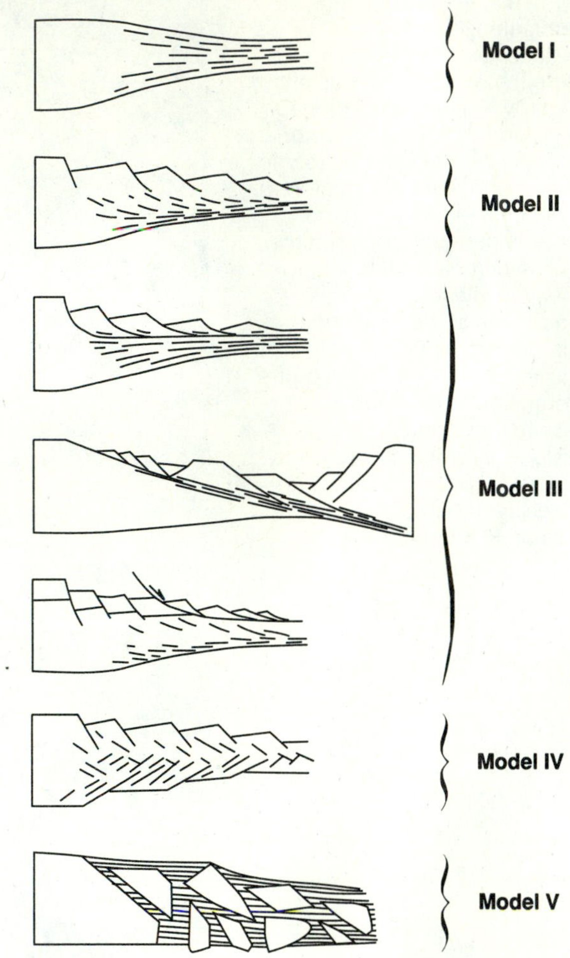

FIGURE 13–19
Models for stretched crust. (After J.-P. Brun and Pierre Choukroune, *Tectonics*, v. 2, p. 345–356, © 1983, American Geophysical Union.)

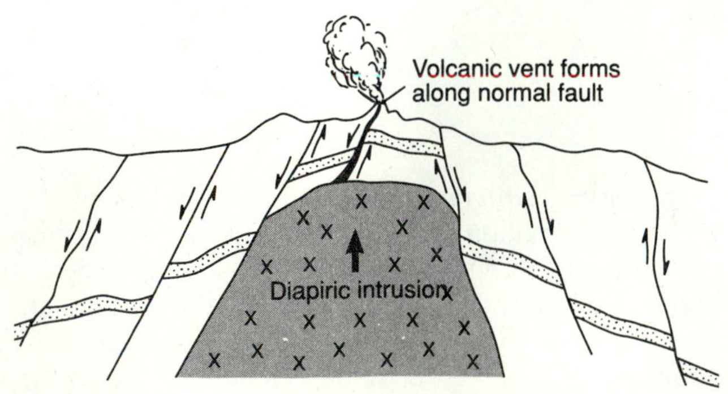

FIGURE 13–20
Normal faults developed in association with volcanoes and plutons.

ESSAY

Inverted Faults and Tectonic Inheritance

An *inverted fault* formed with one sense of motion and was later affected by a differently oriented deformation that reversed the original sense of slip. It may be that any fault not folded can be affected by a later stress field that reverses the original motion sense. Fewer faults undergo any such history, because reversal requires that the fault surface be suitably oriented with respect to the new stress field. Later crustal deformation, however, appears to occur perpendicular to continental margins and older zones of crustal weakness, so that perhaps 40 percent of existing faults may have recurrent motion in the same (original) sense. Consequently, the tendency for reactivation of old faults must be related to the alignment of newly formed plate boundaries in a later tectonic cycle.

Normal faults formed by rifting of continental margins may be inverted and reactivated as thrusts (McClay and Buchanan, 1992). Robert W. H. Butler (1989) concluded that preexisting basin structure influenced the geometry of later thrusts in the Alps, and Gautam Mitra and M. T. Lukert (1982) suggested that the border faults of the very narrow Meechum River basin in the central Appalachian Blue Ridge were later inverted as thrusts.

Some geologists have speculated that the crust may undergo later extension and normal faulting in a direction exactly opposite the direction of compression that earlier produced thrusting. One example of such a reversal is in the Viking graben, in the North Sea (Figure 13E–1). The Viking graben is a complex of extensional structures formed during the late Mesozoic and early Cenozoic (Gibbs, 1983). Where high-quality crustal seismic reflection data have been acquired, the later normal faults appear to have reactivated older thrusts originally formed during the early Paleozoic Caledonian orogeny. Some thrusts had been reactivated earlier, during the Devonian, and formed the extensional Old Red Sandstone basins (McGeary, 1987).

Most of the exposed Triassic-Jurassic basins of the East Coast of the United States trend parallel to the strikes of major contractional Appalachian faults. Some basins are localized along the outcrop traces of older faults, indicating a further genetic connection. The Ramapo fault that crosses the New Jersey–New York border is the northwestern border fault for the Triassic-Jurassic Newark basin (Figure 13E–2). After detailed structural studies along the Ramapo fault, Nicholas Ratcliffe (1971) showed that it formed as a contractional fault during the Paleozoic. He showed that it was reactivated during the Mesozoic as an extensional feature and that it also underwent strike slip during the same period. Historical earthquake epicenters and present-day seismic activity are also located along the fault, strongly suggesting that the Ramapo may once again be undergoing reactivation as a thrust. If so, it may be one of the best-documented examples of both inversion and tectonic inheritance, with a history spanning more than 350 m.y. The Ramapo fault has been interpreted to have weakened the crust to provide a conduit system for mafic intrusions (Figure 3E–2).

The phenomenon of fault inversion leads to both interesting and useful consequences. Inverted faults are important for accumulation of hydrocarbons in the North Sea (Hardman and Booth, 1991) and elsewhere. Formation contacts that were displaced during growth normal faulting must pass through *null points* of zero displacement upon being inverted as thrusts. If the youngest contact was not displaced before inversion, it is a null point before inversion begins but is moved from the null condition as initial movement occurs (Figure 13E–3a). The next lower contact moves toward and through the null condition as displacement continues, then the next lower contact moves as the null points migrate down dip along the fault (Williams and others, 1989). Except for the growth aspects, the opposite occurs if a thrust is inverted as a normal fault (Figure 13E–3b). Shankar Mitra (1993), using both experimental models and natural examples, has suggested that

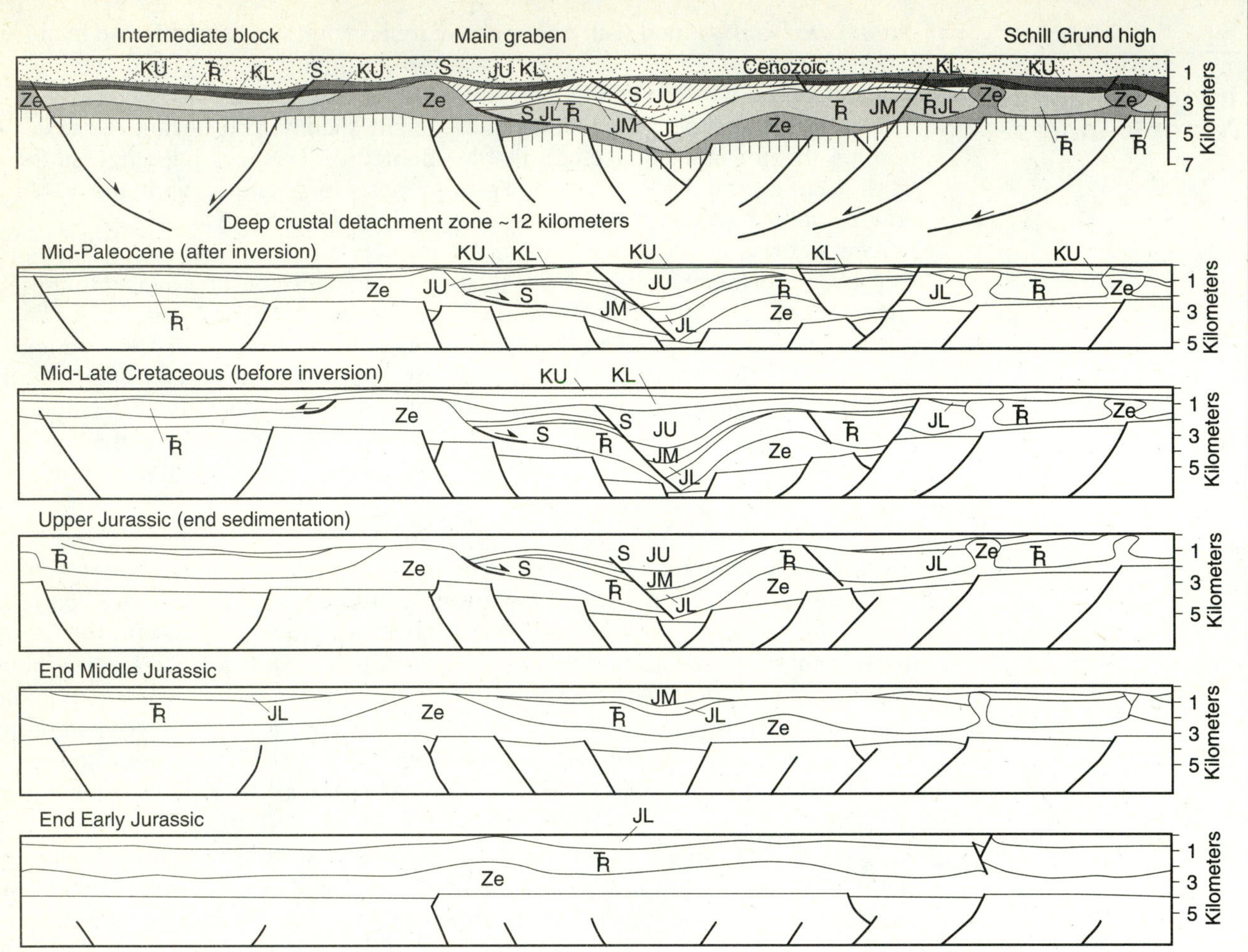

FIGURE 13E–1
Balanced regional cross section through part of the North Sea with successive back-stripping of sediments to the end of Early Jurassic (JL) time, showing older faults that remain beneath the Zechstein (Ze) evaporite sequence. These faults were reactivated contractionally near the end of the Mesozoic and beginning of the Tertiary, producing uplift and erosion of some of the Cretaceous sediments, before sedimentation resumed during the Tertiary. The Viking graben thus formed during Jurassic extension and then was subsequently inverted. TR—Triassic; K—Cretaceous; L—lower; M—middle; U—upper; S—fault active during sedimentation. (Reprinted from *Journal of Structural Geology,* v. 5, A. D. Gibbs, p. 153–160, © 1983, with kind permission from Elsevier Science, Ltd., Kidlington, United Kingdom.)

contractional reactivation of earlier extensional structures occurs by fault-propagation folding, by inversion of planar faults, and by fault-bend folding by inversion of listric faults.

References Cited

Butler, R. W. H., 1989, The influence of preexisting basin structure on thrust system evolution in the western Alps, *in* Cooper, M. A., and Williams, G. D., eds., Inversion tectonics: Oxford, England, Blackwell Scientific Publications, Geological Society of London Special Publication 44, p. 105–122.

Davies, B., 1984, Strain analysis of wrench faults and collision tectonics of the Arabian shield: Journal of Geology, v. 82, p. 37–53.

Gibbs, A. D., 1983, Balanced cross-sections from seismic sections in areas of extensional tectonics: Journal of Structural Geology, v. 5, p. 153–160.

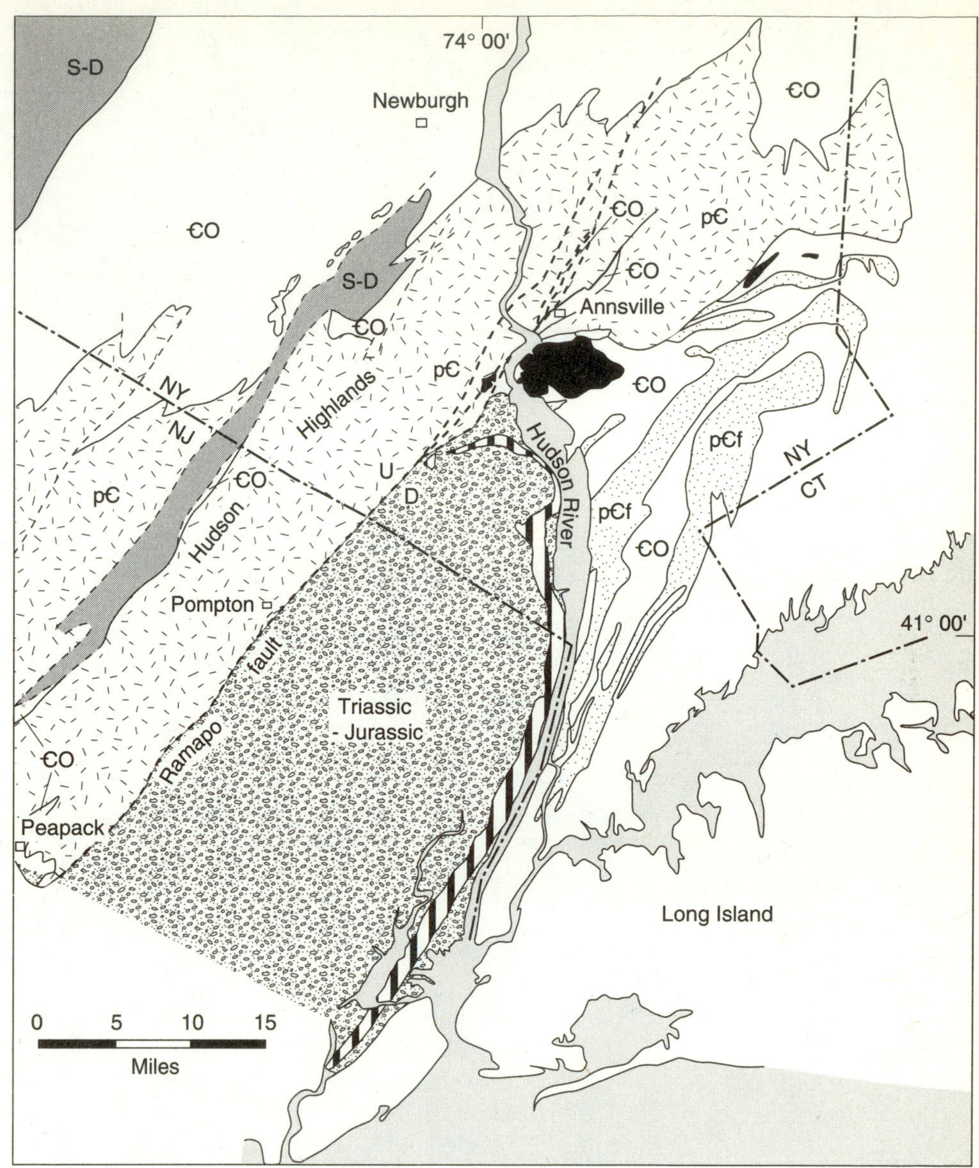

FIGURE 13E–2
Geologic map of part of northern New Jersey and southern New York showing the northern end of the Triassic-Jurassic Newark basin (striped) and the Ramapo fault and its extension northeast into the Hudson Highlands. pꞒ and pꞒf—Precambrian rocks; ꞒO—Cambrian and Ordovician rocks; S-D—Silurian-Devonian. (From N. M. Ratcliffe, Geological Society of America *Bulletin,* v. 82, 1971.)

Hardman, R. P. F., and Booth, J. E., 1991, The significance of normal faults in the exploration and production of North Sea hydrocarbons, *in,* Roberts, A. M., Yielding, G., and Freeman, B., eds., The geometry of normal faults: London, Geological Society of London, Special Publication 56, p. 1–13.

McClay, K. R., and Buchanan, P. G., 1992, Thrust faults in inverted extensional basins, *in* McClay, K. R., ed., Thrust tectonics: London, Chapman and Hall, p. 93–104.

McGeary, S., 1987, Nontypical BIRPS on the margin of the North Sea: The SHET survey: Geophysical Journal of the Royal Astronomical Society, v. 89, p. 231–238.

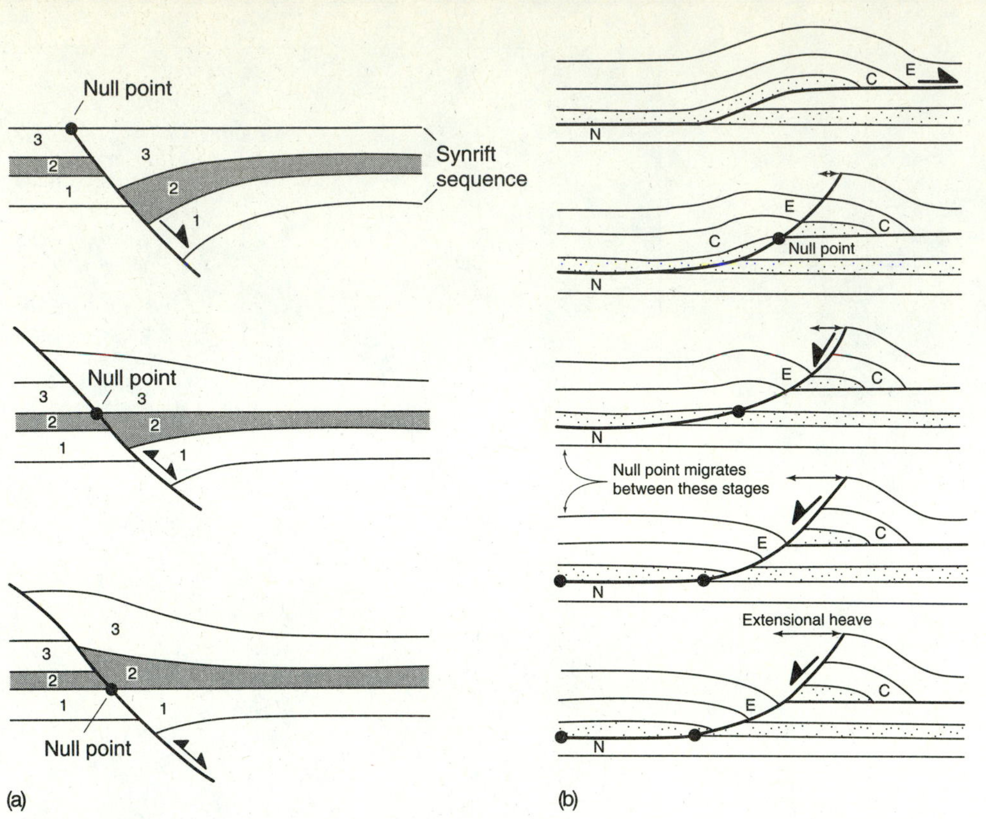

FIGURE 13E–3
Consequences of inversion of (a) a normal fault by thrusting, and (b) a thrust fault by normal faulting. (From Williams, G. D., Powell, C. M., and Cooper, M. A., Geometry and kinematics of inversion tectonics, *in* Cooper, M. A., and Williams, G. D., eds., Inversion tectonics: Oxford, England, Blackwell Scientific Publications, Geological Society of London Special Publication 44, 1989. Used by permission.)

Mitra, G., and Lukert, M. T., 1982, Geology of the Catoctin-Blue Ridge anticlinorium in northern Virginia, *in* Lyttle, P. T., ed., Central Appalachian geology: Northeast-Southeast Geological Society of America '82 Field Trip Guidebook: Falls Church, Virginia, American Geological Institute, p. 83–108.

Mitra, S., 1993, Geometry and kinematic evolution of inversion structures: American Association of Petroleum Geologists Bulletin, v. 77, p. 1159–1191.

Ratcliffe, N. M., 1971, The Ramapo fault system in New York and adjacent northern New Jersey: A case of tectonic heredity: Geological Society of America Bulletin, v. 82, p. 125–142.

Williams, G. D., Powell, C. M., and Cooper, M. A., 1989, Geometry and kinematics of inversion tectonics, *in* Cooper, M. A., and Williams, G. D., eds., Inversion tectonics: Oxford, England, Blackwell Scientific Publications, Geological Society of London Special Publication 44, p. 3–16.

or even asteroids, colliding with the Earth. Some, such as Meteor Crater in Arizona, were formed as recently as the Quaternary. Others (for example, the Wells Creek and Flynn Creek structures in Tennessee) are Paleozoic or Mesozoic structures, and the Vredefort structure in South Africa was probably formed by an impact during the Precambrian.

Relationship to Strike-Slip Faults

The geometry and sense of motion of normal faults was at one time thought to be unique and invariably related to gravity or to crustal extension, but detailed study of fault systems has revealed that normal faults are related both to folds and compressional faults. Strike-slip faults

(Chapter 12) have been shown to be intimately related to formation of isolated normal faults and of horsts and grabens (rhomb grabens and rhombochasms). Resolution of strike-slip faulting into an extensional sense results from the stepping over of strike-slip motion from one fault to another across a zone of "pull-apart" extension.

We have seen that the formation of some normal faults may be linked with formation of other fault types at all scales and that their geometry is similar to that of thrusts, but with an opposite movement sense. With that, we close our discussion of fractures and faults and turn to a discussion of folds and folding.

Questions

1. Why are normal faults also called gravity faults?
2. Why do normal faults splay and form antithetic and synthetic faults?
3. How do reverse drag folds form? Why are they commonly associated with growth faults?
4. How do normal faults terminate?
5. If a listric normal fault zone flattens downward into a detachment in shale, and a listric normal fault forms in the upper crust and flattens into a detachment in the ductile-brittle transition zone, how would you expect them to differ in appearance and in fault rocks (if any)?
6. How would you attempt to document the timing of motion on a growth fault?
7. Photocopy the geologic cross section of the Corsair fault complex in Figure 13–10, then, using a piece of tracing paper, trace the main fault and restore all faults to an undeformed state. What can you conclude about (1) the growth rate through time and (2) the mechanical process of movement along the fault? Can movement on the fault be accounted for by rotation of a block, or must a component of simple shear (parallel to the fault surface or parallel to bedding) be involved, or both? See the article by Hongbin Xiou and John Suppe in the *American Association of Petroleum Geologists Bulletin* (1992, v. 76, p. 509–529), and the abstract by Richard H. Groshong (1993, AAPG New Orleans Annual Meeting Abstracts, p. 111).
8. What kinds of evidence would help you to decide whether or not a particular fault has been reactivated with a sense of motion different from its original motion sense?
9. What kinds of data would you need to determine if rifting occurred by a symmetrical or asymmetric mechanism in an area such as the East African Rift or the continental margins around the present Atlantic Ocean?

Further Reading

Brun, J.-P., and Choukroune, P., 1983, Normal faulting, block tilting and décollement in a stretched crust: Tectonics, v. 2, p. 345–356.
Outlines a series of models for crustal thinning—ductile, ductile-brittle transition, and mechanical boundaries, such as shear zones, brittle faulting, and magma intrusion. All relate to the normal faulting process.

Cloos, E., 1968, Experimental analysis of Gulf Coast fracture patterns: American Association of Petroleum Geologists Bulletin, v. 52, p. 420–444.
Models the normal fault system of the Gulf Coast by subjecting clay blocks to extensional deformation. The listric geometry was successfully modeled, as was the major regional pattern of down-to-basin faults.

Cooper, M. A., and Williams, G. D., eds., 1989, Inversion tectonics: Oxford, England, Blackwell Scientific Publications, Geological Society of London Special Publication 44, p. 105–122.
Contains a spectrum of papers that discuss various aspects of the theoretical, experimental, and field setting of fault inversion.

Coward, M. P., Dewey, J. F., and Hancock, P. L., eds., 1987, Continental extensional tectonics: Oxford, England, Blackwell Scientific Publications, Geological Society of London Special Publication 28, 637 p.

An extensive array of papers dealing with extensional processes, geometry, and mechanics that presents numerous examples from a variety of settings.

Gans, P. B., 1987, An open-system, two-layer crustal stretching model for the eastern Great Basin: Tectonics, v. 6, p. 1–12.

Employs data from structural reconstructions of exposed ranges along with considerations of kinds and volumes of magma that have been erupted in the northern Great Basin to estimate the amount of crustal thinning and extension here.

Jackson, M. P. A., and Vendeville, B. C., 1994, Regional extension as a geologic trigger for diapirism: Geological Society of America Bulletin, v. 106, p. 57–93.

Most of the traditional ideas about the nature and origin of diapirs focus on the diapirs as nontectonic features and not on tectonic processes that may form them (see Chapter 2). This paper provides documentation that regional extension serves not only to trigger diapirs but provides the means to localize and track their evolution.

Keen, C. E., Stockmal, G. S., Welsink, H., Quinlan, G., and Mudford, B., 1987, Deep crustal structure and evolution of the rifted margin northeast of Newfoundland: Results from LITHOPROBE East: Canadian Journal of Earth Sciences, v. 24, p. 1537–1549.

Discusses the crustal structure of part of the present-day Atlantic extensional margin and attempts to evaluate the applicability of both symmetrical and asymmetric spreading models.

Roberts, A. M., Yielding, G., and Freeman, B., eds., 1991, The geometry of normal faults: London, Geological Society of London, Special Publication 56, p. 1–13.

A compendium of papers summarizing the mechanics and applications to hydrocarbon accumulation related to normal faulting with examples from different parts of the world.

Wernicke, B., 1985, Uniform-sense normal simple shear of the continental lithosphere: Canadian Journal of Earth Sciences, v. 22, p. 108–125.

Examines models of pure and simple shear for crustal extension in the Basin and Range and elsewhere. Wernicke concluded that simple shear oriented in a single direction is the most common mechanism for crustal extension.

PART FOUR

Folds and Folding

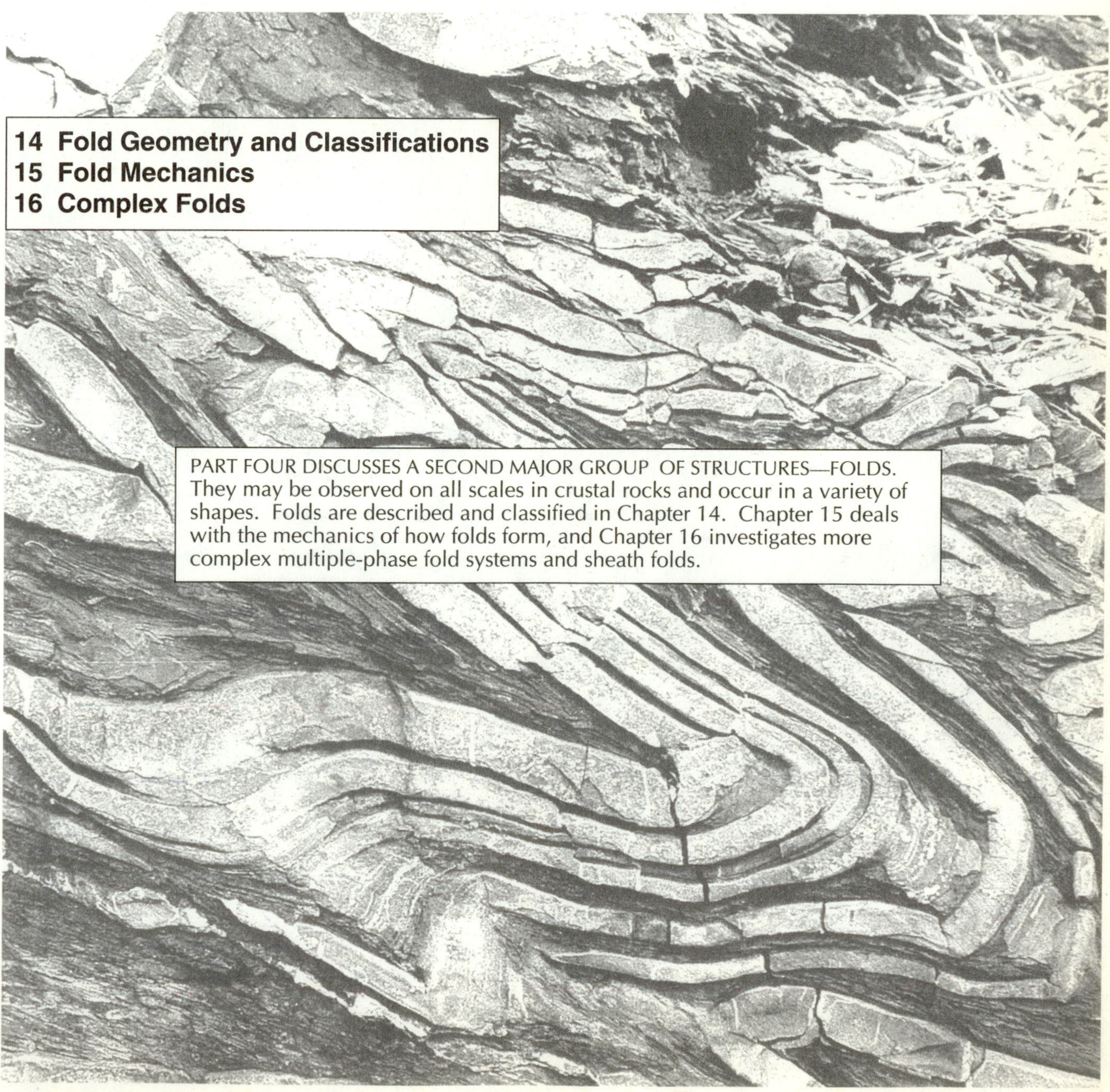

PART FOUR DISCUSSES A SECOND MAJOR GROUP OF STRUCTURES—FOLDS. They may be observed on all scales in crustal rocks and occur in a variety of shapes. Folds are described and classified in Chapter 14. Chapter 15 deals with the mechanics of how folds form, and Chapter 16 investigates more complex multiple-phase fold systems and sheath folds.

14

Fold Geometry and Classifications

Folds in layered rock commonly are nearly perfect geometric figures, approaching segments of circles in parallel or concentric folds, sinusoidal waves in some similar folds and sharp angles in some kink folds. Partly because of their geometric beauty, folds have attracted the attention of geologists almost from the time of the birth of geological science.

ARVID M. JOHNSON and STEPHENSON D. ELLEN, 1974, *Tectonophysics*

FOLDS ARE WAVE LIKE STRUCTURES THAT RESULT FROM deformation of bedding, foliation, or other originally planar surfaces in rocks (Figure 14–1). They occur on all scales, ranging from those visible only under a microscope to those extending for kilometers. Folds form in all deformational environments in the Earth's crust, from near-surface brittle conditions to lower-crust ductile conditions, and from conditions of simple shear to pure shear. They range from very broad and gentle structures to tightly compressed and attenuated structures. They occur singly as isolated folds and in extensive *fold trains* of different sizes. Rocks may be affected by a single folding event or by multiple events leading to serially overprinted generations of folds.

FIGURE 14–1
Disharmonic flexural-flow folds in shale and thin limestone beds of the Poultney Formation (Lower Ordovician) near Whitehall, New York. (RDH photo.)

Folds were the first structures recognized as hydrocarbon traps, leading to the *anticlinal theory* of petroleum accumulation and the development of a major new industry near the end of the nineteenth century. Folds also play an important role in concentrating valuable minerals. For example, the widely known ***saddle-reef deposits***, first identified in Australia, contain minable concentrations of sulfide minerals localized in the hinges of folds.

In this chapter, we introduce folds, describe their properties, and discuss some of the classifications devised to help us to understand folds and communicate about them. Chapter 15 will present current views on the mechanics of folding, and Chapter 16 will discuss development of complex fold systems.

It is important at this point to reintroduce the concept of scale: structures small enough to require magnification to be seen are ***microscopic***. Those whose sizes range from hand specimen to outcrop scale are ***mesoscopic***, and those of map scale or larger are ***macroscopic***. Most observations and measurements that provide information about orientations of the various fold elements (limbs, axes, and axial surfaces) and allow inferences about the folding process are made on the mesoscopic scale, yet microscopic- and map-scale studies also yield useful results. Conclusions regarding the orientation and shape of folds apply on all scales.

Recall ***Pumpelly's rule*** (Chapter 1), which states that *small-scale structures generally mimic larger-scale structures formed at the same time*. If Pumpelly's rule holds true, mesoscopic folds should have the same style and orientation as macroscopic folds (Figure 14–2), enabling study of small folds to enhance our understanding of large folds and the structural history of an area.

DESCRIPTIVE ANATOMY OF SIMPLE FOLDS

Anatomy of Folds

Consider the simple upright fold pair in Figure 14–3a. Any fold of this type has a ***crest*** at the highest point (elevation) on a cross section of the fold, and a ***trough*** at the lowest point on the cross section. The straighter or least-curved segments are called ***limbs*** and connect the parts of the fold exhibiting the greatest curvature—the ***hinge zones***. The line within the folded surface across which layering changes orientation from dip in one direction to dip in the opposite direction across a crest or trough is called the ***crest*** or ***trough line***. For ideal (cylindrical) reversal of dip, the hinge line is parallel to the ***fold axis***, or ***axial line***. The fold axis is the line that is moved parallel to itself to generate the folded surface, and so in a strict sense only cylindrical folds have fold axes. The ***hinge line*** is the line joining points of greatest curvature on a folded surface or midline of arc of greatest curvature, hinge zone. If several hinge lines are connected on successive folded surfaces of the same fold, they form the ***axial surface*** (***axial plane***, if the surface is planar). The hinge and crest lines may not coincide in symmetrical folds that have non-vertical axial surfaces. Fold hinges may be horizontal, but more often they are inclined to the horizontal and are said to ***plunge*** (Figure 14–3b); folds with nonhorizontal hinges are called ***plunging folds***. An *inflection line* is a line on a folded surface that separates concave curvature in one direction from concave curvature in the opposite direction. An *inflection point* is a point separating concave curvature in one direction from concave curvature in the opposite direction identified on a cross section of a fold (Figure 14–3a).

The distance from crest to crest of adjacent anticlines—or trough to trough of adjacent synclines—is the *wavelength*. Half the distance from the crest of an anticline to the trough of an adjacent syncline, measured parallel to the axial plane, is the fold *amplitude*.

The direction of leaning (the opposite of dip direction) of the axial surface or the direction of overturning (producing an inverted limb) is called the ***vergence*** of a fold (Figure 14–4a). This property applies only to folds having one limb that dips more steeply and is shorter than the other—an *asymmetric fold*. Vergence is not a property of *symmetrical folds*, but small folds may be asymmetric on the limbs of a larger symmetrical fold. A related property of folds is *slip lines*, which are generally another indication of vergence. Slip lines are produced by relative motion of originally adjacent reference points on either side of a slip surface during folding or other deformation. They

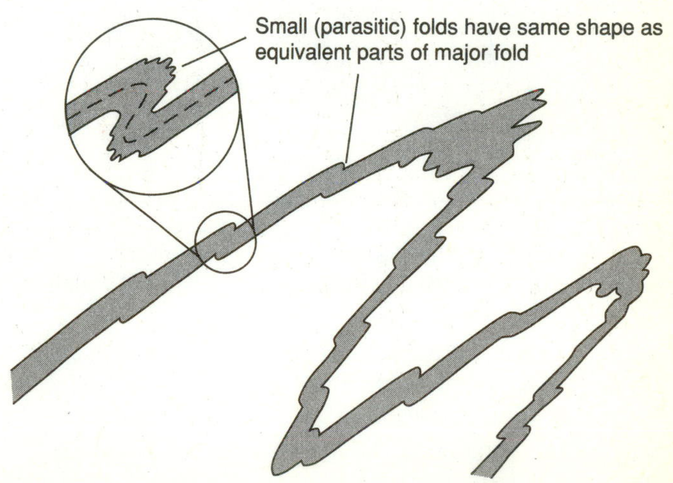

FIGURE 14–2
Pumpelly's rule relating small- and large-scale folds.

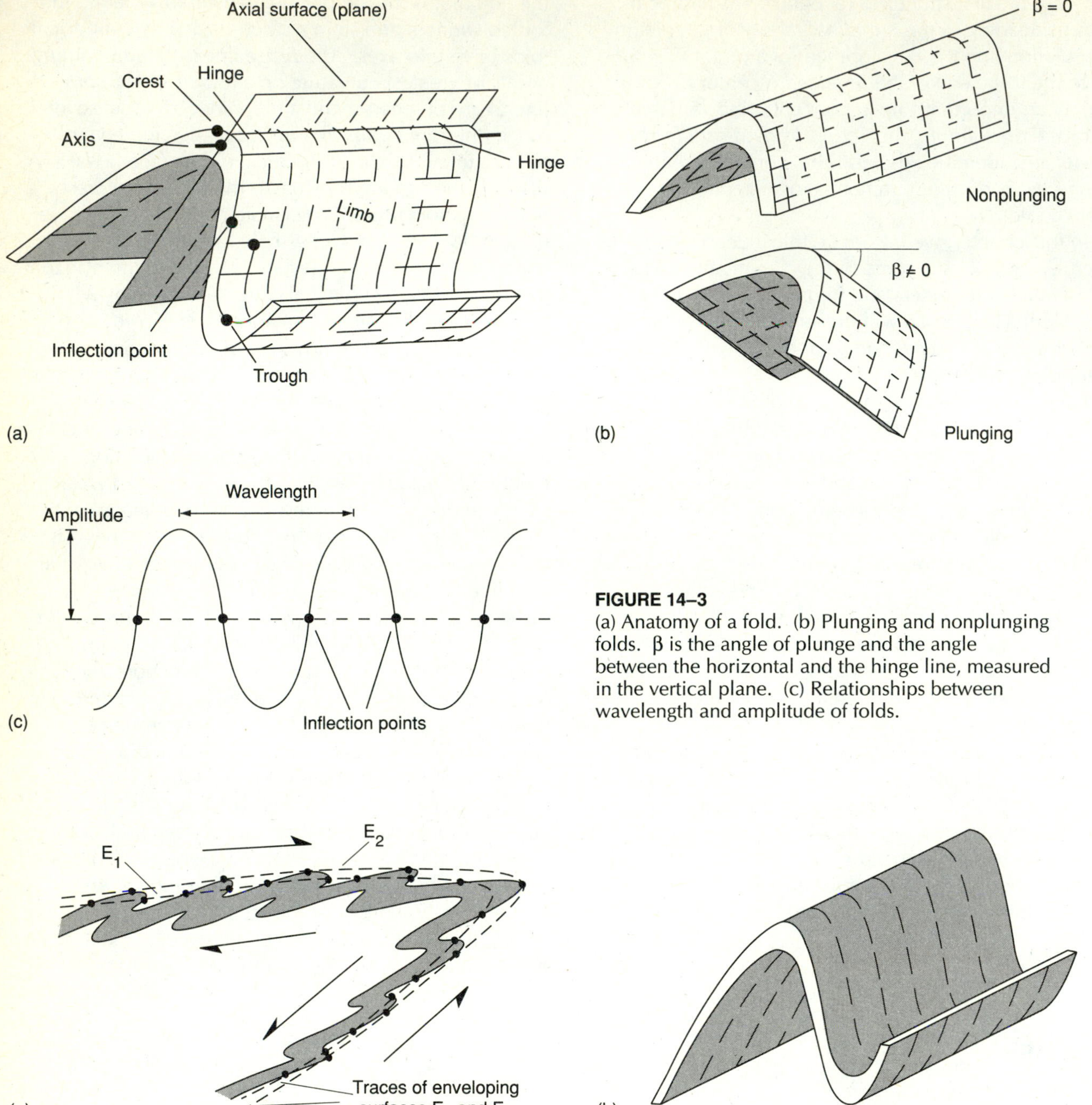

FIGURE 14–3
(a) Anatomy of a fold. (b) Plunging and nonplunging folds. β is the angle of plunge and the angle between the horizontal and the hinge line, measured in the vertical plane. (c) Relationships between wavelength and amplitude of folds.

FIGURE 14–4
(a) Vergence (direction of overturning) of large (first-order) and small (second-order parasitic) folds (sense of overturning, or vergence, indicated by arrows) and relationships to an enveloping surface E_1—by connecting inflection points—or E_2—by connecting a tangent to the crests or troughs of second-order folds. (b) Slip lines illustrated by lines such as fibers or slickensides on a layer surface that indicate the direction of motion of one layer past another.

may occur on real movement surfaces and produce slickensides, fibers, or other visible motion indicators (Figure 14–4b), or they may be imaginary lines deduced from vergence. Determination of vergence may be useful in working out the overall direction of tectonic transport of all structures in an area, in addition to helping to fix an observer's location on a large fold.

The largest folds in a given area are often called *first-order folds;* smaller folds on the flanks are *second-order folds.* Folds of successively higher order are also possible. It may be feasible to relate the geometry of small- to large-scale folds by using an *enveloping surface* (Figure 14–4a), which is a smoothly varying surface tangent to the crests or troughs of a folded

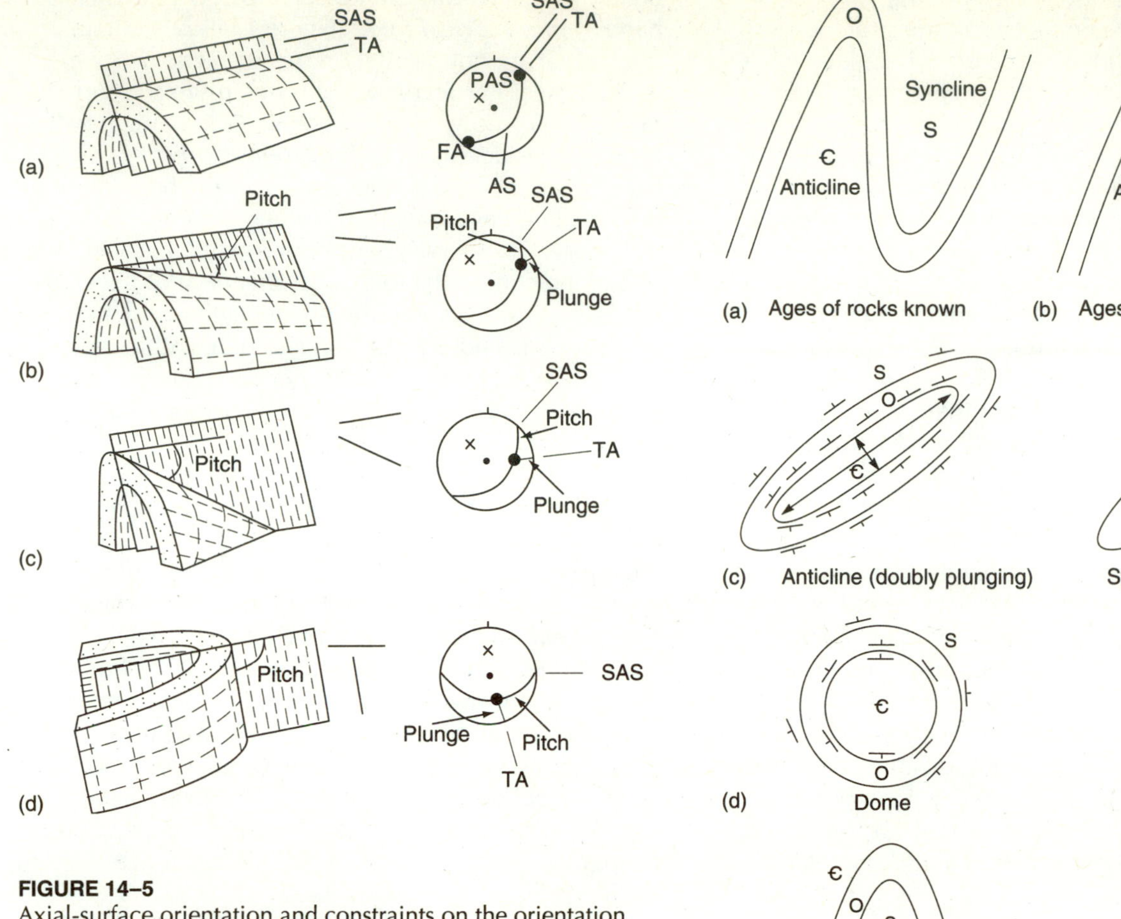

FIGURE 14–5
Axial-surface orientation and constraints on the orientation of the fold axis at zero (a), intermediate (b and c), and steep plunges (d). Note that the plunge of the fold can be read directly from the fabric diagram as the value derived from the line connecting the center of the diagram and the location of the point of emergence of the axis along the great circle trace of the axial surface, with the plunge angle being the amount read along the line from the primitive circle to the fold axis. The pitch is obtained by reading the angle along the great circle from the primitive circle to the fold axis. PAS—pole to axial surface indicated by an **x** in all of the fabric diagrams; AS—axial surface; FA—fold axis; TA—trend of axis; SAS—strike of axial surface. Note that the strike and trend of the axial surface remain constant.

FIGURE 14–6
(a) Anticline and syncline (cross section). (b) Antiform and synform (cross section). (c) Doubly plunging anticline and syncline (map view). (d) Dome and basin (map view). (e) Antiformal syncline and synformal anticline in cross section. Є, O, and S indicate Cambrian, Ordovician, and Silurian rocks.

surface, where crests and troughs are measured perpendicular to the smooth surface joining the inflection points. Enveloping surfaces are useful for studying folds at outcrop scale or in cross section where many small folds occur on limbs of larger folds, but the geometry of the larger folds is not clear.

While deciphering a fold, a geologist must keep in mind the relationship between the axial plane and variations in the plunge of the hinge (Figure 14–5) as the dip of the axial surface also varies. The axis of a fold must lie within the axial surface, but the orientation may vary within the plane. The trend of the fold axis may vary considerably from the strike of the axial surface.

Kinds of Folds

In this section, we will consider simple folds, and you will see that many of the terms are closely interrelated as opposites or as synonyms.

The term ***anticline*** applies to folds that are concave toward older rocks in the structure; an anticline contains older rocks in the center (Figure 14–6a). A fold that is concave downward is called an ***antiform***; the rocks may not be older in the middle, or the age of the rocks may be unknown. In the opposite sense, a ***syncline*** is a structure wherein layering is concave toward the younger rocks, and as a result, younger rocks are found in the central part of the structure (Figure 14–6a). Similarly, a structure where layering is concave up and dips toward the center is a ***synform*** (Figure 14–6b). A ***dome*** is a special kind of antiform wherein layering dips in all directions away from a central point. A ***basin*** is a unique synform in which layering dips inward toward a central point (Figure 14–6d).

Where the sequence can be worked out such that the ages of the rocks are determinable, it may be possible to categorize antiforms and synforms as antiformal synclines or synformal anticlines, depending on the relative ages of rocks found in the centers of the structures (Figure 14–6e). An ***antiformal syncline*** (*downward-facing syncline*) is a structure in which layering dips away from the axis, but the rocks in the center are younger. In contrast, a ***synformal anticline*** (*upward-facing anticline*) is a structure wherein layering dips inward as in a syncline, but the rocks in the center of the structure are older rather than younger. These structures may be produced during multiple episodes of folding (Figure 14–7).

Rocks that dip uniformly in one direction may be described as a ***homocline***. A slight complication of this structure, involving a local steepening of an otherwise uniform regional dip, is called a ***monocline***. A ***structural terrace*** is a local flattening of a uniform regional dip (Figure 14–8).

A fold is ***cylindrical*** (properly *cylindroidal,* or cylinder-like) if the fold can be generated by moving a line—the fold axis—parallel to itself. Generally, cylindrical folds are those in which the hinges are everywhere parallel on successive folds (Figure 14–9a1). *Aberrant folds* deviate slightly from ideal cylindricity (Ramsay and Sturt, 1973a). Most folds probably fall into this class because the deviation from ideal cylindricity is only slight or may be detected only by tracing the fold a great distance along the trend. Aberrant folds are commonly not considered separately but are grouped with cylindrical folds. On the other hand, ***noncylindrical folds*** (Figure 14–9a2) contain hinges that are not parallel on successive folds, or in which hinges in successive folds of the same group may converge toward a point rather than being parallel, and are considered a separate group. *Conical folds* are noncylindrical folds generated by moving one end of a line in a circular arc while the other end remains stationary. These folds have convergent hinges and the axis of the cone is the fold axis. ***Sheath folds*** (Figures 14–9a3 and 14–10) are strongly noncylindrical and closed at one end; the fold hinges curve within the axial surfaces. They commonly occur in shear zones where rocks have been deformed by a strong component of inhomogeneous simple shear.

Upright folds have vertical axial surfaces (Figure 14–11). ***Overturned folds*** have one inverted limb. In ***reclined folds***, axes plunge at nearly the same angle as the dip of the axial surfaces; the axis plunges normal—or at a high angle—to the strike of the axial surface. ***Recumbent folds*** have horizontal axes and axial surfaces. Folds may be classed as ***open***, if the limbs dip very *gently* away from or toward one another (large interlimb angle), or ***tight***, if the fold limbs dip steeply toward or away from one another (small interlimb angle). ***Isoclinal folds*** are tight folds wherein axial surfaces and limbs are parallel (interlimb angle equals zero); they are common in rocks that have been ductilely deformed (Figure 14–12).

To distinguish among many possible combinations of plunging, reclined, inclined, and recumbent folds, Michael Fleuty (1964) has constructed a diagram plotting the plunge of a fold hinge versus the dip of the axial surface (Figure 14–13). M. J. Rickard (1971)

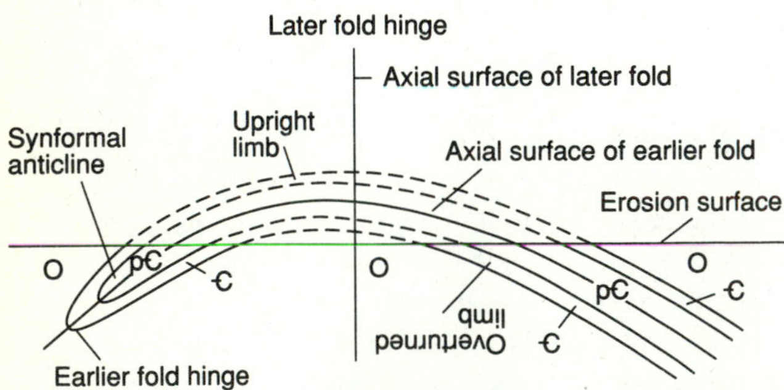

FIGURE 14–7
Synformal anticline as a product of multiple folding. Note axial surfaces of both earlier and later folds. p€, €, and O indicate Precambrian, Cambrian, and Ordovician rocks, respectively.

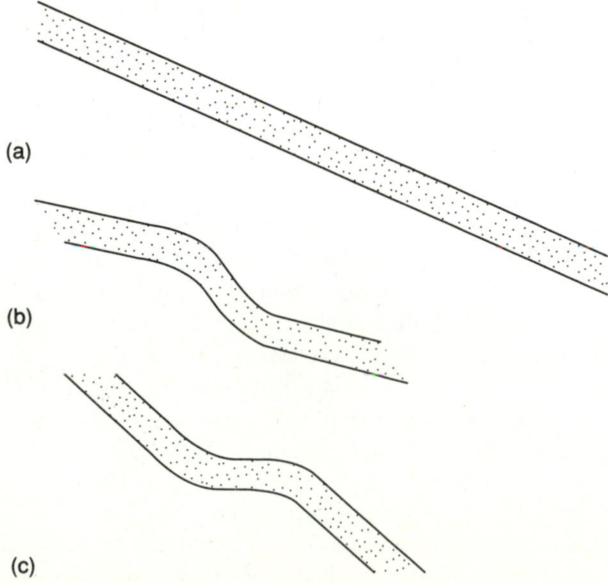

FIGURE 14–8
(a) Homocline, (b) monocline, and (c) structural terrace.

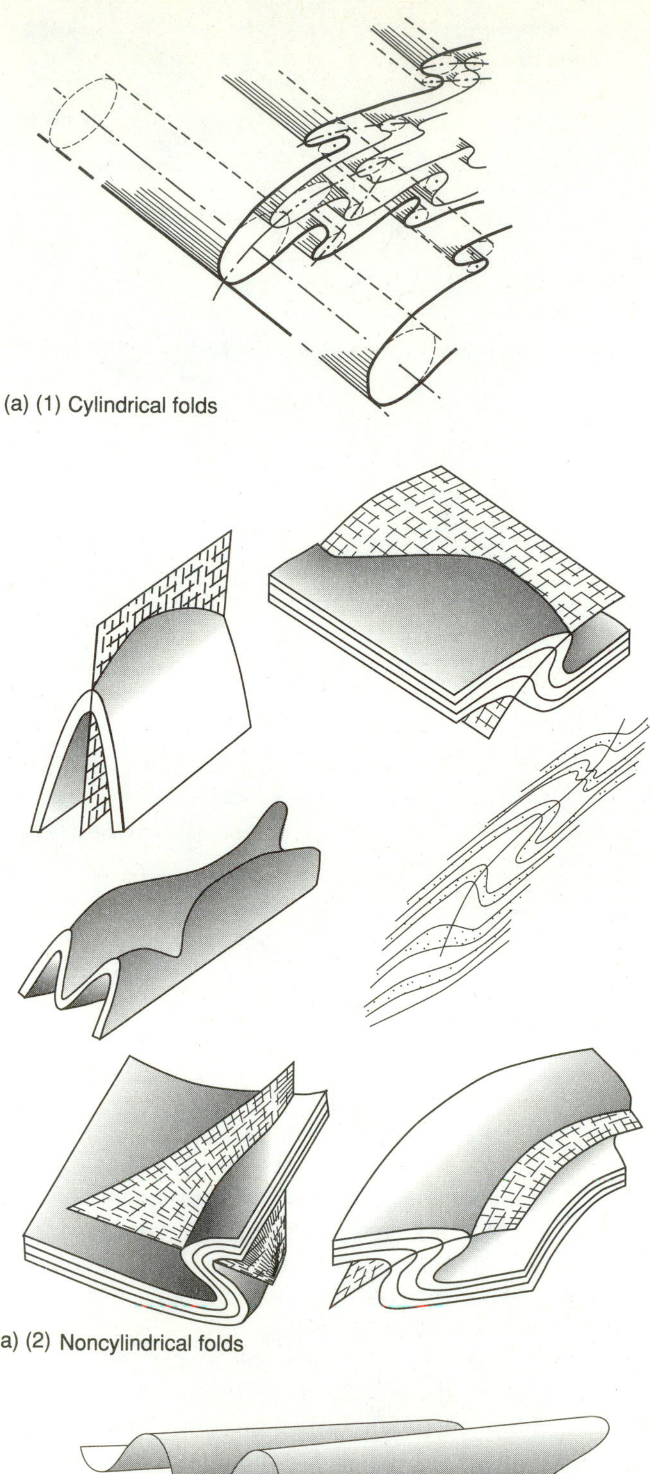

(a) (1) Cylindrical folds

(a) (2) Noncylindrical folds

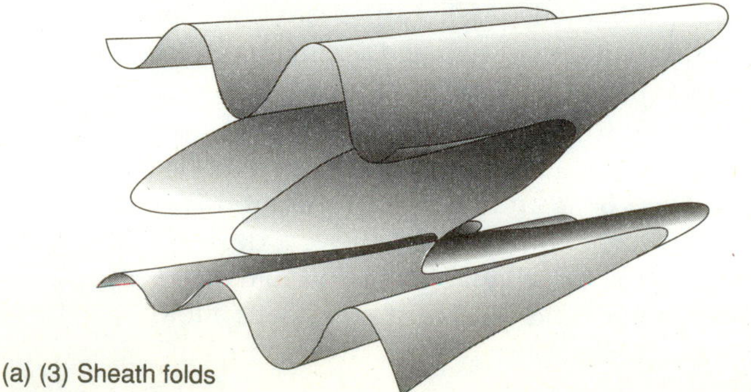

(a) (3) Sheath folds

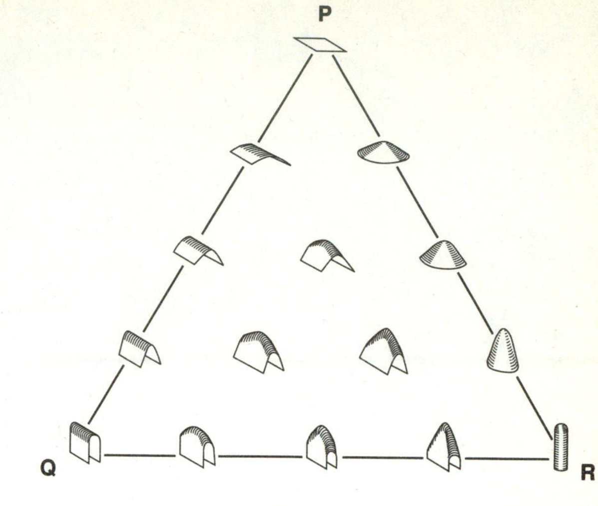

(b) (1)

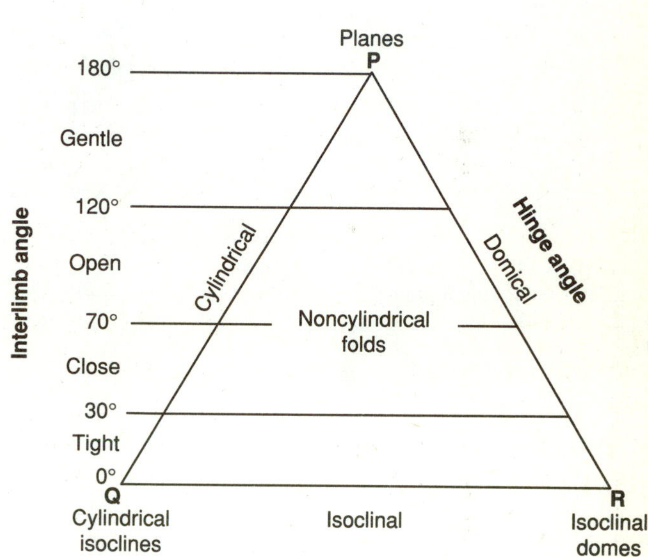

(b) (2)

FIGURE 14–9
(a) Cylindrical (1), noncylindrical (2), and sheath (3), a variety of noncylindrical) folds. Compare the shapes of the hinges from top to bottom: the ideal cylindrical folds have very linear hinges, and the sheath folds have tightly curved hinges. (After Gilbert Wilson, 1982, *Introduction to Small Scale Geological Structures,* Unwin-Hyman Ltd.; *Tectono-physics,* v. 18, D. M. Ramsay and B. A. Sturt, p. 81–107, © 1973, with kind permission from Elsevier Science Ltd., Kidlington, United Kingdom; and *Journal of Structural Geology,* v. 3, J. R. Henderson, p. 203–210, © 1981, with kind permission from Elsevier Science Ltd., Kidlington, United Kingdom.) (b) End-member fold classification (PQR diagram) showing fold shapes and terminology. This classification scheme illustrates the gradual changes in shape from unfolded surfaces (planes, P) through ideal cylindrical isoclinal folds (Q) to ideal isoclinal domes (sheath folds, R). (Reprinted from *Journal of Structural Geology,* v. 1, G. D. Williams and T. J. Chapman, p. 181–185, © 1979, with kind permission from Elsevier Science, Ltd., Kidlington, United Kingdom.)

FIGURE 14–10
Sheath fold of calcsilicate layers in Archean gneiss, Melville Peninsula, District of Franklin, Canada. (J. R. Henderson, Geological Survey of Canada.)

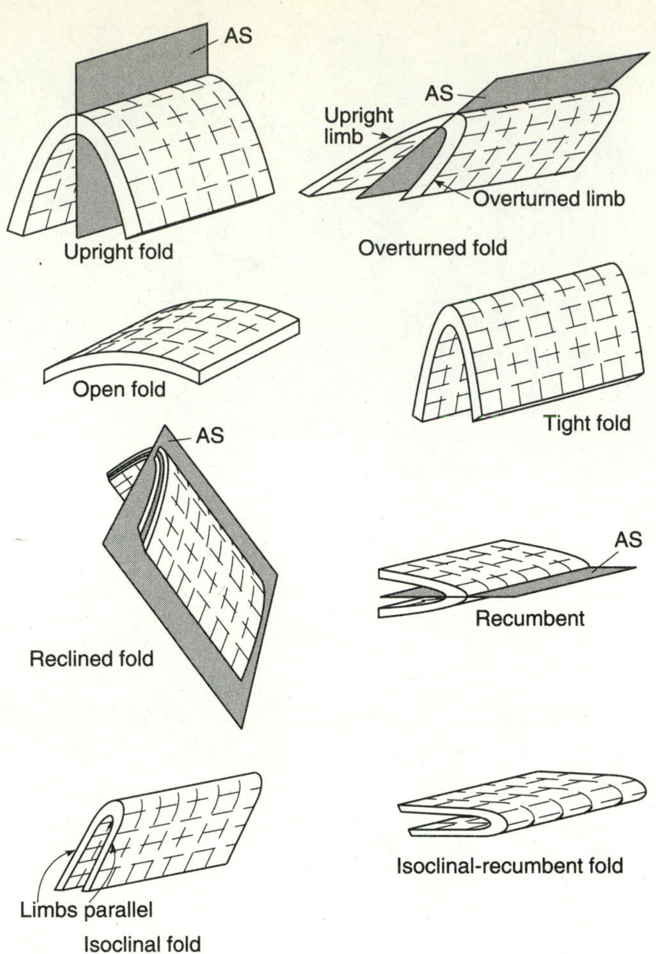

FIGURE 14–11
Upright, overturned, open, tight, reclined, recumbent, isoclinal, and isoclinal recumbent folds. AS—axial surface.

constructed a similar plot with end members—vertical, horizontal, upright, and recumbent folds—that vary by pitch (or rake) of the axis, plunge of the axis, and dip of the axial surface.

Folds that maintain constant layer thickness are called ***parallel folds*** (Figure 14–14a). ***Concentric folds*** are parallel folds in which folded surfaces define circular arcs and maintain the same center of curvature. These characteristics restrict fold shape and require that both anticlines and synclines die out upward and downward from the zone of greatest deformation (Figures 14–14a and 14–15a). ***Ptygmatic folds*** have a lobate and nearly concentric shape, and attenuated limbs—an intestinal appearance. Quartz or quartz-feldspar veins in weak schist frequently produce ptygmatic folds (Figure 14–15c). ***Similar folds*** maintain their shape throughout a section; they do not die out upward or downward but maintain the same curvature in the hinge zones (Figures 14–14b and 14–15d). Layer thickness changes (measured perpendicular to layering) in similar folds, from thicker in the axial zones to thinner on the limbs. Folds with straight limbs and sharp angular hinges are ***chevron*** and ***kink folds*** (Figure 14–14c).

Folds in which shape or wavelength changes from one layer to another are ***disharmonic*** (Figures 14–1 and 14–14d). Disharmonic folds can form during parallel folding by the overtightening and crumpling of thinner layers between thicker layers (Figure 14–15c).

Folds wherein the synclines are thickened and the anticlines thinned are called ***supratenuous folds*** (Figures 14–14e and 14–14f). Many supratenuous folds form in unconsolidated sediments and so are nontectonic; others form where uplift occurs during deposition so that more sediment accumulates in structural lows and less accumulates over highs.

Fault-bend and ***fault-propagation folds*** that form in association with thrust faults were described in Chapter 11. Recall that fault-bend folds are generally open parallel folds, but fault-propagation folds may be tighter parallel folds.

FIGURE 14–12
Isoclinal-recumbent folds in amphibolite and granitic gneiss, near Walhalla (a) and Clemson (b) South Carolina, in the Appalachian Inner Piedmont. Note that the recumbent fold in (a) has been refolded by open folds with moderately to steeply dipping axial surfaces, whereas the axial surface in the fold in (b) remains planar, and the fold has developed an axial-plane foliation. (RDH photos.)

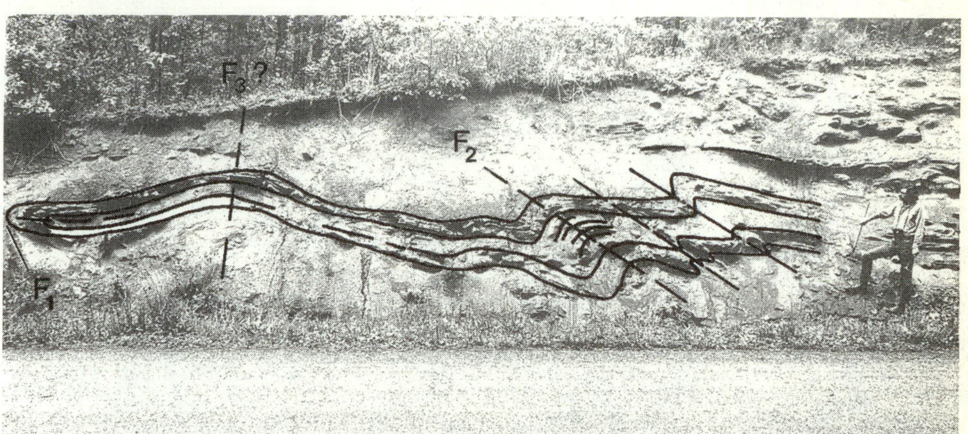

(a)

(b)

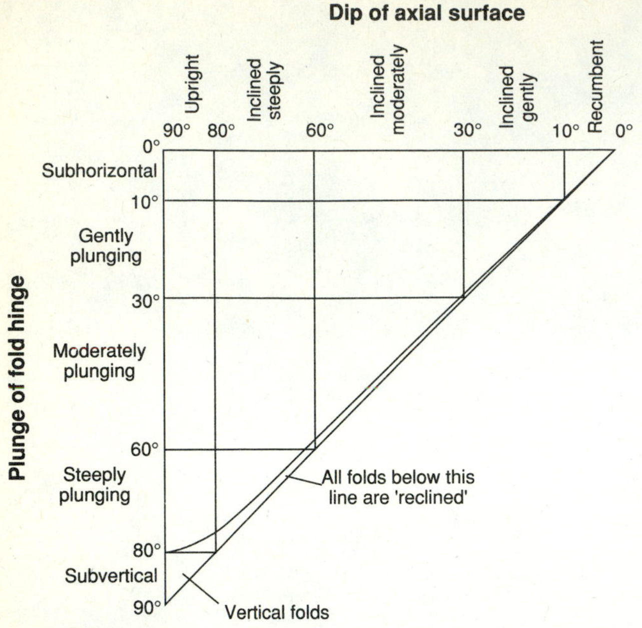

FIGURE 14–13
Plot for distinguishing among several fold types on the basis of relative plunge of the fold hinge and dip of the axial surface. (From M. J. Fleuty, 1964, *Proceedings of the Geologists' Association*, v. 75, p. 461–492.)

Use of Parasitic Folds in Determining Position in a Fold

Pumpelly's rule notes similarities between the orientations of fold hinges and axial surfaces, along with the shape of folds of different sizes on the same major structure (Figures 14–2, 14–4a, and 14–16a), but does not take into account a useful difference. The sense of asymmetry (or vergence) of higher-order folds depends on which limb of the next-lower-order fold they occur. Shear couples commonly form as a result of flow of material out of fold hinges or relative motion of oppositely moving layers in the fold limbs. Figure 14–16a shows that on one limb, small or ***parasitic*** folds have a clockwise (Z) sense of rotation, but on the opposite limb, they have a counterclockwise (S) sense. Small folds in the hinge are symmetrical and have an M or W shape. In the field, this observation helps to locate hinges of the next larger set of folds, but only if both sets of folds were formed at the same time.

Edward Hansen has devised a method for determining the vergence of a large fold by plotting measurements of orientation and vergence of parasitic folds on a stereonet (or equal-area net; Figure 14–17). The *Hansen method* was worked out by noting small folds

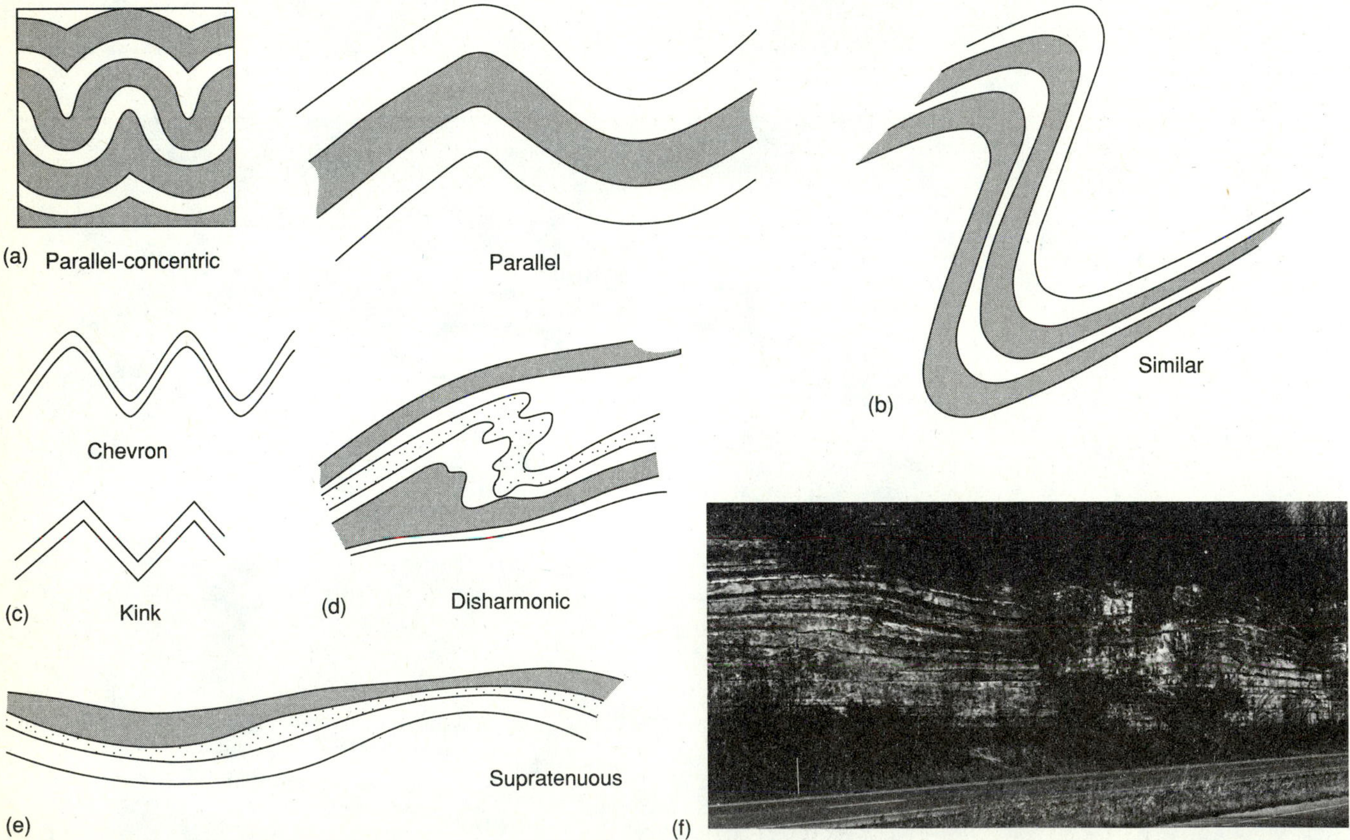

FIGURE 14–14
(a) Ideal parallel-concentric folds and parallel folds. (b) Ideal similar folds. (c) Chevron and kink folds. (d) Disharmonic folds. (e) Supratenuous folds. All are shown in cross-section view. (f) Supratenuous folds in Fort Payne chert and limestone formed by compaction of beds over reefs in the anticlinal hinges. Located on Interstate 40 at Buffalo Valley, Tennessee. (RDH photo.)

(a)

(b)

(c)

(d)

FIGURE 14–15
(a) Parallel-concentric folds in Upper Cambrian Conasauga shale and limestone near Kingston, Tennessee. Note that the folds die out upward in the exposure. (RDH photo.) (b) Similar-like folds in Walden Creek Group Wilhite Slate (Late Proterozoic?) near Walland, Tennessee. Amplitude of folds is about 5 m. (RDH photo.) (c) Folding of a quartz-feldspar vein produced ptygmatic folds in biotite gneiss (metasandstone) near Asheville, North Carolina. (Arthur Keith, U.S. Geological Survey.) (d) Strongly disharmonic folds in Pennsylvanian sandstone and shale near Rockwood, Tennessee. Note fault in left center. (RDH photo.)

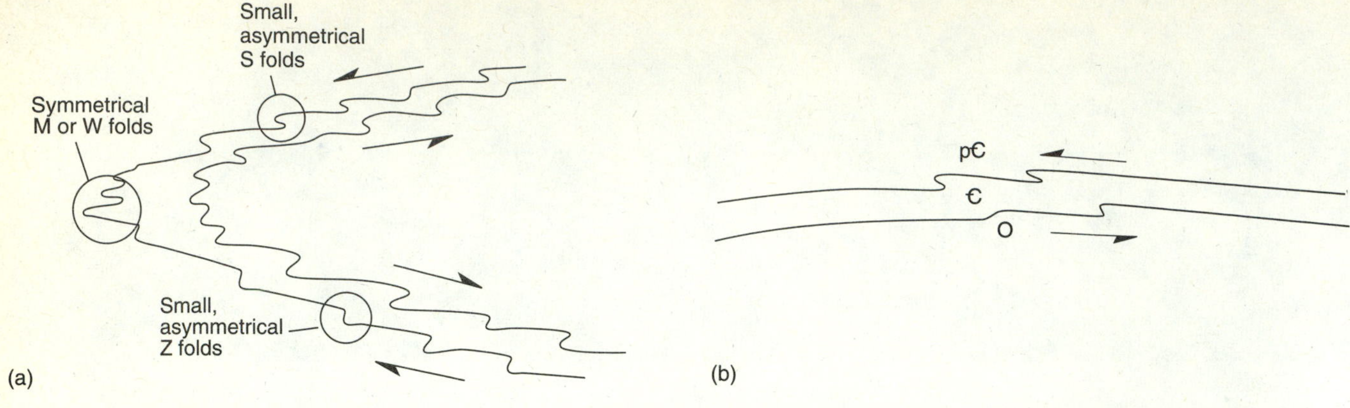

FIGURE 14–16
(a) Relationships of small (second-order) S, Z, and M folds to the location of a large (first-order) fold hinge. (b) Multiple deformation and Pumpelly's rule. Small folds are interpreted to lie on the upright limb of an anticline, but stratigraphy (pЄ—oldest; O—youngest) indicates that the sequence is overturned, and small folds were superposed after the sequence was overturned. Where axial surfaces dip more gently than a fold limb, overturning is commonly assumed. (c) Recumbent fold in Chilhowee Group metasandstone in the Table Rock thrust sheet at Linville Falls, North Carolina. Note parasitic S and Z folds on the limbs. (RDH photo.)

in tundra sod (solifluction lobes) moving downslope on the upper slopes of Blåhø (Norwegian for Blue Mountain), a mountain in Trollheimen, southern Norway, and then was applied to the Caledonian folds in bedrock of the same area (Hansen, 1971). Small folds having one vergence are separated from those having the opposite vergence on a fabric diagram. The line on the diagram (Figure 14–17c) separating the two fields of fold hinges of opposite vergence is the transport direction (*slip line*). Hansen's model for working out this method involved formation of folds above a detachment surface as the entire mass moved downslope, essentially a ductile shear zone. The method therefore might be employed to determine the transport

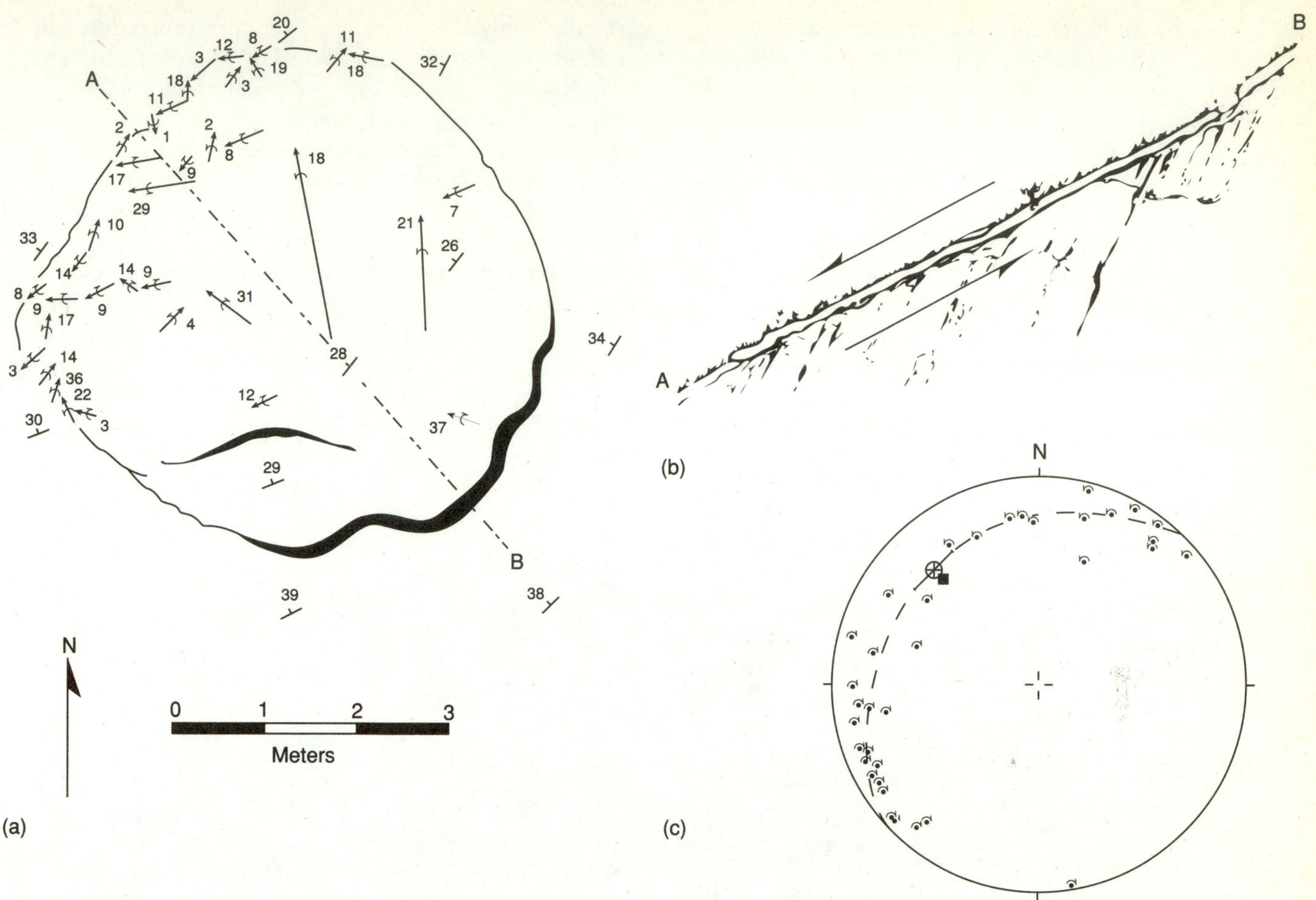

FIGURE 14–17
(a) Map of fold-axis orientations on a lobe of tundra sod moving downslope on Blåhø, Trollheimen, southern Norway, that served as part of the model for development of Hansen's method. Each arrow represents a fold axis; the length of each arrow is proportional to the length of the hinge line measured. Semicircular arrows on each fold axis indicate the sense, or asymmetry, or vergence of each fold in a downplunge direction. Dip-strike symbols indicate the dip of the tundra surface where unaffected by the slide. (b) Cross section A–B; location indicated in (a). (c) Stereoplot of 36 fold axes from map in (a), each point showing the sense of vergence. Counterclockwise arrows are S-folds; clockwise arrows are Z-folds; dark square indicates transport direction. (From Edward Hansen, 1971, *Strain Facies,* Springer-Verlag.)

direction of faults if folds formed next to the detachment surface during movement. J. T. Dillon, Gordon Haxel, and Richard Tosdal (1990) have used this method to determine the transport direction on the Chocolate Mountains thrust in southeastern California.

Observations of overturning and vergence should always be made while viewing fold shape in the direction of plunge (Figure 14–3b). A simple experiment with a piece of folded paper will show that the sense of vergence appears to change when the same fold is viewed from the opposite direction.

Employing vergence to determine the position in a structure is a useful tool, but is risky where it is employed to determine the top of a sequence. In areas with only one generation of folds, it works very well. Where more than one fold generation exists, the geologist must make sure that observations and conclusions are related to folds of the same generation.

Vergence may be used to determine whether the geologist is on an upright or an overturned limb of a fold—if the vergence is related to the fold generation being considered. In many areas, the stratigraphic section may be overturned, but a set of later folds may *falsely* indicate that the sequence is upright (Figure 14–16b). Careful study of both stratigraphy and structure will usually reveal the correct relationships.

MAP-SCALE PARALLEL FOLDS AND SIMILAR FOLDS

Parallel and similar folds are common groupings of folds; they form two classes of contrasting fold shape, but the separation may not be immediately evident in map view (Figures 14–18a and 14–19a). They are clearly illustrated in sections perpendicular to fold axes or in vertical sections constructed through regions with horizontal folds of these different styles (Figures 14–18b and 14–19b). Determination of fold style is made by combining field observations of mesoscopic folds (and related structures) and careful plotting of dip-strike data on sections as they are constructed.

Several techniques have been devised for constructing geologic sections through regions where parallel folds have formed. H. G. Busk (1929) invented a method for section construction, assuming that parallel folds also are concentric. He realized that in parallel-concentric folds, dip of bedding is parallel to circular arcs of folds. He extrapolated folded layers to depth using dip measurements made on the surface and, after identifying the center of curvature of each fold segment, constructed lines perpendicular to bedding to locate the centers of the fold segments and then connected them with circular arcs (Figure 14–20a). He then eliminated overlapping segments of the arcs, producing smooth curves connecting anticlines and synclines. Busk's method works well if parallel-

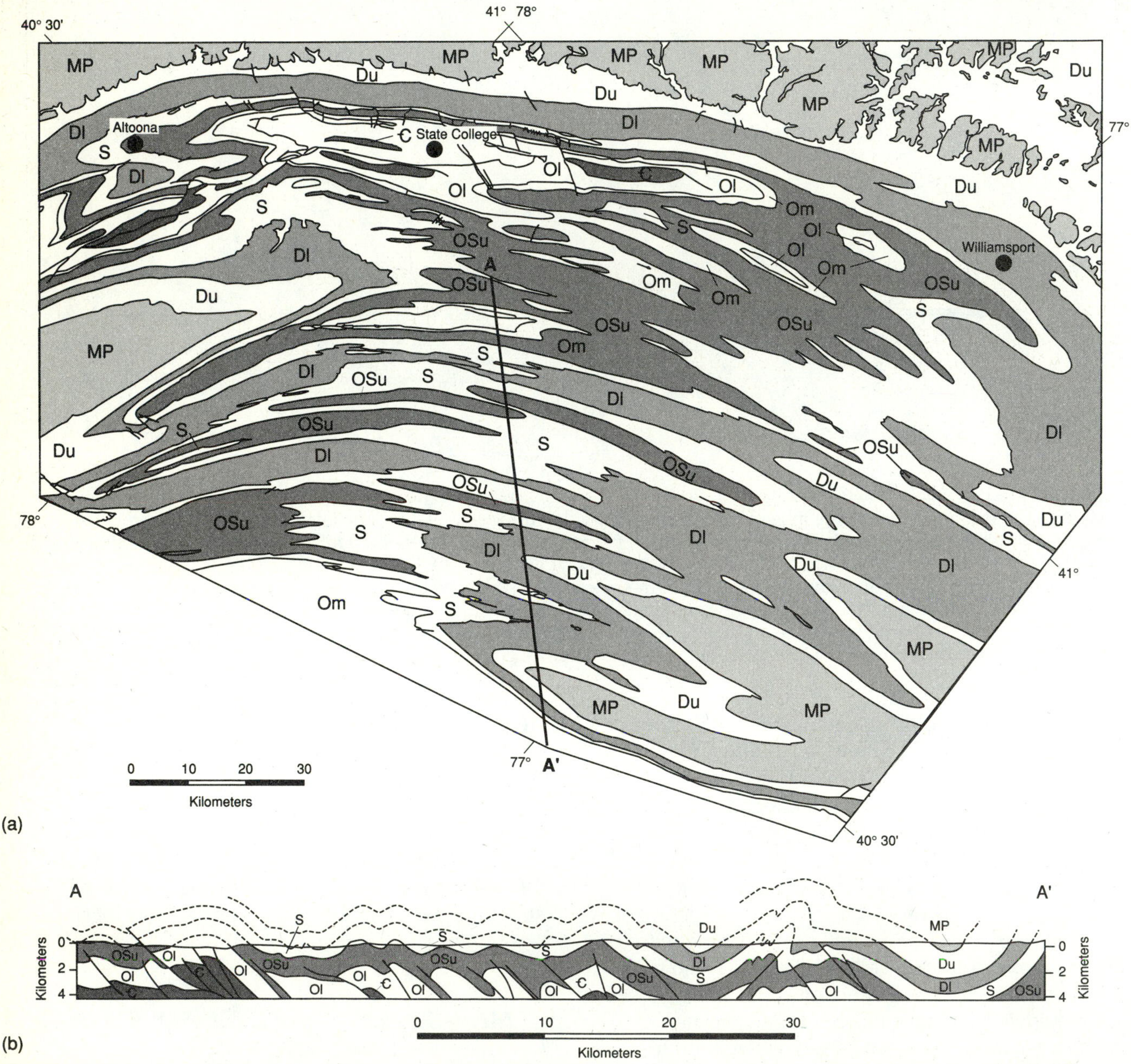

FIGURE 14–18
(a) Map-scale parallel folds in part of the Appalachian Valley and Ridge of Pennsylvania. (b) Cross section along line A–A′ showing the parallel style of folding. Note that the layers remain appropriately parallel and maintain nearly constant thickness and that folds are faulted at depth. Ꞓ—Cambrian; Ol—Lower Ordovician; Om—Middle Ordovician; Osu—Upper Ordovician and Lower Silurian; S—Middle and Upper Silurian; Dl—Lower Devonian; Du—Upper Devonian; MP—Mississippian and Pennsylvanian. (From Pennsylvania Geologic and Topographic Survey, 1980, *Geologic Map of Pennsylvania*.)

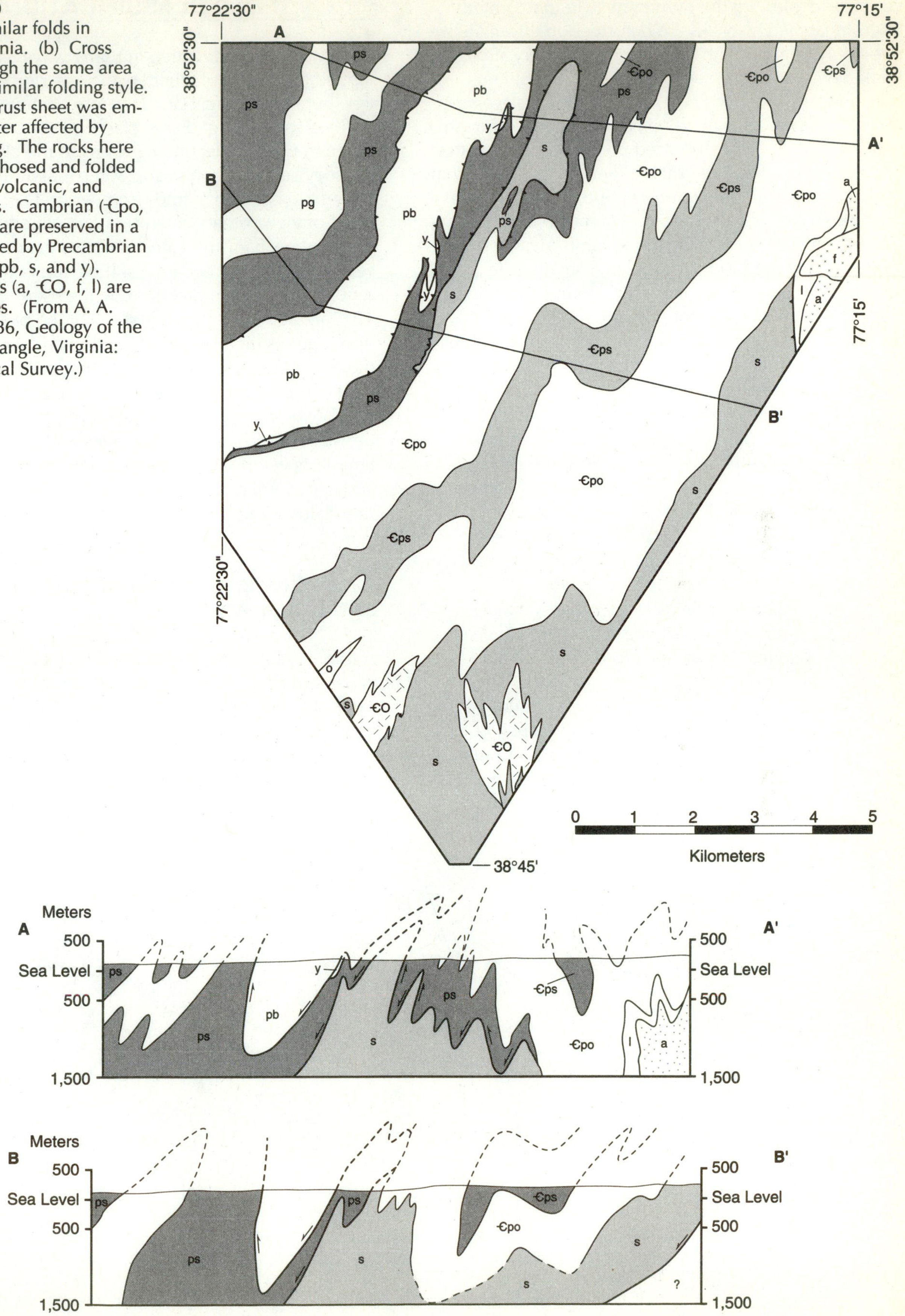

FIGURE 14–19
(a) Map of similar folds in northern Virginia. (b) Cross sections through the same area showing the similar folding style. Note that a thrust sheet was emplaced and later affected by similar folding. The rocks here are metamorphosed and folded sedimentary, volcanic, and plutonic rocks. Cambrian (Єpo, Єps, o) rocks are preserved in a syncline flanked by Precambrian rocks (ps, pg, pb, s, and y). Patterned units (a, ЄO, f, l) are igneous bodies. (From A. A. Drake, Jr., 1986, Geology of the Fairfax Quadrangle, Virginia: U.S. Geological Survey.)

concentric folds are the dominant style that formed. Unfortunately, few rocks are deformed in this manner—only in areas with thick sections of very strong, thickly bedded rocks (such as massive sandstone, dolostone, or limestone) with few weak interlayers.

An alternative way to construct sections through parallel-folded regions is based on the assumption that the rocks were deformed by parallel kinking rather than by concentric folding (Figure 14–20b). Kink folds have been observed for many years in layered rocks, and Roger Faill (1973), working in the Pennsylvania Appalachians, was the first to suggest that small-scale kinks may be extrapolated to map scale. John Suppe (1983, 1985) has used this assumption and the kink style extensively because after a section is constructed, this method makes the section easier to balance (Chapter 11). Results essentially the same as Busk's can be achieved by using many small kinks. Kink folds do exist in foreland fold-thrust belts but are no more common than parallel-concentric folds and probably less common than parallel folds with curved (rather than angular) hinges. Therefore, anyone constructing sections through regions of parallel folding should try to combine the two techniques so that the structure can be interpreted more realistically.

Deformation that yields similar folds involves thinning of limbs and thickening of hinges. Layers no longer are parallel, but the shapes of folds do not change in the section; these are fundamental differences between parallel and similar folds (Figure 14–19). Cross sections are still a useful way to depict the structure but are harder to construct because similar folds also have a greater tendency to be noncylindrical, and extrapolation of layering to depth is less certain.

FOLD CLASSIFICATIONS

Folds have been classified in many ways, each intended to pigeonhole them using one or more properties, but no classification takes into account all aspects of folds and folding. Every classification suffers from inherent defects, depending on what the classification was devised to accomplish. Some classifications attempt to categorize folds from a purely descriptive point of view; others determine and classify the fold-forming mechanism. Genetic classifications have been severely criticized because of their subjectivity and the requirement of interpretation by the user (Hudleston, 1973a). Some geologists argue that the best classification requires the least interpretation and is therefore the most objective, but drawbacks remain. Every classification, regardless of its basis, has some merit. In the next few sections, we consider several fold classifications and discuss their advantages and disadvantages. Emphasis will be placed on learning John Ramsay's geometric and descriptive classification, along with the rationale for the others.

Classifications Based on Interlimb Angle and Hinge Area

A fold classification devised by Michael Fleuty (1964) is based on the angle between the limbs of a fold—the *interlimb angle* (Figure 14–21). The interlimb angle is best measured as the angle between the inflection points on opposite limbs of the same fold. His classification is purely descriptive: the interlimb angle is measured and the tightness of the fold identified as

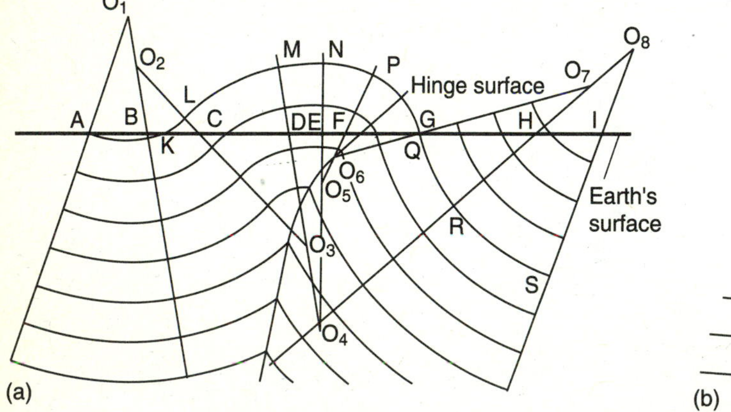

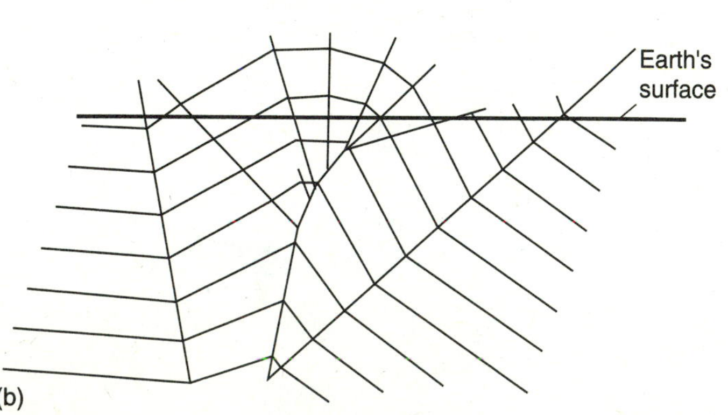

FIGURE 14–20
(a) Busk method for constructing sections through parallel-concentric folds. Normals are constructed through each dip measurement (labeled A through I). Normals intersect at points O_1 through O_8. Using O_1 as a center and O_1A as a radius, an arc is drawn connecting O_1A and O_1B. The process is repeated using O_2 as center and O_2K as radius, creating point L, and the process is continued until the fold system is constructed. The next-deeper layer is constructed by adding or subtracting thickness to the radii, and the process repeated. The overlapping arcs are rounded. (From H. G. Busk, 1929, *Earth Flexures,* Cambridge University Press.) (b) Kink method for reconstructing folds, using the same example as in (a).

gentle, having an interlimb angle of 180° to 120°; *open,* 120° to 70°; *closed,* 70° to 30°; *tight,* 30° to 0°; *isoclinal,* 0°; and *elastica,* wherein the angle is negative. Elasticas are rare in natural rocks but do occur in ductilely folded stronger veins in a weaker matrix (ptygmatic folds; Figure 14–15b). John Ramsay (1967) has pointed out, however, that interlimb angle as a measure of fold tightness, as Fleuty defined it, depends not only on this angle but also on the distribution of curvature around the fold hinge and limbs. Therefore, a working descriptive fold classification suitable for use with most folds must take into account the relative extent of the hinge zone compared with the extent of the limbs of a fold.

One such classification (by Ramsay, 1967) relates variables of hinge area, or the length of the hinge, and the interlimb angle. An equation describes fold shape involving a parameter P_1 (Figure 14–21b):

$$P_1 = \frac{L_{i_1 i_2}}{H_{i_1 i_2}}, \tag{14–1}$$

where $L_{i_1 i_2}$ is the length of projection of the fold limbs on the join of $i_1 i_2$, $H_{i_1 i_2}$ is the length of the hinge zone on the join of $i_1 i_2$, and i_1 and i_2 are inflection points on adjacent limbs of the same fold, and the hinge zone is where maximum curvature occurs (curvature greater than an arc diameter of $i_1 i_2$). Ramsay's P_1 ratio is useful in describing folds where a quantitative representation of the hinge and interlimb angle is important. Ramsay's modified hinge-area classification is purely descriptive; an observer can make measurements on a fold, calculate the parameter P_1, and categorize it without making any interpretation. Therefore, even a person knowing very little about structural geology or the nature of folds can use this classification if he or she can make the required measurements and divide one by the other.

G. D. Williams and T. J. Chapman (1979) have devised a classification of noncylindrical folds using a diagram with three end members: planes, cylindrical isoclines, and isoclinal domes (Figure 14–9b). The classification is based on measurements of both interlimb angle and hinge angle. It permits classification of the entire spectrum of variations between the three end members.

Hudleston (1973a) has devised a simple classification based on the assumption that an observer can recognize basic fold shapes and estimate amplitude (Figure 14–22). His scheme, called *visual harmonic analysis,* involves assigning to a given fold one basic shape (A to F) and one amplitude number (1 to 5). He recommended that diagrams of the fold shape and amplitude be taken to the field for direct comparison with natural folds in rocks.

Ramsay's Standard Classification

A geometric and descriptive classification used by many structural geologists is another devised by John Ramsay (1962, 1967) based on the profile perpendicular to the hinge of a fold. Several measurements enable the fold to be categorized without bias by any observer, even one uninitiated in fold mechanics. Thus, it is one of the most useful of all fold classifications.

Ramsay's standard classification involves an indirect relationship between layer thickness, both normal to layering (t_α) and parallel to the fold axial surface (T_α), and angle of dip (α) at different points on successive folded surfaces (Figure 14–23). The parameters are related by

$$T_\alpha \cos \alpha = t_\alpha. \tag{14–2}$$

At the fold hinge, the thickness $t_0 = T_0$. The ratio

$$t'_\alpha = \frac{t_\alpha}{t_0} \quad or \quad T'_\alpha = \frac{T_\alpha}{T_0} \tag{14–3}$$

may then be calculated and plotted versus the angle of dip α (Ramsay, 1967). Equation 14–3 expresses change in thickness (t_α or T_α) with change in dip.

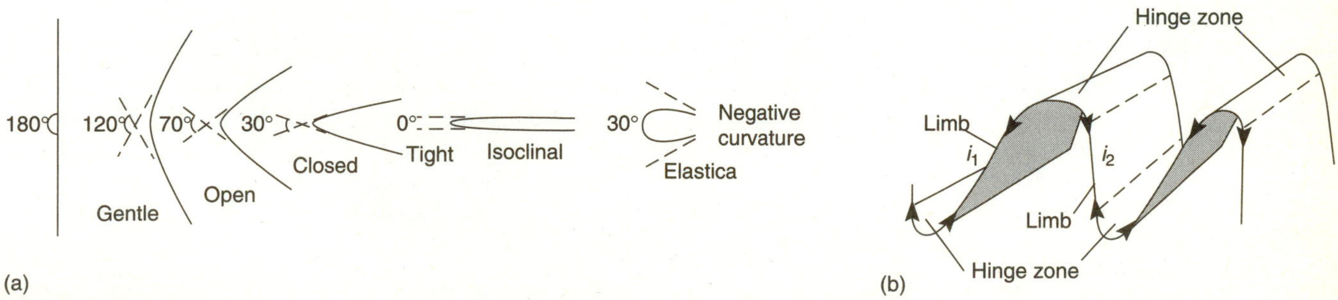

FIGURE 14– 21
(a) Interlimb angle classification and categories. (After M. J. Fleuty, 1964, *Proceedings of the Geologists' Association,* v. 75, p. 461–492.) (b) Illustration of the parameters measured for calculation of Ramsay's P_1 (equation 14–1).

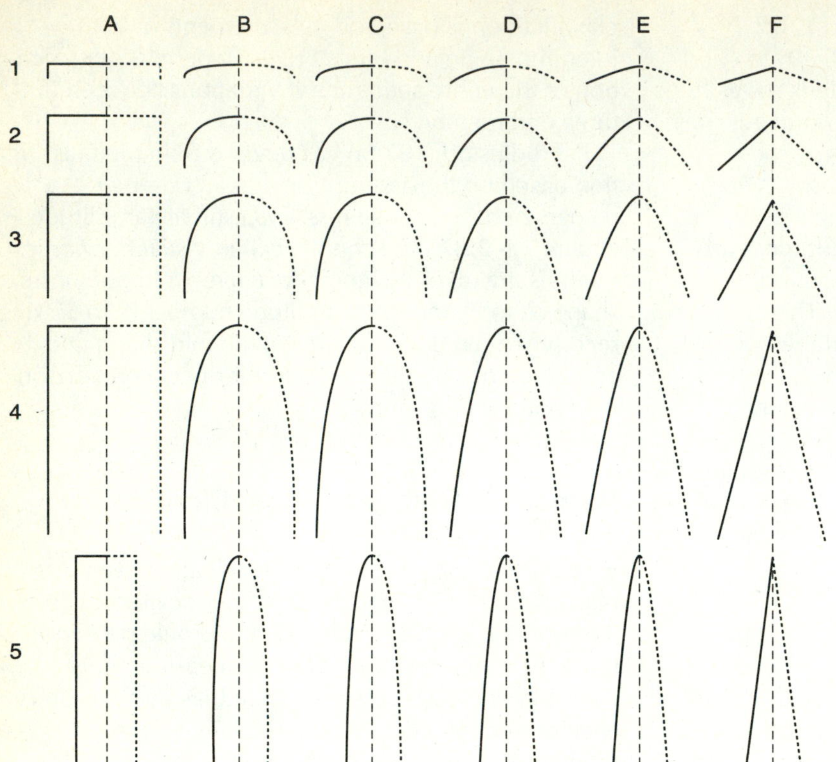

FIGURE 14–22
Hudleston's 30 folds involving different combinations of the ratios of amplitude to wavelength (A–F) and hinge zone to wavelength (1–5). (From *Tectonophysics*, v. 16, P. J. Hudleston, p. 1–46, © 1973, with kind permission from Elsevier Science, Ltd., Kidlington, United Kingdom.)

Lines connecting points of equal slope, or dip, are ***dip isogons***. They may be connected between successive layers and, in Ramsay's technique, are constructed at regular intervals (10° is convenient) on successive layers within a fold (Figures 14–23a and 14–23d). The relative convergence, divergence, or parallelism of the dip isogons is a key to Ramsay's classification, with the degree of convergence of isogons directly related to fold tightening. Folds where the isogons converge toward the concave part of the fold are Class 1 folds. Folds with parallel isogons belong to Class 2, and folds with isogons that diverge toward the concave part of the fold are in Class 3 (Figure 14–24). Class 1 is subdivided into three groups: Class 1A folds have strongly convergent isogons; Class 1B folds correspond to parallel–concentric folds with convergent isogons; Class 1C folds are modified similar or parallel folds that have weakly convergent isogons. Class 1C folds may correspond approximately to the flexural flow folds of Donath and Parker (1964) (see the next section). Class 2 folds are ideal similar folds with parallel isogons, and Class 3 folds are folds with extremely thickened hinges or extremely thinned limbs.

Ramsay's classification has a distinct advantage, being both descriptive and objective, but it classifies many folds—perhaps the majority—as 1C folds, and, as Ramsay (1962) pointed out, similar folds as Class 2 and parallel folds as Class 1B. Ramsay concluded that this is an attribute of the folding process rather than a defect in the scheme; others have suggested that this may be a flaw in the classification. His classification cannot be used readily in the field; either a photograph must be made facing parallel to the hinge of the fold, or the fold must be collected, sawed perpendicular to the hinge, and traced or photocopied so that the isogons can be constructed.

Peter Hudleston (1973a) modified Ramsay's scheme by defining another measurable quantity, ϕ_α, to be used in conjunction with dip isogons. He defined ϕ_α for a folded layer as the angle between the normal to the tangents drawn to either fold surface at an angle of apparent dip α and the isogon (Figure 14–25). The quantity ϕ_α is either positive or negative, depending on whether the isogon is deflected clockwise (–) or counterclockwise (+) relative to the normals to the tangents to the folded surfaces as the observer traces the isogon from the inner to the outer arc. The dip α of the right limbs of antiforms and left limbs of synforms is assumed positive and all other limbs negative. The fold in Figure 14–25a is plotted in Figure 14–26b as a plot of ϕ_α versus α. Figure 14–26a is a plot of t'_α versus α for the same fold to illustrate the standard Ramsay classifi-

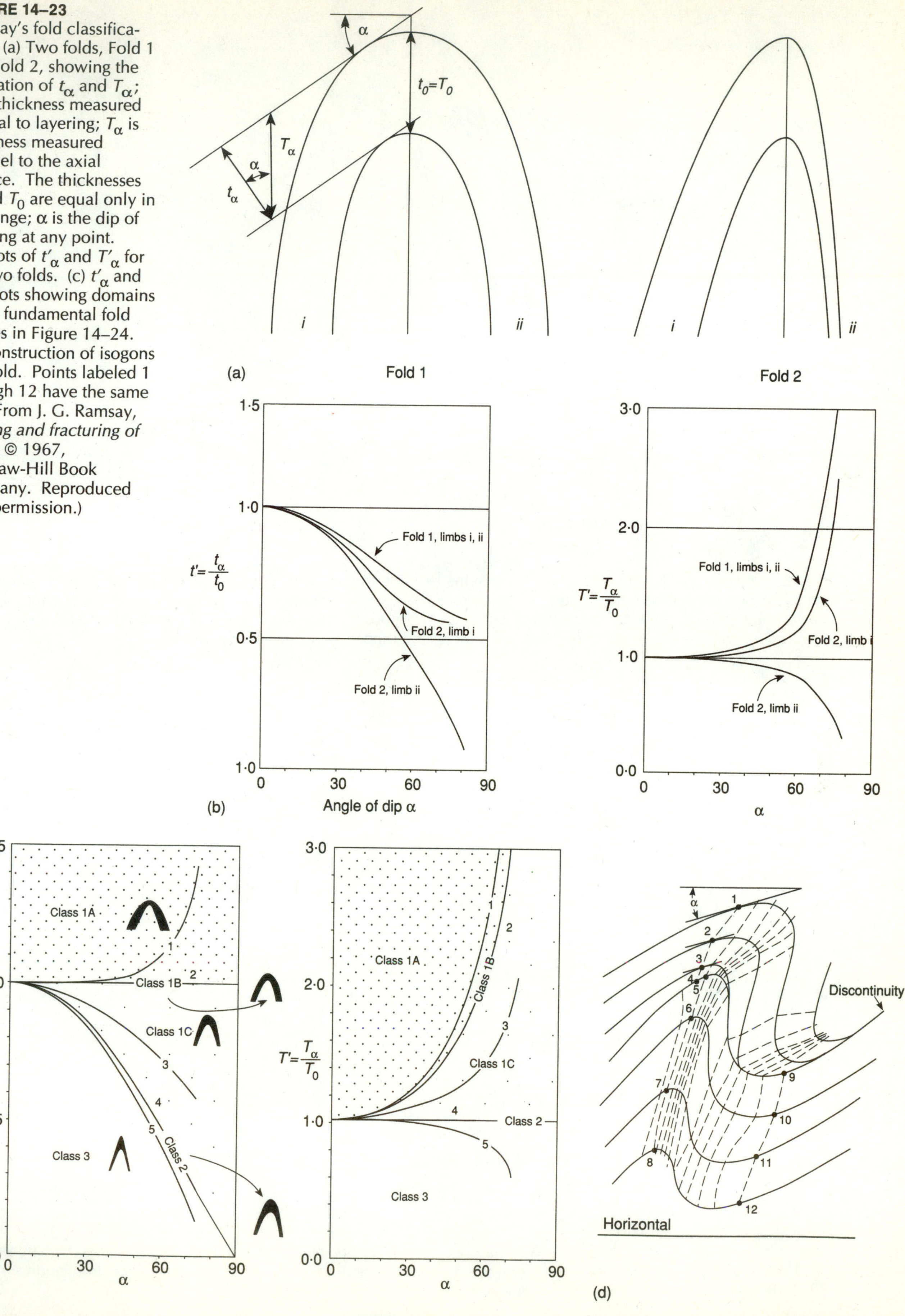

FIGURE 14–23 Ramsay's fold classification. (a) Two folds, Fold 1 and Fold 2, showing the derivation of t_α and T_α; t_α is thickness measured normal to layering; T_α is thickness measured parallel to the axial surface. The thicknesses t_0 and T_0 are equal only in the hinge; α is the dip of layering at any point. (b) Plots of t'_α and T'_α for the two folds. (c) t'_α and T'_α plots showing domains of the fundamental fold classes in Figure 14–24. (d) Construction of isogons in a fold. Points labeled 1 through 12 have the same dip. (From J. G. Ramsay, *Folding and fracturing of rocks*, © 1967, McGraw-Hill Book Company. Reproduced with permission.)

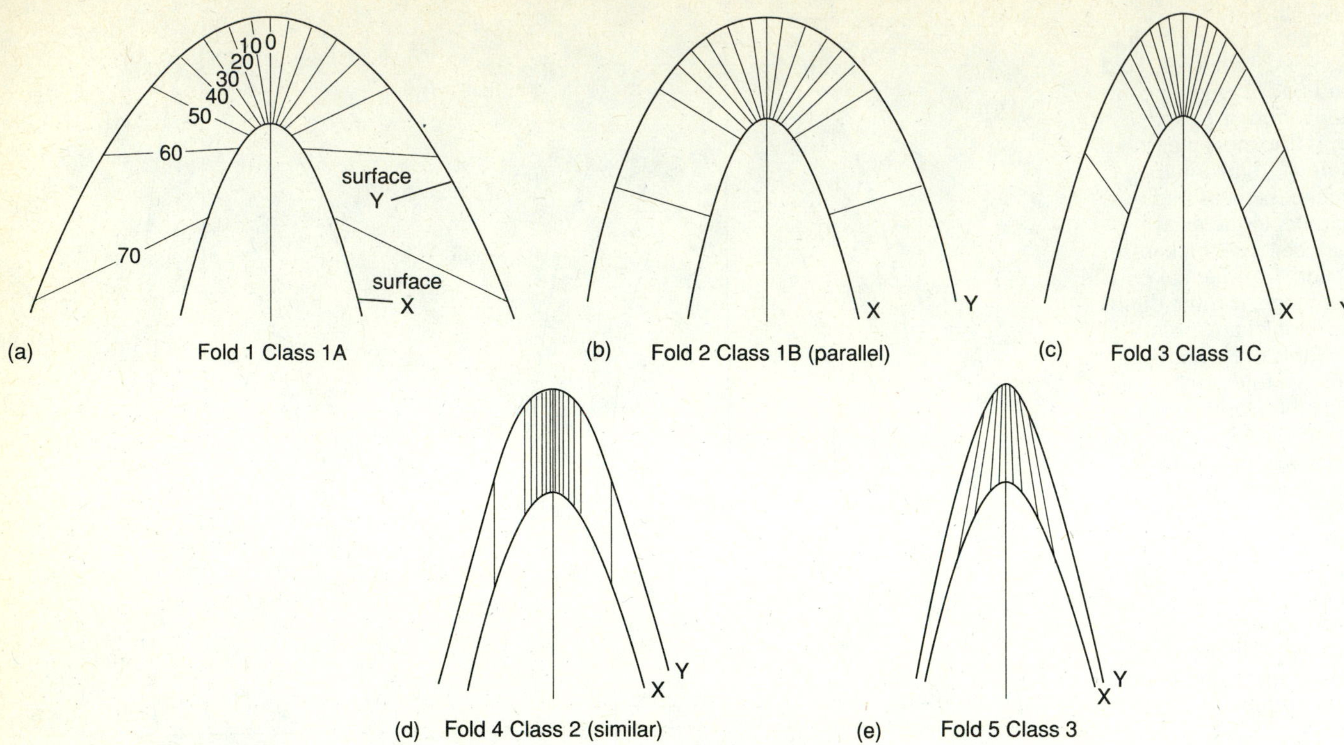

FIGURE 14–24
Ramsay fold classes. (a) Isogons change direction faster than the bedding surfaces they connect. (b) Isogons change direction at the same rate as the bedding surfaces they connect. (c) Isogons change direction more slowly than the bedding surfaces they connect. (d) Isogons are parallel. (e) Isogons change direction in the opposite sense to the surfaces they connect. (From J. G. Ramsay, *Folding and fracturing of rocks,* © 1967, McGraw-Hill Book Company. Reproduced with permission.)

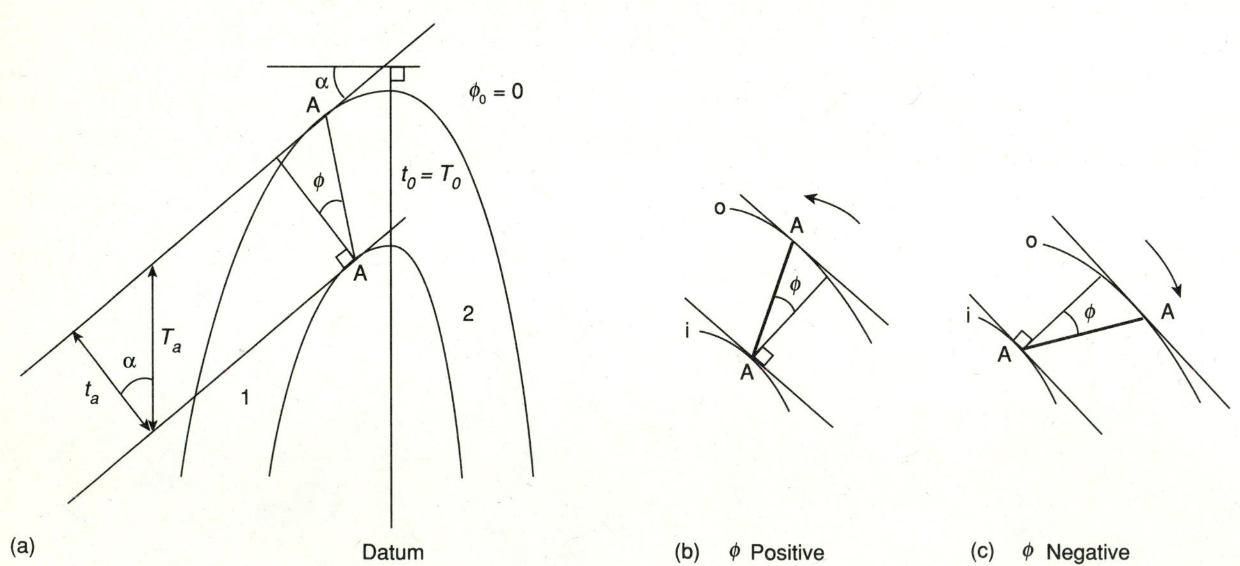

FIGURE 14–25
(a) Derivation of the parameters t_α, T_α, and ϕ_α. A–A is the isogon at dip α. The reference line here is the axial-surface trace. (b) and (c) Sign convention for ϕ: i—inner arc; o—outer arc. A–A is the isogon. (From *Tectonophysics,* v. 16, P. J. Hudleston, p. 1–46, © 1973, with kind permission from Elsevier Science, Ltd., Kidlington, United Kingdom.)

cation. Hudleston's modification of Ramsay's scheme permits limiting each class to a distinct field on the plot of ϕ_α versus α (Figure 14–26b). Table 14–1 summarizes relationships between Ramsay's fold classes and these parameters.

Donath and Parker Classification

A refinement of several older classifications (Knopf and Ingerson, 1938; Turner and Weiss, 1963) is one devised by Fred Donath and Ronald Parker (1964). It is a generic-mechanical scheme based on *mean ductility* and *ductility contrast* within the folded sequence (Figure 14–27), and thus some geologists argue that it does not belong in a discussion of fold properties and classifications. We will consider it here because it contrasts with the descriptive classifications and because it provides another option that can be used in the field.

According to the Donath and Parker scheme, two broad groups of folds exist. One, where fold shape is controlled by the layering in the rocks, is called ***flexural folds***. The other, where layering serves only as a displacement marker during folding, is called ***passive folds***. A second broad, twofold subdivision within the Donath and Parker scheme is fundamentally a separation of brittle and ductile behavior. ***Slip*** along bedding, cleavage, or foliation planes is important in forming brittle folds. The process of *ductile flow* dominates in

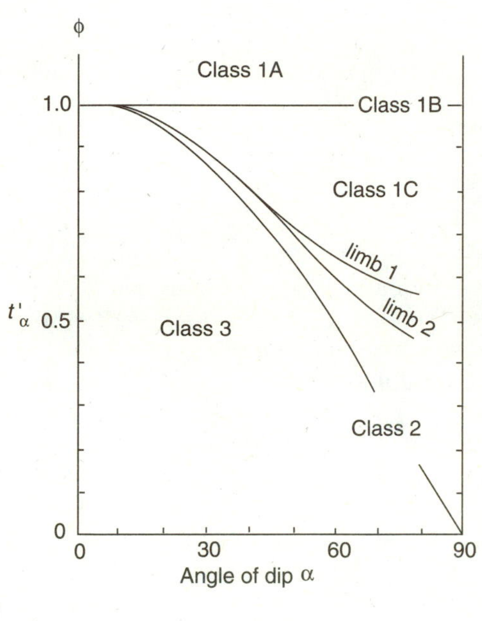

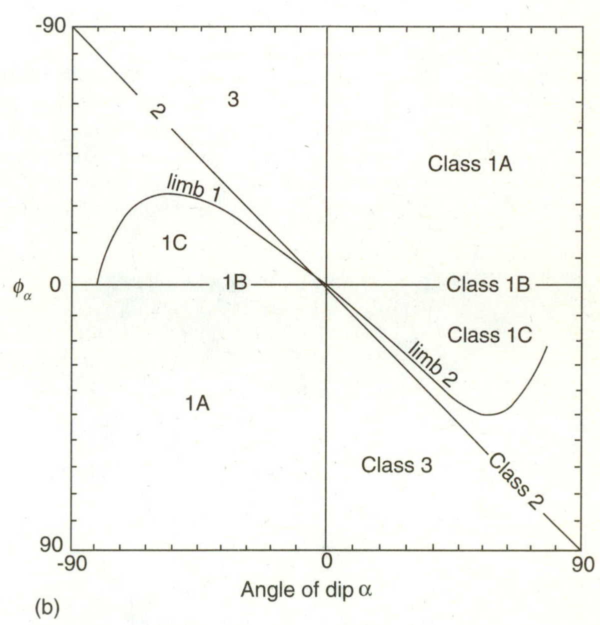

FIGURE 14–26
Plots of t'_α versus dip (α) (a) and ϕ_α versus α (b) for the fold in Figure 14–25a, showing the fields where Ramsay's fold classes plot. (From *Tectonophysics,* v. 16, P. J. Hudleston, p. 1–46, © 1973, with kind permission from Elsevier Science, Ltd., Kidlington, United Kingdom.)

TABLE 14–1
RELATIONSHIPS BETWEEN HUDLESTON'S PARAMETERS AND RAMSAY'S FOLD CLASSES

Class	t'_α	ϕ_α
1A	> 1.0	< 0
1B (parallel)	1.0	0
1C	$\cos\alpha < t'_\alpha < 1.0$	$\alpha > \phi_\alpha > 0$
2 (similar)	$\cos\alpha$	α
3	$< \cos\alpha$	$> \alpha$

Values of ϕ_α are plotted for a positive α. (From P. J. Hudleston, 1972, *Tectonophysics,* v. 16, p. 1–46.)

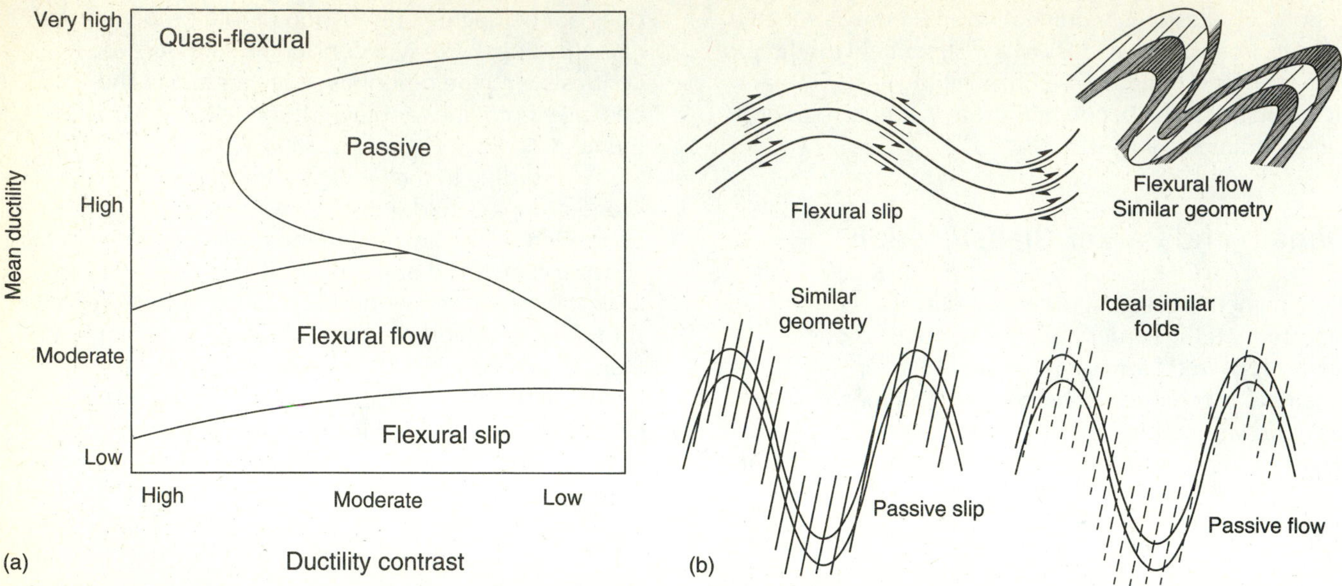

FIGURE 14–27
(a) Basis of Donath and Parker classification. (From F. A. Donath and R. B. Parker, 1964, Geological Society of America *Bulletin.*) (b) Types and mechanisms of Donath and Parker fold types. In flexural-slip folds, layer thicknesses remain constant, and folding is accomplished by slip along layers. In flexural flow, strong layers change thickness little or not at all, weak layers undergo appreciable thickness changes, and cleavage is strong in weak layers, but poorly developed in strong layers. Passive slip folds are ideally developed by movement parallel to a strong cleavage, a mechanism that may not exist in nature. Passive-flow folds develop by ductile flow with limbs thinned (or relatively thickened) equally in all rock types.

FIGURE 14–28
Flexural-slip folds in calcareous sandstone in Middle Ordovician Tellico Formation near Tallassee, southeastern Tennessee. (RDH photo.)

passive folds. Another category of folds that fits into neither twofold subdivision is called ***quasi-flexural*** and corresponds to disharmonic folds (Figure 14–1).

Flexural-slip folds correspond in shape and mode of formation to parallel or parallel-concentric folds (Figure 14–28). Flexural-slip folds form by buckling (being pushed from the ends of layers) or bending (flexing across layering), and slip parallel to layering (Chapter 15). Generally, these folds are not very tight except in weak rocks or in rocks with a strong contrast in competence (ductility). Flexural-slip folds are easily recognized by slickensides, fibers, or other movement indicators (such as slip lines) on layer surfaces and by constant layer thickness.

Chevron folds (Figure 14–29) may form by buckling. They are folds in which the curvature is confined to narrow hinges, and have straight limbs. The temperature and pressure at which these folds form are low. Such folds may overprint earlier folds in rocks of high metamorphic grade, forming after the high-grade rocks have cooled and pressure has dropped.

Passive-slip folds (Figure 14–30) are a type of similar folds thought to form by shearing along planes inclined to the layering (Figure 14–27b), much as playing cards move past each other when pushed at one end of the deck. Folds of this kind would not form by pure shear. This mechanism is difficult to demonstrate in rocks, and apparent displacements that result from pressure solution are easy to mistakenly identify as products of passive slip (see Chapter 15).

FIGURE 14–29
Chevron folds in Early Carboniferous sandstone and shale near Oued Zem, central Morocco. (RDH photo.)

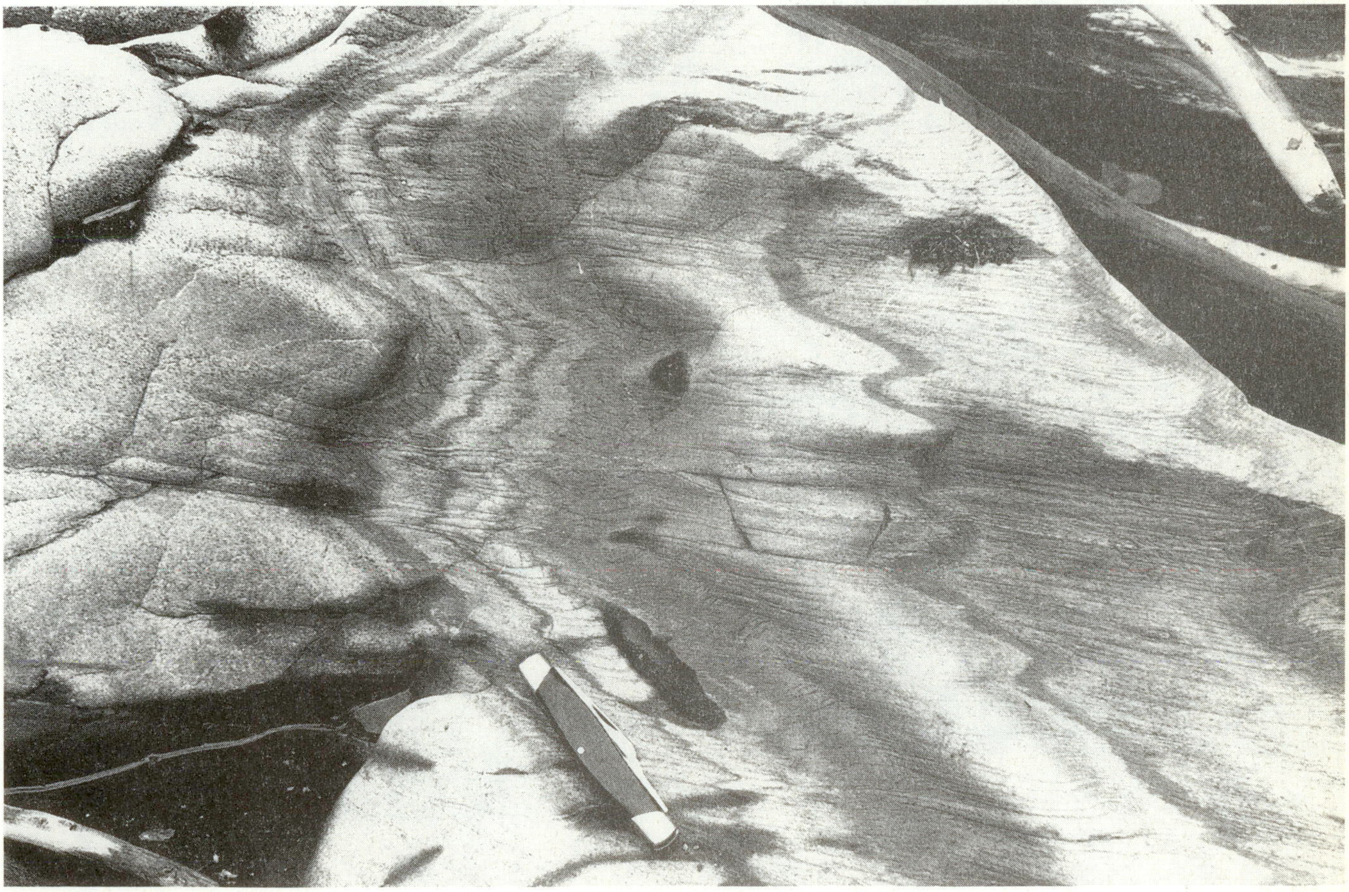

FIGURE 14–30
Folds in cleaved Lardeau Group phyllitic metasiltstone along the Illecillewaet River in southern British Columbia. Displacement of layers parallel to cleavage could result from passive slip or from pressure solution and buckling. (RDH photo.)

Flexural-flow folds form in rocks from low to moderate metamorphic grade, depending on the nature of the rocks (Figure 14–31). Flexural-flow folds are mostly similar-like folds, but may also include some parallel folds. Some layers in single flexural-flow folds maintain constant thickness; others are thickened into axial zones and thinned into limbs as folding proceeds, indicating a higher contrast in internal ductility. As a result, flexural flow may be observed in shales in an interbedded limestone and shale sequence deformed at very low temperature (Figure 14–1). The shale layers change thickness appreciably as they undergo strongly plastic or semiplastic flow, and the limestone layers undergo brittle deformation and maintain constant thickness (Figure 14–31a). In much the same way, flexural-flow folds may occur in moderate to high-grade metamorphic rocks in which a layer, such as quartzite or amphibolite with a relatively low mean ductility (high competence), is interlayered with a weak material such as schist (Figure 14–31b). The schist will flow ductilely, but the quartzite will deform semi-brittlely or have a slower rate of ductile deformation. As a result, the quartzite may not change thickness much during folding. In flexural-flow folding, the layers that do not undergo appreciable thickness changes control the overall shape of the folds.

Ideal ***passive-flow folds*** are similar folds (Figure 14–32) that involve plastic deformation. The layering acts only as a displacement marker to record the effects of deformation. Layers are thinned or thickened in the flow direction equally in all rock types. Where they die out, passive-flow folds become nonsimilar and plot in the Class 1C and Class 3 fields. Passive-flow folds form in metamorphic rocks with low mean ductility and low ductility contrast regardless of grade, and in salt, glacial ice, and water-saturated unconsolidated sediments wherein properties are uniformly ductile (Figures 2–2a, 2–30a, 2–31, and 15–2).

The Donath and Parker classification has one important shortcoming: it is genetic and is meant to explain fold mechanics, thus requiring the user to make a judgment each time a fold is categorized. The user must have more than a cursory understanding of fold mechanics and must be able to apply impartially the criteria for mechanical processes. The principal advantage of the classification lies in its usefulness in the field. For example, to conclude that a fold is a flexural-slip fold, a geologist must observe the fold shape and bed thickness (whether thickness changes or remains constant), make a judgment about ductility contrast and mean ductility, and show that slip occurred parallel to bedding. That the observations can be made in the field is an advantage, but interpretations required by the observer may prove a disadvantage.

Which Classification to Use?

The brief comparison in this chapter points out advantages and disadvantages of descriptive and genetic classifications. No classification is ideal. One reason is that few folds are ideal, and therefore few can be

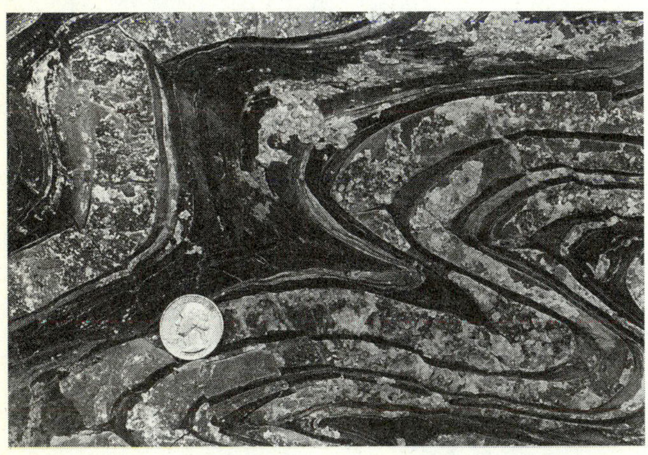

(a)

(b)

FIGURE 14–31
(a) Flexural-flow folds in Poultney Formation shale and limestone near Whitehall, New York. Weak shale layers (dark) thicken and thin by flow; stronger limestone layers change thickness less in the short limbs and hinges. They tend to fracture in the long limbs of folds and where limestone beds are thick. These are also quasi-flexural (disharmonic) folds. (b) Flexural-flow folds in Lower Cambrian Hamill Group quartzite and schist, northern Selkirk Mountains, British Columbia. Darker schist layers undergo greater thickness changes than do the lighter quartzite layers. (RDH photos.)

pigeonholed readily. The descriptive classifications may be too rigid and may not describe real folds, while genetic classifications may be too subjective. Perhaps the ideal classification would be usable in the field and would yield information about the nature of the folding process without demanding much judgment by the observer. Folding is a complex process, and it is often difficult to determine the boundary between brittle and ductile behavior at the time of deformation; some rocks may exhibit brittle behavior, while others in the same fold may exhibit ductile behavior. (The interbedded shale and limestone producing flexural flow is an example.) Rock composition, texture, and overall physical properties involve many more variables than can be accounted for in any simple classification.

Which fold classification should be used? It depends partly on what the observer needs to determine about the folds in an area. Probably the best scheme is Ramsay's standard classification—given an ability to distinguish between parallel folds and similar folds in the field, but Huddleston's is also a very good descriptive classification based on fold shape and it can be easily used in the field. The Donath and Parker classification is still used because of its ready application in the field.

In the next chapter, we will examine the mechanics of fold formation, further expanding our ability to understand the differences between the mechanisms that produce folds.

FIGURE 14–32
Passive-flow and quasi-flexural folds in Middle Proterozoic migmatitic biotite gneiss, Thor-Odin dome, Shuswap complex, British Columbia. These folds have a similar or near-similar geometry. Quartz-feldspar layers (light-colored) provide minimal control of the shape of folds. (RDH photo.)

ESSAY

Folds in the Development of the Petroleum Industry

The first petroleum reservoir trap recognized was an anticlinal fold (Howell, 1934). Petroleum had been known from seeps since ancient times, and tar had been recovered by mining, but actually drilling into the Earth to recover oil was not thought of until holes were drilled in the first half of the 19th century in Germany, Canada, and Kentucky. Edwin L. Drake drilled his first well in August of 1859 in Pennsylvania. Drake drilled into an anticlinal fold in Paleozoic sedimentary rocks of the Appalachian basin, but he did not formulate the anticlinal theory. It is uncertain who did originate the theory, but in 1846, Sir William Logan suggested a relationship between oil seeps and anticlinal folds in rocks of the Gaspé Peninsula in Québec near the mouth of the St. Lawrence River.

Encouraged by Drake's discovery and Logan's suggestion of a link between petroleum and anticlines, geologists began exploring for petroleum under conditions thought to be similar to those responsible for Drake's success. Scientific papers proclaimed the merits of the anticlinal theory and its relationship to the accumulation of both oil and natural gas. Two papers by I. C. White (1885, 1892) on the occurrence of hydrocarbons in West Virginia and Pennsylvania are classic statements and applications. White's use of the anticlinal theory to find hydrocarbons, in his opinion, helped to dispel the notion that geologists were really not very helpful in oil prospecting. He had actually used scientific methods to predict the occurrence and location of economically recoverable quantities of hydrocarbons, and inadvertently he became known as the "father of the anticlinal theory" (Levorsen, 1954).

Years after the petroleum industry became a major factor in the United States and the world economy, geologists realized that many structures trapping petroleum were not anticlines and that an anticline cannot trap petroleum except under special stratigraphic conditions. A. W. McCoy and W. R. Keyte (1934) recommended that anticlines be considered part of a group of structural traps. Others proposed that a parallel group of stratigraphic traps also be recognized. By the 1930s, geologists applying scientific concepts had helped to establish the petroleum industry, many structural and stratigraphic traps had been identified, and geologists had become an essential part of petroleum exploration.

References Cited

Howell, J. V., 1934, Historical development of the structural theory of accumulation of oil and gas, *in* Problems in petroleum geology: Tulsa, Oklahoma, American Association of Petroleum Geologists, p. 1–23.

Levorsen, A. I., 1954, Geology of petroleum: San Francisco, W. H. Freeman, 703 p.

Logan, Sir William, 1846, Report of progress, 1844: Ottawa, Geological Survey of Canada.

McCoy, A. W., and Keyte, W. R., 1934, Present interpretations of the structural theory for oil and gas migration and accumulation, *in* Problems of petroleum geology: Tulsa, Oklahoma, American Association of Petroleum Geologists, p. 253–307.

White, I. C., 1885, The geology of natural gas: Science, v. 5, p. 521–522.

White, I. C., 1892, The Mannington oil field and the history of its development: Geological Society of America Bulletin, v. 3, p. 187–216. (An appendix describes the anticlinal theory for natural gas.)

Questions

1. What is the main pitfall in relating small-scale and large-scale folds using Pumpelly's rule?
2. If a fold axial surface has a strike of N 35° E (035) and a dip of 49° SE, what is the range of possible orientations for the trend and maximum magnitude of plunge of the fold axis?
3. What is meant by the statement, "These folds have a northeast vergence?"
4. Determine the positions and nature of major fold-hinge zones by reconstruction from the small structures represented in the outcrop sketches shown here (cross section view). The layers in the second and fourth exposures are not of the same layer; those in the first, third, and fifth exposures are the same layer.

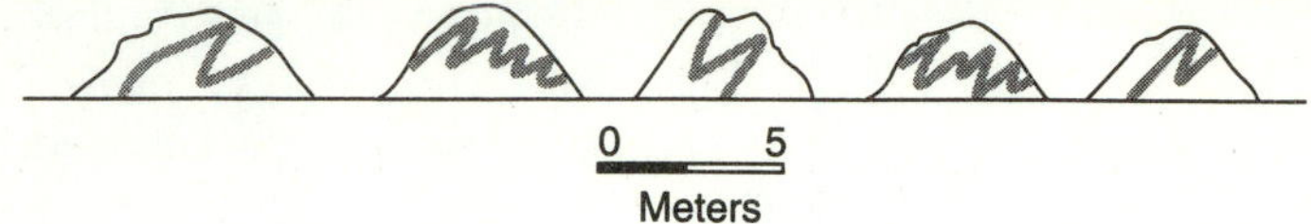

5. What kind of hinge does an ideal dome or basin have?
6. How can a large fold structure be defined using the enveloping surface?
7. What are the relative merits of genetic and descriptive classifications? Shortcomings?
8. What is the basis of the Donath and Parker fold classification? Ramsay's classification?
9. What information would you need to determine whether a fold is a passive-flow fold? A Class 1C fold? How and where would you obtain this information?
10. The following orientations (trend, plunge, sense of rotation—S or Z) of mesoscopic folds above the Chocolate Mountains thrust were collected in southeastern California by Gordon Haxel of the U.S. Geological Survey (see Dillon and others, 1990, Journal of Geophysical Research, v. 95, p. 19,953–19,971). Plot these data on a stereonet, indicating sense of rotation on each point, and then determine the transport direction of the thrust.

N 8 E 19 NE (Z)	N 32 E 4 NE (S)	N 54 W 27 NW (Z)	N 22 E 7 NE (Z)
N 33 E 7 NE (Z)	N 39 W 14 NW (Z)	N 17 W 8 NW (Z)	N 28 E 22 NE (Z)
N 79 E 3 NE (S)	N 1 W 12 NW (Z)	N 68 E 4 NE (S)	N 82 E 7 NE (S)
N 28 E 31 NE (Z)	N 48 E 34 NE (S)	N 86 E 1 NE (S)	N 52 W 4 NW (Z)
N 86 W 4 NW (Z)	N 89 W 27 NW (Z)	N 19 E 22 NE (Z)	N 31W17 NW (Z)
N 88 E 4 NE (S)	N 36 E 19 NE (S)	N 72 W 19 NW (Z)	N 89 E 12 NE (S)
N 37 E 53 NE (S)	N 69 E 12 NE (S)	N 61 W 27 NW (Z)	N 23 W 11 NW (Z)
N 34 E 7 NE (Z)	N 63 W 31 NW (Z)	N 22 E 24 NE (Z)	N 32 E 34 NE (Z)
N 66 W 3 NW (Z)	N 32 E 17 NE (S)	N 24 E 33 NE (Z)	N 70 W 3 NW (Z)
N 53 W 32 NW (Z)	N 24 E 3 NE (Z)	N 73 W 5 NW (Z)	N 87 E 28 NE (S)
N 73 E 8 NE (S)	N 66 W 8 NW (Z)	N 63 E 12 NE (S)	N 76 E 6 NE (S)
N 59 W 6 NW (Z)	N 82 E 33 NE (S)	N 75 E 9 NE (S)	N 56 W 7 NW (Z)
N 79 W 18 SE (S)	N 78 E 22 SW (Z)	N 58 E 33 SW (Z)	N 68 W 8 SE (S)
N 58 W 27 SE (S)	N 57 W 23 SE (S)	N 74 W 23 SE (S)	N 82 E 24 SW (Z)
N 87 W 4 SE (Z)	N 81 E 4 SW (Z)	N 58 W 9 SE (S)	N 47 W 28 SE (S)
N 49 W 17 SE (S)	N 43 W 8 SE (S)	N 28 W 16 SE (S)	N 32 W 11 SE (S)
N 62 E 7 NE (S)	N 69 W 7 NW (Z)		

Further Reading

Donath, F. A., and Parker, R. B., 1964, Folds and folding: Geological Society of America Bulletin, v. 75, p. 45–62.
A useful genetic synthesis and classification of folds based on mean ductility and ductility contrast. The classification is easy to follow and use in the field.

Fleuty, M. J., 1964, The description of folds: Proceedings of the Geological Association: v. 75, p. 461–492.
Critically reviews the descriptive terminology of folds and provides worthwhile comments about the terms Fleuty considers most useful.

Hansen, E., 1971, Strain facies: New York, Springer-Verlag, 207 p.
Describes a useful method for determining transport direction of a large fold from sense of overturning and orientation of parasitic folds.

Hobbs, B. E., Means, W. D., and Williams, P. F., 1976, An outline of structural geology: New York, John Wiley & Sons, 571 p.
The chapter on folds is especially good and contains a useful discussion of fold mechanics.

Hudleston, P. J., 1973a, Fold morphology and some geometrical implications of theories of fold development: Tectonophysics, v. 16, p. 1–46.
Takes the geometric classifications to a higher level with the modification of Ramsay's classification. It also presents a useful discussion of fold mechanisms.

Ramsay, J. G., 1967, Folding and fracturing of rocks: New York, McGraw-Hill, 568 p.
Discusses several of the fold classifications outlined in this chapter. It is an advanced text, but elementary structure students should find the mathematics simple.

Williams, G. D., and Chapman, T., 1979, The geometrical classification of noncylindrical folds: Journal of Structural Geology, v. 1, p. 181–185.
An easy-to-follow classification that should prove useful in the field, as it requires only observation of relationships between axes, hinges, and limbs and measurement of interlimb and hinge angles.

15

Fold Mechanics

One of the intriguing features of layered rocks deformed during natural orogenic processes is that their surfaces are often curved to form folds. The methods of classification of these folds and the determination of their mechanisms of formation have long been controversial subjects in the study of rock structure.

JOHN G. RAMSAY, 1967, *Folding and Fracturing of Rocks*

FOLDED LAYERS IN ROCK HAVE BEEN OBSERVED FOR MANY centuries. The ancient Greeks reported folded rocks in their exploration of the Mediterranean region. Leonardo da Vinci (1452–1519) was impressed by contorted strata preserving marine fossils high in the Alps, as indicated in Vallisnieri's early eighteenth-century sketch (Figure 15–1). Sir James Hall (1761–1832) in 1788 described folds in the sea cliffs on the coast of Scotland (Johnson and Ellen, 1974).

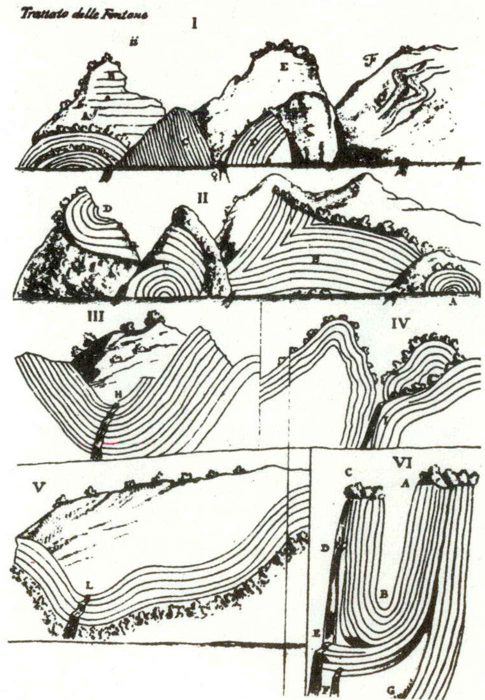

FIGURE 15–1
Folded rocks from sketches by Antonio Vallisnieri of folds on mountainsides in the Alps of Switzerland and Germany. (From *The Birth and Development of the Geological Sciences* by Frank Dawson Adams, © 1938, 1966, Dover Publications, Inc.)

The geologists who first described folds doubtlessly speculated about their origin. The phenomenon is readily observed in glacial ice (Figure 15–2) and in soft sediment deformed by gravity (Figure 2–2). The close association of folds and cleavage was first noted by several geologists before 1850, directly linking forces and folding (Chapter 17). Da Vinci's uplifted and contorted fossiliferous limestones, high above sea level, required tectonic forces to elevate them—forces unexplained for centuries.

The widespread distribution, occurrence on all scales, and great variety of folds provide convincing evidence that they are truly a major class of geologic structures—but *how* do these structures form? In Chapter 14, we described the extensive fold terminology. In this chapter, we will explore the mechanics of folding, which requires us to consider the physical and chemical environment of the rock body being deformed.

The fold mechanism is strongly influenced by the factors that govern other kinds of deformation: temperature, pressure, fluid, and overall properties of the rock body as determined by the composition, texture, and character of individual layers. The relative thickness and contrasting properties with direction—***anisotropy***—greatly influence the kind of fold mechanism affecting a rock body. Anisotropy is in turn affected by changes in temperature and pressure.

Attempts by Bailey Willis (1857–1949) to model folds and fold mechanisms date back nearly a century. His goal was to discover the relationships between folds and thrusts in the Appalachian Valley and Ridge province by duplicating them in a pressure box (Willis, 1893; Figure 10–1). H. M. Cadell, a British geologist of the early twentieth century, used a pressure box to produce folds in sediment with layers having contrasting properties. The studies by Cadell and Willis, although primitive by present standards, laid the foundation for our current understanding of fold

FIGURE 15–2
Similar to similar-like folds several kilometers in amplitude produced by ductile flow in ice and moraine in the Malaspina Glacier, southeastern Alaska. The St. Elias Mountains are in the background. (Austin Post, U.S. Geological Survey.)

mechanics. For example, Willis's break thrusts and shear thrusts of 1893 are very similar to John Suppe's fault-propagation folds and fault-bend folds of 1985.

As we know, folds bear a remarkable similarity to sine waves and other curves that can be expressed using simple mathematical relationships (Figure 15–3), and some mathematical studies of folding are based on wave mechanics. In recent years, modeling of folds has also been done both experimentally and by computer. These methods provide ways to model strain distributions in folds and to otherwise simulate natural fold mechanisms.

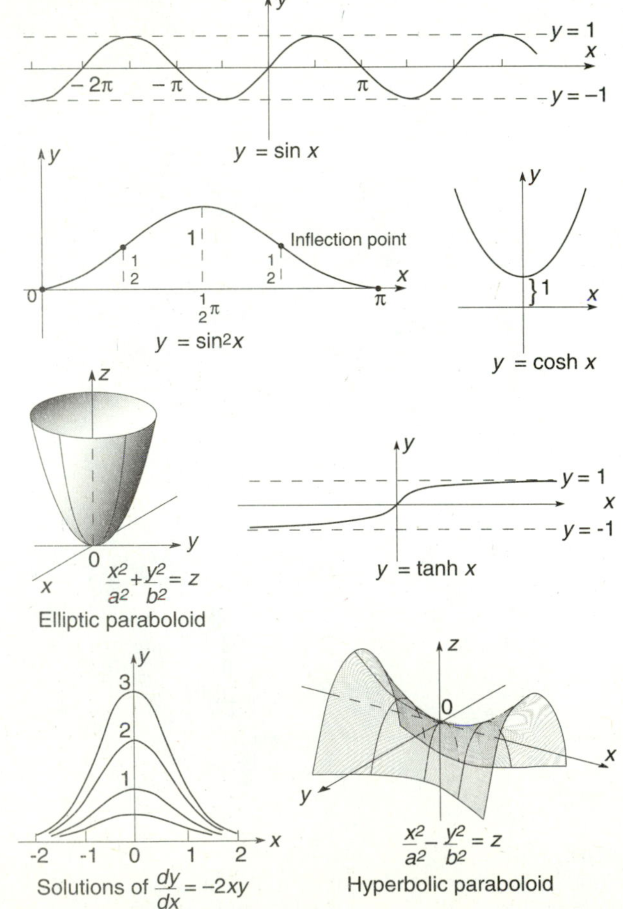

FIGURE 15–3
Simple mathematical representations of various fold shapes, along with their graphs.

FOLD MECHANISMS AND ACCOMPANYING PHENOMENA

Several fold mechanisms have been identified (Figure 15–4a), including ***buckling, bending, and passive (ductile) flow***; the last is sometimes called ***passive strain amplification.*** These mechanisms may be accompanied by the kinematic phenomena of *flexural slip* and *flexural flow.* Richard H. Groshong (1975a) has discussed each and evaluated their relationships to deformation mechanisms involved in a rock mass during folding in unmetamorphosed sedimentary rocks in the Appalachians of Pennsylvania. There Groshong recognized that pressure solution was an important deformation mechanism and that a model including twin gliding, translation gliding, and adjustments along

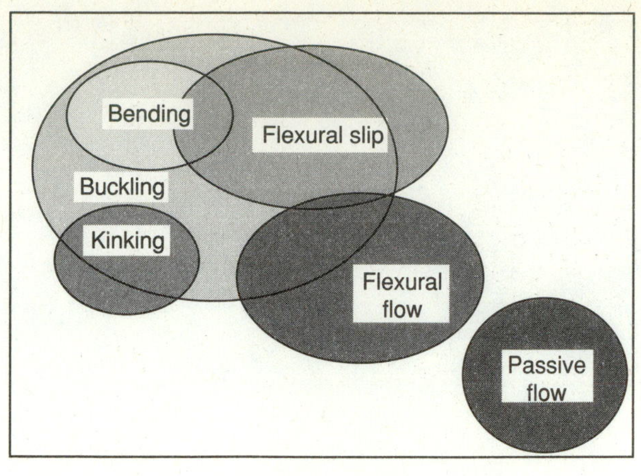

(a)

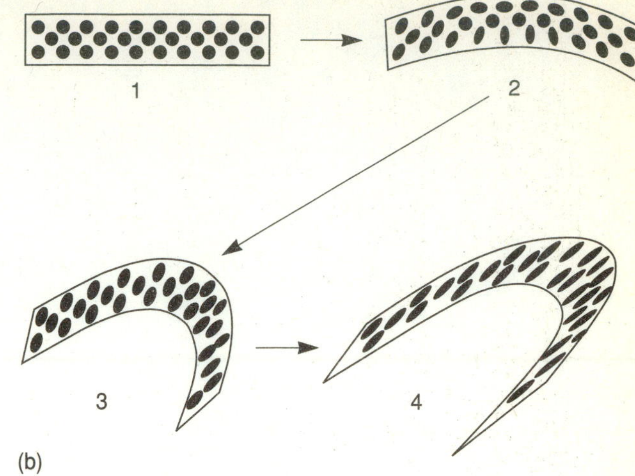

(b)

FIGURE 15–4
(a) Possible relationships of fold mechanisms. The horizontal axis relates increasing temperature and pressure to fold mechanism. The vertical axis reflects contrasting mechanisms under different conditions. Bending and buckling occur wherever layering influences fold shape. Kinking occurs similarly but is independent of buckling and bending. Buckling and bending probably occur at early and later stages over the same temperature range. (b) Stages in the development of a folded layer showing several different processes occurring at early and later stages during the progressive deformation of the layer by folding. Stage 1 is the undeformed condition. Stage 2 involves slight buckling of the layer accompanied by internal heterogeneous strain involving extension of the outer arc and shortening of the inner arc (tangential-longitudinal strain; see Figure 15–15). Stage 3 involves progressive deformation of the layer by heterogeneous simple-shear strain. In Stage 4, this is later superposed by homogeneous strain and pure shear. (b is from J. G. Ramsay, *Folding and fracturing of rocks,* © 1967, McGraw-Hill Book Company. Reproduced with permission).

grain boundaries accounted for some of the deformation accompanying folding; he also concluded that much of the deformation during folding is accomplished by displacement along fractures. As we saw in Chapter 11, fault-bend and fault-propagation folds involve bending and flexural slip associated with thrust faulting.

Parallel-concentric folds and similar-type folds may be created by more than one mechanism. Some folds are the product of only one mechanism; others, of more than one. Thus, the geometric form of a particular fold may be the end product of one or more fold mechanisms—whether the mechanism is parallel (Ramsay's Class 1B), concentric, parallel-concentric, modified concentric (Class 1C), similar (Class 2), or similar-like (Class 1C or Class 3). Each process operates according to the nature of the deforming rock mass and the physical conditions imposed on it.

One fold mechanism, such as buckling accompanied by flexural slip, may operate during the early history of a fold, and then buckling accompanied by flexural flow may dominate as the fold tightens and pressure increases during progressive deformation (Figure 15–4b). Under high temperature and pressure, layers may no longer control the shapes of the folds but may serve only as strain markers. Folds formed by ductile flow result in differential simple shear across layers and displacement of layers. Only seldom can we prove that a sequence of multiple fold mechanisms (as just described) affected a rock mass undergoing progressive deformation. Relict structures formed by several mechanisms have been observed in rocks, indicating that a sequence of mechanisms operated during folding.

Flexural Slip

Layers play a dominant role in folding of most rocks at the low temperature and pressure found at shallow depths in the Earth. For layers to maintain constant thickness during folding of a mass of uniformly layered strong rocks—such as thickly bedded carbonate or sandstone—they must slip past one another (Figures 15–4a and 15–5). (For an instant demonstration of flexural slip, simply flex the pages of this book and watch the pages slip past each other.) Where the mechanical properties of successive layers differ, as with interlayered limestone and shale, flexural slip still occurs, but the shale may become crumpled into the hinges of folds; this produces thickening in the shale layers without ductile flow of material (Figure 15–6a). The phenomenon of flexural slip commonly accompanies the bending (Figures 15–7 and 15–8) and buckling (Figure 15–9) mechanisms and is recognized by slickensides or fibers on bedding surfaces (Figure 15–6b). If flexural slip does occur, the fibers may be oriented perpendicular to fold hinge lines but may not be perpendicular because of variations in the direction of motion between layers (Ramsay, 1967).

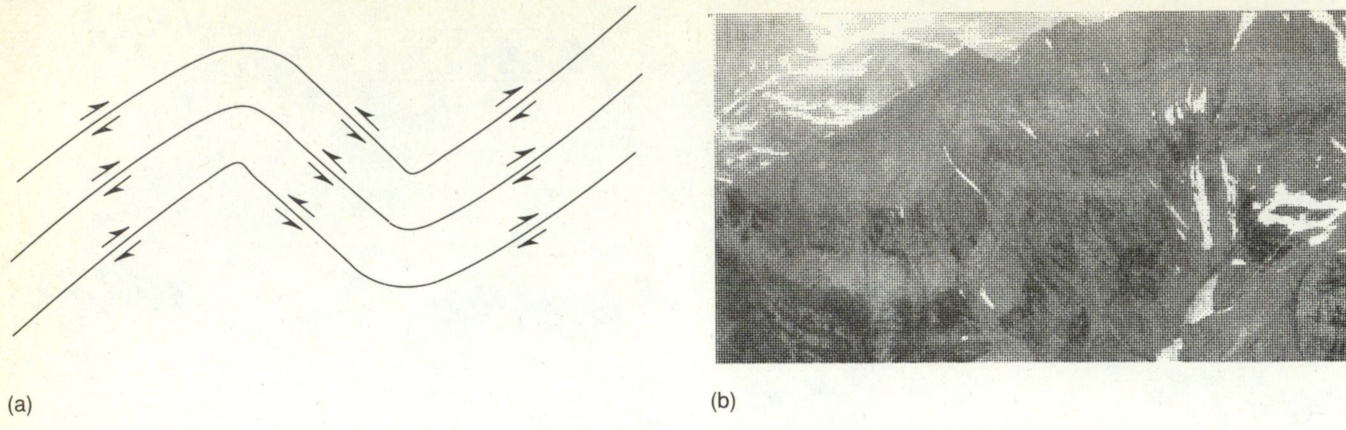

FIGURE 15–5
(a) Flexural-slip folds. Arrows indicate relative motion on each layer surface during folding. (b) Flexural-slip buckle folds in Upper Proterozoic Hamill Group sandstone and shale, Selkirk Mountains, British Columbia. (RDH photo.)

(a)  (b)

FIGURE 15–6
(a) Abandoned barite mine west of Glenwood, Arkansas, exposing interlayered sandstone and shale near the base of the upper Mississippian Stanley Shale. Shale layers were thickened by crumpling into hinges of folds rather than by ductile flow.
(b) Slickenfibers on folded bedding surfaces in calcareous sandstone produced by flexural slip, Middle Ordovician Tellico Sandstone near Tallassee, Tennessee. Note that the dip of bedding decreases toward the top background of the photo into the hinge of the fold. (RDH photos.)

Bending

Rock layers may be subjected to ***bending*** (Figure 15–7 and 15–8), a mechanism that involves application of force across layers. Generally, it produces folds that are very gentle with large interlimb angles. The term has also been used by Hans Ramberg (1963) and John Ramsay (1967) to describe bending in a different way, one that involves passive flow across layers in response to transverse forces and the way folds are generated. Folds produced by bending, in the sense intended here, are common in continental interiors—cratons—where vertical forces may be directed at high angles to originally horizontal bedding, producing the broad domes, basins, swells, and arches so common in cratons. Bending occurs during duplex formation as the rocks forming the roof are arched from below by the growing duplex (Chapter 11; Figure 11–14c). Subduction of rigid objects, such as a seamount, produces bending in the overriding plate. Other examples of bending include arching of cover rocks above a basement uplift, as in the Colorado and Wyoming Rockies (Chapter 11), and above an intruding pluton. Flexural bending of lithospheric plates also occurs at subduction zones and adjacent to oceanic ridges, at smaller accumulations of volcanic rock in the oceans (such as guyots, seamount chains, and aseismic ridges), and where loading by ice or sediment causes the lithosphere to undergo flexural readjustment (Turcotte and Schubert, 1982).

In rocks subjected to bending, layers are bent like an elastic beam that has been supported at the ends and loaded in the middle (Figure 15–8). The layers undergo flexural slip, sliding past each other just as if they were buckled (Figure 15–9), and the layers are deflected into gentle folds. Mathematically, modeling the bending of layers as elastic beams may work well for low-amplitude folds but becomes inexact for large segments of the lithosphere (and also smaller folds) because of inelastic behavior.

Fault-bend folds (Suppe, 1983; Figure 11–11a) form by the bending of a thrust sheet as it passes over a ramp. Such folds, including those produced during duplex formation, basement uplifts, and subduction of rigid objects, are all examples of bending folds formed by tectonic forces. Can you think of others?

Buckling

Folds form by ***buckling*** where force is applied parallel to layering in rocks (as at the ends of an elastic beam) and the easiest direction of relief is normal to the direction of force application (Figure 15–9). The layers are said to buckle, and the structures formed are called ***buckle folds***. Flexural slip commonly accompanies buckling at low temperature and pressure (Figure 15–10), whereas buckling of a stiff layer (with high viscosity, strength, and competence) embedded in a less-stiff medium may occur at higher temperature and pressure without flexural slip (Figure 15–11). The result at low temperature is sinusoidal parallel and parallel-concentric folds. At high temperature, other mechanisms such as flexural flow (discussed below) may accompany buckling to produce similar-like folds. Buckling and formation of a thrust fault between an anticline and syncline can produce fault-propagation folds at low temperatures (Chapter 11).

Buckling is commonly preceded by *layer-parallel shortening.* The layers are shortened laterally and thickened perpendicular to layering and to the direction of applied stress. Buckling accompanied by inelastic strain within strong layers may modify the shape of the fold and may accelerate strain rate. For example, folds formed by a combination of buckling and pressure-solution strain (Figure 15–12) maintain the shapes of buckle folds but may develop a strong cleavage because of the associated flattening. This process may produce strain-softening if strain within the strong layers increases at constant or decreasing stress during progressive deformation. Pressure solution accompanying buckling, however, produces a linear increase in strain with increasing stress.

Numerous theoretical models of buckle folds were proposed in the 1960s and 1970s. Some of them (Biot, 1961; Biot, Odé, and Roever, 1961; Chapple, 1968, 1969; Hudleston and Stephansson, 1973; Ramberg, 1960, 1961; Parrish and others, 1976) involved buckling of a single elastic layer suspended in a medium of lower viscosity. Mathematical models have also been devised by M. A. Biot (1963, 1965a, 1965b) and William Chapple (1970) for buckling of multilayer sequences. Viscosity ratios are commonly assumed to be 100:1, or higher, to simulate the contrast between strong rocks, such as dolomite or sandstone, and weak rocks, such as shale, evaporite, or coal. This ratio is actually geologically indeterminable for most situations where such contrasts occur, because both physical and chemical conditions change during and after folding.

These theoretical studies have clarified the relationship between the thickness of the dominant (strong) member in the sequence being folded and the wavelength of buckle folds produced. Biot (1961) and Ramberg (1961) showed independently that an isolated strong (high-viscosity) layer embedded in a weaker (lower-viscosity) material will form buckle folds wherein wavelength is given by

$$\lambda_d = 2\pi t \sqrt[3]{\frac{\mu_1}{6\mu_2}}, \qquad \textbf{(15–1)}$$

FIGURE 15–7
Candidate for a bending flexural-slip fold in Middle to Upper Ordovician limestone on the Cumberland River at Old Hickory (near Nashville), Tennessee. This could be a buckle fold, but it is located in the "stable interior" where large domes and basins, and possibly smaller folds, are thought to be related to vertical tectonic movement caused by loading of the crust during the late Paleozoic by thrust faulting originating farther east in the Appalachians. (RDH photo.)

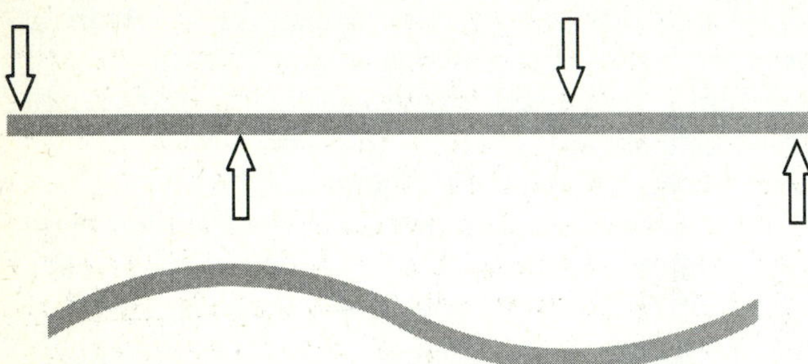

FIGURE 15–8
Formation of folds by bending. The layer is folded by deflection in the same directions as the applied stresses. This produces the same shape as is produced by buckling, but the actual mechanism involves bending of the ends of a beam that is buttressed in the middle. Compare with Figure 15–9.

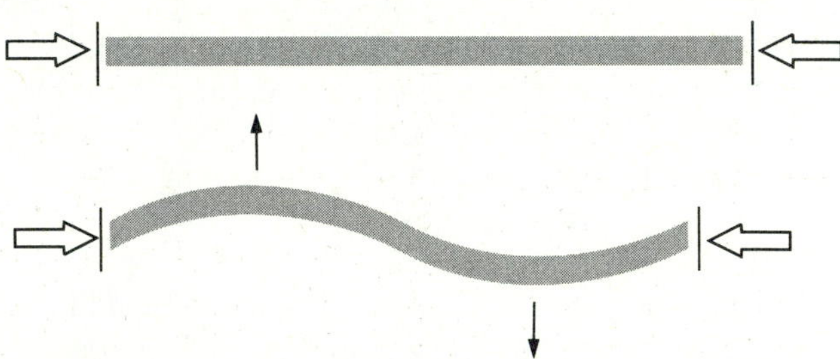

FIGURE 15–9
Formation of buckle folds. The layer is folded by deflection (small black arrows) normal to the shortening direction (large open arrows).

where λ_d is the dominant wavelength (of folds in the strong layer), t the thickness of the strong layer, μ_1 the viscosity of the strong layer, and μ_2 the viscosity of the weak supporting medium or matrix. Biot (1961) recognized that where viscosity ratios are 100:1 or less, layer-parallel shortening becomes important and modifies the shapes of folds produced. Using Biot's observation, J. A. Sherwin and Chapple (1968), and then Peter Hudleston (1973), modified equation 15–1 to

$$\lambda_d = 2\pi t \sqrt[3]{\frac{\mu_1(S+1)}{6\mu_2\, 2S^2}}, \qquad (15\text{–}2)$$

where S is the square root of the ratio of quadratic elongations $(\lambda_1/\lambda_2)^{1/2}$ of the homogeneous deformation on which buckling is superposed. Equations 15–1 and 15–2 approximately depict the geometry of low-amplitude ptygmatic folds (Chapter 14; Figure 15–13), providing reasonable estimates of dominant wavelength only if μ_1/μ_2 is large (>100), that is, if $\lambda_d/t > 16$ (Fletcher, 1974).

Buckling theory enables us to predict the geometry of simple folds and demonstrates the utility of the relationship between the thickness of the dominant (strong) member and the wavelength. J. B. Currie, H. W. Patnode, and R. P. Trump (1962) used theory, field observations, and photoelastic and other rock mechanics laboratory experiments to predict fold shapes, and related fold wavelength to the thickness of the dominant member in buckle folds (Figure 15–13). They studied folds produced in stiff layers in a multilayer package (not isolated stiff layers in a less stiff matrix), and so equations 15–1 and 15–2 could not be applied here. That most folds measured fall on or near the line

FIGURE 15–7 (continued)

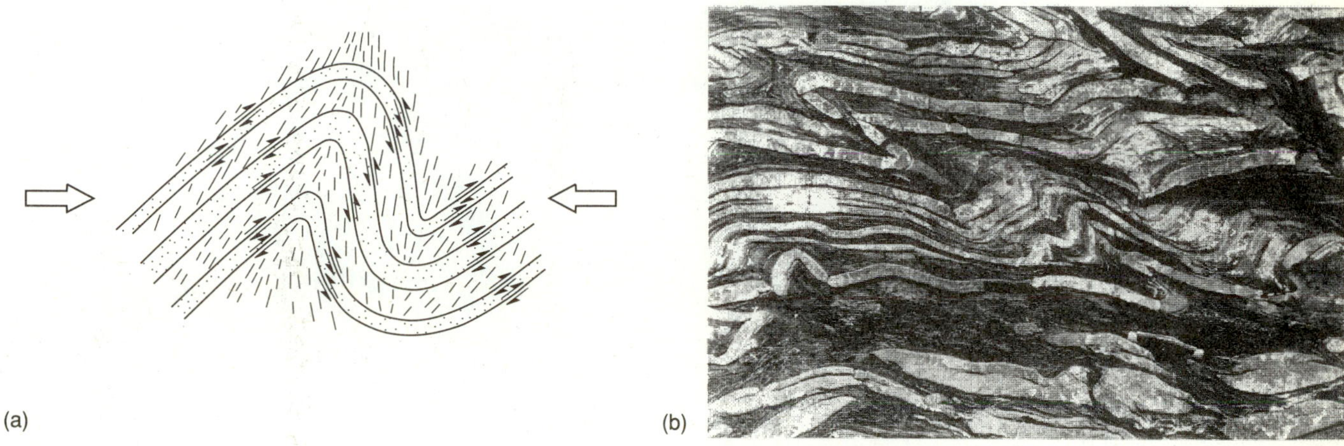

(a) (b)

FIGURE 15–10
(a) Buckling and flexural slip (small black arrows) at low temperature and pressure. The strong layers are stippled. The weak layers contain a cleavage subparallel to fold axial surfaces. (b) Strongly disharmonic buckle folds of limestone (light) in shale (dark) near Whitehall, New York. The differences in thickness between strong limestone layers interlayered with weak shale and the room problem during folding prevent the folds from attaining the same shape throughout. (RDH photo.)

in Figure 15–14 suggests that the rocks have been folded to the maximum curvature possible for the strong unit without internal ductile flow. Remarkably, this relationship holds without regard to rock type.

A characteristic strain geometry related both to bending and buckling mechanisms was termed ***concentric-longitudinal strain*** by Ramberg (1961) and ***tangential-longitudinal strain*** by Ramsay (1967). It was first suggested by L. Ulbo de Sitter (1959) as a fold mechanism involving internal deformation of layers of finite thickness. The inner arcs of the buckled layers are subjected to layer-parallel shortening, whereas the outer arcs of each layer are subjected to layer-parallel extension (Figure 15–15a). This results in a ***neutral surface*** of no finite strain that in each layer separates the zone of extension from the zone of shortening, and flexural slip must occur between layers. During bending the outer arc is extended but the inner arc is *not* shortened, because the ends remain fixed. Folds exhibiting tangential-longitudinal strain may be identified by small structures that indicate shortening in the inner arcs of folds (Figure 15–15b); examples are crenulations and pressure-solution cleavage. In the outer arcs, the layers may be broken by extension fractures or contain other structures (such as boudinage) that indicate extension.

Ramberg (1961) has shown that a simple relationship exists between strain and the ratio of wavelength to thickness in a fold: the maximum concentric shear strain (γ_{max})—shear strains parallel to layering—and maximum concentric strains (ε_{max}) in the crests of folded layers of this type with small ratios of amplitude to wavelength may be related to the thickness of the dominant layer (t) and length of arc of one fold (L) as

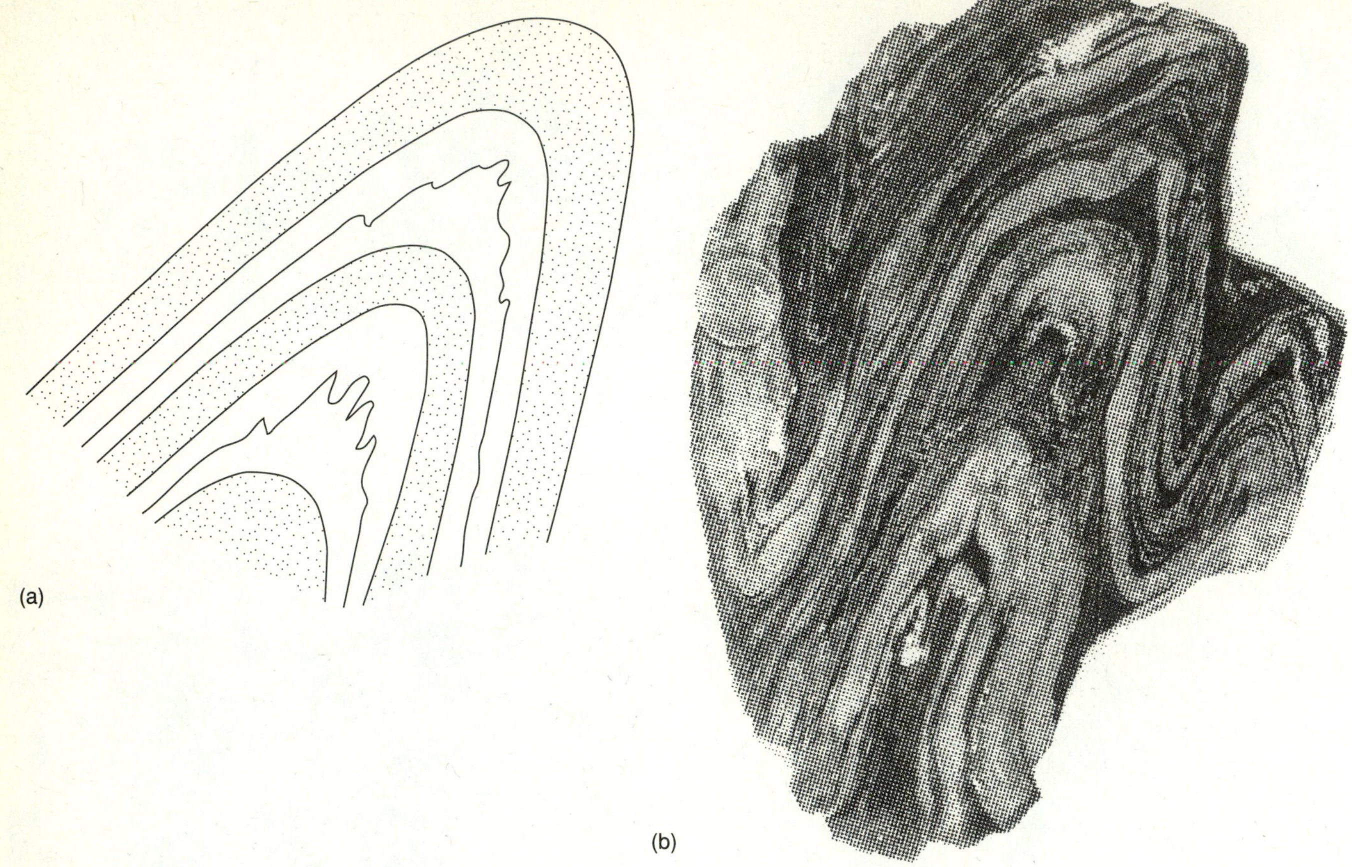

FIGURE 15–11
(a) Buckling and flexural flow at high temperatures and pressures, or in weak materials (salt or glacial ice interlayered with anhydrite or moraine) at low temperatures and pressures, produce similar-like folds without flexural slip. Strong layers are stippled. (b) Sawed specimen of Poor Mountain Amphibolite from near Hendersonville, North Carolina. Light layers are quartz- and feldspar-rich; dark layers are amphibole-rich and flowed more during folding than did the light layers. These similar-like flexural-flow buckle folds were produced under middle- to upper-amphibolite-facies (sillimanite grade) conditions. (Specimen courtesy of Timothy L. Davis, North Carolina Geological Survey.)

$$\frac{\gamma_{max}}{\varepsilon_{max}} = 2\pi\frac{t}{L}. \qquad \textbf{(15–3)}$$

This equation may be useful in determining the ratio of shear to longitudinal strains in layers that have undergone tangential-longitudinal strain, if a strain marker is present, and may be applied only to low-amplitude folds. In deriving it, Ramberg assumed that the layers composing the folds behave as viscous materials.

Edward C. Beutner and Frederick A. Diegel (1985) studied fibers that developed in pressure shadows and veins during formation of buckle folds in the Martinsburg Slate in New Jersey. They found that the hinges in anticlines and synclines actually migrated away from each other by lengthening the connecting common limb (Figure 15–16). The folds were then flattened during the later stages of folding, bringing the fold axial surfaces—but not the hinges—closer again.

FIGURE 15–12
Pressure solution on the inner arcs of a fold may weaken the layer and produce strain-softening during buckling.

Passive Slip—A Problematical Mechanism

Passive slip has been described (Knopf and Ingerson, 1938; Donath and Parker, 1964; Ramsay, 1967) as slip at an angle to layering that results in a new cleavage or schistosity that accommodates movement parallel to the new surface (Figure 15–17). Related folds are called *passive-slip folds* (Donath and Parker, 1964) or *shear folds* (Ramsay, 1967). Bedding, or compositional layering, serves only as a strain marker that records displacement parallel to the cleavage or schistosity. Even so, passive slip is considered by many geologists

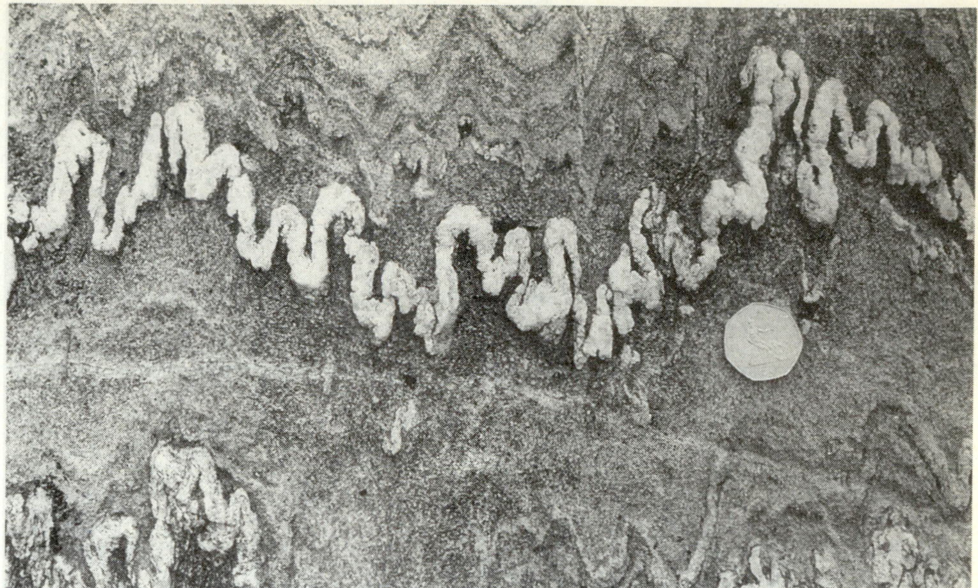

FIGURE 15–13
Buckling producing ptygmatic folds of a pegmatite layer in Middle Proterozoic Moine Series clastic rocks, Scottish Highlands. (Anthony L. Harris, University of Liverpool.)

to not be a viable fold mechanism because cleavage and foliation planes have been shown in many places to form parallel to the *XY* plane of the strain ellipsoid (Chapter 17). Paul Williams (1976) reviewed the evidence for shear parallel to cleavage and found that cleavage may not always form parallel to the *XY* plane, and movement may occur parallel to cleavage if the strain is not coaxial, as during folding. *Coaxial strain* is commonly thought of as strain that accumulates without rotation of the principal strain axes with respect to reference lines in the deforming rock mass. It is difficult to understand how a structure formed parallel to the *XY* plane of the strain ellipsoid and to the axial surfaces of folds can be immediately transformed into shear planes and take part in the folding process. Consequently, passive slip should *not* be considered a fold mechanism under conditions where cleavage forms parallel to the *XY* plane, but it may be possible under conditions of noncoaxial shear deformation. Associated folds probably form by buckling, with or without flexural slip, and the cleavage forms as buckling produces a component of pure shear in the rock mass. Yet displacement parallel to axial-plane foliations has been observed in rocks of higher metamorphic grade (Figure 15–17), as well as in low-grade rocks. A different orientation of the strain ellipsoid that permits the foliation, after it has been shear plane in the strain ellipsoid may require a two-stage mechanism (Figure 15–18). Ramsay (1962) has pointed out that a variable amount of flattening in a rock mass can lead to shear that parallels the axial surfaces of the folds. At large shear strains, a small angle exists between the *XY* plane of finite strain and a shear plane, and so slip on foliation may become more likely if shear strains are large (Chapter 10).

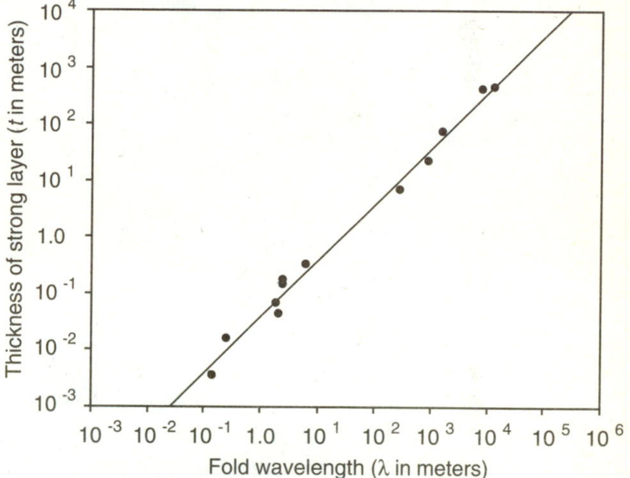

FIGURE 15–14
Log-log plot of thickness of strong layer versus wavelength (ranging from a few centimeters to more than 1 km) of multilayer buckle folds from West Virginia, New York, and Alberta. The slope of the curve (at 45°) indicates a linear relationship. The magnitude of the λ coordinate where $t = 1$ indicates wavelength is about 27 times the thickness of the strong unit. (From J. B. Currie, H. W. Patnode, and R. P. Trump, 1962, Geological Society of America *Bulletin*, v. 73.)

The passive-slip mechanism was originally intended to explain the apparent displacement of a marker layer parallel to a cleavage or foliation surface. The apparent displacement can also be produced by removing part of the layer by pressure solution (Figure 15–19a), but displacement of bedding has been observed in low-grade slates unaffected by pressure solution (Onasch, 1983). Actual passive slip may exist in shear zones (Chapter 10) if foliations develop parallel to shear directions, and folds related to the foliations can be passive-slip folds. Folds of this kind are rarely observed, however, because shear zones are dynamic, producing penetrative structures that shear out (or *transpose,* see Chapter 17) the original marker layers until they parallel the shear planes.

Kink Folding

As defined above, kink and chevron folds have straight limbs and narrow angular hinges. They form in minerals and rocks and occur on any scale from crystal lattices to map scale (Figure 15–20). Kink bands and folds form in regular geometric forms, either as isolated structures or as *conjugate* (paired) sets. Paterson and Weiss (1966) showed that confined samples of thinly layered rocks shortened by 10 percent (or less) parallel to the layers will develop conjugate kink bands at 55° to 65° to the shortening direction. As strain increases from 10 to 30 percent, kinks propagate through the entire mass at the same angle, forming two dominant crossing sets. At strain of 45 percent, the two sets are almost joined, with one set becoming dominant, and,

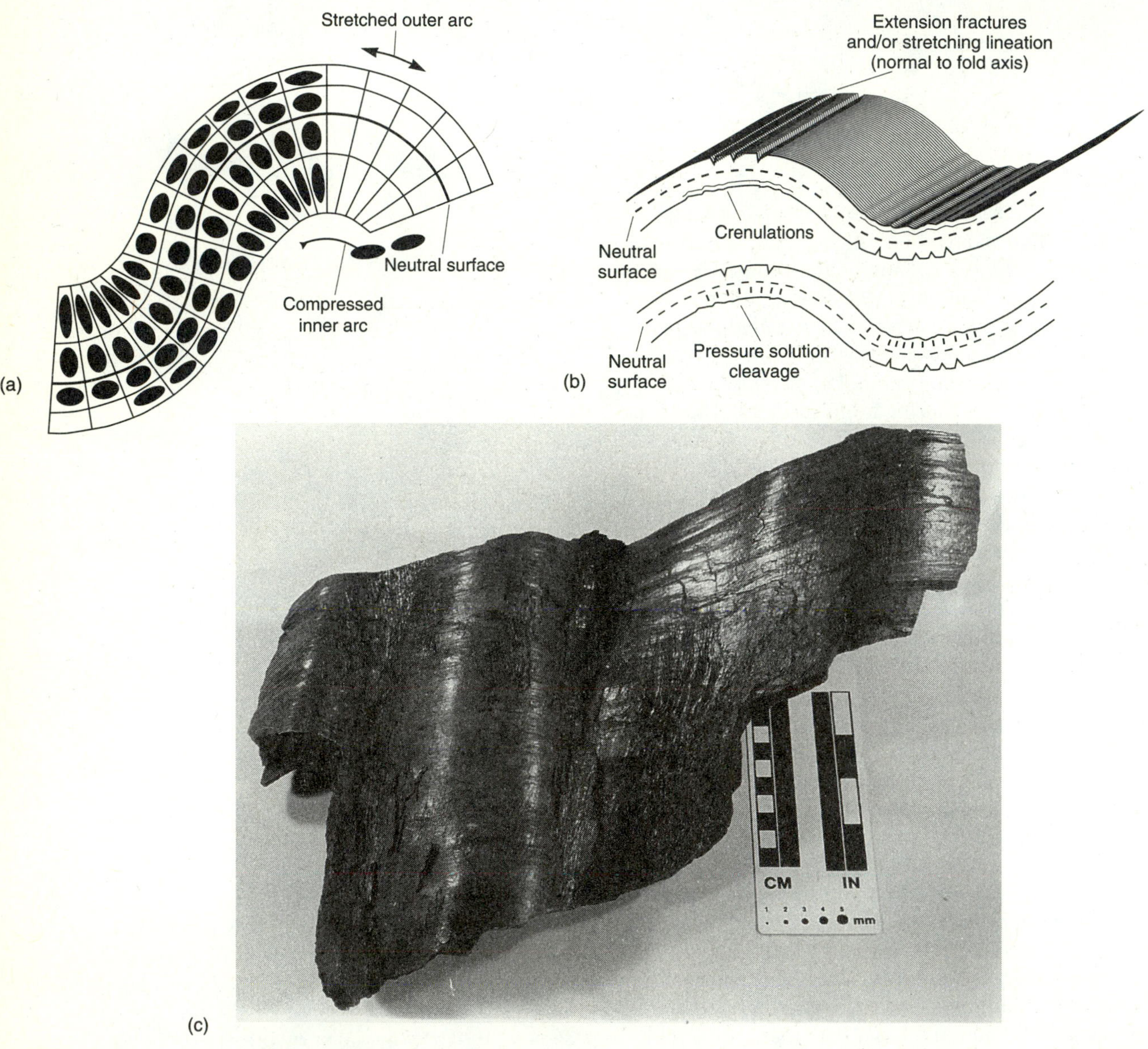

FIGURE 15–15
(a) Strain within a layer folded by buckling and tangential-longitudinal strain. Note that layer-parallel extension and shortening are separated by a neutral surface of no finite strain. (From J. G. Ramsay, *Folding and fracturing of rocks,* © 1967, McGraw-Hill Book Company. Reproduced with permission.) (b) Small structures produced by compression of inner arcs and extension of outer arcs of folds during tangential-longitudinal strain. (c) Folds formed by buckling and tangential-longitudinal strain from Ordovician Walloomsac phyllite near Hoosick Falls, New York. Crenulations in the synform and the stretching lineation oriented normal to the fold hinges suggest extensional motion normal to the antiformal hinges and shortening associated with the synformal hinges.

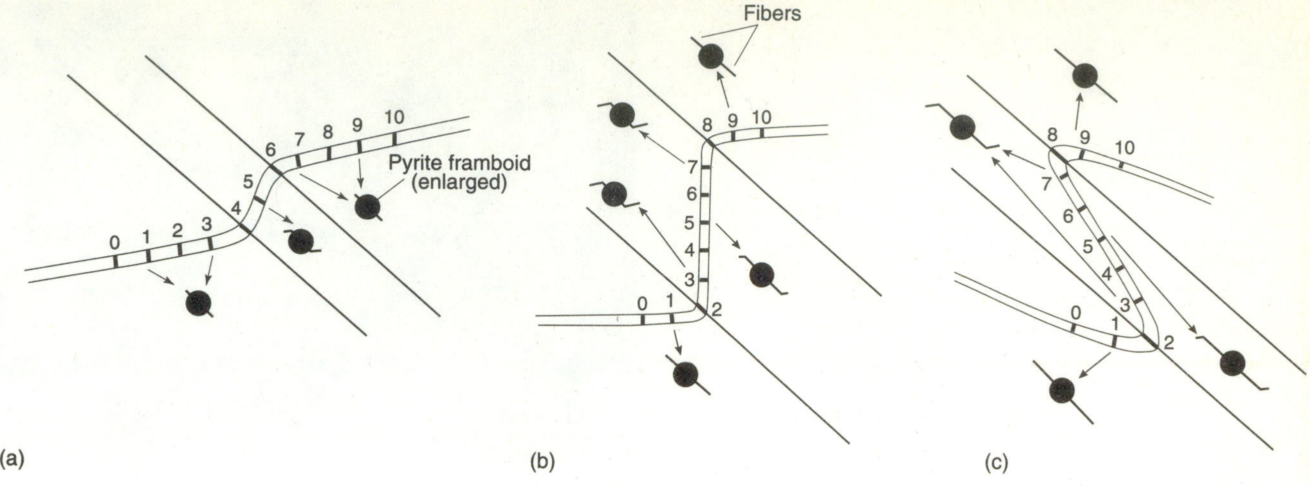

FIGURE 15–16
Beutner and Diegel migrating-hinges model. Fold is initiated (a); the common limb lengthened (b), then flattened and overturned (c). Fibrous pressure shadows on pyrite grains (black circles) are the strain and shear-sense indicators. (From E. C. Beutner and F. A. Diegel, Determination of fold kinematics from syntectonic fibers in pressure shadows, Martinsburg Slate, New Jersey, January 1985, *American Journal of Science*, v. 285, p. 16–50. Reprinted by permission of *American Journal of Science*.)

FIGURE 15–17
Possible passive-slip fold mechanism in Precambrian Moine Series metasandstone at Loch Monar, Scottish Highlands. Note that layering is displaced parallel to fold-axial surfaces, but buckling probably also occurred here. (RDH photo.)

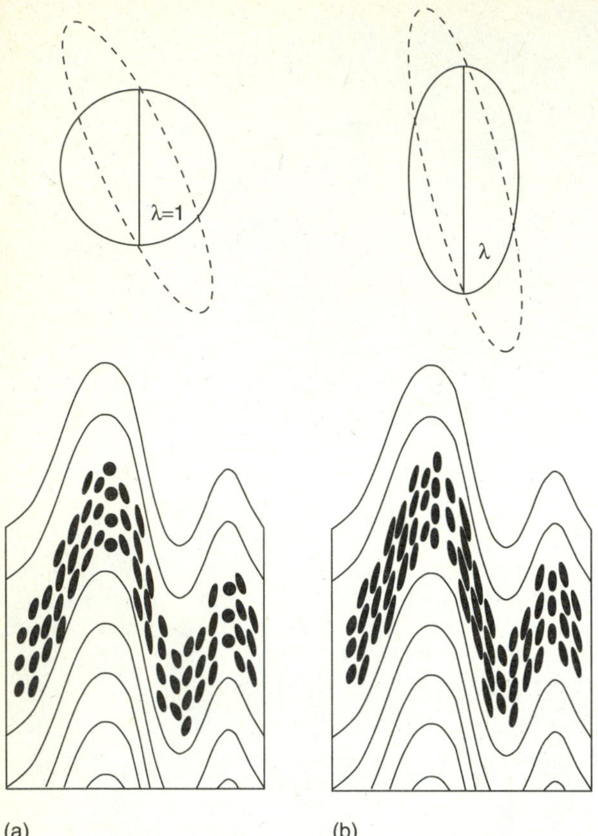

FIGURE 15–18
(a) Heterogeneous simple shear, or (b) heterogeneous simple shear followed by progressive pure shear (flattening) as a possible explanation for the shape of passive slip folds. Upper parts of (a) and (b) show strain in the hinges (solid lines) and limbs (dashed lines) for each case. (From J. G. Ramsay, *Folding and fracturing of rocks,* © 1967, McGraw-Hill Book Company. Reproduced with permission.)

(a)

(b)

FIGURE 15–19
(a) Folding of layering by axial-planar pressure-solution cleavage in Middle Ordovician Womble Shale, near Little Rock, Arkansas, suggesting passive slip; in reality, folds were formed by buckling accompanying cleavage formation. (b) Crenulation folds formed in talc schist in a shear zone at the base of the Karmøy ophiolite, near Haugesund, Island of Karmøy, southern Norway. (RDH photos.)

at strains of 50 percent or more, there is a single set of crenulation folds (Figure 15–21). Ramsay (1967) has suggested two other models for kink formation involving simple shear and rotation of constant-length segments of kinked layers. A third model by Ramsay, like one proposed by M. S. Paterson and L. E. Weiss (1966), involves migration of axial surfaces.

Why do kinks form in some rocks and buckle folds form in others? Each is a product of an instability in the layering as compressive stress is applied parallel to the layering. Arvid Johnson (1977) has formulated two ways to predict whether kinks or sinusoidal folds will form. For his first criterion (K_1), he assumed that initiation of kink folds requires local slippage (flexural slip) between layers within the kink after a threshold shear strength is exceeded; if free slippage can occur on all layers throughout the rock mass, sinusoidal buckle folds will form. Johnson also assumed that the rock mass has the axial load needed to induce slippage between layers. This would occur at critical inflection points on layers where the cohesion (contact strength) approximates the critical load required to nucleate kinks. As a result, his first criterion may be stated as

$$b \cong \tau / G_a, \qquad \textbf{(15–4)}$$

where b is the local increased dip of layering where folding is initiated, τ is cohesion between layers, and G_a is average effective shear modulus of the entire

(a)

(b)

(c)

FIGURE 15–20
(a) Kinked kyanite crystal in graphitic schist thin section from near Black Mountain, North Carolina. Note that the kinks are concentrated in the light area; unkinked kyanite forms the dark portion to the right. Width of field is approximately 1 mm. Crossed polars. (b) Kink folds in phyllite from Tête Jaune Cache, British Columbia. Knife is approximately 10 cm long. (c) Kink folds in thin-bedded late Precambrian quartzite near Oppdal, southern Norway. Exposure is approximately 3 m high. (RDH photos.)

layer sequence. If the shear stress is the same in all layers (both strong and weak), and the cohesion decreases to zero, equation 15–4 becomes

$$G_a = 0 \text{ at } \tau = 0\,. \qquad \textbf{(15–5)}$$

This holds true regardless of the thicknesses or elastic properties of the layers; that is, because shear stress cannot be transmitted, the effective shear modulus of the rock is zero.

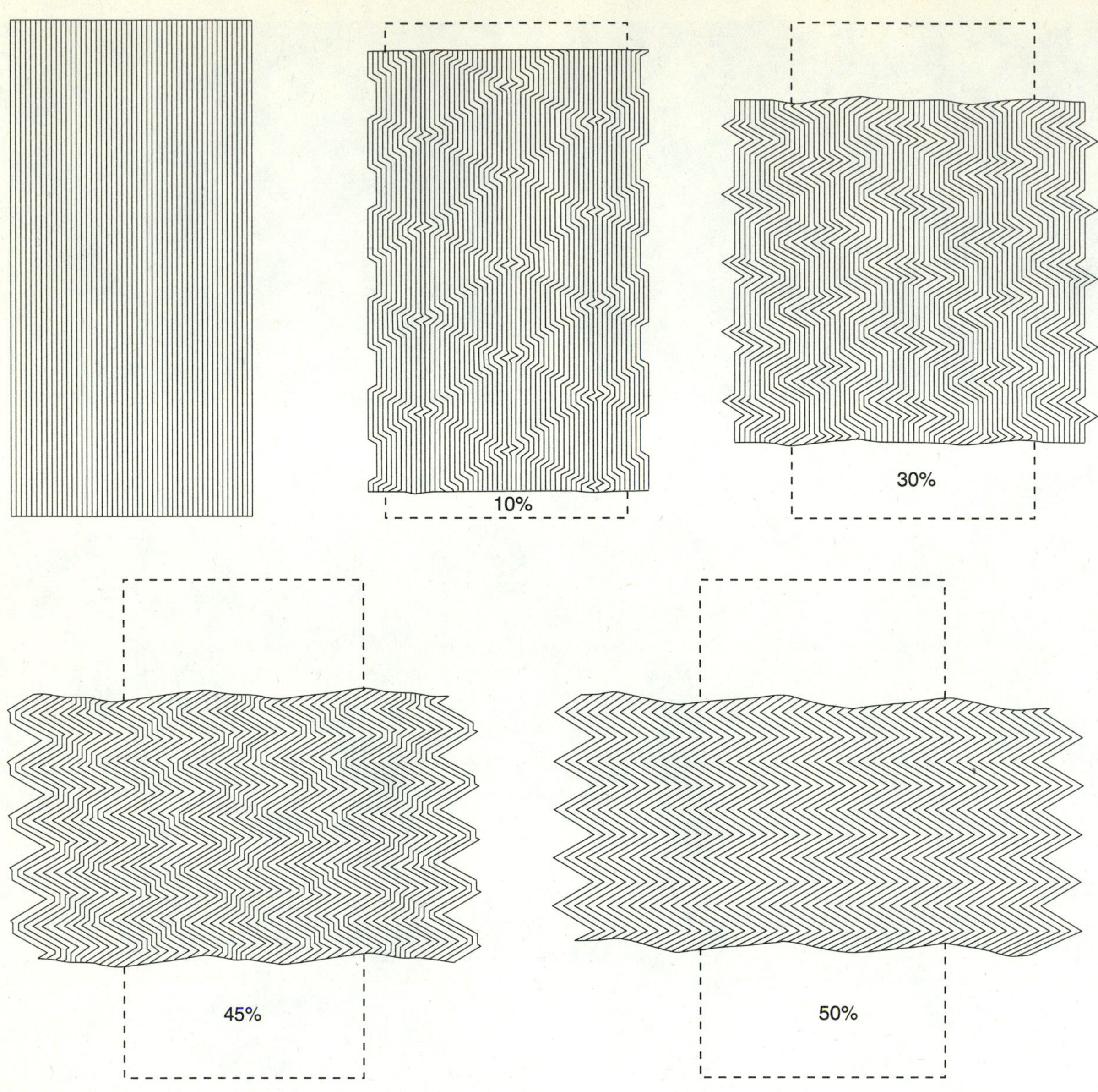

FIGURE 15–21
Kinks that evolve into chevron folds in experiments involving 0, 10, 30, 45, and 50 percent shortening strain in phyllite samples. (From M. S. Paterson and L. E. Weiss, 1966, Geological Society of America *Bulletin,* v. 77.)

If the ratio τ/G_a is small, the slopes of the layers are small when slippage occurs, and the buckle folds may not be clearly distinguishable from kinks formed at the same time. If the ratio is zero, slippage between layers occurs everywhere, and only buckle folds are formed. If the ratio is large, no slippage may occur between layers, and the buckle folds may be transformed into concentric and then chevron shape, rather than into tight kinks. If $G_a >> \tau$ ($\tau \approx 0$), kinks should be prominent. If $\tau = 0 = G_a$, or if $G_a \leq \tau$, sinusoidal buckle folds should form.

Johnson based his second criterion (K_2) on the assumption that an estimate for the critical load for kink folding is the load required to overcome the contact strength between layers all along the waveform at the midpoint of the sequence (Johnson, 1977). Calculations of the critical load, P_B, indicate that for kink folds to be favored over sinusoidal buckle folds,

$$P_0 < P < P_B,$$

or

$$1 < (P/P_0) < (P_B/P_0) = K_2, \qquad \textbf{(15–6)}$$

where P is axial load (parallel to layering), and P_0 is minimum required axial load for slippage to occur all along the layer (taken as minimum load for kinks to develop). Using this criterion, conditions favor kink folding where $K_2 >> 1$, but sinusoidal buckle folds form where $K_2 \sim 1$.

Many real folds represent some intermediate shape between ideal kink and ideal sinusoidal buckle folds, but most folds approach ideal end-member behavior. This model, therefore, serves a practical purpose and is not merely a theoretically and experimentally derived model for imaginary structures.

Before Johnson's work, Biot (1965a) and Peter Cobbold, John Cosgrove, and John Summers (1971) formulated criteria based on the anisotropy of a deforming multilayer rock body to determine whether kink or buckle folds would form, but Johnson's work indicates that anisotropy *per se* is not a factor. His criteria are no doubt applicable at lower temperatures and pressures where the contrasts between kink and buckle folds are commonly observed in crustal rocks. At higher temperature, flexural slip need not accompany buckling because ductile flow occurs in the weaker layers; thus, Johnson's first criterion would not hold under these conditions. The exception seems intended.

Flexural Flow

A layered sequence is said to deform by ***flexural flow*** if some layers flow ductilely while others remain brittle and buckle. Flexural flow requires moderate-to-high ductility contrast between layers. The entire rock mass may be in a state of ductile flow, but some rocks, such as amphibolite and calcsilicate, have greater viscosities under moderate temperature (300 to 500° C) and high pressure than interlayered weaker rocks, such as schist, carbonate rocks, impure sandstone, and felsic volcanic rocks. As a result, some strong layers may not undergo appreciable thickness changes, but weak layers may undergo extreme thickness changes (Figure 15–22). Because of the ductility contrast, the stronger layers control buckling and fold geometry, although not so much as they do in flexural-slip folds. (For an instant demonstration of flexural flow, have someone hold this

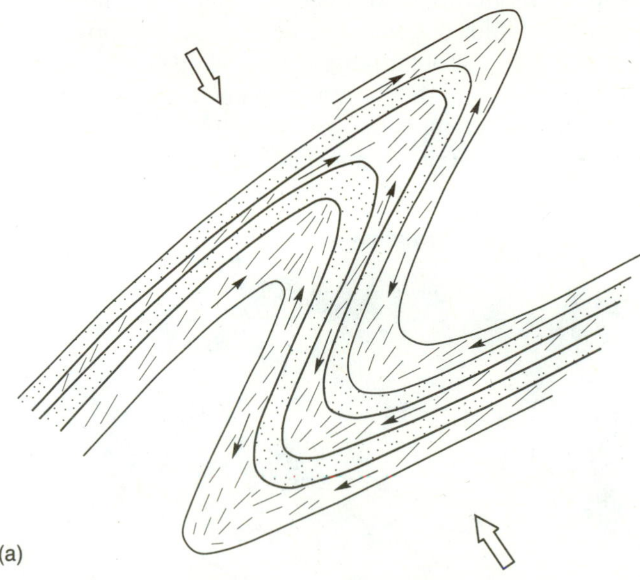

(a)

(b)

FIGURE 15–22
(a) Sequence of stiff (stippled) and weak layers deformed by flexural flow, producing similar-type folds. Note that the strong layers have undergone minimal thickness changes, but the thicknesses of weak layers have changed drastically from axial zones to fold limbs and have developed axial-plane cleavage. Movement (flow) of material from limbs into hinge regions—indicated by small black arrows—occurs in the weak layers, although the magnitude of displacement is small compared to the amount of shortening (indicated by large open arrows); in the strong layers, buckling and a lesser amount of displacement of material (ductile flow) occurs. (b) Flexural-flow buckle fold in interlayered quartzite (light) and amphibolite (dark) in Poor Mountain Amphibolite from near Salem, South Carolina. Quartzite layers were undergoing some ductile flow, but layering still controlled shapes of folds sufficiently that the stronger layers separated during folding, providing a void where the coarse crystalline quartz was deposited (center of photograph). (Photo by D. G. McClanahan, University of Tennessee–Knoxville.)

book several meters away and flex the pages of this book. The pages will fold as a single layer, but the layer [the pages] will readjust internally by apparent flow.) The products of flexural flow are mostly similar-type folds (Class 1C and 3) and rarely ideal similar folds (Class 2).

In flexural flow, fold amplitude and wavelength may be controlled by the original thickness, spacing, and strength of the strong layers, but they will also be modified by the amount of ductile strain parallel to the axial surfaces. The relationships between thickness of the stiff layer and the wavelength of the buckle folds, however, still hold, although this relationship will be modified by ductile strain (Equation 15–2).

Passive Flow (Passive Amplification)

Passive flow involves uniform ductile flow of an entire rock mass, with layering (bedding, foliation, or gneissic banding) serving only as a strain marker (Figures 15–23 and 15–24; see also Figure 2–31b). If passive-flow folds are to form, there must be little or no ductility contrast between layers, even if the composition differs markedly, and there must be flow across the layering. Ramsay's suggestion (1962) of a variable amount of flattening producing simple shear across layers may prove important here. This mechanism grades into flexural flow as ductility contrast increases and the stronger layers begin to influence the shape of the folds. Passive flow is the most likely mechanism for ideal similar folds (Class 2), preserving identical shapes throughout the layered sequence, because of the lack of contrast of properties from layer to layer. Similar-type (Class 3 and 1C) folds are also produced by passive flow.

The actual mechanism by which passive strain markers are folded in passive flow is nonuniform laminar ductile flow produced by heterogeneous pure shear or simple shear (Wynne-Edwards, 1963). Heterogeneous simple shear directed parallel to fold axial surfaces and homogeneous pure shear—whether parallel, oblique, or normal to layering—accentuates passive folding (Figure 15–25). This process also leads to formation of Type F crystalline thrusts (Chapter 11).

Combined Mechanisms

Fold mechanisms, like deformation mechanisms, compete with one another at various temperatures and pressures and in various rock types. The mechanisms of passive flow and flexural flow, which require less or more ductility contrast, may compete at higher temperature in normal silicate rocks. In a rock mass being folded under near-surface conditions, flexural slip and buckling may be initiated; as the folds tighten, frictional resistance between layers increases, thus reducing layer slip. Cleavage-forming mechanisms may take over and produce shortening and flow parallel to the axial surfaces of folds (Figure 15–26). This occurs most frequently in the outer parts of the metamorphic cores of orogens and in the innermost parts of foreland fold-thrust belts. Attempts to pigeonhole folds by geometric or even generic criteria may fail because competing

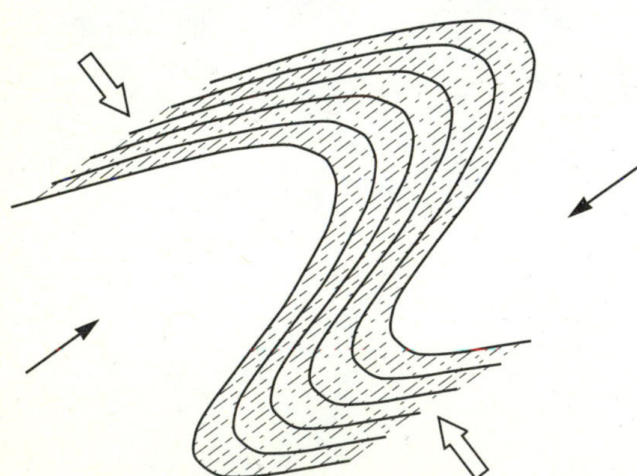

FIGURE 15–23
Passive-flow mechanism. Note that the thickness of each layer changes by the same amount and produces ideal (Class 2, or nearly ideal) similar folds by ductile flow; all hinge zones have the same curvature. Flow can occur in any direction with respect to layering, but it must cross layering to produce folding. Large open arrows indicate shortening direction. Smaller black arrows indicate movement in one axial surface with respect to the next. Axial-plane foliation suggests a strong component of flattening.

FIGURE 15–24
Passive-flow and strongly disharmonic flexural-flow folds in migmatitic biotite gneiss, Thor-Odin dome, Shuswap metamorphic complex, southern British Columbia. Folding here is ductile, and many layers have changed thickness to the same degree along any given flow line. (RDH photo.)

mechanisms produce folds that do not reveal recognizable patterns (Figure 15–27). Strain analysis sometimes helps to resolve problems with fold mechanisms.

Buckling plus flexural slip is probably the most usual combination of folding processes (Figure 15–28). Here, rocks yield by forming open parallel or parallel-concentric folds, but as the fold tightens, the room problem (Chapter 11) arises in the core of the fold. Then disharmonic folding of weaker layers takes place, with or without fracturing, to solve the room problem. In ductile rocks, buckling in some layers may combine with ductile flow in others and produce flexural-flow folds. Disharmonic folds result from stiff layers of different thickness that are widely spaced with respect to each other in a lower viscosity medium, allowing each layer to buckle independently according to its own dominant wavelength.

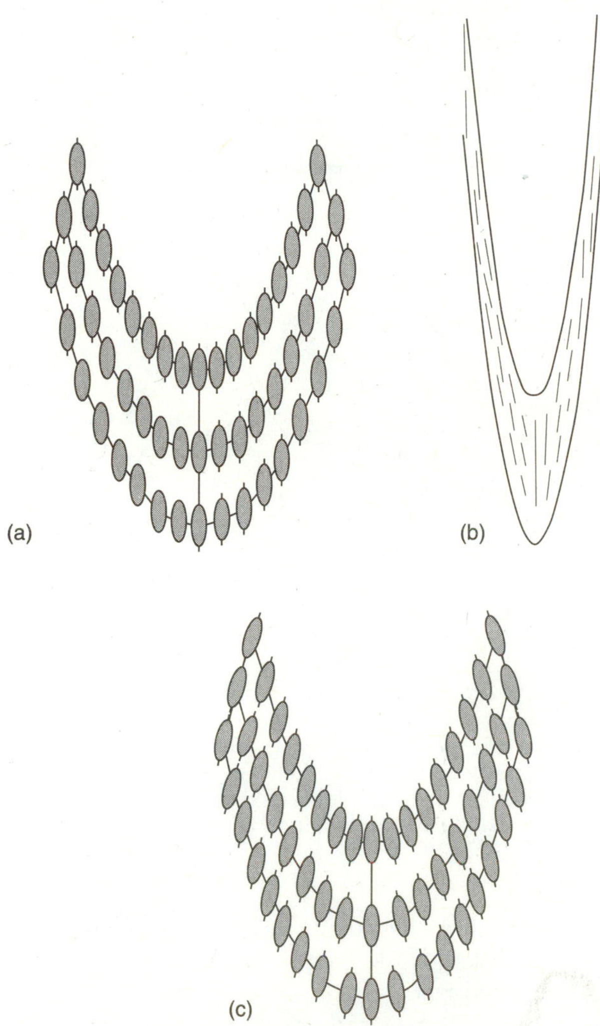

FIGURE 15–25
Strain (a) and changes in layer thickness (b) in passive flow produced by homogeneous strain and greater than 50 percent shortening of an existing fold; (a) shows strain ellipses; (b) lines indicating *X* axes of the strain ellipse, suggesting maximum flattening, producing a similar fold. (c) Strain ellipse orientations involving 50 percent shortening and heterogeneous strain that would prtoduce the same shape. (b and c reprinted from *Tectonophysics,* v. 11, B. E. Hobbs, p. 329–375, © 1971, with kind permission from Elsevier Science Ltd., Kidlington, United Kingdom.)

(a)

(b)

FIGURE 15–26
(a) Buckle fold in Ocoee Series (late Precambrian) rocks near Walland, Tennessee, containing evidence of both buckling and flexural slip. (b) Bedding surface on the southeast (left) limb of the fold in (a) showing slickenlines truncating cleavage. Knife is parallel to cleavage. Note slickenlines on bedding surfaces and axial-plane cleavage in (b). Cleavage formed during homogeneous strain of the rock mass as the layers became unable to slip past each other, but some flexural slip occurred after the cleavage formed. (RDH photos.)

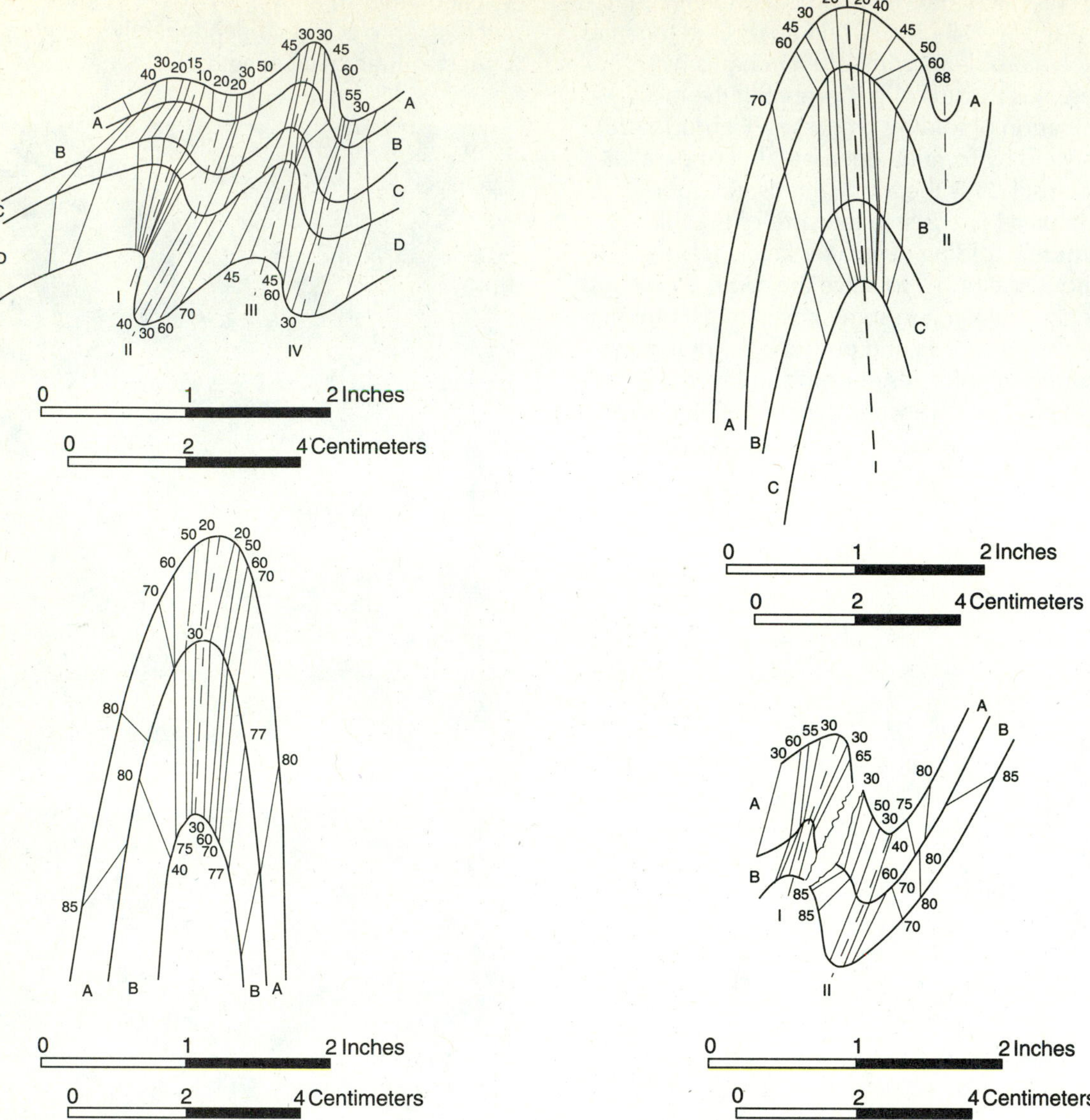

FIGURE 15–27
Isogon analysis of several mesoscopic folds from the Milton area of the Appalachian Piedmont in Virginia and North Carolina. Note that the folds show variations in convergent and divergent characteristics of isogons from layer to layer (of different composition); they would fit more than one of Ramsay's fold classes. Roman numerals identify particular fold sets. Note that the reference frame for the isogons changes from layer to layer (as dictated by changes in axial plane orientation.) (From O. T. Tobisch and Lynn Glover, III, 1971, Geological Society of America *Bulletin,* v. 82, p. 2209–2230.)

DEFORMATION MECHANISMS AND STRAIN

Fold mechanisms in rocks are determined largely by physical conditions during deformation. Consequently, the behavior modes and deformation mechanisms we discussed in Chapters 5 and 6 also play an important role in determining the mechanisms and ultimate shapes of folds. At the low-temperature end, stiff layers bend or buckle and yield brittlely or undergo pressure solution; at the high-temperature end, flow processes dominate and layers undergo ductile flow and recrystallize. At intermediate temperatures, some layers reach the threshold for ductile flow, and others remain strong and undergo little ductile flow. In an interlayered sequence of schist (weak) and amphibolite (strong), the schist may undergo ductile flow with recrystallization, and the amphibolite may still undergo brittle deformation. At still higher temperature, the amphibolite may flow, but less than the schist.

Attempts have been made to determine the amount and distribution of strain in folds by using structures observed within layers or at layer boundaries. These include natural strain markers, such as oöids (Cloos, 1971) and reduction spots (Wood, 1973). Theoretical studies (Ramberg, 1961; Chapple, 1969), experimental rubber and clay models, photoelastic studies (Roberts and Strömgård, 1971), and computer modeling (Dieterich, 1970) have increased our understanding of the folding process.

David Roberts and K.–E. Strömgård (1971) have studied the relationships of cleavage to weak and strong layers in buckle folds, then attempted to model observed strain patterns using rubber, plasticene, and putty and photoelastic models. The stiffer rubber layers exhibited considerable rigidity and clearly displayed tangential-longitudinal strain (Figure 15–29). Multilayer putty and plasticene models provided somewhat different results in different layers (Figure 15–30). The plasticene layers, being stronger, exhibited tangential-longitudinal strain; the weaker putty layers exhibited more- nearly uniform flattening in the axial zones of folds (parallel to the axial surfaces) and thinning of limbs—a flexural-flow mechanism. Strain was, however, refracted in the putty at layer boundaries, particularly on the outer arcs of the strong layers. Photoelastic models yielded similar results (Figure 15–31).

Chapple (1970) has studied strain variability in theoretical folds involving a single layer of high viscosity embedded in a layer of much lower viscosity. He found considerable variation in the nature of strain at layer boundaries as the dip on the fold limb increases from about 23° to about 89°.

Among the first computer models of strain distribution in folds are those of James Dieterich (1970), who based them on the finite-element method. His models of strain distribution (Figure 15–32) reveal patterns like some of the experimental results obtained by Roberts and Strömgård (1972) and even indicate some of the variations they reported at layer boundaries.

Most of the models just described clearly illustrate the effects of various combinations of homogeneous and inhomogeneous strain on fold shape and contrasts in strain at layer boundaries. Homogeneous strain before, during, or after folding by heterogeneous strain (such as by buckling) profoundly affects the shapes of folds and determines whether the folds end up as modified concentric (Class 1C) or similar (Class 2) folds, or some other type. Likewise, layer-parallel shortening before folding also modifies the shapes of folds. Thus, the sequence of homogeneous and heterogeneous strain is also an important factor in fold mechanisms.

DISCUSSION

A number of interesting aspects of fold geometry and mechanisms have been addressed in Chapters 14 and 15. As a part of understanding folds, it is probably useful to entertain some basic questions about folds and how they form, and how different kinds of folds are interrelated. Our purpose here is to undertake such a discussion with the goal of rounding out our investigation of fold geometry and mechanics.

Parallel and Similar Folds

Layering plays a pivotal role in parallel folding; therefore, buckling or bending is the dominant mechanism. Parallel folds formed by buckling or bending of massive sandstone or limestone layers represent an ideal example. Strong sandstone or carbonate rocks interlayered with shale or other weak rocks that change thickness during folding yield buckle folds with similar-like geometry that are not ideal parallel folds. Ductile layers having identical mechanical properties that are folded by passive flow produce ideal similar folds. As layering begins to participate in folding, ideal similar folds are no longer produced, and buckling increases in importance. Buckling *decreases* in importance as metamorphic grade increases and rock becomes more ductile.

Nucleation and Growth of Folds

Most of our discussion of fold geometry and mechanisms has been two-dimensional, visualizing folds in either cross-section or map view (see Figures 14–6, 14–14, 14–18, and 14–19). The questions about how single buckle folds, kinks, and buckle-fold "trains" (series of folds) develop and grow within a sequence of originally flat-lying, undeformed sedimentary rocks (multilayers) have been addressed by a number of geologists. These questions include: Do fold hinges, after being formed, migrate or remain stationary? Do hinges migrate or remain stationary in particular fold mechanisms? Do all folds in which layering participates start out as noncylindrical folds and grow into long, cylindrical folds or larger noncylindrical folds, depending on fold mechanism and the rheology of the sequence being deformed?

Experimental work by A. K. Dubey and Peter Cobbold (1977) produced noncylindrical folds that had chevron-shaped hinges along the crest and rounded hinges at the ends. These folds also had locally strongly curved axial surfaces. The folds were produced by layer-parallel shortening of as much as 46 percent of a plasticene multilayer sequence of alternating weak and strong layers. The fold mechanism was buckling accompanied by flexural slip. Cobbold (1975) earlier had conducted experiments with single layers of stronger material imbedded in material about 0.1 times as strong. He found that the strong layer begins to fold at about 15 percent shortening, forming

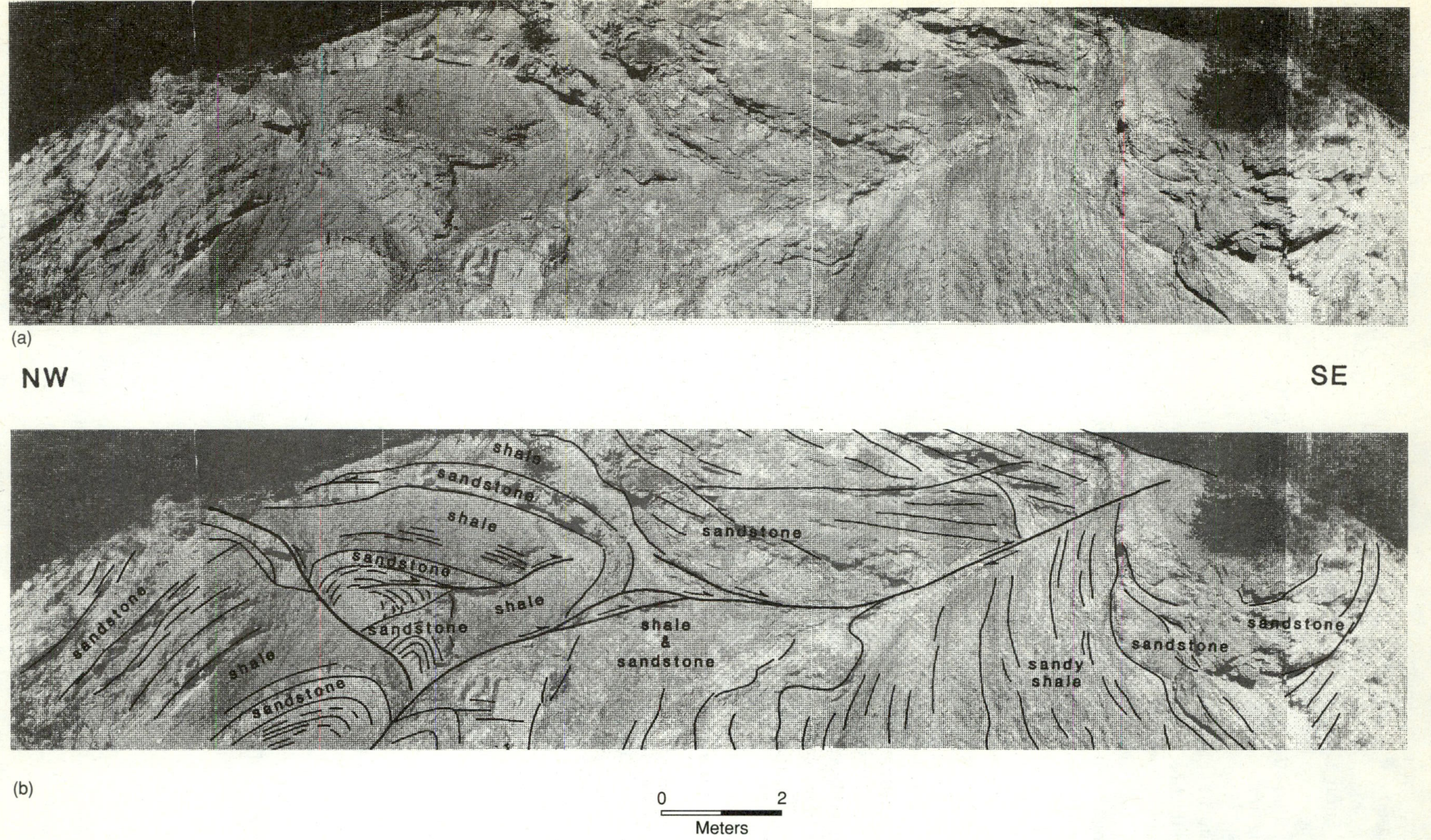

FIGURE 15–28
A fold (a) and sketch (b) combining buckling and flexural slip in Pennsylvanian shale and sandstone at Ozone, Tennessee. Note that the room problem in the hinge of the fold was solved by brittle deformation (faulting) of sandstone and tight folding of shale. (Photograph by Alice L. Stieve, Westinghouse Savannah River Company.)

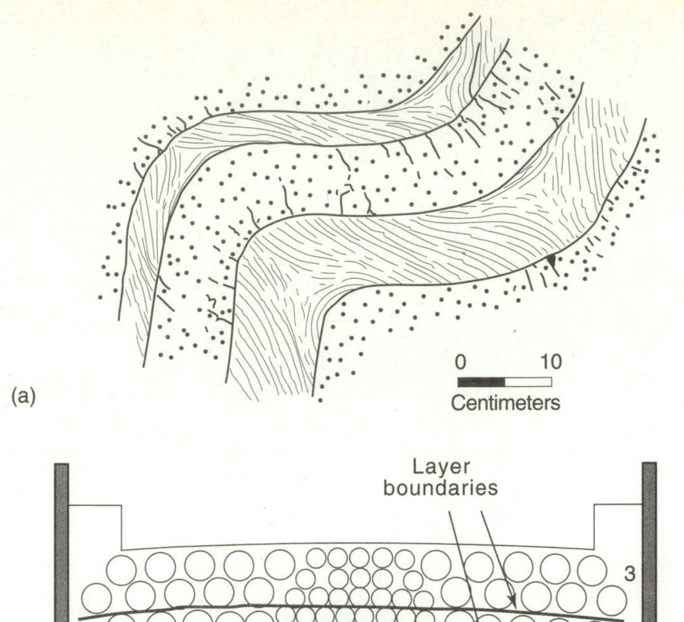

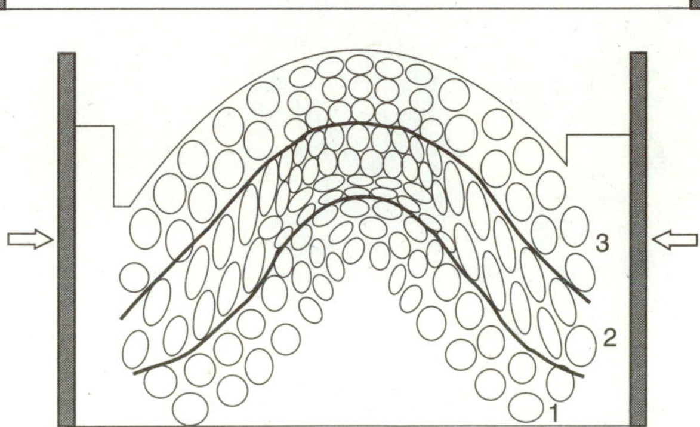

3
2
1
(b)

FIGURE 15–29
(a) Sketch of bedding and cleavage traces (light lines) on a buckle fold from the Varanger Peninsula of Norway. The models in Figures 15–29, 15–30, and 15–31 attempt to duplicate the strain, cleavage, and geometry associated with this fold. (b) Rubber-layer model produced by buckling of a package consisting of layers of soft rubber (2) between layers of stiff rubber (1 and 3, each 13 mm thick). Top, undeformed. Bottom, deformed, showing distinct differences in strain from soft layer to stiff. (Reprinted from *Tectonophysics*, v. 14, David Roberts and K.-E. Strömgård, p. 105–120, © 1971, with kind permission from Elsevier Science Ltd., Kidlington, United Kingdom.)

an anticline flanked by two synclines. With greater shortening, two additional anticlines develop symmetrically outside of the two synclines, and the initial anticline continues to grow by increasing amplitude as the two new anticlines grow, so that it is larger than the two. With continued shortening (to about 45 percent), the process is repeated, so that new flanking anticlines and

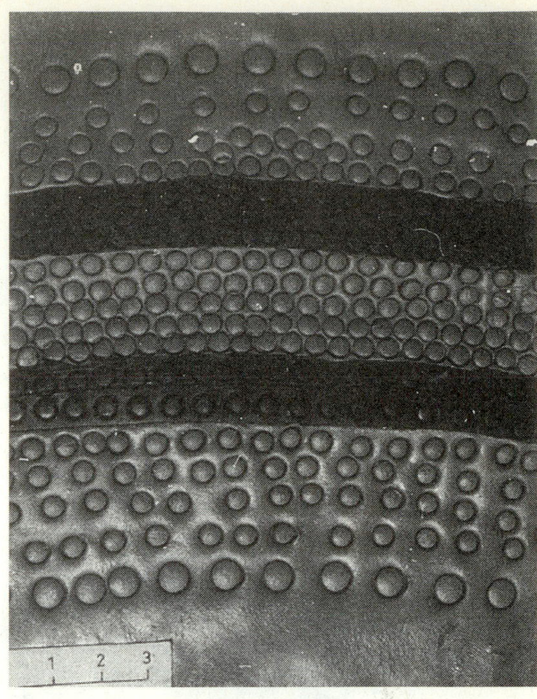

(a)

(b)

FIGURE 15–30
Model made of putty enclosing two stiff layers of plasticene. (a) Undeformed. (b) Deformed, showing strain variations produced by heterogeneous strain and variations near layer boundaries. (Reprinted from *Tectonophysics,* v. 14, David Roberts and K.-E. Strömgård, p. 105–120, © 1971, with kind permission from Elsevier Science Ltd., Kidlington, United Kingdom.)

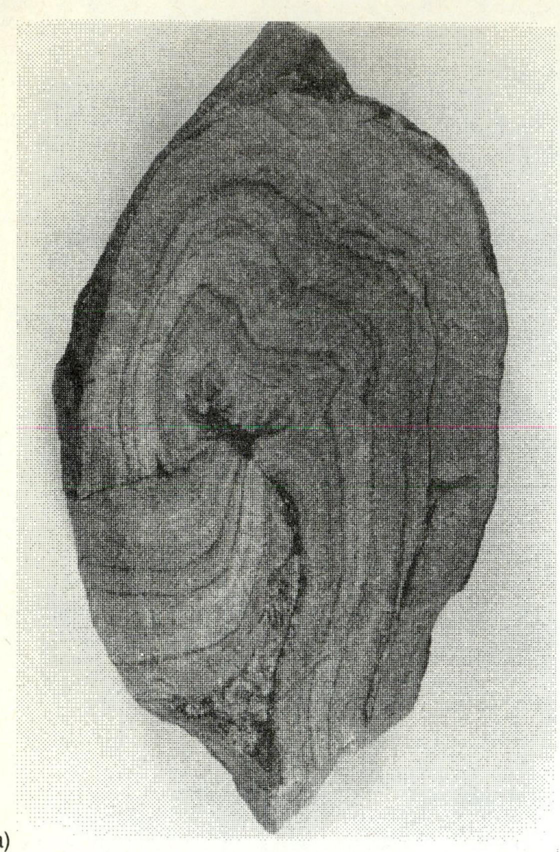

(a)

FIGURE 15–31
(a) Folded quartzite from Poor Mountain Formation near Salem, northwestern South Carolina; layers have separated, and gash zones filled with new quartz. See also Figure 15–22(b) on page 313. (b) Sketch of fold in (a). New (vein) quartz is stippled. (c) Photoelastic model of 18 percent shortening of a multilayer. Isochromatic bands indicate strain pattern in the weaker material. (d) Finite-strain trajectories constructed using the model in (c). They suggest that cleavage may undergo refraction according to these patterns. The dark area near the center of the model is a gash that developed as the layers separated. (Parts c and d reprinted from *Tectonophysics,* v. 14, D. Roberts and K.-E. Strömgård, p. 105–120, © 1971, with kind permission from Elsevier Science Ltd., Kidlington, United Kingdom.) Note similarities in strain trajectories and cleavage with the fold in Figure 15–30a.

(c)

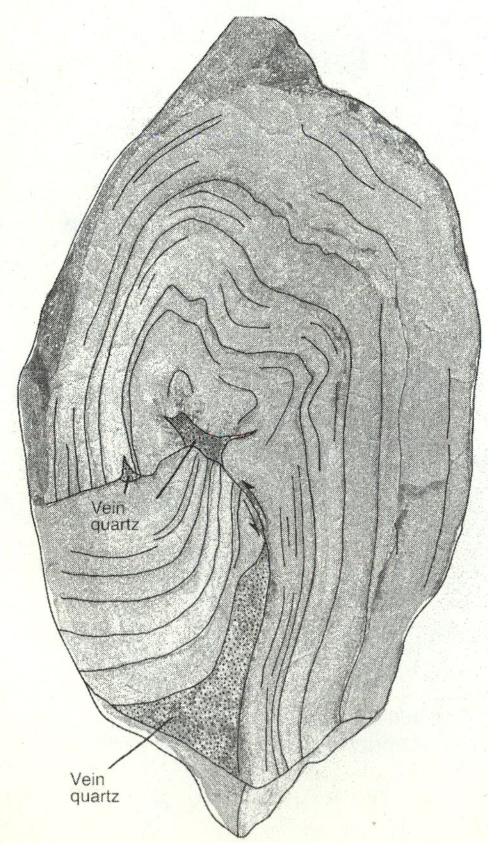

(b)

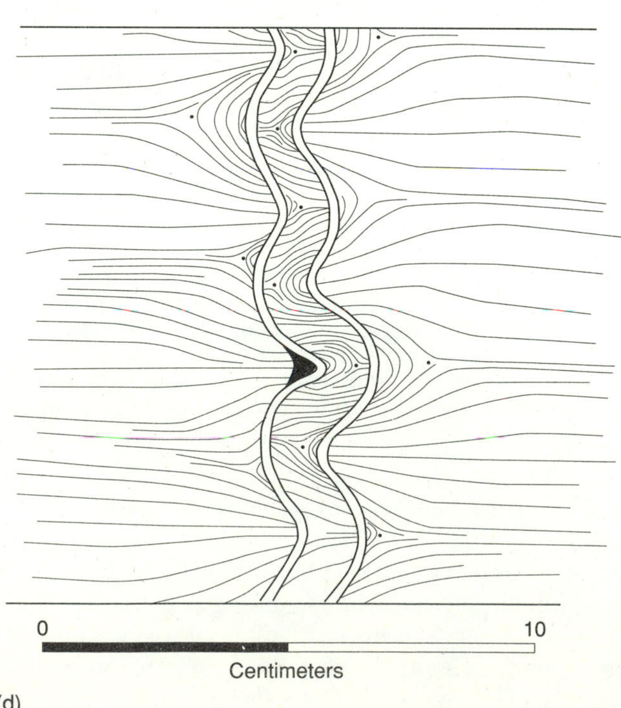

(d)

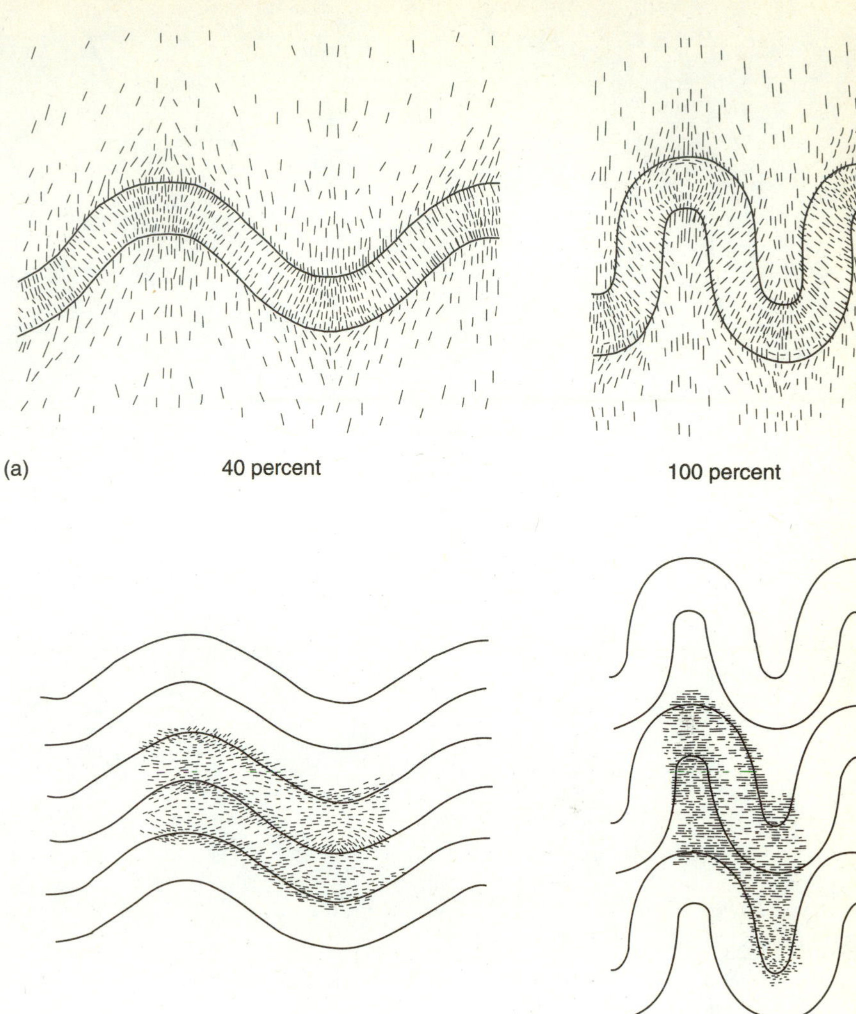

FIGURE 15–32
(a) Computer models of buckle folds. Short lines indicate orientation of the *X* axis of the strain ellipsoid for a single layer shortened by 40 percent and 100 percent. (b) Model buckled multilayer. Short lines are parallel to σ_1. Viscosity contrast between strong and weak layers is 42:1. (From J. H. Dieterich, 1970, *Canadian Journal of Earth Sciences,* v. 7.)

synclines continue to form and grow. Dubey and Cobbold (1977) determined that growth of individual folds involved lengthening of the hinge line by propagation of the terminations parallel to the hinge, and older single folds triggered growth of new folds on both flanks, eventually producing a train of folds with regular wavelength (Figure 15–33). There was some overprinting and interference between folds where hinges propagated into each other, or fold trains attempted to overlap. Some folds also exhibited branching of one hinge into two.

Although all models contained small irregularities before being deformed, Dubey and Cobbold intentionally introduced irregularities with different shape and orientation into several of their models. These produced folds that grew at different rates, depending on the shape and size of the irregularity. Although they concluded that all their folds were noncylindrical, a number of folds were produced that had very linear hinges compared with the width of the fold and became noncylindrical only at the ends. The structures produced resemble the fold patterns in foreland fold-thrust belts, such as the Appalachian Valley and Ridge, the Jura Mountains, and the Zagros Mountains. (Also see Figure 11E–1 for a similar small-scale version of a foreland fold-thrust belt.) All folds must become noncylindrical at the ends unless they are cut off by faults. Many of the very long linear folds that are present in mountain chains may thus have begun as small noncylindrical irregularities.

Kevin Stewart and Walter Alvarez (1991) addressed the problem of the fixed- or mobile-hinge mechanisms for kink and chevron fold formation by employing a combination of natural folds and experimentally produced folds. In the experiments, they deformed a multilayer sequence composed of layers of lead separated by layers of wax-impregnated cloth. Most earlier studies of kink and chevron folds had concluded that these structures formed by a stationary (fixed) hinge, although mobile-hinge kinks were well known in experiments and in deformed single crystals (Paterson and Weiss, 1966; Weiss, 1980). They demon-

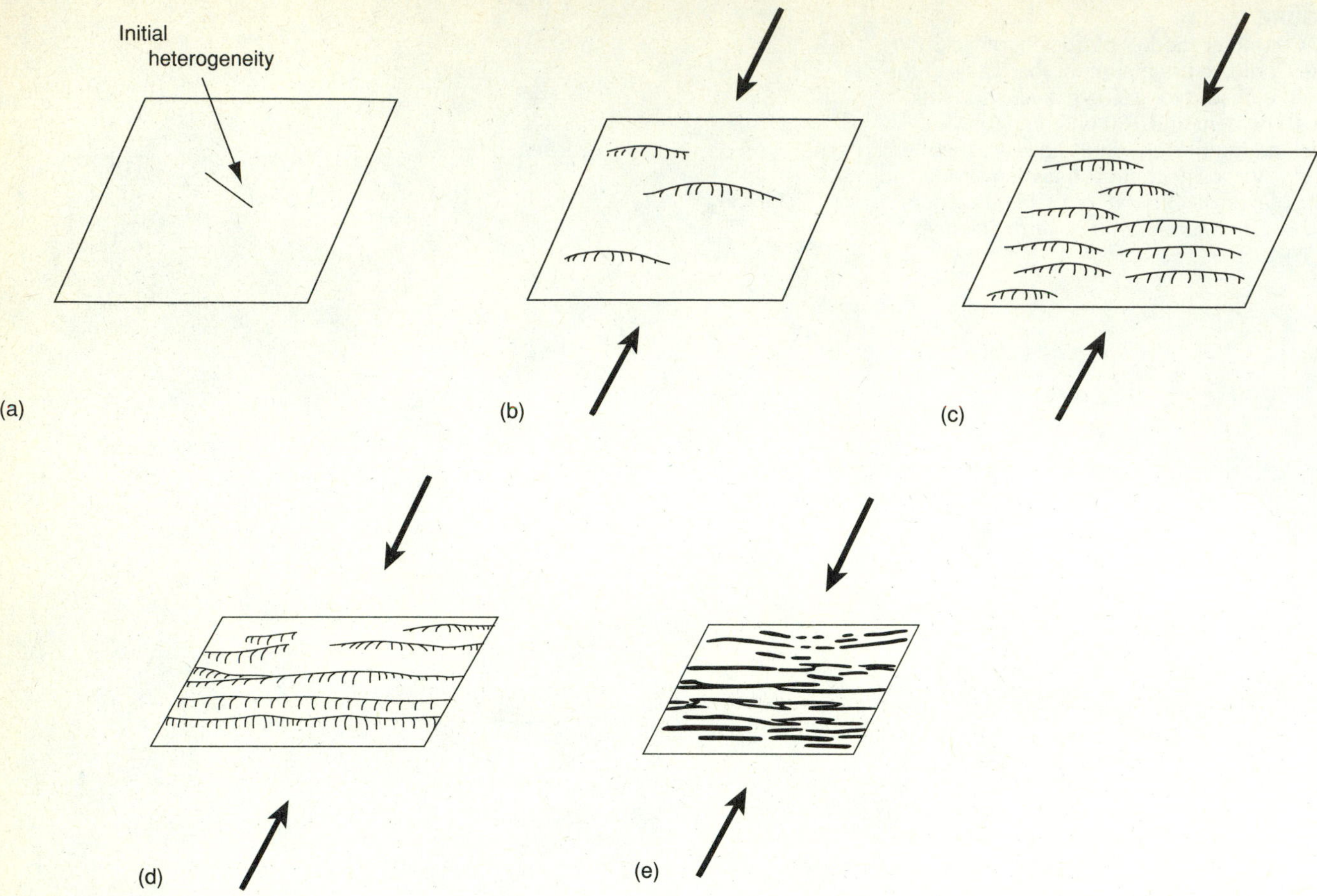

FIGURE 15–33
Stages in development of buckle folds in experiments that involved shortening of a previously undeformed clay layer (a) that contained an initial heterogeneity. Shortening in (b) is 3 to 19 percent, (c) 19 percent, (d) 36 percent, and (e) 46 percent. Compare the pattern that developed with the model shortened by 46 percent with the fold pattern in Figure 14–18, which involves a similar amount of shortening. Arrows indicate orientation of σ_1. (From *Tectonophysics*, v. 38, A. K. Dubey and P. R. Cobbold, p. 223–239, © 1977, with kind permission from Elsevier Science Ltd., Kidlington, United Kingdom.)

strated that chevron folds in a multilayer Cretaceous sequence of stronger limestone and weaker marl in northeastern Italy formed by a migra-ting-hinge mechanism. Their evidence included earlier folds superposed by later kink (or chevron) folds, resulting in rekinking of earlier kinks. They also concluded that mobile-hinge kinking produces no volume change, whereas fixed-hinge kinking produces an initial volume increase as layers sepa-rate during early kinking, but this volume change is lost during the later stages of deformation.

Strongly Noncylindrical and Sheath Folds

The importance of ***noncylindrical and sheath folds*** as products of inhomogeneous strain has been recognized onlyduring the 1980s and 1990s. In a strict sense, all folds are noncylindrical if traced far enough because their hinges become curved as they terminate, unless truncated by faulting; the hinges of some folds, however, are strongly curved over short distances and so they are easily recognized as noncylindrical.

Studies by Donald Ramsay and Brian Sturt (1973a) of the mechanisms and the occurrences of noncylindrical and aberrant folds, and experiments by Peter Cobbold and H. Quinquis (1980) on the formation of sheath folds, have led to explanations of how they form. In most cases, mildly to strongly noncylindrical folds are produced if the fold hinge forms initially as part of a wholly noncylindrical fold in a strain field involving very strong homogeneous simple shear; the strain perturbation that produces the fold either diminishes or becomes inhomogeneous, and the strong component of simple shear is maintained. Inhomogeneous simple shear deforms the preexisting fold hinges so that they become curved within the fold axial surfaces (Figure 15–34). Parts of the same fold hinge are transported differentially, greatly distorting the original folds. If the process continues long enough,

ESSAY

Fold Mechanisms, Space, Time, and Orogenic Belts

Fold mechanisms differ from place to place in mountain chains and are superposed in time. The inner parts of a mountain chain subjected to high-temperature/high-pressure metamorphism generally preserve an early set of ductile structures overprinted by later brittle structures, formed after the mass cooled. Earlier folds formed by flexural-slip or buckle mechanisms may have formed before ductile deformation (that is, before reaching maximum pressure and temperature), but would have been modified or destroyed at high temperature and pressure during penetrative ductile strain and passive or flexural-flow folding (Figure 15E1–1). Generally, once the thermal peak is past, cooling occurs over millions of years, and the rocks become progressively more brittle. Newly applied stresses superpose open buckle folds, accompanied by flexural slip, and kink folds.

(a)

(b)

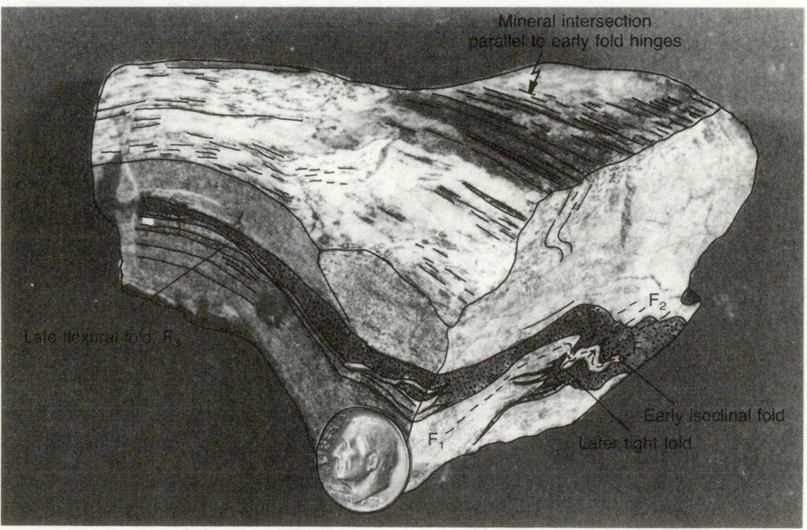

FIGURE 15E1–1
(a) Overprinting of early ductile isoclinal folds by later brittle flexural folds in Poor Mountain amphibolite and feldspathic quartzite from near Toccoa, Georgia. (b) Early (F_1) ductile isoclinal recumbent passive-flow folds on the right side of the specimen were overprinted by more upright (F_2) ductile flexural-flow folds with similar axial trend to the earlier folds, then by flexural (buckle?) folds visible on the left side of the specimen that trend perpendicular to the trend of the earlier folds. Note the strong mineral lineation on top of the specimen parallel to the earlier folds, probably an intersection lineation (Chapter 18) formed by intersection of axial-planar F_1 foliations (Chapter 17) of F_1 and F_2 folds.

Transitions in deformation style from the outer to the inner zones of mountain chains have been described by Richard B. Campbell (1970), Arthur Snoke (1980), David Sanderson (1979), and D. C. Murphy (1987)—specifically, the Cariboo Mountains of east-central British Columbia, the Ruby Mountains of northeastern Nevada, and the Variscan fold belt in southwestern England. Remarkable parallels are present, despite differences in plate-tectonic settings and time of deformation. Each chain exhibits a similar transition from more ductile structures in its inner parts to more brittle structures in the outer parts (Figure 15E1–2). The Cariboo and Ruby Mountains underwent deformation from middle to late Mesozoic and early Tertiary times, and the Variscan chain in the late Paleozoic.

Folds in the outer zones of all chains were produced by flexural slip and buckling. Folds are generally open, and cleavage is weak or nonexistent. Axial-plane cleavage becomes more prominent as the folds tighten toward the cores. In the inner zones of each chain, folds become similar-like and tight-to-isoclinal; passive flow becomes dominant. This transition is most obvious in the Cariboos.

In the Variscan chain, fold interlimb angles systematically decrease from an average of about 80° in the external zone to less than 45° in the innermost zone. Correspondingly, the dip of axial surfaces decreases from about 80° in the outer zones to less than 15° in the core. Sanderson has attributed these changes to increased shear strain toward the inner parts of the Variscan chain.

Superposed folding becomes prominent in the inner zone of the Cariboos. We do not know whether superposition is related to deformational events widely separated in time or to superposition of ductile folds during one event. It seems likely that superposition occurred both during the major thermal event and afterward.

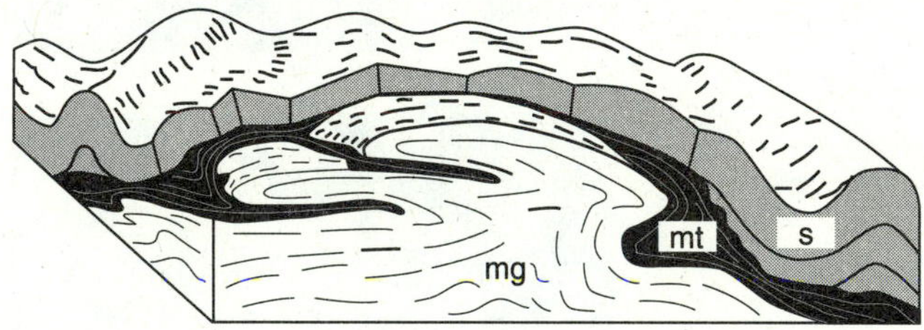

FIGURE 15E1–2
Contrasting and progressive changes occurring in fold style from the outer zone (top of section) to inner zone (bottom); s—sedimentary rocks; mt—metamorphic rocks. mg—migmatite. (From John Haller, 1956, *Geologische Rundschau*, v. 45.)

References Cited

Campbell, R. B., 1970, Structural and metamorphic transitions from infrastructure to suprastructure, Cariboo Mountains, British Columbia: Geological Association of Canada Special Paper 6, p. 67–72.

Murphy, D. C., 1987, Suprastructure/infrastructure transition, east-central Cariboo Mountains, British Columbia: Geometry, kinematics and tectonic implications: Journal of Structural Geology, v. 9, p. 13–29.

Sanderson, D. J., 1979, The transition from upright to recumbent folding in the Variscan fold belt of southwest England: A model based on the kinematics of simple shear: Journal of Structural Geology, v. 1, p. 171–180.

Snoke, A. W., 1980, Transition from infrastructure to suprastructure in the northern Ruby Mountains, Nevada, *in* Crittenden, M. D., Jr., Coney, P. J., and Davis, G. H., eds., Metamorphic core complexes: Geological Society of America Memoir 153, p. 287–333.

fold hinges end up subparallel to the transport direction (Figure 15–34c), and the folds are sheath (or tubular) folds (Figure 15–35). If the process ends before the hinges are rotated into parallelism with the transport direction, the folds are noncylindrical. Sheath folds commonly form in ductile shear zones (Chapter 10).

Lineations and Fold Mechanisms

Lineations (Chapter 18) may indicate the nature of fold mechanisms. Mineral lineations commonly form parallel to the direction of maximum finite extension in the fold axial surfaces perpendicular to the hinges, and

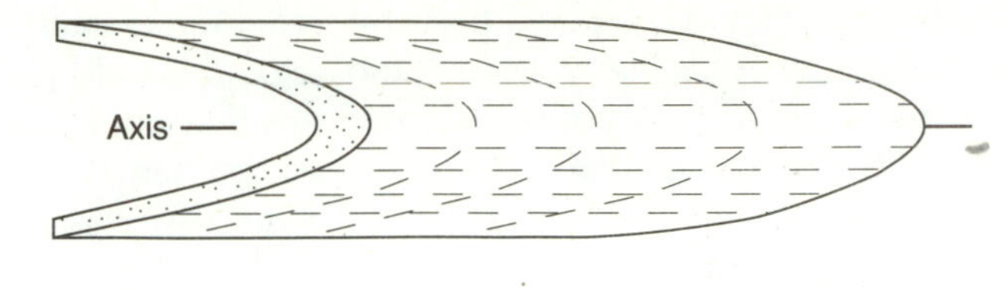

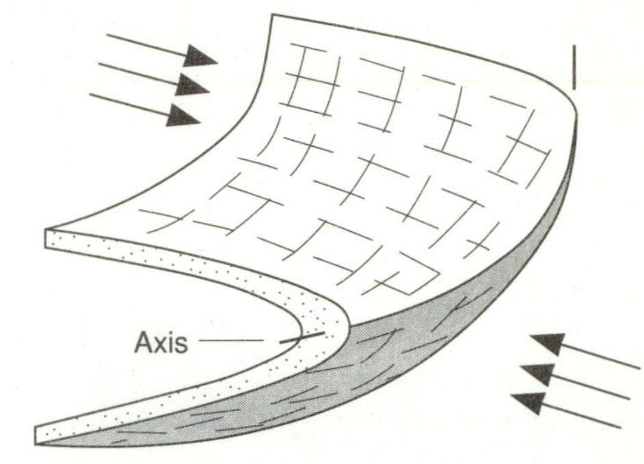

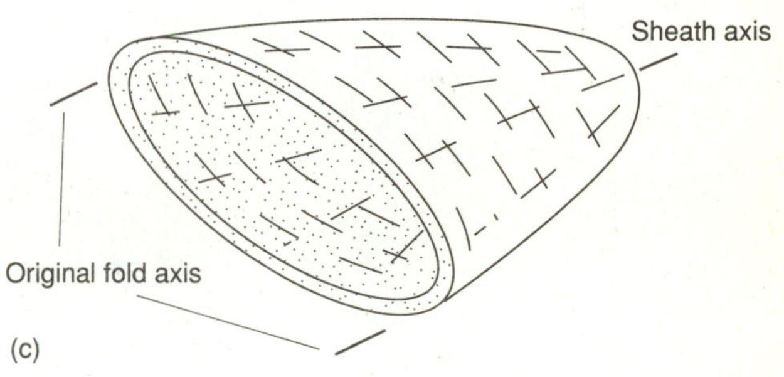

FIGURE 15–34
Formation of noncylindrical and sheath folds through progressive deformation and inhomogeneous simple shear. (a) "Normal" cylindrical fold. (b) Moderately noncylindrical fold produced by a component of inhomogeneous simple shear. (c) Sheath fold produced as an extreme example of inhomogeneous simple shear.

FIGURE 15–35
Map view of a sheath fold in metasandstone in the Middle Proterozoic Moine Series, Scottish Highlands. (Anthony L. Harris, University of Liverpool, England.)

hence are frequently referred to as "stretching lineations" where strain can be measured or sense of movement can be determined independently. Such lineations parallel the *X* axis of the finite strain ellipsoid. Mineral lineations may also form parallel to fold hinges but often become discordant to the fold hinge as noncylindrical folding begins (Figure 15–36). As a result, the lineation may be parallel along the fold hinge but diverge toward a limb as it is traced to another part of the fold (Ramsay and Sturt, 1973b). Similarly, in sheath folds, mineral lineations commonly parallel the sheath-fold axes (with orientations that vary from place to place within the shear zone), and lineations are oriented at various angles, depending on the orientation of the shear zone.

This concluding discussion on the initiation and growth of folds should provide a basis for better understanding of ancient folds that we see in the field. It should also make the diagnosis of the mechanisms that form natural folds easier. Our examination of fold mechanics should provide the background for a discussion of complex folds in the next chapter.

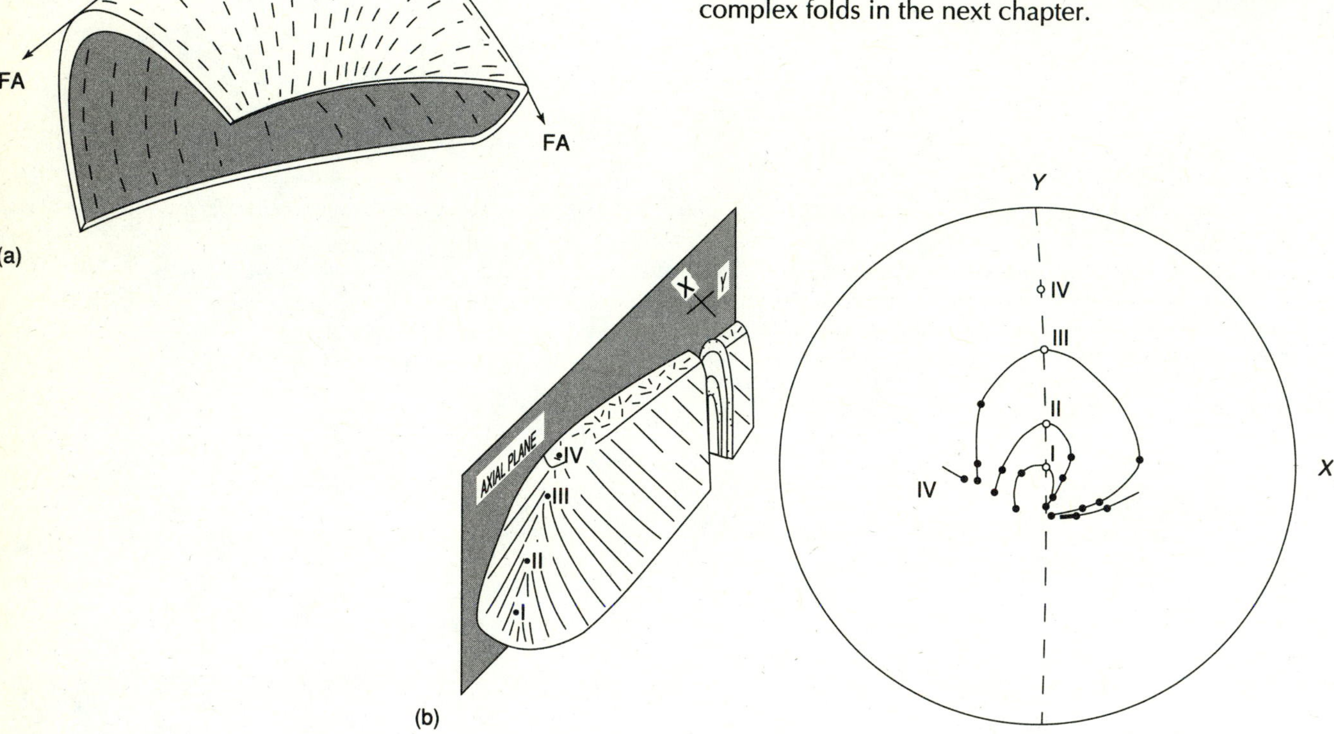

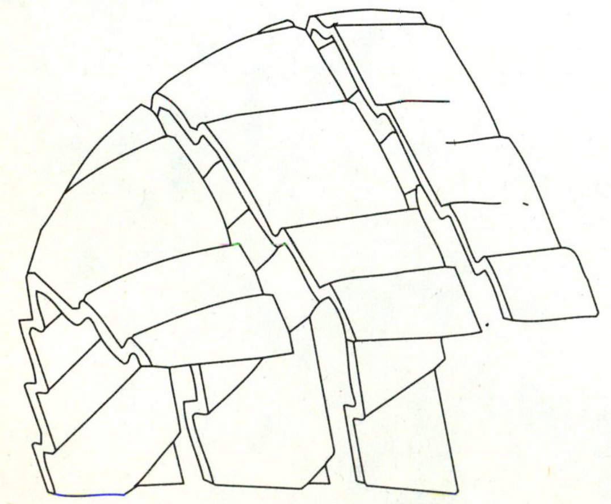

FIGURE 15–36
(a and b) Folded mineral lineation (dark lines on a and b) showing changes in orientation of long axes of minerals in a noncylindrical fold. Some remain parallel to the fold axes (FA); others are rotated into the transport (simple-shear) direction. Fabric diagram in (b) shows the changes in orientation of lineations in different parts of the fold. (c) Small folds that are both parallel to, and divergent from, the hinge of a noncylindrical fold. (Reprinted from *Tectonophysics*, v. 18, D. M. Ramsay and B. A. Sturt, p. 81–107, © 1973, with kind permission from Elsevier Science Ltd., Kidlington, United Kingdom.)

ESSAY

Deciphering the Fold Mechanisms of Two Small Folds

A folded rock (Figure 15E2–1) composed of a layer of mostly garnet (codicule?–a metamorphosed manganese-rich mud) interlayered with micaceous quartzite was collected from the Trondheim nappe complex in southern Norway by Elizabeth A. McClellan (1993) during her dissertation research. A thin section was made of the specimen that revealed the intriguing structure in Figure 15E2–1 and on the cover of this book. Another fold was collected by Darthmouth undergraduate Kurt Larson (Larson and others, 1989) from folded Middle Ordovician shale and graded silt-stone and sandstone in southeastern Tennessee as part of a summer undergraduate research project at the University of Tennessee (Figure 15E2–2). It too, was sectioned, and it revealed an interesting fold formed under physical conditions very different from those in the deeper crust where the garnetiferous quartzite was formed. Despite the differences in physical conditions, both folds have strongly curved hinges (parallel to the hinge) that indicate they are noncylindrical, and both change profile shape in very short distances along the hinge lines. Isogon analysis, however, indicates that both are Class 1C folds.

The Middle Ordovician fold probably formed during a single event, because there is no evidence of refolding or overprinting by younger folds. This initially appears to be a very simple structure, but small fractures that propagated from the inner to the outer hinge of this fold as it was tightened may provide clues to the folding process (Figure 15E2–2b). The fracture probably formed in extension because it was partially filled with calcite, but tightening of the fold appeared to reactivate the fracture, this time in compression. The vein was fractured, and fragments rotated sinistrally (counterclockwise) so that the hinge was thrust over one limb, producing displacements of from 1 mm in the outer hinge to a maximum of 3 mm in the inner hinge. Weaker layers exhibit minor thickening in the inner hinge of the fold, and so the fold mechanism probably was buckling accompanied by flexural flow.

The fold from southern Norway is clearly a refold, but a Type 3 (Ramsay, 1967; Chapter 16) that involved continued folding of an already-formed fold on the same axis, producing a jellyroll-like structure (Figure 15E2–1). There also is an indication of noncylindrical to sheath-fold development in the specimen. The subsequent deformation by layer-parallel simple shear of late veins filled with coarser quartz in the garnet layer indicates that flexural slip occurred after the veins formed, probably during the second stage of folding, but these veins cut smaller earlier fractures that are fanned about the hinges of the late folds. The late veins were folded in the hinge of one of the late folds and possibly also on the limbs. Quartz-mica layers are thickened into the hinges of the last folds (Figure 15E2–1b and 15E2–1c), but anomalously thickened regions of quartz and quartz-mica are also present close to hinges (Figure 15E2–1d). The anomalously thickened quartz-mica layer could represent either the locations of older fold hinges or the locations of sheath-fold hinges out of the plane of the section. The second folding event produced classic flexural-flow behavior. The garnet layer behaved as a stiff material and buckled, and even fractured during folding, whereas the quartz-mica layer deformed by ductile flow, producing a strong preferred orientation of muscovite, followed by static recrystallization of the quartz. Grain size in the quartz-mica layer remains small, a few millimeters away from the boundary with the garnet layer. The earlier folding event probably also followed a flexural flow and buckling mechanism, because the early fractures in the garnet layer indicate layer-parallel extension. Other fractures that formed parallel to layering indicate extension perpendicular to layering. These appear to have formed in several generations.

Folds such as these provide a great deal of insight into the mechanisms of buckling and flexural flow. It should be clear that these mechanisms can operate over a wide range of physical conditions, from near-surface to those found within

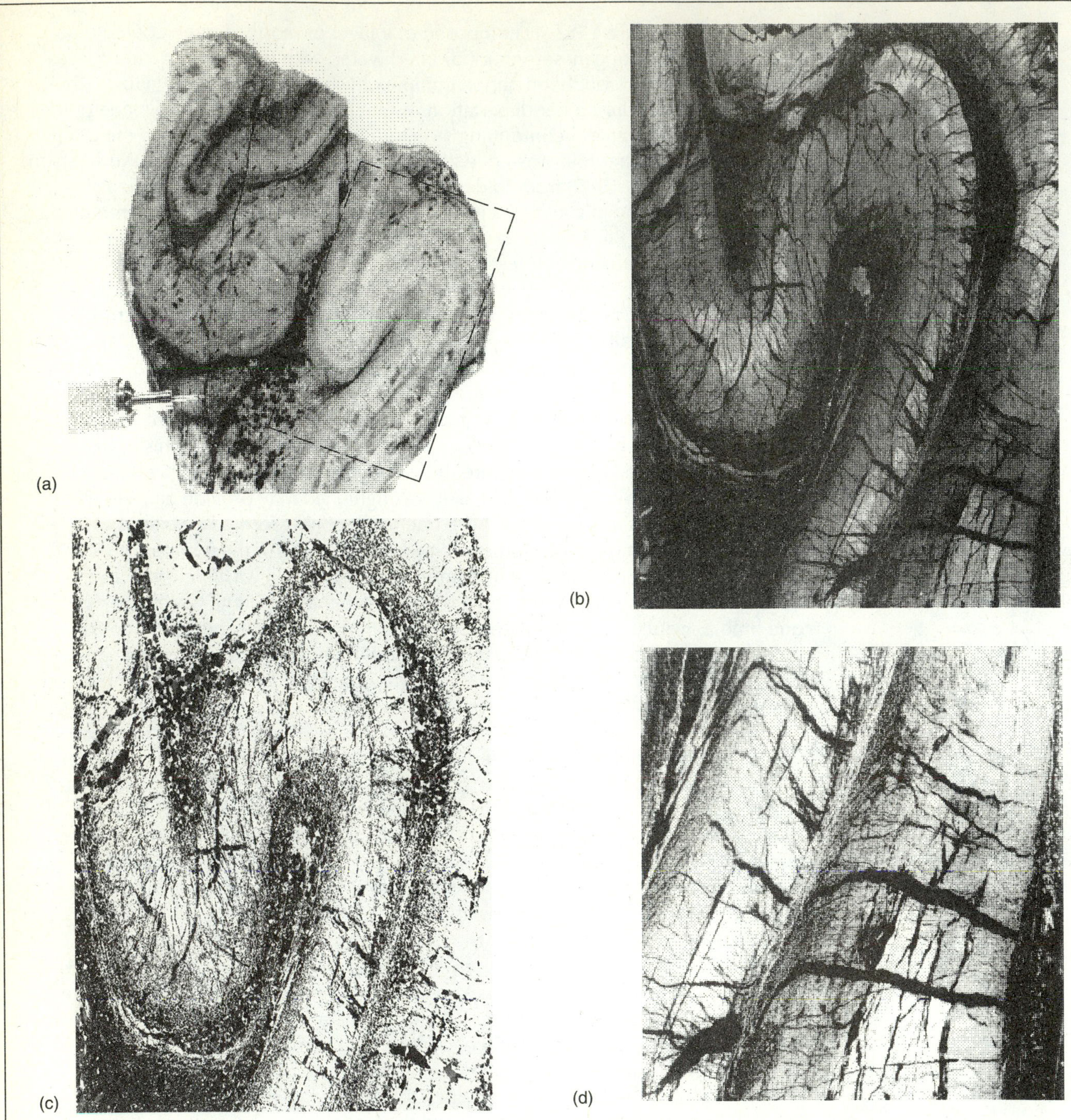

FIGURE 15E2–1
Folded multilayer garnet-quartz (metamorphosed manganiferous mud—codicule?) in a sequence of metasedimentary and metavolcanic rocks in the Trondheim nappe complex from Einunnfjellet near Foldal, southern Norway. (a) Sawed specimen of garnet- and quartz-rich layers. The continuity of the garnet layer throughout the specimen suggests that only one garnet-rich layer is involved. (Dashed rectangle indicates the location of the thin section in b and c.) Negative prints of the section in plane (b) and polarized (c) light indicate the lack of unannealed strain and complete recrystallization of the quartz, along with the more stiff garnet layer that fractured during folding. The maximum long dimension of the thin section is 7.3 cm. (d) The boundary between the garnet- and quartz-rich layers reveals a series of veins that formed early and were later deformed by flexural slip at the layer boundary and were folded inside the garnet layer as folding progressed. Width of field is 7 mm. (Specimen courtesy of Elizabeth A. McClellan, Western Kentucky University.)

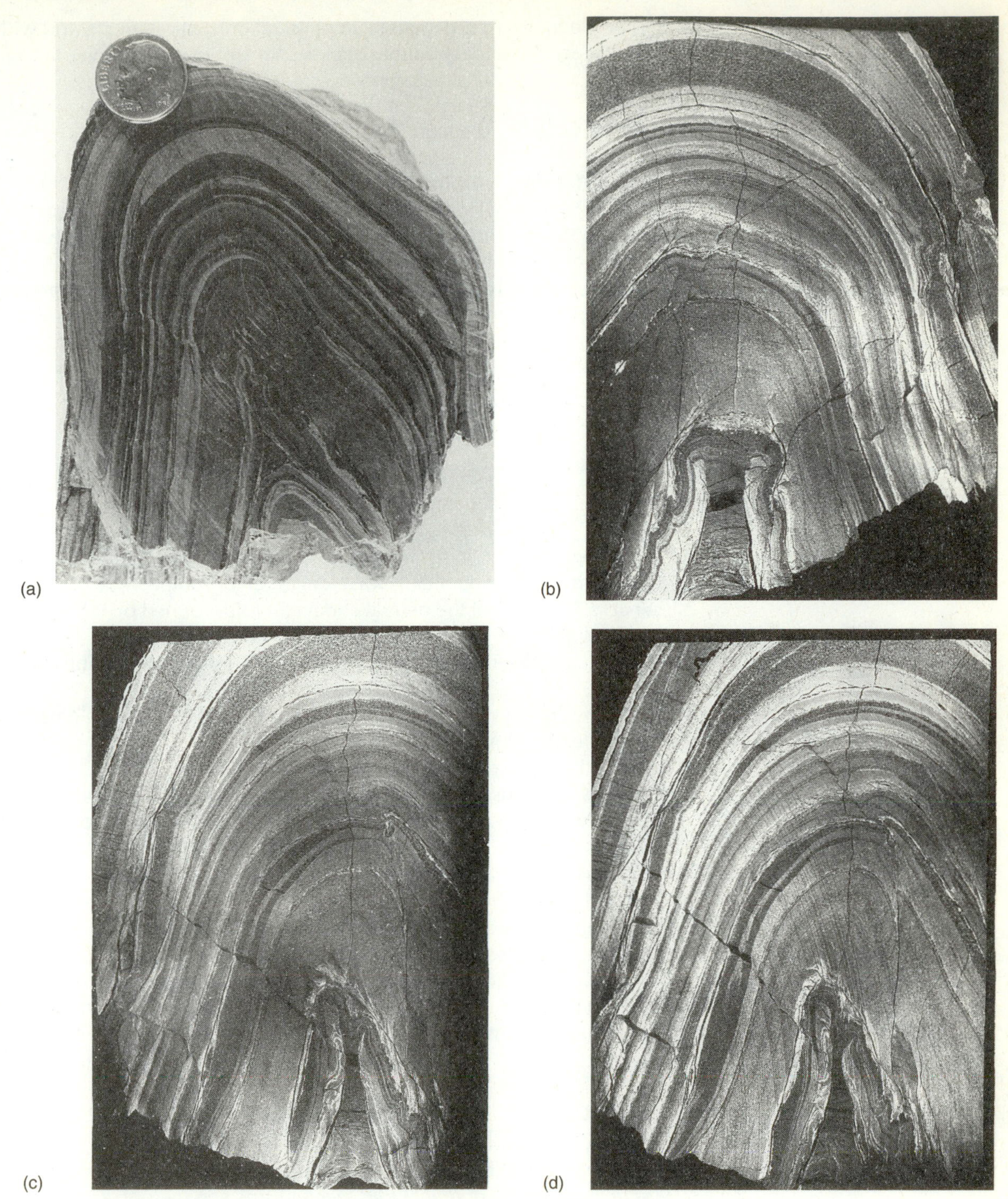

FIGURE 15E2–2
Fold in graded and cross-bedded Middle Ordovician siltstone and shale from near Walland, Tennessee. (a) Sawed specimen reveals external details of the fold. As the fold tightened, some thickening occurred in the weak, thinly bedded shale, and a small fault with displacement decreasing away from the hinge zone formed to compensate for lack of space in the inner hinge. Minor thickening of beds occurs elsewhere in the fold, but several layers exhibit original variations in thickness inherited from the environment of deposition. In both the hand specimen and in thin sections (b, c, and d; all in plane light) cut about 1 cm apart, variations in the shape of the fold occur along the hinge. The small calcite-filled fault zone(s) in thin sections clearly reveal the rotation sense during tightening of the fold, with tightened hinge layers thrust over limbs. The long dimension of the thin sections is approximately 7 cm.

the ductile-brittle transition. Variations in physical properties among and within different rock types doubtlessly enable this mechanism to operate over this range of conditions.

References Cited

Larson, K. W., Hatcher, R. D., Jr., Neuman, R. B., and Finney, S. C., 1989, Structure of theGuess Creek fault and Fair Garden anticline, southeastern Tennessee: Relatives of the Great Smoky fault or unrelated structures?: Geological Society of America Abstracts with Programs, v. 21, p. 46.

McClellan, E. A., 1993, Tectonic evolution of the Einunnfjellet-Savalen area, Hedmark Fylke, central-southern Norwegian Caledonides [unpublished Ph.D. dissertation]: Knoxville, University of Tennessee, 209 p.

Ramsay, J. G., 1967, Folding and fracturing of rocks: New York, McGraw-Hill, 568 p.

Questions

1. Why does variation in strong rock type not influence the thickness of the dominant member-dominant wavelength relationship?
2. What is the origin of the neutral surface in folds formed by tangential-longitudinal strain?
3. What factors determine which mechanism dominates in folding a mass of rocks?
4. What are the physical requirements for flexural-slip buckle folding? Passive flow? Flexural flow?
5. Why is the passive-slip mechanism problematical?
6. Considering Johnson's criteria for the formation of kink and buckle folds, explain why both kinks and sinusoidal folds occur in some deformed single crystals of kyanite, but in other crystals, only kinks occur?
7. Why do occasional mineral lineations parallel a fold axis in one part of the fold and diverge from it in another?
8. Where in an orogenic belt would you expect passive-flow folds? Why do we also see them in soft sediment and salt?
9. What was the sequence of application of homogeneous and heterogeneous strain in the folds in Figure 15–17? The rubber-layer model in Figure 15–28?
10. Could passive-flow folds form by the mechanism proposed by Beutner and Diegel?

Further Reading

Hudleston, P. J., 1986, Extracting information from folds in rocks: Journal of Geological Education, v. 34, p. 237–245.
Discusses the three principal fold mechanisms and their environments of formation.

Hudleston, P. J., and Lan, L., 1993, Information from fold shapes: Journal of Structural Geology, v. 15, p. 253–264.
Outlines the options for determining strains and rheological state of the rock mass during folding.

Johnson, A. M., 1977, Styles of folding: Mechanics and mechanisms of folding of natural elastic materials: Amsterdam, Elsevier, 406 p.
Johnson's introduction summarizes the state of knowledge about fold mechanics in the late 1970s and describes rigorously developed ideas on the formation of buckle and kink folds.

Ramsay, J. G., 1967, Folding and fracturing of rocks: New York, McGraw-Hill, 568p.
The first half of Ramsay's chapter on folds and folding deals with fold mechanics and discusses buckling, shear, and compressive strain and mechanisms that lead to similar folds.

Ramsay, J. G., 1983, Rock ductility and its influence on the development of tectonic structures in mountain belts, *in* Hsü, K. J., Mountain building processes: New York, Academic Press, p. 111–128.
Discusses the influence of ductility contrast on folding style (and other structures), with especially good illustrations.

Tanner, P. W. G., 1989, The flexural-slip mechanism: Journal of Structural Geology, v. 11, p. 635–655.
Summarizes the evidence for flexural slip in layered rocks and the relationships between bedding slip and the development of hinge-perpendicular lineations, displacement of dikes and veins, and the development of other flexural-slip related structures.

16

Complex Folds

Thus the Loch Monar synform has folds which are superimposed upon its limbs, which cut across the major structure and which appear to be unrelated to it. Individual second folds can be traced across the first synformal structure from the north-east, where they are found on the gently dipping limb of the synform to the south-west, where they contort the steeply dipping limb of the synform . . . . A remarkable feature created by the second-fold structures crossing the first folds is that any one stratigraphical horizon is folded at least twice by each second fold.

JOHN G. RAMSAY, 1958, *Quarterly Journal of the Geological Society of London*

SO FAR IN THIS BOOK, WE HAVE CONCENTRATED MORE ON discussions of single folds than on fold systems. Yet, isolated folds are less common than groups of folds. Most folds form in sets with a common orientation of hinges and axial surfaces, but the vergence of second-order folds changes across the axial surfaces of first-order folds of the same generation. The orientation of the strain field can change, or the orientation of principal stresses can change, and either can produce a new set of folds in the same area; the new set may overprint existing folds or extensively modify their geometry. Systems of ***superposed***, or ***interference***, ***folds*** result where one set of folds overprints another. The shapes of interference folds are related to the orientation of the two fold sets and to the physical conditions of the rock body being deformed. Fold interference can result in patterns such as those produced in water when waves cross and interfere with one another (Figure 16–1).

Noncylindrical and sheath folds, too, form geometrically complex systems. Like other complex folds, they usually involve either continued heterogeneous deformation or changes in the strain field through time.

FIGURE 16–1
Interference pattern produced by water waves crossing in the lee of a headland on the coast of Brazil. The waves radiating from the headland are interfering with linear waves to produce domes and basins, particularly well developed southeast of the headland. (U.S. Geological Survey.)

OCCURRENCE AND RECOGNITION

Complex folds occur in a variety of settings. Two folding events may be separated by minutes or by millions of years, or a continuum may exist; in any case, two events are likely to yield fold-interference patterns and thus complex folds. Ductile shear zones and strain variation within ductile rock masses make possible the formation of complex folds. Such conditions arise most commonly in the cores of mountain chains where a continuum of deformation and metamorphism occurs. The same conditions may also exist at plate boundaries—in subduction zones, spreading ridges, or transform fault zones—where varying amounts of thermal energy combine with a strong component of simple shear or pure shear.

FOLD-INTERFERENCE PATTERNS

Near Loch Monar in the northwest Scottish Highlands, studies by John Ramsay (1958, 1962) revealed multiple fold episodes that produced three fundamental types of fold-interference patterns (Figure 16–2). In Australia and elsewhere, S. Warren Carey (1962) also recognized

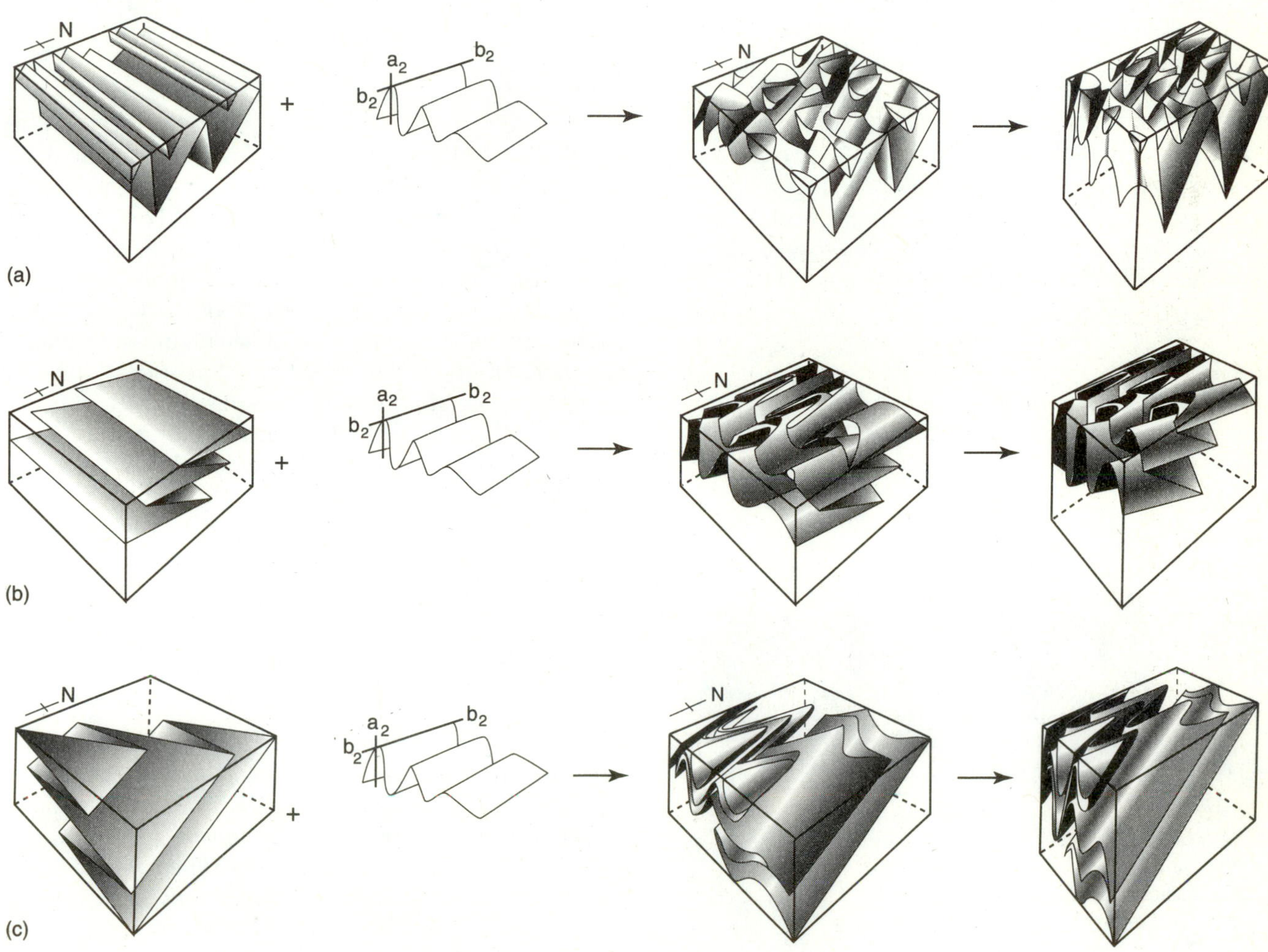

FIGURE 16–2
Ramsay's three fundamental types of fold-interference patterns, showing basic geometry and outcrop patterns. (a) Type 1. Dome-and-basin pattern: early folds with steeply dipping axial surfaces are overprinted by a second set, also having steeply dipping axial surfaces but with an orientation perpendicular to the first set (fold axes also are near-horizontal in each set). (b) Type 2. Early tight folds with gently dipping axial surfaces are overprinted by a set of folds with near-vertical axial surfaces oriented perpendicular to the first set. (c) Type 3. Early tight reclined folds are overprinted by a later set with the same axial trend but with vertical axial surfaces normal to the first set. (Modified from J. G. Ramsay, *Folding and fracturing of rocks*, 1967, McGraw-Hill Book Company. Reproduced with permission.)

the importance of fold interference and interference patterns. Here we will consider these fundamental types, along with intermediate types that often form, as well as the ideal end members.

Type 1

If two sets of folds have upright axial surfaces that intersect at high angles, they interfere and produce a structure called ***egg-carton structure*** or ***dome–and-basin pattern*** (Figures 16–1, 16–2a, and 16–3). Ideally, both fold generations have the same style and the same fold mechanism. Axes of the first fold set are oriented at high angles (70° to 90°) to those of the later set, and both sets of axial surfaces dip steeply with respect to each other. Tight folds interfere and produce high-amplitude dome-and-basin structures; open folds produce gentle, low-amplitude dome-and-basin structures. They may occur on all scales and in any part of an orogen. Othmar Tobisch (1966) reported a large-scale dome-and-basin pattern in the Glen Cannich area of the Scottish Highlands (Figure 16–4). Tobisch and others (1970) have suggested that similar patterns are widespread in that region.

Type 2

If folds having inclined axial surfaces are superposed by folds with steeply dipping axial surfaces and axes at high angles to those of the first set, a ***boomerang pattern*** results (Figures 16–2b, 16–5). Generally, the first folds are inclined isoclinal folds; the second are upright similar-type folds. The shape of the interference pattern—the boomerang—depends on the angle between the first and second folds, the dip of the axial surfaces of the early folds, and the plunge of the axes of both fold sets. The boomerang may become highly asymmetric if the angle between the two axial surfaces is not 90°. A boomerang may be greatly extended and opened to a "tongue" shape if the axial surfaces of the earlier folds dip gently. As the dip of the axial surfaces of the early folds steepens relative to the later folds, boomerangs also grade into dome-and-basin (Type 1) interference folds. A large-scale boomerang pattern exists in the core of the Toxaway dome in South Carolina (Hatcher, 1977; Figure 16–6), where several episodes of folds on all scales are superposed.

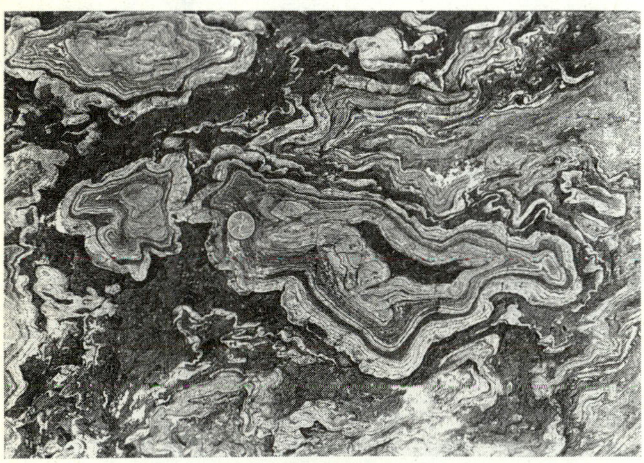

FIGURE 16–3
Dome-and-basin (Type 1) interference pattern in Archean grunerite chert, banded iron formation, Opapimiskan Lake area, northwestern Ontario (North Caribou Lake greenstone belt, Superior Province). (Fred W. Breaks, Ontario Geological Survey.)

Type 3

If early-formed tight to isoclinal folds are refolded about the same axis, they produce an interference pattern called a ***hook*** (Figures 16–2a and 16–7) and may form on small scales as well as large (Figure 16–8). A hook pattern may result from continued folding during the same event (Figure 16–9) or from overprinting by a second generation of folds with the same axial orientation as the first set. To form Type 3 interference patterns, the two fold sets must not have coplanar axial surfaces. It may be difficult to resolve the conditions that form Type 3 folds because we need to know whether folding ended at one time and then resumed later under the same stress conditions—or whether it was continuous within a single event, producing coaxial refolding (Wynne-Edwards, 1963).

Modifications

Combinations of all three types of interference patterns have been found in nature. As noted above, they grade into one another where fold axial surfaces steepen, or become gentler, or approach parallelism. Type 2 interference patterns may grade into Type 3 where the axial trends of early- and late-formed folds become more nearly parallel. Outcrop patterns reflect these gradational fold patterns. Boomerangs appear as tongue-shaped folds in map view or appear more elongate, with one end open in a hooked outcrop pattern (Figure 16–10). Various combinations of outcrop patterns of interference fold shapes have been worked out by Richard Thiessen and W. D. Means (1980), using cuts at different angles through a hypothetical cube of repeatedly deformed crust. Their work also provides better understanding of the three-dimensional aspects of superposed folding.

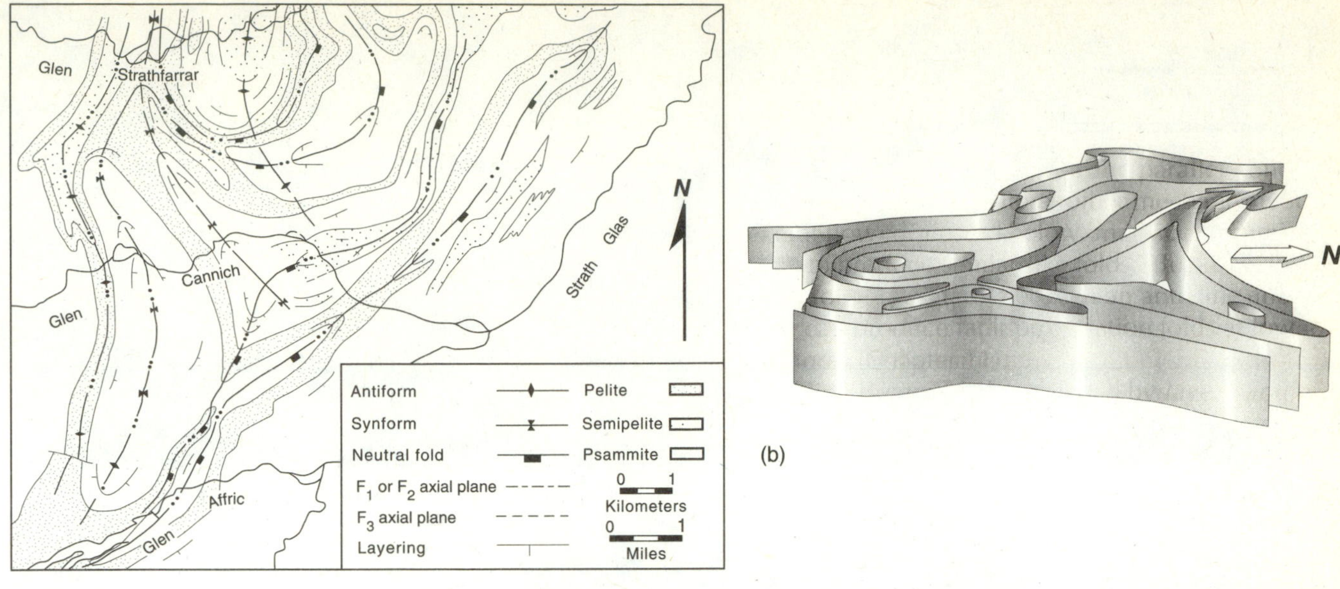

(a)

(b)

FIGURE 16–4
(a) Tectonic sketch map of the Glen Cannich and adjacent areas in the Scottish Highlands, showing the outcrop pattern of the dome-and-basin interference pattern. (b) Simplified three-dimensional diagram of Type 1 folds in the Glen Cannich area. (From O. T. Tobisch, 1966, Geological Society of America *Bulletin*, v. 77.)

RECOGNITION OF MULTIPLE FOLD PHASES

Several characteristics help to identify multiple fold phases. The geometry of outcrop patterns on geologic maps or on the mesoscopic scale is the best key to recognizing superposed folds (Figure 16–10), but in some areas exposures of superposed folds are rare, and other clues are needed. Multiple fold orientations and styles of folds in the same area can be used, with caution, for determining whether superposed folds exist, but alone they do not suffice; exposures of superposed folds must be found, too. Multiple foliations, folded foliations, and folded lineations also indicate multiple deformation and probably superposed folds as well.

FIGURE 16–5
Boomerang (Type 2) interference pattern in Upper Proterozoic Tallulah Falls Formation metasedimentary rocks near Rainy Mountain, northeastern Georgia. (RDH photo.)

NONCYLINDRICAL AND SHEATH FOLDS

All folds become noncylindrical if traced far enough and if they are not truncated by faults. Folds that are noncylindrical over short distances are products of localized strain variations parallel to the hinge line, in addition to the normal circumstance of strain variations in the plane perpendicular to the hinge. The transition from aberrant folds through noncylindrical to sheath folds (see Figures 15–34 and 15–36) is related to increase in magnitude of strain. Homogeneous strain may be responsible for formation of many sheath folds, as indicated by the associated foliations and lineations (Chapters 17 and 18), but some are also the products of heterogeneous strain.

Sheath folds resemble dome-and-basin interference folds (Figure 16–11a), but they can be more isolated. They may be strongly asymmetric, flattened into a C-surface foliation, and elongate rather than symmetrical. Many sheath folds, however, yield highly symmetrical *eyed* or *bulls-eye* sections in plan view. Antiformal and synformal sheath folds may be strongly

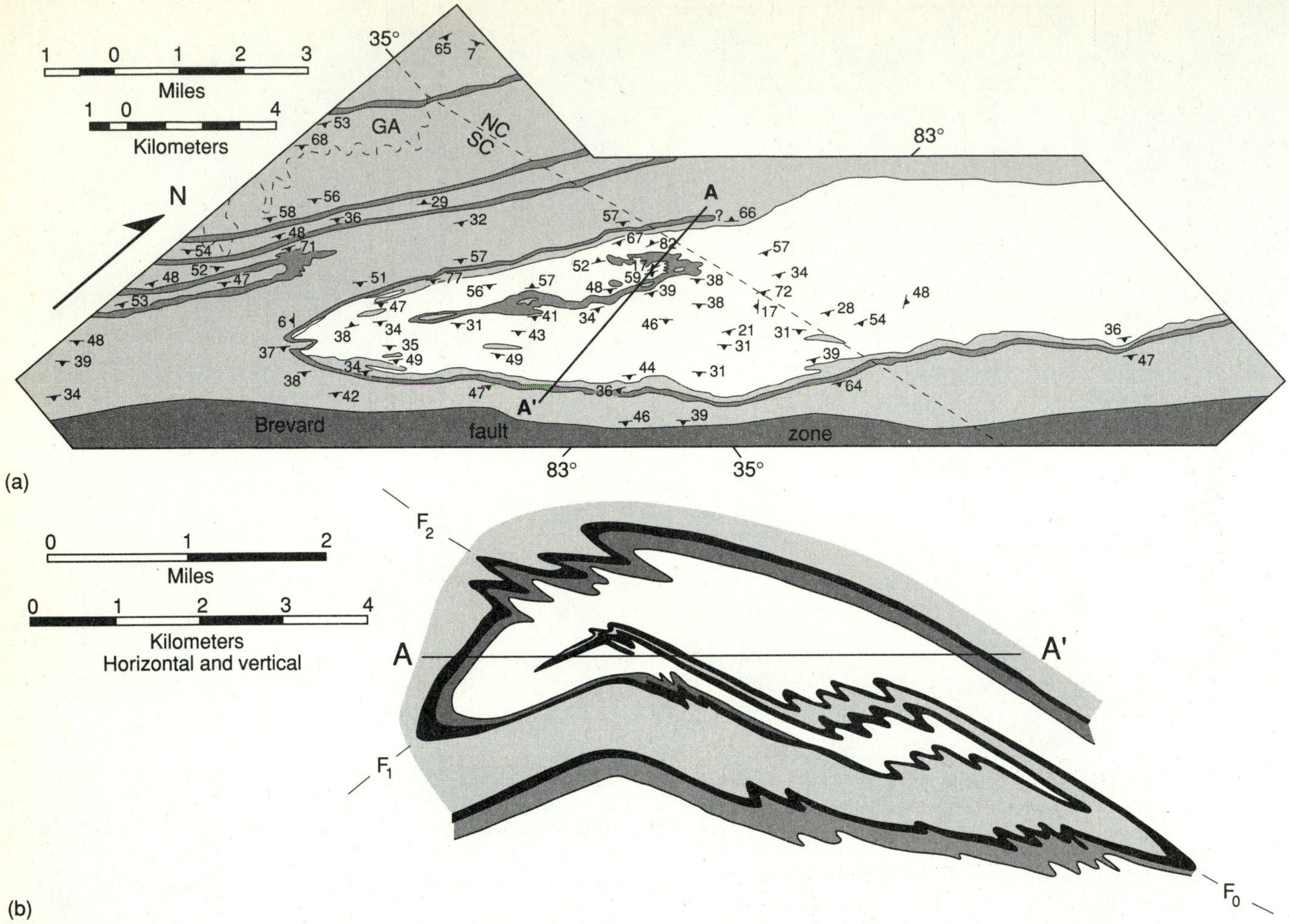

FIGURE 16–6
(a) Geologic map of part of the Toxaway dome, South Carolina–North Carolina, showing the map-scale boomerang (Type 2) interference pattern developed by superposed folding of 1.1 Ga Toxaway Gneiss basement (unpatterned) and cover rocks. (b) North-south geologic section constructed normal to the hinges of the early folds. Note that an earlier ("pre-F_1") set of folds, here called F_0 was resolved after recognizing that the oldest rocks (basement) must have been beneath the cover before deformation. Dark screen pattern is metagraywacke and amphibolite; black screen is aluminous schist, and the light screen is metagrawacke. (From R. D. Hatcher, Jr., 1977, Geological Society of America *Bulletin*, v. 88.)

aligned in shear zones, whereas Type 1 interference folds may be developed over a broad area, although the distinction, however, is not always clear. Most interference folds, on close examination, are also noncylindrical, particularly those of Types 1 and 2.

Many folds originally identified as dome-and-basin interference folds may upon reexamination turn out to be sheath folds. An area where polyphase folding occurs must be studied carefully to ascertain that fold-interference patterns are not really sheath or strongly noncylindrical folds formed during a single deformational event. It is therefore necessary to work out the fold mechanisms as well as the timing of the different fold phases. Sheath folds may be confined to a planar zone of ductile shear or to a large fault zone; fold-interference patterns may be distributed regionally and not confined to particular deformation zones. These generalizations are not always correct: B. Gos-

FIGURE 16–7
Hook (Type 3) interference pattern (left of hammer) in Upper Proterozoic Tallulah Falls Formation metasandstone, Holcombe Creek Falls, near Clayton, Georgia. (RDH photo.)

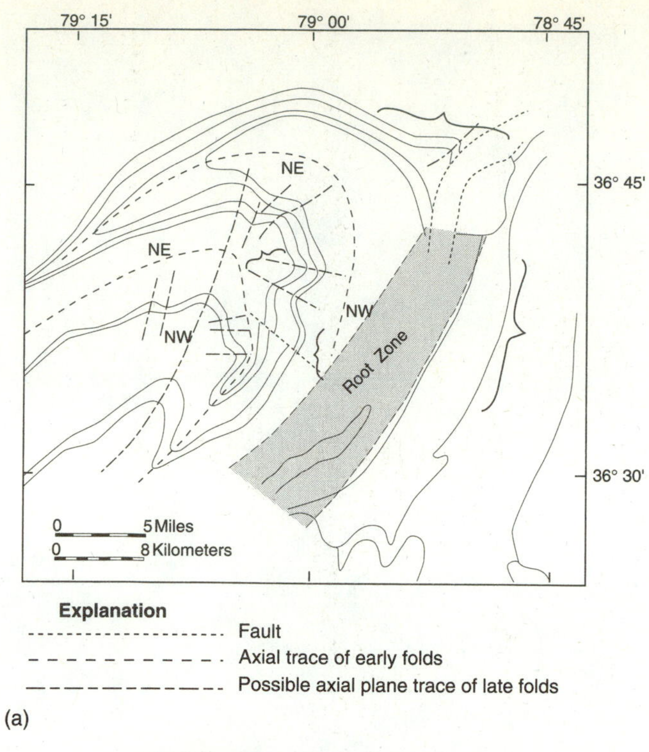

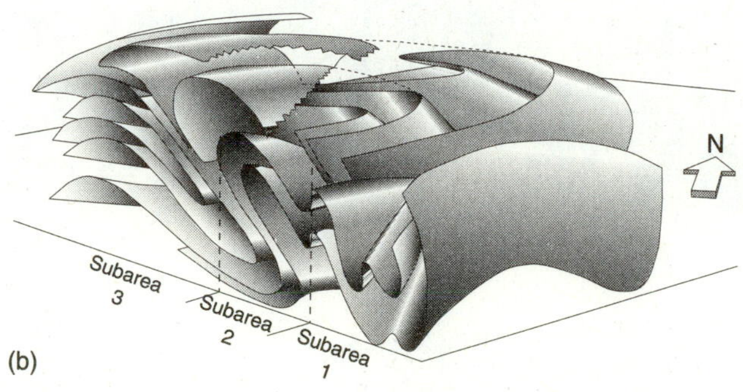

FIGURE 16–8
(a) Map-scale Type 3 interference pattern in the Milton area north of Greensboro, North Carolina. (b) Sketch of the three-dimensional relationships of folds in the Milton area. Subarea 1 is the root zone of a large recumbent fold; subarea 2 is a zone of moderate dip of foliation and dominantly northeast-trending folds; subarea 3 is the gently plunging hinge area of the recumbent fold. (From O. T. Tobisch and Lynn Glover, III, 1971, Geological Society of America *Bulletin*, v. 82.)

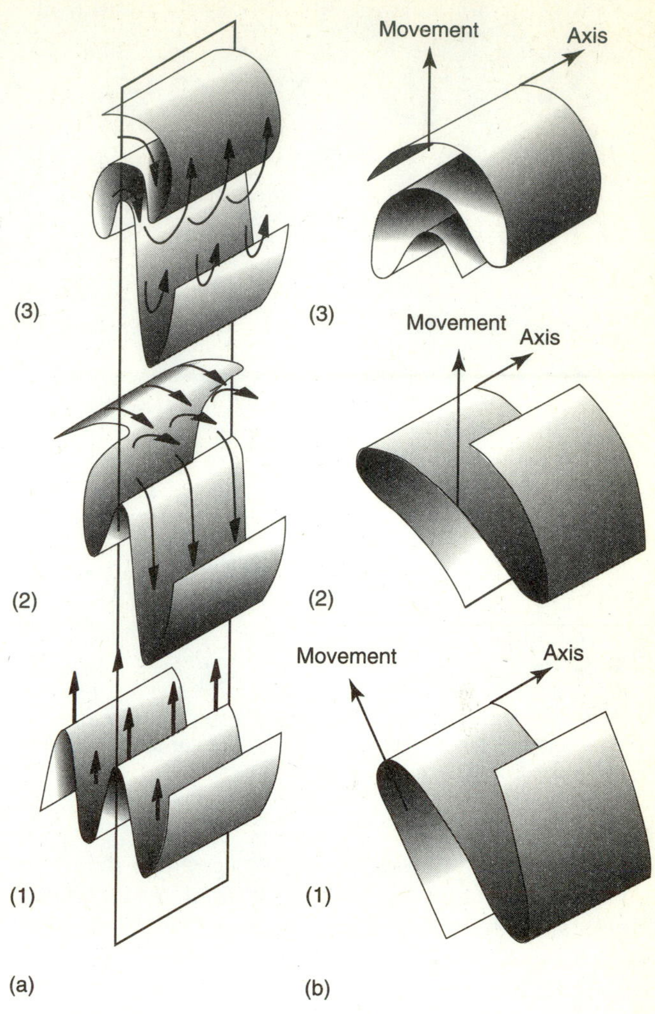

FIGURE 16–9
Formation of Type 3 interference folds by progressive deformation of an already-formed fold without varying axial orientation. (a) Formation by migration of the hinge of an advancing antiform next to a stationary or more slowly advancing synform. (b) Formation by change in direction of transport. (From H. R. Wynne-Edwards, 1963, Flow folding: *American Journal of Science*, v. 261, p. 793–814. Reprinted by permission of the *American Journal of Science*.)

combe (1991) has described a large area of map-scale sheath folds and superposed folding in the Strangways orogen in the Middle Proterozoic rocks of central Australia, and Nicholas Culshaw and others (1991) have described similar structures in the internal parts of the Grenville orogen (Britt and Parry Sound domains) in southern Ontario.

Lilian Skjernaa (1989) attempted to quantitatively separate weakly noncylindrical folds from sheath and *tubular folds* (very tight sheath folds) by employing a ratio of geometric axes x:y or z:y, where x is the cone axis of the fold (Figure 16–12), along with the angle ω, the hinge angle. Using these parameters, she noted that sheath folds should have x:y axes of greater than 4:1 with a hinge angle ω of 20° to 90°. Tubular folds have z:y axes of between 1:1 and 1:10, with most greater than 1:2.5 with hinge angles of < 20°; x:y is commonly greater than 1.

FORMATION OF COMPLEX FOLDS

If superposed folds are to form, strain states must vary in both space and time so as to initiate folding in new orientations in already-folded layers. Analysis of the

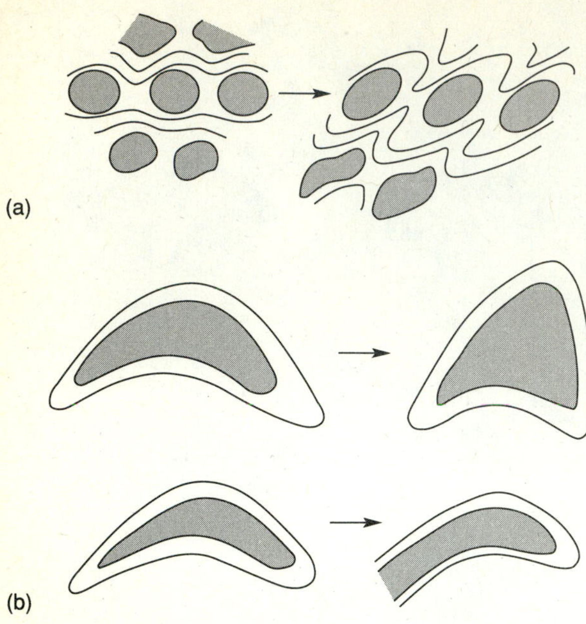

FIGURE 16–10
Changing outcrop pattern with modification of Type 1 domes (a) to elongate domes, (b) ideal Type 2 boomerang folds to tongue-shaped outcrop patterns, and (c) opening up one end of a Type 2 fold to become a Type 3.

(a)

(b)

FIGURE 16–11
(a) Sheath folds in Archean gneiss near Cape Jermain, Foxe fold belt, Melville Peninsula, Canada. View is of a vertical rock face. (From *Journal of Structural Geology*, v. 3, J. R. Henderson, Structural analysis of sheath folds with horizontal *X* axes, Northeast Canada, © 1981, with kind permission from Elsevier Science, Ltd., Kidlington, United Kingdom.) (b) Sheath folds in Late Proterozoic quartzite at Abisko, Sweden. View is up plunge parallel to the sheath axis (x) with the sheath hinge in the exposure. Note "arrowhead" shape in the *YZ* section in the exposure (see Figure 16–15). (RDH photo.)

geometry of superposed folding may begin with consideration of *coaxial* and *noncoaxial deformation.* Noncoaxial deformation is required to form Type 1 and Type 2 interference patterns. Where Type 1 folds develop, the directions of overall extension and shortening are reoriented after the formation of the early folds so that the second folds trend normal to the first. Likewise, with Type 2 and Type 3 interference patterns there must be some reorientation of the directions of overall extension and shortening. Formation of Type 1 and Type 2 patterns, however, requires reorientation of the directions of overall extension and shortening rapid enough that either the mechanical properties of the rocks are not changed by cooling or by diminishing pressure, or the material properties of the rocks change with changing P–T conditions. In either case, this is a more difficult set of circumstances to achieve, especially considering the amount of reorientation required, but it could occur during plate collision if a promontory, a seamount, or other irregularity were encountered on either plate boundary—or possibly by migration of a triple junction. In contrast, Type 3 interference patterns form with only minor reorientation of the directions of overall extension and shortening as a continuation of the same folding process that affected the rocks earlier. Wynne-Edwards (1963; Figure 16–9) has described the circumstances under which Type 3 superposed folds can form during the same event by more or less uninterrupted deformation; the process forms the first folds, then superposes the second set coaxially (here referring to parallel-oriented axes), but not along the original axial surfaces. Both sets of folds in the Wynne-Edwards model, however, form under conditions of simple shear—noncoaxial deformation—in the normal usage.

Skjernaa (1989) considered various combinations of original orientation of cylindrical folds and the application of simple shear to produce noncylindrical and sheath folds. The original variations along the hinges of cylindrical folds, coupled with formation of sheath folds at different angles, produce several variations in outcrop pattern in different sections that should provide clues about the way simple shear strains were superposed (Figure 16–13). Jonathan Mies (1991) made a kinematic analysis of the planar dispersion of sheath folds in shear zones and in equal-area plots of

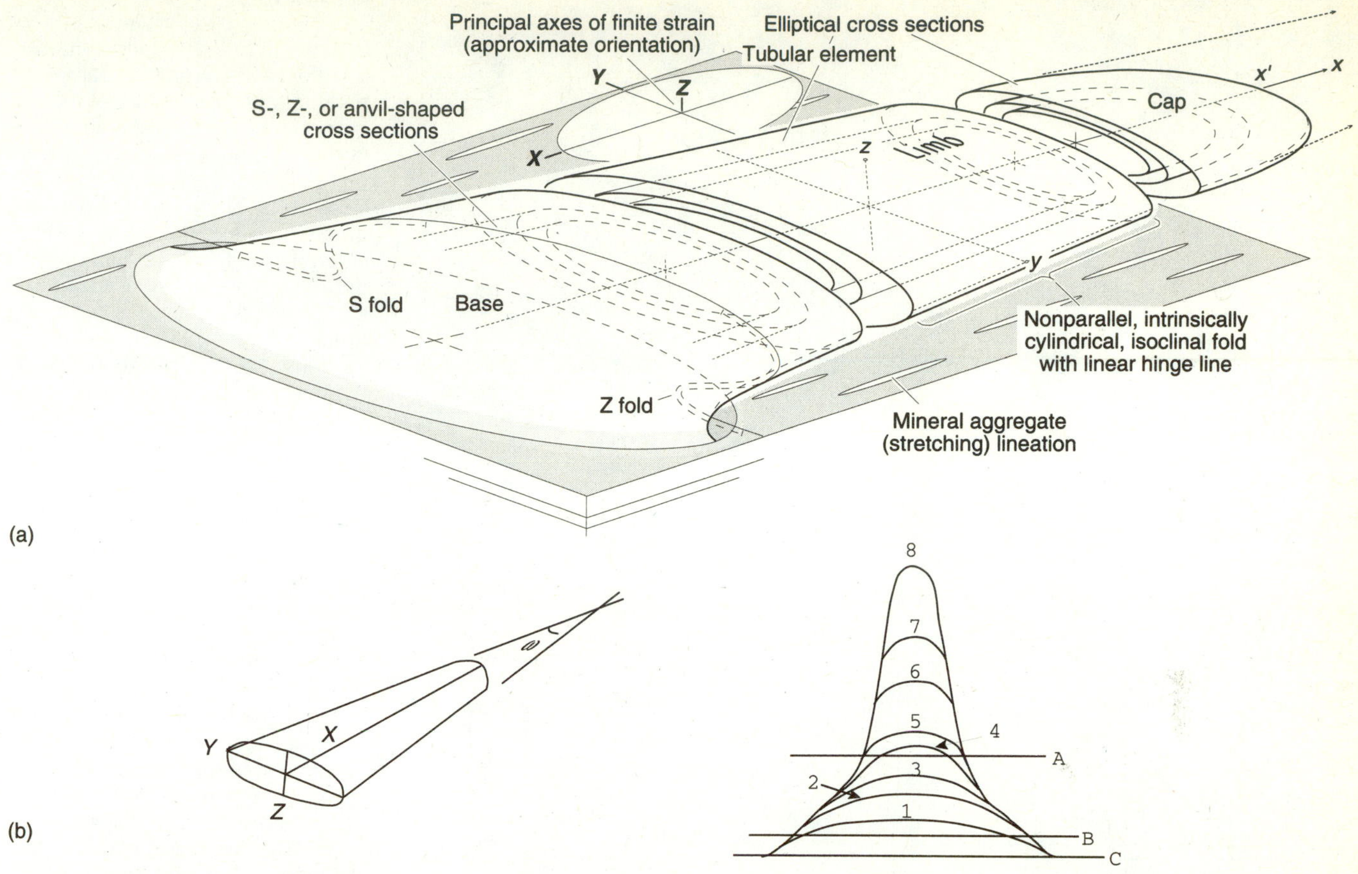

FIGURE 16–12
(a) Geometric and angular components of a sheath fold. ω is the opening angle of Skjernaa (1989). The dimensions *x, y*, and *z* occur at a low angle (decreasing with increasing shear strain) to the principal strain axes *X, Y*, and *Z*. Note how the cross section changes from an S, Z, anvil, or arrowhead shape near the base to elliptical cross sections farther out on the limb of the sheath fold. (From *Journal of Structural Geology*, v. 15, J. W. Mies, p. 983–993, © 1993, with kind permission from Elsevier Science, Ltd., Kidlington, United Kingdom.) (b) Noncylindrical folds 1 through 8, with folds 1 through 4 having ω >90° measured along *x*-*y* section C. Folds 5, 6, and 7 have ω < 90°, measured at section B. Fold 8 has an ω <90° along section A. (From *Journal of Structural Geology*, v. 11, L. Skjernaa, p. 689–703, © 1989, with kind permission from Elsevier Science, Ltd., Kidlington, United Kingdom.)

orientation data measured in shear zones (Figure 16–14). He concluded that hinge lines are rotated into the shear direction and that the angle between the plane of mylonitic foliation and the best-fit girdle to hinge lines may provide an indicator of shear sense, using the Hansen method (Chapter 14).

Studies by Mies (1993) of sheath folds in low-grade marble in the Alabama Appalachians led him to suggest that these structures formed by passive flow under noncoaxial (simple shear) conditions, producing large shear strains ($\gamma \geq 9$). He also determined that the geometry of the sheath folds is governed by the original shape of initial structures required to initiate simple-shear deformation into sheath folds (Figure 16–15a). Symmetrical (circular) initial folds yield sheath folds that are flattened, but horizontal sections remain circular. Transverse initial folds yield sheath folds with elliptical horizontal sections that have maximum elongation in the *Y* direction, whereas longitudinal folds produce sheath folds with elliptical horizontal sections with maximum elongation in the *X* direction. Mies also determined that symmetrical initial structures produce sheath folds with marked limb-thickness differences at high shear strains (Figure 16–15b).

MECHANICAL IMPLICATIONS OF COMPLEX FOLDING

The geometry of superposed and noncylindrical folds implies certain things about the mechanical processes that form them. That they are readily observed as interfering water waves demonstrates that they form in all fluid media. Interference patterns and strongly asymmetric shears resembling folds also appear in

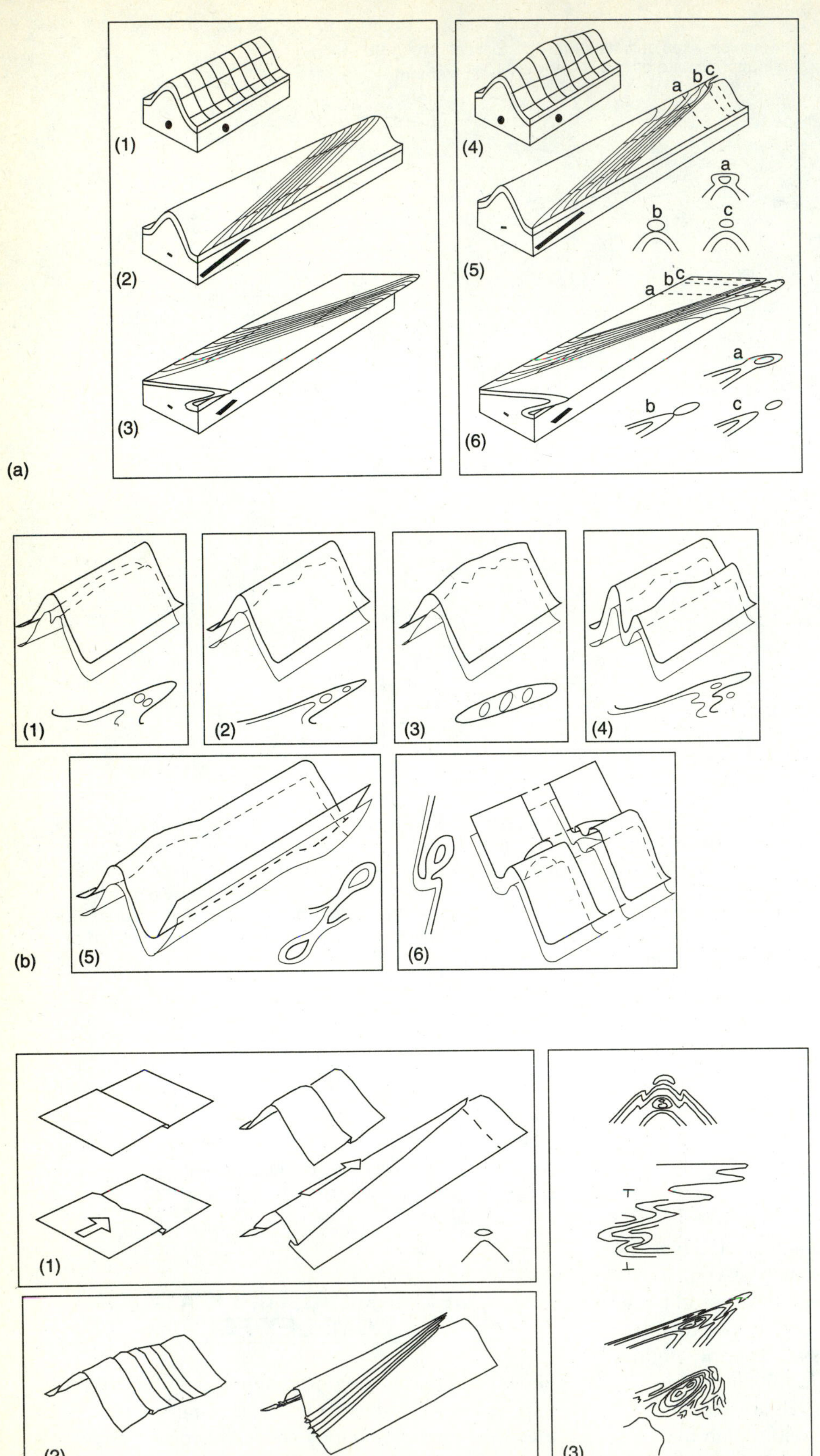

FIGURE 16–13
(a) Progressive deformation of a cylindrical fold by superposition of a dextral horizontal shear parallel to the hinge line of the cylindrical fold (1) and after a horizontal shear at 15° to the axis ($\gamma = 10$) (2). The dots on the sides of the blocks for the cylindrical folds, and the lines on the sides of the blocks for the noncylindrical folds represent strain ellipses for each stage of deformation. 1, 2, and 3 in (2) are cross sections. (b) Possible geometries (with a cross section for each) that may produce tubular folds where horizontal shears are superposed parallel to, or at a small angle to, main axial orientation of the folds. (c) (1) Sequence of folding in which a cylindrical to slightly noncylindrical fold is superposed by a second cylindrical fold with an orientation normal to the first; a tubular fold is produced by renewed shear parallel to the hinge of the second fold, shown in cross section in the lower right. (2) Sequence similar to that in (1) involving several folds. (3) Upper two are sections perpendicular to and parallel to the principal hinge line in the final stage in (2). Dip-strike symbols indicate the location of the upper section in the lower. The third section involves slightly more oblique shear, and the bottom is a sketch of a real fold from the Grapesvare area in northern Sweden. (From *Journal of Structural Geology*, v. 11, L. Skjernaa, p. 689–703, © 1989, with kind permission from Elsevier Science, Ltd., Kidlington, United Kingdom.)

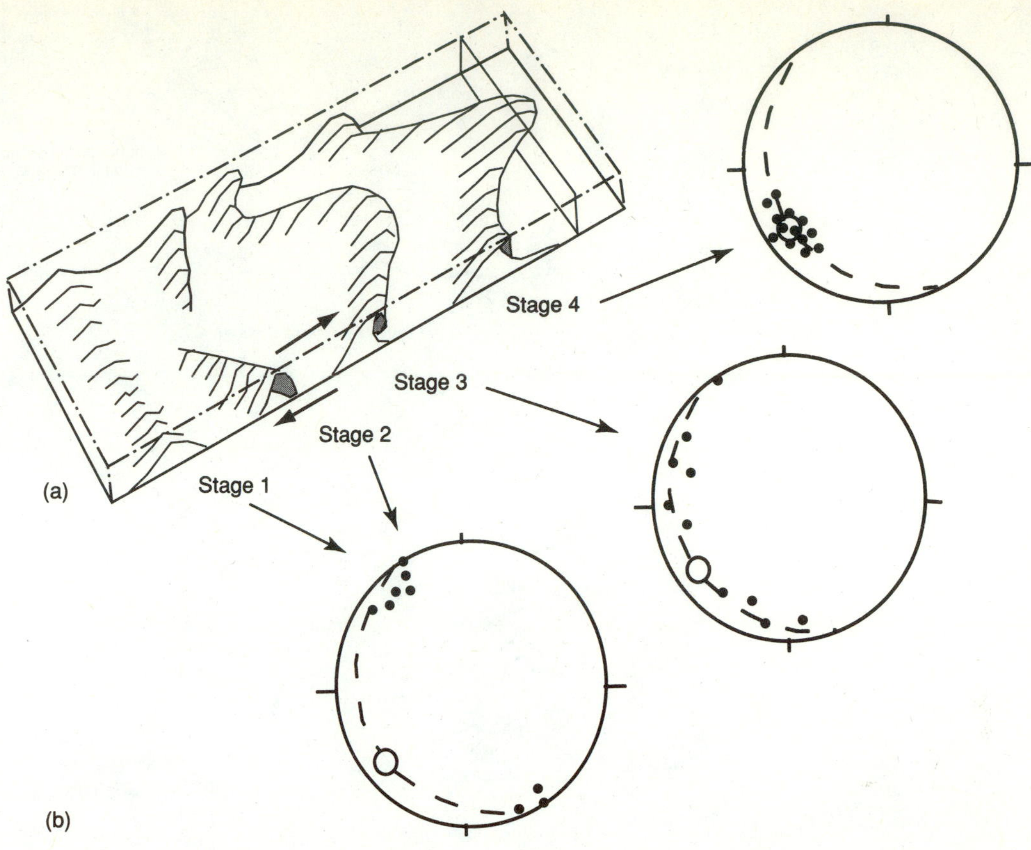

FIGURE 16–14
(a) Progressive amplification of folds (stages 1–4) from cylindrical to sheath geometry in a shear zone, with the corresponding fabric diagrams (b) for each fold. Note the clustering of hinges of the noncylindrical folds about great-circle girdles (dashed lines) that approximate the shear plane. The open circles represent the sheath-fold axes. (Modified from P. J. Hudleston, 1986, *Journal of Geological Education*, v. 34. Used with permission.)

clouds, particularly with the passage of strong frontal systems. Their appearance in ductilely deformed rocks further indicates the similarity of flow processes in all media.

Fold-interference patterns can form by buckling and flexural slip, buckling and flexural flow, or by ductile flow. The standard Type 1, 2, and 3 interference patterns may be associated with a single rheological state or multiple states, either brittle or ductile. Time is required to cool originally ductile rock to a brittle condition. Early buckle folds formed by flexural flow may be superposed by later buckle folds that were formed by flexural slip. Other combinations of passive flow, flexural flow, and flexural slip with buckling are also possible. Deformation mechanisms within individual layers again control fold geometry, as with single folds.

If folds are nucleated by symmetrical buckling and later are deformed by strain variations parallel to the hinge line, as well as variations perpendicular to the hinge, they may be transformed to noncylindrical folds and then to tubular sheath folds. The process may occur under a variety of conditions and rock types. Ductile shear zones are probably the best places for formation of noncylindrical and sheath folds, but they may also form in a purely ductile mass of rocks being deformed by homogeneous strain, which produces locally different rates of ductile flow.

Here we end our discussion of folds and folding, but we will use this knowledge as we now turn to cleavage and foliations, lineations, structural analysis, and plutons.

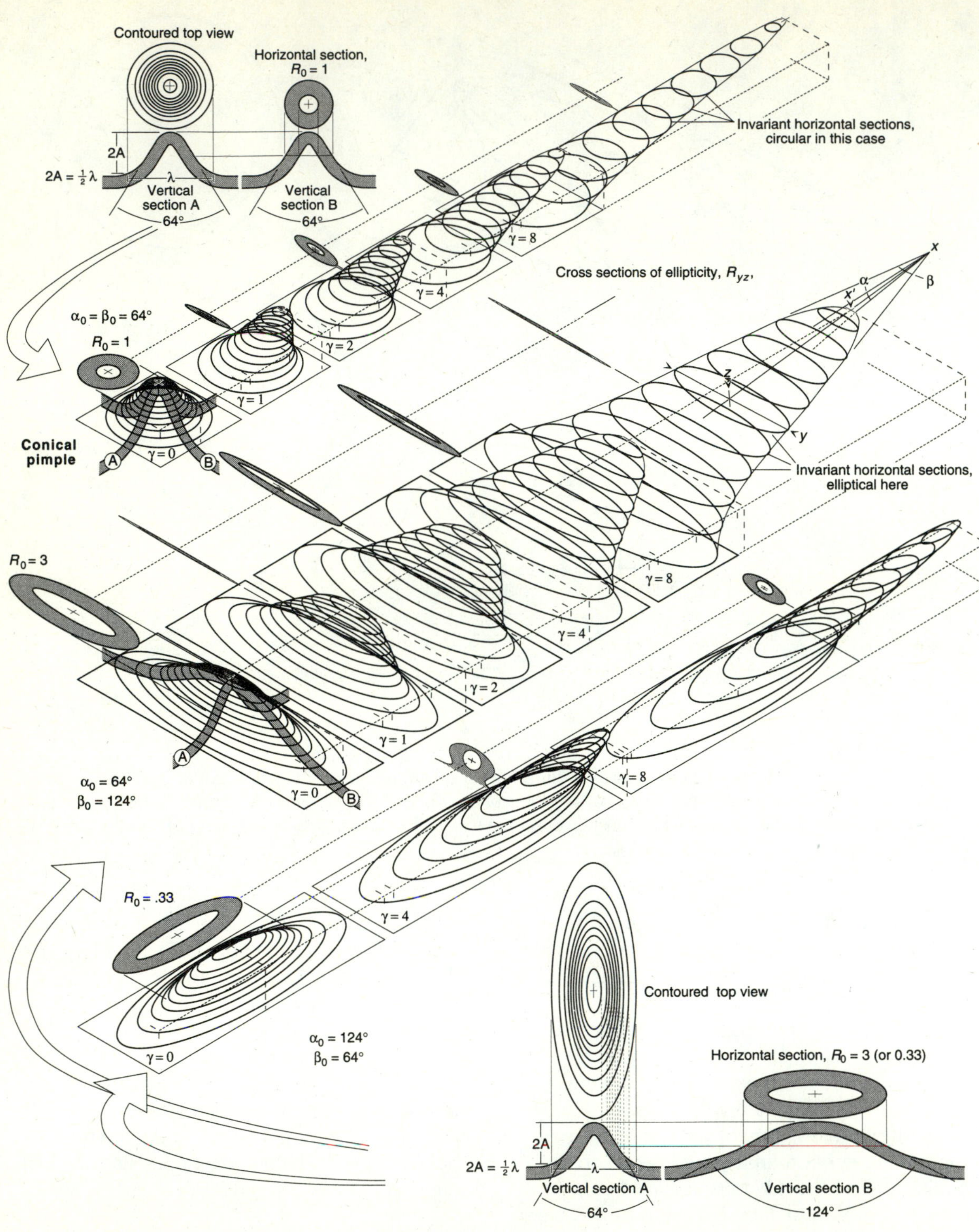

(a)

FIGURE 16–15
(a) Models for formation of sheath folds of different shape by progressive simple shear deformation of initial symmetrical dome (upper), symmetrical transverse fold (middle), and symmetrical longitudinal fold (lower). R_0 is initial ellipticity; λ is wavelength; A is amplitude; γ is shear strain. (b) (Following page) Simple shear deformation of symmetrical and asymmetric (monoclinal) initial folds with different wavelength and amplitude produces sheath folds with different limb-thickness relationships at different stages of deformation that become amplified at large values of γ. (From *Journal of Structural Geology*, v. 15, J. W. Mies, p. 983–993, © 1993, with kind permission from Elsevier Science, Ltd., Kidlington, United Kingdom.)

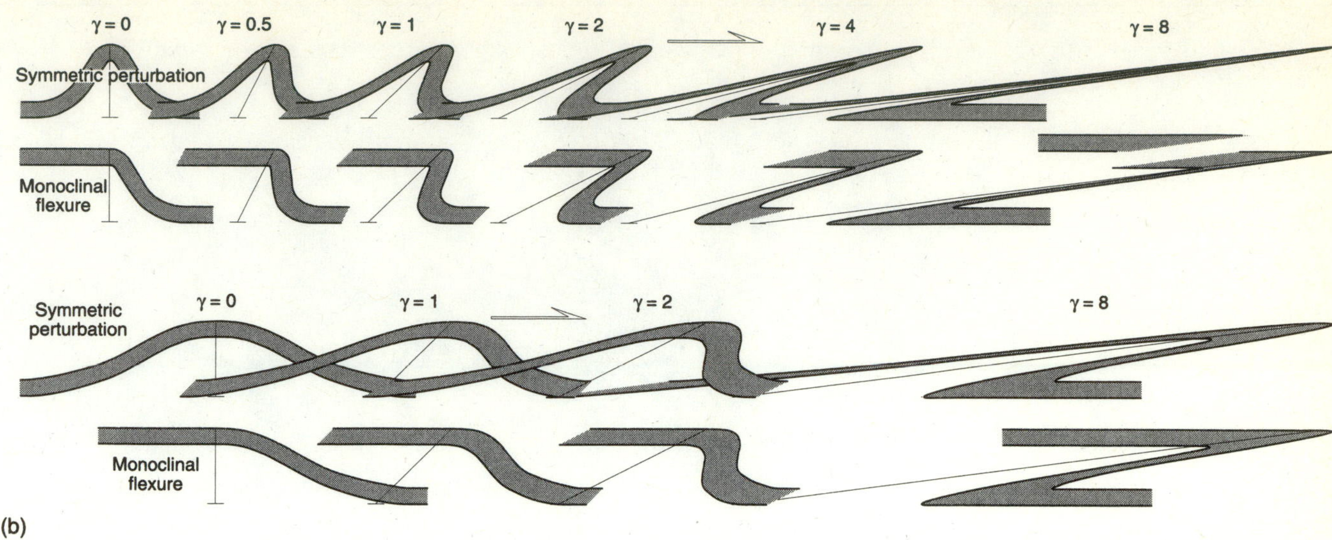

FIGURE 16–15 (continued)

ESSAY

The Value of Rosetta Stones

Fold-interference patterns have been reported in many places in deformed crustal rocks. No doubt, Sam Carey (1962) and many other geologists had recognized the characteristics of superposed folds and considered multiple deformation in a single area before John Ramsay (1962, 1967) described the geometry of superposed folds. Ramsay based his early work largely on exposures of Middle Proterozoic metasedimentary rocks of the Moine Series around Loch Monar in the Scottish Highlands. Impoundment of a reservoir there in the early 1960s resulted in greater variations in the lake level and washing off a superb bedrock exposure along the shore to provide a unique place for observation of fold-interference patterns. All three fundamental interference types are exposed at Loch Monar (Figure 16E–1). Finding all three in a single exposure must have profoundly impressed Ramsay and—because of his ability to synthesize structural geometry—encouraged many structural geologists to read his papers, helping them to understand this phenomenon.

Ramsay originally worked out the fundamental interference types even without a single exposure of such quality, but the Loch Monar exposure certainly made his task much easier. Exposures of this kind are useful in any complexly deformed terrane for observing overprinting relationships and avoiding the pitfalls described by Paul Williams (1970), in which fold generations are recognized and correlated throughout an area by their style alone. As Williams pointed out, changes from one rock type to another, in metamorphic grade and in fluid pressure, introduce many variables and thus produce style changes during a single event.

Along the Chattooga River on the border of South Carolina and Georgia is a large flat exposure called "Woodall Shoals". This exposure is like a Rosetta stone for the deformational history of the eastern Blue Ridge of the southern Appalachians: it preserves several generations of folds and enables us to reconstruct the map-scale deformational history of much of the eastern Blue Ridge (Figure 16E–2; also see the Chapter 20 Essay). The rocks are predominantly sillimanite-grade migmatitic biotite gneiss, amphibolite, anatectic granitoids, and pegmatite belonging to the Late Proterozoic Tallulah Falls Formation. Six or seven separate fold generations are preserved here, along with Types 1 and 2, and possibly Type 3, fold-interference patterns, as well as several folds that may be sheath folds. This does not mean that six or seven deformational events that were separated by large increments of time affected these rocks, but it does mean that favorable conditions

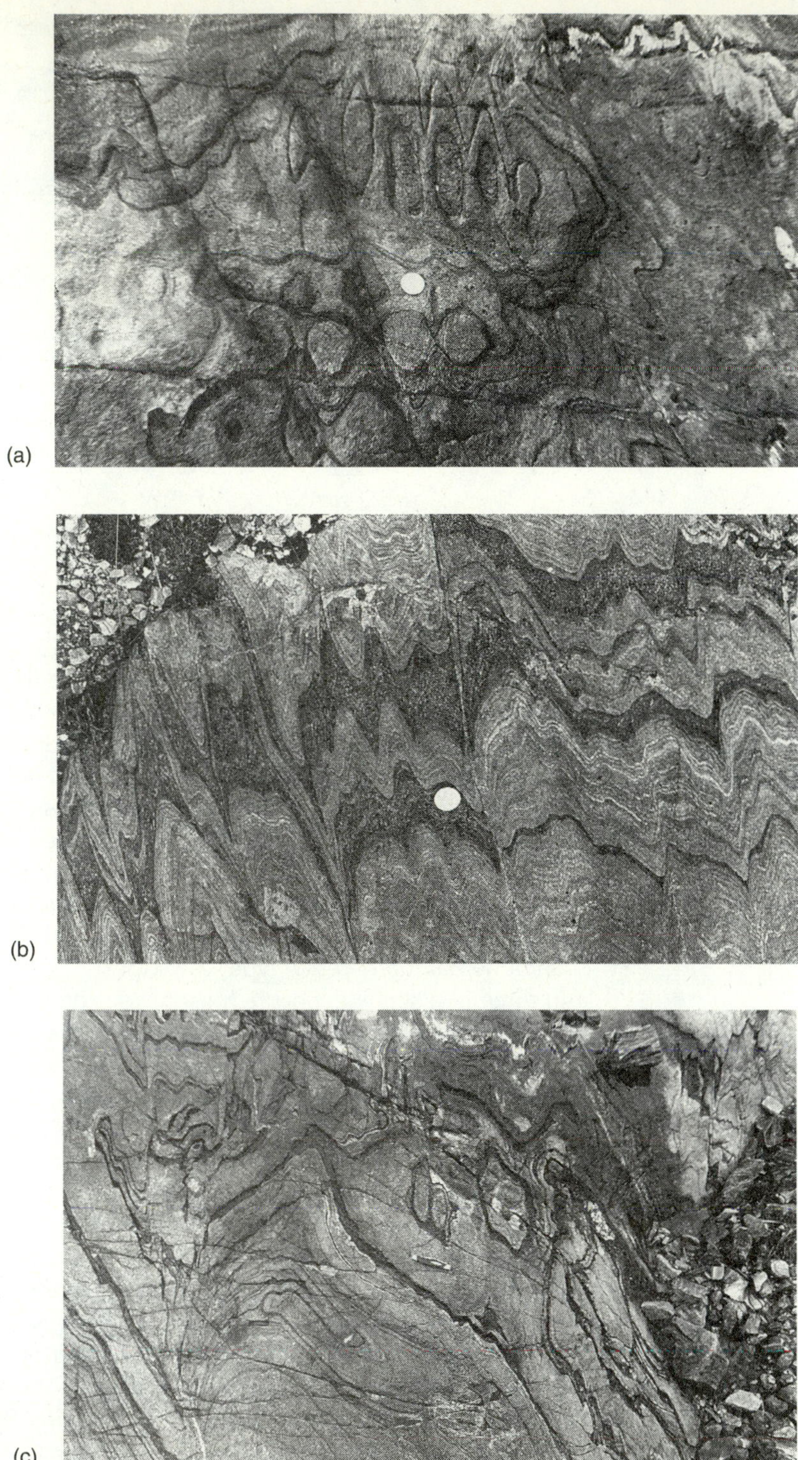

FIGURE 16E–1
Interference-fold patterns originally identified by John Ramsay in Middle Proterozoic Moine Series metasandstone and schist at Loch Monar in the Scottish Highlands. (a) Type 1. Dome-and-basin pattern is in center of photo. (b) Type 2. Boomerang pattern is in schistose (dark) layer beneath coin. (c) Type 3. Hook pattern is to the left and below knife. Type 1 pattern is present above knife. (RDH photos.)

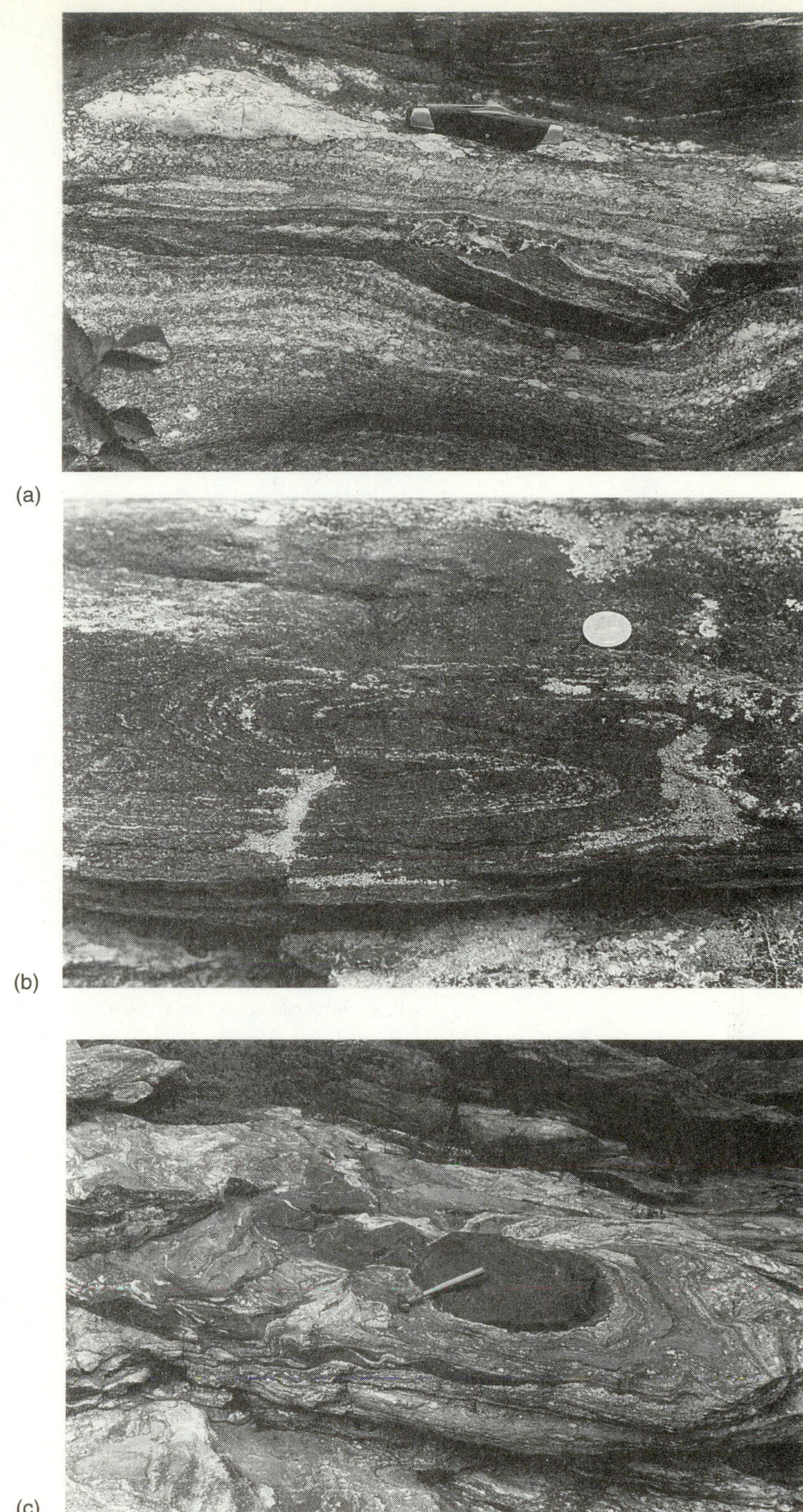

FIGURE 16E–2
Early structures at Woodall Shoals in South Carolina and Georgia. (a) Amphibolite boudin showing foliation in dark amphibolite truncated by enclosing dominant foliation in biotite gneiss. Note small intrafolial fold to the left. (b) Fold in amphibolite boudin is earlier than the earliest folds in the enclosing biotite gneiss. (c) Deformed amphibolite boudin at Woodall Shoals showing structure in enclosing rocks. (RDH photos.)

existed to superpose several sets of folds during perhaps two events (or at most, three) separated by thousands to millions of years.

Before Woodall Shoals was studied carefully, geologists were uncertain which fold generation here was the earliest—the same uncertainty that exists in other complexly deformed areas. It had been assumed that the earliest folds had sheared-out limbs and lay within the dominant foliation *(intrafolial folds*, see Chapter 17)—commonly designated F_1 folds; those that overprint the F_1 folds were designated F_2, F_3, etc. The map-scale (macroscopic) outcrop pattern on the Toxaway dome several kilometers to the northeast (Figure 16–6) strongly suggested that a pre-F_1 fold set existed in the eastern Blue Ridge (Hatcher, 1977). At Woodall Shoals, similar-like folds were preserved in pods (boudins) of less-ductile amphibolite that were wrapped by the dominant regional foliation in the biotite gneiss (Figure 16E–2). This occurrence confirmed that one—possibly two—earlier sets of passive or flexural flow folds were formed before the intrafolial folds lying within the dominant foliation.

This example demonstrates the usefulness of places such as Loch Monar and Woodall Shoals in deciphering both deformational processes and the sequence of overprinting of deformations.

References Cited

Carey S. W., 1962, Folding: Journal of the Alberta Society of Petroleum Geologists, v. 10, p. 95–144.

Hatcher, R. D., Jr., 1977, Macroscopic polyphase folding illustrated by the Toxaway dome, South Carolina–North Carolina: Geological Society of America Bulletin, v. 88, p. 1678–1688.

Ramsay, J. G., 1962, Interference patterns produced by the superposition of folds of "similar" type: Journal of Geology, v. 60, p. 466–481.

Ramsay, J. G., 1967, Folding and fracturing of rocks: New York, McGraw-Hill, 568 p.

Williams, P. F., 1970, A criticism of the use of style in the study of deformed rocks: Geological Society of America Bulletin, v. 81, p. 3283–3296.

Questions

1. How do Type 2 interference folds form?
2. How would you distinguish Type 1 interference patterns from sheath folds?
3. Why do sheath folds commonly occur in ductile shear zones and fault zones?
4. Why is it difficult to say that the early and late folds making up Type 3 interference folds formed at different times?
5. The most common interference fold types are Type 3; next are Type 2, and Type 1 folds are least common. Why?
6. How could you distinguish between complex tectonic folds and complex folds formed by soft-sediment deformation (Figure 2–2b)?
7. Describe conditions in which the late folds in a Type 3 pattern could be separated from the earlier folds.

8. Below is a geologic map of a boomerang (Type 2) interference fold. Photocopy the figure below and sketch in the traces of early and late fold axial surfaces with arrows indicating direction of plunge, using standard symbols for fold hinges, and show which fold axial trace is older or younger. The cross-hatch pattern is Precambrian gneiss; the stipple pattern, Cambrian quartzite; the innermost patterned area is Ordovician marble.

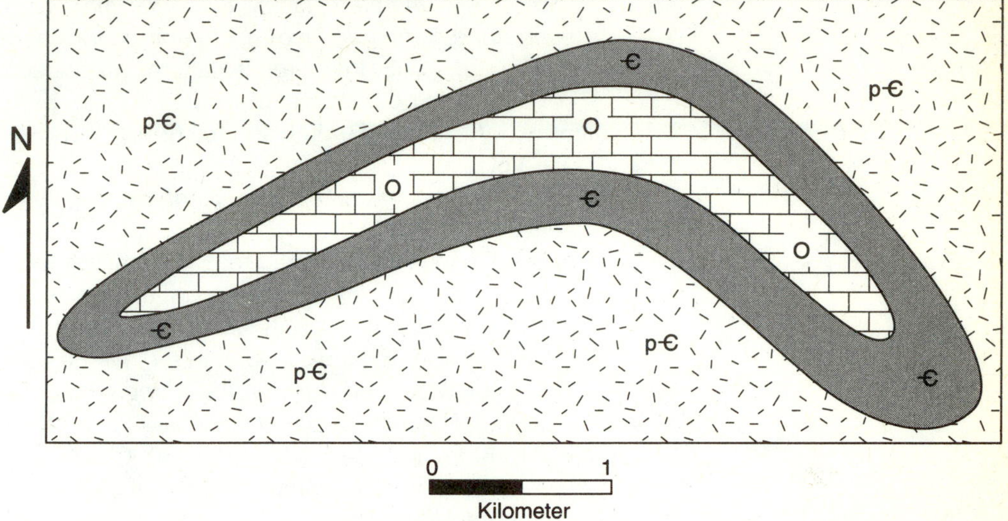

9. Below is a geologic map of a hook (Type 3) interference fold. Photocopy the figure and sketch in the traces of early and late fold axial surfaces with arrows indicating direction of plunge, using standard symbols for fold hinges, and show which fold axial trace is older or younger. The cross-hatch pattern is Precambrian gneiss; the dark pattern, Cambrian quartzite; the brick area is Ordovician marble.

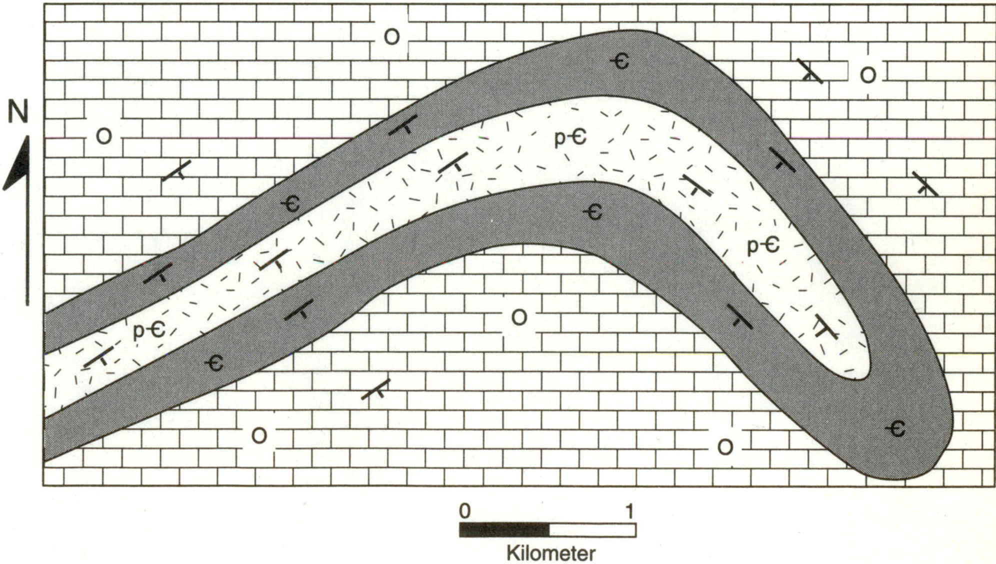

Further Reading

Cobbold, P. R., and Quinquis, H., 1980, Development of sheath folds in shear regimes: Journal of Structural Geology, v. 2, p. 119–126.
Discusses the mechanics of formation of sheath folds and establishes that they form in zones dominated by simple shear.

Mies, J. W., 1993, Structural analysis of sheath folds in the Sylacauga Marble Group, Talladega slate belt, southern Appalachians: Journal of Structural Geology, v. 15, p. 983–993.
Employs both field data and computer models to document a well-illustrated and understandable investigation of the formation of sheath folds.

Ramsay, J. G., 1962, Interference patterns produced by the superposition of folds of "similar type": Journal of Geology, v. 60, p. 466–481.

Ramsay, J. G., 1967, Folding and fracturing of rocks: New York, McGraw-Hill, 568 p.
The 1962 paper and Chapter 10 of the 1967 book first systematized interference fold types.

Skjernaa, Lilian, 1989, Tubular folds and sheath folds: Definitions and conceptual models for their development, with examples from the Grapesvare area, northern Sweden: Journal of Structural Geology, v. 11, p. 689–703.
Provides additional discussion of the mechanisms of formation of noncylindrical and sheath folds, along with defining parameters ω *and the ratio x:y to more quantitatively characterize these structures.*

Thiessen, R. L., and Means, W. D., 1980, Classification of fold-interference patterns: A reexamination: Journal of Structural Geology, v. 2, p. 311–316.
Redefines Ramsay's interference fold types, using the angular relationships between the axes of the first folds and the pole to the axial planes of the second folds. Thiessen and Means realized that a greater variety of interference patterns was possible within Ramsay's basic types.

PART FIVE

Fabrics, Plutons, and Structural Analysis

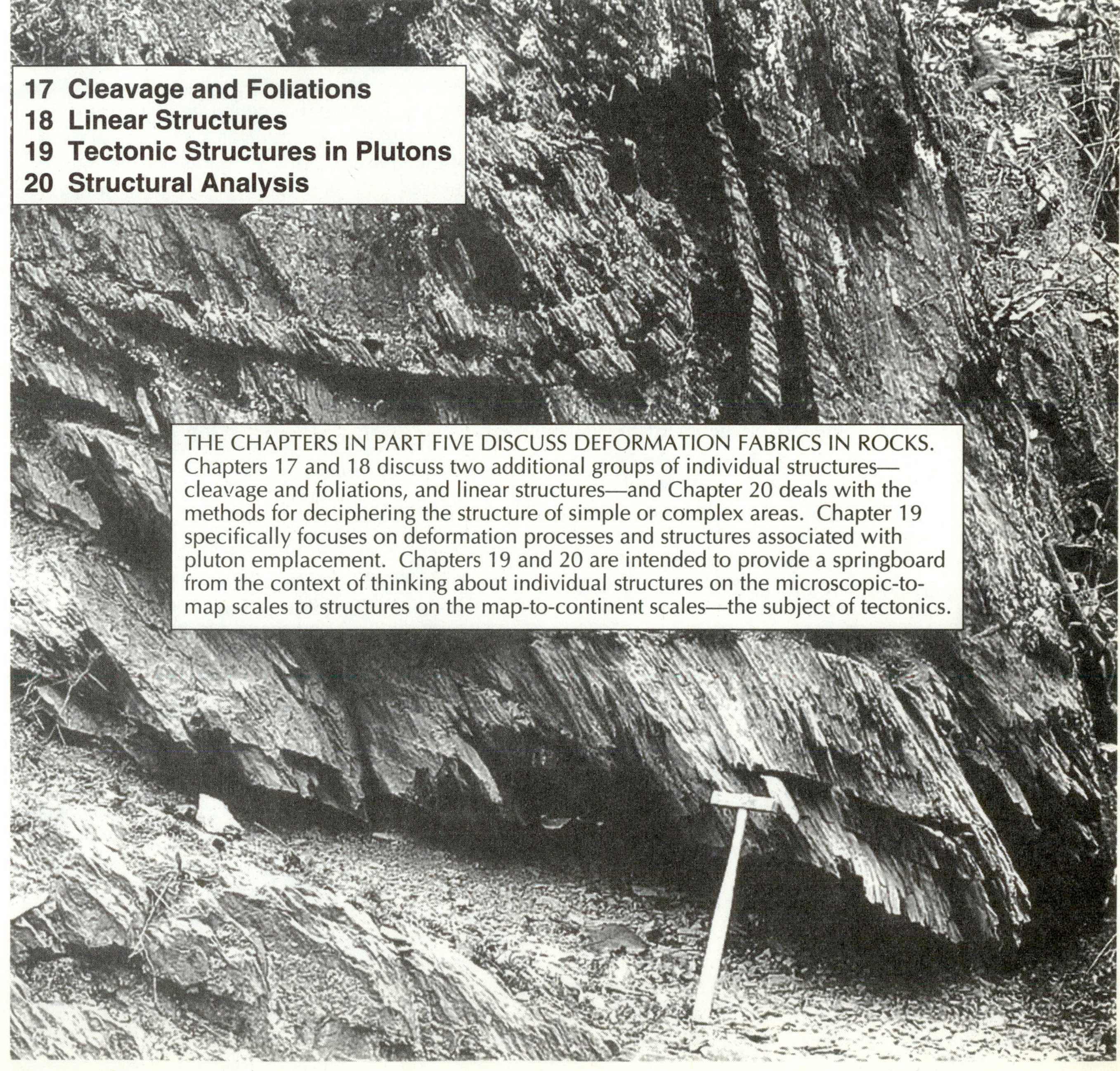

THE CHAPTERS IN PART FIVE DISCUSS DEFORMATION FABRICS IN ROCKS. Chapters 17 and 18 discuss two additional groups of individual structures—cleavage and foliations, and linear structures—and Chapter 20 deals with the methods for deciphering the structure of simple or complex areas. Chapter 19 specifically focuses on deformation processes and structures associated with pluton emplacement. Chapters 19 and 20 are intended to provide a springboard from the context of thinking about individual structures on the microscopic-to-map scales to structures on the map-to-continent scales—the subject of tectonics.

17

Cleavage and Foliations

Rocks affected by slaty cleavage have suffered a compression of their mass in a direction everywhere perpendicular to the plane of cleavage and an expansion of their mass in the direction of cleavage dip.

DANIEL SHARPE, 1847, *Quarterly Journal of the Geological Society of London*

A TENDENCY TO SPLIT ALONG PLANES OTHER THAN BEDDING was recognized more than a century ago as a fundamental property of many deformed rocks, although the significance was probably not known at first. Early geologists did observe bedding cut at a high angle by a prominent planar structure and soon realized that it was a major structure—cleavage—related to both deformation and metamorphism. In this chapter we will trace the early ideas on the origin of slaty cleavage and foliation and will again see the fundamental relationship between small- and large-scale structures first noted by Raphael Pumpelly in the late nineteenth century.

We have several reasons to understand cleavage in rock. Cleavage is directly linked to other deformation processes—especially folding—and to metamorphism. It can help to determine the fold geometry in an area, and if we know the cleavage-forming mechanism in a rock mass, it will lead to better understanding of the deformation processes and also will help to reconstruct the physical conditions during deformation. Cleavage may also serve as a conduit for ground water, particularly in slightly weathered rock.

DEFINITIONS

The term ***fabric*** is used to describe the spatial and geometric relationships among all components that make up a rock. In deformed rocks, this includes planar and linear structures—bedding, cleavage, and the orientation of minerals—and their relationships to texture.

Structures that pervade the rock mass at the scale of observation (Figures 17–1 and 17–2) are ***penetrative*** (Chapter 1). Ideally, penetrative means present on all scales of observation, but some structures, such as joints, are not present on all scales. For example, we can cut thin sections or even break large specimens from a jointed mass and not intersect a joint; but represented on a 1:10,000 or 1:24,000 scale map in the same body of rock, joints may be so closely spaced that they must be considered penetrative. *Slaty cleavage* is generally regarded as penetrative; it consists of parallel grains of thin-layer silicates (clay minerals or micas) or thin anastomosing subparallel zones of insoluble residues produced by pressure solution. Slaty cleavage is observable on the hand-specimen scale and in thin section; it also may be represented on maps and so is a good example of a penetrative structure.

On the other hand, ***nonpenetrative*** structures (Figure 17–2b) include unique nonrecurring structures such as fault planes, so an isolated fault cannot be penetrative. Likewise, an isolated fold hinge or axial surface is nonpenetrative, but folds of the same generation visible on all scales are penetrative structures.

In an early work on fabrics and structures in deformed rocks, Bruno Sander (1930), an Austrian geologist, suggested that planar and some curved structures in deformed rocks be called ***S-surfaces*** (Figure 17–3). S-surfaces include all cleavages and foliations commonly thought of as penetrative structures. They also include one nontectonic planar structure, bedding. A shorthand system has been devised to aid the study of S-surfaces. In areas of multiple S-surfaces, a series of subscripts is assigned: bedding, being oldest, is designated S_0; S_1 is the oldest cleavage (or foliation), and any later structures are given numerically higher subscripts in chronological order (Figure 17–3). Corresponding multiple-fold sets are designated F_1, F_2, and so on; other linear structures, L_1, L_2, . . . and deformations D_1, D_2, ...

Cleavage (or foliation) in fine-grained rocks may be ***continuous*** or ***spaced*** (Figure 17–4). Continuous cleavage pervades the rock mass, but spaced cleavage can be resolved into domains (regions) of uncleaved rock separated by cleavage planes spaced from less than a millimeter to several centimeters (Figure 17–5). Uncleaved zones between cleavage surfaces are called ***microlithons*** (Figure 17–4).

(a)

(b)

FIGURE 17–1
(a) Folded slate in Wilhite Formation (Ocoee Supergroup, Upper Proterozoic?) near Walland, Tennessee, showing bedding and penetrative axial-plane slaty cleavage. Note relationships of dip of cleavage and bedding on the upright and overturned limbs of the fold—cleavage dips more steeply than bedding on the upright limb, less steeply than bedding on the overturned limb. (Arthur Keith, U.S. Geological Survey.) (b) Negative print of a thin section of rhythmically graded bedded slate and metasiltstone from the Wilhite slate near Tellico Plains, Tennessee. Note the differences in spacing of cleavage (white lines nearly perpendicular to bedding) in finer-grained layers, and the near absence of cleavage in coarser-grained layers. Also note the cuspate-lobate structure (cleavage mullions, see Chapter 18) on the bases of several of the dark (actually almost clear sandy) layers. Long axis of the thin section is 7 cm.

Most cleavage in rocks, according to Christopher Powell (1979), is *domainal,* or spaced, if the rock is examined closely enough, but continuous cleavage may be more widespread than Powell realized. He used the following characteristics to distinguish types of cleavage (Figure 17–6): spacing of cleavage domains, shape of cleavage domains, microlithon fabric, and proportion of the rock occupied by cleavage domains. Thus, he created two basic subdivisions, spaced and continuous cleavage, and divided spaced cleavage into ***disjunctive*** (cross-cutting, and not related to original layering) and ***crenulation cleavage*** (which crenulates preexisting layering). Crenulation cleavage may be further divided into *discrete* and *zonal* types (see p. 369); disjunctive cleavage may be divided into *stylolitic, anastomosing, rough,* and *smooth,* depending on the shape of the cleavage domains. Spacing of true slaty cleavage may range from less than 0.01 mm (continuous cleavage) to less than 1.0 mm, that of crenulation cleavage from 0.1 mm to 3 cm, and that of stylolitic and anastomosing cleavage from a few millimeters to several centimeters. Different rock types in the same body, such as interlayered sandstone (or pure limestone) and shale, may display marked differences in cleavage fabrics. Shale may display closely spaced slaty cleavage produced by pressure solution, or recrystallization; the sandstone (or limestone) may contain more widely spaced pressure-solution cleavage (Figures 17–4b and 17–7).

Pressure solution produces spaced cleavage by dissolving the most soluble parts of a rock mass, leaving behind discrete insoluble residues in irregular planar zones that define the cleavage (Figure 17–8). Here, the terminology is a bit confusing because (as we

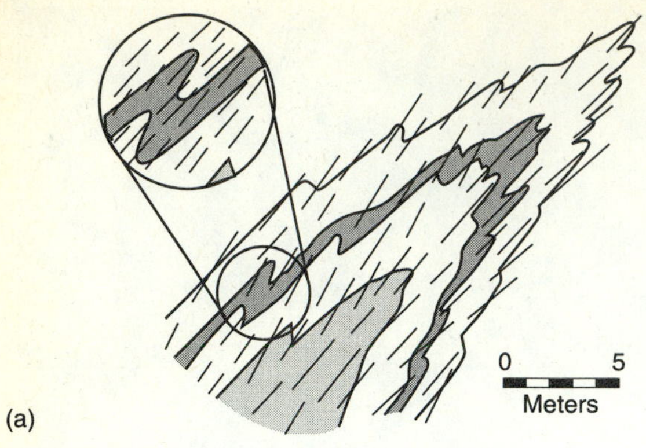

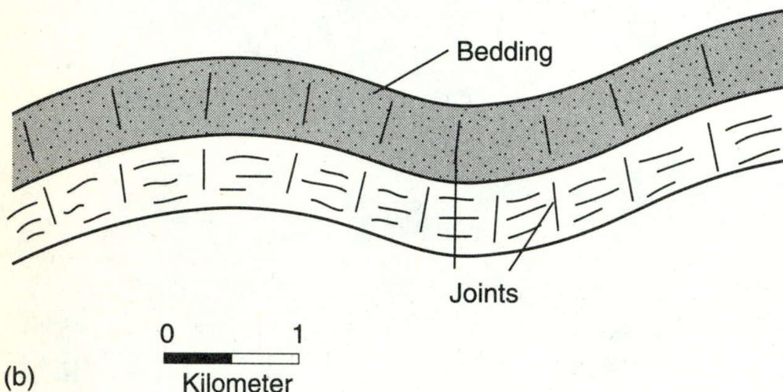

FIGURE 17–2
Penetrative and nonpenetrative structures. (a) Mesoscopic fold with small parasitic folds and an axial-plane foliation. Small folds and foliation are penetrative structures, for they occur throughout the larger fold. (b) Unique fold hinges and widely spaced joints which occur on only one scale and are not repeated on other smaller or larger scales are nonpenetrative.

saw in Chapter 7) pressure solution is a deformation mechanism, but pressure-solution cleavage is recognized today as a cleavage-forming process as well. Spacing of pressure-solution surfaces may range from less than a millimeter to more than a centimeter. They may be somewhat irregular (stylolitic to anastomosing to rough) where poorly developed, to smooth, where the rock mass is more severely deformed.

Slaty cleavage is a planar tectonic structure resulting from parallel orientation of clays or other layer silicates (such as chlorite and muscovite), or seams of insoluble residues (or both) in a fine-grained rock. It is penetrative and thus visible at all scales in the rock mass (Figure 17–1). At microscopic to submicroscopic scale (Figure 17–9), slaty cleavage may be defined by planes made up of aligned layer-silicate minerals that are separated by *microlithons* containing layer or nonlayer silicates that are not aligned. Rocks that commonly develop slaty cleavage are fine-grained sedimentary and volcanic rocks such as shale, mudstone, siltstone, and tuff, and their equivalents at low metamorphic grade (anchizone to chlorite zone).

A *scaly foliation* marked by anastomosing polished surfaces has been recognized in semiconsolidated fine-grained sediment in ancient and modern accretionary complexes and has also been duplicated in the laboratory. This foliation is closely associated with fault zones and appears to develop subparallel to parallel to faults. Scaly foliation surfaces form as small faults—surfaces of simple shear—and are quickly abandoned without attaining an appreciable displacement (Moore and others, 1986).

Cleavage marked by small-scale crinkling or crenulations is called ***crenulation cleavage*** (Figure 17–10). Most crinkles are spaced and asymmetric, and the short limb commonly becomes the cleavage plane. Crenulation cleavage commonly forms by deformation of an earlier cleavage or (much less frequently) bedding. The axial zones or limbs of the crenulations may be aligned with, or transected by, cleavage planes, but the limbs of the crenulations generally remain intact. Microlithons between the crenulation cleavage planes preserve the earlier cleavage or foliation (Figure 17–11).

The various cleavage types, including slaty and crenulation cleavage (some geologists even include bedding), are collectively termed ***foliations***. The term is also used for the planar structure in coarser-grained metamorphic rocks, such as phyllite, schist, and gneiss, where the parallel orientation of at least one mineral dominates the fabric (Figure 17–12). Foliation in coarser-grained rocks (such as schist and gneiss) may be produced by parallel orientation of micas (mostly biotite and muscovite), amphiboles, and flattened quartz grains, and even by alternating layers of aligned quartz, feldspar, and micas. *Schistosity* refers to the foliation in schistose and, occasionally, in gneissic rocks (Figure 17–12b). Banded foliation in gneissic rocks results partly from original differences in composition and partly from the orientation of platy or elongate minerals. Secondarily developed layering in these rocks is called *gneissic banding* (Figure 17–12c). This foliation is easily recognized and may be found where quartz and feldspar layers alternate with layers dominated by micas or amphiboles.

Metamorphic differentiation involves formation of new layering by recrystallization or pressure solution. It reorganizes the chemical components of a rock on the scale of layer thickness to produce new minerals with new orientations. Foliation produced during progressive deformation and metamorphism, largely through recrystallization, is termed ***differentiated layering*** (Figure 17–13). Layering in the original rock mass may be enhanced by this process at high temperature and pressure to produce gneissic banding; formation of spaced slaty and crenulation cleavages may produce differentiated layering at low temperature and pressure.

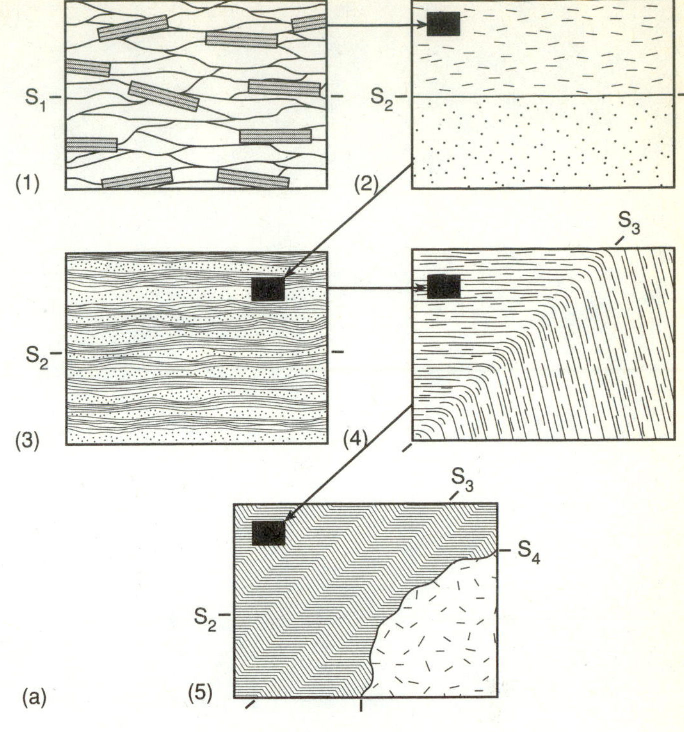

FIGURE 17–3
(a) Planar discontinuities in the same body of rocks at several different scales. Microscopic scale (1) showing preferred orientation of grain boundaries and minerals defining a weakly penetrative planar structure S_1. Grains (2) define a penetrative planar structure S_1 in the upper layer on the mesoscopic scale. The compositional boundary S_2 is nonpenetrative on this scale. Alternating layers (3) parallel to S_2 make S_2 a penetrative structure at this scale (larger mesoscopic). Smaller map scale (4) in which S_3 becomes a series of closely spaced kink surfaces and is penetrative. Map (macroscopic)-scale structures (5) in which a kink S_3 divides the body into two sectors or domains having different strikes. Another nonpenetrative compositional boundary, an intrusive contact (S_4), appears at this scale. (From F. J. Turner and L. E. Weiss, *Structural Analysis of Metamorphic Tectonites*, © 1963, McGraw-Hill Book Company. Reproduced with permission.) (b) Interlayered metasandstone and muscovite-biotite schist at Sill Vinson's Rock near Otto, North Carolina. Original bedding (now transposed) makes up the earliest foliation, S_1. A later foliation, S_2, dips toward the left (west) parallel to the axial surfaces of the folds. A later foliation—a crenulation cleavage, S_3—dips steeply to the right (east, visible above the small ledge) and parallels the axial surfaces of folds at the bottom of the photo. Crenulations occur here only in the schist because the sandstone layers are too quartz-rich. White layers are quartz-feldspar veins. Note that thin quartz-feldspar veins in schist form open to isoclinal ptygmatic folds. Sandstone layers are light gray; schist layers are dark gray. (RDH photo.)

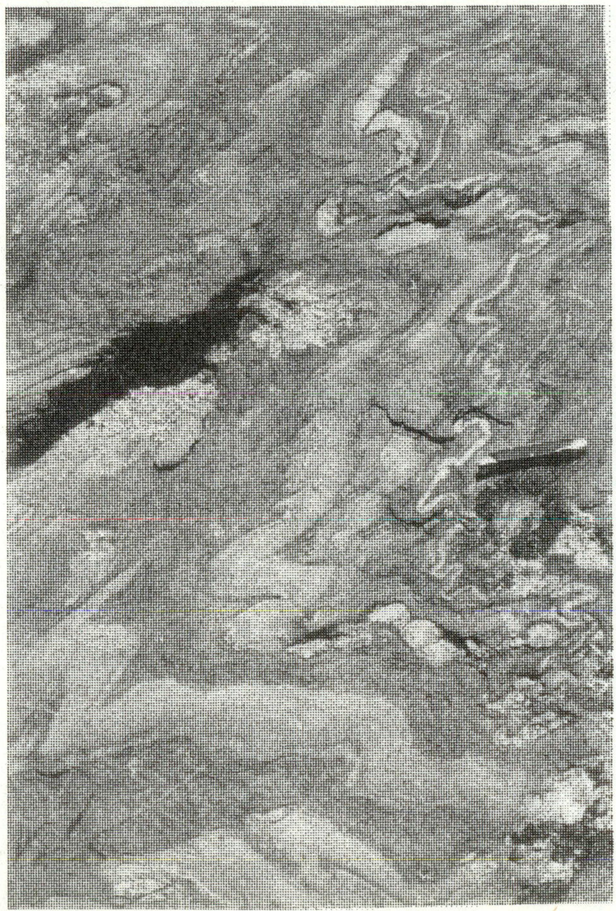

(b)

(a)

(b)

FIGURE 17–4
(a) Continuous cleavage in slate in the Upper Proterozoic(?) Wilhite Formation on U.S. 129 along Chilhowee Lake, southeastern Tennessee. (b) Spaced pressure-solution cleavage in Ordovician Stony Point Limestone, Lessor's Quarry, South Hero Island, Vermont. Hammer head rests on a small thrust. (Note drag of cleavage.) (c) Microlithons of crenulated earlier foliation between nearly vertical crenulation cleavage planes in phyllite near Enosburg Falls, Vermont. (RDH photos.)

Fracture cleavage consists of parallel to subparallel fractures, generally spaced 1 to 3 cm apart (Figure 17–14). As the name implies, this structure is formed by a brittle mechanism, but rocks may also fracture along pressure-solution seams. The latter was originally called "fracture cleavage" because the rock fractures along planes of weakness that appear different from those in slaty cleavage. Fracture cleavage (as used here) tends to develop in slightly coarser-grained rocks than in rocks with slaty cleavage, but it has also been observed in fine-grained rocks. If the origin is traceable to fracturing, many structural geologists call it "jointing." In any case, the term "fracture cleavage" is entrenched in the literature of structural geology. Probably it should be used only as a last resort to describe closely spaced fractures not clearly related genetically to jointing.

(c)

FIGURE 17–4 (continued)

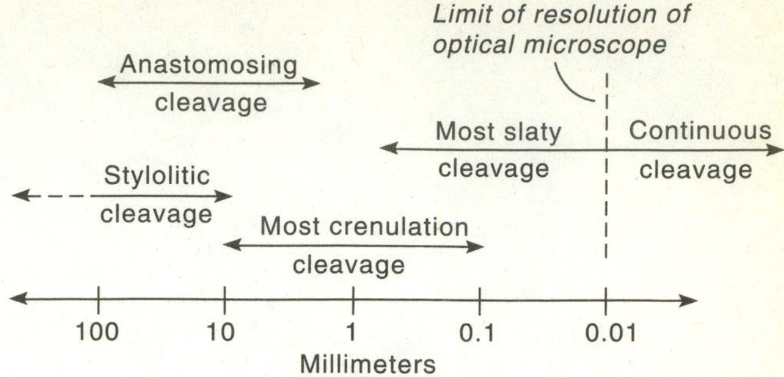

FIGURE 17–5
Type and spacing of cleavages. (From *Tectonophysics,* v. 58, C. McA. Powell, p. 21–34, © 1979, with kind permission from Elsevier Science Ltd., Kidlington, United Kingdom.)

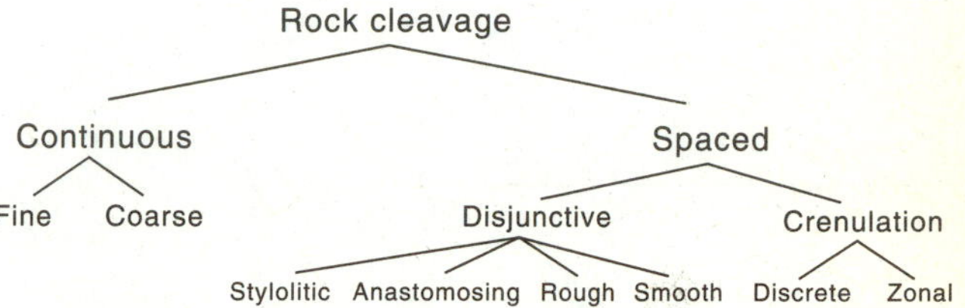

FIGURE 17–6
Powell's classification of cleavages based on morphology. (From *Tectonophysics,* v. 58, C. McA. Powell, p. 21–34, © 1979, with kind permission from Elsevier Science Ltd., Kidlington, United Kingdom.)

CLEAVAGE-BEDDING RELATIONSHIPS

Geologists have long known that slaty cleavage within folds is related to the folding process and also that these structures form synchronously. The angular relationship between cleavage and bedding can be used to determine whether one is observing the upright or the overturned limb of a fold that is incompletely exposed: *If bedding dips less steeply (at a lower angle) but in the same direction as cleavage, the rocks are on upright limbs of folds. Conversely, if bedding dips more steeply than the cleavage (that is to say, the cleavage dips less than bedding) the rocks are on the overturned limb of a fold (see Figure 17–1).* This important rule may be used in areas where the geologist cannot observe complete folds and must project outcrop data to map scale. Conversely, cleavage-bedding relationships can be used to locate the observer's place on a known structure.

Care must be taken to determine whether or not the cleavage being used for determining position within a structure is the only cleavage present and if the cleavage formed at the same time as the structure of interest. Also, if the axial surface of a fold has either rotated through the horizontal or has formed on a small fold on the overturned limb of a recumbent fold (both formed during the same event), cleavage will dip more steeply than does bedding, thus incorrectly indicating that the rocks are upright (Figure 17–15a). If more than one cleavage is found, the cleavage-bedding relationship in each generation of cleavage and folds must be separated before drawing any conclusion about position in the structure (Figure 17–15). In complexly deformed areas with several generations of cleavages, independent information—such as the stratigraphic order based on primary sedimentary structures—should be obtained and used routinely as a check.

CLEAVAGE REFRACTION

A phenomenon frequently observed in cleaved rocks is refraction of cleavage from layer to layer and (as in graded beds) within layers. ***Cleavage refraction*** occurs where the texture and composition—ductility—vary from layer to layer in rocks. The angle between cleavage and bedding changes, or refracts, as the cleavage passes from one layer to another (Figure 17–16). Apparent refraction sometimes results from flexural slip on bedding, producing drag of cleavage surfaces.

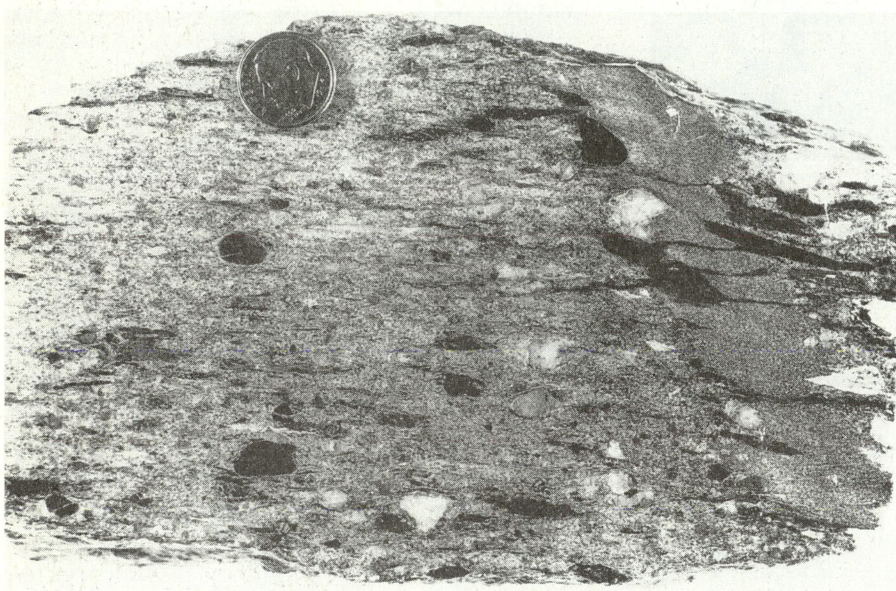

(b)

FIGURE 17–7
(a) Contact between Middle Ordovician slate (below) and sandstone (above) near Smiths Basin, south of Whitehall, New York, showing closely spaced cleavage in the slate and widely spaced pressure-solution cleavage in the sandstone. (RDH photo.) (b) Contacts between a 2.6 cm-thick layer of medium-grained sandstone in coarse-grained conglomeratic arkose of the arkose unit in the Grandfather Mountain Formation (Upper Proterozoic), near Boone, North Carolina. Pressure-solution cleavage is spaced more widely in the thin sandstone layer than in the arkose and is not well developed in either lithology. (Specimen courtesy of Michael J. Neton, GES, Inc., Chattanooga, Tennessee.)

Cleavage refraction produces contrasting effects during folding. Most slaty cleavage forms parallel to axial surfaces in folds but may be displaced or fanned with respect to the hinge as folding proceeds; thus, it is no longer parallel to the axial surface (Figure 17–17). In a different approach, Henderson and others (1986) considered the cleavage-refraction phenomenon as a product of inhomogeneous strain between layers, rotating cleavage—after it had formed by pure shear—by a component of differential simple shear parallel to bedding but by an amount that varied from bed to bed (Figure 17–18).

Refracting cleavage has been confused with cross bedding where the cleavage is traced as a continuous curve from coarser- to finer-grained parts of the same graded bed: the steeper-dipping cleavage (relative to dip of bedding) occurs in the coarser (basal) part of the bed; the gentler-dipping cleavage is in the finer-grained part at the top of the bed (Ramsay, 1967). This indicates a facing direction opposite the direction indicated by graded bedding. In such cases, careful observation of relationships between primary sedimentary and tectonic structures should be sought to solve the problem.

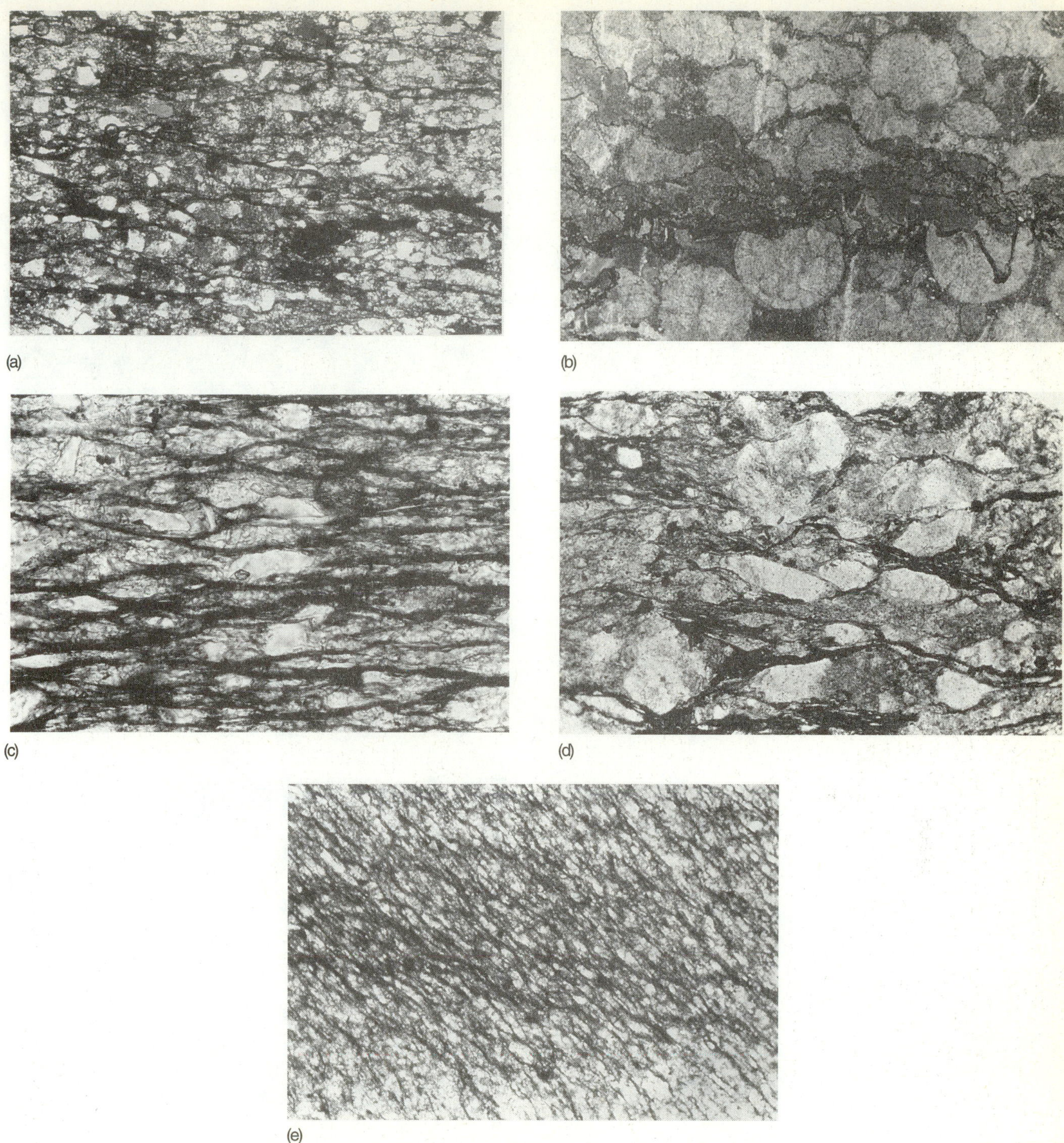

FIGURE 17–8
Pressure-solution cleavage and residues. (a) Residues of insoluble minerals truncating quartz grains along cleavage surfaces in Upper Ordovician Martinsburg Slate, Delaware Water Gap, New Jersey. Plane light. Field is approximately 2 mm wide. (Thin section courtesy of Timothy L. Davis, North Carolina Geological Survey.) (b) Stylolitic seam truncating oöids, producing apparent displacement of an earlier calcite vein in limestone in Upper Cambrian Nolichucky Shale, Oak Ridge, Tennessee. Plane light. Field is approximately 7 mm wide. (Thin section courtesy of Peter J. Lemiszki, Oak Ridge National Laboratory.) (c) Anastomosing slaty cleavage in Devonian Marcellus Shale near Cumberland, Maryland. Plane light. Width of field is approximately 2 mm. (Charles M. Onasch, Bowling Green State University.) (d) Rough cleavage in graywacke in Upper Ordovician Martinsburg Slate near Front Royal, Virginia. Pressure-solution seams have a very irregular shape because of the different amounts of insoluble and soluble constituents present. Plane light. Width of field is approximately 2 mm. (Charles M. Onasch, Bowling Green State University). (e) Penetrative pressure-solution "smooth" cleavage in Upper Ordovician Martinsburg Slate, Delaware Water Gap, New Jersey. Plane light. Width of field is approximately 7 mm. (Thin section courtesy of Timothy L. Davis, North Carolina Geological Survey.)

FIGURE 17–9
Scanning electron micrograph of domainal slaty cleavage from Upper Ordovician Martinsburg Slate near the Delaware River, Pennsylvania. Spacing between cleavage planes (strongly oriented narrow zones) is about 20 μm. (From *Tectonophysics,* v. 82, B. G. Woodland, p. 89–124, © 1982, with kind permission from Elsevier Science Ltd., Kidlington, United Kingdom.)

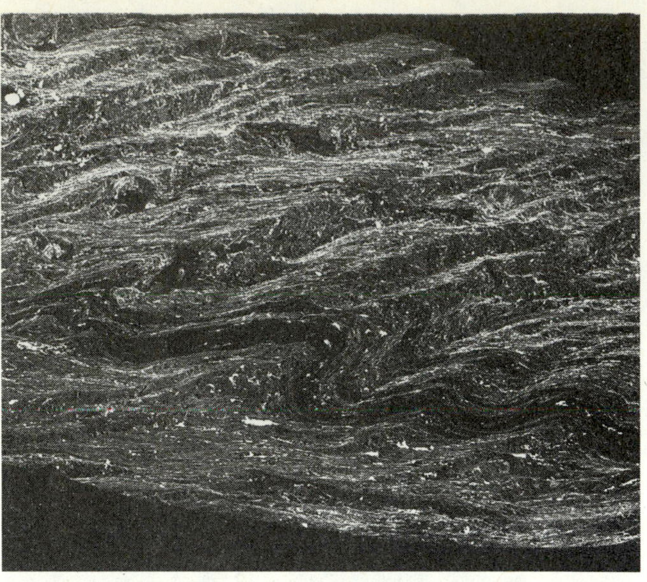

FIGURE 17–10
Negative print of a thin section of garnet-muscovite schist from Einunnfjellet, southern Norway, containing crenulations superposed onto an older foliation that is parallel to bedding. Thin section is 3.8 cm long. (Specimen courtesy of Elizabeth A. McClellan, Western Kentucky University.)

(a)

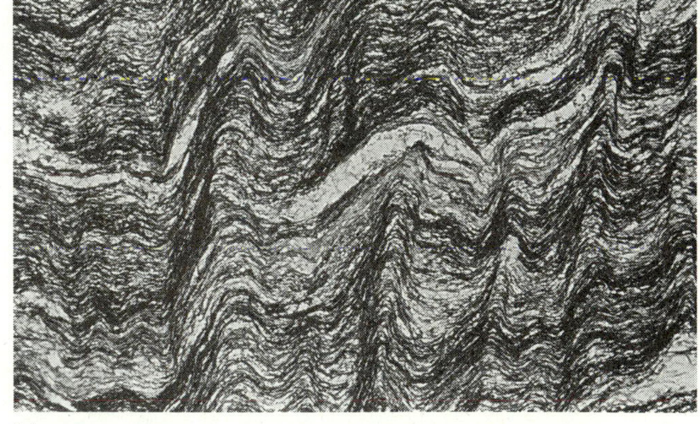

(b)

FIGURE 17–11
Microlithons of earlier deformed material between crenulations at both meso- (a) and micro- (b) scales: (a) Crenulated siltstone in the Upper Proterozoic Hamill Group near Golden, southern British Columbia. (RDH photo.) (b) Chlorite schist from the Wissahickon Group near Westminster, Maryland. Plane light. Width of field is approximately 7 mm. (Charles M. Onasch, Bowling Green State University.)

(a)

(b)

(c)

FIGURE 17–12
Foliations in three types of metamorphic rocks. (a) Phyllite from the Ordovician Walloomsac phyllite near Hoosick Falls, New York. Note the fine grain size and lustrous foliation. (b) Schistosity in crenulated amphibole-biotite schist containing several quartz-feldspar layers on Blahø, Trollheimen region, south-central Norway. (c) Gneissic banding composed of quartz-feldspar–rich layers alternating with biotite-rich layers in migmatitic biotite gneiss, Thor-Odin dome, Shuswap metamorphic complex, south-central British Columbia. (RDH photos.)

EARLY IDEAS ON THE ORIGIN OF SLATY CLEAVAGE

Now we turn to the history of ideas and controversies related to the origin of slaty cleavage. Development of the concepts was presented in greater detail by Andrew Siddans (1972) and also by Dennis Wood (1974). This history bears directly on our present-day concepts on the development of foliation and related structures.

The origin of slaty cleavage has been debated for a century and a half. Some concepts that we now consider fundamental evolved very early but by the turn of the century had been almost forgotten—only to be revived in the 1970s.

Among the first fundamental deductions about the nature of slaty cleavage was that cleavage is related to folding and that cleavage frequently parallels the axial planes of folds. This was recognized independently in the mid-nineteenth century by Adam Sedgwick (1835), Charles Darwin (1846), and Henry D. Rogers (1856).

In the 1830s and '40s, Sedgwick (1835), J. Phillips (1844), and D. Sharpe (1849) thought slaty cleavage developed after folding. Sedgwick understood the axial-plane relationship of cleavage and folding but not that they are contemporaneous. Apparently, J. Tyndall (1856) was the first to see that slaty cleavage and folding develop at the same time.

Several early geologists recognized the relationship between formation of slaty cleavage and the process of metamorphism. Darwin (1846) considered slaty cleavage the end point of metamorphism, again probably including the idea that slaty cleavage is related to a late stage of folding.

Another fundamental observation was that strain markers, such as fossils, are distorted (commonly flattened) when cleavage forms. Phillips (1844) saw that fossils in Welsh slates might be flattened parallel to cleavage planes. Sharpe (1849) noted that shells are most distorted in layers that are most slaty and concluded that the maximum principal shortening direction is perpendicular to cleavage planes.

Among the most important contributions made during the mid-nineteenth century was that by a British geologist, Henry C. Sorby. He used deformed reduction spots in the Cambrian slates of Wales to conclude that the *XY* plane of the strain ellipsoid parallels cleavage planes and estimated that there was as much as 75 percent shortening perpendicular to cleavage

(a)

(b)

FIGURE 17–13
(a) Differentiated layering in gneiss, Toxaway Gneiss near Whitewater Falls on the North Carolina–South Carolina line. (RDH photo.) (b) Pressure-solution–deformed fine-grained sandstone of the Lower Ordovician Halifax Formation, Little Harbor, Tor Bay, Nova Scotia. Earlier cleavage is steeply dipping, but the later cleavage is anastomosing and nearly horizontal. (J. Duncan Keppie, Nova Scotia Department of Mines.)

FIGURE 17–14
Weak fracture cleavage (most visible lower left) that fans around folds in Cambrian limestone and dolomite near Danby, Vermont. Is this a true cleavage? (Arthur Keith, U.S. Geological Survey.)

planes. (Sorby also pioneered the use of petrographic techniques: he is credited with making the first thin section, and he advanced the idea that rotation of inequant grains tends to align their long axes in the plane of cleavage.) Sorby also observed that preferred orientation could be enhanced by recrystallization in the plane of cleavage.

Sorby (1863) was probably the first to recognize pressure solution (see page 120 and Figures 7–18 and 7–19; see also Figure17–4b) and suggest that it is an important mechanism of material transfer in rocks. Ironically, he felt that all other processes in formation of slaty cleavage were secondary to rotation produced by compression.

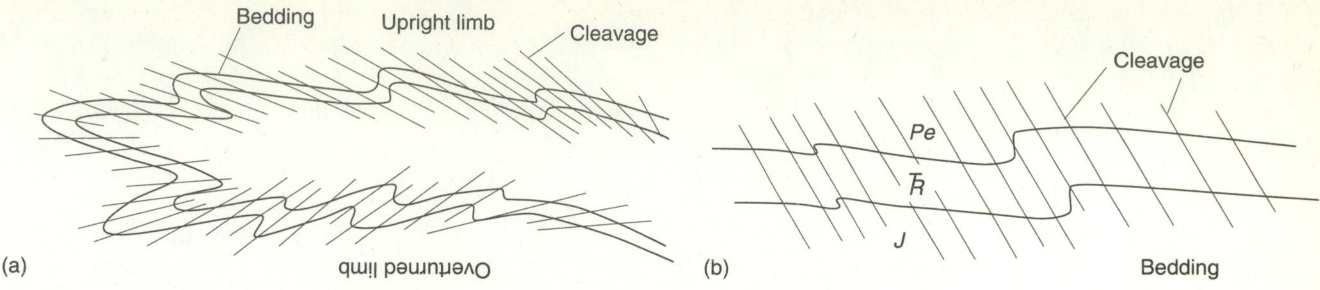

FIGURE 17–15
Cleavage-bedding relationship and its use in determining relative position in a fold. (a) Relationships between orientation of cleavage surfaces and the position on an upright or overturned limb. (b) Cleavage-bedding relationships indicating that the rocks are upright in a sequence that is really overturned, thus showing that the cleavage was emplaced after overturning. J—Jurassic; Ћ—Triassic; Pe—Permian.

FIGURE 17–16
Refracting cleavage in layers of slate (dark), fine sandstone, and coarse sandstone in the Upper Proterozoic(?) Walden Creek Group near Maddens Branch, Ocoee Gorge, southeastern Tennessee. The closely spaced, gently dipping penetrative cleavage in the slate at the bottom of the photo refracted into the fine sandstone at a steeper dip and coarser spacing, then into the thickly bedded sandstone as a spaced cleavage. The clear lithologic control of cleavage here spans much of the range in cleavage spacing—from fine to coarse—in Powell's (1979) classification (Figures 17–5 and 17–6). (RDH photo.)

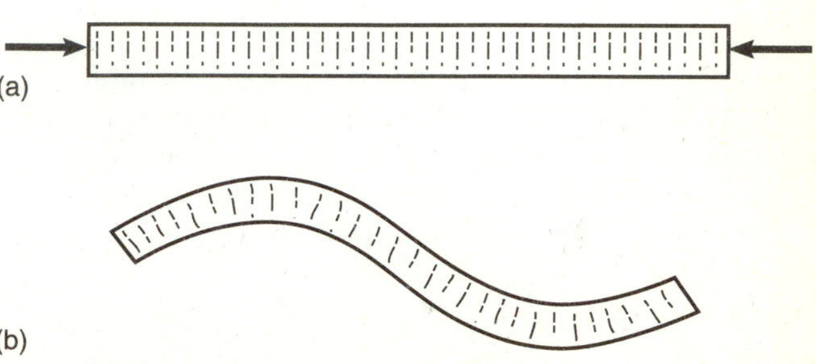

FIGURE 17–17
Formation of a fanned cleavage (b) by the cleavage forming by layer-parallel shortening before or during the early stages of folding (a).

Around the end of the nineteenth century, Charles R. Van Hise suggested—as did Sorby—that slaty cleavage developed perpendicular to the direction of principal shortening (the minimum principal elongation, *Z*). Unlike Sorby, Van Hise believed that new mineral growth in the plane of flattening was the principal cleavage-forming factor.

By the beginning of the twentieth century, all the alternative hypotheses about the origin of slaty cleavage that we recognize today had been outlined. The dispute about rotation versus recrystallization has continued throughout much of the twentieth century. Oddly, the idea that pressure solution is important in the formation of slaty cleavage was almost forgotten until recently and was not mentioned as a viable cleavage-forming mechanism in elementary textbooks of structural geology until the 1980s. As a potential solution to the problem of rotation versus recrystallization, Ramsay (1967) had suggested that rotation dominates at low temperature, but as temperature increases, recrystallization becomes dominant. His notion is no doubt partly correct, but the important role of pressure solution remained in the shadows until recently.

Another controversy focused on whether slaty cleavage formed by simple shear or by pure shear. The question arose in the nineteenth century, being argued by Phillips (1844), A. Laugel (1855), O. Fisher (1884), and G. F. Becker (1896, 1904). Becker advocated development by simple shear parallel to one of the shear planes of the strain ellipsoid. In this century, several geologists, including Bruno Sander (1930), F. J. Turner (1948), R. Hoeppener (1956), and G. Voll (1960), also thought this was the case. Early advocates of the pure shear or flattening mechanism include Sharpe (1847, 1849), Sorby (1853, 1856), Tyndall (1856), S. Haughton (1856), Alfred Harker (1885, 1886), and Van Hise (1896). Still later, structural geologists such as Ramsay (1967), James Dieterich (1969), Andrew Siddans (1972), and Dennis Wood (1973) concluded that pure shear is more likely the dominant cleavage-forming mechanism.

Pure shear as the dominant cleavage-forming mechanism is supported by studies of relationships between finite-strain markers and the orientation of cleavage planes, and the relationships to fold axes or axial surfaces. Even so, it has been shown that in many places slaty cleavage started forming at a high angle to bedding as the rocks were folded and was later sheared to lower angles by flexural slip along bedding. This may imply at first glance that cleavage did not form parallel to the *XY* plane of the strain ellipsoid. In fact, the change in orientation is related to more than one fold mechanism operating at the same time, rather than to the cleavage not being parallel to the *XY* plane.

MECHANICS OF SLATY CLEAVAGE FORMATION

Thus, the origin of all types of slaty cleavage has been controversial. The concepts of rotation versus recrystallization, pure shear versus simple shear, and, more recently, the presence or absence of pressure-solution have been debated throughout much of two centuries. Today, most structural geologists accept Daniel Sharpe's conclusion (1847) that slaty cleavage initially forms by compression perpendicular to the *XY* plane, but most who work with cleaved rocks also conclude that cleavage does not remain parallel to the *XY* plane throughout the folding process. Paul Williams (1976) considered these questions to be fundamental: Does cleavage develop parallel to principal planes and remain so throughout the history of deformation? Or does cleavage develop in some other (shear?) orientation and rotate toward parallelism during progressive deformation? He concluded that if the strain is coaxial (principal planes of the total and incremental strain ellipsoids were always), the cleavage will develop parallel to the *XY* plane and remain so. If, however, the strain is noncoaxial (for example, simple shear strain), cleavage will not develop parallel to the *XY* plane. Formation of cleavage during folding generally involves noncoaxial strain. We do know that the scaly fabric described in subduction complexes forms by simple shear, because of its association with faults, as well as from experimental data (Moore and others, 1986). Debate continues about whether or not cleavage refraction actually exists or wheather it occurs as a product of later inhomogeneous simple shear between layers (Figure 17–18; Williams, 1976; Henderson and others, 1986; Treagus, 1983, 1988). Both appear possible in light of available data.

An important piece of evidence bears directly on the role of pressure solution in formation of slaty cleavage. In calcareous rocks, the relative solubility of carbonate and noncarbonate minerals favors carbonate pressure dissolution in acidic or alkaline conditions. This has long been known, but the concept of a large volume of material—both carbonate and silica—being removed from fine-grained clay or silty rocks did not gain acceptance until recently. (One reason is that the solubility of silicate minerals, mostly quartz, under alkaline conditions was not fully appreciated.) We also know today that during orogenesis a huge volume of water is fluxed out of the deforming rock mass (Figure 11–2; Oliver, 1986). Several factors—the density of shale and slate, the potential for rearranging minerals during compaction, and the loss of soluble minerals (such as quartz in fine-grained rocks)—all suggest an upper limit for volume loss through rearrangement and dissolution. The limit is about 20 percent, but, as we will see, greater loss is possible. Studies by Ramsay and Wood (1973) revealed that a 10 percent loss in volume may occur by reorientation of minerals without actual loss of material. Pressure solution along cleavage planes does remove much material and reduces volume, but with little change in density. Material dissolved during pressure solution need not leave the rock body. As chemical conditions, temperature, and pressure change locally (such as from the inner to the outer parts of layers undergoing tangential-longitudinal strain), the dissolved minerals may be precipitated again nearby in veins or other voids produced by extension.

Loss of 20 to 40 percent volume has been estimated by Ramsay and Wood (1973) for slates in Wales, and by Richard A. Groshong (1975b) in Pennsylvania. A maximum of 60 percent loss during formation of pressure-solution cleavage has been estimated by several geologists in interbedded graywacke and shale in the Martinsburg Formation of Pennsylvania (Wright and Platt, 1982; Onasch, 1983, 1984), and by Edward Beutner and Emmanuel Charles (1985) in the Hamburg sequence in Pennsylvania and New York. In the

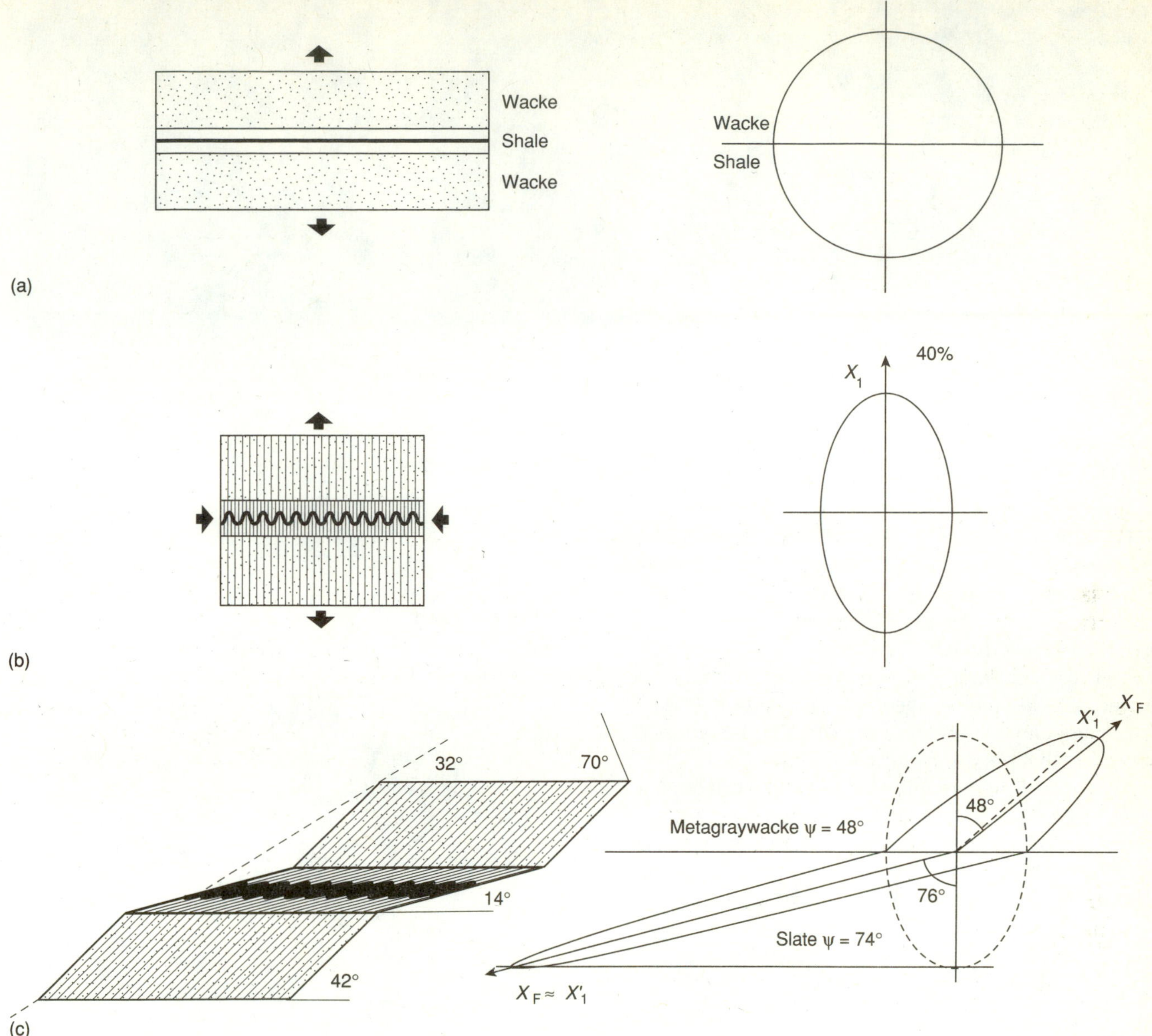

FIGURE 17–18
Formation of a "refracted" cleavage in Cambrian Goldenville Formation sandstone and shale in Nova Scotia. Diagrams on the left represent the original rock mass and structures that developed within it; to the right is a strain ellipse for each stage of deformation. (a) Extension normal to layering, producing a vein (black layer) in the shale. (b) Layer-parallel shortening of 40 percent occurred, buckling the vein and producing an axial-planar slaty cleavage. (c) Displacement of cleavage by differential simple shear parallel to bedding. X_F—final X axis of the strain ellipse following simple shear; X'_1—orientation and elongation of the cleavage relative to the strain ellipse; ψ—simple shear strain. (Modified from J. R. Henderson, T. O. Wright, and M. N. Henderson, 1986, Geological Society of America *Bulletin*, v. 97; and S. H. Treagus, 1988, Geological Society of America *Bulletin*, v. 100.)

Borrowdale volcanic rocks of the Lake District in England, Timothy Bell (1985) could not account for all strain accompanying formation of slaty cleavage by invoking pressure solution alone. He traced strain paths by using strain markers and concluded that a combination of plane strain—such as initial layer-parallel shortening—and volume loss were probably responsible for formation of slaty cleavage.

Determining the actual mechanism that formed slaty cleavage at a particular locality may be difficult because the evidence is often consistent with more than one mechanism. (If an easy solution existed, structural geologists would not have debated the question for more than a century.) Slaty cleavage is frequently accompanied by some recrystallization and reorientation of clays in shale and siltstone. Studies of

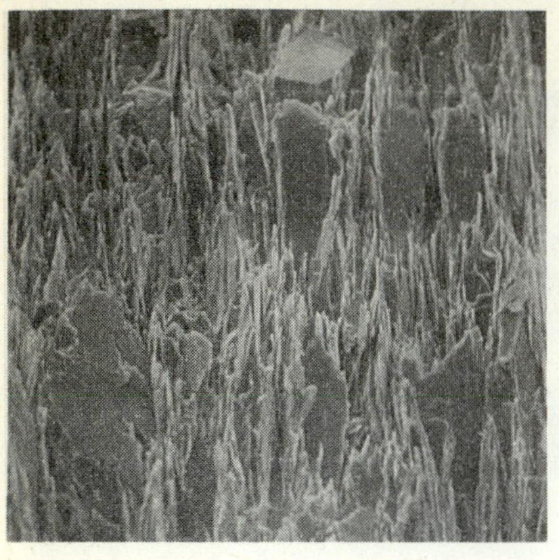

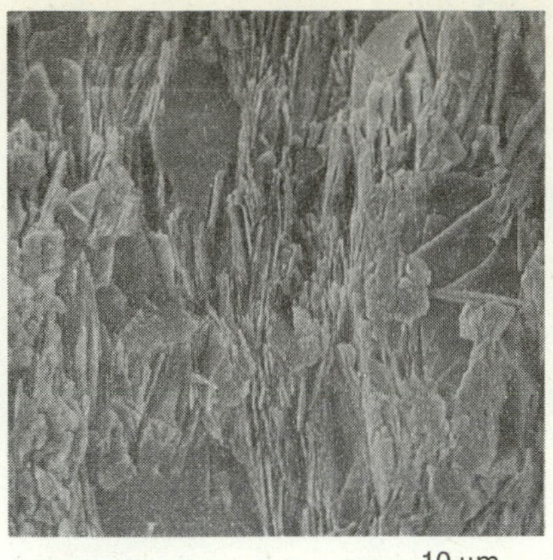

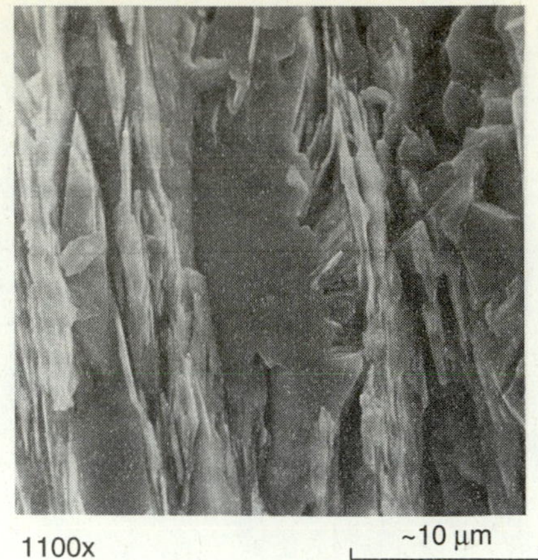

FIGURE 17–19
Scanning electron micrographs of reoriented and recrystallized micas in mica beards in Ordovician slate from the south shore of New World Island, Newfoundland. (From B. A. Van der Pluijm, 1984, *Geologische Rundschau*, v. 73.)

the structure of slates by X-ray diffraction and optical and scanning electron microscopes have revealed that reorientation of grains does occur and that reorientation is usually accompanied by recrystallization (Figures 17–9 and 17–19). A paper by J. H. Lee and others (1986) concluded that layer-silicate grains oriented parallel to cleavage in the Martinsburg Slate in Pennsylvania and New Jersey are virtually undeformed. They noted that pressure solution also occurred, dissolving phyllosilicate minerals originally oriented parallel to bedding. Beutner (1978) suggested that dissolution of even the layer-silicate minerals occurs in proportion to the original orientation of grains relative to the orientation of rock cleavage surfaces (Figure 17–20). Therefore, Beutner concluded that mechanical reorientation of grains that must be rotated through large angles does not occur during formation of slaty cleavage.

Microscopic study of thin sections cut perpendicular to cleavage reveals either that cleavage planes are spaced and contain residues of insoluble minerals (clays, micas, and iron oxides) and a relative lack of soluble minerals (quartz and calcite) or that they consist of a continuous oriented fabric of recrystallized grains (Figure 17–21). Relict partial grains of quartz or calcite truncated at cleavage planes are often found in pressure-dissolved rocks. Slaty cleavage formed by recrystallization (with or without reorientation) does not contain residues or partial grains along cleavage planes. All grains are thus oriented parallel to cleavage planes, and new unstrained minerals are evident. These mechanisms may compete and overlap (Lee and others, 1986). Pressure solution at lower temperatures of the anchizone gives way to recrystallization at the higher temperatures of the greenschist facies where chlorite, micas, and other minerals begin to form in fine-grained rocks.

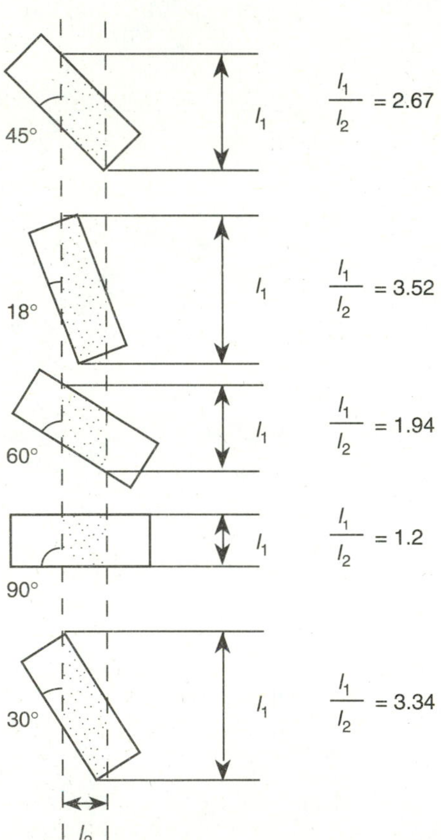

FIGURE 17–20
Corrosion of grains in different orientations to produce new shapes (stippled) proportional to grain orientation. Dashed lines indicate the initial locations of rock cleavage surfaces, and all material outside those lines is removed. New axial ratios are shown as l_1/l_2 from an initial ratio for all grains of 2.64. This mechanism would thus not produce any crystallographic orientation. (From E. C. Beutner, Slaty cleavage and related strain in Martinsburg Slate, Delaware Water Gap, New Jersey, January 1978, *American Journal of Science*, v. 278, p. 1–23. Reprinted by permission of *American Journal of Science*.)

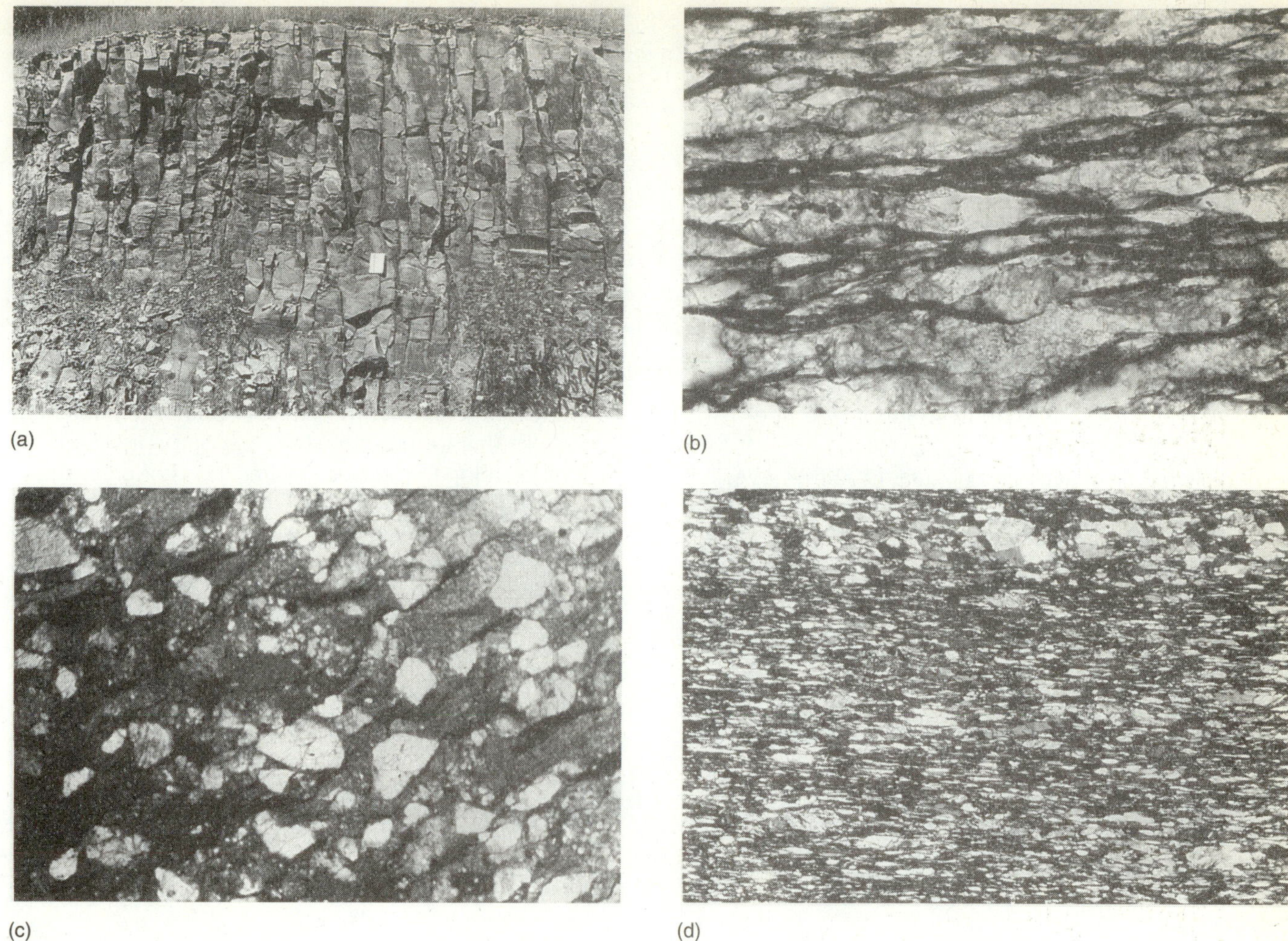

FIGURE 17–21
(a) Pressure-solution spaced cleavage in Middle Ordovician silty carbonate rocks, South Hero Island, Vermont. Cleavage is near vertical; bedding (faint color banding) is nearly horizontal. (RDH photo.) (b) Residues of insoluble minerals along cleavage planes in Devonian Marcellus Shale near Cumberland, Maryland, produced by pressure solution. Plane light. Width of field is approximately 3 mm. (Charles M. Onasch, Bowling Green State University.) (c) Partly dissolved quartz grains next to pressure-solution cleavage planes in Upper Ordovician Martinsburg Slate, Delaware Water Gap, New Jersey. Plane light. Width of field is approximately 2 mm. (Thin section courtesy of Timothy L. Davis, North Carolina Geological Survey.) (d) Slaty cleavage in Upper Ordovician Arvonia Slate, Arvonia, Virginia, produced by recrystallization or rotation of grains. Note the obvious absence of residues parallel to cleavage. Dark areas are biotite grains. Plane light. Width of field is approximately 2 mm. (RDH photo.)

Dissolution of calcite under stress in fine-grained carbonate rocks is directly related to the amount of impurities therein. Stephen Marshak and Terry Engelder (1985) determined that pressure-solution cleavage develops more readily in fine-grained carbonate rocks (such as marl) that contain more silt than does pure limestone. The same relationship probably holds for pressure-solution effects in pure quartzite as compared with less-pure argillite and fine-grained sandstone. Such argillite and sandstone also contain more water. Impurities in large amounts may provide more points to nucleate dissolution, thereby making the process more efficient. Clays, too, may catalyze pressure solution because of their small grain size, and they may serve as both a source of, and a conduit for, movement of water.

Progressive Cleavage Development in Fine-Grained Sediment

Even as sediment is being deposited, it begins readjusting to the new environment. As the overlying sediment weighs down clay particles, water is expelled and the particles begin aligning perpendicular to the vertical maximum principal lithostatic stress to form a bedding fissility. Much of the change comes during expulsion of water and initial compaction, which is the first stage in the transformation of sediment into sedimentary rock. Additional lithification may occur during the early stages of folding. Later, tectonism may transform the rock into a strongly cleaved or foliated mass (Figure 17–22). The entire process has been divided by Ramsay and Huber (1983) into six stages of increasing tectonic strain. Those stages, slightly modified, are as follows:

1. *Undeformed condition*—Bedding fissility present.
2. *Earliest deformation stage*—Volume loss, from reorientation of grains and expulsion of water, enhances bedding fissility.
3. *Pencil structure* (Figures 17–22a and 17–23)—Elongate pencil-like fragments are produced by intersection of bedding and cleavage, with no continuously organized planar cleavage. Such structure is best illustrated by homogeneous silty shale. Pencils generally parallel fold axes. This stage may be found in unmetamorphosed fine-grained rocks.
4. *Embryonic cleavage stage* (Figure 17–22b)—Weak, poorly developed cleavage parallel to fold axial surfaces results from pressure solution of silty carbonate or argillaceous rocks. Minor recrystallization of new minerals—illite, quartz, or calcite—may parallel the cleavage, and an intersection lineation (Chapter 18) of pencil cleavage and bedding may develop parallel to the axes or hinges of folds. This stage is most likely to be found in unmetamorphosed rocks. Layer-parallel shortening may produce cleavage before folding (Engelder and Geiser, 1979).
5. *Cleavage stage* (Figure 17–22c)—A strong planar fabric results from pressure solution, reorientation of platy minerals, or incipient recrystallization of clays. Generally, cleavage has an axial-planar relationship to folds that form at this stage, accompanied by well-developed intersection lineations of cleavage and bedding. Rocks displaying this cleavage stage

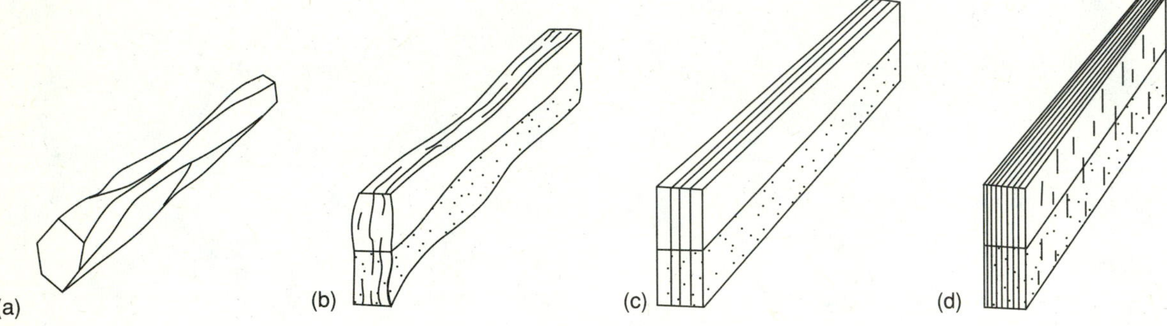

FIGURE 17–22
Stages of cleavage development: (a) Pencil structure. (b) Embryonic cleavage stage. (c) Cleavage stage with accompanying lineation formed by intersection of cleavage and bedding. (d) Well-developed cleavage with mineral lineation.
(After J. G. Ramsay and Martin Huber, 1983, *The Techniques of Modern Structural Geology: Volume 1: Strain Analysis,* Academic Press.)

FIGURE 17–23
Large pencils in siltstone in the Upper Proterozoic(?) Sandsuck Formation near Reliance, Tennessee. (Locality courtesy of J. O. Costello, Georgia Marble Company. RDH photo.)

range in metamorphic grade from anchizone to lower-greenschist facies (chlorite zone).

6. *Strong cleavage with mineral (stretching?) lineation* (Figure 17–22d)—The planar character of slaty cleavage is better developed than in earlier stages, and a faint mineral-elongation lineation (quarry workers' *grain)* appears on the cleavage surface. This lineation is commonly oriented perpendicular to fold hinges, parallel to the X direction of the strain ellipsoid (down the dip of cleavage planes with horizontal fold axes), and lies in the XY plane, but the orientation of the mineral lineation can vary widely in the cleavage plane. The bedding-cleavage intersection lineation is usually well developed. This is the highest stage of cleavage development and may occur throughout the chlorite zone of regional metamorphism; it occasionally persists into the biotite zone. At the upper extreme of this stage, the rock mass is totally recrystallized, with a dominant fabric that is a product of both deformation and metamorphism. Recrystallization processes (dislocation and diffusion creep) and formation of other new minerals become dominant as the metamorphic temperature increases.

Pressure solution as a cleavage-forming mechanism dominates in the early stages of cleavage formation, but the development of cleavage planes in the later stages is probably related to increased recrystallization. Many strongly cleaved or foliated rocks are phyllonites with well-developed S-C structure, indicating that the foliation formed by noncoaxial (simple shear) deformation (Chapter 10).

Progressive development of cleavage in fine-grained rocks is a means by which planar fabrics are organized with increasing deformation (Figure 17–24). Beyond the last stage in Ramsay and Huber's scheme of cleavage and foliation development is the transition to coarser-grain-size recrystallization. A similar scheme was devised by Powell (1979) for slaty-cleavage formation.

Pressure-solution cleavage may be classified under a system devised by Walter Alvarez, Terry Engelder, and Peter Geiser (1978), who applied the terms "weak," "moderate," "strong," and "very strong." Classification depends on the character of cleavage surfaces, spacing of cleavage, and percentage of shortening. Cleavage classified as weak is generally stylolitic, with no preferred orientation; spacing is more than 5 cm, and shortening is 0 to 4 percent. Moderate cleavage has discrete surfaces, and with planes oriented from parallel to 120° to each other; spacing is 1 to 5 cm, and shortening is 4 to 25 percent. Strong cleavage is wispy to locally anastomosing, concentrating in major surfaces; spacing is 0.5 to 1 cm, and shortening is 24 to 35 percent. Very strong cleavage is sigmoidal with transposed bedding; spacing is less than 0.5 cm, and shortening is more than 35 percent. The rock mass may contain abundant calcite veins perpendicular to bedding.

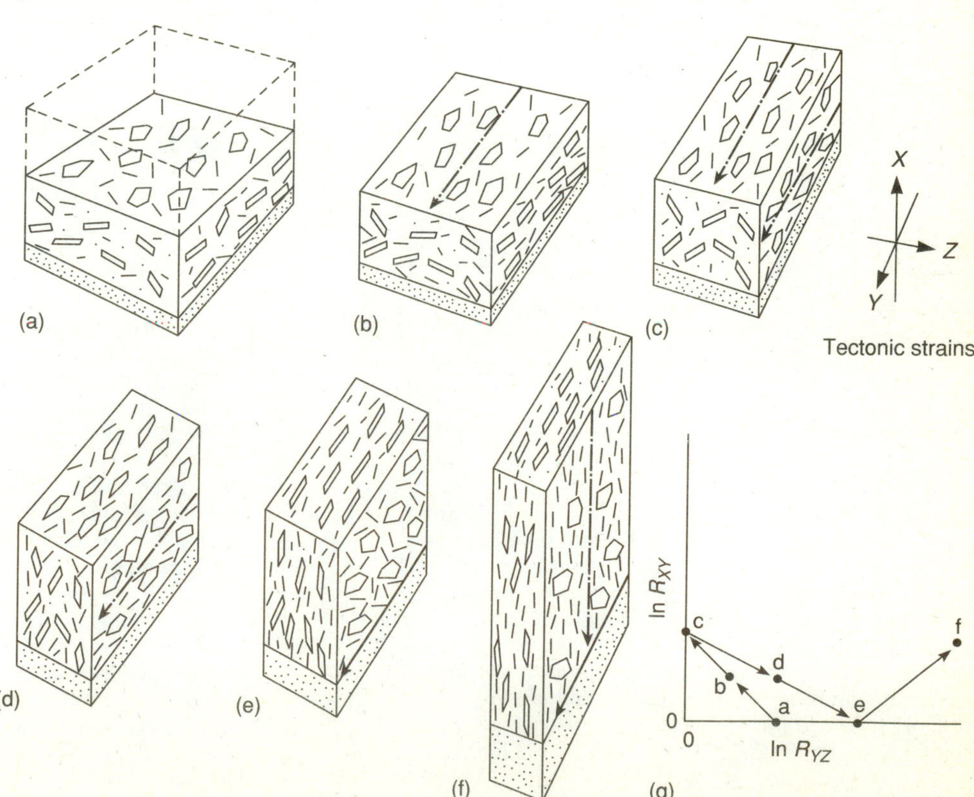

FIGURE 17–24
Relationships between two-dimensional strains and the organization of planar fabrics in rocks; (a) represents the compaction stage; (b) involves the first deformation and minor reorganization of grains; (c) involves development of pencils; (d) is the embryonic cleavage development stage; (e) is the stage of development of a well-developed cleavage; (f) is the stage of further cleavage development where a prominent mineral lineation forms. The Flinn diagram (g) indicates the deformation path from one stage to another. R_{xy} is x/y, R_{yz} is y/z. (From J. G. Ramsay and Martin Huber, 1983, *The Techniques of Modern Structural Geology: Volume 1: Strain Analysis,* Academic Press.)

Strain and Formation of Slaty Cleavage

The amount of strain accompanying slaty cleavage directly indicates the amount of deformation that accompanies folding. As already noted, cleavage-forming mechanisms combine recrystallization and pressure solution. Earlier compaction may account for as much as 30 percent decrease in volume, and the remainder of the deformation involves flattening and extension in the plane of the cleavage (the *XY* plane). The *strain path* during cleavage formation changes with position in folds during progressive deformation as a result of noncoaxial strain. In their study of the Hamburg sequence, Beutner and Charles (1985) found that differences in volume loss by pressure solution occurred in the hinges of folds (up to 59 percent) and limbs (up to 29 percent). Measurements of strain in deformed reduction spots and conodonts indicate that the limbs and hinges were deformed along different strain paths. Fold limbs were flattened throughout deformation; hinges were flattened at the beginning, then constricted, and later flattened again (Figure 17–25).

Assuming that the dominant process in the formation of slaty cleavage was pure shear, Dennis Wood (1973) compared strain values calculated from deformed reduction spots in slates in the Taconic klippes of New York and Vermont, and slates of Wales. In both cases, he found from 50 to 75 percent shortening perpendicular to the cleavage and from 100 to 400 percent extension as a principal elongation in the plane of cleavage.

Graptolites were used by Thomas Wright and Lucian Platt (1982) as finite-strain markers to determine strain in the Martinsburg Slate in Pennsylvania, Maryland, Virginia, and West Virginia; they found from 50 to 70 percent shortening perpendicular to cleavage where the cleavage and bedding are at high angles in the hinges of folds (Figure 17–26). Apparent strain (or the components of strain in the *XY* plane) decreases to nearly zero where bedding and cleavage become parallel in the limbs of folds.

In graywackes of the Martinsburg Formation in the Massanutten synclinorium of Maryland, Charles Onasch (1983) determined the finite strain of rough cleavage using shapes of detrital grains, fibrous minerals, and veins that had been deformed before the cleavage formed. He reported that shortening ranged from 29 to 55 percent normal to the *XY* (cleavage) plane in well-cleaved samples, the variation being caused by an unknown loss of volume by dissolution. Pressure solution was the dominant cleavage-forming mechanism.

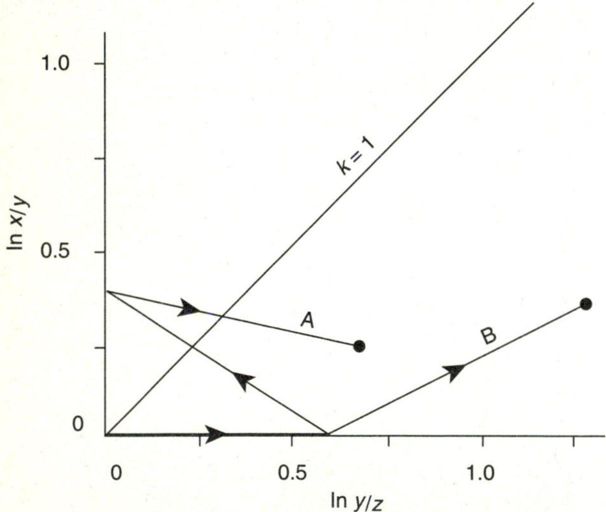

FIGURE 17–25
Flinn diagram of strain paths derived from study of strain indicators in hinges (curve A) and limbs (curve B) of near-isoclinal folds in the Hamburg sequence near Shartlesville, Pennsylvania. Note that the strain in fold hinges (A) was traced from the field of increasing dominant flattening strain to the dominant extension strain field, back into dominant flattening, but with a strong component of triaxial strain. Strain in the limbs of the fold remains in the dominant flattening field and traces toward a greater component of triaxial strain. (From E. C. Beutner and E. G. Charles, 1985, *Geology*, v. 13.)

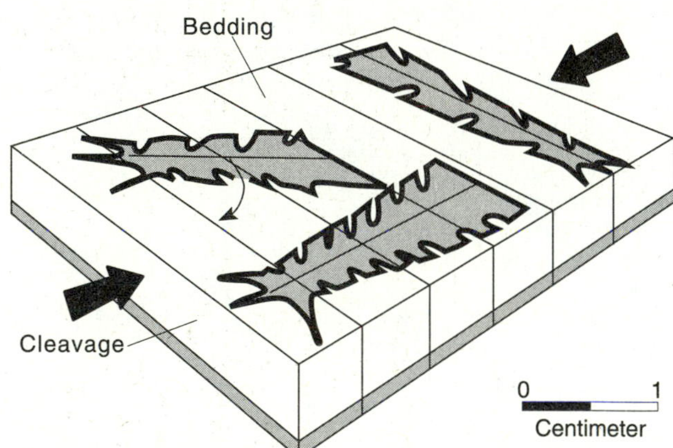

FIGURE 17–26
Relationship of graptolites on a bedding plane to shortening across cleavage. Graptolites parallel to the trace of cleavage on bedding are narrower than usual; those perpendicular to the cleavage trace are shorter. The original spacing of indentations (thecae) on each graptolite is constant for both adults and juveniles of the same species, and so the amount of shortening in any direction can be attributed to tectonic strain. Large arrows on ends of diagram indicate shortening direction; small arow on obliquely oriented graptolite indicates that it would undergo clockwise (dextral) rotation with the shortening indicated, while the other two graptolites, because of their orinetation parallel and normal to the shortening directions, undergo no rotation. (From T. O. Wright and Lucian Platt, 1982, Pressure dissolution and cleavage in the Martinsburg shale, *American Journal of Science*, v. 282, p. 122–135. Reprinted by permission of American Journal of Science.)

I. J. Reks and David Gray (1983) calculated the amount of strain related to cleavage formation in the Pulaski thrust sheet in the Appalachian foreland of southwestern Virginia. Their study of orientation and growth of chlorite pressure shadows on framboidal pyrite nodules revealed from 17 to 35 percent strain. After reconstructing a state of no strain, Reks and Gray said that the cleavage had been superposed parallel to the axial surfaces of already-formed Class 3 buckle folds. Deformation, they concluded, was largely by homogeneous strain, with folds and cleavage forming at different stages of the same event.

The amount of shortening during cleavage formation decreases from the high values just cited to much lower values toward the outer edges of orogenic belts. Under proper conditions, fabrics that are related to pressure solution and accompany layer-parallel shortening can extend to the outer edges of the deformed region. In the central Appalachians, strain was accommodated by layer-parallel shortening above a detachment zone in the rocks beneath (T. Engelder, and Engelder, R., 1977; Engelder, 1979a, 1979b; and Engelder and Geiser, 1979).

CRENULATION CLEAVAGE

Development of crenulation cleavage generally involves formation of a new structure that overprints an existing cleavage or foliation (Figure 17–27), although crenulations may also form on bedding and have even been observed in ice-deformed glacial silt. *Crenulation cleavage,* as defined by David Gray (1977a), occurs in zones of mineral differentiation coincident with the limbs of microfolds in crenulated rock fabrics. Crenulations generally overprint an existing cleavage by crinkling and crenulating the earlier structure. Differentiated layering frequently begins with development of a crenulation structure and continues with pressure-solution removal of quartz from the limbs of crenulations and redeposition in the hinges.

The origin of crenulation cleavage has long been debated. Many structural geologists support its origin by pure shear, but others have found evidence for a simple-shear mechanism. In shear zones, simple shear best explains crenulations (Chapter 10), but the widespread regional crenulation cleavage may best be explained by pure shear, in the sense of an "average" strain ellipsoid.

Gray (1977a) recognized two classes of crenulation cleavage: ***discrete*** and ***zonal***, as used in Powell's classification discussed earlier. *Discrete crenulations* involve sharply defined cleavages that truncate the older fabric in the rock mass. *Zonal crenulations* occur as wide diffuse zones, and the original rock fabric continues uninterrupted (Figure 17–28).

Differentiated crenulation cleavages in Australia and Switzerland led Gray (1977b), and Gray and David Durney (1979) to suggest that the crenulations formed by buckling and fluid-induced mobility. The process was driven by stress, and the crenulations were related to the wavelengths of microfolds and grain-boundary contacts, for grain-boundary diffusion (Chapter 6) was identified as the deformation mechanism. They also concluded that the crenulations formed by pure shear.

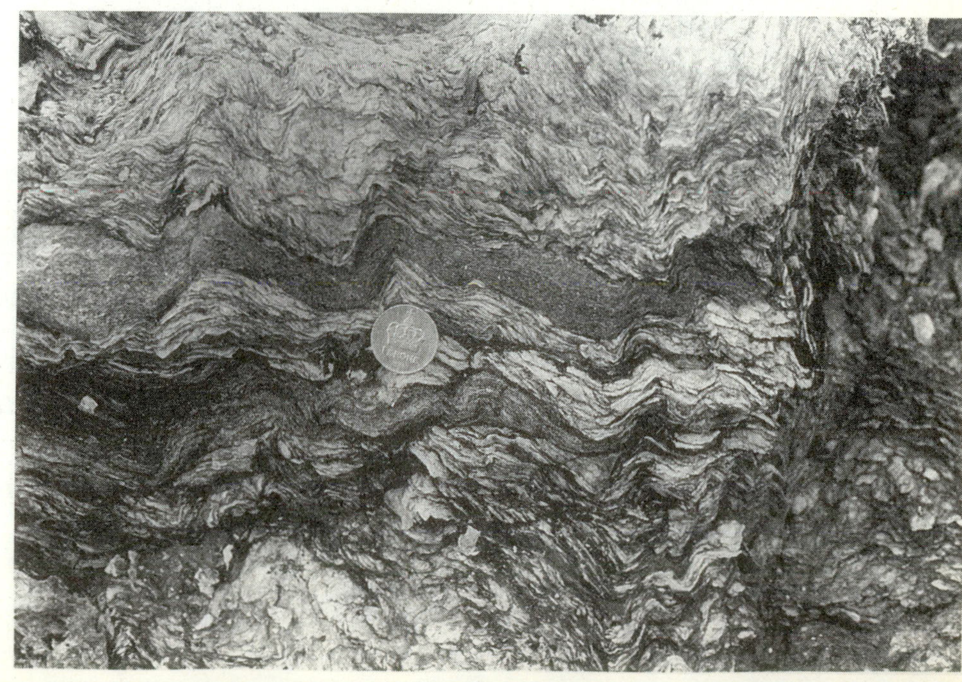

FIGURE 17–27
Crenulation folds developing a cleavage on some long limbs in deformed talc-rich schist (light-colored) near the base of the Karmøy ophiolite near Hagesund, Island of Karmøy, southern Norway. Note that dark (more feldspathic) layers are not crenulated. Also, see Figure 17–11b. (Locality courtesy of Brian A. Sturt, Norwegian Geological Survey; RDH photo.)

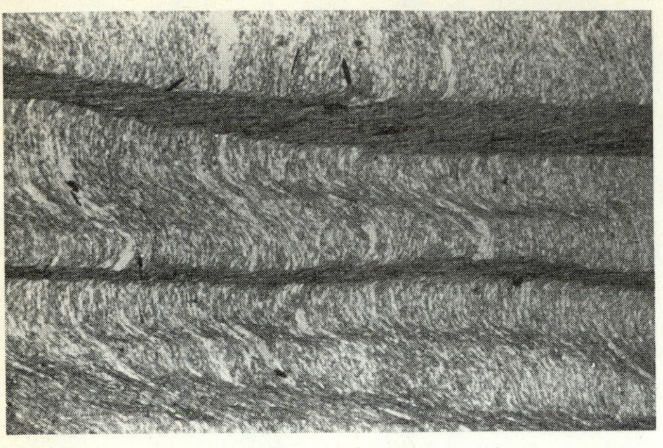

(a)

(b)

FIGURE 17–28
(a) Discrete crenulations in Lower Silurian Vakdal Formation phyllite near Ulven, southeastern Norway. Plane light. Width of field is approximately 16 mm. (b) Zonal crenulations in (Ordovician?) Mineral Bluff Formation phyllite near Murphy, North Carolina. Plane light. Width of field is approximately 16 mm. (RDH photos.)

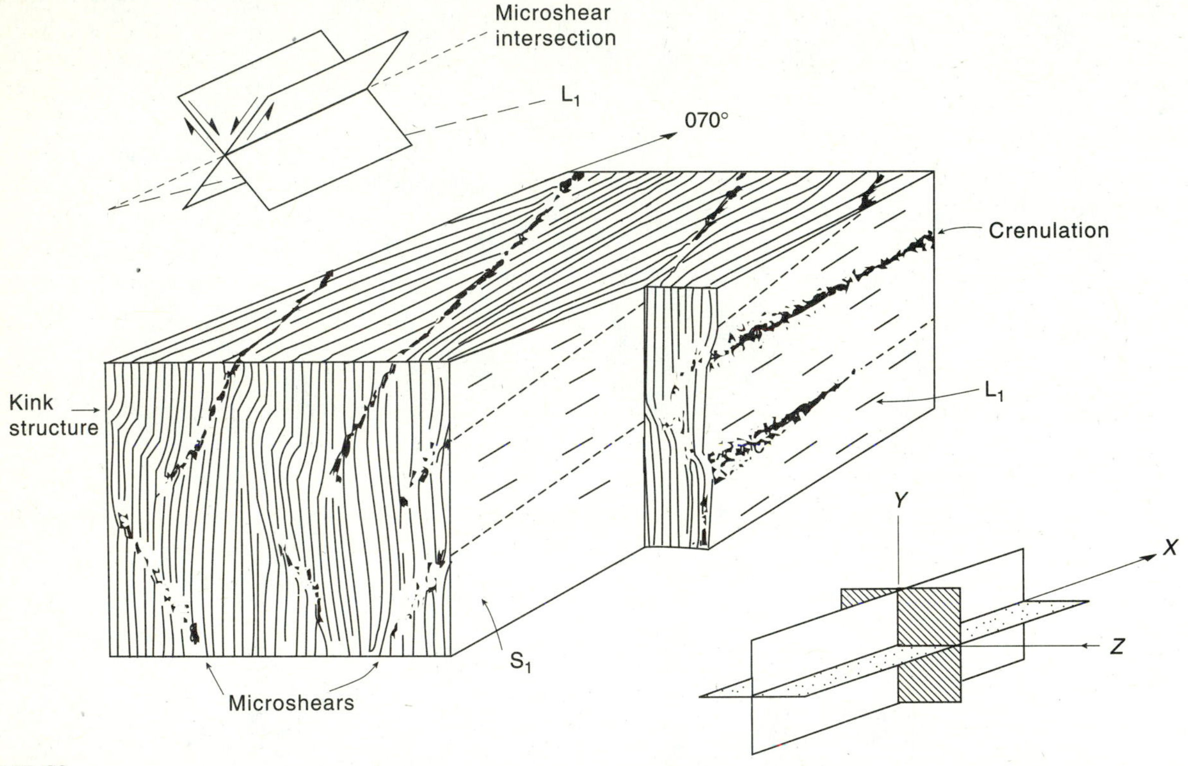

FIGURE 17–29
Vertical S_1 cleavage surfaces, arbitrarily oriented ENE, containing a subhorizontal mineral lineation (L_1). Sections cut normal to *X, Y,* and *Z* show relationships of microshears, crenulations, and kinks to principal planes of the strain ellipsoid. (Reprinted from *Journal of Structural Geology*, v. 1, S. K. Hanmer, p. 81–91, © 1979, with kind permission from Elsevier Science, Ltd., Kidlington, United Kingdom.)

By way of contrast, consider a study by Simon Hanmer (1979), who suggested that crenulation cleavage forms on microshears developed by heterogeneous strain parallel to shear planes (Figure 17–29). He concluded that the character of the existing planar and linear fabrics and porphyroblasts that form during recrystallization all affect initiation and propagation of crenulations. Hanmer's study showed that crenulations also form by simple shear in fine-grained rocks, not as part of a shear zone.

Evidence can thus be cited to show that crenulations can form by pure shear or by simple shear. Paul Williams (1972) observed that crenulations and slaty cleavage grade into one another in the Bermagui region of Australia, implying a common origin by pure shear. We also know that crenulations form in shear zones (Chapter 10) as both *extensional crenulations* (Platt and Vissers, 1979) and as reverse-*sense crenulations* (Dennis and Secor, 1987). They, too, may form by simple shear.

CLEAVAGE FANS AND TRANSECTING CLEAVAGES

A fundamental property of most slaty cleavage is the parallelism (or a fanning relationship) between the orientations of fold axial surfaces and slaty cleavage, providing a genetic link between the two structures (Figure 17–1); yet examples can be found that lack parallelism between fold axial surfaces and cleavage (Figures 17–30 and 17–31). Folds where the cleavage and fold axial surfaces are not coincident—or where the fold axis does not lie in the cleavage—are ***transected folds***.

Convergent and divergent cleavage fans may indicate discordant timing between folding and cleavage formation. Causes range from the obvious noncontemporaneity of formation to subtler relationships resulting from the sequencing of homogeneous and inhomogeneous strain, as well as discordance in strain rate and strain variation from bed to bed.

FIGURE 17–30
Transected fold in Devonian slate from near Zell, Mosel Valley, Germany. The transecting cleavage (S_2) forms an intersection lineation that obliquely crosses the F_2 fold shown here. (From G. J. Borradaile, M. B. Bayly, and C. McA. Powell, eds., *Atlas of Metamorphic and Deformational Rock Fabrics,* © 1982, Springer–Verlag. Reproduced with permission.)

In the simplest relationship, transected folds and cleavage formed at different times with differently oriented principal strain axes. The differences in timing and orientation can be explained if the folds formed as soft-sediment structures, if the folds formed tectonically (earlier than the cleavage), or if the folds formed later than cleavage. Powell (1974) has suggested that slaty cleavage forms as a principal-plane cleavage parallel to the *XY* plane of an incremental strain ellipsoid; but because folding occurred later in rocks he studied, Graham Borradaile (1978) suggested that cleavage does not parallel a principal plane in the finite strain ellipsoid. Susan Treagus and John Treagus (1992) suggested that there may be a relationship between transpression and formation of transecting cleavage. Specifically, they suggested that clockwise transection may represent sinistral transpression, but without independent data on the incremental and finite strain history of a region, such a conclusion is unwarranted. For this hypothesis to be correct would require that cleavage not track the *XY* plane of the finite strain ellipsoid. They suggested alternatively that initial pure-shear, dominated deformation may give way to later simple shear, producing strike-slip motion. They also concluded that transection could not occur in horizontal layering in a transpressive environment without either the folds or the cleavage forming not parallel to the *XY* plane of the strain ellipsoid. If the layering had an initial inclination, transection paralle to the XY plane in the finite strain ellipsoid would be possible in transpression.

In Ramsay's view (1963), cleavage and folds commonly form about the same time, and cleavage forms parallel to the *XY* plane. Noncoaxial strain is superposed during the same deformation, however, so

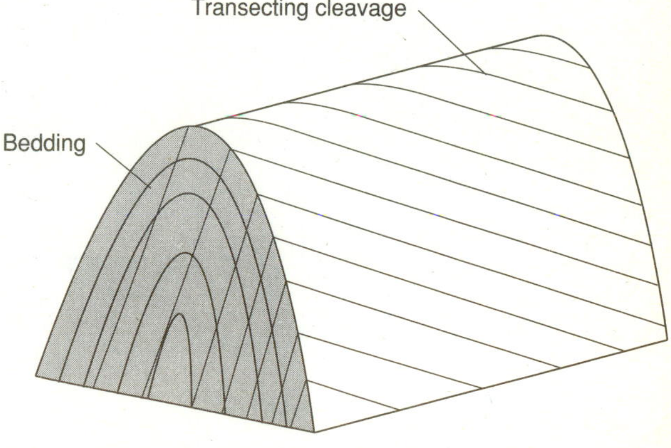

FIGURE 17–31
Transected fold showing relationships between the hinge and parallel cleavage.

that the principal plane of the strain ellipsoid and the *XY* cleavage plane rotate to positions that cut through both fold limbs. Later, Borradaile (1978) concluded that it was unlikely that the strain ellipsoid can be rotated so that cleavage will transect folds.

Both synchronous and nonsynchronous behavior may lead to transected folds, as recognized by Borradaile (1979). For synchronous formation of transected folds, he concluded that a noncoaxial strain history may be possible in which cleavage formation is delayed slightly relative to fold formation within the same deformational event. Borradaile invoked dominance by a competing mechanism, such as dewatering, which prevents cleavage from forming initially during folding; when cleavage does begin to form, it transects the existing folds. Transected folds may also form in shear zones. Borradaile (1981) later suggested that transected folds may result from deformation and rotation of grains during folding, producing a noncoaxial strain history for folds and cleavage.

The most likely cause of transected cleavage may be a combination of differences in timing of formation of folds and cleavage, within a regime of noncoaxial deformation.

TRANSPOSITION

For many years, geologists have recognized the progressive dominance of cleavage and foliation with increasing deformation and metamorphic grade. We also know that layering in rocks of high metamorphic grade is probably not original bedding (Figure 17–32). Transformation of original bedding by folding, ductile shear, or other process into parallelism with cleavage or foliation through progressive deformation is called ***transposition***. Bruno Sander (1911) may have been the first to consider transposition as a separate process. Its importance was recognized by Anna Jonas (1927) during her field studies in the central Appalachians and by Knopf and Ingerson (1938) in their book on structural petrology.

Transposition of bedding, or of any layering, occurs progressively. It begins during the initial stages of slaty-cleavage formation. The cleavage thus formed cuts through the axial zones of folds at a high angle, but through the limbs at a low angle, where cleavage and bedding may be almost parallel (Figure 17–33). As deformation continues, the short limbs of the fold are progressively flattened and thinned, and layering in the

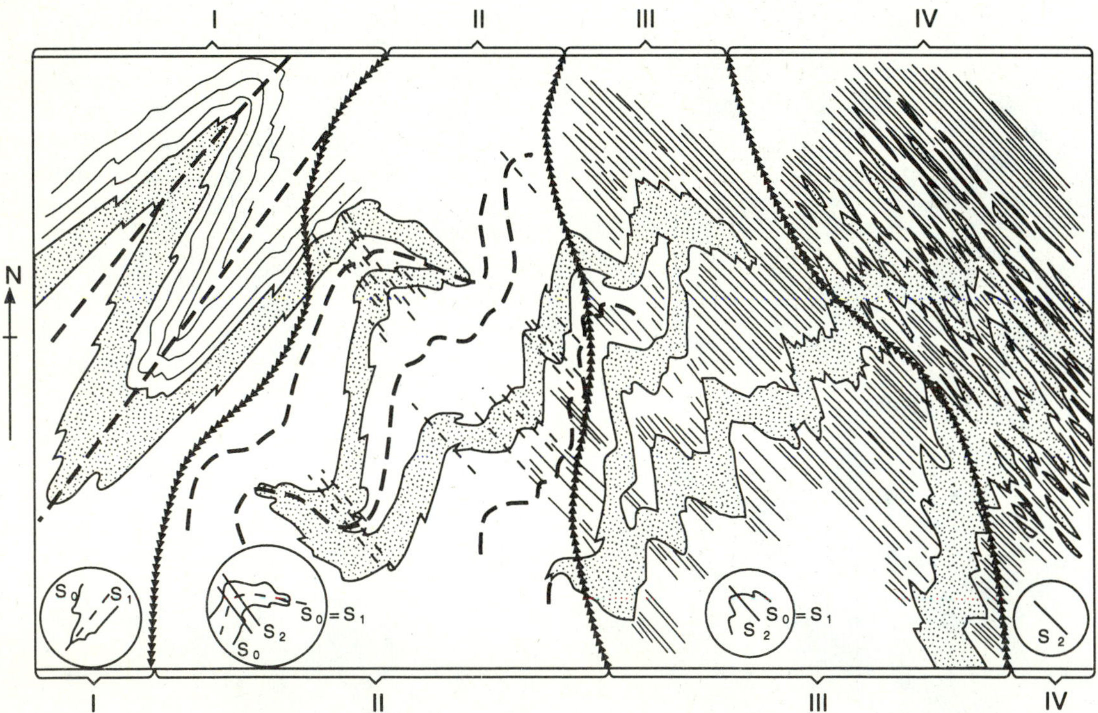

FIGURE 17–32
Progressive deformation and transposition of earlier bedding and cleavage (left side), producing a rock mass containing layers that are all parallel (right side). Roman numerals are stages of progressive deformation. Heavy dashed lines, some labeled S_1 or S_2, represent axial surfaces of folds with parallel development of a foliation. The first development of parallelism between between S_0 and S_1 —transposiiton of bedding into the first foliation—occurs during the first isoclinal folding event . S_1 is then reorineted by porogressive deofrmation into parallelism with S_2. (From F. J. Turner and L. E. Weiss, *Structural Analysis of Metamorphic Tectonites,* © 1963, McGraw-Hill Book Company. Reproduced with permission.)

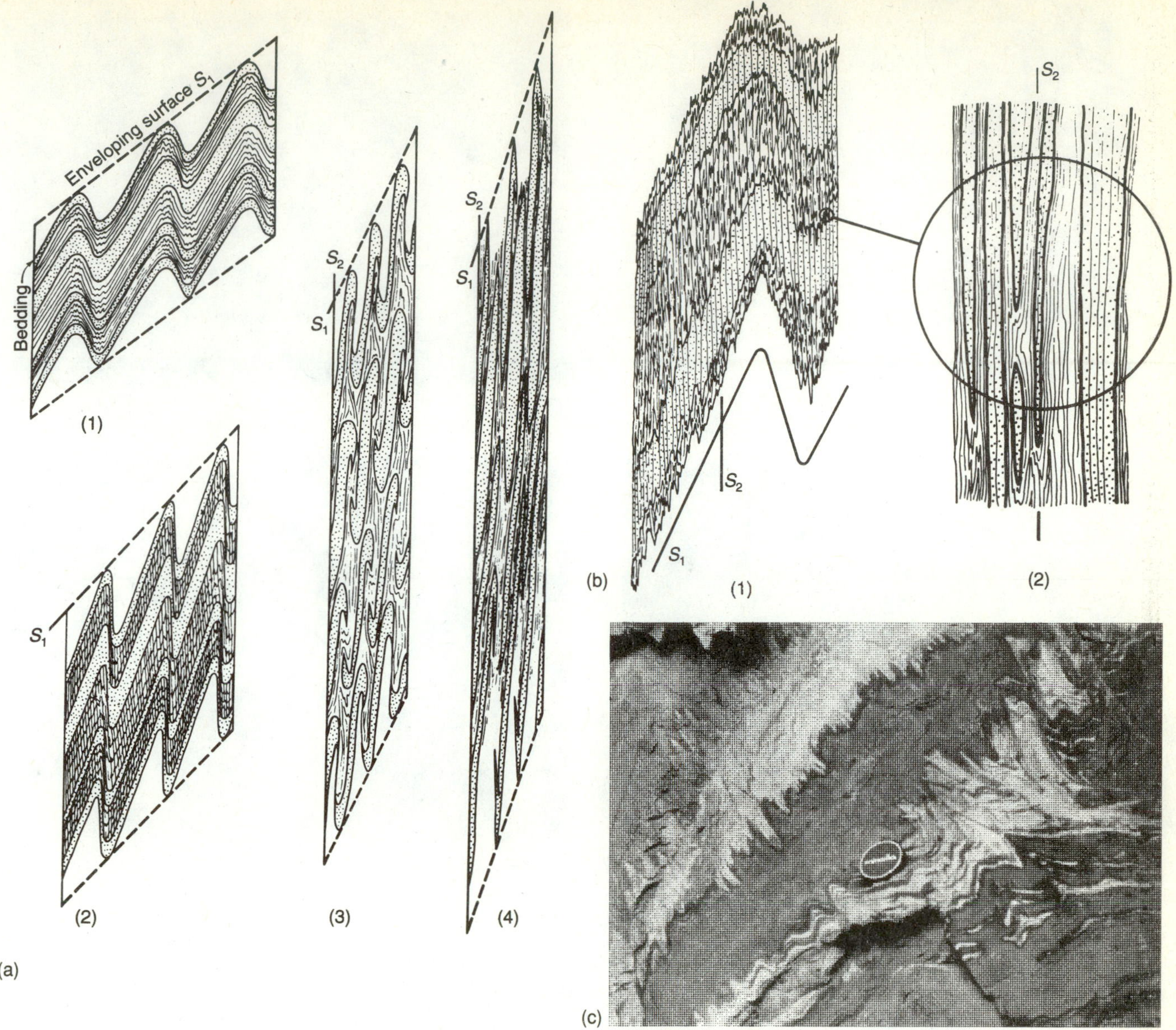

FIGURE 17–33
(a) Relationship between transposition of layering and hinges on the overturned (short) limbs of small folds that have been progressively removed by ductile flow or pressure solution. (b) Incipient transposition in a cleavage-dominated rock mass with closeup of bedding showing reorientation by deformation. (a and b from F. J. Turner and L. E. Weiss, *Structural Analysis of Metamorphic Tectonites,* © 1963, McGraw-Hill Book Company. Reproduced with permission.) (c) Incipient transposition of bedding in interlayered slate (dark) and fine-grained metasiltstone (light-colored) in Upper Proterozoic(?) Wilhite Formation slate, Ocoee Gorge, southeastern Tennessee. (RDH photo.)

axial zones that originally crossed foliation at a high angle is flattened into increased parallelism with the cleavage (or foliation). Recrystallization is favored as deformation continues and temperature rises. Flattening proceeds to an even greater degree, and layering that initially was at a high angle to cleavage planes is flattened into subparallel orientation to cleavage, even in the axial zones of folds (Figures 17–32 and 17–34). Finally, all layering is parallel to the foliation, and original bedding is no longer recognizable—unless primary structures such as graded bedding or cross bedding survive within the layers. Transposition is then complete. An analogous parallel scenario involving transposition of layering in ductile shear zones (Chapter 10) by simple shear can also be described.

Earlier cleavages or foliations may be transposed in much the same way. An initial foliation or cleavage may be overprinted by a new cleavage, possibly crenulation cleavage. With continued deformation, the older S-surface is reoriented into subparallel orientation with

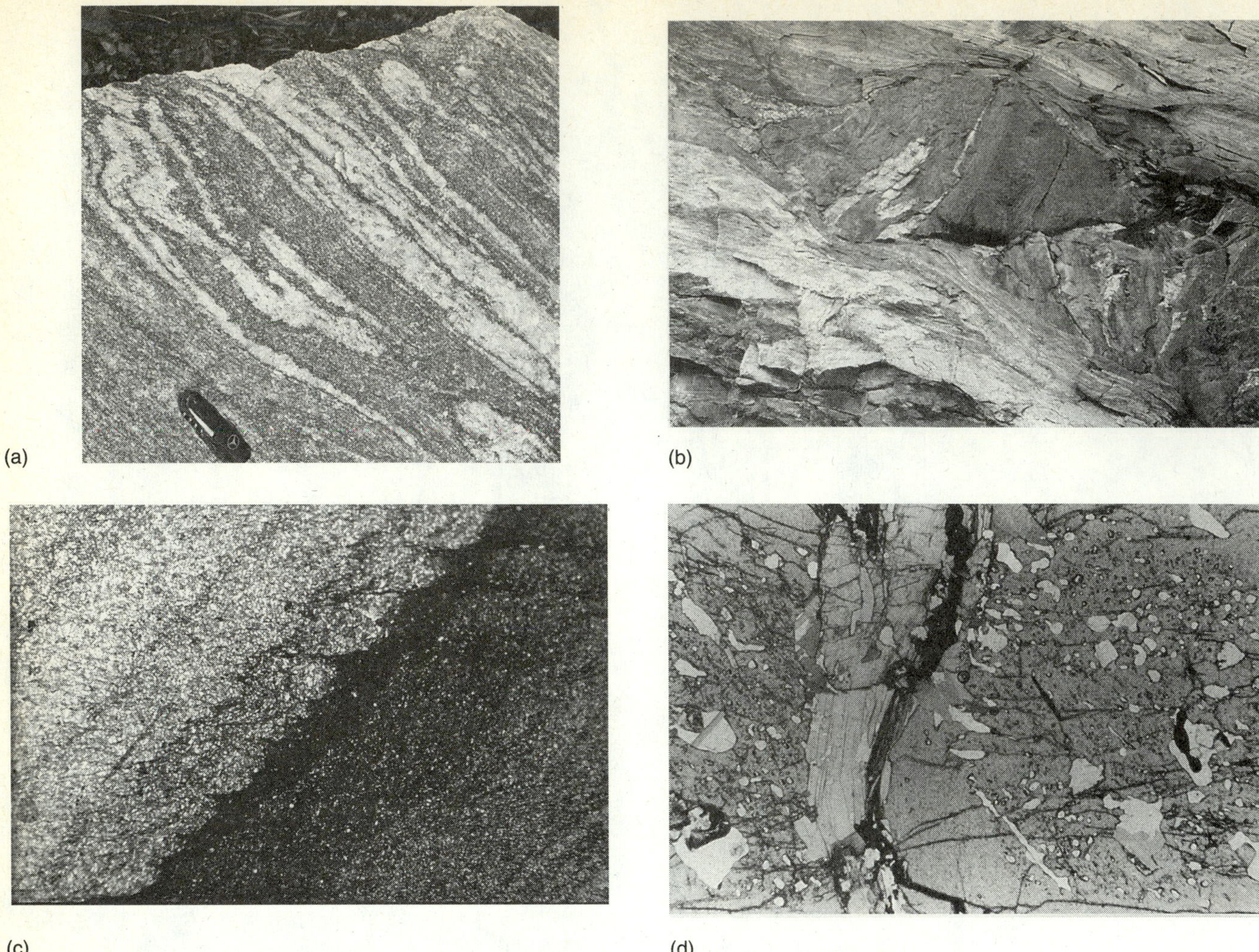

(a) (b) (c) (d)

FIGURE 17–34
(a) Isoclinal intrafolial folds in Middle Proterozoic Toxaway Gneiss, near Whitewater Falls on the North Carolina–South Carolina border. (b) Boudins of amphibolite preserving an earlier foliation truncated and wrapped by the foliation in the enclosing rocks, Shuswap metamorphic complex near Revelstoke, British Columbia. (c) Microlithons of pressure-solution cleavage (dark) preserving bedding (light and dark layers) between the cleavage planes in Upper Proterozoic(?) Wilhite Formation, near Tellico Plains, southeastern Tennessee. Note the cuspate nature of the contact between the light and dark layers. Plane light. Field is approximately 16 mm wide. (d) Garnet porphyroblast from Upper Proterozoic metasedimentary gneisses near Franklin, North Carolina, containing inclusion trails truncated by the enclosing foliation. Plane light. Field is approximately 7 mm wide. (RDH photos.)

the new foliation (Figure 17–32). S_0, the original bedding in the sequence of rocks, may be transposed parallel to S_1, and so the designation $S_0 = S_1$ may be appropriate. Foliation in rocks with no original layering, such as massive plutonic rocks, is designated S_1. A second and younger foliation, S_2, may overprint the older one, then transpose both S_0 and S_1 into subparallelism with the younger structure. The ability to recognize this overprinting relationship helps to unravel the history of complex areas and identify structures that indicate transposition. Othmar Tobisch and Scott Paterson (1988) described the process of multiple transposition during progressive deformation in orogenic belts. They recognized repeated transposition cycles and composite foliations as parts of this process.

Several indicators of transposition occur in rocks. Small to large isolated relict isoclinal folds lying within a foliation, ***intrafolial folds***, are common indicators (Figure 17–34a). They commonly take the form of isolated hinges where the limbs have been attenuated and sheared off through deformation. Boudins (Chapter 18) of resistant rocks, such as amphibolite or calc-silicate, may preserve earlier folds and foliations.

Microlithons may preserve earlier structures on the microscopic scale (Figure 17–34c). They may provide evidence of an earlier foliation in a deformed and metamorphosed rock mass. Additional indicators of earlier deformation are *inclusion trails in porphyroblasts* or other coarse crystals as seen in thin section (Figure 17–34d). Garnets are particularly useful in

ESSAY

Cleavage Formation and the Identification of Elephants

It was six men of Indostan
To learning much inclined,
Who went to see the Elephant
(though all of them were blind),
That each by observation
Might satisfy his mind.

. . .

And so these men of Indostan
Disputed loud and strong
Each in his own opinion
Exceeding stiff and strong
Though each was partly right
And all were in the wrong!

from John Godfrey Saxe,
The Blind Men and the
Elephant—A Hindoo Fable

For a century and a half, geologists have recognized cleavage as an important geologic structure, and they have debated its origin almost as long. The processes of cleavage formation were all known in the 1800s, and pressure solution was ignored for almost a century before being resurrected as a major cleavage-forming mechanism.

Why did we ignore the obvious for so long? Why have we not been able to resolve the question of its origin? The answer is not simple, and so—possibly—either is the solution. Consider the problem of the six blind men examining an elephant. Similarly, independent evidence can be found to favor each suggested mechanism of cleavage formation, but each cleavage examined may also contain evidence that supports another solution.

Carefully examine a rock that contains both slaty cleavage and bedding and also contains at least two rock types with contrasting properties—such as slate and sandstone or carbonate (Figures 17–1 and 17E–1). It is highly likely that you will find evidence for both pressure solution and also for recrystallization. What about rotation of individual minerals? Beutner (1978) argued convincingly that mechanical rotation of layer-silicate minerals is very difficult and that they, too, must be partially dissolved. Borradaile (1981), however, has argued that grain-boundary sliding may be an important factor in reorientation of minerals during cleavage formation. After a century and a half, therefore, we may have to accept two mechanisms, with pressure solution dominating at low temperature. This points again to the complexity of nature and our weak attempts to simplify and pigeonhole it. The dispute over the relationships of cleavage planes to the strain ellipsoid is largely resolved; most structural geologists today agree with Williams (1976) that cleavage does not form parallel to the *XY* plane of the finite strain ellipsoid, except under conditions of coaxial strain.

If your rock is also crenulated, you may find evidence favoring either a pure-shear or simple-shear mechanism. Crenulations and kinks may have a similar origin, but most geologists believe that crenulations develop by pure shear parallel to the *XY* plane (Gray, 1980). Still, some crenulations may form by simple shear (Hanmer, 1979).

The origin of the different cleavages is now partly understood. There probably is more evidence for continuous cleavage than Powell (1979) believed. There probably are some crenulations that originated by simple shear. There may be some rotation of layer-silicate grains during formation of slaty cleavage, although wholesale reorientation of most grains in a rock seems unlikely. Recent facts about

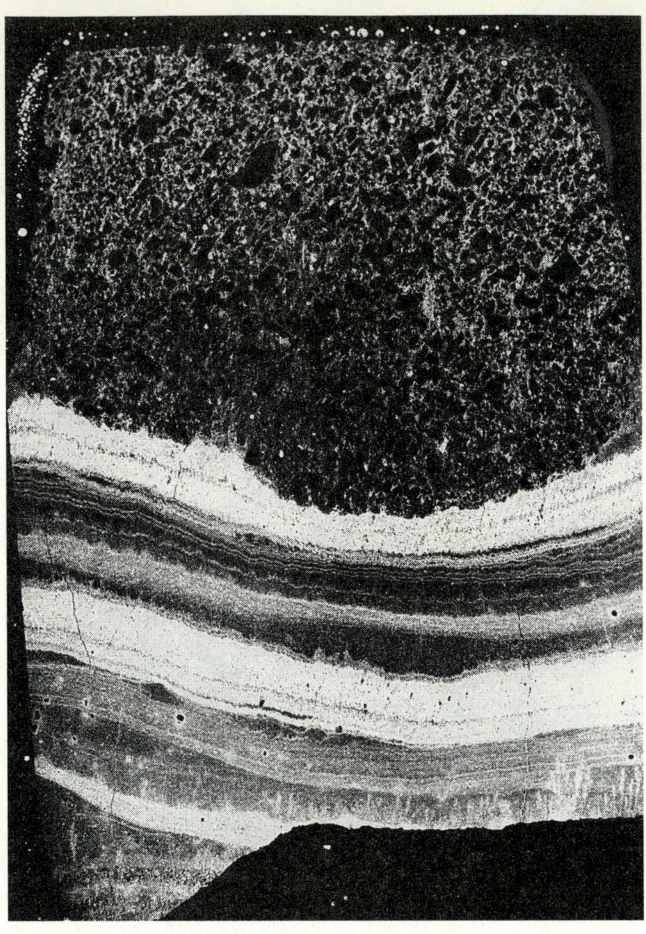

FIGURE 17E–1
Negative print of a contact between sandstone and slate in chlorite-grade Wilhite Formation (Upper Proterozoic?) near Bald River Falls, southeastern Tennessee. Plane light. Long axis of the thin section is 7 cm.

the chemistry of pressure solution and the nature of recrystallization processes have helped enormously in answering some of the questions, but no doubt the problem of cleavage formation will be debated for many decades. We may in time, however, learn enough to enable one geologist to see more than one part of the elephant.

References Cited

Beutner, E. C., 1978, Slaty cleavage and related strain in Martinsburg slate, Delaware Water Gap, New Jersey: American Journal of Science, v. 278, p. 1–23.

Borradaile, G. J., 1981, Particulate flow of rock and the formation of cleavage: Tectonophysics, v. 72, p. 305–321.

Gray, D. R., 1980, Geometry of crenulation-folds and their relationship to crenulation cleavage: Journal of Structural Geology, v. 2, p. 187–204.

Hanmer, S. K., 1979, The role of discrete heterogeneities and linear fabrics in the formation of crenulations: Journal of Structural Geology, v. 1, p. 81–91.

Powell, C. McA., 1979, A morphological classification of rock cleavage: Tectonophysics, v. 58, p. 21–34.

Williams, P. F., 1976, Relationships between axial-plane foliations and strain: Tectonophysics, v. 30, p. 181–196.

preserving inclusion trails that represent earlier foliations, but garnets may be rotated several times during the same deformation. The presence of inclusion trails at a high angle to the dominant foliation therefore may or may not identify transposed layering. Other minerals such as cordierite, staurolite, feldspar, and kyanite may contain inclusion trails that can be used similarly.

We will briefly return to a discussion of cleavage in Chapter 20 after discussing linear structures in Chapter 18. The role played by cleavage and foliations in the formation of certain types of linear structures and their importance in structural analysis should then become clear.

Questions

1. Several long-standing controversies concern the origin of slaty cleavage. Why?
2. How can you distinguish slaty cleavage from "fracture cleavage"?
3. What is cleavage refraction?
4. Explain how slaty cleavage that starts out as a pressure-solution cleavage could evolve into cleavage formed by recrystallization.
5. How can you estimate cleavage-related strain in a rock mass?
6. Purity or lack of purity of a rock affects its ability to exhibit strain when pressure solution forms slaty cleavage. How?
7. How has it been determined that the amount of strain in slate is more than 100 percent in some directions?
8. Why does pencil structure form as a precursor to cleavage?
9. How can you prove transposition has occurred in a rock mass?
10. Suggest two mechanisms for forming strongly discordant cleavage and fold axial orientation.
11. The drawing shows discontinuous exposures of cleaved rock. The heavy solid lines are bedding; light dashed lines are slaty cleavage. Photocopy this page and connect the bedding surfaces both above and below the ground surface to show the larger structure.

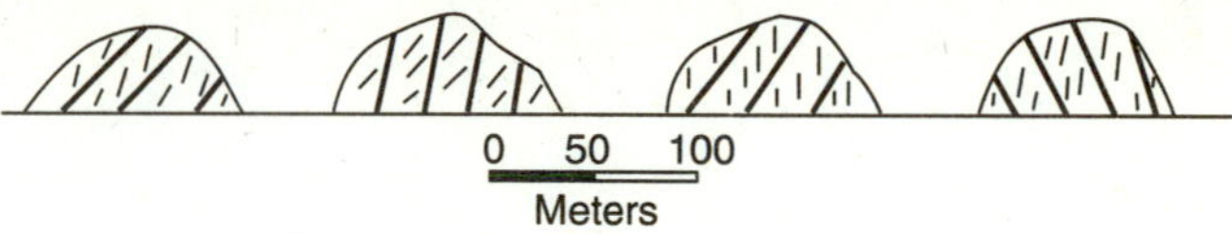

12. How can cleavage that formed parallel to the *XY* plane contain rotated porphyroblasts?

Further Reading

Borradaile, G. J., Bayly, M. B., and Powell, C. McA., eds., 1982, Atlas of deformation and metamorphic rock fabrics: New York, Springer-Verlag, 551 p.
Contains excellent photographs of most kinds of cleavage and foliations.

Gray, D. R., 1977a, Morphological classification of crenulation cleavage: Journal of Geology, v. 85, p. 229–235.
Classifies crenulation cleavage and pigeonholes the different types.

Johnson, T. E., 1991, Nomenclature and geometric classification of cleavage-transected folds: Journal of Structural Geology, v. 13, p. 261–274.
Provides a somewhat modified terminology for characterizing the relationships between the geometric elements of folds and cleavage in transected folds, and a geometric classification of cleavage-transected folds based on the angular relationships between profile transection and axial transection.

Platt, L. B., 1976, A Penrose Conference on cleavage in rocks: Geotimes, v. 21, p. 19–20.
Summarizes a research conference that considered some of the latest ideas on the origin and nature of cleavages. Several papers, such as Powell's (listed below), were later published.

Powell, C. McA., 1979, A morphological classification of rock cleavage: Tectonophysics, v. 58, p. 21–34.
Clarifies the descriptive classification of cleavage types.

Ramsay, J. G., and Huber, M. I., 1983, The techniques of modern structural geology, Volume 1: Strain analysis: New York, Academic Press, 307 p.
Contains useful discussions of cleavage/foliation development and summarizes Sorby's contributions to our knowledge of the nature of slaty cleavage.

Siddans, A. B., 1972, Slaty cleavage, a review of research since 1818: Earth Science Reviews, v. 8, p. 205–232.

Tobisch, O. T., and Paterson, S. R., 1988, Analysis and interpretation of progressive deformation: Journal of Structural Geology, v. 10, p. 745–754.
A useful summary of the sequence of formation of multiple foliations and transposition during progressive deformation in an orogen. The discussion of the development of composite foliations is particularly worthwhile.

Williams, P. F., 1976, Relationships between axial-plane foliations and strain: Tectonophysics, v. 30, p. 181–196.
Clarifies the evidence and theoretical basis for formation of cleavage whether parallel or nearly parallel to a principal plane of the strain ellipsoid.

Wood, D. S., 1974, Current views of the development of slaty cleavage: Annual Reviews of Earth and Planetary Sciences, v. 2, p. 369–501.
Each reviews fundamental concepts and ideas that led to our present state of knowledge. Reference lists are quite useful.

18

Linear Structures

Linear structures are important traces of movement and of the direction of tectonic transport. They are more useful than planes because movements along lines are restricted to two directions, whereas movements in planes can occur in all directions.

ERNST CLOOS, 1957, *Geological Society of America Memoir 18*

LINEAR STRUCTURES WERE FIRST DESCRIBED IN 1833, AND IN 1888 first appeared as symbols on a published geologic map. The map was made by a Norwegian geologist named Hans Reusch, who measured and plotted orientations of the long axes of deformed fossils in Karmøy, in southern Norway. As Reusch's map shows, linear structures are natural vectors: they indicate orientation of fold axes, sense and direction of movement along fault surfaces and, in a mass of rock, the direction of penetrative ductile flow. Interpreting them may be easy and straightforward, or difficult and controversial, or anything in between—depending on the kind of linear structure and its setting.

Any structure that can be expressed as a real or imaginary line is a ***linear structure***, or ***lineation*** (Figure 18–1). Hinge lines of folds (Chapters 14 through 16) are linear structures. Linear structures constitute one kind of ***fabric element***. A *lineament* is a topographic feature consisting of straight or aligned surficial features such as valleys and ridges. Many lineaments in aerial photographs and satellite imagery have been interpreted as having a tectonic origin, but on further examination have turned out to be nontectonic or even artificial. In this chapter, we will discuss only tectonic lineations, from mesoscopic to microscopic scale.

DEFINITIONS

Nonpenetrative Linear Structures

Like foliations, linear structures are either penetrative or nonpenetrative. The latter kind are confined to isolated surfaces in a rock mass. Probably the most common mesoscale nonpenetrative linear structure is ***slickenlines***. Slickenlines are the direct result of frictional sliding and flexural slip, processes described

FIGURE 18–1
Pulled-apart layer of calcsilicate quartzite more competent than the groundmass, producing boudinage in Upper Proterozoic Coleman River Formation micaceous metasandstone, Chunky Gal Mountain, North Carolina. Garnets formed in pressure shadows at ends of boudins, and the boudins have been rotated. The lineation here consists of the boudin axes or the boudin neck lines (both perpendicular to the photo). (RDH photo.)

in Chapters 8 and 16. The term ***slickensides*** refers to the entire movement surface. Slickensides may develop on the surfaces of faults, bedding, and foliations (Figure 18–2). Each involves relative movement parallel to the surface. As a result, the surface may be grooved and polished or may be covered with ***slickenfibers,*** fibrous crystals of calcite, quartz, chlorite, iron oxides, or other minerals where their long axes are oriented in the direction of movement. The slickenside surface is commonly stepped, with the down sides of the steps indicating the direction of motion (Figure 18–3). Means (1987), however, has described a very small-scale (less than a millimeter) slickenside striation consisting of a series of valleys and ridges that nest exactly (ridges in valleys) in those of the opposing block. M. S. Paterson (1958) showed experimentally that for a very small displacement (less than a few millimeters), slickenlines may indicate motion in either direction, but for large displacements, the slickenlines always step down in the relative direction of motion of the opposing block. If slickenlines occur on many layers in a flexural-slip folded sequence, it is penetrative on map scale and even on mesoscopic scale.

FIGURE 18–2
Slickenfibers of calcite on a movement surface in limestone, Romani Road near Rabat, Morocco. (RDH photo.)

At the appropriate scale, an isolated fold hinge is a nonpenetrative linear structure. Large folds, such as the Wills Mountain, Burning Springs, and Sequatchie anticlines, in the Appalachian Plateau from Pennsylvania to Alabama, are nonpenetrative folds. If the fold hinge belongs to a family of large and small folds, however, it is penetrative on map scale.

Penetrative Linear Structures

The most common linear structures are penetrative. They are usually penetrative because of their distribution on all scales throughout the rock mass; thus, they are also fabric elements. Most lineations are mesoscopic structures.

The line formed where two surfaces intersect is called an ***intersection lineation***—sets of such planes may yield a penetrative linear structure (Figure 18–4). The most common intersection involves bedding (S_0) and cleavage (S_1), sometimes designated by the shorthand notation L_{1x0} ("L one cross zero") to show that it is an intersection lineation involving bedding and the earliest cleavage. The convention is to denote lineations associated with a particular foliation or deformation as L_1, L_2, etc., the subscript indicating relative age. The intersection of slaty cleavage and bedding is a structure often measured in low-grade metamorphic rocks. Generally, the intersection of cleavage and bedding parallels a fold axis as a direct consequence of cleavage paralleling the fold axial surface.

Another intersection lineation is defined by two cleavages or other foliations. For example, L_{2x1} denotes the intersection of the earliest-formed foliation by the next earliest. The genetic relationship between the two may not be immediately obvious, but both cleavages may be related to folding, thereby producing a lineation parallel to a fold axis. The intersection of two cleavages or foliations produces a line that can be measured—either directly or by measuring the orienta-

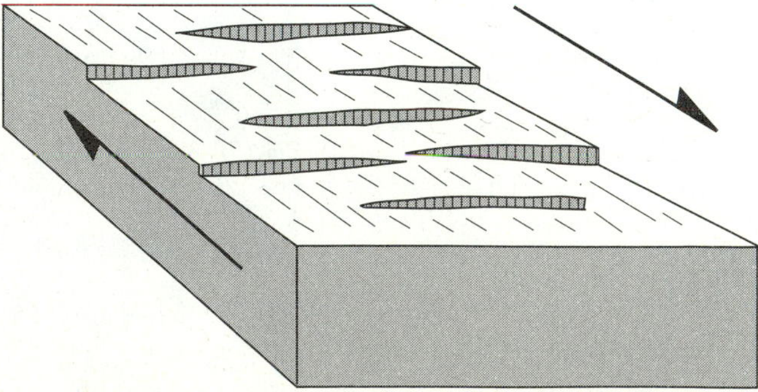

FIGURE 18–3
Formation of slickensides by relative movement. Steps are formed facing the direction of movement (arrow) of the opposite wall. Lines form in the direction of movement.

tions of the two cleavages and plotting them on an equal-area net— the orientation of the intersection can be determined by directly reading the trend and plunge of the intersection from the diagram. This way intersections may be related to larger structures.

Mineral lineations consist of aligned elongate mineral grains and grain aggregates. Amphiboles, micas, feldspars, and even quartz are among minerals that may be aligned (Figure 18–5). ***Pressure shadows*** of quartz, muscovite, chlorite, magnetite, or other minerals on either side of a single crystal of pyrite, garnet, or other, may yield an axis that constitutes a mineral lineation (Figure 18–6). Minerals grow in pressure shadows because the pressure near rigid grains is lower. The ductile matrix deforms around the grain.

Hidemi Tanaka (1992) has identified a stretching lineation along the Akaishi tectonic line in central Japan that appears to result from homogeneous laminar

(a)

(b)

FIGURE 18–4
Intersection lineations. (a) Cleavage-bedding intersection in Upper Proterozoic Kaza Group phyllite, Cariboo Mountains, British Columbia. (b) Intersection of two foliations in Ordovician(?) Mineral Bluff Formation phyllite at Murphy, North Carolina. View looks down on the earliest foliation that was cut by a younger domainal crenulation. Note that the prominent domainal crenulation transects the longer-wavelength folds. (RDH photos.)

FIGURE 18–5
Mineral lineation of quartz, feldspar, and micas in multiply deformed Middle Proterozoic Cranberry Gneiss from near Boone, North Carolina. (Photo by D. G. McClanahan.)

flow of fine-grained cataclastic material. He called this structure a *cataclastic lineation* because it occurs in a shallow zone of brittle deformation. Despite the brittle nature of the deformation, this zone also contains a foliation, aligned porphyroclasts and grains, smeared and aligned aggregates of clays, and other fine-grained minerals analogous to many of the linear structures that appear in more deep-seated ductile shear zones. Polycrystalline quartz porphyroclasts have also been fragmented to produce a lineation.

The axes of rotation of ***rotated minerals*** may compose a mineral lineation, although not a true lineation in the same sense as the others described here. Garnets frequently contain a "snowball" structure or an S-shaped alignment of inclusion trails, which indicates that simple shear and rotation occurred during garnet growth (Figure 18–7). The lineation is the rotation axis.

Some mineral lineations are deformed into elongate ***rods***—the group noun is ***rodding***—or grain aggregates of one or more minerals, such as quartz, feldspars, and micas (Figure 18–8). They also occur at intersections of two foliation planes, thereby obscuring their origin. Rodding of minerals occurs in many places, but it is common in ductile shear zones (Chapter 10). James McLelland (1984) recognized a mineral lineation in the Grenville rocks of the southern

FIGURE 18–6
Pressure shadows of muscovite and chlorite on pyrite crystals in metasiltstone of the Wilhite Formation (Late Proterozoic?) near Servilla, southeastern Tennessee. The lineation passes through the center of the pyrite crystal and is perpendicular to the photo. Plane light. Width of field is approximately 1 mm. (Specimen courtesy of Mark W. Carter, University of Tennessee–Knoxville.)

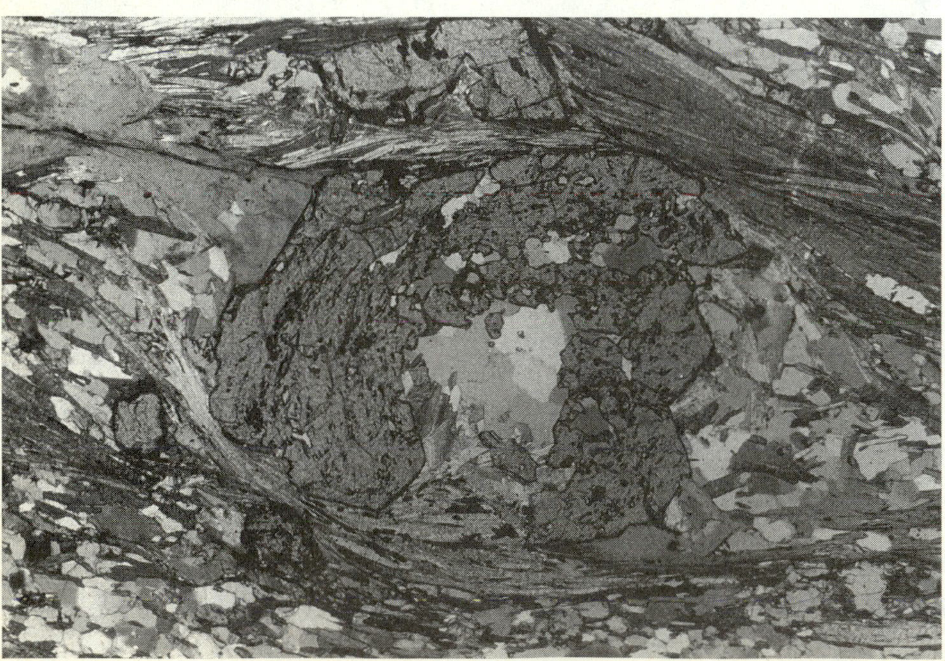

FIGURE 18–7
Garnet in metasiltstone of lower Paleozoic(?) Poor Mountain Formation near Salem, South Carolina, containing concentric inclusion trails in a porphyroblast that has undergone sinistral rotation after formation (as a σ porphyroclast)—indicated by pressure shadows of micas and quartz. Plane light. Garnet is approximately 2.5 mm in diameter. (RDH photo.)

Adirondacks in New York that he called a "ribbon lineation." The lineation is composed of quartz, feldspar, and mafic minerals, with one mineral composing most of the linear structure. McLelland concluded that the lineation formed parallel to the *X* axis of the strain ellipsoid and is parallel to early folds, probably sheath fold axes.

Long axes of objects that are deformed into ***natural strain ellipsoids*** make up another category of lineations. The long axes of pebbles, boulders, vesicles and amygdules, reduction spots, oöids, and pisolites all become natural strain ellipsoids and form measurable lineations during deformation, depending on strain state (Figure 18–9). They were discussed in Chapter 7 as finite-strain markers, where we noted that most were not originally spherical and thus cannot be ideal strain ellipsoids. They do, however, constitute a linear structure, making it possible to measure the approximate orientation of the principal axes of ellipsoids. The lineation to be measured parallels the *X* axis of the strain ellipsoid.

Mullions form at boundaries between differing rock types and consist of corrugated or scalloped surfaces (Figure 18–10). They result from contrast in competence or ductility from layer to layer. The smooth convex sides of a mullion are directed toward

FIGURE 18–8
Rodding of a quartz-feldspar layer in migmatitic Upper Proterozoic Tallulah Falls Formation metagraywacke at Woodall Shoals, South Carolina–Georgia. (RDH photo.)

FIGURE 18–9
Natural strain ellipsoids lineation—deformed quartz pebbles—in Upper Proterozoic Bygdin conglomerate, Bygdin, Norway. Note that the length-to-maximum-width ratios of the pebbles are 10:1 or greater. (RDH photo.)

(a)

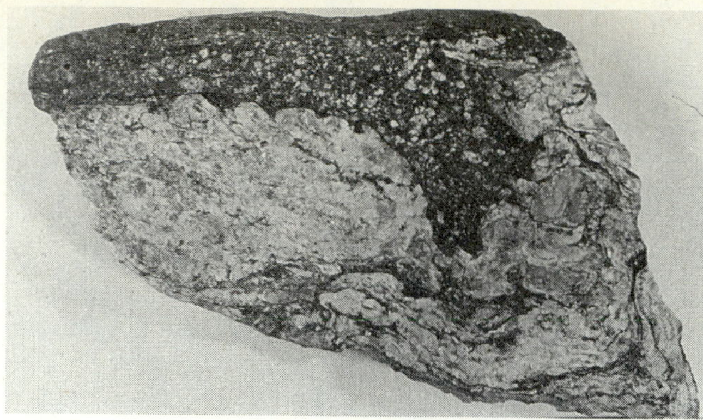

(b)

(c)

FIGURE 18–10
Mullions. (a) Cleavage-bedding–related mullions on a corrugated bedding surface in Cambrian shale near Bouznika, Morocco. Mullions are formed by flattening perpendicular to cleavage (perpendicular to the long axes of mullions) and parallel to bedding. (RDH photo.) (b) Ductility (viscosity) contrast-related mullions without cleavage along a quartz-feldspar (light-colored)-mica-quartz (dark) boundary in mylonite from the Towaliga fault zone from near Indian Springs, Georgia. Specimen is approximately 10 cm long. (Specimen courtesy of Robert J. Hooper, Conoco Research.) (c) Ductility contrast-related mullions along a boundary between amphibolite (dark) and marble (light) in lower Paleozoic(?) Gaffney Marble near Blacksburg, South Carolina. (Photo by D. G. McClanahan.)

the layer with the lower viscosity; the cusps, toward the layer with the higher viscosity. A structure such as slaty cleavage may be well developed in one layer even though the layer above contains a poorly defined and widely spaced cleavage. Mullions form a corrugated boundary between layers. They may also form in uncleaved rocks by ductile deformation at the boundary of layers in which competence differs.

Boudinage consists of lenticular segments of a layer that has been pulled apart and flattened in such a way that the layer is segmented (Figure 18–11). The layer being segmented is less ductile than the enclosing material, and the degree of contrast in competence affects the shape of boudins: large contrast produces boudins with sharp edges; small contrast, boudins with rounded edges. Boudins can develop under conditions either ductile or brittle. Under brittle conditions, most boudins are angular, and the space between them is filled with less-competent rock or fibrous minerals. ***Ordinary boudinage*** consists of segmented, sausage-shaped pieces of a single layer in which the lenticular segments parallel one another. It results from extension of the layer in a single direction. If layer-parallel extension has occurred in two directions, the resulting boudinage consists of a series of three-dimensional blocks called ***chocolate-block*** (or ***chocolate tablet***) ***boudinage***. Boudins may be the most useful of all lineations because they yield information about strain, shear sense, and differences in competence. The "neck line" of the boudin is the lineation, and is commonly oriented parallel to fold axes (Figures 18–11 and 18–12).

If flattening is accompanied by simple shear after the blocks have formed, or if the extended layer is not perpendicular to the direction of shortening (Figure 18–12), boudins commonly become rotated (see Figure 10–20b). Boudinage may be related to the finite and incremental strain ellipsoids, because it forms parallel to the *XY* plane, normal to *Z*. Boudinage is frequently seen in the limbs of folds, where most flattening and layer-parallel extension occurs.

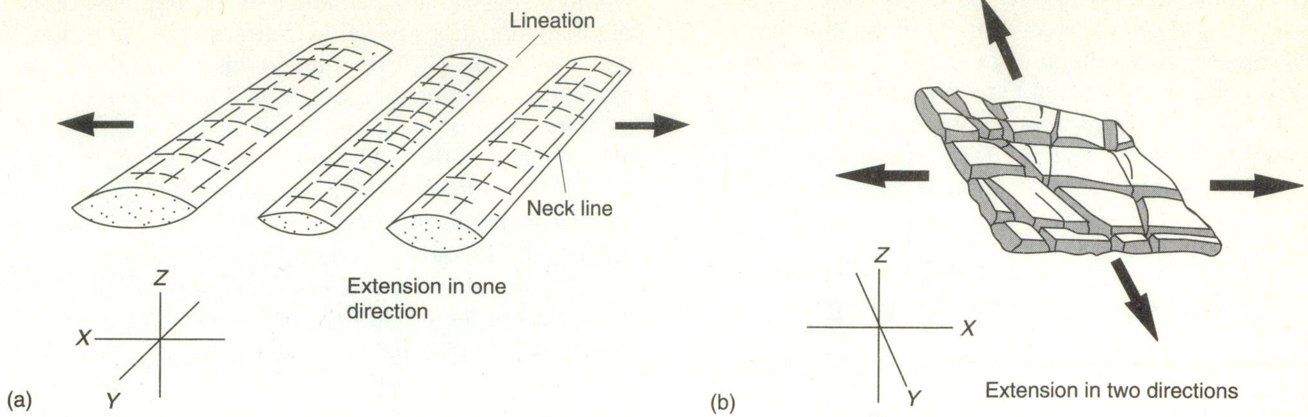

FIGURE 18–11
Boudinage. (a) Ordinary boudinage involving extension in one direction, producing plane strain ($X > Y = 1 > Z$) and segmentation of a layer. (b) Chocolate-block boudinage involving extension in two directions yielding $X > Y > 1 > Z$, or flattening with $X = Y > 1 > Z$.

FIGURE 18–12
Boudinage of competent sandy layers in less-competent limestone in the Berge limestone near Ange, Sweden. (Grooves and striae were produced by Quaternary glaciation.) (RDH photo.)

LINEATIONS AS SHEAR-SENSE INDICATORS

Several of the lineations just described may be used as shear-sense indicators. Slickensides directly indicate movement sense by the direction of their lines and steps. Boudins indicate the extension direction. Mineral lineations yield sense of shear if the linear mineral (e.g., a feldspar) is segmented in the movement direction. Rotated minerals are shear-sense indicators—the direction of movement is perpendicular to the lineation (rotation axis).

FOLDS AND LINEATIONS

Lineations are directly related to folds and can be helpful in deciphering a folded area (Figure 18–13). A close relationship can frequently be proved between fold hinges and intersection lineations, and between poles to axial surfaces and poles to slaty cleavage (Figure 18–13b). Intersection lineations tend to parallel fold axes because of the nature of the lineation—the intersection of an axial-planar cleavage (or foliation) with bedding or earlier compositional layering. Mullions and boudin necks generally parallel the fold axes.

Intersections, mullions, and boudins all carry information about fold orientation. Mineral-elongation lineations sometimes parallel fold axes and are sometimes oriented oblique to normal to fold axes.

Flexural-slip folds frequently produce slickenlines on bedding surfaces (Figure 18–2). These lineations are commonly oriented perpendicular to the fold axis.

DEFORMED LINEATIONS

After they are formed, lineations become strain markers that can be used to decipher later deformation. This deformation yields results both surprising and revealing. We might think that folding of linear structures would be easy to analyze and explain, but most plots of folded lineations in fabric diagrams yield complex patterns (Figure 18–15). If lineations are gently folded by flexural-slip buckling or bending so that the fold axial trend is normal to the lineations, they will plot on a great circle of the *equal-angle net.* A lineation *not* normal to the fold axis will yield a small circle on the equal-area net for flexural-slip folding (with horizontal hinges). If the fold axis is not perpendicular to the trend of the lineation and a different fold mechanism is involved, the deformed lines plot as small circles, spirals, or complex patterns that are difficult to interpret. Mineral lineations should be the only ones used for this kind of analysis because a cleavage superposed on a surface already folded will produce an apparently folded intersection lineation—one that is actually *not*

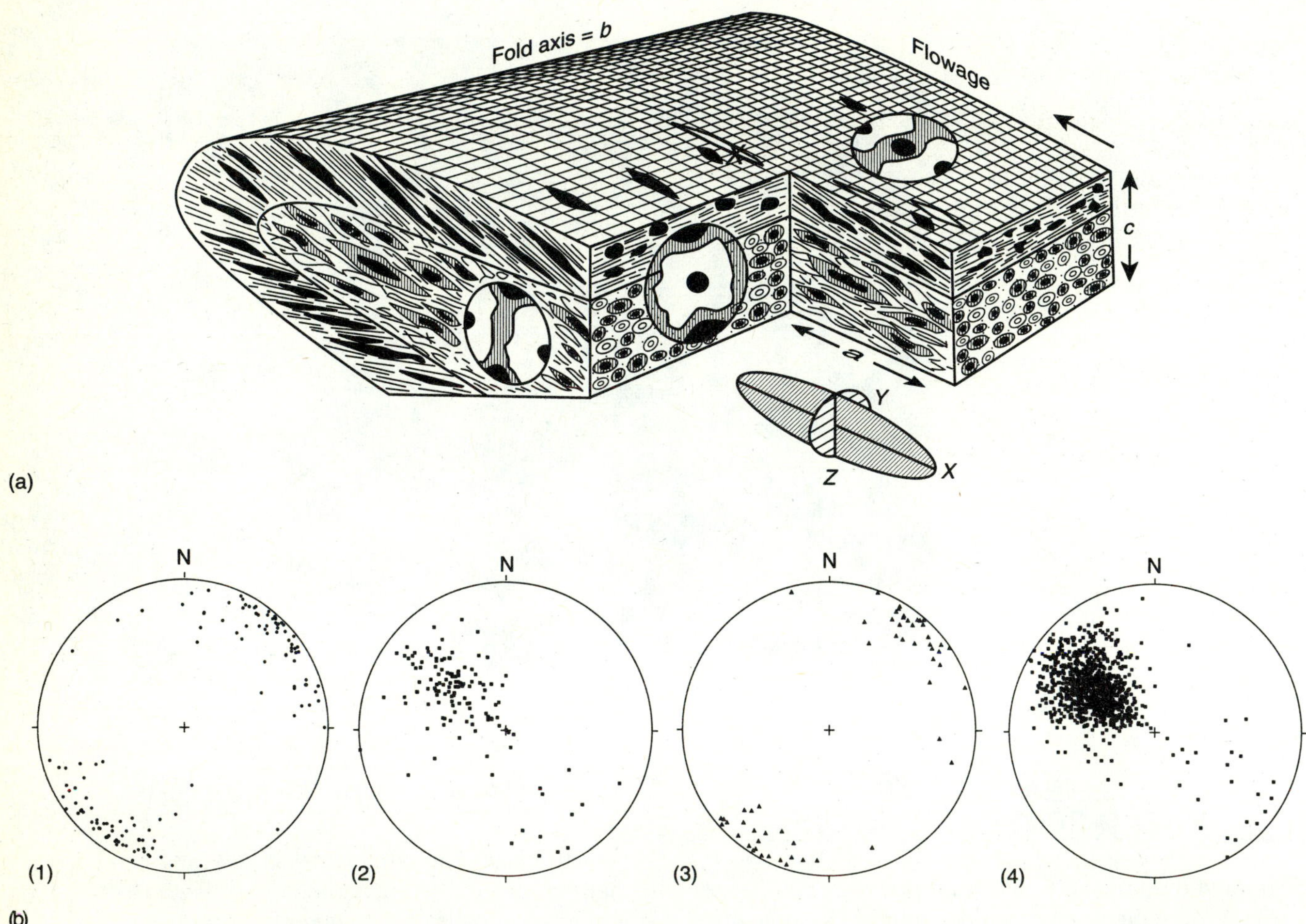

FIGURE 18–13
(a) Relationships between a fold, intersection, oöid, and mineral lineations, and preferred orientations parallel to different planes in the strain ellipsoid. Fabric diagrams in different parts of the fold indicate orientation of dominant structures. (After Ernst Cloos, 1957, Geological Society of America *Memoir* 18.) (b) Relationships between orientations of 140 fold axes (F_1) (1) and poles to axial surfaces (2), and 63 intersection lineations (L_{1x0}) (3) and 980 poles to slaty cleavage (S_1) (4) in low-grade rocks of the Blue Ridge Foothills of southeastern Tennessee. (Unpublished data courtesy of Mark W. Carter, Steven L. Martin, and Donald J. Geddes, University of Tennessee–Knoxville, and were plotted by Mark Carter.) Note the relationships between poles to axial surfaces and slaty cleavage, and fold axes and intersection lineations.

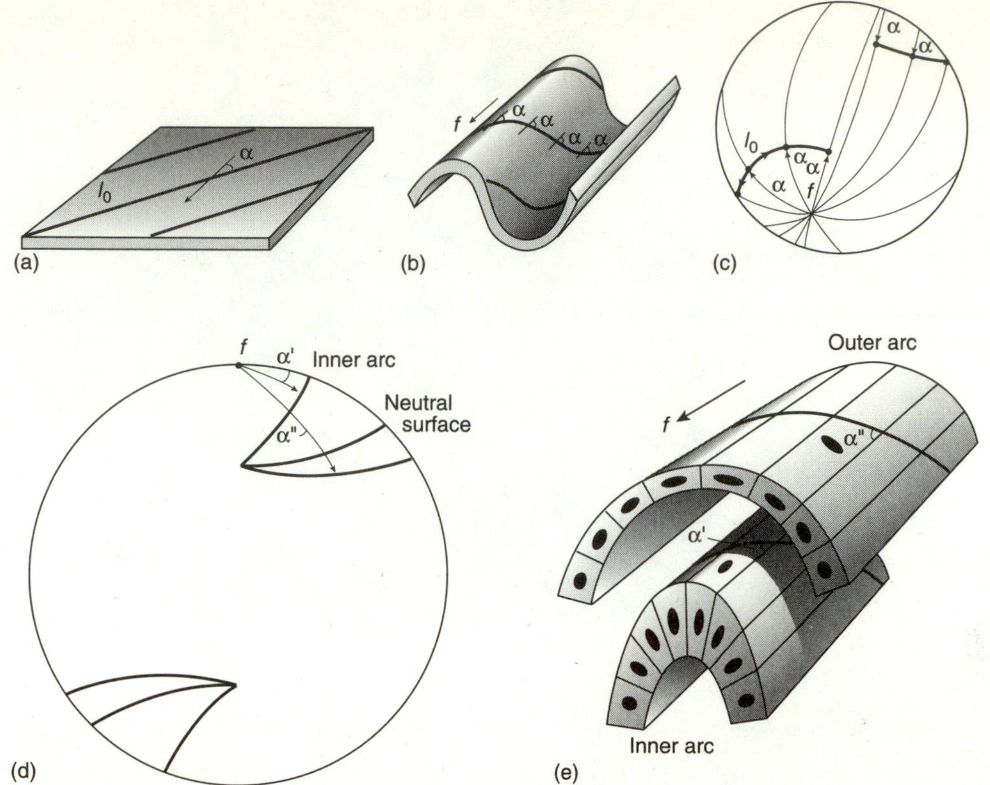

FIGURE 18–14
Deformed linear structures showing fabric diagrams and different potential orientations of lineations and fold axes. Folding about the fold axis *f* of layers (a) containing l_0 by flexural slip (b) results in l_0 plotting as the heavy line in the fabric diagram (c). α represents successive measurements of the angle between the lineation and fold axis in (b) plotted as a small circle in (c); (d) is a fabric diagram showing a plot of lineations folded by tangential longitudinal strain in (e). (From J. G. Ramsay, 1967, *Folding and Fracturing of Rocks:* New York: McGraw-Hill. ©1967. Reproduced with permission.)

deformed. A cleavage superposed on ripple marks would likewise produce a lineation that is only apparently deformed.

The reasons for this apparent complexity of folded lineations are many. One is that standard plotting nets cannot accommodate features that have been deformed into noncircular patterns. Most folded lineations are deformed into helixes, parabolas, or hyperbolas that may not be immediately evident in the field. Another is that different fold mechanisms deform lines differently. Small-circle patterns suggest a flexural-slip mechanism; great-circle patterns indicate passive flow. Whether or not the folds are tight or open is also significant. John Ramsay (1967) and Ramsay and Huber (1987) have discussed the geometry of deformed lineations in greater detail.

INTERPRETATION OF LINEAR STRUCTURES

Linear structures may help to reveal the deformation sequence of an area. Consequently, measuring and interpreting linear structures and associating them with other structures in the field is extremely important in deciphering their history. Generally, it is not difficult to recognize mineral lineations, deformed pebbles, rods, boudins, mullions, and cleavage-bedding intersections. Interpreting most lineations is more difficult. In the past, some geologists have assumed that mineral alignment indicated direction of tectonic transport. It is true that slip parallel to quartz *c* axes frequently parallel the quartz mineral lineations in deformed rocks (Figure 18–15), particularly in large fault zones (McIntyre, 1950; Christie, 1963, 1964; M. R. W. Johnson, 1960, 1964; Reed and Bryant, 1964). Thus, many structural geologists assume that all mineral lineations yield transport direction, but field studies indicate a strong parallelism between quartz mineral elongation lineations and fold axes—thus a paradox. The quartz mineral lineation appears to be both an intersection and an elongation lineation (parallel to the *X* axis of the finite strain ellipsoid). In some cases, where independent evidence exists (such as boudins, rotated porphyroclasts, and other shear-sense indicators), quartz mineral lineations *do* indicate transport direction; in others, they do not. Many boudin and mullion axes parallel fold axes, indicating that a direct relationship exists between folding and formation of boudinage and mullions. Rodding

FIGURE 18–15
Parallelism of quartz-feldspar mineral lineation and fold axes in feldspathic quartzite of the Sarv nappe along Bøfjorden at Staknes Point Lighthouse, Birlandet, south-central Norway. (RDH photo.)

(and McLelland's ribbon lineation) may result from stretching parallel to a major transport direction, especially in ductile shear zones, but rods also frequently parallel fold axes, accompanying formation of sheath folds, which *do* involve stretching parallel to their axes. The difficulty in interpreting mineral lineations, rods, and other linear structures may indicate that they developed by small-scale (infinitesimal) stretching parallel to fold axes during flattening or by simple shear directed perpendicular to fold axes. Another possibility is that mineral lineations developed parallel both to fold axes and to transport directions in zones of heterogeneous simple shear where sheath folds were forming, rotating fold axes into the transport direction. In many places, fold axes and mineral lineations (mostly quartz or micas) are parallel, and no evidence has been found for reorientation of fold axes by any mechanism. Many lineations parallel fold axes; others form high angles to fold axes. Curved fold hinges present another problem in interpretation. Interpreting those mineral lineations parallel to fold axes as elongation (stretching) lineations is difficult without independent evidence of movement sense.

Now that we have examined each major type of geologic structure, we can proceed to Chapters 19 and 20. There we will learn more about the deformation processes and tectonic effects of pluton emplacement and how to decipher combinations of structures in regions that vary in structural complexity.

ESSAY

Pitfalls in Interpreting Linear Structures

Lineations have been studied for many years, and geologists routinely measure them during field study. Many lineations are associated with large fault zones, and their interpretation has provoked controversy. Two examples involve lineations as movement indicators: the Brevard fault zone in the southern Appalachians, studied by Jack Reed and Bruce Bryant (1964); and the Moine thrust zone in the Scottish Highlands, studied by John Christie (1963). In the Brevard fault zone (Figure 18E–1), Reed and Bryant interpreted systematic changes in orientation of lineations as indicating shear sense and strike-slip movement and classified them as "a" lineations (indicating direction of movement). The shallow-plunging—northeast-to-southwest—mineral lineations (mostly quartz) were taken to indicate strike slip. Reed and Bryant also identified lineations trending northwest in the Blue Ridge northwest of the fault zone and interpreted them as a change of orientation of the

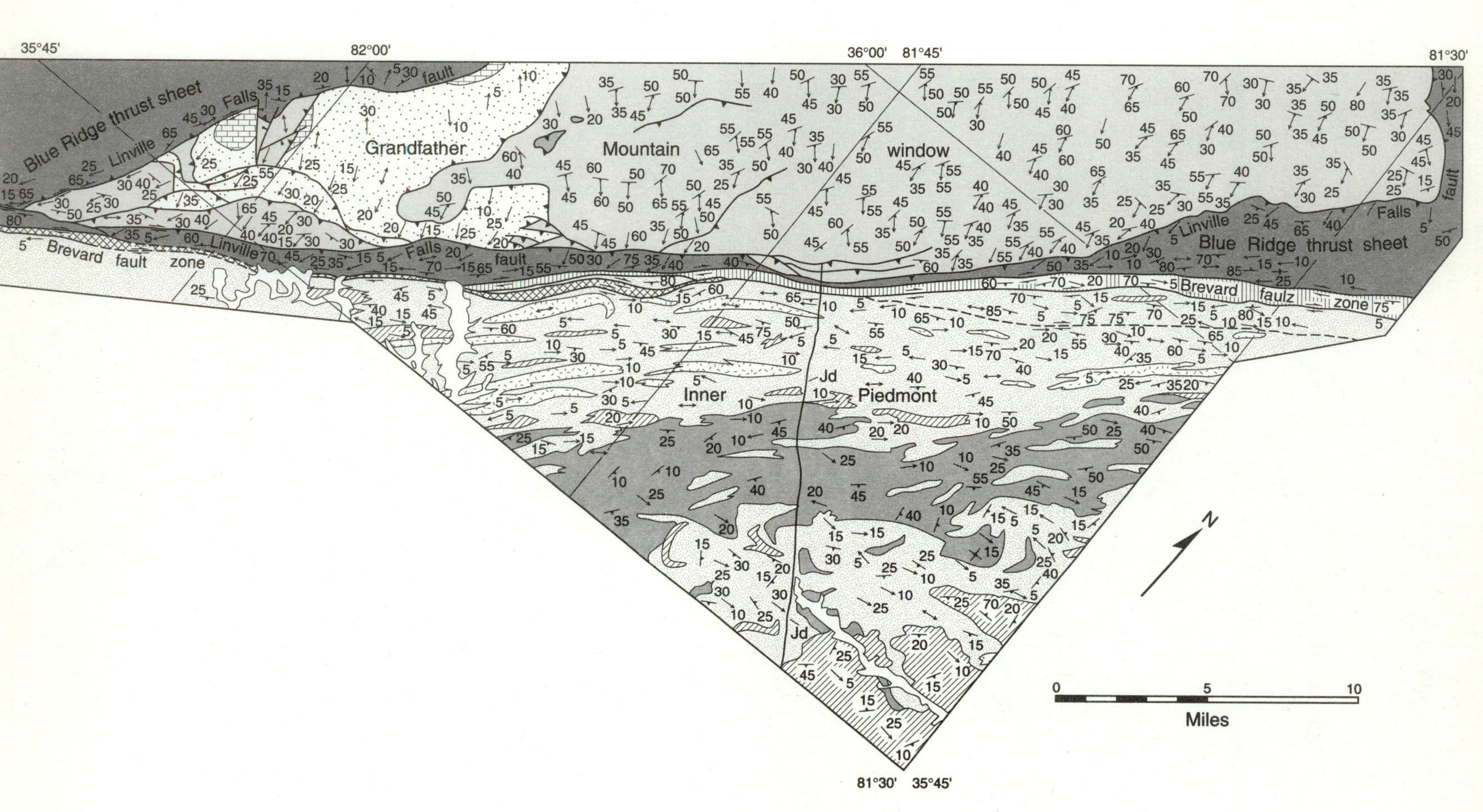

FIGURE 18E–1
Map showing trend and plunge of lineations (arrows) in the Inner Piedmont (light and medium gray screens, diagonal line, and crosshatch patterns), Brevard fault zone (vertical-line and cross-line patterns), Blue Ridge (dark gray screen pattern), and Grandfather Mountain window (medium-light gray screen, stipple, and limestone patterns), North Carolina. Faults are indicated by heavy toothed lines and heavy dashed lines. Heavy northwest-trending solid line in southeast part of map is a Jurassic diabase dike (Jd) that cuts all earlier structures. (From J. C. Reed, Jr., and Bruce Bryant, 1964, Geological Society of America *Bulletin,* v. 75, p. 1177–1195.)

same lineations into a northwest transport direction. The northwest-trending lineations in the Blue Ridge, however, parallel fold axes and are not obvious mineral-elongation lineations. They probably result from intersection of two foliations. Independent evidence for dip slip along the fault zone in South Carolina, based on geologic map patterns, critical rock units, and parallelism of cylindrical folds with the strong northeast-trending mineral lineation (Hatcher, 1971, 1978b), cast doubt on Reed and Bryant's interpretation of strike-slip. Thereafter, the Brevard fault zone was presumed to be a large ductile dip-slip fault until Andy Bobyarchick (1984) used S-C fabrics to indicate a major component of dextral motion. Subsequently, Steven Edelman, Angang Liu, and Hatcher (1987) have suggested that both dip slip and strike slip occurred in the Brevard fault zone and that the strong quartz lineation is associated with the strike-slip events—as Reed and Bryant (1964) originally concluded. As we now understand it, the Brevard fault zone underwent early ductile dip slip, then ductile strike slip, and finally brittle dip slip.

A similar problem existed in the Moine thrust zone. First, H. H. Read (1931) interpreted some of the linear structures as slickenlines, indicating a component of strike-slip motion along the fault. Then John Christie (1963) concluded that the lineations were parallel to fold axes ("b" lineations) and the transport direction 90° from that in Read's interpretation. Later still, Michael Johnson (1964) questioned Christie's interpretation of lineations and the basis for his conclusions.

The debate about interpretation of linear structures continues. Shofa Lin and Paul Williams (1992) have pointed out that assuming the mineral stretching lineations in a shear zone is the transport direction. They noted that the true movement direction is parallel to the lineation (a striation) that lies within the C–surface, or the orthogonal projection of the stretching lineation onto the shear zone boundary. This technique works only if the structures being measured in the shear zone and the boundaries formed during the same event. If the shear zone is the product of more than one event and the boundaries formed during an earlier event, care must be exercised in using this relationship. So, despite a wealth of experimental and field data, and computers to speed calculations and aid modeling, many difficulties in interpreting lineations remain. It is always possible to misidentify and misinterpret a structure, and so use of all available data is necessary.

References Cited

Bobyarchick, A. R., 1984, A late Paleozoic component of strike slip in the Brevard zone, southern Appalachians: Geological Society of America Abstracts with Programs, v. 16, p. 126.

Christie, J. M., 1963, The Moine thrust zone in the Assynt region, Northwest Scotland: University of California Publications in the Geological Sciences, v. 40, p. 335–440.

Edelman, S. H., Liu, A., and Hatcher, R. D., Jr., 1987, The Brevard zone in South Carolina and adjacent areas: An Alleghanian orogen-scale dextral shear zone reactivated as a thrust fault: Journal of Geology, v. 95, p. 793–806.

Hatcher, R. D., Jr., 1971, Structural, petrologic, and stratigraphic evidence favoring a thrust solution to the Brevard problem: American Journal of Science, v. 270, p. 177–202.

Hatcher, R. D., Jr., 1978, Tectonics of the western Piedmont and Blue Ridge, southern Appalachians: Review and speculation: American Journal of Science, v. 278, p. 276–304.

Johnson, M. R. W., 1964, Discussion: Journal of Geology, v. 72, p. 672–676.

Lin, S., and Williams, P. F., 1992, The geometrical relationship between stretching lineation and the movement direction of shear zones: Journal of Structural Geology, v.14, p. 491–497.

Read, H. H., 1931, Geology of central Sutherland: Geological Survey of Great Britain Memoir, 238 p.

Reed, J. C., Jr., and Bryant, B. 1964, Evidence for strike-slip faulting along the Brevard fault zone in North Carolina: Geological Society of America Bulletin, v. 75, p. 1177–1195.

Questions

1. How does a lineation differ from a lineament?
2. What are intersection lineations? How do they originate?
3. Under what conditions can slickenlines be penetrative structures?
4. If all intersections, mineral lineations, and boudin axes are parallel to each other and to the dominant fold trend in an area, what can you say about their origin?
5. How do boudins form? Mullions?
6. Why do pressure shadows form? Could they form in the same stress/strain environment as boudins? If so, how?
7. The data in the following table were collected from a set of folds (axial orientation 024, 3° NE, axial surface 034, 62° NW) that folded a mineral lineation in the Ordovician Walloomsac phyllite just east of the Taconic allochthons near Hoosick Falls, New York (Figures 15–15c and 17–12a). Plot these data on an equal-area net and comment on the nature of the folding process, the effects on the orientation of the lineation and the fold axes, and the pattern observed in the equal-area plot.

119 29°SE	106 36°SE	102 36°SE	288 54°NW	292 52°NW	108 39°SE	144 54°SE
112 35°SE	136 63°SE	038 57°NE	138 44°SE	156 46°SE	146 36°SE	131 61°SE
134 49°SE	123 54°SE	303 69°NW	111 34°SE	127 46°SE	298 72°NW	

8. Identify the type(s) of lineation(s) in the fold below.

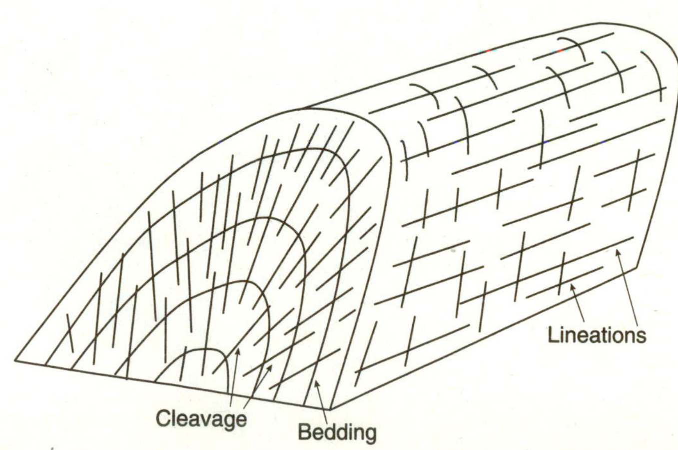

9. Sketch in the strain axes in the rectangular prisms, which represent boudinage. Was the contrast of competence during deformation large or small?

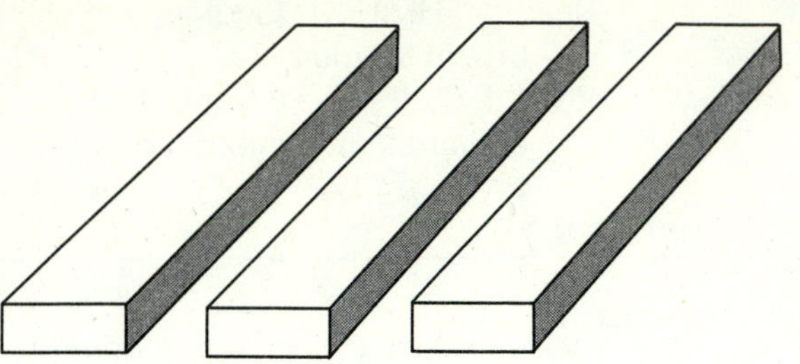

Further Reading

Cloos, E., 1957, Lineation: A critical review and annotated bibliography: Geological Society of America Memoir 18, 136 p.
Comprehensively reviews lineations and discusses various kinds and interpretations of linear structures.

McLelland, J. M., 1984, The origin of ribbon lineation within the southern Adirondacks, U.S.A.: Journal of Structural Geology, v. 6, p. 147–157.
A useful study of lineations in some highly deformed rocks where it is possible to demonstrate the parallelism between lineation and early fold hinges.

Wilson, G., and Cosgrove, J. W., 1982, Introduction to small-scale geological structures: London, Allen & Unwin, 128 p.
A readable compendium of mesoscopic structures, considered singly and in combination. Several sections discuss linear structures.

19

Tectonic Structures in Plutons

Whoever begins to study the structural geology of an igneous mass should be aware that structural problems require close attention in the field. . . . . Geologists whose time in the field is occupied by petrographic studies, or mapping, and are obliged to cover large areas in a short time, will find difficulty in solving structural problems. There is hardly any area where the structural geology is so simple that a reconnaissance survey, or petrological mapping, would furnish enough pertinent data to cover the special problems of the structure; and it may be said without exaggeration that the slowest field work will, in many instances, prove to be the quickest.

ROBERT BALK, 1937, *Geological Society of America Memoir 5*

WE DISCUSSED PRIMARY IGNEOUS STRUCTURES IN CHAPTER 2 with the assumption that they form during emplacement and crystallization of magma without any accompanying tectonic effects. As we shall see, magma is almost never emplaced into an unstressed and benign crust, and the structural geologist should heed the advice of Robert Balk during the investigation of any igneous body. The crust in the environments surrounding intrusions of magma is commonly in a state of compression within and adjacent to subduction zones or is in a state of tension in back-arc regions, in the near-surface parts of magmatic arcs, or in rift environments where tension dominates (Figure 19–1).

The melting process involves several phenomena that can create differences in pressure related solely to the properties of the mineral assemblage in the rock

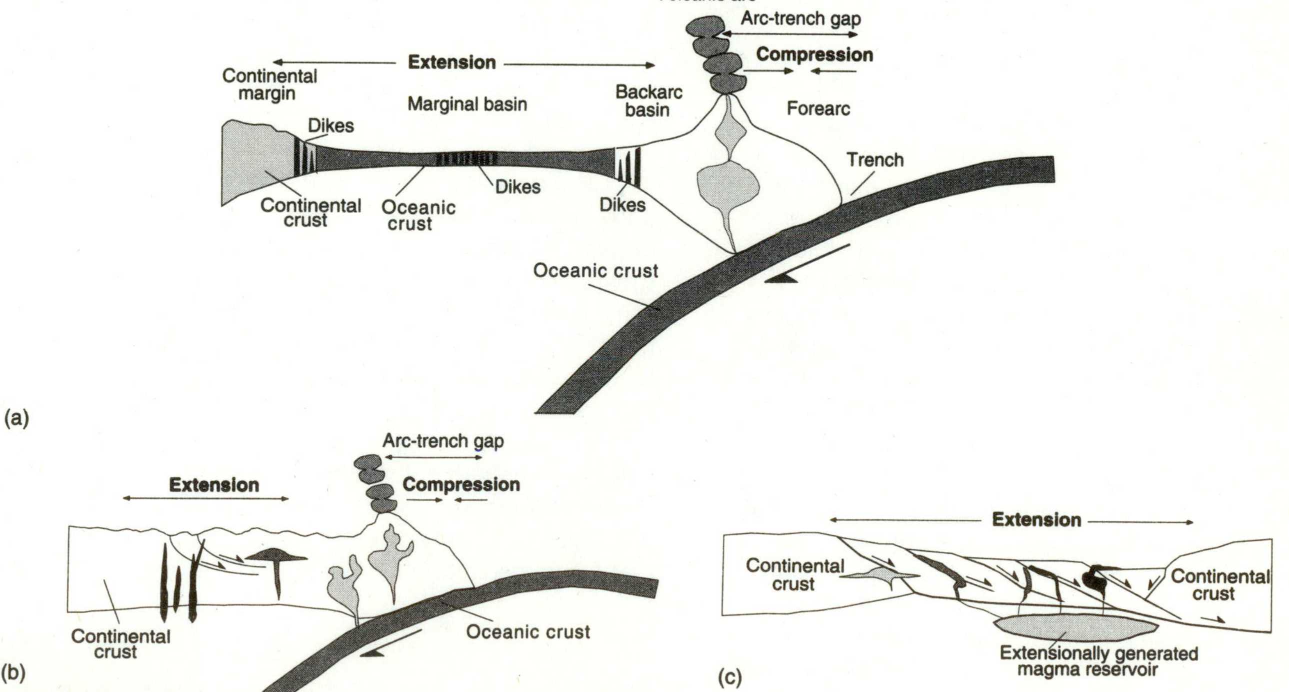

FIGURE 19–1
Several environments for emplacement of plutons. (a) Subduction and island-arc formation involving both compression in the forearc and arc regions, and tension in the backarc region and marginal basin. (b) Similar environments develop by subduction beneath a continental margin. (c) Asymmetric rifting of continental crust produces several conduits and geometries for pluton emplacement and formation of volcanic edifices.

being melted (Suppe, 1985). Melting begins as an intergranular phenomenon that produces small masses of liquid that coalesce into larger magma bodies. It commonly produces a volume increase for most rock-forming minerals ranging from 3 to 20 percent during the complete transformation from solid to melt (Yoder, 1976). This increased volume must be accompanied by an increase in pressure on the enclosing rocks. Increased volume that occurs with melting is also accompanied by a decrease in density of the magma relative to the original rock from which it is derived. This decrease in density commonly produces a density differential between the magma and the country rocks, providing the opportunity for diapiric intrusion into the upper crust, analogous to the diapiric intrusion of salt (Chapter 2).

Turcotte and Emerman (1985) concluded that *magma fracture* (analogous to hydraulic fracturing) is the most efficient mechanism for intrusion into a cold crust, because of the heat-exchange requirements of other mechanisms. They also suggested that magma ascent velocities of 0.1 m sec^{-1} are required to prevent the magma from transferring all of its heat to the country rocks and to permit rapid downward removal of blocks of wallrock into the magma.

DISTINGUISHING MAGMATIC FROM TECTONIC STRUCTURES

The problem of separating magmatic from tectonic structures in an igneous body has been the subject of debate for many years. This problem has no easy solution, because structures form through a range of conditions from high-temperature flow of magma that aligns crystals during intrusion to low-temperature crystal-plastic deformation of a largely solid pluton that results in formation of metamorphic minerals and a deformation fabric. In between is a broad area where the separation of magmatic from tectonic structures is difficult.

Scott Paterson, Ron Vernon, and Othmar Tobisch (1989) have attempted to address the problem, and have listed several means by which foliations are thought to develop in granitoid plutons. These include foliations that develop during flow accompanying ascent of magma, flow accompanying diapiric emplacement and expansion (sometimes referred to as ballooning) of magma, regional deformation that occurs during emplacement, regional deformation that postdates emplacement, and combinations of all these.

Paterson and others formulated a number of criteria that would help to distinguish these different types of foliations but emphasized that, in most cases, no single criterion is sufficient. Primary flow in an igneous body is most commonly identified by recognition of a foliation consisting of primary igneous minerals that exhibit no evidence of plastic deformation or recrystallization. Other criteria include (1) aligned crystals surrounded by randomly oriented anhedral quartz, (2) imbrication or tilting of crystals, (3) alignment of magnetic fields of minerals (magnetic anisotropy) parallel to the presumed flow direction, (4) schlieren layering in the absence of plastic (solid-state) deformation, (5) igneous flow texture in an igneous body that intrudes undeformed rocks, and (6) where flow layering is parallel to the walls of a pluton, phenocrysts may be rare because of the Bagnold effect. This phenomenon causes denser crystals in flowing magma to move away from the walls. Criteria for recognition of solid-state (tectonically produced) flow includes evidence of plastic deformation, recovery and recrystallization of individual grains and aggregates, formation of flattened ribbons of quartz and lens-shaped aggregates of micas resulting in a new compositional banding in the rock mass or local mylonitic (ductile shear) zones, formation of S-C fabrics (Chapter 10), and foliation that passes through xenoliths and other enclaves. Paterson and others cited as one of the most straightforward criteria for recognizing tectonic foliations in granitoids the occurrence of a foliation that without any change in orientation superposes earlier structures in the pluton and the contacts with the host rocks and is a principal foliation in the country rocks. Unfortunately, variations from one rock mass to another in the composition of the rocks, metamorphic conditions, anisotropy, fluid content, and other factors may prevent the formation of a common orientation of these structures, even though they may have been superposed at the same time.

The separation of magmatic from tectonic foliations in plutons may seem an impossible task, but a remarkable amount of success has been achieved in solving this problem for individual plutons and regions, some of which will be brought out in the discussion below. Paterson and others (1989) recommended the following approach to solving this problem employing their criteria: the microstructures and microtextures in a pluton should be examined, along with the foliation patterns in both the pluton and host rocks; relationships between porphyroblasts and foliations should be investigated, and the age relationships between magmatic and metamorphic minerals determined.

EMPLACEMENT OF TABULAR PLUTONS

The emplacement into the crust of tabular plutons—dikes and sills—is traditionally thought of in terms of intrusion parallel to some weakness in the country rocks, such as bedding, foliation, or preexisting fractures (including faults). Perhaps more frequently, however, fractures are produced by the pressure of the magma as it is intruded into a stressed crust by the same mechanism as joints formed by hydraulic fracturing (Chapter 8). Pressure on magma at the source frequently influences the formation of *"magma fractures"* in the crust, which then serve as conduits for magma to form dikes, sills, and even larger plutons. Magma behaves like any pressurized fluid, although it is more viscous than water, by decreasing normal stress by an amount equal to the fluid pressure. It thus obeys the Terzaghi modification of the Coulomb-Mohr equation discussed in Chapters 8 and 10 (Equations 8–20 and 10–3). Radially and concentrically oriented dikes commonly form as a result of tensional forces that develop about a cylindrical plug that connects to a larger magma body beneath it but are also intruded into a regional stress field (Odé, 1957). Dike systems such as Ship Rock in New Mexico or the Spanish Peaks area in Colorado (Figure 19–2) are good examples of radially oriented dikes.

Ernest Anderson (1951) suggested that magma can produce fractures by "*magmatic wedging*" as it forces its way into the crust (Figure 19–3). As long as the magma stress is even slightly greater than the tectonic stress σ_3 in that region of the crust, magma wedging is likely to occur because of the greatly magnified stress at the tip of the wedge (Johnson and Pollard, 1973; Pollard and Johnson, 1973). Anderson concluded that dikes thus intruded would form parallel to regional σ_1 and would probably maintain this orientation at all levels in the crust, especially in the lower to middle crust where the overburden stress is more uniform. In the upper crust, where the overburden decreases and σ_3 may become vertical, tabular plutons are more likely to become horizontally oriented. Secor and Pollard (1975) concluded that if the rate of change of σ_3 with respect to depth (vertical gradient of σ_3) exceeds the unit weight of the fluid (such as either magma or water) by ~100 dyne cm^{-3}, a stable open fracture in granite cannot be longer than ~200 m.

An intriguing corollary to the magma-fracture mechanism has been suggested by Tom Parsons and George Thompson (1991, 1993a), Parsons, Norman Sleep, and Thompson (1992), and Parsons, Thompson, and Sleep (1994) that relates dike intrusion, normal

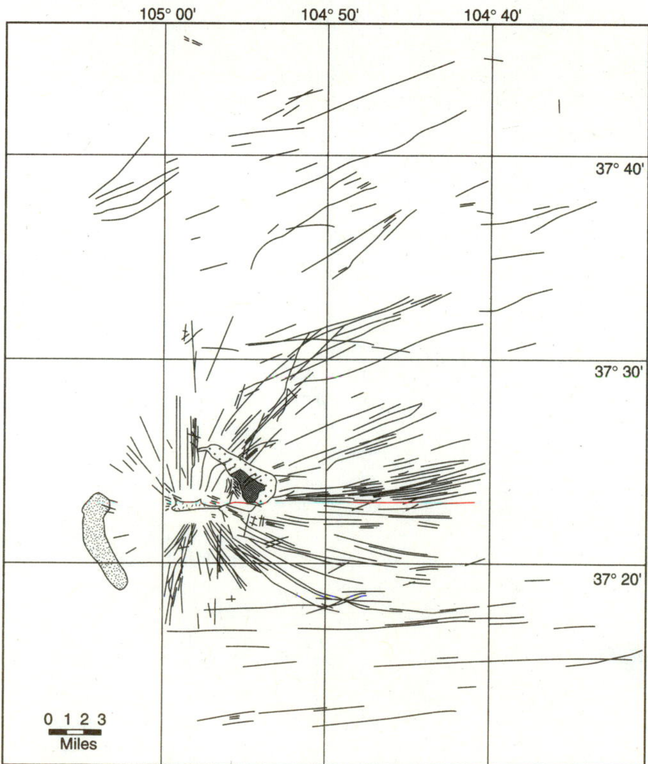

FIGURE 19–2
Dike pattern around Spanish Peaks, southwestern Colorado. Large, irregularly shaped bodies are plutons. (From A. Knopf, Geological Society of America *Bulletin*, v. 47, 1936.)

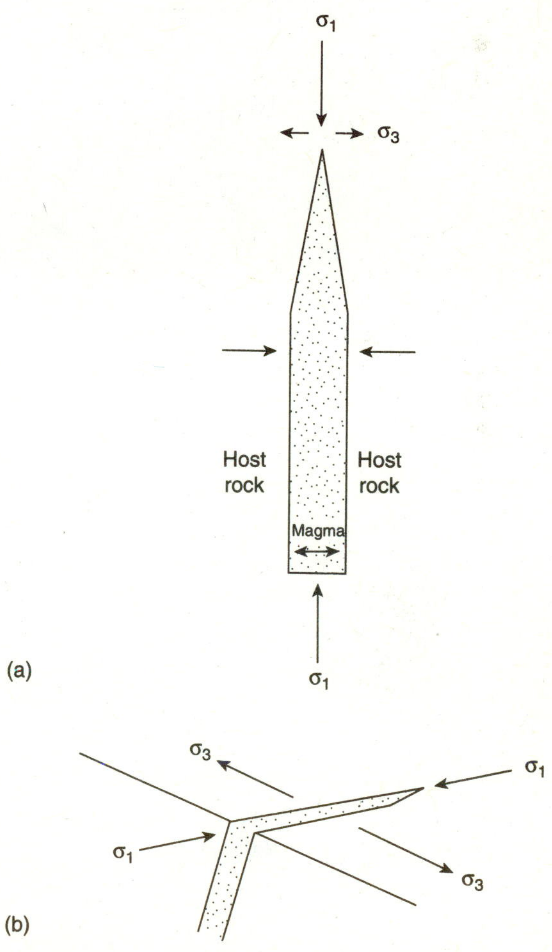

FIGURE 19–3
Anderson's magmatic wedging mechanism. Magma under pressure is forced into the crust by fracturing the host rocks (a) in the same way as water forms joints, either laterally (b) or vertically.

faulting, and regional crustal extension (Figure 19–4). In a segment of the brittle crust under tension (Figure 19–1c), the stress may be relieved by either dike intrusion or normal faulting. If the regional horizontal compressive stress field is reduced slightly in the *X* direction producing a deviatoric stress in the horizontal stress components, the crust will fail by normal faulting if the stress difference is sufficient to produce large enough Mohr circles to intersect the envelope (Figure 19–4b). If, on the other hand, magma is available or forms by melting under reduced stress at the base of the ductile-brittle transition, the magma may fracture the crust in the *YZ* plane, produce a series of dikes that intrude the crust, lower the tensional stress component, and thus prevent the crust from failing by normal faulting (Figure 19–4c). The intrusion process, once complete, would cause both the deviatoric and shear stress components to be reoriented, and, if a sufficiently high deviatoric stress exists above the intrusions, that part of the crust may fail by normal faulting (Figure 19–4d). So, while intrusion of dikes—or other plutons—may on the one hand initially reduce the overall stress in the crust, the process may on the other hand later cause the crust to fail by normal faulting. Parsons and Thompson (1993b) and Parsons and others (1992, 1994) have suggested that, in addition to applicability of the Parsons and Thompson (1991, 1993a) mechanism to dike emplacement and possible attenuation of earthquakes in zones of active extension, the entire process of crustal extension and formation of rifts and zones of extended crust like the Basin and Range (Chapter 13), particularly the formation of "metamorphic core complexes," may be driven by intrusion of both dikes and larger plutons from the base of the lithosphere to the upper crust.

Other modifications of the Andersonian relationships must be considered where nearly tabular plutons, such as laccoliths and lopoliths, are involved. Laccoliths, and their many aberrations—sphenoliths, bysma-

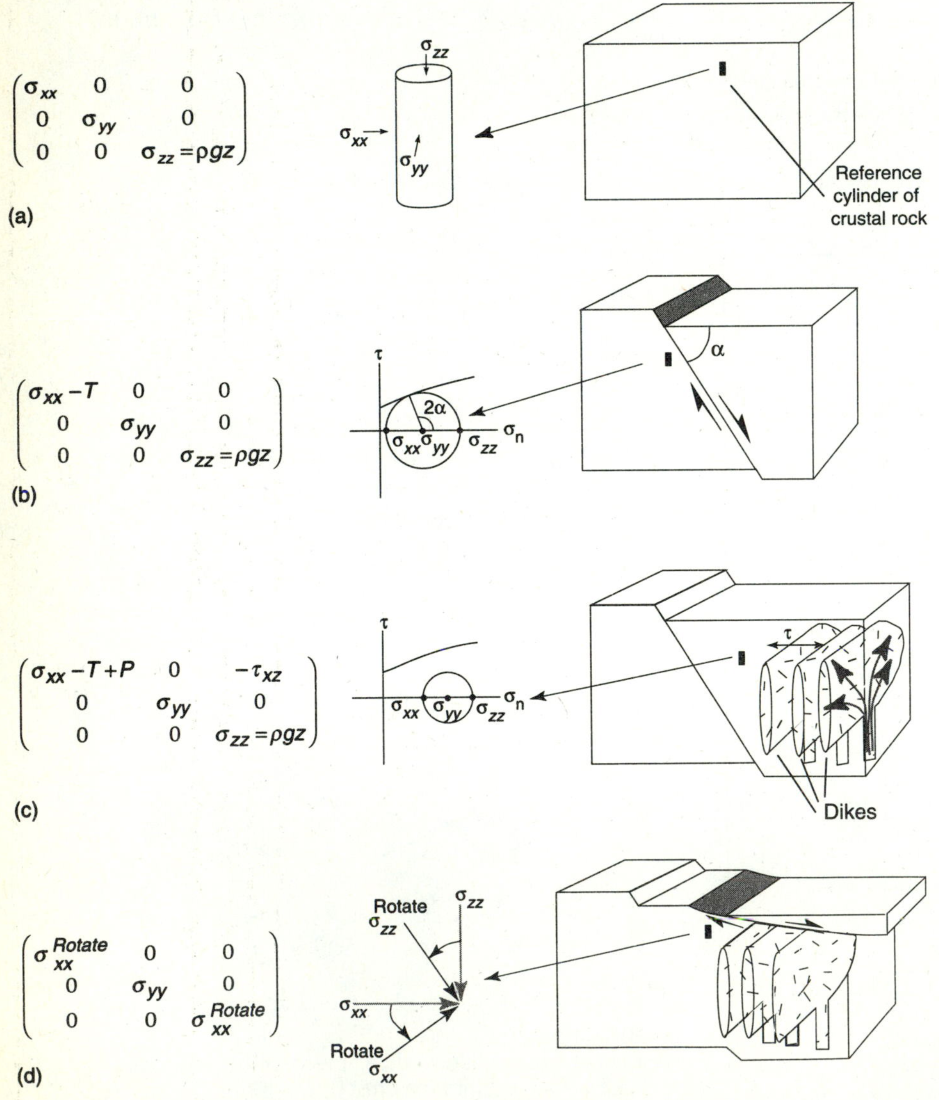

FIGURE 19–4
Parsons and Thompson relationship. Stress tensors on the left show the relationships between the principal stress components, the Mohr diagrams in (b) and (c), and the mechanical models to the right. In the tensors, the vertical component σ_{zz} is the same as ρgz, which is density times the acceleration due to gravity times the height of the column of rock above the reference cylinder. (a) State of stress on a reference cylinder of rock related only to lithostatic load. The stress field can be described in terms of three principal (compressive) normal stresses σ_{xx}, σ_{yy}, and σ_{zz}, with σ_{zz} vertical and the other two in a plane parallel to the surface and perpendicular to σ_{zz}, the same stress system defined by Anderson for normal faulting (Chapter 10). (b) Reduction of the horizontal stress in the *X* direction in an amount *T* resulting in a deviatoric stress large enough to produce normal faulting. (c) Dike intrusion parallel to *Y* increases the horizontal compressive stress in the *X* direction by an amount *P*, and also the shear stress τ on the *XZ* and *ZX* planes. (d) Rotation of the compressive and shear stress directions occurs because of the intrusions, yielding a new deviatoric stress that may induce low-angle normal faulting in the upper crust. σ_{zz} Rotate > σ_{yy} > σ_{xx} Rotate. (From T. Parsons and G. A. Thompson, 1993, *Geology*, v. 21.)

liths, phacoliths, cactoliths, etc., chronicled by Hunt (1953)—are mushroom-shaped or modified in cross section (Figure 19–5a). In addition, the arched roof of the laccolith is formed by bending of the overburden layers. Johnson and Pollard (1973) and Pollard and Johnson (1973) concluded from experimental and field studies of laccoliths in the Henry Mountains in southeastern Utah that the magma is intruded parallel to layering in the host rocks by wedging and that the upward bending of the overburden above the magma increases by the fourth power of the diameter of the laccolith, as

$$w_{c=\infty} = \frac{p_d}{24R}\left(a^4 - 2a^2x^2 + x^4\right) \qquad \textbf{(19–1)}$$

for a laccolith with an anticlinal outline in map view, and

$$w_{a=c} = \frac{p_d}{64R}\left(a^4 - 2a^2x^2 + x^4\right) \qquad \textbf{(19–2)}$$

for a laccolith with a circular outline in map view. w_c and w_a are displacements in the XZ plane (with $y = 0$), R is flexural rigidity of the overburden (elastic modulus, B, times t^3 divided by 12), t is thickness of the overburden, a is the half length of the intrusion in the x direction, and x is the distance along the x axis for half the layer. Thus, if the diameter of a laccolith with a circular plan is doubled, the displacement increases by a factor of 16. Displacement is even greater for a laccolith with an anticlinal (elliptical) plan (Figure 19–5b). Moreover, they concluded from

$$\sigma_{xx} = \frac{p_d{}^z}{t^3}\left(6x^2 - 2a^2\right) \qquad \textbf{(19–3)}$$

that the overburden layer resists bending as a function of the third power of the thickness, where σ_{xx} is the normal stress component in the x direction, and $p_d{}^z$ is the driving pressure of the magma (total pressure minus overburden pressure). Crystallization of part of the magma along the thin edges of the intrusion, without removal of the stress, can cause the remaining magma to increase the vertical displacement of the overburden, without increasing the lateral extent of the intrusion. John Lister and Ross Kerr (1991) derived analogous equations for dike intrusion. They also observed that because of the buoyancy of magma, underplating of magma along the base of the crust is unlikely; magma, once formed, is more likely to move into the upper crust by a magma-fracturing mechanism.

An intriguing modification of sheet intrusions, called "fingers," has been described in sills associated with the Shonkin Sag laccolith in Montana by Pollard, Otto Muller, and David Dockstader (1975), where the edges of a sheet intrusion separate into elongate finger-like projections into the host rocks (Figure 19–6a). Many of the fingers develop offsets perpendicular to the propagation direction of the fingers, and curves, apparently related to the intrusion mechanism. Country rocks are commonly wedged aside and are compacted. Pollard and others demonstrated that finger intrusions require more mechanical energy than intrusion of sheets because of the larger numbers, smaller size, and greater surface area of fingers required for intrusion of an equivalent amount of magma in a sill. The rate of heat transfer to the host rocks is also greater because of the greater total surface area of a series of finger intrusions compared with the equivalent volume of magma intruded as a sheet. Therefore, sills and other sheet intrusions are favored over fingers and are much more common. Pollard and others suggested that fingers are initiated as instabilities that develop where part of the magma fringe with lower viscosity is intruded ahead of other parts and accelerates into the higher-viscosity host. Fingers appear to grow in a surface perpendicular to σ_3, and Pollard and others speculated that offsets develop as the intrusion propagates into a region where the stress ellipsoid changes orientation. Both offsets and grooves in the main sheet (Figure 19–6b) form parallel to the long axes of fingers, and so both can be used to determine the propagation direction and the orientation of the stress field into which the magma was intruded parallel to σ_2. Fingers thus are rare compared to sheets because of the greater mechanical and thermal efficiencies associated with formation of sheet intrusions.

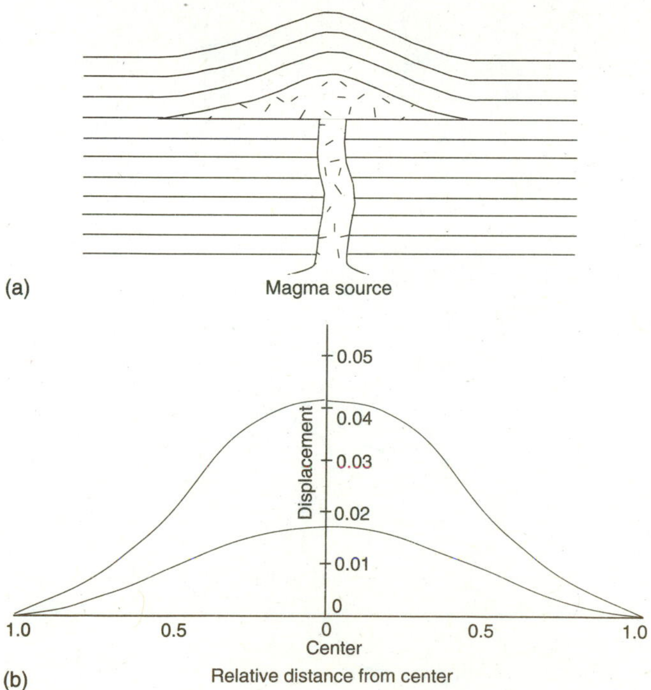

FIGURE 19–5
(a) Geometry of a hypothetical laccolith. (b) Displacement curves for laccoliths having circular (lower) and anticlinal (elliptical, upper) plans. (b is from Pollard and Johnson, *Tectonophysics*, v. 18,© 1973, with kind permission from Elsevier Sciences, Ltd., Kidlington, United Kingdom.)

Sills and other large tabular plutons may also be intruded into fault zones oblique to the principal stress axes. Donald Hutton (1982, 1988, 1992) and Hutton and others (1990) have described the conditions under which magma may be intruded into the extensional regions of thrust, strike-slip, and normal fault zones (Figure 19–7). He distinguished these intrusions from Anderson's magma fractures by calling them *non-Andersonian intrusions.* Hutton has suggested that the Main Donegal Granite in Ireland and the Strontian Granite in Scotland were intruded into active shear zones, and thus were deformed during emplacement. These intrusions exploit zones of weakness in the crust and may thereby require expenditure of less work than intrusions that employ the Andersonian magma-fracturing mechanism.

Maria Luisa Crawford and Lincoln Hollister (1982); Hollister and Crawford (1986); Cameron Davidson, Hollister, and Stefan Schmid (1992); and Hutton and Gary Ingram (1992) have suggested that the "great tonalite sill" in the Coast Mountains of British Columbia and southeastern Alaska was intruded into an environment of active faulting and that faulting facilitated the movement of magma from the lower into the upper crust. Hollister and Crawford (1986) suggested that thermal weakening of the thickened lower crust by either anatectic or mantle-derived magma led to fractures that propagated through the entire crust. Movement during faulting was lubricated by magma. Davidson and others (1992) measured shear-sense indicators and observed metamorphic assemblages in the Valdez Creek shear zone along part of the tonalite sill (Figure 19–8). They concluded that deformation in the tonalite ceased prior to crystallization of some late phases, such as anhedral K–spar along grain boundaries. Also, from the metamorphic assemblage in the country rocks, they observed that the higher grade assemblages are located above the northwest-dipping sills, whereas the lower grade country rocks are located below the sills. They were able to employ these and other data to derive two thermal models—one for the upper crust and another for the lower crust—for generation and emplacement of the Valdez Creek sills. One model suggests that emplacement of the tonalite magma into the upper crust during deformation along the shear zone involved greater heat transfer into the overlying rocks and a more rapid cooling rate to the final crystallization temperature of 700° C. A 2-km-

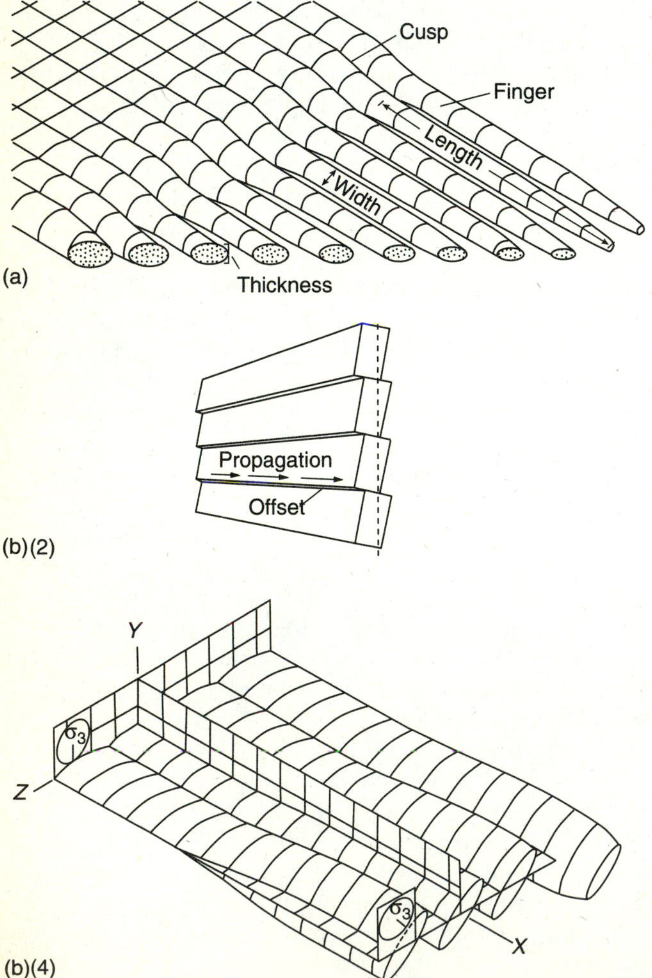

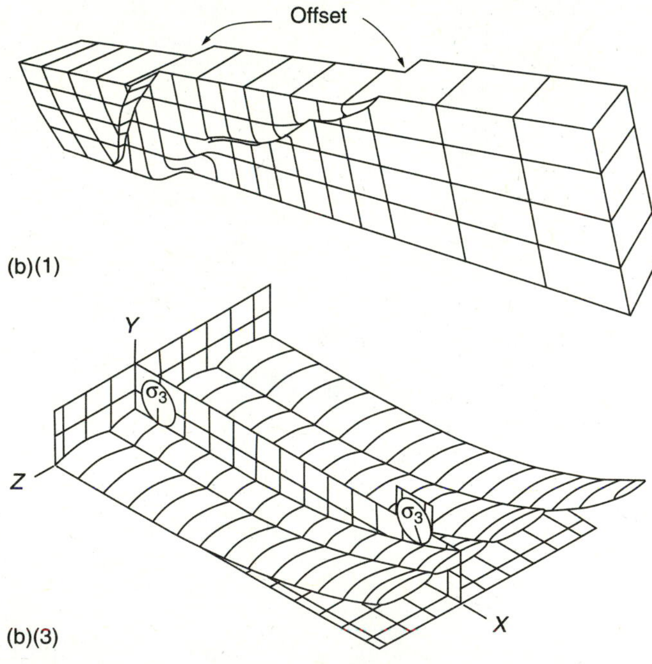

FIGURE 19–6
(a) Geometric relationships between the parent sheet intrusion and fingers that develop along the distal edge of the intrusion. (b) Offsets and curved segments in fingers. (From D. D. Pollard, O. H. Muller, and D. R. Dockstader, 1975, The form and growth of fingered sheet intrusions: Geological Society of America *Bulletin*, v. 86, p. 351–363.)

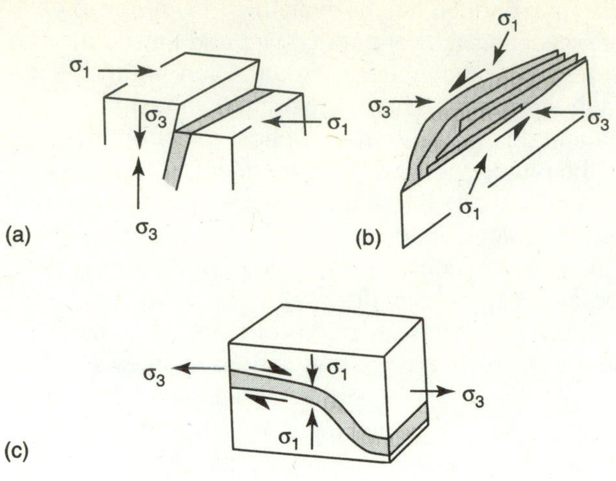

FIGURE 19–7
Hutton's non-Andersonian intrusions into fault zones producing net extension. (a) Contraction producing thrust faulting. (b) Strike-slip faulting. (c) Extension. (From D. H. W. Hutton, Geological Society of America *Special Paper 272*, and *Transactions of the Royal Society of Edinburgh: Earth Sciences*, v. 83, 1992.)

thick sill that is emplaced into the upper crust (500° C country rocks) will cool in 200,000 y, whereas a sill of the same thickness that is emplaced into the deep crust (700° C country rocks) will take 8.37 m.y. to cool. In their analysis, Davidson and others reinforced the conclusion of Hollister and Crawford (1986) by suggesting that faults in the lower crust that were being lubricated by liquid magma would thus take much longer to become locked by crystallized magma than would similar faults in the upper crust.

EMPLACEMENT OF STOCKS AND BATHOLITHS

Many large plutons—stocks and batholiths—have irregular but approximately circular outcrop patterns in map view (Figure 2–22), while others have elliptical to very linear outcrop patterns parallel to the trend of the mountain chain, or frequently parallel to fold axes. Moreover, despite the irregular shape of many indi-

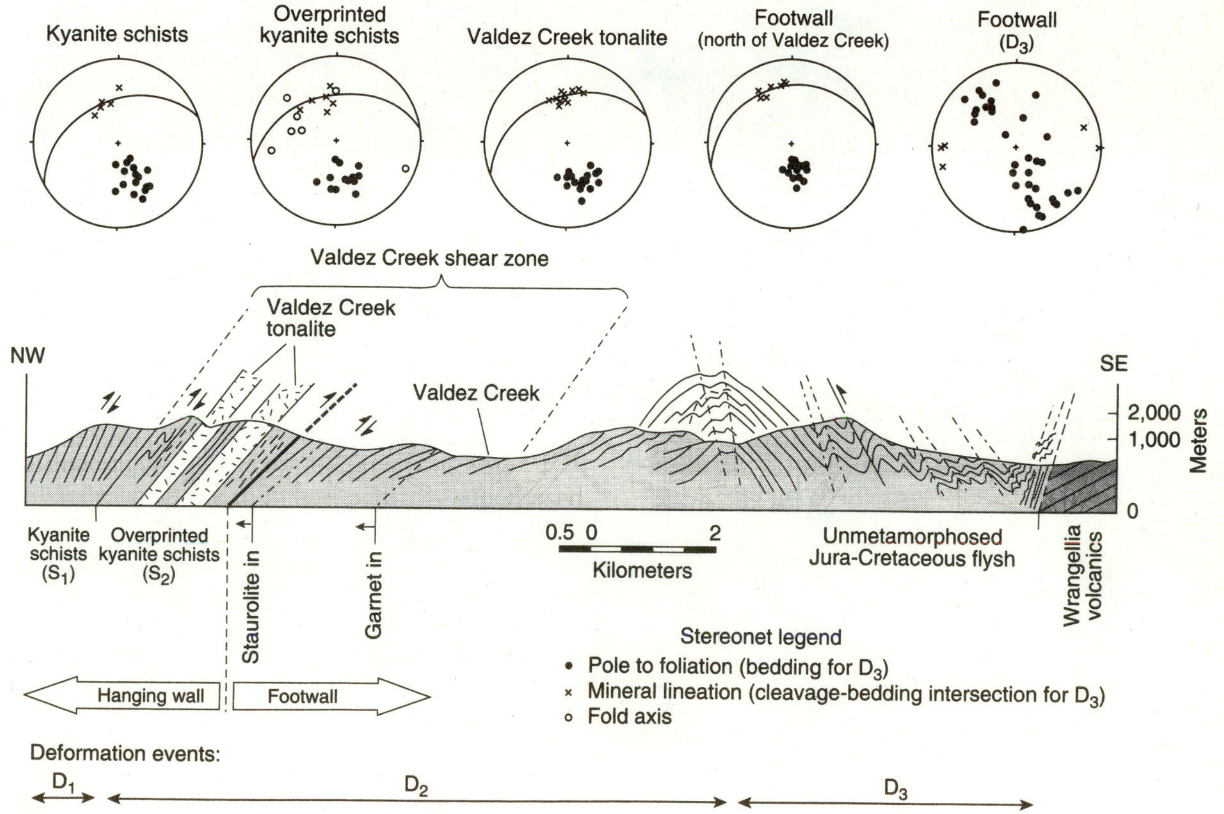

FIGURE 19–8
Cross section showing structure in country rocks, summary fabric diagrams in different domains, locations of metamorphic isograds, and tonalite sill intrusions into the Valdez Creek shear zone, southeastern Alaska. (From C. Davidson, L. S. Hollister, and S. M. Schmid, *Tectonics*, v. 11, 1992.)

vidual stocks, the assemblage of stocks of a particular magmatic suite, such as the Sierra Nevada Mountains in California and Nevada, may collectively occur along a linear trend within a mountain chain. Some, perhaps most, of this linearity may be explained in terms of the ***arc-trench gap*** (Figure 19–1), where intrusions and volcanic activity in andesitic arcs are commonly localized within 80 to 290 km of the surface outcrop of the subduction zone (Hatherton and Dickinson, 1969). Intrusions in an arc system may also be localized by extension fractures in the cooler upper parts of the arc above the subduction zone. These features were created by the diapiric pressure of the rising magma, or bending strains related to a change in dip of the subduction zone. Warren Hamilton (1979) recognized a number of almost concentric arcs that were produced at different times in Indonesia as new subduction zones formed south and west of older ones (Figure 19–9).

We suggested in the previous section that heat transfer and mechanical considerations indicate that the magma-fracturing mechanism for dike intrusion probably is more efficient (Lister and Kerr, 1991), but the abundance of xenoliths in stocks and batholiths indicates that enough heat is available for *stoping* to be a viable mechanism. Kenneth Fowler and Paterson (1993) suggested that material balance and strain compatibilities require both upward movement of magma and downward transfer of host rock from the roof of the pluton, processes independent of emplacement depth in shallow to intermediate-depth plutons in the western United States.

The emplacement of domed granitoid plutons (composed mostly of tonalite, granodiorite, and adamellite) in the Archean greenstone belts and mafic-granitoid terranes in Proterozoic regions of shields (Figure 19–10), along with similar structures in the Phanerozoic crust, has been the subject of debate for several decades. Emplacement has been suggested to be the product of the diapiric rise of lower-density magma in response to thermal convection and gravitational instability in the Archean crust (Talbot, 1968), or solid-state diapiric rise of granitoid bodies in granite-greenstone belts (Ramberg, 1981). Alternatively, domes with granitoid cores have been suggested to be related to Type 1 (dome-and-basin) fold interference, faulting, and buckle folding of these rocks in the solid

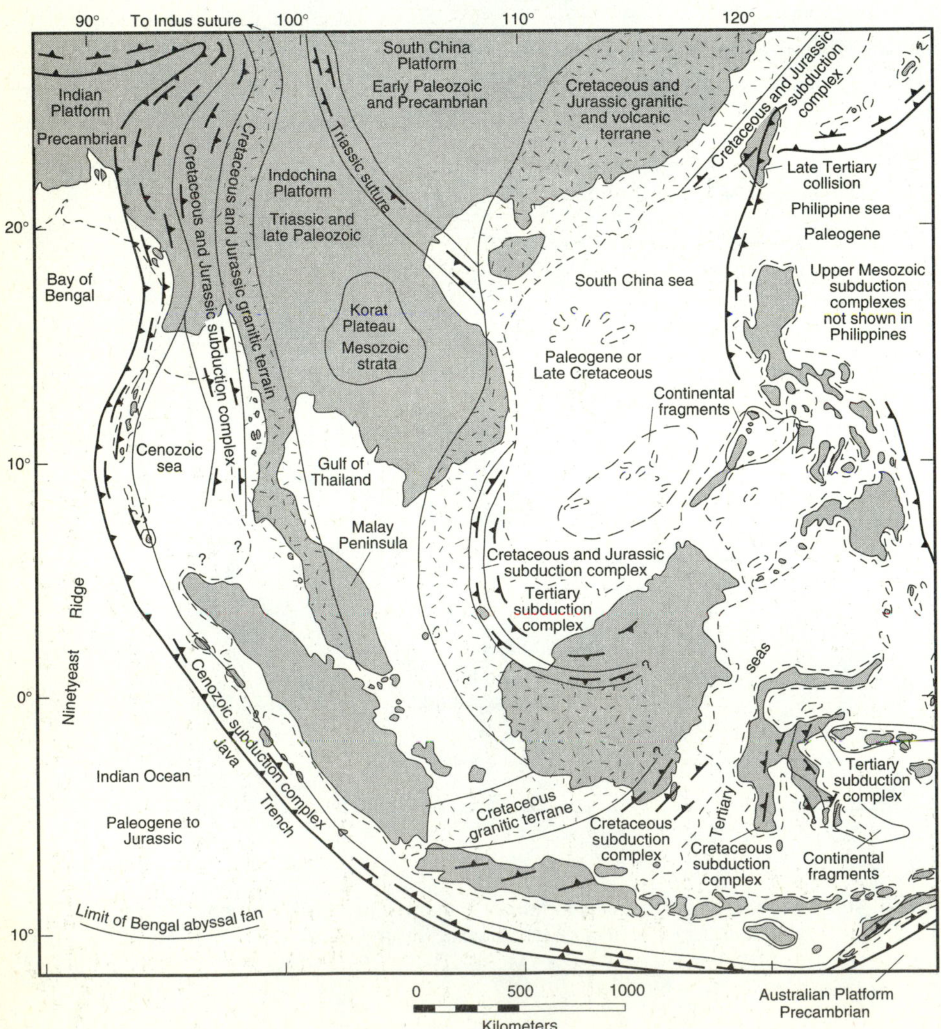

FIGURE 19–9
Mesozoic and Cenozoic arcs related to other tectonic features of southeast Asia and Indonesia. Note the part of Australia being subducted beneath Indonesia in the southeastern corner of the map. (From W. B. Hamilton, *Tectonics of the Indonesian Region*, U. S. Geological Survey Professional Paper 1078, 1979).

(a)

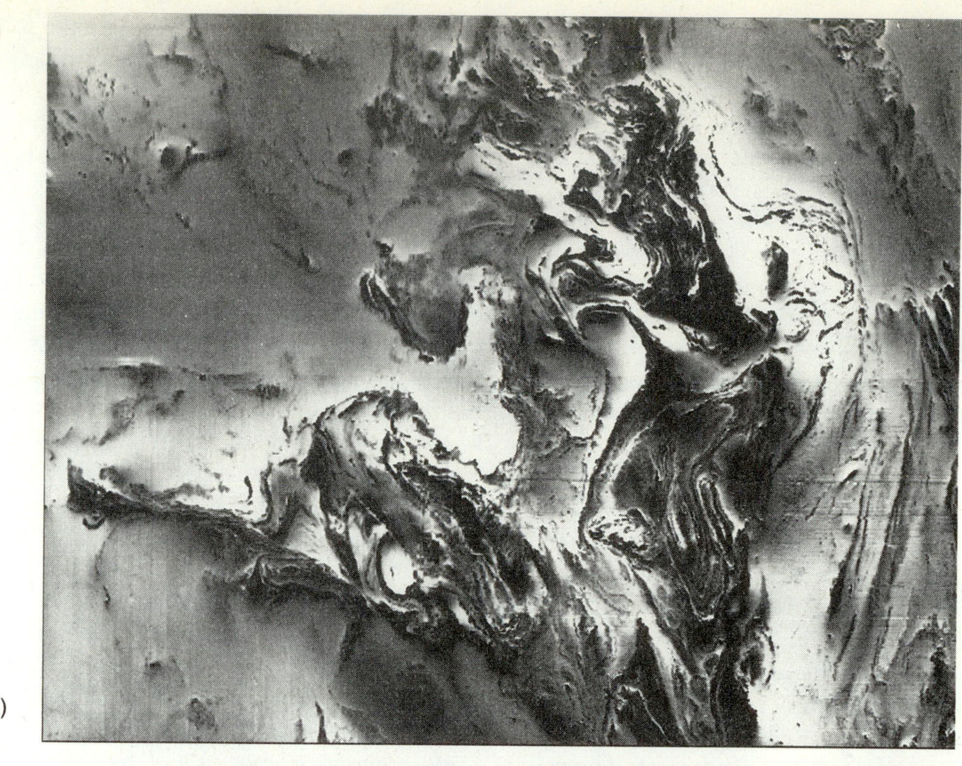

(b)

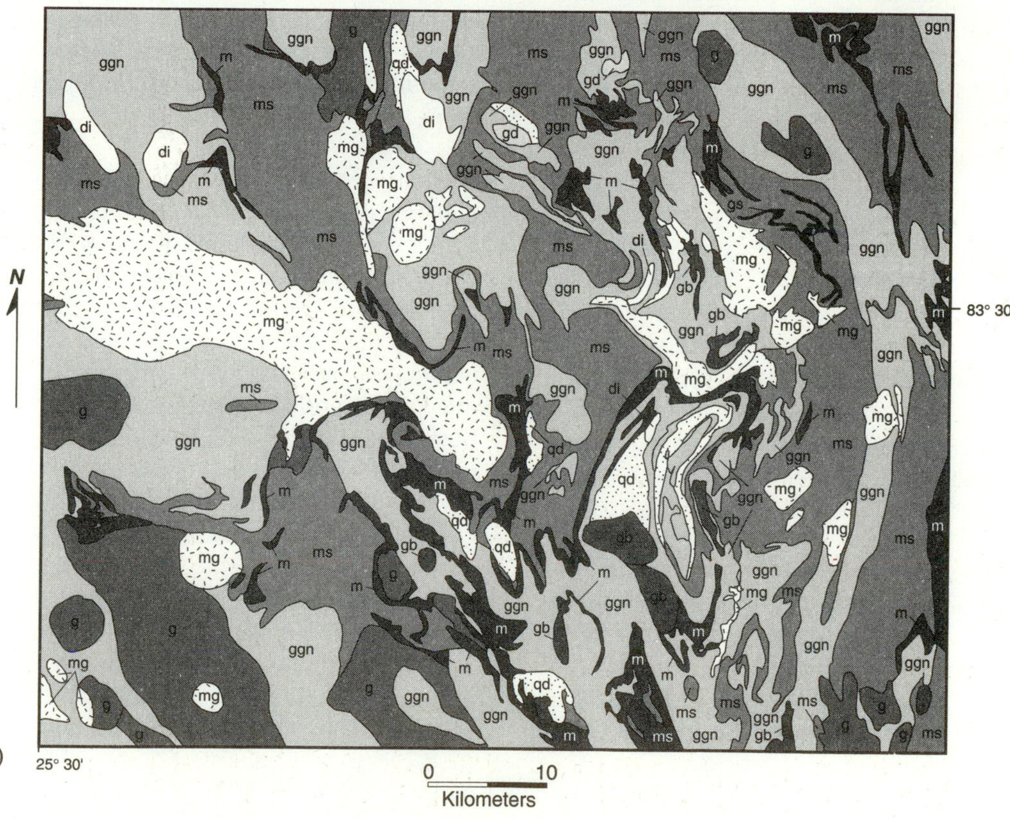

FIGURE 19–10

(a) Aeromagnetic map of part of the Baltic shield of north-central Finland. Granitic plutons produce no magnetic signature, whereas highly magnetic metavolcanic and metasedimentary rocks outline the contacts of plutons with the host rocks, revealing some of the complex structure of that region. (b) Simplified geologic map of the area shown in (a). ggn—granitic gneiss; g—granite; mg—megacrystic granitoid; ms—metasedimentary rocks; mi—metaigneous rocks; m—metavolcanic rocks (mostly mafic) and amphibolite; gb—gabbro; d—diorite; gs—graphitic schist; gd—gabbro and diorite. (From G. Gaál, editor, Geological Survey of Finland *Report of Investigations 80,* 1988. Reproduced with permission.)

state, and not to solid-state diapirism (Snowden and Bickle, 1976; Dixon and Summers, 1983; Myers and Watkins, 1985; Hudleston and others, 1988).

Paterson and Fowler (1993) examined the process of diapiric emplacement of plutons in an attempt to account for the means by which a large volume of magma can be intruded into the middle or upper crust. They noted that there may be only two ways to create additional space in the crust to accommodate this increased volume: lower the Moho or raise the Earth's surface. They suggested that most elliptical diapiric plutons (stocks and batholiths) are probably emplaced through a combination of mechanisms that involve different material transfer processes as a magma body intrudes different parts of the crust: assimilation, rigid translation, and recoil ductile flow of wall rock occur in the deeper crust; stoping and doming of roof rocks dominate in the upper crust. Roberto Weinberg (1994) pointed out, however, that it was difficult to measure the emplacement-related strains in the wall rocks far from the pluton, because they are very small and also because the three-dimensional flow paths of the magma are rarely, if ever, known. Both pointed out that the largest amounts of strain are concentrated near the pluton contacts and that strain decreases exponentially away from the contact into the host rocks. Weinberg suggested that this raises the possibility that other variables, such as strain rate and temperature variations producing variations in strain softening, ductile flow, and volume readjustments far from the pluton, could be even more important than the mechanisms proposed by Paterson and Fowler to explain diapiric emplacement. Paterson and Fowler (1994) in their response cited evidence, such as the concentration of high strain in narrow concordant strain aureoles and the sharply discordant roof contacts of many plutons, that indicates that large-scale ductile flow of wall rocks in the middle to upper crust is unlikely as a mechanism for large-scale volume displacement in response to diapiric magma ascent.

Rainy Lake and Irene-Eltrut Lakes Plutons. The diapir mechanism was employed by Walfried Schwerdtner and S. B. Lumbers (1980) to explain intrusion of Archean plutons in the Canadian Shield, and they suggested that diapiric intrusion will occur where little density contrast exists between magma or ductile solid-state granitoid material and the host rocks. Schwerdtner (1990) formulated a test of the diapir hypothesis using the Rainy Lakes and Irene-Eltrut Lakes elliptically shaped domal granitoid complexes in the western Wabigoon subprovince, located in southwestern Ontario and northern Minnesota. The outcrop patterns and features within these complexes (Figure 19–11) could be explained by either folding or diapirism. Deformation in xenoliths and larger enclaves produced strains that conform with the foliation in the granitoid that occupies the domes, indicating that the foliation and related structures were acquired prior to doming. Schwerdtner concluded that the large granitoid complexes are localized in first-order fold culminations and that smaller gneiss domes occur in second-order fold culminations. Several lines of evidence preclude that these are solid-state diapirs, including the absence of radial structures from later deformational events, presence of large basins adjacent to gneiss domes, and the presence of mafic rocks, (amphibolite-free) zones of low-density rocks (potential parasitic diapirs) at the contacts of gneiss domes. He also found several lines of evidence that suggest magmatic diapirism is unlikely. These include (1) the absence of deformation aureoles that would reveal forced intrusion, (2) identification of foliation patterns that suggest many of the plutons are phacoliths that probably formed by filling voids between layers formed during flexural folding, (3) presence of high-order dome-and-basin fold patterns throughout the subprovince, (4) absence of mesoscopic or small map-scale structures that would indicate late horizontal extension within the cores of gneiss domes as well as upright antiforms that may be underlain by concordant magmatic diapirs, and (5) absence of small-scale structures that would indicate late down-dip extension in a smaller satellite dome.

Chindamora Batholith. John Ramsay (1989) studied the almost circular Chindamora batholith, a small (~200 km^2) granitoid pluton in an area of much larger circular to elliptical granitoid plutons in the Archean granitoid-greenstone crust in Zimbabwe. This pluton consists of several rock types (Figure 19–12a): (1) a central core of megacrystic adamellite that makes up most of the body is largely unfoliated, except along the outer margins of the adamellite where K-spar megacrysts are aligned, (2) a western and northern homogeneous largely unfoliated fine- to medium-grained nonmegacrystic adamellitic granite, (3) a body of strongly foliated granodiorite located in the northeastern part of the batholith that contains abundant foliated xenoliths of both greenstone host rock and the tonalite that make up the outermost zone, and (4) an outer zone of strongly foliated to banded tonalite containing abundant xenoliths of the greenstone host rock. Although the largest bodies of tonalite occur along the pluton margins (Figure 19–12a), some smaller (~100-m diameter) bodies that Ramsay interpreted as roof pendants occur near the center of the batholith. Foliation and schistosity are commonly parallel where they occur in the same rock, but in a few localities schistosity (mica foliation) crosscuts the gneissic foliation, indicating that the two fabrics formed independently of each other. Intensity of schistosity deve-

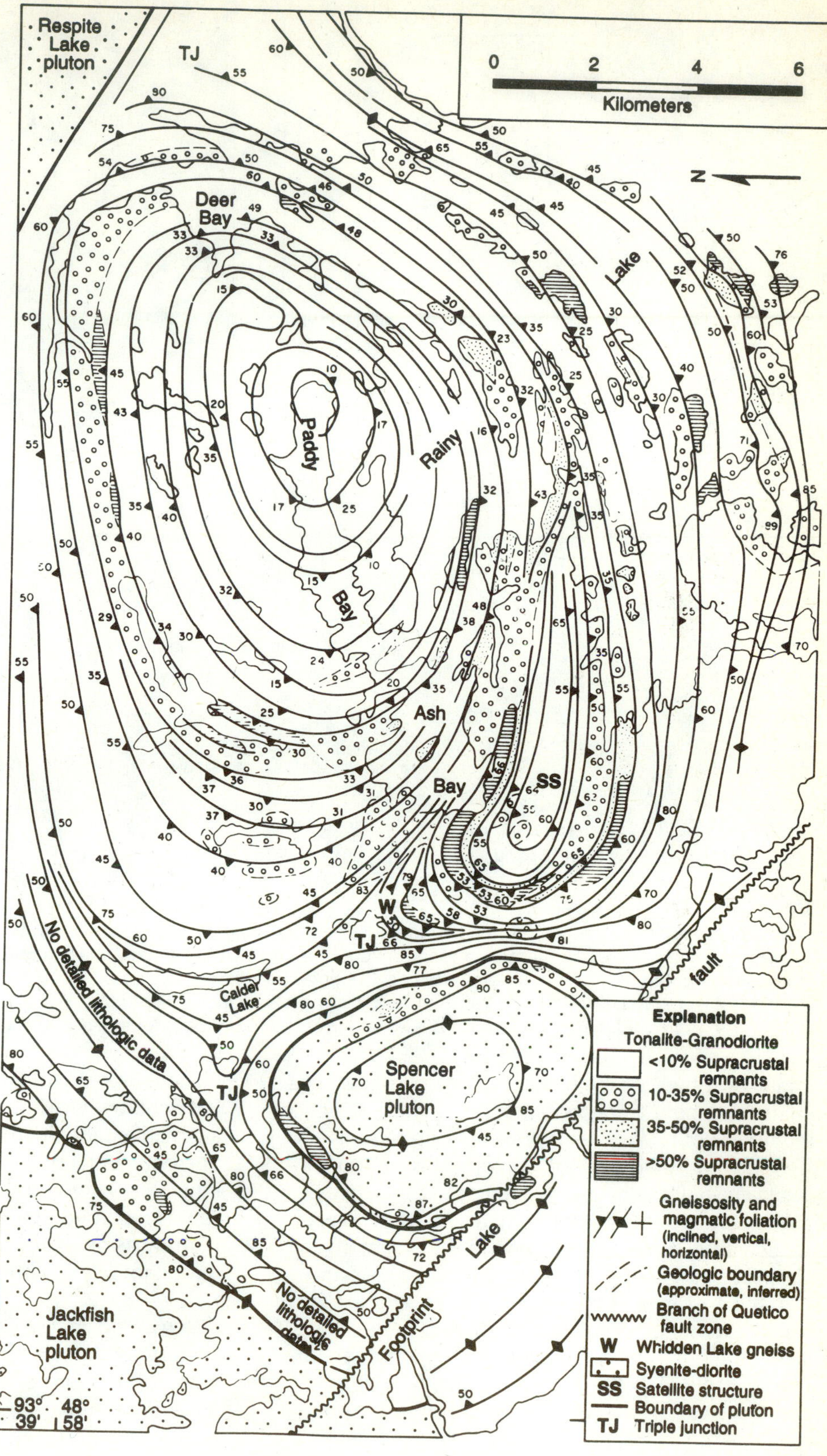

FIGURE 19–11
Structure of the Rainy Lake complex of gneiss domes and basins located in the southwestern part of the Wabigoon subprovince of the Superior province in southwestern Ontario and northeastern Minnesota. The solid lines represent the strike of foliation; triangles attached to the lines, the dip direction. The numbers beside each triangle represent the average dip in that region. Zones containing significant amounts of enclaves in the Rainy Lake complex are patterned. (From W. M. Schwerdtner, *Canadian Journal of Earth Sciences*, v. 27, 1990.)

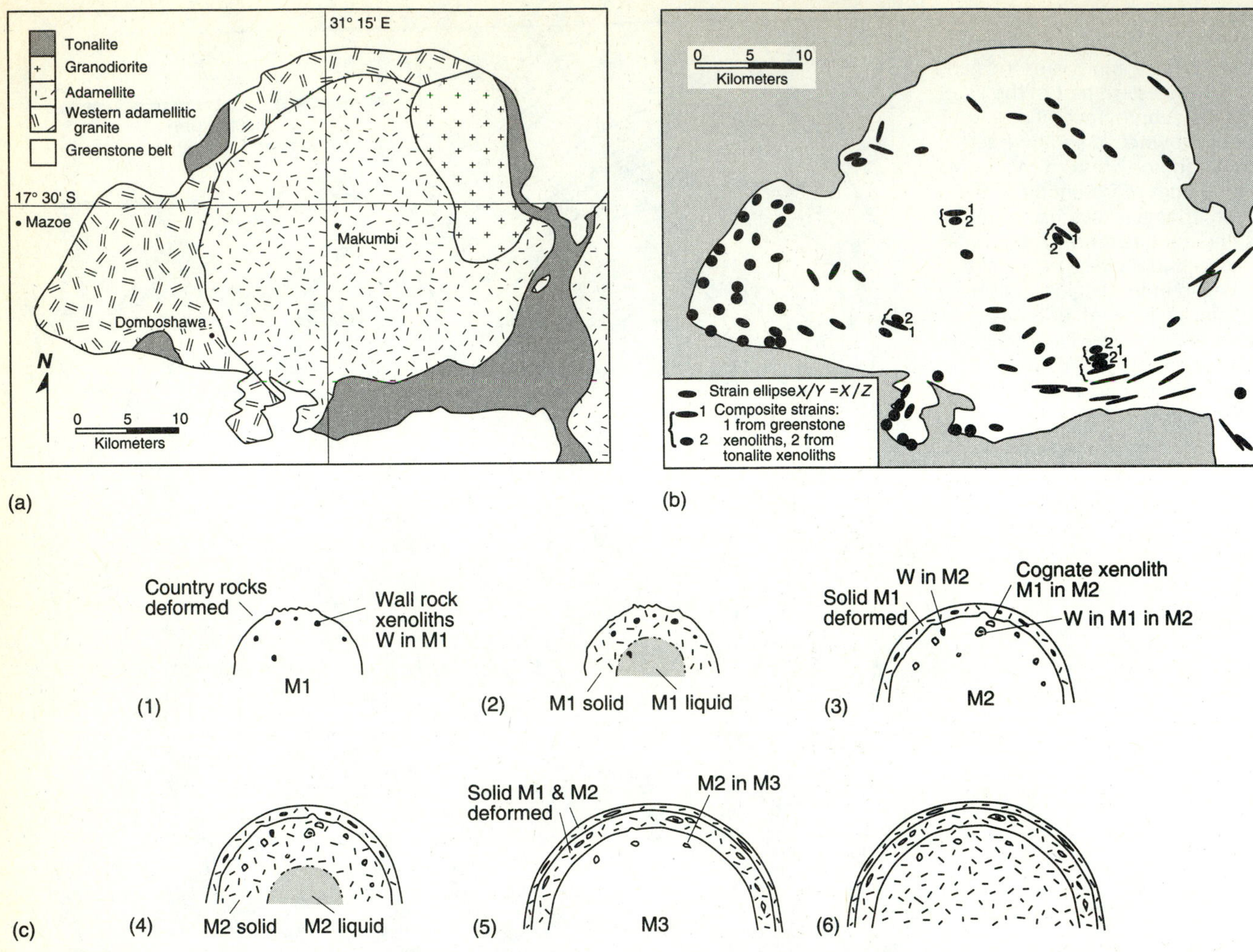

FIGURE 19–12
(a) Geologic map of the 2,750–2,550 Ma Chindamora batholith in the Archean craton of Zimbabwe. (b) Finite strain ellipses determined from the shapes and fabrics present in xenoliths. (c) Ramsay's interpretation of the three-stage development of the Chindamora batholith. (1) First magma phase. (2) Cooling. (3) Second magma pulse. (4) Cooling. (5) Third magma pulse. (6) Cooling. Magma M1 is tonalite, M2 is granodiorite, and M3 is adamellite. (From J. G. Ramsay, *Journal of Structural Geology*, v. 11, © 1989, with kind permission from Elsevier Sciences, Ltd., Kidlington, United Kingdom.)

lopment varies within the batholith but is strongest near the margins where a lineation parallel to the X direction of the finite strain ellipsoid is present. Dip of schistosity is generally steep in the center of the pluton and varies from shallow toward the center to steep away from the center of the batholith; near the walls, schistosity generally dips steeply away from the center. Schistosity is always oriented perpendicular to Z and thus in the XY plane of the finite strain ellipsoid. Ramsay determined from analysis of deformed xenoliths that the finite strain ellipsoid in the batholith is an oblate spheroid ($X = Y > Z$) that records flattening strains that increase toward the margins of the batholith (Figure 19–12b). This pattern is also reflected in the constrictional (prolate) strain recorded in xenoliths that occur in the adamellite in the inner parts of the batholith where xenoliths are highly strained, whereas strains in the enclosing adamellite are not only small but are flattening strains. Ramsay interpreted the emplacement sequence as intrusion of the tonalite first, followed by intrusion of the granodiorite, and then the adamellite (and adamellitic granite) (Figure 19–12c). The outer parts of the pluton were still plastic when the inner parts were being intruded, and so the flattening strains recorded in the outer zones were interpreted by Ramsay as being related to ballooning of the pluton during intrusion of the adamellite.

Discussion. Ramsay's observations and conclusions from study of the Chindamora batholith are consistent with observations of fabrics in plutons in the Sierra Nevada Mountains by Paterson, Elizabeth Yuan, and Fowler (1993). They observed folds, lineations, and foliations in several plutons that have the same orientations as similar structures in the host rocks but formed at high angles to flow directions, schlieren, and other primary structures formed in the magma. They concluded that magmatic foliations probably form late in the crystallization history of a pluton, when the rock mass is plastic, any remaining magma has a high viscosity, and solid crystals are abundant.

The ballooning mechanism employed by Ramsay (1981) has also been suggested by Jean-Pierre Brun and J. Pons (1981) to explain the emplacement of several Paleozoic plutons into a region in Spain undergoing deformation. They concluded that ballooning occurred beneath a barrier formed by a Cambrian limestone unit and that strains recorded by pluton shape (elliptical versus circular) reflect the noncoaxial nature of the deformation environment in which the plutons were being emplaced as well as the position of the pluton in a specific structure at the time of emplacement: greater ellipticity is recorded by the youngest of the plutons, possibly because of the advanced development of the anticline into which it was emplaced. They also noted that the intersection of three foliation directions, "foliation triple points," near the plutons are related to interference between a strain field created by the local effects of ballooning and deformation being produced in the regional strain field.

Giovanni Guglielmo (1993) has employed a computer simulation technique to conduct three-dimensional modeling of the relationships between regional deformation and deformation associated with pluton emplacement where the pluton "balloons" during emplacement into an environment of noncoaxial (simple shear) deformation (Figure 19–13). The computer simulations permit simultaneous evaluation of several variables, including strain-ellipsoid shape and intensity and orientation of mineral-elongation (stretching) lineations and foliations, that affect the shape of the strain ellipsoid through time. They also permit separation of pluton-related and tectonic-related strain paths and the effects of each of the above variables on the finite strain ellipsoid. This is an important factor because the interaction between pluton-generated strains and tectonic strains produces a new three-dimensional strain pattern. Guglielmo's simulation assumed that a spherical pluton expands radially and produces a strain gradient in the host rocks. Models he devised were constrained with data from natural plutons and other published models. Flattening strains are oriented concentrically around both real (Figure 19–12b) and model plutons, and pluton-derived strains decrease gradually away from the pluton but may affect the country rocks as far away as half the pluton radius from the contact with the host rocks. The time of pluton emplacement relative to the peak of regional deformation is also an important variable in Guglielmo's analysis; additional variables include compositions and anisotropies of host rocks and the influence of strains from nearby plutons. He assumed that the regional deformation is homogeneous, ductile, and dominated by simple shear. Guglielmo appreciated the importance of the foliation triple points discussed by Brun and Pons (1981) in their two-dimensional analysis, and he was able to differentiate in three dimensions between triple points that form at the ends of plutons from those that form along the top. Foliation triple points are small and form equilateral triangles at the ends of a pluton, but triple points on the tops of plutons are larger and form isosceles triangles that flatten (opening the apical angle) closer to the pluton, finally becoming parallel to the contact at the pluton margin. The models reveal that foliation, mineral-stretching lineation, and regions of constrictional strain (prolate ellipsoids) form irregular three-dimensional ellipsoid-shaped rings around the pluton (Figure 19–13). Moreover, if the pluton expands rapidly relative to tectonic deformation rate, or if a higher expansion strain occurs, the rings are displaced farther from the pluton. Higher total tectonic strain, or higher tectonic strain rate, forces the rings closer to the pluton. Model simulations such as Guglielmo's permit better interpretation of both geologic map patterns of foliations, lineations, and other structures and meso-fabrics within and surrounding plutons. They also help to predict where to look for particular patterns, such as foliation triangles, and permit better interpretations of the kinematics of pluton emplacement in environments of noncoaxial strain.

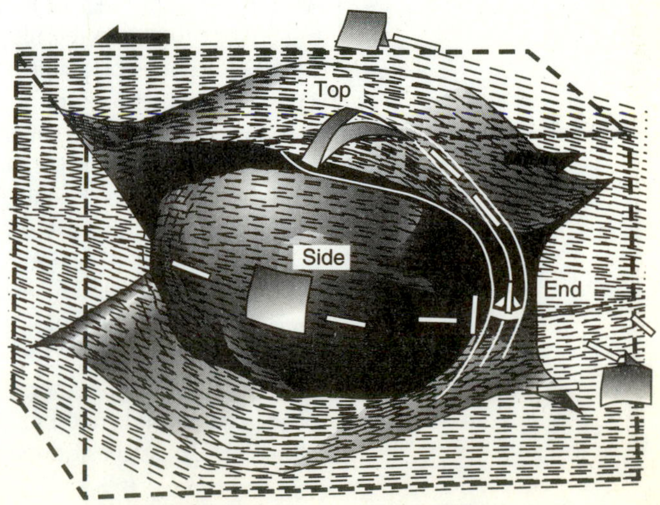

FIGURE 19–13
Relationships between an expanding ellipsoidal pluton and the foliation and lineation rings that develop by interaction of strains introduced by the pluton and regional tectonic strains, indicated by the large arrows here as sinistral motion. Note the large isosceles triangle shape of the foliation triple junction on top of the pluton contrasted with the smaller equilateral triangle shape of that at the end. (From *Journal of Structural Geology*, v. 15, Giovanni Guglielmo, p. 593–608, © 1993, with kind permission from Elsevier Sciences, Ltd., Kidlington, United Kingdom.)

ESSAY

A Tale of Two Plutons

We commonly think of plutons as having simple concordant or discordant contacts along their margins and that these relationships remain mostly unchanged in three dimensions. Several plutons have been described that have margins that change from strongly discordant in one part to partially concordant and even mylonitic in another. These relationships indicate major variations in the mechanism of intrusion, in external tectonic processes that affected part of the pluton, or in the level of erosion. Both the Papoose Flat pluton in California and the Bergell pluton in the Alps have contacts that vary with position along them.

The Papoose Flat pluton is one of several Mesozoic granitoid plutons in the White and Inyo Mountains in California (Figure 19E–1a). These are all shallow (epizonal) plutons that intruded a sequence of weakly deformed Late Proterozoic to Cambrian and younger Paleozoic sedimentary rocks that have been metamorphosed only to lower-greenschist facies (chlorite-grade) assemblages. Papoose Flat is not unique among the plutons of this area because several contain an extensive mylonite zone along the northwestern and western margins. Several also appear nearly concordant with the enclosing sedimentary rocks (Figure 19E–1b). Along the eastern and southeastern margins, however, these plutons are strongly discordant and have produced a rather spectacular but narrow contact aureole with virtually no evidence of accompanying deformation, a characteristic of shallow intrusions.

Detailed investigations of the Papoose Flat pluton by Arthur Sylvester, Gerhard Oertel, Clement Nelson, and John Christie (1978), and more recently by Scott Paterson, Tom Brudos, Ken Fowler, Chris Carlson, Kim Bishop, and Ron Vernon (1991) and by Richard Law, S. S. Morgan, A. G. Sylvester and M. W. Nyman (1992), have helped to unravel some of the complex history of this pluton. Sylvester and others (1978) found that the three Lower Cambrian stratigraphic units that are in contact with the pluton at the surface are thinned from a regional thickness of 1,180 m far from the pluton, to 910 m some 7 km from the pluton, to only 112 m along the western margin of the pluton. All investigations have noted the parallelism of micas and quartz forming a weak foliation in the entire pluton, along with the strong mylonitic foliations developed predominantly in the pluton but also to some extent in the host rocks along the western margin. Minimum temperatures during metamorphism were estimated using a variety of mineral thermometers to range from 380 to 540° C (Law and others, 1992). Foliation in the pluton, interpreted by Paterson and others (1991) to be tectonic, exhibits a wide range of dip, but the strike of foliation ranges from near concentric to subparallel to the pluton margins; mineral (stretching) lineations in both the pluton and host rocks generally trend northward and plunge moderately to shallowly (Figure 19E–2). Shear-sense indicators in the gneissic border facies and in the more internal western parts of the pluton indicate a consistent top-to-the-southeast transport of the cover relative to the pluton (Paterson and others, 1991; Law and others, 1992).

Differences of opinion have arisen among the different researchers about the roles played by either pluton expansion (ballooning) or externally derived top-to-the-southeast-directed shearing of the cover, affecting the western part of this and other plutons in the Inyo and White Mountains. Paterson and others (1991) concluded that the crystal-plastic deformation in the western part of the Papoose Flat pluton and others in the region is low-temperature deformation and that it is primarily related to regional shear, not to earlier emplacement processes. Using the same data as above, along with quartz c-axis data, Law and others (1992) concluded that the crystal-plastic deformation must have occurred at a high temperature (>450° C), because most of the quartz c axes are oriented in the mylonitic foliation perpendicular to the dominant trend of the lineation, indicative of at least amphibolite-facies conditions of deformation. As a result, Law and others (1992) suggested that difficulties with the temperature of deformation, quartz

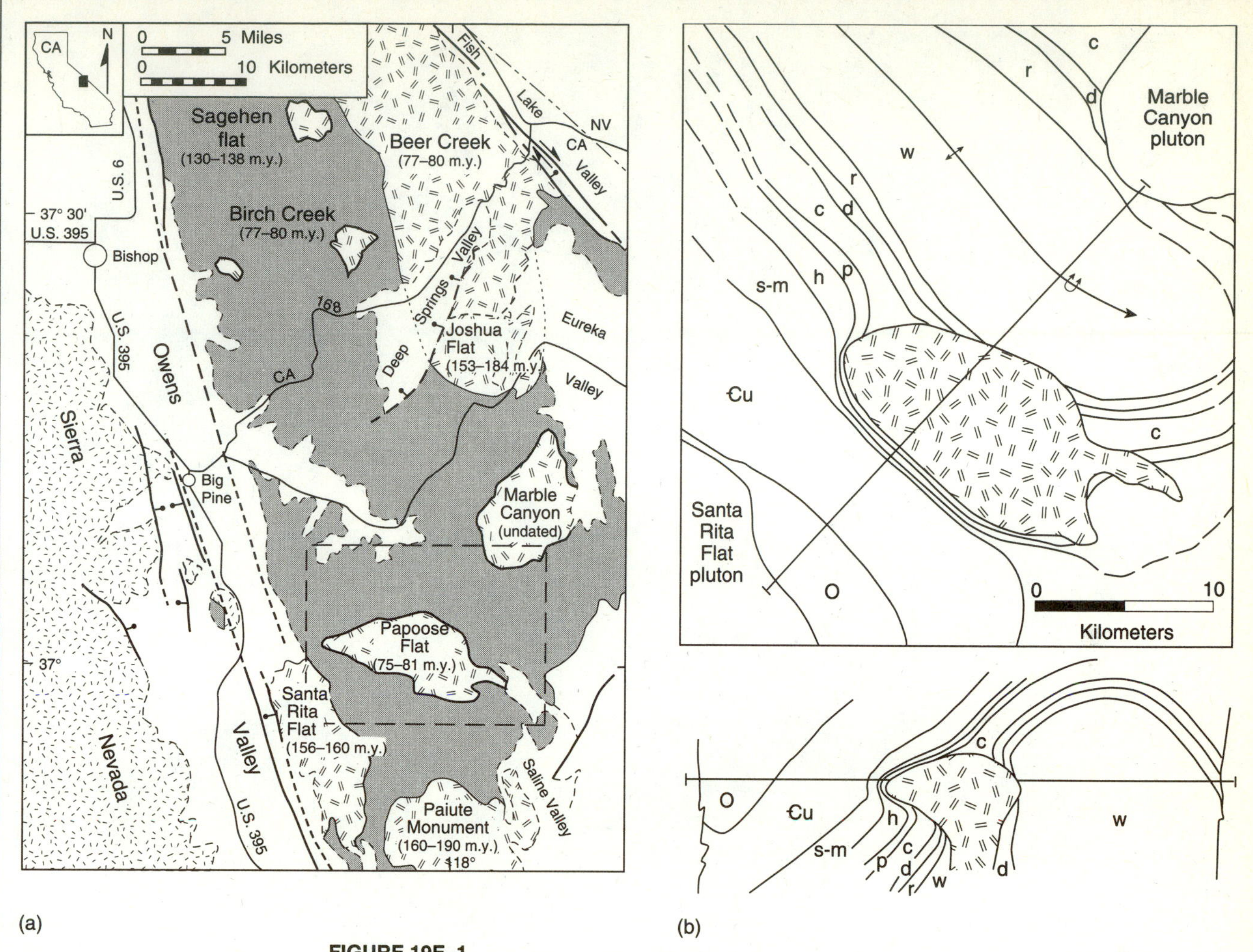

FIGURE 19E–1
(a) Cretaceous plutons in the White and Inyo Mountains, California. (b) Geologic map and cross section of Papoose Flat pluton. Note the abrupt cross-cutting nature of the contact with the country rocks on the eastern side in contrast with the layer-parallel (quasi-concordant) contact and thinned country-rock units along the western margin. w—Wyman Formation; r—Reed dolomite; d—Deep Spring Formation; c—Campito Formation; p—Poleta Formation; h—Harkless Formation; s-m—Saline Valley, Mule Spring, and Monola Formations; u—undifferentiated formations of Middle and Late Cambrian age, including Bonanza King Formation; O—undifferentiated Ordovician formations; patterned area—Papoose Flat pluton. (From A. G. Sylvester, G. Oertel, C. A. Nelson, and J. M. Christie, Geological Society of America *Bulletin*, v. 89, 1978.)

c-axis fabrics, and other data taken together preclude an independent role favoring one mechanism over another. They proposed three different models that would satisfy the most critical observations (Figure 19E–3a-19E–3c): (1) the intense plastic deformation restricted to the western part of the pluton, (2) mylonites indicating a plane strain deformation yielding a top-to-the-southeast shear sense, and (3) the quartz c-axis patterns are indicative of high-temperature deformation. These models are as follows: (1) northwest-directed forced intrusion of the western part of the pluton into the host rocks as an almost solidified wedge of granitoid was driven by the magma that remained in the pluton, (2) the mylonitic deformation around the western part of the pluton was associated with southeast-directed thrusting of the cover rocks over the pluton (the Paterson and others, 1991, model), and (3) both magmatic wedging and southeast-directed thrusting occurred at the same time. A fourth model, attributed by Law and others (1993) to Clifford

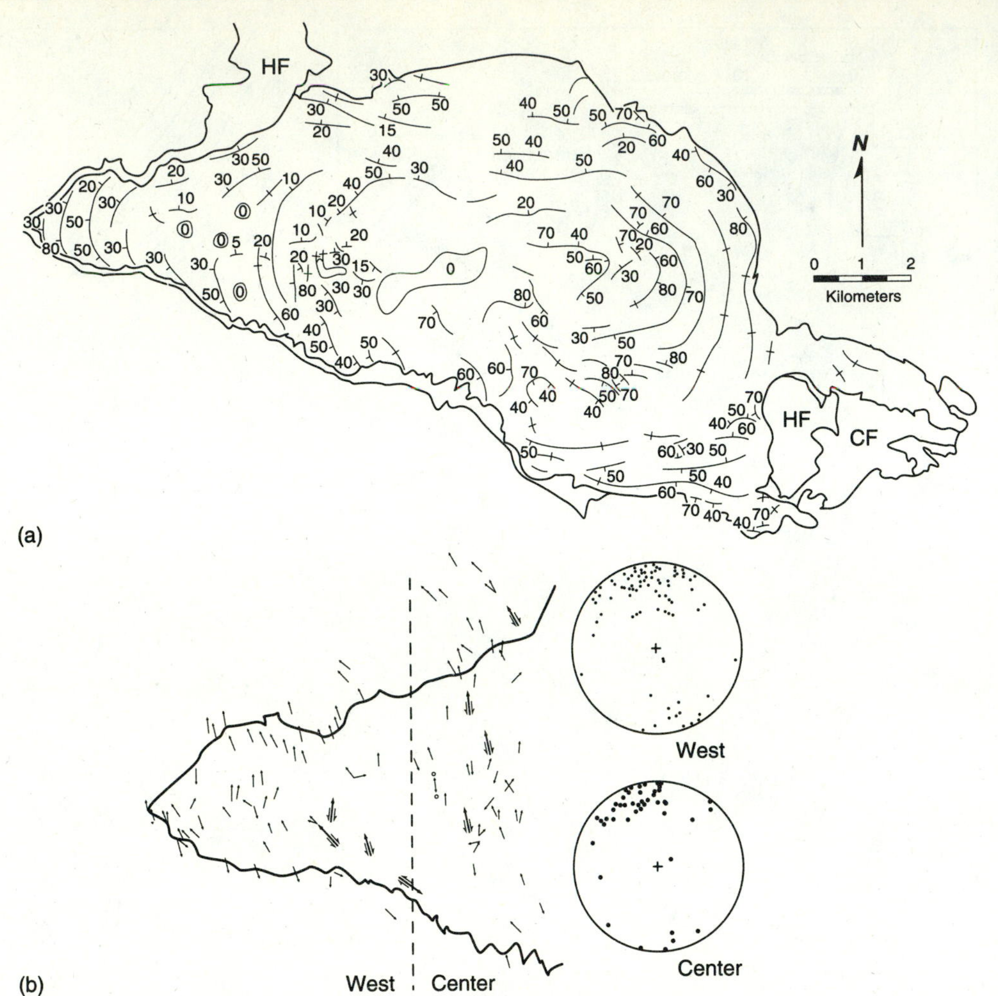

FIGURE 19E–2
(a) Trend lines showing variations of strike and dip of foliation in the Papoose Flat pluton. These were interpreted by Paterson and others (1991) as foliations that developed after the pluton was emplaced but was still in a warm, plastic condition. HF—Harkless Formation. CF—Campito Formation. (b) Mineral lineations (small arrows) and shear-sense determinations (half arrows, with open half arrows indicating motion of upper plate in the Paterson and others interpretation). (From S. R. Paterson, T. Brudos, K. Fowler, C. Carlson, and K. Bishop, *Geology*, v. 19, 1991.)

Hopson, suggests that northwest-directed magmatic wedging could be either accompanied or postdated by localized magma chamber expansion (Figures 19E–3d and 19E–3e). Later intrusion rate was not accompanied by a sufficient rate of lateral opening of a space to accommodate the magma. This precipitated ballooning of the magma chamber and upward arching of the cover rocks. During both stages, southeast-directed shear occurred, but the cover stratigraphy was thinned mainly during the latter stage in the Hopson model.

The Bergell pluton in the Pennine Alps in Switzerland is an Oligocene (30–32 Ma) granitic pluton that was tilted and eroded following intrusion, so that most of a complete section from the bottom to the top of the pluton can be studied (Figure 19E–4). It is one of only two large granitoid plutons generated during the Alpine orogenies in the eastern, central, western, and Italian Alps. The other is the Adamello pluton located in the southern Alps. The Bergell is thus the only major pluton located north of the Insubric line (Figure 19E–4a). It is actually located in the southeastern corner of the Pennine Alps along the boundary with the Austro-

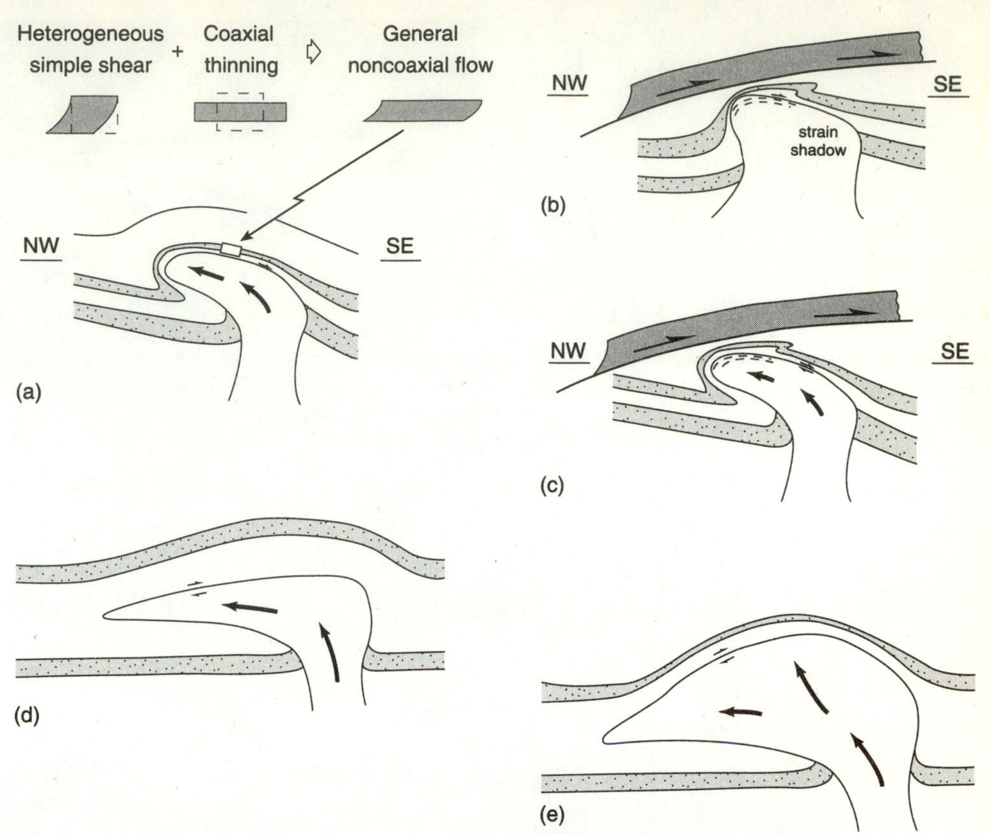

FIGURE 19E–3
Models for emplacement of the Papoose Flat pluton. (a) Magmatic wedging. (b) Regional southeast-directed thrusting of the host rocks over the pluton. (c) Combination of (a) and (b). (d) and (e) Model for emplacement of the Papoose Flat pluton attributable to C. A. Hopson, involving (d) magmatic wedging with sufficient volume to accommodate the magma as it is intruded, followed by a later phase of intrusion (e) in which there is insufficient volume in the magma chamber, causing the pluton to inflate laterally and upward. (a, b, and c from R. D. Law, S. S. Morgan, A. G. Sylvester, and M. Nyman, *Transactions of the Royal Society of Edinburgh: Earth Sciences,* v. 83, 1992. d and e from R. D. Law, A. G. Sylvester, C. A. Nelson, S. S. Morgan, and M. W. Nyman, Deformation associated with emplacement of the Papoose Flat pluton, Inyo Mountains, eastern California: Geologic overview and field guide, *in* M. M. Lahre, J. H. Trexler, Jr., and C. Spinosa, eds., *Crustal evolution of the Great Basin and the Sierra Nevada*: Reno, Nevada, Department of Geological Sciences and McKay School of Mines Field Trip Guidebook for 1993 Joint Meeting of the Cordilleran/Rocky Mountain Sections of the Geological Society of America, 1993.)

alpine thrust sheets. The Austroalpine sheets are interpreted to be part of Africa that were thrust over the Pennine zone rocks, which belong to either Europe or to the intervening closed Tethys ocean (Trümpy, 1975), and so the Bergell pluton was intruded along the suture between Africa and Europe.

The lower (southwest) and eastern margins and tail-like portion of the Bergell pluton consist of mylonitic tonalite (31.8 Ma), whereas the internal parts are composed of less deformed and slightly younger (30 Ma) granodiorite (Rosenberg and others, in press 1994). The pluton was intruded into part of the crust undergoing active deformation at a depth of several kilometers. Crustal deformation at the base of the pluton during the magmatic phase forced the pluton to expand laterally, so that when the pluton reached a particular level in the crust, pressure in the country rocks decreased sufficiently that they yielded plastically and the magma could expand. This enabled the pluton to become locally discordant along part of the margin (Figure 19E–4b). The Bergell pluton was initially intruded at depth into

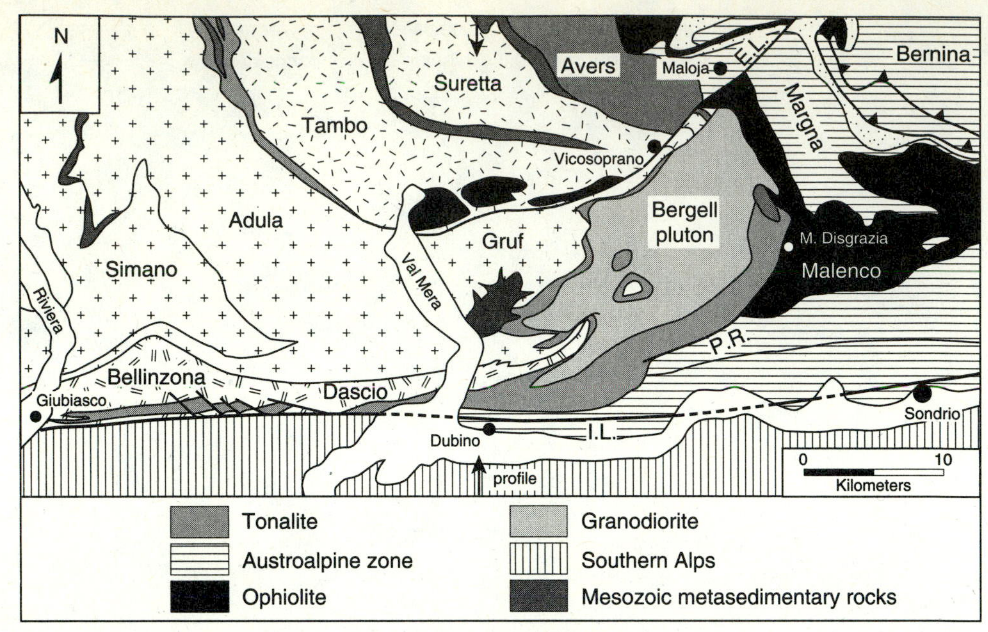

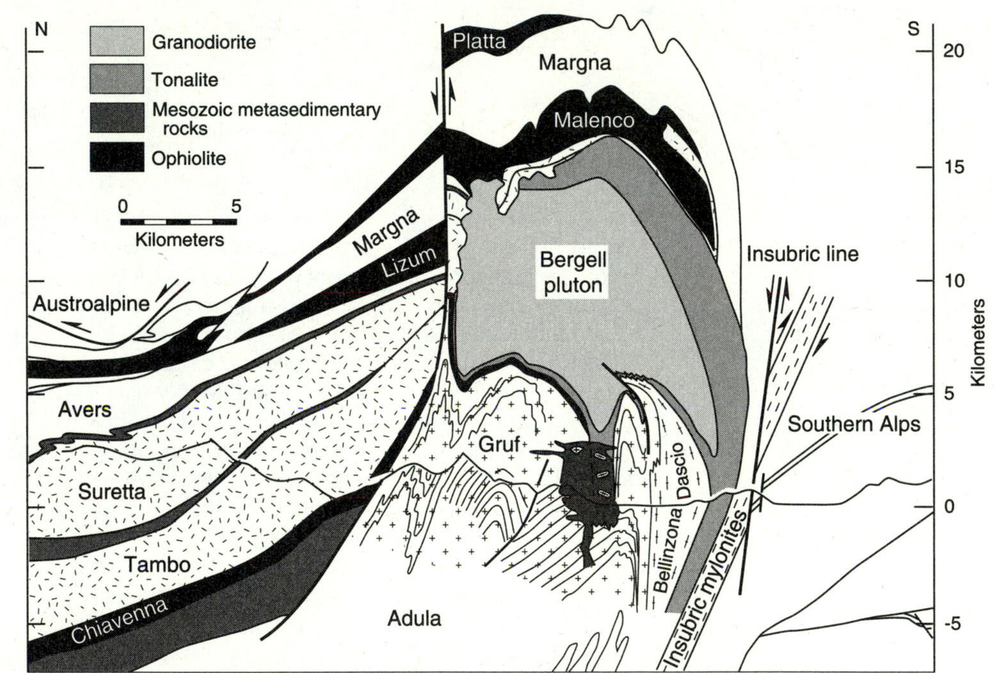

FIGURE 19E–4

(a) Simplified map of part of the western Alps in southern Switzerland and northern Italy showing the location of the Bergell pluton. E.L.—Engadine line; I.L.—Insubor line. Names in patterned areas are major tectonic units. (b) Cross section of the Bergell pluton. (From C. Rosenburg, A. Berger, C. Davidson, and S. M. Schmid, At*ti Ticinensi di Scienze della Terra*, in press, 1994.)

rocks undergoing amphibolite-facies metamorphism, and there was sufficient water present to produce partial melting (migmatization) of the host rocks near the base of the granitoid. Study of kinematic indicators revealed that two different phases of deformation were recorded in the pluton during intrusion: an early phase that involved top-to-the-east shearing along the base of the intrusion, and a later phase of north-south folding and shortening of the early-formed shear zone. This later phase produced shortening along the base of the pluton, causing lateral expansion of the pluton and deformation of the eastern margin.

References Cited

Law, R., Morgan, S. S., Sylvester, A. G., and Nyman, M., 1992, The Papoose Flat pluton of eastern California: A reassessment of its emplacement history in the light of new microstructural and crystallographic fabric observations: Transactions of the Royal Society of Edinburgh: Earth Sciences, v. 83, p. 361–375.

Law, R. D., Sylvester, A. G., Nelson, C. A., Morgan, S. S., and Nyman, M. W., 1993, Deformation associated with emplacement of the Papoose Flat pluton, Inyo Mountains, eastern California: Geologic overview and field guide, *in* Lahre, M. M., Trexler, J. H., Jr., and Spinosa, C., eds., Crustal evolution of the Great Basin and the Sierra Nevada: Reno, Nevada, Department of Geological Sciences and McKay School of Mines Field Trip Guidebook for 1993 Joint Meeting of the Cordilleran/Rocky Mountain Sections of the Geological Society of America, p. 231–261.

Paterson, S., Brudos, T., Fowler, K., Carlson, C., Bishop, K., and Vernon, R. H., 1991, Papoose Flat pluton: Forceful expansion or post emplacement deformation?: Geology, v. 19, p. 324–327.

Rosenburg, C., Berger, A., Davidson, C., and Schmid, S. M., 1994, Messa in Posto del plutone di Masino-Bregaglia (Bergell), Alpi Centrale: Atti Ticinensi di Scienze della Terra (in press).

Sylvester, A. G., Oertel, G., Nelson, C. A., and John Christie, J. M., 1978, Papoose Flat pluton: A granitic blister in the Inyo Mountains, California: Geological Society of America Bulletin, v. 89, p. 1205–1219.

Trümpy, R., 1975, Penninic-Austroalpine boundary in the Swiss Alps: A presumed former continental margin and its problems: American Journal of Science, v. 275–A, p. 209–238.

Questions

1. Among tabular plutons, why are sheet intrusions much more common than finger intrusions?
2. How could you distinguish tectonic from magmatic effects in a foliated pluton?
3. How does the mechanism of "magma fracturing" work?
4. Which is a more efficient mechanism for intrusion of tabular plutons, magma fracturing or stoping and assimilation? Why?
5. Explain the Parsons and Thompson relationship between dike intrusion and normal faulting. Using available geologic and tectonic maps, evaluate the value—or lack of value—of applying this mechanism to the Late Triassic–Jurassic rifting of eastern North America and formation of the present Atlantic Ocean.
6. What are the principal attributes of the diapir mechanism for emplacement of granitic magmas into the crust.
7. What is magmatic inflation, or ballooning, and how does it work?
8. How do magmas that are being intruded into a deforming rock mass interact with the host rocks?
9. Why are the early phases of a deformed pluton commonly more deformed than the late phases?
10. How can plutons be intruded into fault zones?
11. How can magma facilitate the movement on a fault?

Further Reading

Guglielmo, G., Jr., 1993, Interference between pluton expansion and non-coaxial tectonic deformation: Three-dimensional computer model and field implications: Journal of Structural Geology, v. 15, p. 593–608.
Presents a computer model constrained by field data to simulate the interaction between a pluton and host rocks that are undergoing active deformation in an attempt to predict the structures that may form by both processes and make it easier to separate them in the field.

Law, R., Morgan, S. S., Sylvester, A. G., and Nyman, M., 1992, The Papoose Flat pluton of eastern California: A reassessment of its emplacement history in the light of new microstructural and crystallographic fabric observations: Transactions of the Royal Society of Edinburgh: Earth Sciences, v. 83, p. 361–375.

Parsons, T., and Thompson, G. A., 1993, Does magmatism influence low-angle normal faulting?: Geology, v. 21, p. 247–250.
A relationship probably exists between emplacement of magma under a regional tensional stress field and normal faulting. This paper is a concise statement of the relationship.

Paterson, S. R., Brudos, T., Fowler, K., Carlson, C., Bishop, K., and Vernon, R. H., 1991, Papoose Flat pluton: Forceful expansion or post emplacement deformation?: Geology, v. 19, p. 324–327.
Presents several facets of the controversy about the intrusion history of the Papoose Flat pluton in the Inyo Mountains in southeastern California.

Paterson, S. R., and Fowler, K. T., 1993, Re-examining pluton emplacement processes: Journal of Structural Geology, v. 15, p. 191–206.

Weinberg, R. F., 1994, Re-examining pluton emplacement processes: Discussion: Journal of Structural Geology, v. 16, p. 743–746.

Paterson, S. R., and Fowler, K. T., 1994, Reexamining pluton emplacement processes: Reply: Journal of Structural Geology, v. 16, p. 747–748.
Examines the problem of the possible mechanisms that would permit diapiric emplacement of large volumes of magma into the middle to upper crust. Paterson and Fowler suggested that a combination of mechanisms must be responsible for emplacement of large volumes of magma, but Weinberg suggested that the process may occur by a combination of variations in strain rate and temperature, along with far-field flow of host rocks. This is an interesting exchange of ideas illustrating how science progresses through formulation of hypotheses, weighing a wide range of data, and reducing the number of hypotheses based on the available data.

Paterson, S. R., Vernon, R. H., and Tobisch, O. T., 1989, A review of criteria for the identification of magmatic and tectonic foliations in granitoids: Journal of Structural Geology, v. 11, p. 349–363.
Outlines the means by which penetrative structures, mainly foliations, that are products of magmatic processes can be distinguished from those formed by tectonic overprinting. They emphasize that no single criterion can be employed to distinguish these structures and that both structures in the pluton and in the adjacent host rock must also be considered.

Ramsay, J. G., 1989, Emplacement kinematics of a granitic diapir: The Chindamora batholith, Zimbabwe: Journal of Structural Geology, v. 11, p. 191–209.
Explores the evidence from field and mesoscopic data for a multiple emplacement history of a well-exposed pluton, providing a straightforward interpretation about separable early and late magmatic phases.

Schwerdtner, W. M., 1990, Structural tests of diapir hypotheses in Archean crust of Ontario: Canadian Journal of Earth Sciences, v. 27, p. 387–402.
Investigates the mechanism of intrusion of several large plutons in the southern Canadian shield to test the diapir hypothesis.

20

Structural Analysis

On this profile I platted [sic] all the data of structure at every outcrop of rock. I realized then how valuable to me had been the close attention I had given to minute details of structure in my youthful studies in the mountains of Corsica, for this had led me to infer that the structure in a specimen might repeat in miniature that of the great rock masses from which the small piece had come.

RAPHAEL PUMPELLY, 1918, *My Reminiscences*

STRUCTURAL ANALYSIS CONSISTS OF DESCRIBING AND interpreting all the structures in an area, on all scales (Figure 20–1). The technique complements the standard techniques of field geology—geologic mapping and stratigraphic analysis—and should be incorporated in any investigation of any deformed region. *Structural analysis* involves study of the structures described previously in this book and provides a systematic framework for relating them to one another spatially, to the mechanical processes that formed them, to their position in the crust, and to time. That is to say, it involves the observation, description, analysis, and interpretation of the kinds and orientations of folds and other linear structures, foliations and other planar structures, strain and displacement indicators, fractures and faults in any body of rocks of the Earth's crust, as well as in the crust of the other planets. There are several approaches to structural analysis, but each has the common goal described here.

As first developed, structural analysis was a technique for geometric and kinematic analysis of the minerals in a penetratively deformed rock mass. The technique was developed for the microscopic scale, but attempts were later made to relate microscopic fabrics to the large-scale deformation plan of the rock mass.

FIGURE 20–1
Strongly deformed and multiply transposed biotite gneiss in the Thor-Odin dome, Shuswap metamorphic complex, British Columbia. (RDH photo.)

Much structural analysis in the early twentieth century was developed by Bruno Sander, an Austrian structural geologist and metamorphic petrologist. His great work (1930) *Gefügekunde der Gesteine* (the English title is *An Introduction to the Study of the Fabrics of Geological Bodies)* formed the basis for structural analysis at that time and summarized advances that had been made in Germany. Later, Sander's concept of structural analysis was expanded to include all aspects of the deformational history of rock bodies. Further developments were reviewed by E. G. Knopf and E. Ingerson (1938) in *Structural Petrology;* their book enabled English-speaking geologists to apply the technique without extensive familiarity with the European literature. Later, Francis Turner and Lionel Weiss (1963) published *Structural Analysis of Metamorphic Tectonites,* which summarized structural analysis on all scales. During the 1950s, knowledge of multiple deformation, and particularly of mesoscopic and macroscopic polyphase folding (Chapter 16), and how various fold generations are related to foliations, lineations, and other fabric elements, was proliferated by a number of British geologists, including Basil King, Robert Shackleton, John Ramsay, Michael Fleuty, Lionel Weiss, Donald McIntyre, Nicholas Rast, Derek Flinn, and Brian Sturt. Their work, which dealt with specific areas in Scotland and pioneered modern techniques of geometric analysis, spawned many structural analyses in complexly deformed regions all over the world.

Microscopic structural studies require oriented thin sections, a petrographic microscope, and universal stage, and so they are much slower than mesoscopic studies. They are, however, the principal means of determining the orientations of crystallographic axes and other microscopic structures. They also provide a way to determine shear sense (Chapter 10) by measuring and relating the orientations of quartz c axes and glide planes (Simpson and Schmid, 1983), as well as to make an assessment of the temperature of deformation of certain minerals by determining the slip system involved. Microtextural studies, preferably with oriented thin sections, provide useful information about vergence, chronology of deformation, and relative times of deformation and metamorphism.

In this chapter we emphasize mesoscopic analysis, where measurements are made in the field with a compass and clinometer (Appendix 2). The discussion focuses on a scheme for analyzing deformed areas, whether simple or complex. Regardless of complexity, structural analysis should begin with the most fundamental of geologic data—the geologic map. If a detailed map does not exist, one must be made, and existing maps must be revised as new data are gathered. Initial mapping and mesoscopic structural analysis of an area frequently lead to other and more specialized studies, but making all kinds of specialized studies at one time may not be possible, or desirable.

DEFINITIONS

The term *fabric* encompasses the relationships between texture and structure in a rock. It relates the minerals, their orientations, and relative sizes to the structures in the rock.

We use the term ***tectonite*** to describe any rock containing a prominent tectonic fabric that is penetrative (Figure 20–2). The element may be either linear or planar. *S-tectonites* are dominated by foliation or cleavage; *L-tectonites,* by linear elements; *B-tectonites,* by fold axes. Combinations of linear and planar elements are possible—*LS-tectonites,* if linear elements dominate, and *SL–tectonites,* if planar elements dominate. This terminology is falling into disuse, but it conveys useful information and continues to appear in the literature.

RESOLVING STRUCTURES IN MULTIPLY DEFORMED ROCKS

Most structural-analysis techniques were devised in complexly deformed areas—in Phanerozoic orogens or in the exposed roots of Precambrian mountain chains of the continental shields—where they may still be best applied. Below, we will outline the general deformation plan for an orogen, explore options for implementing mesoscopic analysis, and examine the problems of recognizing separate structures. In all cases, the goal is to resolve the history of multiply deformed rocks, with ready application to those less deformed.

Deformation Plan in the Core of an Orogen

Mountain belts have many similarities in spatial arrangement of structures, sequence of development, and other properties. These similarities are repeated in orogenic belts wherever formed, from the Precambrian to the Tertiary (Figure 20–3). A *foreland fold-thrust belt* usually consists of a former trailing continental margin (formed at the beginning of a Wilson cycle; see Chapters 1 and 11) that was altered by thin-skinned deformation into a series of rootless folds and thrust faults directed toward the continent away from the deforming orogen. Foreland fold-thrust belts vary greatly in width—even in the same orogenic belt. They have

(a)

(b)

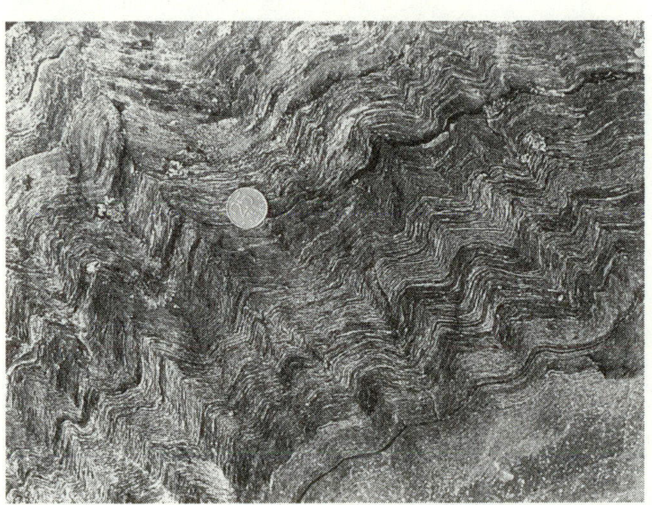

(c)

(d)

(e)

FIGURE 20–2
Different types of tectonites. (a) S-tectonite dominated by a single planar fabric; Middle Proterozoic Toxaway Gneiss near Whitewater Falls on the North Carolina–South Carolina border. (b) L-tectonite—an exposure dominated by mineral and crenulation lineations in the Middle Proterozoic Toxaway Gneiss at Whitewater Falls, North Carolina–South Carolina. The curved outline of the surface of the exposure is the enveloping surface of the smaller crenulation folds that define a larger fold. (c) B-tectonite—strongly crenulated amphibolite near Trondheim, west-central Norway. (d) LS-tectonite—quartzite layer dominated by linear fabrics but containing a strong foliation; near Burrells Ford, Oconee County, South Carolina. (e) SL-tectonite—Lower Cambrian Chilhowee quartzite at Linville Falls, North Carolina, dominated by a foliation but containing two lineations (an intersection and a mineral stretching lineation at 90°). (RDH photos.)

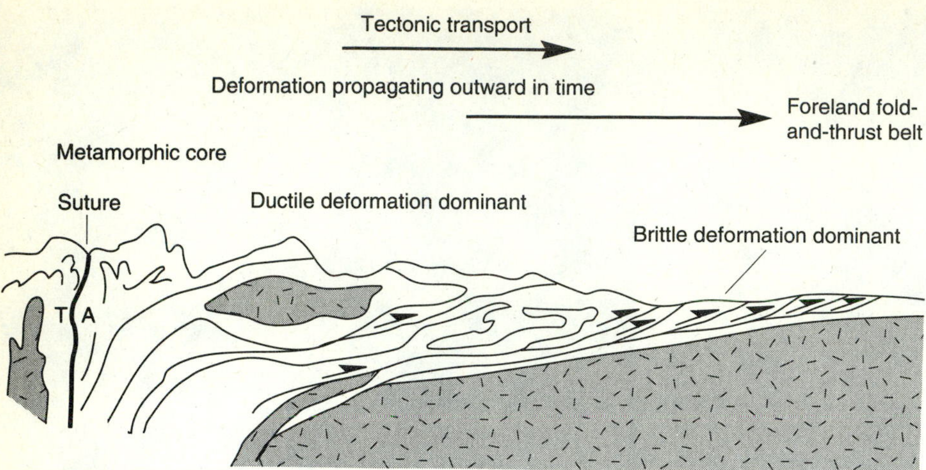

FIGURE 20–3
Sequence, style, and rheology of deformation in the inner and outer parts of an orogen.

been found in most orogens, including the Alps, the Andes, the Appalachians, the British and Scandinavian Caledonides, the North American Cordillera, Taiwan, and the Proterozoic Wopmay orogen of Canada. Toward the inner parts of most orogens, but adjacent to the foreland fold-thrust belt, is the *metamorphic core,* consisting of folded, faulted, cleaved, foliated, and intruded rocks ranging in metamorphic grade from low to high (Figure 11–3). The metamorphic core overlaps the foreland fold-thrust belt and may overlap any *accreted terranes* that have been sutured to the deforming mass.

Generally, the earliest recognizable structures in the metamorphic cores are produced by penetrative ductile deformation, which yields several generations of ductile faults (both dip-slip and strike-slip), folds, foliations, and linear structures. Earlier brittle or ductile structures rarely survive the main thermal-deformational event, but a few indicators of earlier events may be preserved in boudins, the cores of early porphyroblasts, intrafolial folds, or other structures. Later, crenulation cleavage and kinks may deform the early ductile structures, and then the crenulations may be overprinted by open flexural-slip buckle folds. Development of ductile faults, folds, tectonic slides (Chapter 11), and ductile-shear zones accompanies the early phase, which may then be overprinted by brittle structures. Deformation in the orogenic core is generally, but not always, earlier than that in the foreland. For example, the Alps and southern and central Appalachians were deformed from the inside outward in time and space, whereas the New England and Canadian Maritime Appalachians were deformed from west to east, with the oldest deformation in the foreland fold-thrust belt (Hatcher and Odom, 1980; Trümpy, 1980; Laubscher, 1988). Deformation that formed the early ductile structures in the deeper inner parts of the orogen propagates outward to the shallower outer parts and ultimately forms brittle structures. Later structures—open flexural-slip folds and brittle faults—primarily affect the outer parts of foreland fold-thrust belts but may develop throughout the entire orogen. Ductile deformation may occur in the core during the late stage and propagate to the outer zones as brittle structures.

Complex deformation in orogens requires an analytical scheme to resolve the succession of multiple events and to understand the mechanical/chemical processes that form structures at different times. One commonly used scheme begins with fundamentals: cross-cutting and overprinting relationships (Chapter 1). For example, if a foliation is folded, the foliation must have existed before the deformation that folded it; the radiometric age of a singly foliated pluton that cuts a sequence of multiply foliated metasedimentary rocks provides age constraints for both pre- and post-intrusion foliations.

Pitfalls in Using Style and Orientation in Polyphase-Deformed Rocks

Style and orientation of structures are frequently correlated to resolve the sequence of complex deformation. Differences in style and orientation generally show that the structures formed at different times, but they may not have done so; similarities in style and orientation do not necessarily mean that the structures are contemporaneous. Obviously, the best means of deciphering the structure of an area is to find one place where the entire deformational sequence is preserved—a "Rosetta stone" such as that discussed in Chapter 16. Such a place reveals overprinting and crosscutting relationships of superposed structures without forcing the observer to correlate structures from one small exposure to another. It permits resolution of representative styles and orientations of structures of different generations, so that—where exposure is less complete—the observer can confidently separate individual structures and arrange them chronologically. Any area of 100 km^2 or more should contain at least one exposure revealing most mesoscopic structures. Some structural elements may be absent because of the rock type or other factors. At best, correlating particular events, and even resolving the questions of style, may still prove difficult.

In a critique of reasoning processes used in structural analysis, Paul Williams (1970) demonstrated that fold style or orientation may not be confined to one particular generation of folds. Williams' example was the well-exposed coastal section at Bermagui, New South Wales, where he concluded that fold generations in a complexly deformed area can be resolved only by use of overprinting relationships. Thus, an exposure where overprinting relationships may be observed is necessary to resolve the structural chronology and relationships between deformational episodes.

Generally, where contrasting behavior is observed, as in a ductile fabric overprinted by brittle structures, it can usually be assumed that enough time had elapsed after development of the ductile structure that the rock mass cooled before overprinting by brittle deformation. An exception occurs, however, where an abrupt increase in strain rate produces a brittle structure that otherwise would form ductilely at a slower rate.

Problems of Multiply Transposed Rocks

In analyzing mesoscopic fabrics and working out the structural history of an area containing polydeformed rocks and rocks of moderate to high metamorphic grade, identification of the earliest structures may be difficult or impossible because of subsequent transposition. In such cases, the rock mass must remain sufficiently ductile so that penetrative deformation can overprint and mask earlier structures. Such earlier structures are sometimes preserved in boudins formed from more competent layers—amphibolite, quartzite, or calcsilicate—in a mass of more ductile rock (Figure 20–4). Structures preserved in boudins may represent only an earlier pulse of a major deformation phase, and

(a)

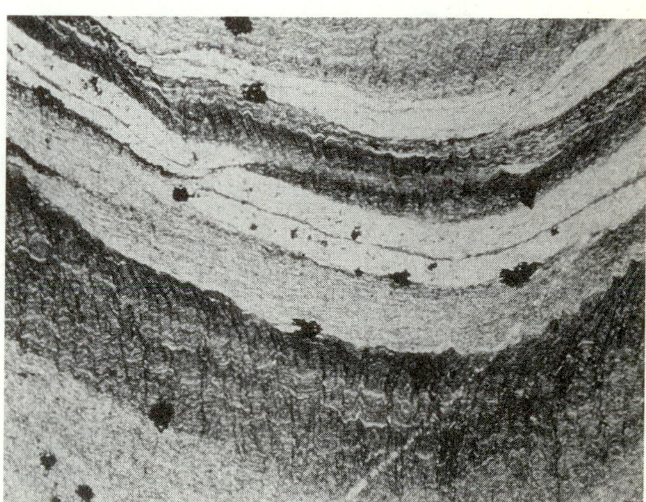

(b)

(c)

FIGURE 20–4
(a) Preservation of early folds and foliation in a boudin of amphibolite about 1 m long, Trans-Canada Highway west of Revelstoke, British Columbia. The enclosing foliation truncates the foliation and folds in boudins. The structures could have formed at different times during the same event or during separate events. (RDH photo.) (b) Earlier foliation (transposed bedding) preserved as microlithons between later foliation developed as crenulation surfaces in the dark layers in Lower Devonian Seboomook Formation schist and metasandstone, Somerset County, Maine. (Arden L. Albee, California Institute of Technology.) (c) Earlier foliation (pseudo cross beds) preserved between later foliation in Middle Proterozoic Toxaway Gneiss near Whitewater Falls, North Carolina–South Carolina. (RDH photo.)

not a phase distinctly different. Formation during another phase of deformation is also possible. Multiple transposition may be deduced from intrafolial folds and microlithons containing an earlier foliation.

MESOSCOPIC ANALYSIS

We will now outline the terminology and fabric elements useful in conducting mesoscopic analysis, which involves resolution of the spatial-temporal relationships among mesoscopic structures. In any study area, all fabric elements should be examined for geometric style, orientation, and overprinting relationships. Fabric data may be plotted on maps, equal-area (Schmid) diagrams, and equal-angle (Wulff) diagrams using techniques described in Appendix 1. Oriented samples should be collected for analysis of microtextures and determination of strain.

Few areas are homogeneous with respect to the structures they contain. In one place, some fabric elements—folds and other lineations, fold axial surfaces, and cleavage or foliation—may have a common orientation, but in another part nearby, the same elements may be oriented differently. A smaller part of the area where one fabric element is oriented the same throughout is called a ***homogeneous domain*** (Figure 20–5). Foliation or bedding may have the same dip and strike, the same kinds of lineations may have a similar trend and plunge, and all fold axes and axial surfaces of the same generation may have similar orientations. Folds may also have the same style and sequence of overprinting, and each set may have

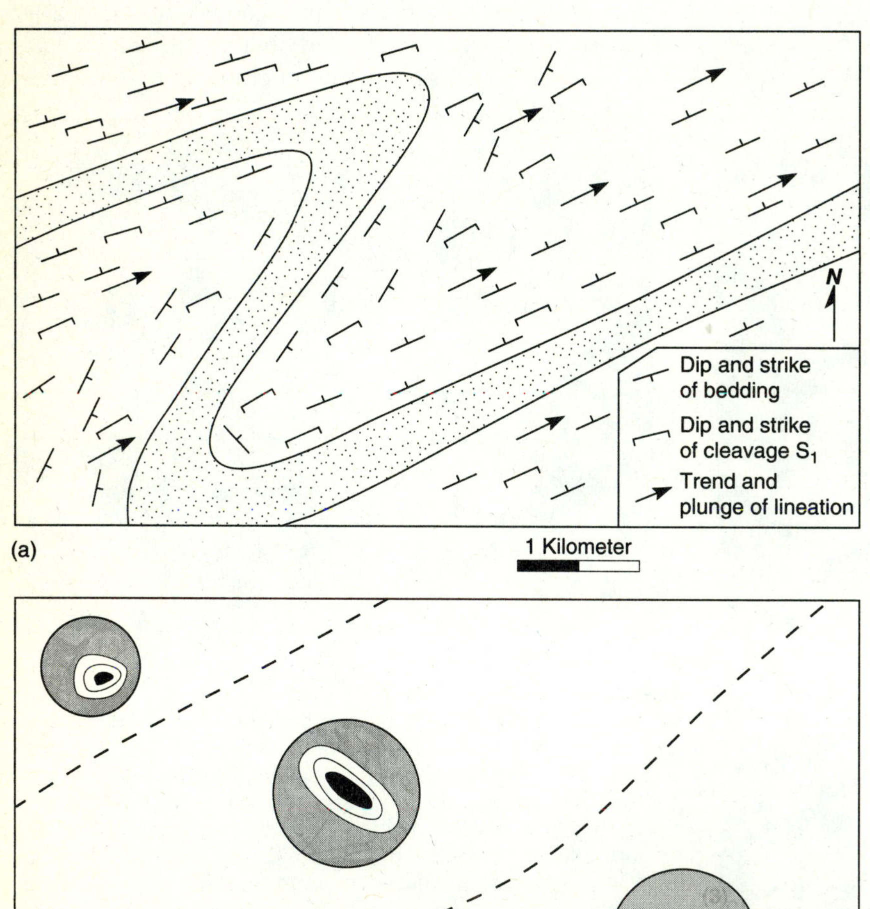

FIGURE 20–5
(a) Geologic map of a hypothetical area with structural data (bedding, S_1, $L_{0\times1}$).
(b) Homogeneous domains of poles to bedding. (c) Fabric diagrams of $L_{0\times1}$ axes and poles to S_1.

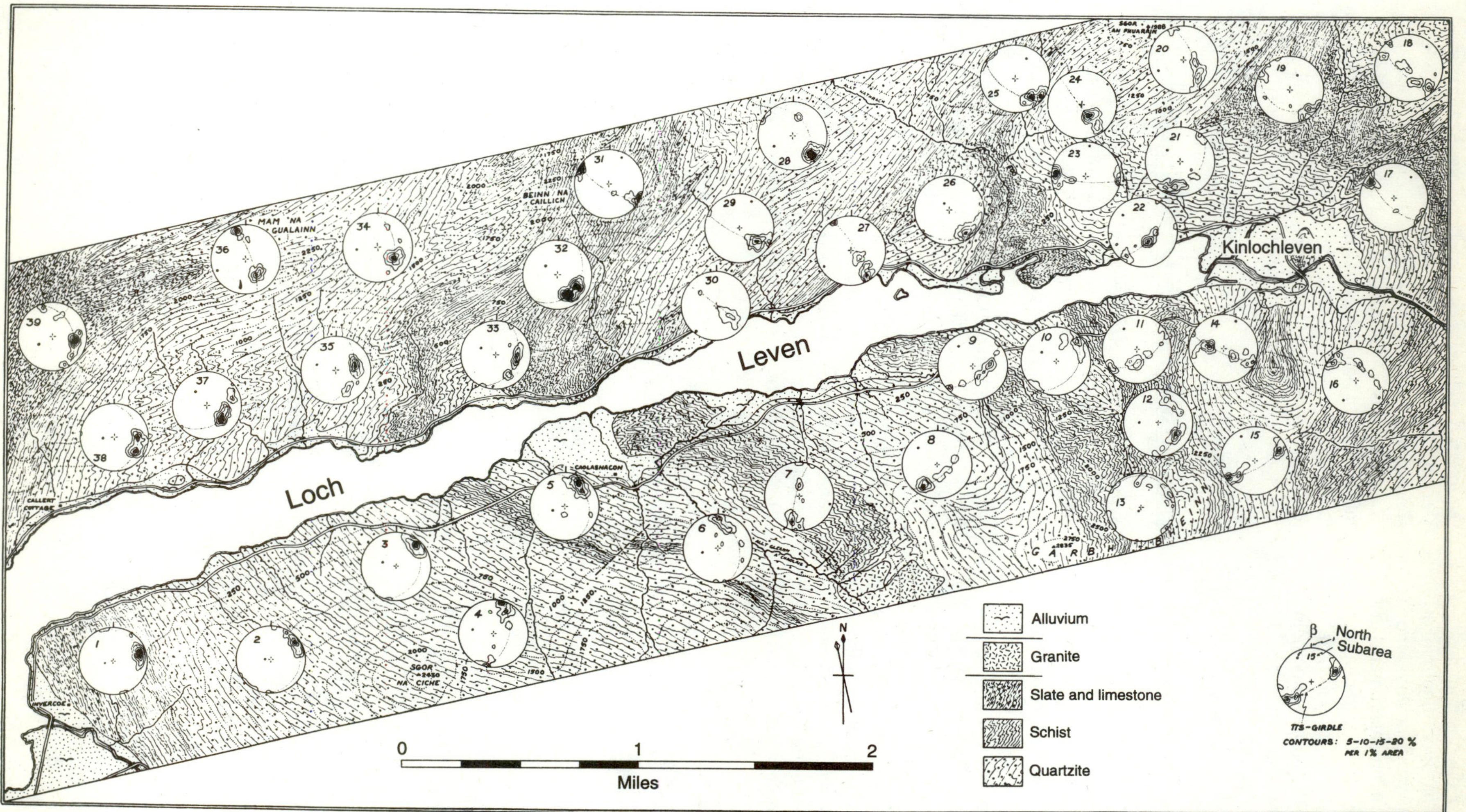

FIGURE 20–6
Homogeneous domains of the dominant foliation in metasedimentary rocks from the Scottish Highlands. The following numbers refer to the subareas (homogeneous domains) and the number of measurements of dip and strike of foliation (plotted as poles) made in each. 1—40. 2—57. 3—85. 4—42. 5—29. 6—52. 7—68. 8—71. 9—31. 10—70. 11—37. 12—15. 13—63. 14—36. 15—33. 16—86. 17—28. 18—64. 19—28. 20—58. 21—42. 22—73. 23—30. 24—86. 25—25. 26—89. 27—37. 28—43. 29—87. 30—98. 31—31. 32—13. 33—38. 34—65. 35—36. 36—38. 37—47. 38—53. 39—18. (From L. E. Weiss and D. B. McIntyre, 1957, Structural geometry of the Dalradian rocks at Loch Leven, Scottish Highlands, *Journal of Geology,* v. 65, p. 575–602. Published by permission of University of Chicago Press.)

formed at the same time. Homogeneous domains can most easily be defined by bedding or dominant foliation data, because that is the structure most commonly measured (Figure 20–6). If enough measurements are available, folds and other linear fabrics also help to distinguish homogeneous domains. It may also be useful to define ***heterogeneous domains***, which help one to recognize fold-hinge zones, superposed folds, ductile shears, and other structures; for a practical example, see the Essay accompanying this chapter.

Structural Measurements

Folds are only one of the fabric elements studied in mesoscopic fabric analysis. When a fold is observed in the field, the geologist should first measure the trend and plunge of the axis, the strike and dip of the axial surface, and the interlimb angles in the fold limbs (Appendix 2). Furthermore, the geologist should do the following: record the fold style and inferred mechanism, along with the vergence of the fold; measure the trend and plunge of other lineations; record relationships of all linear elements to folds; measure strike and dip of planar fabrics (bedding, joints, foliations, and cleavage), and note the character of each foliation for resolution of the various kinds of S-surfaces. Finally, the inferred relative ages of all structures should be recorded. Presence and motion sense of observable shear-sense indicators should be recorded. Rotated porphyroclasts, shear bands (C-surfaces), and other indicators will help to determine sense of movement on faults (Chapter 10). Rotation sense on the limbs of folds can be recorded by observing shear-sense indicators and the asymmetry of small parasitic folds.

SYMMETRY OF FABRICS

Symmetry of structures observed in the field and plots of structural data frequently yield useful information for structural analysis. In rocks with no preferred orientation of constituent grains—most igneous rocks, many sedimentary rocks, and some hornfels—a fabric diagram of orientation of poles to platy minerals (such as biotite) displays *spherical symmetry* (Figure 20–7a). A single, preferred orientation—a mineral lineation—results in the symmetry of a cylinder, *axial symmetry* (Figure 20–7b), produced in an L-tectonite.

Orthorhombic symmetry with two (or three) planes of symmetry is observed in plots of poles to bedding from symmetrical folds (Figures 20–7c and 20–7d). Plots of poles to two sets of joints oriented

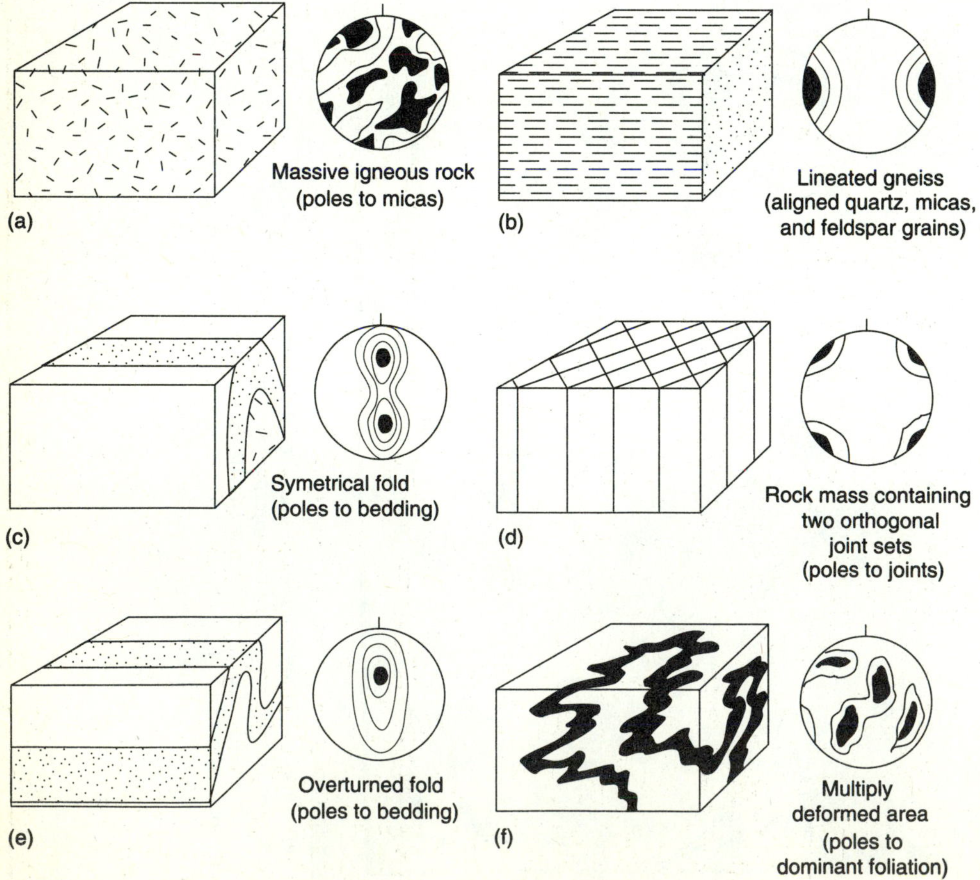

FIGURE 20–7
Fabric symmetries.
(a) Spherical. (b) Axial.
(c) and (d) Orthorhombic.
(e) Monoclinic. (f) Triclinic.
The side of each block nearest the reader faces south. Fabric diagrams plot data from top (map view) of each block.

normal to each other also reveal orthorhombic symmetry (Figure 20–7d). *Monoclinic symmetry* with one symmetry plane is found in asymmetric folds (Figure 20–7e). This lower symmetry is probably more common in nature than the other kinds described above. *Triclinic symmetry* may indicate multiple deformation (Figure 20–7f).

Fabric Axes

Early in the development of structural analysis, geologists found it possible to relate structure and fabrics to coordinate (fabric) axes. The discussion here of fabric axes is included largely for its historical value, because most structural geologists today try to relate structures to the principal axes and planes of the strain ellipsoid, which, as we have seen, is a different set of coordinate axes. ***Geometric axes*** (Figure 20–8) describe the spatial relationships between particular structures and a set of axes *a, b,* and *c.* Here is one scheme, recommended by Turner and Weiss (1963) for orienting fabric axes:

1. The *ab* plane is oriented parallel to the most prominent planar structure in the rock mass, and a prominent lineation lying in *ab* is *b*, particularly if normal to a plane of symmetry of the fabric. If no lineation is found, the orientation of *b* is fixed arbitrarily.
2. In a rock mass containing two or more planar structures that intersect in a common axis, the most prominent of the two planar structures is designated *ab,* and the intersection designated *b.*
3. If more than two planar structures occur and do not intersect in a common axis, the most prominent structure is designated *ab,* and its intersection with the next most prominent structure is *b.*
4. In a rock body dominated by a strong lineation, the orientation of the lineation is *b,* and any direction normal to *b* is *a.* The *a* axis is best oriented in a planar structure.

These rules permit relating fabric axes to fabric symmetry. In rocks dominated by linear fabrics (axial symmetry), *b* or *c* is the principal axis of symmetry. The *a, b,* and *c* axes are perpendicular to symmetry planes in structures (such as symmetrical folds) with orthorhombic symmetry. The *b axis* is perpendicular to the one plane of symmetry in structures (such as asymmetric folds) having monoclinic symmetry, and it parallels the fold axis; the *a* and *c* axes then lie in the plane perpendicular to *b* (*a* may lie in the axial surface of the fold). In *kinematic (movement plan) axes,* as Sander (1930) defined them, the *a* axis is the tectonic transport direction, *b* is perpendicular to *a* in the transport (fold axial) plane, and *c* is perpendicular to the *ab* plane. Realistically, determining true movement directions in multiply deformed rocks is extremely difficult, and so is resolution of kinematic axes. Thus, at the present time, kinematic axes are seldom identified, except to determine the primary transport direction(s). Emphasis instead is on relating structures to the strain ellipsoid and determining orientations of strain axes—a more efficient way of describing structures and determining deformation paths.

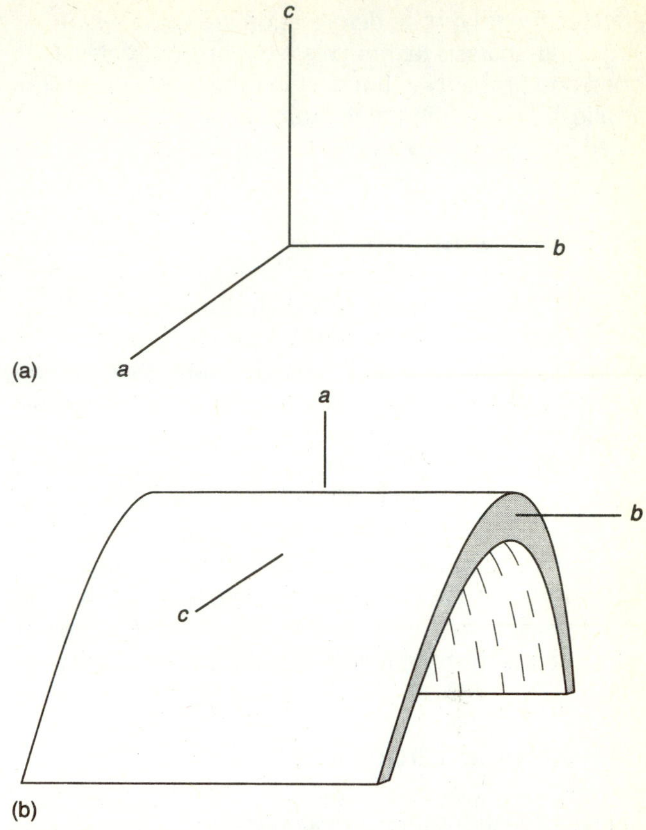

FIGURE 20–8
Geometric fabric axes (a) and conventional use related to a fold (b).

STRUCTURAL-ANALYSIS PROCEDURES

Our goal in this section is to formulate a plan for analysis in any structural setting, whether the area has been deformed singly or multiply. Each setting dictates a slightly different approach to fabric analysis, but as a whole, the techniques are similar. An area deformed by a single event poses fewer problems for structural analysis than an area deformed more than once, but useful results may still be gained by applying these techniques to singly deformed areas. Applying them makes it possible to determine aspects of the structural history that would otherwise be impossible. Areas affected by polyphase deformation are those where structural-analysis techniques were originally developed and probably still have their greatest utility. The applicability of different techniques to different areas is listed in Figure 20–9 and is discussed below.

Geologic Mapping

Detailed geologic mapping—both structural and stratigraphic—should be undertaken to provide the database essential for any geologic study. Geometric distribution of map-scale structures and their constituents must be shown precisely on a map of reasonably large scale (1:24,000, 1:12,000, 1:10,000, or larger). It may be useful to map particular exposures at scales of 1:120 or 1:60, using a measured or surveyed grid laid out on the exposure itself and transferred to graph paper for a base map. Such mapping is the basis of all structural studies and is necessary for understanding of the structural history, regardless of the kind of analysis.

Mesoscopic Structural Analysis

Study of foliations, lineations, folds, and other mesoscopic structural elements is undertaken by first measuring them in the field and then plotting each element separately on a map overlay to display the spatial distribution. The data should also be plotted on fabric diagrams (Appendix 1) to show the relationships between structural elements of different kinds. Relationships of linear fabrics to shear-sense indicators can reveal important information about the kinematics of deformation. Domain analysis (Figures 20–5 and 20–6) should also be undertaken if there are variations in orientations of major fabric element data sets in different parts of the area. Fold axes and other linear structures, plotted on geologic maps or overlays, will reveal the spatial distribution, changes in plunge, and relationships to macroscopic folds. Cleavage may or may not be uniformly present in singly deformed areas but may occur locally in the hinges of mesoscopic folds as an incipient domainal pressure-solution cleavage, as penciling, or even as a well-developed cleavage.

Microtextural Studies

Study of thin sections should reveal relationships between various fabric elements. Techniques involving cross-cutting and overprinting relationships may be applied in thin section more clearly than in the field. Deformation mechanisms, recrystallization phenomena, and many fault fabrics—both brittle and ductile—are best studied on the microscale. Early structures preserved in porphyroblasts of garnet, feldspar, staurolite, and other minerals (Figure 20–10) and in boudins may be a key to understanding multiple deformation and thus the structural history. Many clues to shear sense (Chapter 10) are also easy to find in thin section but are of minimal value unless the section is oriented.

Fracture Analysis

Formation of systematic joints and other fractures may be related to development of folds and faults within the area, and so their measurement and analysis is worthwhile. Fracture analysis is commonly undertaken separately from standard fabric analysis: the goals of fracture analysis are frequently related to interpretation of brittle events affecting the rocks much later than those producing penetrative ductile deformation (Wise and others, 1979). Fracture patterns may be related to late folds and therefore may elucidate fold mechanisms.

Finite- and Incremental-Strain Studies

Rocks in an area deformed penetratively make possible the study of finite and incremental strain (Chapter 5), using folds or other strain markers. The sequence of opening and filling of veins (crack-seal relationships), along with vein orientations, should be observed and measured. Study of finite and incremental strain may begin with measurement of internal strain in veins and any other strain markers; the goal is to determine the magnitude and orientation of strain as it affected the rocks at different times. Study of fibers in veins or pressure shadows (or both) is one of the best ways to measure noncoaxial progressive deformation.

Chronology of Development of Structures

Measurements of all structural elements, on all scales, and results of microtextural studies must be integrated into a complete structural analysis so that the chronology of development can be worked out. When completed, it will provide a reasonably complete picture of the kinematics, mechanics, and sequence of deformation in the area. Such is the principal goal of structural analysis.

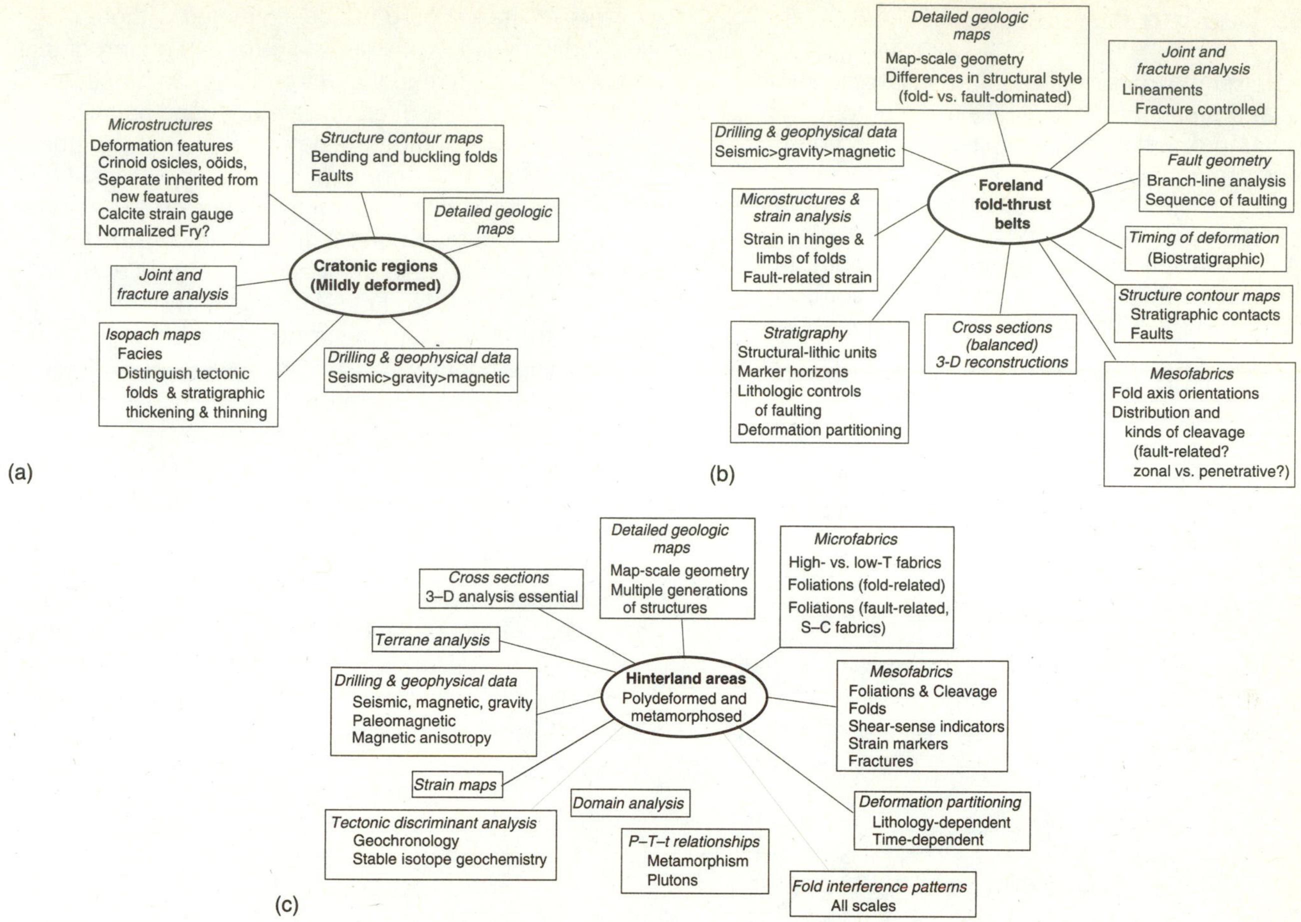

Microstructures
Deformation features
Crinoid osicles, oöids,
Separate inherited from new features
Calcite strain gauge
Normalized Fry?

Structure contour maps
Bending and buckling folds
Faults

Detailed geologic maps

Cratonic regions (Mildly deformed)

Joint and fracture analysis

Isopach maps
Facies
Distinguish tectonic folds & stratigraphic thickening & thinning

Drilling & geophysical data
Seismic>gravity>magnetic

(a)

Detailed geologic maps
Map-scale geometry
Differences in structural style (fold- vs. fault-dominated)

Joint and fracture analysis
Lineaments
Fracture controlled

Drilling & geophysical data
Seismic>gravity>magnetic

Foreland fold-thrust belts

Fault geometry
Branch-line analysis
Sequence of faulting

Microstructures & strain analysis
Strain in hinges & limbs of folds
Fault-related strain

Timing of deformation
(Biostratigraphic)

Structure contour maps
Stratigraphic contacts
Faults

Stratigraphy
Structural-lithic units
Marker horizons
Lithologic controls of faulting
Deformation partitioning

Cross sections (balanced)
3-D reconstructions

Mesofabrics
Fold axis orientations
Distribution and kinds of cleavage (fault-related? zonal vs. penetrative?)

(b)

Detailed geologic maps
Map-scale geometry
Multiple generations of structures

Microfabrics
High- vs. low-T fabrics
Foliations (fold-related)
Foliations (fault-related, S–C fabrics)

Cross sections
3–D analysis essential

Terrane analysis

Hinterland areas
Polydeformed and metamorphosed

Drilling & geophysical data
Seismic, magnetic, gravity
Paleomagnetic
Magnetic anisotropy

Mesofabrics
Foliations & Cleavage
Folds
Shear-sense indicators
Strain markers
Fractures

Strain maps

Domain analysis

Deformation partitioning
Lithology-dependent
Time-dependent

Tectonic discriminant analysis
Geochronology
Stable isotope geochemistry

P–T–t relationships
Metamorphism
Plutons

Fold interference patterns
All scales

(c)

FIGURE 20–9
Structural analysis of areas in different settings requires many of the same techniques, but a greater variety and more data are required in more complexly deformed areas. (a) Cratonic regions. (b) Foreland fold-thrust belts. (c) Internal parts (hinterland regions) of mountain ranges.

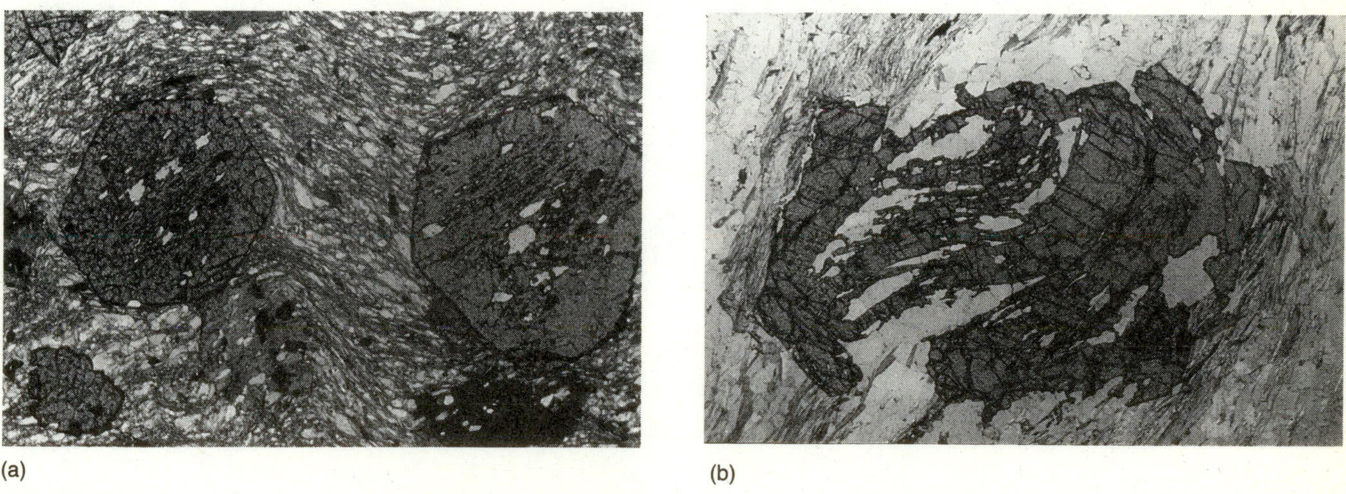

(a) (b)

FIGURE 20–10
Earlier foliation preserved as aligned inclusions in garnet porphyroblasts, suggesting multiple deformation in schist from Upper Proterozoic structural Ammons Formation near Topton, North Carolina. The garnet has a younger rim devoid of inclusions, indicating additional growth after the event that formed the core of the crystal. Could this grain be detrital, and the overgrowth formed in a later thermal event? If so, is it still evidence of multiple deformation? (Thin section courtesy of M. P. Ausburn.) (b) Multiply deformed rotated "skeletal" garnet in pelitic schist from the Einunnfjellet region , south-central Norway. (Specimen courtesy of Elizabeth A. McClellan, Western Kentucky University.) Plane light. Width of field in both is approximately 7 mm.

Cross-Section Analysis

If a single set of structures is dominant, cross sections normal and parallel to strike make a useful tool. Construction of a standard geologic cross section in a singly deformed area requires careful transfer of contacts from geologic maps to the section line (preferably also showing topography) and systematic projection of the contacts between rock units to depth (Chapter 11). If this is done properly and folding is parallel, thicknesses of rock units will remain constant. Vertical and horizontal scales should be the same to depict structural geometry without artificial distortion. Balancing of cross sections involves retrodeforming the section to its undeformed state to see whether the section can be taken apart and then reconstructed (see Figures 11–36 and 11–37). It is always desirable to balance cross sections, an easy task if the rocks are not penetratively deformed. Balancing sections by retrodeforming them should be carried out because it tests the interpretation: if a section will not balance, it cannot be correct; if it does balance, it may be correct (Hossack, 1979; Price, 1981; Elliott, 1983). In a complexly deformed area, construction and balancing of cross sections should not be attempted without detailed knowledge of strain states or without proof that plane strain dominated the deformation. In practice, cross sections in these regions are not often balanced, because in areas of multiple deformation, the structures have more than one dominant orientation. Probably the better course is to construct accurate cross sections along lines in different orientations—which is better than not constructing them at all. In a classic example, Ramsay (1981) resolved the differences in strain between three internally deformed thrust sheets in the Helvetic Alps (Figure 20–11). The lowest thrust sheet (the Morcles) is deformed more than the two above (the Diablerets and the Wildhorn), with the lowest part of the Morcles and the down-dip parts of the two higher thrust sheets internally deformed more than other parts of the sheets.

This completes our survey of geologic structures toward the understanding of large-scale and structural analysis. The remaining chapter discusses geophysical techniques that provide useful information—on the map to continent scale—the subject of tectonics.

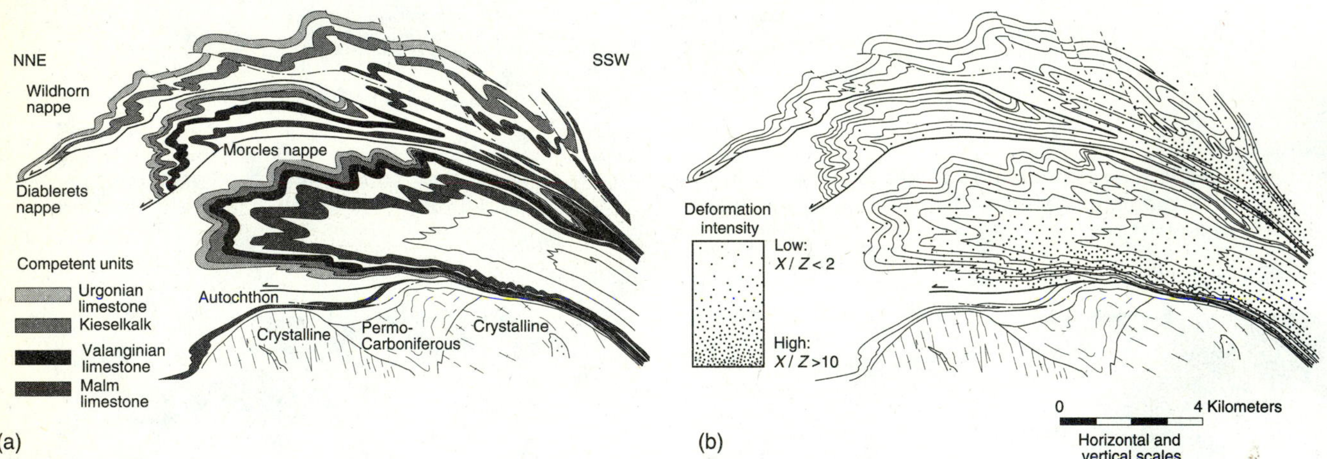

FIGURE 20–11
Cross sections through three internally deformed thrust sheets in the Helvetic Alps in Switzerland. (a) Geologic section showing relationships between rock and tectonic units and internal deformation in each thrust sheet. (b) Intensity of deformation in the cross section (indicated by density of stipple pattern), determined by study of strain markers within each sheet. (From J. G. Ramsay, 1981, *in* N. J. Price and K. R. McClay, eds., *Thrust and Nappe Tectonics:* Geological Society of London Special Publication 9, Blackwell Scientific Publications Limited.)

ESSAY

Structural Analysis at Woodall Shoals

In this essay we present a problem in structural analysis and attempt to solve it. The area chosen is the "Rosetta stone" at Woodall Shoals, described previously (Chapter 16) as a key to the structural history of a large area in the eastern Blue Ridge of the Carolinas and Georgia. (For background, review the Chapter 16 Essay.) Woodall Shoals is a large rapids and rock exposure of about 700 m^2 along the Chattooga River at the border of Georgia and South Carolina (Figure 20E–1). A geologic map of Woodall Shoals is published as part of another map by the South Carolina Geological Survey (Hatcher and others, 1995, in press). The Woodall Shoals exposure was mapped by geology students from Clemson University (Steven H. Poe) and the University of South Carolina (Jack P. Horkowitz, Daniel B. Wynne, Arlene C. Burns) and myself as a part-time project that I coordinated from 1976

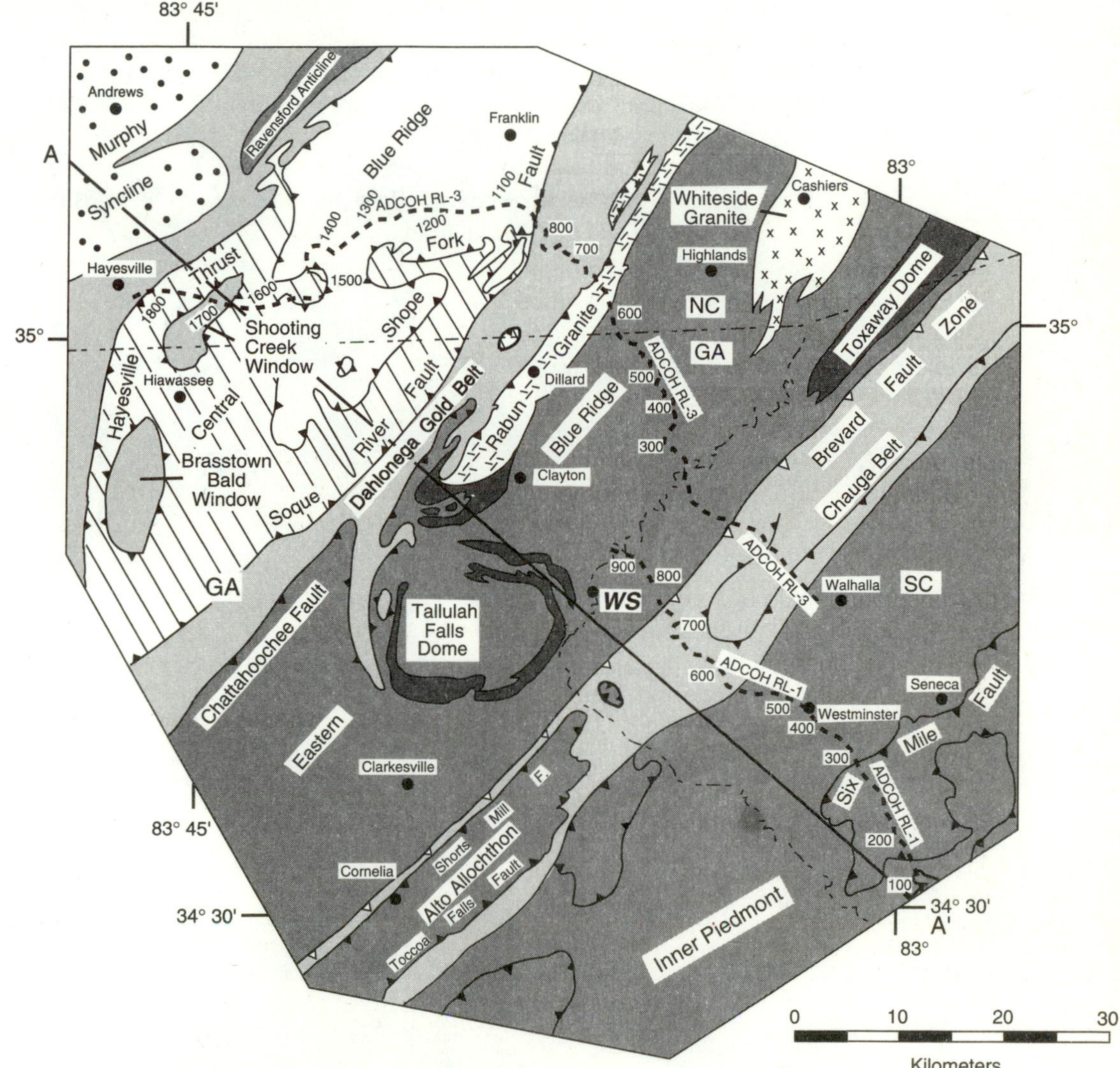

FIGURE 20E–1

Simplified tectonic map of northwestern South Carolina, northeastern Georgia, and adjacent North Carolina in the vicinity of Woodall Shoals *(WS)*. Many of the outcrop patterns in this small-scale map are shown on much larger scale in the geologic map of Woodall Shoals (Figure 20E–2). The medium-screen pattern in the eastern Blue Ridge and Inner Piedmont where Woodall Shoals occurs represents Upper Proterozoic(?) metasedimentary and metavolcanic rocks, mostly metagraywacke, pelitic sctist, and amphibolite. Dark-screened areas in the eastern Blue Ridge represent Grenville (1.0–1.2 Ga old) basement. The dashed lines labeled ADCOH RL–1 and ADCOH RL–3 are seismic reflection profiles, with ADCOH RL–1 presented in Figure 21–14(b). Part of section A–A' to the South Carolina–Georgia state line is Figure 21–14(c). Numbers on the dashed lines are shot-point numbers.

until 1983. We conducted the structural analysis of this exposure using techniques recommended in Figure 20–9(c). First, we marked a series of north-south and east-west grid lines on the exposure at 5 ft intervals. We then used graph paper as a base to construct a geologic map showing the distributions of all rock units (Figure 20E–2) at a scale of 1 in to 5 ft. We also measured mesoscopic fabrics, joints (Figure 20E–3a) and both early and late veins (Figure 20E–3b).

We measured foliation and compositional layering at more than 850 stations, enough for domain analysis of the exposure (Figure 20E–4). In addition, we identified several generations of folds and crenulations on the mesoscopic and map scales and measured the trends of numerous folds, crenulations, and other lineations (Figure 20E–5). All of our data are compiled in Appendix 4.

Our tectonic and geologic maps (Figure 20E–2) reveal the complexity of this exposure and the region, despite the small size of Woodall Shoals. The exposure is dominated by sillimanite-grade migmatitic metasedimentary rocks, mostly biotite gneiss, intruded by both early and late granites. Closer examination of the exposure reveals boudins of amphibolite, garnetiferous amphibolite, calcsilicate quartzite, and granitic gneiss—some preserving earlier foliation truncated by the dominant foliation in the biotite gneiss. Pressure shadows at the ends of some boudins contain quartz, feldspar, and garnet.

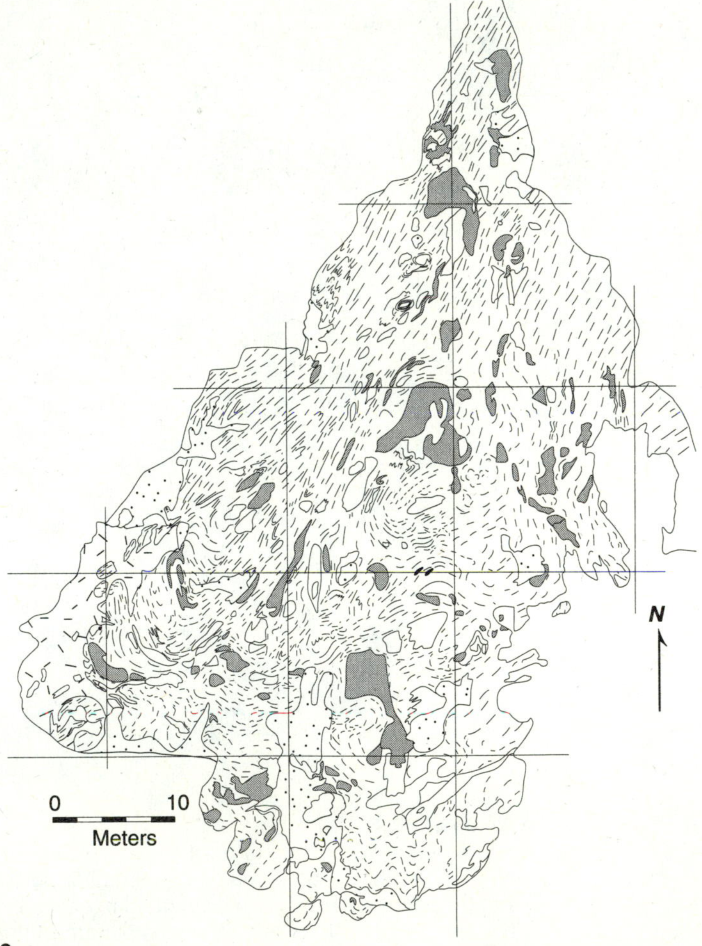

FIGURE 20E–2
Simplified geologic map of Woodall Shoals. The crosshatch pattern is a late pegmatite intrusion; the short-line pattern is migmatitic biotite gneiss; dark screen indicates amphibolite boudins; the unpatterned bodies are foliated granitic gneiss and pegmatite; the fine stipple pattern is recently deposited sand. (The stipple [sand] and cross-hatch [late pegmatite] patterns indicate the same units in all the other maps in this Essay.)

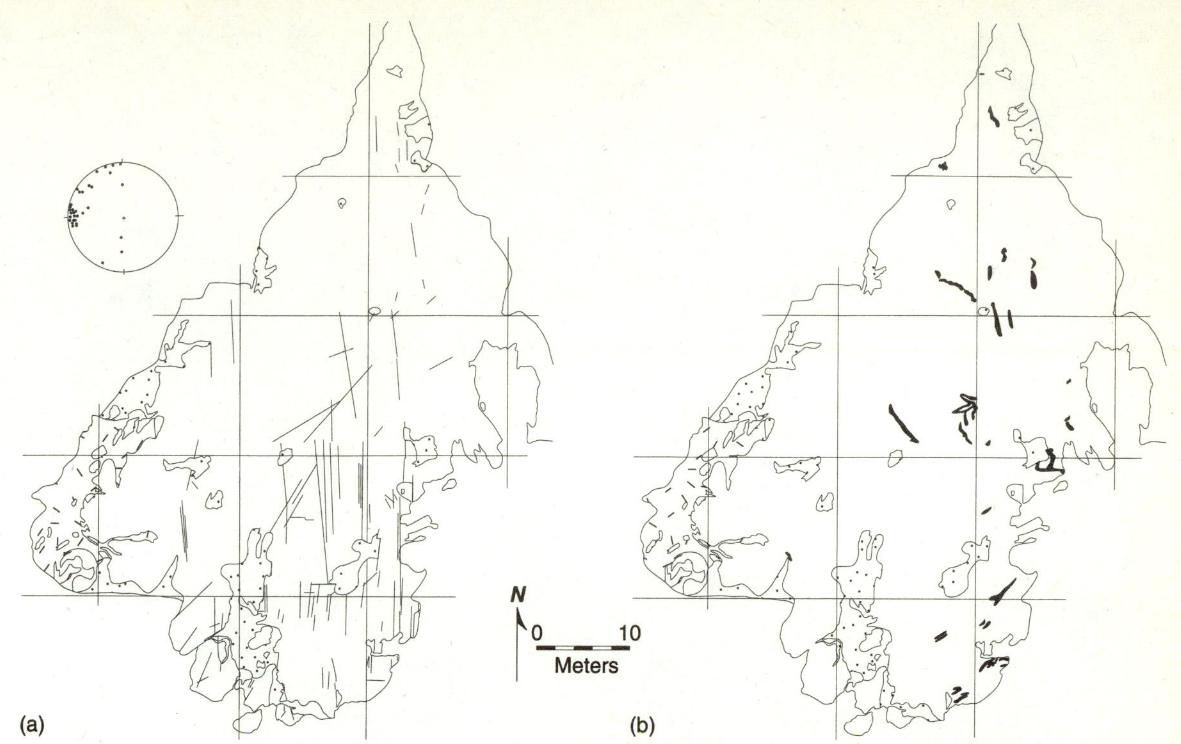

FIGURE 20E–3
(a) Joint distribution at Woodall Shoals and fabric diagram of poles to 34 joints. This and all other fabric diagrams in the figures in this Essay were plotted using a Macintosh™ computer and the Stereonet program written by Richard W. Allmendinger, Cornell University.
(b) Distribution of deformed granitoids and pegmatite veins at Woodall Shoals.

By constructing form lines from strike of dominant foliation in the biotite gneiss (Figure 20E–4), we were able to obtain an indication of the geometry of early folds and the extent of refolding. Foliation strikes dominantly northeast, but, as the map shows in many places, the strike may vary widely over a small area. We found this to be related to folding and fold interference.

The completed geologic map (with plotted structural data) was used to identify boundaries between areas of the same direction of strike and dip (homogeneous domains) and multiple orientations of strike and dip (heterogeneous domains). Thus, we determined the locations of fold hinges and limbs, noting that homogeneous domains will locate limbs. (Most heterogeneous domains are probably early fold hinges or zones of fold interference.) After we plotted strike and dip of the dominant foliation (S_2) on fabric diagrams (Figure 20E–4), the distinction between homogeneous and heterogeneous domains became even more striking. Domains II, III, VII, and XII we identified as homogeneous domains because the diagrams contain clusters of points. Domains IV, V, VI, and IX we judged heterogeneous because points are scattered in the diagrams. Domains I, X, and XI contain enough diversity of orientations that they may be considered heterogeneous, but some clustering of points also occurs in the diagrams. It might have been possible to determine whether domains X and XI are homogeneous, but we did not have enough data. Inspection of the form line patterns in each of these domains indicates that domain I is homogeneous and that domains X and XI are heterogeneous.

Folds here range from north- to northwest-vergent, early passive-flow and flexural-flow, buckle folds with small (0° to 40°) interlimb angles to west-vergent to late upright, symmetrical, flexural-slip, buckle folds with large (90° to 150°)

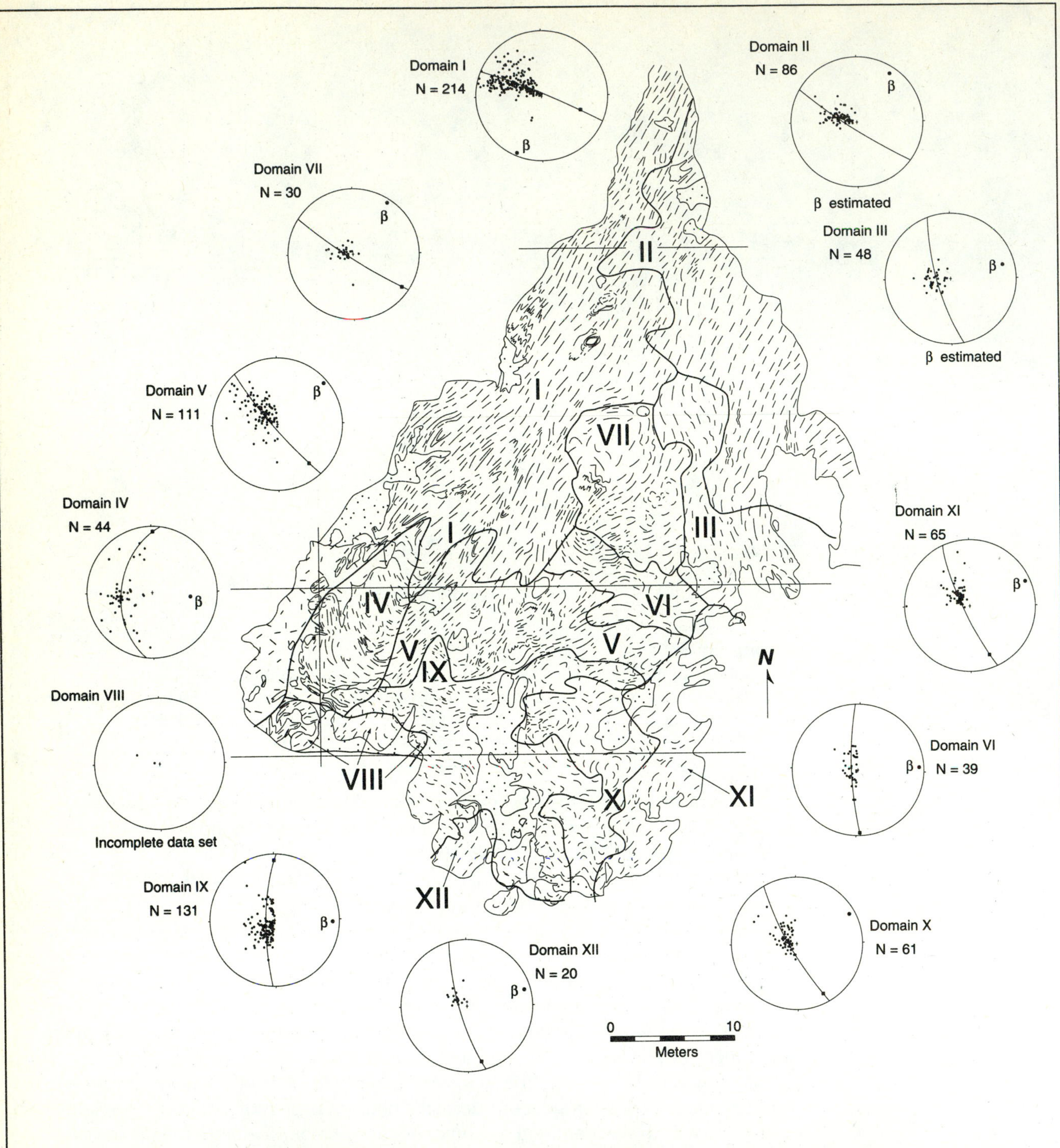

FIGURE 20E–4
Map showing form (strike) lines and domains of similar orientation and dip direction or multiple directions at Woodall Shoals. Fabric diagrams are equal-area plots of poles to the dominant foliation, S_2. Locations of the great circle and β were determined as a best fit by the computer program that plotted the data. All βs indicate a gentle plunge for the fold axis determined, but the trends of some may be biased by lack of scatter in the plot. Carefully examine each plot and position of the great circle and β relative to distribution of points.

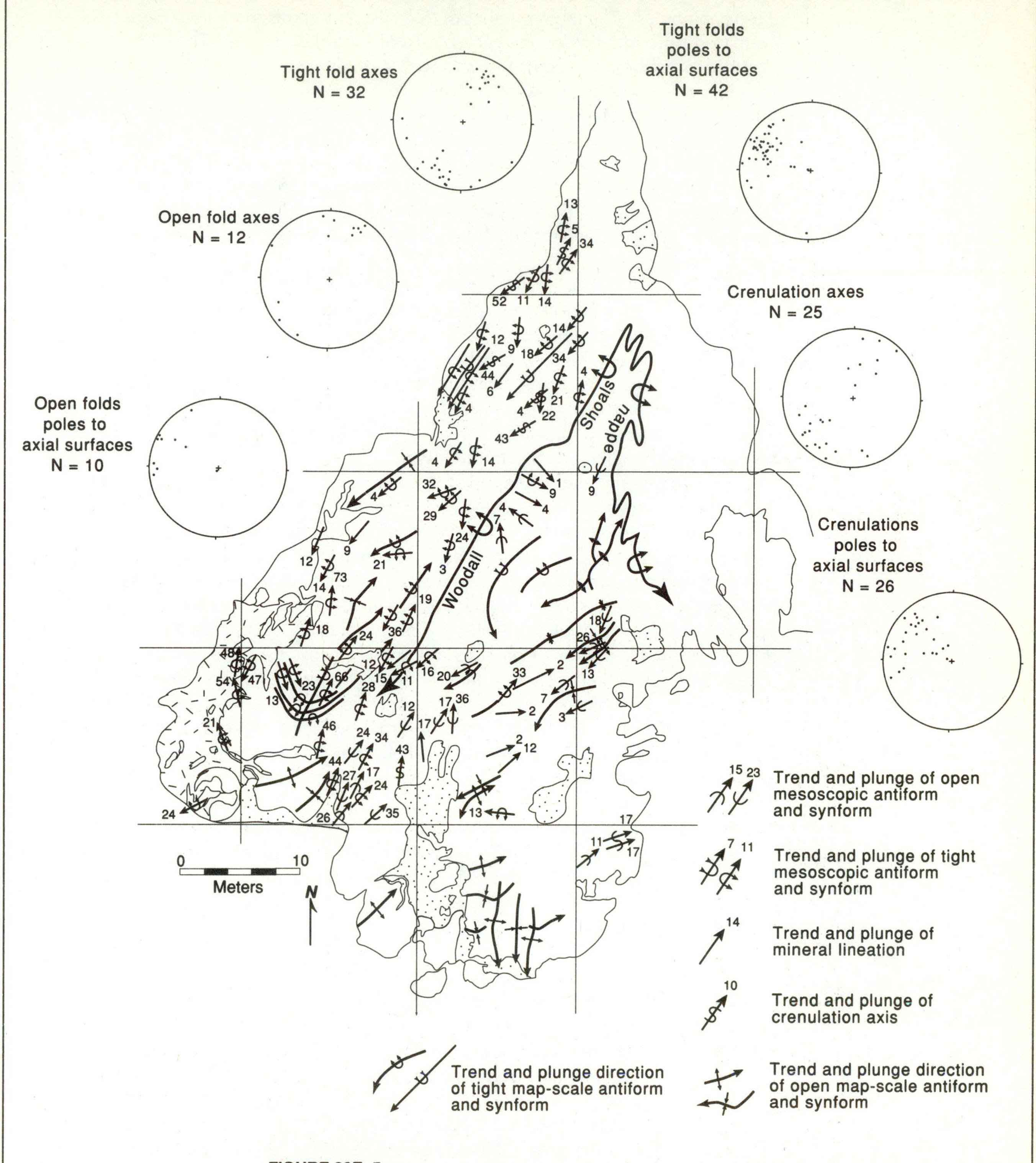

FIGURE 20E–5
Map showing mesoscopic folds and mineral lineations (small lines and symbols with numbers) measured at Woodall Shoals, and map-scale folds (heavier lines) determined from a combination of fabric data and direct observation. Equal-area plots of trend and plunge of fold axes and mineral lineations and poles to axial surfaces of different classes of folds are shown. Also note that the axial surface of the structure labeled "Woodall Shoals nappe" reverses dip direction on opposite sides of the later antiform that refolded the axial surface of the nappe.

interlimb angles. The large number of folds present in this map area (Figure 20E–4) permitted us to identify several generations of superposed folds. The earliest folds occur in amphibolite boudins that survived strong transposition of the earliest foliation in the biotite gneiss, preserving a relict foliation in the boudins (see Figure 16E–2b). Later tight-to-isoclinal folds in the biotite gneiss may be responsible for dismemberment of the amphibolite layer (or layers) and formation of the boudins. Geologic mapping provided us with a perspective heretofore unavailable through studies that I began here in the 1960s: it permitted mapping of a previously unrecognized map-scale recumbent fold, revealed by foliation measurements and the distribution of large amphibolite boudins (Figure 20E–2), throughout the central part of the outcrop.

The map of fold axial traces (Figure 20E–5) revealed several crossing fold sets. Most folds and crenulations here plunge gently northeast or southwest and verge northwest, as indicated in the fabric diagrams. A set of crossing folds in the southwestern part of the map consists of earlier isoclinal folds (F_2?) crossed by a large (F_3?), northwest-verging, tight fold (Figure 20E–5; see also Figure 16E–2). Limbs of small earlier folds produce an apparent plunge both northeast and southwest in the hinge of the later fold. Tracing the later fold a few meters northeast revealed that it really plunges northeastward.

The youngest structures found at Woodall Shoals were joints. Most of the joints that we measured have a steep dip. The majority define a set striking nearly north-south and dipping steeply southeast. A smaller number defines another set that strikes northeast and dips southeast. We measured few gentle dips among joints, although both sets recognized were regional sets (Acker and Hatcher, 1970; Schaeffer and others, 1979). Most joints are unfilled, but several contain quartz, epidote, and pink feldspar, indicating hydrothermal alteration after the fractures formed and while the rock mass was still under greenschist-facies pressure and temperature conditions. These filled joints are crossed by younger unfilled fractures. A late unfoliated pegmatite body along the west side of the exposure (Figures 20E–2 and 20E–3b) contains xenoliths and roof pendants of biotite gneiss. We thought that xenoliths could be distinguished from roof pendants by comparing the degree of parallelism of foliation in the gneiss with that in the enclosing rocks. Lack of parallelism led us to conclude that an inclusion is probably a xenolith.

Analysis of a data set of this type, compiled as maps and fabric data, is a useful exercise; applying the technique will produce valid conclusions in areas of any size and provide immediate insight into the structural history of the area.

References Cited

Acker, L. L., and Hatcher, R. D., Jr., 1970, Relationships between structure and topography in northwest South Carolina: South Carolina Geologic Notes, v. 14, p. 35–48.

Hatcher, R. D., Jr., Acker, L. L., Liu, A., Zupan, A. -J., and Mittwede, S., 1995, Geology and mineral resources of the Whetstone, Holly Springs, Rainy Mountain, and Tugaloo Lake quadrangles, South Carolina–Georgia: South Carolina Geological Survey, scale 1:24,000, (in press).

Schaeffer, M. F., Steffens, R. E., and Hatcher, R. D., Jr., 1979, *In situ* stress and its relationships to joint formation in the Toxaway Gneiss, northwestern South Carolina: Southeastern Geology, v. 20, p. 129–143.

Questions

1. Why does the deformation plan of most orogenic belts follow that of early ductile structures overprinted by late brittle structures? Why not the reverse order?
2. How does use of similar styles and orientations in interpretation of multiple deformation sometimes lead to erroneous results?
3. What evidence is needed to resolve multiple transposition in rocks?
4. Here is a set of fabric diagrams of poles to foliation in several homogeneous domains in an area in the Blue Ridge. What kind of structure is represented? Photocopy this page and sketch in appropriate structural symbols (fold axes, showing plunge and faults).

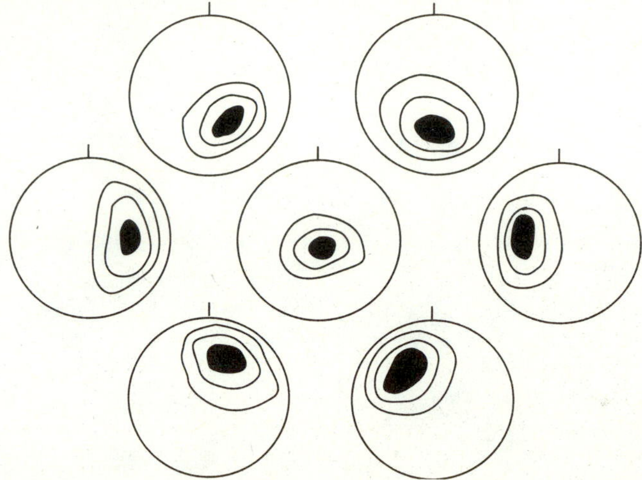

5. What are two interpretations of triclinic symmetry of a fabric diagram of poles to foliation?
6. Why is identification of kinematic axes in most areas considered useless or even impossible?
7. How would you undertake structural study of an area deformed by one event and containing penetratively deformed (cleaved) rocks, crack-seal veins, and mesoscopic- and map-scale folds and faults?
8. Sketch in major fold axes (showing plunge) on a photocopy of Figure 20–6 using only the outcrop patterns and form surfaces on the map, then examine the orientation of β in each of the domain diagrams. Assuming that β axes are parallel to folds, what do they tell you about the small and large structures in that area?

Further Reading

Hossack, J. R., 1979, The use of balanced cross-sections in the calculation of orogenic contraction: A review: Journal of the Geological Society of London, v. 136, p. 705–711.

Reviews principles of cross section balancing and techniques of line and area balancing.

Ramsay, J. G., and Huber, M. I., 1983, The techniques of modern structural geology, Volume 1: Strain analysis: New York, Academic Press, 307 p.

Ramsay, J. G., and Huber, M. I., 1987, The techniques of modern structural geology, Volume 2: Folds and fractures: New York, Academic Press, 392 p.

This two-volume work covers the full spectrum of modern structural geology. Strongly recommended as a standard reference for anyone undertaking structural analysis.

Turner, F. J., and Weiss, L. E., 1963, Structural analysis of metamorphic tectonites: New York, McGraw-Hill, 545 p.

Contains a step-by-step approach to plotting fabric diagrams, analyzing rock fabrics on all scales, and interpreting complex structure.

Whitten, E. H. T., 1966, Structural geology of folded rocks: Chicago, Rand McNally, 678 p.

A different approach to study of structure and fabrics in penetratively deformed rocks. Assumes familiarity with fabric diagrams.

Williams, P. F., 1985, Multiply deformed terrains—problems of correlation: Journal of Structural Geology, v. 7, p. 269–280.

Deals with the problem of correlating foliations in complexly deformed regions; discusses structures of microscopic scale through map scale.

PART SIX

Geophysical Techniques in Structural Geology

SECTION SIX CONTAINS A SINGLE CHAPTER ON GEOPHYSICAL TECHNIQUES commonly used to address structural problems. A large amount of high-quality crustal seismic reflection, refraction, gravity, magnetic, and other geophysical data is available to the structural geologist today. It is important that we know something about the techniques in order to employ them to help constrain cross sections, project structures and plutons, and otherwise make better interpretations of the crust. These techniques are also very useful in working out the structure of areas of poor exposure and areas of complex structure. They provide information on the generally inaccessible third dimension of the crust.

21

Geophysical Techniques

It is the time in history for thorough exploration of the entire continental crust. The seismic reflection profiling technique . . . will almost certainly be the principal tool for probing the deep continental basement. . . . One can safely predict an era in which the deep crustal features of all continents are discovered, mapped, named, understood, and made familiar to all earth scientists.

JACK OLIVER, 1986, American Geophysical Union *Geodynamics Series*

GEOPHYSICAL TECHNIQUES AND CONCEPTS ARE WIDELY USED to resolve individual geologic structures and to explore the crustal- or larger-scale structure of the Earth. They help to resolve near-surface structure and to extract information about the nature of the deeper crust and mantle. Techniques useful in delineating geologic structure include gravity, magnetism, paleomagnetism, electrical properties, seismic reflection, seismic refraction, and earthquake seismology. The resolution and imaging capabilities of each technique are quite variable. Seismic reflection profiling is probably the most precise technique (because it produces a vertical section that resembles a geologic cross section) followed by magnetics, gravity, and seismic refraction techniques. These techniques will be discussed below with examples of applications to structural geology.

This chapter is not intended to provide in-depth background in geophysics but is intended to demonstrate the importance of geophysical techniques in modern structural geology and kindle interest in the subject (Figure 21–1). The techniques to be discussed here help most in interpreting large structures—faults and folds of map scale or larger, plutons, and crustal boundaries—and, as such, provide a useful springboard to tectonics.

POTENTIAL FIELD METHODS

Gravity and magnetism—called *potential fields*—provide useful insights into the structure of the crust and upper mantle. Each has its own limitations and degree of precision and resolution, but of the two, magnetism has the greater resolution in the crust.

Terrestrial Magnetism

Earth magnetism was known to the Chinese at least as early as the eleventh century, and in approximately 1295 Marco Polo brought magnetite (lodestone) from China to Europe. In the sixteenth century, William Gilbert (1540–1603), an English physicist, noted that a sliver of magnetite hung by a string would orient itself more or less north-south.

The Earth's magnetic field is believed to originate in the core, yet temperatures in the core must be well above the Curie temperatures for most materials, including metallic iron, which probably is the dominant element (90% iron, 10% nickel). The *Curie temperature* (after Pierre Curie, 1859–1906, a French physicist) is the temperature above which strongly magnetic (ferromagnetic) materials lose their ability to interact with a magnetic field (Table 21–1); a weaker (paramagnetic) property, like that of most materials, remains. Although it is not exactly certain why the Earth has a magnetic field, the best model is the *self-exciting dynamo,* which was first proposed by a British mathematician, Joseph Larmor (1857–1904), to explain the magnetic field of our sun. His idea was refined and applied by Walter M. Elsasser and Sir Edward Bullard to explain Earth magnetism (Rikitaki,

TABLE 21–1
CURIE TEMPERATURES FOR COMMON MAGNETIC MINERALS

Mineral	Curie Temperature °C
Magnetite	578
Hematite	670
Iron	770
Pyrrhotite	316

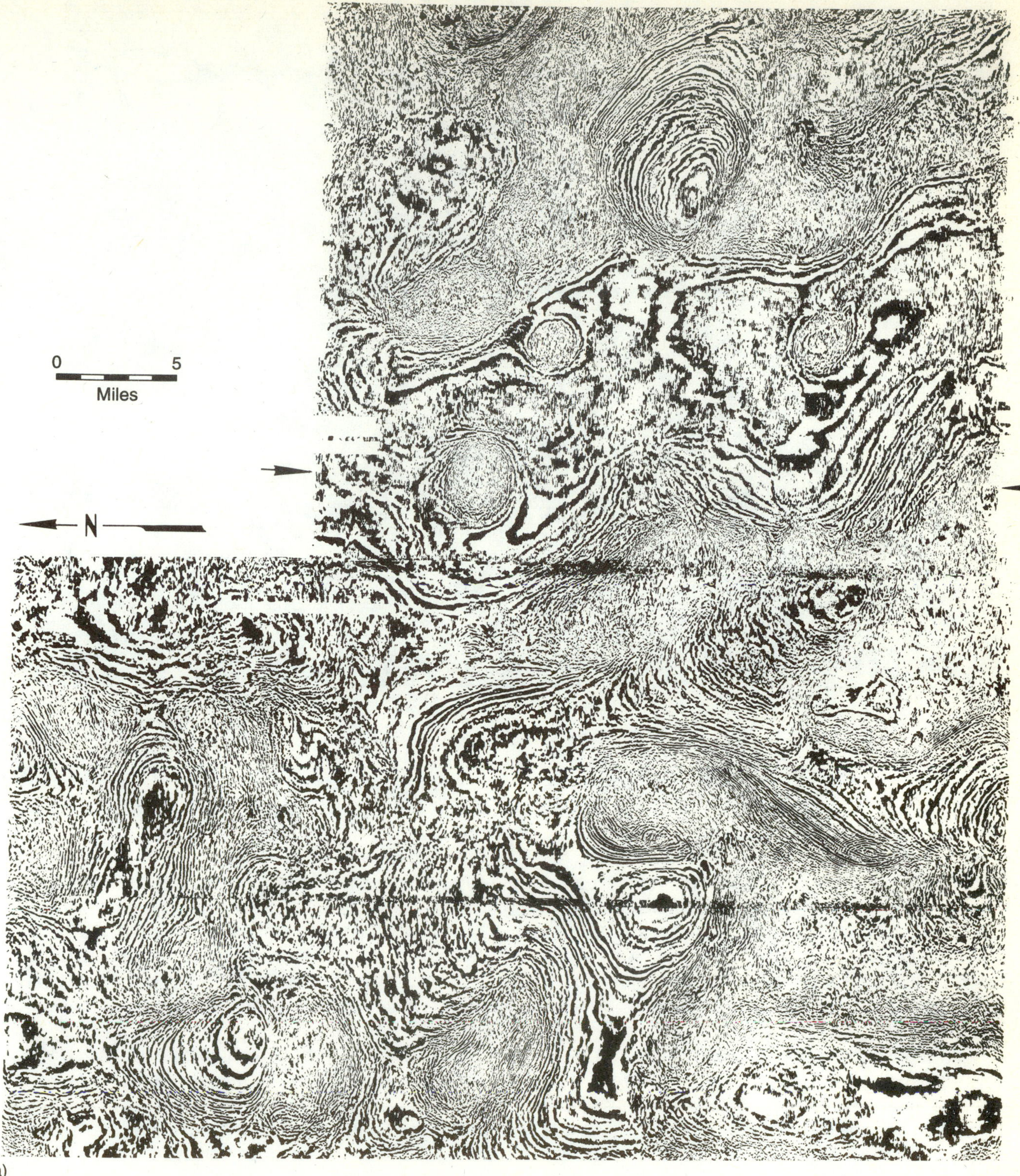

FIGURE 21–1
(a) Map of seismic reflectors at 2.8 s (approximately 3,000 m below sea level) depth in Tracts 1 and 5 in the Green Canyon area in the Gulf of Mexico south of New Orleans, Louisiana, showing nearly circular salt diapirs that have intruded the sedimentary cover. These data were processed through three-dimensional migration. Note the similarity to map patterns in a region of multiple deformation in the core of a mountain chain (see Figures 2–22 and 19E–2). Arrows locate the vertical section in (b) (following page). (b) North-south seismic reflection profile showing a large salt dome near the north end and smaller structures to the south. Subhorizontal layering in sediments is folded upward by diapiric intrusion of salt. (Used by permission and courtesy of Geophysical Services, Inc.)

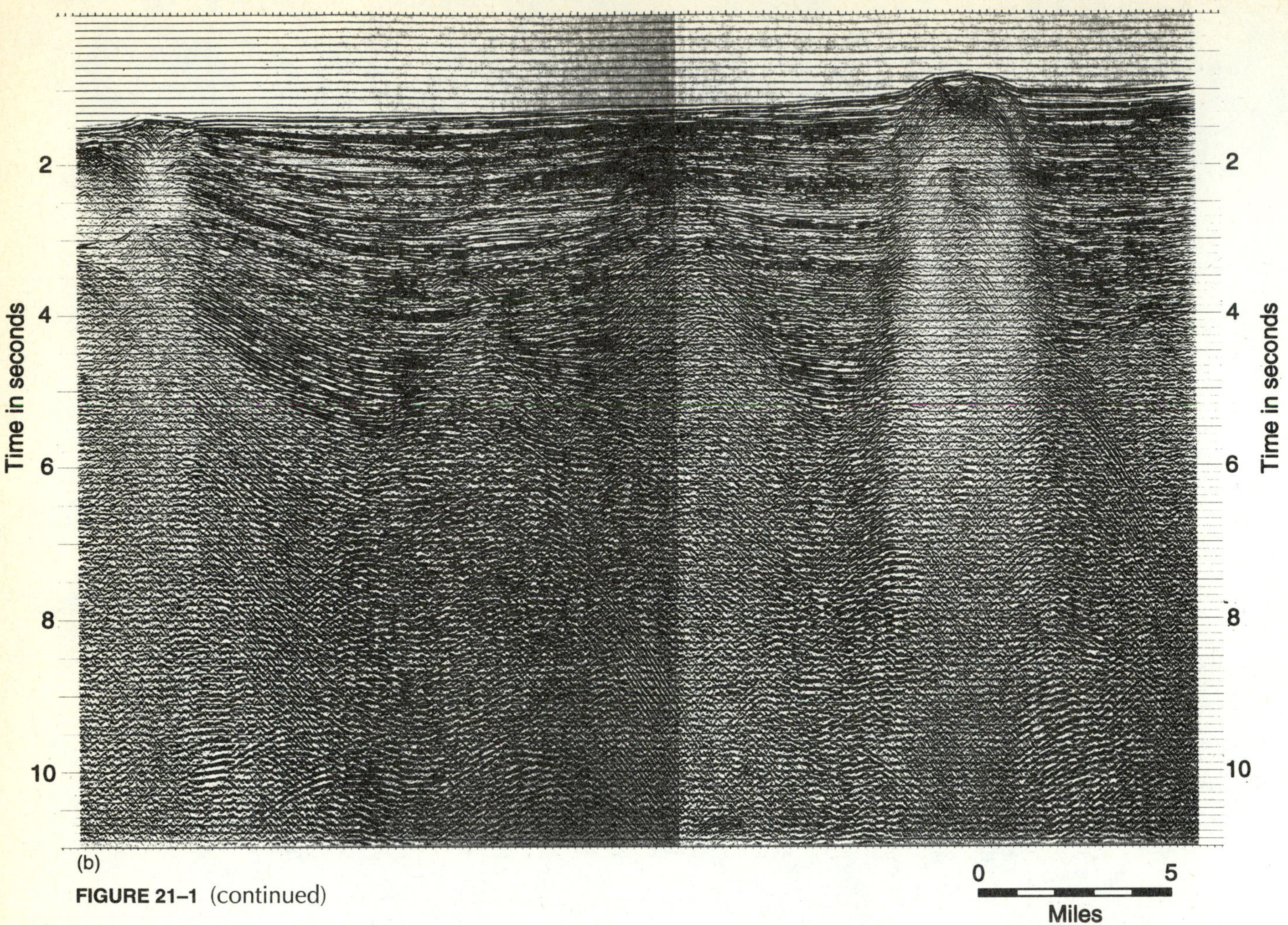

FIGURE 21–1 (continued)

1966). The theory is based on the premise that convection currents exist in the liquid core, transporting hotter material upward and returning cooler material to the interior. The convection cells are assumed to be oriented by the Earth's rotation. The convective transfer of liquid iron-nickel through a thermal gradient generates an electrical potential difference with an associated magnetic field. After the magnetic field is generated, the convective motion enhances and perpetuates it, producing a self-exciting dynamo.

A major difficulty is that the self-exciting dynamo theory does not account for polarity reversals: at intervals since the Paleozoic, the Earth's magnetic field has spontaneously reversed polarity many times (Figure 21–2). The intervals are not uniform, and the time it took to actually reverse polarity was probably on the order of a few thousand years—perhaps only hundreds of years (Fuller and others, 1979). The dipole magnetic field may be lost during the reversal, leaving a weaker nondipole field. During reversals, the strength of the magnetic field decreases to about 15 percent of its normal strength, remaining there until the reversal process is complete (Watkins, 1969). K. A. Hoffman (1988) has summarized evidence showing that the strength of the present-day magnetic field has been decreasing for a century—if it continues at the same rate, it will decrease to zero in 1,500 years. Thus, a polarity reversal may be imminent.

The reversal phenomenon raises many questions about the effect on life forms (among other things) because the magnetic field acts as a radiation shield. Removal of the shield will allow more radiation to reach the Earth's surface, increasing the mutation rate.

Intensity of the magnetic field may be measured using a *magnetometer* on the ground, at sea, or from an airplane. Vertical *magnetic-intensity profiles* may be plotted along the data contoured between successive traverse lines to produce a ***magnetic-anomaly map*** (Figure 21–3). An ***anomaly*** is a deviation from an assumed average value for magnetic-field intensity. Rocks that contain a high percentage of magnetically susceptible minerals (such as magnetite) commonly produce *positive anomalies,* and rocks that contain a low percentage of magnetically susceptible minerals produce *negative anomalies.* The anomalies may also be enhanced, cancelled, or reversed if strong remanent magnetization is present.

The total intensity of a magnetic field measured by a magnetometer is the vector sum of the Earth's field and that produced by the *induced magnetization* of the

FIGURE 21–2
Magnetic polarity reversals from the Mesozoic to the present. Black intervals are normal polarity—as at the present time—and white are reversed. (From A. R. Palmer, Geological Society of America DNAG Time Scale; used by permission.)

rock. These components consist of the primary field intensity, plus the induced magnetic susceptibilities of all crustal rocks (down to the Curie isotherm), plus the *remanent field* component for the entire rock body. In addition, the inclination of the lines of force of the magnetic field varies with latitude. This produces a separation of high and low anomalies (dipoles) at lower latitudes (below 25°), where the inclination is low; at high latitudes (above 70°), a single anomaly occurs directly above the rock body that produced it (Figure 21–4). A modeling technique called *reduction to pole* may be used to correct for the low-latitude separation property and produce single anomalies above the rock bodies that caused them.

The *amplitude* of the magnetic anomaly produced by a rock body varies directly with the susceptibility of the body and with its shape, and inversely with the square of the distance from the body to the magnetometer (Figure 21–5). The slope (or gradient) shown by the spacing of contours on the flank of the anomaly is actually used to determine depth to the anomaly (Vacquier and others, 1951). The technique should be used only for estimating depth; more accurate determinations can be made by other techniques. Magnetic rock bodies, such as plutons, produce high-amplitude anomalies with steep gradients wherein amplitudes (and gradients) decrease as the distance to the magnetometer increases. Dikes only a few meters thick, which generate sharp, high-amplitude anomalies when measured with a ground magnetometer or during a low-altitude (150 m) airborne survey, would not be resolved by high-altitude airborne surveys (300 to 800 m). Therefore, low-altitude airborne surveys are more useful in resolving near-surface geology and structure. Surveys flown at high altitude are more useful in resolving structure in the deep crust and upper mantle.

Suppose contrasting sets of anomalies are surveyed from the same altitude (Figure 21–6): generally, broad, long-wavelength anomalies will be produced by deeper crustal features, and high-frequency, short-wavelength anomalies will be produced by near-surface features. Major boundaries and crustal blocks may be identified by studying magnetic anomaly maps.

Problems arise when working with anomalies, even those reduced to the pole, because the total magnetic field is the vector sum of the induced and remanent fields. The remanent-field component affects the amplitude and slope of contours on the flanks of anomalies by vectorially adding to (or subtracting from) the induced field. The ***Königsberger ratio*, Q**, is represented by

$$\mathbf{Q} = \frac{J_R}{\kappa_b F}, \qquad (21\text{–}1)$$

where J_R is remanent magnetism, κ_b is the magnetic susceptibility of the rock body producing an anomaly, and F is the induced magnetic-field intensity. Values of Q for older rocks are generally less than values of Q in younger rocks. Oceanic rocks generally have very high Q values relative to continental rocks, partly reflecting age but also reflecting a fundamental difference in properties. High Q is one reason the polarity-reversal pattern is strongly expressed in ocean crust. That the Königsberger ratios for continental rocks are low—about 0.1 to 1—indicates that a significant remanent

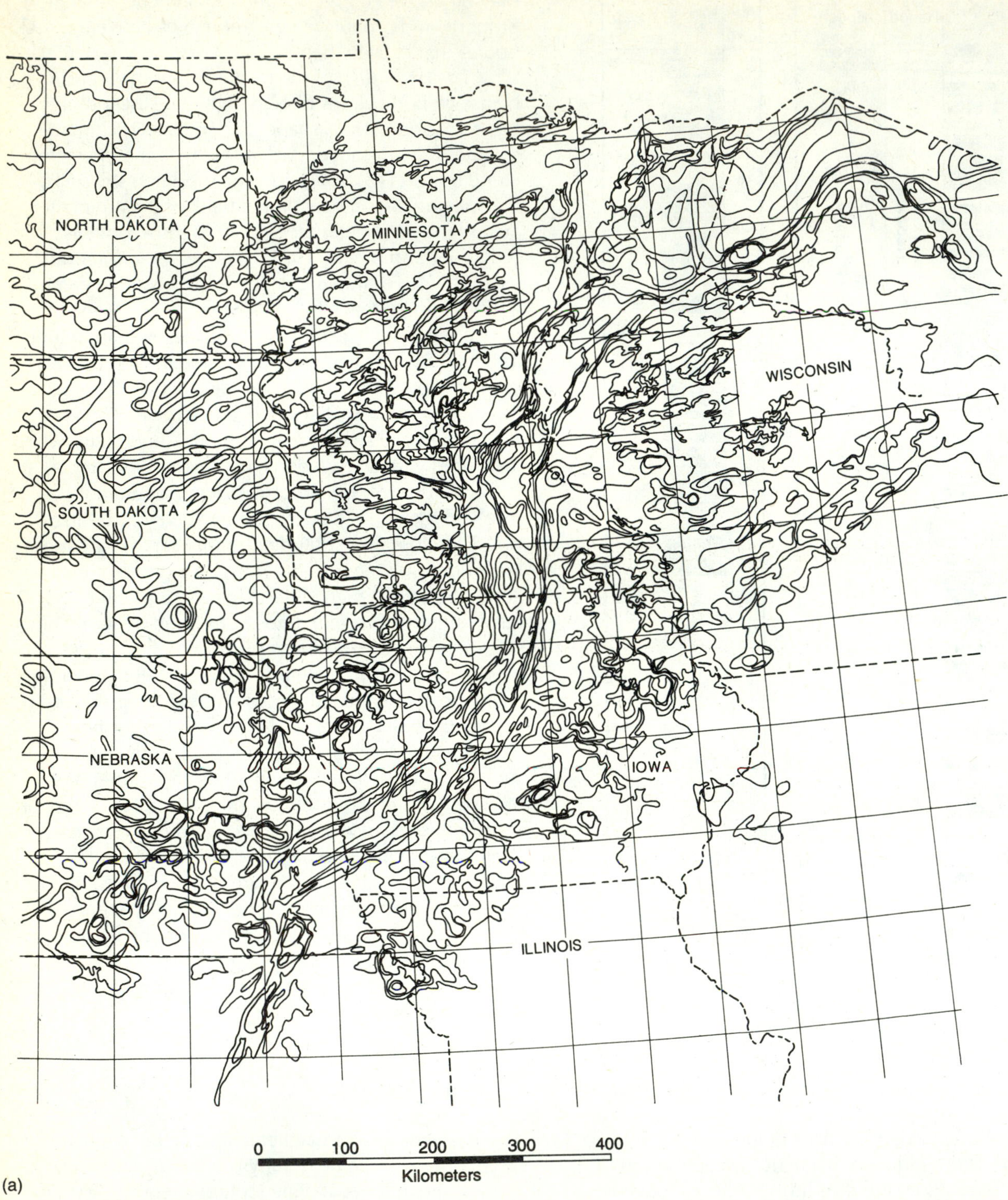

FIGURE 21–3
Magnetic-anomaly (a) and Bouguer gravity-anomaly maps (b) (next page) of part of the same area of the north-central United States, showing distribution of anomalies in the Precambrian basement. The structure that produced the prominent northeast-southwest linear gravity and magnetic high in the center of (a) and (b) is the Mid-Continent rift, formed during the Late Proterozoic. The high is interpreted (c) (following page) as having been produced by mafic igneous rocks that have been traced to the surface as the Keweenawan basalts of northern Michigan. The flanking gravity low is interpreted to have been produced by rift-related sedimentary rocks. Magnetic field contours in (a) are in gammas; gravity contours are in milligals. (Magnetic data from I. Zietz, 1982, Composite magnetic-anomaly map of the conterminous United States: U.S. Geological Survey. Gravity data are from G. P. Wollard and H. R. Joesting, 1964, Bouguer gravity-anomaly map of the United States: U.S. Geological Survey.)

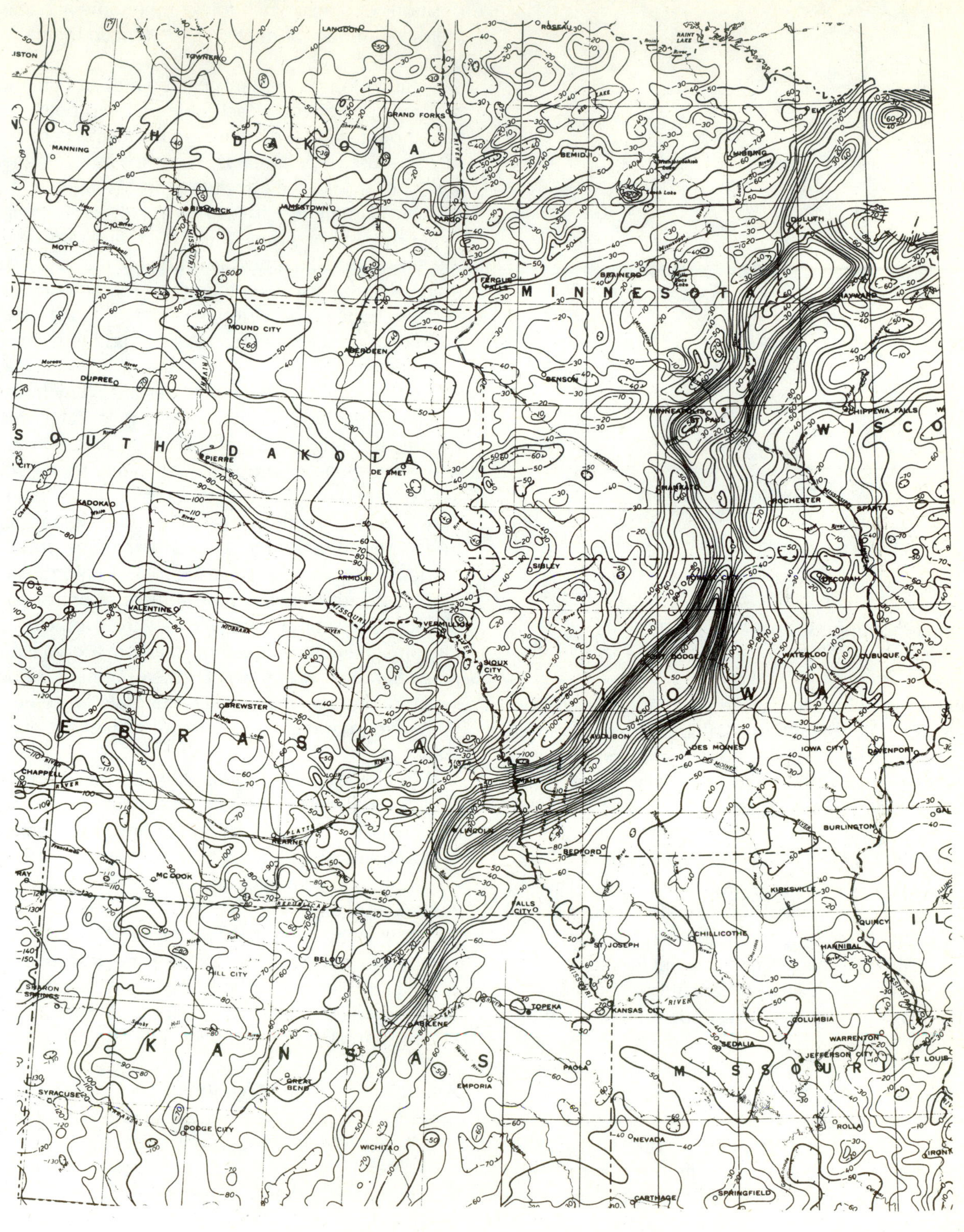

0 100 200 300 400
Kilometers

(b)

FIGURE 21–3 (continued)

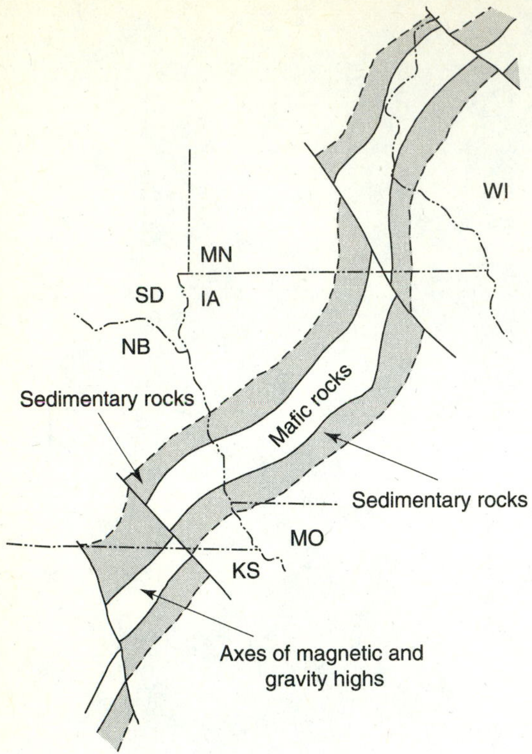

FIGURE 21–3 (continued)

component still exists. It may strongly influence the total field anomaly, but reliable calculations of depth to magnetic anomalies using the slope technique of Victor Vacquier and others (1951) have been made where independent checks are available.

Remanent Magnetism. Remanent magnetism develops in rocks when they crystallize or are deposited. The magnetic axes of magnetically susceptible minerals, such as magnetite, are aligned in the magma chamber or the depositional environment parallel to the magnetic field that exists during crystallization or deposition. In a magma, alignment does not occur until the temperature falls below the Curie temperature for that mineral (Table 21–1). Some minerals may crystallize completely and cool below their Curie temperatures before the entire magma congeals. Thus, the orientation of the Earth's magnetic field at the time of crystallization or deposition is locked in. The study of ***paleomagnetism*** involves measurement of these remanent fields and attempts to reconstruct the positions of the ancient magnetic poles, continents, and oceans.

Remanent magnetism in igneous rocks is termed ***thermo-remanent magnetism (TRM).*** It is measured by collecting an oriented sample and shielding it from the present-day magnetic field so that the orientation of the ancient magnetic field can be measured. The measurement provides an azimuth toward the magnetic pole—the ancient pole position. Comparison with pole positions of other rock bodies of the same age may yield the amount of rotation experienced by the rock body.

Similar measurements can be made of paleomagnetism in sedimentary rocks, where the property is called *depositional remanent magnetism.* Red beds are the most useful sedimentary rocks for paleomagnetic study because of the concentration of magnetic minerals. Measurements can also be made—with greater difficulty—in fine-grained limestone and other sedimentary rocks. Sedimentary and volcanic rocks are more desirable for study than most plutonic rocks because bedding provides an originally horizontal reference surface to which rotations and translations may be related.

Many rocks undergo changes after they form that can alter and even erase paleomagnetic signatures. Chemical changes in composition or oxidation state may alter remanent properties. Treating rocks with acid or complexing agents to remove altered material may sometimes sufficiently restore the original signature to permit measurements. Thermal overprinting, particularly on some Paleozoic rocks in North America, hinders paleomagnetic measurement. Controlled reheating of samples may alleviate the problem. Paleomagnetic measurements made in a *cryogenic magnetometer,* which permits measurement inside a superconducting ring, may compensate for some difficulties, particularly where paleomagnetic signatures are weak.

A check called a *fold test* is performed on all samples to determine if paleomagnetic measurements can be used. The orientation of samples collected from opposite limbs of folds, or from otherwise tilted rocks, are plotted on an equal-area net or stereonet (see Appendix 1) and then rotated back to the horizontal. Paleomagnetic measurements for each sample are compared, and if the rotation does not produce a consistent orientation, the locality may be abandoned or the results discarded as spurious. Further thermal or chemical treatment of samples may help.

Applications Using Magnetism. The concepts of terrestrial magnetism are widely applied in solving structural problems. Faults may be recognized by associated linear trends and by truncation of magnetic trends in rock bodies (Figure 21–7). Folds may be identified by noting the curvature of contrasting magnetically susceptible units. Plutons may be recognized by the shape of the magnetic anomaly. The bulk mafic or felsic composition of plutons may be estimated by noting the kind of associated magnetic anomaly (positive or negative) in conjunction with the corresponding gravity anomaly (positive or negative). Magnetic-anomaly maps may be used to identify

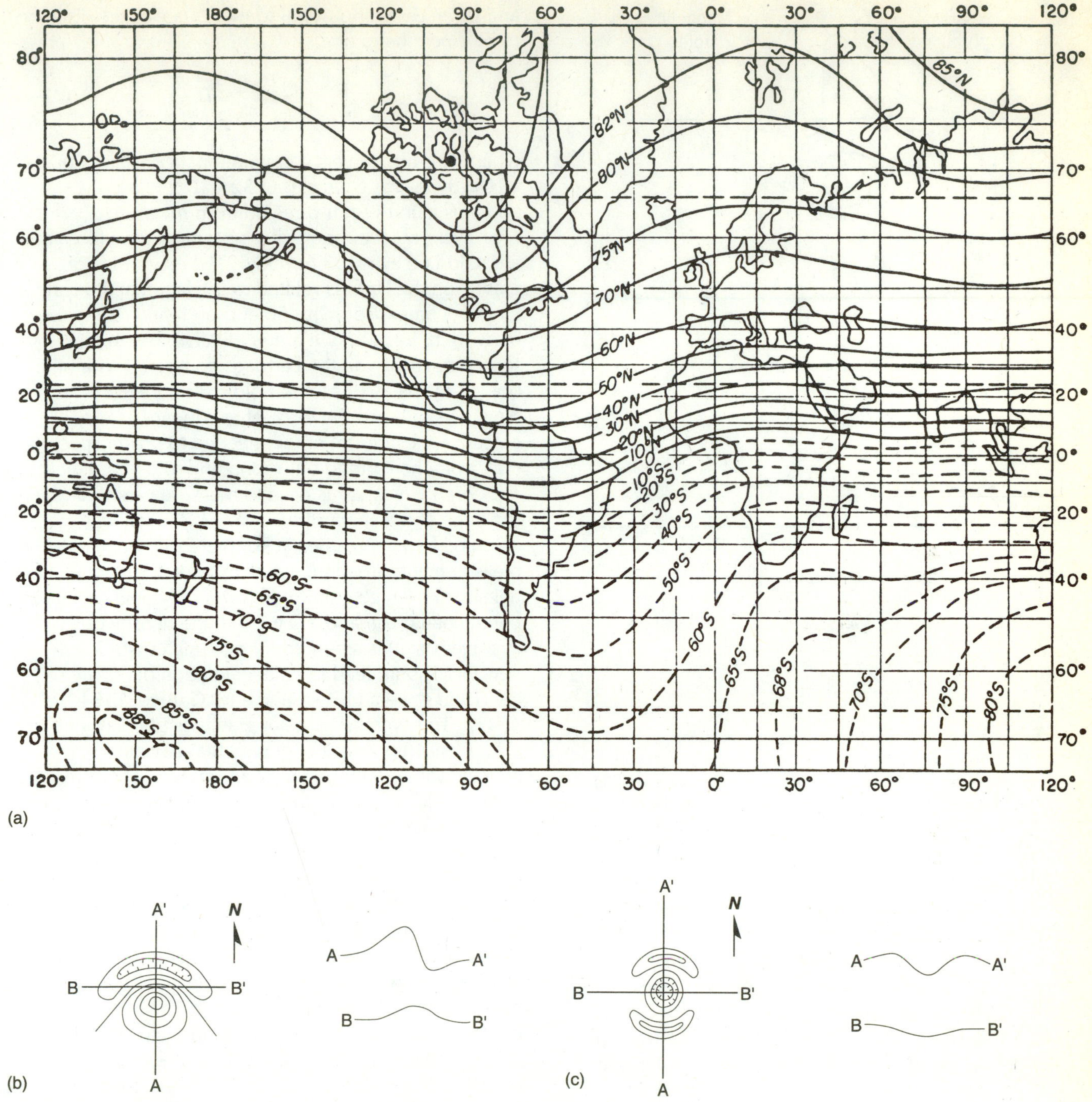

(a)

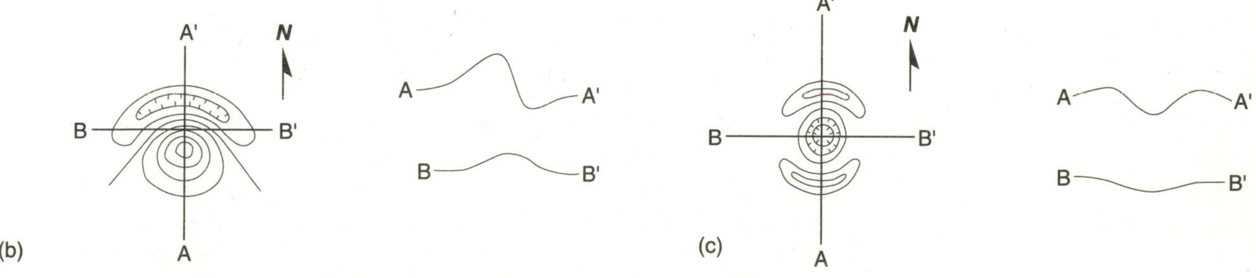

FIGURE 21–4
(a) Latitudinal variation in the inclination of the Earth's magnetic field. (From U.S. Naval Hydrographic Office.) (b) Magnetic-anomaly map and profiles showing a dipole anomaly over a pluton with a field inclination of 60° at 60° S. (c) Map and profiles from low latitude showing separation of anomalies from a single magnetic source into a dipolar pair. [(b) and (c) from S. Breiner, 1973, Applications manual for portable magnetometers, E. G. & G. Geometrics, Inc.; used by permission.]

boundaries and assess the crustal character of suspect terranes. Linear magnetic anomalies in the sea floor have been dated and correlated to provide a better understanding of the evolution of oceanic crust during the past 200 m.y.

Data from paleomagnetic studies are useful in determining the paleolatitude of a rock body or a large crustal block but cannot be used directly to determine longitude. Local rotation of a rock body due to faulting may be measured. *Apparent paleo-polar wandering*

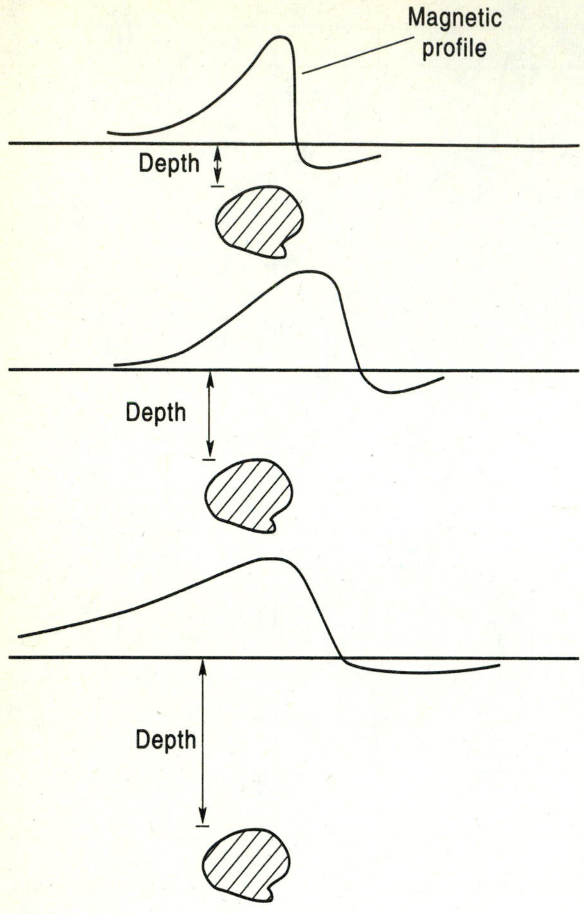

FIGURE 21–5
Depth dependence of the width of magnetic anomalies. As the depth to the source of the anomaly increases, the amplitude decreases and the anomaly widens.

(APW) curves derived from paleomagnetic studies are useful in reproducing ancient positions of a continent. An important assumption made in paleomagnetic studies is that magnetic north, on average, corresponds with the Earth's spin axis (geographic North or South). Such studies have been used to determine the original positions of cratonic blocks in Precambrian shields and also the positions of larger continents and suspect terranes during later geologic periods.

Gravity

Gravity is the force of mutual attraction among all bodies in the universe. It is exerted on and by all components of the Earth, including rock bodies. The more dense and voluminous the body, the stronger the force of gravity, and so denser rocks such as gabbro possess a stronger gravitational force. *Newton's law of gravitation* states that the force of gravity (**F**) is directly proportional to the masses *(m)* of the objects involved and inversely proportional to the square of the distance *(r)* between them, or

$$\mathbf{F} = G\frac{m_1 m_2}{r^2}\gamma_1 , \qquad \mathbf{(21\text{–}2)}$$

where G is the universal gravitational constant, 6.6732 x 10^{-8} dyne cm^2 g^{-2} (or, 6.6732 x 10^{-11} Nm^2 kg^{-2}), and γ_1 is a unit vector directed from m_1 to m_2.

Every object in our solar system exerts a force of attraction on every other body. The magnitude of gravity on the Earth is related to the fact that the Earth is large in comparison with objects on it and to any rock body in it. Because we are relatively close to the source of gravity on the Earth, as compared with other objects in space, this force holds us on the Earth. Newton's law and the force of gravity are used to calculate the amount of fuel needed to power spacecraft from the Earth to our moon and to other planets.

A *gravity meter* is used to measure variations in the force of gravity on the Earth. First, a value is assumed for average density of the crust where the measurements are made. A density of 2.65 g cm^{-3} (the density of average granite) is commonly assumed for continental crust, and 3.0 g cm^{-3} (the density of average basalt) is assumed for oceanic crust. These densities should lead to an acceleration of gravity of 1 gal (1 cm s^{-2})—a unit named after Galileo (1564–1642). The average value of gravity is 980 gals (9.8 m s^{-2}). Measured values of gravity that deviate from average values define *gravity anomalies.* Measurements of differences in the gravity field are made in units of mgal (10^{-3} gal), and modern gravity meters (Figure 21–8) are capable of measurements within $\pm$ 0.04 milligals. If the measured value of gravity at a station is greater than predicted by using the average density assumed, a *positive gravity anomaly* exists; if it is less, a *negative gravity anomaly* exists. Gravity profiles may be constructed or, if the gravity measurements are distributed widely enough, the data may be contoured as a gravity anomaly map (Figures 21–3 and 21–9). If the map is corrected only for elevation above sea level, it is called a *free-air gravity map*; if data are corrected for differences in both elevation and density between sea level and the elevation where measurements were made, a *Bouguer gravity anomaly map* is produced (the name honors a French surveyor and hydrographer of the eighteenth century, Pierre Bouguer). Free-air measurements are most useful for determining how close to isostatic equilibrium is a portion of the crust. If a large free-air anomaly is present, the mass is out of isostatic equilibrium. A *terrain correction* is frequently made for Bouguer data, and, if such corrections are made and plotted as a map, the result is a *complete Bouguer anomaly map.* The magnitude of the gravity field varies with latitude,

FIGURE 21–6

Magnetic-anomaly map showing contrasts in deep (broad anomalies) and shallow (high-frequency anomalies) magnetic sources in Georgia and South Carolina. Contours in units of 200 gammas. (Data from I. Zietz, 1982, Composite magnetic anomaly map of the conterminous United States: U.S. Geological Survey.) The east-west linear anomalies crossing the entire northern part of the map area and passing through the middle of South Carolina record a large, late Paleozoic, dextral fault zone. Circular anomalies are plutons at various depths.

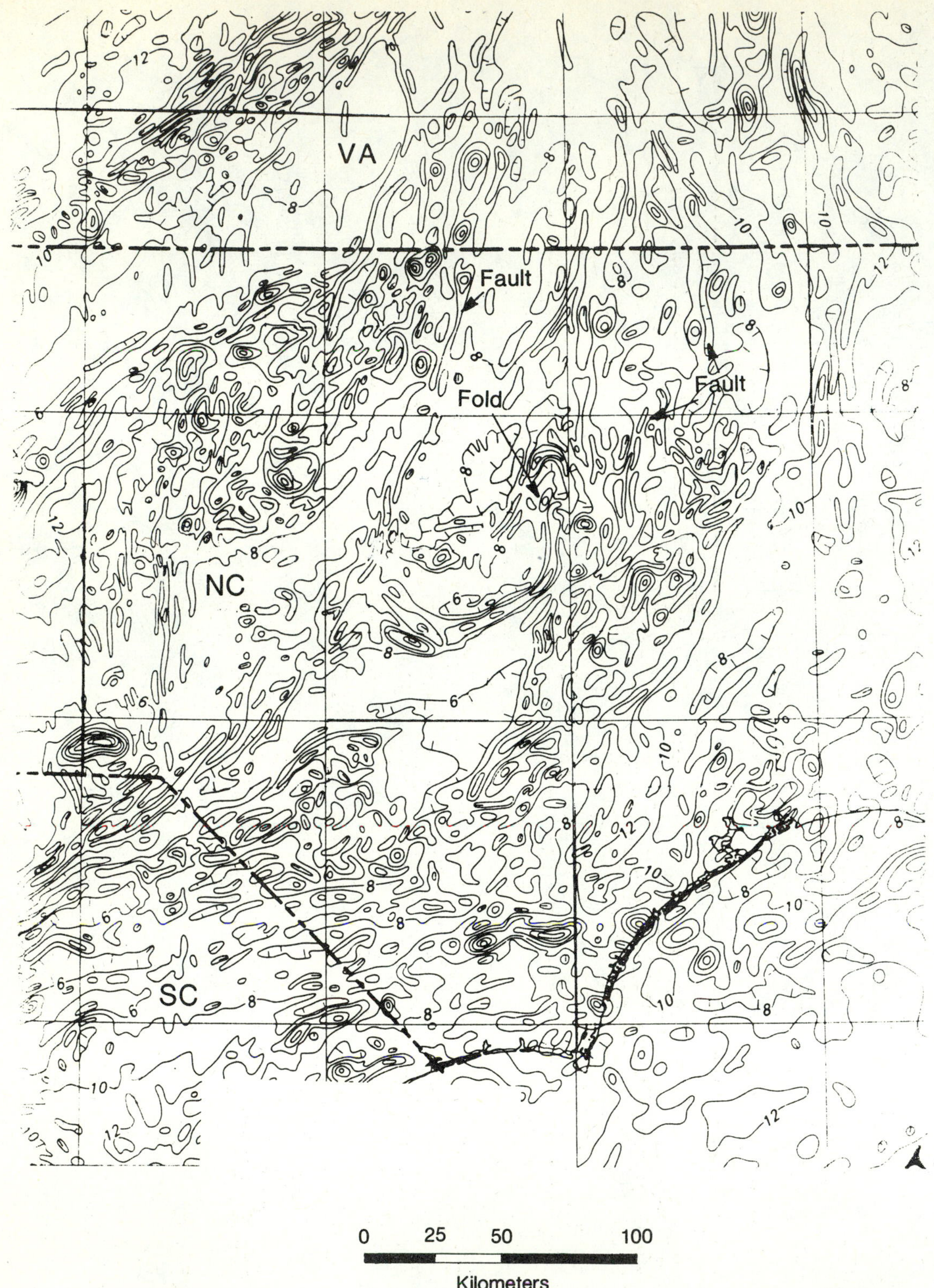

(a)

FIGURE 21–7
Magnetic-anomaly map of part of the North Carolina Piedmont (a) showing truncation of folds by faults (b) (following page). The fold is defined by highly magnetic amphibolites; a swarm of Jurassic diabase dikes is suggested by subparallel linear anomalies. Contour interval is 200 gammas. (From I. Zietz, F. E. Riggle, and F. P. Gilbert, 1984, U.S. Geological Survey Map GP-958.) (c) (page 446) Magnetic-anomaly map showing several plutons in the Piedmont of North Carolina and South Carolina. Contours interval is 200 gammas. (From I. Zietz and F. P. Gilbert, 1980, U.S. Geological Survey Map GP-936.) (d) (page 447) Sea-floor magnetic anomalies in the Pacific off the Pacific Northwest coast. (From W. J. Morgan, *Journal of Geophysical Research,* v. 73, p. 1959–1982, 1968, © American Geophysical Union.)

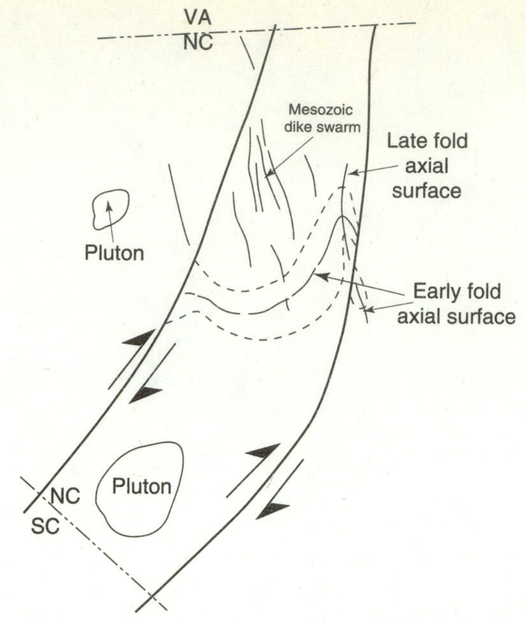

FIGURE 21–7 (continued)

elevation, and Earth tides. Gravity measurements are commonly made from ships, and may be made from aircraft, but with greater difficulty and much less reliability.

Depth to a mass producing a gravity anomaly may be calculated after assuming a particular shape for the body causing the anomaly. The calculations remain simple if the shape is assumed to be a sphere or a cylinder. If the source of an anomaly of amplitude g_0 is assumed to be a spherical body, the depth to the body (center of the sphere) may be calculated from the half width of a profile (at 1/2 the maximum amplitude) through the anomaly (Figure 21–10a) as

$$g(x) = g_0 \left(1 + \frac{x_{1/2}^2}{z^2}\right)^{-3/2}, \qquad \textbf{(21–3)}$$

where x is the width of the gravity anomaly and z is the depth to the anomaly. From the half-width of the anomaly,

$$x_{1/2} = \left(1 + \frac{x_{1/2}^2}{z^2}\right)^{-3/2}, \qquad \textbf{(21–4)}$$

$z = 1.305x^{1/2}$.

Calculation of the depth of a cylindrical body (Figure 21–10b) is even simpler:

$$g(x) = g_0 \left(1 + \frac{x_{1/2}^2}{z^2}\right)^{-1}. \qquad \textbf{(21–5)}$$

Again, using the half-width of the anomaly,

$$x_{1/2} = \left(1 + \frac{x_{1/2}^2}{z^2}\right)^{-1}. \qquad \textbf{(21–6)}$$

If the body is assumed to be cylindrical, the half-width equals the depth to the anomaly.

Applications Using Gravity. Gravity may be used to characterize large regions of the crust. Faults may be recognized by contrasting amplitudes and patterns on adjacent blocks. The linearity of the actual fault is sometimes obvious, but more frequently the truncation, deflection, or offset of anomalies at faults is more readily observed, a result of fault displacement of the masses that produced the anomaly (Figures21–3b and 21–11). Gravity data represent integrated values for the acceleration due to gravity for the whole Earth at a single point. As a result, the data are less specific for near-surface features unless large density contrasts exist in surface rocks and the gravity survey is very detailed. Generally, with a station spacing of 2 to 5 km, the data yield only regional patterns, gradients, crustal boundaries, very large faults, and plutons. Gravity data are useful for determining the shape and broad composition of plutons. Granitic plutons produce gravity lows because of their lower density, and mafic plutons yield gravity highs because of their higher density. Granitic rocks commonly produce both gravity lows and magnetic lows, but if a granite contains enough magnetite, it will yield a gravity low and a magnetic high. Mafic plutons tend to produce both gravity highs and magnetic highs. Neither gravity nor magnetic data alone yield precise identification of individual features, although both used together are a powerful tool for delineating faults, plutons, and other structures (Figures 21–7a and 21–7b).

Gravity data can be used to calculate a *gravity model* for a part of the crust (Figure 21–12). Magnetic modeling is more difficult, because it requires full knowledge or accurate assumptions about susceptibility of the material being modeled. John McBride and Douglas Nelson (1988) have used a combination of seismic reflection, gravity, and magnetic data to produce a magnetic model suggesting that the north-south East Coast magnetic (and gravity) anomaly (ECMA) and the east-west Brunswick magnetic anomaly in Georgia are the same. The model also

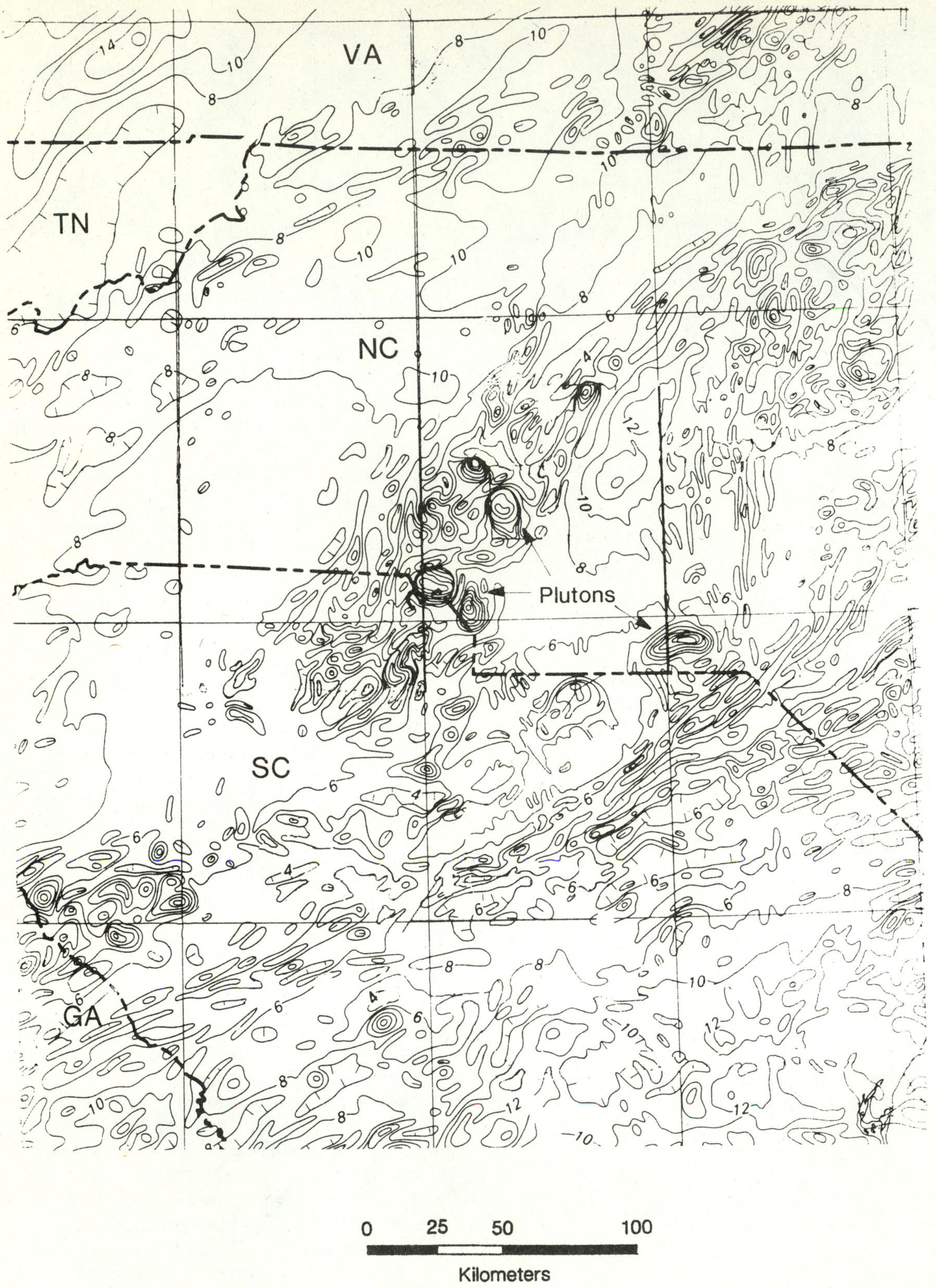

(c)

FIGURE 21–7 (continued)

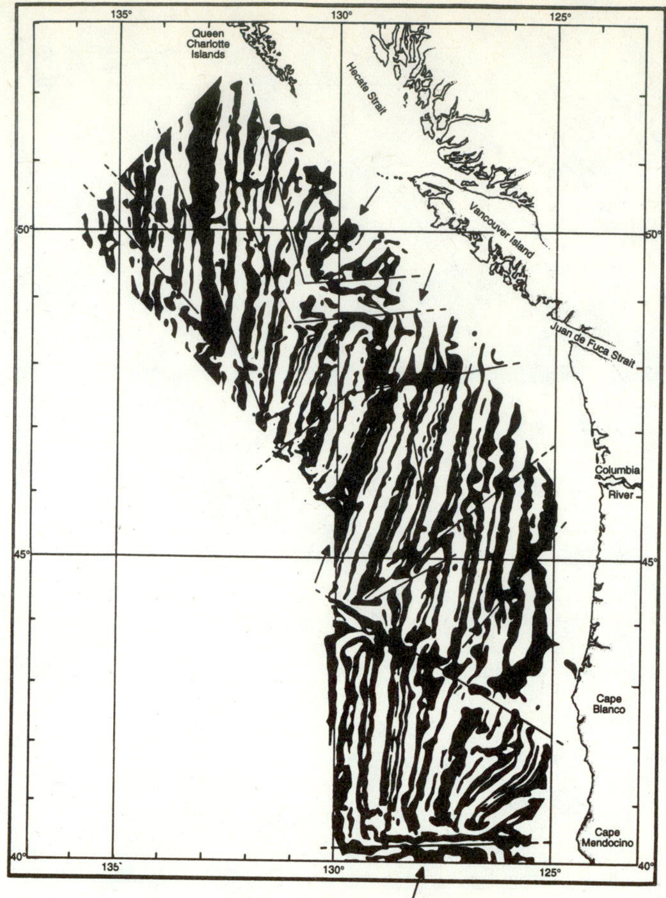

(d)

FIGURE 21–7 (continued)

FIGURE 21–8
Gravity measurement being made with a LaCoste and Romberg Model G gravity meter. This instrument combines the principles of the long-period seismograph and a "zero-length" spring—with tension proportional to the length of the spring. (RDH photo.)

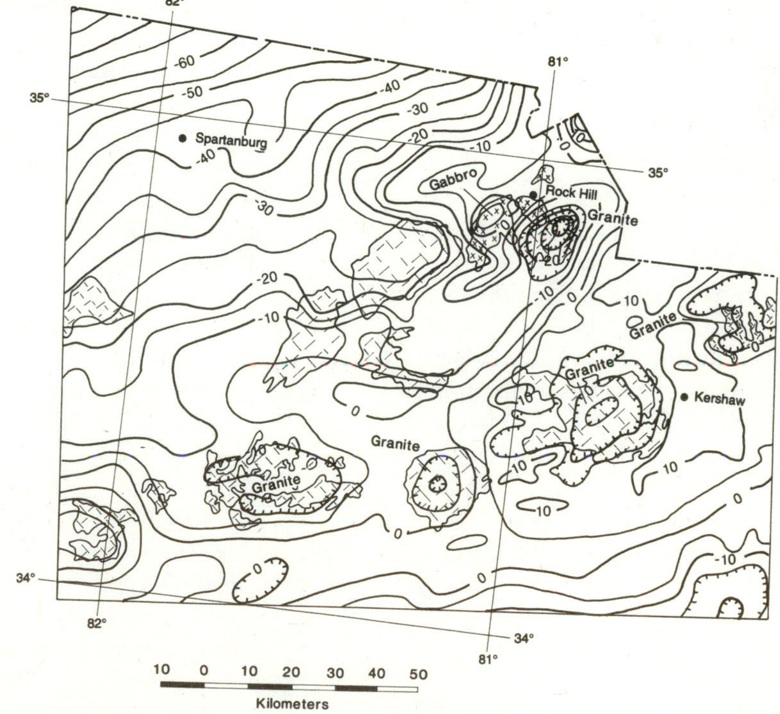

FIGURE 21–9
Bouguer gravity-anomaly map of part of the South Carolina Piedmont shown in Figure 21–7b. Circular anomalies indicate plutons. Gravity lows are granitic plutons; highs are mostly gabbros. Contours are in milligals. (After P. Talwani, L. T. Long, and S. R. Bridges, 1975, Simple Bouguer Map of South Carolina, South Carolina Geological Survey.)

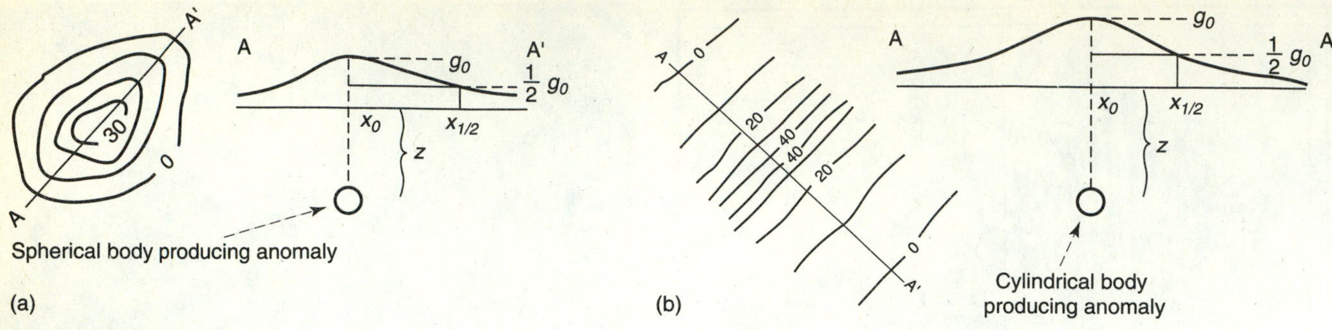

FIGURE 21–10
Calculation from gravity data of the depth to spherical (a) and cylindrical (b) bodies.

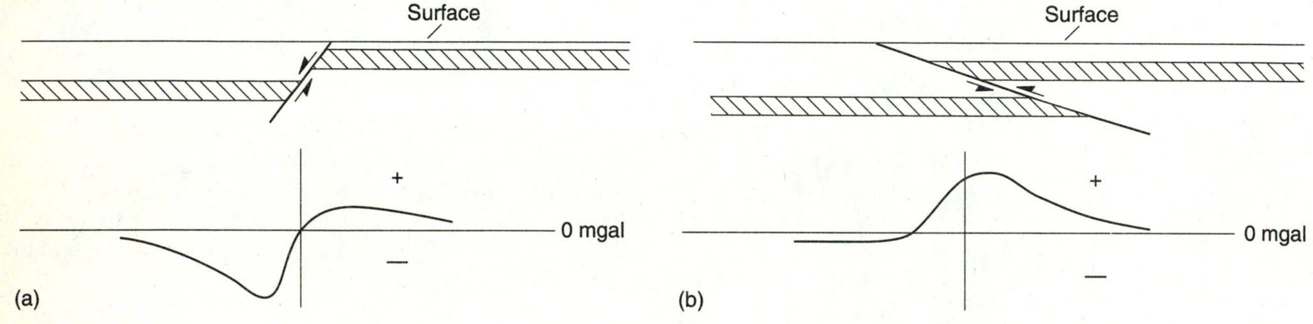

FIGURE 21–11
Geologic cross sections and gravity profiles showing kinds of anomalies that ideally develop along normal faults (a) and thrust faults (b). The striped layer has a higher density.

suggests that differences in amplitude and width of the anomalies are related to the orientations of the bodies producing them.

SEISMIC REFLECTION

The seismic reflection technique has developed rapidly during recent decades because computers and digital techniques have speeded data processing. Seismic reflection was originally developed using explosives; a modification is the vibroseis technique. Vibroseis uses a truck (or more than one) that has a vibrating steel plate mounted under the body (Figure 21–13); at regular intervals, the plate is lowered to the ground, where vibration energy in the form of sound waves is passed into the ground over a specific frequency range. If more than one vibrator truck is used, the spacing and vibration frequencies are synchronized. The waves are transmitted into the crust and some are reflected back from subsurface structures and are recorded at the surface. To record transmitted and reflected waves, an array of *geophones* is laid out as in conventional seismic surveys (which use explosives as a seismic source). The waves are recorded similarly, but with greater control over frequency and recording time. Multiple traces may be recorded and processed over the same structures, producing sharper records. This in turn provides greater latitude in reprocessing data to aid interpretation of crustal structure (Figure 21–14). An airgun may be used in much the same way, but it transmits less energy into the Earth from land than through water, so airguns are mainly used for marine surveys. For maximum transmission of energy, and therefore a maximum return of reflected energy, explosives are still the best energy source, but airguns are less destructive to the marine environment.

Seismic reflection is best suited to recognition of nearly horizontal structures (such as faults and packages of layers) that dip less than 30°. This is entirely due to the way the technique was designed—to transmit acoustic waves more or less vertically into the Earth and to receive almost vertical reflection of waves back to the surface from rock layers of contrasting acoustic properties (Figure 21–14). Theoretically, it is possible to record or process reflections from steeply dipping surfaces, but in practice, few reflectors that dip more than 30° are observed, and they may be selectively removed during processing. This probably introduces bias or even significant error. Steeply dipping structures, such as some faults, may be imaged only by displaced gently dipping layers on both sides of the structure (see Figure 11–34).

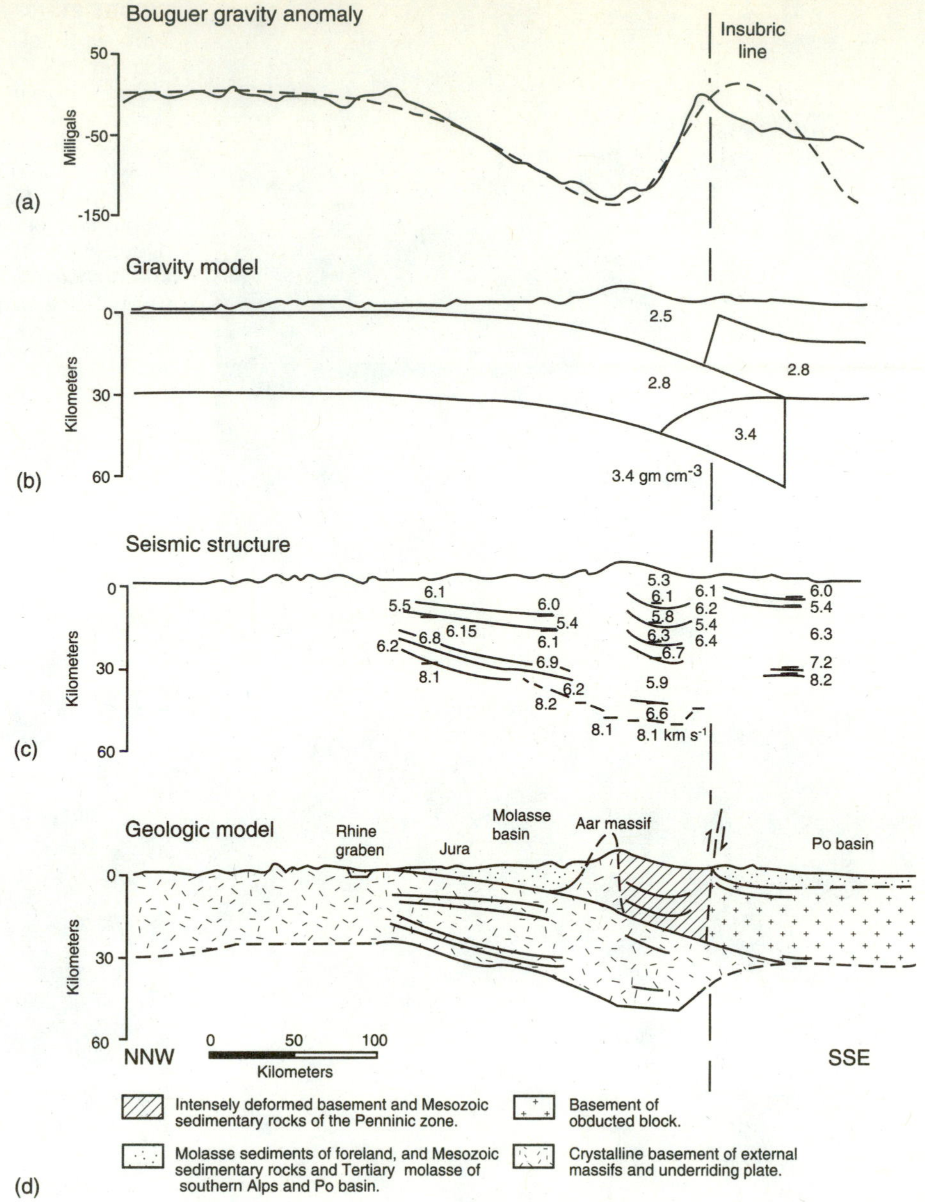

FIGURE 21–12
Gravity profiles observed (solid line in part a) and calculated (dashed line in part a) for the Alps were employed to construct the gravity model in (b). Both gravity and seismic refraction data (c) were used to construct the geologic section in (d). (From G. D. Karner and A. B. Watts, *Journal of Geophysical Research,* v. 88, p. 10,449–10,477, 1983, © by the American Geophysical Union.)

A processed seismic reflection profile consists of a two-dimensional section through part or all of the crust (see Figure 11–34) and resembles a geologic cross section. The vertical axis of the profile is commonly two-way travel time of the acoustic waves in seconds. Knowledge of velocity in different parts of the section permits conversion of a time section to a depth section, with a vertical scale in kilometers or feet. The seismic section may be used (with care) to construct a well-constrained structural cross section along the line of the seismic profile (see Figure 11–34). Many artifacts (unnatural flaws in the data introduced during data acquisition and processing) are only partly removed during processing. Construction of high-quality geologic cross sections (Figure 21–15; see also Figure 13–3) requires knowledge of seismic velocities, field and processing parameters, careful search of the profile for artifacts, and maximum use of surface geologic data and other geophysical data. High-quality seismic reflection profiles are probably the best tool available for imaging the structural geometry of the crust and mantle and for projecting surface structure to depth.

Seismic reflection profiling was originally designed as a tool for petroleum exploration. Since the mid–1970s it has been adapted for study of the deep crust of the continents and the interiors of mountain

FIGURE 21–13
Vibroseis trucks working in tandem on a COCORP line in the House Range in the Basin and Range of western Utah. The recording truck (silhouette) is shown on the right edge of the photograph. Geophones would be buried along the roadside. (Douglas Von Tish, Larry Brown, and Richard W. Allmendinger, Cornell University.)

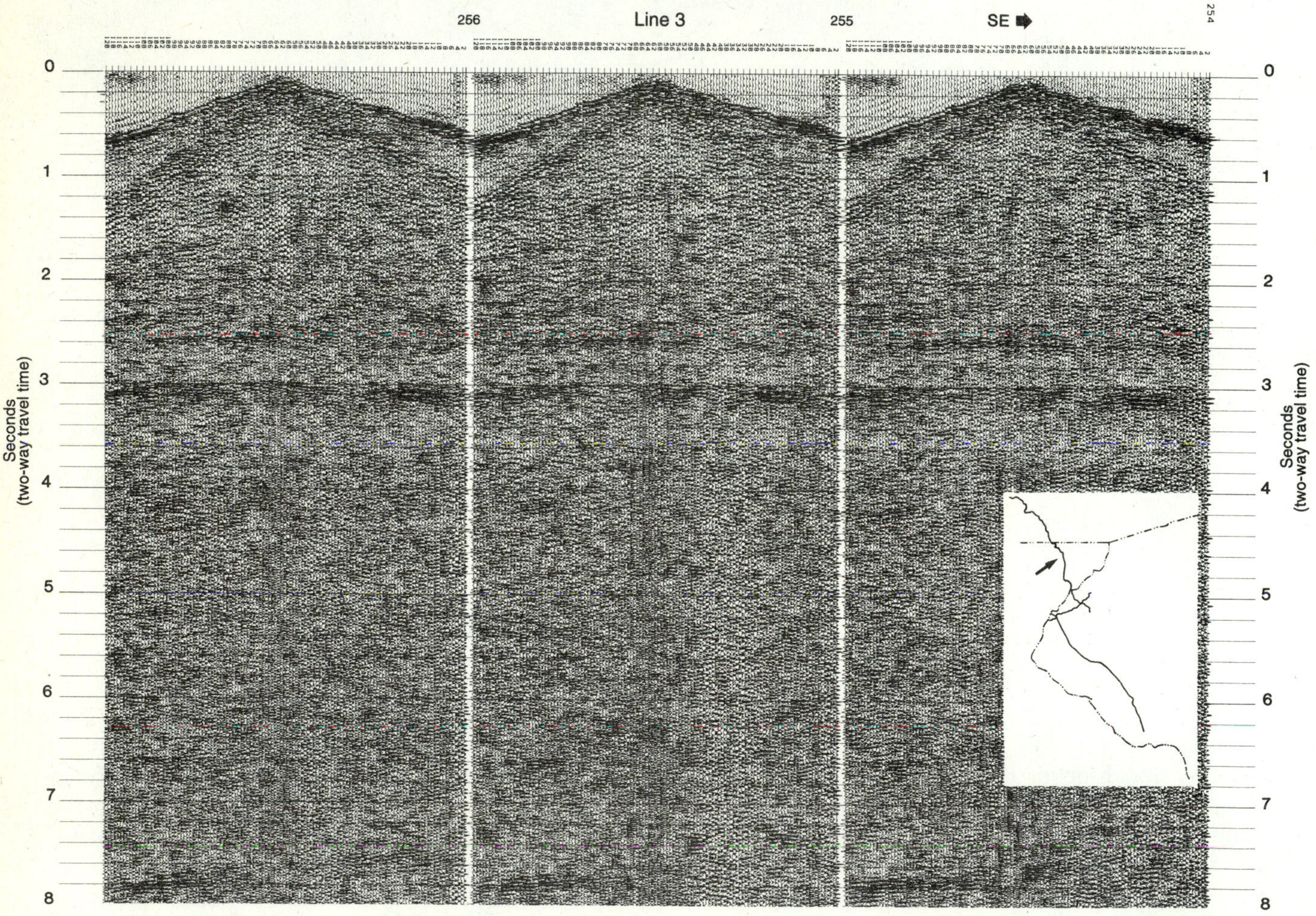

(a)

FIGURE 21–14
(a) Field record for three vibrator shot points collected along part of a vibroseis line as part of the Appalachian Ultradeep Core Hole (ADCOH) site investigation in northeastern Georgia (arrow on inset map), showing major reflectors at 2.6-, 3.1-, and 8-s two-way travel time. Field records for each shot point are correlated and stacked using the computer to construct the initial section—called a "brute stack."

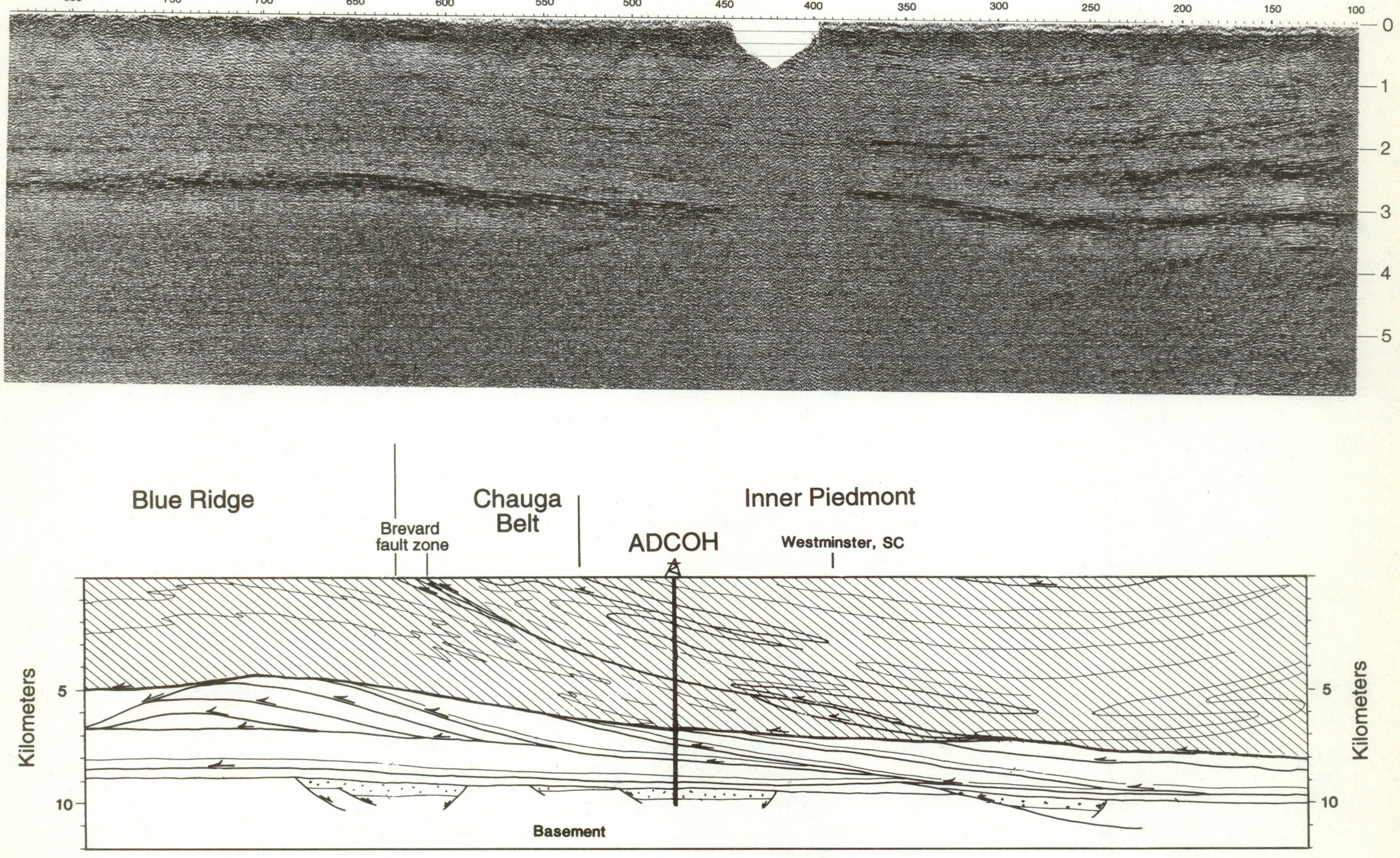

FIGURE 21–14 (continued)
(b) Processed section constructed from field records such as those in (a). Data collected as Line 1 (location shown on map in a) from northwestern South Carolina as part of the ADCOH Project site investigation. (Data processed at Virginia Tech by J. K. Costain and C. Çoruh.) (c) Geologic interpretation of the line shown in (b). Striped area is the Blue Ridge–Piedmont crystalline thrust sheet. Reflectors in the right third of the seismic reflection profile image recumbent folds that are mappable on the surface. Reflectors are produced by strong acoustic contrasts of the folded granitic rocks and amphibolite. Strong reflectors around 3-s two-way travel time are derived from flat-lying sedimentary rocks on top of crystalline basement. Inclined reflectors immediately beneath the 3-s reflectors are interpreted as small Cambrian rift basins. The Brevard fault zone is the distinct set of inclined reflectors that project to the surface near shot-point 700.

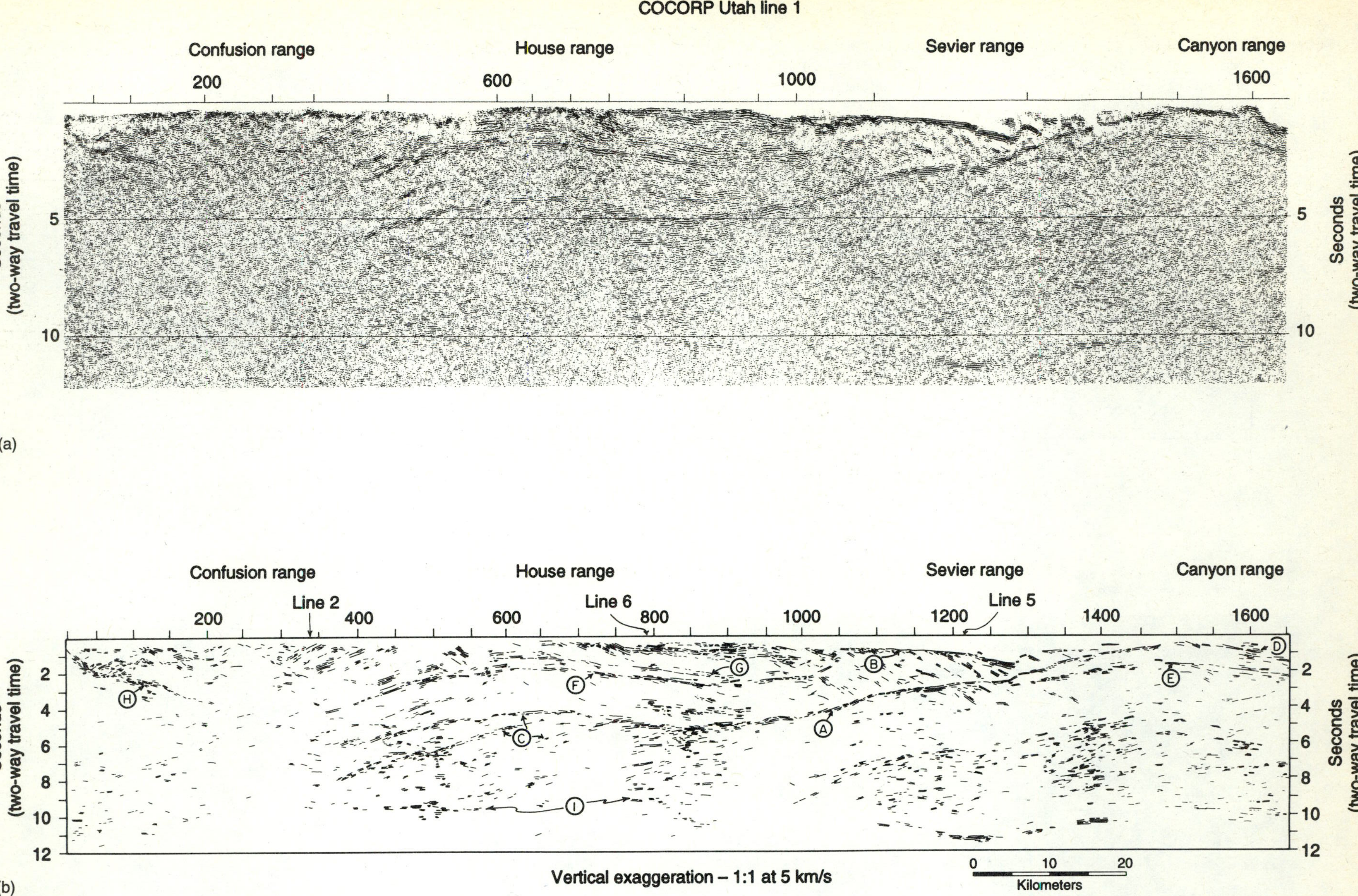

FIGURE 21–15
COCORP seismic reflection profile and tracing of reflectors in the section from the Basin and Range Province of Utah. (From Allmendinger and others, 1983, *Geology,* v. 11.) Structural interpretation of this section is shown in Figure 13–3.

chains (Figure 21–16) by the COCORP (Consortium for Continental Reflection Profiling) group at Cornell University, the U.S. Geological Survey, LITHOPROBE (Canada), BIRPS (British Institutions Reflection Profiling Studies), ECORPS in France, DECORP in West Germany, and others, including the CALCRUST consortium in California. As a result, several thousand kilometers of new crustal data are available to help to confirm, constrain, or disprove hypotheses about the structure of the crust. For example, the hypothesis of large-scale, thin-skinned thrusting of crystalline rocks in the southern Appalachians had been suggested from surface geologic studies (Bryant and Reed, 1970; Hatcher, 1971, 1972). Despite the successes of the COCORP seismic reflection profile in the southern Appalachians that gave the hypothesis greater credibility, it also raised new unanswered questions related to this and other regions in the crust. Some of these unsolved problems include determination of the nature of layered reflectors deep in the crust, the nature of curved reflectors, deep-seated "bright spots" (exceptionally strong reflectors originally identified in the shallow crust as hydrocarbons, but within the deeper crust they may be magma; de Voogd and others, 1988), and the zones of no reflectors, called "transparent zones."

The Mohorovičíc discontinuity (M–discontinuity or Moho, named in honor of Andrija Mohorovičíc, 1857–1936, a Croatian seismologist) is visible on reflection profiles as a series of nearly horizontal layered reflectors. Among related questions: What does the Moho actually represent? Why are there often several Moho reflections rather than a single reflector? Does the Moho represent a zone of laminar flow at a major rheological discontinuity? When did the present Moho form?

Jack Oliver (1988) has suggested a number of alternative interpretations of the Moho—interpretations that we were incapable of discussing without the abundant crustal geophysical data we now have. Reflection profiling has brought to light many aspects of crustal structure previously unknown, has made possible the confirmation and description of several types of structures, and has permitted formulation of new models for the structure of the continental crust. Compendia of papers (such as Barazangi and Brown, 1986; Matthews and Smith, 1987) that interpret the crust on the basis of these data and present new ideas continue to appear.

SEISMIC REFRACTION

The principle of refraction of waves as they pass from one layer to another is used in *seismic refraction* studies. Sound or other waves are detected after they are produced by an explosion or an earthquake. The waves pass into the Earth and are refracted at boundaries between particular layers where acoustic velocity increases downward; then the waves return to the surface, where they are detected by geophones (Figure 21–16a). The technique generally involves angles from the source (called "offsets") to the geophone that are much wider than in seismic reflection. Distances from the source to the detector are known, and so they may be calibrated with the travel times of acoustic waves of known frequency. Refraction is one of the best techniques for calculating seismic velocity and exploring the layered structure of the crust and mantle (Figure 21–16b). It is particularly useful in conjunction with seismic reflection studies, which require detailed knowledge of crustal velocities.

ELECTRICAL METHODS

Measurements of *electrical conductivity* and *resistivity* are useful for determining certain crustal properties. Conductivity studies have been used for many years in the search for sulfide ore bodies, because metallic sulfides are better electrical conductors than most other rock minerals. Resistivity and conductivity are usually measured by setting electrodes in the ground and recording for various lengths of time. Airborne techniques also exist.

Rock layers of greater conductivity than neighboring layers are detectable, but depth to the conducting layer may only approximately be determined. Graphite-rich layers and water containing small quantities of dissolved solids are also conductive and detectable. Electrical-resistivity anomalies may be exhibited by bodies composed of silicate rocks in a terrane otherwise dominated by carbonate, or by evaporites in a sedimentary terrane dominated by clastic rocks.

Here we end our survey of elementary structural geology. We hope the survey will have sowed the seeds of continued interest in those who will not return to the subject, and increased the interest and enthusiasm among those who plan to become structural geologists.

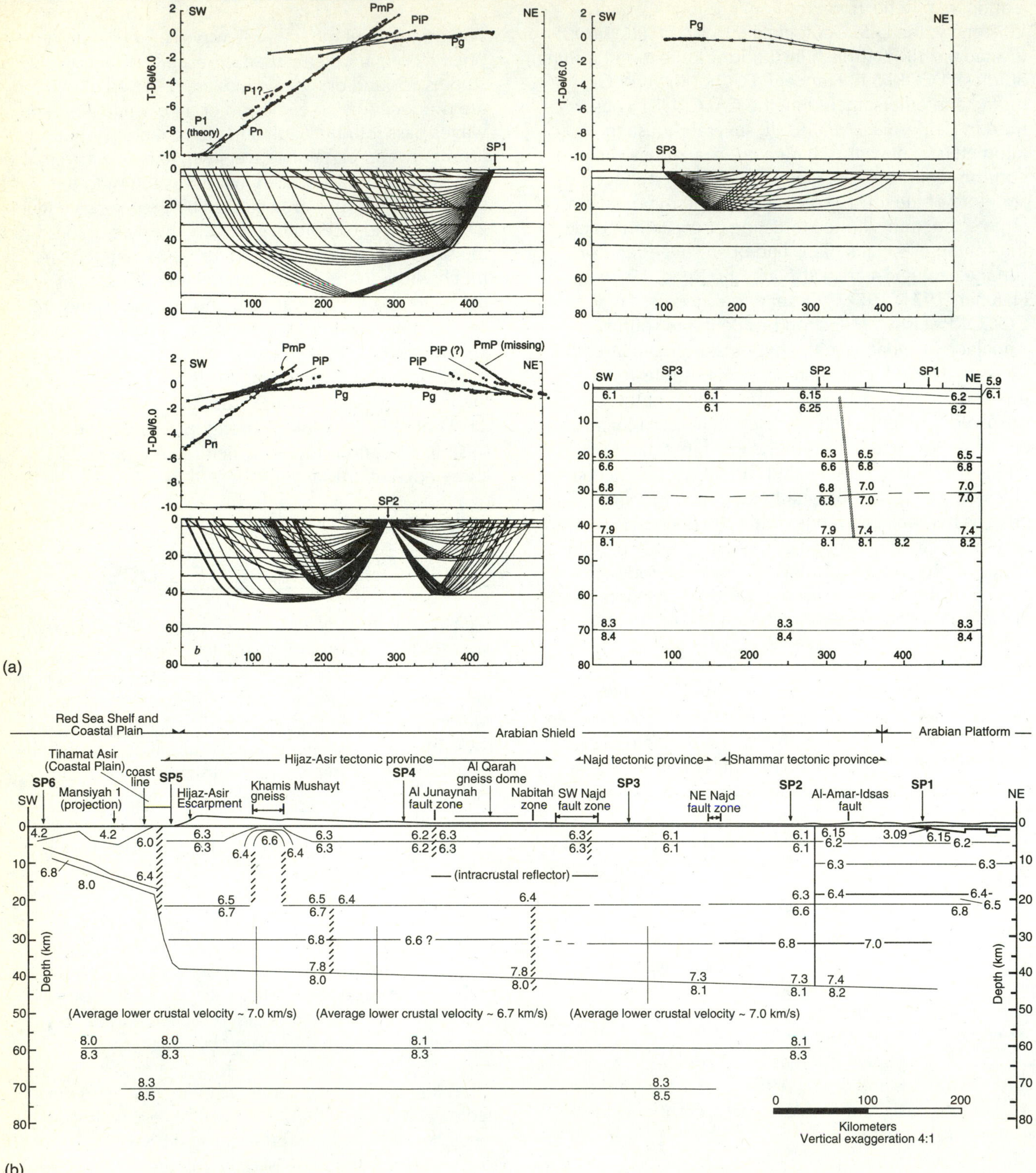

FIGURE 21–16

(a) Ray traces from source to receiver array in a seismic refraction experiment in Saudi Arabia, and a velocity model (lower right). (b) Seismic velocities for various parts of the crust in Saudi Arabia, determined by seismic refraction. (From *Tectonophysics,* v. 111, W. D. Mooney, M. G. Gettings, H. R. Blank, and J. H. Healy, p. 173–246, © 1985, with kind permission from Elsevier Science, Ltd., Kidlington, United Kingdom.)

ESSAY

Enhancing Structural Interpretation with Geophysical Data

The proliferation of high-quality geophysical data that began in the 1970s has led to improved interpretations of geologic structure. Examples can be drawn from all of the continents. Structural trends in the Precambrian basement imaged by aeromagnetic data continue uninterrupted many kilometers westward beneath the foreland fold-thrust belt of the Canadian Rockies, helping to demonstrate the lack of basement involvement in the thrust belt. Continuity of the thin-skinned fold-thrust belt exists farther west beneath the crystalline core of the orogen is more extensive than could be proved by surface geologic data alone (Price, 1981). Earlier, seismic reflection data (Bally, Gordy, and Stewart, 1966) from the outer parts of the foreland had revealed the thin-skinned style of deformation here. Earlier still, its existence had been only speculative (but correct), based on the work of John Rich (Chapter 11) in the Appalachians.

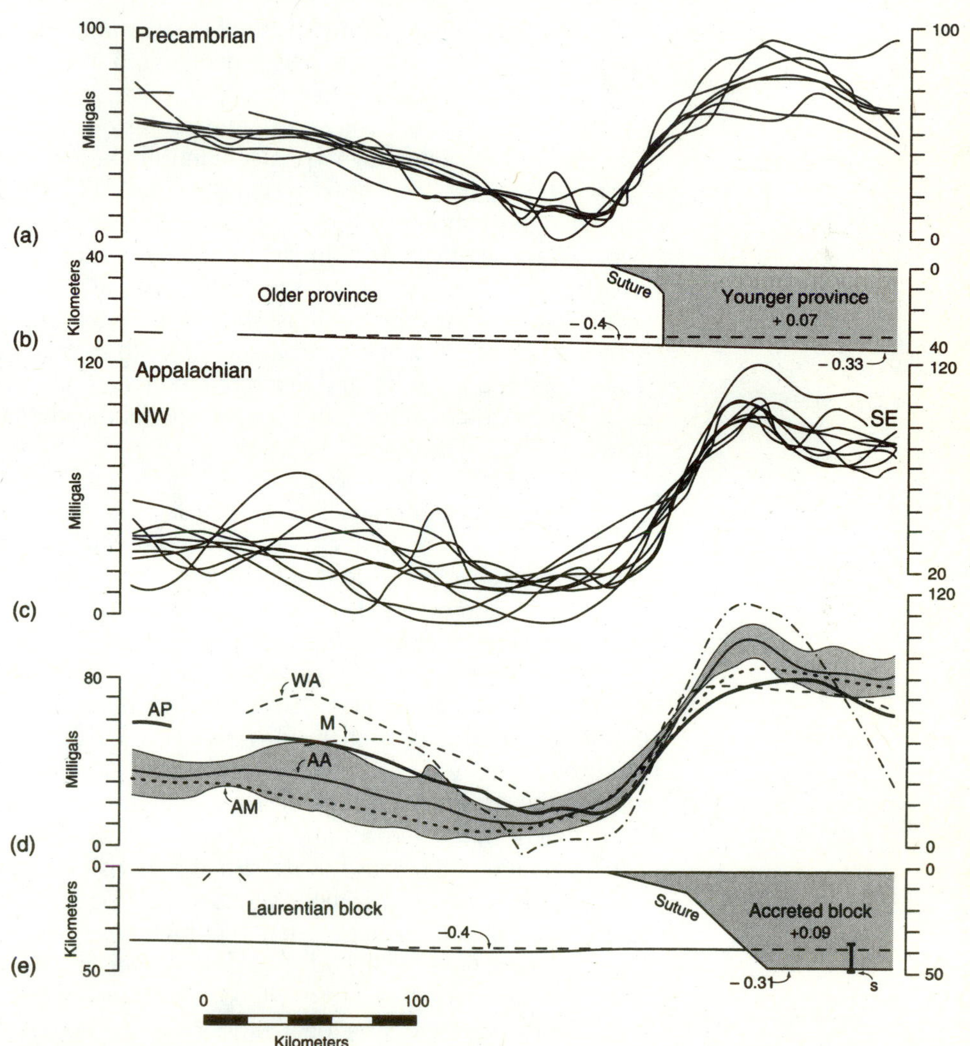

FIGURE 21E–1

Gravity profiles (a) and a gravity-based crustal model (b) across several Precambrian orogens in the Canadian shield, and similar profiles (c) and (d), and a model (e) for the southern and central Appalachians. The gravity low is always found closer to the older province (older core of the continent) and the high closer to the internal parts of the mountain chain. AP—Average Precambrian profile. AA—Average Appalachians profile. AM—curve corresponding to the type Appalachian model. WA—West African craton curve. M—Curve for the Musgrave block in Australia. (After M. D. Thomas, 1983, Geological Society of America *Memoir* 158.)

An integral part of many mountain chains, from the Precambrian to the Tertiary (Figure 21E–1), is a gradient from gravity low toward the foreland to gravity high toward the closed ocean. There, the position of the gradient frequently coincides with the surface position of a suture, and on that basis Michael D. Thomas (1983) concluded that the gradient localized a former collision zone. Changes in the nature of aeromagnetic anomaly patterns also coincide in many places with gravity gradients and sutures, but in the same orogen the gravity gradient may follow a suture for some distance and then diverge from it. The Appalachians is one example: the gravity gradient coincides with the suture between an exotic terrane and North America from Alabama to Virginia, then diverges from the suture to the north. Thus, Deborah Hutchinson, John Grow, and Kim Klitgord (1983) concluded that the Appalachian gravity gradient, and perhaps others, did not have a unique origin. Also, some geologists have speculated that the Appalachian gradient formed during Mesozoic extension and crustal thinning related to the opening of the Atlantic Ocean and not at all to contraction processes. Gary Karner and Alan Watts (1983) declared, however, that the gravity field over both ancient and modern mountain chains was consistent with the processes related to the construction of mountain chains by compression.

The extent of the huge Appalachian Blue Ridge–Piedmont Type C thrust sheet has been better delineated by geophysical data (Figures 21–14b and 21–14c). Its existence throughout the Blue Ridge and part of the Piedmont had been deduced from surface geologic data (Bryant and Reed, 1970; Hatcher, 1972).

Aeromagnetic data from the southern Appalachians reveal very little of the surface structure within the thrust sheet in both the Blue Ridge and much of the Piedmont, leading Hatcher and Zietz (1978) to suggest that the thin sheet extends across all the Blue Ridge and much of the Piedmont. We incorporated the gravity data into the interpretation and confirmed that the changes in magnetic and gravity patterns occur at the same place—the gravity gradient just discussed.

In 1978, COCORP collected the data for a crustal seismic reflection line across the southern Appalachians. A strong set of subhorizontal reflectors was detected at depths of 2 km at the northwest end to 12 km or more at the southeast end. These reflectors were initially interpreted as the Appalachian–Blue Ridge thrust (Cook and others, 1979) but were later said to be derived from sedimentary rocks beneath the thrust sheet (Ando and others, 1983; Cook and others, 1983).

Seismic reflection data acquired in 1984 in the Blue Ridge and western Piedmont (Çoruh and others, 1987) showed that the crystalline thrust sheet is much thinner in the Blue Ridge than suggested by the COCORP data and that the platform sedimentary sequence beneath is deformed into a series of duplexes that, in turn, have arched the crystalline sheet above the duplexes. Large recumbent folds have also been imaged within the crystalline sheet in the Piedmont (Figure 21–14b), but little of the near-surface structure is visible in the Blue Ridge, probably because the surface dip is moderate to steep. Geophysical data should provide more useful data on crustal structure in the future—and imaging of structures heretofore unknown will lead to new controversies. The greatest strides in our knowledge of crustal structure will continue to be made by structural geologists and geophysicists working together to interpret geologic and geophysical data.

References Cited

Ando, C. J., Cook, F. A., Oliver, J. E., Brown, L. D., and Kaufman, S., 1983, Crustal geometry of the Appalachian orogen from seismic reflection studies, *in* Hatcher, R. D., Jr., Williams, H., and Zietz, I., eds., Contributions to the tectonics and geophysics of mountain chains: Geological Society of America Memoir 158, p. 83–102.

Bally, A. W., Gordy, P. L., and Stewart, G. A., 1966, Structure, seismic data, and orogenic evolution of the southern Canadian Rockies: Canadian Petroleum Geology Bulletin, v. 14, p. 337–381.

Cook, F. A., Albaugh, D. S., Brown, L. D., Kaufman, S., Oliver, J. E., and Hatcher, R. D., Jr., 1979, Thin-skinned tectonics in the crystalline southern Appalachians; COCORP seismic reflection profiling of the Blue Ridge and Piedmont: Geology, v. 7, p. 563–567.

Cook, F. A., Brown, L. D., Kaufman, S., and Oliver, J. E., 1983, The COCORP seismic reflection traverse across the southern Appalachians: American Association of Petroleum Geologists Studies in Geology, v. 14, 61 p.

Çoruh, C., Costain, J. K., Hatcher, R. D., Jr., Pratt, T. L., Williams, R. T., and Phinney, R. A., 1987, Results from regional Vibroseis profiling: Appalachian ultradeep core hole site study: Geophysical Journal of the Royal Astronomical Society, v. 89, p. 473–474.

Hatcher, R. D., Jr., and Zietz, I., 1978, Thin crystalline thrust sheets in the southern Appalachian Inner Piedmont and Blue Ridge: Interpretation based upon regional aeromagnetic data: Geological Society of America Abstracts with Programs, v. 10, p. 417.

Hatcher, R. D., Jr., and Zietz, I., 1980, Tectonic implications of regional aeromagnetic and gravity data from the southern Appalachians, *in* Wones, D. R., ed., The Caledonides in the USA: Virginia Tech Geological Sciences Memoir 2, p. 235–244.

Hutchinson, D. R., Grow, J. A., and Klitgord, K. D., 1983, Crustal structure beneath the southern Appalachians: Nonuniqueness of gravity modeling: Geology, v. 11, p. 611–615.

Karner, G. D., and Watts, A. B., 1983, Gravity anomalies and flexure of the lithosphere at mountain ranges: Journal of Geophysical Research, v. 88, p. 10,449–10,477.

Price, R. A., 1981, The Cordilleran foreland thrust and fold belt in the southern Canadian Rockies, *in* McClay, K. R., and Price, N. J., eds., Thrust and nappe tectonics: Geological Society of London Special Publication 9, p. 427–448.

Thomas, M. D., 1983, Tectonic significance of paired gravity anomalies in the southern and central Appalachians, *in* Hatcher, R. D., Jr., Williams, Harold, and Zietz, Isidore, eds., Contributions to the tectonics and geophysics of mountain chains: Geological Society of America Memoir 158, p. 113–124.

Questions

1. What is the basis of the self-exciting dynamo theory?
2. We assume that the Earth's core consists of 90 percent iron and 10 percent nickel and that the mantle is peridotite. On what do we base our assumption?
3. Why is gravity less precise than seismic reflection as an imaging technique?
4. Calculate the depth at which most common magnetically susceptible minerals will no longer interact with the Earth's magnetic field.
5. Why are magnetic anomalies over a single source separated into high and low pairs at low latitudes, but at high latitudes consist of a single anomaly?
6. How can paleomagnetic measurements that do not meet a fold test be rendered usable?
7. What are the differences between free-air and Bouguer gravity maps?
8. What does a gravity model for part of the crust or a body represent?
9. Why does seismic reflection not detect steeply dipping features?
10. Why are paleomagnetic measurements unable to determine longitude?
11. Calculate the depth to a 7-km-wide gravity anomaly thought to be produced by a nearly spherical pluton.

12. A number of Russian geologists and geophysicists have suggested that a 55- to 60-km-deep root of thickened continental crust exists beneath the Ural Mountains, yet the maximum elevations anywhere in the Urals range only slightly above 2,000 m. From your knowledge of gravity and isostasy, do you think this is a reasonable hypothesis? Why?

Further Reading

Bally, A. W., 1983, Seismic expression of structural styles: American Association of Petroleum Geologists, Studies in Geology 15, 3 volumes.
A compendium of seismic reflection profiles showing various kinds of structures and providing examples of artifacts.

Barazangi, M., and Brown, L. D., eds., 1986a, Reflection seismology: A global perspective: American Geophysical Union Geodynamics Series, v. 13, 311 p.

Barazangi, M., and Brown, L. D., eds., 1986b, Reflection seismology: The continental crust: American Geophysical Union Geodynamics Series, v. 14, 339 p.
Each volume applies reflection seismology to solving structural and tectonics problems and recognizes new unsolved problems. Some papers present new data not previously published; others interpret existing data.

Breiner, S., 1973, Applications manual for portable magnetometers: Sunnyvale, California, E. G. & G. Geometrics, 58 p.
Primarily for use of Geometrics™ magnetometers, but summarizes interpretation of magnetic data and adds some elementary theory. May not be widely available.

Dobrin, M. B., 1976, Introduction to geophysical prospecting: New York, McGraw-Hill, 630 p.
Compiles seismic reflection theory and applications; less emphasis on other techniques.

Hoffman, K. A., 1988, Ancient magnetic reversals: Clues to the geodynamo: Scientific American, v. 256, no. 5, p. 76–83.
Discusses the Earth's magnetic field, polarity reversals, and evidence that decrease in magnetic-field intensity during the past century may presage a reversal of polarity in the next 1,500 years or so.

Sharma, P. V., 1976, Geophysical methods in geology: Amsterdam, Elsevier, 428 p.
Surveys the various geophysical methods, probably treating gravity and magnetism best.

Telford, W. M., Geldart, L. P., Sheriff, R. E., and Keys, D. A., 1990, Applied geophysics, 2nd ed.: Cambridge University Press, 770 p.
A balanced treatment of geophysical methods, giving the limitations of each.

Appendices

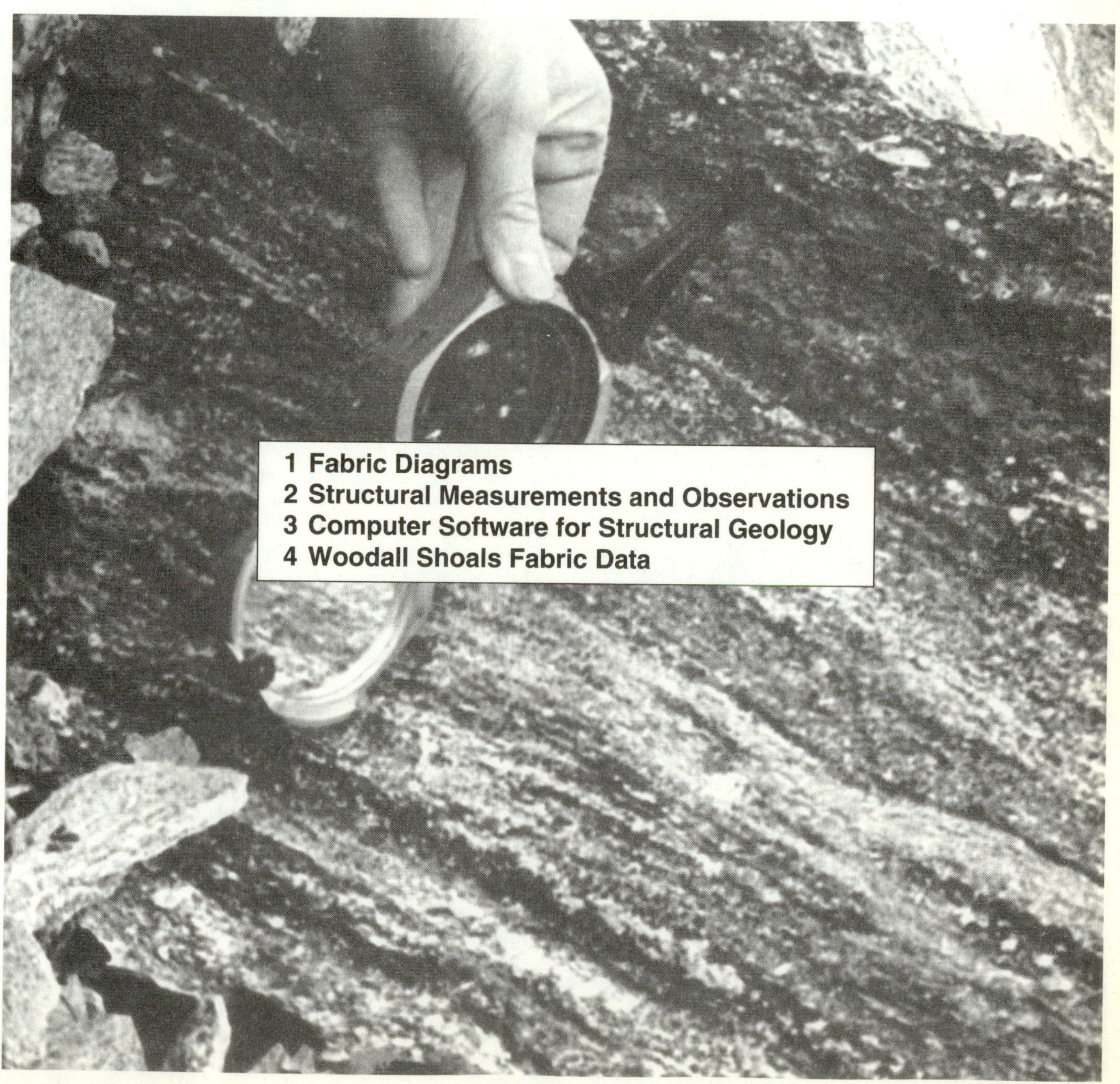

Appendix 1

Fabric Diagrams

In structural analysis, the necessity to evaluate preferred orientations of fabric elements imposes a peculiar limitation on graphic procedure. All equal areas on the surface of the reference sphere must remain equal on the projection itself. This is not true of the stereographic projection, in which centrally situated areas are diminished relative to peripheral areas of equivalent size on the reference sphere. To obviate this difficulty it is customary in structural geology to use a type of equal-area projection.

F. J. TURNER AND L. E. WEISS, 1963, *Structural Analysis of Metamorphic Tectonites*

TWO PLOTTING NETS ARE COMMONLY USED IN GEOLOGY FOR plotting planes and lines: the equal-angle stereographic (Wulff) net and the equal-area (Schmidt, or Lambert) net (Figures A1–1 and A1–2). Each consists of a circular protractor that has been divided, using circular or elliptical arcs, into a number of smaller angular segments. The principal difference between the two projections is the means by which they are constructed. In the case of the stereographic projection, plotting a point is accomplished using the radius of the reference sphere, whereas in the equal-area projection, a point is plotted using twice the radius of the reference sphere (Figure A1–3). It should also be noted that the subdivisions of the equal-area projection are proportional to the areas they represent on a sphere throughout the entire protractor—hence the name—whereas this proportionality differs in different parts of the stereographic projection, but the angular relations are conserved. With either protractor, the projection attempts to portray in two dimensions planes or lines that may otherwise be expressed on the surface of a sphere. A common use of the stereographic projection in mineralogy is to express in two dimensions the relative positions and symmetry of crystal faces. Plotting crystal faces, or other planar or linear features, may be accomplished using either stereographic or equal-area projections. All of the discussion to follow involves the equal-area projection as the standard reference for plotting. It is possible, however, to plot any line or plane on a stereographic projection. It should be noted that, following the convention in structural geology, the lower hemisphere is employed for plotting using either net, whereas the upper hemisphere is employed for plotting in mineralogy. When data are to be contoured, the equal-area projection should be used to make the quasi-statistical principles of contouring meaningful. The stereonet should be used for determination of the angular relationships of small amounts of data where contouring is not needed.

HOW TO BEGIN

An equal-area net should be obtained. The standard net has a 10-cm radius. Prepare the net by attaching a rigid backing of cardboard or plastic, placing a transparent cover over the front of the protractor, and inserting a thumbtack through the exact center of the protractor from the back. The tack is used so that overlays may be placed on the protractor and rotated to conveniently plot data without sliding the overlay out of alignment (Figure A1–4). It is common practice to begin plotting by taking a piece of tracing paper, Mylar™, or acetate film, and placing it over the protractor. The first step in plotting any element on an equal-area (or stereographic) net is to mark off the *primitive circle*, the perimeter of the protractor, on the overlay. Next, mark a convenient reference point so that you can always return to the starting position on the overlay for additional plotting of points or planes. This common reference point in structural geology is the position of north on the primitive circle located at the top at 0°.

PLANAR STRUCTURES

Plotting of planar structures, including bedding, foliation, cleavage, fault surfaces, and joints can be accomplished by two means on an equal-area or a stereographic projection. Planes may be plotted either as the trace of the line of intersection between the plane and the lower hemisphere projected to the equatorial plane, or as *poles* (perpendiculars) to planes that plot as points. Traces of the plane are *great circles* that intercept the reference, or *primitive,* circle of the projection. Because *poles* are perpendicular to the planes of interest, they plot as points on the opposite side of the diagram (actually 90°) from the great circle representing the corresponding plane.

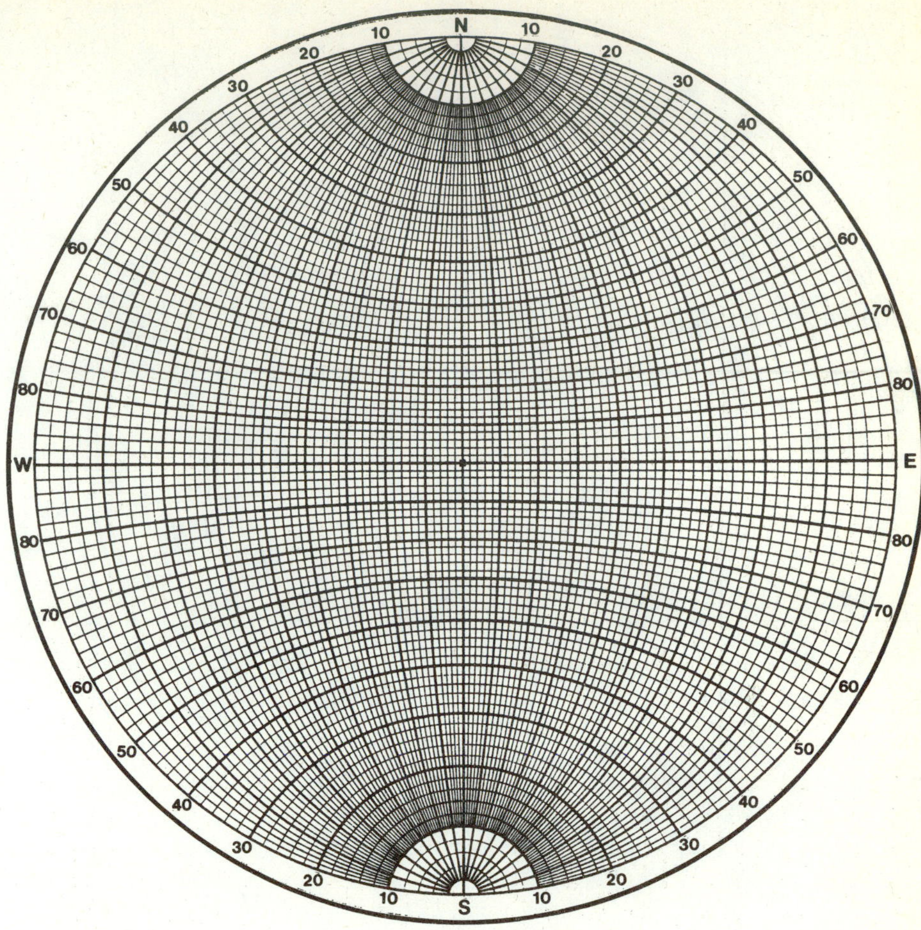

FIGURE A1–1
Stereographic (Wulff) net.

Let's plot a dip-strike measurement on the equal-area net. It will be plotted both as a plane and as a pole to the plane. The measurement has a strike of N 37° W and a dip of 28° SW (alternatively 323, 28° SW). The first step is to count 37° counterclockwise (left) along the edge of the net—to the west of north—along the margin of the net and mark that point with a pencil on the overlay (Figure A1–5). This represents the strike of the plane. Next, rotate the overlay 37° clockwise (right) so that the point that was just marked on the overlay lines up with the north pole of the net. Then, count along the equatorial line 28° from the west (left) edge of the protractor. Trace out on the overlay the great circle connecting the north and south poles of the protractor. The circle corresponds to the trace of the plane in the projection, and so the plane has been plotted as a line with a strike of N 37° W and a dip of 28° SW. Finally, without moving the overlay, plot the pole to the plane by counting 28° from the center outward on the opposite (east) side along the east-west (equatorial) line across the protractor. This point is 90° from the trace of the plane. Then place a point representing the pole to the plane N 37° W, 28° SW on the overlay.

LINES

An overlay is prepared for the equal-area net in exactly the same way as for plotting a plane. The primitive circle is sketched on the overlay, and a mark is made to identify north.

Let's plot a linear structure on the equal-area net. A fold axis plunges 39° toward N 63° E (alternatively expressed 063, 39°). The first step in plotting the line is to mark the trend along the primitive circle (Figure A1–6), as you did in plotting the trace of a plane. In this case, because the trend is N 63° E, you must count 63° to the east (right) of north and place a pencil mark at that point on the primitive circle. Next, rotate the mark on the primitive circle until it coincides with the east-west (equatorial) diameter of the protractor. Now count in 39° from the east edge of the primitive circle and mark the point. This point corresponds to the projection of the intersection of the line with the hemisphere "reflected" back to the plane of the projection.

This method of plotting lines allows the orientations of many structural features to be expressed on equal-area diagrams. Fold axes, boudin axes, intersec-

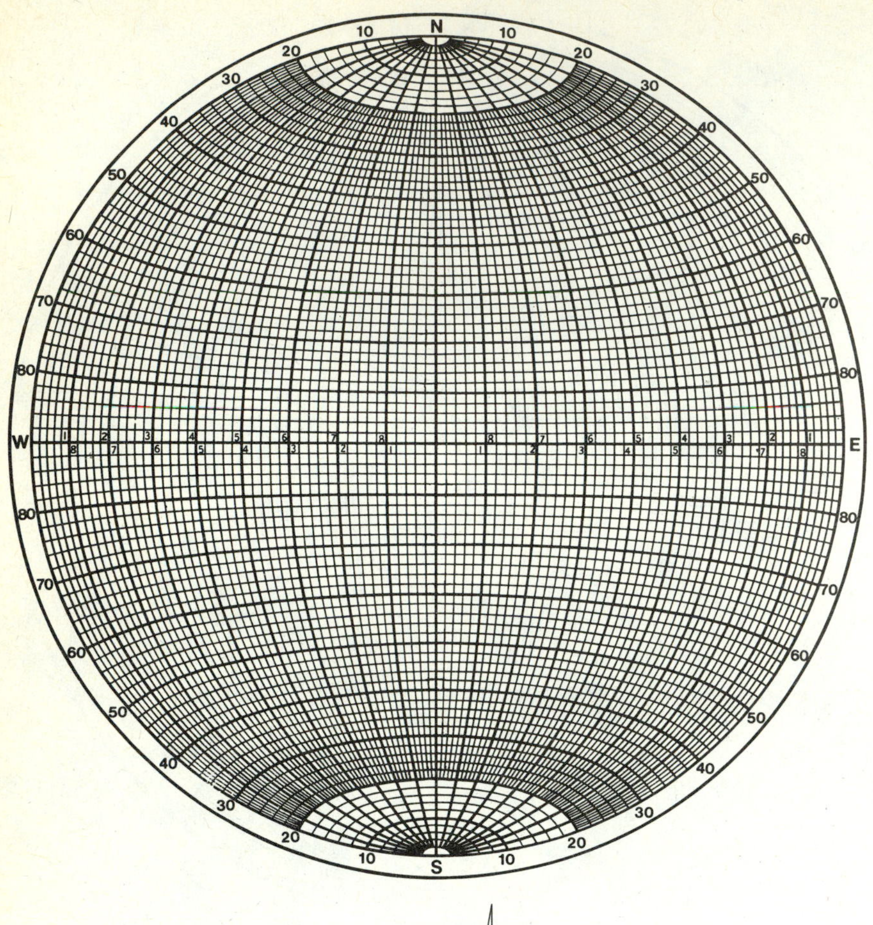

FIGURE A1–2
Equal-area (Schmidt, or Lambert) net. This protractor may be photocopied and enlarged to 10 cm radius for use with the template in Figure A1–9.

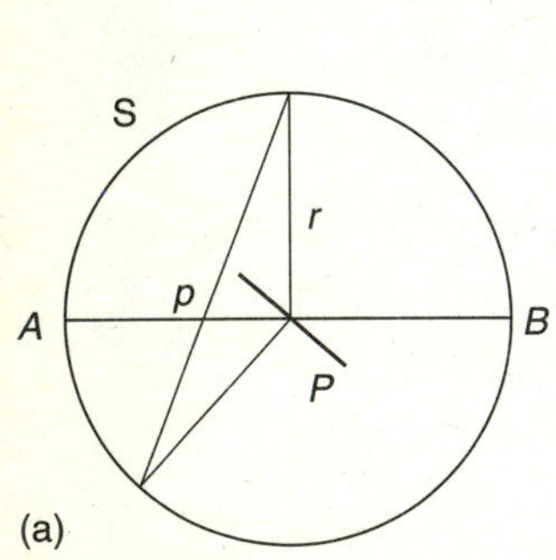

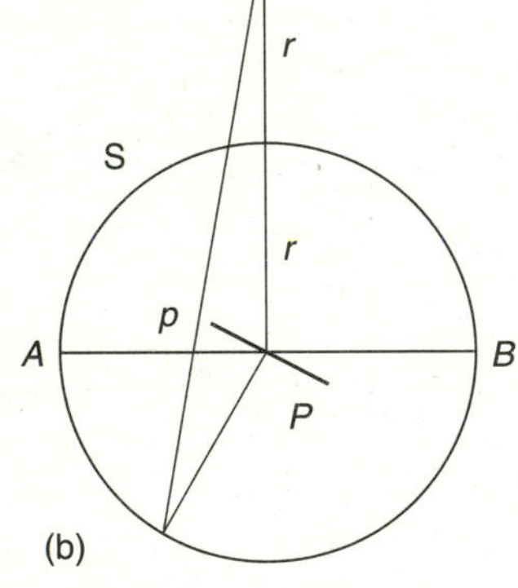

FIGURE A1–3
Basis for plotting a pole (p) to a plane (P) on a stereographic (a) or equal-area net (b). S is the reference sphere of radius r. AB is the equatorial plane of the sphere, and so we are looking at a vertical section through the sphere. Note that a pole to a plane is plotted on the stereographic net (a) by reflecting a normal downward from plane P to the boundary of the sphere and then back to the north pole of the sphere. The point of intersection of the reflection of the pole with the equatorial plane is the point where the pole will plot in the diagram. In contrast, a pole to plane P plotted on the equal-area net (b) is reflected to the boundary of the sphere and then is reflected to a point at a distance $2r$ from the center of the sphere that passes through the north pole. Point p, again, is the point of intersection of the reflection of the pole with the equatorial plane. Note that both (a) and (b) are plotted on and employ the lower hemisphere, as is the convention in structural geology.

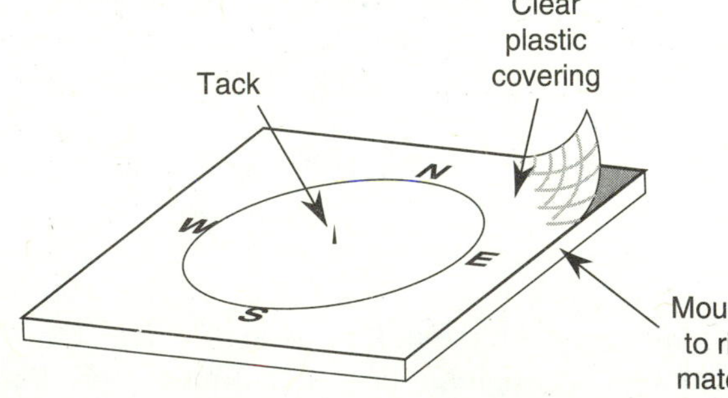

FIGURE A1–4
Mounting a plotting net. After mounting the net on a piece of cardboard or other rigid material, cover the net with a piece of clear, durable plastic, and push a tack through the center of the protractor from the back

tions, and mineral lineations may be plotted, as well as the orientations of crystallographic axes of minerals in thin sections of deformed rocks measured under the microscope using a universal stage.

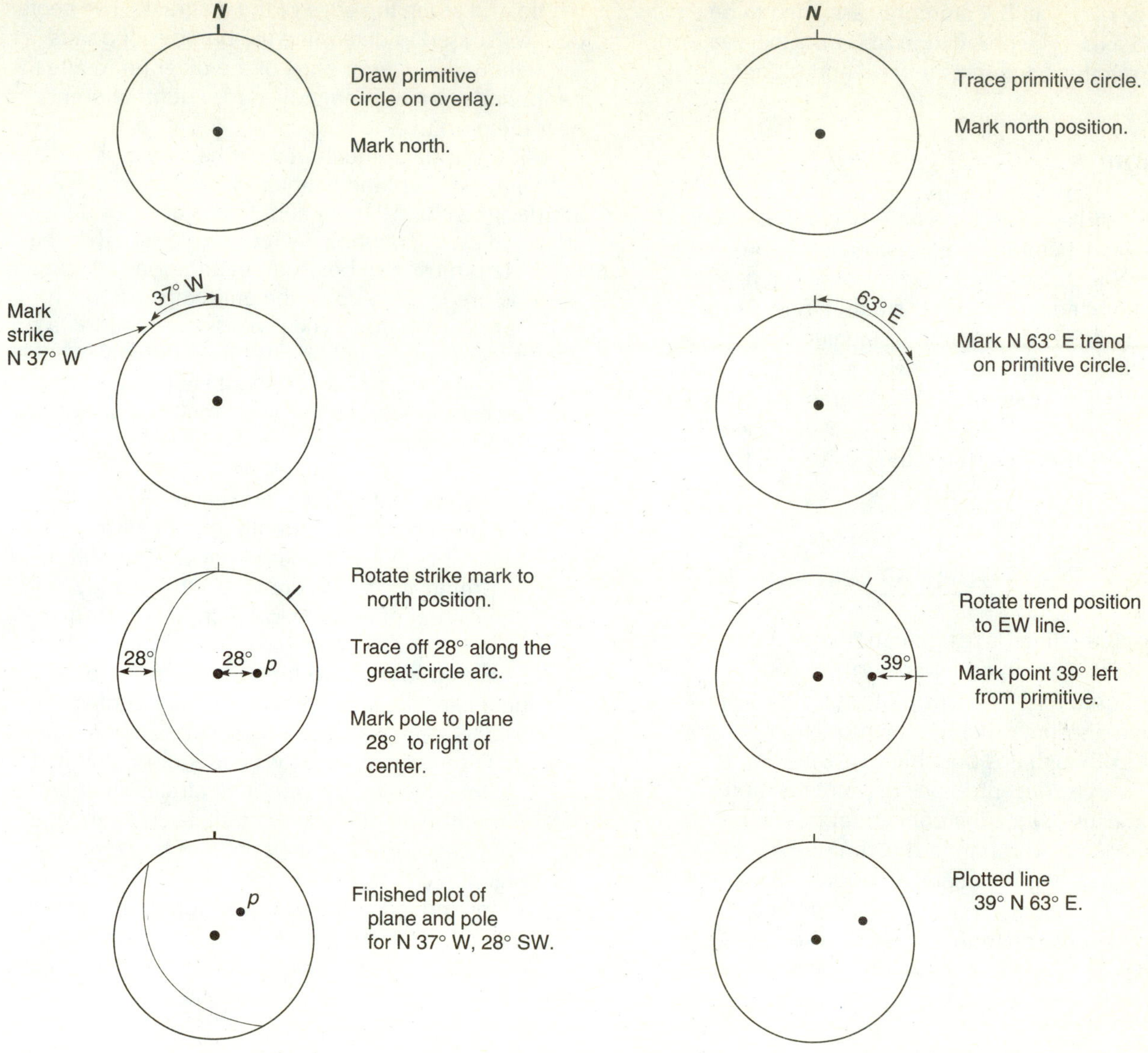

FIGURE A1–5
Plotting a plane with a strike of N 37° W and a dip of 28° SW on the equal-area net. The plane plots as a great-circle arc, and the pole (*p*) plots as a point.

FIGURE A1–6
Plotting a line that has a trend of N 63° E and a plunge of 39° NE on the equal-area net. The line plots as a point.

LOCATING FOLD AXES USING EQUAL-AREA PLOTS: β AND π DIAGRAMS

It is often desirable to determine the orientations of fold axes using dip-strike data where measurements of small fold axes cannot be made in sufficient numbers to accurately determine the orientation of major axes of first-order folds in an area. Two techniques for accomplishing this are the construction of β and π diagrams. Each will be explained below.

β Diagrams

Bedding or foliation planes plotted as traces on an equal-area projection will yield a family of great-circle curves in which the intersections approximate the orientation of the fold axis (Figure A1–7). Rarely will all the traces intersect at a common point, but the traces will form many intersections in a small area on the equal-area diagram, so that the center of mass (best-fit point) of the intersections will approximate the trend and plunge of a line, β, commonly assumed to be the major fold axis. The orientation of β may then be read directly from the diagram as a trend and plunge.

π Diagrams

A set of dip-strike measurements—of either bedding or foliation—may be plotted on an equal-area projection as poles to the planes. If the points plot in a pattern that may be fitted to a great (or small) circle, the pole to the great circle yields a point, designated by β on the diagram, that corresponds to the trend and plunge of the fold axis (Figure A1–8). This diagram is called a π diagram because poles to planes are used instead of the traces of a plane to determine the position of β.

CONTOURING DATA

A set of points—lines (axes, lineations, etc.) or poles to planes—plotted on an equal-area projection is a *point diagram,* or *scatter plot.* This plot is useful for showing the relative distribution of points throughout the diagram. Although it is possible by inspection to readily observe concentrations of points, a more efficient means of locating concentrations is by contouring the data. The standard technique of contouring fabric data involves the use of an equal-area plot of the data set and counting devices called *center* and *peripheral counters* (Figure A1–9). These counting devices are employed in conjunction with a grid to determine the number of points per 1 percent area within and along the edge of the diagram. The center counter is used to determine the number of points throughout the interior parts of the diagram, while the peripheral counter is employed for counting points near the primitive.

The standard equal-area net has a radius of 10 cm. The dimensions of the counters shown in Figure A1–9 are designed for a 10-cm net. If nets of other dimensions are used, counting devices proportional to the other sizes must also be used. In addition, a 1-cm grid must be prepared to cover the entire area of the 10-cm equal-area projection. The process of preparing a contoured diagram will be described stepwise. If the grid is to be used over and over, it should be constructed on a piece of translucent mylar or acetate film, and the lines of the grid inked in a fine-line pattern.

1. The translucent grid is placed over the point diagram, and a transparent paper or Mylar™ overlay placed over both of these. The outline of the primitive circle is drawn on the overlay, and the position of north marked on it. The center counter is placed over each intersection of the 1-cm grid so that two lines of the grid intersect at the center of the circle. The center counter must be used for only for those intersections where the circle on the counter does not enclose any part of the area out-side the primitive circle. The number of points lying within the circle is counted, and the number is written in the center of that circle (Figures A1–10a and A1–10b). The center counter is then moved to the next position of intersecting lines on the 1-cm grid, and the process is repeated. The center counter is moved again and again until a number appears in every intersection position in-

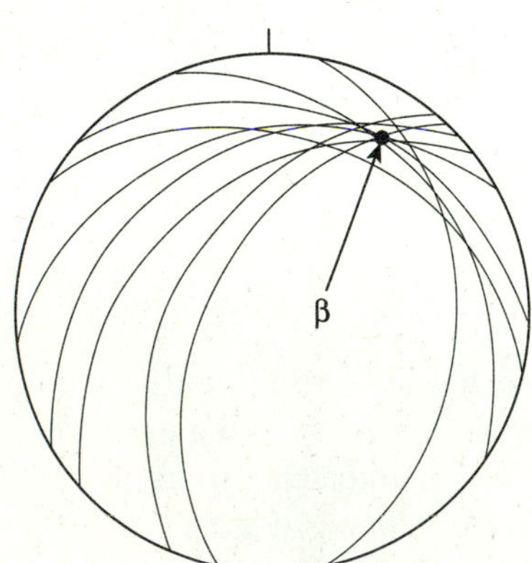

FIGURE A1–7
A β diagram consisting of a set of dip-strike data for a group of planar structures that plot as a series of great-circle arcs that intersect in one particular area of the fabric diagram. The center of this area of intersection is designated by β, and is assumed to approximate the trend and plunge of a fold axis.

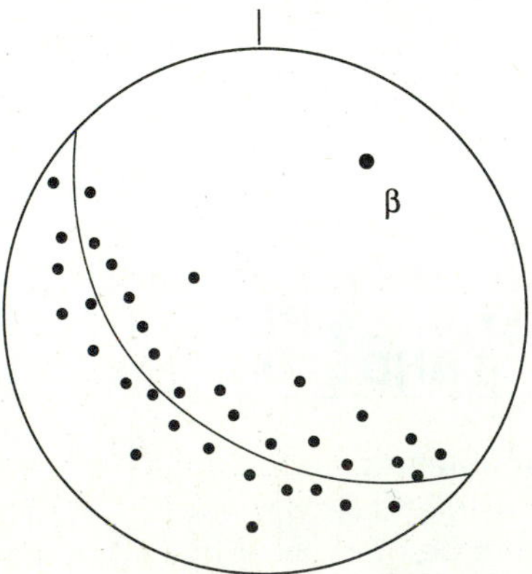

FIGURE A1–8
A π diagram consisting of poles to a set of dip-strike data for a group of planar structures. The resulting fabric diagram consists of a series of points that scatter about a great-circle arc constructed as the best-fit line for the points. The pole to the great circle is designated by β and is assumed to approximate the trend and plunge of a fold axis—here, the same fold axis as in Figure A1–7.

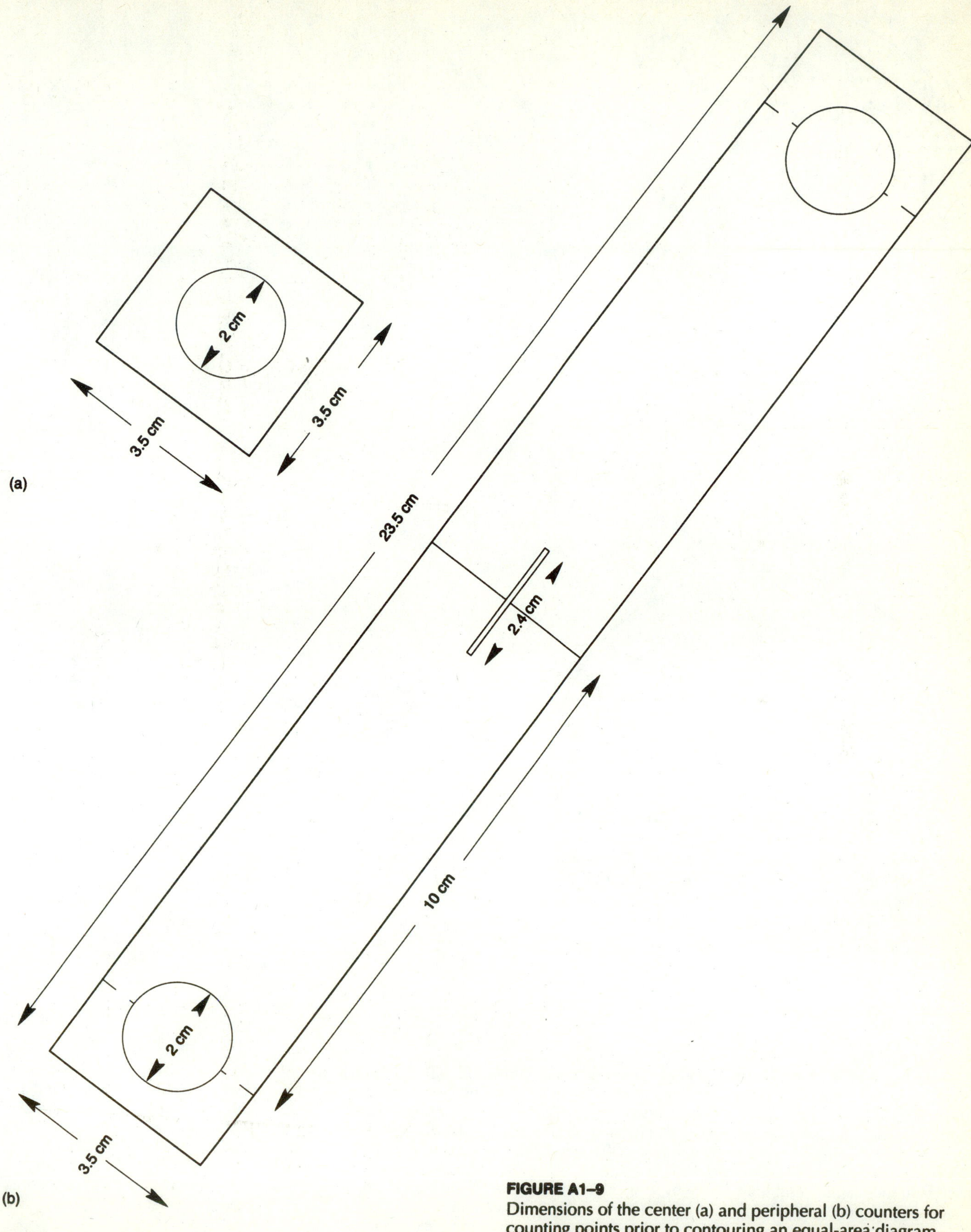

FIGURE A1–9
Dimensions of the center (a) and peripheral (b) counters for counting points prior to contouring an equal-area diagram. This figure can be used as a template to make center and peripheral counters from cardboard or other rigid material.

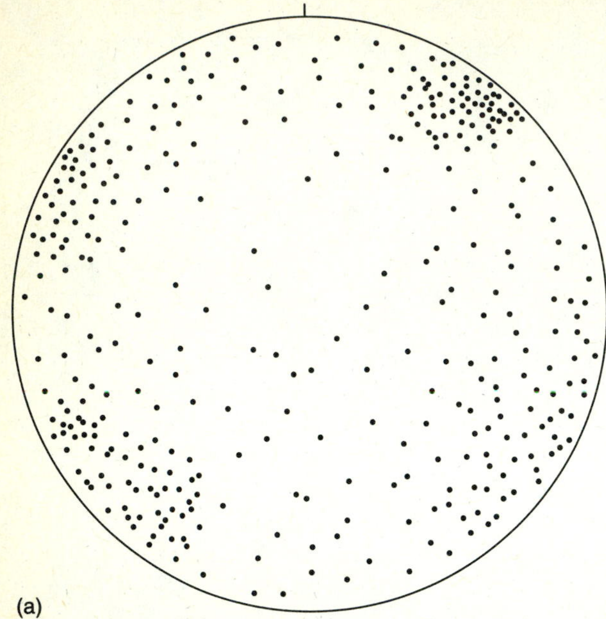

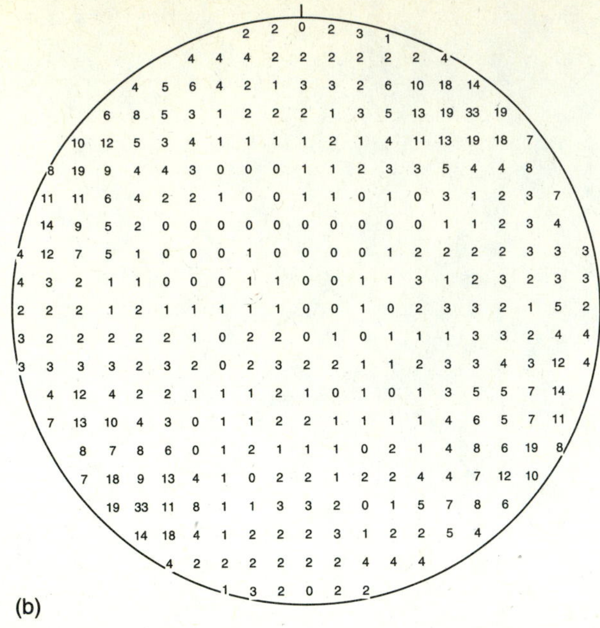

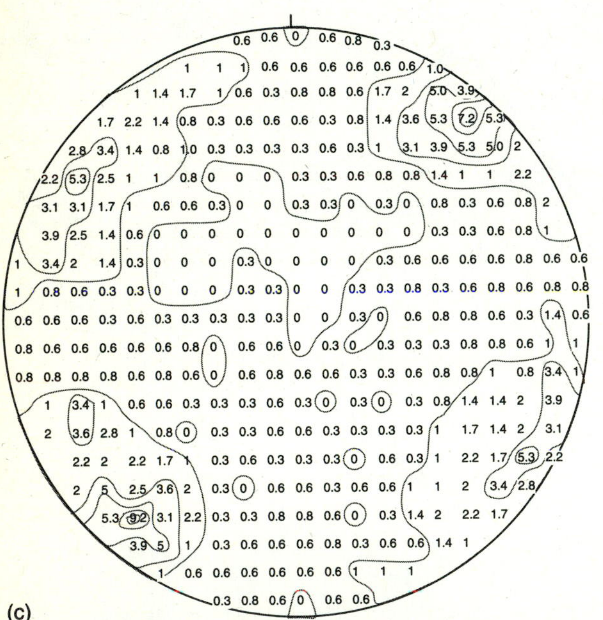

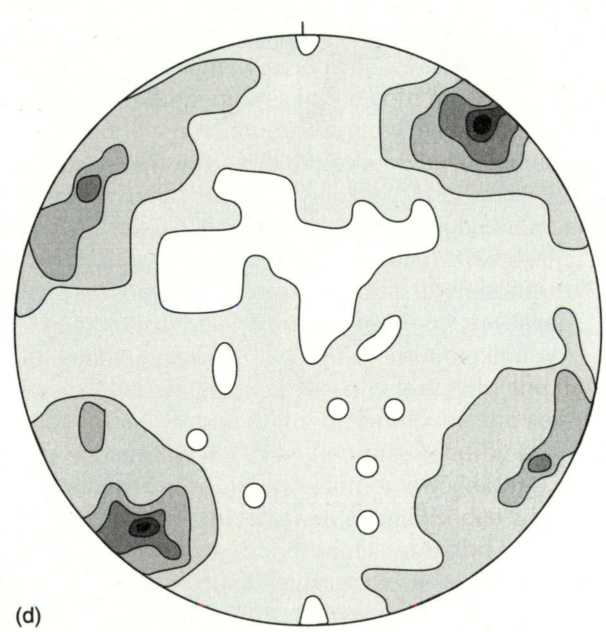

FIGURE A1–10
Contouring a set of points using the percent per percent area method. (a) Point diagram consisting of 357 poles to joints measured in basement rocks in central Colorado. (b) Results of counting points in (a) using the center and peripheral counters and a 1-cm grid. The numbers on this diagram represent the number of points inside the 1-cm circle, with a grid point as the center of the circle for areas inside the diagram. Numbers along the edge of the diagram represent the sum of points counted in both circles of the peripheral counter for grid points that cause the circle to overlap the edge of the diagram. (c) Percent diagram. Numbers here represent points counted in (b) converted to percent per 1 percent area by dividing each by the total number of points in the diagram and multiplying by 100. Contours are sketched here for transfer to the final contour diagram. (d) Contoured diagram with contours 0.3, 1, 3, 5, 7, and 9 percent per 1 percent area. Note that the noninterval contour, 0.3 percent, was drawn to show the boundary between no data and the smallest number of data.

side the primitive circle. After all the interior grid intersections have a number on them, the peripheral counter is used for intersection positions along the edge of the circle, where the intersection on the grid causes part of the center counter to fall outside the primitive circle. Here, the number of points is counted and summed in each partial circle on both ends of the peripheral counter, and the same number is written on both sides of the diagram in the center of each circle of the peripheral counter (Figure A1–10b).

2. The total number of points in the point diagram must be counted. A second overlay is then prepared, on which the numbers previously written at the grid points are converted to percentages (by dividing each number by the total number of points and multiplying by 100), and each percent written on the new overlay at the appropriate grid point (Figure A1–10c). These percentages represent percent per 1 percent of the area of the 10-cm circle.
3. The values of 1 percent per percent area are contoured using an interval chosen by inspection that best fits the numbers on the diagram. Rather than always choosing an equal increment from the lowest to the highest, it might be useful to draw a contour that separates the lowest percentage—such as 0.3—from zero, thus showing the absolute distribution of data over the diagram (Figure A1–10d). An equal increment is then chosen for the remainder of the contours—2, 4, 6, 8, or 5, 10, 15, 20, or whatever is appropriate. The final contouring is then carried out on the percent overlay and the contour interval indicated. If it is desirable to reproduce the contour diagram, this may be accomplished by transferring the contours to another overlay.

The method described here is slightly modified from that described by Turner and Weiss (1963). A number of other contouring methods have been devised to accomplish the same result. One of these is the use of a counting net divided into small triangles, six of which form a hexagon equal to 1 percent of the area of the equal-area net (Kalsbeek, 1963). Some have suggested that the standard free-counter method (described above) for contouring introduces a bias in the shape of contours.

Mellis (1942) invented a contouring method involving the use of overlapping circles that does work well for diagrams containing fewer than 150 points, and particularly for drawing the contour of lowest point density. The center counter is placed above each plotted point, and a circle is drawn around the inside of the counting circle on an overlay, producing a mesh of overlapping circles. The contour representing 2 percent per 1 percent area outlines the area where 2 circles overlap; 4 percent per 1 percent area is outlined by drawing a line through the places where 4 circles overlap, and so on. This contouring technique is less subjective; it is described in greater detail in Turner and Weiss (1963).

A statistical method devised by Barclay Kamb (1959) makes use of contoured multiples of σ (standard deviation). This method contours data sets containing greater than 100 to 150 points better than using percent per percent area contours. Smaller data sets (50 to 150 points) are better contoured using the percent per percent area method. Compare Figure 8–9(c) and 8–9(d) where the same data set of 256 poles to joints was contoured on the computer using the same contouring program and both methods.

The mechanical process of conversion of a point diagram to a contour diagram may be easily accomplished today using a number of computer programs available in the literature (such as Jeran and Mashey, 1970; Cooper and Nutall, 1981) or easily obtained commercially (Appendix 3). Several have been written specifically for IBM and Macintosh personal computers. The diagrams in the Chapter 20 Essay were plotted using a program written for the Macintosh computer by R. W. Allmendinger at Cornell University.

Appendix 2

Structural Measurements and Observations

. . . As the cleavage foliation in some places coincides with the stratification foliation both in strike and dip, in others agrees in strike while differing in angle of dip, and in still others differs from it in the direction of both strike and dip, and . . . the whole matter is attended with much difficulty. This is enhanced by the absence of all outcrops over considerable areas. Satisfactory results can be reached only by accumulating a great number of observations. . . .

T. NELSON DALE, 1894, *U.S. Geological Survey Annual Report*

A WIDE VARIETY OF STRUCTURAL MEASUREMENTS AND observations are commonly made in the field. These range from the orientation of planar and linear structures, which involve making two measurements, to the orientation of folds, thus requiring four to five measurements for a complete characterization of orientation. Determination and recording of shear sense (vergence) of structures is also important. Accurate sketches of the geometry of structures in your field book will prove useful. Photographs of representative and critical mesoscopic structures should also be made to augment—not to replace—sketches. Accurate sketches require more observation than taking a photograph; details of a structure will become better understood if you take the time (sometimes an hour or more) to accurately sketch it in your notebook.

Accompanying the requirement that measurements be made correctly is that the location where measurements are being made be accurately known so that they are reproducible and can be related spatially to other measurements. Accurate location of each measurement is needed if you are constructing a geologic map, making measurements on a particular map-scale structure to investigate the mechanics by which it formed, or making detailed measurements of different kinds of structures in a roadcut, stream, or other small exposure. Such data remain the highest quality, most fundamental, and most quantitative of any data in geological science.

ORIENTATION OF PLANES: STRIKE AND DIP

Two measurements are used to describe the orientation of a plane in the Earth: strike and dip. The *strike* is the trend (compass direction, azimuth) of a horizontal line lying in an inclined plane (Figure A2–1)—the intersection of horizontal and inclined planes. The *dip* of an inclined plane is the largest acute angle made by the plane with the horizontal and is measured perpendicular to strike in the vertical plane. Angles measured within the vertical plane that are less than the maximum are not perpendicular to strike and are measurements of *apparent dip*. Strike is measured using a compass, and dip is measured with a clinometer, commonly built into compasses designed for making these measurements in the field (Figure A2–1).

Measurement of strike and dip can be made in several ways by using a compass. A common way is to place a smooth board (or thin sheet of aluminum) on the inclined surface to be measured to even out any irregularities. The edge of the compass is then placed against the inclined surface. When the compass is leveled, the strike is directly read. The direction perpendicular to the strike, down the inclined plane, is the dip direction; the angle of dip is then measured in this direction using the built-in clinometer in the compass. This method is very good for learning how to measure the strike of planar surfaces, but it can easily

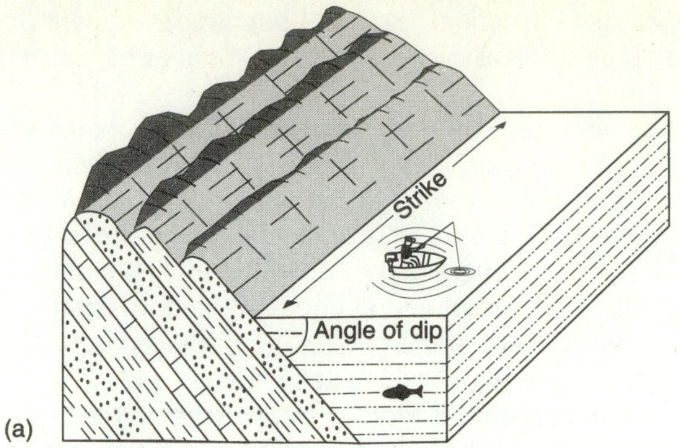

(a)

(b)

(c)

FIGURE A2–1
(a) Strike and dip of inclined bedding. Note that the strike is the line of intersection of the inclined bedding surface with the water surface, a horizontal plane; the dip is the acute angle between the inclined bedding and the water surface. (b) Measurement of the strike of compositional layering in 1.2-billion-year-old Toxaway Gneiss at Whitewater Falls in western North Carolina. (c) Measurement of dip on an inclined crenulated foliation surface at the same locality as (b). Compositional layering is visible on the end of the exposure. (RDH photos.)

yield erroneous measurements if irregularities on the rock surface are allowed to tilt the board out of parallelism with the true strike and dip of the surface. By making hundreds of measurements and developing an ability to visualize the correct orientation of an inclined plane to be measured, more accurate measurements of strike may be obtained by sighting along the strike line of the surface to be measured, and then aligning the clinometer parallel to the dip direction of the inclined surface to obtain the dip. Recording the strike and dip in your field book, then plotting it on the map using a protractor provides a good means of checking your measurement.

Direct measurement of strike can be made by aligning the compass parallel to the line formed by the intersection of an inclined surface with a water surface (Figure A2–1a). The dip is the acute angle the inclined planar rock surface makes with the water surface—a natural horizontal reference surface. Inclined surfaces also produce mesoscopic *cuestas* and *hogbacks,* the crests of which form natural strike ridges. To accurately measure the strike of an inclined layer producing such a feature, the compass is aligned parallel to the crest of the hogback and the strike is directly read. Dip is measured by one of the methods described above. Strike may be recorded in either azimuthal or quadrant format. Quadrant format is commonly measured and recorded relative to north. Azimuthal format records strike direction clockwise from 0° to 360°. For example, a plane with a northwest strike is recorded as N 45° W in quadrant format, or 315 in azimuthal format, not as S 45° E. Dip is usually expressed as an angle and a direction—here, 37° NE. The dip and dip direction may also be expressed as 37° N 45° E (or 37° 045), thereby also providing the strike orientation 90° from the dip direction.

Measurement of strike and dip may be made on any inclined surface, including bedding, foliation, joints, and faults, as well as on surfaces not of primary interest in structural studies, such as roads, topographic slopes, artificial cuts in rocks, and erosion surfaces. Measurement accuracy decreases where the surface being measured has a dip of less than 15°, because of small surface irregularities and the difficulty of making measurements on gently dipping surfaces. Measurements of dip should be accurate to within $\pm$ 2°–3°. At angles less than 15°, even if this level of accuracy can be maintained, $\pm$ 2°–3° error is a significantly greater proportion of the angle being measured than at steeper dip, and so the effective accuracy of measurements decreases where the dip is gentle. Measuring the dip direction and then the perpendicular to get the strike sometimes helps to improve accuracy of strike measurements where the dip is less than 15°; it is easier to visualize the dip direction than the strike direction.

ORIENTATION OF LINES: TREND AND PLUNGE; RAKE

The orientation of linear structures, such as mineral lineations, intersections, boudin axes, and fold axes, is expressed by the trend (azimuth) and plunge (Figure A2–2). The trend is the compass direction measured in the down-plunge direction of the structure. For example, if a lineation plunges southwest, the trend is recorded as S 45° W, or 225. Plunge is the angle made by the axis of the structure with the horizontal, in a vertical plane, and is measured in this plane by using a clinometer. Always try to view the linear structure in at least two nonparallel surfaces so that you do not record apparent orientations. Measurement of linear structures may be facilitated by aligning a pencil parallel to the structure.

Complete description of the orientation of folds requires two additional measurements: the dip and strike of the axial surface and the direction of overturning (vergence). The fold axial surface is a virtual surface, unless an axial-planar cleavage or foliation is present. It is also used to record the symmetry and approximate interlimb angle of folds (either a number or open, tight, isoclinal, etc.), although measurement of enveloping surfaces along with axial surfaces may also be useful.

An additional measurement occasionally made is the *rake* (Figure A2–3). The rake is an angle between a line lying in a plane and a horizontal line that also lies in the plane. Therefore, it is an angle not measured in a horizontal or vertical plane but in the inclined plane containing both the plunging line and the horizontal line. Measurement of the rake of the displacement vector on a fault plane (see next section) is one example of the use of this kind of measurement. Measurement of the rake of the displacement implies that all motion is translational and does not involve rotation, but rotation of large and small blocks has been documented in many areas. A protractor can be used for making these measurements.

SHEAR-SENSE INDICATORS

S-C structures, rotated porphyroclasts, asymmetric folds, and other structures that indicate shear sense may be observed in the field.

The orientations of S- and C-surfaces can be measured as any other planes, and the intersections of S- and C-surfaces are lines that can also be measured or determined using an equal-area net. The shear direction can be recorded perpendicular to the intersection formed between the acute angle between S- and C-surfaces.

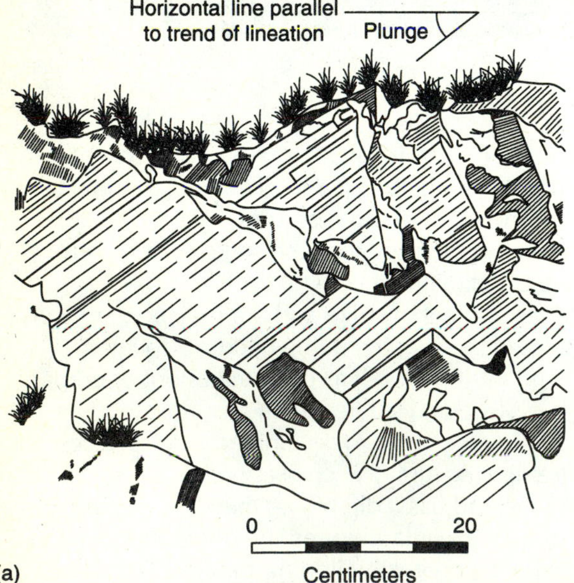

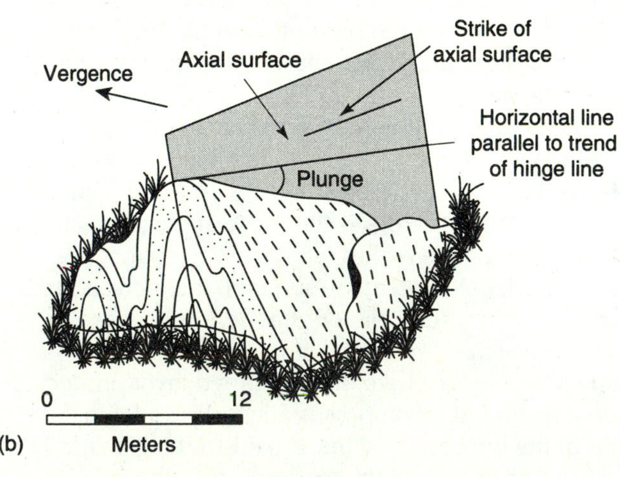

FIGURE A2–2
(a) Sketch of intersection lineation in Figure 18–4a showing the trend and plunge of the lineation. (b) Determination of the orientation of a mesoscopic fold. Note that the strike of the axial surface need not necessarily coincide with the fold hinge line, or axis. Measurements to be made are the trend and plunge of the axis, strike and dip of the axial surface, and the direction of vergence.

Recording the shear direction and sense of rotation of porphyroclasts and asymmetric folds provides the needed information to characterize these structures. Similar measurements should be made of lineations (intersections, mineral lineations, boudins).

RECORDING DATA

There are several ways to accurately record field data. The simplest is to write out descriptions of all stratigraphic and other lithologic features (as needed), write out structural measurements, and formulate a means to relate your data to a location on a map or photograph. Alternatively, field data may be tabulated as it is collected in the field. Data may also be preserved using a tape recorder, to be transcribed later, but it is more difficult to search back through a tape in the field to review a note or measurement than to flip back several pages in a notebook.

It may be useful to invent a shorthand system for recording structural and other data. The system should be efficient and understandable to someone other than yourself. I have used a system for many years that involves the abbreviation of rock type and a detailed record of structural data for a particular station. Each page in the field book (Figure A2–4) has the date and location of the data set—by specific locality, name of a quadrangle, a locality in a quadrangle, or some other means of immediately determining the site where the data set on that page is located. Each place data collected is assigned a separate station number—keyed to a station map (Figure A2–5). The first entry after the station number is the orientation of the dominant planar fabric element at the station—bedding, compositional layering, cleavage, foliation, schistosity—the most likely structure to be plotted on a geologic map at that station. The map symbol for the structure may be written beside the measurement— shorthand for which of the planar elements is represented by the measurement. Following that are the other structural measurements made at that station. A small Roman numeral (or other symbol, e.g., D_2) may be written next to any measurement to indicate the deformational event that might have produced it.

Extensive notes may be written about rock units in an area that has never before been described or one that is unfamiliar to the structural geologist. After the rock units become familiar and recognition becomes routine, a shorthand system may be devised for describing particular rock units or rock types. Abbreviations such as qzt. for quartzite, ss. for sandstone, sh. for shale, amph. for amphibolite, ls. for limestone, and standard map unit abbreviations, such as Dm for Marcellus Shale, IPf for Fountain Sandstone, Mk for Kaibab Limestone, may be used. Alternatively, the number code for the colored pencils used for particular rock types or units is a shorthand I have used; e.g., 745 is the number of the carmine red color of the Berol Verithin™ I have used for metagraywacke, and 742 1/2 is the lavender Berol Verithin™ color I use for pelitic (muscovite-biotite) schist. Rather than writing metagraywacke or pelitic schist in my field book for the dominant lithology at a station, I will write 745 or 742 1/2 opposite the dominant planar structure, or if both are present, 745–742 1/2. Following this scheme, only notes about geologic oddities, variations in rock type, structural relationships, or other nonroutine observations need be extensively described.

Sometimes on a sunny day in the field—sometimes even on a cloudy day—a magnificent idea will reveal itself, demanding space in the field notebook. Some of the best detailed notes I can recall describing a nonroutine observation were written one summer by a Clemson University senior undergraduate student working with me in the thickly forested southern Blue Ridge in North Carolina. Unfortunately, these notes did not describe a new idea about the geology but the appearance and proximity of a large black bear that walked to within less than 5 m of the student as he was sitting on an exposure making structural measurements!

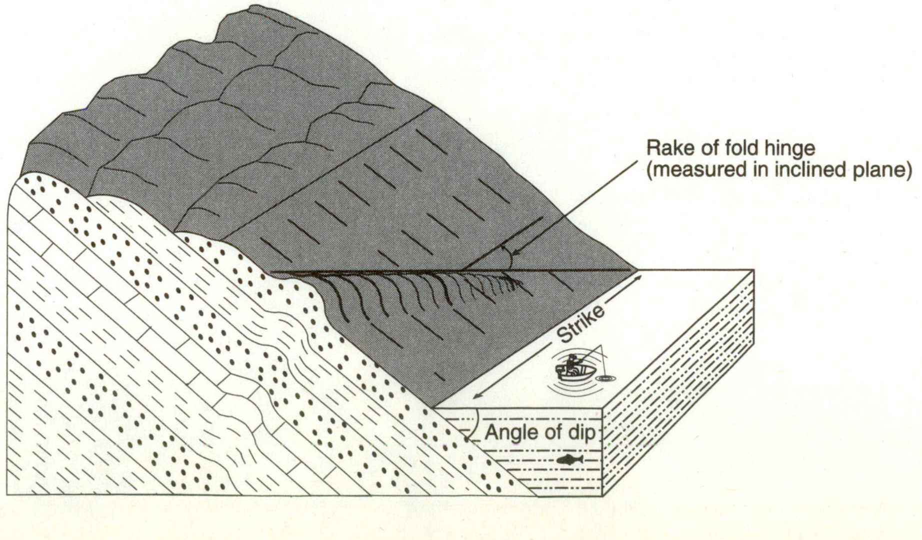

FIGURE A2–3
Rake of a fold hinge line is the angle made by the hinge line measured in the inclined bedding surface.

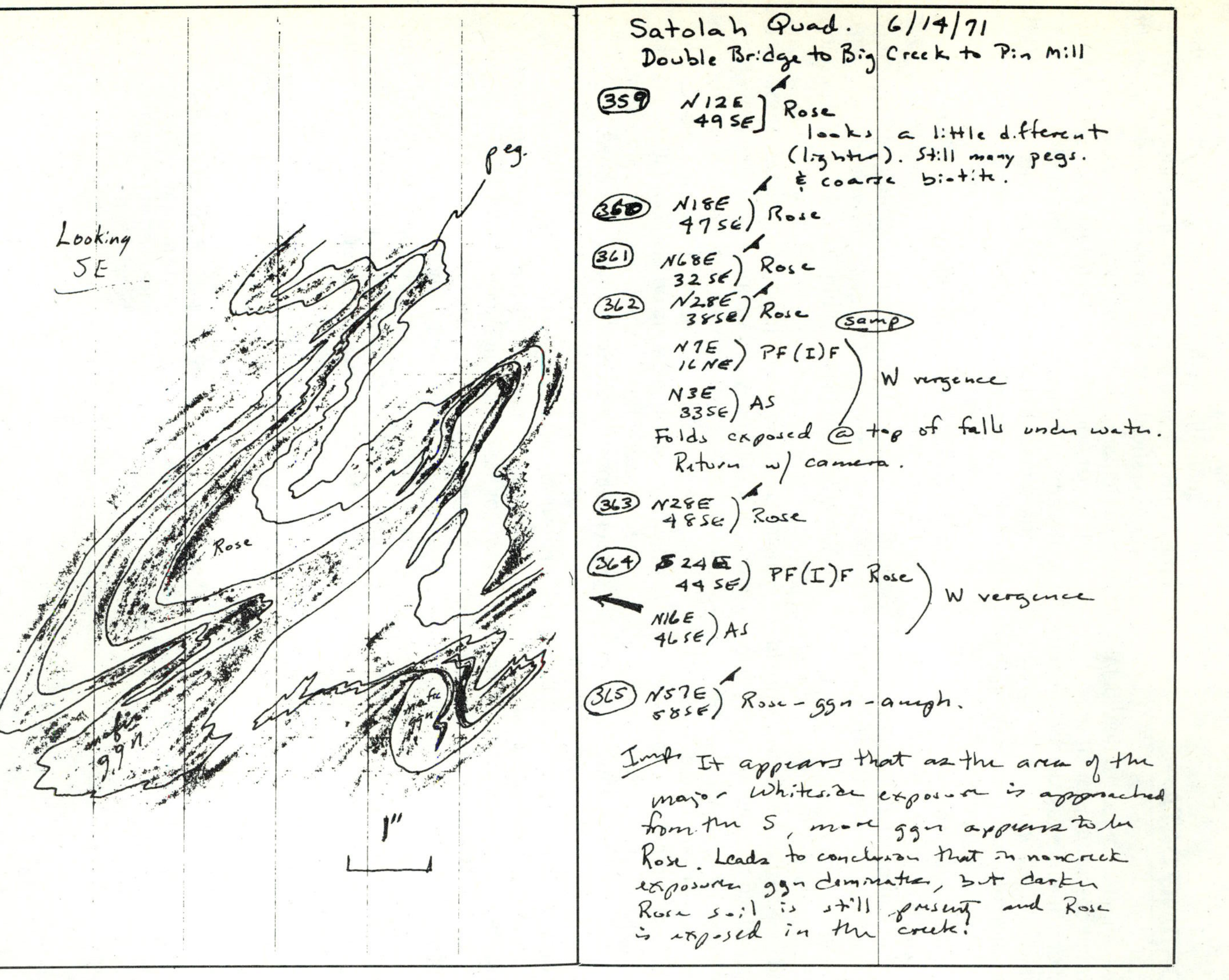

FIGURE A2–4
Two facing pages from a field book for part of the Satolah quadrangle, in the Blue Ridge of northeastern Georgia, showing one way that structural measurements may be recorded and a sketch of a passive-flow fold. At the top of the page is the name of the quadrangle, the date, and the location of the traverse. Note that rock types are abbreviated. Rose is the pencil color used for biotite gneiss; ggn is granitic gneiss; amph. is amphibolite; peg. is pegmatite. A measurement followed by a rock type is a planar structure; the symbol beside the measurement indicates the kind of planar structure. PF(I)F indicates the orientation (trend and plunge) of the axis of a passive-flow isoclinal fold; AS is the orientation (strike and dip) of the axial surface of the fold. The sketch was carefully drawn to make sure all parts of the fold were correctly shown and in the right proportions; rock types were labeled, and the shading indicates that the layers of different composition were colored using the colors chosen for the specific rock types involved. The scale and approximate orientation of the sketch are also indicated.

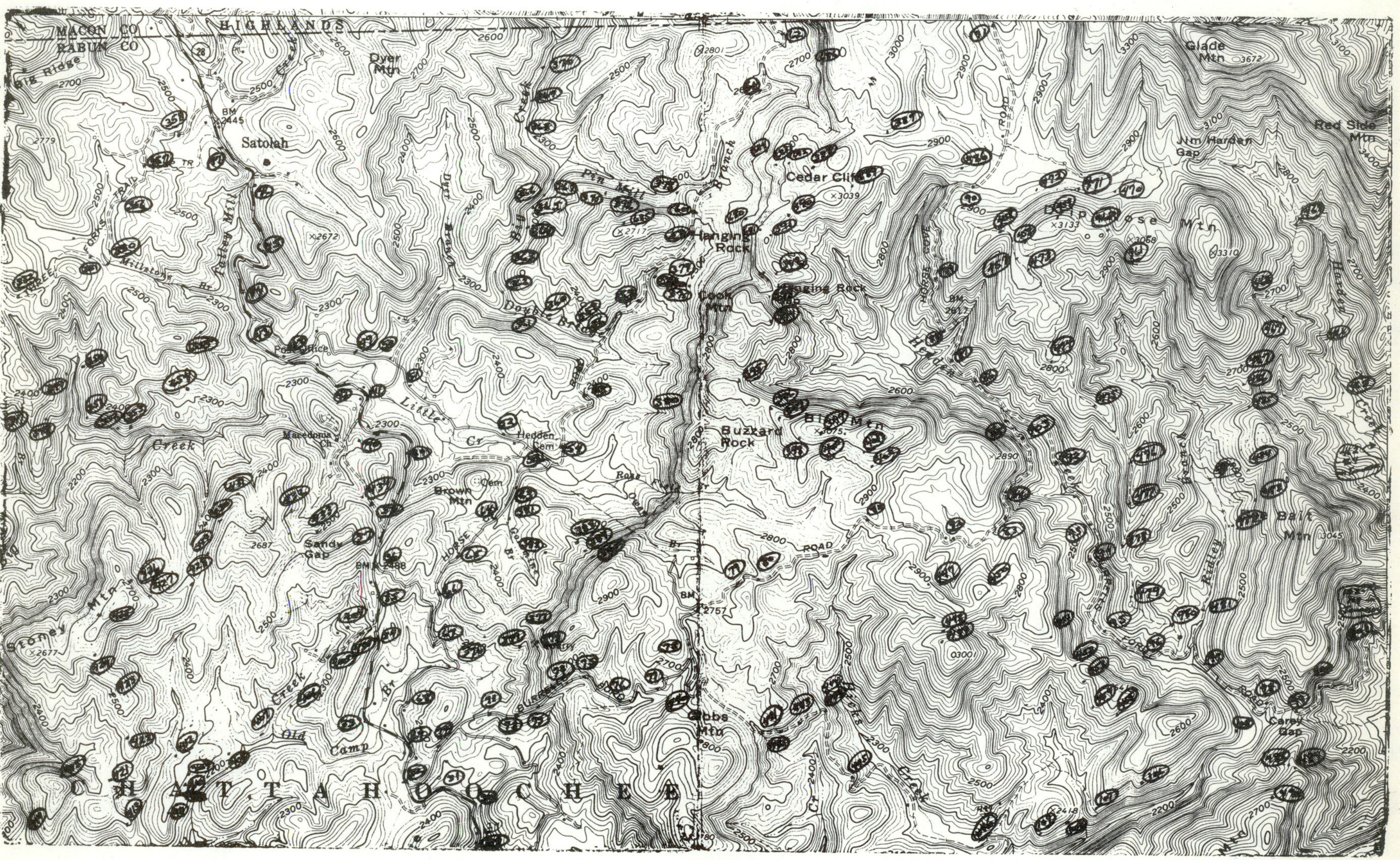

FIGURE A2–5
Part of a station map for the Satolah, Georgia, quadrangle. Circled numbers are keyed to numbers in the field notebook. The stations for the page shown in Figure A2–4 may be found, as indicated at the top of the page in the field notebook, beginning on Double Branch and traversing upstream into Big Creek to Pin Mill Branch, in the upper left center of the figure.

Appendix 3

Computer Software for Structural Geology

A NUMBER OF COMPUTER SOFTWARE PROGRAMS PRESENTLY exist that either are designed directly, or may readily be adapted, for use in structural geology. Many of the routine data compilations and manipulations that we had to do by hand in the past can now be done by the computer. The first programs that were written in the 1970s for plotting structural data on fabric diagrams had to be run on mainframe computers. With the advances in computer design, inexpensive personal and workstation computers have been produced that are as powerful as the mainframes of the 1970s and early 1980s. As a result, software development has become a major undertaking, not only by commercial organizations, but also by individuals who are more interested in providing a professional service than in making a profit. We owe a large debt to individuals who have invested their time to write and upgrade software as a service to the profession. More software is appearing each year for personal computers, and many of the available applications are frequently upgraded.

The following lists are by no means exhaustive but are intended to serve only as examples of the variety of software available to structural geologists. The lists were compiled from several sources and are separated into those available essentially at cost, or at no cost, and commercial software.

TABLE A3–1
PUBLIC-DOMAIN SOFTWARE

Title	Use	Type(s) of Computer(s)	Source	Address	Fee
Stereonet	Plotting Schmidt and Wulff net diagrams of planes and lines, contouring of fabric data.	Macintosh	Richard W. Allmendinger	Dept. of Geological Sciences Snee Hall Cornell University Ithaca, NY 14853–1504	$20 if sending disk; no cost by e-mail.
NETPROG	Plotting Schmidt and Wulff net diagrams of planes and lines, contouring of fabric data.	IBM PC	David Allison	usadta01@asnusa.asn.net (request NETPROG to receive instructions) or Department of Geology and Geography University of South Alabama Mobile, AL 36688–0002	Send 3.5 in. disk or by e-mail $50 (government & commercial only).
Stereoplot	Plotting Schmidt and Wulff net diagrams of planes and lines, contouring of fabric data.	Macintosh	Neil D. Mancktelow	Geologisches Institut ETH–Zentrum CH–8092 Zürich Switzerland	SF100 (~$60).
Restore	Retrodeforming and balancing section in extensional terranes, designed particularly for use in areas with evaporite detachments and salt structures.	Macintosh	Ken Duncan	Applied Geodynamics Laboratory Bureau of Economic Geology University of Texas at Austin Austin, TX 78713	
Calcite Strain	Calculation of calcite strains from data gathered using a universal stage.	IBM PC	Mark A. Evans	Department of Geology & Geography Georgia Southern University Statesboro, GA 30460–8149	None. Send disk.
Stress Analysis 2.1.5A	Manipulation of a variety of structural data.	Macintosh	D. G. De Paor Earth'n Ware, Inc.	4514 Dresden Road Baltimore, MD 21208 [FAX (410) 655–3741]	Single/Site $200/$899
Strain	Digitizing object shape (elliptical or polygonal) and object centers; strain analysis using Fry, normalized Fry, R_f/ϕ, and destraining methods.	Macintosh SE or Plus	W. Adolph Yonkee	Department of Geology Weber State University Ogden, UT 84408–2507	None (& no warranty). Send disk.

TABLE A3–2
COMMERCIAL SOFTWARE

Title	Use	Type(s) of Computer(s)	Source	Address	Fee
Geostructure	Demonstration of fault and fold geometry.	Macintosh	Intellimation, Inc.	Intellimation Library for the Macintosh Department YAS P.O. Box 1922 Santa Barbara, CA 93116–1922	$12.95
					Single/Site License
Boudinage 2.0	Manipulation of a variety of structural data	Macintosh	D. G. De Paor Earth'n Ware, Inc.	4515 Dresden Road Baltimore, MD 21208 [FAX (410) 655–3741]	$33/$66
CarDec 2.0					$69/$199
Contour File Maker 1.0					$39/n.a.
Flinn Plot 2.0					$39/$99
Fry 5.1					$69/$99
Hypercard Structures					$19/$59
Microstructure Folder 1.1					$39/$99
Porphyroblast 1.0					$19/$59
R_f/ϕ 2.02-Fry 5.1 Bundle					$99/$299
Seismac 1.0					$39/$99
Short Course Notes 1993 version					$49/n.a.
Solve 3					$?
Stretches 2.0					$19/$59
Stereo Reader 1.0					$49/$149
Strain Grid 2.02					$99/$299
Stress Analysis 2.1.51					$999/$2,999
Structure Lab 1.2					$99/$299
U4 U-stage driver (requires Futron hardware)					$299/n.a.
Wellman 3.0					$39/$99
Plate Tracker	Interactive plate reconstruction of Phanerozoic paleopositions.	IBM Compatible	David B. Walsh Plate Tracker Software	dbwalsh@albert..ta.edu$ (817) 577–0191 5621 Santa Fe Trail Haltom City, TX 76148	$200 ($100 educational)
STEREO™ PC & Stereo ™ for the Mac	Plotting planes and lines. Contouring.	IBM Compatible DOS & Macintosh	RockWare, Inc.	4251 Kipling Street Suite 595 Wheat Ridge Colorado 80033	$299 $275
Instrain™ & MacINSTRAIN	Analyzes and plots digital strain data to determine strain using normalized Fry, enhanced normalized Fry, and R_f/ϕ methods.	IBM Compatible DOS & Macintosh			$200 (either)
MacMohr	Plots Mohr diagrams from triaxial test data.	Macintosh			$100
Rose PC	Orientation analysis, Rose diagrams, and statistics.	IBM Compatible DOS			$199
Rosy	Orientation analysis, Rose diagrams, and statistics.	Macintosh			$175
Rose/W	Orientation analysis, Rose diagrams, and statistics.	IBM Compatible with Windows™			$129
MacSection II	Plotting cross sections, fence diagrams, plotting lithologic logs.	Macintosh			$399

TABLE A3–2 (continued)

Fault II™	Forward modeling of area-balanced cross sections of fault-bend and fault-propagation folds.	Macintosh IBM	Mac $349 (industry); $149 (education) IBM $199 (industry); $99 (education)
GMM/Geological Map Maker	Plotting structural data on a map.	IBM Compatible with Windows™	$99

Appendix 4

Woodall Shoals Fabric Data

THE FOLLOWING DATA SET IS INTENDED FOR USE IN EXERCISES that will reproduce the fabric diagrams in the Chapter 20 Essay, or that the instructor would like to employ for other uses. These are the actual data gathered during the process of constructing the 1:60-scale geologic map. At least 1,000 professionals and as many geology students have visited this locality since it became known in the early 1970s.

TABLE A4–1
WOODALL SHOALS DOMAIN I—MEASUREMENTS OF S_2

16 54NW	39 33NW	51 55SE	67 38SE	58 38SE	14 64NW	64 69NW	48 24NW	38 06NW
25 12SE	24 49SE	42 64SE	68 16NW	29 57NW	22 39NW	24 39NW	28 56NW	39 12NW
01 18NW	17 67SE	12 52NW	38 39NW	52 34NW	03 53NW	11 62NW	22 69NW	34 22NW
16 69NW	16 13NW	52 48NW	56 41NW	26 68NW	21 33NW	58 23NW	62 55SE	22 63NW
16 40NW	79 51NW	63 27NW	24 l9NW	14 52NW	23 33SE	39 44NW	14 68NW	19 51NW
24 26SE	52 39SE	07 67NW	44 63NW	43 28SE	63 23SE	38 24NW	29 41NW	29 54NW
48 14SE	24 61NW	43 03NW	22 42NW	11 49NW	58 54NW	35332SW	17 67NW	24 54NW
12 59NW	54 73NW	12 46NW	27 39NW	11 66NW	24 46NW	41 75SE	27 33NW	23 29NW
16 38NW	26 41NW	28 47SE	13 47NW	37 36NW	30 14SE	23 06SE	52 36SE	52 27NW
27 48NW	28 33NW	65 19SE	18 27NW	32 52SE	42 72NW	22 58SE	28 18NW	30 24SE
19 67NW	27 41SE	62 28NW	23 47SE	20 17SE	25 17SE	23 62NW	11 68SE	293 39SW
19 55NW	08 07NW	35 28SE	16 73NW	27 52SE	294 34SW	27 63NW	25 34NW	32 12SE
21 56NW	68 56SE	41 29NW	30 23NW	21 34SE	38 62NW	47 23NW	09 48NW	23 25NW
295 30SW	12 28SE	43 62NW	26 48SE	19 51NW	22 25NW	20 10NW	24 7NW	24 70NW
66 22SE	21 68NW	27 34NW	20 03NW	29 25NW	07 47NW	47 22SE	13 73NW	20 15NW
22 15NW	32 46NW	27 21NW	47 16NW	23 44NW	23 21NW	00 10E	34 47NW	32 64NW
48 42NW	19 63NW	23 19NW	00 03W	22 37NW	52 34SE	51 21NW	23 72NW	25 11NW
15 10NW	12 62NW	21 43SE	26 39NW	38 24NW	22 16NW	25 01NW	19 57NW	22 68NW
61 24NW	30 52NW	320 01SW	16 32SE	19 71NW	51 61NW	27 26NW	02 29NW	16 11NW
19 02NW	07 51SE	49 73NW	28 24NW	17 68NW	25 14NW	30 36NW	29 36NW	43 82NW
42 46NW	12 79NW	25 18NW	42 54NW	36 32NW	74 51NW	08 69NW	33 11SE	10 15NW
39 11NW	14 43NW	54 53NW	17 74NW	22 41NW	18 19NW	47 36NW	10 84NW	20 17SE
12 51NW	22 43NW	49 64NW	43 42NW	31 27SE	16 44NW	18 89NW	37 24NW	13 88NW
56 20SE	24 16NW	14 44SE	48 54NW	27 37NW	22 19SE	18 52NW	64 69NW	48 24NW
38 06NW	25 12SE							

TABLE A4–2
WOODALL SHOALS DOMAIN II—MEASUREMENTS OF S_2

34 64SE	06 11SE	00 46E	26 49SE	15 18SE	55 27SE	50 30SE	04 14SE	79 12SE
70 18SE	17 20SE	78 12SE	14 16SE	10 07SE	30 25SE	31 18NW	25 16SE	07 10SE
30 20SE	16 12SE	20 42SE	15 22SE	11 23NW	15 06SE	15 09NW	16 17SE	00 29E
16 16SE	21 34SE	40 15SE	352 25SW	25 10SE	05 31SE	59 26SE	05 39SE	24 15NW
340 22NE	19 17SE	350 10SW	05 38SE	350 6NE	35 31SE	354 14NE	19 19SE	359 21NE
12 34SE	07 22SE	23 21SE	15 24SE	56 41SE	14 28SE	23 29SE	35 30SE	25 21SE
20 25SE	12 25SE	15 10SE	19 21SE	11 12SE	14 25SE	20 25SE	24 27SE	11 09SE
24 28SE	16 30SE	20 29SE	30 15SE	12 31SE	35 12SE	07 40SE	28 14SE	16 47SE
25 28SE	16 42SE	05 24SE	12 37SE	10 24SE	05 23SE	15 25SE	09 35SE	16 11SE
356 24NE	04 10SE	10 36SE	05 09SE	01 42SE				

TABLE A4–3
WOODALL SHOALS DOMAIN III—MEASUREMENTS OF S_2

39 18SE	10 18SE	345 34SW	00 15E	301 21NE	343 26NE	02 47SE	341 27NE	06 30SE
301 20NE	333 26NE	320 23NE	26 18SE	60 28SE	02 32SE	17 29SE	08 29SE	340 27NE
340 24NE	346 22NE	26 30SE	23 26SE	21 29SE	16 16SE	02 26SE	350 24NE	12 04NW
336 27NE	330 18NE	349 24NE	00 00	339 08SW	05 20SE	12 17SE	75 12SE	00 00
02 07SE	07 07SE	18 30SE	330 35NE	340 16NE	317 23NE	331 36NE	334 15NE	00 30NE
305 12NE	350 25NE	348 22NE						

TABLE A4–4
WOODALL SHOALS DOMAIN IV—MEASUREMENTS OF S_2

73 52NW	02 47SE	314 41NE	02 37SE	326 28NE	326 27NE	339 47NE	346 32NE	354 51SE
08 29SE	347 34NE	347 57NE	333 79SW	298 56NE	78 43SE	333 56NE	352 48SE	30 33SE
319 78SW	04 43E	348 12NE	348 40NE	348 43NE	00 69E	357 53NE	14 47SE	356 46NE
293 57NE	08 47SE	357 58NE	64 52SE	82 59SE	344 44NE	274 74NE	354 44NE	03 28SE
01 32SE	54 73NW	341 12NE	347 54NE	348 45NE	39 73SE	77 67NW	287 67NE	38 14SE*
04 37SE*	49 73NW*	319 57NE*	273 42NE	354 45NE*	03 28SE*	353 42NE*		

*Not plotted on fabric diagrams.

TABLE A4–5
WOODALL SHOALS DOMAIN V—MEASUREMENTS OF S_2

52 22SE	33 05SE	27 07NW	303 27SW	24 10SE	61 14NW	62 34SE	62 28NW	40 39SE
75 37SE	46 36NW	68 24SE	13 28SE	47 21NW	00 00	52 17SE	31 47SE	23 24NW
42 19SE	24 52NW	41 37NW	358 27NE	06 37NW	43 22NW	13 24SE	24 12SE	00 00
57 35SE	67 36NW	36 78SE	41 21SE	24 21NW	44 53SE	42 36NW	346 16NE	08 68SE
44 28SE	322 39NE	36 67NW	12 20SE	86 14NW	29 69SE	32 47SE	09 24NW	05 07SE
42 71NW	317 27SW	53 07SE	52 48NW	272 18NE	63 08SE	44 32SE	46 29NW	35 25SE
56 24NW	27 38NW	10 13SE	58 67SE	84 03SE	03 19SE	51 53SE	82 23NW	65 14NW
49 51SE	69 34NW	61 14NW	48 64SE	38 18NW	57 07NW	46 31SE	282 03NE	317 16NE
43 82NW	82 08SE	14 19NW	72 43SE	37 52NW	48 17NW	48 43NW	31 32NW	281 21NE
33 54SE	351 28SW	43 78NW	44 21NW	57 57SE	29 19NW	32 34SE	18 40NW	39 48SE
81 19NW	53 32SE	11 14SE	46 18NW	69 27NW	33 11SE	32 16NW	49 21NW	23 14NW
36 12NW	63 29NW	358 13NE	42 23NW	56 17NW	56 44NW	24 10NW	73 21NW	22 19NW
69 22NW								

TABLE A4–6
WOODALL SHOALS DOMAIN VI—MEASUREMENTS OF S_2

319 09SW	283 12NE	303 03SW	84 24SE	84 30SE	40 39SE	288 22SW	273 38SW	277 39SW
78 7NW	79 30SE	357 18SW	344 16SW	295 14SW	75 19SE	32 20NW	275 21SW	301 12SW
321 13SW	87 20SE	65 25SE	80 16SE	329 15SW	313 14SW	330 08SW	75 17SE	319 12SW
85 21SE	66 20SE	59 23SE	287 13SW	279 13SW	50 26SE	85 20SE	16 16SE	86 16SE
275 16SW	05 11SE							

TABLE A4–7
WOODALL SHOALS DOMAIN VII—MEASUREMENTS OF S_2

271 39NE	46 16NW	351 28NE	11 34SE	57 04SE	343 16NE	317 07NE	353 18NE	09 17SE
329 11NE	03 26SE	14 14SE	48 11NW	20 18NW	20 10SE	358 10NE	14 31SE	22 07SE
24 21NW	357 05SW	39 10SE	27 12SE	00 12E	00 00	89 04SE	345 07NE	74 18SE
89 15NW	330 08NE							

TABLE A4–8
WOODALL SHOALS DOMAIN VIII—MEASUREMENTS OF S_2*

14 6SE	23 30SE

*Incomplete data set.

TABLE A4–9

WOODALL SHOALS DOMAIN IX—MEASUREMENTS OF S_2

83 24SE	352 11NE	333 37NE	334 14NE	72 16SE	308 43NE	09 29SE	88 22NW	277 24SW
355 39NE	303 23NE	65 89SE	317 07SW	356 31NE	351 38NE	333 06NE	88 06SE	325 05NE
05 33SE	72 08SE	63 31NW	321 05NE	335 33NE	302 23NE	277 34SW	282 04NE	344 27NE
82 13NW	300 15NE	329 26NE	317 25NE	305 20NE	283 31NE	290 07NE	327 30NE	315 8NE
82 27NW	337 03NE	313 11NE	330 12NE	334 17NE	337 12NE	320 23NE	279 11NE	303 09NE
344 30NE	357 01NE	83 12SE	307 09NE	346 18NE	336 16SW	280 15NE	350 06NE	318 52SW
87 10NW	318 15NE	83 21NW	85 11NW	311 41NE	277 18NE	23 23SE	323 04NE	277 12NE
317 34NE	312 19NE	60 38NW	284 17NE	282 25NE	68 22NW	307 16NE	326 12NE	55 22NW
300 15NE	281 23NE	72 13NW	325 20NE	321 19NE	82 24NW	307 04SW	311 18NE	308 22NE
310 18NE	311 24NE	02 18NE	283 13NE	272 11NE	293 19NE	275 28SW	281 22NE	342 17NE
295 17NE	305 27NE	292 26NE	294 16NE	00 09E	09 29SE	329 13NE	355 06NE	283 22NE
03 10SE	303 10SW	326 18NE	285 12NE	291 10SW	309 06NE	87 26NW	288 08SW	293 29NE
282 27NE	326 07SW	289 28NE	78 30NW	276 25SW	318 19NE	346 11NE	309 34SW	353 21NE
280 52NE	297 19SW	293 23NE	346 20NE	69 29SE	291 18NE	305 25NE	293 37SW	53 58NW
349 11NE	79 15SE	288 32NE	311 29NE	10 12SE				

TABLE A4–10

WOODALL SHOALS DOMAIN X—MEASUREMENTS OF S_2

353 14NE	25 11SE	353 20NE	77 27NW	358 08NE	15 17SE	68 16SE	15 18SE	44 25NW
09 22NW	07 09NW	315 18NE	62 12NW	28 62SE	19 31SE	75 17NW	63 23SE	87 29SE
00 05W	357 25NE	06 14NW	51 62SE	16 12SE	36 61SE	11 27SE	323 14NE	343 18SW
282 15NE	38 33NW	275 20NE	72 21SE	303 08NE	57 31SE	35 20SE	51 32NW	20 36SE
07 11SE	341 11NE	340 11NE	26 08NW	342 15NE	45 21SE	44 11SE	61 20SE	63 25NW
55 27SE	310 11NE	46 16SE	300 15NE	01 15SE	336 06NE	19 19SE	07 14SE	24 19SE
75 02SE	38 19SE	61 20SE	275 14NE	70 38NW	50 28NW	08 12NW		

TABLE A4–11

WOODALL SHOALS DOMAIN XI—MEASUREMENTS OF S_2

54 23SE	51 21SE	49 22SE	55 05SE	56 18SE	85 55SE	25 15SE	37 44SE	60 15SE
62 36SE	43 27SE	43 22SE	28 27SE	43 16SE	01 86NW	79 73SE	38 35SE	10 16SE
36 39SE	04 28SE	26 20SE	55 33SE	15 24SE	62 25SE	40 20SE	40 11SE	69 20SE
57 34SE	27 30SE	24 24SE	77 24SE	24 13SE	69 24SE	41 25SE	53 16SE	48 13SE
55 13SE	12 08SE	42 22SE	34 15SE	77 33SE	39 13SE	48 22SE	336 06NE	48 23SE
65 39SE	70 08SE	51 19SE	64 15SE	49 21SE	54 19SE	54 19SE	48 24SE	64 25SE
21 17SE	50 22SE	38 14SE	48 15SE	14 14SE	75 20SE	46 13SE	10 11SE	38 12SE
26 20SE	53 15SE							

TABLE A4–12

WOODALL SHOALS DOMAIN XII—MEASUREMENTS OF S_2

84 32NW	72 09NW	30 27SE	38 20SE	25 13SE	34 15SE	28 16SE	52 28SE	01 18SE
17 18SE	89 05SE	28 25SE	52 21SE	33 12SE	27 18SE	30 08SE	35 07SE	09 11NE
30 13SE								

TABLE A4–13

WOODALL SHOALS—ORIENTATIONS OF JOINTS

12 88NW	358 81NE	70 88NW	32 79SE	353 78NE	02 73	276 30NE	88 51NW	05 85SE
01 87SE	04 87NW	07 88SE	06 79SE	07 89SE	02 89SE	274 55NE	05 82SE	06 88SE
87 89SE	01 87SE	09 87SE	15 58NW	351 78SW	43 78NW	73 77SE	02 79NW	07 65NW
51 89SE	358 89NE	358 88SW	44 72SE	294 82NE	351 84SW	31 89SE	31 83SE	354 85NE
358 83NE	07 84SE	61 89SE						

TABLE A4–14
WOODALL SHOALS—ORIENTATIONS OF CRENULATION AXIAL SURFACES*

351 61NE (IV)	344 82NE (IV)	348 68NE (IV)	04 79SE (IV)	12 85SE (IV)	52 78SE (IV)	28 21NW (IV)
42 70NW (IV)	16 35NW (IV)	52 35NW (IV)	52 51NW (IV)	24 25SE (IV)	54 64SE (I)	22 66SE (I)
51 72SE (I)	37 56NW (I)	49 64SE (I)	44 63SE (I)	12 75SE (I)	9 58SE (I)	32 6SE (I)
32 69NW (I)	42 84N W (I)	72 59SE (I)	79 20NW (I)	57 19NW (I)		

*Domain number is indicated in parentheses.

TABLE A4–15
WOODALL SHOALS—TREND AND PLUNGE OF CRENULATION AXES

234 43	202 04	231 22	217 14	229 44	224 52	12 48	09 13	32 34
212 14	222 04	251 21	259 32	237 29	351 54	164 13	168 23	43 49
12 75	232 12	208 24	42 06	196 19	232 16	232 11	24 08	

TABLE A4–16
WOODALL SHOALS—ORIENTATIONS OFAXIAL SURFACES OF TIGHT FOLDS*

24 59NW (I)	14 59NW (I)	33 63NW (I)	31 72SE (I)	21 62SE (I)	18 58NW (I)	13 23SE (I)
13 49NW (I)	42 71SE (I)	09 52SE (I)	43 49SE (I)	12 69SE (I)	32 69SE (II)	18 4SE (VII)
272 82SW (X)	352 51NE (IX)	06 88SE (IX)	27 71SE (IX)	22 77SE (IX)	28 81SE (IX)	44 58SE (IX)
32 81SE (IX)	26 58SE (IX)	11 79SE (IX)	39 75SE (VIII)	62 74NW (V)	48 76SE (V)	28 88NW (V)
06 88SE (V)	47 79NW (V)	33 55NW (V)	16 54SE (V)	26 68SE (V)	350 81SW (IV)	51 77NW (IV)
32 48SE (IV)	23 74NW (IV)	23 77SE (IV)	22 71SE (IV)	31 58SE (IV)	72 37NW (IV)	13 l9SE (IV)
343 77NE (IV)						

*Numbers in parentheses are domain where each measurement was made.

TABLE A4–17
WOODALL SHOALS—TREND AND PLUNGE OF TIGHT FOLDS

204 21	212 19	350 21	194 04	129 07	51 50	213 34	272 13	212 47
211 24	172 03	23 18	201 11	06 43	23 66	198 14	29 34	22 24
193 03	22 28	31 31	193 24	28 24	252 16	222 04	44 27	13 19
198 14	32 19	163 10	223 18	26 44	192 12	11 46		

TABLE A4–18
WOODALL SHOALS—TREND AND PLUNGE OF MINERAL LINEATIONS

92 12	311 15	139 04	03 07	68 02	83 2	226 15

TABLE A4–19
WOODALL SHOALS—ORIENTATIONS OFAXIAL SURFACES OF TIGHT FOLDS

28 89SE	28 89SE	32 65SE	358 78NE	64 84SE	03 83SE	24 88SE	26 83NW	299 89NE
32 48SE	353 84SW							

TABLE A4–20
WOODALL SHOALS—TREND AND PLUNGE OF OPEN FOLDS

28 14	208 07	32 17	358 36	244 20	229 07	0313	24 17	26 27	299 03
32 05	353 07								

Glossary

Aberrant folds. Folds that deviate slightly from ideal cylindrical folds.

Accident. Large, steeply dipping fault with uncertain motion sense.

Accreted terrane. Crustal-scale rock mass that has been tectonically joined to another mass; identified by stratigraphic, metamorphic, or deformational history that contrasts with that in the adjacent block.

Accretionary tectonics. Process whereby suspect and exotic terranes ranging in size from less than continental proportions to masses the size of continents are moved by plate motion to collision with each other or with other continents.

Accretionary wedge. Sediments in a subduction zone along an active margin that are scraped off a descending slab; wedge consists of an imbricate stack of thrusts with transport out of the subduction zone.

Active-roof duplex. Roof thrust in a duplex (thrust or strike-slip fault system) having considerable displacement, in contrast to a passive-roof duplex.

Adamellite. Coarse-grained igneous rock composed of K-spar, sodic plagioclase, quartz, and biotite and/or hornblende. Quartz monzonite.

Allochthon. Large remnant of a far-traveled rock mass, generally a sheet, moved by thrust faulting. Commonly rocks of an allochthon contrast in stratigraphy, metamorphic history, and structural history with rocks upon which they presently rest.

Amontons' laws. *Amontons' first law:* Tangential force parallel to a movement surface necessary to initiate slip is directly proportional to the force normal to the movement surface. *Amontons' second law:* The frictional resistance to motion is independent of the contact area on a movement surface.

Amplitude. Half the crest-to-trough height of any wave. Half the distance from the crest of an anticline to the trough of an adjacent syncline, measured parallel to the axial plane.

Amygdules. Vesicles filled with secondary minerals.

Anatectic. *In situ* melting produced by metamorphism at high temperature and pressure, commonly in the presence of water.

Angular shear. Relative angle of rotation (ψ) of reference lines in a rock mass subjected to simple shear. The tangent of this angle is the shear strain (γ).

Angular unconformity. Angular relationship that exists in rocks where a sequence has been deposited and later tilted, followed by erosion (or nondeposition), then renewed deposition. Produces nonparallel layering above and below the unconformity.

Anisotropy. Characteristic of material in which properties vary with direction.

Annealing. Thermal or mechanical process wherein the deviatoric stress in a crystal is removed and a crystal (or crystals) with less elastic stress results.

Anomaly. Deviation from an average value of magnetic- or gravity-field intensity.

Anticlinal theory. Theory stating that hydrocarbons tend to accumulate in anticlines.

Anticline. Fold in which the layering is concave toward older rocks in the structure. An anticline contains older rocks in the center.

Antiform. Fold that is concave downward and has the shape of an anticline.

Antiformal stack duplex. Duplex (in a thrust-faulted terrane) in which the imbricate thrust sheets have moved forward to overlap each other, producing an antiformal arch of the imbricates and the roof thrust.

Antiformal syncline. Concave-downward fold wherein the rocks in the center are younger. Also called downward-facing syncline.

Anti-Riedel shears (R′). Secondary brittle fractures that form at a high angle (75° to 80°) to the primary fractures in a brittle shear zone and have a motion sense opposite the dominant motion sense of the fracture zone.

Antithetic normal fault. Normal fault that dips oppositely to join a larger (master) normal fault.

Apparent polar wandering (APW) curve. Curve that plots the positions through time of ancient magnetic poles for a continent; provides a record of "polar wandering" when the continent is actually moving.

Arc-trench gap. Distance from the trench axis to the volcanic arc in a subduction complex. Commonly is 80 to 250 km.

Arrest line. Diffuse line on a joint surface that marks a place where joint propagation hesitated or stopped. Several arrest lines would be concentric to the point where propagation began.

Asperities. Microscopic irregularities and imperfections on a fault surface.

Astrobleme. Structure produced by extraterrestrial impact, frequently consisting of concentrically and radially arranged horsts and grabens about a central chaotic zone. Impact structure.

Asymmetric fold. Fold having one limb that is longer than the other.

Aulocogen. Tectonic trough, bounded by normal faults and formed within continental crust at a high angle to a nearby continental margin. May consist of the failed arm of a triple junction.

Axial elongation. Strain resulting from elongation parallel to a unique axis (hexagonal, tetragonal, or cylindrical symmetry); produces deformed objects (e.g., pebbles or oöids) resembling hot dogs (prolate spheroids).

Axial flattening. Strain resulting from shortening parallel to a unique axis, producing deformed objects resembling hamburgers (oblate spheroids).

Axial line. Line on a fold that separates dip in one direction from dip in the opposite direction.

Axial plane or axial surface. Plane or surface that results by connecting hinge lines on successive folded surfaces of the same fold.

Axial symmetry. Symmetry of a cylinder; generally refers to deformation dominated by linear structures, such as folds or a mineral lineation.

Balanced cross section. Cross section that can be restored to an undeformed condition.

Basalt. Fine-grained igneous rock composed of calcic plagioclase and pyroxene, sometimes with lesser amounts of olivine, amphibole, and biotite.

Basement. Rock that is the product of an earlier orogenic cycle; underlies a less deformed and commonly less metamorphosed cover. The cover records only the younger orogenic cycle(s).

Basin. Unique bowl-shaped synform in which layering dips inward toward a central point; generally a concave-up fold.

Bedding. Primary layering in sediments and sedimentary and some volcanic rocks.

Bedding (bedding-plane) fault. Fault that follows bedding or occurs parallel to the orientation of bedding planes.

Bedding fissility. Paper-thin layering in fine-grained sediment wherein platy minerals (such as clays) are aligned horizontally, perpendicular to the vertical maximum principal lithostatic stress.

Bedding plane. Zone of mechanical weakness that forms in sediment because of slight compositional or textural differences at the interface between adjacent beds.

Bedding thrust. Thrust faulting locally parallel to bedding; commonly occurs in a weak rock unit, such as shale, coal, or evaporite, but may also occur parallel to bedding in a strong unit.

Bending. Fold mechanism that involves application of force across rock layers.

Blind thrust. Thrust in which displacement decreases upward within the sedimentary section; blind thrusts never reach the surface.

Blocking temperature. Temperature at which radiogenic isotopes are locked into a crystal lattice.

Body force. Force that acts equally on all parts of a body.

Bolide. Body from outer space (e.g., asteroid or meteor) that explodes on impact with the Earth or other planet.

Boomerang pattern. Fold pattern produced by interference of folds with gently dipping axial surfaces superposed by folds trending perpendicular to the first set that have near-vertical axial surfaces. Ramsay Type 2 fold interference pattern.

Boudinage. Boudins. Sausage-shaped segments of a layer that has been extended. The layer being extended is much less ductile than the enclosing material, and the degree of contrast in competence affects the shape of boudins. Shapes of boudins range from angular (brittle layer segmented) to rounded (ductile layer segmented).

Bouguer gravity-anomaly map. Gravity-anomaly map (generally contoured) that incorporates corrections for both elevation and density between sea level and the elevation where gravity measurements were made.

Bouma sequence. Sequence in sedimentary rocks composed of five intervals that make up a complete turbidite succession. It develops as a graded interval (A), a lower interval of parallel laminations (B), an interval of current ripple laminations (C), an upper interval of parallel laminations (D), and a fine-grained shaly (pelitic) interval (E). Forms a sequence that grades upward from one part into another (A to B). Partial sequences, particularly A and B, are most commonly preserved.

Branch line. Line formed by the intersection of two fault surfaces.

Break thrust. Thrust formed during folding of the connecting limb of an anticline-syncline pair; similar to fault-propagation fold.

Brittle behavior. Failure in the elastic range.

Brittle deformation. Discontinuous deformation.

Broken formation. Zone of broken rock that develops beneath a detachment fault, perhaps by hydrofracturing under high pore pressure.

B-tectonite. Foliated rock dominated by fold axes.

Buckle fold. Fold formed by buckling.

Buckling. Fold mechanism involving application of stress nearly parallel or parallel to layering in rocks.

Bulls-eye fold. See *Eyed fold.*

Burgers circuit. Loop traverse that will not close at the starting point, as in a perfect crystal, because the traverse crosses a dislocation.

Burgers vector (b). Magnitude of the failure of a loop traverse to close in a crystal containing a dislocation.

Burrow. Tunnel produced in sediment by organisms. Indirect indicator of previous organism activity.

Byerlee's law of rock friction. The coefficient of friction in Amontons' first law is independent of rock type and depends solely on values of shear and normal stress.

Calcsilicate. Metamorphic rock composed almost entirely of Ca-bearing silicate minerals, most commonly calcic plagioclase, Ca-garnet, Ca-amphibole, diopside, epidote-clinozoisite, and quartz in different proportions, with or without calcite or dolomite.

Cataclasis. Brittle granulation of rock at low temperature and low to moderate confining pressure along a fault; may be accelerated by high fluid pressure. Is generally a friction-dependent process involving fracturing and rigid-body grain rotation and generally produces grain-size reduction.

Cataclasite. Internally undeformed rock fragments formed by brittle deformation; composed of meter- to micrometer-size original rock or minerals.

Cataclastic flow. Ductile-like flow of very fine-grained (microscopic to submicroscopic) particles along a fault; produced by brittle deformation.

Cataclastic lineation. Alignment of minerals and rock fragments by brittle deformation along a fault.

Cataclastic rock. Rock formed by brittle deformation along a fault. Breccia, cataclasite, and gouge are cataclastic rocks.

Center-to-center method. Method of determination of the shape and orientation of the strain ellipsoid in a rock mass by using the distance and angular relationships of strain-oriented, closely packed objects in the deformed aggregate. Involves measurement of the distances and angles between a reference grain and the nearest neighbors.

Chevron folds. Folds with straight limbs of similar length and sharp angular hinges. Kink folds.

Chocolate-block (chocolate-tablet) boudinage. Square to rectangular boudinage resulting from layer-parallel extension in two directions.

Chronostratigraphic. Time stratigraphic.

Class 1 folds. Folds where dip isogons converge toward the concave part of the fold in Ramsay's standard fold classification.

Class 1A folds. Class 1 folds that have isogons strongly convergent toward the hinge. Isogons change orientation during folding more than the folded layer.

Class 1B folds. Class 1 folds with convergent isogons corresponding to parallel-concentric folds.

Class 1C folds. Class 1 folds having weakly convergent isogons; may correspond to flexural-flow folds. Isogons change orientation less than the folded layer.

Class 2 folds. Folds with parallel isogons; ideal-similar folds.

Class 3 folds. Similar-like folds with isogons that diverge from the tightly curved part of the fold.

Cleavage refraction. Change in the angle between cleavage and bedding as the cleavage passes from one layer to another.

Climb. Movement of a dislocation out of a glide surface normal to Burgers vector. Reorganization of dislocations resulting in formation of subgrain boundaries.

Closed fold. Fold in which the limbs make an interlimb angle of 70° to 30°

Coaxial deformation (coaxial strain). Principal axes of the strain ellipsoid coincide with the same material lines throughout deformation.

Coble creep. A low- to moderate-temperature (<400° C) temperature-dependent atomic-scale creep process that competes (unsuccessfully) with pressure solution at the low-temperature end but becomes dominant at higher temperatures where the physical state or absence of water makes pressure solution inefficient.

Coefficient of internal friction (μ). Material property expressed as the ratio of the shear strength to the normal stress of the material.

Competent. Strong or rigid material. Material of high yield strength.

Complete Bouguer anomaly map. Gravity-anomaly map (generally contoured) that includes corrections in the gravity data for elevation, density, and terrain.

Compositional banding. (a) Layering in an igneous body that results from crystal settling, differentiation, fractional crystallization, multiple parallel intrusions, or flow processes that flatten xenoliths. (b) Alternating layers of different minerals in a metamorphic rock mass.

Compressional stress. Stress that decreases the volume of or shortens a body.

Concentric faults. Arcuate faults that form concentric about a point.

Concentric folds. Parallel folds in which folded surfaces define circular arcs and maintain the same center of curvature.

Concentric-longitudinal strain. See *Tangential-longitudinal strain.*

Concordant mineral. Zircon or other mineral that yields ages that plot on a concordia curve.

Concordant pluton. Pluton intruded parallel (or subparallel) to the dominant layering (bedding, foliation) in the host rock.

Concordia curve. Curve that shows variation in abundance of lead-isotope ratios through time.

Conical folds. Noncylindrical folds with convergent hinge zones or lines.

Conjugate. Structures that form as interactive pairs.

Conjugate shear angle. Angle between maximum principal stress (σ_1) and a fracture that forms as a result of the application of stress.

Contact metamorphic zone (aureole). Zone of thermally induced recrystallization around a pluton.

Continuous cleavage. Foliation in fine-grained rocks that pervades the rock mass.

Continuous creep. Stable sliding along a fault.

Contractional fault. Faults along which bedding or some other reference in any kind of rock is shortened. Also called a wedge.

Coulomb criterion of failure. The absolute value of shear strength (τ_s) is the sum of the inherent shear strength (S_0) and the product of the coefficient of internal friction (μ) and normal stress (σ_n).

Coulomb wedge. Wedge of sedimentary rocks that is undergoing brittle deformation, generally by imbricate thrusting.

Creep process. Rate-dependent lattice-scale deformation mechanism that involves mass transport or diffusion of atoms or ions at grain boundaries, glide and climb of dislocations within a lattice, and diffusion of point defects through lattices.

Crenulation cleavage. Spaced cleavage that overprints bedding, or an existing cleavage or foliation, by crinkling the earlier structure, forming a series of straight-limb-dominant, small folds with narrow, tightly curved hinges.

Crest. Highest point of a fold (or wave).

Critical taper. In a thrust-faulted terrane, the angle between the surface slope and the basal detachment. Once attained, the angle is maintained by erosion, internal deformation of the wedge, and/or displacement of the wedge.

Cross bedding. Layering inclined to the principal bedding planes in sedimentary rocks composed of sand-sized particles.

Cross-section analysis. Study of the structure of an area by construction of vertical cross sections from surface geologic data, down-plunge projection, and geophysical data.

Cryogenic magnetometer. Instrument that measures paleomagnetism inside a superconducting ring, allowing detection of weak signatures.

Crystal lattice. Skeletal pattern of the internal array of atoms, ions, or molecules in a compound.

Crystalline solid. Regular geometric arrangement of atoms, ions, or groups to form a repeat unit in a compound. Crystalline solids need not be single homogeneous crystals but can be crystal aggregates or native element.

Crystalline rocks. Igneous and metamorphic rocks.

Crystalline thrust. Thrust involving crystalline (metamorphic or igneous) rocks.

C-surface. Foliation in shear zones that develops parallel to shear-zone boundaries. As shearing continues, the C-surfaces may be rotated to an angle of 18°–25° in the displacement direction.

Culmination. Structural high.

Curie temperature. Temperature above which strongly magnetic (ferromagnetic) materials lose their ability to interact with a magnetic field—become non-magnetic.

Current ripple marks. Translational ripples that form where a prevailing direction of transport and deposition of sediment occur. They present the same shape whether upright or overturned.

Cylindrical (cylindroidal) fold. Fold that can be generated by moving the fold axis parallel to itself.

Décollement. Movement surface (detachment) produced by flexural-slip folding or faulting; zone of weakness (shale, coal, or other weak rock, or a strong unit susceptible to strain softening) through which a fault may propagate. Separate zones of contrasting deformation above and below the décollement. French for "ungluing."

Defect. Imperfection in a crystal lattice.

Deformation map. Graph that shows the experimentally determined range of physical conditions of several deformation mechanisms for particular rock types or minerals.

Deformation path. Stages of progressive deformation affecting a rock mass.

Deformed state. Strained state.

δ porphyroclast. δ, σ, and θ porphyroclasts have been recognized as products of simple shear deformation. Winged asymmetric relict of a larger crystal, commonly feldspar, produced in ductile shear zones having a δ shape, but most commonly with a tail or wing on diagonally opposite ends.

Delta structure. See *Triangle structure.*
Depositional remanent magnetism. Paleomagnetism in sedimentary rocks; recorded by orientation of grains of magnetite (or other magnetically susceptible minerals) in the Earth's magnetic field at the time of deposition.
Desiccation cracks. See *Mud cracks.*
Detachment. See *Décollement.*
Deviatoric stress. Nonhydrostatic component of stress.
Dextral strike-slip fault. See *Right-lateral strike-slip fault.*
Diamictite. Rock composed of angular, unsorted to poorly sorted, generally noncalcareous material of a wide range of particle sizes.
Diapir. Salt, mud, magma, or other material of lower density than that of rocks enclosing it that moves upward gravitationally and intrudes the overlying sediments or the crust. Most are cylindrical domes.
Differential stress. Difference between the maximum and minimum principal normal stresses ($\sigma_1 - \sigma_3$).
Differentiated layering. Foliation involving segregation of minerals into layers of different composition (such as quartz-feldspar-rich and biotite-rich layers); produced largely through recrystallization during progressive deformation and metamorphism.
Dike. Discordant tabular pluton.
Dilation (Δ). Positive or negative change in volume.
Diorite. Coarse-grained igneous rock composed of plagioclase and amphibole, with minor to zero amounts of biotite, olivine, and quartz.
Dip. Angle made by an inclined surface with the horizontal, measured perpendicular to strike.
Dip isogon. Line connecting points of equal slope, or dip, on successive layers in a fold as seen parallel to the fold axis.
Dip-slip movement. Movement down or up parallel to the dip direction of a fault.
Disconformity. Unconformity produced by deposition of a sequence of sedimentary rocks followed by uplift or a drop in sea level. Uplift and erosion produce a surface having topographic relief without tilting or other deformation of the sequence, followed by subsidence and renewed deposition. Bedding in the rocks above and below the unconformity remains parallel.
Discordant mineral. Zircon or other mineral having lead-isotope ratios that do not plot on the concordia curve.
Discordant pluton. Igneous body in which the contacts cross the layering (bedding, foliation) in the host rock at a high angle.
Discrete crenulations. Sharply defined domainal crenulation cleavage that truncates and preserves the older fabric in microlithons.
Disharmonic folds. Folds that change shape or orientation from layer to layer. Involves moderate to high ductility contrast. Same as quasi-flexural folds.
Disjunctive cleavage. Cross-cutting spaced cleavage, not related to original layering.
Dislocation. Line defect.
Dislocation climb. Thermally driven reorganization of dislocations producing a decrease in dislocation density with the formation of low-angle (subgrain) boundaries.
Dislocation creep. Combination of glide-and-climb motion of dislocations through a crystal lattice; occurs at moderate to high stress and/or temperature (depending on the material).
Dislocation density. Relative concentration of dislocations in a crystal, indicating the degree of deformation or recovery/recrystallization in the lattice.
Dislocation glide. Mechanism in which dislocations (and deformation) in a crystal may result in formation of tangles of dislocations, deformation lamellae, and undulatory, patchy, or sweeping extinction under the microscope.
Dislocation map. Map showing distribution, density, and kinds of dislocations in a crystalline solid.
Displacement. Total amount of motion measured parallel to the movement direction. Amount of relative motion measured on opposite sides of a fault. Net slip.
Distortion. Strain involving change in shape.
Doctrine of uniformitarianism. Processes occurring today upon and within the Earth have probably gone on similarly in the past and will continue in the future.
Domainal cleavage. Cleavage localized in particular zones. A type of spaced cleavage.
Dome. Unique antiform wherein layering dips in all directions away from a central point.
Dome-and-basin pattern. Fold pattern consisting of systematically related domes and basins. May be produced by two generations of folds that overprint each other at high angles, producing a Type 1 fold interference pattern, or one event with shortening from two directions.
Down-to-basin fault. Listric normal fault in which the downthrown side is toward an adjacent basin. Commonly forms during crustal extension or filling of a basin.
Downward-facing syncline. See *Antiformal syncline.*
Drag. Frictional bending of layering in rocks next to a fault; bending occurs during movement.
Drag fold. Fold that forms during movement because of friction on a fault surface. It yields motion sense on the fault by fold asymmetry along the fault.
Drape fold. Fold in cover produced by high-angle normal or thrust faults in the basement that may only partly break the sedimentary cover.
Ductile-brittle transition. Transition from brittle to ductile behavior in the crust and mantle where brittle behavior is inhibited because strain softening occurs as a result of increases in temperature and pressure with depth. The location of the transition is dependent both on composition (ice in glaciers, quartz in the crust, olivine in the mantle) and the geothermal gradient.
Ductile deformation. Continuous deformation by plastic or viscous flow.
Ductility. Permanent strain as viscous or plastic deformation that reflects the capacity of rocks to accommodate large strains homogeneously (affecting all of a rock mass) or heterogeneously (localized in ductile shear zones).
Ductility contrast. Contrast in ability to flow between layers in a layered rock sequence or between a dike or sill and the host rocks.
Duplex. Most commonly, a system of two subparallel master thrust faults separated by a deformed interval dominated by smaller imbricate thrusts that connect the two master faults. The imbricates have the same sense of movement as the master faults. The upper fault is the roof thrust; the lower is the floor thrust. Duplexes may involve either crystalline or sedimentary rocks and also form in normal and strike-slip fault zones.
Dynamic recovery. Deformation processes that increase strain, dislocation density, and dislocation interaction and increase the rate of glide of dislocations.
Dynamic recrystallization. Recrystallization that reduces dislocation density and dislocation interaction; this process occurs at low strain rate and high temperature and produce new grain boundaries and strain-free grains.

Edge dislocation. Edge of an extra partial layer (extra half plane) in a crystal lattice parallel to existing layers. Edge dislocations may propagate in a lattice subjected to simple shear and are normal to Burgers vectors.

Effective normal stress. Normal stress minus fluid pressure ($\sigma_n - P$).

Egg-carton structure. Dome-and-basin fold pattern produced by Type 1 fold interference.

Elasticas. Fold in which the interlimb angle is negative.

Elastic limit. Point on the stress-strain curve beyond which the material begins to undergo permanent deformation or ruptures (or both), initiating plastic or brittle behavior.

Elasticoviscous behavior. Combination of elastic and viscous behavior.

Elastic-plastic behavior. Combination of elastic and plastic end-member behavior.

Elastic rebound. Rapid movement commonly accompanied by earthquakes on a fault after long accumulation of elastic strain.

Elastic strain. Strain that is recovered instantaneously on removal of applied stress, so that the object returns to its original undeformed shape.

Electrical conductivity. Ability of a material to conduct electric current.

Electrical resistivity. Resistance of a material to the flow of electric current.

Elongation (ε). Type of strain measured by the ratio of the length of lines in a deformed mass minus the original length to the original length.

Emergent thrust. See *Erosion thrust.*

Enclave. Inclusion of wall rock or other material in an igneous body. See also *Xenolith.*

***En echelon* faults.** Faults that strike approximately parallel to one another; they occur in short, locally overlapping segments.

Entropy. Measure of the amount of energy not available for useful work in a system.

Enveloping surface. Surface constructed by connecting the inflection points on the limbs of small folds that occur on a larger fold, or by drawing a tangent surface to a fold train.

Episodic lead loss. Loss of lead during one or more thermal events subsequent to crystallization of zircon, monazite, sphene, or other U-bearing mineral; produces discordant lead-isotope ratios that plot on a chord rather than the concordia curve.

Equilibrium. State of rest or balance. Excess energy in one part of a system is compensated by work performed in another, or flow of energy to the lower-energy state, until a state of balance is achieved.

Erosional outlier. Remnant of a formerly more extensive rock mass isolated by erosion. Contacts are mostly gently dipping and are stratigraphic contacts.

Erosion thrust. Thrust that ruptures the ground surface; an emergent thrust.

Exotic terrane. Block of crustal dimensions that bears no resemblance (structural or metamorphic history, faunal assemblage) to the mass to which it is presently attached; commonly derived from the opposite side of an ocean. Boundaries of exotic terranes with the masses to which they are attached are always tectonic.

Extensional crenulation cleavage. Crenulations in a shear zone having movement sense opposite to that of the shear zone and C-surfaces formed by apparent extension and rotation as the shear zone moves. Structures resembling intrafolial folds (folds lying within a foliation) appear between conveniently oriented active S-surfaces inclined opposite the shear direction and the C-surface orientation. They may result from shear motion on an existing S-surface that becomes locked, then is forced to "ramp" to the next higher detachment S-surface. Formation of reverse-sense crenulations requires a preexisting layering or layering produced during motion of the shear zone. An earlier foliation (S-surface) or new C-surfaces provide(s) the layering for formation of extensional crenulations.

Extensional fault. Fault in which bedding undergoes layer-parallel extension.

Extra half plane. Layer forming an edge dislocation produced by deformation of a crystal, so that a discontinuous layer of atoms is produced from existing atoms in the crystal.

Eyed fold. Sheath fold that yields a highly symmetric section viewed perpendicular to the hinge. Also known as a bulls-eye fold.

Eyelid window. Window formed where a thrust sheet is arched into an antiformal fold during emplacement, then broken along the antiform; erosion produces a map pattern exposing younger rocks in the window, but faults are connected to the fault framing the window within the thrust sheet. Instead of having a near-circular shape, an eyelid window has a more elongate eye shape.

Fabric. Relationships between planar (S) and linear (L) structures (including bedding, cleavage, and the orientation of minerals) to texture in rocks.

Fabric element. Any structure, such as foliations, lineations, or small folds, that contributes to the fabric (texture plus structure) in rocks.

Facing direction. Direction toward the original top of a sequence determined using primary sedimentary or volcanic structures, faunal succession, or, less commonly, radiometric age data.

Failed arm. Initial rift in a triple junction that does not continue to develop and spread apart. Also called a failed rift.

Failed rift. See *Failed arm.*

Fault. Fracture having appreciable movement that produces measurable displacement parallel to the plane of the fracture on the scale of observation. Mode II and III fractures.

Fault-bend fold. Fold formed above a thrust as the fault surface changes from steeper dip to shallower in an up-dip direction as it passes over a ramp, forming generally parallel bending folds. A shear thrust.

Fault-line scarp. Fault scarp produced by differential erosion along a fault, lowering the level of the topographic surface and removing a resistant layer in the upthrown side, so that the original displacement may or may not be indicated by offset in the present-day topography.

Fault mechanics. Study of the physical processes and physics that produce faults and faulting.

Fault plane. Actual movement surface along a fault.

Fault-propagation fold. Fold formed as a fold tightens and a thrust fault begins to propagate through the common tightened limb between the anticline and syncline. A break thrust. Alternatively, a fold that develops by deformation at the distal end of a propagation fault.

Fault scarp. Scarp formed where a topographic surface is offset by dip-slip motion along a fault. The present topographic displacement indicates the movement sense on the fault.

Fault zone. Series of interleaving anastomosing brittle faults near the surface, or a ductile shear zone produced by faulting at great depth. Yields a zone of measurable thickness, not a discrete plane.

Fenster. See *Simple window.*

Field relations. Structural measurements made at outcrop scale that permit deciphering outcrop- to map-scale structures.

Filled joint. Joint that has been filled with minerals during or after formation of the joint.

Finite-element method. Numerical method for dividing a continuum into a grid that can be used to understand how structures form by studying the changes that occur in the grid as it is deformed.

Finite strain. Comparison of the difference in the present state of strain with some previous less deformed state. Relates to the instantaneous shape of a rock mass at one time relative to the initial undeformed shape.

First-order folds. Largest folds in an area.

First-rank tensor. Tensor that has three components and can be represented as a vector.

Flexural flow. Fold mechanism in which some layers in a sequence flow ductilely during folding, whereas stronger layers remain brittle (or are less ductile) and buckle. Moderate to high ductility contrast between layers is required.

Flexural-flow folds. Mostly similar-like folds that form in low to moderate metamorphic-grade rocks but may include some parallel folds. Some layers maintain constant thickness; others are thickened into the axial zones and thinned in the limbs as a result of moderate to high ductility contrast.

Flexural folds. Folds in which shape is controlled by the layering in rocks.

Flexural slip. Slip parallel to layering during folding.

Flexural-slip folds. Folds that form by buckling, bending, and slip parallel to layering. They correspond in shape and mode of formation to parallel or parallel-concentric folds.

Flinn diagram. Plot of elongation *(X/Y)* vs. flattening *(Y/Z)* strain.

Floor thrust. Lower of the two master faults in a duplex.

Flow law. Mathematical expressions for different types of viscous or plastic flow.

Flower structure. Upward and outward branching faults (cross-section view) in a strike-slip fault system that have the shape of a bouquet of flowers. Same as palm (-tree) structure.

Flute casts. Spoon-shaped structures that form in sedimentary environments where currents scour and erode a surface.

Fluxion structure. Laminations with or without folding developed in very fine-grained cataclasite, mylonite, and related rocks along faults.

Fold axis. A line that is moved parallel to itself to generate a folded surface (cylindrical folds).

Fold train. A series of folds of the same generation having similar amplitude and wavelength.

Fold test. Test of paleomagnetic data whereby the measurements made on oriented samples collected from opposite limbs of a fold are plotted on a stereonet and then rotated back to the horizontal. Paleomagnetic measurements for samples from opposite limbs treated the same are then compared for consistent orientation of paleomagnetic poles. A valid fold test is necessary to evaluate the usefulness of paleomagnetic measurements.

Foliation. Texture in deformed and metamorphosed rocks imparted by the parallel orientation of platy or elongate minerals. Schistosity, gneissic banding, slaty cleavage, and crenulation cleavage are all foliations.

Footwall. Rock mass beneath a fault plane.

Force (F). Vector that produces a change in the velocity, direction, or acceleration of a body.

Foreland fold-thrust belt. Belt of thrust faults and related folds—generally thin-skinned—that occurs in sedimentary and metamorphic rocks between the undeformed craton (platform) and the metamorphic core of nearly every mountain chain.

Form lines. Lines—commonly strike lines—on a map that outline the trend and shape of a structure.

Fossil. Evidence of an organism preserved in the geologic record.

Fracture cleavage. Cleavage formed by parallel to subparallel fractures, generally spaced 1–3 cm apart.

Frank-Read source. Mechanism whereby dislocations are multiplied during deformation: a segment of a dislocation bows outward as shear stress increases, then expands into a loop that closes back on itself to allow a new dislocation.

Free-air gravity map. Gravity-anomaly map corrected only for elevation above sea level.

Fry method. Simpler version of the center-to-center method for determining strain. It produces a diagram containing a set of points with a central circular to elliptical blank area wherein relative shape and orientation are proportional to the shape and orientation of the strain ellipse. A circular area indicates that there is no strain.

Gabbro. Coarse-grained igneous rock consisting of calcic plagioclase and pyroxene, sometimes with minor amounts of olivine, amphibole, or biotite.

General strain. Strain that produces the shape of a triaxial ellipsoid. Plots in the vicinity of the $k = 1$ line in the Flinn diagram. Plane strain if $k = 1$.

Gentle fold. Fold with an interlimb angle of 180° to 120°.

Geochemical cycle. See *Rock cycle.*

Geometric axes. Spatial relationships between structures and a set of mutually perpendicular axes.

Geophone. Acoustic wave detector used in seismic exploration that produces an electrical signal proportional to the energy (velocity, amplitude) of waves detected. Can be designed to detect waves over a narrow or wide frequency range.

Glide. Motion of a dislocation within a surface that contains both the line of the dislocation and Burgers vector. May produce tangles, deformation lamellae, and undulatory, patchy, or sweeping extinction. The most common indication of plastic deformation in minerals.

Gneissic banding. Foliation in gneissic rocks; commonly alternating layers of different texture, composition, and color.

Graben. Block that has been dropped down between two subparallel normal faults that dip toward each other.

Graded bedding. Beds consisting of coarse particles at the bottom gradually fining upward into smaller particles at the top. Form where sediment of widely different particle sizes is deposited rapidly in the same depositional environment, so that the largest particles settle to the bottom first.

Grain-boundary diffusion creep (Coble creep). Deformation mechanism involving mass transport through diffusion at grain boundaries, occurring at low to moderate temperature and overlapping the pressure-solution realm (<300° C) up to about 400° C.

Grain-boundary sliding. Deformation mechanism where grains slip past one another at moderate temperature in association with Coble creep, or at low temperature under a strong component of simple shear.

Granitoid. Light-colored, quartz-bearing, coarse-grained, igneous rock having a composition in the range from tonalite (quartz diorite) to granite. Includes granite, adamellite (quartz monzonite), granodiorite, and tonalite.
Granodiorite. Coarse-grained, igneous rock composed of plagioclase, K–spar, and quartz, often with minor amounts of biotite, hornblende, or muscovite.
Gravitational spreading. Theory that an orogen compensates for uplift derived from horizontal compression by flowing laterally under the influence of gravity.
Gravity anomaly. Measured value of gravity that deviates from estimated average or assumed value.
Gravity fault. Normal fault, so called by implication that gravity is the primary motive force. Also called extensional fault.
Gravity model. Model used to determine the shape and broad composition of plutons, or other rock bodies, constructed by plotting matching the observed anomaly with a gravity anomaly produced by different assumed densities and volumes.
Griffith cracks. Elliptical microscopic cracks that exist in glass; they suggest that stress is concentrated and magnified by several orders of magnitude at crack tips.
Growth fault. Fault—generally a normal fault—that involves simultaneous deposition and fault motion.
Hackle marks. Corrugations on a joint surface indicating a zone where a joint propagated rapidly. Twist hackle forms in response to changing stress orientation or possibly a component of simple shear.
Half-graben. Block bounded by a normal fault on one side; the other side passes into a gentle bending fold.
Hanging wall. Rock mass resting on (above) a fault plane.
Hartman's rule. Maximum principal stress (σ_1) bisects the acute angle between conjugate shear planes of the stress ellipsoid.
Heave. Horizontal component of fault displacement.
Heterogeneous domain. Domain of a single kind of fabric data (e.g., foliation) containing multiple orientations.
Heterogeneous deformation. Deformation that varies in magnitude with direction.
High-angle boundary. Boundary involving new grains formed to relieve stess in an elastically strained crystal. New grains contain lattices oriented at a high angle to the lattice in the parent crystal and to each other.
Hinge. Part of a fold that has the greatest curvature.
Hinge line. Line where layering changes orientation most rapidly.
Hinterland-dipping duplex. Duplex in which the imbricates dip toward the interior of a mountain chain.
Homocline. Structure with uniform dip of layering in one direction.
Homogeneous domain. Area where one fabric element is oriented the same throughout.
Homogeneous material. Material in which properties are the same throughout.
Homogeneous strain. Strain in which lines originally straight and parallel before deformation remain straight and parallel after deformation.
Hook fold pattern. Pattern produced by refolding of an isoclinal fold by folds having axes parallel to the first folds. Type 3 fold interference pattern.
Horizontal compression. Force oriented in the horizontal that tends to push a mass together.
Horse. Fragment of rock surrounded on all sides by faults that is transported beneath a thrust sheet or within a strike-slip or normal fault zone. A slice.
Horse-tail structure. Branching, frequently anastomosing, array of faults.
Horst. Structurally high block between two oppositely dipping normal faults.
Hummocky laminations. See *Ripple cross laminations.*
Hydraulic joints. Joints produced by abnormally high pore pressure during burial and vertical compaction of sediment at depths greater than 5 km.
Hydrostatic state of stress. State in which normal stress is the same in all directions, and no shear stress exists.
Imbricate stack. Assemblage of imbricate faults, generally thrusts, arranged so that they overlap one another and do not coalesce upward as in a duplex.
Imbricate thrust. Smaller thrust that converges down-dip into a master fault.
Imbricate zone. Zone of imbricate faults. May refer to the entire series of thrusts making up a foreland fold-and-thrust belt.
Impact structure. Kilometer- or larger-scale structure commonly made up of a complex of radial and concentric faults that have a circular or elliptical outline but are not obviously related to tectonic processes; probably the result of impact of an extraterrestrial mass of rock. Astrobleme.
Incompetent. Weak material; material of low viscosity or rigidity and low yield strength.
Incremental strain. Strain that occurs in events small steps during progressive deformation.
Induced magnetization. Magnetization in a rock mass produced by interaction of the magnetically susceptible minerals and an external—commonly the Earth's—magnetic field.
Infinitesimal strain. Strain that occurs in infinitesimally small amounts. Restricted to very small strains relative to an initial condition.
Inherited zircons. Zircons assimilated by magma from another rock mass.
Inhomogeneous property. Property that varies spatially with location at the scale of reference, either microscopic, in a hand specimen, or in a region.
Inhomogeneous strain. Strain in which lines that are straight or parallel before deformation do not remain so after deformation; angular relationships change nonuniformly.
Inlier. Erosionally exposed rock mass surrounded laterally by the unit above it and with a normal (stratigraphic) or unconformable contact with the unit below.
Interference folds. See *Superposed folds.*
Interlimb angle. Angle between the limbs of a fold.
Intersection lineation. Linear structure formed by the intersection of two planar structures; usually penetrative. The most common intersection lineation involves bedding (S_0) and cleavage (S_1), sometimes designated L_{1x0}.
Interstitial atom. Atom present between ideal atomic sites in a crystal.
Interstitial defect. Defect that disturbs the regular arrangement of atoms or ions in a crystal at a point.
Intracontinental transform. Transform fault that separates two continental terranes.
Intrafolial folds. Small to large isolated folds (commonly relict isoclinal folds) lying within a later foliation.
Inverted fault structure. Structure in which original movement sense has been reversed.
Inverted rift structure. Series of normal faults that originally formed by rifting (extension) but were later subjected to compression, reactivating the faults as thrust or strike-slip faults.

Isochron. Line obtained by plotting analyses of the components of some isotopic system, such as $^{87}Sr/^{86}Sr$ versus $^{87}Rb/^{86}Sr$. The slope of the isochron defines the ^{87}Rb-^{86}Sr whole-rock or mineral age.

Isoclinal fold. Fold in which the axial surface and limbs are parallel.

Isostatic equilibrium. State of equilibrium between blocks of different density and volume within the continents or sea floors and between continents and adjacent sea floors.

Isotropy. Lack of contrast in physical properties with direction between and within individual layers of a rock mass. Isotropic materials have the same properties in all directions.

Joint. Fracture along which there has been no appreciable movement parallel to the fracture and only slight movement perpendicular to the fracture plane. A Mode I fracture.

Joint set. Joints that share similar orientation in the same area.

Joint system. Two or more joint sets in the same area.

Laccolith. Confordant pluton that has a mushroom or bell shape in vertical cross section.

Kinematic (movement) axes. Fabric axes in which the *a* axis is the tectonic transport direction, *b* is normal to *a* in the transport (fold axial) plane, and *c* is normal to the *a-b* plane.

Kink folds. Asymmetric folds with straight limbs and angular hinge zones.

Klippe. A small erosional outlier (remnant) of a thrust sheet.

Königsberger ratio (Q). Remanent magnetism divided by the product of magnetic susceptibility of the rock material that produced a magnetic anomaly and the induced magnetic-field intensity.

Lateral ramp. Along-strike change in detachment level where a thrust is forced by stratigraphic pinch-out or other means to rise to a higher level.

Law of cross-cutting relationships. A body of rock is older than structures or igneous bodies that cut through it. See also *Law of igneous cross-cutting relationships; Law of structural relationships.*

Law of faunal succession. Fossil organisms in a sequence evolve toward the top of the sequence.

Law of igneous cross-cutting relationships. An igneous body must be younger than the rocks it intrudes.

Law of original horizontality. Bedding planes within sediments or sedimentary rocks form at the time of deposition in a horizontal to nearly horizontal orientation.

Law of structural relationships. A structure, such as a fold or fault, must be younger than the rocks it deforms or cuts through.

Law of superposition. Within a layered sequence, commonly sedimentary rocks, the oldest rocks will occur at the base of the sequence, and successively younger rocks will occur toward the top, unless the sequence has been inverted through tectonic activity.

Layer-parallel shortening. Deformation resulting in shortening of a layered sequence parallel to bedding. Pressure solution is a common mechanism.

Left-handed screw dislocation. Screw dislocation circumscribed by a counterclockwise spiral of an atomic plane in the distal direction.

Left-lateral strike-slip fault. Strike-slip fault where the left side has moved toward an observer looking along the fault; also called sinistral strike-slip fault.

Limbs. Straighter or least-curved segments of a fold.

Lineament. Topographic feature consisting of aligned surficial features such as valleys and ridges.

Linear structure; lineation. Structure on any scale that can be expressed as a real or imaginary line.

Line defect. Type of dislocation produced by lack of registry of atoms along a line in a crystal. Edge and screw dislocations are both line dislocations.

Listric fault. Thrust or normal fault with concave-up geometry. These faults have steep dip near the surface but flatten with depth.

Lithosphere. The outer shell (commonly ~100 km) of the Earth; includes all the crust and uppermost rigid part of the mantle.

Lithostat. Weight of a column of rock per unit area above a depth in the crust.

Load casts. Depressions in sediment beneath a sand (or gravel) bed. Result from gravitational instability at the interface between a layer of water-saturated sand and underlying mud after deposition and dewatering. May be used to determine facing direction.

Low-angle boundary. Microscopic boundary of a subgrain within a strained grain, indicating a slight (low-angle) lattice misorientation across the boundary.

LS-tectonite. Deformed rock wherein linear structures dominate over foliation or cleavage.

L-tectonite. Rock dominated by a linear fabric element.

Macroscopic scale. Mountainside to map scale.

Macroscopic structure. Structure of mountainside or larger scale.

Magma fracture. Tension fracture produced by pressurized magma. Analogous to hydrofracture.

Magnetic anisotropy. Variation of magnetic properties with direction in a rock.

Magnetic-anomaly map. Map produced by plotting vertical magnetic-intensity profiles along the line of traverse or the data contoured between successive traverse lines.

Magnetic-intensity profile. Plot of vertical magnetic intensity along the line of a profile.

Magnetometer. Instrument used to measure intensity of the Earth's magnetic field.

Master fault. Fault, commonly with large displacement, from which smaller faults may propagate. See also *Imbricate thrust; Antithetic fault; Synthetic fault.*

McKenzie model of rift formation. Model depicting the idea that spreading is symmetrical about a rift axis, decreasing to zero along the axis at a pole of rotation.

Mean ductility. Ductility of an entire rock mass. Rocks of low mean ductility are strong; rocks of high mean ductility are weak. Average ductility or ease of flowage.

Mean stress. The mean of the principal stresses: $(\sigma_1 + \sigma_2 + \sigma_3)/3$.

Megacryst. Large crystal in a finer-grained groundmass. Large phenocryst or porphyroblast.

Megacrystic. Igneous texture involving large crystals embedded in a finer-grained or metamorphic groundmass. Porphyritic or porphyroblastic.

Megascopic. Visible without the aid of magnification. Opposite of microscopic.

Mélange. Mixture of sheared, tectonically mixed weak and strong rock materials. Fragments of strong material may measure up to several kilometers.

Mesoscopic scale. Scale of structures in the size range from hand specimen to outcrop.

Metamorphic core. Internal part of most compressional mountain chains, consisting of folded, faulted, cleaved, foliated, and intruded rocks ranging from low to high metamorphic grade. Also used (principally in the western United States) to describe the internal parts of metamorphic core complexes.

Metamorphic core complex. Domal uplift of deformed metamorphic and igneous rocks exposed beneath a tectonically detached unmetamorphosed cover of sedimentary and volcanic rocks, commonly associated with regions of crustal extension. The Basin and Range Province is a well-described region of large-scale crustal extension that contains numerous metamorphic core complexes.

Metamorphic differentiation. Formation of new layering by pressure solution or recrystallization.

Mica beards. Microscopic texture of pressure shadows composed of micas (commonly muscovite and chlorite) on quartz or other strong grains in clastic sedimentary rocks, volcanic rocks, and others undergoing deformation at lower-greenschist–facies metamorphic conditions.

Microcontinent; microplate. Crustal mass of thrust sheet to subcontinental proportions.

Microcracks. Microscopic-scale fractures in individual crystals and mineral aggregates.

Microlithon. Relatively undeformed zone between cleavage surfaces.

Microscopic scale. Scale of observation requiring magnification.

Migmatite. Mixed igneous-metamorphic-looking rock produced by partial melting or metamorphic differentiation. New material *(neosome)* is commonly more feldspar- and quartz-rich; older material *(paleosome)* is commonly rich in mafic components (hornblende, biotite).

Mineral lineation. Lineation consisting of aligned elongate mineral grains and grain aggregates.

Mixing line. Best-fit line produced by plotting $^{87}Sr/^{86}Sr$ versus $^{87}Rb/^{86}Sr$ (or data from another isotopic system), but in which the points do not plot close to the line to define an isochron and accurately determine rock or mineral age. Scatterchron.

Mode I fracture. Fracture formed by extension; a joint.

Mode II fracture. Fracture formed by sliding; perpendicular to the tip line (line along which shear dies out).

Mode III fracture. Fracture formed by a tearing motion; parallel to the tip line.

Mohr circle for strain. Circle produced by plotting on a set of coordinate axes values of reciprocal quadratic elongation on the horizontal axis versus modified shear strain on the vertical axis.

Mohr construction. Circles produced by a plot of values of σ_1 and σ_3 as diameter on the σ_n axis of a plot of σ_n versus τ. Mohr circles for stress produced. Similar construction can be made for strain.

Mohr envelope. Line separating stable from unstable regions on a Mohr diagram. Tangent to Mohr circles.

Mohr's hypothesis. Shear strength is a function of normal stress.

Monocline. Structure in which an otherwise uniform regional dip is locally steepened.

Monoclinic symmetry. One plane or axis of symmetry. Asymmetric folds have monoclinic symmetry.

Mud cracks. Polygonal cracks that taper downward and terminate in unconsolidated fine-grained sediment produced by drying and shrinkage of the sediment. Useful in determining facing direction. Desiccation cracks.

Mullions. Linear corrugated or scalloped surfaces resulting from folding, cleavage formation, or differential flow; form at boundaries between rock types of different relative ductility. Type of lineation.

Multiple working hypotheses. Principle of formulation of more than one possible explanation from the same data, evaluation of each, and selection of the most likely hypothesis.

Mylonite. Strongly foliated fault rock that exhibits characteristics of high ductile (mostly simple shear) strain. Decrease of grain size from the original rock is characteristic of mylonite, along with ribbon quartz and rotated porphyroclasts. S–C fabrics are commonly present. Microfabric is commonly only partially recovered or recrystallized but may be totally recrystallized if deformation ceased during or before a thermal event.

Nabarro-Herring creep. See *Volume-diffusion creep.*

Natural strain ellipsoids. Features in rocks (oöids, pebbles, vesicles) that are nearly spherical before deformation and deform into ellipsoid shapes.

Negative anomaly. Gravity or magnetic feature that has a less than average intensity.

Net slip. See *Displacement.*

Neutral surface. Surface of no finite strain in a layer undergoing tangential-longitudinal strain during buckling or bending; separates a zone of tension in the outer arcs from a zone of compression in the inner arcs.

Newtonian fluid. Fluid in which stress and shear strain rate are proportional.

Newton's law of gravitation. The force of gravity is directly proportional to the masses of the objects involved and inversely proportional to the square of the distance between them.

Node. Point where three or more dislocations meet.

Noncoaxial deformation. Principal axes of incremental strain ellipsoids do not coincide with the same material lines throughout deformation.

Nonconformity. Unconformity in which igneous or metamorphic rocks—or both—occur below the erosion surface, and sedimentary rocks occur above.

Noncylindrical folds. Folds in which hinges are curved and are commonly not parallel on successive folds.

Nonpenetrative structure. Isolated structure, such as a fault or isolated fold. Some structures, such as joints, may be nonpenetrative at outcrop scale but penetrative at map scale.

Nonsystematic joints. Irregular fractures that do not share a common orientation; fracture surfaces are commonly highly curved, irregular, and exhibit no displacement.

Normal fault. Dip-slip fault in which the hanging wall has moved down relative to the footwall.

Normal-sequence thrust. Thrust fault in a foreland fold-thrust belt that formed by propagation outward from the internal core of the mountain chain.

Normal stress (σ_n). Stress that acts perpendicular to a surface. A vector.

Oblique-slip motion. Combined dip-slip and strike-slip motion on a fault.

Obsequent fault-line scarp. Fault-line scarp that through erosion of a resistant layer faces opposite the direction of the original fault scarp. An incorrect motion sense would be inferred from the topography alone.

Olistostrome. Deposit without internal order made up of rock fragments of diverse size and composition that accumulated in finer sediment by submarine slumping or gravity sliding from an unstable slope. Blocks range from a few millimeters to several kilometers.

Oöid. Millimeter-size, concentrically layered, spheroidal to ellipsoidal, concretionary body composed most frequently of $CaCO_3$. Commonly formed by precipitation around a nucleus (shell fragment, sand grain) in a high-energy (wave) zone. Good finite-strain indicator.

Open folds. Folds in which the limbs dip gently away from or toward one another, producing a large interlimb angle (>70°).

Ordinary boudinage. Boudinage formed by extension in one direction; consists of parallel, segmented, sausage-shaped (in cross section) remnants of a former continuous layer in a more ductile matrix.

Orogeny. Process of mountain building, accompanied by metamorphism, plutonism, and associated deformation, resulting from subduction, terrane accretion, and/or continent-continent collision.

Orphan. A complex of far-traveled horses and duplex imbricates along a thrust derived from a footwall syncline probably located in a former thrust ramp. Orphans contain rocks that are stratigraphically unrelated to the rocks in either the hanging wall or the footwall of the thrust system of which they are presently a part.

Orthorhombic symmetry. Symmetry involving three mutually perpendicular planes. Symmetrical folds and conjugate joints have orthorhombic symmetry.

Oscillatory ripple marks. Symmetrical ripple marks consisting of linear, sharp crests and broad, rounded troughs formed by back-and-forth motion of water. They may be used for determining facing direction.

Out-of-sequence thrust. Younger, higher thrust fault in a foreland fold-thrust belt that formed by breaking an older thrust sheet.

Outrageous hypothesis. Initially unacceptable hypothesis formulated to explain a phenomenon; the value of the hypothesis is not in being right or wrong but in focusing attention on an unsolved problem.

Overburden pressure. Pressure of overlying material.

Overturned folds. Folds that have one tilted past the vertical limb.

Paleomagnetism. Remanent ancient magnetic fields preserved in rocks. Measurement allows reconstruction of the positions of the ancient continents, and oceans.

Palm (-tree) structure. See *Flower structure.*

Paraconformity. Unconformity involving only sedimentary rocks with little relief on the erosion surface; bedding remains parallel on both sides.

Parallel folds. Folds in which the layering maintains thickness, and layers remain parallel.

Parasitic fold. Higher-order fold on the limbs or hinge of a lower-order fold.

Passive flow. Fold mechanism involving uniform ductile flow of the entire rock mass, with layering serving only as a strain marker; little ductility contrast exists between layers.

Passive flow folds. Ideal similar folds that involve plastic deformation. The layering acts only as a strain marker to record the deformation.

Passive folds. Folds in which layering serves only as a strain marker during folding.

Passive-roof duplex. Duplex in which the roof thrust has only minor displacement.

Passive slip. Slip (simple shear) at an angle to layering along a cleavage or schistosity. Associated folds are called passive-slip or shear folds.

Passive-slip folds. Similar-like folds that are thought to form by shearing along planes (cleavage, foliation) inclined to the layering. Problematic fold mechanism.

Pencil structure. Elongate pencil-shaped structure in fine-grained rocks; produced by intersection of bedding and cleavage.

Penetrative structure. Structure—including cleavage, foliation, and some folds—that occurs pervasively. Some structures, such as joints, may not be penetrative at outcrop scale but are penetrative at map scale.

Perfect crystal. Crystal lattice where all the sites are filled with the correct atoms, ions, or groups for a particular mineral; contains no dislocations or interstitial atoms between lattice sites. Stack of orderly repeat units of identical atoms or ions.

Perfect fluid. Stationary fluid that will not transmit shear stress.

Phacolith. Tabular pluton intruded into the hinge of a fold.

Pileup. Accumulation of dislocations in a crystal lattice.

Pillow structures. Clustered volcanic structures with rounded tops and sharply pointed bases resembling pillows; form where lava is erupted beneath or flows into water. Exteriors of pillows are fine-grained and scoriaceous to glassy, with grain size increasing toward the interior.

Pisolite. Concentrically layered spheroidal to ellipsoidal concretionary body that forms under conditions similar to those for formation of oöids. Similar in shape to an oöid but larger and more irregular. Good finite-strain indicator.

Planar (tabular) cross bed. Layering inclined to normal bedding planes that is planar and does not become tangent to the bottom of a bed; it is truncated at both the top and bottom of the bed. Facing direction cannot be determined using planar cross beds because they provide the same perspective upright or overturned.

Plane strain. Strain that plots on the $k = 1$ line in the Flinn diagram. Involves no change in the Y axis in the strain ellipsoid before or after strain, resulting in a triaxial ellipsoid (general strain).

Plastic strain. Permanent, nonrecoverable strain occurring without loss of cohesion. Results from rearranging chemical bonds in crystal lattices and may affect an entire rock mass. Ductile shear zones are narrow zones of plastic strain.

Plate tectonics. Concept of global tectonics in which the Earth's surface is divisible into seven major, rigid plates and several smaller ones that contain all the continents and oceans. Plates are formed at the oceanic ridges as new oceanic crust forms and are consumed by subduction in the trenches. Involves plate generation, motion and destruction, and interaction.

Plumose joint. Joint that has a feathered surface texture.

Plunge. Bearing and amount of inclination from the horizontal of a fold axis or other linear structure.

Plunging fold. Fold with nonhorizontal axis.

Point defect. Substitution, interstitial defect involving atoms, ions, or molecules or a vacancy.

Poisson's ratio (υ). Measure of compressibility defined by the negative of the ratio of extension normal to an applied compressive stress to the extension parallel to the applied compression.

Polygon. Strain-free crystal formed by recovery or recrystallization. Generally has high-angle boundaries but forms initially along low-angle boundaries.

Porphyroblasts. Large grains that have grown in a rock mass during deformation or metamorphism as products of dynamic or static recrystallization.

Porphyroclasts. Relict earlier large grains of one or more minerals that remain as a rock mass is deformed. δ, σ, and θ porphyroclasts have been recognized as products of simple shear deformation.

Positive anomaly. Gravity or magnetic feature that has a greater than average value.

Potential field. Gravity, magnetic, or electric field.

Pressure shadow. Mineral lineation or foliation composed of quartz, muscovite, chlorite, magnetite, or other mineral that forms in the extensional strain field on either side of a larger crystal or detrital grain. Pressure shadows are commonly oriented parallel to a prominent foliation or lineation.

Pressure solution. Deformation mechanism that involves dissolution at grain boundaries under stress of soluble constituents such as calcite or quartz. Generally active at low to moderate temperature in the presence of fluid.

Principal axes of the strain ellipsoid. Three mutually perpendicular symmetry axes of unequal length that define the strain ellipsoid: *X* (greatest), *Y* (intermediate), and *Z* (least).

Principal strain. Strain parallel to one of the principal axes of the strain ellipsoid.

Principal stresses. Three mutually perpendicular normal stresses, σ_1, σ_2, and σ_3. σ_1 is the greatest principal normal stress; σ_3, the least.

Progressive deformation (strain). Strain that results when a series of deformational events produces an increase in the amount of strain in a rock body through time.

Protolith. Original rock type from which a metamorphic rock is derived.

Pseudotachylite. Glass produced by local shear heating along a fault. Very fine-grained material generated by motion may partially melt by buildup of frictional heat and sudden decrease of pressure. Literally, false glass.

Ptygmatic folds. Near-parallel folds formed by buckling of strong layers in a much more ductile matrix.

Pull-apart basin. Rhomb-graben (rhombochasm) basin, produced by *en echelon* dextral strike-slip faults with right step-overs or sinistral strike-slip faults with left step-overs.

Pumpelly's rule. Small structures are a key to, and mimic the style and orientation of, larger structures of the same generation.

Pure shear. Distortion involving homogeneous deformation. Principal strain axes are not rotated.

Push-up range. Rhomb horst, produced by dextral strike-slip faults with left step-overs or sinistral faults with right step-overs.

Quadratic elongation (λ). Square of the ratio of deformed length to original length.

Quartz diorite. See *Tonalite*.

Quartz monzonite. See *Adamellite*.

Quasi-flexural folds. See *Disharmonic folds*.

Radial faults. Faults that converge toward a single point.

Rain imprints. Small crater-like structures produced where rain falls on fine-grained sediment that is intermittently exposed to the atmosphere. Can be used to determine facing direction.

Rake of net slip. Angle from the horizontal of a line in an inclined fault plane measured in the plane. If the dip of the plane is vertical, the rake and plunge of the line are the same; otherwise they are not.

Ramp. High-angle segment along a thrust or normal fault.

Reclined folds. Have fold axes that plunge at the same angle as the dip of the axial surfaces.

Recovery. Processes that reduce dislocation density and increase the rate of dislocation climb in crystals.

Recrystallization. Process that may be initiated independent of recovery that completes the process of removal of accumulated stress from a deformed crystal and produces high-angle (new) grain boundaries.

Recumbent folds. Folds that have horizontal axes and axial surfaces.

Reduction spot. Spherical color alteration in sediment produced by a small grain or fragment that is chemically different from the surrounding mass of sediment. The chemical difference may produce a nearly spherical area of reduction expressed as a color change in the immediate vicinity of the grain in the otherwise oxidized sediment. Reduction spots are useful strain indicators if the time of formation can be resolved.

Reduction to pole. Modeling technique for aeromagnetic data used to correct for the low-latitude field effects on magnetic data where a high and low pair of anomalies for the same body are separated. Reduction to pole produces a single anomaly above the rock body that caused the anomaly analogous to the anomaly that would have occurred for the body at magnetic north.

Regional structural geology. Study of parts of mountain ranges, small parts of continents, trenches, and island arcs by geologic mapping, mesoscopic and microscopic structural studies, and the relationships of the resolved structural history to stresses and tectonic plates.

Release joints. Joints that form near the surface as erosion removes overburden, and thermal-elastic contraction occurs.

Release spectrum. Release of argon from biotite, muscovite, or hornblende as a sample is heated through a range (spectrum) of temperatures, commonly from room temperature to the melting point of the mineral. Used to identify excess argon that introduces an error into K-Ar ($^{40}Ar/^{39}Ar$) age determinations.

Remanent field. Magnetic field remaining in a rock after the present-day field has been removed.

Resequent fault-line scarp. Fault-line scarp along which erosion preserves the original facing direction of the fault scarp.

Reversed graded bedding. Metamorphosed normal graded bedding where the fine-grained top of a bed is recrystallized to coarse micas (or other minerals) that are larger than the granular material at the original bottom of the bed. Also a primary structure formed during deposition of pumice fragments in water, where the smallest fragments are deposited first, followed by successively larger fragments.

Reverse drag. Drag that occurs where layers appear to have been dragged down-dip in the same direction as the relative movement of the hanging wall of a normal fault.

Reverse drag folds. Folds that form along growth (normal) faults where the part of the downthrown block close to the fault is displaced downward more than the parts farther away.

Reverse fault. Fault with moderate to steep dip (45° or more) in which the hanging wall has moved up relative to the footwall. Mechanically the same as a thrust fault.

Reverse-sense crenulations. See *Extensional crenulation cleavage*.

R_f/ϕ method. Method of determining homogeneous strain in a rock mass containing deformed initially elliptical objects; generally results in objects that remain elliptical. The shape of the final ellipse is determined by the initial shape and orientation (ϕ) of the starting ellipse relative to the shape and orientation of the strain ellipse.

Rheology. The study of flow.

Rhomb-graben basin. Basin formed along *en echelon* dextral strike-slip faults with right step-overs, or along *en echelon* sinistral strike-slip faults with left step-overs.

Rhomb horst. Push-up range produced by dextral strike-slip faults with left step-overs, or sinistral faults with right step-overs.

Rhombochasm basin. Rhomb-graben basin or pull-apart basin.

Riedel shears (R). *En echelon* shear fractures formed at 10° to 15° to the principal shear (fracture orientation); they have the same displacement sense as the primary shear fractures.

Right-handed screw dislocation. Screw dislocation circumscribed by a "plane" of atoms that spirals clockwise distally.

Right-lateral strike-slip fault. Fault where the right side moves toward an observer looking parallel to the trace of the fault. Also called dextral strike-slip fault.

Rigid indenter. Promontory on a continental margin that is transported by plate motion into collision with another continent; the promontory produces localized deformation in the collided continent.

Rigidity modulus (G). Elastic constant determined from the ratio of shear stress to shear strain. Shear modulus.

Ripple cross laminations. Small-scale trough cross beds in fine-grained sediment that form under low-velocity conditions. Can be used to determine facing direction. Also known as hummocky laminations.

Ripple marks. Ripple marks formed where sediment finer than 0.6 mm is moved by a current, or where the bottom sediment surface is otherwise disturbed by water moving above a threshold velocity. May be symmetrical or asymmetrical.

Rock cycle. Cyclic changes of energy fluxes ranging from the crystallization of magma to conversion of sedimentary or igneous rocks into metamorphic rocks. Geochemical cycle.

Rock mechanics. Application of principles of mechanics to the study of minerals and rocks.

Rodding. Lineation produced by linear alignment of coarse mineral aggregates, pebbles, and other objects to give the rock an appearance of being composed of rods.

Rods. Aggregates of one or more minerals (quartz, feldspars, micas), pebbles, or other features that have been deformed into elongate, rodlike form. Rods form at intersections of two foliation planes, or by elongation (stretching), and are common in ductile shear zones.

Rollover anticline. Anticline that forms along growth faults where the part of the downthrown block close to the fault is displaced downward more than the parts farther away.

Roof thrust. Upper of the two master faults that bound a duplex.

Room problem. Unfilled voids resulting from improper construction of a cross section that suggest too much room (or volume) in the core of a fold or beneath a fault block.

Rotated mineral. Mineral that has been rotated during deformation.

Rotation. Angular displacement of a mass without distortion.

Rupture. Point at which material loses cohesiveness (breaks).

Saddle-reef deposit. Mineral deposit formed in a void along a fold hinge where strong layers have separated. May be found in successive layers in the hinge of the same fold.

Saint-Venant behavior. Ideal plastic behavior.

Scalar. Quantity having only magnitude. (A tensor rank.)

Scale. Relationship of the dimensions of a feature in the field to the same feature in a photograph or on a map. Map scale commonly expressed as a representative fraction (e.g., 1:50,000) or a bar scale. Also, scale is used to describe the relative size of features as microscopic, mesoscopic, and macroscopic.

Scatterchron. See *Mixing line.*

Schistosity. Foliation in schist and some gneisses.

Schlieren. Concentrations of dark minerals in a granitoid pluton that represent either a partially absorbed xenolith or early crystallization products of the magma. Boundaries with the enclosing rock are gradational.

Scour marks. Marks formed in sediment as currents scour a bedding surface subjected to simple shear.

Screw dislocation. Line dislocation in which the slip sense across the crystallographic slip plane is parallel to the dislocation.

Second law of motion (Newton). Force equals mass times acceleration.

Second-order folds. Smaller folds on the flanks of the largest (first-order) folds in an area.

Second-rank tensor. Tensor consisting of nine components in three-dimensional space.

Sedimentary facies. Lateral changes in the kind of sediment as the environment of deposition changes.

Seismic reflection. Technique in which acoustic (sound) waves, generated by explosives or a mechanical vibrating truck (vibroseis), are transmitted through the Earth, reflected where velocity changes, and recorded by an array of geophones located a short distance from the source.

Seismic refraction. Technique involving measurement of the change of velocity of acoustic waves as they pass from one layer to another, recorded a large distance (tens to hundreds of kilometers) from the source of the waves.

Self-exciting dynamo. Theory that convection currents oriented by the Earth's rotation exist in the liquid iron-nickel core, transport hotter material upward, then return cooler material to the interior through a thermal gradient, thus generating an electric current and a magnetic field as a self-exciting dynamo.

Separation. Amount of apparent offset of a faulted surface measured in a specified direction. May be described as strike separation, dip separation, or stratigraphic separation, etc.

Shatter cone. Centimeter- to meter-size, striated, cone-shaped structure formed as a shock wave propagated down and away from an explosion or high-velocity impact. Common in many meteorite impact craters.

Shear-band foliation. C-surface foliation developed at 18° to 25° to the boundaries of a ductile shear zone that, with continued motion, evolves into parallelism with the shear zone walls, and new C-surfaces form.

Shear fold. See *Passive-slip folds.*

Shear fractures. Modes II and III fractures.

Shear modulus (G). An elastic constant determined from the ratio of shear stress to shear strain. Rigidity modulus.

Shear-sense indicator. An object or texture that provides information on movement direction in a rock.

Shear strain (γ). Strain that results when parts of a rock body are deformed so that angles between originally orthogonal reference lines are rotated from the original orientation. The tangent of the change from a right angle.

Shear strength. Maximum shear stress a material is able to withstand and not undergo permanent strain.

Shear stress (τ). Stress that acts parallel to a surface.
Shear thrust. Thrust fault formed by shearing parallel to and across layering independent of folding. Associated fault-bend folds form as dip of the fault flattens over a ramp.
Shear zone. Zone of closely spaced, interleaving, anastomosing brittle faults and crushed rocks near the surface, or zone of ductile faults and associated mylonitic rocks at great depth.
Sheath folds. Noncylindrical tubular folds that are closed at one end and with fold hinges tightly curved within the axial surfaces formed as a product of simple shear.
Sheeting. Jointing, more or less parallel to surface topography,that forms by unloading , generally in massive rocks. See *Unloading joints.*
σ porphyroclast. Winged asymmetric relict of a larger crystal, commonly feldspar, produced in ductile shear zones having a σ shape, but most commonly with a tail or wing on diagonally opposite ends.
Sill. Tabular pluton intruded parallel to bedding or the dominant foliation.
Similar folds. Folds that maintain the same shape throughout a section normal to the hinge, so that they do not die out upward or downward but maintain the same curvature in the hinge. Layer thickness (measured perpendicular to layering) changes uniformly at the same position in all layers.
Simple shear. Rotational constant-volume homogeneous plane strain in which some elements of a body are rotated, but others that were straight and parallel remain straight and parallel after deformation. Inhomogeneous simple shear involves rotational strain, but the elements of the body that were straight and parallel before deformation do not remain straight and parallel afterward. Also known as noncoaxial deformation.
Simple window. Hole in a thrust sheet produced by erosion that results in footwall rocks being completely surrounded by hanging-wall rocks in map view. Fenster.
Sinistral strike-slip fault. See *Left-lateral strike-slip fault.*
Sinusoidal fold. Fold with cross-section geometry that resembles a sine wave.
Slaty cleavage. Penetrative planar tectonic structure consisting of parallel grains of phyllo-silicates (clay minerals or micas) or thin anastomosing subparallel zones of insoluble residues in a fine-grained rock, generally of low metamorphic grade.
Slice. See *Horse.*
Slickenfibers. Lineation (mostly nonpenetrative) formed by fibrous minerals that have grown in the direction of movement on a fault or bedding surface. Frequently form stepped surfaces.
Slickenlines. Striations on movement surfaces. The direct result of frictional sliding and flexural slip. A nonpenetrative linear structure if planes are isolated.
Slickensides. Striated stepped surfaces along a fault or bedding surface that have undergone movement (flexural slip). Refers to entire movement surface.
Slide. See *Tectonic slide.*
Slip. (a) Movement on a fault plane. (b) Dislocation glide; manifestation of plastic deformation in minerals.
Slip lines. Lines produced by relative motion of reference points in successive layers during folding, faulting, or other deformation. They may produce slickensides, fibers, or other visible lines indicating motion or may be projected lines based on motion sense.
Slip systems. Specific planes and directions in these planes in crystals along which slip may occur.
SL-tectonite. Deformed rock in which planar elements dominate over linear elements.
Snake-head structure. Asymmetric folds visible in seismic sections or in vertical sections. Form as a thrust sheet passes over a ramp and the hanging-wall anticline has not moved very far past the ramp, or that result from fault propagation along the common limb of an anticline-syncline, and the hanging-wall anticline is not displaced very far.
Snowball (Helicitic) structure. Spiral structure in some porphyroblasts that has the appearance of a rolled snowball (or jellyroll).
Sole marks. Marks formed during deposition on the undersides of sandstone beds that are deposited on shale or siltstone; produced by currents, readjustments of the sandstone-shale bedding interface by the weight of deposited sand, dewatering of sediment following deposition, or a combination of these processes. Flute casts, drag marks, channels, load casts, and others are sole marks.
Spaced cleavage. Foliation in fine-grained rocks that can be resolved into domains of uncleaved rock separated by cleavage planes with a spacing ranging from less than a millimeter to several centimeters.
Spherical symmetry. The symmetry of a sphere—a center and an infinite number of planes and axes. A fabric diagram that has an infinite number of symmetry axes.
Splay. Smaller fault that branches from a larger fault.
S-surface. Penetrative planar tectonic structure (including curved surfaces) in rocks. Bedding is commonly included and designated S_0 despite having a nontectonic origin.
Stacking faults. Planar lattice defects consisting of irregularities in the repeat order in a series of layers in a close-packed lattice.
Static recrystallization. Recrystallization, commonly thermally driven, that occurs in a rock mass no longer undergoing deformation.
Steady-state deformation mechanism. Rate-dependent deformation mechanism; generally includes the creep mechanisms and pressure solution.
S-tectonite. Deformed rock in which foliation or cleavage is the dominant structure.
Step-over. Mechanism in faulting by which motion is transferred from one segment of a strike-slip fault to another through zones of extension or compression.
Stick-slip movement. Intermittent movement along a fault; produced by unstable frictional sliding.
Stoping. Igneous process whereby magma either forms fractures or is forced into existing fractures and removes blocks of country rock that fall into the magma.
Strain. Permanent deformation in the form of distortion, translation, and rotation.
Strain compatibility. Geometric features, such as lines or ellipses, that permit relating initial to final (finite) strain states in a heterogeneously deformed rock. If the geometric features indicate that there are no holes or other discontinuities, the strain is compatible; if they indicate that holes or discontinuities are present, the strain is incompatible.
Strain ellipsoid. Triaxial ellipsoid with three mutually perpendicular symmetry axes. Axis magnitudes are $X > Y > Z$; X is the axis of greatest principal strain, Y is the axis of intermediate principal strain, and Z is the axis of least principal strain.

Strain hardening. Increased stress resistance to strain with increasing amount of plastic deformation.
Strain marker. Any deformed object in rocks where the original shape can be quantitatively inferred from the present deformed shape.
Strain partitioning. Separation of strain into different mechanisms (or movement domains) in a rock mass or crystal.
Strain path. Path (e.g., as described on a Flinn diagram) during progressive deformation. May involve initial flattening or elongation, followed by transformation into the other field and possibly later return to the original field.
Strain softening. Decreased stress resistance to strain with increasing amount of plastic deformation.
Strength. Stress required to cause permanent deformation.
Stress (σ). Force applied per unit area (**F**/*A*).
Stress ellipsoid. Triaxial ellipsoid with three mutually perpendicular symmetry axes—the three principal stresses σ_1, σ_2, and σ_3 ($\sigma_1 > \sigma_2 > \sigma_3$).
Stretch (S). The ratio of the deformed line length to the original length of a reference line.
Strike. Compass direction (azimuth) of a horizontal line in an inclined surface.
Strike-slip fault. Fault in which movement is parallel to the strike of the fault plane.
^{87}Sr/^{86}Sr initial ratio. ^{87}Sr/^{86}Sr at the time the rock formed. An initial ratio <0.706 indicates formation in the lower crust, ocean crust, or upper mantle; initial ratio >0.706 indicates formation in the middle to upper crust.
Structural analysis. Description and interpretation of structures on all scales in an area. This involves the observation, description, analysis, and interpretation of the kinds and orientations of folds and other linear structures, foliations and other planar structures, strain and displacement indicators, and faults.
Structural-lithic unit. A rock sequence, comonly sedimentary, that shares common mechanical properties, such as strong or weak, in response to deformation.
Structural terrace. Local flattening of a uniform regional dip.
Strut. Strong unit, such as massive limestone, dolomite, or sandstone, in a sequence of sedimentary rocks; serves as the dominant unit in determining the shape of folds or the size of fault blocks.
Stylolites. Irregular surfaces coated with insoluble minerals or organic matter in limestone, sandstone, or other rock type; results from partial dissolution of the host rock by pressure solution.
Subgrains. Small parts of a grain with lattice orientation that differs by small angles from adjacent parts of the same grain.
Subgrain boundaries. Low-angle boundaries in which the crystal lattice has a misorientation of less than 10° across the boundary, causing a change in optical properties such as extinction angle that are readily detectable with a petrographic microscope.
Substitution defect. Defect in which a foreign atom or ion is present in a crystal lattice, replacing an atom or ion ideally at that site.
Superplastic flow. Gliding along grain boundaries, involving both Coble creep and grain-boundary sliding at temperatures greater than half the absolute melting temperature of the rock. Thought to also occur at low temperature in fine-grained rocks being deformed in fault zones by inhomogeneous simple shear.
Superposed folds. One set of folds that overprints another during the same or a different deformational event. Also called polyphase or interference folds.
Supratenuous folds. Folds commonly produced during deposition by differential thickness accumulation and compaction in which the troughs of synclines are thickened, and the crests of anticlines are thinned.
Surface force. Force that acts on a surface.
Suspect terrane. Mass of fault block of crustal dimensions in which the original position is questionable with respect to the adjacent terrane or stable continental land mass to which it is presently attached. Boundaries of suspect terranes are always faults.
Symmetrical folds. Folds in which the limbs have the same length and are inclined at the same angle away from the hinge surface.
Syncline. Fold in which layering is concave toward the younger rocks; as a result, younger rocks are found in the central part of the structure.
Synform. Fold in which layering is concave-up; age of the rocks is unknown.
Synformal anticline. Fold in which layering is concave-up, but the rocks in the center of the structure are older rather than younger. Upward-opening anticline.
Syntectonic recrystallization. Recrystallization that occurs while the rock mass is undergoing deformation. Dynamic recrystallization.
Synthetic normal fault. Normal fault that dips in the same direction and joins a larger (master) normal fault.
Systematic joints. Planar Mode I fractures that have an approximately parallel orientation and regular spacing.
Tabular pluton. Intrusive igneous body with roughly parallel walls that has a length in two dimensions that greatly exceeds the thickness.
Tangential-longitudinal strain. Buckle or bending (-fold) mechanism in which the inner arcs of layers are subjected to layer-parallel compression, and the outer arcs of each layer are subjected to layer-parallel tension. Zones of tension and compression are separated by a neutral surface of no strain. Same as concentric-longitudinal strain.
Tangle. Accumulation of dislocations in a crystal lattice.
Tear fault. Strike-slip fault that terminates a thrust sheet along strike. Tear faults flatten and merge with the thrust at depth and define the ends of the thrust sheet.
Tectogene concept. Concept involving a subsiding, crumpling crust that forms mountain chains on the surface and is driven by mantle convection.
Tectonic cycle. See *Wilson cycle.*
Tectonic joints. Joints that form at depths of less than 3 km in response to abnormal fluid pressure and involve hydrofracturing.
Tectonics. Study of large crustal or mantle features— mountain ranges, parts of continents, trenches and island arcs, oceanic ridges, mantle plumes, and entire continents and ocean basins—and their relationships to stresses and tectonic plates.
Tectonic slide. Thrust produced by ductile attenuation of the overturned limb between an antiform and synform. Slide. See also *Type F thrusts.*
Tectonic structures. Structures produced in rocks in response to stresses generated, for the most part, by plate motion within the Earth; include faults, folds, cleavage, and other geologic structures.
Tectonic stylolite. Stylolite produced by tectonic processes, commonly oriented at a moderate-to-high angle to bedding.
Tectonite. Rock containing and generally dominated by a penetrative tectonic fabric element. A foliated or lineated rock.

Tensional stress. Stress that tends to pull apart an object or rock mass.

Tensor. Mathematical way of representing a physical quantity by referring it to an appropriate coordinate system. Tensors have the same value in any coordinate system, but the magnitude of the components depends on the choice of coordinate system.

Terrain correction. Correction of gravity data for differences in elevation and proximity to areas of different (usually high) relief.

Thecae. Tubular indentations where individual organisms live along the jagged edge of a branch (stipe) in a graptolite colony.

Thermo-remanent magnetism (TRM). Remanent magnetism that results from orientation in magnetic domains in minerals in the Earth's magnetic field as igneous rocks cool below the Curie temperature for each mineral.

θ porphyroclast. Asymmetric relict of a larger crystal, commonly feldspar, produced in ductile shear zones and having a θ shape, with tails or wings on diagonally opposite ends. Naked inclusion.

Thick-skinned structure. Structure that incorporates basement. Refers primarily to thrust faults, but can refer to other fault types and folds.

Thin-skinned structure. Structure that does not involve basement.

Third law of motion (Newton). All forces acting to move an object in one direction are countered by equal and opposite forces acting to move the object in the opposite direction.

Throw. Vertical component of fault displacement.

Thrust fault. Fault with a low angle of dip (30° or less) in which the hanging wall has moved up relative to the footwall. Thrust faults are produced by horizontal compression.

Tight fold. Fold in which the limbs dip steeply toward or away from one another, with an interlimb angle ranging from 0 to 30°.

Tip line. Line of zero displacement along a fault that separates displaced from not displaced rocks.

Tonalite. Coarse-grained igneous rock composed of plagioclase and quartz, often with minor K–spar, biotite, or hornblende.

Torrential cross bed. Cross bed inclined to primary bedding planes. Provides the same perspective in overturned or upright position, and so cannot be used for determination of facing direction.

Tracks and trails. Fossil impressions left by organisms as they moved across soft sediment. Ichnofossils.

Transcurrent fault. Strike-slip fault that cuts across the dominant regional trend of the structures in an area.

Transected fold. Fold wherein cleavage and fold axial surfaces are not coincident, or where the fold axis does not lie in the cleavage.

Transform fault. Strike-slip fault forming a plate boundary along spreading ridges, arcs, or within continents. Kinds include ridge-ridge, ridge-arc, arc-arc, and intracontinental transforms.

Translation gliding. Process by which deformation occurs in a crystal lattice by slip along an existing crystallographic plane.

Transposition. Transformation of original bedding through progressive deformation— folding, cleavage formation, ductile shearing, or other process—into parallelism with cleavage or foliation.

Transpression. Resolution of strike-slip movement into components of significant compressional dip-slip motion that commonly results from oblique plate convergence. Opposite of transtension.

Transtension. Resolution of strike-slip movement into components of significant tensional dip-slip motion that commonly results from oblique plate divergence. Opposite of transpression.

Trend. Compass direction (azimuth) of the orientation of a linear structure or other feature.

Tresca's criterion. Plastic yield will begin when the maximum shear stress reaches a critical value.

Triangle structure. Structure formed by two thrusts that dip in the same direction as the limbs of a fold and opposite each other; commonly formed to solve the room problem. Also called a delta structure.

Triboplastic behavior. Apparent megascopic ductile behavior shown to be brittle on the microscopic scale.

Triclinic symmetry. Lack of symmetry, except for a center; in a fabric diagram is usually the result of multiple deformation, such as two superposed foliations.

Triple junction. Plate boundary where three tectonic plates meet.

Trough. Lowest point in a syncline or synformal fold.

Trough (festoon) cross beds. Layering that is truncated at the top by a primary bedding plane but curves (concave-up) to tangency with the bedding plane at the bottom of the bed. Can be used for determination of facing direction.

Tubular fold. Sheath fold with a hinge-line angle (ω) of less than 20°.

Turbidite. Deposit produced by rapid (turbulent) flow of a sediment-laden turbidity current down a slope on the sea floor or in a lake. These deposits may develop as graded, channeled, massive, and cross-bedded sequences that grade upward into one another in a Bouma sequence.

Twin gliding. Deformation of a crystal lattice by slip of a segment of the lattice along crystallographic planes, producing stress-induced twinning of the lattice.

Type I S-C mylonite. S-C mylonite in quartzo-feldspathic rocks.

Type II S-C mylonite. S-C mylonite resulting from mylonitization of quartz-mica–rich rocks.

Type 1 (Ramsay) fold interference pattern. Pattern of superposed folds that have upright axial surfaces and horizontal axes that intersect at high angles and interfere to produce an egg-carton or dome-and-basin pattern.

Type 2 (Ramsay) fold interference pattern. Pattern of superposed folds having inclined or reclined axial surfaces that are superposed by folds with steeply dipping axial surfaces and axes at high angles to those of the first set; once eroded, result in a boomerang outcrop pattern.

Type 3 (Ramsay) fold interference pattern. Superposed fold pattern in which early-formed tight to isoclinal folds are folded about the same axes but with different axial surfaces, producing a hook pattern, once eroded.

Type C (composite) thrust sheets. Crystalline thrusts that are detached along the ductile-brittle transition and cut through all earlier structures and rock units to form the largest of thrust sheets.

Type F (fold-related) thrusts. Thrusts that form on all scales as a product of plastic folding by cutting out of the common limb between the anticline and syncline. As they cool and move into higher levels of the crust, they may evolve into type C sheets.

Ultimate strength. Highest point on a stress-strain curve.

Unconformity. Break in the stratigraphic sequence where some portion of geologic history is missing. Produced by nondeposition, erosion, or both, resulting in a loss of strata for part of the geologic record.

Undeformed state. No particles in the body have been displaced since their deposition or magmatic emplacement.

Undulatory extinction. Variable extinction in a thin section in cross-polarized light related to slightly misaligned parts of crystals, produced by unrecovered strain. Most commonly observed in quartz.

Unfilled joint. Joint that has not been filled with minerals.

Unit cell. Building block in a crystal lattice.

Unloading joints. Joints that occur near the surface as erosion removes overburden and thermal-elastic contraction occurs.

Upright folds. Folds with vertical axial surfaces.

Upward-facing anticline. See *Synformal anticline.*

Vacancy. Unoccupied site in a crystal lattice.

Vector. Quantity that has both magnitude and direction.

Velocity discontinuity. Boundary across which occurs a difference in velocity.

Vergence. Direction of leaning or overturning (the opposite of dip direction) of the axial surface, or sense of shear of a fold. Also used to indicate transport direction of thrust faults.

Vesicles. Cavities in volcanic rocks left by gas bubbles that form in magma as pressure decreases.

Viscoelastic behavior. Combination of viscous and elastic behavior.

Viscosity (η). Material constant involving the relative resistance of a liquid to flow, determined as a ratio of stress to strain rate.

Viscous. Pertaining to fluid behavior. Viscous deformation is permanent and involves dependence of stress on strain rate.

Viscous behavior. Fluid behavior. Stress is proportional to strain rate.

Visual harmonic analysis. Visual comparison of the profile shape of a fold with 30 ideal fold forms based on combinations of 6 possible shapes and 5 possible amplitudes.

Volume-diffusion creep. Deformation mechanism, occurring at high temperature and low stress, whereby point defects diffuse through crystals. Also called Nabarro-Herring creep.

Von Mises criterion. Model for plastic deformation that assumes the strain energy of distortion is a constant determined from the sums of the squares of the principal stress differences.

Wavelength. Distance from crest to crest of adjacent anticlines, or trough to trough of adjacent synclines.

Wedge. See *Contractional fault.*

Whole-rock age. Radiometric age determined by measuring the Rb and Sr (or K and Ar, or Pb and U) isotopes in an entire rock sample rather than by analyzing individual minerals (which yields a mineral age).

Wildflysch. Heterogeneous deep-water accumulation of blocks and finer sediment resulting from tectonic or nontectonic processes.

Wilson cycle. Tectonic cycle of opening and closing of ocean basins.

Window. Erosional hole in a thrust sheet. Fenster. See also *Eyelid window; Simple window.*

Wrench fault. See *Strike-slip fault.*

Xenolith. Mass of wall rock or other material incorporated into an igneous body.

Yield point. Point where Hooke's law no longer holds (nonelastic behavior begins). Marked by a change in slope on a stress-strain curve.

Younging direction. Facing direction of the original top of a sequence.

Young's modulus (E). Elastic constant determined from the ratio of normal stress to elongation strain.

Zero-rank tensor. Scalar, with only one component.

Zonal crenulations. Crenulations occurring as wide diffuse zones, wherein the original rock fabric continues uninterrupted outside the crenulated zone.

References Cited

Abbate, E., Bortolotti, V., and Passerini, P., 1970, Olistostromes and olistoliths: Sedimentary Geology, v. 4, p. 521–557.

Acker, L. L., and Hatcher, R. D., Jr., 1970, Relationships between structure and topography in northwest South Carolina: South Carolina Geologic Notes, v. 14, p. 35–48.

Adams, F. D., 1954, Birth and development of the geological sciences: New York, Dover Publications, 506 p.

Aggarwal, Y. P., and Sykes, L. R., 1978, Earthquakes, faults and nuclear power plants in southern New York and northern New Jersey: Science, v. 200, p. 425–429.

Aki, K., 1967, Scaling law of seismic spectra: Journal of Geophysical Research, v. 72, p. 1217–1231.

Alavi, M., 1994, Tectonics of the Zagros orogenic belt of Iran: New data and interpretations: Tectonophysics, v. 229, p. 211–238.

Allard, G. O.,1976, Dore Lake complex and its importance to Chibougamau geology and metallogeny: Quebec Ministry of Natural Resources DP-368, 446 p.

Allmendinger, R. W., Sharp, J. W., Von Tish, D., Serpa, L., Brown, L., Kaufman, S., and Oliver, J., 1983, Cenozoic and Mesozoic structure of the eastern Basin and Range Province, Utah, from COCORP seismic-reflection data: Geology, v. 11, p. 532–536.

Alsop, L. E., and Talwani, M., 1984, The East Coast magnetic anomaly: Science, v. 226, p. 1189–1191.

Alvarez, W., Engelder, T., and Lowrie, W., 1976, Formation of spaced cleavage and folds in brittle limestone by dissolution: Geology, v. 4, p. 698–701.

Alvarez, W., Engelder, T., and Geiser, P. A., 1978, Classification of solution cleavage in pelagic limestones: Geology, v. 6, p. 263–266.

Anderson, E. M., 1942, 1951, The dynamics of faulting and dyke formation, with applications to Britain: Edinburgh, Oliver and Boyd, 191 p.

Ando, C. J., Cook, F. A., Oliver, J. E., Brown, L. D., and Kaufman, Sidney, 1983, Crustal geometry of the Appalachian orogen from seismic reflection studies, *in* Hatcher, R. D., Jr., Williams, Harold, and Zietz, Isidore, eds., Contributions to the tectonics and geophysics of mountain chains: Geological Society of America Memoir 158, p. 83–102.

Argand, E., 1925 (Carozzi, A. V., trans., 1977), Tectonics of Asia: New York, Hafner Press, 218 p.

Armstrong, R. L., and Dick, H. J. B., 1974, A model for the development of thin overthrust sheets of crystalline rock: Geology, v. 1, p. 35–40.

Atwater, T., 1970, Implications of plate tectonics for the Cenozoic tectonic evolution of western North America: Geological Society of America Bulletin, v. 81, p. 3513–3536.

Austrheim, H., and Boundy, T. M., 1994, Pseudotachylytes generated during seismic faulting and eclogitation of the deep crust: Science, v. 265, p. 82–83.

Aydin, A., and Nur, A., 1982, Evolution of pull-apart basins and their scale independence: Tectonics, v. 1, p. 91–105.

Aydin, A., and Page, B. M., 1984, Diverse Pliocene Quaternary tectonics in a transform environment, San Francisco Bay region, California: Geological Society of America Bulletin, v. 95, p. 1303–1317.

Bailey, E. B., 1910, Recumbent folds in the schists of the Scottish Highlands: Quarterly Journal of the Geological Society of London, v. 66, p. 586–618.

Bailey, E. B., 1934, West Highland tectonics: Loch Leven to Glen Roy: Quarterly Journal of the Geological Society of London, v. 90, p. 462–522.

Bailey, E. B., 1935, Tectonic essays, mainly Alpine: Oxford, England, Clarendon Press, 191 p.

Bak, J., Sorensen, K., Grocott, J., Korstgard, N. D., and Waterson, J., 1975, Tectonic implications of Precambrian shear belts in western Greenland: Nature, v. 254, p. 566–569.

Balk, R., 1937, Structural behavior of igneous rocks: Geological Society of America Memoir 5, 177 p.

Bally, A. W., 1983, Seismic expression of structural styles: American Association of Petroleum Geologists, Studies in Geology 15, 3 volumes.

Bally, A. W., Gordy, P. L., and Stewart, G. A., 1966, Structure, seismic data, and orogenic evolution of the southern Canadian Rockies: Canadian Petroleum Geology Bulletin, v. 14, p. 337–381.

Banks, C. J., and Warburton, J., 1986, 'Passive-roof' duplex geometry in the frontal structures of the Kirthar and Sulaiman mountain belts, Pakistan: Journal of Structural Geology, v. 8, p. 229–238.

Barazangi, M., and Brown, L. D., eds., 1986a, Reflection seismology: A global perspective: American Geophysical Union Geodynamics Series, v. 13, 311 p.

Barazangi, M., and Brown, L. D., eds., 1986b, Reflection seismology: The continental crust: American Geophysical Union Geodynamics Series, v. 14, 339 p.

Bartholomew, M. J., Henika, W., and Lewis, S., 1994, Geologic and structural transect of the New River Valley: Valley and Ridge and Blue Ridge provinces, southwestern Virginia, *in* Schultz, A., and Henika, W., eds., Fieldguides to southern Appalachian structure, stratigraphy, and engineering geology: Virginia Tech Department of Geological Sciences Guidebook Number 10, p. 177–228.

Bayly, B., 1992, Mechanics in structural geology: New York, Springer-Verlag, 253 p.

Beach, A., 1980, Numerical models of hydraulic fracturing and the interpretations of syntectonic veins: Journal of Structural Geology, v. 2, p. 425–438.

Becker, G. F., 1896, Schistosity and slaty cleavage: Journal of Geology, v. 4, p. 429–448.

Becker, G. F., 1904, Experiments on schistosity and slaty cleavage: U.S. Geological Survey Bulletin 241, p. 1–34.

Bell, T. H., 1985, Deformation partitioning and porphyroblast rotation in metamorphic rocks: A radical reinterpretation: Journal of Metamorphic Geology, v. 3, p. 109–118.

Bell, T. H., and Etheridge, M. A., 1973, Microstructure of mylonites and their descriptive terminology: Lithos, v. 6, p. 337–348.

Berthe, D., Choukroune, P., and Jegouzo, P., 1979, Orthogneiss, mylonite, and noncoaxial deformation of granites: The example of the South Armorican shear zone: Journal of Structural Geology, v. 2, p. 31–42.

Beutner, E. C., 1978, Slaty cleavage and related strain in Martinsburg Slate, Delaware Water Gap, New Jersey: American Journal of Science, v. 278, p. 1–23.

Beutner, E. C., and Charles, E. G., 1985, Large volume loss during cleavage formation, Hamburg sequence, Pennsylvania: Geology, v. 13, p. 803–805.

Beutner, E. C., and Diegel, F. A., 1985, Determination of fold kinematics from syntectonic fibers in pressure shadows, Martinsburg Slate, New Jersey: American Journal of Science, v. 285, p. 16–50.

Biddle, K. T., and Christie-Blick, N., 1985, Strike-slip deformation, basin formation, and sedimentation: Society of Economic Paleontologists and Mineralogists Special Publication 37, 386 p.

Billings, M. P., 1973, Structural geology, 3rd ed.: New York, Prentice-Hall, 606 p.

Biot, M. A., 1961, Theory of folding of stratified viscoelastic media and its implications in tectonics and orogenesis: Geological Society of America Bulletin, v. 72, p. 1595–1620.

Biot, M. A., 1963, Internal buckling under initial stress in finite elasticity: Royal Society of London Proceedings, v. 273, p. 306–328.

Biot, M. A., 1965a, Theory of viscous buckling and gravity instability of multilayers with large deformation: Geological Society of America Bulletin, v. 76, p. 371–378.

Biot, M. A., 1965b, Further development of the theory of internal buckling of multilayers: Geological Society of America Bulletin, v. 76, p. 833–840.

Biot, M. A., Odé, H., and Roever, W. L., 1961, Experimental verification of the theory of folding of stratified viscoelastic media: Geological Society of America Bulletin, v. 72, p. 1621–1632.

Birkhead, P. K., 1973, Some flinty crush rock exposures in northwest South Carolina and adjoining areas of North Carolina: South Carolina Geologic Notes, v. 17, p. 19–25.

Bjornerud, M., 1989, Mathematical model for folding of layering near rigid objects in shear deformation: Journal of Structural Geology, v. 11, p. 245–254.

Bobyarchick, A. R., 1984, A late Paleozoic component of strike-slip in the Brevard zone, southern Appalachians: Geological Society of America Abstracts with Programs, v. 16, p. 126.

Boggs, S., Jr., 1987, Principles of sedimentology and stratigraphy: New York, Macmillan, 784 p.

Borg, I. Y., and Handin, J., 1966, Experimental deformation of crystalline rocks: Tectonophysics, v. 3, p. 249–368.

Borg, I. Y., and Heard, H. C., 1970, Experimental deformation of plagioclases, *in* Paulitsch, P., ed., Experimental and natural rock deformation: Berlin, Springer-Verlag, p. 375–403.

Borradaile, G. J., 1978, Transected folds: A study illustrated with examples from Canada and Scotland: Geological Society of America Bulletin, v. 89, p. 481–493.

Borradaile, G. J., 1979, Strain study of the Caledonides in the Islay region, S.W. Scotland: Implications for strain histories and deformation mechanisms in greenschist: Journal of the Geological Society of London, v. 136, p. 77–88.

Borradaile, G. J., 1981, Particulate flow of rock and the formation of cleavage: Tectonophysics, v. 72, p. 305–321.

Borradaile, G. J., and Poulsen, K. H., 1981, Tectonic deformation of pillow lava: Tectonophysics, v. 79, p. T17–T26.

Borradaile, G. J., Bayly, M. B., and Powell, C. McA., eds., 1982, Atlas of deformation and metamorphic rock fabrics: New York, Springer-Verlag, 551 p.

Bosworth, W., Lambiase, J., and Keisler, R., 1986, A new look at Gregory's Rift: The structural style of continental rifting: Eos, v. 67, p. 577–583.

Boullier, A. M., and Gueguen, Y., 1975, SP-mylonites: Origin of some mylonites by superplastic flow: Contributions to Mineralogy and Petrology, v. 50, p. 93–104.

Bouma, A. H., 1962, Sedimentology of some flysch deposits: Amsterdam, Elsevier Publishing, 169 p.

Boyer, S. E., and Elliott, D., 1982, Thrust systems: American Association of Petroleum Geologists Bulletin, v. 66, p. 1196–1230.

Brace, W. F., Paulding, B. W., and Scholz, C. H., 1966, Dilatancy in the fracturing of crystalline rocks: Journal of Geophysical Research, v. 71, p. 3939–3954.

Breiner, S., 1973, Applications manual for portable magnetometers: Sunnyvale, California, E. G. & G. Geometrics, 58 p.

Brock, W. G., and Engelder, T., 1977, Deformation associated with the movement of the Muddy Mountains overthrust in the Buffington window, southeastern Nevada: Geological Society of America Bulletin, v. 88, p. 1667–1677.

Brun, J.-P., and Choukroune, P., 1983, Normal faulting, block tilting and décollement in a stretched crust: Tectonics, v. 2, p. 345–356.

Brun, J.-P., and Pons, J., 1981, Strain patterns pf pluton emplacement in a crust undergoing non-coaxial deformation, Sierra Morena, southern Spain: Journal of Structural Geology, v. 3, p. 219–229.

Bryant, B., and Reed, J. C., Jr., 1970, Geology of the Grandfather Mountain window and vicinity, North Carolina and Tennessee: U.S. Geological Survey Professional Paper 615, 190 p.

Bucher, W. H., 1936, Cryptovolcanic structures in the United States: 16th International Geological Congress, v. 2, p. 1055–1084.

Burchfiel, B. C., and Livingston, J. L., 1967, Brevard zone compared to Alpine root zones: American Journal of Science, v. 265, p. 241–256.

Burchfiel, B. C., and Stewart, J. H., 1966, Pull-apart origin of the central segment of Death Valley, California: Geological Society of America Bulletin, v. 77, p. 439–442.

Burchfiel, B. C., Fleck, R. J., Secor, D. T., Vincelette, R. R., and Davis, G. A., 1974, Geology of the Spring Mountains, Nevada: Geological Society of America Bulletin, v. 85, p. 1013–1022.

Burke, K., and Dewey, J. F., 1973, Plume generated triple junctions: Key indicators in applying plate tectonics to old rocks: Journal of Geology, v. 81, p. 406–433.

Busk, H. G., 1929, Earth flexures: Cambridge, England, Cambridge University Press, 106 p.

Butler, R. W. H., 1989, The influence of pre-existing basin structure on thrust system evolution in the western Alps, *in* Cooper, M. A., and Williams, G. D., eds., Inversion tectonics: Oxford, England, Blackwell Scientific Publications, Geological Society of London Special Publication 44, p. 105–122.

Butler, R. W. H., and Coward, M. P., 1984, Geological constraints, structural evolution, and deep geology of the northwest Scottish Caledonides: Tectonics, v. 3, p. 347–365.

Buxtorf, A., 1916, Prognosen and Befunde beim Hauensteinbasis- und Grenchenburg-tunnel und die Bedeutung der letzteren für die Geologie des Juragebirges: Naturforschende Gesellschaft Basel Verhandlungen, v. 27, p. 184–254.

Byerlee, J. D., 1977, Friction of rocks, *in* Evernden, J. F., ed., Experimental studies of rock friction with application to earthquake prediction: Menlo Park, California, U.S. Geological Survey, p. 55–77.

Byerlee, J. D., 1978, Friction of rocks: Pure and Applied Geophysics, v. 116, p. 615–626.

Cadell, H. M., 1890, Experimental researches in mountain-building: Transactions of the Royal Society of Edinburgh, v. 35, p. 337–357.

Campbell, R. B., 1970, Structural and metamorphic transitions from infrastructure to suprastructure, Cariboo Mountains, British Columbia: Geological Association of Canada Special Paper 6, p. 67–72.

Carey, S. W., 1953, The rheid concept in geotectonics: Journal of the Geological Society of Australia, v. 1, p. 67–117.

Carey, S. W., 1962, Folding: Canadian Petroleum Geology Bulletin, v. 10, p. 95–144.

Carreras, J., and White, S. H., 1980, Preface to "Shear zones in rocks": Journal of Structural Geology, v. 2, p. 1.

Carter, N. L., 1971, Static deformation of silica and silicates: Journal of Geophysical Research, v. 76, p. 5514–5540.

Carter, N. L., and Avé Lallemant, H. G., 1970, High temperature flow of dunite and peridotite: Geological Society of America Bulletin, v. 81, p. 2181–2202.

Chapple, W. M., 1968, A mathematical theory of finite amplitude rock-folding: Geological Society of America Bulletin, v. 79, p. 47–68.

Chapple, W. M., 1969, Fold shape and rheology: The folding of an isolated viscous-plastic layer: Tectonophysics, v. 7, p. 97–116.

Chapple, W. M., 1970, The finite-amplitude instability in the folding of layered rocks: Canadian Journal of Earth Sciences, v. 7, p. 457–466.

Chapple, W. M., 1978, Mechanics of thin-skinned fold-and-thrust belts: Geological Society of America Bulletin, v. 89, p. 1189–1198.

Christie, J. M., 1963, The Moine thrust zone in the Assynt region, Northwest Scotland: University of California Publications in the Geological Sciences, v. 40, p. 335–440.

Christie, J. M., 1964, Moine thrust: A reply: Journal of Geology, v. 72, p. 677–681.

Christie, J. M., Griggs, D. T., and Carter, N. L., 1964, Experimental evidence for basal slip in quartz: Journal of Geology, v. 72, p. 734–756.

Cloos, E., 1957, Lineation: A critical review and annotated bibliography: Geological Society of America Memoir 18, 136 p.

Cloos, E., 1964, Wedging, bedding-plane slips, and gravity tectonics in the Appalachians, in Lowry, W. D., ed., Tectonics of the southern Appalachians: Virginia Polytechnic Institute and State University Department of Geological Sciences, Memoir 1, p. 63–70.

Cloos, E., 1947, Oölite deformation in the South Mountain Fold, Maryland: Geological Society of America Bulletin, v. 58, p. 843–918.

Cloos, E., 1968, Experimental analysis of Gulf Coast fracture patterns: American Association of Petroleum Geologists Bulletin, v. 52, p. 420–444.

Cloos, E., 1971, Microtectonics along the western edge of the Blue Ridge, Maryland and Virginia: Baltimore, The Johns Hopkins Press, 234 p.

Cloud, P. E., 1970, Adventures in Earth history: San Francisco, W. H. Freeman and Company, 992 p.

Cobbold, P. R., 1975, Fold propagation in single embedded layers: Tectonophysics, v. 27, p. 333–351.

Cobbold, P. R., and Quinquis, H., 1980, Development of sheath folds in shear regimes: Journal of Structural Geology, v. 2, p. 119–126.

Cobbold, P. R., Cosgrove, J. W., and Summers, J. M., 1971, Development of internal structures in deformed anisotropic rocks: Tectonophysics: v. 12, p. 23–53.

Collinson, J. D., and Thompson, D. B., 1989, Sedimentary structures, 2nd ed.: London, Chapman & Hall, 224 p.

Coney, P. J., 1980, Cordilleran metamorphic core complexes: An overview, *in* Crittenden, M. D., Jr., Coney, P. J., and Davis, G. H., eds., Cordilleran metamorphic core complexes: Geological Society of America Memoir 153, p. 7–34.

Coney, P. J., Jones, D. L., and Monger, J. W. H., 1980, Cordilleran suspect terranes: Nature, v. 288, p. 329–333.

Conley, J. F., and Drummond, K. M., 1965, Ultramylonite zones in the western Carolinas: Southeastern Geology, v. 6, p. 201–211.

Cook, F. A., Albaugh, D. S., Brown, L. D., Kaufman, S., Oliver, J. E., and Hatcher, R. D., Jr., 1979, Thin-skinned tectonics in the crystalline southern Appalachians; COCORP seismic-reflection profiling of the Blue Ridge and Piedmont: Geology, v. 7, p. 563–567.

Cook, F. A., Brown, L. D., Kaufman, S., and Oliver, J. E., 1983, The COCORP seismic reflection traverse across the southern Appalachians: American Association of Petroleum Geologists Studies in Geology 14, 61 p.

Cooper, B. N., 1964, Relations of stratigraphy to structure in the southern Appalachians, *in* Lowry, W. D., ed., Tectonics of the southern Appalachians: Department of Geological Sciences, Virginia Tech, Memoir 1, p. 81–114.

Cooper, M. A., and Nutall, D. J. H., 1981, GODPP: Programs for presentation and analysis of structural data: Computers and Geosciences, v. 7, p. 267–285.

Cooper, M. A., and Williams, G. D., eds., 1989, Inversion tectonics: Oxford, England, Blackwell Scientific Publications, Geological Society of London Special Publication 44, p. 105–122.

Çoruh, C., Costain, J. K., Hatcher, R. D., Jr., Pratt, T. L., Williams, R. T., and Phinney, R. A., 1987, Results from regional Vibroseis profiling: Appalachian ultradeep core hole site study: Geophysical Journal of the Royal Astronomical Society, v. 89, p. 473–474.

Cosgrove, J. W., 1993, The interplay between fluids, folds and thrusts during the deformation of a sedimentary succession: Journal of Structural Geology, v. 15, p. 491–500.

Costello, J. O., and Hatcher, R. D., Jr., 1985, Southern Appalachian foreland-hinterland transition and the boundary conditions of duplex nucleation: Geological Society of America Abstracts with Programs, v. 17, p. 554.

Coward, M. P., and Kim, J. H., 1981, Strain within thrust sheets, *in* McClay, K. R., and Price, N. J., eds., Thrust and nappe tectonics: Geological Society of London Special Publication 9, p. 275–292.

Coward, M. P., Dewey, J. F., and Hancock, P. L., eds., 1987, Continental extensional tectonics: Oxford, England, Blackwell Scientific Publications, Geological Society of London Special Publication 28, 637 p.

Crawford, M. L., and Hollister, L. S., 1982, Contrast of metamorphic and structural histories across the Work Channel Lineament, Coast Plutonic Complex, British Columbia: Journal of Geophysical Research, v. 87, p. 3849–3860.

Crowell, J. C., 1974, Origin of late Cenozoic basins in southern California, in Dickinson, W. R., ed., Tectonics and sedimentation: Society of Economic Paleontologists and Mineralogists Special Publication No. 22, p. 190–204.

Culshaw, N., Corrigan, D., Jamieson, R. A., Ketchum, J., Wallace, P., and Wodicka, N., 1991, Traverse of the central gneiss belt, Grenville Province, Georgian Bay: Geological Association of Canada Field Trip B3: Guidebook, 32 p.

Currie, J. B., Patnode, H. W., and Trump, R. P., 1962, Development of folds in sedimentary strata: Geological Society of America Bulletin, v. 73, p. 655–674.

Dahlen, F. A., Suppe, J., and Davis, D., 1984, Mechanics of fold-and-thrust belts and accretionary wedges: Cohesive Coulomb theory: Journal of Geophysical Research, v. 89, p. 10,087–10,101.

Dahlstrom, C. D. A., 1969, Balanced cross sections: Canadian Journal of Earth Sciences, v. 6, p. 743–757.

Darwin, C., 1846, Geological observations in South America: London, Smith-Elder.

Daubrée, A., 1879, Études synthétiques de géologie expérimentale: Paris, p. 306–314.

Davidson, C., Hollister, L. S., and Schmid, S. M., 1992, Role of melt in the formation of a deep-crustal compressive shear zone: The Maclaren Glacier metamorphic belt, South Central Alaska: Tectonics, v. 11, p. 348–359.

Davies, B., 1984, Strain analysis of wrench faults and collision tectonics of the Arabian shield: Journal of Geology, v. 82, p. 37–53.

Davis, D., Suppe, J., and Dahlen, F. A., 1983, Mechanics of fold-and-thrust belts and accretionary wedges: Journal of Geophysical Research, v. 88, p. 1153–1172.

Davis, G. A., and Burchfiel, B. C., 1973, Garlock fault: An intracontinental transform structure, southern California: Geological Society of America Bulletin, v. 84, p. 1407–1422.

Davis, G. A., and Lister, G. S., 1988, Detachment faulting in continental extension; Perspectives from the southwestern U.S. Cordillera: Geological Society of America Special Paper 218, p. 133–159.

Davis, G. H., 1975, Gravity-induced folding off a gneiss dome complex, Rincon Mountains, Arizona: Geological Society of America Bulletin, v. 86, p. 979–990.

Davis, G. H., 1980, Structural characteristics of metamorphic core complexes, southern Arizona, *in* Crittenden, M. D., Jr., Coney, P. J., and Davis, G. H., eds., Cordilleran metamorphic core complexes: Geological Society of America Memoir 153, p. 35–78.

Davis, G. H., 1984, Structural geology of rocks and regions: New York, John Wiley & Sons, 492 p.

Davis, W. M., 1926, The value of the outrageous geological hypothesis: Science, v. 63, p. 463–468.

de Boer, J., and Snider, F. G., 1979, Magnetic and chemical variations of Mesozoic diabase dikes from eastern North America: Evidence for a hotspot in the Carolinas?: Geological Society of America Bulletin, v. 90, p. 185–198.

de Charpal, O., Montadert, L., Guennoc, P., and Roberts, D. G., 1978, Rifting, crustal attenuation and subsidence in the Bay of Biscay: Nature, v. 275, p. 706–711.

de Sitter, L. U., 1949, Le style structural Nord-Pyrénéen dans les Alpes Bergamasques: Société Géologique de France Bulletin, v. 19, p. 617–621.

de Voogd, B., Serpa, L., and Brown, L., 1988, Crustal extension and magmatic processes: COCORP profiles from Death Valley and the Rio Grande rift: Geological Society of America Bulletin, v. 100, p. 1550–1567.

Dennis, A. J., and Secor, D. T., 1987, A model for the development of crenulations in shear zones with applications from the southern Appalachian Piedmont: Journal of Structural Geology, v. 9, p. 809–817.

Dennis, A. J., and Secor, D. T., 1990, On resolving shear direction in foliated rocks deformed by simple shear: Geological Society of America Bulletin, v. 102, p. 1257–1267.

DePaor, D. G., 1986, A graphical approach to quantitative structural geology: Journal of Geological Education, v. 34, p. 231–236.

DePaor, D. G., 1986, Orthographic analysis of geological structures—II. Practical applications: Journal of Structural Geology, v. 8, p. 87–100.

DePaor, D. G., and Anastasio, D. J., 1987, The Spanish external Sierra: A case history in the advance and retreat of mountains: National Geographic Research, v. 3, p. 199–209.

Dewey, J. F., and Bird, J. M., 1970, Mountain belts and the new global tectonics: Journal of Geophysical Research, v. 75, p. 2615–2647.

Dewey, J. F., Pittman, W., III, Ryan, W. B. F., and Bonnin, J., 1973, Plate tectonics and the evolution of the Alpine system: Geological Society of America Bulletin, v. 84, p. 3137–3180.

Dibblee, T. W., Jr., 1977, Strike-slip tectonics of the San Andreas fault and its role in Cenozoic basin evolvement, *in* Nilsen, T. H., ed., Late Mesozoic and Cenozoic sedimentation and tectonics in California: Bakersfield, California, San Joaquin Geological Society, p. 26–38.

Diegel, F. A., 1986, Topological constraints on imbricate thrust networks, examples from the Mountain City window, Tennessee, U.S.A.: Journal of Structural Geology, v. 8, p. 269–280.

Dieterich, J. H., 1969, Origin of cleavage in folded rocks: American Journal of Science, v. 267, p. 155–165.

Dieterich, J. H., 1970, Computer experiments on mechanics of finite amplitude folds: Canadian Journal of Earth Sciences, v. 7, p. 467–476.

Dietz, R. S., 1960, Shatter cones in cryptoexplosion structures (meteorite impact?): Journal of Geology, v. 67, p. 496–505.

Dillon, J. T., Haxel, G. B., and Tosdal, R. M., 1990, Structural evidence for northeastward movement on the Chocolate Mountains thrust, southeasternmost California: Journal of Geophysical Research, v. 95, p. 19,953–19,971.

Dixon, H. R., and Lundgren, L., 1968, The structure of eastern Connecticut, *in* Zen, E-an, White, W. S., Hadley, J. B., and Thompson, J. B., Jr., Studies of Appalachian geology: Northern and maritime: New York, Wiley-Interscience, p. 261–270.

Dixon, J. M., and Summers, J. M., 1983, Patterns of total and incremental strain in subsiding troughs: Experimental centrifuged models of inter-diapir synclines: Canadian Journal of Earth Sciences, v. 272, p. 20, p. 1843–1861.

Dobrin, M. B., 1976, Introduction to geophysical prospecting: New York, McGraw-Hill, 630 p.

Donath, F. A., 1970, Some information squeezed out of rock: American Scientist, v. 58, p. 54–72.

Donath, F. A., and Parker, R. B., 1964, Folds and folding: Geological Society of America Bulletin, v. 75, p. 45–62.

Dott, R. H., and Batten, R. L., 1981, Evolution of the Earth: New York, McGraw-Hill Book Company, 113 p.

Dubey, A. K., and Cobbold, P. R., 1977, Noncylindrical flexural slip folds in nature and experiment: Tectonophysics, v. 38, p. 223–239.

Duval, B., Cramez, C., and Jackson, M. P. A., 1992, Raft tectonics in the Kwanza Basin, Angola: Marine and Petroleum Geology, v. 9, p. 389–404.

Eaton, G. P., 1979, Regional geophysics, Cenozoic tectonics, and geologic resources of the Basin and Range Province and adjoining regions, *in* Newman, G. W., and Goode, H. D., eds., Basin and Range symposium: Denver, Rocky Mountain Association of Geologists, p. 11–40.

Edelman, S. H., Liu, A., and Hatcher, R. D., Jr., 1987, The Brevard zone in South Carolina and adjacent areas: An Alleghanian orogen-scale dextral shear zone reactivated as a thrust fault: Journal of Geology, v. 95, p. 793–806.

Elliott, D., 1976a, The motion of thrust sheets: Journal of Geophysical Research, v. 81, p. 949–963.

Elliott, D., 1976b, The energy balance and deformation mechanisms of thrust sheets: Philosophical Transactions of the Royal Society of London, v. 283, p. 289–312.

Elliott, D., 1983, The construction of balanced cross sections: Journal of Structural Geology, v. 5, p. 101.

Elliott, D., and Johnson, M. R. W., 1980, Structural evolution in the northern part of the Moine thrust belt, N.W. Scotland: Transactions of the Royal Society of Edinburgh, v. 71, p. 69–96.

Engelder, T., 1979a, Mechanisms for strain within the Upper Devonian clastic sequence of the Appalachian Plateau, western New York: American Journal of Science, v. 279, p. 527–542.

Engelder, T., 1979b, The nature of deformation within the outer limits of the central Appalachian foreland fold and thrust belt in New York State: Tectonophysics, v. 55, p. 289–310.

Engelder, T., 1982, Is there a genetic relationship between selected regional joints and contemporary stress within the lithosphere of North America?: Tectonics, v. 1, p. 161–178.

Engelder, T., 1985, Loading paths to joint propagation during a tectonic cycle: An example from the Appalachian Plateau, USA: Journal of Structural Geology, v. 7, p. 459–476.

Engelder, T., 1987, Joints and shear fractures in rock, *in* Atkinson, B. K., ed., Fracture mechanics of rock: London, Academic Press, p. 27–69.

Engelder, T., and Engelder, R., 1977, Fossil distortion and decollement tectonics on the Appalachian Plateau: Geology, v. 5, p. 457–460.

Engelder, T., and Geiser, P., 1979, The relationship between pencil cleavage and lateral shortening within the Devonian section of the Appalachian Plateau, New York: Geology, v. 7, p. 460–464.

Erslev, E. A., 1988, Normalized center-to-center strain analysis of packed aggregates: Journal of Structural Geology, v. 10, p. 201–209.

Etheridge, M. A., Hobbs, B. E., and Paterson, M. S., 1973, Experimental deformation of single crystals of biotite: Contributions to Mineralogy and Petrology, v. 38, p. 21–36.

Evans, D. M., 1966, Man-made earthquakes in Denver: Geotimes, v. 10, no. 9, p. 11–18.

Evans, K., Oertel, G., and Engelder, T., 1989, Appalachian stress study 2: Analysis of Devonian shale core: Some implications for the nature of contemporary stress variations and Alleghanian deformation in Devonian rocks: Journal of Geophysical Research, v. 94, p. 1755–1770.

Faill, R. T., 1973, Kink-band folding, Valley and Ridge Province, Pennsylvania: Geological Society of America Bulletin, v. 84, p. 1289–1314.

Faure, G., 1986, Principles of isotope geology, 2nd ed.: New York, John Wiley & Sons, 464 p.

Fischer, M. W., and Coward, M. P., 1982, Strains and folds within thrust sheets: An analysis of the Heilan sheet, Northwest Scotland: Tectonophysics, v. 88, p. 291–312.

Fisher, O., 1884, On cleavage and distortion: Geological Magazine, v. 1, p. 268–276 and 396–406.

Fletcher, R. C., 1974, Wavelength selection in the folding of a single layer with power law rheology: American Journal of Science, v. 274, p. 1029–1043.

Fleuty, M. J., 1964, The description of folds: Proceedings of the Geologists' Association: v. 75, p. 461–492.

Flinn, D., 1962, On folding during three-dimensional progressive deformation: Geological Society of London Quarterly Journal, v. 118, p. 385–433.

Fowler, T. K., Jr., and Paterson, S. R., 1993, Reexamining the significance of concordance vs. discordance for pluton emplacement models: Geological Society of America Abstracts with Programs, v. 25, p. 38.

Freund, R., 1965, A model of the structural development of Israel and adjacent areas since Upper Cretaceous times: Geological Magazine, v. 102, p. 188–204.

Freund, R., Zak, I., and Garfunkel, Z., 1968, Age and rate of the sinistral movement along the Dead Sea Rift: Nature, v. 220, p. 253–255.

Fry, N., 1979, Density distribution techniques and strained length methods for determination of finite strains: Journal of Structural Geology, v. 1, p. 221–229.

Fuller, M., Williams, I., and Hoffman, K. A., 1979, Paleomagnetic records of geomagnetic field records and the morphology of the transitional fields: Reviews of Geophysics and Space Physics, v. 17, p. 179–203.

Gans, P. B., 1987, An open-system, two-layer crustal stretching model for the eastern Great Basin: Tectonics, v. 6, p. 1–12.

Garihan, J. M., Preddy, M. S., and Ranson, W. A., 1993, *in* Hatcher, R. D., Jr., and Davis, T. L., eds., Studies of Inner Piedmont geology with a focus on the Columbus Promontory: Carolina Geological Society Annual Field Trip November 6–7, 1993, Field Guide, p. 55–65.

Garihan, J. M., Ransom, W. A., Preddy, M., and Hallman, T. D., 1988, Brittle faults, lineaments and cataclastic rocks in the Slater, Zirconia, and part of the Saluda 71/2 minute quadrangles, northern Greenville County, SC, and adjacent Henderson and Polk Counties, NC, *in* Secor, D. T., Jr., ed., Southeastern geological excursions: Columbia, South Carolina, Southeastern Section, Geological Society of America, p. 266–312.

Gee, D. G., 1978, Nappe displacement in the Scandinavian Caledonides: Tectonophysics, v. 47, p. 393–419.

Geiser, P., and Engelder, T., 1983, The distribution of layer parallel shortening fabrics in the Appalachian foreland of New York and Pennsylvania: Evidence for two noncoaxial phases of the Alleghanian orogeny *in* Hatcher, R. D., Jr., Williams, H., and Zietz, I., Contributions to the tectonics and geophysics of mountain chains: Geological Society of America Memoir 158, p. 161–175.

Gibbs, A. D., 1983, Balanced cross-sections from seismic sections in areas of extensional tectonics: Journal of Structural Geology, v. 5, p. 153–160.

Gilluly, J., 1960, A folded thrust in Nevada—Inferences as to time relations between folding and faulting: American Journal of Science, v. 258-A, p. 68–79.

Glen, W., 1982, The road to Jaramillo: Critical years of the revolution in Earth science: Stanford, California, Stanford University Press, 459 p.

Goguel, J., 1965, Traite de Tectonique, 2nd ed.: Paris, Masson, 457 p.

Goldstein, A., 1982, Geometry and kinematics of ductile faulting in a portion of the Lake Char mylonite zone, Massachusetts and Connecticut: American Journal of Science, v. 282, p. 378–405.

Goscombe, B., 1991, Intense non-coaxial shear and the development of mega-scale sheath folds in the Arunta Block, Central Australia: Journal of Structural Geology, v. 13, p. 299–318.

Gray, D. R., 1977a, Morphological classification of crenulation cleavage: Journal of Geology, v. 85, p. 229–235.

Gray, D. R., 1977b, Differentiation associated with discrete crenulation cleavages: Lithos, v. 10, p. 89–101.

Gray, D. R., 1980, Geometry of crenulation-folds and their relationship to crenulation cleavage: Journal of Structural Geology, v. 2, p. 187–204.

Gray, D. R., and Durney, D. W., 1979, Crenulation cleavage differentiation: Implications of solution deposition processes: Journal of Structural Geology, v. 1, p. 75–80.

Greenly, E., 1919, The geology of Anglesey: Great Britain Geological Survey Memoir, 980 p.

Griffith, A. A., 1920, The phenomena of rupture and flow in solids: Philosophical Transactions Royal Society, series A, v. 221, p. 163–198.

Grocott, J., and Waterson, J., 1980, Strain profile of a boundary within a large shear zone: Journal of Structural Geology, v. 2, p. 111–117.

Groshong, R. H., Jr., 1975a, Strain, fractures, and pressure solution in natural single-layer folds: Geological Society of America Bulletin, v. 86, p. 1363–1376.

Groshong, R. H., Jr., 1975b, "Slip" cleavage caused by pressure solution in a buckle fold: Geology, v. 3, p. 411–413.

Groshong, R. H., Jr., 1988, Low-temperature deformation mechanisms and their interpretation: Geological Society of America Bulletin, v. 100, p. 1329–1360.

Groshong, R. H., Jr., 1993, The importance of layer-parrellel strain in restoration of the hangingwall of the Corsair Fault, Offshore Texas: American Association of Petroleum Geologists New Orleans Annual Meeting Abstracts, p. 111.

Groshong, R. H., Jr., Teufel, W., and Gasteiger, C., 1984, Precision and accuracy of the calcite strain-gage technique: Geological Society of America Bulletin, v. 95, p. 357–363.

Guglielmo, G., Jr., 1993, Interference between pluton expansion and non-coaxial tectonic deformation: Three-dimensional computer model and field implications: Journal of Structural Geology, v. 15, p. 593–608.

Gwinn, V. E., 1970, Kinematic patterns and estimates of lateral shortening, Valley and Ridge and Great Valley provinces, central Appalachians, south-central Pennsylvania, *in* Fisher, G. W., Pettijohn, F. J., Reed, J. C., Jr., and Weaver, K. N., eds., Studies of Appalachian geology: Central and southern: New York, Wiley Interscience, p. 127–146.

Hafner, W., 1951, Stress distributions and faulting: Geological Society of America Bulletin, v. 62, p. 373–398.

Hamilton, W. B., 1979, Tectonics of the Indonesian region: U.S. Geological Survey Professional Paper 1078, 345 p.

Hamilton, W. B., 1982, Structural evolution of the Big Maria Mountains, northeastern Riverside County, southeastern California, *in* Frost, E. G., and Martin, D. L., eds., Mesozoic-Cenozoic tectonic evolution of the Colorado River region, California, Arizona, and Nevada: San Diego, Cordilleran Publishers, p. 1–21.

Hancock, P. L., 1985, Brittle microtectonics: Principles and practice: Journal of Structural Geology, v. 7, p. 437–457.

Hancock, P. L., and Engelder, T., 1989, Neotectonic joints: Geological Society of America Bulletin, v. 101, p. 1197–1208.

Handin, J., and Hager, R. V., 1957, Experimental deformation of sedimentary rocks under confining pressure: Tests at room temperature on dry samples: American Association of Petroleum Geologists Bulletin, v. 41, p. 1–50.

Handin, J., Hager, R. V., Friedman, M., and Feather, J. N., 1963, Experimental deformation of sedimentary rocks under confining pressure: Pore pressure tests: American Association of Petroleum Geologists Bulletin, v. 47, p. 717–755.

Hanmer, S. K., 1979, The role of discrete heterogeneities and linear fabrics in the formation of crenulations: Journal of Structural Geology, v. 1, p. 81–91.

Hanmer, S. K., and Passchier, C., 1991, Shear-sense indicators: A review: Geological Survey of Canada Paper 90–17, 72 p.

Hansen, E., 1971, Strain facies: New York, Springer-Verlag, 207 p.

Harding, T. P., 1974, Petroleum traps associated with wrench faults: American Association of Petroleum Geologists Bulletin, v. 58, p. 1290–1304.

Hardman, R. P. F., and Booth, J. E., 1991, The significance of normal faults in the exploration and production of North Sea hydrocarbons, *in* Roberts, A. M., Yielding, G., and Freeman, B., eds., The geometry of normal faults: London, Geological Society of London Special Publication 56, p. 1–13.

Harker, A., 1885, The cause of slaty cleavage: Geological Magazine, v. 2, p. 15–17.

Harker, A., 1886, On slaty cleavage and allied rock structures: Report to the British Association for the Advancement of Science 1885, p. 813–852.

Harland, W. B., 1971, Tectonic transpression in Caledonian Spitsbergen: Geological Magazine, v. 108, p. 27–42.

Harms, J. C., Southland, J. B., and Walker, R. G., 1982, Structures and sequences in clastic rocks: Society of Economic Paleontologists and Mineralogists Short Course 9, 249 p.

Hatcher, R. D., Jr., 1971, Stratigraphic, petrologic, and structural evidence favoring a thrust solution to the Brevard problem: American Journal of Science, v. 270, p. 177–202.

Hatcher, R. D., Jr., 1972, Developmental model for the southern Appalachians: Geological Society of America Bulletin, v. 83, p. 2735–2760.

Hatcher, R. D., Jr., 1977, Macroscopic polyphase folding illustrated by the Toxaway dome, South Carolina–North Carolina: Geological Society of America Bulletin, v. 88, p. 1678–1688.

Hatcher, R. D., Jr., 1978a, Eastern Piedmont fault system: Reply: Geology, v. 6, p. 580–582.

Hatcher, R. D., Jr., 1978b, Tectonics of the western Piedmont and Blue Ridge, southern Appalachians: Review and speculation: American Journal of Science, v. 278, p. 276–304.

Hatcher, R. D., Jr., 1981, Thrusts and nappes in the North American Appalachian orogen, in McClay, K. R., and Price, N. J., eds., Thrust and nappe tectonics: Geological Society of London Special Publication 9, p. 491–499.

Hatcher, R. D., Jr., 1991, Interactive property of large thrust sheets with footwall rocks—the Subthrust Interactive Duplex Hypothesis: A mechanism of dome formation in thrust sheets: Tectonophysics, v. 191, p. 237–242.

Hatcher, R. D., Jr., and Hooper, R. J., 1981, Controls of mylonitization processes: Relationships to orogenic thermal/metamorphic peaks: Geological Society of America Abstracts with Programs, v. 13, p. 469.

Hatcher, R. D., Jr., and Hooper, R. J., 1992, Evolution of crystalline thrust sheets in the internal parts of mountain chains, *in* McClay, K. R., ed., Thrust tectonics: London, Chapman and Hall, p. 217–233.

Hatcher, R. D., Jr., and Odom, A. L., 1980, Timing of thrusting in the southern Appalachians, USA: Model for orogeny?: Geological Society of London Quarterly Journal, v. 137, p. 321–327.

Hatcher, R. D., Jr., and Williams, R. T., 1986, Mechanical model for single thrust sheets Part 1: Crystalline thrust sheets and their relationships to the mechanical/thermal behavior of orogenic belts: Geological Society of America Bulletin, v. 97, p. 975–985.

Hatcher, R. D., Jr., and Zietz, I., 1978, Thin crystalline thrust sheets in the southern Appalachian Inner Piedmont and Blue Ridge: Interpretation based upon regional aeromagnetic data: Geological Society of America Abstracts with Programs, v. 10, p. 417.

Hatcher, R. D., Jr., and Zietz, I., 1979, Interpretation of regional aeromagnetic and gravity data from the southeastern United States, Part II—Tectonic implications for the southern Appalachians: Geological Society of America Abstracts with Programs, v. 11, p. 181–182.

Hatcher, R. D., Jr., and Zietz, I., 1980, Tectonic implications of regional aeromagnetic and gravity data from the southern Appalachians, *in* Wones, D. R., ed., The Caledonides in the USA: Virginia Polytechnic Institute Department of Geological Sciences, Memoir 2, p. 235–244.

Hatcher, R. D., Jr., Acker, L. L., Liu, A., Zupan, A. -J., and Mittwede, S., 1995, Geology and mineral resources of the Whetstone, Holly Springs, Rainy Mountain, and Tugaloo Lake quadrangles, South Carolina–Georgia: South Carolina Geological Survey, scale 1:24,000 (in press).

Hatherton, T., and Dickinson, W. R., 1969, The relationship between andesitic volcanism and seismicity in Indonesia, the Lesser Antilles, and other island arcs: Journal of Geophysical Research, v. 74, p. 5301–5310.

Haughton, S., 1856, On slaty cleavage and the distortion of fossils: Philosophical Magazine, v. 12, p. 409–421.

Hauksson, E., Jones, L. M., Hutton, K., and Eberhart-Phillips, D., 1993, The 1992 Landers earthquake sequence: Seismological observations: Journal of Geophysical Research, v. 98, p. 19,835–19,858.

Henderson, J. R., Wright, T. O., and Henderson, M. N., 1986, A history of cleavage and folding: An example from the Goldenville Formation, Nova Scotia: Geological Society of America Bulletin, v. 97, p. 1354–1366.

Hess, H. H., 1962, History of ocean basins: Geological Society of America Buddington Volume, p. 599–620.

Hills, E. S., 1963, Elements of structural geology: New York, John Wiley & Sons, 483 p.

Hirth, G., and Tullis, J., 1992, Dislocation creep regimes in quartz aggregates: Journal of Structural Geology, v. 14, p. 145–159.

Hobbs, B. E., Means, W. D., and Williams, P. E., 1976, An outline of structural geology: New York, John Wiley & Sons, 571 p.

Hodgson, R. A., 1961, Regional study of jointing in Comb Ridge Navajo mountain area, Arizona and Utah: American Association of Petroleum Geologists Bulletin, v. 45, p. 1–38.

Hoeppener, R., 1956, Zum Problem der Bruchbildung, Schieferung und Faltung: Geologische Rundschau, v. 45, p. 247–283.

Hoffman, K. A., 1988, Ancient magnetic reversals: Clues to the geodynamo: Scientific American, v. 258, no. 5, p. 76–83.

Hoffman, P., 1973, Evolution of an Early Proterozoic continental margin: The Coronation geosyncline and associated aulocogens of the northwestern Canadian shield: Philosophical Transactions of the Royal Society of London, v. 273, p. 547–581.

Hollister, L. S., and Crawford, M. L., 1986, Melt-enhanced deformation: A major tectonic process: Geology, v. 14, p. 558–561.

Hooker, V. E., and Bickel, L. D., 1974, Overcoring equipment and techniques used in rock stress determination: U.S. Bureau of Mines Information Circular 8618, 32 p.

Hooper, R. J., 1989, Tectonic implications of a regionally extensive brittle fault system in the Piedmont: Evidence from central Georgia: Geological Society of America Abstracts with Programs, v. 21, p. 22.

Hooper, R. J., and Hatcher, R. D., Jr., 1988, Mylonites from the Towaliga fault zone, central Georgia: Products of heterogeneous non-coaxial deformation: Tectonophysics, v. 152, p. 1–17.

Hossack, J. R., 1968, Pebble deformation in the Bygdin area (southern Norway): Tectonophysics, v. 5, p. 315–339.

Hossack, J. R., 1979, The use of balanced cross sections in the calculation of orogenic contraction, A review: Journal of the Geological Society of London, v. 136, p. 705–711.

Hossack, J. R., and Hancock, P. L., eds., 1983, Balanced cross sections and their geological significance: Journal of Structural Geology, v. 5, p. 98–223.

Howell, D. G., 1985a, Terranes: Scientific American, v. 253, no. 5, p. 116–125.

Howell, D. G., ed., 1985b, Tectonostratigraphic terranes of the circum-Pacific region: Houston, Texas Circum-Pacific Council for Energy and Mineral Resources, 581 p.

Howell, J. V., 1934, Historical development of the structural theory of accumulation of oil and gas, *in* Problems in petroleum geology: Tulsa, Oklahoma, American Association of Petroleum Geologists, p. 1–23.

Hsü, K. J., 1968, Principles of mélanges and their bearing on the Franciscan-Knoxville paradox: Geological Society of America Bulletin, v. 79, p. 1063–1074.

Hsü, K. J., 1969, Role of cohesive strength in the mechanics of overthrust faulting and of landsliding: Geological Society of America Bulletin, v. 80, p. 927–952.

Hubbert, M. K., 1945, Strength of the Earth: American Association of Petroleum Geologists Bulletin, v. 29, p. 1630–1653.

Hubbert, M. K., 1951, Mechanical basis for certain familiar geologic structures: Geological Society of America Bulletin, v. 62, p. 355–372

Hubbert, M. K., and Rubey, W. W., 1959, Role of fluid pressure in mechanics of overthrust faulting: Part 1. Mechanics of fluid-filled porous solids and its application to overthrust faulting: Geological Society of America Bulletin, v. 70, p. 115–166.

Hubbert, M. K., and Willis, D. G., 1957, Mechanics of hydraulic fracturing: Journal of Petroleum Technology, v. 9, p. 153–168.

Hudleston, P. J., 1973, Fold morphology and some geometrical implications of theories of fold development: Tectonophysics: v. 16, p. 1–46.

Hudleston, P. J., 1986, Extracting information from folds in rocks: Journal of Geological Education, v. 34, p. 237–245.

Hudleston, P. J., 1989, The association of folds and veins in shear zones: Journal of Structural Geology, v. 11, p. 949–957.

Hudleston, P. J., and Lan, L., 1993, Information from fold shapes: Journal of Structural Geology, v. 15, p. 253–264.

Hudleston, P. J., and Stephansson, O., 1973, Layer shortening and fold-shape development in the buckling of single layers: Tectonophysics, v. 17, p. 299–321.

Hudleston, P. J., Schultz-Ela, D., and Southwick, D. L., 1988, Transpression in an Archean greenstone belt, northern Minnesota: Canadian Journal of Earth Sciences, v. 25, p. 1060–1068.

Hull, D., 1975, Introduction to dislocations: New York, Pergamon, 271 p.

Hull, D., and Bacon, D. J., 1984, Introduction to dislocations, 3rd ed.: Oxford, England, Pergamon Press, 257 p.

Hunt, C. B., 1953, Geology and geography of the Henry Mountains region, Utah: U. S. Geological Survey Professional Paper 228, 234 p.

Hutchinson, D. R., Grow, J. A., and Klitgord, K. D., 1983, Crustal structure beneath the southern Appalachians: Nonuniqueness of gravity modeling: Geology, v. 11, p. 611–615.

Hutton, D. W., 1982, A tectonic model for the emplacement of the Main Donegal Granite, northwest Ireland: Geological Society of London, v. 139, p. 615–631.

Hutton, D. W., 1988, Igneous emplacement in a shear-zone termination: The biotite granite at Strontian, Scotland: Geological Society of America Bulletin, v. 100, p. 1392–1399.

Hutton, D. W., 1992, Granite sheeted complexes: Evidence for the dyking ascent mechanism, in Brown, P. E., and Chappell, B. W., The second Hutton symposium of the origin of granites and related rocks: Geological Society of America Special Paper 272, p. 377–382.

Hutton, D. W., and Ingram, G. M., 1992, The great Tonalite Sill of southeastern Alaska and British Columbia: emplacement into an active contractional high angle reverse shear zone (extended abstract), *in* Brown, P. E., and Chappell, B. W., The second Hutton symposium of the origin of granites and related rocks: Geological Society of America Special Paper 272, p. 383–386.

Hutton, D. W., Dempster, T. J., Brown, P. E., and Becker, S. D., 1990, A new mechanism of granite emplacement: Intrusion in active extensional shear zones: Nature, v. 343, p. 452–455.

Inglis, C. E., 1913, Stresses in a plate due to the presence of cracks and sharp corners: Transactions of the Institute of Naval Architects, v. 55, p. 219–230.

Isacks, B., Oliver, J., and Sykes, L. R., 1968, Seismology and the new global tectonics: Journal of Geophysical Research, v. 72, p. 5855–5900.

Jackson, M. P. A., and Cramez, C., 1989, Seismic recognition of salt welds in salt tectonics regimes, *in* Gulf Coast Section Society of Economic Paleontologists and Mineralogists Foundation Tenth Annual Research Conference Program and Abstract, Houston, Texas, p. 66–71.

Jackson, M. P. A., and Talbot, C. J., 1986, External shapes, strain rates, and dynamics of salt structures: Geological Society of America Bulletin, v. 97, p. 305–323.

Jackson, M. P. A., and Talbot, C. J., 1989, Anatomy of mushroom-shaped diapirs: Journal of Structural Geology, v. 11, p. 211–230.

Jackson, M. P. A., and Vendeville, B. C., 1994, Regional extension as a geologic trigger for diapirism: Geological Society of America Bulletin, v. 106, p. 57–93.

Jackson, M. P. A., Cornelius, R. R., Craig, J. H., Gansser, A., Stocklin, J., and Talbot, C. J., 1990, Salt diapirs of the Great Kavir, central Iran: Geological Society of America Memoir 177, 139 p.

Jaeger, J. C., and Cook, N. G. W., 1976, Fundamentals of rock mechanics: London, Chapman and Hall, 585 p.

Jamieson, R. A., 1986, P–T paths from high-temperature shear zones beneath ophiolites: Journal of Metamorphic Geology, v.4, p. 3–22.

Jeran, P. W., and Mashey, J. R., 1970, A computer program for stereographic analysis of coal fractures and cleats: U.S. Bureau of Mines Information Circular 8454, 34 p.

Johnson, A. M., 1970, Physical processes in geology: San Francisco, Freeman, Cooper, and Company, 577 p.

Johnson, A. M., 1977, Styles of folding: Mechanics and mechanisms of folding of natural elastic materials: Amsterdam, Elsevier, 406 p.

Johnson, A. M., and Ellen, S. D., 1974, A theory of concentric, kink, and sinusoidal folding and of monoclinal flexuring of compressible, elastic multilayers. I. Introduction: Tectonophysics, v. 21, p. 301–339.

Johnson, A. M., and Pollard, D. D., 1973, Mechanics of growth of some laccolithic intrusions in the Henry Mountains, Utah, I, Field observations, Gilbert's model, Physical properties and flow of the magma: Tectonophysics, v. 18, p. 261–309.

Johnson, M. R. W., 1960, The structural history of the Moine thrust zone at Lochcarron, Wester Ross: Royal Society of Edinburgh Transactions, v. 64, p. 139–168.

Johnson, M. R. W., 1964, The Moine thrust: Discussion: Journal of Geology, v. 72, p. 672–676.

Johnson, T. E., 1991, Nomenclature and geometric classification of cleavage-transected folds: Journal of Structural Geology, v. 13, p. 261–274.

Jonas, A. I., 1927, Geologic reconnaissance in the Piedmont of Virginia: Geological Society of America Bulletin, v. 38, p. 837–846.

Jonas, A. I., 1932, Structure of the metamorphic belt of the southern Appalachians: American Journal of Science, 5th series, v. 24, p. 228–243.

Kalsbeek, F., 1963, A hexagonal net for the counting out and testing of fabric diagrams: Neues Jahrbuch für Mineralogie, Monatshefte, v. 7, p. 173–176.

Kamb, W. B., 1959, Ice petrofabric observations from Blue Glacier, Washington, in relation to theory and experiment: Journal of Geophysical Research, v. 64, p. 1891–1909.

Karner, G. D., and Watts, A. B., 1983, Gravity anomalies and flexure of the lithosphere at mountain ranges: Journal of Geophysical Research, v. 88, p. 10,449–10,477.

Keen, C. E., Stockmal, G. S., Welsink, H., Quinlan, G., and Mudford, B., 1987, Deep crustal structure and evolution of the rifted margin northeast of Newfoundland: Results from LITHOPROBE East: Canadian Journal of Earth Sciences, v. 24, p. 1537–1549.

Keith, A., 1907, Description of the Pisgah quadrangle, North Carolina–South Carolina: U.S. Geological Survey Geologic Atlas Folio 147, 8 p.

Kennedy, W. A., 1946, The Great Glen fault: Quarterly Journal of the Geological Society of London, v. 102, p. 41–76.

Kerrich, R., 1978, An historical review and synthesis of research on pressure solution: Zentralblatt für Geologie und Palaeontologie Teil I, v. 5/6, p. 512–550.

Kerrich, R., and Allison, I., 1978, Flow mechanisms in rocks: Microscopic and mesoscopic structures, and their relation to physical conditions of deformation in the crust: Geoscience Canada, v. 5, p. 109–118.

Knipe, R. J., 1989, Deformation mechanisms—recognition from natural tectonites: Journal of Structural Geology, v. 11, p. 127–146.

Knopf, E. G., and Ingerson, E., 1938, Structural petrology: Geological Society of America Memoir 6, 270 p.

Krogh, T. E., 1973, A low-contamination method for hydrothermal decomposition and extraction of U and Pb for isotopic age determination: Geochimica et Cosmochimica ACTA, v. 37, p. 485–494.

Kulander, B. R., Barton, C. C., and Dean, S. L., 1979, The application of fractography to core and outcrop fracture investigations: Morgantown, West Virginia, DOE METC/SP-79/3, 174 p.

Lakes, R., 1987, Foam structures with a negative Poisson's ratio: Science, v. 235, p. 1038–1040.

Larson, K. W., Hatcher, R. D., Jr., Neuman, R. B., and Finney, S. C., 1989, Structure of the Guess Creek fault and Fair Garden anticline, southeastern Tennessee: Relatives of the Great Smoky fault or unrelated structures?: Geological Society of America Abstracts with Programs, v. 21, p. 46.

Latham, J.-P., 1985, A numerical investigation and geological discussion of the relationship between folding, kinking and faulting: Journal of Structural Geology, v. 7, p. 237–249.

Laubscher, H. P., 1962, Die Zweiphasenhypothese der Jurafaltung: Ecolgae Geologicae Helvetiae, v. 55, p. 1–22.

Laubscher, H. P., 1983, Detachment, shear and compresion in the central Alps, *in* Hatcher, R. D., Jr., Williams, H., and Zietz, I., Contributions to the tectonics and geophysics of mountain chains: Geological Society of America Memoir 158, p. 191–211.

Laubscher, H. P., 1988, Material balance in Alpine orogeny: Geological Society of America Bulletin, v. 100, p. 1313–1328.

Laubscher, H. P., 1990, The problem of the deep structure of the Southern Alps: 3–D material balance considerations and regional consequences: Tectonophysics, v. 176, p. 103–121.

Laubscher, H. P., 1991, The arc of the western Alps today: Ecolgae Geologicae Helvetiae, v. 84, p. 1–29.

Laubscher, H. P., 1992, Jura kinematics and the Molasse Basin: Ecolgae Geologicae Helvetiae, v. 85, p. 653–675.

Laugel, A., 1855, Du clivage de roches: Compte Rendu Academie Science, v. 40, p. 182, 185, and 978–980.

Law, R., Morgan, S. S., Sylvester, A. G., and Nyman, M., 1992, The Papoose Flat pluton of eastern California: A reassessment of its emplacement history in the light of new microstructural and crystallographic fabric observations: Transactions of the Royal Society of Edinburgh: Earth Sciences, v. 83, p. 361–375.

Law, R. D., Sylvester, A. G., Nelson, C. A., Morgan, S. S., and Nyman, M. W., 1993, Deformation associated with emplacement of the Papoose Flat pluton, Inyo Mountains, eastern California: Geologic overview and field guide, in Lahre, M. M., Trexler, J. H., Jr., and Spinosa, C., eds, Crustal evolution of the Great Basin and the Sierra Nevada: Reno, Nevada, Department of Geological Sciences and McKay School of Mines Field Trip Guidebook for 1993 Joint Meeting of the Cordilleran/Rocky Mountain Sections of the Geological Society of America, p. 231–261.

Lee, J. H., Peacor, D. R., Lewis, D. D., and Wintsch, R. P., 1986, Evidence for syntectonic crystallization for the mudstone to slate transition at Lehigh Gap, Pennsylvania, U.S.A.: Journal of Structural Geology, v. 8, p. 767–780.

LeFort, J.-P., 1984, Mise en évidence d'une virgation carbonifère induite par la dorsale Reguibat (Mauritanie) dans les Appalaches du Sud (U.S.A.), Arguments géophysiques: Bulletin Société Géologique de France, v. 26, p. 1293–1303.

Le Pichon, X., 1968, Seafloor spreading and continental drift: Journal of Geophysical Research, v. 73, p. 3661–3697.

Levorsen, A. I., 1954, Geology of petroleum: San Francisco, W. H. Freeman, 703 p.

Lewis, S. E., and Bartholomew, M. J., 1989, Orphans - exotic, detached duplexes within thrust sheets of complex history: Geological Society of America, Abstracts with Programs, v. 21, p. A136.

Lin, S., and Williams, P. F., 1992, The geometrical relationship between the stretching lineation and the movement direction of shear zones: Journal of Structural Geology, v. 14, p. 491–497.

Lister, G. S., and Snoke, A. W., 1984, S-C mylonites: Journal of Structural Geology, v. 6, p. 617–638.

Lister, G. S., and Williams, P. F., 1983, The partitioning of deformation in flowing rock masses: Tectonophysics, v. 92, p. 1–33.

Lister, G. S., Etheridge, M. A., and Symonds, P. A., 1986, Detachment faulting and the evolution of passive continental margins: Geology, v. 14, p. 246–250.

Lister, J. R., and Kerr, R. C., 1991, Fluid-mechanical models of crack propagation and their application to magma transport in dykes: Journal of Geophysical Research, v. 96, p. 10,049–10,077.

Lloyd, G. E., and Knipe, R. J., 1992, Deformation mechanisms accommodating faulting of quartzite under upper crustal conditions: Journal of Structural Geology, v. 14, p. 127–143.

Logan, Sir William, 1846, Report of progress, 1844: Ottawa, Geological Survey of Canada.

Lowell, J. D., 1972, Spitsbergen Tertiary orogenic belt and the Spitsbergen fracture zone: Geological Society of America, v. 83, p. 3091–2102.

Mackwell, S. J., Kohlsted, D. L., and Paterson, M. S., 1985, The role of water in the deformation of olivine single crystals: Journal of Geophysical Research, v. 90, p. 11,319–11,333.

Mandl, G., 1988, Mechanics of tectonic faulting: Amsterdam, Elsevier, 407 p.

Marshak, S., and Engelder, T., 1985, Development of cleavage in limestones of a fold-thrust belt in eastern New York: Journal of Structural Geology, v. 7, p. 345–359.

Matthews, D., and Smith, C., eds., 1987, Deep seismic reflection profiling of the continental lithosphere: Geophysical Journal of the Royal Astronomical Society Special Issue, v. 89, p. 1–494.

May, P. R., 1971, Pattern of Triassic–Jurassic diabase dikes around the North Atlantic in the context of predrift position of the continents: Geological Society of America Bulletin, v. 82, p. 1285–1291.

McBride, J. H., and Nelson, K. D., 1988, Integration of COCORP deep reflection and magnetic anomaly analysis in the southeastern United States: Implications for origin of the Brunswick and East Coast magnetic anomalies: Geological Society of America Bulletin, v. 100, p. 436–445.

McClay, K. R., 1992, ed., Thrust tectonics: London, Chapman and Hall, 447 p.

McClay, K. R., and Buchanan, P. G., 1992, Thrust faults in inverted extensional basins, *in* McClay, K. R., ed., Thrust tectonics: London, Chapman and Hall, p. 93–104.

McClay, K. R., and Price, N. J., eds., 1981, Thrust and nappe tectonics: Geological Society of London Special Publication 9, 539 p.

McClellan, E. A., 1993, Tectonic evolution of the Einunnfjellet-Savalen area, Hedmark Fylke, central-southern Norwegian Caledonides [unpublished Ph.D. dissertation]: Knoxville, University of Tennessee, 209 p.

McCoy, A. W., and Keyte, W. R., 1934, Present interpretations of the structural theory for oil and gas migration and accumulation, *in* Problems of petroleum geology: Tulsa, Oklahoma, American Association of Petroleum Geologists, p. 253–307.

McEachran, D. B., and Marshak, S., 1986, Teaching strain theory in structural geology using graphics programs for the Apple Macintosh computer: Journal of Geological Education, v. 34, p. 191–195.

McGeary, S., 1987, Nontypical BIRPS on the margin of the North Sea: The SHET survey: Geophysical Journal of the Royal Astronomical Society, v. 89, p. 231–238.

McIntyre, D. B., 1950, Note on two lineated tectonites from Strathavon, Banfshire: Geological Magazine, v. 87, p. 331–336.

McKenzie, D. P., 1970, Plate tectonics of the Mediterranean region: Nature, v. 226, p. 239–243.

McKenzie, D. P., and Morgan, W. J., 1969, Evolution of triple junctions: Nature, v. 224, p. 125–133.

McKenzie, D. P., and Parker, R. L., 1967, The North Pacific: An example of tectonics on a sphere: Nature, v. 216, p. 1276–1280.

McKenzie, D. P., Davies, D., and Molnar, P., 1970, Plate tectonics of the Red Sea and east Africa: Nature, v. 226, p. 243–248.

McLelland, J. M., 1984, The origin of ribbon lineation within the southern Adirondacks, U.S.A.: Journal of Structural Geology, v. 6, p. 147–157.

Mead, W. J., 1920, Notes on the mechanics of geologic structures: Journal of Geology, v. 28, p. 505–523.

Means, W. D., 1976, Stress and strain: New York, Springer-Verlag, 339 p.

Means, W. D., 1986, Three microstructural exercises for students: Journal of Geological Education, v. 34, p. 224–230.

Means, W. D., 1987, A newly recognized type of slickenside striation: Journal of Structural Geology, v. 9, p. 585–590.

Means, W. D., 1989, Synkinematic microscopy of transparent crystals: Journal of Structural Geology, v. 11, p. 163–174.

Means, W. D., 1990, Review paper. Kinematics, stress, deformation and material behavior: Journal of Structural Geology, v. 12, p. 953–971.

Means, W. D., 1992, How to do anything with Mohr circles (except fry an egg)—A short course about tensors for structural geologists: Geological Society of America Structural Geology and Tectonics Division Short Course Notes, 66 p.

Means, W. D., 1993, Elementary geometry of deformation processes: Journal of Structural Geology, v. 15, p. 343–349.

Mellis, O., 1942, Gefugediagramme in stereographischer Projecktien: Zeitschrift Mineralogie und Petrographie Mitteilungen, v. 53, p. 330–353.

Mies, J. W., 1991, Planar dispersion of folds in ductile shear zones and kinematic interpretation of fold-hinge girdles: Journal of Structural Geology, v. 13, p. 281–297.

Mies, J. W., 1993, Structural analysis of sheath folds in the Sylacauga Marble Group, Talladega slate belt, southern Appalachians: Journal of Structural Geology, v. 15, p. 983–993.

Milici, R. C., 1975, Structural patterns in the southern Appalachians: Evidence for a gravity slide mechanism for Alleghanian deformation: Geological Society of America Bulletin, v. 86, p. 1316–1320.

Miller, E. L., Gans, P. B., and Garing, J., 1983, The Snake Range décollement: An exhumed mid-Tertiary ductile-brittle transition: Tectonics, v. 2, p. 239–263.

Mitra, G., and Lukert, M. T., 1982, Geology of the Catoctin–Blue Ridge anticlinorium in northern Virginia, *in* Lyttle, P. T., ed., Central Appalachian geology, Northeast-Southeast Geological Society of America '82 Field Trip Guidebooks: Falls Church, Virginia, American Geological Institute, p. 83–108.

Mitra, S., 1986, Duplex structures and imbricate thrust systems: Geometry, structural position, and hydrocarbon potential: American Association of Petroleum Geologists Bulletin, v. 70, p. 1087–1112.

Mitra, S., 1988, Three-dimensional geometry and kinematic evolution of the Pine Mountain thrust system, southern Appalachians: Geological Society of America Bulletin, v. 100, p. 72–95.

Mitra, S., 1993, Geometry and kinematic evolution of inversion structures: American Association of Petroleum Geologists Bulletin, v. 77, p. 1159–1191.

Mitra, S., and Fisher, G. W., eds., 1992, Structural geology of fold and thrust belts: Baltimore, Maryland, The Johns Hopkins University Press, 254 p.

Molnar, P., and Tapponnier, P., 1975, Cenozoic tectonics of Asia: Effects of a continental collision: Science, v. 189, p. 419–426.

Monger, J. W. H., Price, R. A., and Templeman-Kluitt, D. J., 1982, Tectonic accretion and the origin of the two major metamorphic and plutonic welts in the Canadian Cordillera: Geology, v. 10, p. 70–75.

Moore, J. C., Roeske, S., Lundberg, N., Schoonmaker, J., Cowan, D. S., Gonzales, E., and Lucas, S. E., 1986, Scaly fabrics from Deep Sea Drilling Project cores from forearcs: Geological Society of America Memoir 166, p. 55–73.

Moores, E. M., 1982, Origin and significance of ophiolites: Reviews of Geophysics and Space Physics, v. 20, p. 735–760.

Morgan, W. J., 1968, Rises, trenches, great faults, and crustal blocks: Journal of Geophysical Research, v. 73, p. 1959–1982.

Morley, C. K., 1986, A classification of thrust fronts: American Association of Petroleum Geologists, v. 70, p. 12–25.

Mosher, S., 1983, Kinematic history of the Narragansett Basin, Massachusetts and Rhode Island: Constraints on late Paleozoic plate reconstructions: Tectonics, v. 2, p. 327–344.

Mount, V. S., and Suppe, J., 1987, State of stress near the San Andreas fault: Implications for wrench tectonics: Geology, v. 15, p. 1143–1146.

Mount, V. S., and Suppe, J., 1992, Present-day stress orientations adjacent to active strike-slip faults: California and Sumatra: Journal of Geophysical Research, v. 97, p. 11,995–12,013.

Muehlberger, W. R., 1968, Internal structures and mode of uplift of Texas and Louisiana salt domes: Geological Society of America Special Paper 88, p. 359–364.

Murphy, D. C., 1987, Suprastructure/infrastructure transition, east-central Cariboo Mountains, British Columbia: Geometry, kinematics and tectonic implications: Journal of Structural Geology, v. 9, p. 13–29.

Myers, J. S., and Watkins, K. P., 1985, Origin of granite-greenstone patterns, Yilgarn Block, Western Australia: Geology, v. 13, p. 778–780.

Nickelsen, R. P., 1979, Sequence of structural stages of the Alleghany orogeny at the Bear Valley Strip Mine, Shamokin, Pennsylvania: American Journal of Science, v. 279, p. 225–271.

Nicol, A., and Wise, D. U., 1992, Paleostress adjacent to the Alpine fault of New Zealand: Fault, vein, and stylolite data from the Doctors Dome area: Journal of Geophysical Research, v. 97, p. 17,685–17,692.

Nicol, J., 1861, On the structure of the North-Western Highlands and the relations of the gneiss, red sandstone, and quartzite of Sutherland and Ross-Shire: Geological Society of London Quarterly Journal, v. 17, p. 85–113.

Nicolas, A., 1987, Principles of rock deformation: Dordrecht, Holland, D. Reidel Publishing, 201 p.

Nicolas, A., and Poirier, J. P., 1976, Crystalline plasticity and solid state flow in metamorphic rocks: London, Wiley-Interscience, 444 p.

Norris, D. K., 1958, Structural conditions in Canadian coal mines: Geological Survey of Canada Bulletin 44, 54 p.

Nye, J. F., 1957, Physical properties of crystals: London, Oxford University Press, 322 p.

Obermeier, S. F., Gohn, G. S., Weems, R. S., and Gelinas, R. L., 1985, Geologic evidence for recurrent moderate to large earthquakes near Charleston, South Carolina: Science, v. 227, p. 408–410.

Odé, H., 1957, Mechanical analysis of the dike pattern of the Spanish Peaks area, Colorado: Geological Society of America Bulletin, v. 68, p. 567–576.

Oliver, J., 1986, Fluids expelled tectonically from orogenic belts: Their role in hydrocarbon migration and other geologic phenomena: Geology, v. 14, p. 99–102.

Oliver, J., 1988, Opinion: Geology, v. 16, p. 291.

Onasch, C. M., 1983, Dynamic analysis of rough cleavage in the Martinsburg Formation, Maryland: Journal of Structural Geology, v. 5, p. 73–81.

Onasch, C. M., 1984, Application of the R_f/ϕ technique to elliptical markers deformed by pressure-solution: Tectonophysics, v. 110, p. 157–165.

Oriel, S. S., 1950, Geology and mineral resources of the Hot Springs window, Madison County, North Carolina: North Carolina Department of Conservation and Development Bulletin 60, 70 p.

Oxburgh, E. R., 1972, Flake tectonics and continental collision: Nature, v. 239, p. 202–209.

Page, B. M., 1990, Evolution and complexities of the transform system in California, U.S.A.: Annales Tectonicae, v. IV, p. 53–69.

Panza, G. F., and Muller, St., 1979, The plate boundary between Eurasia and Africa in the Alpine area: Memorie di Scienze Geologiche, v. 33, p. 43–50.

Park, R. G., 1983, Foundations of structural geology: London, Chapman & Hall, 135 p.

Parrish, D. K., Krivz, A. L., and Carter, N. L., 1976, Finite-element folds of similar geometry: Tectonophysics, v. 32, p. 183–207.

Parsons, T., and Thompson, G. A., 1991, The role of magma overpressure in suppressing earthquakes and topography: Worldwide examples: Science, v. 253, p. 1399–1402.

Parsons, T., and Thompson, G. A., 1993, Does magmatism influence low-angle normal faulting?: Geology, v. 21, p. 247–250.

Parsons, T., and Thompson, G. A., 1993, Does magmatism influence low-angle normal faulting?: Comment and reply: Geology, v. 21, p. 956–958.

Parsons, T., Sleep, N. H., and Thompson, G. A., 1992, Host rock rheology controls on the emplacement of tabular intrusions: implications for underplating of extending crust: Tectonics, v. 11, p. 1348–1356.

Parsons, T., Thompson, G. A., and Sleep, N. H., 1994, Mantle plume influence on the Neogene uplift and extension of the U.S. western Cordillera?: Geology, v. 22, p. 83–86.

Passchier, C. W., and Simpson, C., 1986, Porphyroclast systems as kinematic indicators: Journal of Structural Geology, v. 8, p. 831–843.

Passchier, C. W., ten Brink, C. E., Bons, P. D., and Sokoutis, D., 1993, δ objects as a gauge for stress sensitivity of strain rate in mylonites: Earth and Planetary Science Letters, v. 120, p. 239–245.

Paterson, M. S., 1958, Experimental deformation and faulting in Wombeyan marble: Geological Society of America Bulletin, v. 69, p. 465–476.

Paterson, M. S., and Weiss, L. E., 1966, Experimental deformation and folding in phyllite: Geological Society of America Bulletin, v. 77, p. 343–374.

Paterson, S. R., and Fowler, K. T., 1993, Re-examining pluton emplacement processes: Journal of Structural Geology, v. 15, p. 191–206.

Paterson, S. R., and Fowler, K. T., 1993, Re-examining pluton emplacement processes: Reply: Journal of Strucutral Geology, v. 16, p. 747–748.

Paterson, S. R., Vernon, R. H., and Tobisch, O. T., 1989, A review of criteria for the identification of magmatic and tectonic foliations in granitoids: Journal of Structural Geology, v. 11, p. 349–363.

Paterson, S. R., Brudos, T., Fowler, K., Carlson, C. Bishop, K., and Vernon, R. H., 1991, Papoose Flat pluton: Forceful expansion or postemplacement deformation?: Geology, v. 19, p. 324–327.

Paterson, S. R., Yuan, E. S., and Fowler, T. K., Jr.,1993, Are we interpreting magmatic foliations, lineations and layering correctly?: Geological Society of America Abstracts with Programs, v. 25, p. 131.

Paterson, W. S. B., 1981, The physics of glaciers, 2nd ed.: New York, Pergamon Press, 380 p.

Pavlis, T. L., 1986, The role of strain heating in the evolution of megathrusts: Journal of Geophysical Research, v. 91, p. 12,407–12,422.

Peach, B. N., Horne, J., Clough, C. T., Cadell, H. M., and Dinhaus, C. H., 1888 (reprinted 1923), Assynt District: Geological Survey of Great Britain, scale 1:63,360.

Peacock, S. M., 1992, Blueschist-facies metamorphism, shear heating, and P–T–t paths in subduction shear zones: Journal of Geophysical Research, v. 97, p. 17,693–17,707.

Phillips, J., 1844, On certain movements in the parts of stratified rocks: Report of the British Association for the Advancement of Science, v. 1843, p. 60–61.

Platt, J.-P., and Vissers, R. L. M., 1979, Extensional structures in anisotropic rocks: Journal of Structural Geology, v. 2, p. 397–410.

Platt, L. B., 1976, A Penrose Conference on cleavage in rocks: Geotimes, v. 21, p. 19–20.

Poirier, J.-P., 1985, Creep of crystals: High-temperature deformation processes in metals, ceramics, and minerals: Cambridge, England, Cambridge University Press, 260 p.

Poldervaart, A., and Walker, K. R., 1962, The Palisade sill: International Mineralogical Association, 3rd General Congress, Washington, D.C., Northern Field Excursion Guidebook, p. 5–7.

Pollard, D. D., and Aydin, A., 1988, Progress in understanding joints over the past century: Geological Society of America Bulletin, v. 100, p. 1181–1204.

Pollard, D. D., and Holzhausen, G., 1979, On the mechanical interaction between a fluid-filled fracture and the Earth's surface: Tectonophysics, v. 53, p. 27–57.

Pollard, D. D., and Johnson, A. M., 1973, Mechanics of growth of some laccolithic intrusions in the Henry Mountains, Utah, II, Bending and failure of overburden layers and sill formation: Tectonophysics, v. 18, p. 311–354.

Pollard, D. D., and Muller, O. H., 1976, The effects of gradients in regional stress and magma pressure on the form of sheet intrusions in cross section: Journal of Geophysical Research, v. 81, p. 975–984.

Pollard, D. D., Muller, O. H., and Dockstader, D. R., 1975, The form and growth of fingered sheet intrusions: Geological Society of America Bulletin, v. 86, p. 351–363.

Powell, C. McA., 1974, Timing of slaty cleavage during folding of Precambrian rocks, northwest Tasmania: Geological Society of America Bulletin, v. 85, p. 1043–1060.

Powell, C. McA., 1979, A morphological classification of rock cleavage: Tectonophysics, v. 58, p. 21–34.

Powell, R. E., Weldon, R. J., II, and Matti, J. C., 1993, The San Andreas fault system: Displacement, palinspastic reconstruction, and geologic evolution: Geological Society of America Memoir 178, 332 p.

Price, N. J., 1977, Aspects of gravity tectonics, and the development of listric faults: Journal of the Geological Society of London, v. 133, p. 311–325.

Price, N. J., and Cosgrove, J. W., 1990, Analysis of geological structures, Cambridge, England, Cambridge University Press, 502 p.

Price, R. A., 1973, The mechanical paradox of large overthrusts: Geological Society of America Abstracts with Programs, v. 5, p. 772.

Price, R. A., 1974, Large-scale gravitational flow of supracrustal rocks, southern Canadian Rockies, *in* DeJong, K. A., and Scholten, R., eds., Gravity and tectonics: New York, John Wiley & Sons, p. 491–502.

Price, R. A., 1981, The Cordilleran foreland thrust and fold belt in the southern Canadian Rockies, *in* McClay, K. R., and Price, N. J., eds., Thrust and nappe tectonics: Geological Society of London Special Publication 9, p. 427–448.

Price, R. A., 1988, The mechanical paradox of large overthrusts: Geological Society of America Bulletin, v. 100, p. 1898–1908.

Price, R. A., and Mountjoy, E. W., 1970, Geologic structure of the Canadian Rocky Mountains between Bow and Athabaska Rivers—A progress report: Geological Association of Canada Special Paper 6, p. 7–25.

Proffett, J. M., Jr., 1977, Cenozoic geology of the Yerington district, Nevada, and implications for the nature and origin of Basin and Range faulting: Geological Society of America Bulletin, v. 88, p. 247–266.

Ragland, P. C., Hatcher, R. D., Jr., and Whittington, D., 1983, Juxtaposed Mesozoic diabase dike sets from the Carolinas: A preliminary assessment: Geology, v. 11, p. 394–399.

Raj, R., 1982, Creep polycrystalline aggregates by matter transport through a liquid phase: Journal of Geophysical Research, v. 87, p. 4731–4741.

Raleigh, C. B., 1965, Glide mechanisms in experimentally deformed minerals: Science, v. 150, p. 739–741.

Raleigh, C. B., Healy, J., and Bredehoeft, J., 1976, An experiment in earthquake control at Rangely, Colorado: Science, v. 191, p. 1230–1237.

Ramberg, H., 1960, Relationships between length of arc and thickness of ptygmatically folded veins: American Journal of Science, v. 258, p. 36–46.

Ramberg, H., 1961, Relationship between concentric-longitudinal strain and concentric-shearing strain during folding of homogeneous sheets of rocks: American Journal of Science, v. 259, p. 382–390.

Ramberg, H., 1963, Evolution of drag folds: Geological Magazine, v. 100, p. 97–106.

Ramberg, H., 1967, Gravity, deformation and the Earth's crust as studied by centrifuged models: New York, Academic Press, 241 p.

Ramberg, H., 1981, Gravity, deformation and the Earth's crust, 2nd ed.: London, Academic Press.

Ramsay, D. M., and Sturt, B. A., 1973a, An analysis of noncylindrical and incongruous fold pattern from the Eo-Cambrian rocks of Sørøy, northern Norway, Part I. Noncylindrical, incongruous and aberrant folding: Tectonophysics, v. 18, p. 81–107.

Ramsay, D. M., and Sturt, B. A., 1973b, An analysis of noncyclindrical and incongruous fold pattern from the Eo-Cambrian rocks of Sørøy, northern Norway, Part II. The significance of synfold stretching lineation in the evolution of noncylindrical folds: Tectonophysics, v. 18, p. 109–121.

Ramsay, J. G., 1958, Superimposed folding at Loch Monar, Inverness-Shire and Ross-Shire: Geological Society of London Quarterly Journal, v. 113, p. 271–308.

Ramsay, J. G., 1962, The geometry and mechanics of formation of "similar" type folds: Journal of Geology, v. 70, p. 309–327.

Ramsay, J. G., 1963, Structure and metamorphism of the Moine and Lewisian rocks of the northwest Caledonides, *in* Johnson, M. R. W., and Steward, F. H., eds., The British Caledonides: Edinburgh, Oliver and Boyd, Ltd., p. 1434–1475.

Ramsay, J. G., 1967, Folding and fracturing of rocks: New York, McGraw-Hill, 568 p.

Ramsay, J. G., 1980, Shear zone geometry, a review: Journal of Structural Geology, v. 2, p. 83–99.

Ramsay, J. G., 1981, Tectonics of the Helvetic nappes, *in* McClay, K. R., and Price, N. J., eds., Thrust and nappe tectonics: Geological Society Special Publication No. 9: Oxford, England, Blackwell Scientific Publications, p. 293–309.

Ramsay, J. G., 1983, Rock ductility and its influence on the development of tectonic structures in mountain belts, *in* Hsü, K. J., Mountain building processes: New York, Academic Press, p. 111–128.

Ramsay, J. G., 1989, Emplacement kinematics of a granite diapir: The Chindamora batholith, Zimbabwe: Journal of Structural Geology, v. 11, p. 191–209.

Ramsay, J. G., and Graham, R. H., 1970, Strain variation in shear belts: Canadian Journal of Earth Sciences, v. 7, p. 786–813.

Ramsay, J. G., and Huber, M. I., 1983, The techniques of modern structural geology, v. 1: Strain analysis: New York, Academic Press, 307 p.

Ramsay, J. G., and Huber, M. I., 1987, The techniques of modern structural geology, v. 2: Folds and fractures: London, Academic Press, 392 p.

Ramsay, J. G., and Wood, D. S., 1973, The geometric effects of volume change during deformation processes: Tectonophysics, v. 16, p. 263–277.

Rankin, D. W., 1975, The continental margin of eastern North America in the southern Appalachians: The opening and closing of the Proto-Atlantic Ocean: American Journal of Science, v. 275-A, p. 298–336.

Ratcliffe, N. M., 1971, The Ramapo fault system in New York and adjacent northern New Jersey: A case of tectonic heredity: Geological Society of America Bulletin, v. 82, p. 125–142.

Raymond, L. A., 1975, Tectonite and melange—A distinction: Geology, v. 3, p. 7–9.

Read, H. H., 1931, Geology of central Sutherland: Geological Survey of Great Britain Memoir, 238 p.

Reed, J. C., Jr., and Bryant, B., 1964, Evidence for strike-slip faulting along the Brevard zone in North Carolina: Geological Society of America Bulletin, v. 75, p. 1177–1196.

Reed, J. C., Jr., Bryant, B., and Myers, W. B., 1970, The Brevard zone: A reinterpretation, *in* Fisher, G. W., Pettijohn, F. J., Reed, J. C., Jr., and Weaver, K. N., eds., Studies of Appalachian geology: Central and southern: New York, Wiley Interscience, p. 241–256.

Reed, J. J., 1964, Mylonites, cataclasites, and associated rocks along the Alpine fault, South Island, New Zealand: New Zealand Journal of Geology and Geophysics, v. 7, p. 645–684.

Reineck, H.-E., and Singh, I. B., 1975, Depositional sedimentary environments: Berlin, Springer-Verlag, 439 p.

Reks, I. J., and Gray, D. R., 1983, Strain patterns and shortening in a folded thrust sheet: An example from the southern Appalachians: Tectonophysics, v. 93, p. 99–128.

Rich, J. L., 1934, Mechanics of low-angle overthrust faulting as illustrated by Cumberland thrust block, Virginia, Kentucky and Tennessee: American Association of Petroleum Geologists Bulletin, v. 18, p. 1584–1596.

Richardson, R. M., Soloman, S. C., and Sleep, N. H., 1979, Tectonic stress in the plates: Reviews of Geophysics and Space Physics, v. 17, p. 981–1019.

Rickard, M. J., 1971, A classification diagram for fold orientations: Geological Magazine, v. 108, p. 23–26.

Riedel, W., 1929, Zur Mechanik geologischer Brucherscheinungen: Zentralblatt für Mineralogie, Geologie und Paleontologie, v. B, p. 354–368.

Rikitaki, T., 1966, Electromagnetism and the earth's interior: New York, Elsevier Publishing Company, 308 p.

Roberts, A. M., Yielding, G., and Freeman, B., eds., 1991, The geometry of normal faults: London, Geological Society of London, Special Publication 56, p. 1–13.

Roberts, D., and Strömgård, K.-E., 1972, A comparison of natural and experimental strain patterns around fold hinge zones: Tectonophysics, v. 14, p. 105–120.

Rodgers, J., 1949, Evolution of thought on structure of middle and southern Appalachians: American Association of Petroleum Geologists, v. 33, p. 1643–1654.

Rodgers, J., 1964, Basement and no-basement hypotheses in the Jura and the Appalachian Valley and Ridge, *in* Lowry, W. D., ed., Tectonics of the southern Appalachians: Virginia Polytechnic Institute Department of Geological Sciences Memoir 1, p. 71–80.

Rogers, H. D., 1856, On the laws of structure of the more disturbed zones of the Earth's crust: Transactions of the Royal Society of Edinburgh, v. 21, p. 431–472.

Rosenburg, C., Berger, A., Davidson, C., and Schmid, S. M., 1994, Messa in Posto del plutone di Masino-Bregaglia (Bergell), Alpi Centrale: Atti Ticinensi di Scienze della Terrai (in press).

Rubey, W. W., and Hubbert, M. K, 1959, Role of fluid pressure in mechanics of overthrust faulting, Part II: Overthrust belt in geosynclinal area of western Wyoming in light of fluid-pressure hypothesis: Geological Society of America Bulletin, v. 70, p. 167–206.

Russ, D. P., 1979, Late Holocene faulting and earthquake recurrence in the Reelfoot Lake area, northwestern Tennessee: Geological Society of America Bulletin, v. 90, p. 1013–1018.

Rutter, E. H., 1976, The kinetics of rock deformation by pressure solution: Philosophical Transactions of the Royal Society, v. A283, p. 203–219.

Rutter, E. H., 1983, Pressure solution in nature, theory and experiment: Journal of Geological Society of London, v. 140, p. 725–740.

Rutter, E. H., 1986, On the nomenclature of mode of failure

transitions in rocks: Tectonophysics, v. 122, p. 381–387.

Safford, J. M., 1856, A geological reconnaissance of the state of Tennessee: Nashville, Mercer, 164 p.

Sander, B., 1911, Über Zusammenhänge Zwischen Teilbewegung und Gefüge in Gesteinen: Tschermaks Mineralogie und Petrographie Mittelungen, v. 30, p. 381–384.

Sander, B., 1930, Gefügekunde der Gesteine: Vienna, Springer (English translation: An introduction to the study of fabrics of geological bodies: Oxford, England, Pergamon Press, 641 p.)

Sanderson, D. J., 1979, The transition from upright to recumbent folding in the Variscan fold belt of southwest England: A model based on the kinematics of simple shear: Journal of Structural Geology, v. 1, p. 171–180.

Schaeffer, M. F., Steffens, R. E., and Hatcher, R. D., Jr., 1979, *In situ* stress and its relationships to joint formation in the Toxaway Gneiss, northwestern South Carolina: Southeastern Geology, v. 20, p. 129–143.

Schedl, A., and van der Pluijm, B. A., 1988, A review of deformation microstructures: Journal of Geological Education, v. 36, p. 111–120.

Schmid, S. M., 1983, Microfabric studies as indicators of deformation mechanisms and flow laws operative in mountain building, *in* Hsü, K. J., ed., Mountain building processes: New York, Academic Press, p. 95–110.

Schmid, S. M., 1992, Role of melt in the formation of a deep-crustal compressive shear zone: The Maclaren Glacier metamorphic belt, south central Alaska: Tectonics, v. 11, p. 348–359.

Scholz, C. H., 1980, Shear heating and the state of stress on faults: Journal of Geophysical Research, v. 85, p. 6174–6184.

Scholz, C. H., 1990, The mechanics of earthquakes and faulting: Cambridge, England, Cambridge University Press, 439 p.

Schönborn, G., 1992a, Alpine tectonics and kinematic models of the central Southern Alps: Memorie Di Scienze Geologiche, v. XLIV, p. 229–393.

Schönborn, G., 1992b, Kinematics of a transverse zone in the Southern Alps, Italy, *in* McClay, K. ed., Thrust tectonics: London, Chapman and Hall, p. 299–310.

Schumacher, M.E., 1990, Alpine basement thrusts in the eastern Seengebirge, Southern Alps (Italy/Switzerland): Eclogae Geologicae Helvetiae, v. 83, p. 645–663.

Schwerdtner, W. M., 1990, Structural test of diapir hypotheses in Archean crust of Ontario: Canadian Journal of Earth Sciences, v. 27, p. 387–402.

Schwerdtner, W. M., and Lumbers, S. B., 1980, Major diapiric structures in the Superior and Grenville provinces of the Canadian Shield, *in* Strangway, D. W. ed., The continental crust and its mineral deposits: Geological Association of Canada Special Paper 20, p. 149–180.

Secor, D. T., Jr., 1963, Geology of the central Spring Mountains, Nevada, [unpublished Ph. D. dissertation]: Stanford, California, Stanford University, 152 p.

Secor, D. T., Jr., 1965, Role of fluid pressure in jointing: American Journal of Science, v. 263, p. 633–646.

Secor, D. T., Jr., 1969, Mechanics of natural extension fracturing at depth in the earth's crust: Geological Survey of Canada Paper 68–52, p. 3–48.

Secor, D. T., Jr., and Pollard, D. D., 1975, The stability of open hydraulic fractures in the earth's crust: Geophysical Research Letters, v. 2, p. 510–513.

Sedgwick, A., 1835, Remarks on the structure of large mineral masses, and especially on the chemical changes produced in the aggregation of stratified rocks during different periods after their deposition: Transactions of the Geological Society of London, Series 2, v. 3, p. 461–486.

Segall, P., and Pollard, D. D., 1983, Joint formation in granitic rock of the Sierra Nevada: Geological Society of America Bulletin, v. 94, p. 563–575.

Seifert, K. W., 1965, Deformation bands in albite: American Mineralogist, v. 50, p. 1469–1472.

Shackleton, R. M., 1958, Downward facing structures of the Highland Border: Quarterly Journal of the Geological Society of London, v. 72, p. 361–392.

Shanmugam, G., Damuth, J. E., and Moiola, R. J., 1985, Is the turbidite facies association scheme valid for interpreting ancient submarine environments?: Geology, v. 13, p. 234–237.

Sharma, P. V., 1976, Geophysical methods in geology: Amsterdam, Elsevier, 428 p.

Sharpe, D., 1847, On slaty cleavage: Quarterly Journal of the Geological Society of London, v. 3, p. 74–105.

Sharpe, D., 1849, On slaty cleavage: Quarterly Journal of the Geological Society of London, v. 5, p. 111–129.

Sherwin, J. A., and Chapple, W. M., 1968, Wavelengths of single layer folds: A comparison between theory and observation: American Journal of Science, v. 266, p. 167–179.

Shumskii, P. A., 1964, Principles of structural glaciology; D. Kraus translation of 1955 edition: New York, Dover Publications, 497 p.

Sibson, R. H., 1973, Interactions between temperature and pore pressure during earthquake faulting and a mechanism for partial or total stress relief: Nature, v. 243, p. 66–68.

Sibson, R. H., 1975, Generation of pseudotachylite by ancient seismic faulting: Geophysical Journal of the Royal Astronomical Society, v. 43, p. 775–794.

Sibson, R. H., 1977, Fault rocks and fault mechanisms: Journal of the Geological Society of London, v. 133, p. 191–213.

Sibson, R. H., 1983, Continental fault structure and the shallow earthquake source: Journal of the Geological Society of London, v. 140, p. 741–767.

Siddans, A. B., 1972, Slaty cleavage, a review of research since 1815: Earth Science Reviews, v. 8, p. 205–232.

Sieh, K. E., 1984, Lateral offsets and revised dates of large prehistoric earthquakes at Pallett Creek, southern California: Journal of Geophysical Research, v. 89, p. 7641–7670.

Sieh, K. E., and Jahns, R. H., 1984, Holocene activity of the San Andreas fault at Wallace Creek, California: Geological Society of America Bulletin, v. 95, p. 883–896.

Severinghaus, J., and Atwater, T., 1990, Cenozoic geometry and thermal state of the subducting slabs beneath western North America: Geological Society of America Memoir 176, p. 1–22.

Silver, L. T., and Schultz, P. H., eds., 1982, Geological implications of impacts of large asteroids and comets on the Earth: Geological Society of America Special Paper 190, 528 p.

Simpson, C., 1984, Borrego Springs–Santa Rosa mylonite zone: A Late Cretaceous west-directed thrust in southern California: Geology, v. 12, p. 8–11.

Simpson, C., 1985, Deformation of granitic rocks across the ductile-brittle transition: Journal of Structural Geology, v. 7, p. 503–512.

Simpson, C., 1986, Determination of movement sense in mylonites: Journal of Geological Education, v. 34, p. 246–261.

Simpson, C., and De Paor, D. G., 1993, Strain and kinematic analysis near shear zones: Journal of Structural Geology, v. 15, p. 1–20.

Simpson, C., and Schmid, S. M., 1983, An evaluation of criteria to deduce the sense of movement in sheared rocks: Geological Society of America Bulletin, v. 94, p. 1281–1288.

Simpson, D. W., 1986, Triggered earthquakes: Annual Review of Earth and Planetary Science, v. 14, p. 21–42.

Sinha, A. K., Hewitt, D. A., and Rimstidt, J. D., 1988, Metamorphic petrology and strontium isotope geochemistry associated with the development of mylonites: An example from the Brevard fault zone, North Carolina: American Journal of Science, v. 288-A, p. 115–147.

Skjernaa, Lilian, 1989, Tubular folds and sheath folds: Definitions and conceptual models for their development, with examples from the Grapesvare area, northern Sweden: Journal of Structural Geology, v. 11, p. 689–703.

Smoluchowski, M. S., 1909, Some remarks on the mechanics of overthrusts: Geological Magazine, v. 6, p. 204–205.

Snoke, A. W., 1980, Transition from infrastructure to suprastructure in the northern Ruby Mountains, Nevada, *in* Crittenden, M. D., Jr., Coney, P. J., and Davis, G. H., eds., Metamorphic core complexes: Geological Society of America Memoir 153, p. 287–333.

Snowden, P. A., and Bickle, M. J., 1976, The Chindamora batholith: Diapiric intrusion or interference fold? Journal of the Geological Society of London, v. 132, p. 131.

Sorby, H. C., 1853, On the origin of slaty cleavage: New Philosophical Journal of Edinburgh, v. 55, p. 137–148.

Sorby, H. C., 1856, On slaty cleavage, as exhibited in the Devonian limestones of Devonshire: Philosophical Magazine, v. 11, p. 20–37.

Sorby, H. C., 1863, On the direct correlation of mechanical and chemical forces: Proceedings of the Royal Society, v. 12, p. 538–550.

Stewart, K. G., and Alvarez, W., 1991, Mobile-hinge kinking in layered rocks and models: Journal of Structural Geology, v. 13, p. 243–259.

Stockmal, G. S., 1983, Modeling of large-scale accretionary wedge deformation: Journal of Geophysical Research, v. 88, p. 8271–8288.

Stone, B. D., 1976, Analysis of slump slip lines and deformation fabrics in slumped Pleistocene lake beds: Journal of Sedimentary Petrology, v. 46, p. 313–325.

Stone, B. D., and Koteff, C., 1979, A late Wisconsin ice readvance near Manchester, New Hampshire: American Journal of Science, v. 279, p. 590–601.

Suppe, J., 1981, Mechanics of mountain building and metamorphism in Taiwan: Geological Society of China Memoir 4, p. 67–89.

Suppe, J., 1983, Geometry and kinematics of fault-bend folding: American Journal of Science, v. 283, p. 648–721.

Suppe, J., 1985, Principles of structural geology: Englewood Cliffs, New Jersey, Prentice-Hall, 537 p.

Swanson, M. T., 1982, Preliminary model for an early transform history in central Atlantic rifting: Geology, v. 10, p. 317–320.

Swisher, C. C., III, and 11 others, 1992, Coeval $^{40}Ar/^{39}Ar$ ages of 65.0 million years ago from Chicxulub Crater melt rock and Cretaceous-Tertiary boundary tectites: Science, v. 257, p. 954–958.

Sykes, L. R., 1967, Mechanisms of earthquakes and nature of faulting on the mid-oceanic ridges: Journal of Geophysical Research, v. 72, p. 2131–2153.

Sylvester, A. G., ed., 1984, Wrench fault tectonics: American Association of Petroleum Geologists Reprints Series 28, 374 p.

Sylvester, A. G., 1988, Strike-slip faults: Geological Society of America Bulletin, v. 100, p. 1666–1703.

Sylvester, A. G., and Smith, R. R., 1976, Tectonic transpression and basement-controlled deformation in San Andreas fault zone, Salton Trough, California: American Association of Petroleum Geologists Bulletin, v. 60, p. 2081–2102.

Sylvester, A. G., Oertel, G., Nelson, C. A., and Christie, J. M., 1978, Papoose Flat pluton: A granitic blister in the Inyo Mountains, California: Geological Society of America Bulletin, v. 89, p. 1205–1219.

Talbot, C. J., 1968, Thermal convection in the Archean crust?: Nature, v. 220, p. 552–556.

Talwani, P., and Cox, J., 1985, Paleoseismic evidence for recurrence of earthquakes near Charleston, South Carolina: Science, v. 229, p. 379–381.

Tanaka, H., 1992, Cataclastic lineations: Journal of Structural Geology, v. 14, p. 1239–1252.

Tanner, P. W. G., 1989, The flexural-slip mechanism: Journal of Structural Geology, v. 11, p. 635–655.

Tapponier, P., and Molnar, P., 1977, Active faulting and Cenozoic tectonics of China: Journal of Geophysical Research, v. 82, p. 2905–2930.

Tapponnier, P., Peltzer, G., Le Dain, A. Y., Armijo, R., and Cobbold, P., 1982, Propagating extrusion tectonics in Asia: New insight from simple experiments with Plasticine: Geology, v. 10, p. 611–616.

Tchalenko, J. S., 1970, Similarities between shear zones of different magnitudes: Geological Society of America Bulletin, v. 81, p. 1625–1640.

Telford, W. M., Geldart, L. P., Sheriff, R. E., and Keys, D. A., 1990, Applied geophysics, 2nd ed.: Cambridge, England, Cambridge University Press, 770 p.

Templeman-Kluitt, D. J., 1979, Transported cataclasite, ophiolite and granodiorite in the Yukon: Evidence of arc-continent collision: Geological Survey of Canada Paper 79–14, 27 p.

Terzaghi, K. V., 1923, Die Berechnung der Durchlassigkeitsziffer des Zones aus dem Verlauf der hydrodynamischen Spannungserscheinungen: Sitzungisbericht der Akademie der Wissenschaftes Wien, v. 132, p. 105.

Teufel, L. W., and Logan, J. M., 1978, Effect of shortening rate on the real area of contact and temperatures generated during frictional sliding: Pure and Applied Geophysics, v. 116, p. 840–865.

Thiessen, R. L., and Means, W. D., 1980, Classification of fold interference patterns: A reexamination: Journal of Structural Geology, v. 2, p. 311–316.

Thomas, M. D., 1983, Tectonic significance of paired gravity anomalies in the southern and central Appalachians, *in* Hatcher, R. D., Jr., Williams, H., and Zietz, I., eds., Contributions to the tectonics and geophysics of mountain chains: Geological Society of America Memoir 158, p. 113–124.

Tobisch, O. T., 1966, Large-scale basin-and-dome pattern resulting from the interference of major folds: Geological Society of America Bulletin, v. 77, p. 393–408.

Tobisch, O. T., and Paterson, S. R., 1988, Analysis and interpretation of progressive deformation: Journal of Structural Geology, v. 10, p. 745–754.

Tobisch, O. T., Fleuty, M. J., Merh, S. S., Mukhopadhyay, D., and Ramsay, J. G., 1970, Deformational and metamorphic history of Moinian and Lewisian rocks between Strathconon and Glen Affric: Scottish Journal of Geology, v. 6, p. 243–265.

Törnebohm, A. E., 1872, En geognostisk profil öfver den skandinaviska fjällryggen mellan Östersund och Levanger: Sveriges Geologiska Undersökning, v. 6, 24 p.

Treagus, S. H., 1983, A theory of finite strain variation through contrasting layers, and its bearing on cleavage refraction: Journal of Structural Geology, v. 5, p. 351–368.

Treagus, S. H., 1988, A history of cleavage and folding: An example from the Goldenville Formation, Nova Scotia: Discussion: Geological Society of America Bulletin, v. 100, p. 152–154.

Treagus, S. H., and Treagus, J. E., 1992, Transected folds and transpression: How are they associated?: Journal of Structural Geology, v. 14, p. 361–367.

Trümpy, R., 1960, Paleotectonic evolution of the central and western Alps: Geological Society of America Bulletin, v. 71, p. 843–908.

Trümpy, R., 1975, Penninic-Austroalpine boundary in the Swiss Alps: A presumed former continental margin and its problems: American Journal of Science, v. 275–A, p. 209–238.

Trümpy, R., 1980, Geology of Switzerland: A guide-book: Part A: An outline of the geology of Switzerland, Wepf & Co. Publishers, Basel, Switzerland, 104 p.

Tullis, J., and Yund, R. A., 1985, Dynamic recrystallization of feldspar: A mechanism for ductile shear zone formation: Geology, v. 13, p. 238–241.

Tullis, J., and Yund, R. A., 1987, Transition from cataclastic flow to dislocation creep of feldspar: Geology, v. 15, p. 606–609.

Tullis, J., Christie, J. M., and Griggs, D. T., 1973, Microstructures and preferred orientations of experimentally deformed quartzites: Geological Society of America Bulletin, v. 84, p. 297–314.

Tullis, T. E., 1971, Experimental development of preferred orientation of mica during recrystallization [unpublished Ph.D. dissertation]: Los Angeles, University of California, 262 p.

Turcotte, D. L., and Emerman, S. H., 1985, Magma fracture as a mechanism for magma migration: EOS, v. 66, p. 361.

Turcotte, D. L., and Schubert, G., 1982, Geodynamics: Applications of continuum physics to geological problems: New York, John Wiley & Sons, 450 p.

Turner, F. J., 1948, Mineralogical and structural evolution of the metamorphic rocks: Geological Society of America Memoir 30, 342 p.

Turner, F. J., and Weiss, L. E., 1963, Structural analysis of metamorphic tectonites: New York, McGraw-Hill, 545 p.

Turner, F. J., Griggs, D. T., and Heard, H. C., 1954, Experimental deformation of calcite crystals: Geological Society of America Bulletin, v. 65, p. 883–934.
Twiss, R. J., and Moores, E. M., 1992, Structural geology: New York, W. H. Freeman & Company, 532 p.
Tyndall, J., 1856, Observations on "The theory of the origin of slaty cleavage" by H. C. Sorby: Philosophical Magazine, v. 12, p. 129–135.
Vacquier, V., Nelson, C. S., Henderson, R. G., and Zietz, I., 1951, Interpretation of aeromagnetic maps: Geological Society of America Memoir 47, 151 p.
Van Hise, C. R., 1896, Principles of North American Precambrian geology: U.S. Geological Survey Annual Report 16th (1894 1895), Part 1, p. 571–843.
Vendeville, B. C., and Jackson, M. P. A., 1992a, The rise of diapirs during thin-skinned extension: Marine and Petroleum Geology, v. 9, p. 331–353.
Vendeville, B. C., and Jackson, M. P. A., 1992b, The fall of diapirs during thin-skinned extension: Marine and Petroleum Geology, v. 9, p. 354–371.
Vernon, R. H., 1988, Sequential growth of cordierite and andalusite porphyroblasts, Cooma complex, Australia: Microstructural evidence of a prograde reaction: Journal of Metamorphic Geology, v. 6, p. 255–269.
Voll, G., 1960, New work on petrofabrics: Geological Journal, v. 2, p. 503–597.
Wager, L. R., and Deer, W. A., 1939, Geological investigations in East Greenland, Part III: The petrology of the Skaergaard intrusion, Kangerdlugsuag, East Greenland: Meddlungen Grønland, v. 105, p. 1–352.
Walker, F., 1940, Different ratios of the Palisade diabase, New Jersey: Geological Society of America Bulletin, v. 51, p. 1059–1106.
Waterson, J., 1975, Mechanism for the persistence of tectonic lineaments: Nature, v. 253, p. 520–522.
Watkins, N. D., 1969, Non-dipole behaviour during an Upper Miocene geomagnetic polarity transition in Oregon: Geophysical Journal of the Royal Astronomical Society, v. 17, p. 121–149.
Weijermars, R., and Rondeel, H. E., 1984, Shear band foliation as an indicator of sense of shear: Field observations in central Spain: Geology, v. 12, p. 603–606.
Weinberg, R. F., 1994, Re-examining pluton emplacement processes: Discussion: Journal of Structural Geology, v. 16, p. 743–746.
Weiss, L. E., 1980, Nucleation and growth of kink bands: Tectonophysics, v. 65, p. 1–38.
Weller, J. M., 1947, Relations of the invertebrate paleontologist to geology: Journal of Paleontology, v. 21, p. 570–575.
Wellman, H. G., 1962, A graphic method for analysing fossil distortion caused by tectonic deformation: Geological Magazine, v. 99, p. 348–352.
Wernicke, B., 1985, Uniform-sense normal simple shear of the continental lithosphere: Canadian Journal of Earth Sciences, v. 22, p. 108–125.
Wernicke, B., and Burchfiel, B. C., 1982, Modes of extensional tectonics: Journal of Structural Geology, v. 4, p. 105–115.
Westbrook, G. K., and Smith, M. J., 1983, Long décollements and mud volcanoes: Evidence from the Barbados Ridge complex for the role of high pore-fluid pressure in the development of an accretionary complex: Geology, v. 11, p. 279–283.
Wheeler, J., 1991, A view of texture dynamics: Terra Nova, v. 3, p. 125–136.
White, I. C., 1885, The geology of natural gas: Science, v. 5, p. 521–522.
White, I. C., 1892, The Mannington oil field and the history of its development: Geological Society of America Bulletin, v. 3, p. 187–216. (An appendix describes the anticlinal theory for natural gas.)
White, J. C., and White, S. H., 1980, High-voltage transmission electron microscopy of naturally deformed polycrystalline dolomite: Tectonophysics, v. 66, p. 35–54.
White, S. H., Burrows, S. E., Carreras, J., Shaw, N. D., and Humphreys, F. J., 1980, On mylonites in ductile shear zones: Journal of Structural Geology, v. 2, p. 175–187.
Whitten, E. H. T., 1966, Structural geology of folded rocks: Chicago, Rand McNally, 678 p.
Williams, G. D., and Chapman, T. J., 1979, The geometrical classification of noncylindrical folds: Journal of Structural Geology, v. 1, p. 181–185.
Williams, G. D., Powell, C. M., and Cooper, M. A., 1989, Geometry and kinematics of inversion tectonics, *in* Cooper, M. A., and Williams, G. D., eds., Inversion tectonics: Oxford, England, Blackwell Scientific Publications, Geological Society of London Special Publication 44, p. 3–16.
Williams, H., and Hatcher, R. D., Jr., 1983, Appalachian suspect terranes, *in* Hatcher, R. D., Jr., Williams, H., and Zietz, I., Contributions to the tectonics and geophysics of mountain chains: Geological Society of America Memoir 158, p. 33–53.
Williams, H., and Smyth, W. R., 1973, Metamorphic aureoles beneath ophiolite suites and Alpine peridotites: Tectonic implications with west Newfoundland examples: American Journal of Science, v. 273, p. 594–621.
Williams, P. F., 1970, A criticism of the use of style in the study of deformed rocks: Geological Society of America Bulletin, v. 81, p. 3283–3296.
Williams, P. F., 1972, Development of metamorphic layering and cleavage in low grade metamorphic rocks at Bermagui, Australia: American Journal of Science, v. 272, p. 1–47.
Williams, P. F., 1976, Relationships between axial-plane foliations and strain: Tectonophysics, v. 30, p. 181–196.
Williams, P. F., 1985, Multiply deformed terrains—problems of correlation: Journal of Structural Geology, v. 7, p. 269–280.
Willis, B., 1893, The mechanics of Appalachian structure: U.S. Geological Survey 13th Annual Report 1891–1892, Part 2, p. 212–281.
Willis, B., 1923, Geologic structures, 1st ed.: New York, McGraw-Hill Book Co., 295 p.
Wilson, C. W., Jr., and Stearns, R. G., 1968, Geology of the Wells Creek structure, Tennessee: Tennessee Division of Geology, Bulletin 68, 236 p.
Wilson, G., and Cosgrove, J. W., 1982, Introduction to small-scale geological structures: London, Allen & Unwin, 128 p.
Wilson, J. T., 1965, A new class of faults and their bearing on continental drift: Nature, v. 207, p. 343–347.
Wilson, J. T., 1966, Did the Atlantic close and then reopen?: Nature, v. 211, p. 676–681.
Wilson, J. T., 1968, Static or mobile earth: The current scientific revolution: Proceedings of the American Philosophical Society, v. 112, p. 309–320.
Wilson, M. E., 1941, Noranda District, Québec: Geological Survey of Canada Memoir 229, 162 p.
Wilson, M. E., 1962, Rouyn-Beauchastel map area, Québec: Geological Survey of Canada Memoir 315, 118 p.
Wintsch, R. P., 1979, The Willimantic fault: A ductile fault in eastern Connecticut: American Journal of Science, v. 279, p. 367–393.
Wise, D. U., 1963, An outrageous hypothesis for the tectonic pattern of the North American Cordillera: Geological Society of America Bulletin, v. 74, p. 357–362.
Wise, D. U., 1982, Linesmanship and the practice of linear geo-art: Geological Society of America Bulletin, v. 93, p. 886–888.
Wise, D. U., and others, 1979, Fault, fracture, and lineament data for western Massachusetts and western Connecticut: Amherst, University of Massachusetts, Department of Geology and Geography, 253 p.
Wise, D. U., Dunn, D. E., Engelder, J. T., Geiser, P. A., Hatcher, R. D., Jr., Kish, S. A., Odom, A. L., and Schamel, S., 1984, Fault-related rocks: Suggestions for terminology: Geology, v. 12, p. 391–394.
Wise, D. U., Funiciello, R., Maurizio, P., and Salvini, F., 1985, Topographic lineament swarms: Clues to their origin from domain analysis of Italy: Geological Society of America Bulletin, v. 96, p. 952–967.

Wojtal, S., 1986, Deformation within foreland thrust sheets by populations of minor faults: Journal of Structural Geology, v. 8, p. 341–360.

Wood, D. S., 1973, Patterns and magnitudes of natural strain in rocks: Philosophical Transactions of the Royal Society, Series A, v. 274, p. 373–382.

Wood, D. S., 1974, Current views of the development of slaty cleavage: Annual Review of Earth and Planetary Sciences, v. 2, p. 369–401.

Woodward, N. B., 1987, Geological applicability of critical-wedge thrust-belt models: Geological Society of America Bulletin, v. 99, p. 827–832.

Woodward, N. B., Boyer, S. E., and Suppe, J., 1985, An outline of balanced cross-sections: Knoxville, University of Tennessee, Department of Geological Sciences, Studies in Geology 11, 170 p.

Woodward, N. B., Gray, D. R., and Spears, D. B., 1986, Including strain data in balanced cross-sections: Journal of Structural Geology, v. 8, p. 313–324.

Wright, T. O., and Platt, L. B., 1982, Pressure dissolution and cleavage in the Martinsburg Shale: American Journal of Science, v. 282, p. 122–135.

Wynne-Edwards, H. R., 1963, Flow folding: American Journal of Science, v. 261, p. 793–814.

Xiao, H.-B., and Suppe, J., 1986, Role of compaction in the listric shape of growth faults: Geological Society of America Abstracts with Programs, v. 18, p. 796.

Xiao, H.-B., and Suppe, J., 1992, Origin of rollover: American Association of Petroleum Geologists Bulletin, v. 76, p. 509–529.

Yoder, H. S., Jr., 1976, Generation of basaltic magma: Washington, D. C., National Academy of Sciences, 265 p.

Zoback, M. D., and Haimson, B. C., 1982, Status of the hydraulic fracturing method for *in situ* stress measurements, *in* Society of Mining Engineers, 23rd Symposium on Rock Mechanics, Proceedings: New York, American Institute of Mining and Metallurgical Engineers, p. 143–156.

Zoback, M. D., Tsukahara, H., and Hickman, S., 1980, Stress measurements at depth in the vicinity of the San Andreas fault: Implications for the magnitude of shear stress at depth: Journal of Geophysical Research, v. 85, p. 6157–6173.

Zoback, M. D., and 12 others, 1987, State of stress of the San Andreas fault system: Science, v. 238, p. 1105–1111.

Zoback, M. L., 1992, First- and second-order patterns of stress in the lithosphere: The world stress map: Journal of Geophysical Research, v. 97, p. 11,703–11,728.

Zoback, M. L., and Zoback, M. D., 1980, State of stress in the continental United States: Journal of Geophysical Research, v. 85, p. 6113–6156.

Zoback, M. L., and Zoback, M. D., 1989, Tectonic stress field of the continental United States, *in* Pakiser, L. C., and Mooney, W. D., Geophysical framework of the continental United States: Geological Society of America Memoir 172, p. 523–539.

AUTHOR INDEX

SUBJECT INDEX